Chemical Process Design and Integration

Chemical Process Design and Integration

Robin Smith

Centre for Process Integration,
School of Chemical Engineering and Analytical Science,
University of Manchester.

John Wiley & Sons, Ltd

Previous edition published by McGraw Hill

Telephone (+44) 1243 779777

Email (for orders and customer service enquiries): cs-books@wiley.co.uk
Visit our Home Page on www.wileyeurope.com or www.wiley.com

Other Wiley Editorial Offices

John Wiley & Sons Inc., 111 River Street, Hoboken, NJ 07030, USA

Jossey-Bass, 989 Market Street, San Francisco, CA 94103-1741, USA

Wiley-VCH Verlag GmbH, Boschstr. 12, D-69469 Weinheim, Germany

John Wiley & Sons Australia Ltd, 33 Park Road, Milton, Queensland 4064, Australia

John Wiley & Sons (Asia) Pte Ltd, 2 Clementi Loop #02-01, Jin Xing Distripark, Singapore 129809

John Wiley & Sons Canada Ltd, 22 Worcester Road, Etobicoke, Ontario, Canada M9W 1L1

Wiley also publishes its books in a variety of electronic formats. Some content that appears in print may not be available in electronic books.

Library of Congress Cataloging-in-Publication Data

Smith, R. (Robin)
Chemical process design and integration / Robin Smith.
p. cm.
Includes bibliographical references and index.
ISBN 0-471-48680-9 (HB) (acid-free paper) – ISBN 0-471-48681-7 (PB) (pbk. : acid-free paper)
1. Chemical processes. I. Title.
TP155.7.S573 2005
660′.2812 – dc22

2004014695

British Library Cataloguing in Publication Data

A catalogue record for this book is available from the British Library

ISBN 0-471-48680-9 (cloth)
0-471-48681-7 (paper)

Typeset in 10/12pt Times by Laserwords Private Limited, Chennai, India

This book is printed on acid-free paper responsibly manufactured from sustainable forestry in which at least two trees are planted for each one used for paper production.

To my family

Contents

Preface

This book deals with the design and integration of chemical processes, emphasizing the conceptual issues that are fundamental to the creation of the process. Chemical process design requires the selection of a series of processing steps and their integration to form a complete manufacturing system. The text emphasizes both the design and selection of the steps as individual operations and their integration to form an efficient process. Also, the process will normally operate as part of an integrated manufacturing site consisting of a number of processes serviced by a common utility system. The design of utility systems has been dealt with so that the interactions between processes and the utility system and the interactions between different processes through the utility system can be exploited to maximize the performance of the site as a whole. Thus, the text integrates equipment, process and utility system design.

Chemical processing should form part of a sustainable industrial activity. For chemical processing, this means that processes should use raw materials as efficiently as is economic and practicable, both to prevent the production of waste that can be environmentally harmful and to preserve the reserves of raw materials as much as possible. Processes should use as little energy as is economic and practicable, both to prevent the buildup of carbon dioxide in the atmosphere from burning fossil fuels and to preserve reserves of fossil fuels. Water must also be consumed in sustainable quantities that do not cause deterioration in the quality of the water source and the long-term quantity of the reserves. Aqueous and atmospheric emissions must not be environmentally harmful, and solid waste to landfill must be avoided. Finally, all aspects of chemical processing must feature good health and safety practice.

It is important for the designer to understand the limitations of the methods used in chemical process design. The best way to understand the limitations is to understand the derivations of the equations used and the assumptions on which the equations are based. Where practical, the derivation of the design equations has been included in the text.

The book is intended to provide a practical guide to chemical process design and integration for undergraduate and postgraduate students of chemical engineering, practicing process designers and chemical engineers and applied chemists working in process development. For undergraduate studies, the text assumes basic knowledge of material and energy balances, fluid mechanics, heat and mass transfer phenomena and thermodynamics, together with basic spreadsheeting skills. Examples have been included throughout the text. Most of these examples do not require specialist software and can be solved using spreadsheet software. Finally, a number of exercises have been added at the end of each chapter to allow the reader to practice the calculation procedures.

Robin Smith

Acknowledgements

The author would like to express gratitude to a number of people who have helped in the preparation and have reviewed parts of the text.

From The University of Manchester: Prof Peter Heggs, Prof Ferda Mavituna, Megan Jobson, Nan Zhang, Constantinos Theodoropoulos, Jin-Kuk Kim, Kah Loong Choong, Dhaval Dave, Frank Del Nogal, Ramona Dragomir, Sungwon Hwang, Santosh Jain, Boondarik Leewongtanawit, Guilian Liu, Vikas Rastogi, Clemente Rodriguez, Ramagopal Uppaluri, Priti Vanage, Pertar Verbanov, Jiaona Wang, Wenling Zhang.

From Alias, UK: David Lott.

From AspenTech: Ian Moore, Eric Petela, Ian Sinclair, Oliver Wahnschafft.

From CANMET, Canada: Alberto Alva-Argaez, Abdelaziz Hammache, Luciana Savulescu, Mikhail Sorin.

From DuPont Taiwan: Janice Kuo.

From Monash University, Australia: David Brennan, Andrew Hoadley.

From UOP, Des Plaines, USA: David Hamm, Greg Maher.

Gratitude is also expressed to Simon Perry, Gareth Maguire, Victoria Woods and Mathew Smith for help in the preparation of the figures.

Finally, gratitude is expressed to all of the member companies of the Process Integration Research Consortium, both past and present. Their support has made a considerable contribution to research in the area, and hence to this text.

Acknowledgements

Nomenclature

a	Activity (–), or constant in cubic equation of state ($N \cdot m^4 \cdot kmol^{-2}$), or correlating coefficient (units depend on application), or cost law coefficient (\$), or order of reaction (–)
A	Absorption factor in absorption (–), or annual cash flow (\$), or constant in vapor pressure correlation ($N \cdot m^{-2}$, bar), or heat exchanger area (m^2)
A_{CF}	Annual cash flow ($\$ \cdot y^{-1}$)
A_{DCF}	Annual discounted cash flow ($\$ \cdot y^{-1}$)
A_I	Heat transfer area on the inside of tubes (m^2), or interfacial area (m^2, $m^2 \cdot m^{-3}$)
A_M	Membrane area (m^2)
$A_{NETWORK}$	Heat exchanger network area (m^2)
A_O	Heat transfer area on the outside of tubes (m^2)
A_{SHELL}	Heat exchanger area for an individual shell (m^2)
AF	Annualization factor for capital cost (–)
b	Capital cost law coefficient (units depend on cost law), or constant in cubic equation of state ($m^3 \cdot kmol^{-1}$), or correlating coefficient (units depend on application), or order of reaction (–)
b_i	Bottoms flowrate of Component i ($kmol \cdot s^{-1}$, $kmol \cdot h^{-1}$)
B	Bottoms flowrate in distillation ($kmol \cdot s^{-1}$, $kmol \cdot h^{-1}$), or breadth of device (m), or constant in vapor pressure correlation ($N \cdot K \cdot m^{-2}$, $bar \cdot K$), or total moles in batch distillation (kmol)
B_C	Baffle cut for shell-and-tube heat exchangers (–)
BOD	Biological oxygen demand ($kg \cdot m^{-3}$, $mg \cdot l^{-1}$)
c	Capital cost law coefficient (–), or order of reaction (–)
c_D	Drag coefficient (–)
c_F	Fanning friction factor (–)
c_L	Loss coefficient for pipe or pipe fitting (–)
C	Concentration ($kg \cdot m^{-3}$, $kmol \cdot m^{-3}$, ppm), or constant in vapor pressure correlation (K), or number of components (separate systems) in network design (–)
C_B	Base capital cost of equipment (\$)
Ce	Environmental discharge concentration (ppm)
C_E	Equipment capital cost (\$), or unit cost of energy ($\$ \cdot kW^{-1}$, $\$ \cdot MW^{-1}$)
C_F	Fixed capital cost of complete installation (\$)
C_P	Specific heat capacity at constant pressure ($kJ \cdot kg^{-1} \cdot K^{-1}$, $kJ \cdot kmol^{-1} \cdot K^{-1}$)
$\overline{C_P}$	Mean heat capacity at constant pressure ($kJ \cdot kg^{-1} \cdot K^{-1}$, $kJ \cdot kmol^{-1} \cdot K^{-1}$)
C_V	Specific heat capacity at constant volume ($kJ \cdot kg^{-1} \cdot K^{-1}$, $kJ \cdot kmol^{-1} \cdot K^{-1}$)
C^*	Solubility of solute in solvent ($kg \cdot kg$ $solvent^{-1}$)
CC	Cycles of concentration for a cooling tower (–)
COD	Chemical oxygen demand ($kg \cdot m^{-3}$, $mg \cdot l^{-1}$)
COP_{HP}	Coefficient of performance of a heat pump (–)
COP_{REF}	Coefficient of performance of a refrigeration system (–)

CP — Capacity parameter in distillation ($m \cdot s^{-1}$) or heat capacity flowrate ($kW \cdot K^{-1}$, $MW \cdot K^{-1}$)

CP_{EX} — Heat capacity flowrate of heat engine exhaust ($kW \cdot K^{-1}$, $MW \cdot K^{-1}$)

CW — Cooling water

d — Diameter (μm, m)

d_i — Distillate flowrate of Component i ($kmol \cdot s^{-1}$, $kmol \cdot h^{-1}$)

d_I — Inside diameter of pipe or tube (m)

D — Distillate flowrate ($kmol \cdot s^{-1}$, $kmol \cdot h^{-1}$)

D_B — Tube bundle diameter for shell-and-tube heat exchangers (m)

D_S — Shell diameter for shell-and-tube heat exchangers (m)

$DCFRR$ — Discounted cash flowrate of return (%)

E — Activation energy of reaction ($kJ \cdot kmol^{-1}$), or entrainer flowrate in azeotropic and extractive distillation ($kg \cdot s^{-1}$, $kmol \cdot s^{-1}$), or extract flowrate in liquid–liquid extraction ($kg \cdot s^{-1}$, $kmol \cdot s^{-1}$), or stage efficiency in separation (–)

E_O — Overall stage efficiency in distillation and absorption (–)

EP — Economic potential ($\$ \cdot y^{-1}$)

f — Fuel-to-air ratio for gas turbine (–)

f_i — Capital cost installation factor for Equipment i (–), or feed flowrate of Component i ($kmol \cdot s^{-1}$, $kmol \cdot h^{-1}$), or fugacity of Component i ($N \cdot m^{-2}$, bar)

f_M — Capital cost factor to allow for material of construction (–)

f_P — Capital cost factor to allow for design pressure (–)

f_T — Capital cost factor to allow for design temperature (–)

F — Feed flowrate ($kg \cdot s^{-1}$, $kg \cdot h^{-1}$, $kmol \cdot s^{-1}$, $kmol \cdot h^{-1}$), or future worth a sum of money allowing for interest rates (\$), or volumetric flowrate ($m^3 \cdot s^{-1}$, $m^3 \cdot h^{-1}$)

F_{LV} — Liquid–vapor flow parameter in distillation (–)

F_T — Correction factor for noncountercurrent flow in shell-and-tube heat exchangers (–)

F_{Tmin} — Minimum acceptable F_T for noncountercurrent heat exchangers (–)

g — Acceleration due to gravity (9.81 $m \cdot s^{-2}$)

g_{ij} — Energy of interaction between Molecules i and j in the NRTL equation ($kJ \cdot kmol^{-1}$)

G — Free energy (kJ), or gas flowrate ($kg \cdot s^{-1}$, $kmol \cdot s^{-1}$)

$\overline{G}_i$ — Partial molar free energy of Component i ($kJ \cdot kmol^{-1}$)

$\overline{G}_i^O$ — Standard partial molar free energy of Component i ($kJ \cdot kmol^{-1}$)

h — Settling distance of particles (m)

h_C — Condensing film heat transfer coefficient ($W \cdot m^{-2} \cdot K^{-1}$, $kW \cdot m^{-2} \cdot K^{-1}$)

h_I — Film heat transfer coefficient for the inside ($W \cdot m^{-2} \cdot K^{-1}$, $kW \cdot m^{-2} \cdot K^{-1}$)

h_{IF} — Fouling heat transfer coefficient for the inside ($W \cdot m^{-2} \cdot K^{-1}$, $kW \cdot m^{-2} \cdot K^{-1}$)

h_L — Head loss in a pipe or pipe fitting (m)

h_{NB} — Nucleate boiling heat transfer coefficient ($W \cdot m^{-2} \cdot K^{-1}$, $kW \cdot m^{-2} \cdot K^{-1}$)

h_O — Film heat transfer coefficient for the outside ($W \cdot m^{-2} \cdot K^{-1}$, $kW \cdot m^{-2} \cdot K^{-1}$)

h_{OF} — Fouling heat transfer coefficient for the outside ($W \cdot m^{-2} \cdot K^{-1}$, $kW \cdot m^{-2} \cdot K^{-1}$)

h_W — Heat transfer coefficient for the tube wall ($W \cdot m^{-2} \cdot K^{-1}$, $kW \cdot m^{-2} \cdot K^{-1}$)

H — Enthalpy (kJ, $kJ \cdot kg^{-1}$, $kJ \cdot kmol^{-1}$), or height (m), or Henry's Law Constant ($N \cdot m^{-2}$, bar, atm), or stream enthalpy ($kJ \cdot s^{-1}$, $MJ \cdot s^{-1}$)

H_T — Tray spacing (m)

$\overline{H}_i^O$	Standard heat of formation of Component i ($kJ \cdot kmol^{-1}$)
ΔH^O	Standard heat of reaction (J, kJ)
ΔH_{COMB}	Heat of combustion ($J \cdot kmol^{-1}$, $kJ \cdot kmol^{-1}$)
ΔH_{COMB}^O	Standard heat of combustion at 298 K ($J \cdot kmol^{-1}$, $kJ \cdot kmol^{-1}$)
ΔH_P	Heat to bring products from standard temperature to the final temperature ($J \cdot kmol^{-1}$, $kJ \cdot kg^{-1}$)
ΔH_R	Heat to bring reactants from their initial temperature to standard temperature ($J \cdot kmol^{-1}$, $kJ \cdot kmol^{-1}$)
ΔH_{STEAM}	Enthalpy difference between generated steam and boiler feedwater (kW, MW)
ΔH_{VAP}	Latent heat of vaporization ($kJ \cdot kg^{-1}$, $kJ \cdot kmol^{-1}$)
HETP	Height equivalent of a theoretical plate (m)
HP	High pressure
HR	Heat rate for gas turbine ($kJ \cdot kWh^{-1}$)
i	Fractional rate of interest on money (–), or number of ions (–)
I	Total number of hot streams (–)
J	Total number of cold streams (–)
k	Reaction rate constant (units depend on order of reaction), or thermal conductivity ($W \cdot m^{-1} \cdot K^{-1}$, $kW \cdot m^{-1} \cdot K^{-1}$)
$k_{G,i}$	Mass transfer coefficient in the gas phase ($kmol \cdot m^{-2} \cdot Pa^{-1} \cdot s^{-1}$)
k_{ij}	Interaction parameter between Components i and j in an equation of state (–)
$k_{L,i}$	Mass transfer coefficient in the liquid phase ($m \cdot s^{-1}$)
k_0	Frequency factor for heat of reaction (units depend on order of reaction)
K	Overall mass transfer coefficient ($kmol \cdot Pa^{-1} \cdot m^{-2} \cdot s^{-1}$) or total number of enthalpy intervals in heat exchanger networks (–)
K_a	Equilibrium constant of reaction based on activity (–)
K_i	Ratio of vapor to liquid composition at equilibrium for Component i (–)
$K_{M,i}$	Equilibrium partition coefficient of membrane for Component i (–)
K_p	Equilibrium constant of reaction based on partial pressure in the vapor phase (–)
K_T	Parameter for terminal settling velocity ($m \cdot s^{-1}$)
K_x	Equilibrium constant of reaction based on mole fraction in the liquid phase (–)
K_y	Equilibrium constant of reaction based on mole fraction in vapor phase (–)
L	Intercept ratio for turbines (–), or length (m), or liquid flowrate ($kg \cdot s^{-1}$, $kmol \cdot s^{-1}$), or number of independent loops in a network (–)
L_B	Distance between baffles in shell-and-tube heat exchangers (m)
LP	Low pressure
m	Mass flowrate ($kg \cdot s^{-1}$), or molar flowrate ($kmol \cdot s^{-1}$), or number of items (–)
M	Constant in capital cost correlations (–), or molar mass ($kg \cdot kmol^{-1}$)
MP	Medium pressure
MC_{STEAM}	Marginal cost of steam ($\$ \cdot t^{-1}$)
n	Number of items (–), or number of years (–), or polytropic coefficient (–), or slope of Willans' Line ($kJ \cdot kg^{-1}$, $MJ \cdot kg^{-1}$)
N	Number of compression stages (–), or number of moles (kmol), or number of theoretical stages (–), or rate of transfer of a component ($kmol \cdot s^{-1} \cdot m^{-3}$)
N_i	Number of moles of Component i (kmol)
N_{i0}	Initial number of moles of Component i (kmol)
N_{PT}	Number of tube passes (–)
N_R	Number of tube rows (–)

N_{SHELLS} — Number of number of 1–2 shells in shell-and-tube heat exchangers (–)

N_T — Number of tubes (–)

N_{UNITS} — Number of units in a heat exchanger network (–)

NC — Number of components in a multicomponent mixture (–)

NPV — Net present value (\$)

p — Partial pressure ($N \cdot m^{-2}$, bar)

p_C — Pitch configuration factor for tube layout (–)

p_T — Tube pitch (m)

P — Present worth of a future sum of money (\$), or
pressure ($N \cdot m^{-2}$, bar), or
probability (–), or
thermal effectiveness of 1–2 shell-and-tube heat exchanger (–)

P_C — Critical pressure ($N \cdot m^{-2}$, bar)

P_{max} — Maximum thermal effectiveness of 1–2 shell-and-tube heat exchangers (–)

$P_{M,i}$ — Permeability of Component i for a membrane ($kmol \cdot m \cdot s^{-1} \cdot m^{-2} \cdot bar^{-1}$, kg solvent $\cdot m^{-1} \cdot s^{-1} \cdot bar^{-1}$)

$\overline{P}_{M,i}$ — Permeance of Component i for a membrane ($m^3 \cdot m^{-2} \cdot s^{-1} \cdot bar^{-1}$)

P_{N-2N} — Thermal effectiveness over N_{SHELLS} number of 1–2 shell-and-tube heat exchangers in series (–)

P_{1-2} — Thermal effectiveness over each 1–2 shell-and-tube heat exchanger in series (–)

P^{SAT} — Saturated liquid–vapor pressure ($N \cdot m^{-2}$, bar)

Pr — Prandtl number (–)

q — Heat flux ($W \cdot m^{-2}$, $kW \cdot m^{-2}$), or
thermal condition of the feed in distillation (–), or
Wegstein acceleration parameter for the convergence of recycle calculations (–)

q_C — Critical heat flux ($W \cdot m^{-2}$, $kW \cdot m^{-2}$)

q_{C1} — Critical heat flux for a single tube ($W \cdot m^{-2}$, $kW \cdot m^{-2}$)

q_i — Individual stream heat duty for Stream i ($kJ \cdot s^{-1}$), or
pure component property measuring the molecular van der Waals surface area for Molecule i in the UNIQUAC Equation (–)

Q — Heat duty (kW, MW)

Q_C — Cooling duty (kW, MW)

$Q_{C\,min}$ — Target for cold utility (kW, MW)

Q_{COND} — Condenser heat duty (kW, MW)

Q_{EVAP} — Evaporator heat duty (kW, MW)

Q_{EX} — Heat duty for heat engine exhaust (kW, MW)

Q_{FEED} — Heat duty to the feed (kW, MW)

Q_{FUEL} — Heat from fuel in a furnace, boiler, or gas turbine (kW, MW)

Q_H — Heating duty (kW, MW)

Q_{Hmin} — Target for hot utility (kW, MW)

Q_{HE} — Heat engine heat duty (kW, MW)

Q_{HEN} — Heat exchanger network heat duty (kW, MW)

Q_{HP} — Heat pump heat duty (kW, MW)

Q_{LOSS} — Stack loss from furnace, boiler, or gas turbine (kW, MW)

Q_{REACT} — Reactor heating or cooling duty (kW, MW)

Q_{REB} — Reboiler heat duty (kW, MW)

Q_{REC} — Heat recovery (kW, MW)

Q_{SITE} — Site heating demand (kW, MW)

Q_{STEAM} — Heat input for steam generation (kW, MW)

r — Molar ratio (–), or
pressure ratio (–), or
radius (m)

r_i — Pure component property measuring the molecular van der Waals volume for Molecule i in the UNIQUAC Equation (–), or
rate of reaction of Component i ($kmol^{-1} \cdot s^{-1}$), or
recovery of Component i in separation (–)

R — Fractional recovery of a component in separation (–), or

	heat capacity ratio of 1–2 shell-and-tube heat exchanger (–), or raffinate flowrate in liquid–liquid extraction ($kg \cdot s^{-1}$, $kmol \cdot s^{-1}$), or ratio of heat capacity flowrates (–), or reflux ratio for distillation (–), or removal ratio in effluent treatment (–), or residual error (units depend on application), or universal gas constant ($8314.5\ N \cdot m \cdot kmol^{-1} K^{-1} = J \cdot kmol^{-1} K^{-1}$, $8.3145\ kJ \cdot kmol^{-1} \cdot K^{-1}$)
R_{min}	Minimum reflux ratio (–)
R_F	Ratio of actual to minimum reflux ratio (–)
R_{SITE}	Site power-to-heat ratio (–)
ROI	Return on investment (%)
Re	Reynolds number (–)
s	Reactor space velocity (s^{-1}, min^{-1}, h^{-1}), or steam-to-air ratio for gas turbine (–)
S	Entropy ($kJ \cdot K^{-1}$, $kJ \cdot kg^{-1} \cdot K^{-1}$, $kJ \cdot kmol^{-1} \cdot K^{-1}$), or number of streams in a heat exchanger network (–), or reactor selectivity (–), or reboil ratio for distillation (–), or selectivity of a reaction (–), or slack variable in optimization (units depend on application), or solvent flowrate ($kg \cdot s^{-1}$, $kmol \cdot s^{-1}$), or stripping factor in absorption (–)
S_C	Number of cold streams (–)
S_H	Number of hot streams (–)
t	Time (s, h)
T	Temperature (°C, K)
T_{BPT}	Normal boiling point (°C, K)
T_C	Critical temperature (K), or temperature of heat sink (°C, K)
T_{COND}	Condenser temperature (°C, K)
T_E	Equilibrium temperature (°C, K)
T_{EVAP}	Evaporation temperature (°C, K)
T_{FEED}	Feed temperature (°C, K)
T_H	Temperature of heat source (°C, K)
T_R	Reduced temperature T/T_C(–)
T_{REB}	Reboiler temperature (°C, K)
T_S	Stream supply temperature (°C)
T_{SAT}	Saturation temperature of boiling liquid (°C, K)
T_T	Stream target temperature (°C)
T_{TFT}	Theoretical flame temperature (°C, K)
T_W	Wall temperature (°C)
T_{WBT}	Wet bulb temperature (°C)
T^*	Interval temperature (°C)
ΔT_{LM}	Logarithmic mean temperature difference (°C, K)
ΔT_{min}	Minimum temperature difference (°C, K)
ΔT_{SAT}	Difference in saturation temperature (°C, K)
$\Delta T_{THRESHOLD}$	Threshold temperature difference (°C, K)
TAC	Total annual cost ($\$ \cdot y^{-1}$)
TOD	Total oxygen demand ($kg \cdot m^{-3}$, $mg \cdot l^{-1}$)
u_{ij}	Interaction parameter between Molecule i and Molecule j in the UNIQUAC Equation ($kJ \cdot kmol^{-1}$)
U	Overall heat transfer coefficient ($W \cdot m^{-2} \cdot K^{-1}$, $kW \cdot m^{-2} \cdot K^{-1}$)
v	Velocity ($m \cdot s^{-1}$)
v_T	Terminal settling velocity ($m \cdot s^{-1}$)
v_V	Superficial vapor velocity in empty column ($m \cdot s^{-1}$)
V	Molar volume ($m^3 \cdot kmol^{-1}$), or vapor flowrate ($kg \cdot s^{-1}$, $kmol \cdot s^{-1}$), or volume (m^3), or volume of gas or vapor adsorbed ($m^3 \cdot kg^{-1}$)
V_{min}	Minimum vapor flow ($kg \cdot s^{-1}$, $kmol \cdot s^{-1}$)
VF	Vapor fraction (–)
w	Mass of adsorbate per mass of adsorbent (–)
W	Shaft power (kW, MW), or shaft work (kJ, MJ)
W_{GEN}	Power generated (kW, MW)
W_{INT}	Intercept of Willans' Line (kW, MW)

W_{SITE} Site power demand (kW, MW)

x Liquid-phase mole fraction (–) or variable in optimization problem (–)

x_F Mole fraction in the feed (–)

x_D Mole fraction in the distillate (–)

X Reactor conversion (–) or wetness fraction of steam (–)

X_E Equilibrium reactor conversion (–)

X_{OPT} Optimal reactor conversion (–)

X_P Fraction of maximum thermal effectiveness P_{max} allowed in a 1–2 shell-and-tube heat exchanger (–)

XP Cross-pinch heat transfer in heat exchanger network (kW, MW)

y Integer variable in optimization (–), or vapor-phase mole fraction (–)

z Elevation (m), or feed mole fraction (–)

Z Compressibility of a fluid (–)

GREEK LETTERS

α Constant in cubic equation of state (–), or constants in vapor pressure correlation (units depend on which constant), or fraction open of a valve (–)

α_{ij} Ideal separation factor or selectivity of membrane between Components i and j (–), or parameter characterizing the tendency of Molecule i and Molecule j to be distributed in a random fashion in the NRTL equation (–), or relative volatility between Components i and j (–)

α_{LH} Relative volatility between light and heavy key components (–)

β_{ij} Separation factor between Components i and j (–)

γ Ratio of heat capacities for gases and vapors (–)

γ_i Activity coefficient for Component i (–)

δ_M Membrane thickness (m)

ε Extraction factor in liquid–liquid extraction (–), or pipe roughness (mm)

η Carnot factor (–), or efficiency (–)

η_{BOILER} Boiler efficiency (–)

η_{COGEN} Cogeneration efficiency (–)

η_{GT} Efficiency of gas turbine (–)

η_{IS} Isentropic efficiency of compression or expansion (–)

η_{MECH} Mechanical efficiency of steam turbine (–)

η_P Polytropic efficiency of compression or expansion (–)

η_{POWER} Power generation efficiency (–)

η_{ST} Efficiency of steam turbine (–)

θ Fraction of feed permeated through membrane (–), or root of the Underwood Equation (–)

λ Ratio of latent heats of vaporization (–)

λ_{ij} Energy parameter characterizing the interaction of Molecule i with Molecule j ($kJ \cdot kmol^{-1}$)

μ Fluid viscosity ($kg \cdot m^{-1} \cdot s^{-1}$, $mN \cdot s \cdot m^{-2} = cP$)

π Osmotic pressure ($N \cdot m^{-2}$, bar)

ρ Density ($kg \cdot m^{-3}$, $kmol \cdot m^{-3}$)

σ Surface tension ($mN \cdot m^{-1} = mJ \cdot m^{-2} = dyne \cdot cm^{-1}$)

τ Reactor space time (s, min, h) or residence time (s, min, h)

ϕ Cost-weighing factor applied to film heat transfer coefficients to allow for mixed materials of construction, pressure rating, and equipment types in heat exchanger networks (–), or fugacity coefficient (–)

ω Acentric factor (–)

SUBSCRIPTS

B Blowdown, or bottoms in distillation

BFW Boiler feedwater

cont	Contribution
C	Cold stream, or contaminant
CN	Condensing
COND	Condensing conditions
CP	Continuous phase
CW	Cooling water
D	Distillate in distillation
DS	De-superheating
e	Enhanced, or end zone on the shell-side of a heat exchanger, or environment
E	Extract in liquid–liquid extraction
EVAP	Evaporator conditions
EX	Exhaust
final	Final conditions in a batch
F	Feed, or fluid
G	Gas phase
H	Hot stream
HP	Heat pump, or high pressure
i	Component number, or stream number
I	Inside
IS	Isentropic
in	Inlet
j	Component number, or stream number
k	Enthalpy interval number in heat exchanger networks
L	Liquid phase
LP	Low pressure
m	Stage number in distillation and absorption
max	Maximum
min	Minimum
M	Makeup
MIX	Mixture
n	Stage number in distillation and absorption
out	Outlet
O	Outside, or standard conditions
p	Stage number in distillation and absorption
prod	Products of reaction
P	Particle, or permeate
react	Reactants
R	Raffinate in liquid–liquid extraction
REACT	Reaction
S	Solvent in liquid–liquid extraction
SAT	Saturated conditions
SF	Supplementary firing
SUP	Superheated conditions
T	Treatment
TW	Treated water
V	Vapor
w	Window section on the shell-side of a heat exchanger
W	Conditions at the tube wall, or water
∞	Conditions at distillate pinch point

SUPERSCRIPTS

I	Phase *I*
II	Phase *II*
III	Phase *III*
L	Liquid
O	Standard conditions
V	Vapor
*	Adjusted parameter

1 The Nature of Chemical Process Design and Integration

1.1 CHEMICAL PRODUCTS

Chemical products are essential to modern living standards. Almost all aspects of everyday life are supported by chemical products in one way or another. Yet, society tends to take these products for granted, even though a high quality of life fundamentally depends on them.

When considering the design of processes for the manufacture of chemical products, the market into which they are being sold fundamentally influences the objectives and priorities in the design. Chemical products can be divided into three broad classes:

1. *Commodity or bulk chemicals*: These are produced in large volumes and purchased on the basis of chemical composition, purity and price. Examples are sulfuric acid, nitrogen, oxygen, ethylene and chlorine.
2. *Fine chemicals*: These are produced in small volumes and purchased on the basis of chemical composition, purity and price. Examples are chloropropylene oxide (used for the manufacture of epoxy resins, ion-exchange resins and other products), dimethyl formamide (used, for example, as a solvent, reaction medium and intermediate in the manufacture of pharmaceuticals), *n*-butyric acid (used in beverages, flavorings, fragrances and other products) and barium titanate powder (used for the manufacture of electronic capacitors).
3. *Specialty or effect or functional chemicals*: These are purchased because of their effect (or function), rather than their chemical composition. Examples are pharmaceuticals, pesticides, dyestuffs, perfumes and flavorings.

Because commodity and fine chemicals tend to be purchased on the basis of their chemical composition alone, they are *undifferentiated*. For example, there is nothing to choose between 99.9% benzene made by one manufacturer and that made by another manufacturer, other than price and delivery issues. On the other hand, specialty chemicals tend to be purchased on the basis of their effect or function and are therefore *differentiated*. For example, competitive pharmaceutical products are differentiated according to the efficacy of the product, rather than chemical composition. An adhesive is purchased on the basis of its ability to stick things together, rather than its chemical composition and so on.

Chemical Process Design and Integration R. Smith
 ISBNs: 0-471-48680-9 (HB); 0-471-48681-7 (PB)

However, undifferentiated and differentiated should be thought of as relative terms rather than absolute terms for chemical products. In practice, chemicals do not tend to be completely undifferentiated or completely differentiated. Commodity and fine chemical products might have impurity specifications as well as purity specifications. Traces of impurities can, in some cases, give some differentiation between different manufacturers of commodity and fine chemicals. For example, 99.9% acrylic acid might be considered to be an undifferentiated product. However, traces of impurities, at concentrations of a few parts per million, can interfere with some of the reactions in which it is used and can have important implications for some of its uses. Such impurities might differ between different manufacturing processes. Not all specialty products are differentiated. For example, pharmaceutical products like aspirin (acetylsalicylic acid) are undifferentiated. Different manufacturers can produce aspirin and there is nothing to choose between these products, other than the price and differentiation created through marketing of the product.

Scale of production also differs between the three classes of chemical products. Fine and specialty chemicals tend to be produced in volumes less than 1000 $t \cdot y^{-1}$. On the other hand, commodity chemicals tend to be produced in much larger volumes than this. However, the distinction is again not so clear. Polymers are differentiated products because they are purchased on the basis of their mechanical properties, but can be produced in quantities significantly higher than 1000 $t \cdot y^{-1}$.

When a new chemical product is first developed, it can often be protected by a patent in the early years of commercial exploitation. For a product to be eligible to be patented, it must be novel, useful and unobvious. If patent protection can be obtained, this effectively gives the producer a monopoly for commercial exploitation of the product until the patent expires. Patent protection lasts for 20 years from the filing date of the patent. Once the patent expires, competitors can join in and manufacture the product. If competitors cannot wait until the patent expires, then alternative competing products must be developed.

Another way to protect a competitive edge for a new product is to protect it by secrecy. The formula for Coca-Cola has been kept a secret for over 100 years. Potentially, there is no time limit on such protection. However, for the protection through secrecy to be viable, competitors must not be able to reproduce the product from chemical analysis. This is likely to be the case only for certain classes of specialty and food products for which the properties of the product depend on both the chemical composition and the method of manufacture.

Figure 1.1 illustrates different product *life cycles*[1,2]. The general trend is that when a new product is introduced into the market, the sales grow slowly until the market is established and then more rapidly once the market is established. If there is patent protection, then competitors will not be able to exploit the same product commercially until the patent expires, when competitors can produce the same product and take market share. It is expected that competitive products will cause sales to diminish later in the product life cycle until sales become so low that a company would be expected to withdraw from the market. In Figure 1.1, Product *A* appears to be a poor product that has a short life with low sales volume. It might be that it cannot compete well with other competitive products, and alternative products quickly force the company out of that business. However, a low sales volume is not the main criterion to withdraw from the market. It might be that a product with low volume finds a market niche and can be sold for a high value. On the other hand, if it were competing with other products with similar functions in the same market sector, which keeps both the sale price and volume low, then it would seem wise to withdraw from the market. Product *B* in Figure 1.1 appears to be a better product, showing a longer life cycle and higher sales volume. This has patent protection but sales decrease rapidly after patent protection is lost, leading to loss of market through competition. Product *C* in Figure 1.1 is a still better product. This shows high sales volume with the life of the product extended through reformulation of the product[1]. Finally, Product *D* in Figure 1.1 shows a product life cycle that is typical of commodity chemicals. Commodity chemicals tend not to exhibit the same kind of life cycles as fine and specialty chemicals. In the early years of the commercial exploitation, the sales volume grows rapidly to a high volume, but then does not decline and enters a mature period of slow growth, or, in some exceptional cases, slow decline. This is because commodity chemicals tend to have a diverse range of uses. Even though competition might take away some end uses, new end uses are introduced, leading to an extended life cycle.

The different classes of chemical products will have very different *added value* (the difference between the selling price of the product and the purchase cost of raw materials). Commodity chemicals tend to have low added value, whereas fine and specialty chemicals tend to have high added value. Commodity chemicals tend to be produced in large volumes with low added value, while fine and specialty chemicals tend to be produced in small volumes with high added value.

Because of this, when designing a process for a commodity chemical, it is usually important to keep operating costs as low as possible. The capital cost of the process will tend to be high relative to a process for fine or specialty chemicals because of the scale of production.

When designing a process for specialty chemicals, priority tends to be given to the product, rather than to the process. This is because the unique function of the product must be protected. The process is likely to be small scale and operating costs tend to be less important than with commodity chemical processes. The capital cost of the process will be low relative to commodity chemical processes because of the scale. The time to

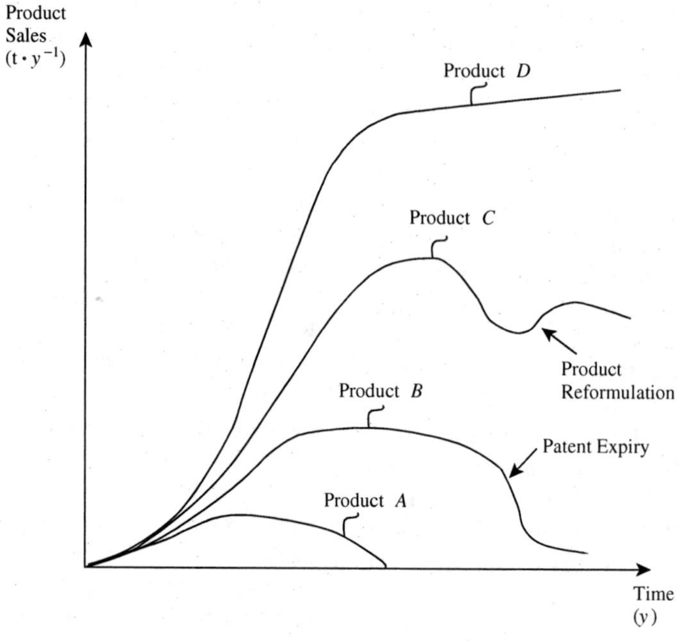

Figure 1.1 Product life cycles. (Adapted from Sharratt PN, 1997, Handbook of Batch Process Design, Blackie Academic and Professional by permission).

market the product is also likely to be important with specialty chemicals, especially if there is patent protection. If this is the case, then anything that shortens the time from basic research, through product testing, pilot plant studies, process design, construction of the plant to product manufacture will have an important influence on the overall project profitability.

All this means that the priorities in process design are likely to differ significantly, depending on whether a process is being designed for the manufacture of a commodity, fine or specialty chemical. In commodity chemicals, there is likely to be relatively little product innovation, but intensive process innovation. Also, equipment will be designed for a specific process step. On the other hand, the manufacture of fine and specialty chemicals might involve:

- selling into a market with low volume,
- short product life cycle,
- a demand for a short time to market, and therefore, less time is available for process development, with product and process development proceeding simultaneously.

Because of this, the manufacture of fine and specialty chemicals is often carried out in multipurpose equipment, perhaps with different chemicals being manufactured in the same equipment at different times during the year. The life of the equipment might greatly exceed the life of the product.

The development of pharmaceutical products is such that high-quality products must be manufactured during the development of the process to allow safety and clinical studies to be carried out before full-scale production. Pharmaceutical production represents an extreme case of process design in which the regulatory framework controlling production makes it difficult to make process changes, even during the development stage. Even if significant improvements to processes for pharmaceuticals can be suggested, it might not be feasible to implement them, as such changes might prevent or delay the process from being licensed for production.

1.2 FORMULATION OF THE DESIGN PROBLEM

Before a process design can be started, the design problem must be formulated. Formulation of the design problem requires a product specification. If a well-defined chemical product is to be manufactured, then the specification of the product might appear straightforward (e.g. a purify specification). However, if a specialty product is to be manufactured, it is the functional properties that are important, rather than the chemical properties, and this might require a *product design* stage in order to specify the product[3]. The initial statement of the design problem is often ill defined. For example, the design team could be asked to expand the production capacity of an existing plant that produces a chemical that is a precursor to a polymer product, which is also produced by the company. This results from an increase in the demand for the polymer product and the plant producing the precursor currently being operated at its maximum capacity. The designer might well be given a specification for the expansion. For example, the marketing department might assess that the market could be expanded by 30% over a two-year period, which would justify a 30% expansion in the process for the precursor. However, the 30% projection can easily be wrong. The economic environment can change, leading to the projected increase being either too large or too small. It might also be possible to sell the polymer precursor in the market to other manufacturers of the polymer and justify an expansion even larger than 30%. If the polymer precursor can be sold in the marketplace, is the current purity specification of the company suitable for the marketplace? Perhaps the marketplace demands a higher purity than what is currently the company specification. Perhaps the current specification is acceptable, but if the specification could be improved, the product could be sold for a higher value and/or at a greater volume. An option might be to not expand the production of the polymer precursor to 30%, but instead to purchase it from the market. If it is purchased from the market, is it likely to be up to the company specifications, or will it need some purification before it is suitable for the company's polymer process? How reliable will the market source be? All these uncertainties are related more to market supply and demand issues than to specific process design issues.

Closer examination of the current process design might lead to the conclusion that the capacity can be expanded by 10% with a very modest capital investment. A further increase to 20% would require a significant capital investment, but an expansion to 30% would require an extremely large capital investment. This opens up further options. Should the plant be expanded by 10% and a market source identified for the balance? Should the plant be expanded to 20% similarly? If a real expansion in the market place is anticipated and expansion to 30% would be very expensive, why not be more aggressive and instead of expanding the existing process, build an entirely new process? If a new process is to be built, then what should be the process technology? New process technology might have been developed since the original plant was built that enables the same product to be manufactured at a much lower cost. If a new process is to be built, where should it be built? It might make more sense to build it in another country that would allow lower operating costs, and the product could be shipped back to be fed to the existing polymer process. At the same time, this might stimulate the development of new markets in other countries, in which case, what should be the capacity of the new plant?

From all of these options, the design team must formulate a number of plausible design options. Thus, from the initial ill-defined problem, the design team must create a series of very specific options and these should then be compared on the basis of a common set of assumptions regarding, for example, raw materials prices and product prices. Having specified an option, this gives the design team a well-defined problem to which the methods of engineering and economic analysis can be applied.

In examining a design option, the design team should start out by examining the problem at the highest level, in terms of its feasibility with the minimum of detail to ensure the design option is worth progressing[4]. Is there a large difference between the value of the product and the cost of the raw materials? If the overall feasibility looks attractive, then more detail can be added, the option re-evaluated, further detail added, and so on. Byproducts might play a particularly important role in the economics. It might be that the current process produces some byproducts that can be sold in small quantities to the market. But, as the process is expanded, there might be market constraints for the new scale of production. If the byproducts cannot be sold, how does this affect the economics?

If the design option appears to be technically and economically feasible, then additional detail can be considered. Material and energy balances can be formulated to give a better definition to the inner workings of the process and a more detailed process design can be developed. The design calculations for this will normally be solved to a high level of precision. However, a high level of precision cannot usually be justified in terms of the operation of the plant after it has been built. The plant will almost never work precisely at its original design flowrates, temperatures, pressures and compositions. This might be because the raw materials are slightly different than what is assumed in the design. The physical properties assumed in the calculations might have been erroneous in some way, or operation at the original design conditions might create corrosion or fouling problems, or perhaps the plant cannot be controlled adequately at the original conditions, and so on, for a multitude of other possible reasons. The instrumentation on the plant will not be able to measure the flowrates, temperatures, pressures and compositions as accurately as the calculations performed. High precision might be required for certain specific parts of the design. For example, the polymer precursor might need certain impurities to be very tightly controlled, perhaps down to the level of parts per million. It might be that some contaminant in a waste stream might be exceptionally environmentally harmful and must be extremely well defined in the design calculations.

Even though a high level of precision cannot be justified in many cases in terms of the plant operation, the design calculations will normally be carried out to a reasonably high level of precision. The value of precision in design calculations is that the consistency of the calculations can be checked to allow errors or poor assumptions to be identified. It also allows the design options to be compared on a valid like-for-like basis.

Because of all the uncertainties in carrying out a design, the specifications are often increased beyond those indicated by the design calculations and the plant is *overdesigned*, or *contingency* is added, through the application of *safety factors* to the design. For example, the designer might calculate the number of distillation plates required for a distillation separation using elaborate calculations to a high degree of precision, only to add an arbitrary extra 10% to the number of plates for contingency. This allows for the feed to the unit not being exactly as specified, errors in the physical properties, upset conditions in the plant, control requirements, and so on. If too little contingency is added, the plant might not work. If too much contingency is added, the plant will not only be unnecessarily expensive, but too much overdesign might make the plant difficult to operate and might lead to a less efficient plant. For example, the designer might calculate the size of a heat exchanger and then add in a large contingency and significantly oversize the heat exchanger. The lower fluid velocities encountered by the oversized heat exchanger can cause it to have a poorer performance and to foul up more readily than a smaller heat exchanger. Thus, a balance must be made between different risks.

In summary, the original problem posed to process design teams is often ill-defined, even though it might appear to be well defined in the original design specification. The design team must then formulate a series of plausible design options to be screened by the methods of engineering and economic analysis. These design options are formulated into very specific design problems. Some design options might be eliminated early by high-level arguments or simple calculations. Others will require more detailed examination. In this way, the design team turns the ill-defined problem into a well-defined one for analysis. To allow for the many unquantifiable uncertainties, overdesign is used. Too little overdesign might lead to the plant not working. Too much overdesign will lead to the plant becoming unnecessarily expensive, and perhaps difficult to operate and less efficient. A balance must be made between different risks.

Consider the basic features of the design of chemical processes now.

1.3 CHEMICAL PROCESS DESIGN AND INTEGRATION

In a chemical process, the transformation of raw materials into desired chemical products usually cannot be achieved in a single step. Instead, the overall transformation is broken down into a number of steps that provide intermediate transformations. These are carried out through reaction, separation, mixing, heating, cooling, pressure change, particle size reduction or enlargement. Once individual steps have been selected, they must be interconnected to carry out the

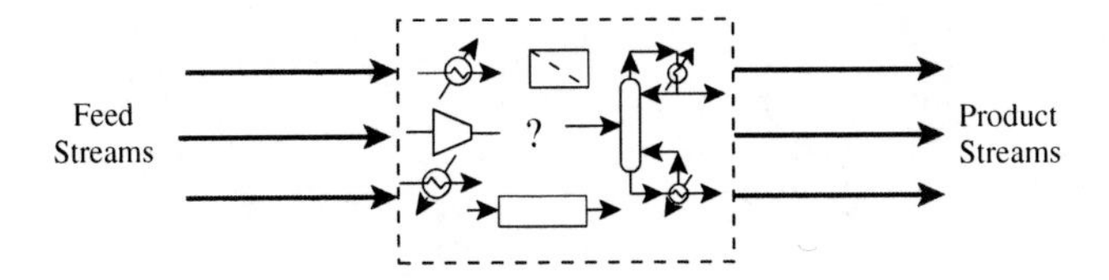

(a) Process design starts with the synthesis of a process to convert raw materials into desired products.

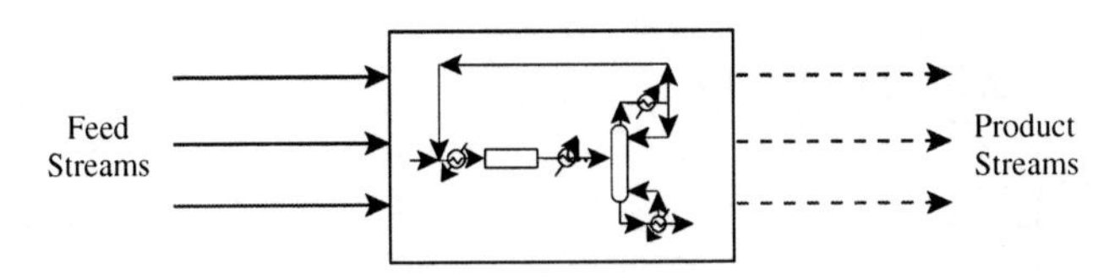

(b) Simulation predicts how a process would behave if it was constructed.

Figure 1.2 Synthesis is the creation of a process to transform feed streams into product streams. Simulation predicts how it would behave if it was constructed.

overall transformation (Figure 1.2a). Thus, the *synthesis* of a chemical process involves two broad activities. First, individual transformation steps are selected. Second, these individual transformations are interconnected to form a complete process that achieves the required overall transformation. A *flowsheet* is a diagrammatic representation of the process steps with their interconnections.

Once the flowsheet structure has been defined, a *simulation* of the process can be carried out. A simulation is a mathematical model of the process that attempts to predict how the process would behave if it were constructed (Figure 1.2b). Having created a model of the process, the flowrates, compositions, temperatures and pressures of the feeds can be assumed. The simulation model then predicts the flowrates, compositions, temperatures, and pressures of the products. It also allows the individual items of equipment in the process to be sized and predicts, for example, how much raw material is being used or how much energy is being consumed. The performance of the design can then be evaluated. There are many facets to the evaluation of performance. Good economic performance is an obvious first criterion, but it is certainly not the only one.

Chemical processes should be designed as part of a sustainable industrial activity that retains the capacity of ecosystems to support both life and industrial activity into the future. Sustainable industrial activity must meet the needs of the present, without compromising the needs of future generations. For chemical process design, this means that processes should use raw materials as efficiently as is economic and practicable, both to prevent the production of waste that can be environmentally harmful and to preserve the reserves of raw materials as much as possible. Processes should use as little energy as is economic and practicable, both to prevent the build-up of carbon dioxide in the atmosphere from burning fossil fuels and to preserve the reserves of fossil fuels. Water must also be consumed in sustainable quantities that do not cause deterioration in the quality of the water source and the long-term quantity of the reserves. Aqueous and atmospheric emissions must not be environmentally harmful, and solid waste to landfill must be avoided.

The process must also meet required health and safety criteria. Start-up, emergency shutdown and ease of control are other important factors. Flexibility, that is, the ability to operate under different conditions, such as differences in feedstock and product specification, may be important. Availability, that is, the number of operating hours per year, may also be critically important. Uncertainty in the design, for example, resulting from poor design data, or uncertainty in the economic data, might guide the design away from certain options. Some of these factors, such as economic performance, can be readily quantified; others, such as safety, often cannot. Evaluation of the factors that are not readily quantifiable, the intangibles, requires the judgment of the design team.

Once the basic performance of the design has been evaluated, changes can be made to improve the performance; the process is *optimized.* These changes might involve the synthesis of alternative structures, that is, *structural optimization*. Thus, the process is simulated and evaluated again, and so on, optimizing the structure. Alternatively, each structure can be subjected to *parameter optimization* by changing operating conditions within that structure.

1.4 THE HIERARCHY OF CHEMICAL PROCESS DESIGN AND INTEGRATION

Consider the process illustrated in Figure 1.3[5]. The process requires a reactor to transform the *FEED* into *PRODUCT* (Figure 1.3a). Unfortunately, not all the *FEED* reacts. Also, part of the *FEED* reacts to form *BYPRODUCT* instead of the desired *PRODUCT.* A separation system is needed to isolate the *PRODUCT* at the required purity. Figure 1.3b shows one possible separation system consisting of two distillation columns. The unreacted *FEED* in Figure 1.3b is recycled, and the *PRODUCT* and *BYPRODUCT* are removed from the process. Figure 1.3b shows a flowsheet where all heating and cooling is provided by external *utilities* (steam and cooling water in this case). This flowsheet is probably too inefficient in its use of energy, and heat would be recovered. Thus, *heat integration* is carried out to exchange heat between those streams that need to be cooled and those that need to be heated. Figure 1.4[5] shows two possible designs for the *heat exchanger network*, but many other heat integration arrangements are possible.

The flowsheets shown in Figure 1.4 feature the same reactor design. It could be useful to explore the changes in reactor design. For example, the size of the reactor could be increased to increase the amount of *FEED* that reacts[5].

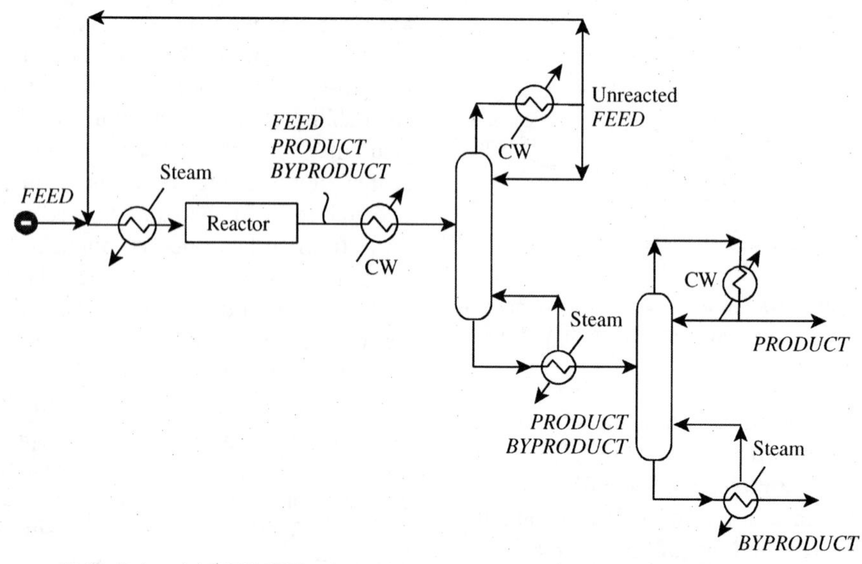

(a) A reactor transforms *FEED* into *PRODUCT* and *BYPRODUCT*.

(b) To isolate the *PRODUCT* and recycle unreacted *FEED* a separation system is needed.

Figure 1.3 Process design starts with the reactor. The reactor design dictates the separation and recycle problem. (From Smith R and Linnhoff B, 1998, *Trans IChemE ChERD*, **66**:195 by permission of the Institution of Chemical Engineers).

Now, there is not only much less *FEED* in the reactor effluent but also more *PRODUCT* and *BYPRODUCT*. However, the increase in *BYPRODUCT* is larger than the increase in *PRODUCT*. Thus, although the reactor has the same three components in its effluent as the reactor in Figure 1.3a, there is less *FEED*, more *PRODUCT* and significantly more *BYPRODUCT*. This change in reactor design generates a different task for the separation system, and it is possible that a separation system different from that shown in Figures 1.3 and 1.4 is now appropriate. Figure 1.5 shows a possible alternative. This also uses two distillation columns, but the separations are carried out in a different order.

Figure 1.5 shows a flowsheet without any heat integration for the different reactor and separation system. As before, this is probably too inefficient in the use of energy, and heat integration schemes can be explored. Figure 1.6[5] shows two of the many possible flowsheets.

Different complete flowsheets can be evaluated by simulation and costing. On this basis, the flowsheet in Figure 1.4b might be more promising than the flowsheets in Figures 1.4a, 1.6a and b. However, the best flowsheet cannot be identified without first optimizing the operating conditions for each. The flowsheet in Figure 1.6b might have greater scope for improvement than that in Figure 1.4b, and so on.

Thus, the complexity of chemical process synthesis is twofold. First, can all possible structures be identified? It might be considered that all the structural options can be found by inspection, at least all of the significant ones. The fact that even long-established processes are still being improved bears evidence to just how difficult this is. Second, can each structure be optimized for a valid comparison? When optimizing the structure, there may be many ways in which each individual task can be performed and many ways in which the individual tasks can be interconnected. This means that the operating conditions for a multitude of structural options must be simulated and optimized. At first sight, this appears to be an overwhelmingly complex problem.

It is helpful when developing a methodology if there is a clearer picture of the nature of the problem. If the process requires a reactor, this is where the design starts. This is likely to be the only place in the process where raw materials are converted into products. The chosen reactor design produces a mixture of unreacted feed materials, products and byproducts that need separating. Unreacted feed material is recycled. The reactor design dictates the separation and recycle problem. Thus, design of the separation and recycle system follows the reactor design. The reactor and separation and recycle system designs together define the process for heating and cooling duties.

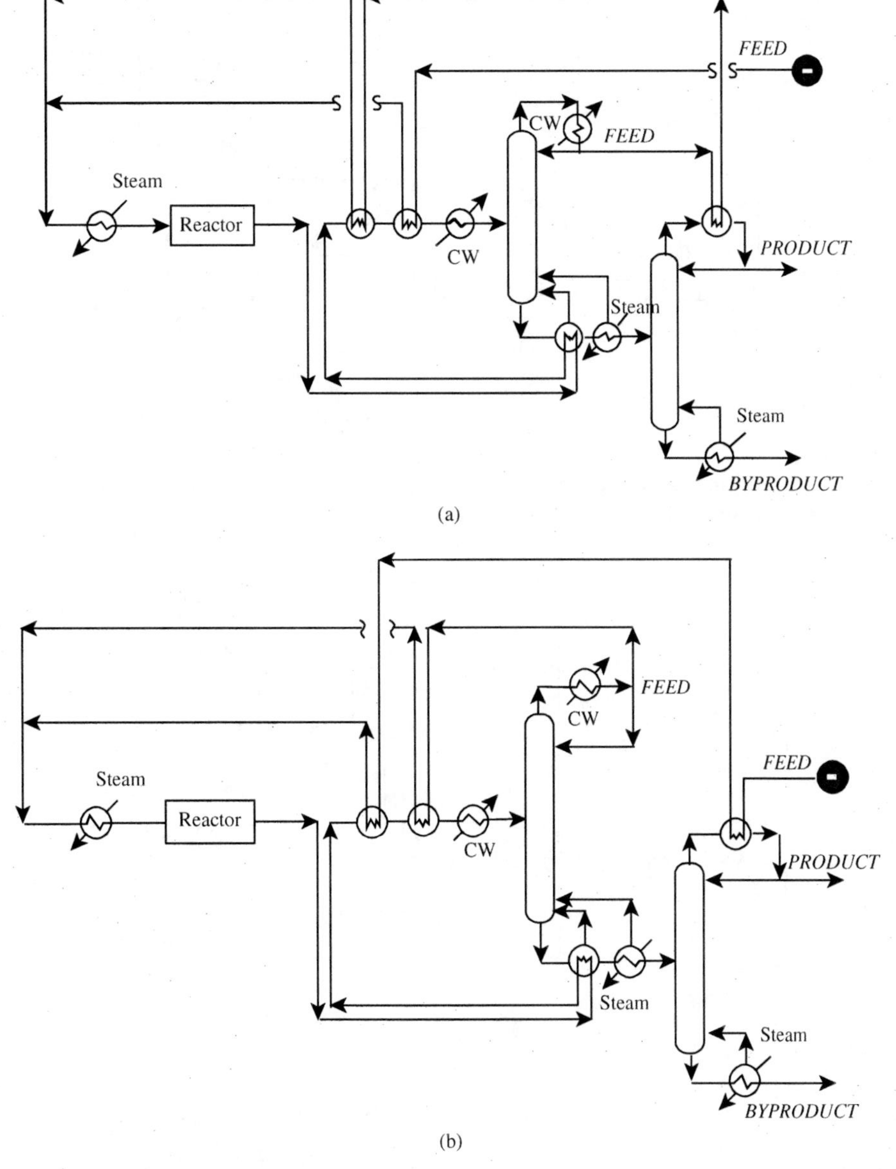

Figure 1.4 For a given reactor and separator design there are different possibilities for heat integration. (From Smith R and Linnhoff B, 1998, *Trans IChemE ChERD*, **66**:195 by permission of the Institution of Chemical Engineers).

Thus, heat exchanger network design comes next. Those heating and cooling duties that cannot be satisfied by heat recovery, dictate the need for external heating and cooling *utilities* (furnace heating, use of steam, steam generation, cooling water, air-cooling or refrigeration). Thus, utility selection and design follows the design of the heat recovery system. The selection and design of the utilities is made more complex by the fact that the process will most likely operate within the context of a site comprising a number of different processes that are all connected to a common utility system. The process and the utility system will both need water, for example, for steam generation, and will also produce aqueous effluents that will have to be brought to a suitable quality for discharge. Thus, the design of the water and aqueous effluent treatment system comes last. Again, the water and effluent treatment system must be considered at the site level as well as the process level.

This hierarchy can be represented symbolically by the layers of the "onion diagram" shown in Figure 1.7[6]. The diagram emphasizes the sequential, or hierarchical, nature of process design. Other ways to represent the hierarchy have also been suggested[4].

Some processes do not require a reactor, for example, some processes just involve separation. Here, the design starts with the separation system and moves outward to the heat exchanger network, utilities and so on. However, the same basic hierarchy prevails.

The synthesis of the correct structure and the optimization of parameters in the design of the reaction and separation systems are often the most important tasks of

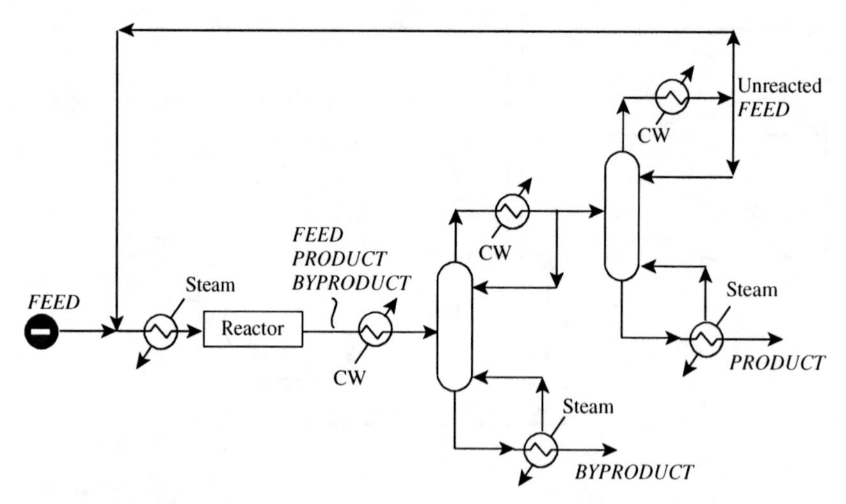

Figure 1.5 Changing the reactor dictates a different separation and recycle problem. (From Smith R and Linnhoff B, 1988, *Trans IChemE ChERD*, **66**: 195 by permission of the Institution of Chemical Engineers).

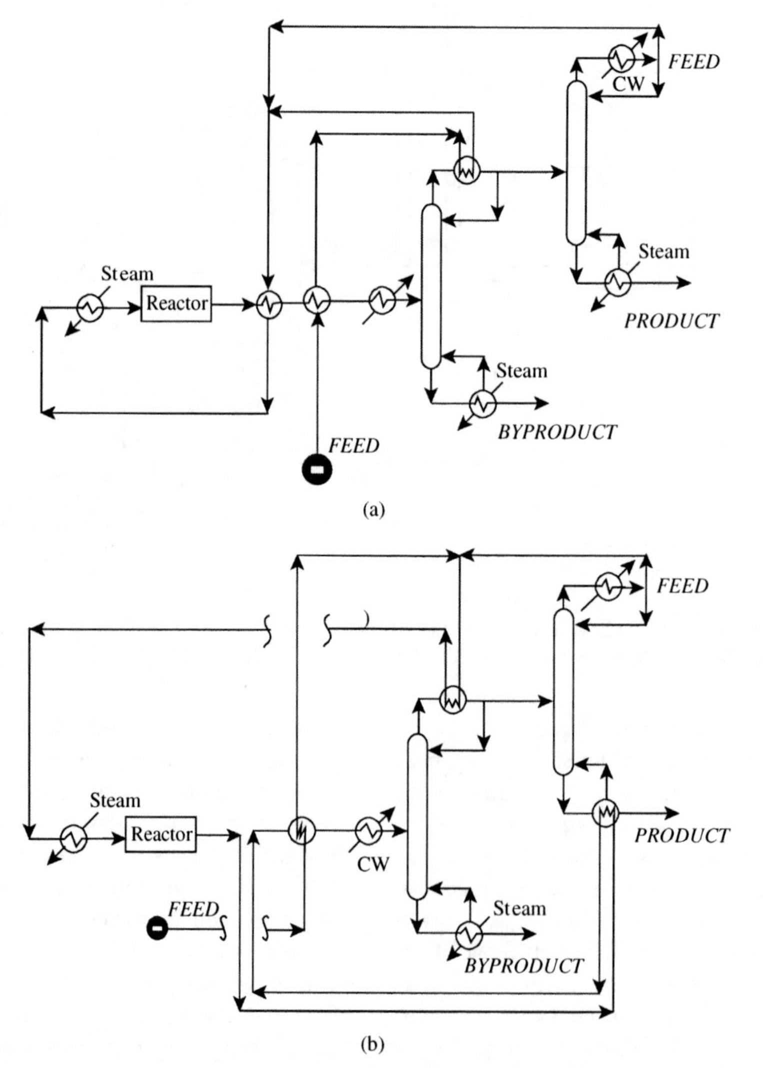

Figure 1.6 A different reactor design not only leads to a different separation system but additional possibilities for heat integration. (From Smith R and Linnhoff B, 1988, *Trans IChemE ChERD*, **66**: 195 by permission of the Institution of Chemical Engineers).

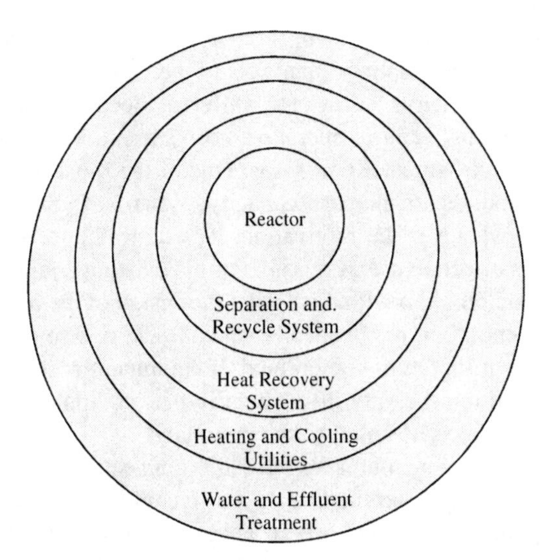

Figure 1.7 The onion model of process design. A reactor is needed before the separation and recycle system can be designed and so on.

process design. Usually, there are many options, and it is impossible to fully evaluate them unless a complete design is furnished for the "outer layers" of the onion. For example, it is not possible to assess which is better, the basic scheme from Figure 1.3b or that from Figure 1.5, without fully evaluating all possible designs, such as those shown in Figures 1.4a and b and Figures 1.6a and b, all completed, including utilities. Such a complete search is normally too time consuming to be practical.

Later, an approach will be presented in which some early decisions (i.e. decisions regarding reactor and separator options) can be evaluated without a complete design for the "outer layers".

1.5 CONTINUOUS AND BATCH PROCESSES

When considering the processes in Figures 1.3 to 1.5, an implicit assumption was made that the processes operated continuously. However, not all processes operate continuously. In a *batch* process, the main steps operate discontinuously. In contrast with a continuous process, a batch process does not deliver its product continuously but in discrete amounts. This means that heat, mass, temperature, concentration and other properties vary with time. In practice, most batch processes are made up of a series of batch and *semicontinuous* steps. A semicontinuous step runs continuously with periodic start-ups and shutdowns.

Consider the simple process shown in Figure 1.8. Feed material is withdrawn from storage using a pump. The feed material is preheated in a heat exchanger before being fed to a batch reactor. Once the reactor is full, further heating takes place inside the reactor by passing steam into the reactor jacket, before the reaction proceeds. During the later stages of the reaction, cooling water is applied to the reactor jacket. Once the reaction is complete, the reactor product is withdrawn using a pump. The reactor product is cooled in a heat exchanger before going to storage.

The first two steps, pumping for reactor filling and feed preheating are both semicontinuous. The heating inside the reactor, the reaction itself and the cooling using the reactor jacket are all batch. The pumping to empty the reactor and the product-cooling step are again semicontinuous.

The hierarchy in batch process design is no different from that in continuous processes and the hierarchy illustrated in Figure 1.7 prevails for batch processes also. However, the time dimension brings constraints that do not present a problem in the design of continuous processes. For example, heat recovery might be considered for the process in Figure 1.8. The reactor effluent (that requires cooling) could be used to preheat the incoming feed to the reactor (that requires heating). Unfortunately, even if the reactor effluent is at a high enough temperature to allow this, the reactor feeding and emptying take place at different times, meaning that this will not be possible without some way to store the heat. Such heat storage is possible but usually uneconomic, especially for small-scale processes.

If a batch process manufactures only a single product, then the equipment can be designed and optimized for

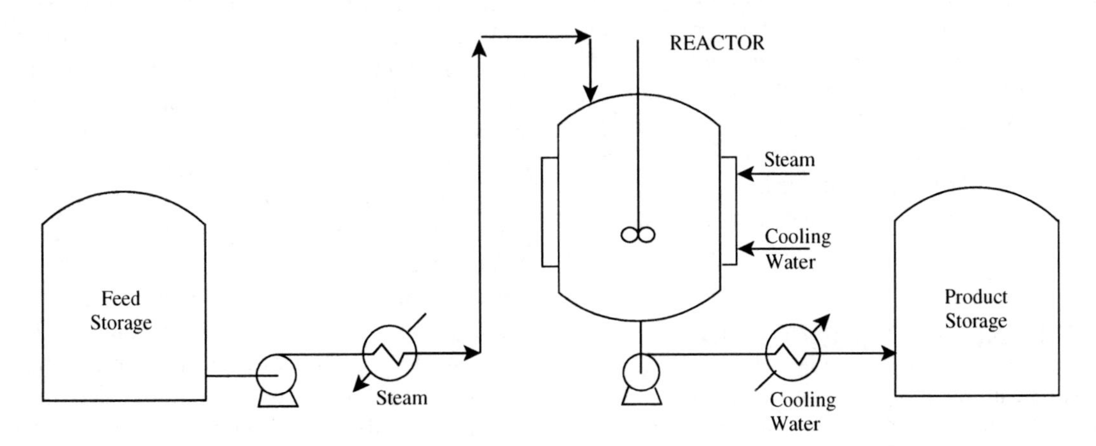

Figure 1.8 A simple batch process.

that product. The dynamic nature of the process creates additional challenges for design and optimization. It might be that the optimization calls for variations in the conditions during the batch through time, according to some *profile*. For example, the temperature in a batch reactor might need to be increased or decreased as the batch progresses.

Multiproduct batch processes, with a number of different products manufactured in the same equipment, present even bigger challenges for design and optimization[7]. Different products will demand different designs, different operating conditions and, perhaps, different trajectories for the operating conditions through time. The design of equipment for multiproduct plants will thus require a compromise to be made across the requirements of a number of different products. The more flexible the equipment and the configuration of the equipment, the more it will be able to adapt to the optimum requirements of each product.

Batch processes

- are economical for small volumes;
- are flexible in accommodating changes in product formulation;
- are flexible in changing production rate by changing the number of batches made in any period of time;
- allow the use of standardized multipurpose equipment for the production of a variety of products from the same plant;
- are best if equipment needs regular cleaning because of fouling or needs regular sterilization;
- are amenable to direct scale-up from the laboratory and
- allow product identification. Each batch of product can be clearly identified in terms of when it was manufactured, the feeds involved and conditions of processing. This is particularly important in industries such as pharmaceuticals and foodstuffs. If a problem arises with a particular batch, then all the products from that batch can be identified and withdrawn from the market. Otherwise, all the products available in the market would have to be withdrawn.

One of the major problems with batch processing is batch-to-batch conformity. Minor changes to the operation can mean slight changes in the product from batch to batch. Fine and specialty chemicals are usually manufactured in batch processes. Yet, these products often have very tight tolerances for impurities in the final product and demand batch-to-batch variation being minimized.

Batch processes will be considered in more detail in Chapter 14.

1.6 NEW DESIGN AND RETROFIT

There are two situations that can be encountered in process design. The first is in the design of *new plant* or *grassroot* design. In the second, the design is carried out to modify an existing plant in *retrofit* or *revamp*. The motivation to retrofit an existing plant could be, for example, to increase capacity, allow for different feed or product specifications, reduce operating costs, improve safety or reduce environmental emissions. One of the most common motivations is to increase capacity. When carrying out a retrofit, whatever the motivation, it is desirable to try and make as effective use as possible of the existing equipment. The basic problem with this is that the design of the existing equipment might not be ideally suited to the new role that it will be put to. On the other hand, if equipment is reused, it will avoid unnecessary investment in new equipment, even if it is not ideally suited to the new duty.

When carrying out a retrofit, the connections between the items of equipment can be reconfigured, perhaps adding new equipment where necessary. Alternatively, if the existing equipment differs significantly from what is required in the retrofit, then in addition to reconfiguring the connections between the equipment, the equipment itself can be modified. Generally, the fewer the modifications to both the connections and the equipment, the better.

The most straightforward design situations are those of grassroot design as it has the most freedom to choose the design options and the size of equipment. In retrofit, the design must try to work within the constraints of existing equipment. Because of this, the ultimate goal of the retrofit design is often not clear. For example, a design objective might be given to increase the capacity of a plant by 50%. At the existing capacity limit of the plant, at least one item of equipment must be at its maximum capacity. Most items of equipment are likely to be below their maximum capacity. The differences in the spare capacity of different items of equipment in the existing design arises from errors in the original design data, different design allowances (or *contingency*) in the original design, changes to the operation of the plant relative to the original design, and so on. An item of equipment at its maximum capacity is the *bottleneck* to prevent increased capacity. Thus, to overcome the bottleneck or *debottleneck*, the item of equipment is modified, or replaced with new equipment with increased capacity, or a new item is placed in parallel or series with the existing item, or the connections between existing equipment are reconfigured, or a combination of all these actions is taken. As the capacity of the plant is increased, different items of equipment will reach their maximum capacity. Thus, there will be thresholds in the plant capacity, created by the limits in different items of equipment. All equipment with capacity less than the threshold must be modified in some way, or the plant reconfigured, to overcome the threshold. To overcome each threshold requires capital investment. As capacity is increased from the existing limit, ultimately, it is likely that it will be prohibitive for the investment to overcome one of the design thresholds. This is likely to become the design limit, as opposed to the original remit of a 50% increase in capacity in the example.

1.7 APPROACHES TO CHEMICAL PROCESS DESIGN AND INTEGRATION

In broad terms, there are two approaches to chemical process design and integration:

1. *Building an irreducible structure*: The first approach follows the "onion logic", starting the design by choosing a reactor and then moving outward by adding a separation and recycle system, and so on. At each layer, decisions must be made on the basis of the information available at that stage. The ability to look ahead to the completed design might lead to different decisions. Unfortunately, this is not possible, and, instead, decisions must be based on an incomplete picture.

This approach to creation of the design involves making a series of best local decisions. This might be based on the use of *heuristics* or *rules of thumb* developed from experience[4] on a more systematic approach. Equipment is added only if it can be justified economically on the basis of the information available, albeit an incomplete picture. This keeps the structure *irreducible*, and features that are technically or economically redundant are not included.

There are two drawbacks to this approach:

(a) Different decisions are possible at each stage of the design. To be sure that the best decisions have been made, the other options must be evaluated. However, each option cannot be evaluated properly without completing the design for that option and optimizing the operating conditions. This means that many designs must be completed and optimized in order to find the best.

(b) Completing and evaluating many options gives no guarantee of ultimately finding the best possible design, as the search is not exhaustive. Also, complex interactions can occur between different parts of a flowsheet. The effort to keep the system simple and not add features in the early stages of design may result in missing the benefit of interactions between different parts of the flowsheet in a more complex system.

The main advantage of this approach is that the design team can keep control of the basic decisions and interact as the design develops. By staying in control of the basic decisions, the intangibles of the design can be included in the decision making.

2. *Creating and optimizing a superstructure*. In this approach, a *reducible* structure, known as a *superstructure*, is first created that has embedded within it all feasible process options and all feasible interconnections that are candidates for an optimal design structure. Initially, redundant features are built into the superstructure. As an example, consider Figure 1.9[8]. This shows one possible structure of a process for the manufacture of benzene from the reaction between toluene and hydrogen. In Figure 1.9, the hydrogen enters the process with a small amount of methane as an impurity. Thus, in Figure 1.9, the option of either purifying the hydrogen feed with a membrane or of passing it directly to the process is embedded. The hydrogen and toluene are mixed and preheated to reaction temperature. Only a furnace has been considered feasible in this case because of the high temperature required. Then, the two alternative reactor options, isothermal and adiabatic reactors, are embedded, and so on. Redundant features have been included in an effort to ensure that all features that could be part of an optimal solution have been included.

The design problem is next formulated as a mathematical model. Some of the design features are continuous, describing the operation of each unit (e.g. flowrate, composition, temperature and pressure), its size (e.g. volume, heat transfer area, etc.) as well as the costs or profits associated with the units. Other features are discrete (e.g. a connection in the flowsheet is included or not, a membrane separator is included or not). Once the problem is formulated mathematically, its solution is carried out through the implementation of an optimization algorithm. An *objective function* is maximized or minimized (e.g. profit is maximized or cost is minimized) in a *structural and parameter* optimization. The optimization justifies the existence of structural features and deletes those features from the structure that cannot be justified economically. In this way, the structure is reduced in complexity. At the same time, the operating conditions and equipment sizes are also optimized. In effect, the discrete decision-making aspects of process design are replaced by a discrete/continuous optimization. Thus, the initial structure in Figure 1.9 is optimized to reduce the structure to the final design shown in Figure 1.10[8]. In Figure 1.10, the membrane separator on the hydrogen feed has been removed by optimization, as has the isothermal reactor and many other features of the initial structure shown in Figure 1.9.

There are a number of difficulties associated with this approach:

(a) The approach will fail to find the optimal structure if the initial structure does not have the optimal structure embedded somewhere within it. The more options included, the more likely it will be that the optimal structure has been included.

(b) If the individual unit operations are represented accurately, the resulting mathematical model will be extremely large and the objective function that must be optimized will be extremely irregular. The profile of the objective function can be like the terrain in a

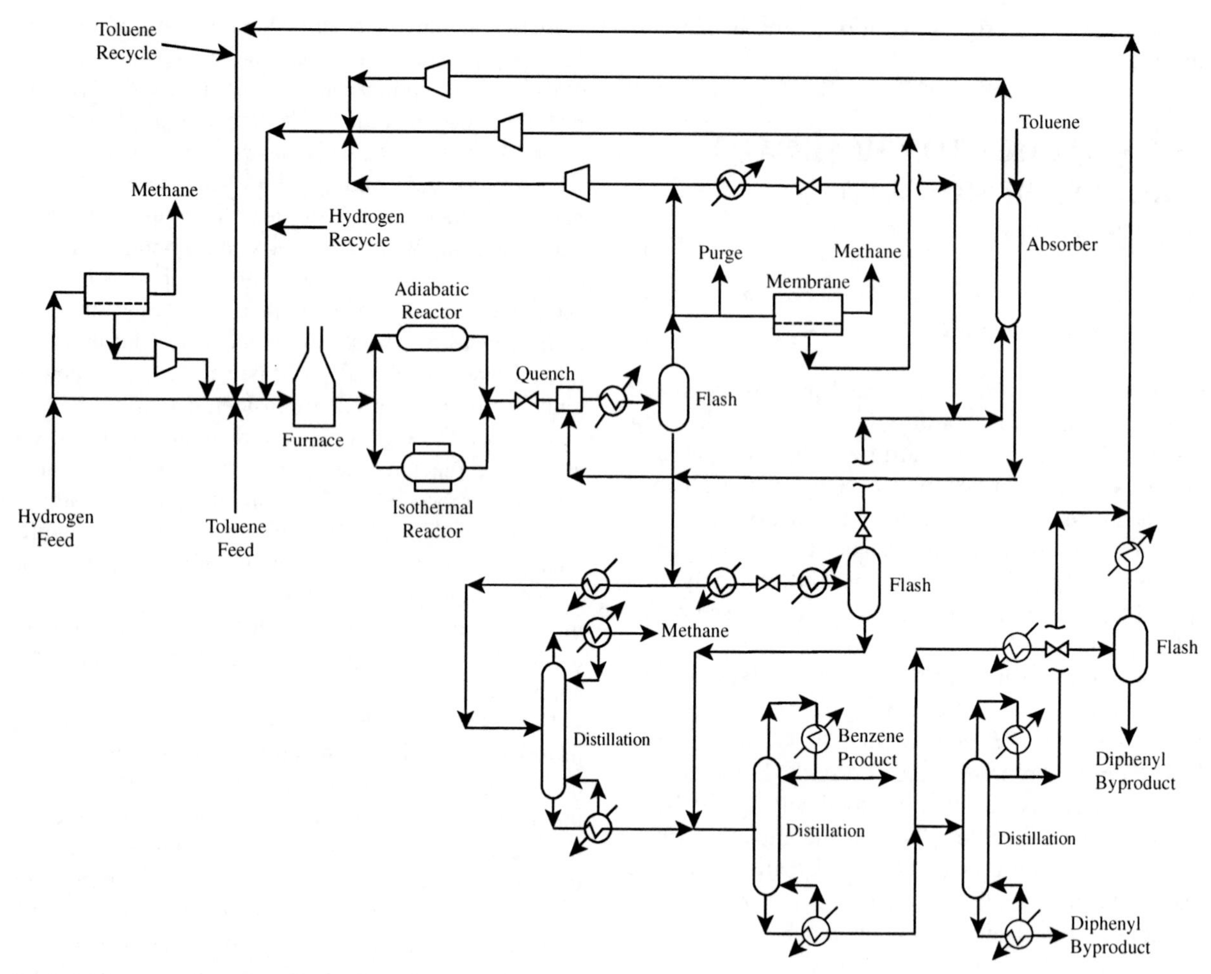

Figure 1.9 A superstructure for the manufacture of benzene from toluene and hydrogen incorporating some redundant features. (From Kocis GR and Grossman IE, 1988, *Comp Chem Eng*, **13**: 797, reproduced by permission.).

range of mountains with many peaks and valleys. If the objective function is to be maximized (e.g. maximize profit), each peak in the mountain range represents a *local optimum* in the objective function. The highest peak represents the *global optimum*. Optimization requires searching around the mountains in a thick fog to find the highest peak, without the benefit of a map and only a compass to tell the direction and an altimeter to show the height. On reaching the top of any peak, there is no way of knowing whether it is the highest peak because of the fog. All peaks must be searched to find the highest. There are crevasses to fall into that might be impossible to climb out of.

Such problems can be overcome in a number of ways. The first way is by changing the model such that the solution space becomes more regular, making the optimization simpler. This most often means simplifying the mathematical model. A second way is by repeating the search many times, but starting each new search from a different initial location. A third way exploits mathematical transformations and bounding techniques for some forms of mathematical expression to allow the global optimum to be found[9]. A fourth way is by allowing the optimization to search the solution space in a series of discrete moves that initially allow the possibility of going downhill, away from an optimum point, as well as uphill. As the search proceeds, the ability of the algorithm to move downhill must be gradually taken away. These problems will be dealt with in more detail in Chapter 3.

(c) The most serious drawback of this approach is that the design engineer is removed from the decision making. Thus, the many intangibles in design, such as safety and layout, which are difficult to include in the mathematical formulation, cannot be taken into account satisfactorily.

On the other hand, this approach has a number of advantages. Many different design options can be considered at the same time. The complex multiple trade-offs usually encountered in chemical process design can be handled by this approach. Also, the entire design procedure can be automated and is capable of producing designs quickly and efficiently.

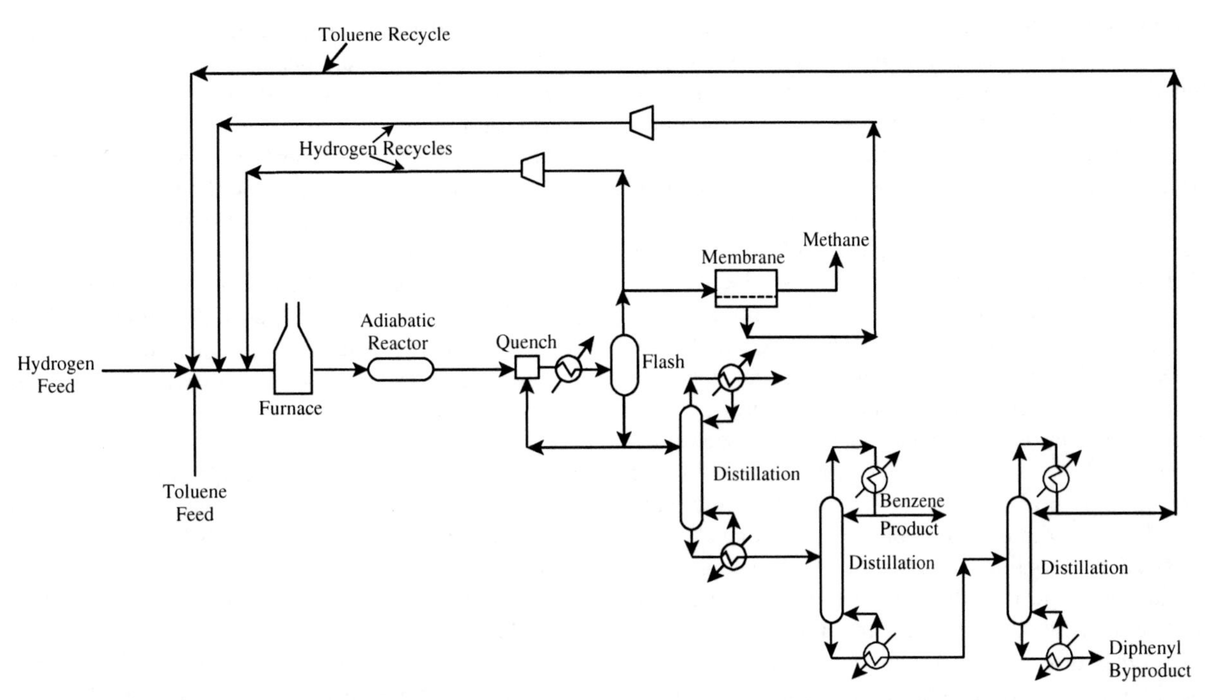

Figure 1.10 Optimization discards many structural features leaving an optimised structure. (From Kocis GR and Grossman IE, 1988, *Comp Chem Eng*, **13**: 797, reproduced by permission.)

In summary, the two general approaches to the chemical process design of building an irreducible structure and creating and optimizing a superstructure have advantages and disadvantages. However, whichever is used in practice, there is no substitute for understanding the problem.

This text concentrates on developing an understanding of the concepts required at each stage of the design. Such an understanding is a vital part of chemical process design and integration, whichever approach is followed.

1.8 PROCESS CONTROL

Once the basic process configuration has been fixed, a *control system* must be added. The control system compensates for the influence of external *disturbances* such as changes in feed flowrate, feed conditions, feed costs, product demand, product specifications, product prices, ambient temperature and so on. Ensuring safe operation is the most important task of a control system. This is achieved by monitoring the process conditions and maintaining them within safe operating limits. While maintaining the operation within safe operating limits, the control system should optimize the process performance under the influence of external disturbances. This involves maintaining product specifications, meeting production targets and making efficient use of raw materials and utilities.

A control mechanism is introduced that makes changes to the process in order to cancel out the negative impact of disturbances. In order to achieve this, instruments must be installed to measure the operational performance of the plant. These *measured variables* could include temperature, pressure, flowrate, composition, level, pH, density and particle size. *Primary measurements* may be made to directly represent the control objectives (e.g. measuring the composition that needs to be controlled). If the control objectives are not measurable, then *secondary measurements* of other variables must be made and these secondary measurements related to the control objective. Having measured the variables that need to be controlled, other variables need to be *manipulated* in order to achieve the control objectives. A control system is then designed, which responds to variations in the measured variables and manipulates variables to control the process.

Having designed a process configuration for a continuous process and having optimized it to achieve some objective (e.g. maximize profit) at steady state, is the influence of the control system likely to render the previously optimized process to now be nonoptimal? Even for a continuous process, the process is always likely to be moving from one state to another in response to the influence of disturbances and control objectives. In the steady-state design and optimization of continuous processes, these different states can be allowed for by considering *multiple operating cases*. Each operating case is assumed to operate for a certain proportion of the year. The contribution of the operating case to the overall steady-state design and optimization is weighted according to the proportion of the time under which the plant operates at that state.

While this takes some account of operation under different conditions, it does not account for the dynamic transition from one state to another. Are these transitory states likely to have a significant influence on the optimality?

If the transitory states were to have a significant effect on the overall process performance in terms of the objective function being optimized, then the process design and control system design would have to be carried out simultaneously. Simultaneous design of the process and the control system presents an extremely complex problem. It is interesting to note that where steady-state optimization for continuous processes has been compared with simultaneous optimization of the process and control system, the two process designs have been found to be almost identical[10–12].

Industrial practice is to first design and optimize the process configuration (taking into account multiple states, if necessary) and then to add the control system. However, there is no guarantee that design decisions made on the basis of steady-state conditions will not lead to control problems once process dynamics are considered. For example, an item of equipment might be oversized for contingency, because of uncertainty in design data or future debottlenecking prospects, based on steady-state considerations. Once the process dynamics are considered, this oversized equipment might make the process difficult to control, because of the large inventory of process materials in the oversized equipment. The approach to process control should adopt an approach that considers the control of the whole process, rather than just the control of the individual process steps in isolation[13].

This text will concentrate on the design and optimization of the process configuration and will not deal with process control. Process control demands expertise in different techniques and will be left to other sources of information[13]. Thus, the text will describe how to develop a *flowsheet* or *process flow diagram*, but will not take the final step of adding the instrumentation, control and auxiliary pipes and valves required for the final engineering design in the *piping and instrumentation diagram* (*P & I D*).

Batch processes are, by their nature, always in a transitory state. This requires the dynamics of the process to be optimized, and will be considered in Chapter 14. However, the control systems required to put this into practice will not be considered.

1.9 THE NATURE OF CHEMICAL PROCESS DESIGN AND INTEGRATION – SUMMARY

Chemical products can be divided into three broad classes: commodity, fine and specialty chemicals. Commodity chemicals are manufactured in large volumes with low added value. Fine and specialty chemicals tend to be manufactured in low volumes with high added value. The priorities in the design of processes for the manufacture of the three classes of chemical products will differ.

The original design problem posed to the design team is often ill-defined, even if it appears on the surface to be well-defined. The design team must formulate well-defined design options from the original ill-defined problem, and these must be compared on the basis of consistent criteria.

The design might be new or a retrofit of an existing process. If the design is a retrofit, then one of the objectives should be to maximize the use of existing equipment, even if it is not ideally suited to its new purpose.

Both continuous and batch process operations can be used. Batch processes are generally preferred for small-scale and specialty chemicals production.

When developing a chemical process design, there are two basic problems:

- Can all possible structures be identified?
- Can each structure be optimized such that all structures can be compared on a valid basis?

Design starts at the reactor, because it is likely to be the only place in the process where raw materials are converted into the desired chemical products. The reactor design dictates the separation and recycle problem. Together, the reactor design and separation and recycle dictate the heating and cooling duties for the heat exchanger network. Those duties that cannot be satisfied by heat recovery dictate the need for external heating and cooling utilities. The process and the utility system both have a demand for water and create aqueous effluents, giving rise to the water system. This hierarchy is represented by the layers in the "onion diagram", Figure 1.7. Both continuous and batch process design follow this hierarchy, even though the time dimension in batch processes brings additional constraints in process design.

There are two general approaches to chemical process design:

- building an irreducible structure;
- creating and optimizing a super structure.

Both of these approaches have advantages and disadvantages.

REFERENCES

1. Sharratt PN (1997) *Handbook of Batch Process Design*, Blackie Academic and Professional.
2. Brennan D (1998) *Process Industry Economics*, IChemE, UK.
3. Cussler EL and Moggridge GD (2001) *Chemical Product Design*, Cambridge University Press.
4. Douglas JM (1985) A Hierarchical Decision Procedure for Process Synthesis, *AIChE J*, **31**: 353.
5. Smith R and Linnhoff B, (1988) The Design of Separators in the Context of Overall Processes, *Trans IChemE ChERD*, **66**: 195.

6. Linnhoff B, Townsend DW, Boland D, Hewitt GF, Thomas BEA, Guy AR and Marsland RH (1982) *A User Guide on Process Integration for the Efficient Use of Energy*, IChemE, Rugby, UK.
7. Biegler LT, Grossmann IE and Westerberg AW (1997), *Systematic Methods of Chemical Process Design*, Prentice Hall.
8. Kocis GR and Grossmann IE (1988) A Modelling and Decomposition Strategy for the MINLP Optimization of Process Flowsheets, *Comput Chem Eng*, **13**: 797.
9. Floudas CA (2000) *Deterministic Global Optimization: Theory, Methods and Applications*, Kluwer Academic Publishers.
10. Bansal V, Perkins JD, Pistikopoulos EN, Ross R and van Shijnedel JMG (2000) Simultaneous Design and Control Optimisation Under Uncertainty, *Comput Chem Eng*, **24**: 261.
11. Bansal V, Ross R, Perkins JD, Pistikopoulos EN (2000) The Interactions of Design and Control: Double Effect Distillation, *J Process Control*, **10**: 219.
12. Kookos IK and Perkins JD (2001) An Algorithm for Simultaneous Process Design and Control, *Ind Eng Chem Res*, **40**: 4079.
13. Luyben WL, Tyreus BD and Luyben ML (1999) *Plant-wide Process Control*, McGraw Hill.

2 Process Economics

2.1 THE ROLE OF PROCESS ECONOMICS

The purpose of chemical processes is to make money. An understanding of process economics is therefore critical in process design. Process economics has three basic roles in process design:

1. *Evaluation of design options.* Costs are required to evaluate process design options; for example, should a membrane or an adsorption process be used for purification?
2. *Process optimization.* The settings of some process variables can have a major influence on the decision-making in developing the flowsheet and on the overall profitability of the process. Optimization of such variables is usually required.
3. *Overall project profitability.* The economics of the overall project should be evaluated at different stages during the design to assess whether the project is economically viable.

Before discussing how to use process economics for decision-making, the most important costs that will be needed to compare options must first be reviewed.

2.2 CAPITAL COST FOR NEW DESIGN

The total investment required for a new design can be broken down into five main parts:

- Battery limits investment
- Utility investment
- Off-site investment
- Engineering fees
- Working capital.

1. *Battery limits investment*: The battery limit is a geographic boundary that defines the manufacturing area of the process. This is that part of the manufacturing system that converts raw materials into products. It includes process equipment and buildings or structures to house it but excludes boiler-house facilities, site storage, pollution control, site infrastructure, and so on. The term battery limit is sometimes used to define the boundary of responsibility for a given project, especially in retrofit projects.

The battery limit investment required is the purchase of the individual plant items and their installation to form a working process. Equipment costs may be obtained from equipment vendors or from published cost data. Care should be taken as to the basis of such cost data. What is required for cost estimates is delivered cost, but cost is often quoted as FOB (Free On Board). Free On Board means the manufacturer pays for loading charges onto a shipping truck, railcar, barge or ship, but not freight or unloading charges. To obtain a delivered cost requires typically 5 to 10% to be added to the FOB cost. The delivery cost depends on location of the equipment supplier, location of site to be delivered, size of the equipment, and so on.

The cost of a specific item of equipment will be a function of:

- size
- materials of construction
- design pressure
- design temperature.

Cost data are often presented as cost versus capacity charts, or expressed as a power law of capacity:

$$C_E = C_B \left(\frac{Q}{Q_B}\right)^M \qquad (2.1)$$

where C_E = equipment cost with capacity Q
C_B = known base cost for equipment with capacity Q_B
M = constant depending on equipment type

A number of sources of such data are available in the open literature[1–8]. Published data are often old, sometimes from a variety of sources, with different ages. Such data can be brought up-to-date and put on a common basis using cost indexes.

$$\frac{C_1}{C_2} = \frac{INDEX_1}{INDEX_2} \qquad (2.2)$$

where C_1 = equipment cost in year 1
C_2 = equipment cost in year 2
$INDEX_1$ = cost index in year 1
$INDEX_2$ = cost index in year 2

Commonly used indices are the Chemical Engineering Indexes (1957–1959 index = 100) and Marshall and Swift (1926 index = 100), published in *Chemical Engineering*

Chemical Process Design and Integration R. Smith
 ISBNs: 0-471-48680-9 (HB); 0-471-48681-7 (PB)

Table 2.1 Typical equipment capacity delivered capital cost correlations.

Equipment	Material of construction	Capacity measure	Base size Q_B	Base cost C_B ($)	Size range	Cost exponent M
Agitated reactor	CS	Volume (m^3)	1	1.15×10^4	1–50	0.45
Pressure vessel	SS	Mass (t)	6	9.84×10^4	6–100	0.82
Distillation column (Empty shell)	CS	Mass (t)	8	6.56×10^4	8–300	0.89
Sieve trays (10 trays)	CS	Column diameter (m)	0.5	6.56×10^3	0.5–4.0	0.91
Valve trays (10 trays)	CS	Column diameter (m)	0.5	1.80×10^4	0.5–4.0	0.97
Structured packing (5 m height)	SS (low grade)	Column diameter (m)	0.5	1.80×10^4	0.5–4.0	1.70
Scrubber (Including random packing)	SS (low grade)	Volume (m^3)	0.1	4.92×10^3	0.1–20	0.53
Cyclone	CS	Diameter (m)	0.4	1.64×10^3	0.4–3.0	1.20
Vacuum filter	CS	Filter area (m^2)	10	8.36×10^4	10–25	0.49
Dryer	SS (low grade)	Evaporation rate (kg $H_2O \cdot h^{-1}$)	700	2.30×10^5	700–3000	0.65
Shell-and-tube heat exchanger	CS	Heat transfer area (m^2)	80	3.28×10^4	80–4000	0.68
Air-cooled heat exchanger	CS	Plain tube heat transfer area (m^2)	200	1.56×10^5	200–2000	0.89
Centrifugal pump (Small, including motor)	SS (high grade)	Power (kW)	1	1.97×10^3	1–10	0.35
Centrifugal pump (Large, including motor)	CS	Power (kW)	4	9.84×10^3	4–700	0.55
Compressor (Including motor)		Power (kW)	250	9.84×10^4	250–10,000	0.46
Fan (Including motor)	CS	Power (kW)	50	1.23×10^4	50–200	0.76
Vacuum pump (Including motor)	CS	Power (kW)	10	1.10×10^4	10–45	0.44
Electric motor		Power (kW)	10	1.48×10^3	10–150	0.85
Storage tank (Small atmospheric)	SS (low grade)	Volume (m^3)	0.1	3.28×10^3	0.1–20	0.57
Storage tank (Large atmospheric)	CS	Volume (m^3)	5	1.15×10^4	5–200	0.53
Silo	CS	Volume (m^3)	60	1.72×10^4	60–150	0.70
Package steam boiler (Fire-tube boiler)	CS	Steam generation ($kg \cdot h^{-1}$)	50,000	4.64×10^5	50,000–350,000	0.96
Field erected steam boiler (Water-tube boiler)	CS	Steam generation ($kg \cdot h^{-1}$)	20,000	3.28×10^5	10,000–800,000	0.81
Cooling tower (Forced draft)		Water flowrate ($m^3 \cdot h^{-1}$)	10	4.43×10^3	10–40	0.63

CS = carbon steel; SS (low grade) = low-grade stainless steel, for example, type 304; SS (high grade) = high-grade stainless steel, for example, type 316

magazine and the Nelson–Farrar Cost Indexes for refinery construction (1946 index = 100) published in the *Oil and Gas Journal*. The Chemical Engineering (CE) Indexes are particularly useful. CE Indexes are available for equipment covering:

- Heat Exchangers and Tanks
- Pipes, Valves and Fittings
- Process Instruments
- Pumps and Compressors
- Electrical Equipment
- Structural Supports and Miscellaneous.

A combined CE Index of Equipment is available. CE Indexes are also available for:

- Construction and Labor Index
- Buildings Index
- Engineering and Supervision Index.

All of the above indexes are combined to produce a CE Composite Index.

Table 2.1 presents data for a number of equipment items on the basis of January 2000 costs[7] (CE Composite Index = 391.1, CE Index of Equipment = 435.8).

Cost correlations for vessels are normally expressed in terms of the mass of the vessel. This means that not only a preliminary sizing of the vessel is required but also a preliminary assessment of the mechanical design[9,10].

Materials of construction have a significant influence on the capital cost of equipment. Table 2.2 gives some approximate average factors to relate the different materials of construction for equipment capital cost.

It should be emphasized that the factors in Table 2.2 are average and only approximate and will vary, amongst other things, according to the type of equipment. As an example, consider the effect of materials of construction on the capital cost of distillation columns. Table 2.3 gives materials of construction cost factors for distillation columns.

The cost factors for shell-and-tube heat exchangers are made more complex by the ability to construct the different components from different materials of construction.

Table 2.2 Typical average equipment materials of construction capital cost factors.

Material	Correction factor f_M
Carbon steel	1.0
Aluminum	1.3
Stainless steel (low grades)	2.4
Stainless steel (high grades)	3.4
Hastelloy C	3.6
Monel	4.1
Nickel and inconel	4.4
Titanium	5.8

Table 2.3 Typical materials of construction capital cost factors for pressure vessels and distillation columns[9,10].

Material	Correction factor f_M
Carbon steel	1.0
Stainless steel (low grades)	2.1
Stainless steel (high grades)	3.2
Monel	3.6
Inconel	3.9
Nickel	5.4
Titanium	7.7

Table 2.4 Typical materials of construction capital cost factors for shell-and-tube heat exchangers[2].

Material	Correction factor f_M
CS shell and tubes	1.0
CS shell, aluminum tubes	1.3
CS shell, monel tubes	2.1
CS shell, SS (low grade) tubes	1.7
SS (low grade) shell and tubes	2.9

Table 2.4 gives typical materials of construction factors for shell-and-tube heat exchangers.

Its operating pressure also influences equipment capital cost as a result of thicker vessel walls to withstand increased pressure. Table 2.5 presents typical factors to account for the pressure rating.

As with materials of construction correction factors, the pressure correction factors in Table 2.5 are average and only approximate and will vary, amongst other things, according to the type of equipment. Finally, its operating temperature also influences equipment capital cost. This is caused by, amongst other factors, a decrease in the allowable stress for materials of construction as the temperature increases. Table 2.6 presents typical factors to account for the operating temperature.

Thus, for a base cost for carbon steel equipment at moderate pressure and temperature, the actual cost can be

Table 2.5 Typical equipment pressure capital cost factors.

Design pressure (bar absolute)	Correction factor f_P
0.01	2.0
0.1	1.3
0.5 to 7	1.0
50	1.5
100	1.9

Table 2.6 Typical equipment temperature capital cost factors.

Design temperature (°C)	Correction factor f_T
0–100	1.0
300	1.6
500	2.1

estimated from:

$$C_E = C_B \left(\frac{Q}{Q_B} \right)^M f_M f_P f_T \qquad (2.3)$$

where C_E = equipment cost for carbon steel at moderate pressure and temperature with capacity Q

C_B = known base cost for equipment with capacity Q_B

M = constant depending on equipment type

f_M = correction factor for materials of construction

f_P = correction factor for design pressure

f_T = correction factor for design temperature

In addition to the purchased cost of the equipment, investment is required to install the equipment. Installation costs include:

- cost of installation
- piping and valves
- control systems
- foundations
- structures
- insulation
- fire proofing
- electrical
- painting
- engineering fees
- contingency.

The total capital cost of the installed battery limits equipment will normally be two to four times the purchased cost of the equipment[11,12].

In addition to the investment within the battery limits, investment is also required for the structures, equipment and services outside of the battery limits that are required to make the process function.

2. *Utility investment*: Capital investment in utility plant could include equipment for:

- electricity generation
- electricity distribution
- steam generation
- steam distribution
- process water
- cooling water
- firewater
- effluent treatment
- refrigeration
- compressed air
- inert gas (nitrogen).

The cost of utilities is considered from their sources within the site to the battery limits of the chemical process served.

3. *Off-site investment*: Off-site investment includes

- auxiliary buildings such as offices, medical, personnel, locker rooms, guardhouses, warehouses and maintenance shops
- roads and paths
- railroads
- fire protection systems
- communication systems
- waste disposal systems
- storage facilities for end product, water and fuel not directly connected with the process
- plant service vehicles, loading and weighing devices.

The cost of the utilities and off-sites (together sometimes referred to as *services*) ranges typically from 20 to 40% of the total installed cost of the battery limits plant[13]. In general terms, the larger the plant, the larger will tend to be the fraction of the total project cost that goes to utilities and off-sites. In other words, a small project will require typically 20% of the total installed cost for utilities and off-sites. For a large project, the figure can be typically up to 40%.

4. *Working capital*: Working capital is what must be invested to get the plant into productive operation. This is money invested before there is a product to sell and includes:

- raw materials for plant start-up (including wasted raw materials)
- raw materials, intermediate and product inventories
- cost of transportation of materials for start-up
- money to carry accounts receivable (i.e. credit extended to customers) less accounts payable (i.e. credit extended by suppliers)
- money to meet payroll when starting up.

Theoretically, in contrast to fixed investment, this money is not lost but can be recovered when the plant is closed down.

Stocks of raw materials, intermediate and product inventories often have a key influence on the working capital and are under the influence of the designer. Issues relating to storage will be discussed in more detail in

Chapters 13 and 14. For an estimate of the working capital requirements, take either[14]:

(a) 30% of annual sales, or
(b) 15% of total capital investment.

5. *Total capital cost*: The total capital cost of the process, services and working capital can be obtained by applying multiplying factors or *installation factors* to the purchase cost of individual items of equipment[11,12]:

$$C_F = \sum_i f_i C_{E,i} \tag{2.4}$$

where C_F = fixed capital cost for the complete system
$C_{E,i}$ = cost of Equipment i
f_i = installation factor for Equipment i

If an average installation factor for all types of equipment is to be used[11],

$$C_F = f_I \sum_i C_{E,i} \tag{2.5}$$

where f_I = overall installation factor for the complete system.

The overall installation factor for new design is broken down in Table 2.7 into component parts according to the dominant phase being processed. The cost of the installation will depend on the balance of gas and liquid processing versus solids processing. If the plant handles only gases and liquids, it can be characterized as fluid processing. A plant can be characterized as solids processing if the bulk of the material handling is solid phase. For example, a solid processing plant could be a coal or an ore preparation plant. Between the two extremes of fluid processing and solids processing are processes that handle a significant amount of both solids and fluids. For example, a shale oil plant involves preparation of the shale oil followed by extraction of fluids from the shale oil and then separation and processing of the fluids. For these types of plant, the contributions to the capital cost can be estimated from the two extreme values in Table 2.7 by interpolation in proportion of the ratio of major processing steps that can be characterized as fluid processing and solid processing.

A number of points should be noted about the various contributions to the capital cost in Table 2.7. The values are:

- based on carbon steel, moderate operating pressure and temperature
- average values for all types of equipment, whereas in practice the values will vary according to the type of equipment
- only guidelines and the individual components will vary from project to project
- applicable to new design only.

Table 2.7 Typical factors for capital cost based on delivered equipment costs.

Item	Type of process	
	Fluid processing	Solid processing
Direct costs		
Equipment delivered cost	1	1
Equipment erection, f_{ER}	0.4	0.5
Piping (installed), f_{PIP}	0.7	0.2
Instrumentation & controls (installed), f_{INST}	0.2	0.1
Electrical (installed), f_{ELEC}	0.1	0.1
Utilities, f_{UTIL}	0.5	0.2
Off-sites, f_{OS}	0.2	0.2
Buildings (including services), f_{BUILD}	0.2	0.3
Site preparation, f_{SP}	0.1	0.1
Total capital cost of installed equipment	3.4	2.7
Indirect costs		
Design, engineering and construction, f_{DEC}	1.0	0.8
Contingency (about 10% of fixed capital costs), f_{CONT}	0.4	0.3
Total fixed capital cost	4.8	3.8
Working capital		
Working capital (15% of total capital cost), f_{WC}	0.7	0.6
Total capital cost, f_I	5.8	4.4

When equipment uses materials of construction other than carbon steel, or operating temperatures are extreme, the capital cost needs to be adjusted accordingly. Whilst the equipment cost and its associated pipework will be changed, the other installation costs will be largely unchanged, whether the equipment is manufactured from carbon steel or exotic materials of construction. Thus, the application of the factors from Tables 2.2 to 2.6 should only be applied to the equipment and pipework:

$$\begin{aligned} C_F = &\sum_i [f_M f_P f_T (1 + f_{PIP})]_i C_{E,i} \\ &+ (f_{ER} + f_{INST} + f_{ELEC} + f_{UTIL} + f_{OS} + f_{BUILD} \\ &+ f_{SP} + f_{DEC} + f_{CONT} + f_{WS}) \sum_i C_{E,i} \end{aligned} \tag{2.6}$$

Thus, to estimate the fixed capital cost:

1. list the main plant items and estimate their size;
2. estimate the equipment cost of the main plant items;

3. adjust the equipment costs to a common time basis using a cost index;
4. convert the cost of the main plant items to carbon steel, moderate pressure and moderate temperature;
5. select the appropriate installation subfactors from Table 2.7 and adjust for individual circumstances;
6. select the appropriate materials of construction, operating pressure and operating temperature correction factors for each of the main plant items;
7. apply Equation 2.6 to estimate the total fixed capital cost.

Equipment cost data used in the early stages of a design will by necessity normally be based on capacity, materials of construction, operating pressure and operating temperature. However, in reality, the equipment cost will depend also on a number of factors that are difficult to quantify[15]:

- multiple purchase discounts
- buyer–seller relationships
- capacity utilization in the fabrication shop (i.e. how busy the fabrication shop is)
- required delivery time
- availability of materials and fabrication labor
- special terms and conditions of purchase, and so on.

Care should also be taken to the geographic location. Costs to build the same plant can differ significantly between different locations, even within the same country. Such differences will result from variations in climate and its effect on the design requirements and construction conditions, transportation costs, local regulations, local taxes, availability and productivity of construction labor, and so on[16]. For example, in the United States of America, Gulf Coast costs tend to be the lowest, with costs in other areas typically 20 to 50% higher, and those in Alaska two or three times higher than US Gulf Coast[16]. In Australia, costs tend to be the lowest in the region of Sydney and the other metropolitan cities, with costs in remote areas such as North Queensland typically 40 to 80% higher[15]. Costs also differ from country to country. For example, relative to costs for a plant located in the US Gulf Coast, costs in India might be expected to be 20% cheaper, in Indonesia 30% cheaper, but in the United Kingdom 15% more expensive, because of labor costs, cost of land, and so on[15].

It should be emphasized that capital cost estimates using installation factors are at best crude and at worst highly misleading. When preparing such an estimate, the designer spends most of the time on the equipment costs, which represents typically 20 to 40% of the total installed cost. The bulk costs (civil engineering, labor, etc.) are factored costs that lack definition. At best, this type of estimate can be expected to be accurate to ±30%. To obtain greater accuracy requires detailed examination of all aspects of the investment. Thus, for example, to estimate the erection cost accurately requires knowledge of how much concrete will be used for foundations, how much structural steelwork is required, and so on. Such detail can only be included from access to a large database of cost information.

The shortcomings of capital cost estimates using installation factors are less serious in preliminary process design if used to compare options on a common basis. If used to compare options, the errors will tend to be less serious as the errors will tend to be consistent across the options.

Example 2.1 A new heat exchanger is to be installed as part of a large project. Preliminary sizing of the heat exchanger has estimated its heat transfer area to be 500 m^2. Its material of construction is low-grade stainless steel, and its pressure rating is 5 bar. Estimate the contribution of the heat exchanger to the total cost of the project (CE Index of Equipment = 441.9).

Solution From Equation 2.1 and Table 2.1, the capital cost of a carbon steel heat exchanger can be estimated from:

$$\begin{aligned} C_E &= C_B\left(\frac{Q}{Q_B}\right)^M \\ &= 3.28 \times 10^4 \left(\frac{500}{80}\right)^{0.68} \\ &= \$11.4 \times 10^4 \end{aligned}$$

The cost can be adjusted to bring it up-to-date using the ratio of cost indexes:

$$\begin{aligned} C_E &= 11.4 \times 10^4 \left(\frac{441.9}{435.8}\right) \\ &= \$11.6 \times 10^4 \end{aligned}$$

The cost of a carbon steel heat exchanger needs to be adjusted for the material of construction. Because of the low pressure rating, no correction for pressure is required (Table 2.5), but the cost needs to be adjusted for the material of construction. From Table 2.4, $f_M = 2.9$, and the total cost of the installed equipment can be estimated from Equation 2.6 and Table 2.7. If the project is a complete new plant, the contribution of the heat exchanger to the total cost can be estimated to be:

$$\begin{aligned} C_F &= f_M(1 + f_{PIP})C_E + (f_{ER} + f_{INST} + f_{ELEC} + f_{UTIL} \\ &\quad + f_{OS} + f_{BUILD} + f_{SP} + f_{DEC} + f_{CONT} + f_{WS})C_E \\ &= 2.9(1 + 0.7)11.6 \times 10^4 + (0.4 + 0.2 + 0.1 + 0.5 \\ &\quad + 0.2 + 0.2 + 0.1 + 1.0 + 0.4 + 0.7)11.6 \times 10^4 \\ &= 8.73 \times 11.6 \times 10^4 \\ &= \$1.01 \times 10^6 \end{aligned}$$

Had the new heat exchanger been an addition to an existing plant that did not require investment in electrical services, utilities, offsites, buildings, site preparation or working capital, then the cost would be estimated from:

$$\begin{aligned} C_F &= f_M(1 + f_{PIP})C_E + (f_{ER} + f_{INST} + f_{DEC} + f_{CONT})C_E \\ &= 2.9(1 + 0.7)11.6 \times 10^4 + (0.4 + 0.2 + 1.0 + 0.4)11.6 \times 10^4 \\ &= 6.93 \times 11.6 \times 10^4 \\ &= \$8.04 \times 10^5 \end{aligned}$$

Installing a new heat exchanger into an existing plant might require additional costs over and above those estimated here. Connecting new equipment to existing equipment, modifying or relocating existing equipment to accommodate the new equipment and downtime might all add to the costs.

2.3 CAPITAL COST FOR RETROFIT

Estimating the capital cost of a retrofit project is much more difficult than for new design. In principle, the cost of individual items of new equipment will usually be the same, whether it is a grassroot design or a retrofit. However, in new design, multiple orders of equipment might lead to a reduction in capital cost from the equipment vendor and lower transportation costs. By contrast, installation factors for equipment in retrofit can be completely different from grassroot design, and could be higher or lower. If the new equipment can take advantage of existing space, foundations, electrical cabling, and so on, the installation factor might in some cases be lower than in new design. This will especially be the case for small items of equipment. However, most often, retrofit installation factors will tend to be higher than in grassroot design and can be very much higher. This is because existing equipment might need to be modified or moved to allow installation of new equipment. Also, access to the area where the installation is required is likely to be much more restricted in retrofit than in the phased installation of new plant. Smaller projects (as the retrofit is likely to be) tend to bring higher cost of installation per unit of installed equipment than larger projects.

As an example, one very common retrofit situation is the replacement of distillation column internals to improve the performance of the column. The improvement in performance sought is often an increase in the throughput. This calls for existing internals to be removed and then to be replaced with the new internals. Table 2.8 gives typical factors for the removal of old internals and the installation of new ones[17].

Table 2.8 Modification costs for distillation column retrofit[17].

Column modification	Cost of modification (multiply factor by cost of new hardware)
Removal of trays to install new trays	0.1 for the same tray spacing 0.2 for different tray spacing
Removal of trays to install packing	0.1
Removal of packing to install new trays	0.07
Installation of new trays	1.0–1.4 for the same tray spacing 1.2–1.5 for different tray spacing 1.3–1.6 when replacing packing
Installation of new structured packing	0.5–0.8

As far as utilities and off-sites are concerned, it is also difficult to generalize. Small retrofit projects are likely not to require any investment in utilities and off-sites. Larger-scale retrofit might demand a major revamp of the utilities and off-sites. Such a revamp of utilities and off-sites can be particularly expensive, because existing equipment might need to be modified or removed to make way for new utilities and off-sites equipment.

Working capital is also difficult to generalize. Most often, there will be no significant working capital associated with a retrofit project. For example, if a few items of equipment are replaced to increase the capacity of a plant, this will not significantly change the raw materials and product inventories, money to carry accounts receivable, money to meet payroll, and so on. On the other hand, if the plant changes function completely, significant new storage capacity is added, and so on, there might be a significant element of working capital.

One of the biggest sources of cost associated with retrofit can be the *downtime* (the period during which the plant will not be productive) required to carry out the modifications. The cost of lost production can be the dominant feature of retrofit projects. The cost of lost production should be added to the capital cost of a retrofit project. To minimize the downtime and cost of lost production requires that as much preparation as possible is carried out whilst the plant is operating. The modifications requiring the plant to be shut down should be minimized. For example, it might be possible for new foundations to be installed and new equipment put into place while the plant is still operating, leaving the final pipework and electrical modifications for the shutdown. Retrofit projects are often arranged such that the preparation is carried out prior to a regular maintenance shutdown, with the final modifications coinciding with the planned maintenance shutdown. Such considerations often dominate the decisions made as to how to modify the process for retrofit.

Because of all of these uncertainties, it is difficult to provide general guidelines for capital cost of retrofit projects. The basis of the capital cost estimate should be to start with the required investment in new equipment. Installation factors for the installation of equipment for grassroot design from Table 2.7 need to be adjusted according to circumstances (usually increased). If old equipment needs to be modified to take up a new role (e.g. move an existing heat exchanger to a new duty), then an installation cost must be applied without the equipment cost. In the absence of better information, the installation cost can be taken to be that for the equivalent piece of new equipment. Some elements of the total cost breakdown in Table 2.7 will not be relevant and should not be included. In general, for the estimation of capital cost for retrofit, a detailed examination of the individual features of retrofit projects is necessary.

Example 2.2 An existing heat exchanger is to be repiped to a new duty in a retrofit project without moving its location. The only significant investment is piping modifications. The heat transfer area of the existing heat exchanger is 500 m^2. The material of construction is low-grade stainless steel, and its design pressure is 5 bar. Estimate the cost of the project (CE Index of Equipment = 441.9).

Solution All retrofit projects have individual characteristics, and it is impossible to generalize the costs. The only way to estimate costs with any certainty is to analyze the costs of all of the modifications in detail. However, in the absence of such detail, a very preliminary estimate can be obtained by estimating the retrofit costs from the appropriate installation costs for a new design. In this case, piping costs can be estimated from those for a new heat exchanger of the same specification, but excluding the equipment cost. For Example 2.1, the cost of a new stainless steel heat exchanger with an area of 500 m^2 was estimated to be $\$11.6 \times 10^4$. The piping costs (stainless steel) can therefore be estimated to be:

$$\begin{aligned}\text{Piping cost} &= f_M f_{PIP} C_E \\ &= 2.9 \times 0.7 \times 11.6 \times 10^4 \\ &= 2.03 \times 11.6 \times 10^4 \\ &= \$2.35 \times 10^5\end{aligned}$$

This estimate should not be treated with any confidence. It will give an idea of the costs and might be used to compare retrofit options on a like-for-like basis, but could be very misleading.

Example 2.3 An existing distillation column is to be revamped to increase its capacity by replacing the existing sieve trays with stainless steel structured packing. The column shell is 46 m tall and 1.5 m diameter and currently fitted with 70 sieve trays with a spacing of 0.61 m. The existing trays are to be replaced with stainless steel structured packing with a total height of 30 m. Estimate the cost of the project (CE Index of Equipment = 441.9).

Solution First, estimate the purchase cost of the new structured packing from Equation 2.1 and Table 2.1, which gives costs for a 5-m height of packing:

$$\begin{aligned}C_E &= C_B\left(\frac{Q}{Q_B}\right)^M \\ &= 1.8 \times 10^4 \times \frac{30}{5}\left(\frac{1.5}{0.5}\right)^{1.7} \\ &= \$6.99 \times 10^5\end{aligned}$$

Adjusting the cost to bring it up-to-date using the ratio of cost indexes:

$$\begin{aligned}C_E &= 6.99 \times 10^5\left(\frac{441.9}{435.8}\right) \\ &= \$7.09 \times 10^5\end{aligned}$$

From Table 2.8, the factor for removing the existing trays is 0.1 and that for installing the new packing is 0.5 to 0.8 (say 0.7). Estimated total cost of the project:

$$\begin{aligned}&= (1 + 0.1 + 0.7)7.09 \times 10^5 \\ &= \$1.28 \times 10^6\end{aligned}$$

2.4 ANNUALIZED CAPITAL COST

Capital for new installations may be obtained from:

a. Loans from banks
b. Issue by the company of common (ordinary) stock, preferred stock or bonds (debenture stock)
c. Accumulated net cash flow arising from company profit over time.

Interest on loans from banks, preferred stock and bonds is paid at a fixed rate of interest. A share of the profit of the company is paid as a dividend on common stock and preferred stock (in addition to the interest paid on preferred stock).

The cost of the capital for a project thus depends on its source. The source of the capital often will not be known during the early stages of a project, and yet there is a need to select between process options and carry out optimization on the basis of both capital and operating costs. This is difficult to do unless both capital and operating costs can be expressed on a common basis. Capital costs can be expressed on an annual basis if it is assumed that the capital has been borrowed over a fixed period (usually 5 to 10 years) at a fixed rate of interest, in which case the capital cost can be annualized according to

$$\text{Annualized capital cost} = \text{capital cost} \times \frac{i(1+i)^n}{(1+i)^n - 1} \qquad (2.7)$$

where i = fractional interest rate per year
n = number of years

The derivation of Equation 2.7 is given in Appendix A.

As stated previously, the source of capital is often not known, and hence it is not known whether Equation 2.7 is appropriate to represent the cost of capital. Equation 2.7 is, strictly speaking, only appropriate if the money for capital expenditure is to be borrowed over a fixed period at a fixed rate of interest. Moreover, if Equation 2.7 is accepted, then the number of years over which the capital is to be annualized is known, as is the rate of interest. However, the most important thing is that, even if the source of capital is not known, and uncertain assumptions are necessary, Equation 2.7 provides a common basis for the comparison of competing projects and design alternatives within a project.

Example 2.4 The purchased cost of a new distillation column installation is \$1 million. Calculate the annual cost of installed capital if the capital is to be annualized over a five-year period at a fixed rate of interest of 5%.

Solution First, calculate the installed capital cost:

$$C_F = f_i C_E$$
$$= 5.8 \times (1{,}000{,}000)$$
$$= \$5{,}800{,}000$$

$$\text{Annualization factor} = \frac{i(1+i)^n}{(1+i)^n - 1}$$
$$= \frac{0.05(1+0.05)^5}{(1+0.05)^5 - 1} = 0.2310$$

$$\text{Annualized capital cost} = 5,800,000 \times 0.2310$$
$$= \$1{,}340{,}000 \text{ y}^{-1}$$

When using annualized capital cost to carry out optimization, the designer should not lose sight of the uncertainties involved in the capital annualization. In particular, changing the annualization period can lead to very different results when, for example, carrying out a trade-off between energy and capital costs. When carrying out optimization, the sensitivity of the result to changes in the assumptions should be tested.

2.5 OPERATING COST

1. *Raw materials cost*: In most processes, the largest individual operating cost is raw materials. The cost of raw materials and the product selling prices tend to have the largest influence on the economic performance of the process. The cost of raw materials and price of products depends on whether the materials in question are being bought and sold under a contractual arrangement (either within or outside the company) or on the open market. Open market prices for some chemical products can fluctuate considerably with time. Raw materials might be purchased and products sold below or above the open market price when under a contractual arrangement, depending on the state of the market. Buying and selling on the open market may give the best purchase and selling prices but give rise to an uncertain economic environment. A long-term contractual agreement may reduce profit per unit of production but gives a degree of certainty over the project life.

The values of raw materials and products can be found in trade journals such as Chemical Marketing Reporter (published by Schnell Publishing Company), European Chemical News and Asian Chemical News (published by Reed Business Information). However, the values reported in such sources will be subject to short-term fluctuations, and long-term forecasts will be required for investment analysis.

2. *Catalysts and chemicals consumed in manufacturing other than raw materials*: Catalysts will need to be replaced or regenerated though the life of a process (see Chapter 7). The replacement of catalysts might be on a continuous basis if homogeneous catalysts are used (see Chapters 5 and 7). Heterogeneous catalysts might also be replaced continuously if they deteriorate rapidly, and regeneration cannot fully reinstate the catalyst activity. More often for heterogeneous catalysts, regeneration or replacement will be carried out on an intermittent basis, depending on the characteristics of the catalyst deactivation.

In addition to the cost of catalysts, there might be significant costs associated with chemicals consumed in manufacturing that do not form part of the final product. For example, acids and alkalis might be consumed to adjust the pH of streams. Such costs might be significant.

3. *Utility operating cost*: Utility operating cost is usually the most significant variable operating cost after the cost of raw materials. This is especially the case for the production of commodity chemicals. Utility operating cost includes:

- fuel
- electricity
- steam
- cooling water
- refrigeration
- compressed air
- inert gas.

Utility costs can vary enormously between different processing sites. This is especially true of fuel and power costs. Not only do fuel costs vary considerably between different fuels (coal, oil, natural gas) but costs also tend to be sensitive to market fluctuations. Contractual relationships also have a significant effect on fuel costs. The price paid for fuel depends very much on how much is purchased and the pattern of usage.

When electricity is bought from centralized power-generation companies under long-term contract, the price tends to be more stable than fuel costs, since power-generation companies tend to negotiate long-term contracts for fuel supply. However, purchased electricity prices (and sales price if excess electricity is generated and exported) are normally subject to tariff variations. Electricity tariffs can depend on the season of the year (winter versus summer), the time of day (night versus day) and the time of the week (weekend versus weekday). In hot countries, electricity is usually more expensive in the summer than in the winter because of the demand from air conditioning systems. In cold countries, electricity is usually more expensive in the winter than in the summer because of the demand from space heating. The price structure for electricity can be complex, but should be predictable if based on contractual arrangements. If electricity is purchased from a spot market in those countries that have such arrangements, then prices can vary wildly.

Steam costs vary with the price of fuel and electricity. If steam is only generated at low pressure and not used for power generation in steam turbines, then the cost can be estimated from fuel costs assuming an efficiency of generation and distribution losses. The efficiency of generation depends on the boiler efficiency and the steam consumed in boiler feedwater production (see Chapter 23). Losses from the steam distribution system include heat losses from steam distribution and condensate return pipework to the environment, steam condensate lost to drain and not returned to the boilers and steam leaks. The efficiency of steam generation (including auxiliary boiler-house requirements, see Chapter 23) is typically around 85 to 90% and distribution losses of perhaps another 10%, giving an overall efficiency for steam generation and distribution of typically around 75 to 80% (based on the net calorific value of the fuel). Care should be exercised when considering boiler efficiency and the efficiency of steam generation. These figures are most often quoted on the basis of *gross calorific value* of the fuel, which includes the latent heat of the water vapor from combustion. This latent heat is rarely recovered through condensation of the water vapor in the combustion gases. The *net calorific value* of the fuel assumes that the latent heat of the water vapor is not recovered and is therefore the most relevant value. Yet, figures are most often quoted on the basis of gross calorific value.

If high-pressure steam mains are used, then the cost of steam should be related in some way to its capacity to generate power in a steam turbine rather than simply to its additional heating value. The high-pressure steam is generated in the utility boilers, and the low-pressure steam is generated by reducing pressure through steam turbines to produce power. This will be discussed in more detail in Chapter 23. One simple way to cost steam is to calculate the cost of the fuel required to generate the high-pressure steam (including any losses), and this fuel cost is then the cost of the high-pressure steam. Low-pressure mains have a value equal to that of the high-pressure mains minus the value of power generated by letting the steam down to the low pressure in a steam turbine. To calculate the cost of steam that has been expanded though a steam turbine, the power generated in such an expansion must be calculated. The simplest way to do this is on the basis of a comparison between an ideal and a real expansion though a steam turbine. Figure 2.1 shows a steam turbine expansion on an enthalpy-entropy plot. In an ideal turbine, steam with an initial pressure P_1 and enthalpy H_1 expands isentropically to pressure P_2 and enthalpy H_2. In such circumstances, the ideal work output is $(H_1 - H_2)$. Because of the frictional effects in the turbine nozzles and blade passages, the exit enthalpy is greater than it would be in an ideal turbine, and the work output is consequently less, given by H_2' in Figure 2.1. The actual work output is given by $(H_1 - H_2')$. The turbine isentropic efficiency η_{IS} measures the ratio of the actual to ideal work obtained:

$$\eta_{IS} = \frac{H_1 - H_2'}{H_1 - H_2} \tag{2.8}$$

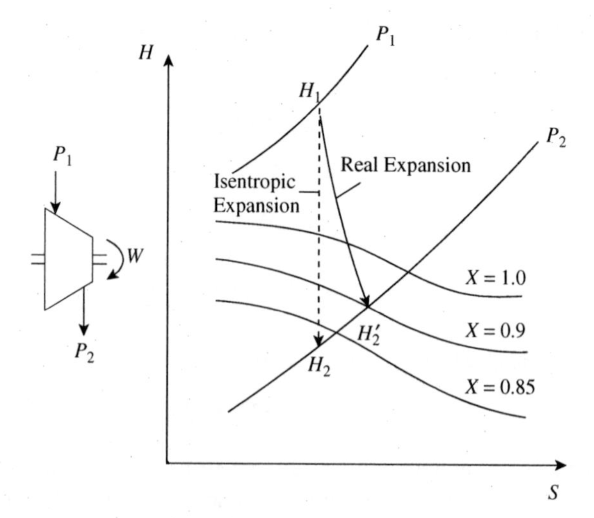

Figure 2.1 Steam turbine expansion.

The output from the turbine might be superheated or partially condensed, as is the case in Figure 2.1. The following example illustrates the approach.

Example 2.5 The pressures of three steam mains have been set to the conditions given in Table 2.9. High-pressure (HP) steam is generated in boilers at 41 barg and superheated to 400°C. Medium-pressure (MP) and low-pressure (LP) steam are generated by expanding high-pressure steam through a steam turbine with an isentropic efficiency of 80%. The cost of fuel is \$4.00 GJ^{-1}, and the cost of electricity is \$0.07 $kW^{-1}{\cdot}h^{-1}$. Boiler feedwater is available at 100°C with a heat capacity of 4.2 $kJ{\cdot}kg^{-1}{\cdot}K^{-1}$. Assuming an efficiency of steam generation of 85% and distribution losses of 10%, estimate the cost of steam for the three levels.

Table 2.9 Steam mains pressure settings.

Mains	Pressure (barg)
HP	41
MP	10
LP	3

Solution *Cost of 41 barg steam.* From steam tables, for 41 barg steam at 400°C:

$$\text{Enthalpy} = 3212 \text{ kJ}{\cdot}\text{kg}^{-1}$$

For boiler feedwater:

$$\text{Enthalpy} = 4.2(100 - 0)(\text{relative to water at } 0^\circ\text{C})$$
$$= 420 \text{ kJ}{\cdot}\text{kg}^{-1}$$

To generate 41 barg steam at 400°C:

$$\text{Heat duty} = 3212 - 420 = 2792 \text{ kJ}{\cdot}\text{kg}^{-1}$$

For 41 barg steam:

$$\text{Cost} = 4.00 \times 10^{-6} \times 2792 \times \frac{1}{0.75}$$
$$= \$0.01489 \text{ kg}^{-1}$$
$$= \$14.89 \text{ t}^{-1}$$

Cost of 10 barg steam. Here 41 barg steam is now expanded to 10 barg in a steam turbine. From steam tables, inlet conditions at 41 barg and 400°C are:

$$H_1 = 3212 \text{ kJ·kg}^{-1}$$
$$S_1 = 6.747 \text{ kJ·kg}^{-1}\text{·K}^{-1}$$

Turbine outlet conditions for isentropic expansion to 10 barg are:

$$H_2 = 2873 \text{ kJ·kg}^{-1}$$
$$S_2 = 6.747 \text{ kJ·kg}^{-1}\text{·K}^{-1}$$

For single-stage expansion with isentropic efficiency of 80%:

$$H_2' = H_1 - \eta_I(H_1 - H_2)$$
$$= 3212 - 0.8(3212 - 2873)$$
$$= 2941 \text{ kJ·kg}^{-1}$$

From steam tables, the outlet temperature is 251°C, which corresponds to a superheat of 67°C. Although steam for process heating is preferred at saturated conditions, it is not desirable in this case to de-superheat by boiler feedwater injection to bring to saturation conditions. If saturated steam is fed to the main, then the heat losses from the main will cause a large amount of condensation in the main, which is undesirable. Hence, it is standard practice to feed steam to the main with a superheat of at least 10°C to avoid condensation in the main.

$$\text{Power generation} = 3212 - 2941$$
$$= 271 \text{ kJ·kg}^{-1}$$
$$\text{Value of power generation} = 271 \times \frac{0.07}{3600}$$
$$= \$0.00527 \text{ kg}^{-1}$$
$$\text{Cost of 10 barg steam} = 0.01489 - 0.00527$$
$$= \$0.00962 \text{ kg}^{-1}$$
$$= \$9.62 \text{ t}^{-1}$$

Cost of 3 barg steam. Here, 10 barg steam from the exit of the first turbine is assumed to be expanded to 3 barg in another turbine.

From steam tables, inlet conditions of 10 barg and 251°C are:

$$H_1 = 2941 \text{ kJ·kg}^{-1}$$
$$S_1 = 6.880 \text{ kJ·kg}^{-1}\text{·K}^{-1}$$

Turbine outlet conditions for isentropic expansion to 3 barg are:

$$H_2 = 2732 \text{ kJ·kg}^{-1}$$
$$S_2 = 6.880 \text{ kJ·kg}^{-1}\text{·K}^{-1}$$

For a single-stage expansion with isentropic efficiency of 80%

$$H_2' = H_1 - \eta_{IS}(H_1 - H_2)$$
$$= 2941 - 0.8(2941 - 2732)$$
$$= 2774 \text{ kJ·kg}^{-1}$$

From steam tables, the outlet temperature is 160°C, which is superheated by 16°C. Again, it is desirable to have some superheat for the steam fed to the low-pressure main.

Power generation

$$= 2941 - 2774$$
$$= 167 \text{ kJ·kg}^{-1}$$

Value of power generation

$$= 167 \times \frac{0.07}{3600}$$
$$= \$0.00325 \text{ kg}^{-1}$$

Cost of 3 barg steam

$$= 0.00962 - 0.00325$$
$$= \$0.00637 \text{ kg}^{-1}$$
$$= \$6.37 \text{ t}^{-1}$$

If the steam generated in the boilers is at a very high pressure and/or the ratio of power to fuel costs is high, then the value of low-pressure steam can be extremely low or even negative.

This simplistic approach to costing steam is often unsatisfactory, especially if the utility system already exists. Steam costs will be considered in more detail in Chapter 23.

The operating cost for cooling water tends to be low relative to the value of both fuel and electricity. The principal operating cost associated with the provision of cooling water is the cost of power to drive the cooling tower fans and cooling water circulation pumps. The cost of cooling duty provided by cooling water is in the order of 1% that of the cost of power. For example, if power costs \$0.07 $\text{kW}^{-1}\text{·h}^{-1}$, then cooling water will typically cost $0.07 \times 0.01/3600 = \$0.19 \times 10^{-6} \text{ kJ}^{-1}$ or \$0.19 GJ^{-1}. Cooling water systems will be discussed in more detail in Chapter 24.

The cost of power required for a refrigeration system can be estimated as a multiple of the power required for an ideal system:

$$\frac{W_{IDEAL}}{Q_C} = \frac{T_H - T_C}{T_C} \tag{2.9}$$

where W_{IDEAL} = ideal power required for the refrigeration cycle
Q_C = the cooling duty
T_C = temperature at which heat is taken into the refrigeration cycle (K)

T_H = temperature at which heat is rejected from the refrigeration cycle (K)

The ratio of ideal to actual power is often around 0.6. Thus

$$W = \frac{Q_C}{0.6}\left(\frac{T_H - T_C}{T_C}\right) \quad (2.10)$$

where W is the actual power required for the refrigeration cycle.

Example 2.6 A process requires 0.5 MW of cooling at −20°C. A refrigeration cycle is required to remove this heat and reject it to cooling water supplied at 25°C and returned at 30°C. Assuming a minimum temperature difference (ΔT_{min}) of 5°C and both vaporization and condensation of the refrigerant occur isothermally, estimate the annual operating cost of refrigeration for an electrically driven system operating 8000 hours per year. The cost of electricity is \$0.07 $kW^{-1} \cdot h^{-1}$.

Solution

$$W = \frac{Q_C}{0.6}\left(\frac{T_H - T_C}{T_C}\right)$$

$$T_H = 30 + 5 = 35^\circ C = 308 \text{ K}$$

$$T_C = -20 - 5 = -25^\circ C = 248 \text{ K}$$

$$W = \frac{0.5}{0.6}\left(\frac{308 - 248}{248}\right)$$

$$= 0.202 \text{ MW}$$

$$\text{Cost of electricity} = 0.202 \times 10^3 \times 0.07 \times 8{,}000$$

$$= \$113{,}120 \text{ y}^{-1}$$

More accurate methods to calculate refrigeration costs will be discussed in Chapter 24.

4. *Labor cost*: The cost of labor is difficult to estimate. It depends on whether the process is batch or continuous, the level of automation, the number of processing steps and the level of production. When synthesizing a process, it is usually only necessary to screen process options that have the same basic character (e.g. continuous), have the same level of automation, have a similar number of processing steps and the same level of production. In this case, labor costs will be common to all options and hence will not affect the comparison.

If, however, options are to be compared that are very different in nature, such as a comparison between batch and continuous operation, some allowance for the difference in the cost of labor must be made. Also, if the location of the plant has not been fixed, the differences in labor costs between different geographical locations can be important.

5. *Maintenance*: The cost of maintenance depends on whether processing materials are solids on the one hand or gas and liquid on the other. Handling solids tends to increase maintenance costs. Highly corrosive process fluids increase maintenance costs. Average maintenance costs tend to be around 6% of the fixed capital investment[8].

2.6 SIMPLE ECONOMIC CRITERIA

To evaluate design options and carry out process optimization, simple economic criteria are needed. Consider what happens to the revenue from product sales after the plant has been commissioned. The sales revenue must pay for both fixed costs that are independent of the rate of production and variable costs, which do depend on the rate of production. After this, taxes are deducted to leave the net profit.

Fixed costs independent of the rate of production include:

- Capital cost repayments
- Routine maintenance
- Overheads (e.g. safety services, laboratories, personnel facilities, administrative services)
- Quality control
- Local taxes
- Labor
- Insurance

Variable costs that depend on the rate of production include:

- Raw materials
- Catalysts and chemicals consumed in manufacturing (other than raw materials)
- Utilities (fuel, steam, electricity, cooling water, process water, compressed air, inert gases, etc.)
- Maintenance costs incurred by operation
- Royalties
- Transportation costs

There can be an element of maintenance that is a fixed and an element that is variable. Fixed maintenance costs cover routine maintenance such as statutory maintenance on safety equipment that must be carried out irrespective of the rate of production. Variable maintenance costs result from certain items of equipment needing more maintenance as the production rate increases. Also, the royalties that cover the cost of purchasing another company's process technology may have different bases. Royalties may be a variable cost, since they can sometimes be paid in proportion to the rate of production or sales revenue. Alternatively, the royalty might be a single-sum payment at the beginning of the project. In this case, the single-sum payment will become part of the project capital investment. As such, it will be included in the annual capital repayment, and this becomes part of the fixed cost.

Two simple economic criteria are useful in process design:

1. *Economic potential (EP)*:

$$EP = \text{value of products} - \text{fixed costs} - \text{variable costs} - \text{taxes} \quad (2.11)$$

2. *Total annual cost (TAC)*:

$$TAC = \text{fixed costs} + \text{variable costs} + \text{taxes} \quad (2.12)$$

When synthesizing a flowsheet, these criteria are applied at various stages when the picture is still incomplete. Hence, it is usually not possible to account for all the fixed and variable costs listed above during the early stages of a project. Also, there is little point in calculating taxes until a complete picture of operating costs and cash flows has been established.

The preceding definitions of economic potential and total annual cost can be simplified if it is accepted that they will be used to compare the relative merits of different structural options in the flowsheet and different settings of the operating parameters. Thus, items that will be common to the options being compared can be neglected.

2.7 PROJECT CASH FLOW AND ECONOMIC EVALUATION

As the design progresses, more information is accumulated. The best methods of assessing the profitability of alternatives are based on projections of the cash flows during the project life[18].

Figure 2.2 shows the cash flow pattern for a typical project. The cash flow is a cumulative cash flow. Consider Curve 1 in Figure 2.2. From the start of the project at Point A, cash is spent without any immediate return. The early stages of the project consist of development, design and other preliminary work, which causes the cumulative curve to dip to Point B. This is followed by the main phase of capital investment in buildings, plant and equipment, and the curve drops more steeply to Point C. Working capital is spent to commission the plant between Points C and D. Production starts at D, where revenue from sales begins. Initially, the rate of production is likely to be below design conditions until full production is achieved at E. At F, the cumulative cash flow is again zero. This is the project breakeven point. Toward the end of the projects life at G, the net rate of cash flow may decrease owing to, for example, increasing maintenance costs, a fall in the market price for the product, and so on.

Ultimately, the plant might be permanently shut down or given a major revamp. This marks the end of the project, H. If the plant is shut down, working capital is recovered,

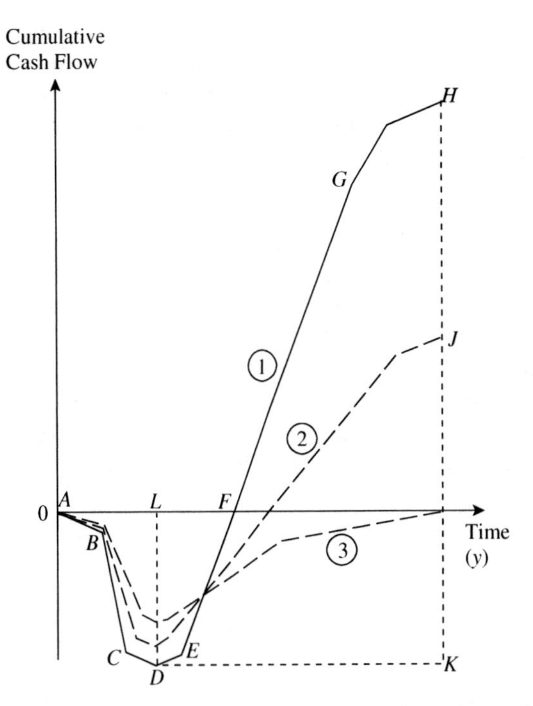

Figure 2.2 Cash flow pattern for a typical project. (From Allen DH, 1980, A Guide to the Economic Evaluation of Projects, IChemE, reproduced by permission of the Institution of Chemical Engineers.)

and there may be salvage value, which would create a final cash inflow at the end of the project.

The predicted cumulative cash flow curve for a project throughout its life forms the basis for more detailed evaluation. Many quantitative measures or indices have been proposed. In each case, important features of the cumulative cash flow curve are identified and transformed into a single numerical measure as an index.

1. *Payback time*: Payback time is the time that elapses from the start of the project (A in Figure 2.2) to the breakeven point (F in Figure 2.2). The shorter the payback time, the more attractive is the project. Payback time is often calculated as the time to recoup the capital investment based on the mean annual cash flow. In retrofit, payback time is usually calculated as the time to recoup the retrofit capital investment from the mean annual improvement in operating costs.

2. *Return on Investment (ROI)*: Return on investment (ROI) is usually defined as the ratio of average yearly income over the productive life of the project to the total initial investment, expressed as a percentage. Thus, from Figure 2.2

$$ROI = \frac{KH}{KD} \times \frac{100}{LD} \%\text{per year} \quad (2.13)$$

Payback and ROI select particular features of the project cumulative cash flow and ignore others. They take no account of the *pattern* of cash flow during a project.

The other indices to be described, net present value and discounted cash flow return, are more comprehensive because they take account of the changing pattern of project net cash flow with time. They also take account of the *time value* of money.

3. *Net present value (NPV)*: Since money can be invested to earn interest, money received now has a greater value than money if received at some time in the future. The net present value of a project is the sum of the present values of each individual cash flow. In this case, the *present* is taken to be the start of a project.

Time is taken into account by discounting the annual cash flow A_{CF} with the rate of interest to obtain the annual discounted cash flow A_{DCF}. Thus at the end of year 1

$$A_{DCF1} = \frac{A_{CF1}}{(1+i)}$$

at the end of year 2,

$$A_{DCF2} = \frac{A_{CF2}}{(1+i)^2}$$

and at the end of year n,

$$A_{DCFn} = \frac{A_{CFn}}{(1+i)^n} \tag{2.14}$$

The sum of the annual discounted cash flows over n years ΣA_{DCF} is known as the *net present value* (NPV) of the project.

$$NPV = \sum A_{DCF} \tag{2.15}$$

The value of NPV is directly dependent on the choice of the fractional interest rate i and project lifetime n.

Returning to the cumulative cash flow curve for a project, the effect of discounting is shown in Figure 2.2. Curve 1 is the original curve with no discounting, that is, $i = 0$, and the project NPV is equal to the final net cash position given by H. Curve 2 shows the effect of discounting at a fixed rate of interest, and the corresponding project NPV is given by J. Curve 3 in Figure 2.2 shows a larger rate of interest, but it is chosen such that the NPV is zero at the end of the project.

The greater the positive NPV for a project, the more economically attractive it is. A project with a negative NPV is not a profitable proposition.

4. *Discounted cash flow rate of return*: Discounted cash flow rate of return is defined as the discount rate i, which makes the NPV of a project to zero (Curve 3 in Figure 2.2):

$$NPV = \sum A_{DCF} = 0 \tag{2.16}$$

The value of i given by this equation is known as the *discounted cash flow rate of return* ($DCFRR$). It may be found graphically or by trial and error.

Example 2.7 A company has the alternative of investing in one of two projects, A or B. The capital cost of both projects is \$10 million. The predicted annual cash flows for both projects are shown in Table 2.10. Capital is restricted, and a choice is to be made on the basis of discounted cash flow rate of return, based on a five-year lifetime.

Table 2.10 Predicted annual cash flows.

	Cash flows (\$10⁶)	
Year	Project A	Project B
0	−10	−10
1	1.6	6.5
2	2.8	5.2
3	4.0	4.0
4	5.2	2.8
5	6.4	1.6

Project *A*

Start with an initial guess for $DCFRR$ of 20% and increase as detailed in Table 2.11.

Table 2.11 Calculation of $DCFRR$ for Project A.

		DCF 20%		DCF 30%		DCF 25%	
Year	A_{CF}	A_{DCF}	ΣA_{DCF}	A_{DCF}	ΣA_{DCF}	A_{DCF}	ΣA_{DCF}
0	−10	−10	−10	−10	−10	−10	−10
1	1.6	1.33	−8.67	1.23	−8.77	1.28	−8.72
2	2.8	1.94	−6.73	1.66	−7.11	1.79	−6.93
3	4.0	2.31	−4.42	1.82	−5.29	2.05	−4.88
4	5.2	2.51	−1.91	1.82	−3.47	2.13	−2.75
5	6.4	2.57	0.66	1.72	−1.75	2.10	−0.65

Twenty percent is too low since ΣA_{DCF} is positive at the end of year 5. Thirty percent is too large since ΣA_{DCF} is negative at the end of year 5, as is the case with 25%. The answer must be between 20 and 25%. Interpolating on the basis of ΣA_{DCF} the $DCFRR \approx 23\%$.

Project *B*

Again, start with an initial guess for $DCFRR$ of 20% and increase as detailed in Table 2.12.

From ΣA_{DCF} at the end of year 5, 20% is to low, 40% too high and 35% also too low. Interpolating on the basis of ΣA_{DCF}, the $DCFRR \approx 38\%$. Project B should therefore be chosen.

2.8 INVESTMENT CRITERIA

Economic analysis should be performed at all stages of an emerging project as more information and detail become available. The decision as to whether to proceed with a project will depend on many factors. There is most often

Table 2.12 Calculation of *DCFRR* for Project *B*.

		DCF 20%		DCF 40%		DCF 35%	
Year	A_{CF}	A_{DCF}	ΣA_{DCF}	A_{DCF}	ΣA_{DCF}	A_{DCF}	ΣA_{DCF}
0	−10	−10	−10	−10	−10	−10	−10
1	6.4	5.42	−4.58	4.64	−5.36	4.81	−5.19
2	5.2	3.61	−0.97	2.65	−2.71	2.85	−2.34
3	4.0	2.31	1.34	1.46	−1.25	1.63	−0.71
4	2.8	1.35	2.69	0.729	−0.521	0.843	0.133
5	1.6	0.643	3.33	0.297	−0.224	0.357	0.490

stiff competition within companies for any capital available for investment in projects. The decision as to where to spend the available capital on a particular project will, in the first instance but not exclusively, depend on the economic criteria discussed in the previous section. Criteria that account for the timing of the cash flows (the *NPV* and *DCFRR*) should be the basis of the decision-making. The higher the value of the *NPV* and *DCFRR* for a project, the more attractive it is. The absolute minimum acceptable value of the *DCFRR* is the market interest rate. If the *DCFRR* is lower than market interest rate, it would be better to put the money in the bank. For a *DCFRR* value greater than this, the project will show a profit, for a lesser value it will show a loss. The essential distinction between *NPV* and *DCFRR* is:

- *Net Present Value* measures profit but does not indicate how efficiently capital is being used.
- *DCFRR* measures how efficiently capital is being used but gives no indication of how large the profits will be.

If the goal is to maximize profit, *NPV* is the more important measure. If the supply of capital is restricted, which is usual, *DCFRR* can be used to decide which projects will use the capital most efficiently. Both measures are therefore important to characterize the economic value of a project.

Predicting future cash flows for a project is extremely difficult. There are many uncertainties, including the project life. Also, the appropriate interest rate will not be known with certainty. The acceptability of the rate of return will depend on the risks associated with the project and the company investment policy. For example, a *DCFRR* of 20% might be acceptable for a low risk project. A higher return of say 30% might be demanded of a project with some risk, whereas a high-risk project with significant uncertainty might demand a 50% *DCFRR*.

The sensitivity of the economic analysis to the underlying assumptions should always be tested. A sensitivity analysis should be carried out to test the sensitivity of the economic analysis to:

- errors in the capital cost estimate
- delays in the start-up of the project after the capital has been invested (particularly important for a high capital cost project)
- changes in the cost of raw materials
- changes in the sales price of the product
- reduction in the market demand for the product, and so on.

When carrying out an economic evaluation, the magnitude and timing of the cash flows, the project life and interest rate are not known with any certainty. However, providing that consistent assumptions are made for projections of cash flows and the assumed rate of interest, the economic analysis can be used to choose between competing projects. It is important to compare different projects and options within projects, on the basis of consistent assumptions. Thus, even though the evaluation will be uncertain in an absolute sense, it can still be meaningful in a relative sense for choosing between options. Because of this, it is important to have a reference against which to judge any project or option within a project.

However, the final decision to proceed with a project will be influenced as much by business strategy as by the economic measures described above. The business strategy might be to gradually withdraw from a particular market, perhaps because of adverse long-term projections of excessive competition, even though there might be short-term attractive investment opportunities. The long-term business strategy might be to move into different business areas, thereby creating investment priorities. Priority might be given to increasing market share in a particular product to establish business dominance in the area and achieve long-term global economies of scale in the business.

2.9 PROCESS ECONOMICS – SUMMARY

Process economics is required to evaluate design options, carry out process optimization and evaluate overall project profitability. Two simple criteria can be used:

- economic potential
- total annual cost.

These criteria can be used at various stages in the design without a complete picture of the process.

The dominant operating cost is usually raw materials. However, other significant operating costs involve catalysts and chemicals consumed other than raw materials, utility costs, labor costs and maintenance.

Capital costs can be estimated by applying installation factors to the purchase costs of individual items of equipment. However, there is considerable uncertainty associated with cost estimates obtained in this way, as equipment costs are typically only 20 to 40% of the total installed costs, with the remainder based on factors. Utility investment, off-site investment and working capital are also needed to complete the capital investment. The capital cost can be annualized by considering it as a loan over a fixed period at a fixed rate of interest.

As a more complete picture of the project emerges, the cash flows through the project life can be projected. This allows more detailed evaluation of project profitability on the basis of cash flows. Net present value can be used to measure the profit taking into account the time value of money. Discounted cash flow rate of return measures how efficiently the capital is being used.

Overall, there are always considerable uncertainties associated with an economic evaluation. In addition to the errors associated with the estimation of capital and operating costs, the project life or interest rates are not known with any certainty. The important thing is that different projects, and options within projects, are compared on the basis of consistent assumptions. Thus, even though the evaluation will be uncertain in an absolute sense, it will still be meaningful in a relative sense for choosing between options.

2.10 EXERCISES

1. The cost of a closed atmospheric cylindrical storage vessels can be considered to be proportional to the mass of steel required. Derive a simple expression for the dimensions of such a storage tank to give minimum capital cost. Assume the top and bottom are both flat.
2. A new agitated reactor with new external shell-and-tube heat exchanger and new centrifugal pump are to be installed in an existing facility. The agitated reactor is to be glass-lined, which can be assumed to have an equipment cost of three times the cost of a carbon steel vessel. The heat exchanger, pump and associated piping are all high-grade stainless steel. The equipment is rated for moderate pressure. The reactor has a volume of 9 m^3, the heat exchanger an area of 50 m^2 and the pump has a power of 5 KW. No significant investment is required in utilities, off-sites, buildings, site preparation or working capital. Using Equation 2.1 and Table 2.1 (extrapolating beyond the range of the correlation if necessary), estimate the cost of the project (CE Index of Equipment = 441.9).
3. Steam is distributed on a site via a high-pressure and low-pressure steam mains. The high-pressure mains is at 40 bar and 350°C. The low-pressure mains is at 4 bar. The high-pressure steam is generated in boilers. The overall efficiency of steam generation and distribution is 75%. The low-pressure steam is generated by expanding the high-pressure stream through steam turbines with an isentropic efficiency of 80%. The cost of fuel in the boilers is 3.5 $\$\cdot GJ^{-1}$, and the cost of electricity is \$0.05 $KW^{-1}\cdot h^{-1}$. The boiler feedwater is available at 100°C with a heat capacity of 4.2 $kJ\cdot kg^{-1}\cdot K^{-1}$. Estimate the cost of the high-pressure and low-pressure steam.
4. A refrigerated distillation condenser has a cooling duty of 0.75 MW. The condensing stream has a temperature of −10°C. The heat from a refrigeration circuit can be rejected to cooling water at a temperature of 30°C. Assuming a temperature difference in the distillation condenser of 5°C and a temperature difference for heat rejection from refrigeration to cooling water of 10°C, estimate the power requirements for the refrigeration.
5. Acetone is to be produced by the dehydrogenation of an aqueous solution of isopropanol according to the reaction:

$$\underset{\text{Isopropanol}}{(CH_3)_2CHOH} \longrightarrow \underset{\text{Acetone}}{CH_3COCH_3} + \underset{\text{Hydrogen}}{H_2}$$

   The effluent from the reactor enters a phase separator that separates vapor from liquid. The liquid contains the bulk of the product, and the vapor is a waste stream. The vapor stream is at a temperature of 30°C and an absolute pressure of 1.1 bar. The component flowrates in the vapor stream are given in Table 2.13, together with their raw material values and fuel values. Three options are to be considered:
   a. Burn the vapor in a furnace
   b. Recover the acetone by absorption in water recycled from elsewhere in the process with the tail gas being burnt in a furnace. It is expected that 99% will be recovered by this method at a cost of 1.8 $\$\cdot kmol^{-1}$ acetone recovered.
   c. Recover the acetone by condensation using refrigerated coolant with the tail gas being burnt in a furnace. It is anticipated that a temperature of −10°C will need to be achieved in the condenser. It can be assumed that the hydrogen is an inert that will not dissolve in the liquid acetone. The vapor pressure of acetone is given by

$$\ln P = 10.031 - \frac{2940.5}{T - 35.93}$$

   where P = pressure (bara)
   T = absolute temperature (K)

   The cost of refrigerant is \$11.5 GJ^{-1}, the mean molal heat capacity of the vapor is 40 $kJ\cdot kmol^{-1}\cdot K^{-1}$, and the latent heat of acetone is 29,100 $kJ\cdot kmol^{-1}$.

   Calculate the economic potential of each option given the data in Table 2.13.

Table 2.13 Data for exercise 5.

Component	Flowrate in vapor ($kmol\cdot h^{-1}$)	Raw material value ($\$\cdot kmol^{-1}$)	Fuel value ($\$\cdot kmol^{-1}$)
Hydrogen	51.1	0	0.99
Acetone	13.5	34.8	6.85

6. A process for the production of cellulose acetate fiber produces a waste stream containing mainly air but with a small quantity of acetone vapor. The flowrate of air is 300 $kmol\cdot h^{-1}$ and that of acetone is 4.5 $kmol\cdot h^{-1}$. It is proposed to recover the acetone from the air by absorption into water followed by distillation of the acetone-water mixture. The absorber requires a flow of water 2.8 times that of the air.
   a. Assuming acetone costs 34.8 $\$\cdot kmol^{-1}$, process water costs \$0.004 $kmol^{-1}$ and the process operates for 8000 $h\cdot y^{-1}$, calculate the maximum economic potential assuming complete recovery of the acetone.
   b. If the absorber and distillation column both operate at 99% recovery of acetone and the product acetone overhead from the distillation column must be 99% pure, sketch the flowsheet for the system and calculate the flows of acetone and water to and from the distillation column.

c. Calculate the revised economic potential to allow for incomplete recovery in the absorption and distillation columns. In addition, the effluent from the bottom of the distillation column will cost $50 for each kmol of acetone plus $0.004 for each kmol of water to treat before it can be disposed of to sewer.

7. A company has the option of investing in one of the two projects *A* or *B*. The capital cost of both projects is $1,000,000. The predicted annual cash flows for both projects are shown in Table 2.14. For each project, calculate the:
   a. payback time for each project in terms of the average annual cash flow
   b. return on investment
   c. discounted cash flow rate of return
   What do you conclude from the result?

Table 2.14 Cash flows for two competing projects.

Year	Cash flows $1000	
	Project *A*	Project *B*
0	−1000	−1000
1	150	500
2	250	450
3	350	300
4	400	200
5	400	100

8. A company is considering the projects given in Table 2.15.

Table 2.15 Cash flow for two competing projects.

	End of year	Project *A*	Project *B*
Investment ($)	0	210,000	50,000
Net cash inflows ($)	1	70,000	20,000
	2	70,000	20,000
	3	70,000	20,000
	4	70,000	20,000
	5	70,000	20,000

   For both projects, calculate the following.
   a. The payback time for each project in terms of the average annual cash flow
   b. The net present value at the current lending interest rate of 10%
   c. The discounted cash flow rate of return.
   On the basis of a comparison of these three measures, which project would you prefer? Explain your decision.

9. A process has been developed for a new product for which the market is uncertain. A plant to produce 50,000 $t \cdot y^{-1}$ requires an investment of $10,000,000, and the expected project life is five years. Fixed operating costs are expected to be $750,000 y^{-1}, and variable operating costs (excluding raw materials) expected to be $40 t^{-1} product. The stoichiometric raw material costs are $80 t^{-1} product. The yield of product per ton of raw material is 80%. Tax is paid in the same year as the relevant profit is made at a rate of 35%. Calculate the selling price of the product to give a minimum acceptable discounted cash flowrate of return of 15% year.

10. How can the concept of simple payback be improved to give a more meaningful measure of project profitability?

11. It is proposed to build a plant to produce 170,000 $t \cdot y^{-1}$ of a commodity chemical. A study of the supply and demand projections for the product indicates that current installed capacity in the industry is 6.8×10^6 $t \cdot y^{-1}$, whereas total production is running at 5.0×10^6 $t \cdot y^{-1}$. Maximum plant utilization is thought to be around 90%. If the demand for the product is expected to grow at 8% per year, and it will take 3 years to commission a new plant from the start of a project, what do you conclude about the prospect for the proposed project?

REFERENCES

1. Guthrie KM (1969) Data and Techniques for Preliminary Capital Cost Estimating, *Chem Eng*, **76**: 114.
2. Anson HA (1977) *A New Guide to Capital Cost Estimating*, IChemE, UK.
3. Hall RS, Matley J and McNaughton KJ (1982) Current Costs of Process Equipment, *Chem Eng*, **89**: 80.
4. Ulrich GD (1984) *A Guide to Chemical Engineering Process Design and Economics*, John Wiley, New York.
5. Hall RS, Vatavuk WM and Matley J (1988) Estimating Process Equipment Costs, *Chem Eng*, **95**: 66.
6. Remer DS and Chai LH (1990) Design Cost Factors for Scaling-up Engineering Equipment, *Chem Eng Prog*, **86**: 77.
7. Gerrard AM (2000) *Guide to Capital Cost Estimating*, 4th Edition, IChemE, UK.
8. Peters MS, Timmerhaus KD and West RE (2003) *Plant Design and Economics for Chemical Engineers*, 5th Edition, McGraw-Hill, New York.
9. Mulet A, Corripio AB and Evans LB (1981) Estimate Costs of Pressure Vessels Via Correlations, *Chem Eng*, **Oct**: 145.
10. Mulet A, Corripio AB and Evans LB (1981) Estimate Costs of Distillation and Absorption Towers Via Correlations, *Chem Eng*, **Dec**: 77.
11. Lang HJ (1947) Cost Relationships in Preliminary Cost Estimation, *Chem Eng*, **54**: 117.
12. Hand WE (1958) From Flowsheet to Cost Estimate, *Petrol Refiner*, **37**: 331.
13. Bauman HC (1955) Estimating Costs of Process Auxiliaries, *Chem Eng Prog*, **51**: 45.
14. Holland FA, Watson FA and Wilkinson JK (1983) *Introduction to Process Economics*, 2nd Edition, John Wiley, New York.
15. Brennan D (1998) *Process Industry Economics*, IChemE, UK.
16. Gary JH and Handwerk GE (2001) *Petroleum Refining Technology and Economics*, 4th Edition, Marcel Dekker.
17. Bravo JL (1997) Select Structured Packings or Trays? *Chem Eng Prog*, **July**: 36.
18. Allen DH, (1980) *A Guide to the Economic Evaluation of Projects*, IChemE, Rugby, UK.

3 Optimization

3.1 OBJECTIVE FUNCTIONS

Optimization will almost always be required at some stage in a process design. It is usually not necessary for a designer to construct an optimization algorithm in order to carry out an optimization, as general-purpose software is usually available for this. However, it is necessary for the designer to have some understanding of how optimization works in order to avoid the pitfalls that can occur. More detailed accounts of optimization can be found elsewhere[1-3].

Optimization problems in process design are usually concerned with maximizing or minimizing an *objective function*. The objective function might typically be to maximize economic potential or minimize cost. For example, consider the recovery of heat from a hot waste stream. A heat exchanger could be installed to recover the waste heat. The heat recovery is illustrated in Figure 3.1a as a plot of temperature versus enthalpy. There is heat available in the hot stream to be recovered to preheat the cold stream. But how much heat should be recovered? Expressions can be written for the recovered heat as:

$$Q_{REC} = m_H C_{P,H}(T_{H,in} - T_{H,out}) \quad (3.1)$$

$$Q_{REC} = m_C C_{P,C}(T_{C,out} - T_{C,in}) \quad (3.2)$$

where Q_{REC} = recovered heat
m_H, m_C = mass flowrates of the hot and cold streams
$C_{P,H}, C_{P,C}$ = specific heat capacity of the hot and cold streams
$T_{H,in}, T_{H,out}$ = hot stream inlet and outlet temperatures
$T_{C,in}, T_{C,out}$ = cold stream inlet and outlet temperatures

The effect of the heat recovery is to decrease the energy requirements. Hence, the energy cost of the process:

$$Energy\ cost = (Q_H - Q_{REC})C_E \quad (3.3)$$

where Q_H = process hot utility requirement prior to heat recovery from the waste stream
C_E = unit cost of energy

There is no change in cost associated with cooling as the hot stream is a waste stream being sent to the environment.

Chemical Process Design and Integration R. Smith
 ISBNs: 0-471-48680-9 (HB); 0-471-48681-7 (PB)

An expression can also be written for the heat transfer area of the recovery exchanger:

$$A = \frac{Q_{REC}}{U \Delta T_{LM}} \quad (3.4)$$

where A = heat transfer area
U = overall heat transfer coefficient
ΔT_{LM} = logarithmic mean temperature difference

$$= \frac{(T_{H,in} - T_{C,out}) - (T_{H,out} - T_{C,in})}{\ln\left[\dfrac{T_{H,in} - T_{C,out}}{T_{H,out} - T_{C,in}}\right]}$$

In turn, the area of the heat exchanger can be used to estimate the annualized capital cost:

$$Annualized\ capital\ cost = (a + bA^c)AF \quad (3.5)$$

where a, b, c = cost coefficients
AF = annualization factor (see Chapter 2)

Suppose that the mass flowrates, heat capacities and inlet temperatures of both streams are fixed and the current hot utility requirement, unit cost of energy, overall heat transfer coefficient, cost coefficients and annualization factor are known. Equations 3.1 to 3.5, together with the specifications for the 13 variables m_H, m_C, $C_{P,H}$, $C_{P,C}$, $T_{H,in}$, $T_{C,in}$, U, a, b, c, AF, Q_H and C_E, constitute 18 *equality constraints*. In addition to these 13 variables, there are a further six unknown variables Q_{REC}, $T_{H,out}$, $T_{C,out}$, *Energy cost, Annualized capital cost* and A. Thus, there are 18 equality constraints and 19 variables, and the problem cannot be solved. For the system of equations (equality constraints) to be solved, the number of variables must be equal to the number of equations (equality constraints). It is underspecified. Another specification (equality constraint) is required to solve the problem; there is one *degree of freedom*. This degree of freedom can be optimized; in this case, it is the sum of the annualized energy and capital costs (i.e. the total cost), as shown in Figure 3.1b.

If the mass flowrate of the cold stream through the exchanger had not been fixed, there would have been one fewer equality constraint, and this would have provided an additional degree of freedom and the optimization would have been a two-dimensional optimization. Each degree of freedom provides an opportunity for optimization.

Figure 3.1b illustrates how the investment in the heat exchanger is optimized. As the amount of recovered heat increases, the cost of energy for the system decreases. On the other hand, the size and capital cost of the heat exchange equipment increase. The increase in size of heat

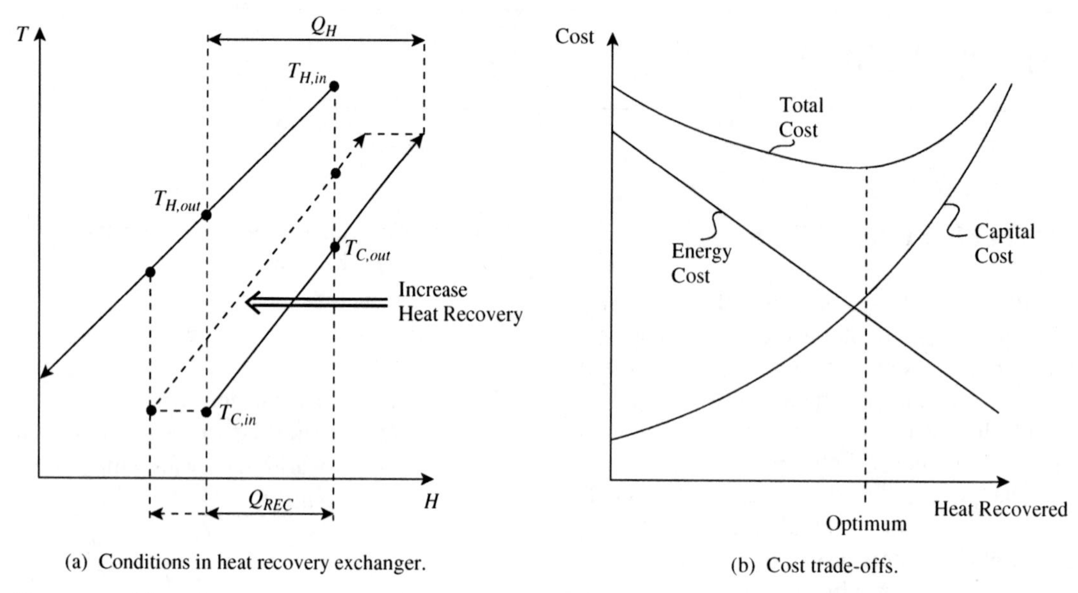

(a) Conditions in heat recovery exchanger. (b) Cost trade-offs.

Figure 3.1 Recovery of heat from a waste steam involves a trade-off between reduced energy cost and increased capital cost of heat exchanger.

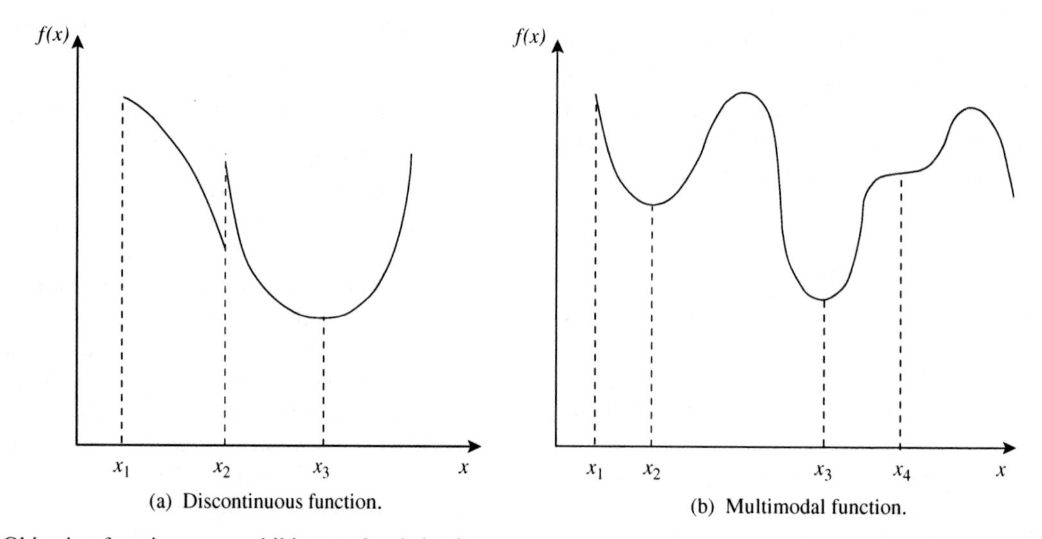

(a) Discontinuous function. (b) Multimodal function.

Figure 3.2 Objective functions can exhibit complex behavior.

exchanger results from the greater amount of heat to be transferred. Also, because the conditions of the waste stream are fixed, as more heat is recovered from the waste stream, the temperature differences in the heat exchanger become lower, causing a sharp rise in the capital cost. In theory, if the recovery was taken to its limit of zero temperature difference, an infinitely large heat exchanger with infinitely large capital cost would be needed. The costs of energy and capital cost can be expressed on an annual basis, as explained in Chapter 2, and combined as shown in Figure 3.1b to obtain the total cost. The total cost shows a minimum, indicating the optimum size of the heat exchanger.

Given a mathematical model for the objective function in Figure 3.1b, finding the minimum point should be straightforward and various strategies could be adopted for this purpose. The objective function is continuous. Also, there is only one extreme point. Functions with only one extreme point (maximum or minimum) are termed *unimodal*. By contrast, consider the objective functions in Figure 3.2. Figure 3.2a shows an objective function to be minimized that is discontinuous. If the search for the minimum is started at point x_1, it could be easily concluded that the optimum point is at x_2, whereas the true optimum is at x_3. Discontinuities can present problems to optimization algorithms searching for the optimum. Figure 3.2b shows an objective function that has a number of points where the gradient is zero. These points where the gradient is zero are known as *stationary points*. Functions that exhibit a number of stationary points are known as *multimodal*. If

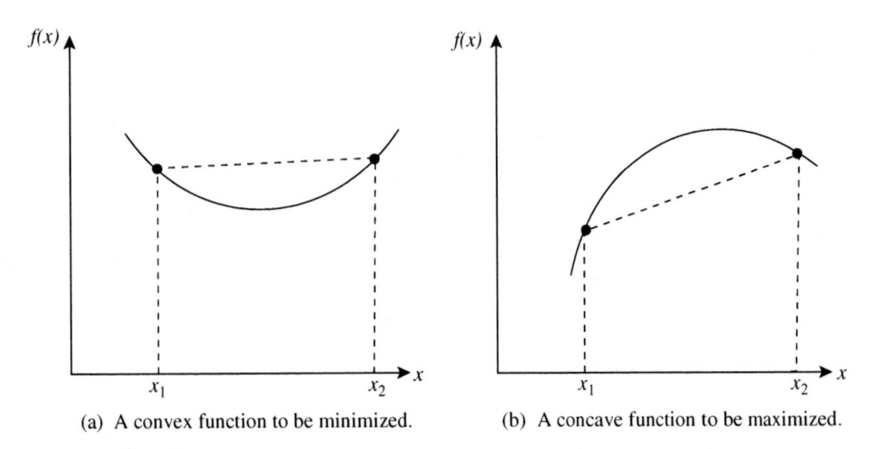

(a) A convex function to be minimized.

(b) A concave function to be maximized.

Figure 3.3 Convex and concave functions.

the function in Figure 3.2b is to be minimized, it has a *local optimum* at x_2. If the search is started at x_1, it could be concluded that the global optimum is at x_2. However, the *global optimum* is at x_3. The gradient is zero at x_4, which is a *saddle point*. By considering the gradient only, this could be confused with a maximum or minimum. Thus, there is potentially another local optimum at x_4. Like discontinuities, multimodality presents problems to optimization algorithms. It is clear that it is not sufficient to find a point of zero gradient in the objective function in order to ensure that the optimum has been reached. Reaching a point of zero gradient is a *necessary condition* for optimality but not a *sufficient condition*.

To establish whether an optimal point is a local or a global optimum, the concepts of *convexity* and *concavity* must be introduced. Figure 3.3a shows a function to be minimized. In Figure 3.3a, if a straight line is drawn between any two points on the function, then the function is *convex* if all the values of this line lie above the curve. Similarly, if an objective function is to be maximized, as shown in Figure 3.3b, and a straight line drawn between any two points on the function, then if all values on this line lie below the curve, the function is *concave*. A convex or concave function provides a single optimum. Thus, if a minimum is found for a function that is to be minimized and is known to be convex, then it is the global optimum. Similarly, if a maximum is found for a function that is to be maximized and known to be concave, then it is the global optimum. On the other hand, a nonconvex or nonconcave function may have multiple local optima.

Searching for the optimum in Figures 3.1 to 3.3 constitutes a one-dimensional search. If the optimization involves two variables, say x_1 and x_2 corresponding to the function $f(x_1, x_2)$, it can be represented as a *contour plot* as shown in Figure 3.4. Figure 3.4 shows a function $f(x_1, x_2)$ to be minimized. The contours represent lines of uniform values of the objective function. The objective function in Figure 3.4 is multimodal, involving a local optimum and a global optimum. The concepts of convexity and concavity can be extended to problems with more than one

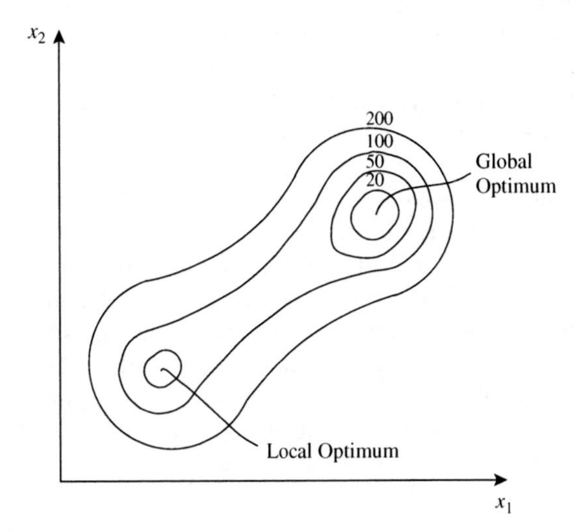

Figure 3.4 A contour plot of a multimodal function to be minimized.

dimension[1]. The objective function in Figure 3.4 is nonconvex. A straight line cannot be drawn between any two points on the surface represented by the contours to ensure that all points on the straight line are below the surface. These concepts can be expressed in a formal way mathematically, but is outside the scope of this text[1-3].

The optimization might well involve more than two variables in a *multivariable optimization*, but in this case, it is difficult to visualize the problem.

3.2 SINGLE-VARIABLE OPTIMIZATION

Searching for the optimum (minimum) for the objective function in Figure 3.1b involves a one-dimensional search across a single variable. In the case of Figure 3.1b, a search is made for the amount of heat to be recovered. An example of a method for single-variable search is *region elimination*. The function is assumed to be unimodal. Figure 3.5 illustrates the approach. In Figure 3.5, the objective function is evaluated at two points. If the function

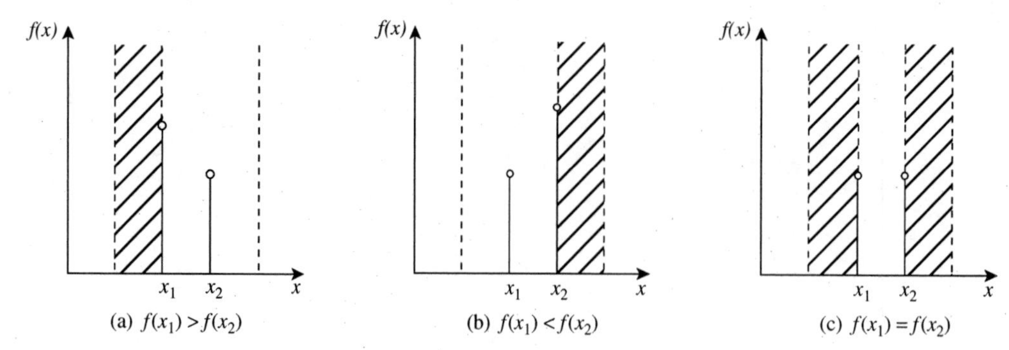

Figure 3.5 Region elimination for the optimization of a single variable.

is being minimized, then in the case of Figure 3.5a, x_2 is less than x_1 and the region to the left of x_1 can be eliminated. This is possible only for the case of a unimodal function. In Figure 3.5b, the objective function is lower at x_1 than at x_2. In this case, the region to the right of x_2 can be eliminated. Finally, in Figure 3.5c, x_1 and x_2 are equal. In this case, the regions to the left of x_1 and to the right of x_2 can be eliminated. If equal intervals are chosen to determine the objective function, as shown in Figure 3.5, then at least one-third of the region can be eliminated. This process can then be repeated in the remaining region in order to identify more precisely the location of the optimum. A number of other methods for region elimination are also possible[1].

A more sophisticated method for optimization of a single variable is Newton's method, which exploits first and second derivatives of the objective function. Newton's method starts by supposing that the following equation needs to be solved:

$$f(x) = 0 \tag{3.6}$$

The solution of this equation needs to start with an initial guess for the solution as x_1. The next guess x_2 can be approximated by:

$$f(x_2) \approx f(x_1) + (x_2 - x_1)f'(x) \tag{3.7}$$

where $f'(x)$ is the derivative (gradient) of $f(x)$. If x_2 is to solve Equation 3.6, then $f(x_2) = 0$. This is substituted into Equation 3.7 and rearranged to give:

$$x_2 = x_1 - \frac{f(x_1)}{f'(x_1)} \quad \text{for } f'(x_1) \neq 0 \tag{3.8}$$

However, x_2 is only an approximation to the solution, as Equation 3.7 is only an approximation. Hence, the application of Equation 3.8 must be repeated such that x_1 in Equation 3.8 is replaced by the new approximation (x_2). Solving Equation 3.8 again provides the next approximation x_3, and so on. Equation 3.8 is applied repeatedly until successive new approximations to the solution change by less than an acceptable tolerance. However, in a single-variable optimization, rather than solving Equation 3.6, the

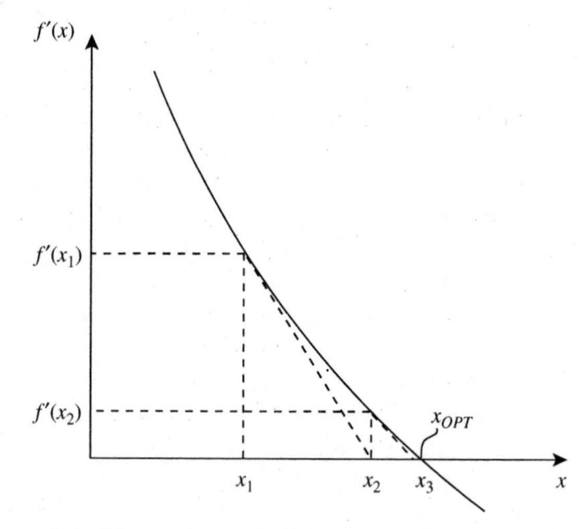

Figure 3.6 Newton's method for the optimization of a single variable.

stationary point needs to be found where the gradient is zero:

$$f'(x) = 0 \tag{3.9}$$

For this, Equation 3.8 takes the form:‘

$$x_{k+1} = x_k - \frac{f'(x_k)}{f''(x_k)} \quad \text{for } f''(x_k) \neq 0 \tag{3.10}$$

where $f''(x_k)$ is the second derivative of $f(x)$ at x_k on iteration k. Equation 3.10 is applied repeatedly until the location of the stationary point is identified within a specified tolerance. This procedure is illustrated in Figure 3.6. Newton's method can be an extremely efficient way to carry out single-variable optimization. However, if the objective function is not unimodal, the method can become unstable. In such situations, the stability depends on the initial guess.

3.3 MULTIVARIABLE OPTIMIZATION

The problem of multivariable optimization is illustrated in Figure 3.4. Search methods used for multivariable optimization can be classified as *deterministic* and *stochastic*.

1. *Deterministic methods.* Deterministic methods follow a predetermined search pattern and do not involve any guessed or random steps. Deterministic methods can be further classified into *direct* and *indirect* search methods. Direct search methods do not require derivatives (gradients) of the function. Indirect methods use derivatives, even though the derivatives might be obtained numerically rather than analytically.

(a) *Direct search methods.* An example of a direct search method is a *univariate search*, as illustrated in Figure 3.7. All of the variables except one are fixed and the remaining variable is optimized. Once a minimum or maximum point has been reached, this variable is fixed and another variable optimized, with the remaining variables being fixed. This is repeated until there is no further improvement in the objective function. Figure 3.7 illustrates a two-dimensional search in which x_1 is first fixed and x_2 optimized. Then x_2 is fixed and x_1 optimized, and so on until no further improvement in the objective function is obtained. In Figure 3.7, the univariate search is able to locate the global optimum. It is easy to see that if the starting point for the search in Figure 3.7 had been at a lower value of x_1, then the search would have located the local optimum, rather than the global optimum. For searching multivariable optimization problems, often the only way to ensure that the global optimum has been reached is to start the optimization from different initial points.

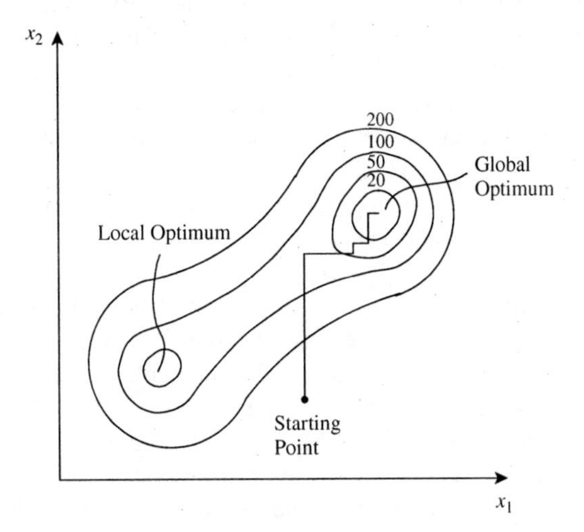

Figure 3.7 A univariate search.

Another example of a direct search is a *sequential simplex* search. The method uses a regular geometric shape (a simplex) to generate search directions. In two dimensions, the simplest shape is an equilateral triangle. In three dimensions, it is a regular tetrahedron. The objective function is evaluated at the vertices of the simplex, as illustrated in Figure 3.8. The objective function must first be evaluated at the Vertices A, B and C. The general direction of search is projected away from the worst vertex (in this case Vertex A) through the centroid of the remaining vertices (B and C), Figure 3.8a. A new simplex is formed by replacing the worst vertex by a new point that is the mirror image of the simplex (Vertex D), as shown in Figure 3.8a. Then Vertex D replaces Vertex A, as Vertex A is an inferior point. The simplex vertices for the next step are B, C and D. This process is repeated for successive moves in a zigzag fashion, as shown in Figure 3.8b. The direction of search can change as illustrated in Figure 3.8c. When the simplex is close to the optimum, there may be some repetition of simplexes, with the search going around in circles. If this is the case, then the size of the simplex should be reduced.

(b) *Indirect search methods.* Indirect search methods use derivatives (gradients) of the objective function. The

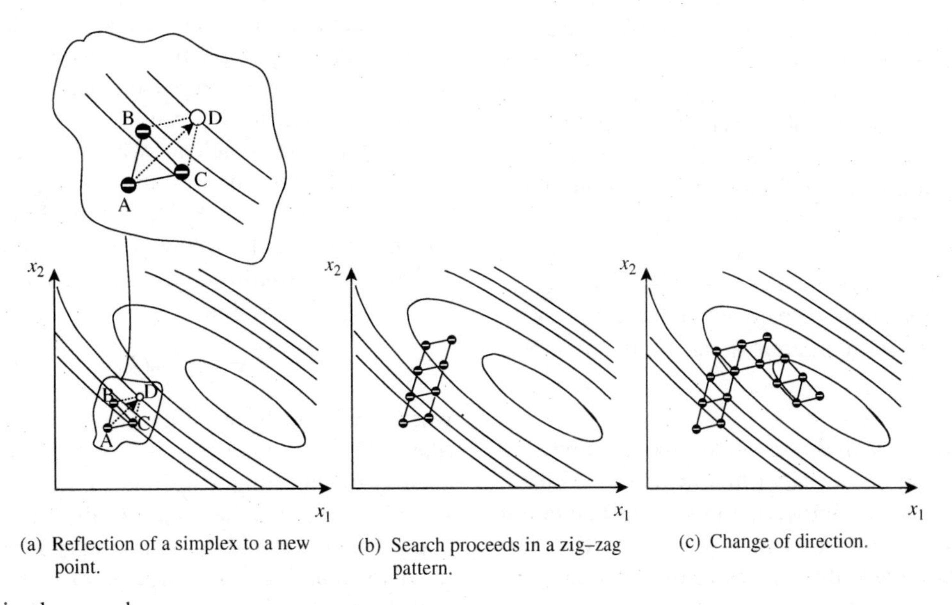

Figure 3.8 The simplex search.

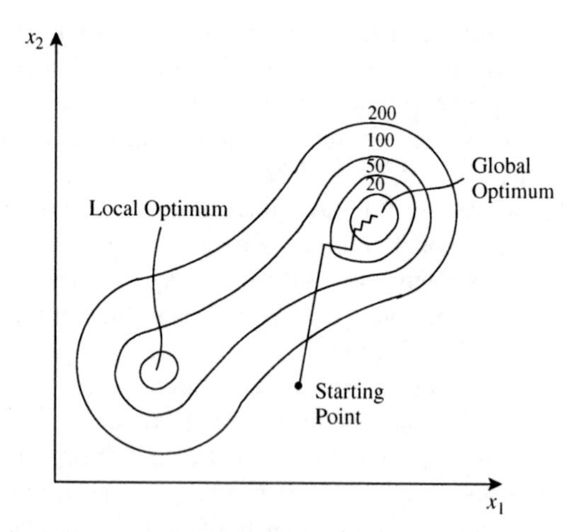

Figure 3.9 Method of steepest descent.

derivatives may be obtained analytically or numerically. Many methods are available for indirect search. An example is the method of *steepest descent* in a minimization problem. The direction of steepest descent is the search direction that gives the maximum rate of change for the objective function from the current point. The method is illustrated in Figure 3.9. One problem with this search method is that the appropriate step size is not known, and this is under circumstances when the gradient might change significantly during the search. Another problem is that the search can slow down significantly as it reaches the optimum point. If a search is made for the maximum in an objective function, then the analogous search is one of *steepest ascent*.

The method of steepest descent uses only first-order derivatives to determine the search direction. Alternatively, Newton's method for single-variable optimization can be adapted to carry out multivariable optimization, taking advantage of both first- and second-order derivatives to obtain better search directions[1]. However, second-order derivatives must be evaluated, either analytically or numerically, and multimodal functions can make the method unstable. Therefore, while this method is potentially very powerful, it also has some practical difficulties.

One fundamental practical difficulty with both the direct and indirect search methods is that, depending on the shape of the solution space, the search can locate local optima, rather than the global optimum. Often, the only way to ensure that the global optimum has been reached is to start the optimization from different initial points and repeat the process.

2. *Stochastic search methods*. In all of the optimization methods discussed so far, the algorithm searches the objective function seeking to improve the objective function at each step, using information such as gradients. Unfortunately, as already noted, this process can mean that the search is attracted towards a local optimum. On the other hand, stochastic search methods use random choice to guide the search and can allow deterioration of the objective function during the search. It is important to recognize that a randomized search does not mean a directionless search. Stochastic search methods generate a randomized path to the solution on the basis of probabilities. Improvement in the objective function becomes the ultimate rather than the immediate goal, and some deterioration in the objective function is tolerated, especially during the early stages of the search. In searching for a minimum in the objective function, rather than the search always attempting to go downhill, stochastic methods allow the search to also sometimes go uphill. Similarly, if the objective function is to be maximized, stochastic methods allow the search to sometimes go downhill. As the optimization progresses, the ability of the algorithm to accept deterioration in the objective function is gradually removed. This helps to reduce the problem of being trapped in a local optimum.

Stochastic methods do not need auxiliary information, such as derivatives, in order to progress. They only require an objective function for the search. This means that stochastic methods can handle problems in which the calculation of the derivatives would be complex and cause deterministic methods to fail.

Two of the most popular stochastic methods are *simulated annealing* and *genetic algorithms*.

(a) *Simulated annealing*. Simulated annealing emulates the physical process of annealing of metals[4,5]. In the physical process, at high temperatures, the molecules of the liquid move freely with respect to one another. If the liquid is cooled slowly, thermal mobility is lost. The atoms are able to line themselves up and form perfect crystals. This crystal state is one of minimum energy for the system. If the liquid metal is cooled quickly, it does not reach this state but rather ends up in a polycrystalline or amorphous state having higher energy. So the essence of the process is slow cooling, allowing ample time for redistribution of the atoms as they lose mobility to reach a state of minimum energy.

With this physical process in mind, an algorithm can be suggested in which a system moves from one point to another, with the resulting change in the objective function from E_1 to E_2. The probability of this change can be assumed to follow a relationship similar to the Boltzmann probability distribution (maintaining the analogy with physical annealing)[4]:

$$P = \exp[-(E_2 - E_1)/T] \tag{3.11}$$

where P is the probability, E is the objective function (the analogy of energy) and T is a control parameter (the analogy of annealing temperature). Relationships other than Equation 3.11 can be used. If the objective function is being minimized and E_2 is less than E_1 (i.e. the objective function improves as a result of the move), then the probability from Equation 3.11 is greater than unity. In such cases, the move is arbitrarily assigned to a probability of unity, and the system should always accept such a move. If the probability in Equation 3.11 is 0, the move is rejected. If the probability is between zero and unity (i.e. the objective function gets worse as a result of the move), then a criterion is needed to dictate whether the move is accepted or rejected. An arbitrary cutoff point (e.g. reject any move with a probability of less than 0.5) could be made, but the most appropriate value is likely to change from problem to problem. Instead, the analogy with physical annealing is maintained, and the probability from Equation 3.11 is compared with the output of a random number generator that creates random numbers between zero and unity. If Equation 3.11 predicts a probability greater than the random number generator, then the move is accepted. If it is less, then the move is rejected and another move is attempted instead. Thus, Equation 3.11 dictates whether a move is accepted or rejected. In this way, the method will always accept a downhill step when minimizing an objective function, while sometimes it will take an uphill step.

However, as the optimization progresses, the possibility of accepting moves that result in a deterioration of the objective function needs to be removed gradually. This is the role of the control parameter (the analogy of temperature) in Equation 3.11. The control parameter is gradually decreased, which results in the probability in Equation 3.11 gradually decreasing for moves in which the objective function deteriorates. When this is compared with the random number, it gradually becomes more likely that the move will be rejected. An *annealing schedule* is required that controls how the control parameter is lowered from high to low values.

Thus, the way the algorithm works is to set an initial value for the control parameter. At this setting of the control parameter, a series of random moves are made. Equation 3.11 dictates whether an individual move is accepted or rejected. The control parameter (annealing temperature) is lowered and a new series of random moves is made, and so on. As the control parameter (annealing temperature) is lowered, the probability of accepting deterioration in the objective function, as dictated by Equation 3.11, decreases. In this way, the acceptability for the search to move uphill in a minimization or downhill during maximization is gradually withdrawn.

Whilst simulated annealing can be extremely powerful in solving difficult optimization problems with many local optima, it has a number of disadvantages. Initial and final values of the control parameter, an annealing schedule and the number of random moves for each setting of the control parameter must be specified. Also, because each search is random, the process can be repeated a number of times to ensure that an adequate search of the solution space has been made. This can be very useful in providing a number of solutions in the region of the global optimum, rather than a single solution. Such multiple solutions can then be screened not just on the basis of cost but also on many other issues, such as design complexity, control, safety, and so on, that are difficult to include in the optimization.

(b) *Genetic algorithms*. Genetic algorithms draw their inspiration from biological evolution[6]. Unlike all of the optimization methods discussed so far, which move from one point to another, a genetic algorithm moves from one set of points (termed a *population*) to another set of points. Populations of *strings* (the analogy of chromosomes) are created to represent an underlying set of parameters (e.g. temperatures, pressures or concentrations). A simple genetic algorithm exploits three basic operators: reproduction, crossover and mutation.

Reproduction is a process in which individual members of a population are copied according to the objective function in order to generate new population sets. The operator is inspired by "natural selection" and the "survival of the fittest". The easiest way to understand reproduction is to make an analogy with a roulette wheel. In a roulette wheel, the probability of selection is proportional to the area of the slots in the wheel. In a genetic algorithm, the probability of reproduction is proportional to the fitness (the objective function). Although the selection procedure is stochastic, fitter strings are given a better chance of selection (survival). The reproduction operator can be implemented in a genetic algorithm in many ways[6].

Crossover involves the combination of genetic material from two successful parents to form two *offspring* (children). Crossover involves cutting two parent strings at random points and combining differently to form new offspring. The crossover point is generated randomly. Crossover works as a local search operator and spreads good properties amongst the population. The fraction of new population generated by crossover is generally large (as observed in nature) and is controlled stochastically.

Mutation creates new strings by randomly changing (mutating) parts of strings, but (as with nature) with a low probability of occurring. A random change is made in one of the genes in order to preserve diversity. Mutation creates a new solution in the neighborhood of a point undergoing mutation.

A genetic algorithm works by first generating an initial population randomly. The population is evaluated according to its fitness (value of the objective function). A reproduction operator then provides an intermediate population using stochastic selection but biased towards survival of the fittest. Crossover and mutation operators are then applied to the intermediate population to create a new generation of population. The new population is evaluated according to its fitness and the search is continued with further reproduction, crossover and mutation until the population meets the required convergence criterion (generally the difference between the average and maximum fitness value or number of generations).

In this way, genetic algorithms create better solutions by reproduction, crossover and mutation such that candidate solutions improve from one *generation* to the next. The algorithm evolves over many generations and eventually converges on the fittest string. A parallel search reduces the chance of being stuck in a local optimum.

The major strengths of stochastic search methods are that they can tackle the most difficult optimization problems and guarantee finding an optimal solution. Another major advantage is that, if the solution space is highly irregular, they can produce a range of solutions with close to optimal performance, rather a single optimal point. This opens up a range of solutions to the designer, rather than having just one option. However, there are disadvantages with stochastic optimization methods also. They can be very slow in converging and usually need to be adapted to solve particular problems. Tailoring the methods to suit specific applications makes them much more efficient.

3.4 CONSTRAINED OPTIMIZATION

Most optimization problems involve constraints. For example, it might be necessary for a maximum temperature or maximum flowrate not to be exceeded. Thus, the general form of an optimization problem involves three basic elements:

1. An objective function to be optimized (e.g. minimize total cost, maximize economic potential, etc.).
2. Equality constraints, which are equations describing the model of the process or equipment.
3. Inequality constraints, expressing minimum or maximum limits on various parameters.

These three elements of the general optimization problem can be expressed mathematically as

$$\begin{aligned} &\text{minimize} && f(x_1, x_2, \ldots\ldots, x_n) \\ &\text{subject to} && h_i(x_1, x_2, \ldots\ldots, x_n) = 0 \; (i = 1, p) \\ & && g_i(x_1, x_2, \ldots\ldots, x_n) \geq 0 \; (i = 1, q) \end{aligned} \tag{3.12}$$

In this case, there are n design variables, with p equality constraints and q inequality constraints. The existence of such constraints can simplify the optimization problem by reducing the size of the problem to be searched or avoiding problematic regions of the objective function. In general though, the existence of the constraints complicates the problem relative to the problem with no constraints.

Now consider the influence of the inequality constraints on the optimization problem. The effect of inequality constraints is to reduce the size of the solution space that must be searched. However, the way in which the constraints bound the feasible region is important. Figure 3.10 illustrates the concept of convex and nonconvex regions.

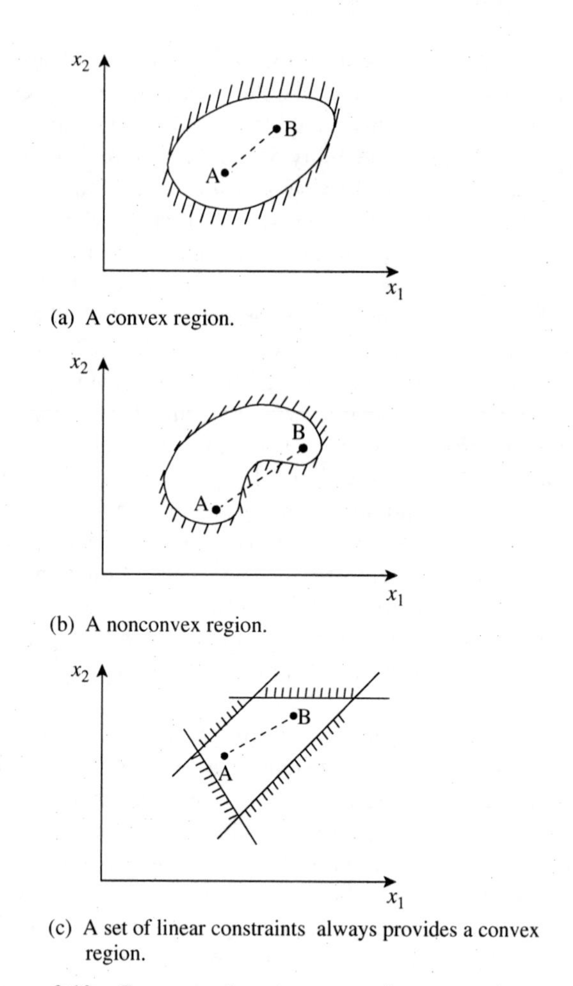

Figure 3.10 Convex and nonconvex regions.

Figure 3.10a shows a convex region. In a convex region, a straight line can be drawn between any two points A and B located within the feasible region and all points on this straight line will also be located within the feasible region. By contrast, Figure 3.10b shows a nonconvex region. This time when a straight line is drawn between two points A and B located within the region, some of the points on the straight line can fall outside the feasible region. Figure 3.10c shows a region that is constrained by a set of linear inequality constraints. The region shown in Figure 3.10c is convex. But it is worth noting that a set of linear inequality constraints will always provide a convex region[1]. These concepts can be represented mathematically for the general problem[1–3].

Now superimpose the constraints onto the objective function. Figure 3.11a shows a contour diagram that has a set of inequality constraints imposed. The feasible region is convex and an appropriate search algorithm should be able to locate the unconstrained optimum. The unconstrained optimum lies inside the feasible region in Figure 3.11a. At the optimum none of the constraints are *active*. By contrast, consider Figure 3.11b. In this case, the region is also convex, but the unconstrained optimum lies outside

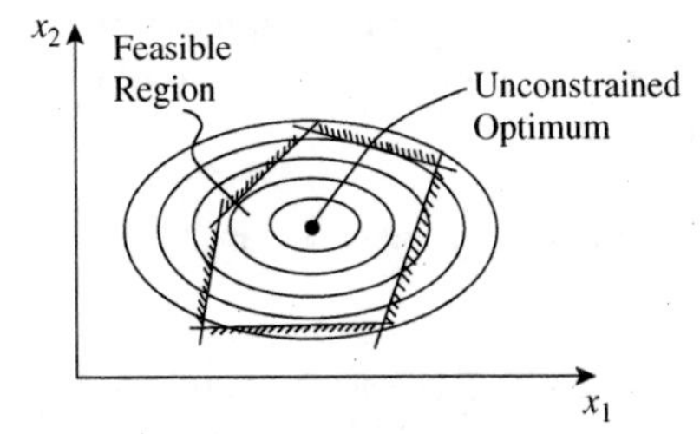

(a) Unconstrained optimum can be reached.

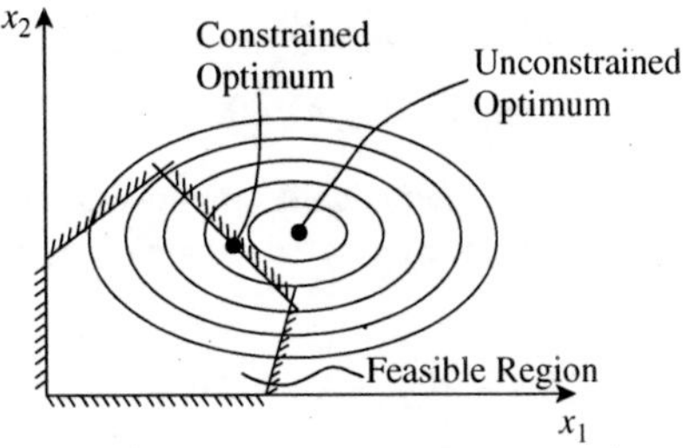

(b) Unconstrained maximum cannot be reached.

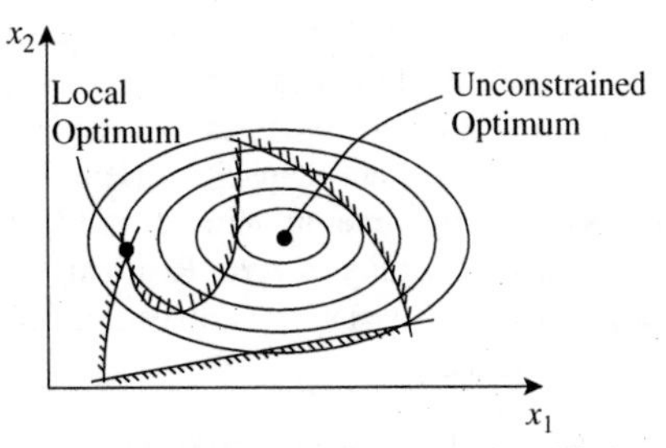

(c) A nonconvex region might prevent the global optimum from being reached.

Figure 3.11 Effects of constraint and optimization.

of the feasible region. This time, the optimum point is on the edge of the feasible region and one of the constraints is active. When the inequality constraint is satisfied, as in Figure 3.11b, it becomes an equality constraint. Figure 3.11c illustrates the potential problem of having a nonconvex region. If the search is initiated in the right-hand part of the diagram at high values of x_1, then it is likely that the search will find the global optimum. However, if the search is initiated to the left of the diagram at low values of x_1, it is likely that the search will locate the local optimum[1].

It should be noted that it is not sufficient to simply have a convex region in order to ensure that a search can locate the global optimum. The objective function must also be convex if it is to be minimized or concave if it is to be maximized.

It is also worth noting that the stochastic optimization methods described previously are readily adapted to the inclusion of constraints. For example, in simulated annealing, if a move suggested at random takes the solution outside of the feasible region, then the algorithm can be constrained to prevent this by simply setting the probability of that move to 0.

3.5 LINEAR PROGRAMMING

An important class of the constrained optimization problems is one in which the objective function, equality constraints and inequality constraints are all linear. A linear function is one in which the dependent variables appear only to the first power. For example, a linear function of two variables x_1 and x_2 would be of the general form:

$$f(x_1, x_2) = a_0 + a_1x_1 + a_2x_2 \tag{3.13}$$

where a_0, a_1 and a_2 are constants. Search methods for such problems are well developed in *linear programming* (LP). Solving such linear optimization problems is best explained through a simple example.

Example 3.1 A company manufactures two products (Product 1 and Product 2) in a batch plant involving two steps (Step I and Step II). The value of Product 1 is 3 $\$\cdot kg^{-1}$ and that of Product 2 is 2 $\$\cdot kg^{-1}$. Each batch has the same capacity of 1000 kg per batch but batch cycle times differ between products. These are given in Table 3.1.

Step I has a maximum operating time of 5000 $h\cdot y^{-1}$ and Step II 6000 $h\cdot y^{-1}$. Determine the operation of the plant to obtain the maximum annual revenue.

Solution For Step I, the maximum operating time dictates that:

$$25n_1 + 10n_2 \leq 5000$$

where n_1 and n_2 are the number of batches per year manufacturing Products 1 and 2 on Step I. For Step II, the corresponding equation is:

$$10n_1 + 20n_2 \leq 6000$$

The feasible solution space can be represented graphically by plotting the above inequality constraints as equality constraints:

$$25n_1 + 10n_2 = 5000$$

$$10n_1 + 20n_2 = 6000$$

This is shown in Figure 3.12. The feasible solution space in Figure 3.12 is given by *ABCD*.

The total annual revenue A is given by

$$A = 3000n_1 + 2000n_2$$

Table 3.1 Times for different steps in the batch process.

	Step I (h)	Step II (h)
Product 1	25	10
Product 2	10	20

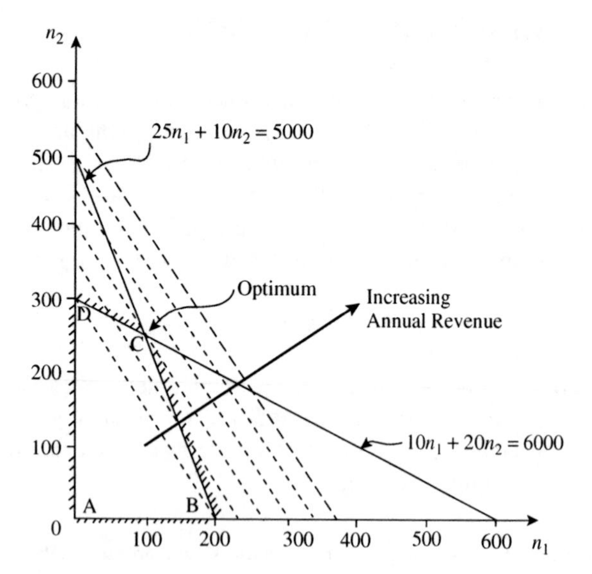

Figure 3.12 Graphical representation of the linear optimization problem from Example 3.1.

On a plot of n_1 versus n_2 as shown in Figure 3.12, lines of constant annual revenue will follow a straight line given by:

$$n_2 = -\frac{3}{2}n_1 + \frac{A}{2000}$$

Lines of constant annual revenue are shown as dotted lines in Figure 3.12, with revenue increasing with increasing distance from the origin. It is clear from Figure 3.12 that the optimum point corresponds with the extreme point at the intersection of the two equality constraints at Point C.

At the intersection of the two constraints:

$$n_1 = 100$$

$$n_2 = 250$$

As the problem involves discrete batches, it is apparently fortunate that the answer turns out to be two whole numbers. However, had the answer turned out not to be a whole number, then the solution would still have been valid because, even though part batches might not be able to be processed, the remaining part of a batch can be processed the following year.

Thus the maximum annual revenue is given by:

$$A = 3000 \times 100 + 2000 \times 250$$

$$= \$800{,}000\ \$\,\mathrm{y}^{-1}$$

Whilst Example 3.1 is an extremely simple example, it illustrates a number of important points. If the optimization problem is completely linear, the solution space is convex and a global optimum solution can be generated. The optimum always occurs at an extreme point, as is illustrated in Figure 3.12. The optimum cannot occur inside the feasible region, it must always be at the boundary. For linear functions, running up the gradient can always increase the objective function until a boundary wall is hit.

Whilst simple two variable problems like the one in Example 3.1 can be solved graphically, more complex problems require a more formal nongraphical approach. This is illustrated by returning to Example 3.1 to solve it in a nongraphical way.

Example 3.2 Solve the problem in Example 3.1 using an analytical approach.

Solution The problem in Example 3.1 was expressed as:

$$A = 3000n_1 + 2000n_2$$

$$25n_1 + 10n_2 \leq 5000$$

$$10n_1 + 20n_2 \leq 6000$$

To solve these equations algebraically, the inequality signs must first be removed by introducing *slack variables* S_1 and S_2 such that:

$$25n_1 + 10n_2 + S_1 = 5000$$

$$10n_1 + 20n_2 + S_2 = 6000$$

In other words, these equations show that if the production of both products does not absorb the full capacities of both steps, then the slack capacities of these two processes can be represented by the variables S_1 and S_2. Since slack capacity means that a certain amount of process capacity remains unused, it follows that the economic value of slack capacity is zero. Realizing that negative production rates and negative slack variables are infeasible, the problem can be formulated as:

$$3000n_1 + 2000n_2 + 0S_1 + 0S_2 = A \quad (3.14)$$

$$25n_1 + 10n_2 + 1S_1 + 0S_2 = 5000 \quad (3.15)$$

$$10n_1 + 20n_2 + 0S_1 + 1S_2 = 6000 \quad (3.16)$$

where $n_1, n_2, S_1, S_2 \geq 0$

Equations 3.15 and 3.16 involve four variables and can therefore not be solved simultaneously. At this stage, the solution can lie anywhere within the feasible area marked $ABCD$ in Figure 3.12. However, providing the values of these variables are not restricted to integer values; two of the four variables will assume zero values at the optimum. In this example, n_1, n_2, S_1 and S_2 are treated as real and not integer variables.

The problem is started with an initial feasible solution that is then improved by a stepwise procedure. The search will be started at the worst possible solution when n_1 and n_2 are both zero. From Equations 3.15 and 3.16:

$$S_1 = 5000 - 25n_1 - 10n_2 \quad (3.17)$$

$$S_2 = 6000 - 10n_1 - 20n_2 \quad (3.18)$$

When n_1 and n_2 are zero:

$$S_1 = 5000 \ S_2 = 6000$$

Substituting in Equation 3.14:

$$A = 3000 \times 0 + 2000 \times 0 + 0 \times 5000 + 0 \times 6000$$

$$= 0 \quad (3.19)$$

This is Point A in Figure 3.12. To improve this initial solution, the value of n_1 and/or the value of n_2 must be increased, because

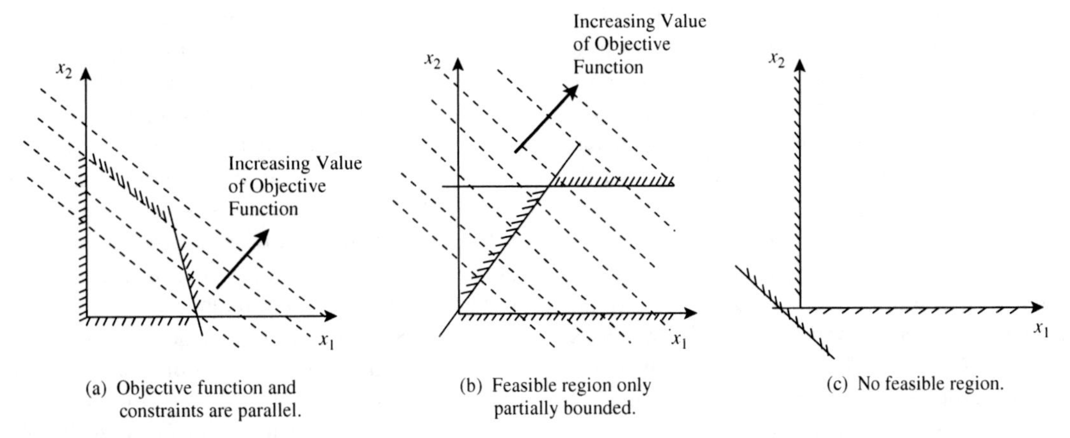

Figure 3.13 Degenerate linear programming problems.

these are the only variables that possess positive coefficients to increase the annual revenue in Equation 3.14. But which variable, n_1 or n_2, should be increased first? The obvious strategy is to increase the variable that makes the greatest increase in the annual revenue, which is n_1. According to Equation 3.18, n_1 can be increased by $6000/10 = 600$ before S_2 becomes negative. If n_1 is assumed to be 600 in Equation 3.17, then S_1 would be negative. Since negative slack variables are infeasible, Equation 3.17 is the dominant constraint on n_1 and it follows that its maximum value is $5000/25 = 200$. Rearranging Equation 3.17:

$$n_1 = 200 - 0.4n_2 - 0.04S_1 \qquad (3.20)$$

which would give a maximum when n_1 and S_1 are zero. Substituting the expression for n_1 in the objective function, Equation 3.14, gives:

$$A = 600{,}000 + 800n_2 - 120S_1 \qquad (3.21)$$

Since n_2 is initially zero, the greatest improvement in the objective function results from making S_1 zero. This is equivalent to making n_1 equal to 200 from Equation 3.20, given n_2 is initially zero. For $n_1 = 200$ and $n_2 = 0$, the annual revenue $A =$ 600,000. This corresponds with Point B in Figure 3.12. However, Equation 3.21 also shows that the profit can be improved further by increasing the value of n_2. Substituting n_1 from Equation 3.20 in Equation 3.18 gives:

$$n_2 = 250 + 0.025S_1 - 0.0625S_2 \qquad (3.22)$$

This means n_2 takes a value of 250 if both S_1 and S_2 are zero. Substituting n_2 from Equation 3.22 in Equation 3.20 gives:

$$n_1 = 100 - 0.05S_1 + 0.025S_2 \qquad (3.23)$$

This means n_1 takes a value of 100 if both S_1 and S_2 are zero. Finally, substituting the expression for n_1 and n_2 in the objective function, Equation 3.14 gives:

$$A = 800{,}000 - 100S_1 - 50S_2 \qquad (3.24)$$

Equations 3.22 to 3.24 show that the maximum annual revenue is \$800,000 y^{-1} when $n_1 = 100$ and $n_2 = 250$. This corresponds with Point C in Figure 3.12.

It is also interesting to note the Equation 3.24 provides some insight into the sensitivity of the solution. The annual revenue would decrease by \$100 for each hour of production lost through poor utilization of Step I. The corresponding effect for Step II would be a reduction of \$50 for each hour of production lost. These values are known as *shadow prices*. If S_1 and S_2 are set to their availabilities of 5000 and 6000 hours respectively, then the revenue from Equation 3.24 becomes zero.

While the method used for the solution of Example 3.1 is not suitable for automation, it gives some insights into the way linear programming problems can be automated. The solution is started by turning the inequality constraints into equality constraints by the use of slack variables. Then the equations are solved to obtain an initial feasible solution. This is improved in steps by searching the extreme points of the solution space. It is not necessary to explore all the extreme points in order to identify the optimum. The method usually used to automate the solution of such linear programming problems is the *simplex algorithm*[1,2]. Note, however, that the simplex algorithm for linear programming should not be confused with the simplex search described previously, which is quite different. Here the term simplex is used to describe the shape of the solution space, which is a convex polyhedron, or simplex.

If the linear programming problem is not formulated properly, it might not have a unique solution, or even any solution at all. Such linear programming problems are termed *degenerate*[1]. Figure 3.13 illustrates some degenerate linear programming problems[1]. In Figure 3.13a, the objective function contours are parallel with one of the boundary constraints. Here there is no unique solution that maximizes the objective function within the feasible region. Figure 3.13b shows a problem in which the feasible region is unbounded. Hence the objective function can increase without bound. A third example is shown in Figure 13.3c, in which there is no feasible region according to the specified constraints.

3.6 NONLINEAR PROGRAMMING

When the objective function, equality or inequality constraints of Equation 3.7 are nonlinear, the optimization becomes a *nonlinear programming* (NLP) problem. The worst case is when all three are nonlinear. Direct and indirect methods that can be used for nonlinear optimization have previously been discussed. Whilst it is possible to include some types of constraints, the methods discussed are not well suited to the inclusion of complex sets of constraints. The stochastic methods discussed previously can readily handle constraints by restricting moves to infeasible solutions, for example in simulated annealing by setting their probability to 0. The other methods discussed are not well suited to the inclusion of complex sets of constraints. It has already been observed in Figure 3.11 that, unlike the linear optimization problem, for the nonlinear optimization problem the optimum may or may not lie on the edge of the feasible region and can, in principle, be anywhere within the feasible region.

One approach that has been adopted for solving the general nonlinear programming problem is *successive linear programming*. These methods linearize the problem and successively apply the linear programming techniques described in the previous section. The procedures involve initializing the problem and linearizing the objective function and all of the constraints about the initial point, so as to fit the linear programming format. Linear programming is then applied to solve the problem. An improved solution is obtained and the procedure repeated. At each successive improved feasible solution, the objective function and constraints are linearized and the linear programming solution repeated, until the objective function does not show any significant improvement. If the solution to the linear programming problem moves to an infeasible point, then the nearest feasible point is located and the procedure applied at this new point.

Another method for solving nonlinear programming problems is based on *quadratic programming* (QP)[1]. Quadratic programming is an optimization procedure that minimizes a quadratic objective function subject to linear inequality or equality (or both types) of constraints. For example, a quadratic function of two variables x_1 and x_2 would be of the general form:

$$f(x_1, x_2) = a_0 + a_1x_1 + a_2x_2 + a_{11}x_1^2 + a_{22}x_2^2 + a_{12}x_1x_2 \tag{3.25}$$

where a_{ij} are constants. Quadratic programming problems are the simplest form of nonlinear programming with inequality constraints. The techniques used for the solution of quadratic programming problems have many similarities with those used for solving linear programming problems[1]. Each inequality constraint must either be satisfied as an equality or it is not involved in the solution of the problem. The quadratic programming technique can thus be reduced to a vertex searching procedure, similar to linear programming[1]. In order to solve the general nonlinear programming problem, quadratic programming can be applied successively, in a similar way to that for successive linear programming, in *successive (or sequential) quadratic programming* (SQP). In this case, the objective function is approximated locally as a quadratic function. For a function of two variables, the function would be approximated by Equation 3.25. By approximating the function as a quadratic and linearizing the constraints, this takes the form of a quadratic programming problem that is solved in each iteration[1]. In general, successive quadratic programming tends to perform better than successive linear programming, because a quadratic rather than a linear approximation is used for the objective function.

It is important to note that neither successive linear nor successive quadratic programming are guaranteed to find the global optimum in a general nonlinear programming problem. The fact that the problem is being turned into a linear or quadratic problem, for which global optimality can be guaranteed, does not change the underlying problem that is being optimized. All of the problems associated with local optima are still a feature of the background problem. When using these methods for the general nonlinear programming problem, it is important to recognize this and to test the optimality of the solution by starting the optimization from different initial conditions.

Stochastic optimization methods described previously, such as simulated annealing, can also be used to solve the general nonlinear programming problem. These have the advantage that the search is sometimes allowed to move uphill in a minimization problem, rather than always searching for a downhill move. Or, in a maximization problem, the search is sometimes allowed to move downhill, rather than always searching for an uphill move. In this way, the technique is less vulnerable to the problems associated with local optima.

3.7 PROFILE OPTIMIZATION

There are many situations in process design when it is necessary to optimize a profile. For example, a reactor design might involve solid catalyst packed into tubes with heat removal via a coolant on the outside of the tubes. Prior to designing the heat transfer arrangement in detail, the designer would like to know the temperature profile along the tube that optimizes the overall reaction conditions[7]. Should it be a constant temperature along the tube? Should it increase or decrease along the tube? Should any increase or decrease be linear, exponential, and so on? Should the profile go through a maximum or minimum? Once the optimum profile has been determined, the catalyst loading and heat transfer arrangements can then be designed to come as close as possible to the optimum temperature profile.

Rather than a parameter varying through space, as in the example of the temperature profile along the reactor, the profile could vary through time. For example, in a batch reactor, the reactants might be loaded into the reactor at the beginning of the batch, the reaction initiated by heating the contents, adding a catalyst, and so on, and the reaction allowed to continue for the required time. The designer would like to know the optimum temperature of the batch through time as the reaction proceeds. Should the temperature be held constant, allowed to increase, decrease, and so on? This presents a *dynamic optimization.*

In continuous processes, parameter profiles might be required to be optimized through space. In batch processes, parameter profiles might need to be optimized through time. How can this be achieved?

A profile generator algorithm can be developed to generate various families of curves that are continuous functions through space or time. In principle, this can be achieved in many ways[7,8]. One way is to exploit two different profiles to generate a wide variety of shapes. Although many mathematical expressions could be used, two types of profile described by the following mathematical equations can be exploited[8].

Type I

$$x = x_F - (x_F - x_0)\left[1 - \frac{t}{t_{total}}\right]^{a_1} \qquad (3.26)$$

Type II

$$x = x_0 - (x_0 - x_F)\left[\frac{t}{t_{total}}\right]^{a_2} \qquad (3.27)$$

In these equations, x is the instantaneous value of any control variable at any space or time t. x_0 is the initial value and x_F is the final value of the control variable. In principle, x can be any control variable such as temperature, reactant feed rate, evaporation rate, heat removed or supplied, and so on. t_{total} is the total distance or time for the profile. The convexity and concavity of the curves are governed by the values of a_1 and a_2. Figures 3.14a and b illustrate typical forms of each curve.

Combining these two profiles across the space or time horizon allows virtually all types of continuous curves to be produced that can be implemented in a practical design. When the two profiles are combined, two additional variables are needed. The value of t_{inter} indicates the point in space or time where the two curves meet and x_{inter} is the corresponding value of the control variable where the curves meet. Figure 3.14c illustrates the form of Type I followed by Type II and Figure 3.14d the form of Type II followed by Type I.

By combining the two curves together, only six variables are needed to generate the various profiles. These six variables are the initial and final values of the control variable x_0 and x_F, two exponential constants a_1 and a_2, the intermediate point in space or time where the two profiles converge t_{inter} and the corresponding intermediate value of the control variable where the profiles converge x_{inter}. Figure 3.15 illustrates the range of shapes that can be produced by putting the two curve types together in different orders and manipulating the six variables. In limiting cases, only one type of profile will be used, rather than two.

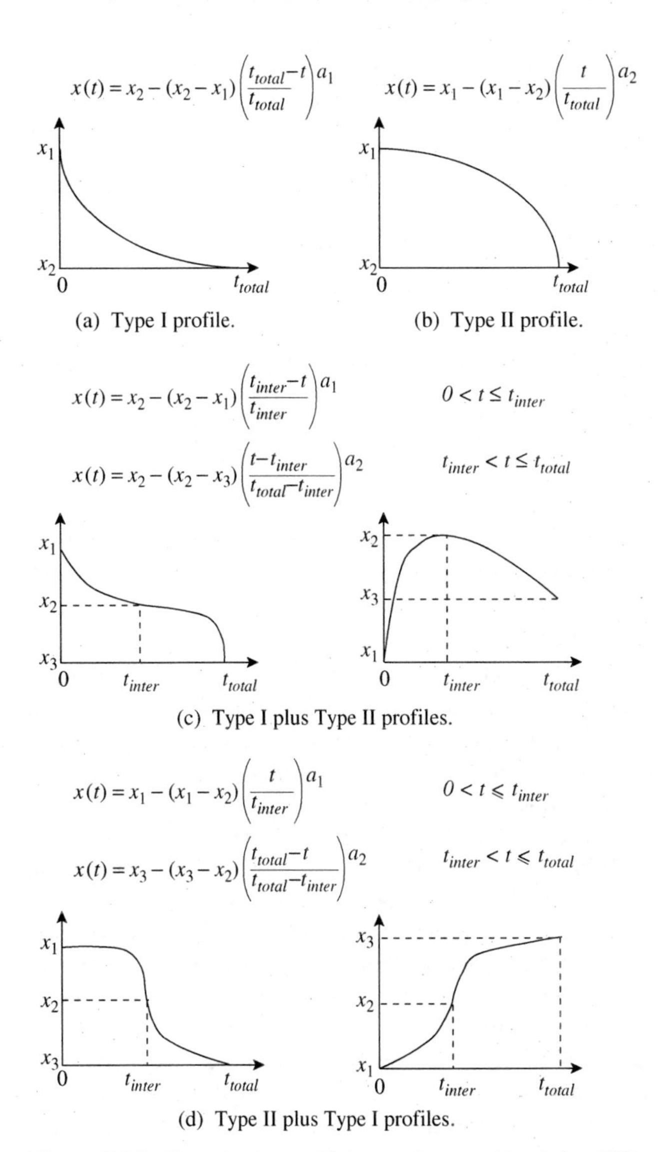

Figure 3.14 Two basic profiles can be combined in different ways.

In formulating profiles, emphasis should be placed on searching for profiles that are continuous and easily implemented in practice. Therefore, curves that include serious discontinuities should normally be avoided. It is meaningless to have a global optimum solution with complex and practically unrealizable profiles. Curve combinations from the above equations such as Type I + Type I or Type II + Type II should not normally be considered as there would be a prominent discontinuity at the intermediate point. The profile generator can be easily extended to combine three or more curves across the space or time horizon instead of two. However, there is little practical use to employ more than two different curves for the majority of problems. The complexity of the profiles increases with the number of curves generated. It should not be forgotten that a way must be found to realize the profile in practice. For a continuous process, the equipment must somehow be

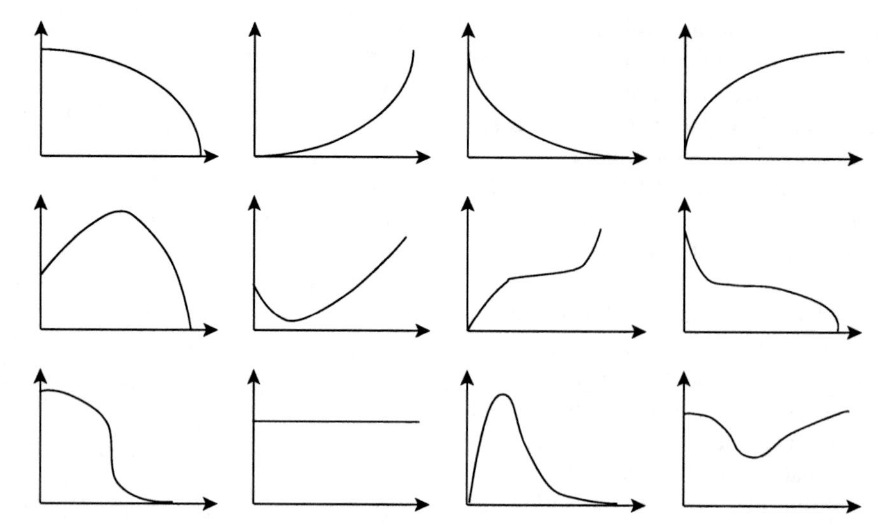

Figure 3.15 Combining the profiles together allows a wide range of profiles to be generated using only seven variables x_1, x_2, x_3, t_{inter}, t_{total}, a_1 and a_2.

designed to follow the profile through space by adjusting rates of reaction, mass transfer, heat transfer, and so on. In a dynamic problem varying through time, a control system must be designed that will allow the profile to be followed through time.

At each point along the profile, the process will have a different performance, depending on the value of the control variable. The process can be modeled in different ways along the profile. The profile can be divided into increments in space, or time in the case of batch processes, and a model developed for each time increment using algebraic equations. The size of the increments must be assessed in such an approach to make sure that the increments are small enough to follow the changes in the profile adequately. Alternatively, the rate of change through space or time can be modeled by differential equations. The profile functions (Equations 3.26 and 3.27) are readily differentiated to obtain gradients for the solution of the differential equations.

Having evaluated the system performance for each setting of the six variables, the variables are optimized simultaneously in a multidimensional optimization, using for example SQP, to maximize or minimize an objective function evaluated at each setting of the variables. However, in practice, many models tend to be nonlinear and hence a stochastic method can be more effective.

The control variables can be constrained to fixed values (e.g. fixed initial temperature in a temperature profile) or constrained to be between certain limits. In addition to the six variables dictating the shape of the profile, t_{total} can also be optimized if required. For example, this can be important in batch processes to optimize the batch cycle time in a batch process, in addition to the other variables.

The approach is readily extended to problems involving multiple profiles. For example, in a batch crystallization process, the temperature profile and evaporation profile can be optimized simultaneously[9]. Each profile optimization would involve six variables using the above profile equations.

Once the optimum profile(s) has been established, its practicality for implementation must be assessed. For a continuous process, the equipment must be able to be designed such that the profile can be followed through space by adjusting rates of reaction, mass transfer, heat transfer, and so on. In a dynamic problem, a control system must be designed that will allow the profile to be followed through time. If the profile is not practical, then the optimization must be repeated with additional constraints added to avoid the impractical features.

3.8 STRUCTURAL OPTIMIZATION

In Chapter 1, two alternative ways were discussed that can be used to develop the structure of a flowsheet. In the first way, an irreducible structure is built by successively adding new features if these can be justified technically and economically. The second way to develop the structure of a flowsheet is to first create a superstructure. This superstructure involves redundant features but includes the structural options that should be considered. This superstructure is then subjected to optimization. The optimization varies the settings of the process parameters (e.g. temperature, flowrate) and also optimizes the structural features. Thus to adopt this approach, both structural and parameter optimization must be carried out. So far, the discussion of optimization has been restricted to parameter optimization. Consider now how structural optimization can be carried out.

The methods discussed for linear and nonlinear programming can be adapted to deal with structural optimization by introducing integer (binary) variables that identify whether a feature exists or not. If a feature exists, its binary variable takes the value 1. If the feature does not exist, then it is set to 0. Consider how different kinds of decisions can be formulated using binary variables[2].

a. *Multiple choice constraints.* It might be required to select only one item from a number of options. This can be represented mathematically by a constraint:

$$\sum_{j=1}^{J} y_j = 1 \qquad (3.28)$$

where y_j is the binary variable to be set to 0 or 1 and the number of options is J. More generally, it might be required to select only m items from a number of options. This can be represented by:

$$\sum_{j=1}^{J} y_j = m \qquad (3.29)$$

Alternatively, it might be required to select at most m items from a number of options, in which case the constraint can be represented by:

$$\sum_{j=1}^{J} y_j \leq m \qquad (3.30)$$

On the other hand, it might be required to select at least m items from a number of options. The constraint can be represented by:

$$\sum_{j=1}^{J} y_j \geq m \qquad (3.31)$$

b. *Implication constraints.* Another type of logical constraint might be that if Item k is selected, Item j must be selected, but not vice versa, then this is represented by the constraint:

$$y_k - y_j \leq 0 \qquad (3.32)$$

A binary variable can be used to set a continuous variable to 0. If a binary variable y is 0, the associated continuous variable x must also be 0 if a constraint is applied such that:

$$x - Uy \leq 0, \quad x \geq 0 \qquad (3.33)$$

where U is an upper limit to x.

c. *Either–or constraints.* Binary variables can also be applied to either–or constraints, known as *disjunctive constraints*. For example, either constraint $g_1(x) \leq 0$ or constraint $g_2(x) \leq 0$ must hold:

$$g_1(x) - My \leq 0 \qquad (3.34)$$

$$g_2(x) - M(1 - y) \leq 0 \qquad (3.35)$$

where M is a large (arbitrary) value that represents an upper limit $g_1(x)$ and $g_2(x)$. If $y = 0$, then $g_1(x) \leq 0$ must be imposed from Equation 3.34. However, if $y = 0$, the left-hand side of Equation 3.35 becomes a large negative number whatever the value of $g_2(x)$, and as a result, Equation 3.35 is always satisfied. If $y = 1$, then the left-hand side of Equation 3.34 is a large negative number whatever the value of $g_1(x)$ and Equation 3.34 is always satisfied. But now $g_2(x) \leq 0$ must be imposed from Equation 3.35.

Some simple examples can be used to illustrate the application of these principles.

Example 3.3 A gaseous waste stream from a process contains valuable hydrogen that can be recovered by separating the hydrogen from the impurities using pressure swing adsorption (PSA), a membrane separator (MS) or a cryogenic condensation (CC). The pressure swing adsorption and membrane separator can in principle be used either individually, or in combination. Write a set of integer equations that would allow one from the three options of pressure swing adsorption, membrane separator or cryogenic condensation to be chosen, but also allow the pressure swing adsorption and membrane separator to be chosen in combination.

Solution Let y_{PSA} represent the selection of pressure swing adsorption, y_{MS} the selection of the membrane separator and y_{CC} the selection of cryogenic condensation. First restrict the choice between pressure swing adsorption and cryogenic condensation.

$$y_{PSA} + y_{CC} \leq 1$$

Now restrict the choice between membrane separator and cryogenic condensation.

$$y_{MS} + y_{CC} \leq 1$$

These two equations restrict the choices, but still allow the pressure swing adsorption and membrane separator to be chosen together.

Example 3.4 The temperature difference in a heat exchanger between the inlet temperature of the hot stream $T_{H,in}$ and the outlet of the cold stream $T_{C,out}$ is to be restricted to be greater than a practical minimum value of ΔT_{min}, but only if the option of having the heat exchanger is chosen. Write a disjunctive constraint in the form of an integer equation to represent this constraint.

Solution The temperature approach constraint can be written as

$$T_{H,in} - T_{C,out} \geq \Delta T_{min}$$

But this should apply only if the heat exchanger is selected. Let y_{HX} represent the option of choosing the heat exchanger.

$$T_{H,in} - T_{C,out} + M(1 - y_{HX}) \geq \Delta T_{min}$$

where M is an arbitrary large number. If $y_{HX} = 0$ (i.e. the heat exchanger is not chosen), then the left-hand side of this equation is bound to be greater than ΔT_{min} no matter what the values of $T_{H,in}$ and $T_{C,out}$ are. If $y_{HX} = 1$ (i.e. the heat exchanger is chosen), then the equation becomes $(T_{H,in} - T_{C,out}) \geq \Delta T_{min}$ and the constraint must apply.

When a linear programming problem is extended to include integer (binary) variables, it becomes a *mixed integer linear programming problem* (MILP). Correspondingly, when a nonlinear programming problem is extended to include integer (binary) variables, it becomes a *mixed integer nonlinear programming problem* (MINLP).

First consider the general strategy for solving an MILP problem. Initially, the binary variables can be treated as continuous variables, such that $0 \leq y_i \leq 1$. The problem can then be solved as an LP. The solution is known as a *relaxed solution*. The most likely outcome is that some of the binary variables will exhibit noninteger values at the optimum LP solution. Because the relaxed solution is less constrained than the true mixed integer solution in which all of the binary variables have integer values, it will in general give a better value for the objective function than the true mixed integer solution. In general, the noninteger values of the binary variables cannot simply be rounded to the nearest integer value, either because the rounding may lead to an infeasible solution (outside the feasible region) or because the rounding may render the solution nonoptimal (not at the edge of the feasible region). However, this relaxed LP solution is useful in providing a *lower bound* to the true mixed integer solution to a minimization problem. For maximization problems, the relaxed LP solutions form the *upper bound* to the solution. The noninteger values can then be set to either 0 or 1 and the LP solution repeated. The setting of the binary variables to be either 0 or 1 creates a solution space in the form of a tree, Figure 3.16[3]. As the solution is stepped through, the number of possibilities increases by virtue of the fact that each binary variable can take a value of 0 or 1, Figure 3.16. At each point in the search, the best relaxed LP solution provides a *lower bound* to the optimum of a minimization problem. Correspondingly, the best true mixed integer solution provides an upper bound. For maximization problems the best relaxed LP solution forms an *upper bound* to the optimum and the best true mixed integer solution provides the *lower bound*. A popular method of solving MILP problems is to use a *branch and bound search*[7]. This will be illustrated by a simple example from Edgar, Himmelblau and Lasdon[1].

Example 3.5 A problem involving three binary variables y_1, y_2 and y_3 has an objective function to be maximized[1].

$$\begin{aligned} \text{maximize:} \quad & f = 86y_1 + 4y_2 + 40y_3 \\ \text{subject to:} \quad & 774y_1 + 76y_2 + 42y_3 \leq 875 \\ & 67y_1 + 27y_2 + 53y_3 \leq 875 \\ & y_1, y_2, y_3 = 0, 1 \end{aligned} \tag{3.36}$$

Solution The solution strategy is illustrated in Figure 3.17a. First the LP problem is solved to obtain the relaxed solution, allowing y_1, y_2 and y_3 to vary continuously between 0 and 1. Both $y_1 = 1$ and $y_3 = 1$ at the optimum of the relaxed solution but the value of y_2 is 0.776, Node 1 in Figure 3.17a. The objective function for this relaxed solution at Node 1 is $f = 129.1$. From this point, y_2 can be set to be either 0 or 1. Various strategies can be adopted to decide which one to choose. A very simple strategy will be adopted here of picking the closest integer to the real number. Given that $y_2 = 0.776$ is closer to 1 than 0, set $y_2 = 1$ and solve the LP at Node 2, Figure 3.17a. Now $y_2 = 1$ and $y_3 = 1$ at the optimum of the relaxed solution but the value of y_1 is 0.978, Node 2 in Figure 3.17a. Given that $y_1 = 0.978$ is closer to 1 than 0, set $y_1 = 1$ and solve the LP at Node 3. This time y_1 and y_2 are integers, but $y_3 = 0.595$ is a noninteger. Setting $y_3 = 1$ yields an infeasible solution at Node 4 in Figure 3.17a as it violates the first inequality constraint in Equation 3.36. Backtracking to Node 3 and setting $y_3 = 0$ yields the first feasible integer solution at Node 5 for which $y_1 = 1$, $y_2 = 1$, $y_3 = 0$ and $f = 90.0$. There is no point in searching further from Node 4 as it is an infeasible solution, or from Node 5 as it is a valid integer solution. When the search is terminated at a node for either reason, it is deemed to be *fathomed*. The search now backtracks to Node 2 and sets $y_1 = 0$. This yields the second feasible integer solution at Node 6 for which $y_1 = 0$, $y_2 = 1$, $y_3 = 1$ and $f = 44.0$. Finally, backtrack to Node 1 and set $y_2 = 0$. This yields the third feasible

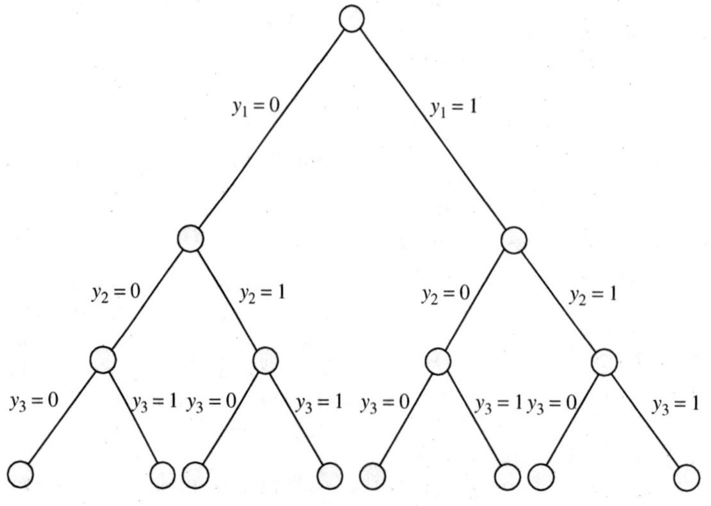

Figure 3.16 Setting the binary variables to 0 or 1 creates a tree structure. (Reproduced from Floudas CA, 1995, Nonlinear and Mixed-Integer Optimization, by permission of Oxford University Press).

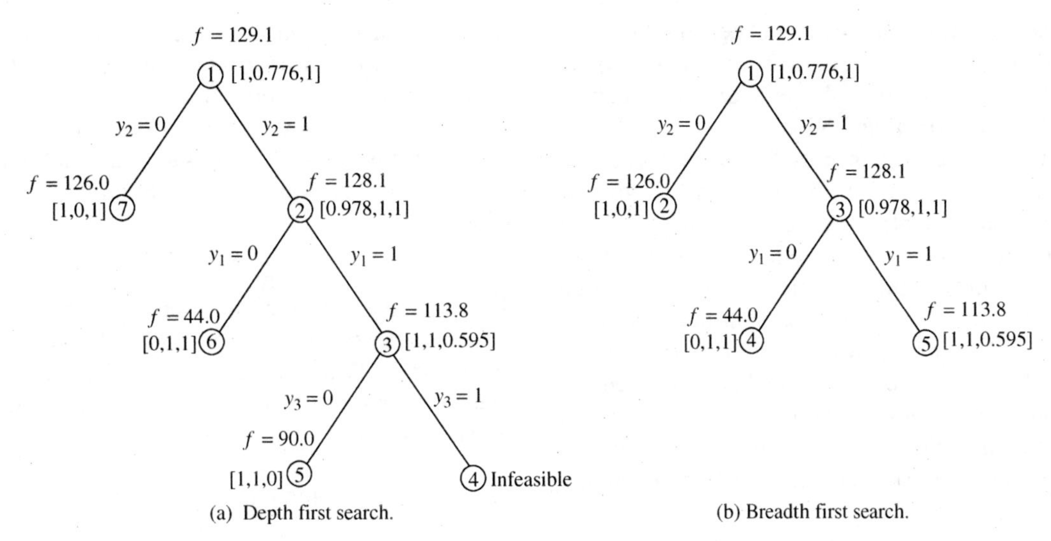

Figure 3.17 Branch and bound search.

integer solution at Node 7 for which $y_1 = 1$, $y_2 = 0$, $y_3 = 1$ and $f = 126.0$. Since the objective function is being maximized, Node 7 is the optimum for the problem.

Searching the tree in the way done in Figure 3.17a is known as a *depth first* or *backtracking* approach. At each node, the branch was followed that appeared to be more promising to solve. Rather than using a depth first approach, a *breadth first* or *jumptracking* approach can be used, as illustrated in Figure 3.17b. Again start at Node 1 and solve the relaxed problem in Figure 3.17b. This gives an upper bound for the maximization problem of $f = 129.1$. However, this time the search goes across the tree with the initial setting of $y_2 = 0$. This yields a valid integer solution at Node 2 with $y_1 = 1$, $y_2 = 0$, $y_3 = 1$ and $f = 126.0$. Node 2 now forms a lower bound and is fathomed because it is an integer solution. In this approach, the search now backtracks to Node 1 whether Node 2 is fathomed or not. From Node 1 now set $y_2 = 1$. The solution at Node 3 in Figure 3.17b gives $y_1 = 0.978$, $y_2 = 1$, $y_3 = 1$ and $f = 128.1$, which is the new upper bound. Setting $y_1 = 0$ and branching to Node 4 gives the second valid integer solution. Now backtrack to Node 3 and set $y_1 = 1$. The solution at Node 5 has $y_1 = 1$, $y_2 = 1$, $y_3 = 0.595$ and $f = 113.8$. At this point, Node 5 is fathomed, even though it is neither infeasible nor a valid integer solution. The upper bound of this branch at Node 5 has a value of the objective function lower than that of the integer solution at Node 2. Setting the values to be integers from Node 5 can only result in an inferior solution. In this way, the search is bounded.

In this case, the breadth first search yields the optimum with a fewer number of nodes to be searched. Different search strategies than the ones used here can readily be used[10]. It is likely that different problems would be suited to different search strategies.

Thus, the solution of the MILP problem is started by solving the first relaxed LP problem. If integer values are obtained for the binary variables, the problem has been solved. However, if integer values are not obtained, the use of bounds is examined to avoid parts of the tree that are known to be suboptimal. The node with the best noninteger solution provides a lower bound for minimization problems and the node with the best feasible mixed integer solution provides an upper bound. In the case of maximization problems, the node with the best noninteger solution provides an upper bound and the node with the best feasible mixed integer solution provides a lower bound. Nodes with noninteger solutions are fathomed when the value of the objective function is inferior to the best integer solution (the lower bound). The tree can be searched by following a depth first approach or a breadth first approach, or a combination of the two. Given a more complex problem than Example 3.5, the search could for example set the values of the noninteger variables to be 0 and 1 in turn and carry out an evaluation of the objective function (rather than an optimization). This would then indicate the best direction in which to go for the next optimization. Many strategies are possible[10].

The series of LP solutions required for MILP problems can be solved efficiently by using one LP to initialize the next. An important point to note is that, in principle, a global optimum solution can be guaranteed in the same way as with LP problems.

The general strategy for solving mixed integer nonlinear programming problems is very similar to that for linear problems[3]. The major difference is that each node requires the solution of a nonlinear program, rather than the solution of a linear program. Unfortunately, searching the tree with a succession of nonlinear optimizations can be extremely expensive in terms of the computation time required, as information cannot be readily carried from one NLP to the next as can be done for LP. Another major problem is that, because a series of nonlinear optimizations is being carried out, there is no guarantee that the optimum will even be close to the global optimum, unless the NLP problem being solved at each node is convex. Of course, different initial points can be tried to overcome this problem, but there can still be no guarantee of global optimality for the general problem.

Another way to deal with such nonlinear problems is to first approximate the solution to be linear and apply MILP, and then apply NLP to the problem. The method then iterates between MILP and NLP[2].

In some cases, the nonlinearity in a problem can be isolated in one or two of the functions. If this is the case, then one simple way to solve the problem is to linearize the nonlinear function by a series of straight-line segments. Integer logic can then be used to ensure that only one of the straight-line segments is chosen at a time and MILP used to carry out the optimization. For some forms of nonlinear mathematical expressions, deterministic optimization methods can be tailored to find the global optimum through the application of mathematical transformations and bounding techniques[11].

Stochastic optimization can be extremely effective for structural optimization if the optimization is nonlinear in character. This does not require a search through a tree as branch and bound methods require. For example, when using simulated annealing, at each setting of the control parameter (annealing temperature), a series of random moves is performed. These moves can be either step changes to continuous variables or equally well be changes in structure (either the addition or removal of a structural feature). Because stochastic optimization allows some deterioration in the objective function, it is not as prone to being trapped by a local optimum as MINLP. However, when optimizing a problem involving both continuous variables (e.g. temperature and pressure) and structural changes, stochastic optimization algorithms take finite steps for the continuous variables and do not necessarily find the exact value for the optimum setting. Because of this, a deterministic method (e.g. SQP) can be applied after stochastic optimization to fine-tune the answer. This uses the stochastic method to provide a good initialization for the deterministic method.

Also, it is possible to combine stochastic and deterministic methods as *hybrid* methods. For example, a stochastic method can be used to control the structural changes and a deterministic method to control the changes in the continuous variables. This can be useful if the problem involves a large number of integer variables, as for such problems, the tree required for branch and bound methods explodes in size.

3.9 SOLUTION OF EQUATIONS USING OPTIMIZATION

It is sometimes convenient to use optimization to solve equations, or sets of simultaneous equations. This arises from the availability of general-purpose optimization software, such as that available in spreadsheets. *Root finding* or *equation solving* is a special case for the optimization, where the objective is to reach a value of 0. For example, suppose it is necessary to solve a function $f(x)$ for the value x that satisfies $f(x) = a$, where a is a constant. As illustrated in Figure 3.18a, this can be solved for

$$f(x) - a = 0 \qquad (3.37)$$

The objective of the optimization would be for the equality constraint given by Equation 3.37 to be satisfied as nearly as possible. There are a number of ways in which this objective can be made specific for optimization. Three possibilities are[12]

1. $$\text{minimize}|f(x) - a| \qquad (3.38)$$
2. $$\text{minimize}[f(x) - a]^2 \qquad (3.39)$$
3. $$\text{minimize}(S_1 + S_2) \qquad (3.40)$$

   subject to

$$f(x) - a + S_1 - S_2 = 0 \qquad (3.41)$$

$$S_1, S_2 \geq 0 \qquad (3.42)$$

where S_1 and S_2 are slack variables

Which of these objectives would be the best to use depends on the nature of the problem, the optimization algorithm being used and the initial point for the solution. For example, minimizing Objective 1 in Equation 3.38 can present problems to optimization methods as a result of the gradient being discontinuous, as illustrated in Figure 3.18b. However, the problem can be transformed such that the objective function for the optimization has no discontinuities in the gradient. One possible transformation is Objective 2 in Equation 3.39. As illustrated in Figure 3.18c, this now has a continuous gradient. However, a fundamental disadvantage in using the transformation in Equations 3.39 is that if the equations to be solved are linear, then the objective function is transformed from linear to nonlinear. Also, if the equations to be solved are nonlinear, then such transformations will increase the nonlinearity. On the other hand, Objective 3 in Equation 3.40 avoids both an increase in the nonlinearity and discontinuities in the gradient, but at the expense of introducing slack variables. Note that two slack variables are needed. Slack variable S_1 in Equation 3.41 for $S_1 \geq 0$ ensures that:

$$f(x) - a \leq 0 \qquad (3.43)$$

Whereas, slack variable S_2 in Equation 3.41 for $S_2 \geq 0$ ensures that:

$$f(x) - a \geq 0 \qquad (3.44)$$

Equations 3.43 and 3.44 are only satisfied simultaneously by Equation 3.37. Thus, x, S_1 and S_2 can be varied

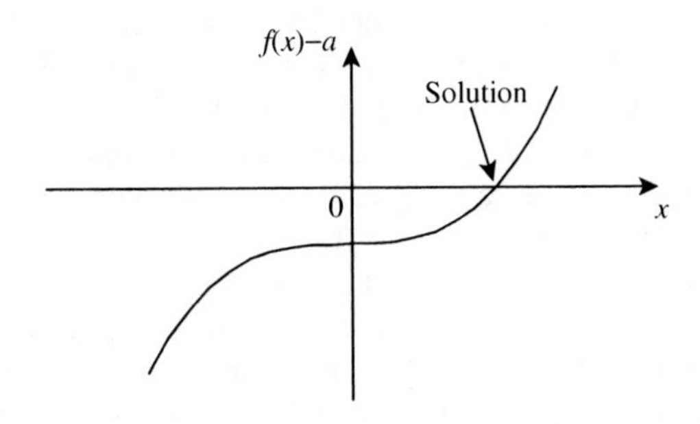

(a) An equation to be solved.

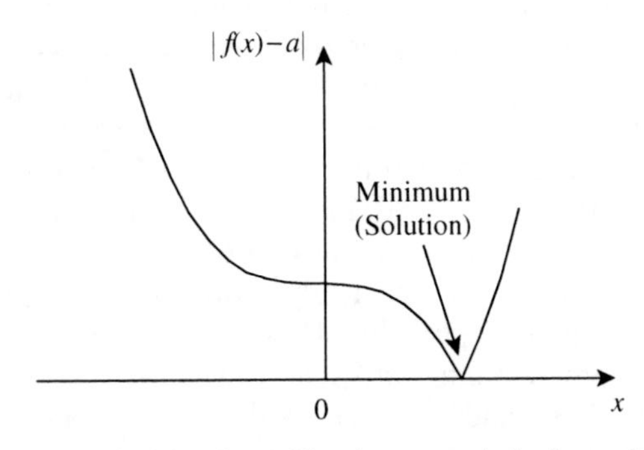

(b) Transforming the problem into an optimization problem.

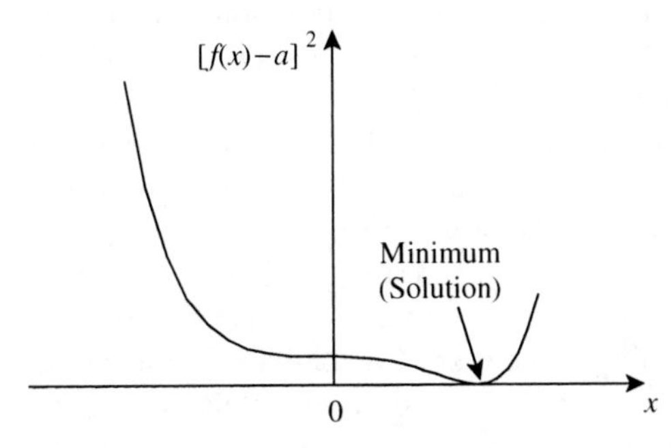

(c) Transformation of the objective function to make the gradient continuous.

Figure 3.18 Solving equations using optimization.

simultaneously to solve Equations 3.40 to 3.43 without increasing the nonlinearity.

The approach can be extended to solve sets of simultaneous equations. For example, suppose a solution is required for x_1 and x_2 such that:

$$f_1(x_1, x_2) = a_1 \text{ and } f_2(x_1, x_2) = a_2 \qquad (3.45)$$

Three possible ways to formulate the objective for optimization are:

1. $\text{minimize}\{|f_1(x_1, x_2) - a_1| + |f_2(x_1, x_2) - a_2|\}$ (3.46)
2. $\text{minimize}\{[f_1(x_1, x_2) - a_1]^2 + [f_2(x_1, x_2) - a_2]^2\}$ (3.47)
3. $\text{minimize}(S_{11} + S_{12} + S_{21} + S_{22})$ (3.48)

subject to

$$f_1(x_1, x_2) - a_1 + S_{11} - S_{12} = 0 \qquad (3.49)$$

$$f_2(x_1, x_2) - a_2 + S_{21} - S_{22} = 0 \qquad (3.50)$$

$$S_{11}, S_{12}, S_{21}, S_{22} \geq 0 \qquad (3.51)$$

where S_{11}, S_{12}, S_{21} and S_{22} are slack variables

As will be seen later, these techniques will prove to be useful when solving design problems in general-purpose software, such as spreadsheets. Many of the numerical problems associated with optimization can be avoided by appropriate formulation of the model. Further details of model building can be found elsewhere[12].

3.10 THE SEARCH FOR GLOBAL OPTIMALITY

From the discussion in this chapter, it is clear that the difficulties associated with optimizing nonlinear problems are far greater than those for optimizing linear problems. For linear problems, finding the global optimum can, in principle, be guaranteed.

Unfortunately, when optimization is applied to most relatively large design problems, the problem usually involves solving nonlinear optimization. In such situations, standard deterministic optimization methods will find only the first local optimum encountered. Starting from different initializations can allow the optimization to explore different routes through the solution space and might help identify alternative solutions. However, there is no guarantee of finding the global optimum. For some forms of nonlinear mathematical expressions, deterministic optimization methods can be tailored to find the global optimum through the application of mathematical transformations and bounding techniques[11].

Alternatively, stochastic optimization (e.g. simulation annealing or genetic algorithms) can be used. These have the advantage of, in principle, being able to locate the global optimum for the most general nonlinear optimization problems. They do not require good initialization and do not require gradients to be defined. However, they involve parameters that are system-dependent and might need to be adjusted for problems that are different in character. Another disadvantage is that they can be extremely slow in solving large complex optimization problems. The relative advantages of deterministic and stochastic methods can be combined using hybrid methods by using stochastic methods to provide a good initial point for a deterministic method. As mentioned previously, stochastic and deterministic methods can also be combined to solve structural optimization problems.

In Chapter 1, an objective function for a nonlinear optimization was likened to the terrain in a range of mountains. If the objective function is to be maximized, each peak in the mountain range represents a local optimum in the objective function. The highest peak represents the global optimum. Optimization requires searching around the mountains in a thick fog to find the highest peak, without the benefit of a map and only a compass to tell direction and an altimeter to show height. On reaching the top of any peak, there is no way of knowing whether it is the highest peak because of the fog.

When solving such nonlinear optimization problems, it is not desirable to terminate the search at a peak that is grossly inferior to the highest peak. The solution can be checked by repeating the search but starting from a different initial point.

However, the shape of the optimum for most optimization problems bears a greater resemblance to Table Mountain in South Africa, rather than to Mount Everest. In other words, for most optimization problems, the region around the optimum is fairly flat. Although on one hand a grossly inferior solution should be avoided, on the other hand the designer should not be preoccupied with improving the solution by tiny amounts in an attempt to locate exactly the global optimum. There will be uncertainty in the design data, especially economic data. Also, there are many issues to be considered other than simply maximizing economic potential or minimizing cost. There could be many reasons why the solution at the exact location of the global optimum might not be preferred, but a slightly suboptimal solution preferred for other reasons, such as safety, ease of control, and so on. Different solutions in the region of the optimum should be examined, rather than considerable effort being expended on finding the solution at the exact location of the global optimum and considering only that solution. In this respect, stochastic optimization has advantages, as it can provide a range of solutions in the region of the optimum.

3.11 SUMMARY – OPTIMIZATION

Most design problems will require optimization to be carried out at some stage. The quality of the design is characterized by an objective function to be maximized (e.g. if economic potential is being maximized) or minimized (e.g. if cost is being minimized). The shape of the objective function is critical in determining the optimization strategy. If the objective function is convex in a minimization problem or concave in a maximization problem, then there is a single optimum point. If this is not the case, there can be local optima as well as the global optimum.

Various search strategies can be used to locate the optimum. Indirect search strategies do not use information on gradients, whereas direct search strategies require this information. These methods always seek to improve the objective function in each step in a search. On the other hand, stochastic search methods, such as simulated annealing and genetic algorithms, allow some deterioration in the objective function, especially during the early stages of the search, in order to reduce the danger of being attracted to a local optimum rather than the global optimum. However, stochastic optimization can be very slow in converging, and usually needs to be adapted to solve particular problems. Tailoring the methods to suit specific applications makes them much more efficient.

The addition of inequality constraints complicates the optimization. These inequality constraints can form convex or nonconvex regions. If the region is nonconvex, then this means that the search can be attracted to a local optimum, even if the objective function is convex in the case of a minimization problem or concave in the case of a maximization problem. In the case that a set of inequality constraints is linear, the resulting region is always convex.

The general case of optimization in which the objective function, the equality and inequality constraints are all linear can be solved as a linear programming problem. This can be solved efficiently with, in principle, a guarantee of global optimality. However, the corresponding nonlinear programming problem cannot, in general, be solved efficiently and with a guarantee of global optimality. Such problems are solved by successive linear or successive quadratic programming. Stochastic optimization methods can be very effective in solving nonlinear optimization, because they are less prone to be stuck in a local optimum than deterministic methods.

One of the approaches that can be used in design is to carry out structural and parameter optimization of a superstructure. The structural optimization required can be carried out using mixed integer linear programming in the case of a linear problem or mixed integer nonlinear programming in the case of a nonlinear problem. Stochastic optimization can also be very effective for structural optimization problems.

3.12 EXERCISES

1. The overhead of vapor of a distillation column is to be condensed in the heat exchanger using cooling water. There is a trade-off involving the flowrate of cooling water and the size of the condenser. As the flowrate of cooling water increases, its cost increases. However, as the flowrate increases, the return temperature of the cooling water to the cooling tower decreases. This decreases the temperature differences in the condenser and increases its heat transfer area, and hence its capital cost. The condenser has a duty of 4.1 MW and the vapor condenses at a constant temperature of 80°C. Cooling water is available at 20°C with a cost of \$ 0.02 t^{-1}. The overall heat transfer coefficient can be assumed to be 500 $W \cdot m^{-1} \cdot K^{-1}$. The cost of the condenser can be assumed to be \$2500 m^{-2} with an installation factor of 3.5. Annual capital charges can be assumed to be 20% of the capital costs. The heat capacity of the cooling water can be assumed constant at 4.2 $kJ \cdot kg^{-1} \cdot K^{-1}$. The distillation column operates for 8000 $h \cdot y^{-1}$. Set up an equation for the heat transfer area of the condenser, and hence the annual capital cost of the condenser, in terms of the cooling

water return temperature. Using this equation, carry out a trade-off between the cost of the cooling water and the cost of the condenser to determine approximately the optimum cooling water return temperature. The maximum return temperature should be 50°C. The heat exchange area required by the condenser is given by:

$$A = \frac{Q}{U \Delta T_{LM}}$$

where A = heat transfer area (m^2)
Q = heat duty (W)
U = overall heat transfer coefficient ($W \cdot m^{-1} \cdot K^{-1}$)
ΔT_{LM} = logarithmic mean temperature difference

$$= \frac{(T_{COND} - T_{CW2}) - (T_{COND} - T_{CW1})}{\ln\left(\frac{T_{COND} - T_{CW2}}{T_{COND} - T_{CW1}}\right)}$$

T_{COND} = condenser temperature (°C)
T_{CW1} = inlet cooling water temperature (°C)
T_{CW2} = outlet cooling water temperature (°C)

2. A vapor stream leaving a styrene production process contains hydrogen, methane, ethylene, benzene, water, toluene, styrene and ethylbenzene and is to be burnt in a furnace. It is proposed to recover as much of the benzene, toluene, styrene and ethylbenzene as possible from the vapor using low-temperature condensation. The low-temperature condensation requires refrigeration. However, the optimum temperature for the condensation needs to be determined. This involves a trade-off in which the amount and value of material recovered increases as the temperature decreases, but the cost of the refrigeration increases as the temperature decreases. The material flows in the vapor leaving the flash drum are given in Table 3.2, together with their values.

Table 3.2 Stream flowrates and component values.

Component	Flowrate ($kmol \cdot s^{-1}$)	Value ($\$ \cdot kmol^{-1}$)
Hydrogen	146.0	0
Methane	3.7	0
Ethylene	3.7	0
Benzene	0.67	21.4
Water	9.4	0
Toluene	0.16	12.2
Ethylbenzene	1.6	40.6
Styrene	2.4	60.1
Total	167.63	

The low-temperature condensation requires refrigeration, for which the cost is given by:

$$\text{Refrigeration cost} = 0.033 Q_{COND} \left(\frac{40 - T_{COND}}{T_{COND} + 268}\right)$$

where Q_{COND} = condenser duty (MW)
T_{COND} = condenser temperature (°C)

The fraction of benzene, toluene, styrene and ethylbenzene condensed can be determined from phase equilibrium calculations. The percent of the various components entering the condenser that leave with the vapor are given in Table 3.3 as a function of temperature. The total enthalpy of the flash drum vapor stream as a function of temperature is given in Table 3.4. Calculate the optimum condenser temperature. What practical difficulties do you foresee in using very low temperatures?

Table 3.3 Condenser performance.

	Percent of Component Entering Condenser that Leaves with the Vapor										
	Condensation temperature (°C)										
	40	30	20	10	0	−10	−20	−30	−40	−50	−60
Benzene	100	93	84	72	58	42	27	15	8	3	1
Toluene	100	80	60	41	25	14	7	3	1	1	0
Ethylbenzene	100	59	33	18	9	4	2	1	0	0	0
Styrene	100	54	29	15	8	4	2	1	0	0	0

Table 3.4 Stream enthalpy data.

Temperature (°C)	Stream enthalpy ($MJ \cdot kmol^{-1}$)
40	0.45
30	−1.44
20	−2.69
10	−3.56
0	−4.21
−10	−4.72
−20	−5.16
−30	−5.56
−40	−5.93
−50	−6.29
−60	−6.63

3. A tank containing 1500 m^3 of naphtha is to be blended with two other hydrocarbon streams to meet the specifications for gasoline. The final product must have a minimum research octane number (RON) of 95, a maximum Reid Vapor Pressure (RVP) of 0.6 bar, a maximum benzene content of 2% vol and maximum total aromatics of 25% vol. The properties and costs of the three streams are given in the Table 3.5.

Table 3.5 Blending streams.

	RON	RVP (bar)	Benzene (% vol)	Total aromatics (% vol)	Cost $\$ \cdot m^{-3}$
Naphtha	92	0.80	1.5	15	275
Reformate	98	0.15	15	50	270
Alkylate	97.5	0.30	0	0	350

Assuming that the properties of the mixture blend are in proportion to the volume of stream used, how much reformate and alkylate should be blended to minimize cost?

4. A petroleum refinery has two crude oil feeds available. The first crude (Crude 1) is high-quality feed and costs \$30 per barrel (1 barrel = 42 US gallons). The second crude (Crude 2) is a low-quality feed and costs \$20 per barrel. The crude oil is separated into gasoline, diesel, jet fuel and fuel oil. The percent yield of each of these products that can be obtained from Crude 1 and Crude 2 are listed in Table 3.6, together with maximum allowable production flowrates of the products in barrels per day and processing costs.

Table 3.6 Refinery data.

	Yield (% volume)		Value of product ($\$\cdot bbl^{-1}$)	Maximum production ($bbl\cdot day^{-1}$)
	Crude 1	Crude 2		
Gasoline	80	47	75	120,000
Jet fuel	4	8	55	8,000
Diesel	10	30	40	30,000
Fuel oil	6	15	30	–
Processing cost ($\$\ bbl^{-1}$)	1.5	3.0		

The economic potential can be taken to be the difference between the selling price of the products and the cost of the crude oil feedstocks. Determine the optimum feed flowrate of the two crude oils from a linear optimization solved graphically.

5. Add a constraint to the specifications for Exercise 4 above such that the production of fuel oil must be greater than 15,000 $bbl\cdot day^{-1}$. What happens to the problem? How would you describe the characteristics of the modified linear programming problem?
6. Devise a superstructure for a distillation design involving a single feed, two products, a reboiler and a condenser that will allow the number of plates in the column itself to be varied between 3 and 10 and at the same time vary the location of the feed tray.
7. A reaction is required to be carried out between a gas and a liquid. Two different types of reactor are to be considered: an agitated vessel (AV) and a packed column (PC). Devise a superstructure that will allow one of the two options to be chosen. Then describe this as integer constraints for the gas and liquid feeds and products.

REFERENCES

1. Edgar TF, Himmelblau DM and Lasdon LS (2001) *Optimization of Chemical Processes*, 2nd Edition, McGraw-Hill.
2. Biegler LT, Grossmann IE and Westerberg AW (1997) *Systematic Methods of Chemical Process Design*, Prentice Hall.
3. Floudas CA (1995) *Nonlinear and Mixed-Integer Optimization*, Oxford University Press.
4. Metropolis N, Rosenbluth AW, Rosenbluth MN, Teller AH and Teller E (1953) Equation of State Calculations by Fast Computing Machines, *J Chem Phys*, **21**: 1087.
5. Kirkpatrick S, Gelatt CD and Vecchi MP (1983) Optimization by Simulated Annealing, *Science*, **220**: 671.
6. Goldberg DE (1989) *Genetic Algorithms in Search Optimization and Machine Learning*, Addison-Wesley.
7. Mehta VL and Kokossis AC (1988) New Generation Tools for Multiphase Reaction Systems: A Validated Systematic Methodology for Novelty and Design Automation, *Comput Chem Eng*, **22S**: 5119.
8. Choong KL and Smith R (2004) Optimization of Batch Cooling Crystallization, *Chem Eng Sci*, **59**: 313.
9. Choong KL and Smith R (2004) Novel Strategies for Optimization of Batch, Semi-batch and Heating/Cooling Evaporative Crystallization, *Chem Eng Sci*, **59**: 329.
10. Taha HA (1975) *Integer Programming Theory and Applications*, Academic Press.
11. Floudas CA (2000) *Deterministic Global Optimization: Theory, Methods and Applications*, Kluwer Academic Publishers.
12. Williams HP (1997) *Model Building in Mathematical Programming*, 3rd Edition, John Wiley.

4 Thermodynamic Properties and Phase Equilibrium

A number of design calculations require a knowledge of thermodynamic properties and phase equilibrium. In practice, the designer most often uses a commercial physical property or a simulation software package to access such data. However, the designer must understand the basis of the methods for thermodynamic properties and phase equilibrium, so that the most appropriate methods can be chosen and their limitations fully understood.

4.1 EQUATIONS OF STATE

The relationship between pressure, volume and temperature for fluids is described by *equations of state*. For example, if a gas is initially at a specified pressure, volume and temperature and two of the three variables are changed, the third variable can be calculated from an equation of state.

Gases at low pressure tend towards *ideal gas* behavior. For a gas to be ideal,

- the volume of the molecules should be small compared with the total volume;
- there should be no intermolecular forces.

The behavior of ideal gases can be described by the ideal gas law[1,2]:

$$PV = NRT \tag{4.1}$$

where P = pressure ($N \cdot m^{-2}$)
V = volume occupied by N kmol of gas (m^3)
N = moles of gas (kmol)
R = gas constant (8314 $N \cdot m \cdot kmol^{-1} \cdot K^{-1}$ or $J \cdot kmol^{-1} \cdot K^{-1}$)
T = absolute temperature (K)

The ideal gas law describes the actual behavior of most gases reasonably well at pressures below 5 bar.

If standard conditions are specified to be 1 atm (101,325 $N \cdot m^{-2}$) and 0°C (273.15 K), then from the ideal gas law, the volume occupied by 1 kmol of gas is 22.4 m^3.

For gas mixtures, the *partial pressure* is defined as the pressure that would be exerted if that component alone occupied the volume of the mixture. Thus, for an ideal gas,

$$p_i V = N_i RT \tag{4.2}$$

where p_i = partial pressure ($N \cdot m^{-2}$)
N_i = moles of Component i (kmol)

Chemical Process Design and Integration R. Smith

ISBNs: 0-471-48680-9 (HB); 0-471-48681-7 (PB)

The mole fraction in the gas phase for an ideal gas is given by a combination of Equations 4.1 and 4.2:

$$y_i = \frac{N_i}{N} = \frac{p_i}{P} \tag{4.3}$$

where y_i = mole fraction of Component i

For a mixture of ideal gases, the sum of the partial pressures equals the total pressure (Dalton's Law):

$$\sum_i^{NC} p_i = P \tag{4.4}$$

where p_i = partial pressure of Component i ($N \cdot m^{-2}$)
NC = number of components (–)

The behavior of real gases and liquids can be accounted for by introducing a compressibility factor (Z), such that[1–3]:

$$PV = ZRT \tag{4.5}$$

where Z = compressibility factor (–)
V = molar volume ($m^3 \cdot kmol^{-1}$)

$Z = 1$ for an ideal gas and is a function of temperature, pressure and composition for mixtures. A model for Z is needed.

In process design calculations, *cubic* equations of state are most commonly used. The most popular of these cubic equations is the Peng–Robinson equation of state given by[3]:

$$Z = \frac{V}{V-b} - \frac{aV}{RT(V^2 + 2bV - b^2)} \tag{4.6}$$

where
$$Z = \frac{PV}{RT} \tag{4.7}$$

V = molar volume

$$a = 0.45724\frac{R^2T_c^2}{P_c}\alpha$$

$$b = 0.0778\frac{RT_c}{P_c}$$

$$\alpha = (1 + \kappa(1 - \sqrt{T_R}))^2$$

$$\kappa = 0.37464 + 1.54226\,\omega - 0.26992\,\omega^2$$

T_C = critical temperature

P_C = critical pressure

ω = acentric factor

$$= \left[-\log\left(\frac{P^{SAT}}{P_C}\right)_{T_R=0.7}\right] - 1$$

R = gas constant

$T_R = T/T_C$

T = absolute temperature

The acentric factor is obtained experimentally. It accounts for differences in molecular shape, increasing with non-sphericity and polarity, and tabulated values are available[3]. Equation 4.6 can be rearranged to give a cubic equation of the form:

$$Z^3 + \beta Z^2 + \gamma Z + \delta = 0 \tag{4.8}$$

where $Z = \dfrac{PV}{RT}$

$\beta = B - 1$

$\gamma = A - 3B^2 - 2B$

$\delta = B^3 + B^2 - AB$

$A = \dfrac{aP}{R^2T^2}$

$B = \dfrac{bP}{RT}$

This is a cubic equation in the compressibility factor Z, which can be solved analytically. The solution of this cubic equation may yield either one or three real roots[4]. To obtain the roots, the following two values are first calculated[4]:

$$q = \frac{\beta^2 - 3\gamma}{9} \tag{4.9}$$

$$r = \frac{2\beta^3 - 9\beta\gamma + 27\delta}{54} \tag{4.10}$$

If $q^3 - r^2 \geq 0$, then the cubic equation has three roots. These roots are calculated by first calculating:

$$\theta = \arccos\left(\frac{r}{q^{3/2}}\right) \tag{4.11}$$

Then, the three roots are given by:

$$Z_1 = -2q^{1/2}\cos\left(\frac{\theta}{3}\right) - \frac{\beta}{3} \tag{4.12}$$

$$Z_2 = -2q^{1/2}\cos\left(\frac{\theta + 2\pi}{3}\right) - \frac{\beta}{3} \tag{4.13}$$

$$Z_3 = -2q^{1/2}\cos\left(\frac{\theta + 4\pi}{3}\right) - \frac{\beta}{3} \tag{4.14}$$

If $q^3 - r^2 < 0$, then the cubic equation has only one root given by:

$$Z_1 = -\mathrm{sign}(r)\{[(r^2 - q^3)^{1/2} + |r|]^{1/3} + \frac{q}{[(r^2 - q^3)^{1/2} + |r|]^{1/3}}\} - \frac{\beta}{3} \tag{4.15}$$

where sign(r) is the sign of r, such that sign(r) = 1 if $r > 0$ and sign(r) = −1 if $r < 0$.

If there is only one root, there is no choice but to take this as the compressibility factor at the specified conditions of temperature and pressure, but if three real values exist, then the largest corresponds to the vapor compressibility factor and the smallest is the liquid compressibility factor. The middle value has no physical meaning. A superheated vapor might provide only one root, corresponding with the compressibility factor of the vapor phase. A subcooled liquid might provide only one root, corresponding with the compressibility factor of the liquid phase. A vapor–liquid system should provide three roots, with only the largest and smallest being significant. Equations of state such as the Peng–Robinson equation are generally more reliable at predicting the vapor compressibility than the liquid compressibility.

For multicomponent systems, mixing rules are needed to determine the values of a and b[3,5]:

$$b = \sum_i^{NC} x_i b_i \tag{4.16}$$

$$a = \sum_i^{NC}\sum_j^{NC} x_i x_j \sqrt{a_i a_j}(1 - k_{ij}) \tag{4.17}$$

where k_{ij} is a binary interaction parameter found by fitting experimental data[5].

Hydrocarbons	: k_{ij} very small, nearly zero
Hydrocarbon/Light gas	: k_{ij} small and constant
Polar mixtures	: k_{ij} large and temperature dependent

Some interaction parameters have been published[5].

Example 4.1 Using the Peng–Robinson equation of state:

a. determine the vapor compressibility of nitrogen at 273.15 K and 1.013 bar, 5 bar and 50 bar, and compare with an ideal gas. For nitrogen, T_C = 126.2 K, P_C = 33.98 bar and ω = 0.037. Take R = 0.08314 bar·m³·kmol⁻¹·K⁻¹.
b. determine the liquid density of benzene at 293.15 K and compare this with the measured value of ρ_L = 876.5 kg·m⁻³. For benzene, T_C = 562.05 K, P_C = 48.95 bar and ω = 0.210.

Solution

a. For nitrogen, Equation 4.8 must be solved for the vapor compressibility factor. In this case, $q^3 - r^2 < 0$ and there is one root given by Equation 4.15. The solution is summarized in Table 4.1.

   From Table 4.1, it can be seen that the nitrogen can be approximated by ideal gas behavior at moderate pressures.
b. For benzene, at 293.15 K and 1.013 bar (1 atm), Equation 4.8 must be solved for the liquid compressibility factor. In

Table 4.1 Solution of the Peng–Robinson equation of state for nitrogen.

	Pressure 1.013 bar	Pressure 5 bar	Pressure 50 bar
κ	0.43133	0.43133	0.43133
α	0.63482	0.63482	0.63482
a	0.94040	0.94040	0.94040
b	2.4023×10^{-2}	2.4023×10^{-2}	2.4023×10^{-2}
A	1.8471×10^{-3}	9.1171×10^{-3}	9.1171×10^{-2}
B	1.0716×10^{-3}	5.2891×10^{-3}	5.2891×10^{-2}
β	−0.99893	0.99471	−0.94711
γ	-2.9947×10^{-4}	-1.5451×10^{-3}	-2.3004×10^{-2}
δ	-8.2984×10^{-7}	-2.0099×10^{-5}	-1.8767×10^{-3}
q	0.11097	0.11045	0.10734
r	-3.6968×10^{-2}	-3.6719×10^{-2}	-3.6035×10^{-2}
$q^3 - r^2$	-2.9848×10^{-8}	-7.1589×10^{-7}	-6.1901×10^{-5}
Z_1	0.9993	0.9963	0.9727
RT/P ($m^3 \cdot kmol^{-1}$)	22.42	4.542	0.4542
Z_1RT/P ($m^3 \cdot kmol^{-1}$)	22.40	4.525	0.4418

this case, $q^3 - r^2 > 0$ and there are three roots given by Equations 4.12 to 4.14. The parameters for the Peng–Robinson equation are given in Table 4.2.

Table 4.2 Solution of the Peng–Robinson equation of state for benzene.

	Pressure 1.013 bar
κ	0.68661
α	1.41786
a	28.920
b	7.4270×10^{-2}
A	4.9318×10^{-2}
B	3.0869×10^{-3}
β	−0.99691
γ	4.3116×10^{-2}
δ	-1.4268×10^{-4}
q	9.6054×10^{-2}
r	-2.9603×10^{-2}
$q^3 - r^2$	9.9193×10^{-6}
Z_1	0.0036094
Z_2	0.95177
Z_3	0.04153

From the three roots, only Z_1 and Z_2 are significant. The smallest root (Z_1) relates to the liquid and the largest root (Z_2) to the vapor. Thus, the density of liquid benzene is given by:

$$\rho_L = \frac{78.11}{Z_1RT/P}$$

$$= \frac{78.11}{0.0036094 \times 24.0597}$$

$$= 899.5 \text{ kg m}^{-3}$$

This compares with an experimental value of $\rho_L = 876.5 \text{ kg} \cdot \text{m}^{-3}$ (an error of 3%).

4.2 PHASE EQUILIBRIUM FOR SINGLE COMPONENTS

The phase equilibrium for pure components is illustrated in Figure 4.1. At low temperatures, the component forms a solid phase. At high temperatures and low pressures, the component forms a vapor phase. At high pressures and high temperatures, the component forms a liquid phase. The phase equilibrium boundaries between each of the phases are illustrated in Figure 4.1. The point where the three phase equilibrium boundaries meet is the *triple point*, where solid, liquid and vapor coexist. The phase equilibrium boundary between liquid and vapor terminates at the *critical point*. Above the critical temperature, no liquid forms, no matter how high the pressure. The phase equilibrium boundary between liquid and vapor connects the triple point and the

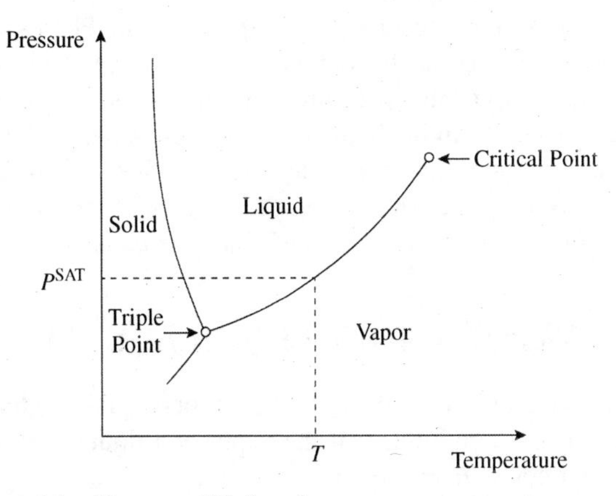

Figure 4.1 Phase equilibrium for a pure component.

critical point, and marks the boundary where vapor and liquid coexist. For a given temperature on this boundary, the pressure is the *vapor pressure*. When the vapor pressure is 1 atm, the corresponding temperature is the *normal boiling point*. If, at any given vapor pressure, the component is at a temperature less than the phase equilibrium, it is *subcooled*. If it is at a temperature above the phase equilibrium, it is *superheated*. Various expressions can be used to represent the vapor pressure curve.

The simplest expression is the Clausius–Clapeyron equation[1,2]:

$$\ln P^{SAT} = A - \frac{B}{T} \tag{4.18}$$

where A and B are constants and T is the absolute temperature. This indicates that a plot of $\ln P^{SAT}$ versus $1/T$ should be a straight line. Equation 4.18 gives a good correlation only over small temperature ranges. Various modifications have been suggested to extend the range of application, for example, the Antoine equation[1-3]:

$$\ln P^{SAT} = A - \frac{B}{C + T} \tag{4.19}$$

where A, B and C are constants determined by correlating experimental data[3]. Extended forms of the Antoine equation have also been proposed[3]. Great care must be taken when using correlated vapor pressure data not to use the correlation coefficients outside the temperature range over which the data has been correlated; otherwise, serious errors can occur.

4.3 FUGACITY AND PHASE EQUILIBRIUM

Having considered single component systems, multicomponent systems need to be addressed now. If a closed system contains more than one phase, the equilibrium condition can be written as:

$$f_i^I = f_i^{II} = f_i^{III} \quad i = 1, 2, \ldots, NC \tag{4.20}$$

where f_i is the fugacity of Component i in Phases I, II and III and NC is the number of components. Fugacity is a thermodynamic pressure, but has no strict physical significance. It can be thought of as an "escaping tendency". Thus, Equation 4.20 states that if a system of different phases is in equilibrium, then the "escaping tendency" of Component i from the different phases is equal.

4.4 VAPOR–LIQUID EQUILIBRIUM

Thermodynamic equilibrium in a vapor–liquid mixture is given by the condition that the vapor and liquid fugacities for each component are equal[2]:

$$f_i^V = f_i^L \tag{4.21}$$

where f_i^V = fugacity of Component i in the vapor phase
f_i^L = fugacity of Component i in the liquid phase

Thus, equilibrium is achieved when the "escaping tendency" from the vapor and liquid phases for Component i are equal. The *vapor-phase fugacity coefficient*, ϕ_i^V, can be defined by the expression:

$$f_i^V = y_i \phi_i^V P \tag{4.22}$$

where y_i = mole fraction of Component i in the vapor phase
ϕ_i^V = vapor-phase fugacity coefficient
P = system pressure

The *liquid-phase fugacity coefficient* ϕ_i^L can be defined by the expression:

$$f_i^L = x_i \phi_i^L P \tag{4.23}$$

The *liquid-phase activity coefficient* γ_i can be defined by the expression:

$$f_i^L = x_i \gamma_i f_i^O \tag{4.24}$$

where x_i = mole fraction of Component i in the liquid phase
ϕ_i^L = liquid-phase fugacity coefficient
γ_i = liquid-phase activity coefficient
f_i^O = fugacity of Component i at standard state

For moderate pressures, f_i^O can be approximated by the saturated vapor pressure, P_i^{SAT}; thus, Equation 4.24 becomes[3]:

$$f_i^L = x_i \gamma_i P_i^{SAT} \tag{4.25}$$

Equations 4.21, 4.22 and 4.23 can be combined to give an expression for the K-value, K_i, that relates the vapor and liquid mole fractions:

$$K_i = \frac{y_i}{x_i} = \frac{\phi_i^L}{\phi_i^V} \tag{4.26}$$

Equation 4.26 defines the relationship between the vapor and liquid mole fractions and provides the basis for vapor–liquid equilibrium calculations on the basis of equations of state. Thermodynamic models are required for ϕ_i^V and ϕ_i^L from an equation of state. Alternatively, Equations 4.21, 4.22 and 4.25 can be combined to give

$$K_i = \frac{y_i}{x_i} = \frac{\gamma_i P_i^{SAT}}{\phi_i^V P} \tag{4.27}$$

This expression provides the basis for vapor–liquid equilibrium calculations on the basis of *liquid-phase activity coefficient models*. In Equation 4.27, thermodynamic models are required for ϕ_i^V (from an equation of state) and γ_i from a liquid-phase activity coefficient model. Some examples will be given later. At moderate pressures, the vapor phase becomes ideal, as discussed previously, and $\phi_i^V = 1$. For an ideal vapor phase, Equation 4.27 simplifies to:

$$K_i = \frac{y_i}{x_i} = \frac{\gamma_i P_i^{SAT}}{P} \tag{4.28}$$

When the liquid phase behaves as an ideal solution,

- all molecules have the same size;
- all intermolecular forces are equal;
- the properties of the mixture depend only on the properties of the pure components comprising the mixture.

Mixtures of isomers, such as *o*-, *m*- and *p*-xylene mixtures, and adjacent members of homologous series, such as *n*-hexane–*n*-heptane and benzene–toluene mixtures, give close to ideal liquid-phase behavior. For this case, $\gamma_i = 1$, and Equation 4.28 simplifies to:

$$K_i = \frac{y_i}{x_i} = \frac{P_i^{SAT}}{P} \tag{4.29}$$

which is Raoult's law and represents both ideal vapor- and liquid-phase behavior. Correlations are available to relate component vapor pressure to temperature, as discussed above.

Comparing Equations 4.28 and 4.29, the liquid-phase nonideality is characterized by the activity coefficient γ_i. When $\gamma_i = 1$, the behavior is ideal. If $\gamma_i \neq 1$, then the value of γ_i can be used to characterize the nonideality:

- $\gamma_i < 1$ represents negative deviations from Raoult's Law;
- $\gamma_i > 1$ represents positive deviations from Raoult's Law.

The vapor–liquid equilibrium for noncondensable gases in equilibrium with liquids can often be approximated by Henry's Law[1–3]:

$$p_i = H_i x_i \tag{4.30}$$

where p_i = partial pressure of Component i
H_i = Henry's Law constant (determined experimentally)
x_i = mole fraction of Component i in the liquid phase

Assuming ideal gas behavior ($p_i = y_i P$):

$$y_i = \frac{H_i x_i}{P} \tag{4.31}$$

Thus, the K-value is given by:

$$K_i = \frac{y_i}{x_i} = \frac{H_i}{P} \tag{4.32}$$

A straight line would be expected from a plot of y_i against x_i.

The ratio of equilibrium K-values for two components measures their *relative volatility*:

$$\alpha_{ij} = \frac{K_i}{K_j} \tag{4.33}$$

where α_{ij} = volatility of Component i relative to Component j

These expressions form the basis for two alternative approaches to vapor–liquid equilibrium calculations:

a. $K_i = \phi_i^L / \phi_i^V$ forms the basis for calculations based entirely on equations of state. Using an equation of state for both the liquid and vapor phase has a number of advantages. Firstly, f_i^O need not be specified. Also, in principle, continuity at the critical point can be guaranteed with all thermodynamic properties derived from the same model. The presence of noncondensable gases, in principle, causes no additional complications. However, the application of equations of state is largely restricted to nonpolar components.
b. $K_i = \gamma_i P_i^{SAT} / \phi_i^V P$ forms the basis for calculations based on liquid-phase activity coefficient models. It is used when polar molecules are present. For most systems at low pressures, ϕ_i^V can be assumed to be unity. If high pressures are involved, then ϕ_i^V must be calculated from an equation of state. However, care should be taken when mixing and matching different models for γ_i and ϕ_i^V for high-pressure systems to ensure that appropriate combinations are taken.

Example 4.2 A gas from a combustion process has a flowrate of 10 $m^3 \cdot s^{-1}$ and contains 200 ppmv of oxides of nitrogen, expressed as nitric oxide (NO) at standard conditions of 0°C and 1 atm. This concentration needs to be reduced to 50 ppmv (expressed at standard conditions) before being discharged to the environment. It can be assumed that all of the oxides of nitrogen are present in the form of NO. One option being considered to remove the NO is by absorption in water at 20°C and 1 atm. The solubility of the NO in water follows Henry's Law, with $H_{NO} = 2.6 \times 10^4$ atm at 20°C. The gas is to be contacted countercurrently with water such that the inlet gas contacts the outlet water. The concentration of the outlet water can be assumed to reach 90% of equilibrium. Estimate the flowrate of water required, assuming the molar mass of gas in kilograms occupies 22.4 m^3 at standard conditions of 0°C and 1 atm.

Solution Molar flowrate of gas

$$= 10 \times \frac{1}{22.4}$$
$$= 0.446 \text{ kmol·s}^{-1}$$

Assuming the molar flowrate of gas remains constant, the amount of NO to be removed

$$= 0.446(200 - 50) \times 10^{-6}$$
$$= 6.69 \times 10^{-5} \text{ kmol·s}^{-1}$$

Assuming the water achieves 90% of equilibrium and contacts countercurrently with the gas, from

Henry's Law (Equation 4.38):

$$x_{NO} = \frac{0.9 y_{NO}^* P}{H_{NO}}$$

where y_{NO}^{*} = equilibrium mole fraction in the gas phase

$$x_{NO} = \frac{0.9 \times 200 \times 10^{-6} \times 1}{2.6 \times 10^4}$$
$$= 6.9 \times 10^{-9}$$

Assuming the liquid flowrate is constant, water flowrate

$$= \frac{6.69 \times 10^{-5}}{6.9 \times 10^{-9} - 0}$$
$$= 9696 \text{ kmol·s}^{-1}$$
$$= 9696 \times 18 \text{ kg·s}^{-1}$$
$$= 174{,}500 \text{ kg·s}^{-1}$$

This is an impractically large flowrate.

4.5 VAPOR–LIQUID EQUILIBRIUM BASED ON ACTIVITY COEFFICIENT MODELS

In order to model liquid-phase nonideality at moderate pressures, the liquid activity coefficient γ_i must be known:

$$K_i = \frac{\gamma_i P_i^{SAT}}{P} \tag{4.34}$$

γ_i varies with composition and temperature. There are three popular activity coefficient models[3],

a. Wilson
b. Nonrandom two liquid (NRTL)
c. Universal quasi-chemical (UNIQUAC)

These models are semiempirical and are based on the concept that intermolecular forces will cause nonrandom arrangement of molecules in the mixture. The models account for the arrangement of molecules of different sizes and the preferred orientation of molecules. In each case, the models are fitted to experimental binary vapor–liquid equilibrium data. This gives binary interaction parameters that can be used to predict multicomponent vapor–liquid equilibrium. In the case of the UNIQUAC equation, if experimentally determined vapor–liquid equilibrium data are not available, the Universal Quasi-chemical Functional Group Activity Coefficients (UNIFAC) method can be used to estimate UNIQUAC parameters from the molecular structures of the components in the mixture[3].

a. *Wilson equation.* The Wilson equation activity coefficient model is given by[3]:

$$\ln \gamma_i = -\ln\left[\sum_j^{NC} x_j \Lambda_{ij}\right] + 1 - \sum_k^{NC}\left[\frac{x_k \Lambda_{ki}}{\sum_j^{N} x_j \Lambda_{kj}}\right] \tag{4.35}$$

where

$$\Lambda_{ij} = \frac{V_j^L}{V_i^L} \exp\left[-\frac{\lambda_{ij} - \lambda_{ii}}{RT}\right] \tag{4.36}$$

where V_i^L = molar volume of pure Liquid i
λ_{ij} = energy parameter characterizing the interaction of Molecule i with Molecule j
R = gas constant
T = absolute temperature
$\Lambda_{ii} = \Lambda_{jj} = \Lambda_{kk} = 1$
$\Lambda_{ij} = 1$ for an ideal solution
$\Lambda_{ij} < 1$ for positive deviation from Raoult's Law
$\Lambda_{ij} > 1$ for negative deviation from Raoult's Law

For each binary pair, there are two adjustable parameters that must be determined from experimental data, that is, $(\lambda_{ij} - \lambda_{ii})$, which are temperature dependent. The ratio of molar volumes V_j^L/V_i^L is a weak function of temperature. For a binary system, the Wilson equation reduces to[3]:

$$\ln \gamma_1 = -\ln[x_1 + x_2\Lambda_{12}] + x_2\left[\frac{\Lambda_{12}}{x_1 + \Lambda_{12}x_2} - \frac{\Lambda_{21}}{x_1\Lambda_{21} + x_2}\right] \tag{4.37}$$

$$\ln \gamma_2 = -\ln[x_2 + x_1\Lambda_{21}] - x_1\left[\frac{\Lambda_{12}}{x_1 + \Lambda_{12}x_2} - \frac{\Lambda_{21}}{x_1\Lambda_{21} + x_2}\right] \tag{4.38}$$

where

$$\Lambda_{12} = \frac{V_2^L}{V_1^L} \exp\left[-\frac{\lambda_{12} - \lambda_{11}}{RT}\right]$$
$$\Lambda_{21} = \frac{V_1^L}{V_2^L} \exp\left[-\frac{\lambda_{21} - \lambda_{22}}{RT}\right] \tag{4.39}$$

The two adjustable parameters, $(\lambda_{12} - \lambda_{11})$ and $(\lambda_{21} - \lambda_{22})$, must be determined experimentally[6].

b. *NRTL equation.* The NRTL equation is given by[3]:

$$\ln \gamma_i = \frac{\sum_j^{NC} \tau_{ji} G_{ji} x_j}{\sum_k^{NC} G_{ki} x_k} + \sum_j^{NC} \frac{x_j G_{ij}}{\sum_k^{NC} G_{kj} x_k} \times \left(\tau_{ij} - \frac{\sum_k^{NC} x_k \tau_{kj} G_{kj}}{\sum_k^{NC} G_{kj} x_k}\right) \tag{4.40}$$

where $G_{ij} = \exp(-\alpha_{ij}\tau_{ij}),\ G_{ji} = \exp(-\alpha_{ij}\tau_{ji})$

$$\tau_{ij} = \frac{g_{ij} - g_{jj}}{RT},\ \tau_{ji} = \frac{g_{ji} - g_{ii}}{RT}$$

$$G_{ij} \neq G_{ji},\ \tau_{ij} \neq \tau_{ji},\ G_{ii} = G_{jj} = 1,\ \tau_{ii} = \tau_{jj} = 0$$

$\tau_{ij} = 0$ for ideal solutions

g_{ij} and g_{ji} are the energies of interactions between Molecules i and j. α_{ij} characterizes the tendency of Molecule i and Molecule j to be distributed in a random fashion, depends on molecular properties and usually lies in the range 0.2 to 0.5. For each binary pair of components, there are three adjustable parameters, $(g_{ij} - g_{jj})$, $(g_{ji} - g_{ii})$ and $\alpha_{ij}(= \alpha_{ji})$, which are temperature dependent. For a binary system, the NRTL equation reduces to[3]:

$$\ln \gamma_1 = x_2^2 \left[\tau_{21} \left(\frac{G_{21}}{x_1 + x_2 G_{21}} \right)^2 + \frac{\tau_{12} G_{12}}{(x_2 + x_1 G_{12})^2} \right] \tag{4.41}$$

$$\ln \gamma_2 = x_1^2 \left[\tau_{12} \left(\frac{G_{12}}{x_2 + x_1 G_{12}} \right)^2 + \frac{\tau_{21} G_{21}}{(x_1 + x_2 G_{21})^2} \right] \tag{4.42}$$

where $G_{12} = \exp(-\alpha_{12}\tau_{12})$, $G_{21} = \exp(-\alpha_{21}\tau_{21})$,

$$\tau_{12} = \frac{g_{12} - g_{22}}{RT}, \qquad \tau_{21} = \frac{g_{21} - g_{11}}{RT}$$

The three adjustable parameters, $(g_{12} - g_{22})$, $(g_{21} - g_{11})$ and $\alpha_{12}(= \alpha_{21})$, must be determined experimentally[6].

c. *UNIQUAC equation.* The UNIQUAC equation is given by[3]:

$$\ln \gamma_i = \ln\left(\frac{\Phi_i}{x_i}\right) + \frac{z}{2} q_i \ln\left(\frac{\theta_i}{\Phi_i}\right) + l_i - \frac{\Phi_i}{x_i}\sum_j^{NC} x_j l_j + q_i \left[1 - \ln\left(\sum_j^{NC} \theta_j \tau_{ji}\right) - \sum_j^{NC} \frac{\theta_j \tau_{ij}}{\sum_k^{NC} \theta_k \tau_{kj}} \right] \tag{4.43}$$

where $\Phi_i = \dfrac{r_i x_i}{\sum_k^{NC} r_k x_k}$, $\theta_i = \dfrac{q_i x_i}{\sum_k^{NC} q_k x_k}$

$l_i = \dfrac{z}{2}(r_i - q_i) - (r_i - 1)$,

$\tau_{ij} = \exp - \left(\dfrac{u_{ij} - u_{jj}}{RT}\right)$

u_{ij} = interaction parameter between Molecule i and Molecule j $(u_{ij} = u_{ji})$

z = coordination number $(z = 10)$

r_i = pure component property, measuring the molecular van der Waals volume for Molecule i

q_i = pure component property, measuring the molecular van der Waals surface area for Molecule i

R = gas constant

T = absolute temperature

$u_{ij} = u_{ji}$, $\tau_{ii} = \tau_{jj} = 1$

For each binary pair, there are two adjustable parameters that must be determined from experimental data, that is, $(u_{ij} - u_{jj})$, which are temperature dependent. Pure component properties r_i and q_i measure molecular van der Waals volumes and surface areas and have been tabulated[6].

For a binary system, the UNIQUAC equation reduces to[3]:

$$\ln \gamma_1 = \ln\left(\frac{\Phi_1}{x_1}\right) + \frac{z}{2} q_1 \ln\left(\frac{\theta_1}{\Phi_1}\right) + \Phi_2\left(l_1 - l_2\frac{r_1}{r_2}\right) - q_1 \ln(\theta_1 + \theta_2\tau_{21}) + \theta_2 q_1 \left(\frac{\tau_{21}}{\theta_1 + \theta_2\tau_{21}} - \frac{\tau_{21}}{\theta_2 + \theta_1\tau_{12}}\right) \tag{4.44}$$

$$\ln \gamma_2 = \ln\left(\frac{\Phi_2}{x_2}\right) + \frac{z}{2} q_2 \ln\left(\frac{\theta_2}{\Phi_2}\right) + \Phi_1\left(l_2 - l_1\frac{r_2}{r_1}\right) - q_2 \ln(\theta_2 + \theta_1\tau_{12}) + \theta_1 q_2 \left(\frac{\tau_{12}}{\theta_2 + \theta_1\tau_{12}} - \frac{\tau_{21}}{\theta_1 + \theta_2\tau_{21}}\right) \tag{4.45}$$

where $\Phi_1 = \dfrac{r_1 x_1}{r_1 x_1 + r_2 x_2}$, $\Phi_2 = \dfrac{r_2 x_2}{r_1 x_1 + r_2 x_2}$

$\theta_1 = \dfrac{q_1 x_1}{q_1 x_1 + q_2 x_2}$, $\theta_2 = \dfrac{q_2 x_2}{q_1 x_1 + q_2 x_2}$

$l_1 = \dfrac{z}{2}(r_1 - q_1) - (r_1 - 1)$,

$l_2 = \dfrac{z}{2}(r_2 - q_2) - (r_2 - 1)$

$\tau_{12} = \exp - \left(\dfrac{u_{12} - u_{22}}{RT}\right)$,

$\tau_{21} = \exp - \left(\dfrac{u_{21} - u_{11}}{RT}\right)$

The two adjustable parameters, $(u_{12} - u_{22})$ and $(u_{21} - u_{11})$, must be determined experimentally[6]. Pure component properties r_1, r_2, q_1 and q_2 have been tabulated[6].

Since all experimental data for vapor–liquid equilibrium have some experimental uncertainty, it follows that the parameters obtained from data reduction are not unique[3]. There are many sets of parameters that can represent the experimental data equally well, within experimental uncertainty. The experimental data used in data reduction are not sufficient to fix a unique set of "best" parameters. Realistic data reduction can determine only a region of parameters[2].

Published interaction parameters are available[6]. However, when more than one set of parameters is available for a binary pair, which should be chosen?

a. Check if the experimental data is thermodynamically consistent. The Gibbs–Duhem equation[1,2] can be applied to experimental binary data to check its thermodynamic consistency and it should be consistent with this equation.
b. Choose parameters fitted at the process pressure.
c. Choose data sets covering the composition range of interest.
d. For multicomponent systems, choose parameters fitted to ternary or higher systems, if possible.

4.6 VAPOR–LIQUID EQUILIBRIUM BASED ON EQUATIONS OF STATE

Before an equation of state can be applied to calculate vapor–liquid equilibrium, the fugacity coefficient ϕ_i for each phase needs to be determined. The relationship between the fugacity coefficient and the volumetric properties can be written as:

$$\ln \phi_i = \frac{1}{RT}\int_V^{\infty}\left[\left(\frac{\partial P}{\partial N_i}\right)_{T,V,N_j} - \frac{RT}{V}\right] dV - RT \ln Z \tag{4.46}$$

For example, the Peng–Robinson equation of state for this integral yields[2]:

$$\ln \phi_i = \frac{b_i}{b}(Z-1) - \ln(Z-B) - \frac{A}{2\sqrt{2}B}\left(\frac{2\sum x_i a_i}{a} + \frac{b_i}{b}\right)\ln\left(\frac{Z+(1+\sqrt{2})B}{Z+(1-\sqrt{2})B}\right) \tag{4.47}$$

where $A = \dfrac{aP}{R^2T^2}$

$B = \dfrac{bP}{RT}$

Thus, given critical temperatures, critical pressures and acentric factors for each component, as well as a phase composition, temperature and pressure, the compressibility factor can be determined and hence component fugacity coefficients for each phase can be calculated. Taking the ratio of liquid to vapor fugacity coefficients for each component gives the vapor–liquid equilibrium K-value for that component. This approach has the advantage of consistency between the vapor- and liquid-phase thermodynamic models. Such models are widely used to predict vapor–liquid equilibrium for hydrocarbon mixtures and mixtures involving light gases.

A vapor–liquid system should provide three roots from the cubic equation of state, with only the largest and smallest being significant. The largest root corresponds to the vapor compressibility factor and the smallest is the liquid compressibility factor. However, some vapor–liquid mixtures can present problems. This is particularly so for mixtures involving light hydrocarbons with significant amounts of hydrogen, which are common in petroleum and petrochemical processes. Under some conditions, such mixtures can provide only one root for vapor–liquid systems, when there should be three. This means that both the vapor and liquid fugacity coefficients cannot be calculated and is a limitation of such cubic equations of state.

If an activity coefficient model is to be used at high pressure (Equation 4.27), then the vapor-phase fugacity coefficient can be predicted from Equation 4.47. However, this approach has the disadvantage that the thermodynamic models for the vapor and liquid phases are inconsistent. Despite this inconsistency, it might be necessary to use an activity coefficient model if there is reasonable liquid-phase nonideality, particularly with polar mixtures.

4.7 CALCULATION OF VAPOR–LIQUID EQUILIBRIUM

In the case of vapor–liquid equilibrium, the vapor and liquid fugacities are equal for all components at the same temperature and pressure, but how can this solution be found? In any phase equilibrium calculation, some of the conditions will be fixed. For example, the temperature, pressure and overall composition might be fixed. The task is to find values for the unknown conditions that satisfy the equilibrium relationships. However, this cannot be achieved directly. First, values of the unknown variables must be guessed and checked to see if the equilibrium relationships are satisfied. If not, then the estimates must be modified in the light of the discrepancy in the equilibrium, and iteration continued until the estimates of the unknown variables satisfy the requirements of equilibrium.

Consider a simple process in which a multicomponent feed is allowed to separate into a vapor and a liquid phase with the phases coming to equilibrium, as shown in Figure 4.2. An overall material balance and component material balances can be written as:

$$F = V + L \tag{4.48}$$

$$Fz_i = Vy_i + Lx_i \tag{4.49}$$

where F = feed flowrate (kmol·s^{-1})
V = vapor flowrate from the separator (kmol·s^{-1})
L = liquid flowrate from the separator (kmol·s^{-1})
z_i = mole fraction of Component i in the feed (–)
y_i = mole fraction of Component i in vapor (–)
x_i = mole fraction of Component i in liquid (–)

The vapor–liquid equilibrium relationship can be defined in terms of K-values by:

$$y_i = K_i x_i \tag{4.50}$$

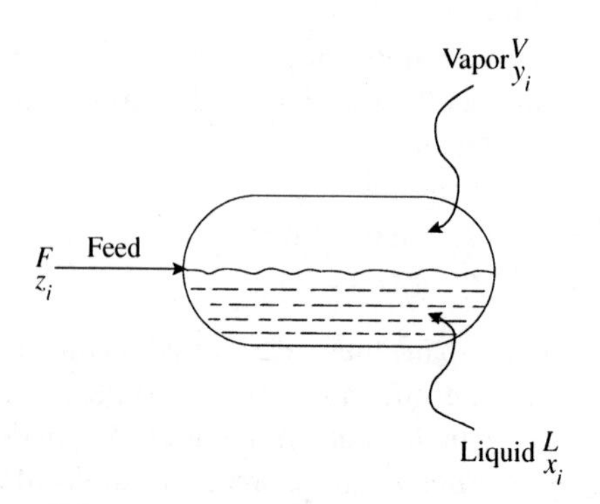

Figure 4.2 Vapor–liquid equilibrium.

Equations 4.48 to 4.50 can now be solved to give expressions for the vapor- and liquid-phase compositions leaving the separator:

$$y_i = \frac{z_i}{\frac{V}{F} + \left(1 - \frac{V}{F}\right)\frac{1}{K_i}} \tag{4.51}$$

$$x_i = \frac{z_i}{(K_i - 1)\frac{V}{F} + 1} \tag{4.52}$$

The vapor fraction (V/F) in Equations 4.51 and 4.52 lies in the range $0 \leq V/F \leq 1$.

For a specified temperature and pressure, Equations 4.51 and 4.52 need to be solved by trial and error. Given that:

$$\sum_i^{NC} y_i = \sum_i^{NC} x_i = 1 \tag{4.53}$$

where NC is the number of components, then:

$$\sum_i^{NC} y_i - \sum_i^{NC} x_i = 0 \tag{4.54}$$

Substituting Equations 4.51 and 4.52 into Equation 4.54, after rearrangement gives[7]:

$$\sum_i^{NC} \frac{z_i(K_i - 1)}{\frac{V}{F}(K_i - 1) + 1} = 0 = f(V/F) \tag{4.55}$$

To solve Equation 4.55, start by assuming a value of V/F and calculate $f(V/F)$ and search for a value of V/F until the function equals zero.

Many variations are possible around the basic *flash* calculation. Pressure and V/F can be specified and T calculated, and so on. Details can be found in King[7]. However, two special cases are of particular interest. If it is necessary to calculate the *bubble point*, then $V/F = 0$ in Equation 4.55, which simplifies to:

$$\sum_i^{NC} z_i(K_i - 1) = 0 \tag{4.56}$$

and given:

$$\sum_i^{NC} z_i = 1 \tag{4.57}$$

This simplifies to the expression for the bubble point:

$$\sum_i^{NC} z_i K_i = 1 \tag{4.58}$$

Thus, to calculate the bubble point for a given mixture and at a specified pressure, a search is made for a temperature to satisfy Equation 4.58. Alternatively, temperature can be specified and a search made for a pressure, the *bubble pressure*, to satisfy Equation 4.58.

Another special case is when it is necessary to calculate the *dew point*. In this case, $V/F = 1$ in Equation 4.55, which simplifies to:

$$\sum_i^{NC} \frac{z_i}{K_i} = 1 \tag{4.59}$$

Again, for a given mixture and pressure, temperature is searched to satisfy Equation 4.59. Alternatively, temperature is specified and pressure searched for the *dew pressure*.

If the K-value requires the composition of both phases to be known, then this introduces additional complications into the calculations. For example, suppose a bubble-point calculation is to be performed on a liquid of known composition using an equation of state for the vapor–liquid equilibrium. To start the calculation, a temperature is assumed. Then, calculation of K-values requires knowledge of the vapor composition to calculate the vapor-phase fugacity coefficient, and that of the liquid composition to calculate the liquid-phase fugacity coefficient. While the liquid composition is known, the vapor composition is unknown and an initial estimate is required for the calculation to proceed. Once the K-value has been estimated from an initial estimate of the vapor composition, the composition of the vapor can be reestimated, and so on.

Figure 4.3 shows, as an example, the vapor–liquid equilibrium behavior for a binary mixture of benzene and toluene[8]. Figure 4.3a shows the behavior of temperature of the saturated liquid and saturated vapor (i.e. equilibrium pairs) as the mole fraction of benzene is varied (the balance being toluene). This can be constructed by calculating the bubble and dew points for different concentrations. Figure 4.3b shows an alternative way of representing the vapor–liquid equilibrium in a composition or *x–y diagram*. The *x–y* diagram can be constructed from the relative volatility. From the definition of relative volatility for a binary mixture of Components A and B:

$$\alpha_{AB} = \frac{y_A/x_A}{y_B/x_B} = \frac{y_A/x_A}{(1 - y_A)/(1 - x_A)} \tag{4.60}$$

Rearranging gives:

$$y_A = \frac{x_A \alpha_{AB}}{1 + x_A(\alpha_{AB} - 1)} \tag{4.61}$$

Thus, by knowing α_{AB} from vapor–liquid equilibrium and by specifying x_A, y_A can be calculated. Figure 4.3a also shows a typical vapor–liquid equilibrium pair, where the mole fraction of benzene in the liquid phase is 0.4 and that in the vapor phase is 0.62. A diagonal line across the *x–y* diagram represents equal vapor and liquid compositions. The phase equilibrium behavior shows a curve above the diagonal line. This indicates that benzene has a higher concentration in the vapor phase than toluene, that is,

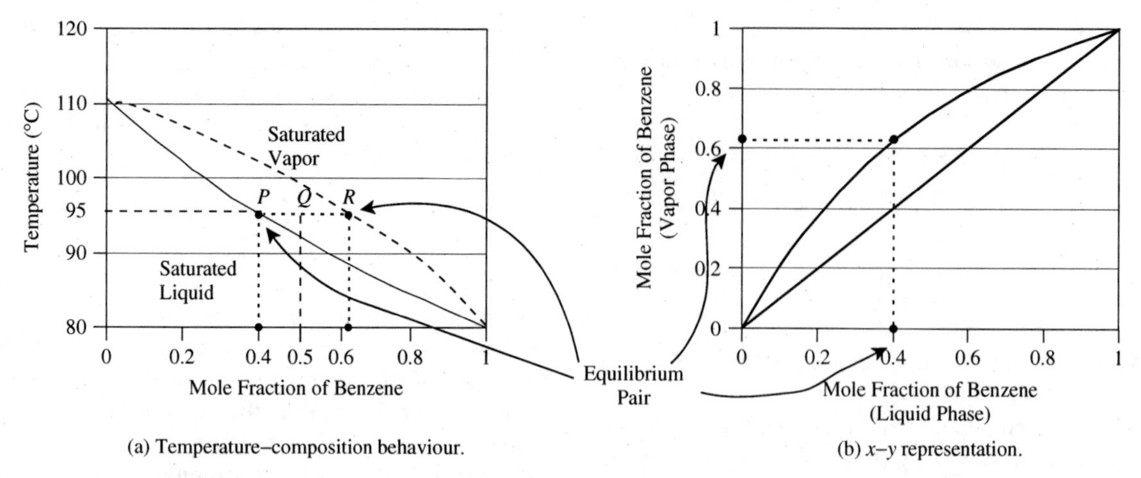

(a) Temperature–composition behaviour. (b) x–y representation.

Figure 4.3 Vapor–liquid equilibrium for a binary mixture of benzene and toluene at a pressure of 1 atm. (From Smith R and Jobson M, 2000, Distillation, Encyclopedia of Separation Science, Academic Press; reproduced by permission).

benzene is the more volatile component. Figure 4.3b shows the same vapor–liquid equilibrium pair as that shown in Figure 4.3a[8].

Figure 4.3a can be used to predict the separation in a single equilibrium stage, given a specified feed to the stage and a stage temperature. For example, suppose the feed is a mixture with equal mole fractions of benzene and toluene of 0.5 and this is brought to equilibrium at 95°C (Point Q in Figure 4.3a). Then, the resulting liquid will have a mole fraction of benzene of 0.4 and the vapor, a mole fraction of 0.62. In addition, the quantity of each phase formed can be determined from the lengths of the lines PQ and QR in Figure 4.3a. An overall material balance across the separator gives:

$$m_Q = m_P + m_R \tag{4.62}$$

A material balance for Component i gives:

$$m_Q x_{i,Q} = m_P x_{i,P} + m_R x_{i,R} \tag{4.63}$$

Substituting Equation 4.62 into Equation 4.63 and rearranging gives:

$$\frac{m_P}{m_R} = \frac{x_{i,R} - x_{i,Q}}{x_{i,Q} - x_{i,P}} \tag{4.64}$$

Thus, in Figure 4.3:

$$\frac{m_P}{m_R} = \frac{PQ}{QR} \tag{4.65}$$

The ratio of molar flowrates of the vapor and liquid phases is thus given by the ratio of the opposite line segments. This is known as the *Lever Rule,* after the analogy with a lever and fulcrum[7].

Consider first, a binary mixture of two Components A and B; the vapor–liquid equilibrium exhibits only a moderate deviation from ideality, as represented in Figure 4.4a. In this case, as pure A boils at a lower temperature than pure B in the temperature–composition diagram in Figure 4.4a, Component A is more volatile than Component B. This is also evident from the vapor–liquid composition diagram (x–y diagram), as it is above the line of $y_A = x_A$. In addition, it is also clear from Figure 4.4a that the order of volatility does not change as the composition changes. By contrast, Figure 4.4b shows a more highly nonideal behavior in which $\gamma_i > 1$ (positive deviation from Raoult's Law) forms a *minimum-boiling azeotrope*. At the azeotropic composition, the vapor and liquid are both at the same composition for the mixture. The lowest boiling temperature is below that of either of the pure components and is at the minimum-boiling azeotrope. It is clear from Figure 4.4b that the order of volatility of Components A and B changes, depending on the composition. Figure 4.4c also shows azeotropic behavior. This time, the mixture shows a behavior in which $\gamma_i < 1$ (negative deviation from Raoult's Law) forms a *maximum-boiling azeotrope*. This maximum-boiling azeotrope boils at a higher temperature than either of the pure components and would be the last fraction to be distilled, rather than the least volatile component, which would be the case with nonazeotropic behavior. Again, from Figure 4.4c, it can be observed that the order of volatility of Components A and B changes depending on the composition. Minimum-boiling azeotropes are much more common than maximum-boiling azeotropes.

Some general guidelines for vapor–liquid mixtures in terms of their nonideality are:

a. Mixtures of isomers usually form ideal solutions.
b. Mixtures of close-boiling aliphatic hydrocarbons are nearly ideal below 10 bar.
c. Mixtures of compounds close in molar mass and structure frequently do not deviate greatly from ideality (e.g. ring compounds, unsaturated compounds, naphthenes etc.).
d. Mixtures of simple aliphatics with aromatic compounds deviate modestly from ideality

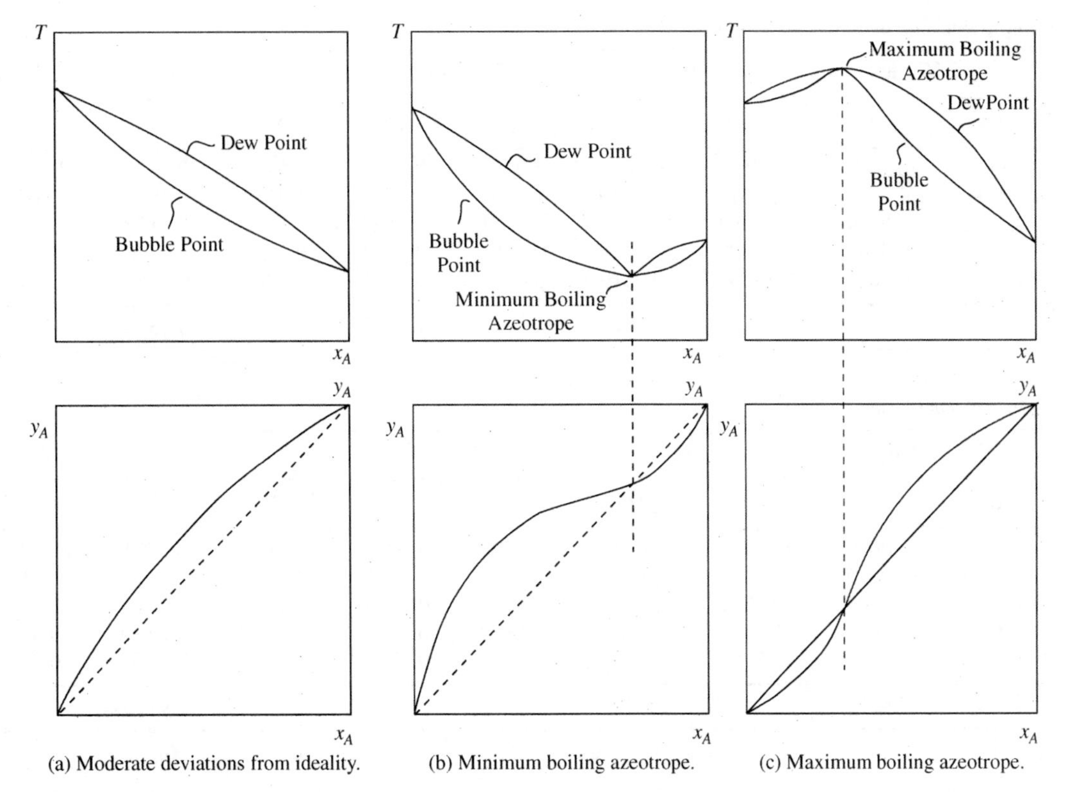

Figure 4.4 Binary vapor-liquid equilibrium behavior.

e. Noncondensables such as CO_2, H_2S, H_2, N_2, and so on, that are present in mixtures involving heavier components tend to behave nonideally with respect to the other compounds.
f. Mixtures of polar and nonpolar compounds are always strongly nonideal.
g. Azeotropes and phase separation into liquid–liquid mixtures represent the ultimate in nonideality.

Moving down the list, the nonideality of the system increases.

Example 4.3 A mixture of ethane, propane, n-butane, n-pentane and n-hexane is given in the Table 4.3. For this calculation, it can be assumed that the K-values are ideal. For the mixture in Table 4.3, an equation of state method might have been a more appropriate choice. However, this makes the calculation of the K-values much more complex. The ideal K-values for the mixture can be expressed in terms of the Antoine Equation as:

$$K_i = \frac{1}{P} \exp\left[A_i - \frac{B_i}{T + C_i}\right] \tag{4.66}$$

where P is the pressure (bar), T the absolute temperature (K) and A_i, B_i and C_i are constants given in the Table 4.3:

a. For a pressure of 5 bar, calculate the bubble point.
b. For a pressure of 5 bar, calculate the dew point.
c. Calculate the pressure needed for total condensation at 313 K.
d. At a pressure of 6 bar and a temperature of 313 K, how much liquid will be condensed?

Table 4.3 Feed and physical property for components.

Component	Formula	Feed (kmol)	A_i	B_i	C_i
1. Ethane	C_2H_6	5	9.0435	1511.4	−17.16
2. Propane	C_3H_8	25	9.1058	1872.5	−25.16
3. n-Butane	C_4H_{10}	30	9.0580	2154.9	−34.42
4. n-Pentane	C_5H_{12}	20	9.2131	2477.1	−39.94
5. n-Hexane	C_6H_{14}	20	9.2164	2697.6	−48.78

Solution

a. The bubble point can be calculated from Equation 4.58 or from Equation 4.55 by specifying $V/F = 0$. The bubble point is calculated from Equation 4.58 in Table 4.4 to be 296.4 K. The search can be readily automated in spreadsheet software.

b. The dew point can be calculated from Equation 4.59 or 4.55 by specifying $V/F = 1$. In Table 4.5, the dew point is calculated from Equation 4.59 to be 359.1 K.

c. To calculate the pressure needed for total condensation at 313 K, the bubble pressure is calculated. The calculation is essentially the same as that in Part a above, with the temperature fixed at 313 K and the pressure varied, instead, in an iterative fashion until Equation 4.64 is satisfied. The resulting bubble pressure at 313 K is 7.4 bar.

If distillation was carried out giving an overhead composition, as given in Table 4.3, and if cooling water was used in the condenser, then total condensation at 313 K (40°C) would require an operating pressure of 7.4 bar.

Table 4.4 Bubble-point calculation.

i	z_i	$T = 275$ K		$T = 350$ K		$T = 300$ K		$T = 290$ K		$T = 296.366$ K	
		K_i	z_iK_i	K_i	z_iK_i	K_i	z_iK_i	K_i	z_iK_i	K_I	z_iK_i
1	0.05	4.8177	0.2409	18.050	0.9025	8.0882	0.4044	6.6496	0.3325	7.5448	0.3772
2	0.25	1.0016	0.2504	5.6519	1.4130	1.9804	0.4951	1.5312	0.3828	1.8076	0.4519
3	0.30	0.2212	0.0664	1.8593	0.5578	0.5141	0.1542	0.3742	0.1123	0.4594	0.1378
4	0.20	0.0532	0.0106	0.6802	0.1360	0.1464	0.0293	0.1000	0.0200	0.1279	0.0256
5	0.20	0.0133	0.0027	0.2596	0.0519	0.0437	0.0087	0.0280	0.0056	0.0373	0.0075
	1.00		0.5710		3.0612		1.0917		0.8532		1.0000

Table 4.5 Dew-point calculation.

i	z_i	$T = 325$ K		$T = 375$ K		$T = 350$ K		$T = 360$ K		$T = 359.105$ K	
		K_i	z_i/K_i	K_i	z_i/K_i	K_i	z_i/K_i	K_i	z_i/K_i	K_i	z_i/K_i
1	0.05	12.483	0.0040	24.789	0.0020	18.050	0.0028	20.606	0.0024	20.370	0.0025
2	0.25	3.4951	0.0715	8.5328	0.0293	5.6519	0.0442	6.7136	0.0372	6.6138	0.0378
3	0.30	1.0332	0.2903	3.0692	0.0977	1.8593	0.1614	2.2931	0.1308	2.2517	0.1332
4	0.20	0.3375	0.5925	1.2345	0.1620	0.6802	0.2941	0.8730	0.2291	0.8543	0.2341
5	0.20	0.1154	1.7328	0.5157	0.3879	0.2596	0.7704	0.3462	0.5778	0.3376	0.5924
	1.00		2.6911		0.6789		1.2729		0.9773		1.0000

Table 4.6 Flash calculation.

i	z_i	$V/F = 0$		$V/F = 0.5$	$V/F = 0.1$	$V/F = 0.05$	$V/F = 0.0951$
		K_i	$\frac{z_i(K_i-1)}{\frac{V}{F}(K_i-1)+1}$	$\frac{z_i(K_i-1)}{\frac{V}{F}(K_i-1)+1}$	$\frac{z_i(K_i-1)}{\frac{V}{F}(K_i-1)+1}$	$\frac{z_i(K_i-1)}{\frac{V}{F}(K_i-1)+1}$	$\frac{z_i(K_i-1)}{\frac{V}{F}(K_i-1)+1}$
1	0.05	8.5241	0.3762	0.0790	0.2147	0.2734	0.2193
2	0.25	2.2450	0.3112	0.1918	0.2768	0.2930	0.2783
3	0.30	0.6256	−0.1123	−0.1382	−0.1167	−0.1145	−0.1165
4	0.20	0.1920	−0.1616	−0.2711	−0.1758	−0.1684	−0.1751
5	0.20	0.0617	−0.1877	−0.3535	−0.2071	−0.1969	−0.2060
	1.00		0.2258	−0.4920	−0.0081	0.0866	0.0000

d. Equation 4.55 is used to determine how much liquid will be condensed at a pressure of 6 bar and 313 K. The calculation is detailed in Table 4.6, giving $V/F = 0.0951$, thus, 90.49% of the feed will be condensed.

Simple phase equilibrium calculations, like the one illustrated here, can be readily implemented in spreadsheet software and automated. In practice, the calculations will most often be carried out in commercial physical property packages, allowing more elaborate methods for calculating the equilibrium K-values to be used.

Example 4.4 Calculate the vapor composition of an equimolar liquid mixture of methanol and water at 1 atm (1.013 bar)

a. assuming ideal vapor- and liquid-phase behavior, that is, using Raoult's Law
b. using the Wilson equation.

Vapor pressure in bar can be predicted for temperature in Kelvin from the Antoine equation using coefficients in Table 4.7[3]. Data for the Wilson equation are given in Table 4.8[6]. Assume the gas constant $R = 8.3145$ kJ·kmol^{-1}·K^{-1}.

Table 4.7 Antoine coefficients for methanol and water[6].

	A_i	B_i	C_i
Methanol	11.9869	3643.32	−33.434
Water	11.9647	3984.93	−39.734

Table 4.8 Data for methanol (1) and water (2) for the Wilson equation at 1 atm[6].

V_1 (m^3·kmol^{-1})	V_2 (m^3·kmol^{-1})	$(\lambda_{12} - \lambda_{11})$ (kJ·kmol^{-1})	$(\lambda_{21} - \lambda_{22})$ (kJ·kmol^{-1})
0.04073	0.01807	347.4525	2179.8398

Table 4.10 Bubble-point calculation for a methanol–water mixture using the Wilson equation.

z_i	$T = 340$ K			$T = 350$ K			$T = 346.13$ K		
	γ_i	K_i	z_iK_i	γ_i	K_i	z_iK_i	γ_i	K_i	z_iK_i
0.5	1.1429	1.2501	0.6251	1.1363	1.8092	0.9046	1.1388	1.5727	0.7863
0.5	1.2307	0.3289	0.1645	1.2227	0.5012	0.2506	1.2258	0.4273	0.2136
1.00			0.7896			1.1552			0.9999

Solution

a. K-values assuming ideal vapor and liquid behavior are given by Raoult's Law (Equation 4.66). The composition of the liquid is specified to be $x_1 = 0.5, x_2 = 0.5$ and the pressure 1 atm, but the temperature is unknown. Therefore, a bubble-point calculation is required to determine the vapor composition. The procedure is the same as that in Example 4.2 and can be carried out in spreadsheet software. Table 4.9 shows the results for Raoult's Law.

Table 4.9 Bubble-point calculation for an ideal methanol–water mixture.

z_i	$T = 340$ K		$T = 360$ K		$T = 350$ K	
	K_i	z_iK_i	K_i	z_iK_i	K_i	z_iK_i
0.5	1.0938	0.5469	2.2649	1.1325	1.5906	0.7953
0.5	0.2673	0.1336	0.6122	0.3061	0.4094	0.2047
1.00		0.6805		1.4386		1.0000

Thus, the composition of the vapor at 1 atm is $y_1 = 0.7953$, $y_2 = 0.2047$, assuming an ideal mixture.

b. The activity coefficients for the methanol and water can be calculated using the Wilson equation (Equations 4.37 to 4.38). The results are summarized in Table 4.10.

Thus, the composition of the vapor phase at 1 atm is $y_1 = 0.7863$, $y_2 = 0.2136$ from the Wilson Equation. For this mixture, at these conditions, there is not much difference between the predictions of Raoult's Law and the Wilson equation, indicating only moderate deviations from ideality at the chosen conditions.

Example 4.5 2-Propanol (isopropanol) and water form an azeotropic mixture at a particular liquid composition that results in the vapor and liquid compositions being equal. Vapor–liquid equilibrium for 2-propanol–water mixtures can be predicted by the Wilson equation. Vapor pressure coefficients in bar with temperature in Kelvin for the Antoine equation are given in Table 4.11[3]. Data for the Wilson equation are given in Table 4.12[6]. Assume the gas constant $R = 8.3145$ kJ·kmol^{-1}·K^{-1}. Determine the azeotropic composition at 1 atm.

Table 4.11 Antoine equation coefficients for 2-propanol and water[6].

	A_i	B_i	C_i
2-propanol	13.8228	4628.96	−20.524
Water	11.9647	3984.93	−39.734

Table 4.12 Data for 2-propanol (1) and water (2) for the Wilson equation at 1 atm[6].

V_1 (m^3·kmol^{-1})	V_2 (m^3·kmol^{-1})	$(\lambda_{12} - \lambda_{11})$ (kJ·kmol^{-1})	$(\lambda_{21} - \lambda_{22})$ (kJ·kmol^{-1})
0.07692	0.01807	3716.4038	5163.0311

Solution To determine the location of the azeotrope for a specified pressure, the liquid composition has to be varied and a bubble-point calculation performed at each liquid composition until a composition is identified, whereby $x_i = y_i$. Alternatively, the vapor composition could be varied and a dew-point calculation performed at each vapor composition. Either way, this requires iteration. Figure 4.5 shows the x–y diagram for the 2-propanol–water system. This was obtained by carrying out a bubble-point calculation at different values of the liquid composition. The point where the x–y plot crosses the diagonal line gives the azeotropic composition. A more direct search for the azeotropic composition can be carried out for such a binary system in a spreadsheet by varying T and x_1 simultaneously and by solving the objective function (see Section 3.9):

$$(x_1K_1 + x_2K_2 - 1)^2 + (x_1 - x_1K_1)^2 = 0$$

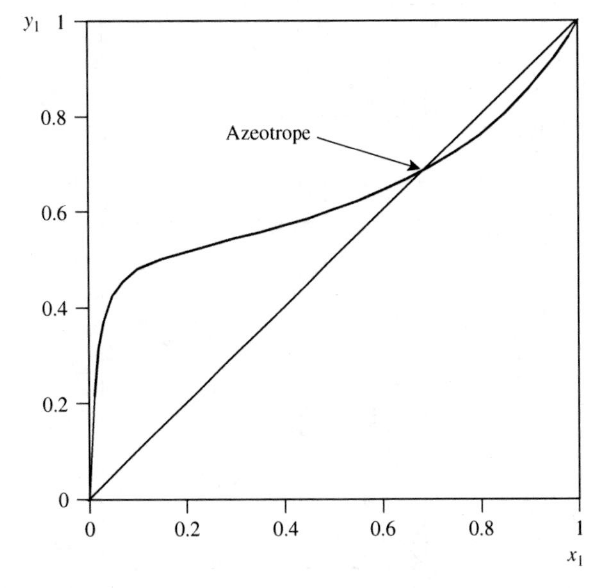

Figure 4.5 x–y plot for the system 2-propanol (1) and water (2) from the Wilson equation at 1 atm.

The first bracket in this equation ensures that the bubble-point criterion is satisfied. The second bracket ensures that the vapor and liquid compositions are equal. The solution of this is given when $x_1 = y_1 = 0.69$ and $x_2 = y_2 = 0.31$ for the system of 2-propanol-water at 1 atm.

4.8 LIQUID–LIQUID EQUILIBRIUM

As the components in a liquid mixture become more chemically dissimilar, their mutual solubility decreases. This is characterized by an increase in their activity coefficients (for positive deviation from Raoult's Law). If the chemical dissimilarity, and the corresponding increase in activity coefficients, become large enough, the solution can separate into two-liquid phases.

Figure 4.6 shows the vapor–liquid equilibrium behavior of a system exhibiting two-liquid phase behavior. Two-liquid phases exist in the areas *abcd* in Figures 4.6a and 4.6b. Liquid mixtures outside of this two-phase region are homogeneous. In the two-liquid phase region, below *ab*, the two-liquid phases are subcooled. Along *ab*, the two-liquid phases are saturated. The area of the two-liquid phase region becomes narrower as the temperature increases. This is because the mutual solubility normally increases with increasing temperature. For a mixture within the two-phase region, say Point Q in Figures 4.6a and 4.6b, at equilibrium, two-liquid phases are formed at Points P and R. The line PR is the *tie line*. The analysis for vapor–liquid separation in Equations 4.56 to 4.59 also applies to a liquid–liquid separation. Thus, in Figures 4.6a and 4.6b, the relative amounts of the two-liquid phases formed from Point Q at P and R follows the Lever Rule given by Equation 4.65.

In Figure 4.6a, the azeotropic composition at Point *e* lies outside the region of two-liquid phases. In Figure 4.6b, the azeotropic composition lies inside the region of two-liquid phases. Any two-phase liquid mixture vaporizing along *ab*, in Figure 4.6b, will vaporize at the same temperature and have a vapor composition corresponding with Point e. This results from the lines of vapor–liquid equilibrium being horizontal in the vapor–liquid region, as shown in Figure 4.3. A liquid mixture of composition e, in Figure 4.6b, produces a vapor of the same composition and is known as a *heteroazeotrope*. The $x-y$ diagrams in Figure 4.6c and 4.6d exhibit a horizontal section, corresponding with the two-phase region.

For liquid–liquid equilibrium, the fugacity of each component in each phase must be equal:

$$(x_i \gamma_i)^I = (x_i \gamma_i)^{II} \tag{4.67}$$

where I and II represent the two-liquid phases in equilibrium. The equilibrium K-value or *distribution coefficient*

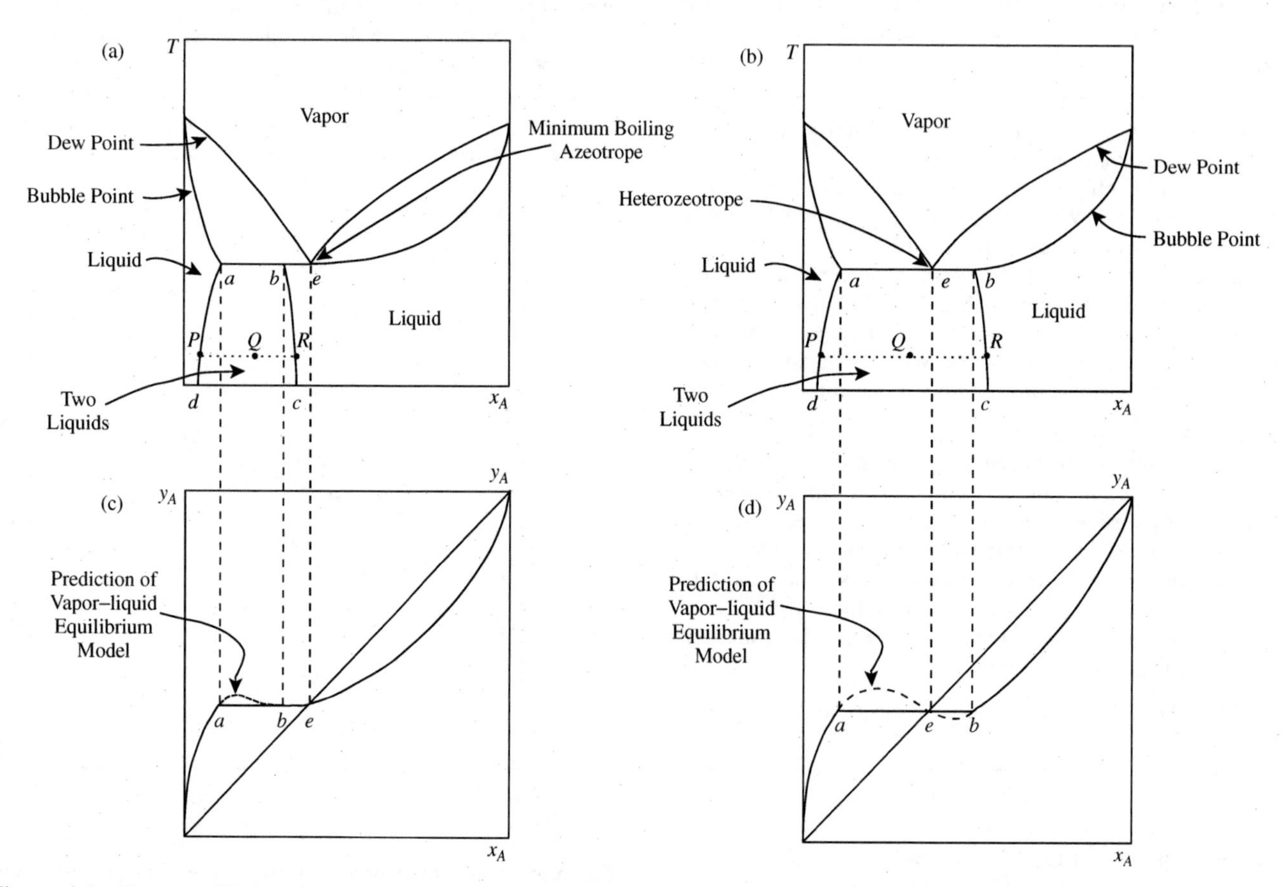

Figure 4.6 Phase equilibrium featuring two-liquid phases.

for Component i can be defined by

$$K_i = \frac{x_i^I}{x_i^{II}} = \frac{\gamma_i^{II}}{\gamma_i^I} \quad (4.68)$$

4.9 LIQUID–LIQUID EQUILIBRIUM ACTIVITY COEFFICIENT MODELS

A model is needed to calculate liquid–liquid equilibrium for the activity coefficient from Equation 4.67. Both the NRTL and UNIQUAC equations can be used to predict liquid–liquid equilibrium. Note that the Wilson equation is not applicable to liquid–liquid equilibrium and, therefore, also not applicable to vapor–liquid–liquid equilibrium. Parameters from the NRTL and UNIQUAC equations can be correlated from vapor–liquid equilibrium data[6] or liquid–liquid equilibrium data[9,10]. The UNIFAC method can be used to predict liquid–liquid equilibrium from the molecular structures of the components in the mixture[3].

4.10 CALCULATION OF LIQUID–LIQUID EQUILIBRIUM

The vapor–liquid x–y diagram in Figures 4.6c and d can be calculated by setting a liquid composition and calculating the corresponding vapor composition in a bubble point calculation. Alternatively, vapor composition can be set and the liquid composition determined by a dew point calculation. If the mixture forms two-liquid phases, the vapor–liquid equilibrium calculation predicts a maximum in the x–y diagram, as shown in Figures 4.6c and d. Note that such a maximum cannot appear with the Wilson equation.

To calculate the compositions of the two coexisting liquid phases for a binary system, the two equations for phase equilibrium need to be solved:

$$(x_1\gamma_1)^I = (x_1\gamma_1)^{II}, (x_2\gamma_2)^I = (x_2\gamma_2)^{II} \quad (4.69)$$

where $x_1^I + x_2^I = 1, x_1^{II} + x_2^{II} = 1 \quad (4.70)$

Given a prediction of the liquid-phase activity coefficients, from say the NRTL or UNIQUAC equations, then Equations 4.69 and 4.70 can be solved simultaneously for x_1^I and x_1^{II}. There are a number of solutions to these equations, including a trivial solution corresponding with $x_1^I = x_1^{II}$. For a solution to be meaningful:

$$0 < x_1^I < 1, \quad 0 < x_1^{II} < 1, \quad x_1^I \neq x_1^{II} \quad (4.71)$$

For a ternary system, the corresponding equations to be solved are:

$$(x_1\gamma_1)^I = (x_1\gamma_1)^{II},$$
$$(x_2\gamma_2)^I = (x_2\gamma_2)^{II}, \quad (x_3\gamma_3)^I = (x_3\gamma_3)^{II} \quad (4.72)$$

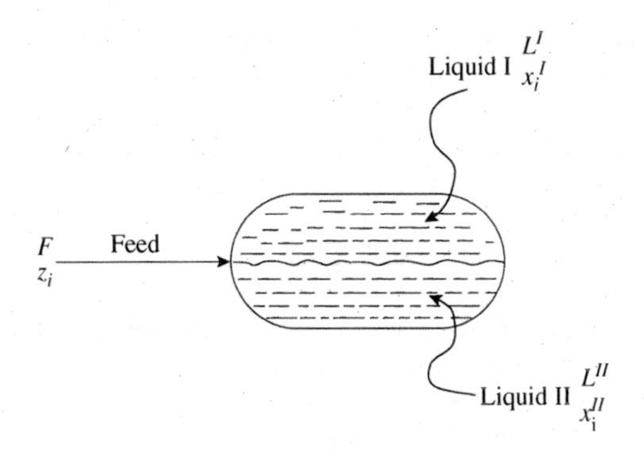

Figure 4.7 Liquid–liquid equilibrium.

These equations can be solved simultaneously with the material balance equations to obtain x_1^I, x_2^I, x_1^{II} and x_2^{II}. For a multicomponent system, the liquid–liquid equilibrium is illustrated in Figure 4.7. The mass balance is basically the same as that for vapor–liquid equilibrium, but is written for two-liquid phases. Liquid I in the liquid–liquid equilibrium corresponds with the vapor in vapor–liquid equilibrium and Liquid II corresponds with the liquid in vapor–liquid equilibrium. The corresponding mass balance is given by the equivalent to Equation 4.55:

$$\sum_i^{NC} \frac{z_i(K_i - 1)}{\dfrac{L^I}{F}(K_i - 1) + 1} = 0 = f(L^I/F) \quad (4.73)$$

where F = feed flowrate (kmol·s^{-1})
L^I = flowrate of Liquid I from the separator (kmol·s^{-1})
L^{II} = flowrate of Liquid II from the separator (kmol·s^{-1})
z_i = mole fraction of Component i in the feed (–)
x_i^I = mole fraction of Component i in Liquid I (–)
x_i^{II} = mole fraction of Component i in Liquid II (–)
K_i = K-value, or distribution coefficient, for Component i (–)

Also, the liquid–liquid equilibrium K-value needs to be defined for equilibrium to be

$$K_i = \frac{x_i^I}{x_i^{II}} = \frac{\gamma_i^{II}}{\gamma_i^I} \quad (4.74)$$

$x_1^I, x_2^I, \ldots, x_{NC-1}^I; x_1^{II}, x_2^{II}, \ldots, x_{NC-1}^{II}$ and L^I/F need to be varied simultaneously to solve Equations 4.73 and 4.74.

Example 4.6 Mixtures of water and 1-butanol (n-butanol) form two-liquid phases. Vapor–liquid equilibrium and liquid–liquid equilibrium for the water–1-butanol system can be predicted by the NRTL equation. Vapor pressure coefficients in bar with temperature in Kelvin for the Antoine equation are given in Table 4.13[6]. Data for the NRTL equation are given in Table 4.14, for a pressure of 1 atm[6]. Assume the gas constant $R = 8.3145$ kJ·kmol^{-1}·K^{-1}.

Table 4.13 Antoine coefficient for water and 1-butanol[6].

	A_i	B_i	C_i
Water	11.9647	3984.93	−39.734
1-butanol	10.3353	3005.33	−99.733

Table 4.14 Data for water (1) and 1-butanol (2) for the NRTL equation at 1 atm[6].

$(g_{12} - g_{22})$ (kJ·kmol^{-1})	$(g_{21} - g_{11})$ (kJ·kmol^{-1})	α_{ij} (–)
11,184.9721	1649.2622	0.4362

a. Plot the x–y diagram at 1 atm.
b. Determine the compositions of the two-liquid phase region for saturated vapor–liquid–liquid equilibrium at 1 atm.

Solution

a. For a binary system, the calculations can be performed in spreadsheet software. As with Example 4.4, a series of bubble-point calculations can be performed at different liquid-phase compositions (or dew-point calculations at different vapor-phase compositions). The NRTL equation is modeled using Equations 4.41 and 4.42. The resulting x–y diagram is shown in Figure 4.8. The x–y diagram displays the characteristic maximum for two-liquid phase behavior.

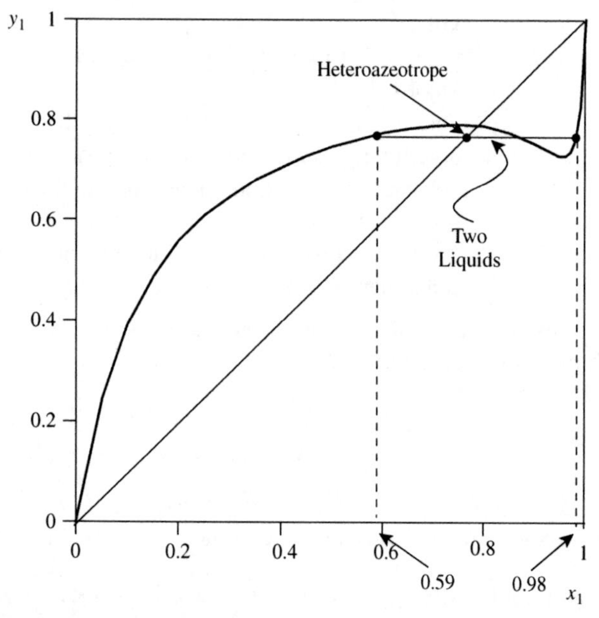

Figure 4.8 x–y plot for the system water (1) and 1-butanol (2) from the NRTL equation at 1 atm.

b. To determine the compositions of the two-liquid phase region, the NRTL equation is set up for two-liquid phases by writing Equations 4.41 and 4.42 for each phase. The constants are given in Table 4.12. Note that if x_1^I and x_1^{II} are specified, then:

$$x_2^I = 1 - x_1^I \text{ and } x_2^{II} = 1 - x_1^{II}$$

A search is then made by varying x_1^I and x_1^{II} simultaneously (e.g. using a spreadsheet solver) to solve the objective function (see Section 3.9):

$$(x_1^I \gamma_1^I - x_1^{II} \gamma_1^{II})^2 + (x_2^I \gamma_2^I - x_2^{II} \gamma_2^{II})^2 = 0$$

This ensures liquid–liquid equilibrium. Trivial solutions whereby $x_1^I = x_1^{II}$ need to be avoided. The results are shown in Figure 4.8 to be $x_1^I = 0.59$, $x_1^{II} = 0.98$. The system forms a heteroazeotrope.

To ensure that the predicted two-phase region corresponds with that for the saturated vapor–liquid–liquid equilibrium, the temperature must be specified to be bubble-point predicted by the NRTL equation at either $x_1 = 0.59$ or $x_1 = 0.98$ (366.4 K in this case).

Care should be exercised in using the coefficients from Table 4.14 to predict two-liquid phase behavior under subcooled conditions. The coefficients in Table 4.14 were determined from vapor–liquid equilibrium data at saturated conditions.

Although the methods developed here can be used to predict liquid–liquid equilibrium, the predictions will only be as good as the coefficients used in the activity coefficient model. Such predictions can be critical when designing liquid–liquid separation systems. When predicting liquid–liquid equilibrium, it is always better to use coefficients correlated from liquid–liquid equilibrium data, rather than coefficients based on the correlation of vapor–liquid equilibrium data. Equally well, when predicting vapor–liquid equilibrium, it is always better to use coefficients correlated to vapor–liquid equilibrium data, rather than coefficients based on the correlation of liquid–liquid equilibrium data. Also, when calculating liquid–liquid equilibrium with multicomponent systems, it is better to use multicomponent experimental data, rather than binary data.

4.11 CALCULATION OF ENTHALPY

The calculation of enthalpy is required for energy balance calculations. It might also be required for the calculation of other derived thermodynamic properties. The absolute value of enthalpy for a substance cannot be measured and only changes in enthalpy are meaningful. Consider the change in enthalpy with pressure. At a given temperature T, the change in enthalpy of a fluid can be determined from the derivative of enthalpy with pressure at fixed temperature, $(\partial H/\partial P)_T$. Thus, the change in enthalpy relative to a reference is given by:

$$[H_P - H_{P_O}]_T = \int_{P_O}^{P} \left(\frac{\partial H}{\partial P}\right)_T dP \qquad (4.75)$$

where H = enthalpy at pressure P (kJ·kmol^{-1})
H_{P_O} = enthalpy at reference pressure P_O (kJ·kmol^{-1})
P = pressure (bar)
P_O = reference pressure (bar)

The change in enthalpy with pressure is given by[1]:

$$\left(\frac{\partial H}{\partial P}\right)_T = V - T\left(\frac{\partial V}{\partial T}\right)_P \quad (4.76)$$

Also:

$$\left(\frac{\partial V}{\partial T}\right)_P = \left[\frac{\partial}{\partial T}\left(\frac{ZRT}{P}\right)\right]_P = \frac{R}{P}\left[Z + T\left(\frac{\partial Z}{\partial T}\right)_P\right] \quad (4.77)$$

where Z = compressibility (–)
R = universal gas constant

Combining Equations 4.75 to 4.77 gives the difference between the enthalpy at pressure P and that at the standard pressure P_O and is known as the *enthalpy departure*.

$$[H_P - H_{P_O}]_T = -\int_{P_O}^{P}\left[\frac{RT^2}{P}\left(\frac{\partial Z}{\partial T}\right)_P\right]_T dP \quad (4.78)$$

Equation 4.78 defines the enthalpy departure from a reference state at temperature T and pressure P_O.

The value of $(\partial Z/\partial T)_P$ can be obtained from an equation of state, such as the Peng–Robinson equation of state, and the integral in Equation 4.78 evaluated[3]. The enthalpy departure for the Peng–Robinson equation of state is given by[3]:

$$[H_P - H_{P_O}]_T = RT(Z-1) + \frac{T(da/dT) - a}{2\sqrt{2}b} \times \ln\left[\frac{Z + (1+\sqrt{2})B}{Z + (1-\sqrt{2})B}\right] \quad (4.79)$$

where
$$da/dT = -0.45724\frac{R^2T_C^2}{P_C}\kappa\sqrt{\frac{\alpha}{T \cdot T_C}}$$

$$B = \frac{bP}{RT}$$

Equations of state, such as the Peng–Robinson equation, are capable of predicting both liquid and vapor behavior. The appropriate root for the equation of a liquid or vapor must be taken, as discussed previously. Equation 4.79 is therefore capable of predicting both liquid and vapor enthalpy[3]. Equations of state such as the Peng–Robinson equation are generally more reliable at predicting the vapor compressibility than the liquid compressibility, and hence, more reliable predicting vapor enthalpy than liquid enthalpy.

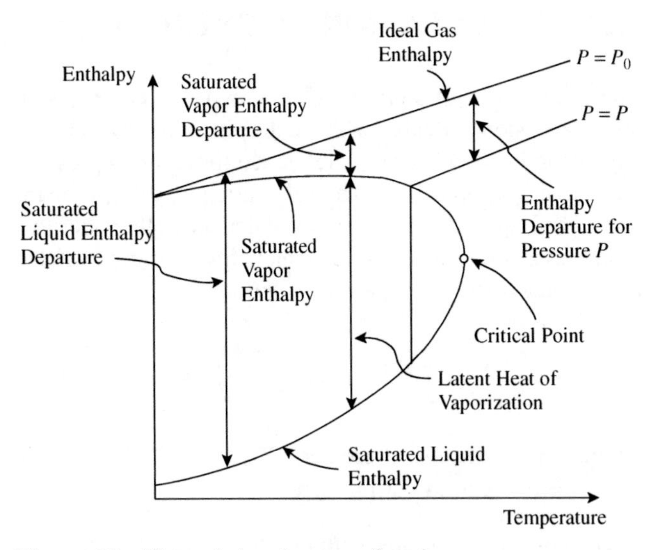

Figure 4.9 The enthalpy departure function.

One problem remains. The reference enthalpy must be defined at temperature T and pressure P_O. The reference state for enthalpy can be taken as an ideal gas. At zero pressure, fluids are in their ideal gaseous state and the enthalpy is independent of pressure. The ideal gas enthalpy can be calculated from ideal gas heat capacity data[3]:

$$H_T^O = H_{T_O}^O + \int_{T_O}^{T} C_P^O\, dT \quad (4.80)$$

where H_T^O = enthalpy at zero pressure and temperature T (kJ kmol^{-1})
$H_{T_O}^O$ = enthalpy at zero pressure and temperature T_O, defined to be zero (kJ·kmol^{-1})
T_O = reference temperature (K)
C_P^O = ideal gas enthalpy (kJ·kmol^{-1}·K^{-1})

The ideal gas enthalpy can be correlated as a function of temperature, for example[3]:

$$\frac{C_P^O}{R} = \alpha_o + \alpha_1 T + \alpha_2 T^2 + \alpha_3 T^3 + \alpha_4 T^4 \quad (4.81)$$

where $\alpha_1, \alpha_2, \alpha_3, \alpha_4$ = constants determined by fitting experimental data

To calculate the enthalpy of liquid or gas at temperature T and pressure P, the enthalpy departure function (Equation 4.78) is evaluated from an equation of state[2]. The ideal gas enthalpy is calculated at temperature T from Equation 4.81. The enthalpy departure is then added to the ideal gas enthalpy to obtain the required enthalpy. Note that the enthalpy departure function calculated from Equation 4.78 will have a negative value. This is illustrated in Figure 4.9. The calculations are complex and usually carried out using physical property or simulation software packages. However, it is important to understand the basis of the calculations and their limitations.

4.12 CALCULATION OF ENTROPY

The calculation of entropy is required for compression and expansion calculations. Isentropic compression and expansion is often used as a reference for real compression and expansion processes. The calculation of entropy might also be required in order to calculate other derived thermodynamic properties. Like enthalpy, entropy can also be calculated from a departure function:

$$[S_P - S_{P_O}]_T = \int_{P_O}^{P} \left(\frac{\partial S}{\partial P}\right)_T dP \tag{4.82}$$

where S_P = entropy at pressure P
S_{P_O} = entropy at pressure P_O

The change in entropy with pressure is given by[2]:

$$\left(\frac{\partial S}{\partial P}\right)_T = -\left(\frac{\partial V}{\partial T}\right)_P \tag{4.83}$$

Combining Equations 4.89 and 4.90 with Equation 4.83 gives:

$$[S_P - S_{P_O}]_T = -\int_{P_O}^{P} \left[\frac{RZ}{P} + \frac{RT}{P}\left(\frac{\partial Z}{\partial T}\right)_P\right]_T dP \tag{4.84}$$

The integral in Equation 4.84 can be evaluated from an equation of state[3]. However, before this *entropy departure function* can be applied to calculate entropy, the reference state must be defined. Unlike enthalpy, the reference state cannot be defined at zero pressure, as the entropy of a gas is infinite at zero pressure. To avoid this difficulty, the standard state can be defined as a reference state at low pressure P_O (usually chosen to be 1 bar or 1 atm) and at the temperature under consideration. Thus,

$$S_T = S_{T_O} + \int_{T_O}^{T} \frac{C_P^O}{T} dT \tag{4.85}$$

where S_T = entropy at temperature T and reference pressure P_O
S_{T_O} = entropy of gas at reference temperature T_O and reference pressure P_O

To calculate the entropy of a liquid or gas at temperature T and pressure P, the entropy departure function (Equation 4.84) is evaluated from an equation of state[3]. The entropy at the reference state is calculated at temperature T from Equation 4.85. The entropy at the reference state is then added to the entropy departure function to obtain the required entropy. The entropy departure function is illustrated in Figure 4.10. As with enthalpy departure, the calculations are complex and are usually carried out in physical property or simulation software packages.

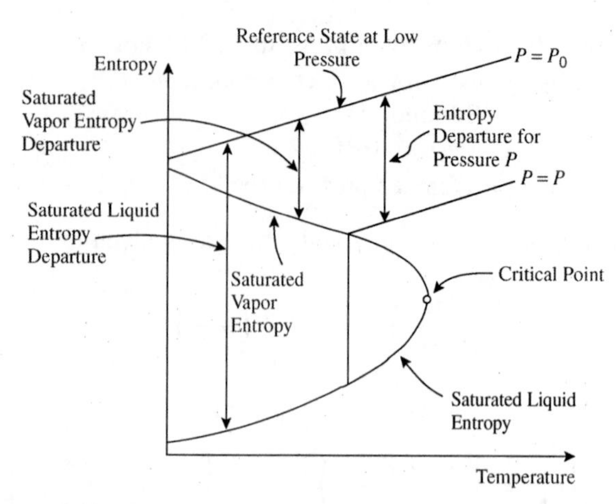

Figure 4.10 The entropy departure function.

4.13 PHASE EQUILIBRIUM AND THERMODYNAMIC PROPERTIES – SUMMARY

Phase equilibrium and thermodynamic properties are required for many design calculations. Vapor and liquid equilibrium can be calculated on the basis of activity coefficient models or equations of state. Activity coefficient models are required when the liquid phase shows a significant deviation from ideality. Equation of state models are required when the vapor phase shows a significant deviation from ideality. Equations of state are normally applied to vapor–liquid equilibrium for hydrocarbon systems and light gases under significant pressure. Such vapor–liquid equilibrium calculations are normally carried out using physical property or simulation software packages.

Prediction of liquid–liquid equilibrium also requires an activity coefficient model. The choice of models of liquid–liquid equilibrium is more restricted than that for vapor–liquid equilibrium, and predictions are particularly sensitive to the model parameters used.

Both enthalpy and entropy can be calculated from an equation of state to predict the deviation from ideal gas behavior. Having calculated the ideal gas enthalpy or entropy from experimentally correlated data, the enthalpy or entropy departure function from the reference state can then be calculated from an equation of state.

Finally, it should be noted that the methods outlined in this chapter are not appropriate for systems in which the components exhibit chemical association in the vapor phase (e.g. acetic acid) or to electrolytic systems.

4.14 EXERCISES

1. For the following mixtures, suggest suitable models for both the liquid and vapor phases to predict vapor–liquid equilibrium.
   a. H_2S and water at 20°C and 1.013 bar.

b. Benzene and toluene (close to ideal liquid-phase behavior) at 1.013 bar.
c. Benzene and toluene (close to ideal liquid-phase behavior) at 20 bar.
d. A mixture of hydrogen, methane, ethylene, ethane, propylene and propane at 20 bar.
e. Acetone and water (nonideal liquid phase) at 1.013 bar.
f. A mixture of 2-propanol, water and diiospropyl ether (nonideal liquid phase forming two-liquid phases) at 1.013 bar.

2. Air needs to be dissolved in water under pressure at 20°C for use in a dissolved-air flotation process (see Chapter 8). The vapor–liquid equilibrium between air and water can be predicted by Henry's Law with a constant of 6.7×10^4 bar. Estimate the mole fraction of air that can be dissolved at 20°C, at a pressure of 10 bar.
3. For the mixture of aromatics in Table 4.15, determine:

Table 4.15 Data for mixture of aromatics.

Component	Feed (kmol)	A_i	B_i	C_i
Benzene	1	9.2806	2789.51	−52.36
Toluene	26	9.3935	3096.52	−53.67
Ethylbenzene	6	9.3993	3279.47	−59.95
Xylene	23	9.5188	3366.99	−59.04

a. The bubble point at a pressure of 1 bar.
b. The bubble point at a pressure of 5 bar.
c. The pressure needed for total condensation at a temperature of 313 K.
d. At a pressure of 1 bar and a temperature of 400 K, how much liquid will be condensed?

Assume that K-values can be correlated by Equation 4.66 with pressure in bar, temperature in Kelvin and constants A_i, B_i and C_i given Table 4.15.

4. Acetone is to be produced by the dehydrogenation of 2-propanol. In this process, the product from the reactor contains hydrogen, acetone, 2-propanol and water, and is cooled before it enters a flash drum. The purpose of the flash drum is to separate hydrogen from the other components. Hydrogen is removed in the vapor stream and sent to a furnace to be burnt. Some acetone is, however, carried over in the vapor steam. The minimum temperature that can be achieved in the flash drum using cooling water is 35°C. The operating pressure is 1.1 bar absolute. The component flowrates to the flash drum are given in Table 4.16. Assume that the K-values can be correlated by Equation 4.66 with pressure in bar, temperature in Kelvin and constants A_i, B_i and C_i given in Table 4.16 for the individual components:

Table 4.16 Flowrate and vapor–liquid equilibrium data for acetone production.

Component	Flowrate (kmol·h^{-1})	A_i	B_i	C_i
Hydrogen	76.95	7.0131	164.90	3.19
Acetone	76.95	10.0311	2940.46	−35.93
2-propanol	9.55	12.0727	3640.20	−53.54
Water	36.6	11.6834	3816.44	−46.13

a. Estimate the flow of acetone and 2-propanol in the vapor stream.
b. Estimate the flow of hydrogen in the liquid stream.
c. Would another method for calculating K-values have been more appropriate?

5. A mixture of benzene and toluene has a relative volatility of 2.34. Sketch the x–y diagram for the mixture, assuming the relative volatility to be constant.
6. The system methanol–cyclohexane can be modeled using the NRTL equation. Vapor pressure coefficients for the Antoine equation for pressure in bar and temperature in Kelvin are given in Table 4.17[6]. Data for the NRTL equation at 1 atm are given in Table 4.18[6]. Assume the gas constant $R = 8.3145$ kJ·kmol^{-1}·K^{-1}. Set up a spreadsheet to calculate the bubble point of liquid mixtures and plot the x–y diagram.

Table 4.17 Antoine coefficients for methanol and cyclohexane at 1 atm[6].

	A_i	B_i	C_i
Methanol	11.9869	3643.32	−33.434
Cyclohexane	9.1559	2778.00	−50.024

Table 4.18 Data for methanol (1) and cyclohexane (2) for the NRTL equation at 1 atm[6].

$(g_{12} - g_{22})$ (kJ·kmol^{-1})	$(g_{21} - g_{11})$ (kJ·kmol^{-1})	α_{ij} (–)
5714.00	6415.36	0.4199

7. At an azeotrope $y_i = x_i$, and Equation 4.34 simplifies to give:

$$\gamma_i = \frac{P}{P_i^{SAT}} \tag{4.86}$$

Thus, if the saturated vapor pressure is known at the azeotropic composition, the activity coefficient can be calculated. If the composition of the azeotrope is known, then the compositions and activity of the coefficients at the azeotrope can be substituted into the Wilson equation to determine the interaction parameters. For the 2-propanol–water system, the azeotropic composition of 2-propanol can be assumed to be at a mole fraction of 0.69 and temperature of 353.4 K at 1 atm. By combining Equation 4.93 with the Wilson equation for a binary system, set up two simultaneous equations and solve Λ_{12} and Λ_{21}. Vapor pressure data can be taken from Table 4.11 and the universal gas constant can be taken to be 8.3145 kJ·kmol^{-1}·K^{-1}. Then, using the values of molar volume in Table 4.12, calculate the interaction parameters for the Wilson equation and compare with the values in Table 4.12.
8. Mixtures of 2-butanol (*sec*-butanol) and water form two-liquid phases. Vapor–liquid equilibrium and liquid–liquid equilibrium for the 2-butanol–water system can be predicted by the NRTL equation. Vapor pressure coefficients for the

Antoine equation for pressure in bar and temperature in Kelvin are given in Table 4.19[6]. Data for the NRTL equation at 1 atm are given in Table 4.20[6]. Assume the gas constant $R = 8.3145$ kJ·kmol^{-1}·K^{-1}.

a. Plot the x–y diagram for the system.
b. Determine the compositions of the two-liquid phase region for saturated vapor–liquid–liquid equilibrium.
c. Does the system form a heteroazeotrope?

Table 4.19 Antoine coefficients for 2-butanol and water[6].

	A_i	B_i	C_i
2-butanol	9.9614	2664.0939	−104.881
Water	11.9647	3984.9273	−39.734

Table 4.20 Data for 2-butanol (1) and water (2) for the NRTL equation at 1 atm[6].

$(g_{12} - g_{22})$ (kJ·kmol^{-1})	$(g_{21} - g_{11})$ (kJ·kmol^{-1})	α_{ij} (–)
1034.30	10,098.50	0.4118

REFERENCES

1. Hougen O.A, Watson K.M and Ragatz R.A (1954) *Chemical Process Principles Part I Material and Energy Balances*, 2nd Edition, John Wiley.
2. Hougen O.A, Watson K.M and Ragatz R.A (1959) *Chemical Process Principles Part II Thermodynamics*, 2nd Edition, John Wiley.
3. Poling B.E, Prausnitz J.M and O'Connell J.P (2001) *The Properties of Gases and Liquids*, McGraw-Hill.
4. Press W.H, Teukolsky S.A, Vetterling W.T and Flannery B.P (1992) *Numerical Recipes in Fortran*, Cambridge University Press.
5. Oellrich L, Plöcker U, Prausnitz J.M and Knapp H (1981) Equation-of-state Methods for Computing Phase Equilibria and Enthalpies, *Int Chem Eng*, **21**(1): 1.
6. Gmehling J, Onken U and Arlt W (1977–1980) *Vapor-Liquid Equilibrium Data Collection*, Dechema Chemistry Data Series.
7. King C.J (1980) *Separation Processes*, McGraw-Hill.
8. Smith R and Jobson M (2000) *Distillation, Encyclopedia of Separation Science*, Academic Press.
9. Sorenson J.M and Arlt W (1980) *Liquid-Liquid Equilibrium Data Collection*, Dechema Chemistry Data Series.
10. Macedo E.A and Rasmussen P (1987) *Liquid-Liquid Equilibrium Data Collection*, Dechema Chemistry Data Series.

5 Choice of Reactor I – Reactor Performance

Since process design starts with the reactor, the first decisions are those that lead to the choice of reactor. These decisions are amongst the most important in the whole process design. Good reactor performance is of paramount importance in determining the economic viability of the overall design and fundamentally important to the environmental impact of the process. In addition to the desired products, reactors produce unwanted byproducts. These unwanted byproducts not only lead to a loss of revenue but can also create environmental problems. As will be discussed later, the best solution to environmental problems is not to employ elaborate treatment methods, but to not produce waste in the first place.

Once the product specifications have been fixed, some decisions need to be made regarding the *reaction path*. There are sometimes different paths to the same product. For example, suppose ethanol is to be manufactured. Ethylene could be used as a raw material and reacted with water to produce ethanol. An alternative would be to start with methanol as a raw material and react it with synthesis gas (a mixture of carbon monoxide and hydrogen) to produce the same product. These two paths employ chemical reactor technology. A third path could employ a *biochemical* reaction (or *fermentation*) that exploits the metabolic processes of *microorganisms* in a biochemical reactor. Ethanol could therefore also be manufactured by fermentation of a carbohydrate.

Reactors can be broadly classified as chemical or biochemical. Most reactors, whether chemical or biochemical, are catalyzed. The strategy will be to choose the catalyst, if one is to be used, and the ideal characteristics and operating conditions needed for the reaction system. The issues that must be addressed for reactor design include:

- Reactor type
- Catalyst
- Size
- Operating conditions (temperature and pressure)
- Phase
- Feed conditions (concentration and temperature).

Once basic decisions have been made regarding these issues, a practical reactor is selected, approaching as nearly as possible the ideal in order that the design can proceed. However, the reactor design cannot be fixed at this stage, since, as will be seen later, it interacts strongly with the rest of the flowsheet. The focus here will be on the choice of reactor and not its detailed sizing. For further details of sizing, see, for example, Levenspiel[1], Denbigh and Turner[2] and Rase[3].

5.1 REACTION PATH

As already noted, given that the objective is to manufacture a certain product, there are often a number of alternative reaction paths to that product. Reaction paths that use the cheapest raw materials and produce the smallest quantities of byproducts are to be preferred. Reaction paths that produce significant quantities of unwanted byproducts should especially be avoided, since they can create significant environmental problems.

However, there are many other factors to be considered in the choice of reaction path. Some are commercial, such as uncertainties regarding future prices of raw materials and byproducts. Others are technical, such as safety and energy consumption.

The lack of suitable catalysts is the most common reason preventing the exploitation of novel reaction paths. At the first stage of design, it is impossible to look ahead and see all of the consequences of choosing one reaction path or another, but some things are clear even at this stage. Consider the following example.

Example 5.1 Given that the objective is to manufacture vinyl chloride, there are at least three reaction paths that can be readily exploited[4].

Path 1

$$\underset{\text{acetylene}}{C_2H_2} + \underset{\substack{\text{hydrogen}\\ \text{chloride}}}{HCl} \longrightarrow \underset{\substack{\text{vinyl}\\ \text{chloride}}}{C_2H_3Cl}$$

Path 2

$$\underset{\text{ethylene}}{C_2H_4} + \underset{\text{chlorine}}{Cl_2} \longrightarrow \underset{\text{dichloroethane}}{C_2H_4Cl_2}$$

$$\underset{\text{dichloroethane}}{C_2H_4Cl_2} \xrightarrow{\text{heat}} \underset{\substack{\text{vinyl}\\ \text{chloride}}}{C_2H_3Cl} + \underset{\substack{\text{hydrogen}\\ \text{chloride}}}{HCl}$$

Path 3

$$\underset{\text{ethylene}}{C_2H_4} + \underset{\text{oxygen}}{1/2O_2} + \underset{\substack{\text{hydrogen}\\ \text{chloride}}}{2HCl} \longrightarrow \underset{\text{dichloroethane}}{C_2H_4Cl_2} + \underset{\text{water}}{H_2O}$$

$$\underset{\text{dichloroethane}}{C_2H_4Cl_2} \xrightarrow{\text{heat}} \underset{\substack{\text{vinyl}\\ \text{chloride}}}{C_2H_3Cl} + \underset{\substack{\text{hydrogen}\\ \text{chloride}}}{HCl}$$

Chemical Process Design and Integration R. Smith
 ISBNs: 0-471-48680-9 (HB); 0-471-48681-7 (PB)

The market values and molar masses of the materials involved are given in Table 5.1.

Table 5.1 Molar masses and values of materials in Example 5.1.

Material	Molar mass ($kg{\cdot}kmol^{-1}$)	Value ($\${\cdot}kg^{-1}$)
Acetylene	26	1.0
Chlorine	71	0.23
Ethylene	28	0.53
Hydrogen chloride	36	0.39
Vinyl chloride	62	0.46

Oxygen is considered to be free at this stage, coming from the atmosphere. Which reaction path makes most sense on the basis of raw material costs, product and byproduct values?

Solution Decisions can be made on the basis of the economic potential of the process. At this stage, the best that can be done is to define the economic potential (*EP*) as (see Chapter 2):

$$EP = (\text{value of products}) - (\text{raw materials costs})$$

Path 1

$$EP = (62 \times 0.46) - (26 \times 1.0 + 36 \times 0.39)$$
$$= -11.52\ \${\cdot}kmol^{-1}\ \text{vinyl chloride product}$$

Path 2

$$EP = (62 \times 0.46 + 36 \times 0.39) - (28 \times 0.58 + 71 \times 0.23)$$
$$= 9.99\ \${\cdot}kmol^{-1}\ \text{vinyl chloride product}$$

This assumes the sale of the byproduct HCl. If it cannot be sold, then:

$$EP = (62 \times 0.46) - (28 \times 0.58 + 71 \times 0.23)$$
$$= -4.05\ \${\cdot}kmol^{-1}\text{vinyl chloride product}$$

Path 3

$$EP = (62 \times 0.46) - (28 \times 0.58 + 36 \times 0.39)$$
$$= -1.76\ \${\cdot}kmol^{-1}\text{vinyl chloride product}$$

Paths 1 and 3 are clearly not viable. Only Path 2 shows a positive economic potential when the byproduct HCl can be sold. In practice, this might be quite difficult, since the market for HCl tends to be limited. In general, projects should not be justified on the basis of the byproduct value.

The preference is for a process based on ethylene rather than the more expensive acetylene, and chlorine rather than the more expensive hydrogen chloride. Electrolytic cells are a much more convenient and cheaper source of chlorine than hydrogen chloride. In addition, it is preferred to produce no byproducts.

Example 5.2 Devise a process from the three reaction paths in Example 5.1 that uses ethylene and chlorine as raw materials and produces no byproducts other than water[4]. Does the process look attractive economically?

Solution A study of the stoichiometry of the three paths shows that this can be achieved by combining Path 2 and Path 3 to obtain a fourth path.

Paths 2 and 3

$$\underset{\text{ethylene}}{C_2H_4} + \underset{\text{chlorine}}{Cl_2} \longrightarrow \underset{\text{dichloroethane}}{C_2H_4Cl_2}$$

$$\underset{\text{ethylene}}{C_2H_4} + \underset{\text{oxygen}}{1/2O_2} + \underset{\text{hydrogen chloride}}{2HCl} \longrightarrow \underset{\text{dichloroethane}}{C_2H_4Cl_2} + \underset{\text{water}}{H_2O}$$

$$\underset{\text{dichloroethane}}{2C_2H_4Cl_2} \xrightarrow{\text{heat}} \underset{\text{vinyl chloride}}{2C_2H_3Cl} + \underset{\text{hydrogen chloride}}{2HCl}$$

These three reactions can be added to obtain the overall stoichiometry.

Path 4

$$\underset{\text{ethylene}}{2C_2H_4} + \underset{\text{chlorine}}{Cl_2} + \underset{\text{oxygen}}{1/2O_2} \longrightarrow \underset{\text{vinyl chloride}}{2C_2H_3Cl} + \underset{\text{water}}{H_2O}$$

or

$$\underset{\text{ethylene}}{C_2H_4} + \underset{\text{chlorine}}{1/2Cl_2} + \underset{\text{oxygen}}{1/4O_2} \longrightarrow \underset{\text{vinyl chloride}}{C_2H_3Cl} + \underset{\text{water}}{1/2H_2O}$$

Now the economic potential is given by:

$$EP = (62 \times 0.46) - (28 \times 0.58 + 1/2 \times 71 \times 0.23)$$
$$= 4.12\ \${\cdot}kmol^{-1}\ \text{vinyl chloride product}$$

In summary, Path 2 from Example 5.1 is the most attractive reaction path if there is a large market for hydrogen chloride. In practice, it tends to be difficult to sell the large quantities of hydrogen chloride produced by such processes. Path 4 is the usual commercial route to vinyl chloride.

5.2 TYPES OF REACTION SYSTEMS

Having made a choice of the reaction path, a choice of reactor type must be made, together with some assessment of the conditions in the reactor. This allows assessment of the reactor performance for the chosen reaction path in order for the design to proceed.

Before proceeding to the choice of reactor and operating conditions, some general classifications must be made regarding the types of reaction systems likely to be encountered. Reaction systems can be classified into six broad types:

1. *Single reactions.* Most reaction systems involve multiple reactions. In practice, the *secondary* reactions can sometimes be neglected, leaving a single *primary* reaction to consider. Single reactions are of the type:

$$FEED \longrightarrow PRODUCT \tag{5.1}$$

or

$$FEED \longrightarrow PRODUCT + BYPRODUCT \tag{5.2}$$

or

$$FEED1 + FEED2 \longrightarrow PRODUCT \tag{5.3}$$

and so on.

An example of this type of reaction that does not produce a byproduct is *isomerization* (the reaction of a feed to a product with the same chemical formula but a different molecular structure). For example, allyl alcohol can be produced from propylene oxide[5]:

$$\underset{\text{propylene oxide}}{CH_3HC\overset{O}{-}CH_2} \longrightarrow \underset{\text{allyl alcohol}}{CH_2{=}CHCH_2OH}$$

An example of a reaction that does produce a byproduct is the production of acetone from isopropyl alcohol, which produces a hydrogen byproduct:

$$\underset{\text{isopropyl alcohol}}{(CH_3)_2CHOH} \longrightarrow \underset{\text{acetone}}{CH_3COCH_3} + H_2$$

2. *Multiple reactions in parallel producing byproducts.* Rather than a single reaction, a system may involve secondary reactions producing (additional) byproducts in *parallel* with the primary reaction. Multiple reactions in parallel are of the type:

$$\begin{aligned} FEED &\longrightarrow PRODUCT \\ FEED &\longrightarrow BYPRODUCT \end{aligned} \tag{5.4}$$

or

$$\begin{aligned} FEED &\longrightarrow PRODUCT + BYPRODUCT1 \\ FEED &\longrightarrow BYPRODUCT2 + BYPRODUCT3 \end{aligned} \tag{5.5}$$

or

$$\begin{aligned} FEED1 + FEED2 &\longrightarrow PRODUCT \\ FEED1 + FEED2 &\longrightarrow BYPRODUCT \end{aligned} \tag{5.6}$$

and so on.

An example of a parallel reaction system occurs in the production of ethylene oxide[5]:

$$\underset{\text{ethylene}}{CH_2 = CH_2} + \underset{\text{oxygen}}{1/2O_2} \longrightarrow \underset{\text{ethylene oxide}}{H_2C\overset{O}{-}CH_2}$$

with the parallel reaction:

$$\underset{\text{ethylene}}{CH_2 = CH_2} + \underset{\text{oxygen}}{3O_2} \longrightarrow \underset{\text{carbon dioxide}}{2CO_2} + \underset{\text{water}}{2H_2O}$$

Multiple reactions might not only lead to a loss of materials and useful product but might also lead to byproducts being deposited on, or poisoning catalysts (see Chapters 6 and 7).

3. *Multiple reactions in series producing byproducts.* Rather than the primary and secondary reactions being in parallel, they can be in *series*. Multiple reactions in series are of the type:

$$\begin{aligned} FEED &\longrightarrow PRODUCT \\ PRODUCT &\longrightarrow BYPRODUCT \end{aligned} \tag{5.7}$$

or

$$\begin{aligned} FEED &\longrightarrow PRODUCT + BYPRODUCT1 \\ PRODUCT &\longrightarrow BYPRODUCT2 + BYPRODUCT3 \end{aligned} \tag{5.8}$$

or

$$\begin{aligned} FEED1 + FEED2 &\longrightarrow PRODUCT \\ PRODUCT &\longrightarrow BYPRODUCT1 + BYPRODUCT2 \end{aligned} \tag{5.9}$$

and so on.

An example of a series reaction system is the production of formaldehyde from methanol:

$$\underset{\text{methanol}}{CH_3OH} + \underset{\text{oxygen}}{1/2O_2} \longrightarrow \underset{\text{formaldehyde}}{HCHO} + \underset{\text{water}}{H_2O}$$

A series reaction of the formaldehyde occurs:

$$\underset{\text{formaldehyde}}{HCHO} \longrightarrow \underset{\text{carbon monoxide}}{CO} + \underset{\text{hydrogen}}{H_2}$$

As with parallel reactions, series reactions might not only lead to a loss of materials and useful products, but might also lead to byproducts being deposited on, or poisoning catalysts (see Chapters 6 and 7).

4. *Mixed parallel and series reactions producing byproducts.* In more complex reaction systems, both parallel and series reactions can occur together. *Mixed parallel and series* reactions are of the type:

$$\begin{aligned} FEED &\longrightarrow PRODUCT \\ FEED &\longrightarrow BYPRODUCT \\ PRODUCT &\longrightarrow BYPRODUCT \end{aligned} \tag{5.10}$$

or

$$\begin{aligned} &FEED \longrightarrow PRODUCT \\ &FEED \longrightarrow BYPRODUCT1 \\ &PRODUCT \longrightarrow BYPRODUCT2 \end{aligned} \quad (5.11)$$

or

$$\begin{aligned} &FEED1 + FEED2 \longrightarrow PRODUCT \\ &FEED1 + FEED2 \longrightarrow BYPRODUCT1 \\ &PRODUCT \longrightarrow BYPRODUCT2 + BYPRODUCT3 \end{aligned} \quad (5.12)$$

An example of mixed parallel and series reactions is the production of ethanolamines by reaction between ethylene oxide and ammonia[5]:

$$\underset{\text{ethylene oxide}}{H_2C{-}CH_2 \;(O)} + \underset{\text{ammonia}}{NH_3} \longrightarrow \underset{\text{monoethanolamine}}{NH_2CH_2CH_2OH}$$

$$\underset{\text{monoethanolamine}}{NH_2CH_2CH_2OH} + \underset{\text{ethylene oxide}}{H_2C - CH_2 \;(O)} \longrightarrow \underset{\text{diethanolamine}}{NH(CH_2CH_2OH)_3}$$

$$\underset{\text{diethanolamine}}{NH(CH_2CH_2OH)_2} + \underset{\text{ethylene oxide}}{H_2C - CH_2 \;(O)} \longrightarrow \underset{\text{triethanolamine}}{N(CH_2CH_2OH)_3}$$

Here the ethylene oxide undergoes parallel reactions, whereas the monoethanolamine undergoes a series reaction to diethanolamine and triethanolamine.

5. *Polymerization reactions.* In polymerization reactions, *monomer* molecules are reacted together to produce a high molar mass *polymer*. Depending on the mechanical properties required of the polymer, a mixture of monomers might be reacted together to produce a high molar mass *copolymer*. There are two broad types of polymerization reactions, those that involve a *termination step* and those that do not[2]. An example that involves a termination step is free-radical polymerization of an alkene molecule, known as *addition polymerization*. A free radical is a free atom or fragment of a stable molecule that contains one or more unpaired electrons. The polymerization requires a free radical from an initiator compound such as a peroxide. The initiator breaks down to form a free radical (e.g. $\bullet CH_3$ or $\bullet OH$), which attaches to a molecule of alkene and in so doing generates another free radical. Consider the polymerization of vinyl chloride from a free-radical initiator $\bullet R$. An *initiation step* first occurs:

$$\underset{\text{initiator}}{\dot{R}} + \underset{\text{vinyl chloride}}{CH_2 = CHCl} \longrightarrow \underset{\text{vinyl chloride free radical}}{RCH_2 - \dot{C}HCl}$$

A *propagation step* involving growth around an *active center* follows:

$$RCH_2 - \dot{C}HCl + CH_2 = CHCl \longrightarrow RCH_2 - CHCl - CH_2 - \dot{C}HCl$$

and so on, leading to molecules of the structure:

$$R - (CH_2 - CHCl)_n - CH_2 - \dot{C}HCl$$

Eventually, the chain is terminated by steps such as the union of two radicals that consume but do not generate radicals:

$$R - (CH_2 - CHCl)_n - CH_2 - \dot{C}HCl + \dot{C}HCl - CH_2 - (CHCl - CH_2)_m - R \longrightarrow R - (CH_2 - CHCl)_n - CH_2 - CHCl - CHCl - CH_2 - (CHCl - CH_2)_m - R$$

This termination step stops the subsequent growth of the polymer chain. The period during which the chain length grows, that is, before termination, is known as the *active life* of the polymer. Other termination steps are possible.

The orientation of the groups along the carbon chain, its *stereochemistry*, is critical to the properties of the product. The stereochemistry of addition polymerization can be controlled by the use of catalysts. A polymer where repeating units have the same relative orientation is termed *stereoregular*.

An example of a polymerization without a termination step is *polycondensation*[2].

$$HO - (CH_2)_n - COOH + HO - (CH_2)_n - COOH \longrightarrow HO - (CH_2)_n - COO - (CH_2)_n - COOH + H_2O \quad \text{etc.}$$

Here the polymer grows by successive esterification with elimination of water and no termination step. Polymers formed by linking monomers with carboxylic acid groups and those that have alcohol groups are known as *polyesters*. Polymers of this type are widely used for the manufacture of artificial fibers. For example, the esterification of terephthalic acid with ethylene glycol produces polyethylene terephthalate.

6. *Biochemical reactions.* Biochemical reactions, often referred to as *fermentations*, can be divided into two broad types. In the first type, the reaction exploits the metabolic pathways in selected microorganisms (especially bacteria,

yeasts, moulds and algae) to convert feed material (often called *substrate* in biochemical reactor design) to the required product. The general form of such reactions is:

$$FEED \xrightarrow{\text{microorganisms}} PRODUCT + [\text{More microorganisms}] \quad (5.13)$$

or

$$FEED1 + FEED2 \xrightarrow{\text{microorganisms}} PRODUCT + [\text{More microorganisms}] \quad (5.14)$$

and so on.

In such reactions, the microorganisms reproduce themselves. In addition to the feed material, it is likely that nutrients (e.g. a mixture containing phosphorus, magnesium, potassium, etc.) will need to be added for the survival of the microorganisms. Reactions involving microorganisms include:

- hydrolysis
- oxidation
- esterification
- reduction.

An example of an oxidation reaction is the production of citric acid from glucose:

$$\underset{\text{glucose}}{C_6H_{12}O_6} + 3/2O_2 \longrightarrow \underset{\text{citric acid}}{HOOCCH_2COH(COOH)CH_2COOH} + 2H_2O$$

In the second group, the reaction is promoted by *enzymes*. Enzymes are the catalyst proteins produced by microorganisms that accelerate chemical reactions in microorganisms. The biochemical reactions employing enzymes are of the general form:

$$FEED \xrightarrow{\text{enzyme}} PRODUCT \quad (5.15)$$

and so on.

Unlike reactions involving microorganisms, in enzyme reactions the catalytic agent (the enzyme) does not reproduce itself. An example in the use of enzymes is the isomerization of glucose to fructose:

$$\underset{\text{glucose}}{CH_2OH(CHOH)_4CHO} \xrightarrow{\text{enzyme}} \underset{\text{fructose}}{CH_2OHCO(CHOH)_3CH_2OH}$$

Although nature provides many useful enzymes, they can also be engineered for improved performance and new applications. Biochemical reactions have the advantage of operating under mild reaction conditions of temperature and pressure and are usually carried out in an aqueous medium rather than using an organic solvent.

5.3 REACTOR PERFORMANCE

Before exploring how reactor conditions can be chosen, some measure of reactor performance is required.

For polymerization reactors, the main concern is the characteristics of the product that relate to the mechanical properties. The distribution of molar masses in the polymer product, orientation of groups along the chain, cross-linking of the polymer chains, copolymerization with a mixture of monomers, and so on, are the main considerations. Ultimately, the main concern is the mechanical properties of the polymer product.

For biochemical reactions, the performance of the reactor will normally be dictated by laboratory results, because of the difficulty of predicting such reactions theoretically[6]. There are likely to be constraints on the reactor performance dictated by the biochemical processes. For example, in the manufacture of ethanol using microorganisms, as the concentration of ethanol rises, the microorganisms multiply more slowly until at a concentration of around 12% it becomes toxic to the microorganisms.

For other types of reactors, three important parameters are used to describe their performance[7]:

$$\text{Conversion} = \frac{(\text{reactant consumed in the reactor})}{(\text{reactant } \textit{fed} \text{ to the reactor})} \quad (5.16)$$

$$\text{Selectivity} = \frac{(\text{desired product produced})}{(\text{reactant } \textit{consumed} \text{ in the reactor})} \times \text{stoichiometric factor} \quad (5.17)$$

$$\text{Reactor yield} = \frac{(\text{desired product produced})}{(\text{reactant } \textit{fed} \text{ to the reactor})} \times \text{stoichiometric factor} \quad (5.18)$$

in which the stoichiometric factor is the stoichiometric moles of reactant required per mole of product. When more than one reactant is required (or more than one desired product produced) Equations 5.16 to 5.18 can be applied to each reactant (or product).

The following example will help clarify the distinctions among these three parameters.

Example 5.3 Benzene is to be produced from toluene according to the reaction[8]:

$$\underset{\text{toluene}}{C_6H_5CH_3} + \underset{\text{hydrogen}}{H_2} \longrightarrow \underset{\text{benzene}}{C_6H_6} + \underset{\text{methane}}{CH_4}$$

Some of the benzene formed undergoes a number of secondary reactions in series to unwanted byproducts that can be characterized by the reaction to diphenyl, according to the reaction:

$$\underset{\text{benzene}}{2C_6H_6} \rightleftharpoons \underset{\text{diphenyl}}{C_{12}H_{10}} + \underset{\text{hydrogen}}{H_2}$$

Table 5.2 gives the compositions of the reactor feed and effluent streams.

Table 5.2 Reactor feed and effluent streams.

Component	Inlet flowrate ($kmol \cdot h^{-1}$)	Outlet flowrate ($kmol \cdot h^{-1}$)
H_2	1858	1583
CH_4	804	1083
C_6H_6	13	282
$C_6H_5CH_3$	372	93
$C_{12}H_{10}$	0	4

Calculate the conversion, selectivity and reactor yield with respect to the:

a. toluene feed
b. hydrogen feed.

Solution

a. Toluene conversion $= \dfrac{\text{(toluene consumed in the reactor)}}{\text{(toulene fed to the reactor)}}$

$$= \frac{372 - 93}{372}$$

$$= 0.75$$

Stoichiometric factor = stoichiometric moles of toluene required per mole of benzene produced

$$= 1$$

Benzene selectivity from toluene

$$= \frac{\text{(benzene produced in the reactor)}}{\text{(toluene consumed in the reactor)}} \times \text{stoichiometric factor}$$

$$= \frac{282 - 13}{372 - 93} \times 1$$

$$= 0.96$$

Reactor yield of benzene from toluene

$$= \frac{\text{(benzene produced in the reactor)}}{\text{(toluene fed to the reactor)}} \times \text{stoichiometric factor}$$

$$= \frac{282 - 13}{372} \times 1$$

$$= 0.72$$

b. Hydrogen conversion

$$= \frac{\text{(hydrogen consumed in the reactor)}}{\text{(hydrogen fed to the reactor)}}$$

$$= \frac{1858 - 1583}{1858}$$

$$= 0.15$$

Stoichiometric factor

= stoichiometric moles of hydrogen required per mole of benzene produced

$$= 1$$

Benzene selectivity from hydrogen

$$= \frac{\text{(benzene produced in the reactor)}}{\text{(hydrogen consumed in the reactor)}} \times \text{stoichiometric factor}$$

$$= \frac{282 - 13}{1858 - 1583} \times 1$$

$$= 0.98$$

Reactor yield of benzene from hydrogen

$$= \frac{\text{(benzene produced in the reactor)}}{\text{(hydrogen fed to the reactor)}} \times \text{stoichiometric factor}$$

$$= \frac{282 - 13}{1858} \times 1$$

$$= 0.14$$

Because there are two feeds to this process, the reactor performance can be calculated with respect to both feeds. However, the principal concern is performance with respect to toluene, since it is more expensive than hydrogen.

As will be discussed in the next chapter, if a reaction is reversible there is a maximum conversion, the *equilibrium conversion*, that can be achieved, which is less than 1.0.

In describing reactor performance, selectivity is often a more meaningful parameter than reactor yield. Reactor yield is based on the reactant fed to the reactor rather than on that which is consumed. Part of the reactant fed might be material that has been recycled rather than fresh feed. Reactor yield takes no account of the ability to separate and recycle unconverted raw materials. Reactor yield is only a meaningful parameter when it is not possible for one reason or another to recycle unconverted raw material to the reactor inlet. However, the yield of the overall process is an extremely important parameter when describing the performance of the overall plant, as will be discussed later.

5.4 RATE OF REACTION

To define the rate of a reaction, one of the components must be selected and the rate defined in terms of that component. The rate of reaction is the number of moles formed with respect to time, per unit volume of reaction mixture:

$$r_i = \frac{1}{V}\left(\frac{dN_i}{dt}\right) \quad (5.19)$$

where r_i = rate of reaction of Component i ($kmol \cdot m^{-3} \cdot s^{-1}$)
N_i = moles of Component i formed (kmol)
V = reaction volume (m^3)
t = time (s)

If the volume of the reactor is constant (V = constant):

$$r_i = \frac{1}{V}\left(\frac{dN_i}{dt}\right) = \frac{dN_i/V}{dt} = \frac{dC_i}{dt} \quad (5.20)$$

where C_i = molar concentration of Component i ($kmol \cdot m^{-3}$)

The rate is negative if the component is a reactant and positive if it is a product. For example, for the general irreversible reaction:

$$bB + cC + \cdots \longrightarrow sS + tT + \cdots \quad (5.21)$$

The rates of reaction are related by:

$$-\frac{r_B}{b} = -\frac{r_C}{c} = - \cdots = \frac{r_S}{s} = \frac{r_T}{t} = \cdots \quad (5.22)$$

If the rate-controlling step in the reaction is the collision of the reacting molecules, then the equation to quantify the reaction rate will often follow the stoichiometry such that:

$$-r_B = k_B C_B^b C_C^c \ldots. \quad (5.23)$$

$$-r_C = k_C C_B^b C_C^c \ldots. \quad (5.24)$$

$$r_S = k_S C_B^b C_C^c \ldots. \quad (5.25)$$

$$r_T = k_T C_B^b C_C^c \ldots. \quad (5.26)$$

where r_i = reaction rate for Component i ($kmol \cdot m^{-3} \cdot s^{-1}$)
k_i = reaction rate constant for Component i ($[kmol \cdot m^{-3}]^{NC-b-c-\ldots} s^{-1}$)
NC = is the number of components in the rate expression
C_i = molar concentration of Component i ($kmol \cdot m^{-3}$)

The exponent for the concentration ($b, c, \ldots$) is known as the *order of reaction*. The reaction rate constant is a function of temperature, as will be discussed in the next chapter.

Thus, from Equations 5.22 to 5.26:

$$\frac{k_B}{b} = \frac{k_C}{c} = \ldots.\frac{k_S}{s} = \frac{k_T}{t} = \ldots. \quad (5.27)$$

Reactions for which the rate equations follow the stoichiometry as given in Equations 5.23 to 5.26 are known as *elementary reactions*. If there is no direct correspondence between the reaction stoichiometry and the reaction rate, these are known as *nonelementary reactions* and are often of the form:

$$-r_B = k_B C_B^\beta C_C^\delta \ldots C_S^\varepsilon C_T^\xi \ldots. \quad (5.28)$$

$$-r_C = k_C C_B^\beta C_C^\delta \ldots. C_S^\varepsilon C_T^\xi \ldots. \quad (5.29)$$

$$r_S = k_S C_B^\beta C_C^\delta \ldots. C_S^\varepsilon C_T^\xi \ldots. \quad (5.30)$$

$$r_T = k_T C_B^\beta C_C^\delta \ldots. C_S^\varepsilon C_T^\xi \ldots. \quad (5.31)$$

where $\beta, \delta, \varepsilon, \xi$ = order of reaction

The reaction rate constant and the orders of reaction must be determined experimentally. If the reaction mechanism involves multiple steps involving chemical intermediates, then the form of the reaction rate equations can be of a more complex form than Equations 5.28 to 5.31.

If the reaction is reversible, such that:

$$bB + cC + \ldots \rightleftharpoons sS + tT + \ldots \quad (5.32)$$

then the rate of reaction is the net rate of the forward and reverse reactions. If the forward and reverse reactions are both elementary, then:

$$-r_B = k_B C_B^b C_C^c \ldots. - k'_B C_S^s C_T^t \ldots. \quad (5.33)$$

$$-r_C = k_C C_B^b C_C^c \ldots. - k'_C C_C^s C_T^t \ldots. \quad (5.34)$$

$$r_S = k_S C_B^b C_C^c \ldots. - k'_S C_S^s C_T^t \ldots. \quad (5.35)$$

$$r_T = k_T C_B^b C_C^c \ldots. - k'_T C_S^s C_T^t \ldots. \quad (5.36)$$

where k_i = reaction rate constant for Component i for the forward reaction
k'_i = reaction rate constant for Component i for the reverse reaction.

If the forward and reverse reactions are nonelementary, perhaps involving the formation of chemical intermediates in multiple steps, then the form of the reaction rate equations can be more complex than Equations 5.33 to 5.36.

5.5 IDEALIZED REACTOR MODELS

Three idealized models are used for the design of reactors[1–3]. In the first (Figure 5.1a), the *ideal-batch* model, the reactants are charged at the beginning of the operation. The contents are subjected to perfect mixing for a certain period, after which the products are discharged. Concentration changes with time, but the perfect mixing ensures that at any instant the composition and temperature throughout the reactor are both uniform.

In the second model (Figure 5.1b), the *mixed-flow* or *continuous well-mixed* or *continuous-stirred-tank (CSTR)* model, feed and product takeoff are both continuous, and the reactor contents are assumed to be perfectly mixed. This leads to uniform composition and temperature throughout the reactor. Because of the perfect mixing, a fluid element can leave the instant it enters the reactor or stay for an extended period. The residence time of individual fluid elements in the reactor varies.

In the third model (Figure 5.1c), the *plug-flow* model, a steady uniform movement of the reactants is assumed,

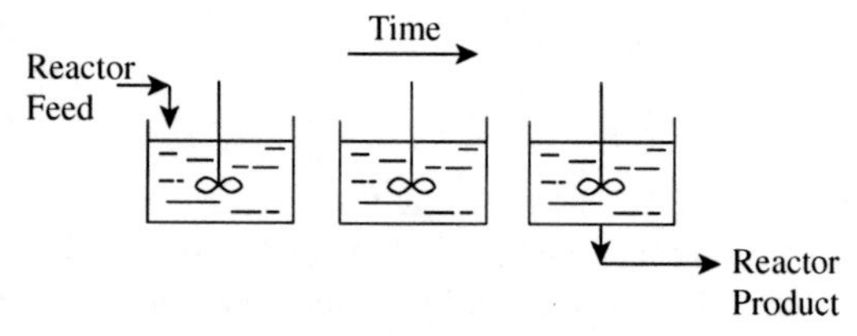

(a) Ideal batch model.

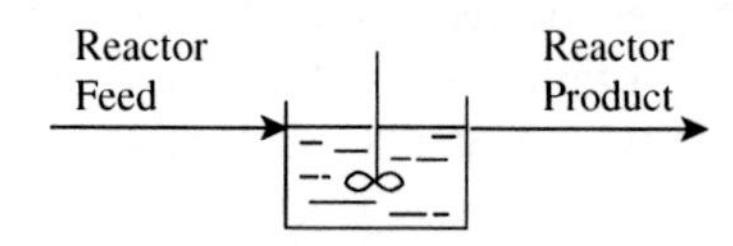

(b) Mixed-flow reactor.

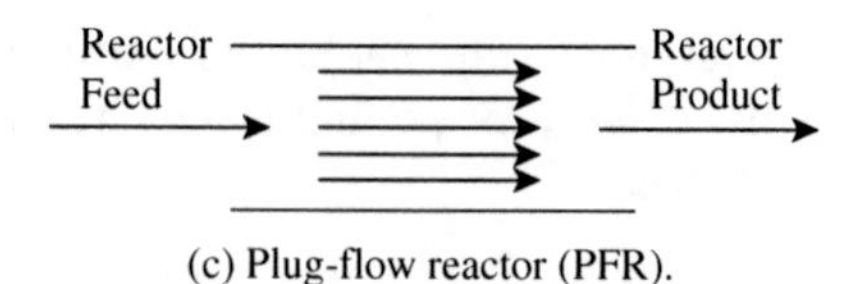

(c) Plug-flow reactor (PFR).

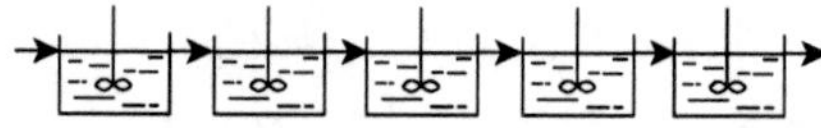
(d) A series of mixed-flow reactors approaches plug-flow

Figure 5.1 The idealized models used for reactor design (From Smith R and Petela EA, 1991, *The Chemical Engineer*, No. 509/510:12, reproduced by permission of the Institution of Chemical Engineers).

with no attempt to induce mixing along the direction of flow. Like the ideal-batch reactor, the residence time in a plug-flow reactor is the same for all fluid elements. Plug-flow operation can be approached by using a number of mixed-flow reactors in series (Figure 5.1d). The greater the number of mixed-flow reactors in series, the closer is the approach to plug-flow operation.

1. *Ideal-batch reactor.* Consider a batch reactor in which the feed is charged at the beginning of the batch and no product is withdrawn until the batch is complete. Given that:

$$\begin{bmatrix}\text{moles of reactant}\\ \text{converted}\end{bmatrix} = -r_i = -\frac{1}{V}\frac{dN_i}{dt} \qquad (5.37)$$

Integration of Equation 5.37 gives:

$$t = \int_{N_{i0}}^{N_{it}} \frac{dN_i}{r_i V} \qquad (5.38)$$

where t = batch time
N_{i0} = initial moles of Component i
N_{it} = final moles of Component i after time t

Alternatively, Equations 5.37 can be written in terms of reactor conversion X_i:

$$\frac{dN_i}{dt} = \frac{d[N_{i0}(1 - X_i)]}{dt} = -N_{i0}\frac{dX_i}{dt} = r_i V \qquad (5.39)$$

Integration of Equation 5.39 gives:

$$t = N_{i0}\int_0^{X_i} \frac{dX_i}{-r_i V} \qquad (5.40)$$

Also, from the definition of reactor conversion, for the special case of a constant density reaction mixture:

$$X_i = \frac{N_{i0} - N_{it}}{N_{i0}} = \frac{C_{i0} - C_{it}}{C_{i0}} \qquad (5.41)$$

where C_i = molar concentration of Component i
C_{i0} = initial molar concentration of Component i
C_{it} = final molar concentration of Component i at time t

Substituting Equation 5.41 into Equation 5.39 and noting that $N_{i0}/V = C_{i0}$ gives:

$$-\frac{dC_i}{dt} = -r_i \qquad (5.42)$$

Integration of Equation 5.42 gives:

$$t = -\int_{C_{i0}}^{C_{it}} \frac{dC_i}{-r_i} \qquad (5.43)$$

2. *Mixed-flow reactor.* Consider now the mixed-flow reactor in Figure 5.2b in which a feed of Component i is reacting. A material balance for Component i per unit time gives:

$$\begin{bmatrix}\text{moles of reactant in}\\ \text{feed per unit time}\end{bmatrix} - \begin{bmatrix}\text{moles of reactant}\\ \text{converted per unit time}\end{bmatrix} = \begin{bmatrix}\text{moles of reactant}\\ \text{in product per unit time}\end{bmatrix} \qquad (5.44)$$

Equation 5.44 can be written per unit time as:

$$N_{i,in} - (-r_i V) = N_{i,out} \qquad (5.45)$$

where $N_{i,in}$ = inlet moles of Component i per unit time
$N_{i,out}$ = outlet moles of Component i per unit time

Rearranging Equation 5.45 gives:

$$N_{i,out} = N_{i,in} + r_i V \qquad (5.46)$$

Substituting $N_{i,out} = N_{i,in}(1 - X_i)$ into Equation 5.46 gives:

$$V = \frac{N_{i,in}X_i}{-r_i} \qquad (5.47)$$

For the special case of a constant density system, Equation 5.41 can be substituted to give:

$$V = \frac{N_{i,in}(C_{i,in} - C_{i,out})}{-r_i C_{i,in}} \qquad (5.48)$$

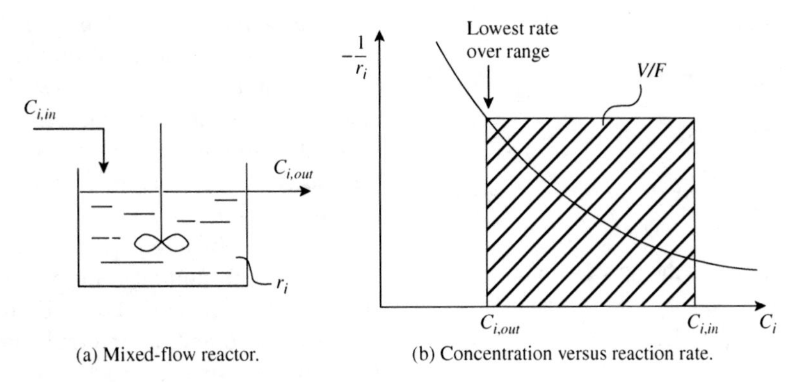

(a) Mixed-flow reactor. (b) Concentration versus reaction rate.

Figure 5.2 Rate of reaction in a mixed-flow reactor.

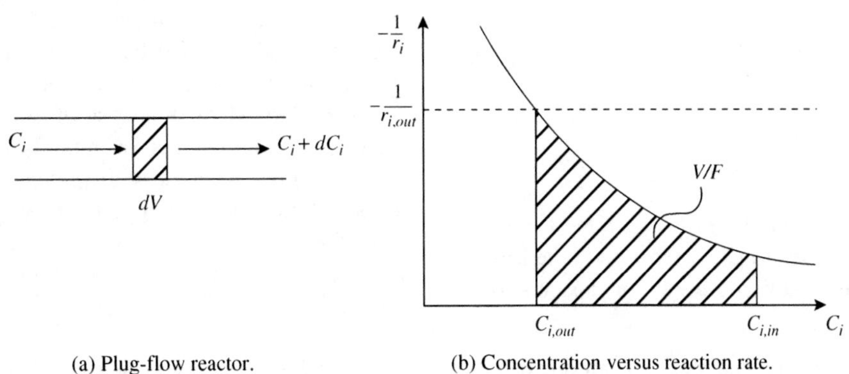

(a) Plug-flow reactor. (b) Concentration versus reaction rate.

Figure 5.3 Rate of reaction in a plug-flow reactor.

Analogous to time as a measure of batch process performance, *space–time* (τ) can be defined for a continuous reactor:

$$\tau = [\text{Time required to process one reactor volume of feed}] \quad (5.49)$$

If the space–time is based on the feed conditions:

$$\tau = \frac{V}{F} = \frac{C_{i,out}V}{N_{i,in}} \quad (5.50)$$

where F = volumetric flowrate of the feed ($m^3 \cdot s^{-1}$)

The reciprocal of space–time is *space–velocity* (s):

$$s = \frac{1}{\tau} = [\text{Number of reactor volumes processed in a unit time}] \quad (5.51)$$

Combining Equations 5.48 for the mixed-flow reactor with constant density and 5.50 gives:

$$\tau = \frac{C_{i,in} - C_{i,out}}{-r_i} \quad (5.52)$$

Figure 5.2b is a plot of Equation 5.52. From $C_{i,in}$ to $C_{i,out}$, the rate of reaction decreases to a minimum at $C_{i,out}$. As the reactor is assumed to be perfectly mixed, $C_{i,out}$ is the concentration throughout the reactor, that is, this gives the lowest rate throughout the reactor. The shaded area in Figure 5.2b represents the space–time (V/F).

3. *Plug-flow reactor.* Consider now the plug-flow reactor in Figure 5.3a, in which Component i is reacting. A material balance can be carried out per unit time across the incremental volume dV in Figure 5.3a:

$$\begin{bmatrix}\text{moles of reactant}\\ \text{entering incremental}\\ \text{volume per unit time}\end{bmatrix} - \begin{bmatrix}\text{moles of reactant}\\ \text{converted per unit}\\ \text{time}\end{bmatrix} = \begin{bmatrix}\text{moles of reactant}\\ \text{leaving incremental}\\ \text{volume per unit time}\end{bmatrix} \quad (5.53)$$

Equation 5.53 can be written per unit time as:

$$N_i - (-r_i dV) = N_i + dN_i \quad (5.54)$$

where N_i = moles of Component i per unit time

Rearranging Equation 5.54 gives:

$$dN_i = r_i\, dV \quad (5.55)$$

Substituting the reactor conversion into Equation 5.55 gives:

$$dN_i = d[N_{i,in}(1 - X_i)] = r_i\, dV \quad (5.56)$$

where $N_{i,in}$ = inlet moles of Component i per unit time

Rearranging Equation 5.56 gives:

$$N_{i,in}dX_i = -r_i\,dV \tag{5.57}$$

Integration of Equation 5.57 gives:

$$V = N_{i,in}\int_0^{X_i}\frac{dX_i}{-r_i} \tag{5.58}$$

Writing Equation 5.58 in terms of the space–time:

$$\tau = C_{i,in}\int_0^{X_i}\frac{dX_i}{-r_i} \tag{5.59}$$

For the special case of constant density systems, substitution of Equation 5.50 gives:

$$V = -\frac{N_{i,in}}{C_{i,in}}\int_{C_{i,in}}^{C_{i,out}}\frac{dC_i}{-r_i} \tag{5.60}$$

$$\tau = -\int_{C_{i,in}}^{C_{i,out}}\frac{dC_i}{-r_i} \tag{5.61}$$

Figure 5.3b is a plot of Equation 5.61. The rate of reaction is high at $C_{i,in}$ and decreases to $C_{i,out}$ where it is the lowest. The area under the curve now represents the space–time.

It should be noted that the analysis for an ideal-batch reactor is the same as that for a plug-flow reactor (compare Equations 5.43 and 5.61). All fluid elements have the same residence time in both cases. Thus

$$t_{Ideal-batch} = \tau_{Plug-flow} \tag{5.62}$$

Figure 5.4a compares the profiles for a mixed-flow and plug-flow reactor between the same inlet and outlet concentrations, from which it can be concluded that the mixed-flow reactor requires a larger volume. The rate of reaction in a mixed-flow reactor is uniformly low as the reactant is instantly diluted by the product that has already been formed. In a plug-flow or ideal-batch reactor, the rate of reaction is high initially and decreases as the concentration of reactant decreases.

By contrast, in an *autocatalytic* reaction, the rate starts low, as little product is present, but increases as product is formed. The rate reaches a maximum and then decreases as the reactant is consumed. This is illustrated in Figure 5.4b. In such a situation, it would be best to use a combination of reactor types, as illustrated in Figure 5.4b, to minimize the volume for a given flowrate. A mixed-flow reactor should be used until the maximum rate is reached and the intermediate product fed to a plug-flow (or ideal-batch) reactor. Alternatively, if separation and recycle of unconverted material is possible, then a mixed-flow reactor could be used up to the point where maximum rate is achieved, then unconverted material separated and recycled back to the reactor inlet. Whether this separation and recycle option is cost-effective or not will depend on the cost of separating and recycling material. Combinations of mixed-flow reactors in series and plug-flow reactors with recycle can also be used for autocatalytic reactions, but their performance will always be inferior to either a combination of mixed and plug-flow reactors or a mixed-flow reactor with separation and recycle[1].

Example 5.4 Benzyl acetate is used in perfumes, soaps, cosmetics and household items where it produces a fruity, jasminelike aroma, and it is used to a minor extent as a flavor. It can be manufactured by the reaction between benzyl chloride and sodium acetate in a solution of xylene in the presence of triethylamine as catalyst[9].

$$C_6H_5CH_2Cl + CH_3COONa \longrightarrow CH_3COOC_6H_5CH_2 + NaCl$$

or

$$A + B \longrightarrow C + D$$

The reaction has been investigated experimentally by Huang and Dauerman[9] in a batch reaction carried out with initial conditions given in Table 5.3[9].

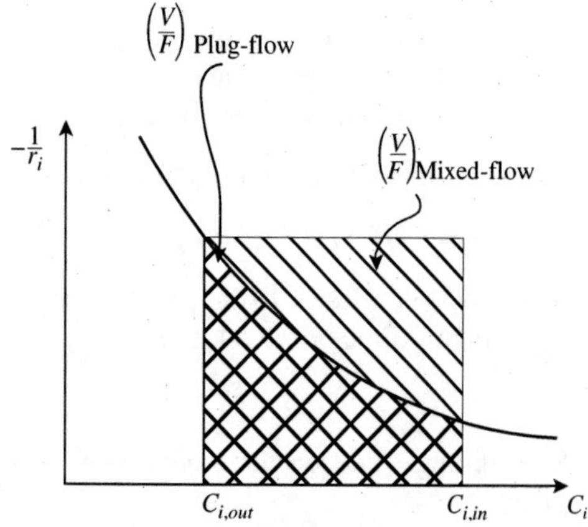

(a) Comparison of mixed-flow and plug-flow reactors.

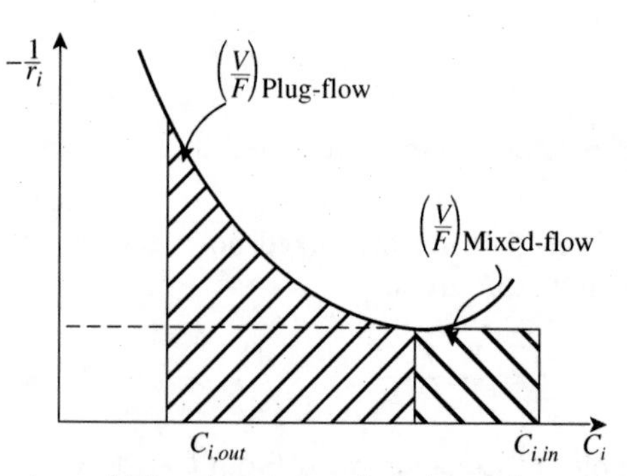

(b) Reaction rate goes through a maximum.

Figure 5.4 Use of mixed-flow and plug-flow reactors.

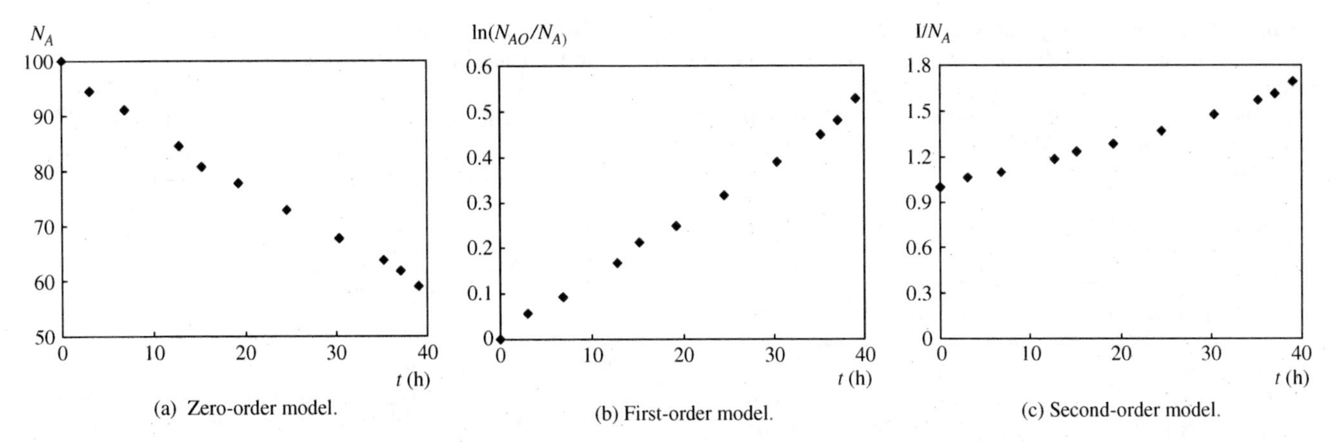

Figure 5.5 Kinetic model for the production of benzyl acetate.

Table 5.3 Initial reaction mixture for the production of benzyl acetate.

Component	Moles
Benzyl chloride	1
Sodium acetate	1
Xylene	10
Triethylamine	0.0508

The solution volume was $1.321 \times 10^{-3}\text{m}^3$ and the temperature maintained to be 102°C. The measured mole per cent benzyl chloride versus time in hours are given in Table 5.4.[9]

Table 5.4 Experimental data for the production of benzyl acetate.

Reaction time (h)	Benzyl chloride (mole%)
3.0	94.5
6.8	91.2
12.8	84.6
15.2	80.9
19.3	77.9
24.6	73.0
30.4	67.8
35.2	63.8
37.15	61.9
39.1	59.0

Derive a kinetic model for the reaction on the basis of the experimental data. Assume the volume of the reactor to be constant.

Solution The equation for a batch reaction is given by Equation 5.38:

$$t = \int_{N_{A0}}^{N_{Af}} \frac{dN_A}{r_A V}$$

Initially, it could be postulated that the reaction could be zero order, first order or second order in the concentration of A and B. However, given that all the reaction stoichiometric coefficients are unity, and the initial reaction mixture has equimolar amounts of A and B, it seems sensible to first try to model the kinetics in terms of the concentration of A. This is because, in this case, the reaction proceeds with the same rate of change of moles for the two reactants. Thus, it could be postulated that the reaction could be zero order, first order or second order in the concentration of A. In principle, there are many other possibilities. Substituting the appropriate kinetic expression into Equation 5.47 and integrating gives the expressions in Table 5.5:

Table 5.5 Expressions for a batch reaction with different kinetic models.

Order of reaction	Kinetic expression	Ideal-batch model
Zero order	$-r_A = k_A$	$\frac{1}{V}(N_{A0} - N_A) = k_A t$
First order	$-r_A = k_A C_A$	$\ln \frac{N_{A0}}{N_A} = k_A t$
Second order	$-r_A = k_A C_A^2$	$V\left(\frac{1}{N_A} - \frac{1}{N_{A0}}\right) = k_A t$

The experimental data have been substituted into the three models and presented graphically in Figure 5.5. From Figure 5.5, all three models seem to give a reasonable representation of the data, as all three give a reasonable straight line. It is difficult to tell from the graph which line gives the best fit. The fit can be better judged by carrying out a least squares fit to the data for the three models. The difference between the values calculated from the model and the experimental values are summed according to:

$$\text{minimize } R^2 = \sum_{j}^{10} [(N_{A,j})_{calc} - (N_{A,j})_{exp}]^2$$

where
R = residual
$(N_{A,j})_{calc}$ = calculated moles of A for Measurement j
$(N_{A,j})_{exp}$ = experimentally measured moles of A for Measurement j

However, the most appropriate value of the rate constant for each model needs to be determined. This can be determined, for example, in a spreadsheet by setting up a function for R^2 in the spreadsheet and then using the spreadsheet solver to minimize R^2 by manipulating the value of k_A. The results are summarized in Table 5.6.

Table 5.6 Results of a least squares fit for the three kinetic models.

Order of reaction	Rate constant	R^2
Zero order	$k_A V = 1.066$	26.62
First order	$k_A = 0.01306$	6.19
Second order	$k_A/V = 0.0001593$	15.65

From Table 5.6, it is clear that the best fit is given by a first-order reaction model: $r_A = k_A C_A$ with $k_A = 0.01306\ \text{h}^{-1}$.

Example 5.5 Ethyl acetate is widely used in formulating printing inks, adhesives, lacquers and used as a solvent in food processing. It can be manufactured from the reaction between ethanol and acetic acid in the liquid phase according to the reaction:

$$\underset{\text{acetic acid}}{CH_3COOH} + \underset{\text{ethanol}}{C_2H_5OH} \rightleftharpoons \underset{\text{ethyl acetate}}{CH_3COOC_2H_5} + \underset{\text{water}}{H_2O}$$

or

$$A + B \rightleftharpoons C + D$$

Experimental data are available using an ion-exchange resin catalyst based on batch experiments at 60°C[10]. These data are presented in Table 5.7[10].

Table 5.7 Experimental data for the production of ethyl acetate.

Sample point	Time (min)	A (mass%)	B (mass%)	C (mass%)	D (mass%)
1	0	56.59	43.41	0.00	0.00
2	5	49.70	38.10	10.00	2.20
3	10	46.30	35.50	15.10	3.10
4	15	42.50	32.50	20.70	4.30
5	30	35.40	27.20	30.90	6.50
6	60	28.10	21.90	41.40	8.60
7	90	24.20	18.60	47.60	9.60
8	120	22.70	17.40	49.80	10.10
9	150	21.20	17.00	51.10	10.70
10	180	20.90	16.50	51.70	10.90
11	210	20.50	16.20	52.30	11.00
12	240	20.30	15.70	52.90	11.10

Molar masses and densities for the components are given in Table 5.8.

Initial conditions are such that the reactants are equimolar with product concentrations of initially zero. From the experimental data, assuming a constant density system:

a. Fit a kinetic model to the experimental data.
b. For a plant producing 10 tons of ethyl acetate per day, calculate the volume required by a mixed-flow reactor and a plug-flow reactor operating at 60°C. Assume no product is recycled to the reactor and the reactor feed is an equimolar mixture of ethanol and acetic acid. Also, assume the reactor conversion to be 95% of the conversion at equilibrium.

Table 5.8 Molar masses for the components involved in the manufacture of ethyl acetate.

Component	Molar mass (kg·kmol^{-1})	Density at 60°C (kg·m^{-3})
Acetic acid	60.05	1018
Ethanol	46.07	754
Ethyl acetate	88.11	847
Water	18.02	980

Solution

a. To fit a model to the data, first convert the mass percent data from Table 5.7 into molar concentrations.

Volume of reaction mixture per kg of reaction mixture (assuming no volume change of solution)

$$= \frac{0.566}{1018} + \frac{0.434}{754}$$

$$= 1.1316 \times 10^{-3}\ \text{m}^3$$

Molar concentrations are given in Table 5.9.

Table 5.9 Molar concentrations for the manufacture of ethyl acetate.

Time (min)	A (kmol·m^{-3})	B (kmol·m^{-3})	C (kmol·m^{-3})	D (kmol·m^{-3})
0	8.3277	8.3266	0.0000	0.0000
5	7.3138	7.3081	1.0029	1.0789
10	6.8134	6.8094	1.5144	1.5202
15	6.2542	6.2339	2.0761	2.1087
30	5.2094	5.2173	3.0991	3.1875
60	4.1352	4.2007	4.1522	4.2174
90	3.5612	3.5677	4.7740	4.7078
120	3.3405	3.3376	4.9946	4.9530
150	3.1198	3.2608	5.1250	5.2472
180	3.0756	3.1649	5.1852	5.3453
210	3.0167	3.1074	5.2454	5.3943
240	2.9873	3.0115	5.3055	5.4434

A reaction rate expression can be assumed of the form:

$$-r_A = k_A C_A^\alpha C_B^\beta - k'_A C_C^\delta C_D^\gamma$$

$\alpha, \beta, \delta, \gamma$ can be 0, 1 or 2 such that $n_1 = \alpha + \beta$ and $n_2 = \delta + \gamma$. Many other forms of model could be postulated. The equation for an ideal-batch reactor is given by Equation 5.38.

$$t = -\int_{C_{A0}}^{C_A} \frac{dC_A}{-r_A}$$

Substituting the kinetic model and integrating give the results in Table 5.10 depending on the values of the order of reaction. The integrals can be found from tables of standard integrals[11].

Table 5.10 Kinetic models used to fit the data for the manufacture of ethyl acetate.

n_1	n_2	Rate equation	Final concentration from the model
1	1	$-r_A = k_A C_A - k' C_C$	
		$-r_A = k_A C_B - k'_A C_C$	$C_A = \dfrac{k_A \exp(-(k_A + k'_A)t) + k'_A}{k_A + k'_A} C_{A0}$
		$-r_A = k_A C_A - k'_A C_D$	
		$-r_A = k_A C_B - k'_A C_D$	
2	1	$-r_A = k_A C_A^2 - k'_A C_C$	
		$-r_A = k_A C_B^2 - k'_A C_C$	
		$-r_A = k_A C_A^2 - k'_A C_D$	$C_A = \dfrac{(k'_A - a)(2k_A C_{A0} + k'_A + a) - (k'_A + a)(2k_A C_{A0} + k'_A - a)\exp(-at)}{2k_A(2k_A C_{A0} + k'_A - a)\exp(-at) - 2k_A(2k_A C_{A0} + k'_A + a)}$
		$-r_A = k_A C_B^2 - k'_A C_D$	
		$-r_A = k_A C_A C_B - k'_A C_C$	$a = \sqrt{k'_A k'_A + 4k_A k'_A C_{A0}}$
		$-r_A = k_A C_A C_B - k'_A C_D$	
1	2	$-r_A = k_A C_A - k'_A C_C^2$	
		$-r_A = k_A C_B - k'_A C_C^2$	$C_A = \dfrac{(k_A + 2k'_A C_{A0} + b)(k_A - b)\exp(-bt) - (k_A + 2k'_A C_{A0} - b)(k_A + b)}{-2k'_A(k_A + b) + 2k'_A(k_A - b)\exp(-bt)}$
		$-r_A = k_A C_A - k'_A C_D^2$	
		$-r_A = k_A C_B - k'_A C_D^2$	$b = \sqrt{k_A^2 + 4k_A k'_A C_{A0}}$
		$-r_A = k_A C_A - k'_A C_C C_D$	
		$-r_A = k_A C_B - k'_A C_C C_D$	
2	2	$-r_A = k_A C_A^2 - k'_A C_C^2$	
		$-r_A = k_A C_B^2 - k'_A C_C^2$	
		$-r_A = k_A C_A^2 - k'_A C_D^2$	
		$-r_A = k_A C_B^2 - k'_A C_D^2$	$C_A = \dfrac{\sqrt{k_A k'_A}(1 + \exp(2C_{A0}\sqrt{k_A k'_A} t))}{(k_A + \sqrt{k_A k'_A})\exp(2C_{A0}\sqrt{k_A k'_A} t) - (k_A - \sqrt{k_A k'_A})} C_{A0}$
		$-r_A = k_A C_A C_B - k'_A C_C^2$	
		$-r_A = k_A C_A C_B - k'_A C_D^2$	
		$-r_A = k C_A^2 - k' C_C C_D$	
		$-r_A = k_A C_B^2 - k'_A C_C C_D$	
		$-r_A = k_A C_A C_B - k'_A C_C C_D$	

Note in Table 5.10 that many of the integrals are common to different kinetic models. This is specific to this reaction where all the stoichiometric coefficients are unity and the initial reaction mixture was equimolar. In other words, the change in the number of moles is the same for all components. Rather than determine the integrals analytically, they could have been determined numerically. Analytical integrals are simply more convenient if they can be obtained, especially if the model is to be fitted in a spreadsheet, rather than purpose-written software. The least squares fit varies the reaction rate constants to minimize the objective function:

$$\text{minimize } R^2 = \sum_{i=1}^{4} \sum_{j=1}^{11} [(C_{i,j})_{calc} - (C_{i,j})_{exp}]^2$$

where R = residual

$(C_{i,j})_{calc}$ = calculated molar concentration of Component i for Measurement j

$(C_{i,j})_{exp}$ = experimentally measured molar concentration of Component i for Measurement j

Again, this can be carried out, for example, in spreadsheet software, and the results are given in Table 5.11.

Table 5.11 Results of model fitting for the manufacture of ethyl acetate.

n_1	n_2	k_A	k'_A	R^2
1	1	0.01950	0.01172	1.01108
2	1	0.002688	0.004644	0.3605
1	2	0.01751	0.002080	1.7368
2	2	0.002547	0.0008616	0.5554

From Table 5.11, there is very little to choose between the best two models. The best fit is given by a second-order model for the forward and a first-order model for the reverse reaction with:

$$k_A = 0.002688 \text{ m}^3\cdot\text{kmol}^{-1}\cdot\text{min}^{-1}$$

$$k'_A = 0.004644 \text{ min}^{-1}$$

However, there is little to choose between this model and a second-order model for both forward and reverse reactions.
b. Now use the kinetic model to size a reactor to produce 10 tons per day of ethyl acetate. First, the conversion at equilibrium needs to be calculated. At equilibrium, the rates of forward and reverse reactions are equal:

$$k_A C_A^2 = k'_A C_C$$

Substituting for the conversion at equilibrium X_E gives:

$$k_A[C_{A0}(1 - X_E)]^2 = k'_A C_{A0} X_E$$

Rearranging gives:

$$\frac{(1 - X_E)^2}{X_E} = \frac{k'_A}{k_A C_{A0}}$$

Substituting $k_A = 0.002688$, $k'_A = 0.004644$ and $C_{A0} = 8.3277$ and solving for X_E by trial and error gives:

$$X_E = 0.6366$$

Assume the actual conversion to be 95% of the equilibrium conversion:

$$X = 0.95 \times 0.6366$$
$$= 0.605$$

A production rate of 10 tons per day equates to 0.0788 kmol·min^{-1}. For a mixed-flow reactor with equimolar feed $C_{A0} = C_{B0} = 8.33$ kmol·m^{-3} at 60°C:

$$V = \frac{N_{A,in} X_A}{-r_A} = \frac{N_{C,out}}{k_A C_{A0}^2 (1 - X_A)^2 - k'_A C_{A0} X_A}$$

$$= \frac{0.0788}{0.002688 \times 8.33^2 \times (1 - 0.605)^2 - 0.004644 \times 8.33 \times 0.605}$$

$$= 13.83 \text{ m}^3$$

For a plug-flow reactor:

$$\tau = \frac{V C_{A0}}{N_{A0}} = \int_0^{X_A} \frac{C_{A0} dX_A}{-k_A C_{A0}{}^2 (1 - X_A)^2 + k'_A C_{A0} X_A}$$

From tables of standard integrals[10], this can be integrated to give:

$$\tau = -\frac{1}{a} \ln \frac{(2k_A C_{A0}(1 - X_A) + k'_A - a)(2k_A C_{A0} + k'_A + a)}{(2k_A C_{A0}(1 - X_A) + k'_A + a)(2k_A C_{A0} + k'_A - a)}$$

where $a = \sqrt{k'_A k'_A + 4k_A k'_A C_{A0}}$
Substituting the values of k_A, k'_A, C_{A0} and X_A gives:

$$\tau = 120.3 \text{ min}$$

The reactor volume is given by:

$$V = \frac{\tau N_{A0}}{C_{A0}} = \frac{\tau N_C}{X_A C_{A0}} = \frac{120.3 \times 0.0788}{0.605 \times 8.33} = 1.88 \text{ m}^3$$

Alternatively, the residence time in the plug-flow reactor could be calculated from the batch equations given in Table 5.10. This results from the residence time being equal for both. Thus the final concentration in a plug-flow reactor is given by (Table 5.10):

$$C_A = \frac{(k'_A - a)(2k_A C_{A0} + k'_A + a) - (k'_A + a)(2k_A C_{A0} + k'_A - a)\exp(-a\tau)}{2k_A(2k_A C_{A0} + k'_A - a)\exp(-a\tau) - 2k_A(2k_A C_{A0} + k'_A + a)}$$

where $a = \sqrt{k'_A k'_A + 4k_A k'_A C_{A0}}$
The final concentration of C_A from the conversion is $8.33 \times 0.395 = 3.29$ kmol·m^3 (assuming no volume change). Thus, the above equation can be solved by trial and error to give the residence time of 120.3 min.

As expected, the result shows that the volume required by a mixed-flow reactor is much larger than that for plug-flow.

A number of points should be noted regarding Examples 5.4 and 5.5:

1. The kinetic models were fitted to experimental data at specific conditions of molar feed ratio and temperature. The models are only valid for these conditions. Use for nonequimolar feeds or at different temperatures will not be valid.
2. Given that kinetic models are only valid for the range of conditions over which they are fitted, it is better that the experimental investigation into the reaction kinetics and the reactor design are carried out in parallel. If this approach is followed, then it can be assured that the range of experimental conditions used in the laboratory cover the range of conditions used in the reactor design. If the experimental programme is carried out and completed prior to the reactor design, then there is no guarantee that the kinetic model will be appropriate for the conditions chosen in the final design.
3. Different models often give very similar predictions over a limited range of conditions. However, the differences between different models are likely to become large if used outside the range over which they were fitted to experimental data.

5.6 CHOICE OF IDEALIZED REACTOR MODEL

Consider now which of the idealized models is preferred for the six categories of reaction systems introduced in Section 5.2.

1. *Single reactions.* Consider the single reaction from Equation 5.1:

$$FEED \longrightarrow PRODUCT \quad r = kC_{FEED}^a \qquad (5.63)$$

where r = rate of reaction
k = reaction rate constant
C_{FEED} = molar concentration of *FEED*
a = order of reaction

Clearly, the highest rate of reaction is maintained by the highest concentration of feed (C_{FEED}, kmol·m^{-3}). As already discussed, in the mixed-flow reactor the incoming feed is instantly diluted by the product that has already been formed. The rate of reaction is thus lower in the mixed-flow reactor than in the ideal-batch and plug-flow reactors, since it operates at the low reaction rate corresponding with the outlet concentration of feed. Thus, a mixed-flow reactor requires a greater volume than an ideal-batch or plug-flow reactor. Consequently, for single reactions, an ideal-batch or plug-flow reactor is preferred.

2. *Multiple reactions in parallel producing byproducts.* Consider the system of parallel reactions from Equation 5.4 with the corresponding rate equations[1–3]:

$$\begin{aligned} &FEED \longrightarrow PRODUCT \qquad r_1 = k_1 C_{FEED}^{a_1} \\ &FEED \longrightarrow BYPRODUCT \qquad r_2 = k_2 C_{FEED}^{a_2} \end{aligned} \tag{5.64}$$

where r_1, r_2 = rates of reaction for primary and secondary reactions
k_1, k_2 = reaction rate constants for primary and secondary reactions
C_{FEED} = molar concentration of *FEED* in the reactor
a_1, a_2 = order of reaction for primary and secondary reactions

The ratio of the rates of the secondary and primary reactions gives [1–3]:

$$\frac{r_2}{r_1} = \frac{k_2}{k_1} C_{FEED}^{a_2 - a_1} \tag{5.65}$$

Maximum selectivity requires a minimum ratio r_2/r_1 in Equation 5.65. A batch or plug-flow reactor maintains higher average concentrations of feed (C_{FEED}) than a mixed-flow reactor, in which the incoming feed is instantly diluted by the *PRODUCT* and *BYPRODUCT*. If $a_1 > a_2$ in Equations 5.64 and 5.65 the primary reaction to *PRODUCT* is favored by a high concentration of *FEED*. If $a_1 < a_2$ the primary reaction to *PRODUCT* is favored by a low concentration of *FEED*. Thus, if

- $a_2 < a_1$, use a batch or plug-flow reactor.
- $a_2 > a_1$, use a mixed-flow reactor.

In general terms, if the reaction to the desired product has a higher order than the byproduct reaction, use a batch or plug-flow reactor. If the reaction to the desired product has a lower order than the byproduct reaction, use a mixed-flow reactor.

If the reaction involves more than one feed, the picture becomes more complex. Consider the reaction system from Equation 5.6 with the corresponding rate equations:

$$\begin{aligned} &FEED1 + FEED2 \longrightarrow PRODUCT \\ &\qquad r_1 = k_1 C_{FEED1}^{a_1} C_{FEED2}^{b_1} \\ &FEED1 + FEED2 \longrightarrow BYPRODUCT \\ &\qquad r_2 = k_2 C_{FEED1}^{a_2} C_{FEED2}^{b_2} \end{aligned} \tag{5.66}$$

where C_{FEED1}, C_{FEED2} = molar concentrations of *FEED 1* and *FEED 2* in the reactor
a_1, b_1 = order of reaction for the primary reaction
a_2, b_2 = order of reaction for the secondary reaction

Now the ratio that needs to be minimized is given by[1–3]:

$$\frac{r_2}{r_1} = \frac{k_2}{k_1} C_{FEED1}^{a_2 - a_1} C_{FEED2}^{b_2 - b_1} \tag{5.67}$$

Given this reaction system, the options are:

- Keep both C_{FEED1} and C_{FEED2} low (i.e. use a mixed-flow reactor).
- Keep both C_{FEED1} and C_{FEED2} high (i.e. use a batch or plug-flow reactor).
- Keep one of the concentrations high while maintaining the other low (this is achieved by charging one of the feeds as the reaction progresses).

Figure 5.6 summarizes these arguments to choose a reactor for systems of multiple reactions in parallel[12].

3. *Multiple reactions in series producing byproducts.* Consider the system of series reactions from Equation 5.7:

$$\begin{aligned} &FEED \longrightarrow PRODUCT \qquad r_1 = k_1 C_{FEED}^{a_1} \\ &PRODUCT \longrightarrow BYPRODUCT \qquad r_2 = k_2 C_{PRODUCT}^{a_2} \end{aligned} \tag{5.68}$$

where r_1, r_2 = rates of reaction for primary and secondary reactions
k_1, k_2 = reaction rate constants
C_{FEED} = molar concentration of *FEED*
$C_{PRODUCT}$ = molar concentration of *PRODUCT*
a_1, a_2 = order of reaction for primary and secondary reactions

For a certain reactor conversion, the *FEED* should have a corresponding residence time in the reactor. In the mixed-flow reactor, *FEED* can leave the instant it enters or remains for an extended period. Similarly, *PRODUCT* can remain for an extended period or leave immediately. Substantial fractions of both *FEED* and *PRODUCT* leave before and after what should be the specific residence time

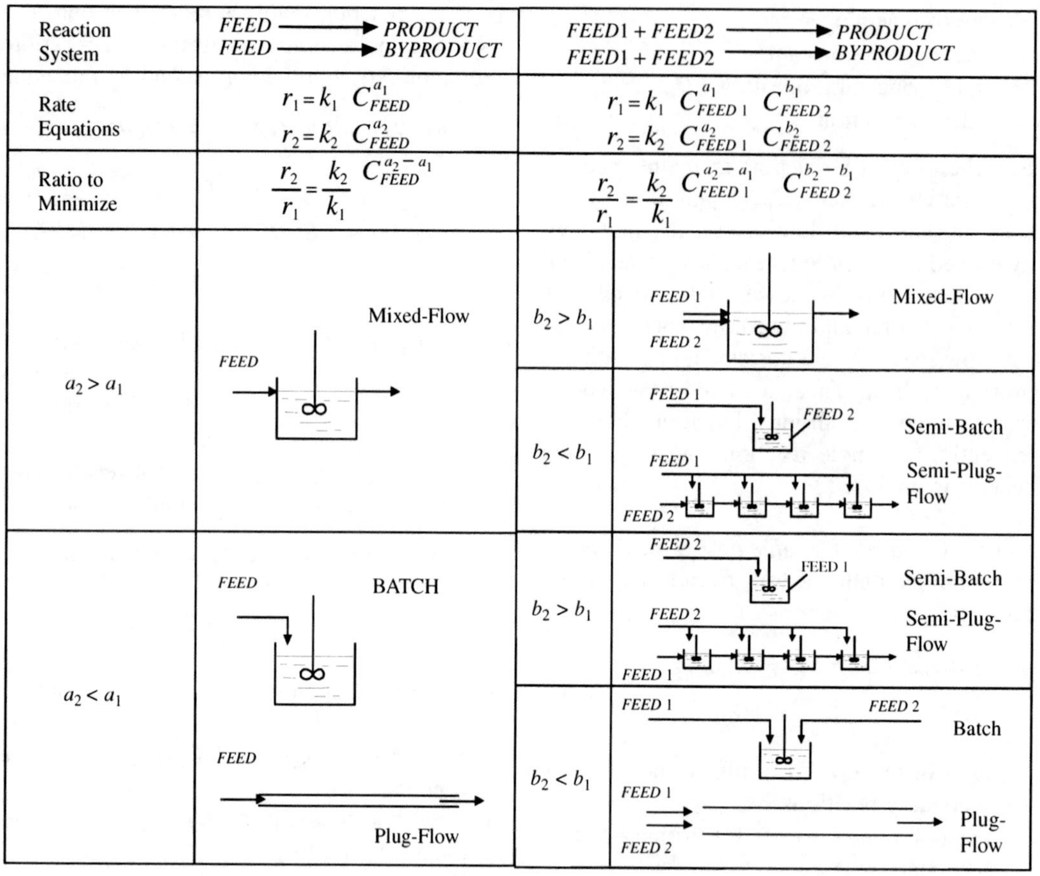

Reaction System	$FEED \longrightarrow PRODUCT$ $FEED \longrightarrow BYPRODUCT$		$FEED1 + FEED2 \longrightarrow PRODUCT$ $FEED1 + FEED2 \longrightarrow BYPRODUCT$
Rate Equations	$r_1 = k_1\ C_{FEED}^{a_1}$ $r_2 = k_2\ C_{FEED}^{a_2}$		$r_1 = k_1\ C_{FEED\,1}^{a_1}\ C_{FEED\,2}^{b_1}$ $r_2 = k_2\ C_{FEED\,1}^{a_2}\ C_{FEED\,2}^{b_2}$
Ratio to Minimize	$\frac{r_2}{r_1} = \frac{k_2}{k_1} C_{FEED}^{a_2 - a_1}$		$\frac{r_2}{r_1} = \frac{k_2}{k_1} C_{FEED\,1}^{a_2 - a_1}\ C_{FEED\,2}^{b_2 - b_1}$
$a_2 > a_1$	FEED Mixed-Flow	$b_2 > b_1$	FEED 1, FEED 2 Mixed-Flow
		$b_2 < b_1$	FEED 1, FEED 2 Semi-Batch FEED 1, FEED 2 Semi-Plug-Flow
$a_2 < a_1$	FEED BATCH FEED Plug-Flow	$b_2 > b_1$	FEED 2, FEED 1 Semi-Batch FEED 2, FEED 1 Semi-Plug-Flow
		$b_2 < b_1$	FEED 1, FEED 2 Batch FEED 1, FEED 2 Plug-Flow

Figure 5.6 Reactor choice for parallel reaction systems. (From Smith R and Petela EA, 1991, *The Chemical Engineer*, No. 509/510:12, reproduced by permission of the Institution of Chemical Engineers).

for a given conversion. Thus, the mixed-flow model would be expected to give a poorer selectivity or yield than a batch or plug-flow reactor for a given conversion.

A batch or plug-flow reactor should be used for multiple reactions in series.

4. *Mixed parallel and series reactions producing byproducts*. Consider the mixed parallel and series reaction system from Equation 5.10 with the corresponding kinetic equations:

$$FEED \longrightarrow PRODUCT$$

$$r_1 = k_1 C_{FEED}^{a_1}$$

$$FEED \longrightarrow BYPRODUCT$$

$$r_2 = k_2 C_{FEED}^{a_2} \qquad (5.69)$$

$$PRODUCT \longrightarrow BYPRODUCT$$

$$r_3 = k_3 C_{PRODUCT}^{a_3}$$

As far as the parallel byproduct reaction is concerned, for high selectivity, if:

- $a_1 > a_2$, use a batch or plug-flow reactor
- $a_1 < a_2$, use a mixed-flow reactor

The series byproduct reaction requires a plug-flow reactor. Thus, for the mixed parallel and series system above, if:

- $a_1 > a_2$, use a batch or plug-flow reactor

But what is the correct choice if $a_1 < a_2$? Now the parallel byproduct reaction calls for a mixed-flow reactor. On the other hand, the byproduct series reaction calls for a plug-flow reactor. It would seem that, given this situation, some level of mixing between a plug-flow and a mixed-flow reactor will give the best overall selectivity[12]. This could be obtained by a:

- series of mixed-flow reactors (Figure 5.7a)
- plug-flow reactor with a recycle (Figure 5.7b)
- series combination of plug-flow and mixed-flow reactors (Figures 5.7c and 5.7d).

The arrangement that gives the highest overall selectivity can only be deduced by a detailed analysis and optimization of the reaction system. This will be dealt with in Chapter 7.

5. *Polymerization reactions*. Polymers are characterized mainly by the distribution of molar mass about the mean

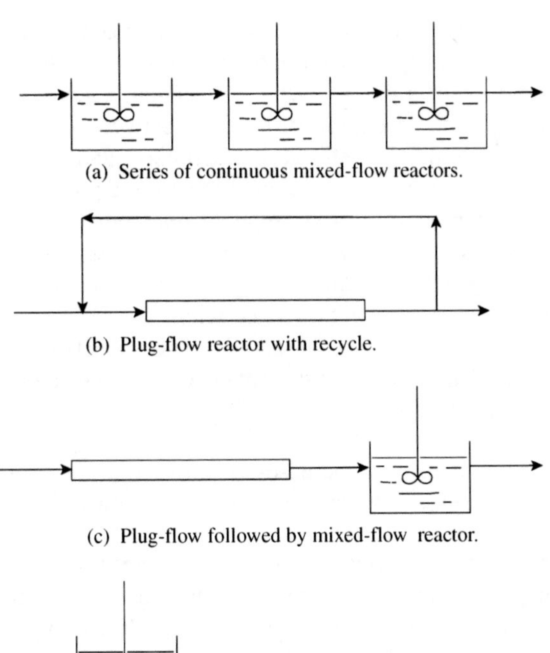

(a) Series of continuous mixed-flow reactors.

(b) Plug-flow reactor with recycle.

(c) Plug-flow followed by mixed-flow reactor.

(d) Mixed-flow followed by plug-flow reactor.

Figure 5.7 Choice of reactor type for mixed parallel and series reactions when the parallel reaction has a higher order than the primary reaction.

as well as by the mean itself. Orientation of groups along chains, cross-linking of the polymer chains, and so on, also affect the properties. The breadth of distribution of molar mass depends on whether a batch or plug-flow reactor or a mixed-flow reactor is used. The breadth has an important influence on the mechanical and other properties of the polymer, and this is an important factor in the choice of reactor.

Two broad classes of polymerization reactions can be identified[2]:

(a) In a batch or plug-flow reactor, all molecules have the same residence time, and without the effect of termination (see Section 5.2) all will grow to approximately equal lengths, producing a narrow distribution of molar masses. By contrast, a mixed-flow reactor will cause a wide distribution because of the distribution of residence times in the reactor.

(b) When polymerization takes place by mechanisms involving free radicals, the life of these actively growing centers may be extremely short because of termination processes such as the union of two free radicals (see Section 5.2). These termination processes are influenced by free-radical concentration, which in turn is proportional to monomer concentration. In batch or plug-flow reactors, the monomer and free-radical concentrations decline. This produces increasing chain lengths with increasing residence time and thus a broad distribution of molar masses. The mixed-flow reactor maintains a uniform concentration of monomer and thus a constant chain-termination rate. This results in a narrow distribution of molar masses. Because the active life of the polymer is short, the variation in residence time does not have a significant effect.

6. *Biochemical reactions.* As discussed previously, there are two broad classes of biochemical reactions: reactions that exploit the metabolic pathways in selected microorganisms and those catalyzed by enzymes. Consider microbial biochemical reactive of the type:

$$A \xrightarrow{C} R + C$$

where A is the feed, R is the product and C represents the cells (microorganisms). The kinetics of such reactions can be described by the Monod Equation[1,6]:

$$-r_A = k\frac{C_C C_A}{C_A + C_M} \tag{5.70}$$

where r_A = rate of reaction
k = rate constant
C_C = concentration of cells (microorganisms)
C_A = concentration of feed
C_M = a constant (Michaeli's constant that is a function of the reaction and conditions)

The rate constant can depend on many factors, such as temperature, the presence of trace elements, vitamins, toxic substances, light intensity, and so on. The rate of reaction depends not only on the availability of feed (food for the microorganisms) but also on the buildup of wastes from the microorganisms that interfere with microorganism multiplication. There comes a point at which their waste inhibits growth. An excess of feed material or microorganisms can slow the rate of reaction. Without poisoning of the kinetics due to the buildup of waste material, Equation 5.70 gives the same characteristic curve as autocatalytic reactions as shown in Figure 5.4b. Thus, depending on the concentration range, mixed-flow, plug-flow, a combination of mixed-flow and plug-flow or mixed-flow with separation and recycle might be appropriate.

For enzyme-catalyzed biochemical reactions of the form

$$A \xrightarrow{\text{enzymes}} R$$

The kinetics can be described by the Monod Equation in the form[1,6]:

$$-r_A = k\frac{C_E C_A}{C_M + C_A} \tag{5.71}$$

where C_E = enzyme concentration

The presence of some substances can cause the reaction to slow down. Such substances are known as *inhibitors*.

This is caused either by the inhibitor competing with the feed material for active sites on the enzyme or by the inhibitor attacking an adjacent site and, in so doing, inhibiting the access of the feed material to the active site.

High reaction rate in Equation 5.71 is favored by a high concentration of enzymes (C_E) and high concentration of feed (C_A). This means that a plug-flow or ideal-batch reactor is favored if both the feed material and enzymes are to be fed to the reactor.

5.7 CHOICE OF REACTOR PERFORMANCE

It is now possible to define the goals for the choice of the reactor at this stage in the design. Unconverted material usually can be separated and recycled later. Because of this, the reactor conversion cannot be fixed finally until the design has progressed much further than just choosing the reactor. As shall be seen later, the choice of reactor conversion has a major influence on the rest of the process. Nevertheless, some decisions must be made regarding the reactor for the design to proceed. Thus, some guess for the reactor conversion must be made in the knowledge that this is likely to change once more detail is added to the total system design.

Unwanted byproducts usually cannot be converted back to useful products or raw materials. The reaction to unwanted byproducts creates both raw materials costs due to the raw materials that are wasted in their formation and environmental costs for their disposal. Thus, maximum selectivity is required for the chosen reactor conversion. The objectives at this stage can be summarized as follows:

1. *Single reactions.* In a single reaction, such as Equation 5.2 that produces a byproduct, there can be no influence on the relative amounts of product and byproduct formed. Thus, with single reactions such as Equations 5.1 to 5.3, the goal is to minimize the reactor capital cost, which often (but not always) means minimizing reactor volume, for a given reactor conversion. Increasing the reactor conversion increases size and hence cost of the reactor but, as will be discussed later, decreases the cost of many other parts of the flowsheet. Because of this, the initial setting for reactor conversion for single irreversible reactions is around 95% and that for a single reversible reaction is around 95% of the equilibrium conversion[12]. Equilibrium conversion will be considered in more detail in the next chapter.

For batch reactors, account must be taken of the time required to achieve a given conversion. Batch cycle time is addressed later.

2. *Multiple reactions in parallel producing byproducts.* Raw materials costs usually will dominate the economics of the process. Because of this, when dealing with multiple reactions, whether parallel, series or mixed, the goal is usually to minimize byproduct formation (maximize selectivity) for a given reactor conversion. Chosen reactor conditions should exploit differences between the kinetics and equilibrium effects in the primary and secondary reactions to favor the formation of the desired product rather than the byproduct, that is, improve selectivity. Making an initial guess for conversion is more difficult than with single reactions, since the factors that affect conversion also can have a significant effect on selectivity.

Consider the system of parallel reactions from Equations 5.64 and 5.65. A high conversion in the reactor tends to decrease C_{FEED}. Thus:

- $a_2 > a_1$ selectivity increases as conversion increases
- $a_2 < a_1$ selectivity decreases as conversion increases

If selectivity increases as conversion increases, the initial setting for reactor conversion should be in the order of 95%, and that for reversible reactions should be in the order of 95% of the equilibrium conversion. If selectivity decreases with increasing conversion, then it is much more difficult to give guidance. An initial setting of 50% for the conversion for irreversible reactions or 50% of the equilibrium conversion for reversible reactions is as reasonable as can be guessed at this stage. However, these are only initial guesses and will almost certainly be changed later.

3. *Multiple reactions in series producing byproduct.* Consider the system of series reactions from Equation 5.68. Selectivity for series reactions of the types given in Equation 5.7 to 5.9 is increased by low concentrations of reactants involved in the secondary reactions. In the preceding example, this means reactor operation with a low concentration of *PRODUCT*, in other words, with low conversion. For series reactions, a significant reduction in selectivity is likely as the conversion increases.

Again, it is difficult to select the initial setting of the reactor conversion with systems of reactions in series. A conversion of 50% for irreversible reactions or 50% of the equilibrium conversion for reversible reactions is as reasonable as can be guessed at this stage.

Multiple reactions also can occur with impurities that enter with the feed and undergo reaction. Again, such reactions should be minimized, but the most effective means of dealing with byproduct reactions caused by feed impurities is not to alter reactor conditions but to carry out feed purification.

5.8 CHOICE OF REACTOR PERFORMANCE – SUMMARY

Some initial guess for the reactor conversion must be made in order that the design can proceed. This is likely to change later in the design because, as will be seen later, there is a strong interaction between the reactor conversion and the rest of the flowsheet.

1. *Single reactions*. For single reactions, a good initial setting is 95% conversion for irreversible reactions and 95% of the equilibrium conversion for reversible reactions.

2. *Multiple reactions*. For multiple reactions where the byproduct is formed in parallel, the selectivity may increase or decrease as conversion increases. If the byproduct reaction is a higher order than the primary reaction, selectivity increases for increasing reactor conversion. In this case, the same initial setting as single reactions should be used. If the byproduct reaction of the parallel system is a lower order than the primary reaction, a lower conversion than that for single reactions is expected to be appropriate. The best initial guess at this stage is to set the conversion to 50% for irreversible reactions or to 50% of the equilibrium conversion for reversible reactions.

For multiple reactions where the byproduct is formed in series, the selectivity decreases as conversion increases. In this case, lower conversion than that for single reactions is expected to be appropriate. Again, the best guess at this stage is to set the conversion to 50% for irreversible reactions or to 50% of the equilibrium conversion for reversible reactions.

It should be emphasized that these recommendations for the initial settings of the reactor conversion will almost certainly change at a later stage, since reactor conversion is an extremely important optimization variable.

When dealing with multiple reactions, selectivity or reactor yield is maximized for the chosen conversion. The choice of mixing pattern in the reactor and feed addition policy should be chosen to this end.

5.9 EXERCISES

1. Acetic anhydride is to be produced from acetone and acetic acid. In the first stage of the process, acetone is decomposed at 700°C and 1.013 bar to ketene via the reaction:

$$\underset{\text{acetone}}{CH_3COCH_3} \longrightarrow \underset{\text{ketene}}{CH_2CO} + \underset{\text{methane}}{CH_4}$$

Unfortunately, some of the ketene formed decomposes further to form unwanted ethylene and carbon monoxide via the reaction:

$$\underset{\text{ketene}}{CH_2CO} \longrightarrow \underset{\text{ethylene}}{1/2C_2H_4} + \underset{\text{carbon monoxide}}{CO}$$

Laboratory studies of these reactions indicate that the ketene selectivity S (kmol ketene formed per kmol acetone converted) varies with conversion X (kmol acetone reacted per kmol acetone fed) follows the relationship[13]:

$$S = 1 - 1.3X$$

The second stage of the process requires the ketene to be reacted with glacial acetic acid at 80°C and 1.013 bar to produce acetic anhydride via the reaction:

$$\underset{\text{ketene}}{CH_2CO} + \underset{\text{acetic acid}}{CH_3COOH} \longrightarrow \underset{\text{acetic anhydride}}{CH_3COOCOCH_3}$$

The values of the chemicals involved, together with their molar masses are given in Table 5.12.

Table 5.12 Data for acetic anhydride production.

Chemical	Molar mass ($kg \cdot kmol^{-1}$)	Value ($\$ \cdot kg^{-1}$)
Acetone	58	0.60
Ketene	42	0
Methane	16	0
Ethylene	28	0
Carbon monoxide	28	0
Acetic acid	60	0.54
Acetic anhydride	102	0.90

Assuming that the plant will produce 15,000 $t \cdot y^{-1}$ acetic anhydride:

a. Calculate the economic potential assuming the side reaction can be suppressed and hence obtain 100% yield.
b. Determine the range of acetone conversions (X) over which the plant will be profitable if the side reaction cannot be suppressed.
c. Published data indicates that the capital cost for the project will be at least \$35 m. If annual fixed charges are assumed to be 15% of the capital cost, revise the range of conversions over which the plant will be profitable.

2. Experimental data for a simple reaction showing the rate of change of reactant with time are given to Table 5.13.

Table 5.13 Experimental data for a simple reaction.

Time (min)	Concentration ($kg \cdot m^{-3}$)
0	16.0
10	13.2
20	11.1
35	8.8
50	7.1

Show that the data gives a kinetic equation of order 1.5 and determine the rate constant.

3. Component A reacts to Component B in an irreversible reaction in the liquid phase. The kinetics are first order with

respect to A with reaction rate constant $k_A = 0.003$ min^{-1}. Find the residence times for 95% conversion of A for:

a. a mixed-flow reactor
b. 3 mixed-flow reactors in series of equal volume
c. a plug-flow reactor

4. Styrene (A) and Butadiene (B) are to be polymerized in a series of mixed-flow reactors, each of volume 25 m^3. The rate of reaction of both A and B are given by:

$$-r = kC_AC_B$$

where $k = 10^{-5}$ m^3·kmol^{-1}·s^{-1}
The initial concentration of styrene is 0.8 kmol·m^{-3} and of butadiene is 3.6 kmol·m^{-3}. The feed rate of reactants is 20 t·h^{-1}. Estimate the total number of reactors required for polymerization of 85% of the styrene. Assume the density of reaction mixture to be 870 kg·m^{-3}.

5. It is proposed to react 1 t·h^{-1} of a pure liquid A to a desired product B. Byproducts C and D are formed through series and parallel reactions:

$$A \xrightarrow{k_1} B \xrightarrow{k_2} C$$

$$B \xrightarrow{k_3} D$$

$$k_1 = k_2 = k_3 = 0.1 \text{ min}^{-1}$$

Assuming rates of reaction to be first order and an average density of 800 kg·m^{-3}, estimate the size of reactor that will give the maximum yield of B for:

a. a mixed-flow reactor
b. 3 equal-sized mixed-flow reactors in series
c. a plug-flow reactor.

6. What reactor configuration would you use to maximize the selectivity in the following parallel reactions:

$$A + B \longrightarrow R \quad r_R = 15C_A^{0.5}\,C_B$$

$$A + B \longrightarrow S \quad r_S = 15C_AC_B$$

where R is the desired product and S is the undesired product.

7. A desired liquid-phase reaction:

$$A + B \xrightarrow{k_1} R \quad r_R = k_1\,C_A^{0.3}\,C_B$$

is accompanied by a parallel reaction:

$$A + B \xrightarrow{k_2} S \quad r_S = k_2\,C_A^{1.5}\,C_B^{0.5}$$

a. A number of reactor configurations are possible. Ideal-batch, semi-batch, plug-flow and semi-plug-flow could all be used. In the case of the semi-batch and semi-plug-flow reactors, the order of addition of A and B can be changed. Order the reactor configurations from the least desirable to the most desirable to maximize production of the desired product.
b. The feed stream to a mixed-flow reactor is an equimolar mixture of pure A and B (each has a density of 20 kmol·m^{-3}). Assuming $k_1 = k_2 = 1.0$ kmol·m^{-3}·min^{-1}, calculate the composition of the exit stream from the mixed-flow reactor for a conversion of 90%.

8. For a free-radical polymerization and a condensation polymerization process, explain why the molar mass distribution of the polymer product will be different depending on whether a mixed-flow or a plug-flow reactor is used. What will be the difference in the distribution of molar mass?

REFERENCES

1. Levenspiel O (1999) *Chemical Reaction Engineering*, 3rd Edition, John Wiley.
2. Denbigh KG and Turner JCR (1984) *Chemical Reactor Theory*, 3rd Edition, Cambridge University Press.
3. Rase HF (1977) *Chemical Reactor Design for Process Plants*, Vol. 1, John Wiley.
4. Rudd DF, Powers GJ and Siirola JJ (1973) *Process Synthesis*, Prentice Hall.
5. Waddams AL (1978) *Chemicals From Petroleum*, John Murray.
6. Shuler ML and Kargi F (2002) *Bioprocess Engineering*, 2nd Edition, Prentice Hall
7. Wells GL and Rose LM (1986) *The Art of Chemical Process Design*, Elsevier.
8. Douglas JM (1985) A Hierarchical Decision Procedure for Process Synthesis, *AIChE J*, **31**: 353.
9. Huang I and Dauerman L (1969) Exploratory Process Study, *Ind Eng Chem Prod Res Dev*, **8**(3): 227.
10. Helminen J, Leppamaki M, Paatero E and Minkkinen (1998) Monitoring the Kinetics of the Ion-exchange Resin Catalysed Esterification of Acetic Acid with Ethanol Using Near Infrared Spectroscopy with Partial Least Squares (PLS) Model, *Chemometr Intell Lab Syst*, **44**: 341.
11. Dwight HB (1961) *Tables of Integrals and Other Mathematical Data*, The Macmillan Company.
12. Smith R and Petela EA (1992) Waste Minimization in the Process Industries Part 2 Reactors, *Chem Eng*, **Dec**(509–510): 17.
13. Jeffreys GV (1964) *A Problem in Chemical Engineering Design – The Manufacture of Acetic Acid*, The Institution of Chemical Engineers, UK.

6 Choice of Reactor II - Reactor Conditions

6.1 REACTION EQUILIBRIUM

In the preceding chapter, the choice of reactor type was made on the basis of the most appropriate concentration profile as the reaction progressed, in order to minimize reactor volume for single reactions or maximize selectivity (or yield) for multiple reactions for a given conversion. However, there are still important effects regarding reaction conditions to be considered. Before considering reaction conditions, some basic principles of chemical equilibrium need to be reviewed.

Reactions can be considered to be either reversible or essentially irreversible. An example of a reaction that is essentially irreversible is:

$$\underset{\text{ethylene}}{C_2H_4} + \underset{\text{chlorine}}{Cl_2} \longrightarrow \underset{\text{dichloroethane}}{C_2H_4Cl_2} \tag{6.1}$$

An example of a reversible reaction is:

$$\underset{\text{hydrogen}}{3H_2} + \underset{\text{nitrogen}}{N_2} \rightleftharpoons \underset{\text{ammonia}}{2NH_3} \tag{6.2}$$

For reversible reactions:

a. For a given mixture of reactants at a given temperature and pressure, there is a maximum conversion (the *equilibrium conversion*) that cannot be exceeded and is independent of the reactor design.
b. The equilibrium conversion can be changed by appropriate changes to the concentrations of reactants, temperature and pressure.

The key to understanding reaction equilibrium is the *Gibbs free energy*, or *free energy*, defined as[1-4]:

$$G = H - TS \tag{6.3}$$

where G = free energy (kJ)
H = enthalpy (kJ)
T = absolute temperature (K)
S = entropy ($kJ \cdot K^{-1}$)

Like enthalpy, the absolute value of G for a substance cannot be measured, and only changes in G are meaningful. For a process occurring at constant temperature, the change in free energy from Equation 6.3 is:

$$\Delta G = \Delta H - T\,\Delta S \tag{6.4}$$

A negative value of ΔG implies a spontaneous reaction of reactants to products. A positive value of ΔG implies the reverse reaction is spontaneous. This is illustrated in Figure 6.1. A system is in equilibrium when[1-4]:

$$\Delta G = 0 \tag{6.5}$$

The free energy can also be expressed in differential form[1-4]:

$$dG = -SdT + VdP \tag{6.6}$$

where V is the system volume. At constant temperature, $dT = 0$ and Equation 6.6 becomes:

$$dG = VdP \tag{6.7}$$

For N moles of an ideal gas $PV = NRT$, where R is the universal gas constant. Substituting this in Equation 6.7 gives:

$$dG = NRT\frac{dP}{P} \tag{6.8}$$

For a change in pressure from P_1 to P_2 Equation 6.8 can be integrated to give:

$$\Delta G = G_2 - G_1 = NRT \int_{P_1}^{P_2} \frac{dP}{P} \tag{6.9}$$

Values for free energy are usually referred to the standard free energy G^O. The standard state is arbitrary and designates the datum level. A gas is considered here to be at a standard state if it is at a pressure of 1 atm or 1 bar for the designated temperature of an isothermal process. Thus, integrating Equation 6.9 from standard pressure P_O to pressure P gives:

$$G = G^O + NRT \ln \frac{P}{P_O} \tag{6.10}$$

For a real system, rather than an ideal gas, Equation 6.10 is written as:

$$\begin{aligned} G &= G^O + NRT \ln \frac{f}{f^O} \\ &= G^O + NRT \ln a \end{aligned} \tag{6.11}$$

where f = fugacity ($N \cdot m^2$ or bar)
f^O = fugacity at standard conditions ($N \cdot m^2$ or bar)
a = activity (–)
$= f/f^O$

The concepts of *fugacity* and *activity* have no strict physical significance but are introduced to transform

Chemical Process Design and Integration R. Smith

ISBNs: 0-471-48680-9 (HB); 0-471-48681-7 (PB)

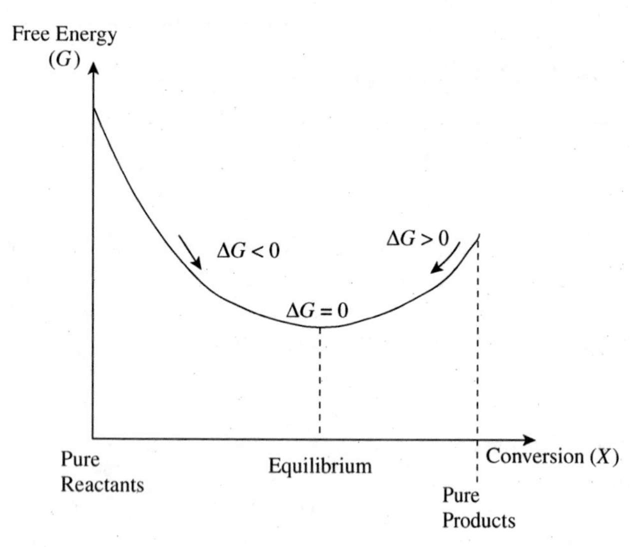

Figure 6.1 Variation of free energy of a reaction mixture.

equations for ideal systems to real systems. However, it does help to relate them to something that does have physical significance. Fugacity can be regarded as an "effective pressure", and activity can be regarded as an "effective concentration" relative to a standard state[4].

Consider the general reaction:

$$bB + cC + \dots \rightleftharpoons sS + tT + \dots \quad (6.12)$$

The free energy change for the reaction is given by:

$$\Delta G = s\overline{G}_S + t\overline{G}_T + \dots - b\overline{G}_B - c\overline{G}_C - \dots \quad (6.13)$$

where ΔG = free energy change of reaction (kJ)
$\overline{G}_i$ = partial molar free energy of Component i ($kJ{\cdot}kmol^{-1}$)

Note that the *partial molar free energy* is used to designate the molar free energy of the individual component *as it exists in the mixture.* This is necessary because, except in ideal systems, the properties of a mixture are not additive properties of the pure components. As a result, $G = \sum_i N_i\overline{G}_i$ for a mixture. The free energy change for reactants and products at standard conditions is given by:

$$\Delta G^O = s\overline{G}_S^O + t\overline{G}_T^O + \dots - b\overline{G}_B^O - c\overline{G}_C^O - \dots \quad (6.14)$$

where ΔG^O = free energy change of reaction when all of the reactants and products are at their respective standard conditions (kJ)
$\overline{G}_i^O$ = partial molar free energy of Component i at standard conditions ($kJ{\cdot}kmol^{-1}$)

Combining Equations 6.13 and 6.14:

$$\Delta G - \Delta G^O = s(\overline{G}_S - \overline{G}_S^O) + t(\overline{G}_T - \overline{G}_T^O) + \dots - b(\overline{G}_B - \overline{G}_B^O) - c(\overline{G}_c - \overline{G}_C^O) - \dots \quad (6.15)$$

Writing Equation 6.11 for the partial molar free energy of Component i in a mixture:

$$\overline{G}_i - \overline{G}_i^O = RT \ln a_i \quad (6.16)$$

where a_i refers to the activity of Component i in the mixture. Substituting Equation 6.16 in Equation 6.15 gives:

$$\Delta G - \Delta G^O = sRT \ln a_S + tRT \ln a_T + \dots - bRT \ln a_B - cRT \ln a_c - \dots \quad (6.17)$$

Since $s \ln a_S = \ln a_S^s$, $t \ln a_T = \ln a_T^t \dots$, Equation 6.17 can be arranged to give:

$$\Delta G - \Delta G^O = RT \ln \left(\frac{a_S^s a_T^t \dots}{a_B^b a_C^c \dots} \right) \quad (6.18)$$

Note that in rearranging $s \ln a_S$ to $\ln a_S^s$, the units of $RT \ln a_S^s$ remain kJ and a_S^s is dimensionless. At equilibrium $\Delta G = 0$ and Equation 6.18 becomes:

$$\Delta G^O = -RT \ln \left(\frac{a_S^s\ a_T^t \dots}{a_B^b\ a_C^c \dots} \right) = -RT \ln K_a \quad (6.19)$$

$$\text{where } K_a = \frac{a_S^s a_T^t \dots}{a_B^b a_C^c \dots} \quad (6.20)$$

K_a is known as the equilibrium constant. It represents the equilibrium activities for a system under standard conditions and is a constant at constant temperature.

1. *Homogeneous gaseous reactions*
Substituting for the fugacity in Equation 6.20 gives:

$$K_a = \left(\frac{f_S^s\ f_T^t \dots}{f_B^b\ f_C^c \dots} \right) \left(\frac{f_B^{Ob}\ f_C^{Oc} \dots}{f_S^{Os}\ f_T^{To} \dots} \right) \quad (6.21)$$

For homogeneous gaseous reactions, the standard state fugacities can be considered to be unity, that is, $f_i^O = 1$. If the fugacity is expressed as the product of partial pressure and fugacity coefficient[1–4]:

$$f_i = \phi_i\ p_i \quad (6.22)$$

where f_i = fugacity of Component i
ϕ_i = fugacity coefficient of Component i
p_i = partial pressure of Component i

Substituting Equation 6.22 into Equation 6.21, assuming $f_i^O = 1$:

$$K_a = \left(\frac{\phi_S^s\ \phi_T^t \dots}{\phi_B^b\ \phi_C^c \dots} \right) \left(\frac{p_S^s\ p_T^t \dots}{p_B^b\ p_C^c \dots} \right) = K_\phi K_P \quad (6.23)$$

Also, by definition (see Chapter 4):

$$p_i = y_i P \tag{6.24}$$

where y_i = mole fraction of Component i
P = system pressure

Substituting Equation 6.24 into Equation 6.23 gives:

$$K_a = \left(\frac{\phi_S^s \, \phi_T^t \cdots}{\phi_B^b \, \phi_C^c \cdots}\right)\left(\frac{y_S^s \, y_T^t \cdots}{y_B^b \, y_C^c \cdots}\right) P^{\Delta N} = K_\varphi K_y P^{\Delta N} \tag{6.25}$$

where $\Delta N = s + t + \ldots - b - c - \ldots$

For an ideal gas $\phi_i = 1$, thus:

$$K_a = K_p = \left(\frac{p_S^s \, p_T^t \cdots}{p_B^b \, p_C^c \cdots}\right) \tag{6.26}$$

$$K_a = K_y P^{\Delta N} = \left(\frac{y_S^s \, y_T^t \cdots}{y_B^b \, y_C^c \cdots}\right) P^{\Delta N} \tag{6.27}$$

2. *Homogeneous liquid reactions*
For homogeneous liquid reactions, the activity can be expressed as[1–4]

$$a_i = \gamma_i x_i \tag{6.28}$$

where a_i = activity of Component i
γ_i = activity coefficient for Component i
x_i = mole fraction of Component i

The activity coefficient in Equation 6.28 ($\gamma_i = a_i/x_i$) can be considered to be the ratio of the effective to actual concentration. Substituting Equation 6.28 into Equation 6.20 gives:

$$K_a = \left(\frac{\gamma_S^s \, \gamma_T^t \cdots}{\gamma_B^b \, \gamma_C^c \cdots}\right)\left(\frac{x_S^s \, x_T^t \cdots}{x_B^b \, x_C^c \cdots}\right) = K_\gamma K_x \tag{6.29}$$

For ideal solutions, the activity coefficients are unity, giving:

$$K_a = \left(\frac{x_S^s x_T^t \cdots\cdots}{x_B^b x_C^c \cdots\cdots}\right) \tag{6.30}$$

3. *Heterogeneous reactions*
For a heterogeneous reaction, the state of all components is not uniform, for example, a reaction between a gas and a liquid. This requires standard states to be defined for each component. The activity of a solid in the equilibrium constant can be taken to be unity.

Consider now an example to illustrate the application of these thermodynamic principles.

Example 6.1 A stoichiometric mixture of nitrogen and hydrogen is to be reacted at 1 bar:

$$\underset{\text{hydrogen}}{3H_2} + \underset{\text{nitrogen}}{N_2} \rightleftharpoons \underset{\text{ammonia}}{2NH_3}$$

Assuming ideal gas behavior ($R = 8.3145$ kJ·K^{-1}·kmol^{-1}), calculate:

a. equilibrium constant
b. equilibrium conversion of hydrogen
c. composition of the reaction products at equilibrium

at 300 K. Standard free energy of formation data are given in Table 6.1[5].

Table 6.1 Standard free energy of formation data for ammonia synthesis.

	$\overline{G}_{300}^{O}$ (kJ·kmol^{-1})
H_2	0
N_2	0
NH_3	−16,223

Solution

a.

$$3H_2 + N_2 \rightleftharpoons 2NH_3$$

$$\Delta G^O = 2\overline{G}_{NH_3}^O - 3\overline{G}_{H_2}^O - \overline{G}_{N_2}^O$$
$$= -RT \ln K_a$$
$$\ln K_a = \frac{-(2\overline{G}_{NH_3}^O - 3\overline{G}_{H_2}^O - \overline{G}_{N_2}^O)}{RT}$$

At 300 K:

$$\ln K_a = \frac{-(-2 \times 16223 - 0 - 0)}{8.3145 \times 300}$$
$$= 13.008$$
$$K_a = 4.4597 \times 10^5$$

Also:

$$K_a = \frac{p_{NH_3}^2}{p_{H_2}^3 p_{N_2}}$$

Note that whilst K_a appears to be dimensional, it is actually dimensionless, since $f_i^O = 1$ bar. Note also that K_a depends on the specification of the number of moles in the stoichiometric equation. For example, if the stoichiometry is written as:

$$\frac{3}{2}H_2 + \frac{1}{2}N_2 \rightleftharpoons NH_3$$

$$K_a' = \frac{p_{NH_3}}{p_{H_2}^{3/2} p_{N_2}^{1/2}}$$

$$K_a' = \sqrt{K_a}$$

It does not matter which specification is adopted as long as one specification is used consistently.

b. The number of moles and mole fractions, initially and at equilibrium, are given in Table 6.2.

$$K_a = \frac{y_{NH_3}^2}{y_{H_2}^3 y_{N_2}} P^{-2}$$

For $P = 1$ bar:

$$K_a = \frac{16X^2(2 - X)^2}{27(1 - X)^4}$$

At 300 K:

$$4.4597 \times 10^5 = \frac{16X^2(2-X)^2}{27(1-X)^4}$$

This can be solved by trial and error or using spreadsheet software (see Section 3.9):

$$X = 0.97$$

c. To calculate the composition of the reaction products at equilibrium, $X = 0.97$ is substituted in the expressions from Table 6.2 at equilibrium:

$$y_{H_2} = 0.0437$$

$$y_{N_2} = 0.0146$$

$$y_{NH_3} = 0.9418$$

Table 6.2 Moles and mole fractions for ammonia synthesis.

	H_2	N_2	NH_3
Moles in initial mixture	3	1	0
Moles at equilibrium	$3-3X$	$1-X$	$2X$
Mole fraction at equilibrium	$\frac{3(1-X)}{4-2X}$	$\frac{1-X}{4-2X}$	$\frac{2X}{4-2X}$

Some points should be noted from this example. Firstly, ideal gas behavior has been assumed. This is an approximation, but it is reasonable for the low pressure assumed in the calculation. Later the calculation will be repeated at higher pressure when the ideal gas approximation will be poor. Also, it should be clear that the calculation is very sensitive to the thermodynamic data. Errors in the thermodynamic data can lead to a significantly different result. Thermodynamic data, even from reputable sources, should be used with caution.

Equation 6.19 can be interpreted qualitatively to give guidance on the equilibrium conversion. If ΔG^O is negative, the position of the equilibrium will correspond to the presence of more products than reactants ($\ln K_a > 0$). If ΔG^O is positive ($\ln K_a < 0$), the reaction will not proceed to such an extent and reactants will predominate in the equilibrium mixture. Table 6.3 presents some guidelines as to the composition of the equilibrium mixture for various values of ΔG^O and the equilibrium constant[4].

Table 6.3 Variation of equilibrium composition with ΔG^O and the equilibrium constant at 298 K.

ΔG^O (kJ)	K_a	Composition of equilibrium mixture
−50,000	6×10^8	Negligible reactants
−10,000	57	Products dominate
−5000	7.5	
0	1.0	
+5000	0.13	
+10,000	0.02	Reactants dominate
+50,000	1.7×10^{-9}	Negligible products

Source: Reproduced from Smith EB, 1982, *Basic Chemical Thermodynamics*, 3rd Edition, by permission of Oxford University Press.

When setting the conditions in chemical reactors, equilibrium conversion will be a major consideration for reversible reactions. The equilibrium constant K_a is only a function of temperature, and Equation 6.19 provides the quantitative relationship. However, pressure change and change in concentration can be used to shift the equilibrium by changing the activities in the equilibrium constant, as will be seen later.

A basic principle that allows the qualitative prediction of the effect of changing reactor conditions on any chemical system in equilibrium is *Le Châtelier's Principle*:

> "If any change in the conditions of a system in equilibrium causes the equilibrium to be displaced, the displacement will be in such a direction as to oppose the effect of the change."

Le Châtelier's Principle allows changes to be directed to increase equilibrium conversion. Now consider the setting of conditions in chemical reactors.

6.2 REACTOR TEMPERATURE

The choice of reactor temperature depends on many factors. Consider first the effect of temperature on equilibrium conversion. A quantitative relationship can be developed as follows. Start by writing Equation 6.6 at constant pressure:

$$dG = -S\,dT \qquad (6.31)$$

Equation 6.31 can be written as:

$$\left(\frac{\partial G}{\partial T}\right)_P = -S \qquad (6.32)$$

Substituting Equation 6.31 into Equation 6.3 gives, after rearranging:

$$\left(\frac{\partial G}{\partial T}\right)_P = \frac{G}{T} - \frac{H}{T} \qquad (6.33)$$

At standard conditions G and H are not functions of pressure, by definition. Thus, Equation 6.33 can be written at standard conditions for finite changes in G^O and H^O as:

$$\frac{d\Delta G^O}{dT} = \frac{\Delta G^O}{T} - \frac{\Delta H^O}{T} \qquad (6.34)$$

Also, because:

$$\frac{d}{dT}\left(\frac{\Delta G^O}{T}\right) = \frac{1}{T}\frac{d\Delta G^O}{dT} + \Delta G^O\frac{d(1/T)}{dT}$$

$$= \frac{1}{T}\frac{d\Delta G^O}{dT} - \frac{\Delta G^O}{T^2} \qquad (6.35)$$

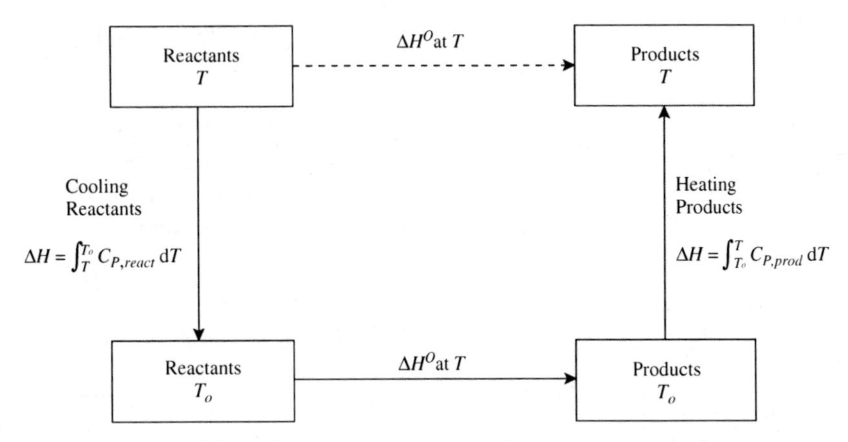

Figure 6.2 Calculation of standard heat of formation at any temperature from data at standard temperature.

Combining Equations 6.34 and 6.35 gives:

$$\frac{d}{dT}\left(\frac{\Delta G^O}{T}\right) = -\frac{\Delta H^O}{T^2} \tag{6.36}$$

Substituting ΔG^O (kJ) from Equation 6.19 into Equation 6.36 gives:

$$\frac{d \ln K_a}{dT} = \frac{\Delta H^O}{RT^2} \tag{6.37}$$

Equation 6.37 can be integrated to give:

$$\ln K_{a2} - \ln K_{a1} = \frac{1}{R}\int_{T_1}^{T_2} \frac{\Delta H^O}{T^2} dT \tag{6.38}$$

If it is assumed that ΔH^O is independent of temperature, Equation 6.38 gives

$$\ln \frac{K_{a2}}{K_{a1}} = -\frac{\Delta H^O}{R}\left(\frac{1}{T_2} - \frac{1}{T_1}\right) \tag{6.39}$$

In this expression, K_{a1} is the equilibrium constant at T_1, and K_{a2} is the equilibrium constant at T_2. ΔH^O is the standard heat of reaction (kJ) when all the reactants and products are at standard state, given by:

$$\Delta H^O = (s\overline{H}_S^O + t\overline{H}_T^O + \ldots) - (b\overline{H}_B^O + c\overline{H}_C^O + \ldots) \tag{6.40}$$

where $\overline{H}_i^O$ is the molar standard enthalpy of formation of Component i (kJ·kmol^{-1}).

Equation 6.39 implies that a plot of ($\ln K_a$) against ($1/T$) should be a straight line with a slope of ($\Delta H^O/R$). For an exothermic reaction, $\Delta H^O < 0$ and K_a decreases as temperature increases. For an endothermic reaction, $\Delta H^O > 0$ and K_a increases with increasing temperature.

Equation 6.39 can be used to estimate the equilibrium constant at the required temperature given enthalpy of formation data at some other temperature. Enthalpy of formation data is usually available at standard temperature, and therefore Equation 6.39 can be used to estimate the equilibrium constant at the required temperature given data at standard temperature. However, Equation 6.39 assumes ΔH^O is constant. If data is available for ΔH^O at standard temperature, together with heat capacity data, the equilibrium constant can be calculated more accurately using the thermodynamic path shown in Figure 6.2. Thus[1–4]:

$$\Delta H_T^O = \Delta H_{T_O}^O + \int_{T_O}^{T} C_{P,prod} dT - \int_{T_O}^{T} C_{P,react} dT \tag{6.41}$$

where $\Delta H_{T_O}^O$ = standard enthalpy of formation at T_O
$C_{P,prod}$ = heat capacity of reaction products as a function of temperature
$C_{P,react}$ = heat capacity of reactants as a function of temperature

Heat capacity data are often available in some form of polynomial as a function of temperature, for example[6]:

$$\frac{C_P}{R} = \alpha_o + \alpha_1 T + \alpha_2 T^2 + \alpha_3 T^3 + \alpha_4 T^4 \tag{6.42}$$

where
C_P = heat capacity (kJ·K^{-1}·kmol^{-1})
R = gas constant (kJ·K^{-1}·kmol^{-1})
T = absolute temperature (K)
$\alpha_0, \alpha_1, \alpha_2, \alpha_3, \alpha_4$ = constants determined by fitting experimental data

Thus, Equation 6.41 can be written as:

$$\Delta H_T^O = \Delta H_{T_O}^O + \int_{T_O}^{T} \Delta C_P dT \tag{6.43}$$

For the general reaction given in Equation 6.12:

$$\frac{\Delta C_P}{R} = \Delta\alpha_0 + \Delta\alpha_1 T + \Delta\alpha_2 T^2 + \Delta\alpha_3 T^3 + \Delta\alpha_4 T^4 \tag{6.44}$$

where

$$\Delta\alpha_0 = s\alpha_0 + t\alpha_0 + \ldots - b\alpha_0 - c\alpha_0 - \ldots$$

$$\Delta\alpha_1 = s\alpha_1 + t\alpha_1 + \ldots - b\alpha_1 - c\alpha_1 - \ldots$$

and so on.

From Equations 6.43 and 6.44:

$$\Delta H_T^O = \Delta H_{T_O}^O + R\int_{T_O}^{T}(\Delta\alpha_0 + \Delta\alpha_1 T + \Delta\alpha_2 T^2 + \Delta\alpha_3 T^3 + \Delta\alpha_4 T^4)dT$$

$$= \Delta H_{T_O}^O + R\left[\Delta\alpha_0 T + \frac{\Delta\alpha_1 T^2}{2} + \frac{\Delta\alpha_2 T^3}{3} + \frac{\Delta\alpha_3 T^4}{4} + \frac{\Delta\alpha_4 T^5}{5}\right]_{T_O}^{T}$$

$$= \Delta H_{T_O}^O + R\left(\Delta\alpha_0 T + \frac{\Delta\alpha_1 T^2}{2} + \frac{\Delta\alpha_2 T^3}{3} + \frac{\Delta\alpha_3 T^4}{4} + \frac{\Delta\alpha_4 T^5}{5}\right) - I \tag{6.45}$$

where

$$I = R\left(\Delta\alpha_0 T_O + \frac{\Delta\alpha_1 T_O^2}{2} + \frac{\Delta\alpha_2 T_O^3}{3} + \frac{\Delta\alpha_3 T_O^4}{4} + \frac{\Delta\alpha_4 T_O^5}{5}\right) \tag{6.46}$$

Substituting Equation 6.45 into Equation 6.38 gives:

$$[\ln K_a]_{K_{aT_O}}^{K_{aT}} = \frac{1}{R}\int_{T_O}^{T}\left(\frac{\Delta H_{T_O}^O}{T^2} + \frac{R}{T^2}\left[\Delta\alpha_0 T + \frac{\Delta\alpha_1 T^2}{2} + \frac{\Delta\alpha_2 T^3}{3} + \frac{\Delta\alpha_3 T^4}{4} + \frac{\Delta\alpha_4 T^5}{5}\right] - \frac{I}{T^2}\right)dT$$

$$\ln K_{aT} - \ln K_{aT_O} = \int_{T_O}^{T}\left(\frac{\Delta H_{T_O}^O}{RT^2} - \frac{I}{RT^2} + \frac{\Delta\alpha_0}{T} + \frac{\Delta\alpha_1}{2} + \frac{\Delta\alpha_2 T}{3} + \frac{\Delta\alpha_3 T^2}{4} + \frac{\Delta\alpha_4 T^3}{5}\right)dT$$

$$\ln\frac{K_{aT}}{K_{aT_O}} = \left[-\frac{\Delta H_{T_O}^O}{RT} + \frac{I}{RT} + \Delta\alpha_0 \ln T + \frac{\Delta\alpha_1 T}{2} + \frac{\Delta\alpha_2 T^2}{6} + \frac{\Delta\alpha_3 T^3}{12} + \frac{\Delta\alpha_4 T^4}{20}\right]_{T_O}^{T} \tag{6.47}$$

Substituting for K_{aT_O}:

$$\ln K_{aT} = \left[-\frac{\Delta H_{T_O}^O}{RT} + \frac{I}{RT} + \Delta\alpha_0 \ln T + \frac{\Delta\alpha_1 T}{2} + \frac{\Delta\alpha_2 T^2}{6} + \frac{\Delta\alpha_3 T^3}{12} + \frac{\Delta\alpha_4 T^4}{20}\right]_{T_O}^{T} - \frac{\Delta G_{T_O}^O}{RT_O} \tag{6.48}$$

Given $\Delta G_{T_O}^O$, $\Delta H_{T_O}^O$ and α_0, α_1, α_2, α_3, α_4, for each component, Equation 6.48 is used to calculate $K_{a,T}$.

Alternatively, Equation 6.48 can be written for standard conditions at temperature T:

$$\Delta G_T^O = \Delta H_T^O - T\Delta S_T^O \tag{6.49}$$

Substituting Equation 6.19 into Equation 6.47 gives:

$$\ln K_{aT} = \frac{\Delta S_T^O}{R} - \frac{\Delta H_T^O}{RT} \tag{6.50}$$

The analogous expression to Equation 6.43 for ΔH_T^O for ΔS_T^O is given by[1–4]:

$$\Delta S_T^O = \Delta S_{T_O}^O + \int_{T_O}^{T}\frac{\Delta Cp}{T}dT \tag{6.51}$$

Substituting Equation 6.44:

$$\Delta S_T^O = \Delta S_{T_O}^O + \int_{T_O}^{T}\frac{R}{T}(\Delta\alpha_0 + \Delta\alpha_1 T + \Delta\alpha_2 T^2 + \Delta\alpha_3 T^3 + \Delta\alpha_4 T^4)dT$$

$$= \Delta S_{T_O}^O + R\int_{T_O}^{T}\left(\frac{\Delta\alpha_0}{T} + \Delta\alpha_1 + \Delta\alpha_2 T + \Delta\alpha_3 T^2 + \Delta\alpha_4 T^3\right)dT$$

$$= \Delta S_{T_O}^O + R\left[\Delta\alpha_0 \ln T + \Delta\alpha_1 T + \frac{\Delta\alpha_2 T^2}{2} + \frac{\Delta\alpha_3 T^3}{3} + \frac{\Delta\alpha_4 T^4}{4}\right]_{T_O}^{T} \tag{6.52}$$

Thus ΔH_T^O can be calculated from Equation 6.45 and ΔS_T^O from Equation 6.52 and the results substituted in Equation 6.50.

Example 6.2 Following Example 6.1:

a. Calculate $\ln(K_a)$ at 300 K, 400 K, 500 K, 600 K and 700 K at 1 bar and test the validity of Equation 6.39. Standard free

Table 6.4 Thermodynamic data for ammonia at various temperatures[5].

T(K)	$\overline{G}^O_{NH_3}$ (kJ·kmol^{-1})	$\overline{H}^O$ (kJ·kmol^{-1})
298.15	−16,407	−45,940
300	−16,223	−45,981
400	−5980	−48,087
500	4764	−49,908
600	15,846	−51,430
700	27,161	−52,682

Table 6.5 Heat capacity data[6].

	C_P (kJ·kmol^{-1}·K^{-1})				
	α_0	$\alpha_1 \times 10^3$	$\alpha_2 \times 10^5$	$\alpha_3 \times 10^8$	$\alpha_4 \times 10^{11}$
H_2	2.883	3.681	−0.772	0.692	−0.213
N_2	3.539	−0.261	0.007	0.157	−0.099
NH_3	4.238	−4.215	2.041	−2.126	0.761

energy of formation and enthalpy of formation data for NH_3 are given in Table 6.4[6]. Free energy of formation data for H_2 and N_2 is zero.

b. Calculate the values of $\ln(K_a)$ from Equation 6.39 and Equation 6.48 from standard data at 298.15 K and compare with values calculated from Table 6.4. Heat capacity coefficients are given in Table 6.5[6].

c. Determine the effect of temperature on equilibrium conversion of hydrogen using the data in Table 6.4.

Again assume ideal gas behavior and $R = 8.3145$ kJ·K^{-1}·kmol^{-1}.

Solution

a. $\ln K_a = -(2\overline{G}^O_{NH_3} - 3\overline{G}^O_{H_2} - \overline{G}^O_{N_2})/RT$

At		
	300 K	$\ln K_a = 13.008$ (from Example 6.1)
	400 K	$\ln K_a = 3.5961$
	500 K	$\ln K_a = -2.2919$
	600 K	$\ln K_a = -6.3528$
	700 K	$\ln K_a = -9.3334$

Figure 6.3a shows a plot of $\ln K_a$ versus $1/T$. This is a straight line and appears to be in good agreement with Equation 6.39. From the slope of the graph it can be deduced that ΔH^O has a value of −97,350 kJ. The standard heat of reaction for ammonia synthesis is given by:

$$\begin{aligned}\Delta H^O &= 2\overline{H}^O_{NH_3} - 3\overline{H}^O_{H_2} - \overline{H}^O_{N_2}\\ &= 2\overline{H}^O_{NH_3} - 0 - 0\end{aligned}$$

This implies a standard enthalpy of formation of −48,675 kJ·kmol^{-1}. From standard enthalpy of formation data, ΔH^O varies from −45,981 kJ·kmol^{-1} at 300 K to −52,682 kJ·kmol^{-1} at 700 K, with an average value over the range of −49,332 kJ·kmol^{-1}. This again appears to be in good agreement with the average value from Figure 6.3a. However, this is on the basis of averages across the range of temperature. In the next calculation, the accuracy of Equation 6.37 will be examined in more detail.

b. As a datum, first calculate $\ln K_{aT}$ from the tabulated data for $\overline{G}^O$ listed in Table 6.4:

$$\ln K_{aT} = -\frac{\Delta G^O_T}{RT}$$

where $\Delta G^O_T = 2\overline{G}^O_{NH_3,T} - 3\overline{G}^O_{H_2,T} - \overline{G}^O_{N_2,T}$

Substituting values of $\overline{G}^O$ and T leads to the results in Table 6.6. Next calculate K_{aT} for a range of temperatures from Equation 6.37:

$$\begin{aligned}\ln K_{aT} &= -\frac{\Delta H^O_{T_O}}{R}\left(\frac{1}{T} - \frac{1}{T_O}\right) + \ln K_{aT_O}\\ &= -\frac{\Delta H^O_{T_O}}{R}\left(\frac{1}{T} - \frac{1}{T_O}\right) - \frac{\Delta G^O_{T_O}}{RT_O}\\ \Delta G^O &= 2\overline{G}^O_{NH_3} - 3\overline{G}^O_{H_2} - \overline{G}^O_{N_2}\\ \Delta G_{T_O} &= 2 \times (-16{,}407) - 0 - 0 = 32{,}814 \text{ kJ}\\ \Delta H^O &= 2\overline{H}^O_{NH_3} - 3\overline{H}^O_{H_2} - \overline{H}^O_{N_2}\\ \Delta H^O_{T_O} &= 2 \times (-45{,}940) - 0 - 0 = -91{,}880 \text{ kJ}\\ \ln K_{aT} &= -\frac{-91{,}880}{8.3145}\left(\frac{1}{T} - \frac{1}{298.15}\right) - \frac{-32{,}814}{8.3145 \times 298.15}\end{aligned}$$

Substituting the values of T leads to the results in Table 6.6. Next calculate K_{aT} from Equation 6.46 using the coefficients in Table 6.5:

$$\Delta\alpha_i = 2\alpha_{iNH_3} - 3\alpha_{iH_2} - \alpha_{iN_2}$$

Thus

$$\Delta\alpha_0 = -3.7120$$

$$\Delta\alpha_1 = -1.9212 \times 10^{-2}$$

(a) Plot of ln*K* versus 1/*T*.

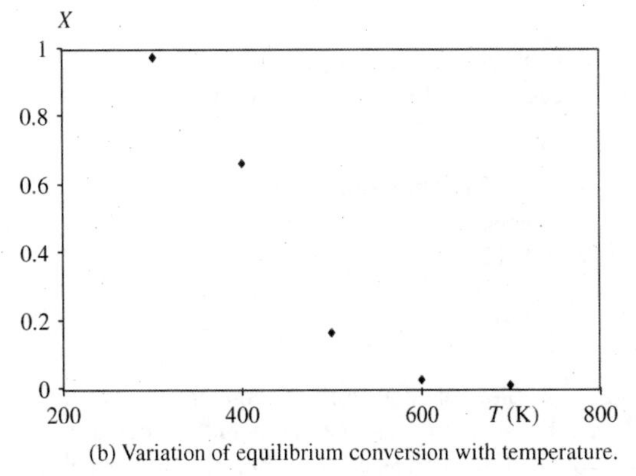

(b) Variation of equilibrium conversion with temperature.

Figure 6.3 The effect of temperature on the ammonia synthesis reaction.

Table 6.6 Comparison of the methods to calculate $\ln K_{aT}$ at various temperatures.

T (K)	$\ln K_{aT}$		
	$-\frac{\Delta G_T^O}{RT}$ G_T^O from Table 6.4	$-\frac{\Delta H_{T_0}}{R} \times \left(\frac{1}{T} - \frac{1}{T_0}\right) - \frac{\Delta G_T^O}{RT_0}$	Equation 6.48
300	13.0078	13.0084	13.0083
400	3.5961	3.7996	3.5977
500	−2.2919	−1.7257	−2.2891
600	−6.3528	−6.4092	−6.3497
700	−9.3334	−8.0403	−9.3300

$$\Delta\alpha_2 = 6.3910 \times 10^{-5}$$

$$\Delta\alpha_3 = -6.4850 \times 10^{-8}$$

$$\Delta\alpha_4 = 2.2600 \times 10^{-11}$$

Substitute these values in Equation 6.46:

$$I = -12{,}583 \text{ kJ·kmol}^{-1}$$

Substituting the values of temperature into Equation 6.48 leads to the results in Table 6.6. From Table 6.6, it can be seen that calculation of $\ln K_{aT}$ from heat capacity data on the basis of $\Delta H_{T_0}^O$ maintains a good agreement with $\ln K_{aT}$ calculated from tabulated values of $\overline{G}_T^O$ even for large ranges in temperature. On the other hand, calculation from Equation 6.39 assuming a constant value of $\Delta H_{T_0}^O$ can lead to significant errors when extrapolated over large ranges of temperature.

c. At 300 K $K_a = 4.4597 \times 10^5$ (from Example 6.1)
400 K $K_a = 36.456$
500 K $K_a = 0.10107$
600 K $K_a = 1.7419 \times 10^{-3}$
700 K $K_a = 8.8421 \times 10^{-5}$

Also from Example 6.1:

$$K_a = \frac{16X^2(2-X)^2}{27(1-X)^4}$$

Substitute K_a and solve for X. This can again be conveniently carried out in spreadsheet software (see Section 3.9):

At 300 K $X = 0.97$ (from Example 6.1)
400 K $X = 0.66$
500 K $X = 0.16$
600 K $X = 0.026$
700 K $X = 0.0061$

Figure 6.3b shows a plot of equilibrium conversion versus temperature. It can be seen that as the temperature increases, the equilibrium conversion decreases (for this reaction). This is consistent with the fact that this is an exothermic reaction. However, it should not necessarily be concluded that the reactor should be operated at low temperature, as the rate of reaction has yet to be considered. Also, catalysts and the deactivation of catalysts have yet to be considered.

It should also be again noted that firstly ideal gas behavior has been assumed, which is reasonable at this pressure, and secondly that small data errors might lead to significant errors in the calculations.

Example 6.2 shows that for an exothermic reaction, the equilibrium conversion decreases with increasing temperature. This is consistent with Le Châtelier's Principle. If the temperature of an exothermic reaction is decreased, the equilibrium will be displaced in a direction to oppose the effect of the change, that is, increase the conversion.

Now consider the effect of temperature on the rate of reaction. A qualitative observation is that most reactions go faster as the temperature increases. An increase in temperature of 10°C from room temperature typically doubles the rate of reaction for organic species in solution. It is found in practice that if the logarithm of the reaction rate constant is plotted against the inverse of absolute temperature, it tends to follow a straight line. Thus, at the same concentration, but at two different temperatures:

$$\ln k = \text{intercept} + \text{slope} \times \frac{1}{T} \qquad (6.53)$$

If the intercept is denoted by $\ln k_O$ and the slope by $-E/R$, where k_O is called the *frequency factor*, E is the *activation energy* of the reaction and R is the universal gas constant, then:

$$\ln k = \ln k_O - \frac{E}{RT} \qquad (6.54)$$

or

$$k = k_O \exp\left[-\frac{E}{RT}\right] \qquad (6.55)$$

At the same concentration, but at two different temperatures T_1 and T_2:

$$\ln \frac{r_2}{r_1} = \ln \frac{k_2}{k_1} = \frac{E}{R}\left(\frac{1}{T_1} - \frac{1}{T_2}\right) \qquad (6.56)$$

This assumes E to be constant.

Generally, the higher the rate of reaction, the smaller is the reactor volume. Practical upper limits are set by safety considerations, materials-of-construction limitations, maximum operating temperature for the catalyst or catalyst life. Whether the reaction system involves single or multiple reactions, and whether the reactions are reversible, also affects the choice of reactor temperature.

1. *Single reactions.*

(a) *Endothermic reactions.* If an endothermic reaction is reversible, then Le Châtelier's Principle dictates that operation at a high temperature increases the maximum conversion. Also, operation at high temperature increases the rate of reaction, allowing reduction of reactor volume. Thus, for endothermic reactions, the temperature should be set as high as possible, consistent with safety, materials-of-construction limitations and catalyst life.

Figure 6.4a shows the behavior of an endothermic reaction as a plot of equilibrium conversion against temperature. The plot can be obtained from values of ΔG^O over a range of temperatures and the equilibrium conversion calculated as illustrated in Examples 6.1 and 6.2. If it is assumed that the reactor is operated adiabatically, a heat balance can be carried out to show the change in temperature with reaction conversion. If the mean molar heat capacity of the reactants and products are assumed constant, then for a given starting temperature for the reaction T_{in}, the temperature of the reaction mixture will be proportional to the reactor conversion X for adiabatic operation, Figure 6.4a. As the conversion increases, the temperature decreases because of the reaction endotherm. If the reaction could proceed as far as equilibrium, then it would reach the equilibrium temperature T_E. Figure 6.4b shows how equilibrium conversion can be increased by dividing the reaction into stages and reheating the reactants between stages. Of course, the equilibrium conversion could also have been increased by operating the reactor nonadiabatically and adding heat as the reaction proceeds so as to maximize conversion within the constraints of feasible heat transfer, materials of construction, catalyst life, safety, and so on.

(b) *Exothermic reactions.* For single exothermic irreversible reactions, the temperature should be set as high as possible, consistent with materials of construction, catalyst life and safety, in order to minimize reactor volume.

For reversible exothermic reactions, the situation is more complex. Figure 6.5a shows the behavior of an exothermic reaction as a plot of equilibrium conversion against temperature. Again, the plot can be obtained from values of ΔG^O over a range of temperatures and the equilibrium conversion calculated as discussed previously. If it is assumed that the reactor is operated adiabatically, and the mean molar heat capacity of the reactants and products is constant, then for a given starting temperature for the reaction T_{in}, the temperature of the reaction mixture will be proportional to the reactor conversion X for adiabatic operation, Figure 6.5a.

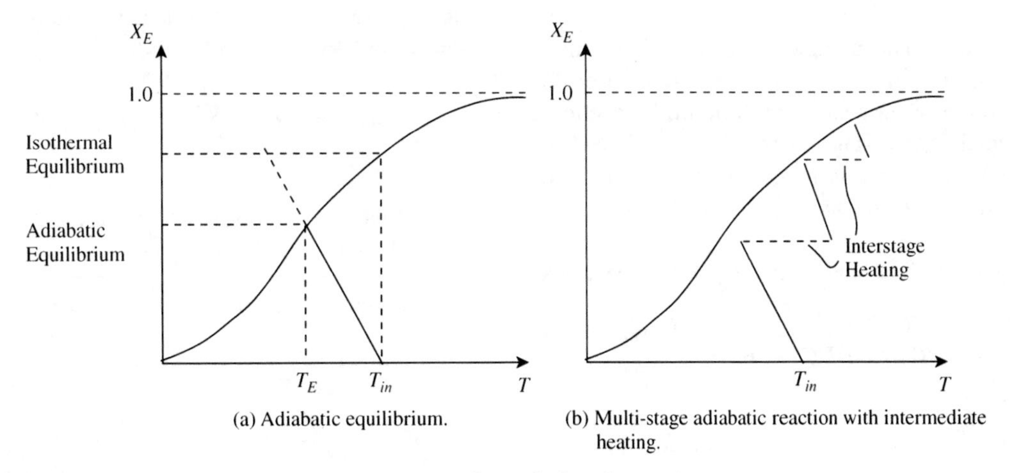

Figure 6.4 Equilibrium behavior with change in temperature for endothermic reactions.

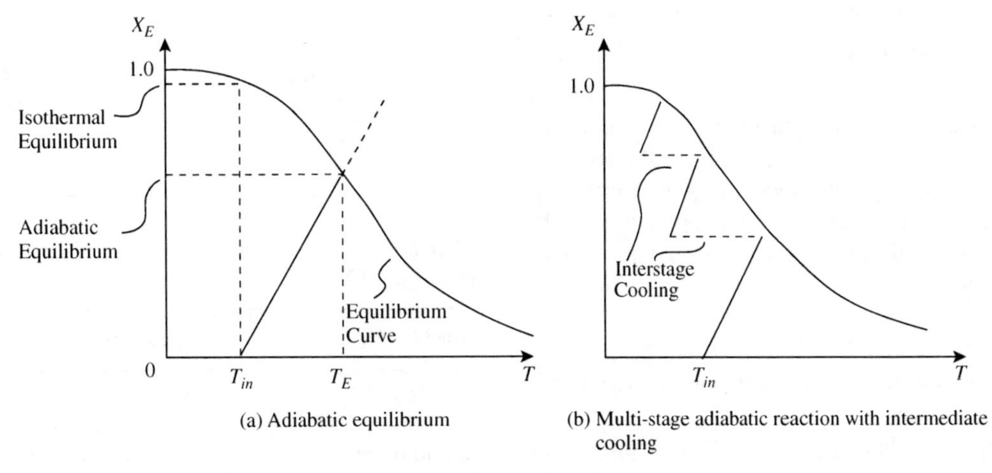

Figure 6.5 Equilibrium behavior with change in temperature for exothermic reactions.

As the conversion increases, the temperature rises because of the reaction exotherm. If the reaction could proceed as far as equilibrium, then it would reach the equilibrium temperature T_E, Figure 6.5a. Figure 6.5b shows how equilibrium conversion can be increased by dividing the reaction into stages and cooling the reactants between stages. Rather than adiabatic operation, the equilibrium conversion could also have been increased by operating the reactor nonadiabatically and removing heat as the reaction proceeds so as to maximize conversion within the constraints of feasible heat transfer, materials of construction, catalyst life, safety, and so on.

Thus, if an exothermic reaction is reversible, then Le Châtelier's principle dictates that operation at a low temperature increases maximum conversion. However, operation at a low temperature decreases the rate of reaction, thereby increasing the reactor volume. Then ideally, when far from equilibrium, it is advantageous to use a high temperature to increase the rate of reaction. As equilibrium is approached, the temperature should be lowered to increase the maximum conversion. For reversible exothermic reactions, the ideal temperature is continuously decreasing as conversion increases.

2. *Multiple reactions*. The arguments presented for minimizing reactor volume for single reactions can be used for the primary reaction when dealing with multiple reactions. However, the goal at this stage of the design, when dealing with multiple reactions, is to maximize selectivity or reactor yield rather than to minimize volume, for a given conversion.

Consider Equations 5.64 and 5.65 for parallel reactions:

$$\begin{array}{lll} FEED \longrightarrow PRODUCT & r_1 = k_1 C_{FEED}^{a_1} \\ FEED \longrightarrow BYPRODUCT & r_2 = k_2 C_{FEED}^{a_2} \end{array} \quad (5.64)$$

$$\frac{r_2}{r_1} = \frac{k_2}{k_1} C_{FEED}^{a_2 - a_1} \quad (5.65)$$

and Equation 5.68 for series reactions:

$$\begin{array}{lll} FEED \longrightarrow PRODUCT & r_1 = k_1 C_{FEED}^{a_1} \\ PRODUCT \longrightarrow BYPRODUCT & r_2 = k_2 C_{PRODUCT}^{a_2} \end{array} \quad (5.68)$$

The rates of reaction for the primary and secondary reactions both change with temperature, since the reaction rate constants k_1 and k_2 both increase with increasing temperature. The rate of change with temperature might be significantly different for the primary and secondary reactions.

- If k_1 increases faster than k_2, operate at high temperature (but beware of safety, catalyst life and materials-of-construction constraints).
- If k_2 increases faster than k_1, operate at low temperature (but beware of capital cost, since low temperature, although increasing selectivity, also increases reactor size). Here there is an economic trade-off between decreasing by product formation and increasing capital cost.

Example 6.3 Example 5.4 developed a kinetic model for the manufacture of benzyl acetate from benzyl chloride and sodium acetate in a solution of xylene in the presence of triethylamine as catalyst, according to:

$$C_6H_5CH_2Cl + CH_3COONa \longrightarrow CH_3COOC_6H_5CH_2 + NaCl$$

or

$$A + B \longrightarrow C + D$$

Example 5.4 developed a kinetic model for an equimolar feed at a temperature of 102°C, such that:

$$-r_A = k_A C_A \text{ with } k_A = 0.01306 \text{ h}^{-1}$$

Further experimental data are available at 117°C. The measured mole per cent benzyl chloride versus time in hours at 117°C are given in Table 6.7.

Again, assume the volume of the reactor to be constant. Determine the activation energy for the reaction.

Solution The same three kinetic models as Example 5.4 can be subjected to a least squares fit given by:

$$\text{minimize } R^2 = \sum_j^{11} [(N_{A,j})_{calc} - (N_{A,j})_{exp}]^2$$

Table 6.7 Experimental data for the production of benzyl acetate at 117°C.

Reaction time (h)	Benzyl chloride (mole %)
0.00	100.0
3.00	94.5
6.27	88.1
7.23	87.0
9.02	83.1
10.02	81.0
12.23	78.8
13.23	76.9
16.60	69.0
18.00	69.4
23.20	63.0
27.20	57.8

Table 6.8 Results of a least squares fit for the three kinetic models at 117°C.

Order of reaction	Rate constant	R^2
Zero order	$k_A V = 1.676$	37.64
First order	$k_A = 0.02034$	7.86
Second order	$k_A / V = 0.0002432$	24.94

where R = residual
$(N_{A,j})_{calc}$ = calculated moles of A for Measurement j
$(N_{A,j})_{exp}$ = experimentally measured moles of A for Measurement j

As in Example 5.4, one way this can be carried out is by setting up a function for R^2 in the spreadsheet, and then using the spreadsheet solver to minimize R^2 by manipulating the value of k_A (see Section 3.9). The results are given in Table 6.8.

Again it is clear from Table 6.8 that the best fit is given by a first-order reaction model:

$$-r_A = k_A C_A \text{ with } k_A = 0.02034 \text{ h}^{-1}$$

Next, substitute the results at 102°C (375 K) and 117°C (390 K) in Equation 6.56:

$$\ln\frac{r_2}{r_1} = \ln\frac{k_2}{k_1} = \frac{E}{R}\left(\frac{1}{T_1} - \frac{1}{T_2}\right)$$

$$\ln\frac{0.02034}{0.01306} = \frac{E}{8.3145}\left(\frac{1}{375} - \frac{1}{390}\right)$$

$$E = 35{,}900 \text{ kJ·kmol}^{-1}$$

6.3 REACTOR PRESSURE

Now consider the effect of pressure. For reversible reactions, pressure can have a significant effect on the equilibrium conversion. Even though the equilibrium constant is only a function of temperature and not a function of pressure, equilibrium conversion can still be influenced through changing the activities (fugacities) of the reactants and products.

Consider again the ammonia synthesis example from Examples 6.1 and 6.2

Example 6.4 Following Example 6.2, the reactor temperature will be set to 700 K. Examine the effect of increasing the reactor pressure by calculating the equilibrium conversion of hydrogen at 1 bar, 10 bar, 100 bar and 300 bar. Assume initially ideal gas behavior.

Solution From Equation 6.26 and 6.27:

$$K_a = \frac{p_{NH_3}{}^2}{p_{H_2}{}^3 p_{N_2}} = \frac{y_{NH_3}{}^2}{y_{H_2}{}^3 y_{N_2}} P^{-2}$$

Note again that K_a is dimensionless and depends on the specification of the number of moles in the stoichiometric equation. From Example 6.2 at 700 K:

$$K_a = 8.8421 \times 10^{-5}$$

Thus:

$$8.8421 \times 10^{-5} = \frac{y_{NH_3}{}^2}{y_{H_2}{}^3 y_{N_2}} P^{-2}$$

$$8.8421 \times 10^{-5} = \frac{16X^2(2-X)^2}{27(1-X)^4} P^{-2}$$

This can be solved by trial and error as before.

At 1 bar $X = 0.0061$ (from Example 6.2)
10 bar $X = 0.056$
100 bar $X = 0.33$
300 bar $X = 0.54$

It is clear that a very significant improvement in the equilibrium conversion for this reaction can be achieved through an increase in pressure.

It should again be noted that ideal gas behavior has been assumed. Carrying out the calculations for real gas behavior using an equation of state and including K_ϕ could change the results significantly, especially at the higher pressures, as will now be investigated.

Example 6.5 Repeat the calculations from Example 6.4 taking into account vapor-phase nonideality. Fugacity coefficients can be calculated from the Peng–Robinson Equation of State (see Poling, Prausnitz and O'Connell[6] and Chapter 4).

Solution The ideal gas equilibrium constants can be corrected for real gas behavior by multiplying the ideal gas equilibrium constant by K_ϕ, as defined by Equation 6.23, which for this problem is:

$$K_\phi = \frac{\phi_{NH_3}{}^2}{\phi_{H_2}{}^3 \phi_{N_2}}$$

The fugacity coefficients ϕ_i can be calculated from the Peng–Robinson Equation of State. The values of ϕ_i are functions of temperature, pressure and composition, and the calculations are complex (see Pohling, Prausnitz and O'Connell[6] and Chapter 4). Interaction parameters between components are here assumed to be zero. The results showing the effect of nonideality are given in Table 6.9:

It can be seen in Table 6.9 that as pressure increases, nonideal behavior changes the equilibrium conversion. Note that like K_a, K_ϕ also depends on the specification of the number of moles in the stoichiometric equation. For example, if the stoichiometry is written as:

$$\frac{3}{2}H_2 + \frac{1}{2}N_2 \rightleftharpoons NH_3$$

then

$$K'_\phi = \frac{\phi_{NH_3}}{\phi_{N_2}{}^{1/2}\phi_{H_2}{}^{3/2}}$$

$$K'_\phi = \sqrt{K_\phi}$$

Table 6.9 Equilibrium conversion versus pressure for real gas behavior.

Pressure (bar)	X_{IDEAL}	K_ϕ	X_{REAL}
1	0.0061	0.9990	0.0061
10	0.056	0.9897	0.056
100	0.33	0.8772	0.34
300	0.54	0.6182	0.58

The selection of reactor pressure for vapor-phase reversible reactions depends on whether there is a decrease or an increase in the number of moles. The value of ΔN in Equation 6.25 dictates whether the equilibrium conversion will increase or decrease with increasing pressure. If ΔN is negative, the equilibrium conversion will increase with increasing pressure. If ΔN is positive, it will decrease. The choice of pressure must also take account of whether the system involves multiple reactions.

Increasing the pressure of vapor-phase reactions increases the rate of reaction and hence decreases reactor volume both by decreasing the residence time required for a given reactor conversion and increasing the vapor density. In general, pressure has little effect on the rate of liquid-phase reactions.

1. *Single reactions.*
(a) *Decrease in the number of moles.* A decrease in the number of moles for vapor-phase reactions decreases the volume as reactants are converted to products. For a fixed reactor volume, this means a decrease in pressure as reactants are converted to products. The effect of an increase in pressure of the system is to cause a shift of the composition of the gaseous mixture toward one occupying a smaller volume. Increasing the reactor pressure increases the equilibrium conversion. Increasing the pressure increases the rate of reaction and also reduces reactor volume. Thus, if the reaction involves a decrease in the number of moles, the pressure should be set as high as practicable, bearing in mind that the high pressure might be costly to obtain through compressor power, mechanical construction might be expensive and high pressure brings safety problems.

(b) *Increase in the number of moles.* An increase in the number of moles for vapor-phase reactions increases the volume as reactants are converted to products. Le Châtelier's principle dictates that a decrease in reactor pressure increases equilibrium conversion. However, operation at a low pressure decreases the rate of reaction in vapor-phase reactions and increases the reactor volume. Thus, initially, when far from equilibrium, it is advantageous to use high pressure to increase the rate of reaction. As equilibrium is approached, the pressure should be lowered to increase the conversion. The ideal pressure would continuously decrease as conversion increases to the desired value. The low pressure required can be obtained by operating the system at reduced absolute pressure or by introducing a diluent to decrease the partial pressure. The diluent is an inert material (e.g. steam) and is simply used to lower the partial pressure in the vapor phase. For example, ethyl benzene can be dehydrogenated to styrene according to the reaction[7]:

$$\underset{\text{ethylbenzene}}{C_6H_5CH_2CH_3} \rightleftharpoons \underset{\text{styrene}}{C_6H_5CH = CH_2} + H_2$$

This is an endothermic reaction accompanied by an increase in the number of moles. High conversion is favored by high temperature and low pressure. The reduction in pressure is achieved in practice by the use of superheated steam as a diluent and by operating the reactor below atmospheric pressure. The steam in this case fulfills a dual purpose by also providing heat for the reaction.

2. *Multiple reactions producing by products.* The arguments presented for the effect of pressure on single vapor-phase reactions can be used for the primary reaction when dealing with multiple reactions. Again, selectivity and reactor yield are likely to be more important than reactor volume for a given conversion.

If there is a significant difference between the effect of pressure on the primary and secondary reactions, the pressure should be chosen to reduce as much as possible the rate of the secondary reactions relative to the primary reaction. Improving the selectivity or reactor yield in this way may require changing the system pressure or perhaps introducing a diluent.

For liquid-phase reactions, the effect of pressure on the selectivity and reactor volume is less pronounced, and the pressure is likely to be chosen to:

- prevent vaporization of the products;
- allow vaporization of liquid in the reactor so that it can be condensed and refluxed back to the reactor as a means of removing the heat of reaction;
- allow vaporization of one of the components in a reversible reaction in order that removal increases maximum conversion (to be discussed in Section 6.5);
- allow vaporization to feed the vapor directly into a distillation operation to combine reactions with separation (to be discussed in Chapter 13).

6.4 REACTOR PHASE

Having considered reactor temperature and pressure, the reactor phase can now be considered. The reactor phase can be gas, liquid or multiphase. Given a free choice between gas and liquid-phase reactions, operation in the liquid phase is usually preferred. Consider the single reaction system:

$$FEED \longrightarrow PRODUCT \quad r = kC_{FEED}^a$$

Clearly, in the liquid phase much higher concentrations of C_{FEED} (kmol·m^{-3}) can be maintained than in the gas phase. This makes liquid-phase reactions in general more rapid and hence leads to smaller reactor volumes for liquid-phase reactors.

However, a note of caution should be added. In many multiphase reaction systems, as will be discussed in the next chapter, rates of mass transfer between different phases can be just as important as, or even more important than, reaction kinetics in determining the reactor volume. Mass transfer rates are generally higher in gas-phase than liquid-phase systems. In such situations, it is not so easy to judge whether gas or liquid phase is preferred.

Very often the choice is not available. For example, if reactor temperature is above the critical temperature of the chemical species, then the reactor must be in the gas phase. Even if the temperature can be lowered below critical, an extremely high pressure may be required to operate in the liquid phase.

The choice of reactor temperature, pressure and hence phase must, in the first instance, take account of the desired equilibrium and selectivity effects. If there is still freedom to choose between gas and liquid phase, operation in the liquid phase is preferred.

6.5 REACTOR CONCENTRATION

When more than one reactant is used, it is often desirable to use an excess of one of the reactants. It is also sometimes desirable to feed an inert material to the reactor or to separate the product partway through the reaction before carrying out further reaction. Sometimes it is desirable to recycle unwanted by products to the reactor. These cases will now be examined.

1. *Single irreversible reactions.* An excess of one feed component can force another component toward complete conversion. As an example, consider the reaction between ethylene and chlorine to dichloroethane:

$$\underset{\text{ethylene}}{C_2H_4} + \underset{\text{chlorine}}{Cl_2} \longrightarrow \underset{\text{dichloroethane}}{C_2H_4Cl_2}$$

An excess of ethylene is used to ensure essentially complete conversion of the chlorine, which is thereby eliminated as a problem for the downstream separation system. In a single irreversible reaction (where selectivity is not a problem), the usual choice of excess reactant is to eliminate the component that is more difficult to separate in the downstream separation system. Alternatively, if one of the components is more hazardous (as is chlorine in this example), complete conversion has advantages for safety.

2. *Single reversible reactions.* The maximum conversion in reversible reactions is limited by the equilibrium conversion, and conditions in the reactor are usually chosen to increase the equilibrium conversion:

(a) *Feed ratio.* If to a system in equilibrium, an excess of one of the feeds is added, then the effect is to shift the equilibrium to decrease the feed concentration. In other words, an excess of one feed can be used to increase the equilibrium conversion. Consider the following examples.

Example 6.6 Ethyl acetate can be produced by the esterification of acetic acid with ethanol in the presence of a catalyst such as sulfuric acid or an ion-exchange resin according to the reaction:

$$\underset{\text{acetic acid}}{CH_3COOH} + \underset{\text{ethanol}}{C_2H_5OH} \rightleftharpoons \underset{\text{ethyl acetate}}{CH_3COOC_2H_5} + H_2O$$

Laboratory studies have been carried out to provide design data on the conversion. A stoichiometric mixture of 60 g acetic acid and 46 g ethanol was reacted and held at constant temperature until equilibrium was achieved. The reaction products were analyzed and found to contain 63.62 g ethyl acetate.

a. Calculate the equilibrium conversion of acetic acid.
b. Estimate the effect of using a 50% and 100% excess of ethanol.

Assume the liquid mixture to be ideal and the molar masses of acetic acid and ethyl acetate to be 60 $kg\cdot kmol^{-1}$ and 88 $kg\cdot kmol^{-1}$ respectively.

Solution

a. Moles of acetic acid in reaction mixture $= 60/60 = 1.0$
   Moles of ethyl acetate formed $= 63.62/88 = 0.723$
   Equilibrium conversion of acetic acid $= 0.723$
b. Let r be the molar ratio of ethanol to acetic acid. Table 6.10 presents the moles initially and at equilibrium and the mole fractions.
   Assuming an ideal solution (i.e. $K_\gamma = 1$ in Equation 6.29):

$$K_a = \frac{X^2}{(1 - X)(r - X)}$$

For the stoichiometric mixture, $r = 1$ and $X = 0.723$ from the laboratory measurements:

$$K_a = \frac{0.723^2}{(1 - 0.723)^2} = 6.813$$

Table 6.10 Mole fractions at equilibrium for the production of ethyl acetate.

	CH_3COOH	C_2H_5OH	$CH_3COOC_2H_5$	H_2O
Moles in initial mixture	1	r	0	0
Moles at equilibrium	$1 - X$	$r - X$	X	X
Mole fraction at equilibrium	$\frac{1-X}{1+r}$	$\frac{r-X}{1+r}$	$\frac{X}{1+r}$	$\frac{X}{1+r}$

Table 6.11 Variation of equilibrium conversion with feed ration for the production of ethyl acetate.

r	X
1.0	0.723
1.5	0.842
2.0	0.894

For excess ethanol:

$$6.813 = \frac{X^2}{(1 - X)(r - X)}$$

Substitute $r = 1.5$ and 2.0 and solve for X by trial and error. The results are presented in Table 6.11.

Increasing the excess of ethanol increases the conversion of acetic acid to ethyl acetate. To carry out the calculation more accurately would require activity coefficients to be calculated for the mixture (see Poling, Prausnitz and O'Connell[6] and Chapter 4). The activity coefficients depend on correlating coefficients between each binary pair in the mixture, the concentrations and temperature.

Example 6.7 Hydrogen can be manufactured from the reaction between methane and steam over a catalyst. Two principal reactions occur:

$$\underset{\text{methane}}{CH_4} + \underset{\text{water}}{H_2O} \rightleftharpoons \underset{\text{hydrogen}}{3H_2} + \underset{\text{carbon monoxide}}{CO}$$

$$\underset{\text{carbon monoxide}}{CO} + \underset{\text{water}}{H_2O} \rightleftharpoons \underset{\text{hydrogen}}{H_2} + \underset{\text{carbon dioxide}}{CO_2}$$

Assuming the reaction takes place at 1100 K and 20 bar, calculate the equilibrium conversion for a molar ratio of steam to methane in the feed of 3, 4, 5 and 6. Assume ideal gas behavior ($K_\phi = 1$, $R = 8.3145\ \text{kJ}\cdot\text{K}^{-1}\cdot\text{kmol}^{-1}$). Thermodynamic data are given in Table 6.12.

Table 6.12 Thermodynamic data for hydrogen manufacture[6].

	$\overline{G}^O_{1100K}$ (kJ·kmol^{-1})
CH_4	30,358
H_2O	−187,052
H_2	0.0
CO	−209,084
CO_2	−359,984

Solution For the first reaction:

$$\Delta G^O_1 = 3\Delta\overline{G}^O_{H_2} + \Delta\overline{G}^O_{CO} - \Delta\overline{G}^O_{CH_4} - \Delta\overline{G}^O_{H_2O}$$
$$= 3 \times 0 + (-209{,}084) - 30{,}358 - (-187{,}052)$$
$$= -52{,}390\ \text{kJ}$$
$$-RT \ln K_{a_1} = -52{,}390\ \text{kJ}$$
$$K_{a1} = 307.42$$

For the second reaction:

$$\Delta G^O_2 = \Delta\overline{G}^O_{H_2} + \Delta\overline{G}^O_{CO_2} - \Delta\overline{G}^O_{CO} - \Delta\overline{G}^O_{H_2O}$$
$$= 0 + (-359{,}984) - (-209{,}084) - (-187{,}052)$$
$$= 36{,}152\ \text{kJ}$$
$$-RT \ln K_{a_2} = 36{,}152$$
$$K_{a2} = 1.9201 \times 10^{-2}$$

Let r be the molar ratio of water to methane in the feed, X_1 be the conversion of the first reaction and X_2 be the conversion of the second reaction.

$$\underset{1 - X_1}{CH_4} + \underset{r - X_1 - X_2}{H_2O} \rightleftharpoons \underset{3X_1}{3H_2} + \underset{X_1 - X_2}{CO}$$

$$\underset{X_1 - X_2}{CO} + \underset{r - X_1 - X_2}{H_2O} \rightleftharpoons \underset{X_2}{H_2} + \underset{X_2}{CO_2}$$

Total moles at equilibrium

$$= (1 - X_1) + (r - X_1 - X_2) + 3X_1 + (X_1 - X_2) + X_2 + X_2$$
$$= r + 1 + 2X_1$$

Mole fractions are presented in Table 6.13.

$$K_{a1} = \frac{y^3_{H_2}\, y_{CO}}{y_{CH_4}\, y_{H_2O}} P^2$$

$$= \frac{\left(\dfrac{3X_1 + X_2}{r + 1 + 2X_1}\right)^3 \left(\dfrac{X_1 - X_2}{r + 1 + 2X_1}\right)}{\left(\dfrac{1 - X_1}{r + 1 + 2X_1}\right)\left(\dfrac{r - X_1 - X_2}{r + 1 + 2X_1}\right)} P^2$$

$$= \frac{(3X_1 + X_2)^3 (X_1 - X_2)}{(1 - X_1)(r - X_1 - X_2)(r + 1 + 2X_1)^2} P^2 \tag{6.57}$$

$$K_{a2} = \frac{y_{H_2}\, y_{CO_2}}{y_{CO}\, y_{H_2O}} P^0$$

$$= \frac{\left(\dfrac{3X_1 + X_2}{r + 1 + 2X_1}\right)\left(\dfrac{X_2}{r + 1 + 2X_1}\right)}{\left(\dfrac{X_1 - X_2}{r + 1 + 2X_1}\right)\left(\dfrac{r - X_1 - X_2}{r + 1 + 2X_1}\right)} P^0$$

$$= \frac{(3X_1 + X_2) X_2}{(X_1 - X_2)(r - X_1 - X_2)} P^0 \tag{6.58}$$

Knowing P, K_{a1} and K_{a2} and setting a value for r, these two equations can be solved simultaneously for X_1 and X_2. However, X_1 or X_2 cannot be eliminated by substitution in this case, and the equations must be solved numerically. This can be done by assuming a value for X_1 and substituting this in both equations,

Table 6.13 Mole fractions at equilibrium for the manufacture of hydrogen.

	CH_4	H_2O	H_2	CO	CO_2
Moles at equilibrium	$1 - X_1$	$r - X_1 - X_2$	$3X_1 + X_2$	$X_1 - X_2$	X_2
Mole fraction at equilibrium	$\dfrac{1 - X_1}{r + 1 + 2X_1}$	$\dfrac{r - X_1 - X_2}{r + 1 + 2X_1}$	$\dfrac{3X_1 + X_2}{r + 1 + 2X_1}$	$\dfrac{X_1 - X_2}{r + 1 + 2X_1}$	$\dfrac{X_2}{r + 1 + 2X_1}$

thus yielding values of X_2 from both equations. X_1 is then varied until the values of X_2 predicted by both equations are equal. Alternatively, X_1 and X_2 can be varied simultaneously in a nonlinear optimization algorithm to minimize the errors in the two equations. If spreadsheet software is used, this can be done by taking K_{a1} to the right-hand side of Equation 6.57 and giving the right-hand side a value of say *Objective 1*, which must ultimately be zero. Also, K_{a2} can be taken to the right-hand side of Equation 6.58 and the right-hand side given a value of say *Objective 2*, which must also ultimately be zero. Values of X_1 and X_2 must then be determined such that *Objective 1* and *Objective 2* are both zero. The spreadsheet solver can then be used to vary X_1 and X_2 simultaneously to search for (see Section 3.9):

$$(Objective\ 1)^2 + (Objective\ 2)^2 = 0 \text{ (within a tolerance)}$$

The results are shown in Table 6.14.

From Table 6.14, the mole fraction of CH_4 and H_2 decrease as the molar ratio of H_2O/CH_4 increases. The picture becomes clearer when the results are presented on a dry basis as shown in Table 6.15.

From Table 6.14, the mole fraction of CH_4 and H_2 decrease as the molar ratio of H_2O/CH_4 increases. The picture becomes clearer when the results are presented on a dry basis as shown in Table 6.15.

From Table 6.15, on a dry basis as the molar ratio H_2O/CH_4 increases, the mole fraction of CH_4 decreases and that of H_2 increases.

The next two stages in the process carry out shift conversion at lower temperatures in which the second reaction above is used to convert CO to H_2 to higher conversion.

(b) *Inert concentration.* Sometimes, an inert material is present in the reactor. This might be a solvent in a liquid-phase reaction or an inert gas in a gas-phase reaction. Consider the reaction

$$A \rightleftharpoons B + C$$

Table 6.14 Equilibrium conversions and product mole fractions for the manufacture of hydrogen.

H_2O/CH_4	X_1	X_2	y_{CH_4}	y_{H_2O}	y_{H_2}	y_{CO}	y_{CO_2}
3	0.80	0.015	0.0357	0.3902	0.4313	0.1402	0.0027
4	0.86	0.019	0.0208	0.4644	0.3868	0.1251	0.0028
5	0.91	0.025	0.0115	0.5198	0.3523	0.1132	0.0032
6	0.93	0.031	0.0079	0.5687	0.3184	0.1015	0.0035

Table 6.15 Product mole fractions for the manufacture of hydrogen on a dry basis.

H_2O/CH_4	y_{CH_4}	y_{H_2}
3	0.0586	0.7072
4	0.0389	0.7221
5	0.0240	0.7337
6	0.0183	0.7383

The effect of the increase in moles can be artificially decreased by adding an inert material. Le Châtelier's Principle dictates that this will increase the equilibrium conversion. For example, if the above reaction is in the ideal gas phase:

$$K_a = \frac{p_B p_C}{p_A} = \frac{y_B y_C}{y_A} P = \frac{N_B N_C}{N_A N_T} P \tag{6.59}$$

where:

N_i = number of moles of Component i

N_T = total number of moles

Thus:

$$\frac{N_B N_C}{N_A} = \frac{K_a N_T}{P} \tag{6.60}$$

Increasing N_T as a result of adding inert material will increase the ratio of products to reactants. Adding an inert material causes the number of moles per unit volume to be decreased, and the equilibrium will be displaced to oppose this by shifting to a higher conversion. If inert material is to be added, then ease of separation is an important consideration. For example, steam is added as an inert to hydrocarbon cracking reactions and is an attractive material in this respect because it is easily separated from the hydrocarbon components by condensation.

Consider the reaction:

$$E + F \rightleftharpoons G$$

For example, if the above reaction is in the ideal gas phase:

$$K_a = \frac{p_G}{p_E p_F} = \frac{y_G}{y_E y_F} \frac{1}{P} = \frac{N_G N_T}{N_E N_F} \frac{1}{P} \tag{6.61}$$

Thus:

$$\frac{N_G}{N_E N_F} = \frac{K_a P}{N_T} \tag{6.62}$$

Decreasing N_T as a result of removing inert material will increase the ratio of products to reactants. Removing inert material causes the number of moles per unit volume to be increased, and the equilibrium will be displaced to oppose this by shifting to a higher conversion.

If the reaction does not involve any change in the number of moles, inert material has no effect on equilibrium conversion.

(c) *Product removal during reaction.* Sometimes the equilibrium conversion can be increased by removing the product (or one of the products) continuously from the reactor as the reaction progresses, for example, by allowing it to vaporize from a liquid-phase reactor. Another way is to carry out the reaction in stages with intermediate separation of the products. As an example of intermediate separation, consider the production of sulfuric acid as illustrated in Figure 6.6. Sulfur dioxide is oxidized to sulfur trioxide:

$$\underset{\text{sulphur dioxide}}{2SO_2} + O_2 \rightleftharpoons \underset{\text{sulphur trioxide}}{2SO_3}$$

This reaction can be forced to effective complete conversion by first carrying out the reaction to approach equilibrium. The sulfur trioxide is then separated (by absorption). Removal of sulfur trioxide shifts the equilibrium, and further reaction of the remaining sulfur dioxide and oxygen allows effective complete conversion of the sulfur dioxide, Figure 6.6.

Intermediate separation followed by further reaction is clearly most appropriate when the intermediate separation is straightforward, as in the case of sulfuric acid production.

3. *Multiple reactions in parallel producing by products.* After the reactor type is chosen for parallel reaction systems in order to maximize selectivity or reactor yield, conditions can be altered further to improve selectivity. Consider the parallel reaction system from Equation 5.66. To maximize selectivity for this system, the ratio given by Equation 5.67 is minimized:

$$\frac{r_2}{r_1} = \frac{k_2}{k_1} C_{FEED1}^{a_2-a_1} C_{FEED2}^{b_2-b_1} \qquad (5.67)$$

Even after the type of reactor is chosen, excess of *FEED 1* or *FEED 2* can be used.

- If $(a_2 - a_1) > (b_2 - b_1)$ use excess *FEED 2*.
- If $(a_2 - a_1) < (b_2 - b_1)$ use excess *FEED 1*.

If the secondary reaction is reversible and involves a decrease in the number of moles, such as:

$$\begin{aligned} FEED1 + FEED2 &\longrightarrow PRODUCT \\ FEED1 + FEED2 &\rightleftharpoons BYPRODUCT \end{aligned} \qquad (6.63)$$

then, if inerts are present, increasing the concentration of inert material will decrease by product formation. If the secondary reaction is reversible and involves an increase in the number of moles, such as:

$$\begin{aligned} FEED1 + FEED2 &\longrightarrow PRODUCT \\ FEED1 &\rightleftharpoons BYPRODUCT1 \\ &\quad + BYPRODUCT2 \end{aligned} \qquad (6.64)$$

then, if inert material is present, decreasing the concentration of inert material will decrease by product formation. If the secondary reaction has no change in the number of moles, then concentration of inert material does not affect it.

For all reversible secondary reactions, deliberately feeding *BY PRODUCT* to the reactor inhibits its formation at source by shifting the equilibrium of the secondary reaction. This is achieved in practice by separating and recycling *BY PRODUCT* rather than separating and disposing of it directly.

An example of such recycling in a parallel reaction system is in the "Oxo" process for the production of C_4 alcohols. Propylene and synthesis gas (a mixture of carbon monoxide and hydrogen) are first reacted to *n*- and iso-butyraldehydes using a cobalt-based catalyst. Two parallel reactions occur[7]:

$$\underset{\text{propylene}}{C_3H_6} + \underset{\substack{\text{carbon}\\ \text{monoxide}}}{CO} + \underset{\text{hydrogen}}{H_2} \rightleftharpoons \underset{\text{n – butyraldehyde}}{CH_3CH_2CH_2CHO}$$

$$\underset{\text{propylene}}{C_3H_6} + \underset{\substack{\text{carbon}\\ \text{monoxide}}}{CO} + \underset{\text{hydrogen}}{H_2} \rightleftharpoons \underset{\text{iso – butyraldehyde}}{CH_3CH(CH_3)CHO}$$

The *n*-isomer is more valuable. Recycling the *iso*-isomer can be used as a means of suppressing its formation[7].

4. *Multiple reactions in series producing by products.* For the series reaction system in Equation 5.68, the series reaction is inhibited by low concentrations of *PRODUCT*. It has been noted already that this can be achieved by operating with a low conversion.

If the reaction involves more than one feed, it is not necessary to operate with the same low conversion on all the feeds. Using an excess of one of the feeds enables operation with a relatively high conversion of other feed material and still inhibits series reactions. Consider again the series reaction system from Example 5.3:

$$\underset{\text{toluene}}{C_6H_5CH_3} + \underset{\text{hydrogen}}{H_2} \longrightarrow \underset{\text{benzene}}{C_6H_6} + \underset{\text{methane}}{CH_4}$$

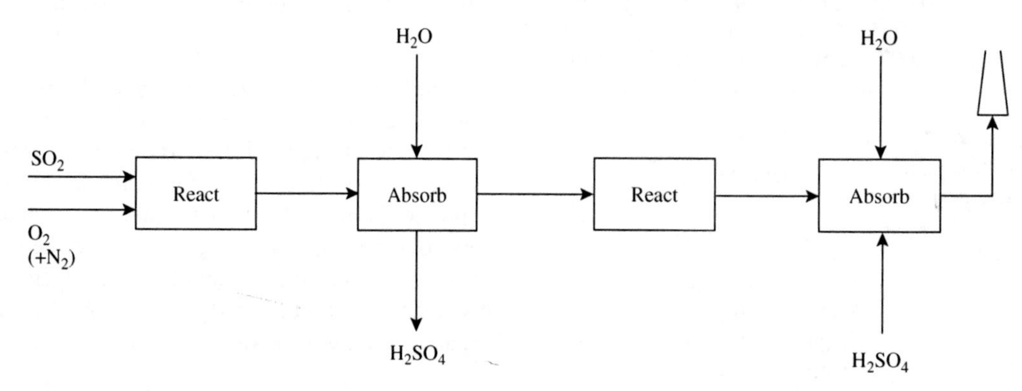

Figure 6.6 The equilibrium conversion for sulfuric acid production can be increased by intermediate separation of the product followed by further reaction.

$$\underset{\text{benzene}}{2C_6H_6} \rightleftharpoons \underset{\text{dyphenyl}}{C_{12}H_{10}} + \underset{\text{hydrogen}}{H_2}$$

It is usual to operate this reactor with a large excess of hydrogen[7]. The molar ratio of hydrogen to toluene entering the reactor is of the order 5 : 1. The excess of hydrogen encourages the primary reaction directly and discourages the secondary reaction by reducing the concentration of the benzene product. Also, in this case, because hydrogen is a by product of the secondary reversible reaction, an excess of hydrogen favors the reverse reaction to benzene. In fact, unless a large excess of hydrogen is used, series reactions that decompose the benzene all the way to carbon become significant, known as *coke formation*.

Another way to keep the concentration of *PRODUCT* low is to remove the product as the reaction progresses, for example, by intermediate separation followed by further reaction. For example, in a reaction system such as Equation 5.68, intermediate separation of the *PRODUCT* followed by further reaction maintains a low concentration of *PRODUCT* as the reaction progresses. Such intermediate separation is most appropriate when separation of the product from the reactants is straightforward.

If the series reaction is also reversible, such as

$$\begin{aligned} &FEED \longrightarrow PRODUCT \\ &PRODUCT \rightleftharpoons BY\ PRODUCT \end{aligned} \qquad (6.65)$$

then, again, removal of the *PRODUCT* as the reaction progresses, for example, by intermediate separation of the *PRODUCT*, maintains a low concentration of *PRODUCT* and at the same time shifts the equilibrium for the secondary reaction toward *PRODUCT* rather than *BY PRODUCT* formation.

If the secondary reaction is reversible and inert material is present, then to improve the selectivity:

- increase the concentration of inert material if the *BY PRODUCT* reaction involves a decrease in the number of moles;
- decrease the concentration of inert material if the *BY PRODUCT* reaction involves an increase in the number of moles.

An alternative way to improve selectivity for the reaction system in Equation 6.65 is again to deliberately feed *BY PRODUCT* to the reactor to shift the equilibrium of the secondary reaction away from *BY PRODUCT* formation.

An example of where recycling can be effective in improving selectivity or reactor yield is in the production of benzene from toluene. The series reaction is reversible. Hence, recycling diphenyl to the reactor can be used to suppress its formation at the source.

5. *Mixed parallel and series reactions producing by products.* As with parallel and series reactions, use of an excess of one of the feeds can be effective in improving selectivity with mixed reactions. As an example, consider the chlorination of methane to produce chloromethane[7]. The primary reaction is:

$$\underset{\text{methane}}{CH_4} + \underset{\text{chlorine}}{Cl_2} \longrightarrow \underset{\text{chloro-methane}}{CH_3Cl} + \underset{\text{hydrogen chloride}}{HCl}$$

Secondary reactions can occur to higher chlorinated compounds:

$$\underset{\text{chloromethane}}{CH_3Cl} + Cl_2 \longrightarrow \underset{\text{dichloro-methane}}{CH_2Cl_2} + HCl$$

$$\underset{\text{dichloromethane}}{CH_2Cl_2} + Cl_2 \longrightarrow \underset{\text{chloroform}}{CHCl_3} + HCl$$

$$\underset{\text{chloroform}}{CHCl_3} + Cl_2 \longrightarrow \underset{\text{carbon tetrachloride}}{CCl_4} + HCl$$

The secondary reactions are series with respect to the chloromethane but parallel with respect to chlorine. A very large excess of methane (molar ratio of methane to chlorine is of the order 10 : 1) is used to suppress selectivity losses[7]. The excess of methane has two effects. First, because it is only involved in the primary reaction, it encourages the primary reaction. Second, by diluting the product chloromethane, it discourages the secondary reactions, which prefer a high concentration of chloromethane.

Removal of the product as the reaction progresses is also effective in suppressing the series element of the by product reactions, providing the separation is straightforward.

If the by product reaction is reversible and inert material is present, then changing the concentration of inert material if there is a change in the number of moles should be considered, as discussed above. Whether or not there is a change in the number of moles, recycling by products can in some cases suppress their formation if the by product–forming reaction is reversible. An example is in the production of ethyl benzene from benzene and ethylene[7]:

$$\underset{\text{benzene}}{C_6H_6} + \underset{\text{ethylene}}{C_2H_4} \rightleftharpoons \underset{\text{ethylbenzene}}{C_6H_5CH_2CH_3}$$

Polyethylbenzenes (diethylbenzene, triethylbenzene, etc.) are also formed as unwanted by products through reversible reactions in series with respect to ethyl benzene but parallel with respect to ethylene. For example:

$$\underset{\text{ethylbenzene}}{C_6H_5CH_2CH_3} + \underset{\text{ethylene}}{C_2H_4} \rightleftharpoons \underset{\text{diethylbenzene}}{C_6H_4(C_2H_5)_2}$$

These polyethylbenzenes are recycled to the reactor to inhibit formation of fresh polyethylbenzenes[7]. However, it should be noted that recycling of by products is by no means always beneficial as by products can break down to cause deterioration in catalyst performance.

6.6 BIOCHEMICAL REACTIONS

Biochemical reactions must cater for living systems and as a result are carried out in an aqueous medium within a narrow range of conditions. Each species of microorganism grows best under certain conditions. Temperature, pH, oxygen levels, concentrations of reactants and products and possibly nutrient levels must be carefully controlled for optimum operation.

1. *Temperature.* Microorganisms can be classified according to the optimum temperature ranges for their growth[8]:

- *Psychrophiles* have an optimum growth temperature of around 15°C and a maximum growth temperature below 20°C.
- *Mesophiles* grow best between 20 and 45°C.
- *Thermophiles* have optimum growth temperatures between 45 and 80°C.
- *Extremphiles* grow under extreme conditions, some species above 100°C, other species as low as 0°C.

Yeasts, moulds and algae typically have a maximum temperature around 60°C.

2. *pH.* The pH requirements of microorganisms depend on the species. Most microorganisms prefer pH values not far from neutrality. However, some microorganisms can grow in extremes of pH.

3. *Oxygen levels.* Reactions can be carried out under *aerobic* conditions in which free oxygen is required or under *anaerobic* conditions in the absence of free oxygen. Bacteria can operate under aerobic or anaerobic conditions. Yeasts, moulds and algae prefer aerobic conditions but can grow with reduced oxygen levels.

4. *Concentration.* The rate of reaction depends on the concentrations of feed, trace elements, vitamins and toxic substances. The rate also depends on the build-up of wastes from the microorganisms that interfere with microorganism multiplication. There comes a point where their waste inhibits growth.

6.7 CATALYSTS

Most processes are catalyzed where catalysts for the reaction are known. The choice of catalyst is crucially important. Catalysts increase the rate of reaction but are ideally unchanged in quantity and chemical composition at the end of the reaction. If the catalyst is used to accelerate a reversible reaction, it does not by itself alter the position of the equilibrium. However, it should be noted if a porous solid catalyst is used, then different rates of diffusion of different species within a catalyst can change the concentration of the reactants at the point where the reaction takes place, thus influencing the equilibrium indirectly.

When systems of multiple reactions are involved, the catalyst may have different effects on the rates of the different reactions. This allows catalysts to be developed that increase the rate of the desired reactions relative to the undesired reactions. Hence the choice of catalyst can have a major influence on selectivity.

The catalytic process can be homogeneous, heterogeneous or biochemical.

1. *Homogeneous catalysts.* With a homogeneous catalyst, the reaction proceeds entirely in either the vapor or liquid phase. The catalyst may modify the reaction mechanism by participation in the reaction but is regenerated in a subsequent step. The catalyst is then free to promote further reaction. An example of such a homogeneous catalytic reaction is the production of acetic anhydride. In the first stage of the process, acetic acid is pyrolyzed to ketene in the gas phase at 700°C.

$$\underset{\text{acetic acid}}{CH_3COOH} \longrightarrow \underset{\text{ketene}}{CH_2 = C = O} + \underset{\text{water}}{H_2O}$$

The reaction uses triethyl phosphate as a homogeneous catalyst[7].

In general, heterogeneous catalysts are preferred to homogeneous catalysts because the separation and recycling of homogeneous catalysts often can be very difficult. Loss of homogeneous catalyst not only creates a direct expense through loss of material but also creates an environmental problem.

2. *Heterogeneous catalysts.* In heterogeneous catalysis, the catalyst is in a different phase from the reacting species. Most often, the heterogeneous catalyst is a solid, acting on species in the liquid or gas phase. The solid catalyst can be either of the following.

- *Bulk catalytic materials*, in which the gross composition does not change significantly through the material, such as platinum wire mesh.
- *Supported catalysts*, in which the active catalytic material is dispersed over the surface of a porous solid.

Catalytic gas-phase reactions play an important role in many bulk chemical processes, such as in the production of methanol, ammonia, sulfuric acid and most petroleum refinery processes. In most processes, the effective area of the catalyst is critically important. Industrial catalysts are usually supported on porous materials, since this results in a much larger active area per unit of reactor volume.

As well as depending on catalyst porosity, the reaction rate is some function of the reactant concentrations, temperature and pressure. However, this function may not be as simple as in the case of uncatalyzed reactions. Before a reaction can take place, the reactants must diffuse through the pores to the solid surface. The overall rate of a heterogeneous gas–solid reaction on a supported catalyst is made up of a series of physical steps as well as the chemical reaction. The steps are as follows.

(a) Mass transfer of reactant from the bulk gas phase to the external solid surface.
(b) Diffusion from the solid surface to the internal active sites.
(c) Adsorption on solid surface.
(d) Activation of the adsorbed reactants.
(e) Chemical reaction.
(f) Desorption of products.
(g) Internal diffusion of products to the external solid surface.
(h) Mass transfer to the bulk gas phase.

All of these steps are rate processes and are temperature dependent. It is important to realize that very large temperature gradients may exist between active sites and the bulk gas phase. Usually, one step is slower than the others, and it is this rate-controlling step. The *effectiveness factor* is the ratio of the observed rate to that which would be obtained if the whole of the internal surface of the pellet were available to the reagents at the same concentrations as they have at the external surface. Generally, the higher the effectiveness factor, the higher the rate of reaction.

The effectiveness factor depends on the size and shape of the catalyst pellet and the distribution of active material within the pellet.

(a) *Size of pellet.* If active material is distributed uniformly throughout the pellet, then the smaller the pellet, the higher the effectiveness. However, smaller pellets can produce an unacceptably high-pressure drop through packed bed reactors. For gas-phase reactions in packed beds, the pressure drop is usually less than 10% of the inlet pressure[9].

(b) *Shape of pellet.* Active material can be distributed on pellets with different shapes. The most commonly used shapes are spheres, cylinders and slabs. If the same amount of active material is distributed uniformly throughout the pellet, then for the same volume of pellet, the effectiveness factor is in the order:

$$\text{slab} > \text{cylinder} > \text{sphere}$$

(c) *Distribution of active material.* During the catalyst preparation, it is possible to control the distribution of the active material within the catalyst pellet[9]. Nonuniform distribution of the active material can increase the conversion, selectivity or resistance to deactivation compared to uniformly active catalysts. It is possible, in principle, to distribute the active material with almost any profile by the use of suitable impregnation techniques[10]. Figure 6.7 illustrates some of the possible distributions through the pellet. In addition to uniform distribution (Figure 6.7a), it is possible to distribute as *egg-shell* in which the active material is located toward the outside of the pellet (Figure 6.7b), or *egg yolk* in which the active material is located toward the core of the pellet (Figure 6.7c). A *middle distribution* locates the active material between the core and the outside of the pellet (Figure 6.7d). For a single reaction involving a fixed amount of active material, it can be shown that for conditions in which there are no catalyst deactivation mechanisms, the effectiveness of a supported catalyst is maximized when the active sites are concentrated at a precise location with zero width as a Dirac Delta Function[11], as illustrated in Figure 6.7e. The optimal performance of the catalyst is obtained by locating the Dirac Delta catalyst distribution so as to maximize the reaction rate by taking advantage of both temperature and concentration gradients within the pellet. The Dirac Delta Function could be located at the surface, in the center or anywhere between. Unfortunately, it is not possible from a practical point of view to locate the catalyst as a Dirac Delta Function. However, it can be approximated as a step function in a *layered catalyst*, as shown in Figure 6.7f. Providing the thickness of the active layer is less than ~5% of the pellet characteristic dimension (e.g. radius for a spherical pellet), the behavior of the Dirac Delta and step distributions is virtually the same[11]. The location of the active material needs to be optimized. However, it should be emphasized that if the catalyst is subject to degradation in its performance because of surface deposits, and so on, the Dirac Delta (and its step function equivalent) can be subject to a sharp deterioration in performance and is not necessarily the best choice under those conditions.

More often than not, solid-catalyzed reactions are multiple reactions. For reactions in parallel, the key to high selectivity is to maintain the appropriate high or low

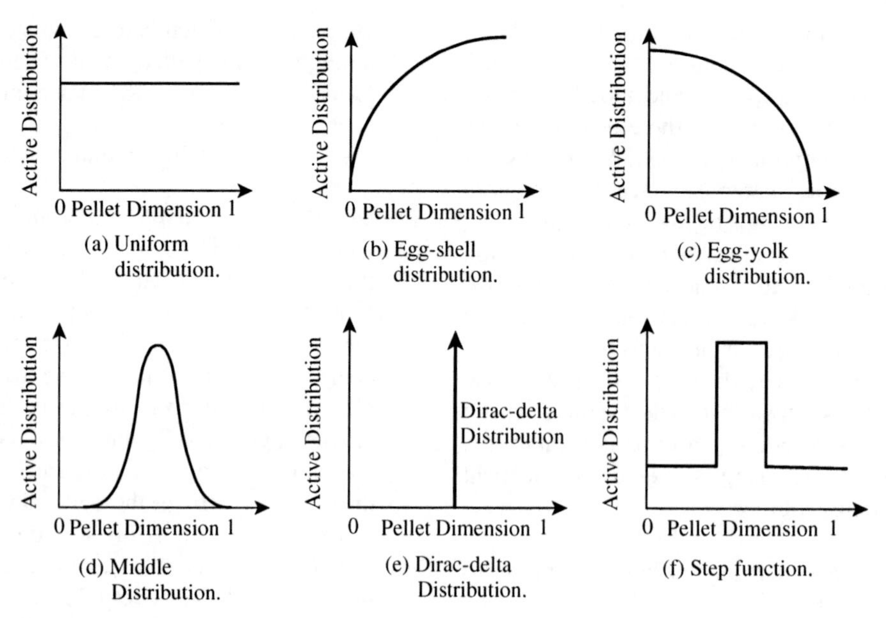

Figure 6.7 Catalyst distribution within a pellet of supported catalyst.

concentration and temperature levels of reactants at the catalyst surface, to encourage the desired reaction and to discourage the byproduct reactions. For reactions in series, the key is to avoid the mixing of fluids of different compositions. These arguments for the gross flow pattern of fluid through any reactor have already been developed.

However, before extrapolating the arguments from the gross patterns through the reactor for homogeneous reactions to solid-catalyzed reactions, it must be recognized that in catalytic reactions the concentration and temperature in the interior of catalyst pellets may differ from the main body of the gas. Local nonhomogeneity caused by lowered reactant concentration or change in temperature within the catalyst pellets results in a product distribution different from what would otherwise be observed for a homogeneous system. Consider two extreme cases:

(a) *Surface reaction controls.* If surface reaction is rate controlling, then the concentrations of reactant within the pellets and in the main gas stream are essentially the same. In this situation, the considerations for the gross flow pattern of fluid through the reactor would apply.

(b). *Diffusion controls.* If diffusional resistance controls, then the concentration of reactant at the catalyst surface would be lower than in the main gas stream. Referring back to Equation 5.64, for example, lowered reactant concentration favors the reaction of lower order. Hence, if the desired reaction is of lower order, operating under conditions of diffusion control would increase selectivity. If the desired reaction is of higher order, the opposite holds.

For multiple reactions, in cases where there is no catalyst deactivation, as with single reactions, the optimal catalyst distribution within the pellet is a Dirac Delta Function[10]. This time the location within the pellet needs to be optimized for maximum selectivity or yield. Again, in practice, a step function approximates the performance of a Dirac Delta Function as long as it is less than ~5% of the pellet characteristic dimension[11]. However, a word of caution must again be added. As will be discussed in the next chapter, the performance of supported catalysts deteriorates through time for a variety of reasons. The optimal location of the step function should take account of this deterioration of performance through time. Indeed, as discussed previously, if the catalyst is subject to degradation in its performance because of surface deposits, and so on, the Dirac Delta (and its step function equivalent) might be subject to a sharp deterioration in performance and is not necessarily the best choice when considering the whole of the catalyst life.

Heterogeneous catalysts can thus have a major influence on selectivity. Changing the catalyst can change the relative influence on the primary and by product reactions. This might result directly from the reaction mechanisms at the active sites or the relative rates of diffusion in the support material or a combination of both.

3. *Biochemical catalysts.* Some reactions can catalyzed be by enzymes. The attraction in using enzymes rather than microorganisms is an enormous rate enhancement that can be obtained in the absence of the microorganisms. This is restricted to situations when the enzyme can be isolated and is also stable. In addition, the chemical reaction does not have to cater for the special requirements of living cells.

However, just like microorganisms, enzymes are sensitive, and care must be exercised in the conditions under which they are used. A disadvantage in using enzymes is that use of the isolated enzyme is frequently more expensive on a single-use basis than the use of propagated microorganisms. Another disadvantage in using enzymes is that the enzymes must be removed from the product once the reaction has been completed, which might involve an expensive separation. Long reaction times may be necessary if a low concentration of enzyme is used to decrease enzyme cost.

Some of these difficulties in using enzymes can be overcome by fixing, or *immobilizing*, the enzyme in some way. A number of methods for enzyme immobilization have been developed. These can be classified as follows.

(a) *Adsorption*. The enzyme can be adsorbed onto an ion-exchange resin, insoluble polymer, porous glass or activated carbon.
(b) *Covalent bonding*. Reactions of side groups or cross-linking of enzymes can be used.
(c) *Entrapment*. Enzymes can be entrapped in a gel that is permeable to both the feeds and products, but not to the enzyme. Alternatively, a membrane can be used within which the enzymes have been immobilized, and feed material is made to flow through the membrane by creating a pressure difference across the membrane.

There are many other possibilities.

Whatever the nature of the reaction, the choice of catalyst and the conditions of reaction can be critical to the performance of the process, because of the resulting influence on the selectivity of the reaction and reactor cost.

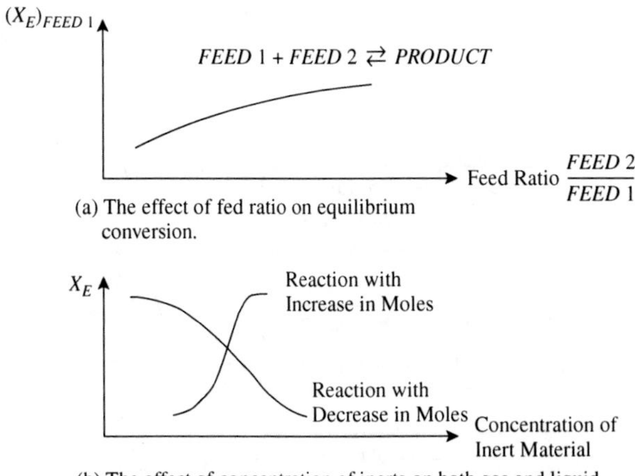

(a) The effect of fed ratio on equilibrium conversion.

(b) The effect of concentration of inerts on both gas and liquid phase reaction.

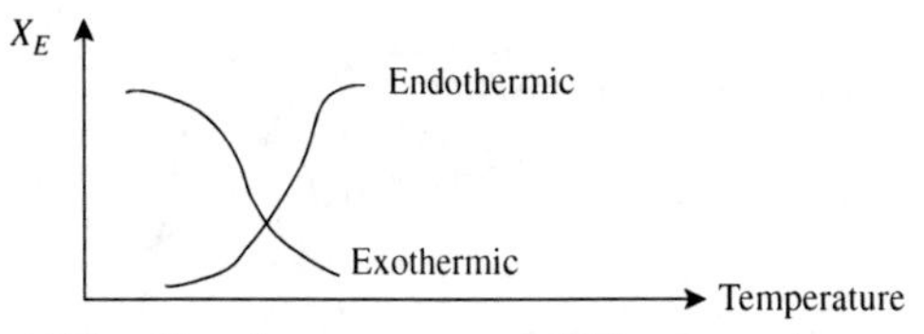

(c) The effect of temperature on equilibrium conversion.

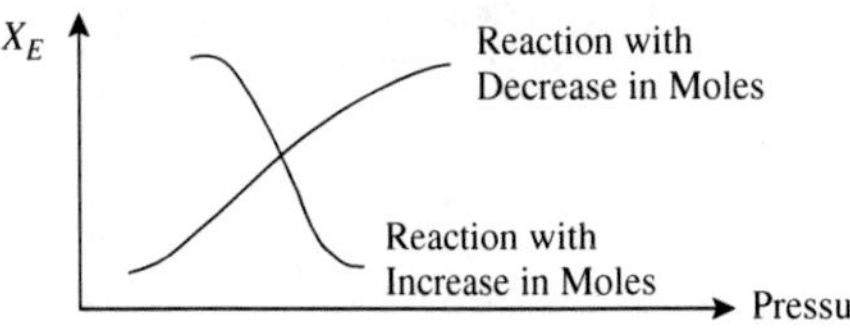

(d) The effect of pressure on equilibrium conversion for gas phase reactants.

Figure 6.8 Various measures can be taken to increase equilibrium conversion in reversible reactions. (From Smith R and Petela EA, 1991, *The Chemical Engineer*, No. 509/510:12, reproduced by permission of the Institution of Chemical Engineers).

6.8 CHOICE OF REACTOR CONDITIONS – SUMMARY

Chemical equilibrium can be predicted from data for the free energy. There are various sources for data on free energy.

1. Tabulated data are available for standard free energy of formation at different temperatures.
2. Tabulated data are available for ΔH^O and ΔS^O at different temperatures and can be used to calculate ΔG^O from Equation 6.4 written at standard conditions:

$$\Delta G^O = \Delta H^O - T\Delta S^O$$

3. Tabulated data are available for ΔG^O and ΔH^O at standard temperature. This can be extrapolated to other temperatures using heat capacity data.
4. Methods are available to allow the thermodynamic properties of compounds to be estimated from their chemical structure[6].

The equilibrium conversion can be calculated from knowledge of the free energy, together with physical properties to account for vapor and liquid-phase nonidealities. The equilibrium conversion can be changed by appropriate changes to the reactor temperature, pressure and concentration. The general trends for reaction equilibrium are summarized in Figure 6.8.

For reaction systems involving multiple reactions producing by products, selectivity and reactor yield can also be enhanced by appropriate changes to the reactor temperature, pressure and concentration. The appropriate choice of catalyst can also influence selectivity and reactor yield. The arguments are summarized in Figure 6.9[12].

For supported layered catalysts, optimizing the location of the active sites within the catalyst pellets maximizes the effectiveness or the selectivity or reactor yield.

Reactions can be catalyzed using the metabolic pathways in microorganisms or the direct use of enzymes in biochemical reactors. The use of enzymes directly can have

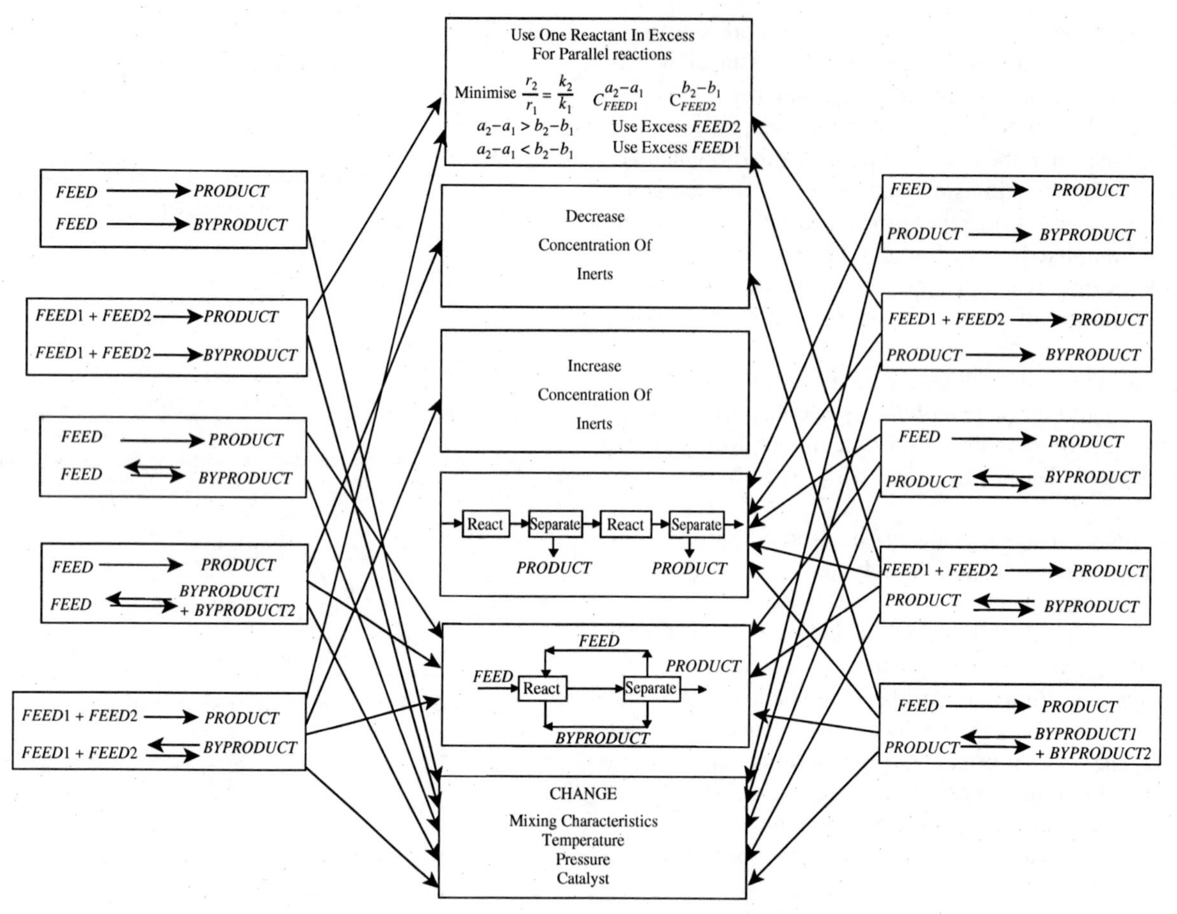

Figure 6.9 Choosing the reactor to maximize security for multiple reactants producing by products. (From Smith R and Petela EA, 1991, *The Chemical Engineer*, No. 509/510:12, reproduced by permission of the Institution of Chemical Engineers).

significant advantages if the enzymes can be isolated and immobilized in some way.

6.9 EXERCISES

1. Ethylene oxide is produced from ethylene by the following reaction:

$$CH_2 = CH_2 + \frac{1}{2}O_2 \longrightarrow H_2C\text{-}CH_2 \text{ (with O bridging)}$$

ethylene   oxygen   ethylene oxide

$$\Delta H^O = -119{,}950 \text{ kJ}$$

A parallel reaction occurs leading to a selectivity loss:

$$CH_2 = CH_2 + 3O_2 \longrightarrow 2CO_2 + 2H_2O$$

ethylene   oxygen   carbon dioxide   water

$$\Delta H^O = -1{,}323{,}950 \text{ kJ}$$

The reactor conversion is 20% and selectivity 80%. Find the heat removal rate from a plant producing 100 tons per day of ethylene oxide (see Table 6.16). Excess oxygen of 10% above stoichiometric is used. The oxygen enters at 150°C, the ethylene at 200°C, and the products leave at 280°C.

Table 6.16 Data for exercise 1.

Component	Molar mass ($kg \cdot kmol^{-1}$)	Mean heat capacity ($kJ \cdot kmol^{-1} \cdot K^{-1}$)
C_2H_4	28	63.7
O_2	32	31.3
C_2H_4O	44	80.2
CO_2	44	45.5
H_2O	18	36.7

2. Ammonia is to be produced by passing a gaseous mixture containing 60% H_2, 20% N_2, and the remainder inert gas is passed over a catalyst at a pressure of 50 bar pressure. Estimate the maximum conversion if $K_P = 0.0125$ for the reaction:

$$N_2 + 3H_2 \rightleftharpoons 2\ NH_3$$

If the system follows ideal gas behavior, what are the equilibrium conversions of nitrogen and hydrogen for a feed containing 65%H_2, 15%N_2 and the balance inert material?

3. A gas mixture of 25 mole% CO_2 and 75% CO is mixed with the stoichiometric H_2 for the following reactions:

$$CO_2 + 3H_2 \rightleftharpoons CH_3OH + H_2O \quad K_P = 2.82 \times 10^{-6}$$

$$CO + 2H_2 \rightleftharpoons CH_3OH \quad K_P = 2.80 \times 10^{-5}$$

The conversion to equilibrium is effected in the presence of a catalyst at a pressure of 300 bar. Assuming the behavior to remain ideal, calculate the fractional conversion for each reaction and hence the volume composition of the equilibrium mixture.

4. For the reaction:

$$2CO + 4H_2 \rightleftharpoons C_2H_5OH + H_2O$$

The feed consists of r moles of H_2 per mole of CO. Find an expression for the fractional conversion of CO in the presence of excess hydrogen (i.e. for $r > 2$) at 20 atm pressure and 573 K for which $\Delta G^\circ = 8,900$ kJ·kmol^{-1}. Develop the corresponding expression for the fractional conversion of H_2 when CO is in excess.

5. The following temperature-time-conversion data was obtained for the batch experiments in the gas phase for the isomerization of reactant A to product B. The equilibrium constant for the reaction is large over the temperature range concerned in Table 6.17.

Table 6.17 Data for isomerization reaction.

Temperature	Conversion at time (s)					
	0	100	500	1000	2000	5000
100°C	0	0.053	0.220	0.363	0.550	0.781
125°C	0	0.101	0.374	0.550	0.737	0.924
150°C	0	0.184	0.550	0.734	0.880	1.000

Show that the reaction is first order with respect to A and deduce the activation energy for the forward reaction.

6. The flow through a plug-flow reactor effecting a first-order forward reaction is increased by 20%, and in order to maintain the fractional conversion at its former value it is decided to increase the reactor operating temperature. If the reaction has an activation energy of 18,000 kJ·kmol^{-1} and the initial temperature is 150°C, find the new operating temperature. Would the required elevation of temperature be different if the reactor was mixed-flow?

7. The flue gas from a combustion process contains oxides of nitrogen NO_x (principally NO and NO_2). The flow of flue gas is 10 Nm^3·s^{-1} and contains 0.1% vol NO_x (expressed as NO_2 at 0°C and 1 atm) and 3% vol oxygen. There is a reversible reaction in the gas phase between the two principal oxides of nitrogen according to:

$$NO + \tfrac{1}{2} O_2 \rightleftharpoons NO_2$$

The equilibrium relationship for the reaction is given by:

$$K_a = \frac{p_{NO_2}}{p_{NO}\, p_{O_2}^{0.5}}$$

where K_a is the equilibrium constant for the reaction and p the partial pressure. At 25°C the equilibrium constant is 1.4×10^6 and at 725°C is 0.14.

a. Calculate the molar ratio of NO_2 to NO assuming chemical equilibrium at 25°C and 725°C.

b. Calculate the mass flowrate of NO_2 to NO assuming chemical equilibrium at 25°C and at 725°C. Assume the kilogram molar mass occupies 22.4 m^3 at standard conditions and the molar masses of NO and NO_2 are 30 and 46 kg·kmol^{-1} respectively.

8. In the manufacture of sulfuric acid, a mixture of sulfur dioxide and air is passed over a series of catalyst beds, and sulfur trioxide is produced according to the reaction:

$$2SO_2 + O_2 \rightleftharpoons 2SO_3$$

The sulfur dioxide enters the reactor with an initial concentration of 10% by volume, the remainder being air. At the exit of the first bed, the temperature is 620°C. Assume ideal gas behavior, the reactor operates at 1 bar and $R =$ 8.3145 kJ·K^{-1}·kmol^{-1}. Assume air to be 21% O_2 and 79% N_2. Thermodynamic data at standard conditions at 298.15 K are given in Table 6.18[6].

Table 6.18 Thermodynamic data at standard conditions and 298.15 K for sulfuric acid production.

	$\overline{H}^O$ (kJ·kmol^{-1})	$\overline{G}^O$ (kJ·kmol^{-1})
O_2	0	0
SO_2	−296,800	−300,100
SO_3	−395,700	−371,100

a. Calculate the equilibrium conversion and equilibrium concentrations for the reactor products after the first catalyst bed at 620°C assuming ΔH^O is constant over temperature range.

b. Is the reaction exothermic or endothermic?

c. From Le Châtelier's Principle, how would you expect the equilibrium conversion to change as the temperature increases?

d. How would you expect the equilibrium conversion to change as the pressure increases?

9. Repeat the calculation in Exercise 8 for equilibrium conversion and equilibrium concentration, but taking into account variation of ΔH^O with temperature. Again assume ideal gas behavior. Heat capacity coefficients for Equation 6.42 are given in Table 6.19[6].
Compare your answer with the result from Exercise 8.

Table 6.19 Heat capacity data for sulfuric acid production.

	C_P (kJ·kmol^{-1}·K^{-1})				
	α_0	$\alpha_1 \times 10^3$	$\alpha_2 \times 10^5$	$\alpha_3 \times 10^8$	$\alpha_4 \times 10^{11}$
O_2	3.630	−1.794	0.658	−0.601	0.179
SO_2	4.417	−2.234	2.344	−3.271	1.393
SO_3	3.426	6.479	1.691	−3.356	1.590

10. In Exercises 8 and 9, it was assumed that the feed to the reactor was 10% SO_2 in air. However, the nitrogen in the air is an inert and takes no part in the reaction. In practice, oxygen-enriched air could be fed.
    a. From Le Châtelier's Principle, what would you expect to happen to the equilibrium conversion if oxygen-enriched air was used for the same O_2 to SO_2 ratio?
    b. Repeat the calculation from Exercise 8 assuming the same O_2 to SO_2 ratio in the feed but using an air feed enriched from 21% purity to 50%, 70%, 90% and 99% purity.
    c. Apart from any potential advantages on equilibrium conversion, what advantages would you expect from feeding enriched air?

REFERENCES

1. Dodge BF (1944) *Chemical Engineering Thermodynamics*, McGraw-Hill.
2. Hougen OA, Watson KM and Ragatz RA (1959) *Chemical Process Principles Part II Thermodynamics*, John Wiley.
3. Coull J and Stuart EB (1964) *Equilibrium Thermodynamics*, John Wiley.
4. Smith EB (1983) *Basic Chemical Thermodynamics*, 3rd Edition, Clarendon Press, Oxford.
5. Lide DR (2003) *CRC Handbook of Chemistry and Physics*, 84th Edition, CRC Press.
6. Poling BE, Prausnitz JM and O'Connell JP (2001) *The Properties of Gases and Liquids*, 5th Edition, McGraw-Hill.
7. Waddams AL (1978) *Chemicals From Petroleum*, John Murray.
8. Madigan MT, Martinko JM and Parker J (2003) *Biology of Microorganisms*, Prentice Hall.
9. Rase HF (1990) *Fixed-Bed Reactor Design and Diagnostics-Gas Phase Reactions*, Butterworths.
10. Shyr Y-S and Ernst WR (1980) Preparation of Nonuniformly Active Catalysts, *J Catal*, **63**: 426.
11. Morbidelli M, Gavriilidis A and Varma A (2001) *Catalyst Design: Optimal Distribution of Catalyst in Pellets, Reactors, and Membranes*, Cambridge University Press.
12. Smith R and Petela EA (1992) Waste Minimisation in the Process Industries, *IChemE Symposium on Integrated Pollution Control Through Clean Technology Wilmslow UK*, Paper 9: 20.

7 Choice of Reactor III – Reactor Configuration

By contrast with ideal models, practical reactors must consider many factors other than variations in temperature, concentration and residence time. Consider the temperature control of the reactor first.

7.1 TEMPERATURE CONTROL

In the first instance, adiabatic operation of the reactor should be considered since this leads to the simplest and cheapest reactor design. If adiabatic operation produces an unacceptable rise in temperature for exothermic reactions or an unacceptable fall in temperature for endothermic reactions, this can be dealt with in a number of ways:

a. *Cold shot and hot shot.* The injection of cold fresh feed directly into the reactor at intermediate points, known as *cold shot*, can be extremely effective for control of temperature in exothermic reactions. This not only controls the temperature by direct contact heat transfer through mixing with cold material but also controls the rate of reaction by controlling the concentration of feed material. If the reaction is endothermic, then fresh feed that has been preheated can be injected at intermediate points, known as *hot shot*. Again the temperature control is through a combination of direct contact heat transfer and control of the concentration.

b. *Indirect heat transfer with the reactor.* Indirect heating or cooling can also be considered. This might be by a heat transfer surface inside the reactor, such as carrying out the reaction inside a tube and providing a heating or cooling medium outside of the tube. Alternatively, material could be taken outside of the reactor at an intermediate point to a heat transfer device to provide the heating or cooling and then returned to the reactor. Different arrangements are possible and will be considered in more detail later.

c. *Heat carrier.* An inert material can be introduced with the reactor feed to increase its heat capacity flowrate (i.e. the product of mass flowrate and specific heat capacity) and to reduce the temperature rise for exothermic reactions or reduce temperature fall for endothermic reactions. Where possible, one of the existing process fluids should be used as heat carrier. For example, an excess of feed material could be used to limit the temperature change, effectively decreasing the conversion, but for temperature-control purposes. Product or by product could be recycled to the reactor to limit the temperature change, but care must be taken to ensure that this does not have a detrimental effect on the selectivity or reactor yield. Alternatively, an extraneous inert material such as steam can be used to limit the temperature rise or fall.

d. *Catalyst profiles.* If the reactor uses a heterogeneous catalyst in which the active material is supported on a porous base, the size, shape and distribution of active material within the catalyst pellets can be varied, as discussed in Chapter 6. For a uniform distribution of active material, smaller pellets increase the effectiveness at the expense of an increased pressure drop in packed beds, and the shape for the same pellet volume generally provides effectiveness in the order

$$\text{slab} > \text{cylinder} > \text{sphere}$$

Cylinders have the advantage that they are cheap to manufacture. In addition to varying the shape, the distribution of the active material within the pellets can be varied, as illustrated in Figure 6.7. For packed-bed reactors, the size and shape of the pellets and the distribution of active material within the pellets can be varied through the length of the reactor to control the rate of heat release (for exothermic reactions) or heat input (for endothermic reactions). This involves creating different zones in the reactor, each with its own catalyst designs.

For example, suppose the temperature of a highly exothermic reaction needs to be controlled by packing the catalyst inside the tubes and passing a cooling medium outside of the tubes. If a uniform distribution of catalyst is used, a high heat release would be expected close to the reactor inlet, where the concentration of feed material is high. The heat release would then gradually decrease through the reactor. This typically manifests itself as an increasing temperature from the reactor inlet, because of a high level of heat release that the cooling medium does not remove completely in the early stages, reaching a peak and then decreasing towards the reactor exit as the rate of heat release decreases. Using a zone with a catalyst design with low effectiveness at the inlet and zones with increasing effectiveness through the reactor would control the rate of reaction to a more even profile through the reactor, allowing better temperature control.

Chemical Process Design and Integration R. Smith
 ISBNs: 0-471-48680-9 (HB); 0-471-48681-7 (PB)

Alternatively, rather than using a different design of pellet in different zones through the reactor, a mixture of catalyst pellets and inert pellets can be used to effectively "dilute" the catalyst. Varying the mixture of active and inert pellets allows the rate of reaction in different parts of the bed to be controlled more easily. Using zones with decreasing amounts of inert pellets through the reactor would control the rate of reaction to a more even profile through the reactor, allowing better temperature control.

As an example, consider the production of ethylene oxide, which uses a silver-supported catalyst[1]:

$$\underset{\text{ethylene}}{CH_2{=}CH_2} + \underset{\text{oxygen}}{1/2O_2} \longrightarrow \underset{\text{ethylene oxide}}{H_2C{-}CH_2\ (\text{bridged by } O)}$$

$$\Delta H^O = -119{,}950 \text{ kJ}$$

A parallel reaction occurs leading to a selectivity loss:

$$\underset{\text{ethylene}}{CH_2 = CH_2} + \underset{\text{oxygen}}{3O_2} \longrightarrow \underset{\text{carbon dioxide}}{2CO_2} + \underset{\text{water}}{2H_2O}$$

$$\Delta H^O = -1{,}323{,}950 \text{ kJ}$$

The reaction system is highly exothermic and is carried out inside tubes packed with catalyst, and a coolant is circulated around the exterior of the tubes to remove the heat of reaction. If a uniform catalyst design is used throughout the tubes, a high peak in temperature occurs close to the inlet. The peak in the temperature promotes the secondary reaction, leading to a high selectivity loss. Using a catalyst with lower effectiveness at the reactor inlet can reduce the temperature peak and increase the selectivity. It is desirable to vary the catalyst design in different zones through the reactor to obtain an even temperature profile along the reactor tubes.

As another example, consider the application of a fixed-bed tubular reactor for the production of methanol. Synthesis gas (a mixture of hydrogen, carbon monoxide and carbon dioxide) is reacted over a copper-based catalyst[2]. The main reactions are

$$\underset{\text{carbon monoxide}}{CO} + \underset{\text{hydrogen}}{2H_2} \rightleftharpoons \underset{\text{methanol}}{CH_3OH}$$

$$\underset{\text{carbon dioxide}}{CO_2} + \underset{\text{hydrogen}}{H_2} \longrightarrow \underset{\text{carbon monoxide}}{CO} + \underset{\text{water}}{H_2O}$$

The first reaction is exothermic, and the second is endothermic. Overall, the reaction evolves considerable heat. Figure 7.1 shows two alternative reactor designs[2]. Figure 7.1a shows a shell-and-tube type of device that generates steam on the shell side. The temperature profile shows a peak shortly after the reactor inlet because of a high rate of reaction close to the inlet. The temperature then comes back into control. The temperature profile through the reactor in Figure 7.1a is seen to be relatively smooth. Figure 7.1b shows an alternative reactor design that uses cold-shot cooling. By contrast with the tubular reactor, the cold-shot reactor in Figure 7.1b experiences significant temperature fluctuations. Such fluctuations can, under some circumstances, cause accidental catalyst overheating and shorten catalyst life.

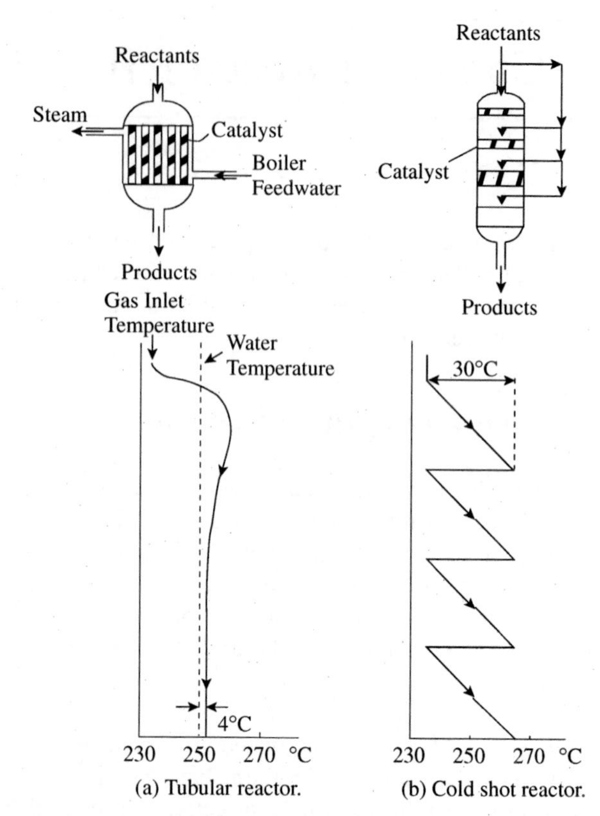

Figure 7.1 Two alternative reactor designs for methanol production give quite different thermal profiles.

Even if the reactor temperature is controlled within acceptable limits, the reactor effluent may need to be cooled rapidly, or *quenched*, for example, to stop the reaction quickly to prevent excessive by product formation. This quench can be accomplished by indirect heat transfer using conventional heat transfer equipment or by direct heat transfer by mixing with another fluid. A commonly encountered situation is one in which gaseous products from a reactor need rapid cooling and this is accomplished by mixing with a liquid that evaporates. The heat required to evaporate the liquid causes the gaseous products to cool rapidly. The quench liquid can be a recycled, cooled product or an inert material such as water.

In fact, cooling of the reactor effluent by direct heat transfer can be used for a variety of reasons:

- The reaction is very rapid and must be stopped quickly to prevent excessive byproduct formation.

- The reactor products are so hot or corrosive that if passed directly to a heat exchanger, special materials of construction or an expensive mechanical design would be required.
- The reactor product cooling would cause excessive fouling in a conventional exchanger.

The liquid used for the direct heat transfer should be chosen such that it can be separated easily from the reactor product and hence recycled with the minimum expense. Use of extraneous materials, that is, materials that do not already exist in the process, should be avoided if possible because of the following reasons:

- Extraneous material can create additional separation problems, requiring new separations that do not exist inherently within the process.
- The introduction of extraneous material can create new problems in achieving product purity specifications.
- It is often difficult to separate and recycle extraneous material with high efficiency. Any material not recycled can become an environmental problem. As shall be discussed later, the best way to deal with effluent problems is not to create them in the first place.

7.2 CATALYST DEGRADATION

The performance of most catalysts deteriorates with time[3–5]. The rate at which the deterioration takes place is not only an important factor in the choice of catalyst and reactor conditions but also the reactor configuration.

Loss of catalyst performance can occur in a number of ways:

a. *Physical loss.* Physical loss is particularly important with homogeneous catalysts, which need to be separated from reaction products and recycled. Unless this can be done with high efficiency, it leads to physical loss (and subsequent environmental problems). However, physical loss, as a problem, is not restricted to homogeneous catalysts. It also can be a problem with heterogeneous catalysts. This is particularly the case when catalytic fluidized-bed reactors (to be discussed later) are employed. Catalyst particles are held in suspension and are mixed by the upward flow of a gas stream that is blown through the catalyst bed. Attrition of the particles causes the catalyst particles to be broken down in size. Particles carried over from the fluidized bed by entrainment are normally separated from the reactor effluent and recycled to the bed. However, the finest particles are not separated and recycled, and are lost.

b. *Surface deposits.* The formation of deposits on the surface of solid catalysts introduces a physical barrier to the reacting species. The deposits are most often insoluble (in liquid-phase reactions) or nonvolatile (in gas-phase reactions) by products of the reaction. An example of this is the formation of carbon deposits (known as *coke*) on the surface of catalysts involved in hydrocarbon reactions. Such coke formation can sometimes be suppressed by suitable adjustment of the feed composition. If coke formation occurs, the catalyst can often be regenerated by air oxidation of the carbon deposits at elevated temperatures.

c. *Sintering.* With high-temperature gas-phase reactions that use solid catalysts, *sintering* of the support or the active material can occur. Sintering is a molecular rearrangement that occurs below the melting point of the material and causes a reduction in the effective surface area of the catalyst. This problem is accelerated if poor heat transfer or poor mixing of reactants leads to local hot spots in the catalyst bed. Sintering can also occur during regeneration of catalysts to remove surface deposits of carbon by oxidation at elevated temperatures. Sintering can begin at temperatures as low as half the melting point of the catalyst.

d. *Poisoning. Poisons* are materials that chemically react with, or form strong chemical bonds, with the catalyst. Such reactions degrade the catalyst and reduce its activity. Poisons are usually impurities in the raw materials or products of corrosion. They can either have a reversible or irreversible effect on the catalyst.

e. *Chemical change.* In theory, a catalyst should not undergo chemical change. However, some catalysts can slowly change chemically, with a consequent reduction in activity.

The rate at which the catalyst is lost or degrades has a major influence on the design of the reactor. Deterioration in performance lowers the rate of reaction, which, for a given reactor design, manifests itself as a lowering of conversion with time. An operating policy of gradually increasing the temperature of the reactor through time can often be used to compensate for this deterioration in performance. However, significant increases in temperature can degrade selectivity considerably and can often accelerate the mechanisms that cause catalyst degradation.

If degradation is rapid, the reactor configuration must make provision either by having standby capacity or by removing the catalyst from the bed on a continuous basis. This will be discussed in more detail later, when considering reactor configuration. In addition to the cost implications, there are also environmental implications, since the lost or degraded catalyst represents waste. While it is often possible to recover useful materials from the degraded catalyst and recycle those materials in the manufacture of new catalyst, this still inevitably creates waste since the recovery of material can never be complete.

7.3 GAS–LIQUID AND LIQUID–LIQUID REACTORS

There are many reactions involving more than one reactant, where the reactants are fed in different phases as gas–liquid or liquid–liquid mixtures. This might be inevitable because the feed material is inherently in different phases at the inlet conditions. Alternatively, it might be desirable to create two-phase behavior in order to remove an unwanted component from one of the phases or to improve the selectivity. If the reaction is two-phase, then it is necessary that the phases be intimately mixed so that mass transfer of the reactants between phases can take place effectively. The overall rate of reaction must take account of the mass transfer resistance in order to bring the reactants together as well as the resistance of the chemical reactions. The three aspects of mixing, mass transfer and reaction can present widely differing difficulties, depending on the problem.

1. *Gas–liquid reactors.* Gas–liquid reactors are quite common. Gas-phase components will normally have a small molar mass. Consider the interface between a gas and a liquid that is assumed to have a flow pattern giving a stagnant film in the liquid and the gas on each side of the interface, as illustrated in Figure 7.2. The bulk of the gas and the liquid are assumed to have a uniform concentration. It will be assumed here that Reactant A must transfer from the gas to the liquid for the reaction to occur. There is diffusional resistance in the gas film and the liquid film.

Consider an extreme case in which there is no resistance to reaction and all of the resistance is due to mass transfer. The rate of mass transfer is proportional to the interfacial area and the concentration of the driving force. An expression can be written for the rate of transfer of Component i from gas to liquid through the gas film per unit volume of reaction mixture:

$$N_{G,i} = k_{G,i}A_I(p_{G,i} - p_{I,i}) \tag{7.1}$$

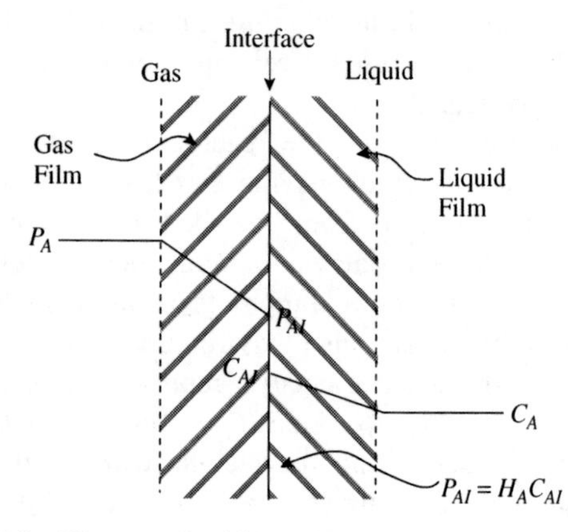

Figure 7.2 The gas–liquid interface.

where $N_{G,i}$ = rate of transfer of Component i in the gas film ($kmol \cdot s^{-1} \cdot m^{-3}$)
$k_{G,i}$ = mass transfer coefficient in the gas film ($kmol \cdot Pa^{-1} \cdot m^{-2} \cdot s^{-1}$)
A_I = interfacial area per unit volume ($m^2 \cdot m^{-3}$)
$p_{G,i}$ = partial pressure of Component i in the bulk gas phase (Pa)
$p_{I,i}$ = partial pressure of Component i at the interface (Pa)

An expression can also be written for the rate of transfer of Component i through the liquid film, per unit volume of reaction mixture:

$$N_{L,i} = k_{L,i}A_I(C_{I,i} - C_{L,i}) \tag{7.2}$$

where $N_{L,i}$ = rate of transfer of Component i in the liquid film ($kmol \cdot s^{-1} \cdot m^{-3}$)
$k_{L,i}$ = mass transfer coefficient in the gas film ($m \cdot s^{-1}$)
A_I = interfacial area per unit volume ($m^2 \cdot m^{-3}$)
$C_{I,i}$ = concentration of Component i at the interface ($kmol \cdot m^{-3}$)
$C_{L,i}$ = concentration of Component i in the bulk liquid phase ($kmol \cdot m^{-3}$)

If equilibrium conditions at the interface are assumed to be described by Henry's Law (see Chapter 4):

$$p_{I,i} = H_i x_{I,i} = \frac{H_i}{\rho_L} C_{I,i} \tag{7.3}$$

where H_A = Henry's Law constant (Pa)
$x_{I,i}$ = mole fraction of Component i in the liquid at the interface (–)
$C_{I,i}$ = concentration of Component i in the liquid at the interface ($kmol \cdot m^{-3}$)
ρ_L = molar density of the liquid phase ($kmol \cdot m^{-3}$)

The Henry's Law constant varies between different gases and must be determined experimentally. If steady state is assumed ($N_{G,i} = N_{L,i} = N_i$), then Equations 7.1, 7.2 and 7.3 can be combined to obtain

$$N_i = \frac{1}{\left[\dfrac{1}{k_{G,i}A_I} + \dfrac{1}{k_{L,i}A_I}\dfrac{H_i}{\rho_L}\right]}\left(p_{G,i} - \frac{H_i}{\rho_L}C_{L,i}\right) \tag{7.4}$$

or

$$N_i = K_{GL,i}A_I\left(p_{G,i} - \frac{H_i}{\rho_L}C_{L,i}\right) \tag{7.5}$$

where
$$\frac{1}{K_{GL,i}A_I} = \frac{1}{k_{G,i}A_I} + \frac{1}{k_{L,i}A_I}\frac{H_i}{\rho_L} \tag{7.6}$$

$K_{GL,i}$ = overall mass transfer coefficient ($kmol \cdot Pa^{-1} \cdot m^{-2} \cdot s^{-1}$)

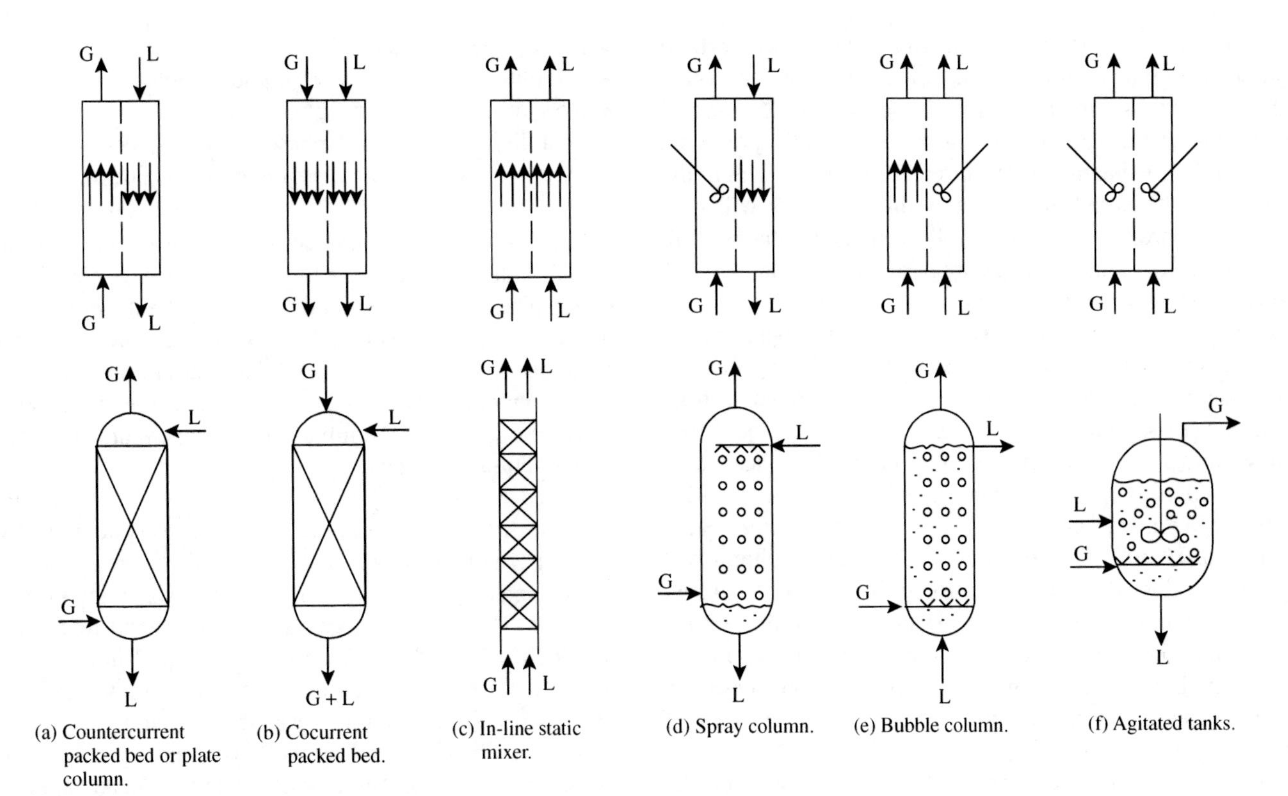

Figure 7.3 Contacting patterns for gas–liquid reactors.

$k_{G,i}$, $k_{L,i}$ and A_I are functions of physical properties and the contacting arrangement. The first term on the right-hand side of Equation 7.6 represents the gas-film resistance and the second term, the liquid-film resistance. If $k_{G,i}$ is large relative to $k_{L,i}/H_i$, the mass transfer is liquid-film controlled. This is the case for low solubility gases (H_i is large). If $k_{L,i}/H_i$ is large relative to $k_{G,i}$, it is gas-film controlled. This is the case for highly soluble gases (H_i is small).

The solubility of gases varies widely. Gases with a low solubility (e.g. N_2, O_2) have large values of the Henry's Law coefficient. This means that the liquid-film resistance in Equation 7.6 is large relative to the gas-film resistance. On the other hand, if the gas is highly soluble (e.g. CO_2, NH_3), the Henry's Law coefficient is small. This leads to the gas-film resistance being large relative to the liquid-film resistance in Equation 7.6. Thus,

- liquid-film resistance controls for gases with low solubility;
- gas-film resistance controls for gases with high solubility.

The capacity of a gas to dissolve in a liquid is determined by the solubility of the gas. The capacity of a liquid to dissolve a gas is increased if it reacts with a species in the liquid.

Now consider the effect of chemical reaction. If the reaction is fast, its effect is to reduce the liquid-film resistance. The result is an effective increase in the overall mass transfer coefficient. The capacity of the liquid is also increased.

If the reaction is slow, there is a small effect on the overall mass transfer coefficient. The driving force for mass transfer will be greater than that for physical absorption alone, as a result of the dissolving gas reacting and not building up in the bulk liquid to the same extent as with pure physical absorption.

Figure 7.3 illustrates some of the arrangements that can be used to carry out gas–liquid reactions. The first arrangement in Figure 7.3a shows a countercurrent arrangement where plug-flow is induced in both the gas and the liquid. Packing material or trays can be used to create interfacial area between the gas and the liquid. In some cases, the packed bed might be a heterogeneous solid catalyst rather than an inert material.

Figure 7.3b shows a packed-bed arrangement where plug-flow is induced in both the gas and the liquid, but both phases are flowing cocurrently. This, in general, will give a poorer performance than the countercurrent arrangement in Figure 7.3a. However, the cocurrent arrangement, known as a *trickle-bed reactor*, might be necessary if the flow of gas is much greater than the flow of liquid. This can be the case for some gas–liquid reactions involving a heterogeneous catalyst with a large excess of the gas phase. A continuous gas phase and a liquid phase in the form of films and rivulets characterize the flow pattern. Countercurrent contacting might well be preferred but a large flow of gas makes this extremely difficult.

Another way to provide cocurrent plug-flow for both phases is to feed both phases to a pipe containing an in-line static mixer, as shown in Figure 7.3c. Various designs of static mixer are available, but mixing is usually promoted by repeatedly changing the direction of flow within the device as the liquid and gas flow through. This will give a good approximation to plug-flow in both phases with cocurrent flow. Static mixers are particularly suitable when a short residence time is required.

Figure 7.3d shows a spray column. This approximates mixed-flow behavior in the gas phase and plug-flow in the liquid phase. The arrangement tends to induce a high gas-film mass transfer coefficient and a low liquid-film mass transfer coefficient. As the liquid film will tend to be controlling, the arrangement should be avoided when reacting a gas that has a low solubility (large values of Henry's Law coefficient). The spray column will generally give a lower driving force than a countercurrent packed column, but might be necessary if, for example, the liquid contains solids or solids are formed in the reaction. If the reaction has a tendency to foul a packed bed, a spray column might be preferred for reasons of practicality.

Figure 7.3e shows a bubble column. This approximates plug-flow in the gas phase and mixed-flow in the liquid phase. The arrangement tends to induce a low gas-film mass transfer coefficient and a high liquid-film mass transfer coefficient. As the gas film will tend to be controlling, the arrangement should be avoided when reacting a gas with a high solubility (small values of Henry's Law coefficient). Although the bubble column will tend have a lower performance than a countercurrent packed bed, the arrangement has two advantages over a packed bed. Firstly, the liquid hold-up per unit reactor volume is higher than a packed bed, which gives greater residence time for a slow reaction for a given liquid flowrate. Secondly, if the liquid contains a dispersed solid (e.g. a biochemical reaction using microorganisms), then a packed bed will rapidly become clogged. A disadvantage is that it will be ineffective if the liquid is highly viscous.

Finally, Figure 7.3f shows an agitated tank in which the gas is sparged through the liquid. This approximates mixed-flow behavior in both phases. The driving force is low relative to a countercurrent packed bed. However, there may be practical reasons to use a sparged agitated vessel. If the liquid is viscous (e.g. a biochemical reaction using microorganisms), the agitator allows the gas to be dispersed as small bubbles and the liquid to be circulated to maintain good contact between the gas and the liquid.

Of the contacting patterns in Figure 7.3, countercurrent packed beds offer the largest mass transfer driving force and agitated tanks the lowest.

The influence of temperature on gas–liquid reactions is more complex than homogeneous reactions. As the temperature increases,

- rate of reaction increases,
- solubility of the gas in the liquid decreases,
- rates of mass transfer increase;
- volatility of the liquid phase increases, decreasing the partial pressure of the dissolving gas in Equation 7.4.

Some of these effects have an enhancing influence on the overall rate of reaction. Others will have a detrimental effect. The relative magnitude of these effects will depend on the system in question. To make matters worse, if multiple reactions are being considered that are reversible and that also produce by products, all of the factors discussed in Chapter 6 regarding the influence of temperature also apply to gas–liquid reactions.

Added to this, the mass transfer can also influence the selectivity. For example, consider a system of two parallel reactions in which the second reaction produces an unwanted by product and is slow relative to the primary reaction. The dissolving gas species will tend to react in the liquid film and not reach the bulk liquid in significant quantity for further reaction to occur there to form the by product. Thus, in this case, the selectivity would be expected to be enhanced by the mass transfer between the phases. In other cases, little or no influence can be expected.

2. *Liquid–liquid reactors.* Examples of liquid–liquid reactions are the nitration and sulfonation of organic liquids. Much of the discussion for gas–liquid reactions also applies to liquid–liquid reactions. In liquid–liquid reactions, mass needs to be transferred between two immiscible liquids for the reaction to take place. However, rather than gas- and liquid-film resistance as shown in Figure 7.2, there are two liquid-film resistances. The reaction may occur in one phase or both phases simultaneously. Generally, the solubility relationships are such that the extent of the reactions in one of the phases is so small that it can be neglected.

For the mass transfer (and hence, reaction) to take place, one liquid phase must be dispersed in the other. A decision must be made as to which phase should be dispersed in a continuous phase of the other. In most cases, the liquid with the smaller volume flowrate will be dispersed in the other. The overall mass transfer coefficient depends on the physical properties of the liquids and the interfacial area. In turn, the size of the liquid droplets and the volume fraction of the dispersed phase in the reactor govern the interfacial area. Dispersion requires the input of power either through an agitator or by pumping of the liquids. The resulting degree of dispersion depends on the power input, interfacial tension between the liquids and their physical properties. While it is generally desirable to have a high interfacial area and, therefore, small droplets, too effective a dispersion might lead to the formation of an emulsion that is difficult to separate after the reactor.

Figure 7.4 illustrates some of the arrangements that can be used for liquid–liquid reactors. The first arrangement

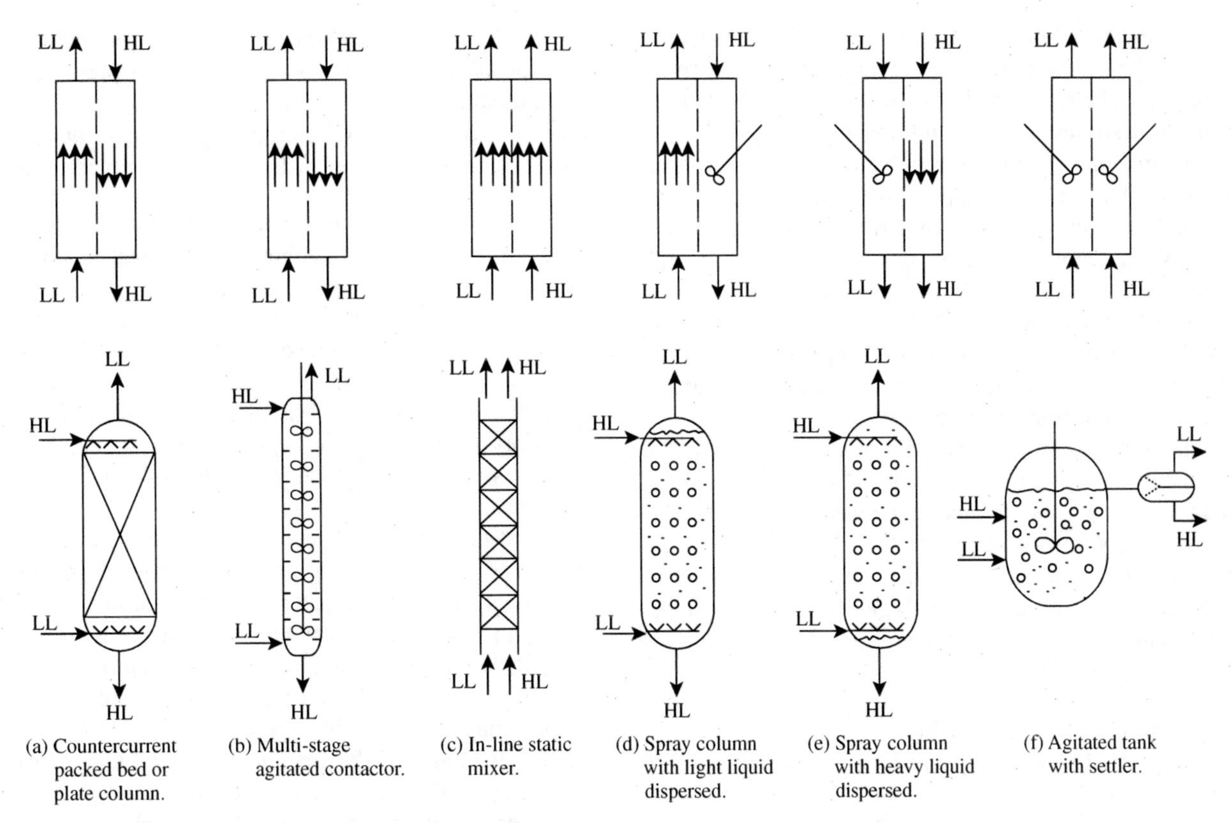

Figure 7.4 Contacting patterns for liquid–liquid reactors.

shown in Figure 7.4a is a packed bed in which the two liquids flow countercurrently. This is similar to Figure 7.3a for gas–liquid reactions. Plates can also be used to create the contact between the two liquid phases, rather than a packed bed. This arrangement will approximate plug-flow in both phases.

Figure 7.4b shows a multistage agitated contactor. A large number of stages and low backmixing will tend to approximate plug-flow in both phases. However, the closeness to plug-flow will depend on the detailed design.

Figure 7.4c shows an in-line static mixer. Dispersion is usually promoted by repeatedly changing the direction of flow locally within the mixing device as the liquids are pumped through. This will give a good approximation to plug-flow in both phases in cocurrent flow. As with gas–liquid reactors, static mixers are particularly suitable when a short residence time is required.

Figure 7.4d shows a spray column in which the light liquid is dispersed. This approximates plug-flow in the light liquid phase and mixed-flow behavior in the heavy liquid phase. Figure 7.4e shows a spray column in which the heavy liquid is dispersed. This approximates mixed-flow in the light liquid phase and plug-flow behavior in the heavy liquid phase. Spray columns will generally give a lower driving force than countercurrent packed columns, multistage agitated contactors and in-line static mixers.

Figure 7.4f shows an agitated tank followed by a settler in a *mixer–settler* arrangement, which will exhibit mixed-flow in both phases. Although Figure 7.4f shows a single-stage agitated tank and settler, a number of agitated tanks, each followed by a settler, can be connected together. For such a cascade of mixer–settler devices, the two liquid phases can be made to flow countercurrently through the cascade. The more stages that are used, the more the cascade will tend to countercurrent plug-flow behavior. Rather than a countercurrent flow arrangement through the cascade, a cross-flow arrangement can be used in which one of the phases is progressively added and removed at different points through the cascade. Such a flow arrangement can be useful if the reaction is limited by chemical equilibrium. If removal of the liquid removes the product that has formed, the reaction can be forced to higher conversion than that of a countercurrent arrangement, as discussed in Chapter 6.

7.4 REACTOR CONFIGURATION

Consider now some of the more common types of reactor configuration and their use:

1. *Tubular reactors.* Although tubular reactors often take the actual form of a tube, they can be any reactor in which there is steady movement only in one direction. If heat needs to be added or removed as the reaction proceeds, the tubes may be arranged in parallel, in a construction similar

to a shell-and-tube heat exchanger. Here, the reactants are fed inside the tubes and a cooling or heating medium is circulated around the outside of the tubes. If a high temperature or high heat flux into the reactor is required, then the tubes are constructed in the radiant zone of a furnace.

Because tubular reactors approximate plug-flow, they are used if careful control of residence time is important, as is the case where there are multiple reactions in series. A high ratio of heat transfer surface area to volume is possible, which is an advantage if high rates of heat transfer are required. It is sometimes possible to approach isothermal conditions or a predetermined temperature profile by careful design of the heat transfer arrangements.

Tubular reactors can be used for multiphase reactions, as discussed in the previous section. However, it is often difficult to achieve good mixing between phases, unless static mixer tube inserts are used.

One mechanical advantage tubular devices have is when high pressure is required. Under high-pressure conditions, a small-diameter cylinder requires a thinner wall than a large-diameter cylinder.

2. *Stirred-tank reactors*. Stirred-tank reactors consist simply of an agitated tank and are used for reactions involving a liquid. Applications include:

- homogeneous liquid-phase reactions
- heterogeneous gas–liquid reactions
- heterogeneous liquid–liquid reactions
- heterogeneous solid–liquid reactions
- heterogeneous gas–solid–liquid reactions.

Stirred-tank reactors can be operated in batch, semi-batch, or continuous mode. In batch or semi-batch mode:

- operation is more flexible for variable production rates or for manufacture of a variety of similar products in the same equipment;
- labor costs tend to be higher (although this can be overcome to some extent by use of computer control).

In continuous operation, automatic control tends to be more straightforward (leading to lower labor costs and greater consistency of operation).

In practice, it is often possible with stirred-tank reactors to come close to the idealized mixed-flow model, providing the fluid phase is not too viscous. For homogenous reactions, such reactors should be avoided for some types of parallel reaction systems (see Figure 5.6) and for all systems in which byproduct formation is via series reactions.

Stirred-tank reactors become unfavorable if the reaction must take place at high pressure. Under high-pressure conditions, a small-diameter cylinder requires a thinner wall than a large-diameter cylinder. Under high-pressure conditions, use of a tubular reactor is preferred; although mixing problems with heterogeneous reactions and other factors may prevent this. Another important factor to the disadvantage of the continuous stirred-tank reactor is that for a given conversion, it requires a large inventory of material relative to a tubular reactor. This is not desirable for safety reasons if the reactants or products are particularly hazardous.

Heat can be added to or removed from stirred-tank reactors via external jackets (Figure 7.5a), internal coils (Figure 7.5b) or separate heat exchangers by means of a flow loop (Figure 7.5c). Figure 7.5d shows vaporization of the contents being condensed and refluxed to remove heat. A variation on Figure 7.5d would not reflux the evaporated

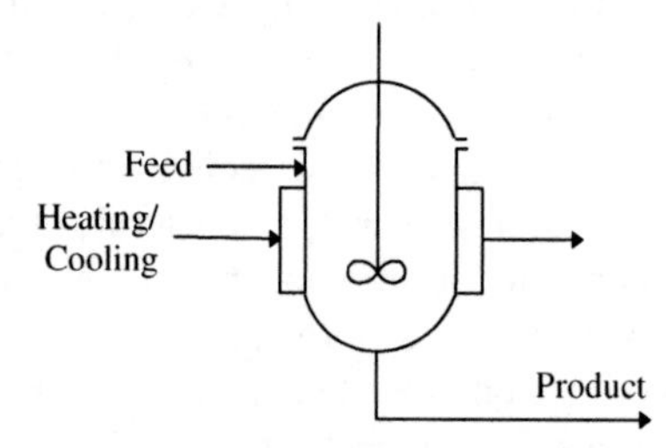

(a) Stirred tank with external jacket.

Feed
Heating/Cooling
Product

(b) Stirred tank with internal coil.

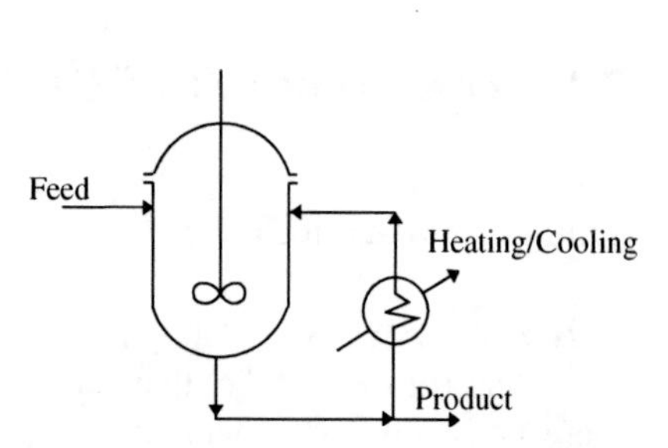

(c) Stirred tank with external heat exchanger.

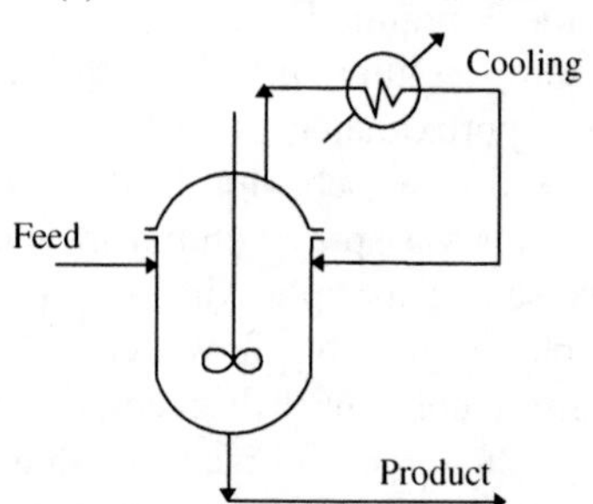

(d) Stirred tank with reflux for heat removal.

Figure 7.5 Heat transfer to and from stirred tanks.

material back to the reactor, but would remove it as a product. Removing evaporated material in this way if it is a product or byproduct of a reversible reaction can be used to increase equilibrium conversion, as discussed in Chapter 6.

If plug-flow is required, but the volume of the reactor is large, then plug-flow operation can be approached by using stirred tanks in series, since large volumes are often more economically arranged in stirred tanks than in tubular devices. This can also offer the advantage of better temperature control than the equivalent tubular reactor arrangement.

3. *Fixed-bed catalytic reactors.* Here, the reactor is packed with particles of solid catalyst. Most designs approximate to plug-flow behavior. The simplest form of fixed-bed catalytic reactor uses an adiabatic arrangement, as shown in Figure 7.6a. If adiabatic operation is not acceptable because of a large temperature rise for an exothermic reaction or a large decrease for an endothermic reaction, then cold shot or hot shot can be used, as shown in Figure 7.6b. Alternatively, a series of adiabatic beds with intermediate cooling or heating can be used to maintain temperature control, as shown in Figure 7.6c. The heating or cooling can be achieved by internal or external heat exchangers. Tubular reactors similar to a shell-and-tube heat exchanger can be used, in which the tubes are packed with catalyst, as shown in Figure 7.6d. The heating or cooling medium circulates around the outside of the tubes.

Generally, temperature control in fixed beds is difficult because heat loads vary through the bed. The temperature inside catalyst pellets can be significantly different from the bulk temperature of the reactants flowing through the bed, due to the diffusion of reactants through the catalyst pores to the active sites for reaction to occur. In exothermic reactors, the temperature in the catalyst can become locally excessive. Such "hot spots" can cause the onset of undesired reactions or catalyst degradation. In tubular devices such as shown in Figure 7.6d, the smaller the diameter of tube, the better is the temperature control. As discussed in Section 7.1, temperature-control problems also can be overcome by using a profile of catalyst through the reactor to even out the rate of reaction and achieve better temperature control.

If the catalyst degrades (e.g. as a result of coke formation on the surface), then a fixed-bed device will have to be taken off-line to regenerate the catalyst. This can either mean shutting down the plant or using a standby reactor. If a standby reactor is to be used, two reactors are periodically switched, keeping one online while the other is taken off-line to regenerate the catalyst. Several reactors might be used in this way to maintain an overall operation that is

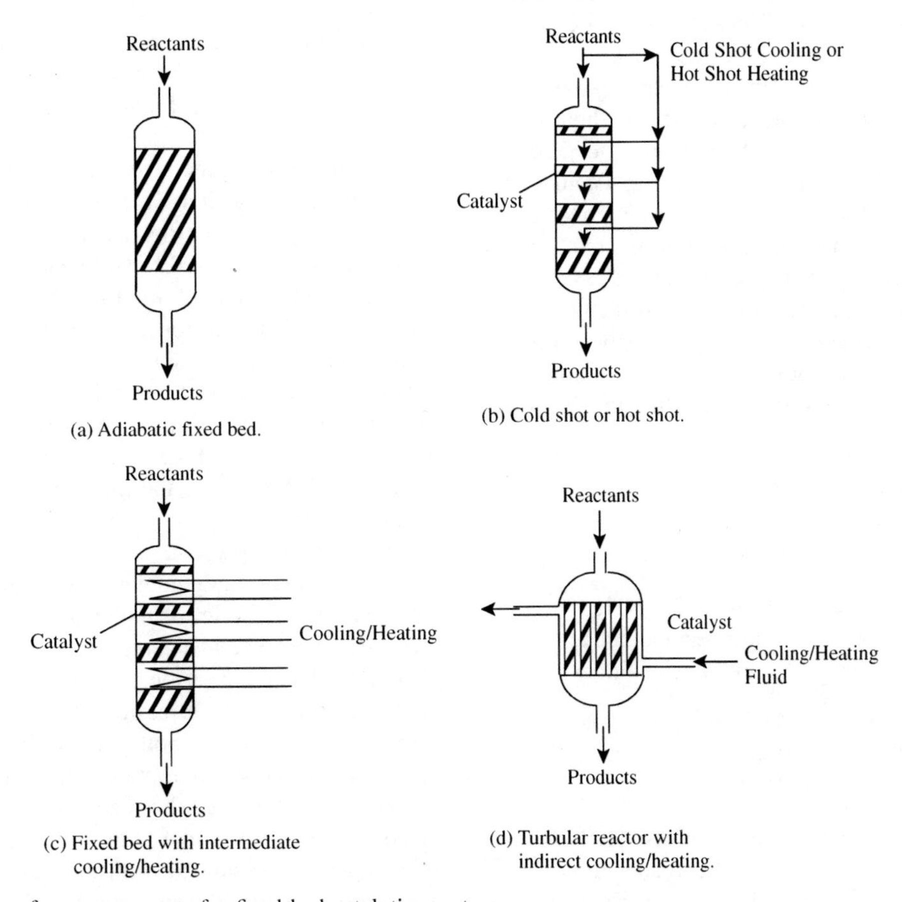

Figure 7.6 Heat transfer arrangements for fixed-bed catalytic reactors.

close to steady state. However, if frequent regeneration is required, then fixed beds are not suitable, and under these circumstances, a moving bed or a fluidized bed is preferred, as will be discussed later.

Gas–liquid mixtures are sometimes reacted in catalytic packed beds. Different contacting methods for gas–liquid reactions have been discussed in Section 7.3.

4. *Fixed-bed noncatalytic reactors.* Fixed-bed noncatalytic reactors can be used to react a gas and a solid. For example, hydrogen sulfide can be removed from fuel gases by reaction with ferric oxide:

$$\underset{\text{ferricoxide}}{Fe_2O_3} + \underset{\text{hydrogen sulfide}}{3H_2S} \longrightarrow \underset{\text{ferricsulfide}}{Fe_2S_3} + 3H_2O$$

The ferric oxide is regenerated using air:

$$2Fe_2S_3 + 3O_2 \longrightarrow 2Fe_2O_3 + 6S$$

Two fixed-bed reactors can be used in parallel, one reacting and the other regenerating. However, there are many disadvantages in carrying out this type of reaction in a packed bed. The operation is not under steady state conditions, and this can present control problems. Eventually, the bed must be taken off line to replace the solid. Fluidized beds (to be discussed later) are usually preferred for gas–solid noncatalytic reactions.

Fixed-bed reactors in the form of gas absorption equipment are used commonly for noncatalytic gas–liquid reactions. Here, the packed bed serves only to give good contact between the gas and liquid. Both cocurrent and countercurrent operation are used. Countercurrent operation gives the highest reaction rates. Cocurrent operation is preferred if a short liquid residence time is required or if the gas flowrate is so high that countercurrent operation is difficult.

For example, hydrogen sulfide and carbon dioxide can be removed from natural gas by reaction with monoethanolamine in an absorber, according to the following reactions[6]

$$\underset{\text{monoethanolamine}}{HOCH_2CH_2NH_2} + \underset{\text{hydrogen sulfide}}{H_2S} \rightleftharpoons \underset{\text{monoethanolamine hydrogen sulfide}}{HOCH_2CH_2NH_3HS}$$

$$\underset{\text{monoethanolamine}}{HOCH_2CH_2NH_2} + \underset{\text{carbon dioxide}}{CO_2} + H_2O \rightleftharpoons \underset{\text{monoethanolamine hydrogen carbonate}}{HOCH_2CH_2NH_3HCO_3}$$

These reactions can be reversed in a stripping column. The input of heat in the stripping column releases the hydrogen sulfide and carbon dioxide for further processing. The monoethanolamine can then be recycled.

5. *Moving-bed catalytic reactors.* If a solid catalyst degrades in performance, the rate of degradation in a fixed bed might be unacceptable. In this case, a moving-bed reactor can be used. Here, the catalyst is kept in motion by the feed to the reactor and the product. This makes it possible to remove the catalyst continuously for regeneration. An example of a refinery hydrocracker reactor is illustrated in Figure 7.7a.

6. *Fluidized-bed catalytic reactors.* In fluidized-bed reactors, solid material in the form of fine particles is held in suspension by the upward flow of the reacting fluid. The effect of the rapid motion of the particles is good heat transfer and temperature uniformity. This prevents the formation of the hot spots that can occur with fixed-bed reactors.

The performance of fluidized-bed reactors is not approximated by either the mixed-flow or plug-flow idealized models. The solid phase tends to be in mixed-flow, but the bubbles lead to the gas phase behaving more like plug-flow. Overall, the performance of a fluidized-bed reactor often lies somewhere between the mixed-flow and plug-flow models.

In addition to the advantage of high heat transfer rates, fluidized beds are also useful in situations where catalyst particles need frequent regeneration. Under these circumstances, particles can be removed continuously from the reactor bed, regenerated and recycled back to the bed. In exothermic reactions, the recycling of catalyst can be used to remove heat from the reactor, or in endothermic reactions, it can be used to add heat.

One disadvantage of fluidized beds, as discussed previously, is that attrition of the catalyst can cause the generation of catalyst fines, which are then carried over from the bed and lost from the system. This carryover of catalyst fines sometimes necessitates cooling the reactor effluent through direct contact heat transfer by mixing with a cold fluid, since the fines tend to foul conventional heat exchangers.

Figure 7.7b shows the essential features of a refinery catalytic cracker. Large molar mass hydrocarbon molecules are made to *crack* into smaller hydrocarbon molecules in the presence of a solid catalyst. The liquid hydrocarbon feed is atomized as it enters the catalytic cracking reactor and is mixed with the catalyst particles being carried by a flow of steam or light hydrocarbon gas. The mixture is carried up the riser and the reaction is essentially complete at the top of the *riser*. However, the reaction is accompanied by the deposition of carbon (coke) on the surface of the catalyst. The catalyst is separated from the gaseous products at the top of the reactor. The gaseous products leave the reactor

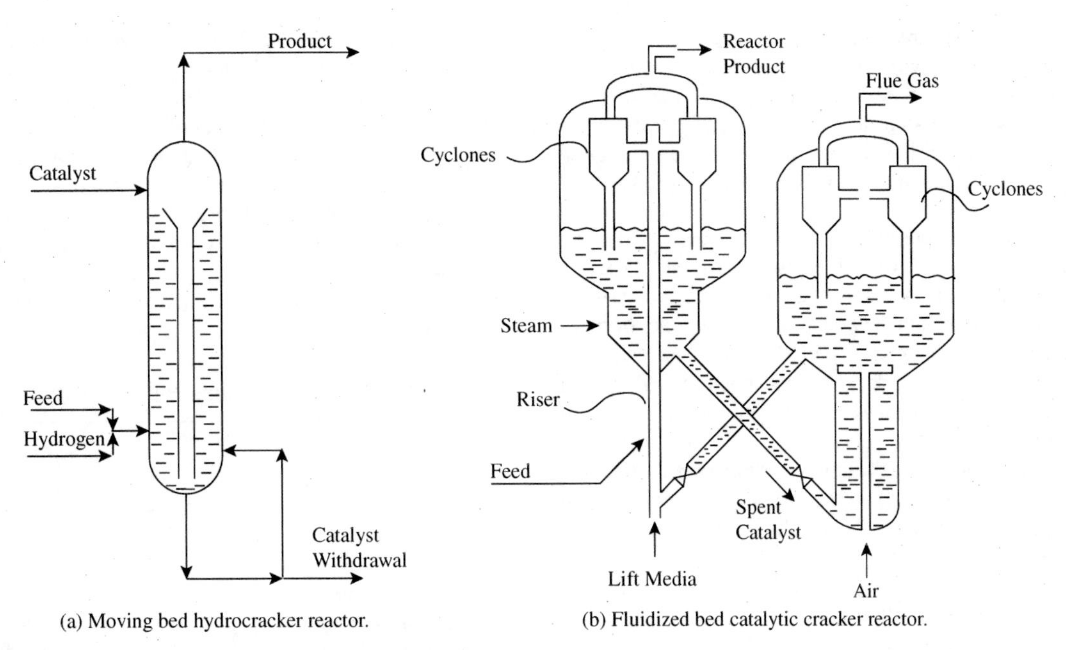

(a) Moving bed hydrocracker reactor. (b) Fluidized bed catalytic cracker reactor.

Figure 7.7 A moving bed or fluidized bed allows the catalyst to be continuously withdrawn and regenerated.

and go on for separation. The catalyst flows to a regenerator in which air is contacted with the catalyst in a fluidized bed. The air oxidizes the carbon that has been deposited on the surface of the catalyst, forming carbon dioxide and carbon monoxide. The regenerated catalyst then flows back to the reactor. The catalytic cracking reaction is endothermic and the catalyst regeneration is exothermic. The hot catalyst leaving the regenerator provides the heat of reaction to the endothermic cracking reactions. The catalyst, in this case, provides a dual function of both catalyzing the reaction and exchanging heat between the reactor and regenerator.

7. *Fluidized-bed noncatalytic reactors.* Fluidized beds are also suited to gas–solid noncatalytic reactions. All the advantages described earlier for gas–solid catalytic reactions apply here. As an example, limestone (principally, calcium carbonate) can be heated to produce calcium oxide in a fluidized-bed reactor according to the reaction

$$CaCO_3 \xrightarrow{\text{heat}} CaO + CO_2$$

Air and fuel fluidize the solid particles, which are fed to the bed and burnt to produce the high temperatures necessary for the reaction.

8. *Kilns.* Reactions involving free-flowing solid, paste and slurry materials can be carried out in kilns. In a rotary kiln, a cylindrical shell is mounted with its axis making a small angle to the horizontal and rotated slowly. The solid material to be reacted is fed to the elevated end of the kiln and it tumbles down the kiln as a result of the rotation. The behavior of the reactor usually approximates plug-flow. High-temperature reactions demand refractory lined steel shells and are usually heated by direct firing. An example of a reaction carried out in such a device is the production of hydrogen fluoride.

$$\underset{\text{calcium flouride}}{CaF_2} + \underset{\text{sulfuric acid}}{H_2SO_4} \longrightarrow \underset{\text{hydrogen flouride}}{2HF} + \underset{\text{calcium sulfate}}{CaSO_4}$$

Other designs of kilns use static shells rather than rotating shells and rely on mechanical rakes to move solid material through the reactor.

Having discussed the choice of reactor type and operating conditions, consider two examples.

Example 7.1 Monoethanolamine is required as a product. This can be produced from the reaction between ethylene oxide and ammonia[1]:

$$\underset{\text{ethylene oxide}}{H_2C\overset{O}{-}CH_2} + \underset{\text{ammonia}}{NH_3} \longrightarrow \underset{\text{monoethanolamine}}{NH_2CH_2CH_2OH}$$

Two principal secondary reactions occur to form diethanolamine and triethanolamine:

$$\underset{\text{monoethanolamine}}{NH_2CH_2CH_2OH} + \underset{\text{ethylene oxide}}{H_2C\overset{O}{-}CH_2} \longrightarrow \underset{\text{diethanolamine}}{NH(CH_2CH_2OH)_2}$$

$$\underset{\text{diethanolamine}}{NH(CH_2CH_2OH)_2} + \underset{\text{ethylene oxide}}{H_2C\overset{O}{-}CH_2} \longrightarrow \underset{\text{triethanolamine}}{N(CH_2CH_2OH)_3}$$

The secondary reactions are parallel with respect to ethylene oxide but are in series with respect to monoethanolamine. Monoethanolamine is more valuable then both the di- and triethanolamine. As a first step in the flowsheet synthesis, make an initial choice of a reactor, which will maximize the production of monoethanolamine relative to di- and triethanolamine.

Solution As much as possible, the production of di- and triethanolamine needs to be avoided. These are formed by series reactions with respect to monoethanolamine. In a mixed-flow reactor, part of the monoethanolamine formed in the primary reaction could stay for extended periods, thus increasing its chances of being converted to di- and triethanolamine. The ideal batch or plug-flow arrangement is preferred to carefully control the residence time in the reactor.

Further consideration of the reaction system reveals that the ammonia feed takes part only in the primary reaction and in neither of the secondary reactions. Consider the rate equation for the primary reaction:

$$r_1 = k_l C_{EO}^{a_1} C_{NH_3}^{b_1}$$

where
r_1 = reaction rate of primary reaction
k_1 = reaction rate constant for the primary reaction
C_{EO} = molar concentration of ethylene oxide in the reactor
C_{NH_3} = molar concentration of ammonia in the reactor
a_1, b_1 = order of primary reaction

Operation with an excess of ammonia in the reactor has the effect of increasing the rate due to the $C_{NH_3}^{b_1}$ term. However, operation with excess ammonia decreases the concentration of ethylene oxide, and the effect is to decrease the rate due to the $C_{EO}^{a_l}$ term. Whether the overall effect is a slight increase or decrease in reaction rate depends on the relative magnitude of a_1 and b_1. Consider now the rate equations for the by product reactions:

$$r_2 = k_2 C_{MEA}^{a_2} C_{EO}^{b_2}$$

$$r_3 = k_3 C_{DEA}^{a_3} C_{EO}^{b_3}$$

where
r_2, r_3 = rates of reaction to diethanolamine and triethanolamine, respectively
k_2, k_3 = reaction rate constants for the diethanolamine and triethanolamine reactions, respectively
C_{MEA} = molar concentration of monoethanolamine
C_{DEA} = molar concentration of diethanolamine
a_2, b_2 = order of reaction for the diethanolamine reaction
a_3, b_3 = order of reaction for the triethanolamine reaction

An excess of ammonia in the reactor decreases the concentrations of monoethanolamine, diethanolamine and ethylene oxide and decreases the rates of reaction for both secondary reactions.

Thus, an excess of ammonia in the reactor has a marginal effect on the primary reaction but significantly decreases the rate of the secondary reactions. Using excess ammonia can also be thought of as operating the reactor with a low conversion with respect to ammonia.

The use of an excess of ammonia is borne out in practice[1]. A mole ratio of ammonia to ethylene oxide of 10:1 yields 75% monoethanolamine, 21% diethanolamine and 4% triethanolamine. Using equimolar proportions, under the same reaction conditions, the respective proportions become 12, 23 and 65%.

Another possibility to improve selectivity is to reduce the concentration of monoethanolamine in the reactor by using more than one reactor with an intermediate separation of the monoethanolamine. Considering the boiling points of the components given in Table 7.1, then separation by distillation is apparently possible. Unfortunately, repeated distillation operations are likely to be very expensive. Also, there is a market to sell both di- and triethanolamine even though their value is lower than monoethanolamine. Thus, in this case, repeated reaction and separation is probably not justified and the choice is a single plug-flow reactor.

Table 7.1 Normal boiling points of the components.

Component	Normal boiling point (K)
Ammonia	240
Ethylene oxide	284
Monoethanolamine	444
Diethanolamine	542
Triethanolamine	609

An initial guess for the reactor conversion is difficult to make. A high conversion increases the concentration of monoethanolamine and increases the rates of the secondary reactions. A low conversion has the effect of decreasing the reactor capital cost but increasing the capital cost of many other items of equipment in the flowsheet. Thus, an initial value of 50% conversion is probably as good a guess as can be made at this stage.

Example 7.2 *Tert*-butyl hydrogen sulfate is required as an intermediate in a reaction sequence. This can be produced by the reaction between isobutylene and moderately concentrated sulfuric acid:

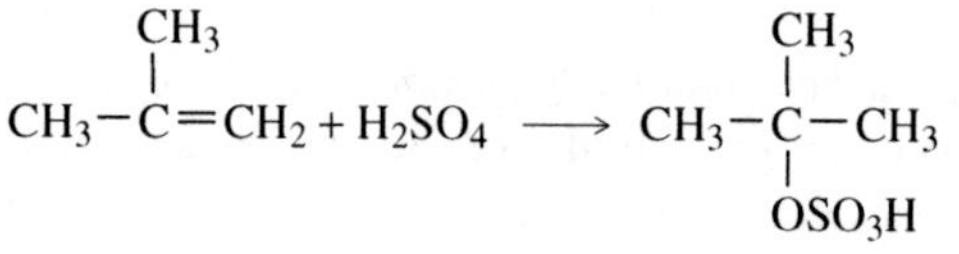

$$CH_3{-}\overset{\displaystyle CH_3}{\overset{|}{C}}{=}CH_2 + H_2SO_4 \longrightarrow CH_3{-}\underset{\displaystyle OSO_3H}{\underset{|}{\overset{\displaystyle CH_3}{\overset{|}{C}}}}{-}CH_3$$

isobutylene, sulfuric acid → *tert*-butyl hydrogen sulfate

Series reactions occur where the *tert*-butyl hydrogen sulfate reacts to form unwanted *tert*-butyl alcohol:

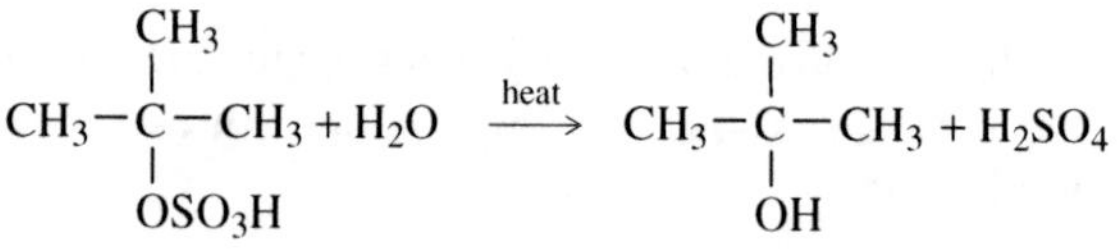

$$CH_3{-}\underset{\displaystyle OSO_3H}{\underset{|}{\overset{\displaystyle CH_3}{\overset{|}{C}}}}{-}CH_3 + H_2O \xrightarrow{\text{heat}} CH_3{-}\underset{\displaystyle OH}{\underset{|}{\overset{\displaystyle CH_3}{\overset{|}{C}}}}{-}CH_3 + H_2SO_4$$

tert-butyl hydrogen sulfate, water → *tert*-butyl alcohol, sulfuric acid

Other series reactions form unwanted polymeric material. Further information on the reaction is

- The primary reaction is rapid and exothermic.
- Laboratory studies indicate that the reactor yield is maximum when the concentration of sulfuric acid is maintained at 63%[7].

- The temperature should be maintained around 0°C or excessive by product formation occurs[7,8].

Make an initial choice of reactor.

Solution The byproduct reactions to avoid are all series in nature. This suggests that a mixed-flow reactor should not be used, rather either a batch or plug-flow reactor should be used.

However, the laboratory data seem to indicate that maintaining a constant concentration in the reactor to maintain 63% sulfuric acid in the reactor would be beneficial. Careful temperature control is also important. These two factors would suggest that a mixed-flow reactor is appropriate. There is a conflict. How can a well-defined residence time and a constant concentration of sulfuric acid be simultaneously maintained?

Using a batch reactor, a constant concentration of sulfuric acid can be maintained by adding concentrated sulfuric acid as the reaction progresses, that is, semi-batch operation. Good temperature control of such systems can be maintained.

By choosing to use a continuous rather than a batch reactor, plug-flow behavior can be approached using a series of mixed-flow reactors. This again allows concentrated sulfuric acid to be added as the reaction progresses, in a similar way as suggested for some parallel systems in Figure 4.7. Breaking the reactor down into a series of mixed-flow reactors also allows good temperature control.

To make an initial guess for the reactor conversion is again difficult. The series nature of the byproduct reactions suggests that a value of 50% is probably as good as can be suggested at this stage.

7.5 REACTOR CONFIGURATION FOR HETEROGENEOUS SOLID-CATALYZED REACTIONS

Heterogeneous reactions involving a solid supported catalyst form an important class of reactors and require special consideration. As discussed in the previous section, such reactors can be configured in different ways:

- fixed-bed adiabatic
- fixed-bed adiabatic with intermediate cold shot or hot shot
- tubular with indirect heating or cooling
- moving bed
- fluidized bed

Of these, fixed-bed adiabatic reactors are the cheapest in terms of capital cost. Tubular reactors are more expensive than fixed-bed adiabatic reactors, with the highest capital costs associated with moving and fluidized beds. The choice of reactor configuration for reactions involving a solid supported catalyst is often dominated by the deactivation characteristics of the catalyst.

If deactivation of the catalyst is very short, then moving- or fluidized-bed reactors are required so that the catalyst can be withdrawn continuously, regenerated and returned to the reactor. The example of refinery catalytic cracking was discussed previously, as illustrated in Figure 7.7b, where the catalyst is moved rapidly from the reaction zone in the riser to regeneration. Here, the catalyst deactivates in a few seconds and must be removed from the reactor rapidly and regenerated. If the deactivation is slower, then a moving bed can be used, as illustrated in Figure 7.7a. This still allows the catalyst to be removed continuously, regenerated and returned to the reactor. If the catalyst deactivation is slower, of the order of a year or longer, then a fixed-bed adiabatic or tubular reactor can be used. Such reactors must be taken off-line for the catalyst to be regenerated. Multiple reactors can be used with standby reactors, such that one of the reactors can be taken off-line to regenerate the catalyst, with the process kept running. However, there are significant capital cost implications associated with standby reactors.

Thus, the objective of the designer should be to use a fixed-bed adiabatic reactor if possible. The reactor conditions and the catalyst design can be manipulated to minimize the deactivation. Reactor inlet temperature, pressure, composition of the reactants in the feed, catalyst shape and size, mixtures of inert catalyst, profiles of active material within the catalyst pellets, hot shot, cold shot and the introduction of inert gases in the feed can all be manipulated with this objective. There are often trade-offs to be considered between reactor size, selectivity and catalyst deactivation, as well as interactions with the rest of the process. If the temperature control requires indirect heating or cooling, then a tubular reactor should be considered, with the same variables being manipulated along with the heat transfer characteristics. If everything else fails, then continuous catalytic regeneration needs to be considered (risers, fluidized beds and moving beds).

7.6 REACTOR CONFIGURATION FROM OPTIMIZATION OF A SUPERSTRUCTURE

The factors influencing the choice of reactor configuration and conditions have been reviewed at length. This has been based on the development of the conceptual issues affecting those decisions. However, there is another approach that can be adopted to make these decisions, based on the optimization of a superstructure. The basis of this approach to process design was discussed in Chapter 1. To apply this approach to reactor design, a superstructure must first be suggested that has all the structural features that might be candidates for the final design. The superstructure will contain redundant features that need to be removed. Subjecting it to a combined structural and parameter optimization carries out the evolution of the superstructure to the final design. Consider the case of isothermal reactors first.

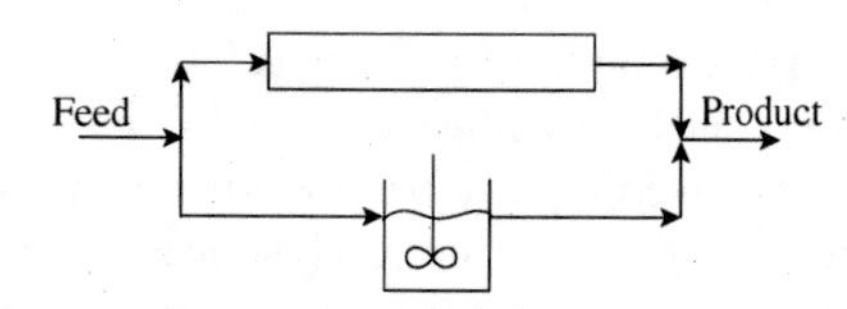

(a) A simple superstructure for a homogeneous reaction.

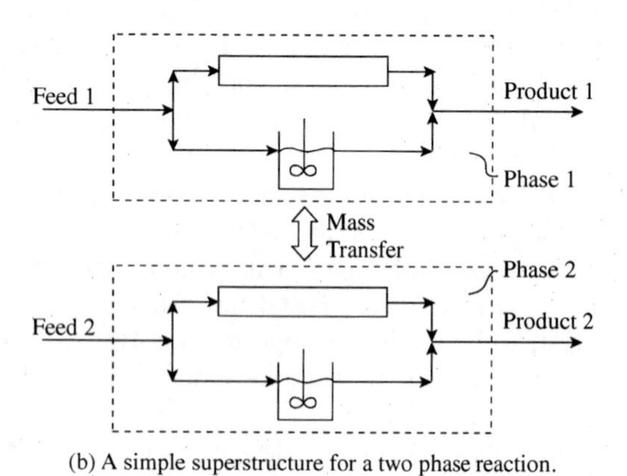

(b) A simple superstructure for a two phase reaction.

Figure 7.8 Simple superstructure for single- and two-phase reaction involving plug-flow and mixed-flow options.

1. *Isothermal reactors.* Figure 7.8a shows a simple case for a homogeneous reaction where only the two options of using either a plug-flow or mixed-flow reactor are considered. Both options are incorporated into a superstructure in which they operate in parallel. A model of the reactors needs to be created and optimized for maximum yield, maximum selectivity, minimum cost, and so on, in a combined structural and parameter optimization. Such optimizations have been discussed in Chapter 3. If it is desirable that only a plug-flow or mixed-flow reactor is chosen, then an integer constraint would have to be introduced to ensure this, as discussed in Chapter 3. Otherwise, the solution might incorporate two reactors in parallel.

The superstructure can be extended to multiphase reactors by introducing separate superstructures for each phase and by allowing mass transfer between the phases, as illustrated in Figure 7.8b. This allows either plug-flow or mixed-flow in each phase, and mass transfer between the phases. If the superstructure in Figure 7.8b was optimized, the various flow arrangements shown in Figure 7.3 for gas–liquid reactors and Figure 7.4 for liquid–liquid reactors might be obtained by the optimization choosing between different combinations of plug-flow and mixed-flow in each phase.

However, the arrangements such as those shown in Figures 7.3 and 7.4 are all conventional reactor designs. To open up the possibility of novel reactor arrangements being developed, further options need to be allowed in the superstructure. For example, a combination of different mixing patterns, as shown in Figure 7.9a, might lead to the novel design in Figure 7.9b. If a greater number of combinations of mixing patterns is allowed, this might result in a much better overall performance of the reactor. Complex reactor designs, such as that shown in Figure 7.9a, should be viewed as the ideal mixing pattern in the reactor that can, in principle, be interpreted in different ways in the final reactor design[9].

A more complex superstructure for single-phase reaction can be developed[10]. A superstructure is shown in Figure 7.10a. This involves a set of mixed-flow reactors with intermediate feed and another single mixed-flow reactor. Now, the options include semi-plug-flow with various feed addition policies, plug-flow (by elimination of intermediate feed points), or a mixed-flow reactor. However, if only one option from the semi-plug-flow, plug-flow or mixed-flow options in the superstructure is allowed on an exclusive basis, complex arrangements cannot be obtained, such as the reactor in Figure 7.9. A more complex

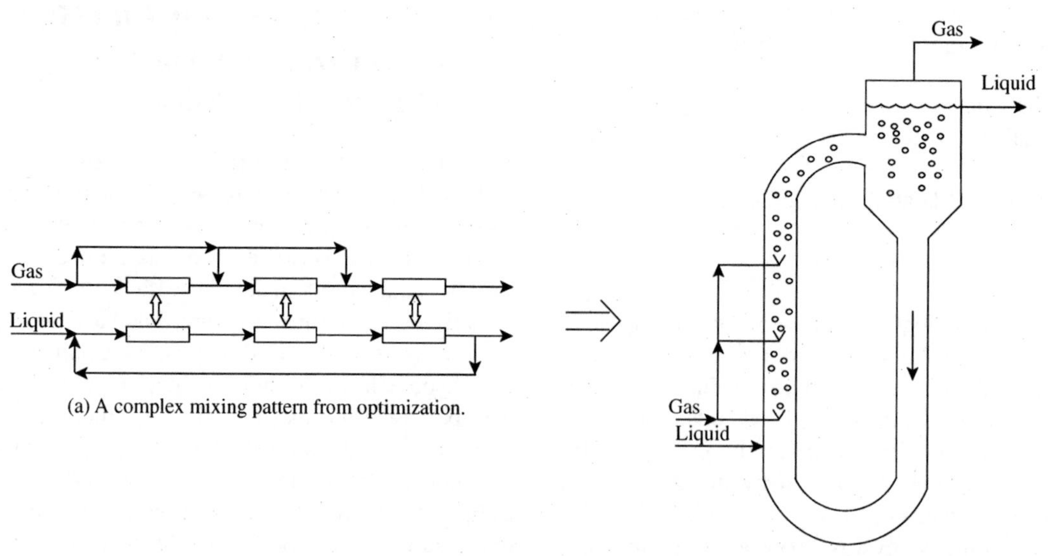

(a) A complex mixing pattern from optimization.

(b) One possible interpretation of the mixing pattern.

Figure 7.9 Complex patterns can be interpreted as novel reactor designs.

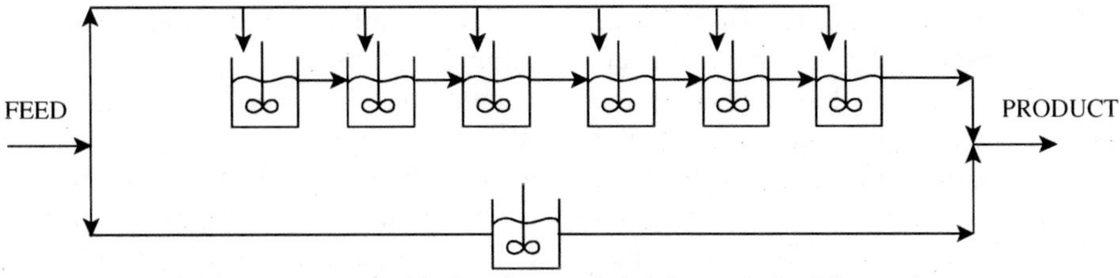

(a) A simple superstructure with plug-flow, semi-plug-flow and mixed-flow options.

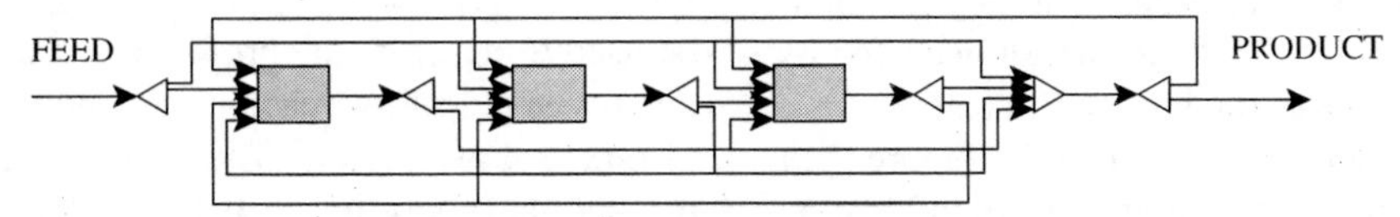

(b) A superstructure of reactor components allows series, parallel, series-parrallel and parralled-series arrangements of plug-flow, semi plug-flow and mixed-flow options.

Figure 7.10 More complex superstructures can lead to more complex mixing arrangements.

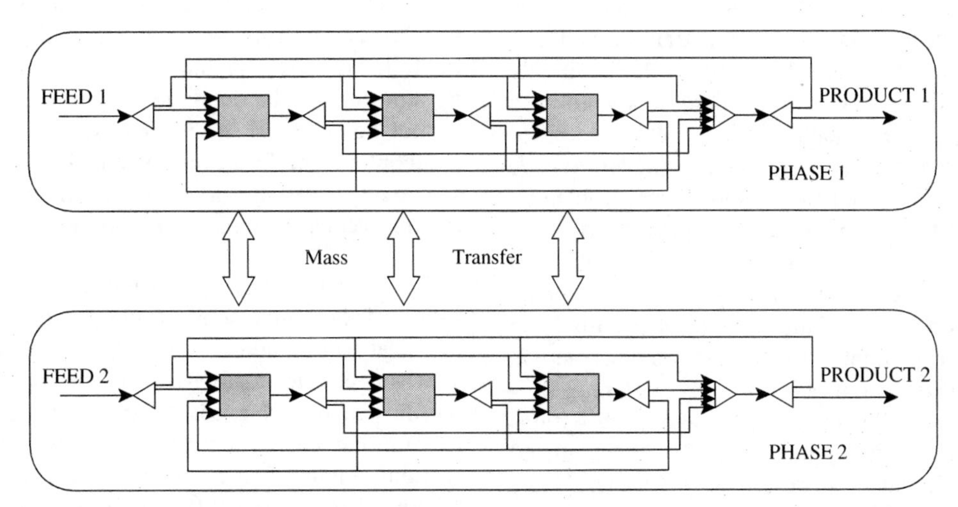

Figure 7.11 Superstructure for two-phase reactions with three reactor compartments in each phase. Mass transfer is only allowed with the corresponding shadow compartment.

superstructure that includes further options for a single phase is shown in Figure 7.10b[9,10]. The superstructure involves three *compartments*[9]. Within each compartment, there is a structure, as shown in Figure 7.10a, leading to one of plug-flow, semi-plug-flow or mixed-flow. The three compartments can then be connected in different possible ways to allow the plug-flow, semi-plug-flow or mixed-flow compartments to be connected in series, parallel, series – parallel or parallel–series arrangements. Recycle streams are also allowed. For a three-compartment superstructure, as shown in Figure 7.10b, up to three different reactor compartments can be connected in different arrangements of mixing patterns. If a greater number of possibilities were desirable, then the superstructure in Figure 7.10b could be extended to include a greater number of reactor compartments. So far, the superstructure shown in Figure 7.10b is restricted to a single phase. Now, consider how this might be extended to multiple phases[9].

In adding a greater number of options to the superstructure, care should be taken to ensure that the complexity does not increase unnecessarily by adding combinations of mixing patterns and mass transfer between phases that can never be possible in a practical reactor. For example, in the mixing pattern in Figure 7.9, it is not possible for every part of the liquid phase throughout the design to exchange mass with every part of the gas phase throughout the design. Mass can only be exchanged where reactor compartments are adjacent. Figure 7.11 shows a superstructure involving two phases, with three compartments in each phase. Mass transfer is limited to be between a reactor compartment in one phase and a single partner reactor compartment in the other phase, termed its *shadow* reactor compartment[9]. The superstructures in Figures 7.10b and 7.11 involve one feed and one product from each phase. It is straightforward to extend the superstructure to include multiple feeds and products.

When optimizing a superstructure for a multiphase reaction, the rate of mass transfer must be specified. This will, to a large extent, be determined by the design of the equipment. Yet, the objective of the superstructure optimization is to design the equipment. Thus, some assumption must be made regarding the mass transfer coefficient for the design to proceed. Once a design emerges, the assumption can be refined and the optimization repeated to ensure that another configuration is not appropriate. Also, the superstructures for multiphase reactions assume mass transfer between shadow compartments. This assumes that the reaction will occur in the bulk phase. This is a reasonable assumption for slow reactions, but fast reactions will occur near the interface. This can be allowed for by artificially enhancing the mass transfer coefficient.

Setting up superstructures like the one in Figure 7.11 is one thing, but carrying out a satisfactory optimization is another. The discussion of optimization techniques in Chapter 3 noted the difficulty of finding a solution close to the global optimum if aspects of the problem to be optimized are nonlinear. In this case, the mathematical models for the reaction kinetics, mass transfer and any hydrodynamic models included are likely to be highly nonlinear. In addition, the optimization is a combined structural and parameter optimization. Methods based on mixed integer nonlinear programming (MINLP) have been found to be inappropriate to solve such problems[9]. However, stochastic optimization has proved to be a reliable method. Simulated annealing, as described in Chapter 3, has been used successfully to solve complex reactor design problems[9,11]. This starts by first initializing the problem. For example, for a gas–liquid reaction, both phases could be initialized arbitrarily with plug-flow behavior. This is a simple configuration, with other options in the superstructure initially switched off. The simulated annealing is then initiated, as described in Chapter 3, and both the structural and parameter settings are changed in a series of random moves. New structural features can be added from the superstructure, and existing structural features can be removed by the optimization moves. The acceptance criterion for a move can be taken from the Metropolis criterion described in Chapter 3. During the course of the optimization, the objective function is allowed to deteriorate during the initial stages of the optimization. As the optimization progresses, the possibility of deterioration in the objective function is gradually removed according to the annealing schedule, as described in Chapter 3. In this way, the optimization reduces the problem of being trapped in a local optimum and allows confidence that solutions close to the global optimum can be obtained. However, it should be noted that there are usually a number of competitive solutions that are close to the global optimum, rather than there being a single global optimum solution far better than other neighboring solutions.

2. *Nonisothermal reactors.* Nonisothermal operation brings additional complexity to the superstructure approach[11,12]. In the first instance, the optimum temperature policy could be explored without consideration of how this might be engineered in practice. For this, the optimum temperature profile through the reactor is determined using profile optimization, as described in Chapter 3. The approach adopted here is to impose a temperature profile and optimize the shape of this profile, its inlet temperature, exit temperature and the maximum or minimum temperature (if the shape has a maximum or minimum) until optimum performance has been achieved[9]. Dimensionless length must be used to describe the profile because of the way the reactors (even plug-flow reactors) are modeled as a combination of mixed-flow reactors. The dimensionless length then represents the proportion of the total reactor volume.

The objective function is maximized or minimized by varying the shape of the temperature profile. This takes no account of whether the optimum profile can be achieved in a practical design. It might be that the heat transfer engineering will not allow the target profile to be followed exactly, but it provides an ultimate target to aim for in the final design. If the optimum profile cannot be achieved in the final design, this will result in a suboptimal performance. Alternatively, the optimization can be repeated with additional constraints imposed to avoid the impractical features of the profile.

3. *Nonisothermal reactors with adiabatic beds.* Optimization of the temperature profile described above assumes that heat can be added or removed wherever required and at whatever rate required so that the optimal temperature profile can be achieved. A superstructure can be set up to examine design options involving adiabatic reaction sections. Figure 7.12 shows a superstructure for a reactor with adiabatic sections[9,12] that allows heat to be transferred indirectly or directly through intermediate feed injection.

As with isothermal reactor design, the optimization of superstructures for nonisothermal reactors can be carried out reliably, using simulated annealing.

One final comment needs to be made regarding the data for the optimization calculations described here. It obviously requires a model of the chemistry and the reaction kinetics. It might also require data on mass transfer. If the models are based on data that was measured under conditions different than those in the region of the optimum, then this introduces uncertainty in the design and might invalidate the conclusions of the optimization. Whether there is considered to be uncertainty in the model or not, the outcome from the optimization needs to be validated by detailed simulation or laboratory work or pilot plant work. The best strategy is to carry out experimental work for the model development and the design and optimization in parallel. In this way, the experimental work and model development will focus on conditions that are appropriate for the final design.

Example 7.3 Monochlorinated carboxylic acids are important intermediates for the chemical industry in the production of

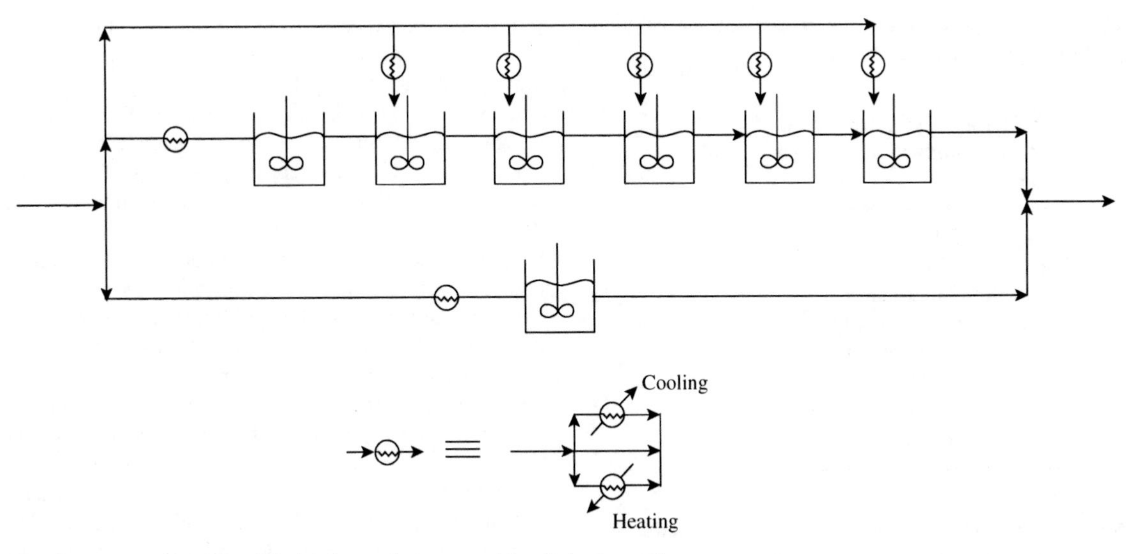

Figure 7.12 Superstructure for a Nonisothermal reactor with adiabatic sections.

pharmaceuticals, dyes, herbicides and cellulose derivatives. The chlorination of butanoic acid (BA) involves two parallel reactions that take place in the liquid phase[13].

$$BA + Cl_2 \rightarrow MBA + HCl$$

$$BA + 2Cl_2 \rightarrow DBA + 2HCl$$

where MBA is α-monochlorobutanoic acid and DBA is α, α-dichlorobutanoic acid. These reactions are catalyzed and carried out in the organic phase but are accomplished by the absorption of Cl_2 and the desorption of the volatile product, HCl. Kinetic equations have been developed[13,14]:

$$r_1 = \beta \frac{k_1 \sqrt{C_{L,MBA}} + k_2}{1 + k_3 C_{L,Cl2}} \quad (7.7)$$

$$r_2 = k_3 r_1 C_{L,Cl2} \quad (7.8)$$

where

$$\beta = \frac{\sqrt{C_{TOT} - (1 + k_3 C_{L,Cl2}) C_{L,MBA}}}{\sqrt{C_{TOT} - (1 + k_3 C_{L,Cl2}) C_{L,MBA}} + k_4 C_{L,MBA}^{1/2}} \quad (7.9)$$

$$C_{TOT} = C_{L,BA} + C_{L,MBA} + C_{L,DBA} \quad (7.10)$$

$C_{L,i}$ = molar concentration of Component i in the liquid phase($kmol \cdot m^{-3}$)

Table 7.2 gives details of the kinetic data. The catalyst mole fraction x_{cat} is assumed to be constant at 0.037[13].

Table 7.2 Kinetic parameters and physical data for the chlorination of butanic acid.

$k_1 = 0.0456 x_{cat}^{1/2} \exp[5.76 - 3260/T]$ $(kmol \cdot m^{-3})^{1/2} \cdot s^{-1}$
$k_2 = 0.4758 x_{cat} \exp[5.34 - 3760/T]$ $kmol \cdot m^{-3} \cdot s^{-1}$
$k_3 = 1.3577$ $m^3 \cdot kmol^{-1}$
$k_4 = 0.01$
T = temperature (K)

Both reactions are slow compared to the film diffusion in the liquid phase[13–15]. Hence, the reactions can be assumed to take place predominantly in the bulk phase of the liquid. The rate of mass transfer can be calculated using Equation 7.2. The interfacial concentration can be calculated using Henry' Law. Mass transfer coefficients, interfacial area and gas hold-up data are required. Gas hold-up is defined as:

$$\text{Gas hold-up} = \frac{\text{volume of gas}}{\text{volume of reaction mixture}}$$

The hydrodynamic data will depend on the mixing characteristics of the reactor. In terms of the modeling of the reactor, the hydrodynamic data will depend on the mixing characteristics of each phase and the combinations of mixing characteristics for the phases. Table 7.3 gives typical data for various combinations of mixing characteristics for gas–liquid reactors[16].

Table 7.3 Typical hydrodynamic data for various types of gas–liquid reactors[16].

Mixing characteristics of phase		Practical device	Gas hold-up (%)	$k_L\ A$ (s^{-1})
Gas	Liquid			
Plug-flow	Plug-flow	Packed column	95	0.005–0.02
Plug-flow	Plug-flow	Plate column	85	0.01–0.05
Plug-flow	Mixed-flow	Bubble column	5	0.005–0.0
Mixed-flow	Mixed-flow	Mechanically agitated vessel	10	0.02–0.2

Table 7.4 gives the hydrodynamic data used for the calculations. Gas hold-ups were taken from Table 7.3.

Table 7.4 Hydrodynamic data for the chlorination of butanic acid.

$H_{Cl_2} = 212$ bar
$H_{HCl} = 212$ bar
$\rho_{L,BA} = 9.750$ kmol·m^{-3}
$\rho_{L,MBA} = 8.857$ kmol·m^{-3}
$\rho_{L,DBA} = 6.824$ kmol·m^{-3}
$A = 255$ m^2·m^{-3}
$k_{L,Cl_2} = 0.666 \times 10^{-4}$ m·s^{-1}
$k_{L,HCl} = 0.845 \times 10^{-4}$ m·s^{-1}

The reactions are assumed to take place under isothermal conditions at 130°C at 10 bar. The liquid feed of *BA* is 0.0133 kmol·s^{-1} and the gaseous feed of chlorine is 0.1 kmol·s^{-1}. The objective is to maximize the fractional yield of α-monochlorobutanoic acid with respect to butanoic acid. Specialized software is required to perform the calculations, in this case, using simulated annealing.

Solution Start by assuming three standard design configurations:

- countercurrent packed bed
- bubble column
- mechanically agitated vessel.

A countercurrent packed bed is modeled with one plug-flow compartment in each phase (Figure 7.3a). A bubble column is modeled as mixed-flow in the liquid phase and plug-flow in the gas phase (Figure 7.3e). A mechanically agitated vessel is modeled as a mixed-flow compartment in the liquid phase and its shadow mixed-flow compartment in the gas phase (Figure 7.3f). The performances of these configurations with initial flowrates are given in Table 7.5 for a 99% conversion of *BA*.

From Table 7.5, the mechanically agitated vessel gives the best performance, not only in terms of selectivity and yield but also in terms of the reactor volume. The reactor volumes in Table 7.5 are only indicative as they are based on an assumed hold-up of the gas in the reactor.

Table 7.5 Initial yields of *MBA* from *BA* for different reactor configurations.

Reactor	Selectivity (%)	Yield (%)	Volume (m^3)
Countercurrent packed bed	65.8	65.1	163.9
Bubble column	77.6	77.9	7.05
Mechanically agitated vessel	78.9	78.2	7.35

It would be expected from the reaction stoichiometry that the concentration of chlorine in the liquid would affect the yield as a high concentration of chlorine will help promote the secondary reaction to α, α-dichlorobutanoic acid. This indicates that the flowrate of chlorine to the reactor should be optimized rather than kept to a fixed flowrate, as assumed so far. A superstructure model can be set up, as illustrated in Figure 7.11. This is then subjected to optimization. In this case, the flowrate of *BA* will be fixed, but the flowrate of chlorine will be allowed to be optimized, in addition to optimization of the reactor network structure. Given the highly nonlinear nature of the model, this is not suited to a deterministic MINLP optimization method. The problem is better suited to a stochastic method. In this case, the superstructure is optimized by simulated annealing to search for the optimum design for continuous operation in the form of a continuous reactor network. Specialist software is required for this. One of the virtues of stochastic methods is that they provide not just a single answer at the optimum point, as is the case with deterministic methods, but, rather, a range of solutions with near-optimum performance. When using simulated annealing, multiple solutions are obtained by carrying out multiple optimizations. Because of the random nature of the search, different solutions are obtained from the multiple runs in the region of the optimum. For this problem, many different reactor networks are possible to give near-optimum answers. However, the preferred answers are always the simplest network structures, as complex network structures will be difficult to engineer. In this case, it turns out that a reactor network with mixed-flow characteristics in both phases gives a good performance, in other words, a mechanically agitated vessel. Table 7.6 presents a summary of the relative performance of three standard configurations after optimization.

Table 7.6 Optimized yields of *MBA* from *BA* for different reactor configurations.

Reactor	Selectivity (%)	Yield (%)	Flowrate of chlorine (kmol·s^{-1})	Volume (m^3)
Countercurrent packed bed	79.6	78.8	0.0230	594
Bubble column	99.8	98.8	0.0146	5.3
Mechanically agitated vessel	99.9	98.9	0.0190	5.6

It can be seen that the bubble column and mechanically agitated vessel, both give a good performance. In this case there is no need to resort to more complex designs. The performance of a countercurrent packed bed after optimization is shown in Table 7.6 for comparison. The performance after optimization is significantly improved relative to a fixed flowrate of chlorine. It is not surprising that the bubble column and mechanically agitated vessel are the same (at least, as far as the model predictions are concerned). Both assume a well-mixed liquid phase. Given that the reaction is assumed to occur in the liquid phase, it therefore makes no difference that one model assumes plug-flow in the gas phase and the other, mixed-flow.

There are many assumptions in the calculations and there are also many uncertainties. Designers should take great care when carrying out such calculations to test the sensitivity of the model to different assumptions. For example, there are major uncertainties regarding the mass transfer coefficients, Henry's Law constants and hold-up, in addition to all of the uncertainties regarding the kinetic model. The designer should repeat the calculations, changing the assumptions to test whether the optimized configuration changes as a result of changing the underlying assumptions. Once the designer is satisfied with the result, the design would need to be investigated further; either using detailed simulation of the multiphase reactor (e.g. using computational fluid dynamics) and experimental investigation in the laboratory or through plant trials.

7.7 CHOICE OF REACTOR CONFIGURATION – SUMMARY

In choosing the reactor, the overriding consideration is usually the raw materials efficiency (bearing in mind materials of construction, safety, etc.). Raw materials costs are usually the most important costs in the whole process. Also, any inefficiency in the use of raw materials is likely to create waste streams that become an environmental problem. The reactor creates inefficiency in the use of raw materials in the following ways.

- If low conversion is obtained and unreacted feed material is difficult to separate and recycle.
- Through the formation of unwanted by products. Sometimes, the by product has value as a product in its own right; sometimes, it simply has value as fuel. Sometimes, it is a liability and requires disposal in expensive waste treatment processes.
- Impurities in the feed can undergo reaction to form additional by products. This is best avoided by purification of the feed before reaction.

Temperature control of the reactor can be achieved through

- cold shot and hot shot
- indirect heat transfer
- heat carriers
- catalyst profiles.

In addition, it is common to have to quench the reactor effluent to stop the reaction quickly or to avoid problems with conventional heat transfer equipment.

Catalyst degradation can be a dominant issue in the choice of reactor configuration, depending on the rate of deactivation. Slow deactivation can be dealt with by periodic shutdown and regeneration or by replacement of the catalyst. If this is not acceptable, then standby reactors can be used to maintain plant operation. If deactivation is rapid, then moving-bed and fluidized-bed reactors, in which catalyst is removed continuously for regeneration, might be the only option.

When dealing with gas–liquid and liquid–liquid reactions, mass transfer can be as equally important a consideration as reaction.

Reactor configurations for conventional designs can be categorized as

- tubular
- stirred-tank
- fixed-bed catalytic
- fixed-bed noncatalytic
- moving-bed catalytic
- fluidized-bed catalytic
- fluidized-bed noncatalytic
- kilns

The choice of reactor configuration and conditions can also be based on the optimization of a superstructure. Combinations of complexities can be included in the optimization. An added advantage of the approach is that it also allows novel configurations to be identified, as well as standard configurations.

The decisions made in the reactor design are often the most important in the whole flowsheet. The design of the reactor usually interacts strongly with the rest of the flowsheet. Hence, a return to the decisions for the reactor must be made when the process design has progressed further to understand the full consequences of those decisions.

7.8 EXERCISES

1. Chlorobenzene is manufactured by the reaction between benzene and chlorine. A number of secondary reactions occur to form undesired by products.

$$C_6H_6 + Cl_2 \longrightarrow C_6H_5Cl + HCl$$

$$C_6H_5Cl + Cl_2 \longrightarrow C_6H_4Cl_2 + HCl$$

$$C_6H_5Cl_2 + Cl_2 \longrightarrow C_6H_3Cl_3 + HCl$$

   Make an initial choice of reactor type.

2. 1000 kmol of A and 2000 kmol of B react at 400 K to form C and D, according to the reversible scheme

$$A + B \rightleftharpoons C + D$$

   The reaction takes place in a gas phase. Component C is the desired product and the equilibrium constant at 400 K is $K_a = 1$. Calculate the equilibrium conversion and explain what advantage may be gained by employing a reactor in which C is continuously removed from the reaction mixture so as to keep its partial pressure low.

3. Components A and G react by three simultaneous reactions to form three products, one that is desired (D) and two that are undesired (W and U). These gas-phase reactions, together with their corresponding rate laws, are given below.
Desired product,

$$A + G \rightarrow D$$

$$r_D = \left\{0.0156 \exp\left[18{,}200\left(\frac{1}{300} - \frac{1}{T}\right)\right]C_A C_G\right.$$

First unwanted byproduct,

$$A + G \rightarrow U$$

$$r_U = \left\{0.0234 \exp\left[17{,}850\left(\frac{1}{300} - \frac{1}{T}\right)\right]C_A^{1.5} C_G\right.$$

Second unwanted byproduct,

$$A + G \rightarrow W$$

$$r_W = \left\{0.0588 \exp\left[3{,}500\left(\frac{1}{300} - \frac{1}{T}\right)\right]C_A^{0.5} C_G\right.$$

T is the temperature (K) and C_A and C_G are the concentrations of A and G.
The objective is to select conditions to maximize the yield of D.
   a. Select the type of the reactor that appears appropriate for the given reaction kinetics. What are the drawbacks of the reactor types you would not recommend?
   b. It has been suggested to introduce inert material. What would be the effect of inerts on the yield?
   c. Assess whether the yield can be improved by operating the reactor at:
      - high or low temperatures
      - high or low pressures

4. Pure reactant A is fed at 330 K into an adiabatic reactor where it converts reversibly to useful product B:

$$A \rightleftharpoons B$$

The reactor brings the reaction mixture to equilibrium at the outlet temperature. The reaction is exothermic and the equilibrium constant K is given by:

$$K = 120{,}000 \exp\left[-20.0\left(\frac{T - 298}{T}\right)\right]$$

T is the temperature in K. The heat of reaction is $-60{,}000\ kJ{\cdot}kmol^{-1}$. The heat capacities of A and B are $190\ kJ{\cdot}kmol^{-1}{\cdot}K^{-1}$.
   a. Use an enthalpy balance to calculate the temperature of the reaction mixture as a function of the conversion. Plot the temperature along the reactor length.
   b. Calculate the exit temperature and the equilibrium conversion.
   c. A choice is required between different reactors of volume V:
      - a single adiabatic reactor
      - two adiabatic reactors
      - four adiabatic reactors

   Which of these choices will lead to a better conversion?

5. In the reaction of ethylene to ethanol:

$$\underset{\text{ethylene}}{CH_2 = CH_2} + \underset{\text{water}}{H_2O} \rightleftharpoons \underset{\text{ethanol}}{CH_3 - CH_2 - OH}$$

A side reaction occurs, where diethyl ether is formed:

$$\underset{\text{ethanol}}{2CH_3{-}CH_2{-}OH} \rightleftharpoons \underset{\text{diethylether}}{CH_3CH_2{-}O{-}CH_2CH_3} + \underset{\text{water}}{H_2O}$$

   a. Make an initial choice of reactor as a first step. How would the reactor be able to maximize selectivity to a desired product?
   b. What operating pressure would be suitable in this reactor?
   c. How would an excess of water (steam) in the reactor feed affect the selectivity of the reactor?

REFERENCES

1. Waddams A.L (1978) *Chemicals From Petroleum*, John Murray.
2. Supp E (1973) Technology of Lurgi's Low Pressure Methanol Process, *Chem Tech*, **3**: 430.
3. Butt J.B and Petersen E.E (1988) *Activation, Deactivation and Poisoning of Catalysts*, Academic Press.
4. Rase H.F (1990) *Fixed-Bed Reactor Design and Diagnostics – Gas Phase Reactions*, Butterworth.
5. Wijngaarden R.J, Kronberg A and Westerterp K.R (1998) *Industrial Catalysis – Optimizing Catalysts and Processes*, Wiley-VCH.
6. Kohl, A.L and Riesenfeld F.C (1979) *Gas Purification*, Gulf Publishing Company.
7. Morrison R.T and Boyd R.N (1992) *Organic Chemistry*, 6th Edition, Prentice-Hall.
8. Albright L.F and Goldsby A.R (1977) Industrial and Laboratory Alkylations, *ACS Symposium Series No. 55* ACS, Washington DC.
9. Mehta V.L and Kokossis A.C (1988) New Generation Tools for Multiphase Reaction Systems: A Validated Systematic Methodology for Novelty and Design Automation, *Comp Chem Eng*, **22S**: 5119.
10. Kokossis A.C and Floudas C.A (1990) Optimization of Complex Reactor Networks – I Isothermal Operation, *Chem Eng Sci*, **45**: 595.
11. Marcoulaki E.C and Kokossis A.C (1999) Scoping and Screening Complex Reaction Networks Using Stochastic Optimization, *AIChE J*, **45**: 1977.
12. Kokossis A.C and Floudas C.A (1994) Optimization of Complex Reactor Networks – II Nonisothermal Operation, *Chem Eng Sci*, **49**: 1977.

13. Romanainen, J.J and Salmi T (1992) The Effect of Reaction Kinetics, Mass Transfer and Flow Pattern on Noncatalytic and Homogeneously Catalyzed Gas-Liquid Reactions in Bubble Columns, *Chem Eng Sci*, **47**: 2493.
14. Salmi T, Paatero E and Fagerstolt K (1993a) Kinetic Model for the Synthesis of α-Chlorocarboxylic Acids, *Chem Eng Sci*, **48**(4): 735–751.
15. Salmi T, Paatero E and Fagerstolt K (1993b) Optimal Degree of Backmixing in Autocatalytic Reactions A Case Study: Chlorination of Dodecanoic Acid, *Chem Eng Res Des*, **71**: 531–542.
16. Nauman E.B (2002) *Chemical Reactor Design, Optimization and Scale-up*, McGraw Hill.

8 Choice of Separator for Heterogeneous Mixtures

8.1 HOMOGENEOUS AND HETEROGENEOUS SEPARATION

Having made an initial specification for the reactor, attention is turned to separation of the reactor effluent. In some circumstances, it might be necessary to carry out separation before the reactor to purify the feed. Whether before or after the reactor, the overall separation task might need to be broken down into a number of intermediate separation tasks. Consider now the choice of separator for the separation tasks. Later in Chapters 11 to 14, consideration will be given as to how separation tasks should be connected together and connected to the reactor. As with reactors, emphasis will be placed on the choice of separator, together with its preliminary specifications, rather than its detailed design.

When choosing between different types of reactors, both continuous and batch reactors were considered from the point of view of the performance of the reactor (continuous plug-flow and ideal batch being equivalent in terms of residence time). If a batch reactor is chosen, it will often lead to a choice of separator for the reactor effluent that also operates in batch mode, although this is not always the case as intermediate storage can be used to overcome the variations with time. Batch separations will be dealt with in Chapter 14.

If the mixture to be separated is homogeneous, separation can only be performed by the creation of another phase within the system or by the addition of a mass separation agent. For example, if a vapor mixture is leaving a reactor, another phase could be created by partial condensation. The vapor resulting from the partial condensation will be rich in the more volatile components and the liquid will be rich in the less volatile components, achieving a separation. Alternatively, rather than creating another phase, a mass separation agent can be added. Returning to the example of a vapor mixture leaving a reactor, a liquid solvent could be contacted with the vapor mixture to act as a mass separation agent to preferentially dissolve one or more of the components from the mixture. Further separation is required to separate the solvent from the process materials so as to recycle the solvent, and so on. A number of physical properties can be exploited to achieve the separation of homogeneous mixtures[1,2].

If a heterogeneous or multiphase mixture needs to be separated, then separation can be done physically by exploiting the differences in density between the phases. Separation of the different phases of a heterogeneous mixture should be carried out before homogeneous separation, taking advantage of what already exists. Phase separation tends to be easier and should be done first. The phase separations likely to be carried out are:

- Gas–liquid (or vapor–liquid)
- Gas–solid (or vapor–solid)
- Liquid–liquid (immiscible)
- Liquid–solid
- Solid–solid.

A fully comprehensive survey is beyond the scope of this text, and many good surveys are already available[1–6].

The principal methods for the separation of heterogeneous mixtures are:

- Settling and sedimentation
- Inertial and centrifugal separation
- Electrostatic precipitation
- Filtration
- Scrubbing
- Flotation
- Drying.

8.2 SETTLING AND SEDIMENTATION

In *settling* processes, particles are separated from a fluid by gravitational forces acting on the particles. The particles can be liquid drops or solid particles. The fluid can be a gas, vapor or liquid.

Figure 8.1a shows a simple device used to separate by gravity a gas–liquid (or vapor–liquid) mixture. The velocity of the gas or vapor through the vessel must be less than the settling velocity of the liquid drops.

When a particle falls under the influence of gravity, it will accelerate until the combination of the frictional drag in the fluid and buoyancy force balances the opposing gravitational force. If the particle is assumed to be a rigid sphere, at this terminal velocity, a force balance gives[3,4,7,8]

$$\underbrace{\rho_P \frac{\pi d^3}{6} g}_{\text{gravitational force}} = \underbrace{\rho_F \frac{\pi d^3}{6} g}_{\text{buoyancy force}} + \underbrace{c_D \frac{\pi d^2}{4} \frac{\rho_F v_T^2}{2}}_{\text{drag force}} \tag{8.1}$$

Chemical Process Design and Integration R. Smith

ISBNs: 0-471-48680-9 (HB); 0-471-48681-7 (PB)

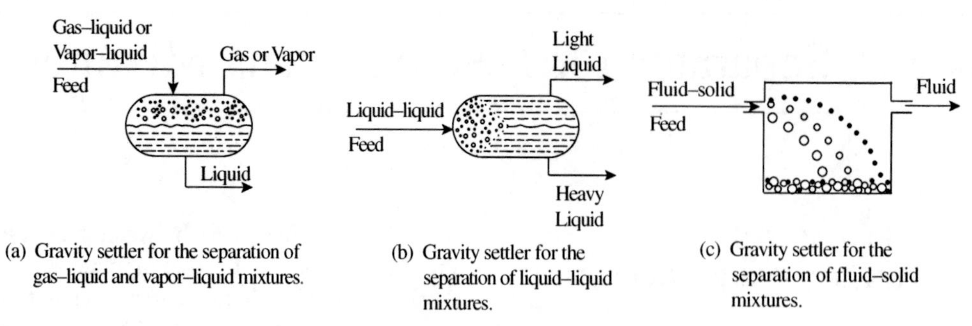

Figure 8.1 Settling processes used for the separation of heterogeneous mixtures.

where ρ_P = density of particles (kg·m^{-3})
ρ_F = density of dispersing fluid (kg·m^{-3})
d = particle diameter (m)
g = the gravitational constant (9.81 m·s^{-2})
c_D = drag coefficient (–)
v_T = terminal settling velocity (m·s^{-1})

Rearranging Equation 8.1 gives:

$$v_T = \sqrt{\left(\frac{4gd}{3c_D}\right)\left(\frac{\rho_P - \rho_F}{\rho_F}\right)} \tag{8.2}$$

More generally, Equation 8.2 can be written as:

$$v_T = K_T\sqrt{\frac{\rho_P - \rho_F}{\rho_F}} \tag{8.3}$$

where K_T = parameter for the terminal velocity (m·s^{-1})

If the particles are assumed to be rigid spheres, then from Equations 8.2 and 8.3[3,4,7,8]:

$$K_T = \left(\frac{4gd}{3c_D}\right)^{1/2} \tag{8.4}$$

However, more general correlations can be found for K_T in Equation 8.3. If in addition to assuming the particles to be rigid spheres, it is also assumed that the flow is in the laminar region, known as the Stoke's Law region, for Reynolds number less than 1 (but can be applied up to a Reynolds number of 2 without much error):

$$c_D = \frac{24}{Re} = \frac{24}{dV_T\rho_F/\mu_F} \qquad 0 < Re < 2 \tag{8.5}$$

where Re = Reynolds Number
μ_F = fluid viscosity (kg·m^{-1}·s^{-1})

Substituting Equation 8.5 into Equation 8.2 gives:

$$v_T = \frac{gd^2(\rho_P - \rho_F)}{18\mu_F} \qquad 0 < Re < 2 \tag{8.6}$$

When applying Equation 8.6, there is a tacit assumption that there is no turbulence in the settler. In practice, any turbulence will mean that a settling device sized on the basis of Equation 8.6 will have a lower efficiency than predicted.

Above a Reynolds number of around 2, Equation 8.5 will underestimate the drag coefficient and hence overestimate the settling velocity. Also, for $Re > 2$, an empirical expression must be used[7]:

$$c_D = \frac{18.5}{Re^{0.6}} \qquad 2 < Re < 500 \tag{8.7}$$

Substituting Equation 8.7 into Equation 8.2 gives:

$$v_T = \left(\frac{gd^{1.6}(\rho_P - \rho_F)}{13.875\rho_F^{0.4}\mu_F^{0.6}}\right)^{0.7143} \qquad 2 < Re < 500 \tag{8.8}$$

For higher values of Re[7]:

$$c_D = 0.44 \qquad 500 < Re < 200{,}000 \tag{8.9}$$

Substituting Equation 8.9 into Equation 8.2 gives:

$$v_T = \sqrt{\frac{gd(\rho_P - \rho_F)}{3.03\rho_F}} \qquad 500 < Re < 200{,}000 \tag{8.10}$$

When designing a settling device of the type in Figure 8.1a, the maximum allowable velocity in the device must be less than the terminal settling velocity. Before Equations 8.6 to 8.10 can be applied, the particle diameter must be known. For gas–liquid and vapor–liquid separations, there will be a range of particle droplet sizes. It is normally not practical to separate droplets less than 100 μm diameter in such a simple device. Thus, the design basis for simple settling devices of the type illustrated in Figure 8.1a is usually taken to be a vessel in which the velocity of the gas (or vapor) is the terminal settling velocity for droplets of 100 μm diameter[3,9].

The separation of gas–liquid (or vapor–liquid) mixtures can be enhanced by installing a mesh pad at the top of the disengagement zone to coalesce the smaller droplets to larger ones. If this is done, then the K_T in Equation 8.3 is normally specified to be 0.11 m·s^{-1}, although this can take lower values down to 0.06 m·s^{-1} for vacuum systems[8].

Figure 8.1b shows a simple gravity settler or *decanter* for removing a dispersed liquid phase from another liquid phase. The horizontal velocity must be low enough to allow the low-density droplets to rise from the bottom of the vessel to the interface and coalesce and for the high-density droplets to settle down to the interface and coalesce. The decanter is sized on the basis that the velocity of the continuous phase should be less than the terminal settling velocity of the droplets of the dispersed phase. The velocity of the continuous phase can be estimated from the area of the interface between the settled phases[3,8,9]:

$$v_{CP} = \frac{F_{CP}}{A_I} \tag{8.11}$$

where v_{CP} = velocity of the continuous phase ($m \cdot s^{-1}$)
F_{CP} = volumetric flowrate of continuous phase ($m^3 \cdot s^{-1}$)
A_I = decanter area of interface (m^2)

The terminal settling velocity is given by Equation 8.6 or 8.8. Decanters are normally designed for a droplet size of 150 μm[3,9], but can be designed for droplets down to 100 μm. Dispersions of droplets smaller than 20 μm tend to be very stable. The band of droplets that collect at the interface before coalescing should not extend to the bottom of the vessel. A minimum of 10% of the decanter height is normally taken for this[3].

An empty vessel may be employed, but horizontal baffles can be used to reduce turbulence and assist the coalescence through preferential wetting of the solid surface by the disperse phase. More elaborate methods to assist the coalescence include the use of mesh pads in the vessel or the use of an electric field to promote coalescence. Chemical additives can also be used to promote coalescence.

In Figure 8.1c is a schematic diagram of a gravity settling chamber. A mixture of gas, vapor or liquid and solid particles enters at one end of a large chamber. Particles settle toward the base. Again the device is specified on the basis of the terminal settling velocity of the particles. For gas–solid particle separations, the size of the solid particles is more likely to be known than the other types discussed so far. The efficiency with which the particles of a given size will be collected from the simple setting devices in Figure 8.1c is given by[10]:

$$\eta = \frac{h}{H} \tag{8.12}$$

where η = efficiency of collection (–)
h = settling distance of the particles during the residence time in the device (m)
H = height of the setting zone in the device (m)

When high concentrations of particles are to be settled, the surrounding particles interfere with individual particles. This is particularly important when settling high concentrations of solid particles in liquids. For such *hindered* settling, the viscosity and fluid density terms in Equation 8.6 can be modified to allow for this. The walls of the vessel can also interfere with settling[4,9].

When separating a mixture of water and fine solid particles in a gravity settling device such as the one shown in Figure 8.1c, it is common in such operations to add a *flocculating agent* to the mixture to assist the settling process. This agent has the effect of neutralizing electric charges on the particles that cause them to repel each other and remain dispersed. The effect is to form *aggregates* or *flocs*, which, because they are larger in size, settle more rapidly.

The separation of suspended solid particles from a liquid by gravity settling into a clear fluid and a slurry of higher solids content is called *sedimentation*. Figure 8.2 shows a sedimentation device known as a *thickener*, the prime function of which is to produce a more concentrated slurry. The feed slurry in Figure 8.2 is fed at the center of the tank below the surface of the liquid. Clear liquid overflows from the top edge of the tank. A slowly revolving rake serves to scrape the thickened slurry sludge toward the center of the base for removal and at the same time assists the formation of a more concentrated sludge. Again, it is common in such operations to add a flocculating agent to the mixture to assist the settling process. When the prime function of the sedimentation is to remove solids from a liquid rather than to produce a more concentrated solid–liquid mixture, the device is known as a *clarifier*. Clarifiers are often similar in design to thickeners.

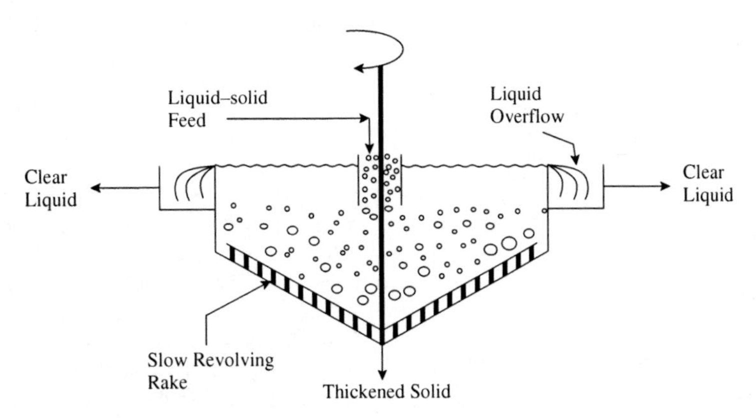

Figure 8.2 A thickener for liquid–solid separation.

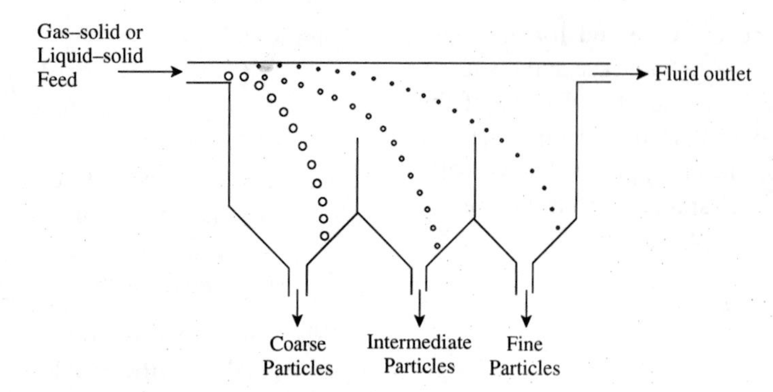

Figure 8.3 Simple gravity settling classifier.

Figure 8.3 shows a simple type of *classifier*. In the device in Figure 8.3, a large tank is subdivided into several sections. A size range of solid particles suspended in gas, vapor or liquid enters the tank. The larger, faster-settling particles settle to the bottom close to the entrance, and the slower-settling particles settle to the bottom close to the exit. The vertical baffles in the tank allow the collection of several fractions.

This type of classification device can be used to carry out solid–solid separation in mixtures of different solids. The mixture of particles is first suspended in a fluid and then separated into fractions of different size or density in a device similar to that in Figure 8.3.

Example 8.1 Solid particles with a size greater than 100 μm are to be separated from larger particles in a settling chamber. The flowrate of gas is 8.5 $m^3 \cdot s^{-1}$. The density of the gas is 0.94 $kg \cdot m^{-3}$ and its viscosity 2.18×10^{-5} $kg \cdot m^{-1} \cdot s^{-1}$. The density of the particles is 2780 $kg \cdot m^{-3}$.

a. Calculate the settling velocity, assuming the particles are spherical.
b. The settling chamber is to be box-shaped, with a rectangular cross section for the gas flow. If the length and breadth of the settling chamber are equal, what should the dimensions of the chamber be for 100% removal of particles greater than 100 μm?

Solution

a. Assume initially that the settling is in the Stoke's Law region

$$v_T = \frac{gd^2(\rho_P - \rho_F)}{18\mu_F}$$

$$= \frac{9.81 \times (100 \times 10^{-6})^2 \times (2780 - 0.94)}{18 \times 2.18 \times 10^{-5}}$$

$$= 0.69 \text{ m} \cdot \text{s}^{-1}$$

Check the Reynolds number

$$Re = \frac{dv_T\rho_F}{\mu_F}$$

$$= \frac{100 \times 10^{-6} \times 0.69 \times 0.94}{2.18 \times 10^{-5}}$$

$$= 3.0$$

This is just outside the range of validity of Stoke's Law. Whilst the errors would not be expected to be too serious, compare with the prediction of Equation 8.8.

$$v_T = \left(\frac{gd^{1.6}(\rho_P - \rho_F)}{13.875\rho_F^{0.4}\mu_F^{0.6}}\right)^{0.7143}$$

$$= \left(\frac{9.81 \times (100 \times 10^{-6})^{1.6} \times (2780 - 0.94)}{13.875 \times 0.94^{0.4} \times (2.18 \times 10^{-5})^{0.6}}\right)^{0.7143}$$

$$= 0.61 \text{ m} \cdot \text{s}^{-1}$$

b. For 100% separation of particles:

$$\tau = \frac{H}{v_T}$$

where τ = mean residence time in the settler
H = height of the settling chamber

Also: $$\tau = \frac{V}{F} = \frac{LBH}{F}$$

where V = volume of settling chamber
F = volumetric flowrate
L = length of settling chamber
B = breadth of settling chamber

Assuming $L = B$, then:

$$\frac{H}{v_T} = \frac{L^2H}{F} \tag{8.13}$$

$$L = \sqrt{\frac{F}{v_T}}$$

$$= \sqrt{\frac{7.5}{0.61}}$$

$$= 3.51 \text{ m}$$

From Equation 8.13, in principle, any height can be chosen. If a large height is chosen, then the particles will have further to settle, but the residence time will be long. If a small height is chosen, then the particles will have a shorter distance to travel, but have a shorter residence time. To keep down the capital cost, a small height should be chosen. However, the bulk velocity through the settling chamber should not be too high; otherwise reentrainment of settled particles will start to occur. The maximum bulk mean velocity of the gas would normally be kept below around 5 m·s^{-1} [10]. Also, for maintenance and access, a minimum height of around 1 m will be needed. If the height is taken to be 1 m, then the bulk mean velocity is F/LH = 2.1 m·s^{-1}, which seems reasonable.

Example 8.2 A liquid–liquid mixture containing 5 kg·s^{-1} hydrocarbon and 0.5 kg·s^{-1} water is to be separated in a decanter. The physical properties are given in Table 8.1.

The water can be assumed to be dispersed in the hydrocarbon. Estimate the size of decanter required to separate the mixture in a horizontal drum with a length to diameter ratio of 3 to 1 and an interface across the center of the drum.

Solution Assuming a droplet size of 150 μm and the flow to be in the Stoke's Law region,

$$v_T = \frac{9.81 \times (150 \times 10^{-6})^2 \times (993 - 730)}{18 \times 8.1 \times 10^{-4}}$$

$$= 0.0040 \text{ m·s}^{-1}$$

Check the Reynolds number

$$Re = \frac{730 \times 150 \times 10^{-6} \times 0.0040}{8.1 \times 10^{-4}}$$

$$= 0.54$$

From Equation 8.10,

$$v_{CP} = \frac{F_{CP}}{A_I} = v_T$$

$$0.0040 = \frac{5.0}{730 \times A_I}$$

$$A_I = 1.712 \text{ m}^2$$

Also: $A_I = DL$

where D = diameter of decanter

L = length of decanter

Assume $3D = L$.

$$A_I = \frac{L^2}{3}$$

$$L = \sqrt{3A_I}$$

$$= 2.27 \text{ m}$$

$$D = L/3$$

$$= 0.76 \text{ m}$$

Check the continuous phase (hydrocarbon) droplets that could be entrained in the dispersed phase (water).

Velocity of the water phase

$$= \frac{0.5}{993 \times 1.712}$$

$$= 2.94 \times 10^{-4} \text{ m·s}^{-1}$$

Table 8.1 Physical property data for Example 8.2.

	Density (kg·m^{-3})	Viscosity (kg·m^{-1}·s^{-1})
Hydrocarbon	730	8.1×10^{-4}
Water	993	8.0×10^{-4}

The diameter of hydrocarbon droplets entrained by the velocity of the water phase

$$v_T = -2.94 \times 10^{-4} \text{ m·s}^{-1} \text{ (i.e. rising)}$$

$$d = \sqrt{\frac{18\, v_T \mu_F}{g(\rho_P - \rho_F)}}$$

$$= \sqrt{\frac{18 \times -2.94 \times 10^{-4} \times 8.0 \times 10^{-4}}{9.81 \times (730 - 993)}}$$

$$= 0.4 \times 10^{-6} \text{ m}$$

Only hydrocarbon droplets smaller than 0.4 μm will be entrained.

8.3 INERTIAL AND CENTRIFUGAL SEPARATION

In the preceding processes, the particles were separated from the fluid by gravitational forces acting on the particles. Sometimes gravity separation may be too slow because of the closeness of the densities of the particles and the fluid, because of small particle size leading to low settling velocity or, in the case of liquid–liquid separations, because of the formation of a stable emulsion.

Inertial or *momentum* separators improve the efficiency of gas–solid settling devices by giving the particles downward momentum, in addition to the gravitational force. Figure 8.4 illustrates three possible types of inertial separator. Many other arrangements are possible[10,11]. The design of inertial separators for the separation of gas–solid separations is usually based on collection efficiency curves, as illustrated schematically in Figure 8.5. The curve is obtained experimentally and shows what proportion of particles of a given size is expected to be collected by the device. As the particle size decreases, the collection efficiency decreases. Collection efficiency curves for standard designs are published[8], but is preferable to use data provided by the equipment supplier.

Centrifugal separators take the idea of an inertial separator a step further and make use of the principle that an object whirled about an axis at a constant radial distance from the point is acted on by a force. Use of centrifugal forces increases the force acting on the particles. Particles that do not settle readily in gravity settlers often can be separated from fluids by centrifugal force.

The simplest type of centrifugal device is the *cyclone* separator for the separation of solid particles or liquid droplets from a gas or vapor (Figure 8.6). This consists of a vertical cylinder with a conical bottom. Centrifugal force is generated by the motion of the fluid. The mixture enters through a tangential inlet near the top, and the rotating motion so created develops centrifugal force that throws the dense particles radially toward the wall. The entering fluid flows downward in a spiral adjacent to the wall. When the fluid reaches the bottom of the cone, it spirals upward in

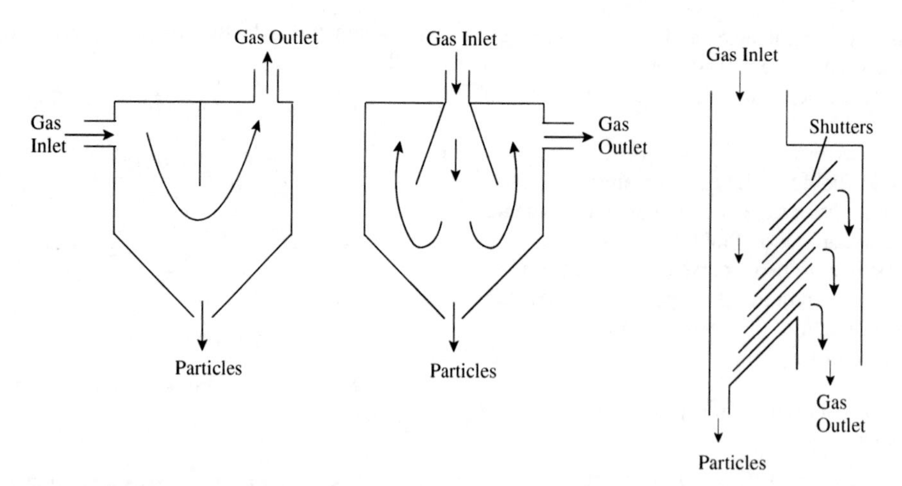

Figure 8.4 Inertial separators increase the efficiency of separation by giving the particles downward momentum.

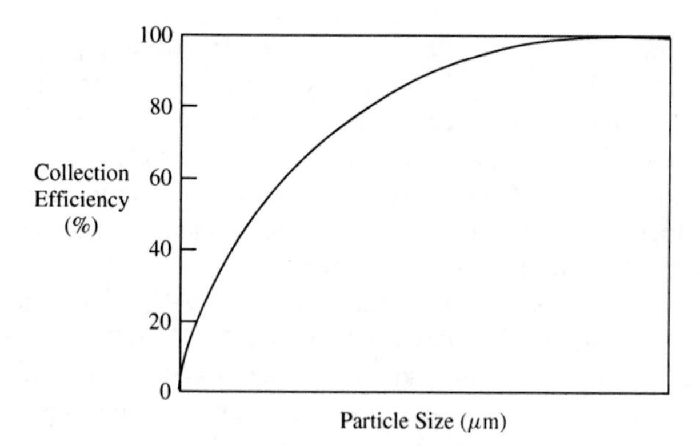

Figure 8.5 A collection efficiency curve shows the fraction of particles of a given particle size that will be collected.

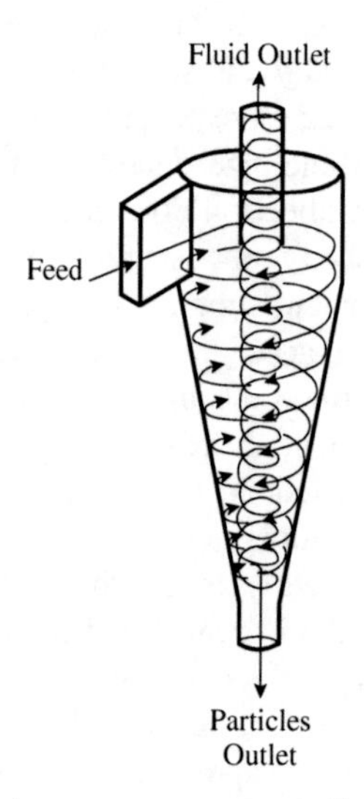

Figure 8.6 A cyclone generates centrifugal force by the fluid motion.

a smaller spiral at the center of the cone and cylinder. The downward and upward spirals are in the same direction. The particles of dense material are thrown toward the wall and fall downward, leaving the bottom of the cone.

The design of cyclones is normally based on collection efficiency curves, such as the one shown in Figure 8.5[10,11]. Curves for cyclones with standard dimensions are published that can be scaled to different dimensions using scaling parameters[10,11]. Again, it is preferable to use curves supplied by equipment manufactures. Standard designs tend to be used in practice and units placed in parallel to process large flowrates.

The same principle can be used for the separation of solids from a liquid in a *hydrocyclone*. Although the principle is the same, whether a gas or vapor is being separated from a liquid, the geometry of the cyclone will change accordingly. Hydrocyclones can also be used to separate mixtures of immiscible liquids, such as mixtures of oil and water. For the separation of oil and water, the water is denser than the oil and is thrown to the wall by the centrifugal force leaving from the conical base. The oil leaves from the top. Again, the design of hydrocyclones is normally based on collection efficiency curves, such as the one shown in Figure 8.5 with standard designs used and units placed in parallel to process large flowrates.

Figure 8.7 shows *centrifuges*, in which a cylindrical bowl is rotated to produce the centrifugal force. In Figure 8.7a, the cylindrical bowl is shown rotating with a feed consisting of a liquid–solid mixture fed at the center. The feed is thrown outward to the walls of the container. The particles settle horizontally outward. Different arrangements are possible to remove the solids from the bowl. In Figure 8.7b, two liquids having different densities are separated by the centrifuge. The more dense fluid occupies the outer periphery, since the centrifugal force is greater on the more dense fluid.

Example 8.3 A dryer vent is to be cleaned using a bank of cyclones. The gas flowrate is 60 $m^3 \cdot s^{-1}$, density of solids 2700 $kg \cdot m^{-3}$ and the concentration of solids is 10 $g \cdot m^{-3}$. The size distribution of the solids is given in Table 8.2:

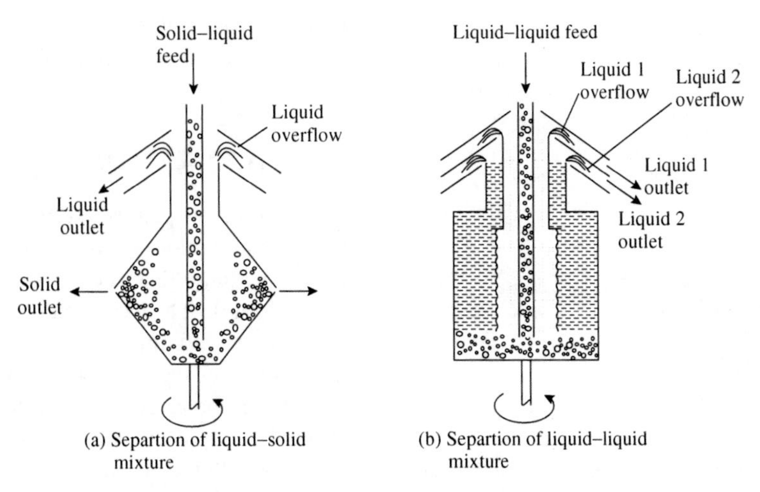

Figure 8.7 A centrifuge uses rotating cylindrical bowl to produce centrifugal force.

Table 8.2 Particle size distribution for Example 8.3.

Particle size (μm)	Percentage by weight less than
50	90
40	86
30	80
20	70
10	45
5	25
2	10

The collection efficiency curve for the design of cyclone used is given in Figure 8.8. Calculate the solids removal and the final outlet concentration.

Solution The particles are first divided into size ranges and the collection efficiency for the average size applied with the size range as shown in Table 8.3:

The overall collection efficiency is 69.5% and the concentration of solids at exit is 3.05 $g \cdot m^{-3}$

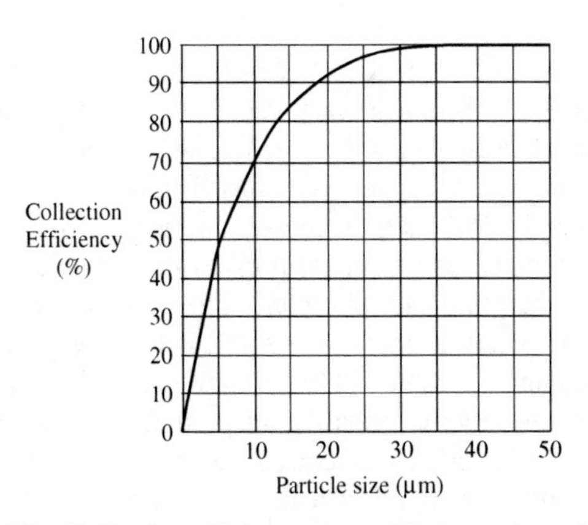

Figure 8.8 Collection efficiency curve for the cyclone design in Example 8.1.

8.4 ELECTROSTATIC PRECIPITATION

Electrostatic precipitators are generally used to separate particulate matter that is easily ionized from a gas stream[3,8,10]. This is accomplished by an electrostatic field produced between wires or grids and collection plates by

Table 8.3 Collection efficiency for Example 8.3.

Particle size range (μm)	Per cent in range (%)	Efficiency at mean size (%)	Overall collection (%)	Outlet	
				Fraction in range	Outlet %
>50	10	100	10	0	0
40 – 50	4	100	4	0	0
30 – 40	6	99	5.9	0.1	0.3
20 – 30	10	96	9.6	0.4	1.3
10 – 20	25	83	20.8	4.2	13.8
5 – 10	2	60	12.0	8.0	26.2
2 – 5	15	41	6.2	8.8	28.9
0 – 2	10	10	1.0	9.0	29.5
			69.5	30.5	100.0

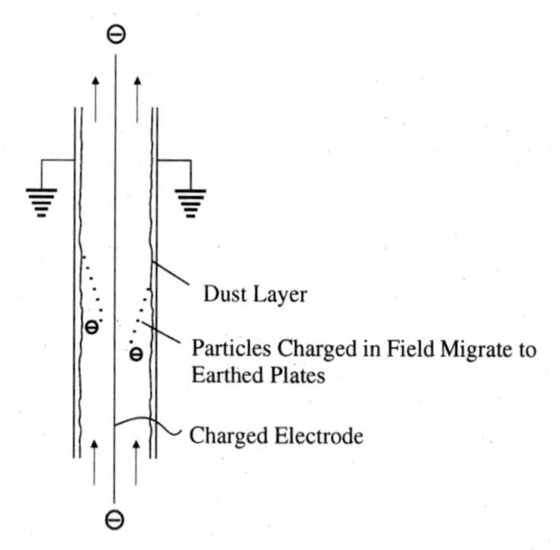

Figure 8.9 Electrostatic precipitation. (Reproduced from Stenhouse JIT, 1981, in Teja (ed) *Chemical Engineering and the Environment*, Blackwell Scientific Publications, Oxford.).

applying a high voltage between the two, as illustrated in Figure 8.9. A corona is established around the negatively charged electrode. As the particulate-laden gas stream passes through the space, the corona ionizes molecules of gases such as O_2 and CO_2 present in the gas stream. These charged molecules attach themselves to particulate matter, thereby charging the particles. The oppositely charged collection plates attract the particles. Particles collect on the plates and are removed by vibrating the collection plates mechanically, thereby dislodging particles that drop to the bottom of the device. Electrostatic precipitation is most effective when separating particles with a high resistivity. The operating voltage typically varies between 25 and 45 kV or more, depending on the design and the operating temperature. The application of electrostatic precipitators is normally restricted to the separation of fine particles of solid or liquid from a large volume of gas. Preliminary designs can be based on collection efficiency curves, as illustrated in Figure 8.5[8].

8.5 FILTRATION

In filtration, suspended solid particles in a gas, vapor or liquid are removed by passing the mixture through a porous medium that retains the particles and passes the fluid (filtrate). The solid can be retained on the surface of the filter medium, which is *cake filtration*, or captured within the filter medium, which is *depth filtration*. The filter medium can be arranged in many ways.

Figure 8.10 shows four examples of cake filtration in which the filter medium is a *cloth* of natural or artificial fibers. In different arrangements, the filter medium might even be ceramic or metal. Figure 8.10a shows the filter cloth arranged between plates in an enclosure in a *plate-and-frame filter* for the separation of solid particles from liquids. Figure 8.10b shows the cloth arranged as a *thimble* or *candle*. This arrangement is common for the separation of solid particles from a gas and is known as a *bag filter*. As the particles build up on the inside of the thimble, the unit is periodically taken off-line and the flow reversed to recover the filtered particles. For the separation of solid particles from gases, conventional filter media can be used up to a temperature of around 250°C. Higher temperatures require ceramic or metallic (e.g. stainless steel sintered fleece) filter media. These can be used for temperatures of 250 to 1000°C and higher. Cleaning of high-temperature media requires the unit to be taken off-line and pressure pulses applied to recover the filtered particles. Figure 8.10c shows a rotating belt for the separation of a slurry of solid particles in a liquid, and Figure 8.10d shows a rotating drum in which

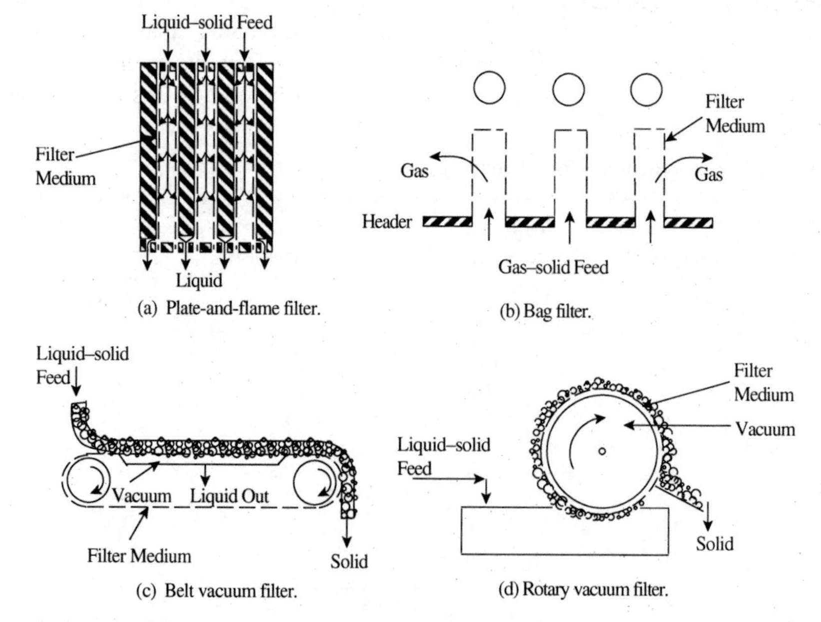

Figure 8.10 Filtration can be arranged in many ways.

the drum rotates through the slurry. In both cases, the flow of liquid is induced by the creation of a vacuum. When filtering solids from liquids, if the purity of the filter cake is not important, *filter aids*, which are particles of porous solid, can be added to the mixture to aid the filtration process. When filtering solids from a liquid, a thickener is often used upstream of filtration to concentrate the mixture prior to filtration.

When separating solid particles from a liquid filtrate, if the solid filter cake is a product, rather than a waste, then it is usual to wash the filter cake to remove the residual filtrate from the filter cake. The washing of the filter cake after filtration takes place by displacement of the filtrate and by diffusion.

Rather than using a cloth, a *granular* medium consisting of layers of particulate solids on a support grid can be used. Downward flow of the mixture causes the solid particles to be captured within the medium. Such *deep-bed* filters are most commonly used to remove small quantities of solids from large quantities of liquids. To release the solid particles captured within the bed, the flow is periodically reversed, causing the bed to expand and release the particles that have been captured. Around 3% of the throughput is needed for this backwashing.

Whereas the liquid–solid filtration processes described so far can separate particles down to a size of around 10 μm, for smaller particles that need to be separated, a porous polymer membrane can be used. This process, known as *microfiltration*, retains particles down to a size of around 0.05 μm. A pressure difference across the membrane of 0.5 to 4 bar is used. The two most common practical arrangements are spiral wound and hollow fiber. In the spiral wound arrangement, flat membrane sheets separated by spacers for the flow of feed and filtrate are wound into a spiral and inserted in a pressure vessel. Hollow fiber arrangements, as the name implies, are cylindrical membranes arranged in series in a pressure vessel in an arrangement similar to a shell-and-tube heat exchanger. The feed enters the shell-side and permeate passes through the membrane to the center of the hollow fibers. The main factor affecting the flux of the filtrate through the membrane is the accumulation of deposits on the surface of the filter. Of course, this is usual in cake filtration systems when the particles are directed normally toward the surface of the filter medium. However, in the case of microfiltration, the surface deposits cause the flux of filtrate to decrease with time. To ameliorate this effect in microfiltration, *cross-flow* can be used, in which a high rate of shear is induced across the surface of the membrane from the feed. The higher the velocity across the surface of the filter medium, the lower is the deterioration in flux of the filtrate. Even with a significant velocity across the surface of the membrane, there is still likely to be a deterioration in the flux caused by fouling of the membrane. Periodic flushing or cleaning of the membrane will be required to correct this. The method of cleaning depends on the type of foulant, the membrane configuration and the membrane's resistance to cleaning agents. In the simplest case, a reversal of the flow (*backflush*) might be able to remove the surface deposits. In the worst case, cleaning agents such as sodium hydroxide might be required. Microfiltration is used for the recovery of paint from coating processes, oil–water separations, separation of biological cells from a liquid, and so on.

8.6 SCRUBBING

Scrubbing with liquid (usually water) can enhance the collection of particles when separating gas–solid mixtures. Figure 8.11 shows three of the many possible designs for

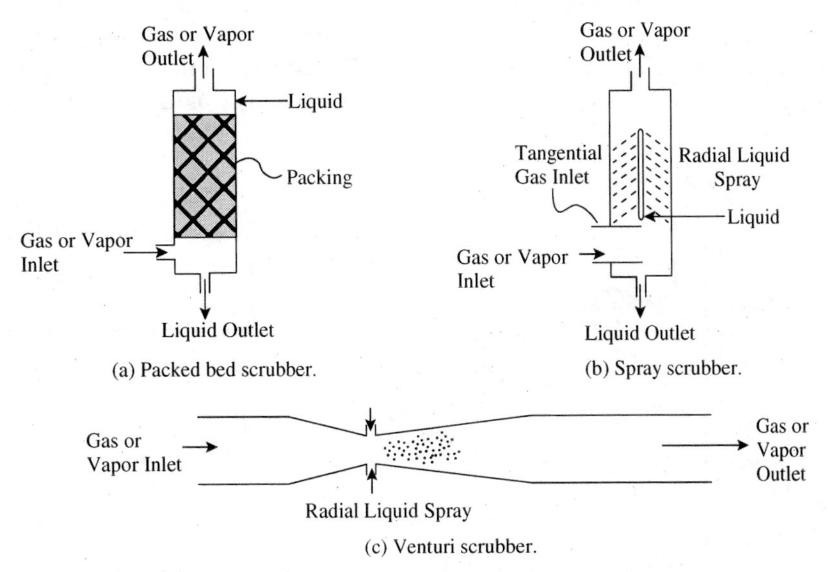

Figure 8.11 Various scrubber designs can be used to separate solid from gas or vapor. (Reproduced from Stenhouse JIT, 1981, in Teja (ed) *Chemical Engineering and the Environment*, Blackwell Scientific Publications, Oxford.).

scrubbers. Figure 8.11a illustrates a packed tower, similar to an absorption tower. Whilst this can be effective, it suffers from the problem that the packing can become clogged with solid particles. Towers using perforated plates similar to a distillation or absorption column can also be used. As with packed columns, plate columns can also encounter problems of clogging. By contrast, Figure 8.11b uses a spray system that will be less prone to fouling. The design in Figure 8.11b uses a tangential inlet to create a swirl to enhance the separation. The design shown in Figure 8.11c uses a *venturi*. Liquid is injected into the throat of the venturi, where the velocity of the gas is highest. The gas accelerates the injected water to the gas velocity, and breaks up the liquid droplets into a relatively fine spray. The particles are then captured by the fine droplets. Very high collection efficiencies are possible with venturi scrubbers. The main problem with venturi scrubbers is the high pressure loss across the device. As with other solids separation devices, preliminary design can be carried out using collection efficiency curves like the one illustrated in Figure 8.5[8].

8.7 FLOTATION

Flotation is a gravity separation process that exploits the differences in the surface properties of particles. Gas bubbles are generated in a liquid and become attached to solid particles or immiscible liquid droplets, causing the particles or droplets to rise to the surface. This is used to separate mixtures of solid–solid particles after dispersion in a liquid, or solid particles already dispersed in a liquid or liquid–liquid mixtures of finely divided immiscible droplets. The liquid used is normally water and the particles of solid or immiscible liquid will attach themselves to the gas bubbles if they are *hydrophobic* (e.g. oil droplets dispersed in water).

The bubbles of gas can be generated by three methods:

a. *dispersion*, in which the bubbles are injected directly by some form of sparging system;
b. *dissolution* in the liquid under pressure and then liberation in the flotation cell by reducing the pressure;
c. *electrolysis* of the liquid.

Flotation is an important technique in mineral processing, where it is used to separate different types of ores. When used to separate solid–solid mixtures, the material is ground to a particle size small enough to *liberate* particles of the chemical species to be recovered. The mixture of solid particles is then dispersed in the flotation medium, which is usually water. The mixture is then fed to a flotation cell, as illustrated in Figure 8.12a. Here, gas is also fed to the cell where gas bubbles become attached to the solid particles, thereby allowing them to float to the surface of the liquid. The solid particles are collected from the surface by an overflow weir or mechanical scraper. The separation of the solid particles depends on the different species having different surface properties such that one species is preferentially attached to the bubbles. A number of chemicals can be added to the flotation medium to meet the various requirements of the flotation process:

a. *Modifiers* are added to control the pH of the separation. These could be acids, lime, sodium hydroxide, and so on.
b. *Collectors* are water-repellent reagents that are added to preferentially adsorb onto the surface of one of the solids. Coating or partially coating the surface of one of the solids renders the solid to be more hydrophobic and increases its tendency to attach to the gas bubbles.
c. *Activators* are used to "activate" the mineral surface for the collector.
d. *Depressants* are used to preferentially attach to one of the solids to make it less hydrophobic and decrease its tendency to attach to the gas bubbles.
e. *Frothers* are surface-active agents added to the flotation medium to create a stable froth and assist the separation.

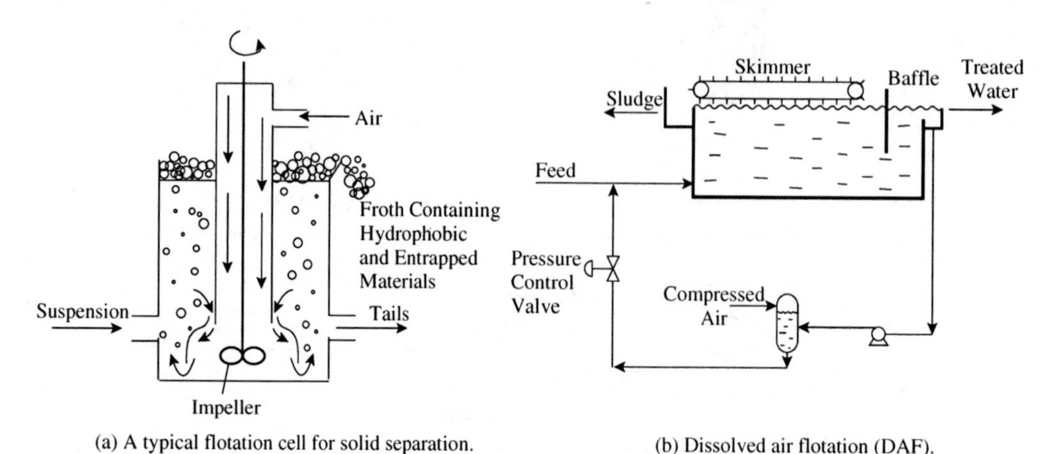

(a) A typical flotation cell for solid separation.

(b) Dissolved air flotation (DAF).

Figure 8.12 Flotation arrangements.

Flotation is also used in applications such as the separation of low-density solid particles (e.g. paper pulp) from water and oil droplets from oil–water mixtures. It is not necessary to add reagents if the particles are naturally hydrophobic, as is the case, for example, with oil–water mixtures, as the oil is naturally hydrophobic.

When separating low-density solid particles or oil droplets from water, the most common method used is *dissolved-air flotation*. A typical arrangement is shown in Figure 8.12b. This shows some of the effluent water from the unit being recycled, and air being dissolved in the recycle under pressure. The pressure of the recycle is then reduced, releasing the air from solution as a mist of fine bubbles. This is then mixed with the incoming feed that enters the cell. Low-density material floats to the surface with the assistance of the air bubbles and is removed.

8.8 DRYING

Drying refers to the removal of water from a substance through a whole range of processes, including distillation, evaporation and even physical separations such as centrifuges. Here, consideration is restricted to the removal of moisture from solids into a gas stream (usually air) by heat, namely, *thermal drying*. Some of the types of equipment for removal of water also can be used for removal of organic liquids from solids.

Four of the more common types of thermal dryers used in the process industries are illustrated in Figure 8.13.

1. *Tunnel dryers* are shown in Figure 8.13a. Wet material on trays or a conveyor belt is passed through a tunnel, and drying takes place by hot air. The airflow can be countercurrent, cocurrent or a mixture of both. This method is usually used when the product is not free flowing.
2. *Rotary dryers* are shown in Figure 8.13b. Here, a cylindrical shell mounted at a small angle to the horizontal is rotated at low speed. Wet material is fed at the higher end and flows under gravity. Drying takes place from a flow of air, which can be countercurrent or cocurrent. The heating may be direct to the dryer gas or indirect through the dryer shell. This method is usually used when the material is free flowing. Rotary dryers are not well suited to materials that are particularly heat sensitive because of the long residence time in the dryer.
3. *Drum dryers* are shown in Figure 8.13c. This consists of a heated metal roll. As the roll rotates, a layer of liquid or slurry is dried. The final dry solid is scraped off the roll. The product comes off in flaked form. Drum dryers are suitable for handling slurries or pastes of solids in fine suspension and are limited to low and moderate throughput.

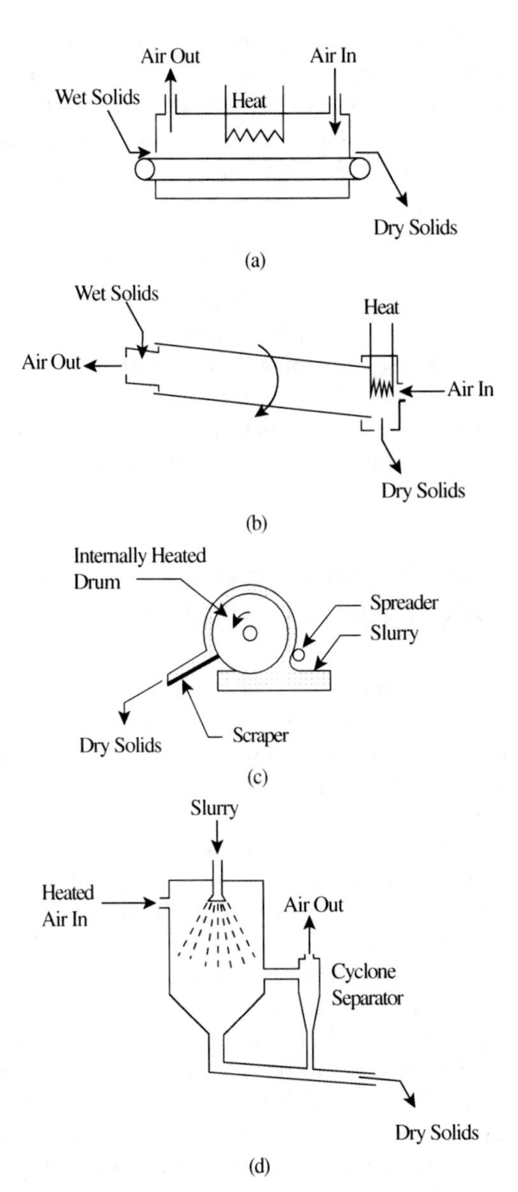

Figure 8.13 Four of the more common types of thermal dryer.

4. *Spray dryers* are shown in Figure 8.13d. Here, a liquid or slurry solution is sprayed as fine droplets into a hot gas stream. The feed to the dryer must be pumpable to obtain the high pressures required by the atomizer. The product tends to be light, porous particles. An important advantage of the spray dryer is that the product is exposed to the hot gas for a short period. Also, the evaporation of the liquid from the spray keeps the product temperature low, even in the presence of hot gases. Spray dryers are thus particularly suited to products that are sensitive to thermal decomposition, such as food products.

Another important class of dryers is the *fluidized-bed* dryer. Some designs combine spray and fluidized-bed dryers. Choice between dryers is usually based on practicalities such as the materials' handling characteristics, product decomposition, product physical form (e.g. if a porous granular material is required), and so on. Also, dryer efficiency can be used to compare the performance of different dryer designs. This is usually defined as follows:

$$\text{Dryer efficiency} = \frac{\text{heat of vaporization}}{\text{total heat consumed}} \quad (8.14)$$

If the total heat consumed is from an external utility (e.g. mains steam), then a high efficiency is desirable, even perhaps at the expense of a high capital cost. However, if the heat consumed is by recovery from elsewhere in the process, as is discussed in Chapter 22, then comparison based on dryer efficiency becomes less meaningful.

8.9 SEPARATION OF HETEROGENEOUS MIXTURES – SUMMARY

For a heterogeneous or multiphase mixture, separation usually can be achieved by phase separation. Such phase separation should be carried out before any homogeneous separation. Phase separation tends to be easier and should be done first.

The simplest devices for separating heterogeneous mixtures exploit gravitational force in settling and sedimentation devices. The velocity of the fluid through such a device must be less than the terminal settling velocity of the particles to be separated. When high concentrations of particles are to be settled, the surrounding particles interfere with individual particles and the settling becomes hindered. Separating a dispersed liquid phase from another liquid phase can be carried out in a decanter. The decanter is sized on the basis that the velocity of the continuous phase should be less than the terminal settling velocity of the droplets of the dispersed phase.

If gravitational forces are inadequate to achieve the separation, the separation can be enhanced by exploiting inertial, centrifugal or electrostatic forces.

In filtration, suspended solid particles in a gas, vapor or liquid are removed by passing the mixture through a porous medium that retains the particles and passes the fluid.

Scrubbing with liquid (usually water) can enhance the collection of particles when separating gas–solid mixtures. Flotation is a gravity separation process that exploits differences in the surface properties of particles. Gas bubbles are generated in a liquid and become attached to solid particles or immiscible liquid droplets, causing the particles or droplets to rise to the surface.

Thermal drying can be used to remove moisture from solids into a gas stream (usually air) by heat. Many types of dryers are available and can be compared on the basis of their thermal efficiency.

For the separation of gas–solid mixtures, preliminary design of inertial separators, cyclones, electrostatic precipitators, scrubbers and some types of filters can be carried out on the basis of collection efficiency curves derived from experimental performance.

No attempt should be made to carry out any optimization at this stage in the design.

8.10 EXERCISES

1. A gravity settler is to be used to separate particles less than 75 μm from a flowrate of gas of 1.6 $m^3 \cdot s^{-1}$. The density of the particles is 2100 $kg \cdot m^{-3}$. The density of the gas is 1.18 $kg \cdot m^{-3}$ and its viscosity is 1.85×10^{-5} $kg \cdot m^{-1} \cdot s^{-1}$ Estimate the dimensions of the settling chamber assuming a rectangular cross section with length to be twice that of the breadth.
2. When separating a liquid–liquid mixture in a cylindrical drum, why is it usually better to mount the drum horizontally than vertically?
3. A liquid–liquid mixture containing 6 $kg \cdot s^{-1}$ of water with 0.5 $kg \cdot s^{-1}$ of entrained oil is to be separated in a decanter. The properties of the fluids are given in Table 8.4 below.

Table 8.4 Fluid properties for exercise 8.3.

	Density ($kg \cdot m^{-3}$)	Viscosity ($kg \cdot m^{-1} \cdot s^{-1}$)
Water	993	0.8×10^{-3}
Oil	890	3.5×10^{-3}

Assuming the water to be the continuous phase, the decanter to be a horizontal drum with a length to diameter ratio of 3 to 1 and an interface across the center of the drum, estimate the dimensions of the decanter to separate water droplets less than 150 μm.

4. In Exercise 3, it was assumed that the interface would be across the center of the drum. How would the design change it if the interface was lower such that length of the interface across the drum was 50% of the diameter?
5. A mixture of vapor and liquid ammonia is to be separated in a cylindrical vessel mounted vertically with a mesh pad to assist separation. The flowrate of vapor is 0.3 $m^3 \cdot s^{-1}$. The density of the liquid is 648 $kg \cdot m^{-3}$ and that of the vapor 2.71 $kg \cdot m^{-3}$. Assuming $K_T = 0.11$ $m \cdot s^{-1}$, estimate the diameter of vessel required.
6. A gravity settler has dimensions of 1 m breadth, 1 m height and length of 3 m. It is to be used to separate solid particles from a gas with a flowrate of 1.6 $m^3 \cdot s^{-1}$. The density of the particles is 2100 $kg \cdot m^{-3}$. The density of the gas is 1.18 $kg \cdot m^{-3}$ and its viscosity is 1.85×10^{-5} $kg \cdot m^{-1} \cdot s^{-1}$. Construct an approximate collection efficiency curve for the settling of particles ranging from 0 to 50 μm.
7. Table 8.5 Shows the size distribution of solid particles in a gas stream.

Table 8.5 Particle size distribution.

Particle size (μm)	Per cent by weight less than
30	95
20	90
10	70
5	30
2	20

The flowrate of the gas is 10 $m^3 \cdot s^{-1}$ and the concentration solids is 12 $g \cdot m^{-3}$. The particles are to be separated in a scrubber with a collection efficiency curve shown in Figure 8.14. Estimate the overall collection efficiency and the concentration of solids in the exit gas.

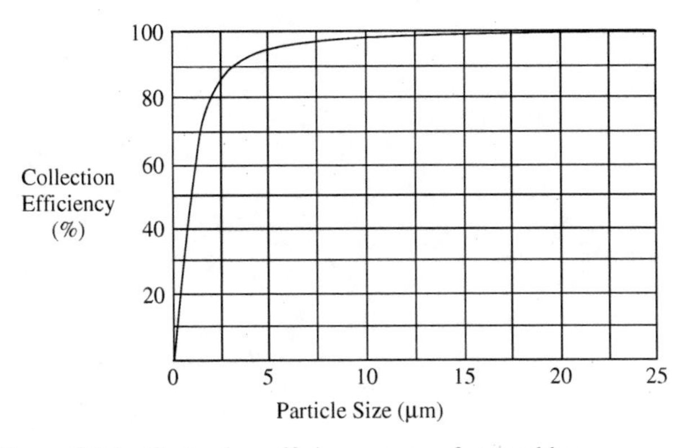

Figure 8.14 Collection efficiency curve for scrubber.

REFERENCES

1. King C.J (1980) *Separation Processes*, 2nd Edition, McGraw-Hill, New York.
2. Rousseau R.W (1987) *Handbook of Separation Process Technology*, Wiley, New York.
3. Schweitzer P.A (1997) *Handbook of Separation Process Techniques for Chemical Engineers*, 3rd Edition, McGraw-Hill, New York.
4. Coulson J.M and Richardson J.F with Backhurst J.R and Harker J.H (1991) *Chemical Engineering*, Vol. 2, 4th Edition, Butterworth Heinemann.
5. Foust A.S, Wenzel L.A, Clump C.W, Maus L and Anderson L.B (1980) *Principles of Unit Operations*, Wiley, New York.
6. Walas S.M (1988) *Chemical Process Equipment Selection and Design*, Butterworths.
7. Geankopolis C.J (1993) *Transport Processes and Unit Operations*, 3rd Edition, Prentice Hall.
8. Ludwig E.E (1977) *Applied Process Design for Chemical and Petrochemical Plants*, 2nd Edition, Gulf Publishing Company, Houston.
9. Woods D.R (1995) *Process Design and Engineering Practice*, Prentice Hall, New Jersey.
10. Dullien F.AL (1989) *Introduction to Industrial Gas Cleaning*, Academic Press.
11. Svarovsky, L (1981) *Solid-Gas Separation*, Elsevier Scientific, New York.

9 Choice of Separator for Homogeneous Fluid Mixtures I – Distillation

9.1 SINGLE-STAGE SEPARATION

As pointed out in the previous chapter, the separation of a homogeneous fluid mixture requires the creation of another phase or the addition of a mass separation agent. Consider a homogeneous liquid mixture. If this liquid mixture is partially vaporized, then another phase is created, and the vapor becomes richer in the more volatile components (i.e. those with the lower boiling points) than the liquid phase. The liquid becomes richer in the less-volatile components (i.e. those with the higher boiling points). If the system is allowed to come to equilibrium conditions, then the distribution of the components between the vapor and liquid phases is dictated by vapor–liquid equilibrium considerations (see Chapter 4). All components can appear in both phases.

On the other hand, rather than partially vaporize a liquid, the starting point could have been a homogeneous mixture of components in the vapor phase and the vapor partially condensed. There would still have been a separation, as the liquid that was formed would be richer in the less-volatile components, while the vapor would have become depleted in the less-volatile components. Again, the distribution of components between the vapor and liquid is dictated by vapor–liquid equilibrium considerations if the system is allowed to come to equilibrium.

When a mixture contains components with large relative volatilities, either a partial condensation from the vapor phase or a partial vaporization from the liquid phase followed by a simple phase split can often produce an effective separation[1].

Figure 4.2 shows a feed being separated into a vapor and liquid phase and being allowed to come to equilibrium. If the feed to the separator and the vapor and liquid products are continuous, then the material balance is described by Equations 4.57, 4.58 and 4.61[1]. If K_i is large relative to V/F (typically $K_i > 10$) in Equation 4.57, then[2]:

$$y_i \approx z_i/(V/F) \quad \text{that is, } Vy_i \approx Fz_i \tag{9.1}$$

This means that all of Component i entering with the feed, Fz_i, leaves in the vapor phase as Vy_i. Thus, if a component is required to leave in the vapor phase, its K-value should be large relative to V/F.

On the other hand, if K_i is small relative to V/F (typically $K_i < 0.1$) in Equation 4.58, then[2]:

$$x_i \approx Fz_i/L \quad \text{that is, } Lx_i \approx Fz_i \tag{9.2}$$

This means that all of Component i entering with the feed, Fz_i, leaves with the liquid phase as Lx_i. Thus, if a component is required to leave in the liquid phase, its K-value should be small relative to V/F.

Ideally, the K-value for the light key component in the phase separation should be greater than 10 and, at the same time, the K-value for the heavy key less than 0.1. Such circumstances can lead to a good separation in a single stage. However, use of phase separators might still be effective in the flow sheet if the K-values for the key components are not so extreme. Under such circumstances, a more crude separation must be accepted.

9.2 DISTILLATION

A single equilibrium stage can only achieve a limited amount of separation. However, the process can be repeated by taking the vapor from the single-stage separation to another separation stage and partially condensing it and taking the liquid to another separation stage and partially vaporizing it, and so on. With each repeated condensation and vaporization, a greater degree of separation will be achieved. In practice, the separation to multiple stages is extended by creating a cascade of stages as shown in Figure 9.1. It is assumed in the cascade that liquid and vapor streams leaving each stage are in equilibrium. Using a cascade of stages in this way allows the more volatile components to be transferred to the vapor phase and the less-volatile components to be transferred to the liquid phase. In principle, by creating a large enough cascade, an almost complete separation can be carried out.

At the top of the cascade in Figure 9.1, liquid is needed to feed the cascade. This is produced by condensing vapor that leaves the top stage and returning this liquid to the first stage of the cascade as *reflux*. All of the vapor leaving the top stage can be condensed in a *total condenser* to produce a liquid top product. Alternatively, only enough of the vapor to provide the reflux can be condensed in a *partial condenser* to produce a vapor top product if a liquid top product is not desired. Vapor is also needed to feed the cascade at the bottom of the column. This is produced

Chemical Process Design and Integration R. Smith
 ISBNs: 0-471-48680-9 (HB); 0-471-48681-7 (PB)

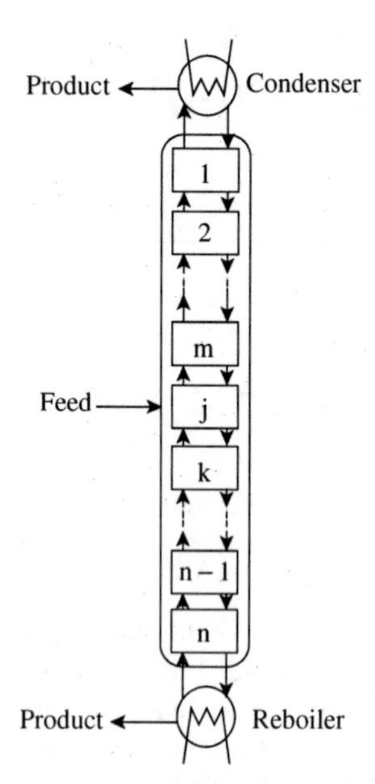

Figure 9.1 A cascade of equilibrium stages with refluxing and reboiling. (From Smith R and Jobson M, 2000, *Distillation, Encyclopedia of Separation Science*, Academic Press; reproduced by permission).

by vaporizing some of the liquid leaving the bottom stage and returning the vapor to the bottom stage of the cascade in a *reboiler*. The feed to the process is introduced at an intermediate stage, and products are removed from the condenser and the reboiler.

The methods by which the vapor and liquid are contacted in each stage of the cascade in distillation fall into two broad categories. Figure 9.2a shows a *plate*, or *tray*, column. Liquid enters the first plate at the top of the column and flows across what is shown in Figure 9.2a as a perforated plate. Liquid is prevented from *weeping* through the holes in the plate by the up-flowing vapor. In this way, the vapor and liquid are contacted. The liquid from each plate flows over a *weir* and down a *downcomer* to the next plate, and so on. The design of plate used in Figure 9.2a involving a plate with simple holes, known as a *sieve plate*, is the most common arrangement used. It is cheap, simple and well understood in terms of its performance. Many other designs of plate are available. For example, valve arrangements in the holes can be used to improve the performance and the flexibility of operation to be able to cope with a wider variety of liquid and vapor flow rates in the column. One particular disadvantage of the conventional plate in Figure 9.2a is that the downcomer arrangement makes a significant proportion of the area within the column shell not available for contacting liquid and vapor. In an attempt to overcome this, *high capacity trays*, with increased *active area*, have been developed. Figure 9.2b illustrates the concept. Again, many different designs are available for high capacity trays. In practice, the column will need more plates than the number of equilibrium stages, as mass transfer limitations and poor contacting efficiency prevent equilibrium being achieved on a plate.

The other broad class of contacting arrangement for the cascade in a distillation column is that of *packed columns*. Here the column is filled with a solid material that has a high voidage. Liquid trickles across the surfaces of the packing, and vapor flows upward through the voids in the packing, contacting the liquid on its way up the column. Many different designs of packing are available. Figure 9.3a shows a traditional design of packing, which is *random*, or *dumped*, packing. The random packing is pieces of preformed ceramic, metal or plastic, which, when dumped in the column, produce a body with a high voidage. Figure 9.3b illustrates *structured packing*. This is manufactured by sheets of metal being preformed with corrugations and holes and then joined together to produce a preformed packing with a high voidage. This is manufactured in slabs and built up in layers within the distillation column. Many types of both random and structured packing are available.

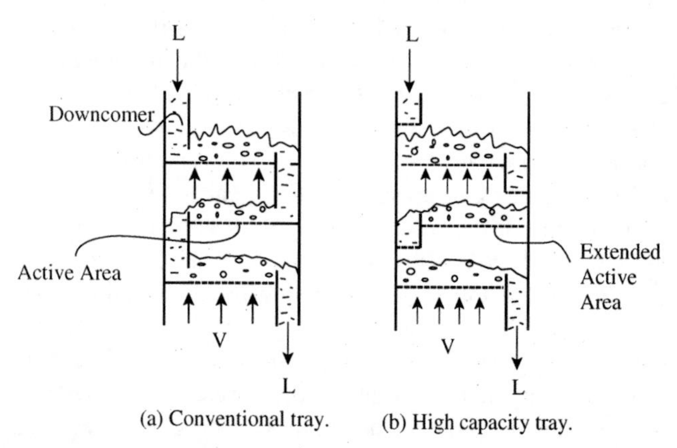

Figure 9.2 Distillation trays. (From Smith R and Jobson M, 2000, *Distillation, Encyclopedia of Separation Science*, Academic Press; reproduced by permission).

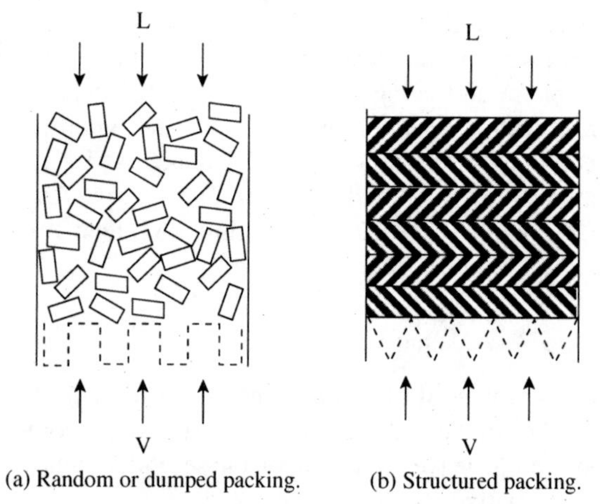

Figure 9.3 Distillation packing. (From Smith R and Jobson M, 2000, *Distillation, Encyclopedia of Separation Science*, Academic Press; reproduced by permission).

Whereas the change in composition in a plate column is finite from stage to stage, the change in composition in a packed column is continuous. To relate the continuous change in composition to that of an equilibrium stage requires the concept of the *height equivalent of a theoretical plate*, *HETP* to be introduced. This is the height of packing required to bring about the same concentration change as an equilibrium stage.

Generally, trays should be used when:

- liquid flowrate is high relative to vapor flowrate (occurs when the separation is difficult);
- diameter of the column is large (packings can suffer from maldistribution of the liquid over the packing);
- there is variation in feed composition (trays can be more flexible to variations in operating conditions);
- the column requires multiple feeds or multiple products (tray columns are simpler to design for multiple feeds or products).

Packings should be used when:

- column diameter is small (relative cost for fabrication of small-diameter trays is high);
- vacuum conditions are used (packings offer lower pressure drop and reduced entrainment tendencies);
- low pressure drop is required (packings have a lower pressure drop than trays);
- the system is corrosive (a greater variety of corrosion resistant materials is available for packings);
- the system is prone to foaming (packings have a lower tendency to promote foaming);
- low liquid hold-up in the column is required (liquid hold-up in packings is lower than that for trays).

To develop a rigorous approach to distillation design, consider Figure 9.4. This shows a general equilibrium stage in the distillation column. It is a general stage and allows for many design options other than simple columns with one feed and two products. Vapor and liquid enter this stage. In principle, an additional product can be withdrawn from the distillation column at an intermediate stage as a liquid or vapor *sidestream*. Feed can enter, and heat can be transferred to or from the stage. By using a general representation of a stage, as shown in Figure 9.4, designs with multiple feeds, multiple products and intermediate heat exchange are possible. Equations can be written to describe the material and energy balance for each stage:

1. Material balance for Component i and Stage j (NC equations for each stage):

$$L_{j-1}x_{i,j-1} + V_{j+1}y_{i,j+1} + F_j z_{i,j} - (L_j + U_j)x_{i,j} - (V_j + W_j)y_{i,j} = 0 \quad (9.3)$$

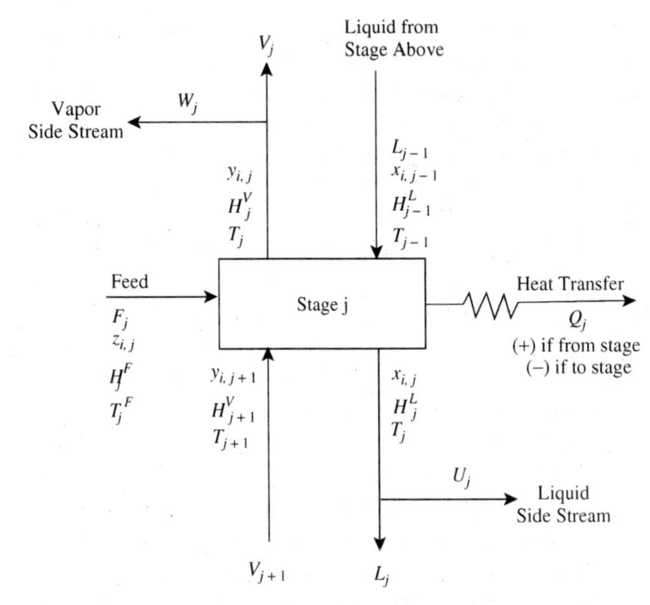

Figure 9.4 A general equilibrium stage for distillation. (From Smith R and Jobson M, 2000, *Distillation, Encyclopedia of Separation Science*, Academic Press; reproduced by permission).

2. Equilibrium relation for each Component i (NC equations for each stage):

$$y_{i,j} - K_{i,j}x_{i,j} = 0 \quad (9.4)$$

3. Summation equations (one for each stage):

$$\sum_{i=1}^{NC} y_{i,j} - 1.0 = 0, \quad \sum_{i=1}^{NC} x_{i,j} - 1.0 = 0 \quad (9.5)$$

4. Energy balance (one for each Stage j):

$$L_{j-1}H_{j-1}^L + V_{j+1}H_{j+1}^V + F_j H_j^F - (L_j + U_j)H_j^L - (V_j + W_j)H_j^V - Q_j = 0 \quad (9.6)$$

where
$z_{i,j} y_{i,j}, x_{i,j}$ = feed, vapor and liquid mole fractions for Stage j
F_j, V_j, L_j = feed, vapor and liquid molar flowrates for Stage j
W_j, U_j = vapor and liquid sidestreams for Stage j
H_F = molar enthalpy of the feed
H_j^V, H_j^L = vapor and liquid molar enthalpies for Stage j
Q_j = heat transfer from Stage j (negative for heat transfer to the stage)
$K_{i,j}$ = vapor–liquid equilibrium constant between x_i and y_i for Stage j
NC = number of components

These equations require physical property data for vapor–liquid equilibrium and enthalpies, and the set of equations must be solved simultaneously. The calculations are complex, but many methods are available to solve the equations[1]. The details of the methods used are outside the scope of this text. In practice, designers most often use commercial computer simulation packages.

However, before the above set of equations can be solved, many important decisions must be made about the distillation column. The thermal condition of the feed, the number of equilibrium stages, feed location, operating pressure, amount of reflux, and so on, all must be chosen.

To explore these decisions in a systematic way, shortcut design methods can be used. These exploit simplifying assumptions to allow many more design options to be explored than would be possible with detailed simulation and allow conceptual insights to be gained. Once the major decisions have been made, a detailed simulation needs to be carried out as described in outline above.

Conceptual insights into the design of distillation are best developed by first considering distillation of binary mixtures.

9.3 BINARY DISTILLATION

Consider the material balance for a simple binary distillation column. A simple column has one feed, two products, one reboiler and one condenser. Such a column is shown in Figure 9.5. An overall material balance can be written as:

$$F = D + B \tag{9.7}$$

A material balance can also be written for Component i as:

$$Fx_{i,F} = Dx_{i,D} + Bx_{i,B} \tag{9.8}$$

However, to fully understand the design of the column, the material balance must be followed through the column. To simplify the analysis, it can be assumed that the molar vapor and liquid flowrates are constant in each column section, which is termed *constant molar overflow*. This is strictly only true if the component molar latent heats of vaporization are the same, there is no heat of mixing between the components, heat capacities are constant and there is no external addition or removal of heat. In fact, this turns out to be a good assumption for many mixtures of organic compounds that exhibit reasonably ideal behavior. But it can also turn out to be a poor assumption for many mixtures, such as many mixtures of alcohol and water.

Start by considering the material balance for the part of the column above the feed, the *rectifying* section. Figure 9.6 shows the rectifying section of a column and the flows and compositions of the liquid and vapor in the rectifying section. First, an overall balance is written for the rectifying section (assuming L and V are constant, i.e. constant molar overflow):

$$V = L + D \tag{9.9}$$

A material balance can also be written for Component i:

$$Vy_{i,n+1} = Lx_{i,n,i} + Dx_{i,D} \tag{9.10}$$

The reflux ratio, R, is defined to be:

$$R = L/D \tag{9.11}$$

Given the reflux ratio, the vapor flow can be expressed in terms of R:

$$V = (R + 1)D \tag{9.12}$$

These expressions can be combined to give an equation that relates the vapor entering and liquid flows leaving Stage n:

$$y_{i,n+1} = \frac{R}{R+1}x_{i,n} + \frac{1}{R+1}x_{i,D} \tag{9.13}$$

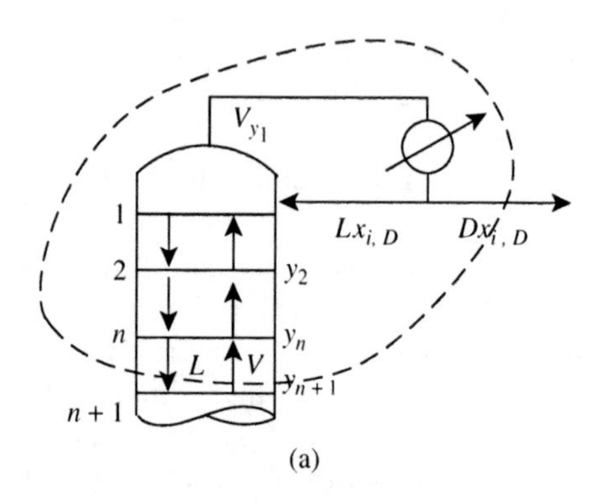

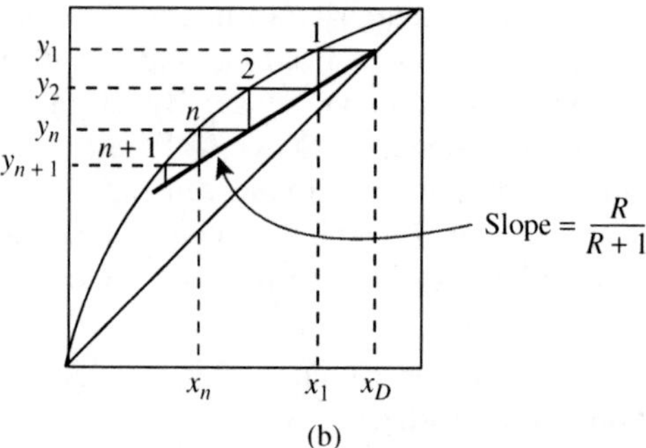

Figure 9.6 Mass balance for the rectifying section. (From Smith R and Jobson M, 2000, *Distillation, Encyclopedia of Separation Science*, Academic Press; reproduced by permission).

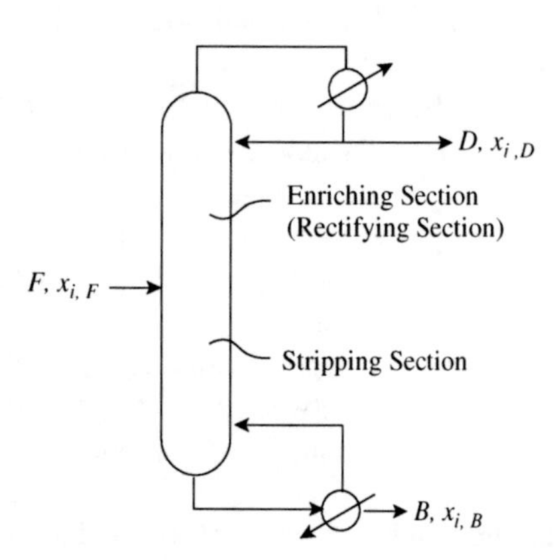

Figure 9.5 Mass balance on a simple distillation column.

On an $x-y$ diagram for Component i this is a straight line starting at the distillate composition with slope $R/(R+1)$ and which intersects the diagonal line at $x_{i,D}$.

Starting at the distillate composition $x_{i,D}$ in Figure 9.6b, a horizontal line across to the equilibrium line gives the composition of the vapor in equilibrium with the distillate (y_1). A vertical step down gives the liquid composition leaving Stage 1, x_1. Another horizontal line across to the equilibrium line gives the composition of the vapor leaving Stage 2 (y_2). A vertical line to the operating line gives the composition of the liquid leaving Stage 2 (x_2), and so on. Thus, stepping between the operating line and equilibrium line in Figure 9.6b, the change in vapor and liquid composition is followed through the rectifying section of the column.

Now consider the corresponding mass balance for the column below the feed, the *stripping* section. Figure 9.7a shows the vapor and liquid flows and compositions through the stripping section of a column. An overall mass balance for the stripping section around Stage $m+1$ gives:

$$L' = V' + B \tag{9.14}$$

A component balance gives:

$$L'x_{i,m} = V'y_{i,m+1} + Bx_{i,B} \tag{9.15}$$

Again, assuming constant molar overflow (L' and V' are constant), these expressions can be combined to give an equation relating the liquid entering and the vapor leaving Stage $m+1$:

$$y_{i,m+1} = \frac{L'}{V'}x_{i,m} - \frac{B}{V'}x_{i,B} \tag{9.16}$$

This line can be plotted in an $x-y$ plot as shown in Figure 9.7b. It is a straight line with slope L'/V', which intersects the diagonal line at $x_{i,B}$. Starting from the bottom composition, $x_{i,B}$, a vertical line up to the equilibrium line gives the composition of the vapor leaving the reboiler (y_B). A horizontal line across to the operating line gives the composition of the liquid leaving Stage N (x_N). A vertical line up to the equilibrium line then gives the vapor leaving Stage N (y_N), and so on.

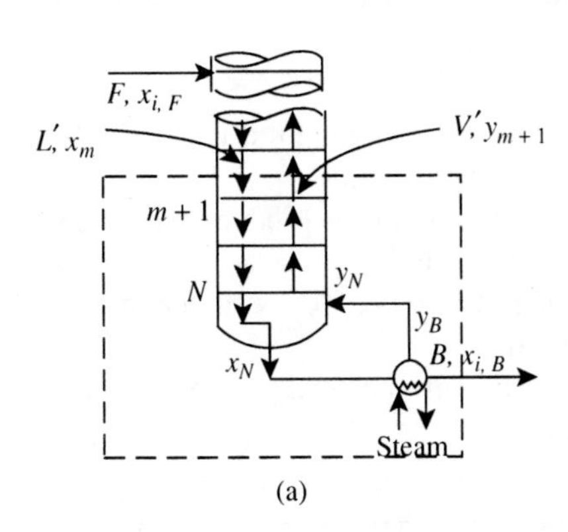

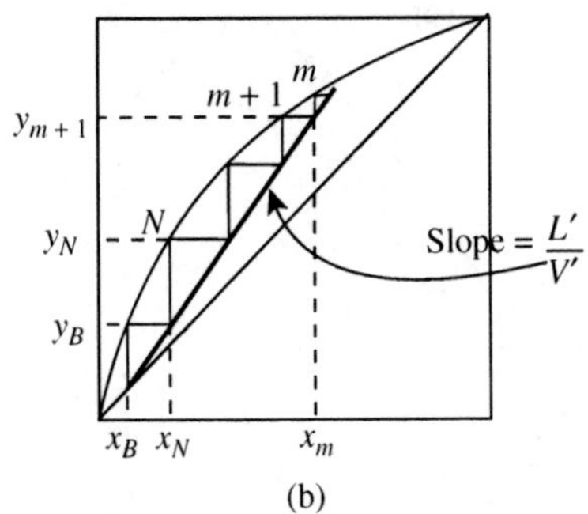

Figure 9.7 Mass balance for the stripping section. (From Smith R and Jobson M, 2000, *Distillation, Encyclopedia of Separation Science*, Academic Press; reproduced by permission).

Now the rectifying and stripping sections can be brought together at the feed stage. Consider the point of intersection of the operating lines for the rectifying and stripping sections. From Equations 9.10 and 9.15:

$$Vy_i = Lx_i + Dx_{i,D} \tag{9.17}$$

$$V'y_i = L'x_i - Bx_{i,B} \tag{9.18}$$

where y_i and x_i are the intersection of the operating lines. Subtracting Equations 9.17 and 9.18 gives:

$$(V - V')y_i = (L - L')x_i + Dx_{i,D} + Bx_{i,B} \tag{9.19}$$

Substituting the overall mass balance, Equation 9.7, gives:

$$(V - V')y_i = (L - L')x_i + Fx_{i,F} \tag{9.20}$$

Now it is necessary to determine how the vapor and liquid flowrates change at the feed stage.

What happens at the feed stage depends on the condition of the feed, whether it is subcooled, saturated liquid, partially vaporized, saturated vapor or superheated vapor. To define the condition of the feed, the variable q is introduced, defined as:

$$q = \frac{\text{heat required to vaporize 1 mole of feed}}{\text{molar latent heat of vaporization of feed}} \tag{9.21}$$

For a saturated liquid feed $q = 1$, and for a saturated vapor feed $q = 0$. The flowrate of feed entering the column as liquid is $q \cdot F$. The flowrate of feed entering the column as vapor is $(1 - q) \cdot F$. Figure 9.8 shows the feed stage. An overall mass balance on the feed stage for the vapor gives:

$$V = V' + (1 - q)F \tag{9.22}$$

An overall mass balance for the liquid on the feed stage gives:

$$L' = L + qF \tag{9.23}$$

Combining Equations 9.20, 9.22 and 9.23 together gives a relationship between the compositions of the feed and the vapor and liquid leaving the feed tray:

$$y_i = \frac{q}{q-1} \cdot x_i - \frac{1}{q-1} \cdot x_{i,F} \tag{9.24}$$

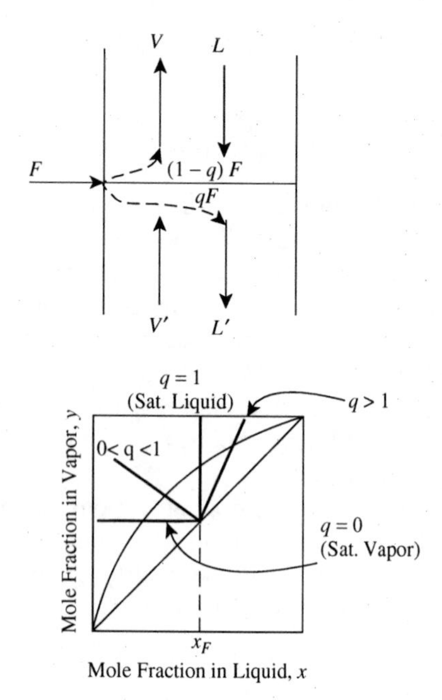

Figure 9.8 Mass balance for the feed stage. (From Smith R and Jobson M, 2000, *Distillation, Encyclopedia of Separation Science*, Academic Press; reproduced by permission).

This equation is known as the *q-line*. On the x–y plot, it is a straight line with slope $q/(q-1)$ and intersects the diagonal line at $x_{i,F}$. It is plotted in Figure 9.8 for various values of q.

The material balances for the rectifying and stripping sections can now be brought together. Figure 9.9a shows the complete design. The construction is started by plotting the operating lines for the rectifying and stripping sections. The q-line intercepts the operating lines at their intersection. The construction steps off between the operating lines and the equilibrium lines. The intersection of the operating lines is the correct feed stage, that is, the feed stage necessary to minimize the overall number of theoretical stages. The stepping procedure changes from one operating line to the other at the intersection with the q-line. The construction can be started either from the overhead composition working down or from the bottoms composition working up. This method of design is known as the *McCabe–Thiele Method*[3].

Figure 9.9b shows an alternative stepping procedure in which the change from the rectifying to the stripping operating lines leads to a feed stage location below the optimum feed stage. The result is an increase in the number of theoretical stages for the same separation. Figure 9.9c shows a stepping procedure in which the feed stage location is above the optimum, again resulting in an increase in the number of theoretical stages.

Figure 9.10 contrasts a total (Figure 9.10a) and partial condenser (Figure 9.10b) in the McCabe–Thiele Diagram. Although a partial condenser can provide a theoretical stage in principle, as shown in Figure 9.11b, in practice the performance of a condenser will not achieve the performance of a theoretical stage.

Figure 9.11 contrasts the two general classes of reboiler. Figure 9.11a is fed with a liquid. This is partially vaporized and discharges liquid and vapor streams that are in equilibrium. This is termed a *partial reboiler* or *once-through reboiler* and, as shown in the McCabe–Thiele diagram in Figure 9.11a, acts as a theoretical stage. In the other class of reboilers shown in Figure 9.11b, known as *thermosyphon reboilers*, part of the liquid flow from the bottom of the column is taken and partially vaporized. Whilst some separation obviously occurs in the thermosyphon reboiler, this will be less than that in a theoretical stage. It is safest to assume that the necessary stages are provided in the column itself, as illustrated in the McCabe–Thiele Diagram in Figure 9.11b. The choice of reboiler depends on[4]:

- the nature of the process fluid (particularly its viscosity and fouling tendencies);
- sensitivity of the bottoms product to thermal degradation;
- operating pressure;
- temperature difference between the process and heating medium;
- equipment layout (particularly the space available for headroom).

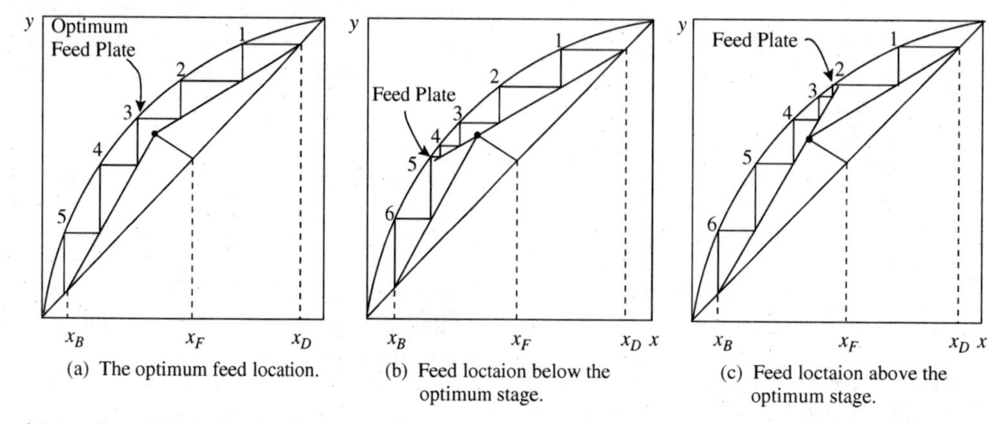

Figure 9.9 Combining the rectifying and stripping sections.

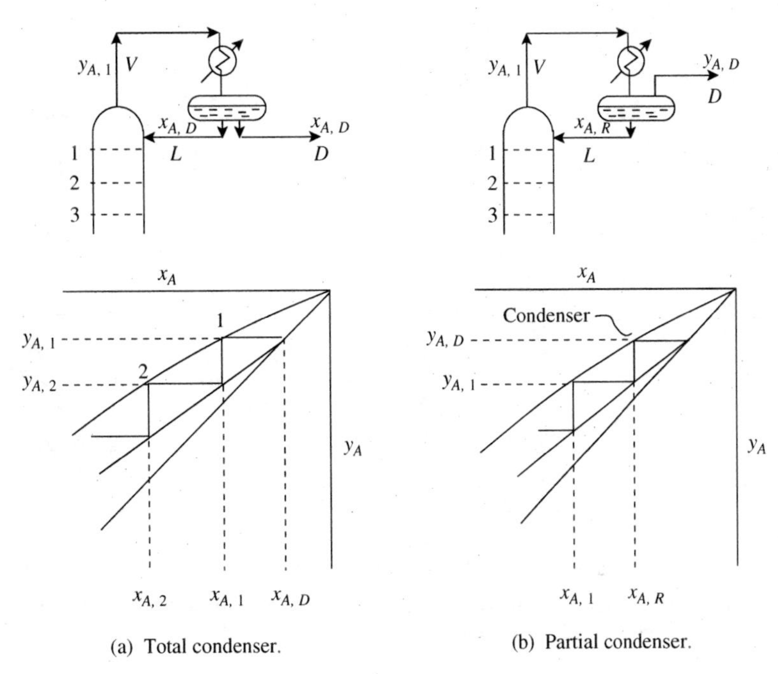

Figure 9.10 Total and partial condensers in the McCabe–Thiele Diagram.

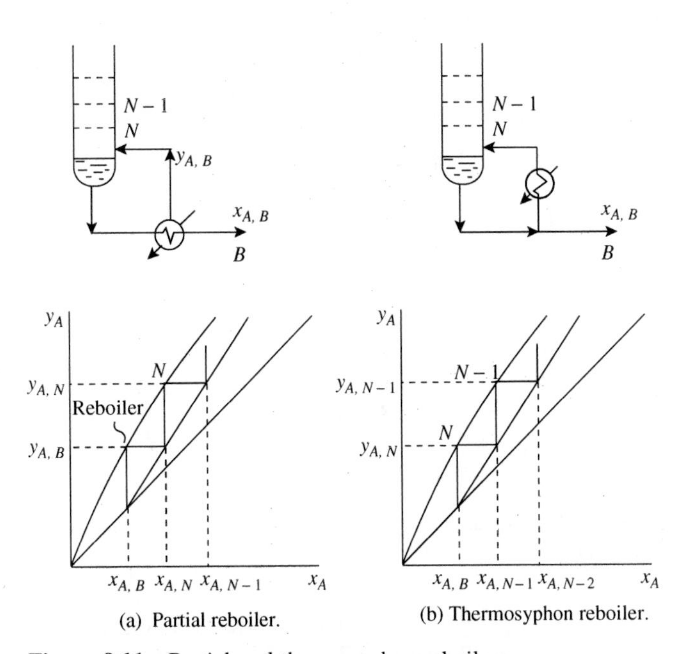

Figure 9.11 Partial and thermosyphon reboilers.

Thermosyphon reboilers are usually cheapest but not suitable for high-viscosity liquids or vacuum operation. Further discussion of reboilers will be deferred until Chapter 15. In the preliminary phases of a design, it is best to assume that enough stages are available in the column itself to achieve the required separation.

There are two important limits that need to be considered for distillation. The first is illustrated in Figure 9.12a. This is total reflux in which no products are taken and there is no feed. At total reflux, the entire overhead vapor is refluxed and all the bottoms liquid reboiled. Figure 9.12a also shows total reflux on an x–y plot. This corresponds with the minimum number of stages required for the separation. The other limiting case, shown in Figure 9.12b, is where the reflux ratio is chosen such that the operating lines intersect at the equilibrium line. As this stepping procedure approaches the q-line from both sides, an infinite number of steps are required to approach the q-line. This is the minimum reflux condition in which there are zones of constant composition (*pinches*) above and below the feed. For binary distillation, the zones of constant composition are usually located at the feed stage, as illustrated in Figure 9.12b. The pinch can also be away from the feed as illustrated in Figure 9.12c. This time, a *tangent pinch* occurs above the feed stage. A tangent pinch can also be obtained below the feed stage, depending on the shape of the x–y diagram.

The McCabe–Thiele Method is restricted in its application because it only applies to binary systems and involves the simplifying assumption of constant molar overflow. However, it is an important method to understand as it gives important conceptual insights into distillation that cannot be obtained in any other way.

9.4 TOTAL AND MINIMUM REFLUX CONDITIONS FOR MULTICOMPONENT MIXTURES

Just as with binary distillation, it is important to understand the operating limits for multicomponent distillation. The two extreme conditions of total and minimum reflux will

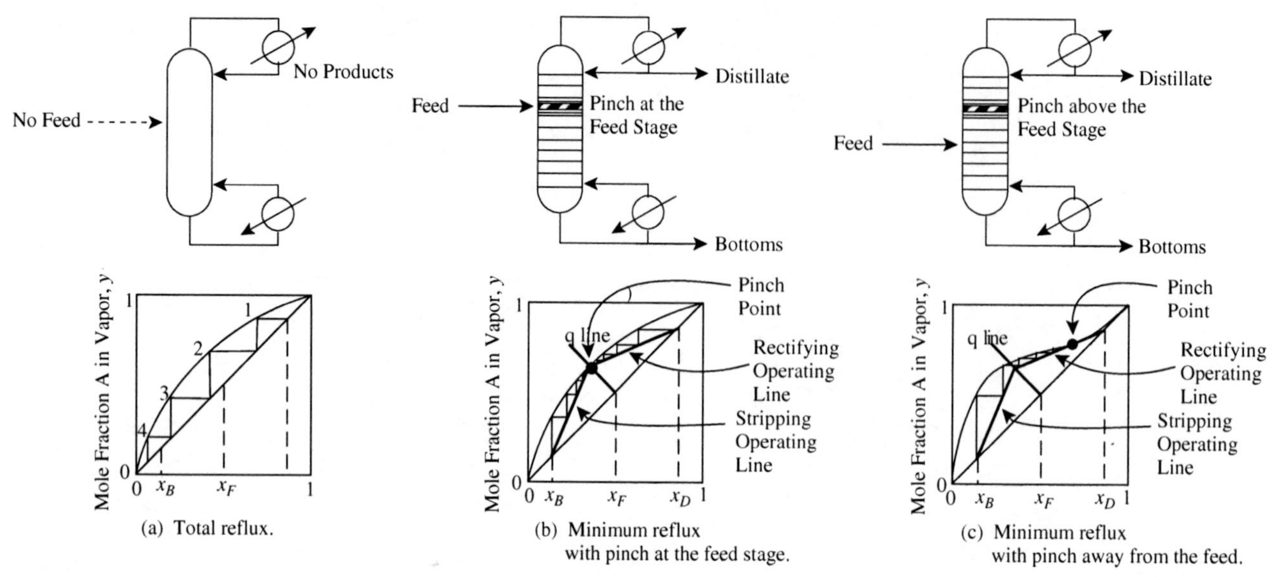

Figure 9.12 Total and maximum reflux in binary distillation. (Adapted from Smith R and Jobson M, 2000, *Distillation, Encyclopedia of Separation Science*, Academic Press; reproduced by permission).

therefore be considered. However, before a multicomponent distillation column can be designed, a decision must be made as to the two *key components* between which it is desired to make the separation. The *light key component* is the one to be kept out of the bottom product according to some specification. The *heavy key component* is the one to be kept out of the top product according to some specification. The separation of the nonkey components cannot be specified. Lighter than light key components will tend to go predominantly with the overhead product, and heavier than heavy key components will tend to go predominantly with the bottoms product. Intermediate-boiling components between the keys will distribute between the products. In preliminary design, the recovery of the light key or the concentration of the light key in the overhead product must be specified, as must the recovery of the heavy key or the concentration of the heavy key in the bottom product.

Consider first total reflux conditions, corresponding with the minimum number of theoretical stages. The bottom of a distillation column at total reflux is illustrated in Figure 9.13.

$$V_n = L_{n-1} \tag{9.25}$$

A component mass balance gives:

$$V_n y_{i,n} = L_{n-1} x_{i,n-1} \tag{9.26}$$

Combining Equations 9.25 and 9.26 gives:

$$y_{i,n} = x_{i,n-1} \tag{9.27}$$

Applying Equation 4.33 to the bottom of the column for a binary separation gives:

$$\left(\frac{y_A}{y_B}\right)_R = \alpha_R \left(\frac{x_A}{x_B}\right)_B \tag{9.28}$$

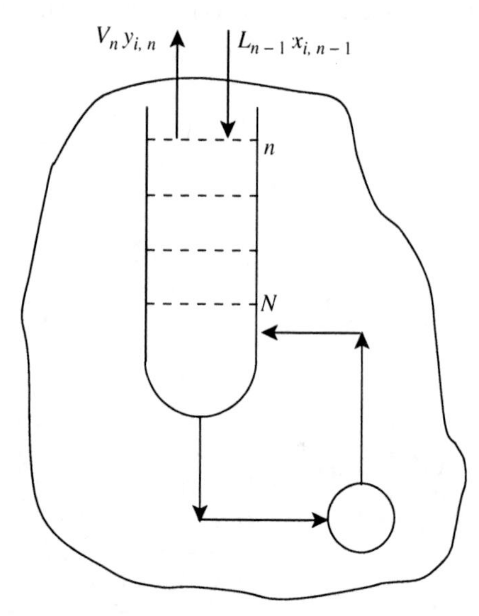

Figure 9.13 Stripping section of a column under total reflux conditions.

where subscript R refers to the reboiler and subscript B to the bottoms. Because reflux conditions are total, the liquid composition in the reboiler is the same as the bottom composition in this case. Combining Equation 9.27 with Equation 9.28 gives:

$$\left(\frac{x_A}{x_B}\right)_N = \alpha_R \left(\frac{x_A}{x_B}\right)_B \tag{9.29}$$

Similarly, the composition at Stage N–1 can be related to Stage N:

$$\left(\frac{x_A}{x_B}\right)_{N-1} = \alpha_N \left(\frac{x_A}{x_B}\right)_N \tag{9.30}$$

Combining this with Equation 9.29 gives:

$$\left(\frac{x_A}{x_B}\right)_N = \alpha_N \alpha_R \left(\frac{x_A}{x_B}\right)_B \qquad (9.31)$$

Continuing up the column to the distillate:

$$\left(\frac{y_A}{y_B}\right)_1 = \alpha_1 \alpha_2 \ldots\ldots \alpha_N \alpha_R \left(\frac{x_A}{x_B}\right)_B \qquad (9.32)$$

Assuming a total condenser and the relative volatility to be constant gives:

$$\left(\frac{x_A}{x_B}\right)_D = \alpha_{AB}^{N_{\min}} \left(\frac{x_A}{x_B}\right)_B \qquad (9.33)$$

where N_{min} = minimum number of theoretical stages
α_{AB} = relative volatility between A and B

Subscript D refers to the distillate. Equation 9.33 predicts the number of theoretical stages for a specified binary separation at total reflux and is known as the *Fenske Equation*[5].

In fact, the development of the equation does not include any assumptions limiting the number of components in the system. It can also be written for a multicomponent system for any two reference Components i and j[1,6–8]:

$$\frac{x_{i,D}}{x_{j,D}} = \alpha_{ij}^{N_{\min}} \frac{x_{i,B}}{x_{j,B}} \qquad (9.34)$$

where $x_{i,D}$ = mole fraction of Component i in the distillate
$x_{i,B}$ = mole fraction of Component i in the bottoms
$x_{j,D}$ = mole fraction of Component j in the distillate
$x_{j,B}$ = mole fraction of Component j in the bottoms
α_{ij} = relative volatility between Components i and j
N_{min} = minimum number of stages

Equation 9.34 can also be written in terms of the molar flowrates of the products:

$$\frac{d_i}{d_j} = \alpha_{ij}^{N_{\min}} \frac{b_i}{b_j} \qquad (9.35)$$

where d_i = molar distillate flow of Component i
b_i = molar bottoms flow of Component i

Alternatively, Equation 9.34 can be written in terms of recoveries of the products:

$$\frac{r_{i,D}}{r_{j,D}} = \alpha_{ij}^{N_{\min}} \frac{r_{i,B}}{r_{j,B}} \qquad (9.36)$$

where $r_{i,D}$ = recovery of Component i in the distillate
$r_{i,B}$ = recovery of Component i in the bottoms
$r_{j,D}$ = recovery of Component j in the distillate
$r_{j,B}$ = recovery of Component j in the bottoms

When Component i is the light key component L, and j is the heavy key component H, this becomes[1,6–8]:

$$N_{min} = \frac{\log\left[\dfrac{d_L}{d_H} \cdot \dfrac{b_H}{b_L}\right]}{\log \alpha_{LH}} \qquad (9.37)$$

$$N_{min} = \frac{\log\left[\dfrac{x_{L,D}}{x_{H,D}} \cdot \dfrac{x_{H,B}}{x_{L,B}}\right]}{\log \alpha_{LH}} \qquad (9.38)$$

$$N_{min} = \frac{\log\left[\dfrac{r_{L,D}}{1 - r_{L,D}} \cdot \dfrac{r_{H,B}}{1 - r_{H,B}}\right]}{\log \alpha_{LH}} \qquad (9.39)$$

The Fenske Equation can be used to estimate the composition of the products. Equation 9.35 can be written in a form:

$$\log\left[\frac{d_i}{b_i}\right] = N_{\min} \log \alpha_{ij} + \log\left[\frac{d_j}{b_j}\right] \qquad (9.40)$$

Equation 9.40 indicates that a plot of log $[d_i/b_i]$ will be a linear function of log α_{ij} with a gradient of N_{min}. This can be rewritten as:

$$\log\left[\frac{d_i}{b_i}\right] = A \log \alpha_{ij} + B \qquad (9.41)$$

where A and B are constants. The parameters A and B are obtained by applying the relationship to the light and heavy key components. This allows the compositions of the nonkey components to be estimated, and this is illustrated in Figure 9.14. Having specified the distribution of the light and heavy key components, knowing the relative volatilities of the other components allows their compositions to be estimated. The method is based on total reflux conditions. It assumes that the component distributions do not depend on reflux ratio.

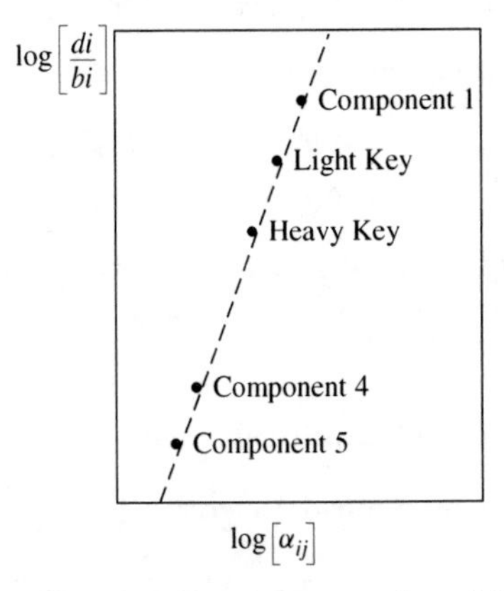

Figure 9.14 A plot of *di/bi* versus *ij* tends to follow a straight line when plotted on logarithmic axes. (From Smith R and Jobson M, 2000, *Distillation, Encyclopedia of Separation Science*, Academic Press; reproduced by permission).

Equation 9.35 can be written in a more convenient form than Equations 9.40 and 9.41. By combining an overall component balance:

$$f_i = d_i + b_i \tag{9.42}$$

with Equation 9.35, the resulting equation is[8]:

$$d_i = \frac{\alpha_{ij}^{N_{min}} f_i \left(\dfrac{d_j}{b_j}\right)}{1 + \alpha_{ij}^{N_{min}} \left(\dfrac{d_j}{b_j}\right)} \tag{9.43}$$

and

$$b_i = \frac{f_i}{1 + \alpha_{ij}^{N_{min}} \left(\dfrac{d_j}{b_j}\right)} \tag{9.44}$$

The Fenske Equation assumes that relative volatility is constant. The relative volatility can be calculated from the feed composition, but this might not be characteristic of the overall column[1]. The relative volatility will in practice vary through the distillation column, and an average needs to be taken. The average would best be taken at the average temperature of the column. To estimate the average, first assume that $\ln P_i^{SAT}$ varies linearly as $1/T$ (see Chapter 4). If it is assumed that the two key components both vary linearly in this way, then the difference ($\ln P_i^{SAT} - \ln P_j^{SAT}$) also varies as $1/T$, which in turn means $\ln(P_i^{SAT}/P_j^{SAT})$ varies as $1/T$. If it is now assumed Raoult's Law applies (i.e. ideal vapor–liquid equilibrium behavior, see Chapter 4), then the relative volatility α_{ij} is the ratio of the saturated vapor pressures P_i^{SAT}/P_j^{SAT}. Therefore, $\ln \alpha_{ij}$ varies as $1/T$. The mean temperature in the column is given by:

$$T_{mean} = 1/2(T_{top} + T_{bottom}) \tag{9.45}$$

Assuming $\ln(\alpha_{ij})$ is proportional to $1/T$:

$$\frac{1}{\ln(\alpha_{ij})_{mean}} = \frac{1}{2}\left(\frac{1}{\ln(\alpha_{ij})_{top}} + \frac{1}{\ln(\alpha_{ij})_{bottom}}\right) \tag{9.46}$$

which on rearranging gives:

$$(\alpha_{ij})_{mean} = \exp\left[\frac{2\ln(\alpha_{ij})_{top}\ln(\alpha_{ij})_{bottom}}{\ln(\alpha_{ij})_{top} + \ln(\alpha_{ij})_{bottom}}\right] \tag{9.47}$$

If, instead of assuming α_{ij} is proportional to $1/T$, it is assumed that α_{ij} is proportional to T, at the mean temperature:

$$\ln(\alpha_{ij})_{mean} = 1/2[\ln(\alpha_{ij})_{top} + \ln(\alpha_{ij})_{bottom}] \tag{9.48}$$

which can be rearranged to give what is the geometric mean:

$$(\alpha_{ij})_{mean} = \sqrt{(\alpha_{ij})_{top} \cdot (\alpha_{ij})_{bottom}} \tag{9.49}$$

Either Equation 9.47 or 9.49 can be used to obtain a more realistic relative volatility for the Fenske Equation. Generally, Equation 9.47 gives a better prediction than Equation 9.49. However, the greater the variation of relative volatility across the column, the greater the error in using any averaging equation. By making some assumption of the product compositions, the relative volatility at the top and bottom of the column can be calculated and a mean taken. Iteration can be carried out for greater accuracy. Start by calculating the relative volatility from the feed conditions. Then the product compositions can be estimated (say from the Fenske Equation). The average relative volatility can then be estimated from the top and bottom products. However, this new relative volatility will lead to revised estimates of the product compositions, and hence the calculation should iterate to converge. How good the approximations turn out to be largely depends on the ideality (or nonideality) of the mixture (see Chapter 4) being distilled. Generally, the more nonideal the mixture, the greater the errors.

In addition to the changes in relative volatility resulting from changes in composition and temperature through the column, there is also a change in pressure through the column due to the pressure drop. The change in pressure affects the relative volatility. In preliminary process design, the effect of the pressure drop through the column is usually neglected.

Total reflux is one extreme condition for distillation and can be approximated by the Fenske Equation. The other extreme condition for distillation is minimum reflux. At minimum reflux, there must be at least one zone of constant composition for all components. If this is not the case, then additional stages would change the separation. In binary distillation, the zones of constant composition are usually adjacent to the feed, Figure 9.15a. If there are no heavy nonkey components in a multicomponent distillation, the zone of constant composition above the feed will still be adjacent to the feed. Similarly, if there are no light nonkey components, the zone of constant composition below the feed will still be adjacent to the feed, Figure 9.15b. If one or more of the heavy nonkey components do not appear in the distillate, the zone of constant composition above the feed will move to a higher position in the column, so that the nondistributing heavy nonkey can diminish to zero mole fraction in the stages immediately above the feed, Figure 9.15c. If one or more of the light nonkey components do not appear in the bottoms, the zone of constant composition below the feed will move to a lower position in the column so that the nondistributing light key components can diminish to zero mole fraction in the stages immediately below the feed, Figure 9.15d. Generally, there will be a zone of constant composition (pinch) above the feed in the rectifying section and one in the stripping section below the feed, Figure 9.15e.

The Underwood Equations can be used to predict the minimum reflux for multicomponent distillation[9]. The derivation of the equations is lengthy, and the reader is

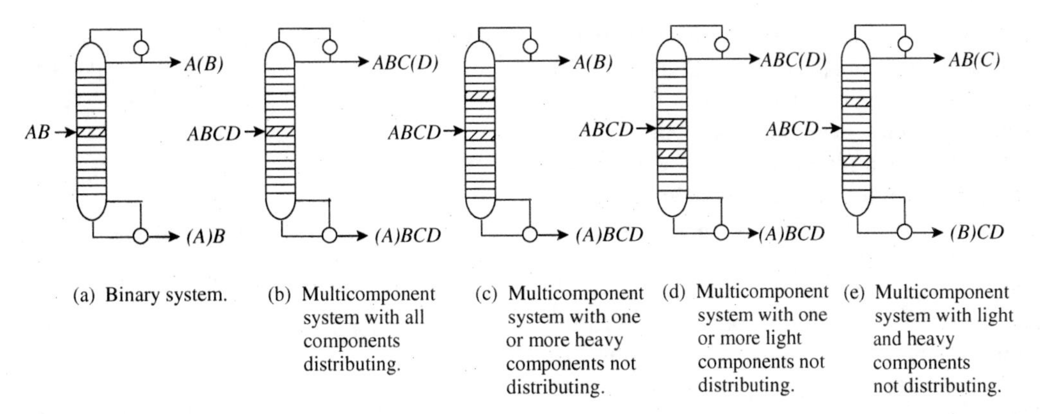

Figure 9.15 Pinch location (zones of constant composition) for binary and multicomponent systems. Brackets indicate key components remaining in a product stream due to incomplete recovery.

referred to other sources for the details of the derivation[1,7,9]. The equations assume that the relative volatility and molar overflow are constant between the zones of constant composition. There are two equations. The first is given by[9]:

$$\sum_{i=1}^{NC} \frac{\alpha_{ij} x_{i,F}}{\alpha_{ij} - \theta} = 1 - q \tag{9.50}$$

where α_{ij} = relative volatility
$x_{i,F}$ = mole fraction of Component i in the feed
θ = root of the equation
q = feed condition
$= \dfrac{\text{heat required to vaporize one mole of feed}}{\text{molar latent heat of vaporization of feed}}$
= 1 for a saturated liquid feed, 0 for a saturated vapor feed
NC = number of components

To solve Equation 9.50, start by assuming a feed condition such that q can be fixed. Saturated liquid feed (i.e. $q = 1$) is normally assumed in an initial design as it tends to decrease the minimum reflux ratio relative to a vaporized feed. Liquid feeds are also preferred because the pressure at which the column operates can easily be increased if required by pumping the liquid to a higher pressure. Increasing the pressure of a vapor feed is much more expensive as it requires a compressor rather than a pump. Feeding a subcooled liquid or a superheated vapor brings inefficiency to the separation as the feed material must first return to saturated conditions before it can participate in the distillation process.

Equation 9.50 can be written for all NC components of the feed and solved for the necessary values of θ. There are $(NC - 1)$ real positive values of θ that satisfy Equation 9.50, and each lies between the α's of the components. The second equation is then written for each value of θ obtained to determine the minimum reflux ratio, R_{min}[9]:

$$R_{min} + 1 = \sum_{i=1}^{NC} \frac{\alpha_{ij} x_{i,D}}{\alpha_{ij} - \theta} \tag{9.51}$$

To solve Equation 9.51, it is necessary to know the values of not only $\alpha_{i,j}$ and θ but also $x_{i,D}$. The values of $x_{i,D}$ for each component in the distillate in Equation 9.51 are the values at the minimum reflux and are unknown. Rigorous solution of the Underwood Equations, without assumptions of component distribution, thus requires Equation 9.50 to be solved for $(NC - 1)$ values of θ lying between the values of $\alpha_{i,j}$ of the different components. Equation 9.51 is then written $(NC - 1)$ times to give a set of equations in which the unknowns are R_{min} and $(NC - 2)$ values of $x_{i,D}$ for the nonkey components. These equations can then be solved simultaneously. In this way, in addition to the calculation of R_{min}, the Underwood Equations can also be used to estimate the distribution of nonkey components at minimum reflux conditions from a specification of the key component separation. This is analogous to the use of the Fenske Equation to determine the distribution at total reflux. Although there is often not too much difference between the estimates at total and minimum reflux, the true distribution is more likely to be between the two estimates.

However, this calculation can be simplified significantly by making some reasonable assumptions regarding the component distributions to approximate $x_{i,D}$. It is often a good approximation to assume all of the lighter than light key components go to the overheads and all of the heavier than heavy key components go to the column bottoms. Also, if the light and heavy key components are adjacent in volatility, there are no components between the keys, and all $x_{i,D}$ are known, as the key component split is specified by definition. Equation 9.50 can then be solved by trial and error for the single value of θ required that lies between the relative volatilities of the key components. This value of θ can then be substituted in Equation 9.51 to solve for R_{min} directly, as all $x_{i,D}$ are known.

If the assumption is maintained that all of the lighter than light key components go to the overheads and all of the heavier than heavy key components go to the column bottoms, for cases where the light and heavy key components are not adjacent in volatility, one more value of θ is required than there are components between the keys.

For this case, $x_{i,D}$ are unknown for the components between the keys. The values of θ obtained from Equation 9.50, where each θ lies between an adjacent pair of relative volatilities. Once the values of θ have been determined, Equation 9.51 can be written for each value of θ with $x_{i,D}$ of the components between the keys as unknowns. This set of equations is then solved simultaneously for R_{min} and $x_{i,D}$ for the components between the keys.

Another approximation that can be made to simplify the solution of the Underwood Equations is to use the Fenske Equation to approximate $x_{i,D}$. These values of $x_{i,D}$ will thus correspond with total reflux rather than minimum reflux.

Example 9.1 A distillation column operating at 14 bar with a saturated liquid feed of 1000 kmol·h^{-1} with composition given in Table 9.1 is to be separated into an overhead product that recovers 99% of the n-butane overhead and 95% of the i-pentane in the bottoms. Relative volatilities are also given in Table 9.1.

Table 9.1 Distillation column feed and relative volatilities.

Component	f_i (kmol·h^{-1})	α_{ij}
Propane	30.3	16.5
i-Butane	90.7	10.5
n-Butane (LK)	151.2	9.04
i-Pentane (HK)	120.9	5.74
n-Pentane	211.7	5.10
n-Hexane	119.3	2.92
n-Heptane	156.3	1.70
n-Octane	119.6	1.00

a. Calculate the minimum number of stages using the Fenske Equation.
b. Estimate the compositions of the overhead and bottoms products using the Fenske Equation.
c. Calculate the minimum reflux ratio using the Underwood Equations.

Solution

a. Substitute $r_{L,D} = 0.99, r_{H,B} = 0.95, \alpha_{LH} = 1.5749$ in Equation 9.39:

$$N_{\min} = \frac{\log\left[\dfrac{0.99}{(1-0.99)}\cdot\dfrac{0.95}{1-0.95}\right]}{\log 1.5749}$$

$$= 16.6$$

b. Using the heavy key component as the reference:

$$\frac{d_H}{b_H} = \frac{f_H(1-r_{H,B})}{f_H r_{H,B}} = \frac{1-r_{H,B}}{r_{H,B}} = \frac{1-0.95}{0.95} = 0.05263$$

Substitute N_{min}, $\alpha_{i,H}$, f_i and (d_H/b_H) in Equation 9.43 to obtain d_i and determine b_i by mass balance. For the first component (propane):

$$d_i = \frac{2.875^{16.6}\times 30.3\times 0.05263}{1+2.875^{16.6}\times 0.05263}$$

$$= 30.30 \text{ kmol·h}^{-1}$$

$$b_i = f_i - d_i$$

$$= 30.30 - 30.30$$

$$= 0 \text{ kmol·h}^{-1}$$

and so on, for the other components to obtain the results in Table 9.2.

Table 9.2 Distribution of components for Example 9.1.

Component	d_i	b_i	$x_{i,D}$	$x_{i,B}$
Propane	30.30	0.0	0.1089	0.0
i-Butane	90.62	0.08	0.3257	0.0001
n-Butane	149.69	1.51	0.5380	0.0021
i-Pentane	6.05	114.86	0.0217	0.1591
n-Pentane	1.55	210.15	0.0056	0.2911
n-Hexane	0.0	119.30	0.0	0.1653
n-Heptane	0.0	156.30	0.0	0.2165
n-Octane	0.0	119.60	0.0	0.1657
Total	278.21	721.80	0.9999	0.9999

c. To calculate minimum reflux ratio, first solve Equation 9.50. A search must be carried out for the root θ that satisfies Equation 9.50. Since there are no components between the key components, there is only one root, and the root will have a value between α_L and α_H. This involves trial and error to satisfy the summation to be equal to zero, as summarized in Table 9.3.

Table 9.3 Solution for the root of the Underwood Equation.

$x_{F,i}$	α_{ij}	$\alpha_{ij}x_{i,F}$	$\dfrac{\alpha_{ij}x_{i,F}}{\alpha_{ij}-\theta}$			
			$\theta = 7.0$	$\theta = 7.3$	$\theta = 7.2$	$\theta = 7.2487$
0.0303	16.5	0.5000	0.0526	0.0543	0.0538	0.0540
0.0907	10.5	0.9524	0.2721	0.2796	0.2886	0.2929
0.1512	9.04	1.3668	0.6700	0.7855	0.7429	0.7630
0.1209	5.74	0.6940	−0.5508	−0.4449	−0.4753	−0.4600
0.2117	5.10	1.0797	−0.5682	−0.4908	−0.5141	−0.5025
0.1193	2.92	0.3484	−0.0854	−0.0795	−0.0814	−0.0805
0.1563	1.70	0.2657	−0.0501	−0.0474	−0.0483	−0.0479
0.1196	1.00	0.1196	−0.0199	−0.0190	−0.0193	−0.0191
			−0.2797	0.0559	−0.0532	0.0000

Now substitute $\theta = 7.2487$ in Equation 9.51, as summarized in Table 9.4.

$$R_{\min} + 1 = 3.866$$

$$R_{\min} = 2.866$$

The calculation in this example can be conveniently carried out in spreadsheet software. However, many implementations are available in commercial flowsheet simulation software.

Table 9.4 Solution of the second Underwood Equation.

$x_{D,i}$	α_{ij}	$\alpha_{ij}x_{i,D}$	$\dfrac{\alpha_{ij}x_{i,D}}{\alpha_{ij}-\theta}$
0.1089	16.5	1.7970	0.1942
0.3257	10.5	3.4202	1.0520
0.5380	9.04	4.8639	2.7153
0.0217	5.74	0.1247	−0.0827
0.0056	5.10	0.0285	−0.0133
0.0	2.92	0.0	0.0
0.0	1.70	0.0	0.0
0.0	1.00	0.0	0.0
			3.8655

The Underwood Equation is based on the assumption that the relative volatilities and molar overflow are constant between the pinches. Given that the relative volatilities change throughout the column, which are the most appropriate values to use in the Underwood Equations? The relative volatilities could be averaged according to Equations 9.47 or 9.49. However, it is generally better to use the ones based on the feed conditions rather than the average values based on the distillate and bottoms compositions. This is because the location of the pinches is often close to the feed.

The Underwood Equations tend to underestimate the true value of the minimum reflux ratio. The most important reason for this is the assumption of constant molar overflow. As mentioned previously, the Underwood Equations assumed constant molar overflow between the pinches. So far, in order to determine the reflux ratio of the column, this assumption has been extended to the whole column. However, some compensation can be made for the variation in molar overflow by carrying out an energy balance around the top pinch for the column, as shown in Figure 9.16. Thus

$$Q_{COND} = V(H_V - H_L) \tag{9.52}$$

where Q_{COND} = heat rejected in the condenser
V = vapor flowrate at the top of the column
H_V, H_L = molar enthalpy of the vapor and liquid at the top of the column

An energy balance gives:

$$V_\infty H_{V\infty} = L_\infty H_{L\infty} + Dh_L + Q_{COND} \tag{9.53}$$

where V_∞, L_∞ = vapor and liquid flowrates at the rectifying pinch
$H_{V\infty}, H_{L\infty}$ = molar enthalpies of the vapor and liquid at the rectifying pinch
D = distillate rate

$$\text{Also } V_\infty = L_\infty + D \tag{9.54}$$

$$V = L + D \tag{9.55}$$

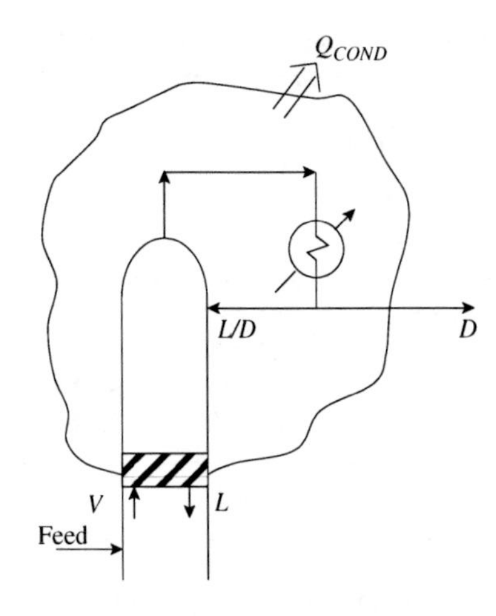

Figure 9.16 Energy balance around the rectifying pinch under minimum reflux conditions.

Combining Equations 9.52 to 9.55 and rearranging gives[8]:

$$\frac{L}{D} = \frac{(L_\infty/D)(H_{V\infty} - H_{L\infty}) + H_{V\infty} - H_V}{H_V - H_L} \tag{9.56}$$

where (L_∞/D) is the reflux ratio given by the Underwood Equations and (L/D) is the reflux ratio compensated for the variation in molar overflow. One problem remains before Equation 9.56 can be applied. The calculation of $H_{V\infty}$ and $H_{L\infty}$ is required, for which the vapor and liquid composition at the top pinch is required. However, these are given by the Underwood Equations[1,8,9]:

$$x_{i,\infty} = \frac{x_{i,D}}{\dfrac{L_\infty}{D}\left(\dfrac{\alpha_{ij}}{\theta} - 1\right)} \tag{9.57}$$

where $x_{i,\infty}$ = liquid mole fraction of Component i at the rectifying pinch
α_{ij} = relative volatility
θ = root of the Underwood Equation that satisfies the second equation (Equation 9.51) at the known minimum reflux ratio. The appropriate root is the one associated with the heavy key, which is the one below α_{HK}. This is given by $\alpha_{HNK} < \theta < \alpha_{HK}$ where α_{HNK} is the relative volatility of the component below the heavy key. If there is no component heavier than the heavy key $0 < \theta < \alpha_{HK}$.

A material balance around the rectifying pinch gives the mole fraction in the vapor phase:

$$y_{i,\infty} = \frac{x_{i,\infty}(L_\infty/D) + x_{i,D}}{(L_\infty/D) + 1} \tag{9.58}$$

where $y_{i,\infty}$ = vapor mole fraction of Component i at the rectifying pinch

Table 9.5 Root of the Underwood Equation.

Component	$x_{i,D}$	α_{ij}	$\alpha_{ij}\ x_{i,D}$	$\sum_{i=1}^{NC} \frac{\alpha_{ij} x_{i,D}}{\alpha_{ij} - \theta} - R_{min} - 1$			
				$\theta = 5.7$	$\theta = 5.6$	$\theta = 5.65$	$\theta = 5.6598$
Propane	0.1089	16.5	1.7970	0.1664	0.1649	0.1656	0.1658
i-Butane	0.3257	10.5	3.4202	0.7126	0.6980	0.7052	0.7066
n-Butane	0.5380	9.04	4.8639	1.4562	1.4139	1.4348	1.4389
i-Pentane	0.0217	5.74	0.1247	3.1180	0.8909	1.3858	1.5551
	0.9943			1.5882	−0.6973	−0.1736	0.0014

Example 9.2 Apply the enthalpy correction to the prediction of minimum reflux ratio for Example 9.1 to obtain a more accurate estimate of the minimum reflux ratio.

Solution Calculate the liquid composition at the rectifying pinch using Equation 9.57. But first, the root of the second Underwood Equation associated with the heavy key component must be calculated. From Table 9.1, this lies in the range:

$$5.74 < \theta < 5.10$$

Assume that components heavier than the heavy key component do not appear in the distillate. The root is determined in Table 9.5.

Now substitute $\theta = 5.6598$ in Equation 9.57 to calculate $x_{i,\infty}$ and solve for $y_{i,\infty}$ from Equation 9.58. The results are given in Table 9.6.

Table 9.6 Vapor and liquid mole fractions at the rectifying pinch.

Component	$x_{i,\infty}$	$y_{i,\infty}$
Propane	0.0198	0.0429
i-Butane	0.1329	0.1828
n-Butane	0.3144	0.3722
i-Pentane	0.5351	0.4023
	1.0022	1.0002

Now calculate the molar enthalpies of the vapor and liquid streams. Enthalpies were calculated here from ideal gas enthalpy data corrected using the Peng–Robinson Equation of State (see Chapter 4):

$$H_V = -122{,}900\ \text{kJ·kmol}^{-1}$$

$$H_L = -138{,}800\ \text{kJ·kmol}^{-1}$$

$$H_{V\infty} = -130{,}700\ \text{kJ·kmol}^{-1}$$

$$H_{L\infty} = -150{,}600\ \text{kJ·kmol}^{-1}$$

Substitute these values and $L_\infty/D = 2.866$ in Equation 9.56:

$$\frac{L}{D} = \frac{2.866(-130{,}700 - (-150{,}600)) - 130{,}700 + 122{,}900}{-122{,}900 - (-138{,}800)}$$

$$= 3.095$$

This is compared with an uncorrected reflux ratio of 2.866.

To obtain a rigorous value for the minimum reflux ratio to assess the results of the Underwood Equation, a rigorous simulation of the column is needed. A model is set up with a large number of stages (say 200 or more) that carry out the separation. The reflux ratio is then gradually turned down and resimulated with each setting. This is repeated until zones of constant composition appear (pinches). For this separation, using the Peng–Robinson Equation of State, it turns out to be $R_{min} = 3.545$. It should be emphasized that this value of 3.545 accounts for both variation in relative volatility and molar overflow.

9.5 FINITE REFLUX CONDITIONS FOR MULTICOMPONENT MIXTURES

Having obtained the minimum number of stages from the Fenske Equation and minimum reflux ratio from the Underwood Equations, the empirical relationship of Gilliland[10] can be used to determine the number of stages. The original correlation was presented in graphical form[10]. Two parameters (X and Y) were used to correlate the data:

$$Y = \frac{N - N_{min}}{N + 1},\ X = \frac{R - R_{min}}{R + 1} \tag{9.59}$$

where X, Y = correlating parameters
N = actual number of theoretical stages
N_{min} = minimum number of theoretical stages
R = actual reflux ratio
R_{min} = minimum reflux ratio

Various attempts have been made to represent the correlation algebraically. For example[11]:

$$Y = 0.2788 - 1.3154X + 0.4114X^{0.2910} + 0.8268 \cdot \ln X + 0.9020 \ln\left(X + \frac{1}{X}\right) \tag{9.60}$$

Equation 9.60 allows an estimate of the number of theoretical stages the column requires.

Example 9.3 Assuming that the distillation column in Examples 9.1 and 9.2 uses a total condenser, estimate the number of theoretical stages required if $R/R_{min} = 1.1$ for the values of R_{min} obtained in Examples 9.1 and 9.2

Solution

$$R_{min} = 2.866$$

Given $R/R_{min} = 1.1$

$$R = 3.153$$

From Equation 9.59

$$X = \frac{R - R_{\min}}{R + 1} = \frac{3.153 - 2.866}{3.153 + 1} = 0.0691$$

Substitute $X = 0.0691$ in Equation 9.60

$$Y = 0.2788 - 1.3154 \times 0.0691 + 0.4114 \times 0.0691^{0.2910} + 0.8268 \ln 0.0691 + 0.9020 \ln \left[0.0691 + \frac{1}{0.0691}\right] = 0.5822$$

Substitute $Y = 0.5822$ in Equation 9.59

$$0.5822 = \frac{N - 16.6}{N + 1}$$

$$N = 41.1$$

Thus, 42 theoretical stages are needed.

Repeat the calculation for $R_{min} = 3.095$

$$R = 3.405$$

$$X = 0.0703$$

$$Y = 0.5805$$

$$N = 41.0$$

Thus, the error in the prediction of the minimum reflux ratio has little effect on the estimate of the number of theoretical stages for this example. The error in the prediction of the energy consumption is likely to be more serious.

The actual number of trays required in the column will be greater than the number of theoretical stages, as mass transfer limitations will prevent equilibrium being achieved on each tray. To estimate the actual number of trays, the number of theoretical stages must be divided by the overall stage efficiency. This is most often in the range between 0.7 and 0.9, depending on the separation being carried out and the design of the distillation tray used. Efficiencies significantly below 0.7 can be encountered (perhaps as low as 0.4 in extreme cases), and efficiencies in excess of 0.9. O'Connell[12] produced a simple graphical plot that can be used to obtain a first estimate of the overall stage efficiency. This graphical plot can be represented by[13]:

$$E_O = 0.542 - 0.285 \log(\alpha_{LH} \mu_L) \qquad (9.61)$$

where E_O = overall stage efficiency ($0 < E_O < 1$)
α_{LH} = relative volatility between the key components
μ_L = viscosity of the feed at average column conditions ($mN \cdot s \cdot m^{-2} = cP$)

Equation 9.61 is only approximate at best. For distillation trays with long flow paths across the active plate area, Equation 9.61 tends to underestimate the overall efficiency.

It should be emphasized that the overall stage efficiency from Equation 9.61 should only be used to derive a first estimate of the actual number of distillation trays. More elaborate methods are available but are outside the scope of this text. In practice, the stage efficiency varies from component to component, and more accurate calculations require much more information on tray type and geometry and physical properties of the fluids.

In practice, the number of trays is often increased by 5 to 10% to allow for uncertainties in the design. The height of the column can then be estimated by multiplying the actual number of trays by the tray spacing. The tray spacing is often taken to be 0.45 m or 0.6 m, but the tendency is to use closer tray spacing given the tendency to use more sophisticated trays than simple sieve trays. Additional height of 1 m to 2 m needs to be added at the top of the column for vapor disengagement and at the bottom of the column for vapor–liquid disengagement for the reboiler return and a liquid sump. However, there is a maximum height for a column, beyond which the column must be split into multiple shells. Large columns are designed to be free standing. This means that the column must be able to mechanically withstand the wind load, as must the foundations. The maximum height depends on the local weather (especially susceptibility to hurricanes and typhoons), ground conditions for foundations and susceptibility to earthquakes. The maximum height is normally restricted to be below 100 m, but might be significantly smaller for adverse conditions. The taller the column, the greater the incentive to decrease the tray spacing to decrease the column height, but the practical minimum tray spacing is of the order of 0.35 m.

Having estimated the number of trays, the column diameter can then be estimated. This is usually estimated with reference to the *flood point* for the column. The flood point occurs when the relative flowrates of the vapor and liquid are such that the liquid can no longer flow down the column in such a way as to allow efficient operation of the column[14]. For plate columns, there are a number of different mechanisms that can create flooding, but in one way or another, either[14]:

- these involve excessive entrainment of liquid up the column with the vapor, or
- the downcomer is no longer able to allow the liquid to flow at the required flowrate.

For preliminary design, liquid entrainment is usually used as a reference. To prevent entrainment, the vapor velocity for tray columns is usually in the range 1.5 to 3.5 m·s^{-1}. However, the entrainment of liquid droplets can be predicted using Equation 8.3 to calculate the settling velocity. To apply Equation 8.3 requires the parameter K_T to be specified. For distillation using tray columns, K_T is correlated in terms of a liquid–vapor flow parameter F_{LV}, defined by:

$$F_{LV} = \left(\frac{M_L L}{M_V V}\right)\left(\frac{\rho_V}{\rho_L}\right)^{0.5} \tag{9.62}$$

where F_{LV} = liquid–vapor flow parameter (–)
L = liquid molar flowrate (kmol·s^{-1})
V = vapor molar flowrate (kmol·s^{-1})
M_L = liquid molar mass (kg·kmol^{-1})
M_V = vapor molar mass (kg·kmol^{-1})
ρ_V = vapor density (kg·m^{-3})
ρ_L = liquid density (kg·m^{-3})

The vapor–liquid flow parameter is related to operating pressure, with low values corresponding to vacuum distillation and high values corresponding to high-pressure distillation. Generally, for values of F_{LV} lower than 0.1, packings are preferred. Fair[15] presented a graphical correlation for K_T that can be used for preliminary design. The original graphical correlation for K_T can be expressed as:

$$K_T = \left(\frac{\sigma}{20}\right)^{0.2} \exp[-2.979 - 0.717 \ln F_{LV} - 0.0865(\ln F_{LV})^2 + 0.997 \ln H_T - 0.07973 \ln F_{LV} \ln H_T + 0.256(\ln H_T)^2] \tag{9.63}$$

where K_T = parameter for terminal velocity (m·s^{-1})
σ = surface tension (mN·m^{-1} = mJ·m^{-2} = dyne·cm^{-1})
H_T = tray spacing (m)

The correlation is valid in the range 0.25 m < H_T < 0.6 m. Equation 9.63 tends to predict low values of K_T and hence can be considered to be conservative, especially at high pressures and high liquid rates. Also, the system might be subject to foaming, in which case K_T must be decreased to allow for this. For a system particularly subject to foaming, the value of K_T from Equation 9.63 should be multiplied by a factor that is less than unity to allow for this[13]:

- Distillation involving light gases typically 0.8 to 0.9
- Crude oil distillation typically 0.85
- Absorbers typically 0.7 to 0.85
- Strippers typically 0.6 to 0.8

The correlation in Equation 9.63 requires the spacing between the trays to be specified. Tray spacing can vary between 0.15 and 1 m, depending on the diameter of the column, tray design, vapor and liquid flowrates and physical properties of the fluids. Tray spacing is more usually in the range 0.45 to 0.6 m. If access is required for maintenance, cleaning or inspection purposes, 0.45 m is a practical minimum spacing. For preliminary design, a value of 0.45 m is usually a reasonable assumption. Having determined the flooding velocity from Equation 8.3, the operation of the column is fixed at some proportion of the flooding velocity. Designs usually lie in the range of 70 to 90% of the flooding velocity. In preliminary design, a value of 80% of the flooding velocity is usually reasonable. The diameter of the column can now be estimated from the vapor flowrate up the column. If conventional trays are to be used, then an allowance must be made for the area of the column cross section that will be taken up by the downcomers. The size of the downcomer depends on the tray geometry, vapor and liquid flowrates, operating pressure and physical properties of the fluids. The area of the column cross section taken up by downcomers is typically in the range of 5 to 15% of the cross-sectional area of the column. Generally, the higher the column operating pressure, the smaller is the downcomer area as a proportion of the total cross-sectional area. A value of 10% is usually a reasonable assumption in preliminary design.

If the column is to be packed, then the height of packing can be estimated from:

$$H = N \times HETP \tag{9.64}$$

where H = packing height
N = number of theoretical stages
$HETP$ = height equivalent of a theoretical plate

HETP is typically in the range 0.3 to 0.9 m for random packings and 0.2 to 0.7 m for structured packings. Various correlations are available for *HETP*, but the designer should use them with great caution, and reliable values can only be obtained from experimental data or packing manufacturers. *HETP* is normally correlated against an *F-Factor*[13]:

$$F_F = v_V \sqrt{\rho_V} \tag{9.65}$$

where F_F = F-factor
v_V = superficial vapor velocity related to empty column (m·s^{-1})
ρ_V = vapor density (kg·m^{-3})

It is best to refer to manufacturer's data as the performance of packing can vary widely, but Kister[13] has presented a considerable amount of data for various packings. Generally, the lower the F-factor, the lower the resulting *HETP* for a given packing and separation.

The height of the column for a packed column needs to allow for liquid distribution and redistribution. As the feed enters the column, the liquid above the feed needs to be collected, combined with the feed liquid and distributed across the packing below the feed using troughs, weirs, drip-pipes, and so on. Also, as the liquid flows down through the packing, flow gradually becomes less effectively distributed over the packing, mainly due to interaction with the column walls. This requires the liquid to be periodically collected and redistributed over the packing below. The maximum height of packed bed is normally taken to be 6 m or 10 column diameters, whichever is smaller. Liquid collection and distribution typically require a height of 0.5 to 1 m. Also, additional height needs to be added of 1 to 2 m at the top of the column for vapor disengagement and at the bottom of the column for vapor–liquid disengagement for the reboiler return and a liquid sump.

The diameter for packed columns is again taken to be that giving a vapor velocity to be some proportion the flooding velocity. In preliminary design, a value of 80% of the flooding velocity is usually reasonable. As with plate columns, flooding occurs in packed columns when the liquid can no longer flow down the column in such a way as to allow efficient operation of the column[13]. This is characterized by a condition in which the pressure drop through the packed bed starts to increase very rapidly as the vapor flowrates are increased with simultaneous loss of mass transfer efficiency. Heavy liquid entrainment will also occur. To prevent entrainment, the vapor velocity for packed columns is often in the range 0.5 to 2 m·s^{-1}. The flooding velocity for different types of packing is usually correlated in terms of the flow parameter used for plate columns, together with a capacity parameter:

$$CP = v_V \left(\frac{\rho_V}{\rho_L - \rho_V} \right)^{0.5} \tag{9.66}$$

where CP = capacity parameter (m·s^{-1})
v_V = vapor velocity in the empty column (m·s^{-1})

For a given liquid–vapor flow parameter F_{LV}, the capacity parameter depends on the design of packing. Figure 9.17 shows a typical plot of CP versus F_{LV}. Data for a variety of packings have been given by Kister[13]. For preliminary design using structured packing, a first estimate of the flooding velocity can be obtained from:

$$CP = \exp[-1.931 - 0.402 \ln F_{LV} - 0.0342(\ln F_{LV})^2]$$

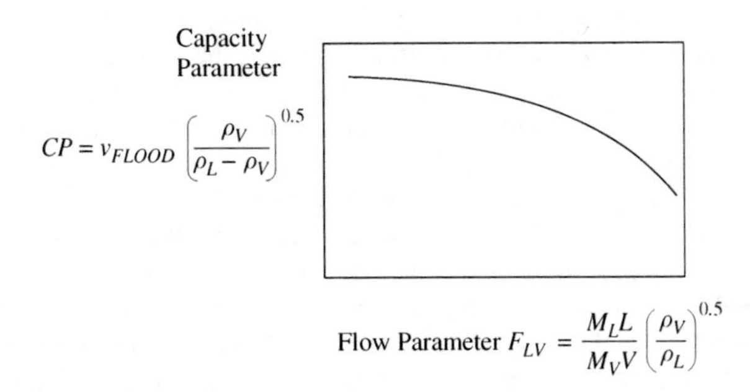

Figure 9.17 Flooding correlations for packings.

However, it should be again emphasized that it is always best to choose a specific packing and use experimental data specific for that packing to ensure a reliable design.

Example 9.4 For the distillation column from Examples 9.1, 9.2 and 9.3, assuming $R_{min} = 3.095$ and $R/R_{min} = 1.1$, estimate:

a. The actual number of trays
b. Height of the column
c. Diameter of the column based on conditions at the top and bottom of the column. The system can be assumed to be susceptible to moderate foaming with a foaming factor of 0.9.

The relative volatility between the key components is 1.57 and the viscosity of the feed is 0.1 mN·s·m^{-2}. The physical properties for the distillate and bottoms compositions are given in Table 9.7.

Table 9.7 Physical properties of distillation and bottoms compositions.

	Distillate	Bottoms
M_L (kg·kmol^{-1})	57.0	87.5
M_V (kg·kmol^{-1})	55.6	80.3
ρ_L (kg·m^{-3})	476	483
ρ_V (kg·m^{-3})	34.9	41.2
σ (mN·m^{-1})	4.6	3.7

Solution

a. From Example 9.3, for $R_{min} = 3.095$, the number of theoretical stages $N = 41.0$. The overall plate efficiency can be estimated from Equation 9.61:

$$E_O = 0.542 - 0.285 \log(\alpha_{LH} \mu_L)$$
$$= 0.542 - 0.285 \log(1.57 \times 0.1)$$
$$= 0.77$$

Number of real trays

$$= \frac{41.0}{0.77}$$
$$= 53.2$$

say 54

b. Assuming a plate spacing of 0.45 m and 4 m allowance at the top of the column for vapor–liquid disengagement and the bottom for a sump:

$$\text{Height} = 0.45(54 - 1) + 4$$
$$= 27.85 \text{ m}$$

c. The diameter of the column will be based on the flooding velocity, which is correlated in terms of the liquid–vapor flow parameter F_{LV}. F_{LV} requires the liquid and vapor rates, which change through the column. Here, the assessment will be based on the flow at the top and bottom of the column assuming constant molar overflow. From Example 9.1, the distillate flow is:

$$D = 278.21 \text{ kmol·h}^{-1}$$

A mass balance around the top of the column gives:

$$V = D + L$$
$$= D(R + 1)$$

where V and L are the vapor and liquid molar flowrates at the top of the column.

$$V = 278.21(3.095 \times 1.1 + 1)$$
$$= 1225.4 \text{ kmol·h}^{-1}$$
$$L = RD$$
$$= 3.095 \times 1.1 \times 278.21$$
$$= 947.2 \text{ kmol·h}^{-1}$$

The feed of 1000 kmol·h^{-1} is saturated liquid. Thus the liquid flowrate below the feed is:

$$L' = 947.2 + 1000$$
$$= 1947.2 \text{ kmol·h}^{-1}$$

From Example 9.1, the bottoms flowrate is:

$$B = 721.8 \text{ kmol·h}^{-1}$$

The vapor flowrate below the feed is:

$$V' = 1947.2 - 721.8$$
$$= 1225.4 \text{ kmol·h}^{-1}$$

The liquid–vapor flow parameter is given by Equation 9.62:

$$F_{LV} = \left(\frac{M_L L}{M_V V}\right)\left(\frac{\rho_V}{\rho_L}\right)^{0.5}$$

At the top of the column:

$$F_{LV} = \left(\frac{57.0 \times 947.2}{55.6 \times 1225.4}\right)\left(\frac{34.9}{476}\right)^{0.5}$$
$$= 0.2146$$

At the bottom of the column:

$$F'_{LV} = \left(\frac{87.5 \times 1947.2}{80.3 \times 1225.4}\right)\left(\frac{41.2}{483}\right)^{0.5}$$
$$= 0.5057$$

The terminal velocity parameter K_T is given by Equation 9.63:

$$K_T = \left(\frac{\sigma}{20}\right)^{0.2} \exp(-2.979 - 0.717 \ln F_{LV} - 0.0865(\ln F_{LV})^2$$
$$+ 0.997 \ln H_T - 0.07973 \ln F_{LV} \ln H_T + 0.256(\ln H_T)^2$$

At the top of the column:

$$\sigma = 4.6 \text{ mN·m}^{-1}$$
$$F_{LV} = 0.2146$$
$$H_T = 0.45 \text{ m}$$

Substituting in Equation 9.63:

$$K_T = 0.0448 \text{ m·s}^{-1}$$

At the bottom of the column:

$$\sigma = 3.7 \text{ mN·m}^{-1}$$
$$F_{LV} = 0.5057$$
$$H_T = 0.45 \text{ m}$$

Substituting in Equation 9.63:

$$K'_T = 0.0289 \text{ m·s}^{-1}$$

The vapor flooding velocity can now be calculated from Equation 8.3 assuming a foaming factor of 0.9:

$$v_T = 0.9K_T\left(\frac{\rho_L - \rho_V}{\rho_V}\right)^{0.5}$$

At the top of the column:

$$v_T = 0.9 \times 0.0448\left(\frac{476 - 34.9}{34.9}\right)^{0.5}$$
$$= 0.143 \text{ m·s}^{-1}$$

At the bottom of the column:

$$v_T = 0.9 \times 0.0289\left(\frac{483 - 41.2}{41.2}\right)^{0.5}$$
$$= 0.0852 \text{ m·s}^{-1}$$

To obtain the column diameter, an allowance must be made for downcomer area (say 10%), and the vapor velocity should be some fraction (say 80%) of the flooding velocity.

$$\text{Diameter} = \left(\frac{4M_V V}{0.9 \times 0.8 \times \pi\rho_V v_T}\right)^{0.5}$$

At the top of the column:

$$\text{Diameter} = \left(\frac{4 \times 55.6 \times 1225.4/3600}{0.9 \times 0.8 \times \pi \times 34.9 \times 0.143}\right)^{0.5}$$
$$= 2.59 \text{ m}$$

At the bottom of the column:

$$\text{Diameter} = \left(\frac{4 \times 80.3 \times 1225.4/3600}{0.9 \times 0.8 \times \pi \times 41.2 \times 0.0852}\right)^{0.5}$$
$$= 3.71 \text{ m}$$

The calculation shows a significant difference between the diameter at the top and bottom. In conceptual design, it is reasonable to take the largest value. Later, when the design is considered in more detail, if different sections of a column require diameters that differ by greater than 20%, it would usually be engineered with different diameters[14]. Such decisions can only be made after a much more detailed analysis of the column design. Also, the design here is based on the flooding velocity at the top and bottom of the column only. In practice, the flooding velocity changes throughout the column, and intermediate points might need to be considered in a more detailed analysis.

9.6 CHOICE OF OPERATING CONDITIONS

The feed composition and flowrate to the distillation are usually specified. Also, the specifications of the products are usually known, although there may be some uncertainty in product specifications. The product specifications may be expressed in terms of product purities or recoveries of certain components. The operating parameters to be selected by the designer include:

- operating pressure
- reflux ratio
- feed condition
- type of condenser.

(a) Pressure. The first decision is operating pressure. As pressure is raised:

- separation becomes more difficult (relative volatility decreases), that is, more stages or reflux are required;
- latent heat of vaporization decreases, that is, reboiler and condenser duties become lower;
- vapor density increases, giving a smaller column diameter;
- reboiler temperature increases with a limit often set by thermal decomposition of the material being vaporized, causing excessive fouling;
- condenser temperature increases.

As pressure is lowered, these effects reverse. The lower limit is often set by the desire to avoid:

- vacuum operation
- refrigeration in the condenser.

Both vacuum operation and the use of refrigeration incur capital and operating cost penalties and increase the complexity of the design. They should be avoided if possible.

For a first pass through the design, it is usually adequate, if process constraints permit, to set distillation pressure to as low a pressure above ambient as allows cooling water or air-cooling to be used in the condenser. If a total condenser is to be used, and a liquid top product taken, the pressure should be fixed such that:

- if cooling water is to be used, the bubble point of the overhead product should be typically 10°C above the summer cooling water temperature, or
- if air-cooling is to be used, the bubble point of the overhead product should be typically 20°C above the summer air temperature, or
- the pressure should be set to atmospheric pressure if either of these conditions would lead to vacuum operation.

If a partial condenser is to be used and a vapor top product taken, then the above criteria should be applied to the dew point of the vapor top product, rather than the bubble point of the liquid top product. Also, if a vapor top product is to be taken, then the operating pressure of the destination for the product might determine the column pressure (e.g. overhead top product being sent to the fuel gas system).

There are two major exceptions to these guidelines:

- If the operating pressure of the distillation column becomes excessive as a result of trying to operate the condenser against cooling water or air-cooling, then a combination of high operating pressure and low-temperature condensation using refrigeration should be used. This is usually the case when separating gases and light hydrocarbons.
- If process constraints restrict the maximum temperature of the distillation, then vacuum operation must be used in order to reduce the boiling temperature of the material to below a value at which product decomposition occurs. This tends to be the case when distilling high molar mass material.

(b) Reflux ratio. Another variable that needs to be set for distillation is reflux ratio. For a stand-alone distillation column (i.e. utility used for both reboiling and condensing), there is a capital–energy trade-off, as illustrated in Figure 9.18. As the reflux ratio is increased from its minimum, the capital cost decreases initially as the number of plates reduces from infinity, but the utility costs increase as more reboiling and condensation are required, (Figure 9.18). If the capital costs of the column, reboiler and condenser are annualized (see Chapter 2) and combined with the annual cost of utilities (see Chapter 2), the optimal reflux ratio is obtained. The optimal ratio of actual to minimum reflux is often less than 1.1. However, most designers are reluctant to design columns closer to minimum reflux than 1.1, except in special circumstances, since a small error in design data or a small change in

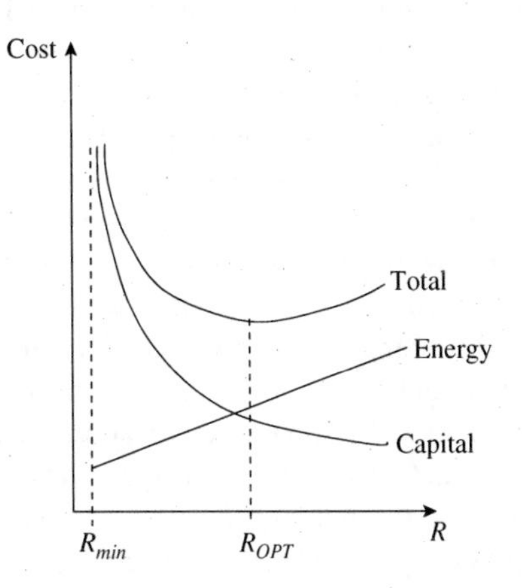

Figure 9.18 The capital–energy trade-off standalone distillation columns.

operating conditions might lead to an infeasible design. Also, the total cost curve is often relatively flat over a range of relative volatility around the optimum. No attempt should be made to do the optimization illustrated in Figure 9.18 until a picture of the overall process design has been established. Later, when heat integration of the column with the rest of the process is considered, the nature of the trade-off changes, and the optimal reflux ratio for the heat-integrated column can be very different from that for a stand-alone column. Any reasonable assumption is adequate in the initial stages of a design, say, a ratio of actual to minimum reflux of 1.1.

(c) Feed condition. Another variable that needs to be fixed is feed condition. For the distillation itself, the optimum feed point should minimize any mixing between the feed and the flows within the column. In theory, for a binary distillation it is possible to get an exact match between a liquid feed and the liquid on a feed stage. In practice, this might not be able to be achieved, as changes from stage to stage are finite. For a multicomponent system, in general it is not possible to achieve an exact match, except under special circumstances. If the feed is partially vaporized, the degree of vaporization provides a degree of freedom to obtain a better match between the composition of the liquid and vapor feed and the liquid and vapor flows in the column. However, minimizing feed mixing does not necessarily minimize the operating costs.

Heating the feed most often:

- increases trays in the rectifying section but decreases trays in the stripping section;
- requires less heat in the reboiler but more cooling in the condenser.

Cooling the feed most often reverses these effects.

Heat added to provide feed preheating may not substitute heat added to the reboiler on an equal basis[16]. The ratio of heat added to preheat the feed divided by the heat saved in the reboiler tends to be less than unity. Although heat added to the feed may not substitute heat added in the reboiler, it changes the minimum reflux ratio as a result of change to the feed condition (q in Equation 9.50). As the condition of the feed is changed from saturated liquid feed ($q = 1$) to saturated vapor feed ($q = 0$), the minimum reflux ratio tends to increase. Thus the ratio of heat added to preheat the feed divided by the heat saved in the reboiler depends on the change in q, the relative volatility between the key components, feed concentration and ratio of actual to minimum reflux. In some circumstances, particularly at high values of actual to minimum reflux ratio, heating the feed can increase the reboiler duty[16].

For a given separation, the feed condition can be optimized. No attempt should be made to do this in the early stages of an overall design, since heat integration is likely to change the optimal settings later in the design. It is usually adequate to set the feed to saturated liquid conditions. This tends to equalize the vapor rate below and above the feed, and having a liquid feed allows the column pressure to be increased if necessary through the feed pump.

(d) Type of condenser. Either a total or partial condenser can be chosen. Most designs use a total condenser. A total condenser is necessary if the top product needs to be sent to intermediate or final product storage. Also, a total condenser is best if the top product is to be fed to another distillation at a higher pressure as the liquid pressure can readily be increased using a pump.

If a partial condenser is chosen, then the partial condenser in theory acts as an additional stage, although in practice the performance tends to be less than a theoretical stage. A partial condenser reduces the condenser duty, which is important if the cooling service to the condenser is expensive, such as low-temperature refrigeration. It is often necessary to use a partial condenser when distilling mixtures with low-boiling components that would require very low-temperature (and expensive) refrigeration for a total condenser. Also, in these circumstances a *mixed condenser* might be used. This condenses what is possible as liquid for reflux and a liquid top product also taken. However, when condensing a mixture containing low-boiling components, a vapor overhead product might be taken, in addition to the liquid top product. In such designs, the uncondensed material often goes to a gas collection system, or *fuel header*, to be used, for example, as fuel gas. If the uncondensed material from a partial condenser is to be sent to a downstream distillation column for further processing, as has already been noted under the discussion on feed condition, there will be implications for the reboiler and condenser duties of the downstream column. A vapor feed will generally increase the condenser duty and decrease the reboiler duty of the downstream column. Whether this is good or bad for the operating costs of the downstream column depends on whether the cold utility used for the condenser or the hot utility used for the reboiler is more expensive. Also, when heat integration is discussed later, there might be important implications resulting from using partial condensers in the design of the overall process.

9.7 LIMITATIONS OF DISTILLATION

The most common method for the separation of homogeneous fluid mixtures with fluid products is distillation. Distillation allows virtually complete separation of most homogeneous fluid mixtures. It is no accident that distillation is the most common method used for the separation of homogeneous fluid mixtures with fluid products. Distillation has the following three principal advantages relative to competitive separation processes.

1. The ability to separate mixtures with a wide range of throughputs; many of the alternatives to distillation can only handle low throughput.
2. The ability to separate mixtures with a wide range of feed concentrations; many of the alternatives to distillation can only handle relatively pure feeds.
3. The ability to produce high-purity products; many of the alternatives to distillation only carry out a partial separation and cannot produce pure products.

However, distillation does have limitations. The principal cases where distillation is not well suited for the separation are as follows.

1. *Separation of materials with low molar mass.* Low molar mass materials are distilled at high pressure to increase their condensing temperature and to allow, if possible, the use of cooling water or air-cooling in the column condenser. Very low molar mass materials require refrigeration in the condenser in conjunction with high pressure. This significantly increases the cost of the separation since refrigeration is expensive. Absorption, adsorption and membrane gas separators are the most commonly used alternatives to distillation for the separation of low molar mass materials.
2. *Separation of heat-sensitive materials.* High molar mass material is often heat sensitive and will decompose if distilled at high temperature. Low molar mass material can also be heat sensitive, particularly when its nature is highly reactive. Such material will normally be distilled under vacuum to reduce the boiling temperature. Crystallization and liquid–liquid extraction can be used as alternatives to the separation of high molar mass heat-sensitive materials.
3. *Separation of components with a low concentration.* Distillation is not well suited to the separation of products that form a low concentration in the feed mixture. Adsorption and absorption are both effective alternative means of separation in this case.
4. *Separation of classes of components.* If a class of components is to be separated (e.g. a mixture of aromatic components from a mixture of aliphatic components), then distillation can only separate according to boiling points, irrespective of the class of component. In a complex mixture where classes of components need to be separated, this might mean isolating many components unnecessarily. Liquid–liquid extraction and adsorption can be applied to the separation of classes of components.
5. *Mixtures with low relative volatility or which exhibit azeotropic behavior.* Some homogeneous liquid mixtures exhibit highly nonideal behavior that form constant boiling azeotropes. At an azeotropic composition, the vapor and liquid are both at the same composition for the mixture. Thus, separation cannot be carried out beyond an azeotropic composition using conventional distillation. The most common method used to deal with such problems is to add a mass separation agent to the distillation to alter the relative volatility of the key separation in a favorable way and make the separation feasible. If the separation is possible but extremely difficult because of low relative volatility, then a mass separation agent can also be used in these circumstances, in a similar way to that used for azeotropic systems. These processes are considered in detail later in Chapter 12. Crystallization, liquid–liquid extraction and membrane processes can be used as alternatives to distillation for the separation of mixtures with low relative volatility or which exhibit azeotropic behavior.
6. *Separation of mixtures of condensable and noncondensable components.* If a vapor mixture contains both condensable and noncondensable components, then a partial condensation followed by a simple phase separator often can give a good separation. This is essentially a single-stage distillation operation. It is a special case that again deserves attention in some detail later in Chapter 13.

In summary, distillation is not well suited for separating either low molar mass materials or high molar mass heat-sensitive materials. However, distillation might still be the best method for these cases, since the basic advantages of distillation (potential for high throughput, any feed concentration and high purity) still prevail.

9.8 SEPARATION OF HOMOGENEOUS FLUID MIXTURES BY DISTILLATION – SUMMARY

Even though choices must be made for separators at this stage in the design, the assessment of separation processes ideally should be done in the context of the total system. As is discussed later, separators such as distillation that use an input of heat to carry out the separation often can be run at effectively zero energy cost if they are appropriately heat integrated with the rest of the process. Although energy intensive, heat-driven separators can be energy efficient in terms of the overall process if they are properly heat integrated. It is not worth expending any effort optimizing pressure, feed condition or reflux ratio until the overall heat integration picture has been established. These parameters very often change later in the design.

The design of a distillation involves a number of steps as follows.

a. Set the product specifications.
b. Set the operating pressure.

c. Determine the number of theoretical stages required and the energy requirements.
d. Determine the actual number of trays or height of packing needed and the column diameter.
e. Design the column internals, which involve determining the dimensions of the trays, packing, liquid and vapor distribution systems, and so on.
f. Carry out the mechanical design to determine the thickness of the vessel walls, internal fittings, and so on.

9.9 EXERCISES

1. In a mixture of methanol and water, methanol is the most volatile component. At a pressure of 1 atm, the relative volatility can be assumed to be constant and equal to 3.60. Construct the x–y diagram.
2. A feed mixture of methanol and water containing a mole fraction of methanol of 0.4 is to be separated by distillation at a pressure of 1 atm. The overhead product should achieve a purity of 95 mole % methanol and the bottoms product a purity of 95 mole% water. Assume the feed to be saturated liquid. Using the x–y diagram constructed in Exercise 1 and the McCabe–Thiele construction:
   a. determine the minimum reflux ratio
   b. for a reflux ratio of 1.1 times the minimum reflux, determine the number of theoretical stages for the separation.
3. A binary mixture is to be separated by distillation into relatively pure products. Where in the distillation column is the vapor–liquid equilibrium data required at the highest accuracy?
4. A distillation calculation is to be performed on a multicomponent mixture. The vapor–liquid equilibrium for this mixture is likely to exhibit significant departures from ideality, but a complete set of binary interaction parameters is not available. What factors would you consider in assessing whether the missing interaction parameters are likely to have an important effect on the calculations?
5. The two components, α and β, are to be separated in a distillation column for which the physical properties are largely unknown. The mole fractions of α and β in the overheads are 0.96 and 0.04 respectively. The overhead vapor can be assumed to be condensed at a uniform temperature corresponding with the bubble point temperature of the overhead mixture. Cooling water at an assumed temperature of 30°C is to be used in the condenser. It can be assumed also that the minimum allowable temperature difference in the condenser is 10°C. No vapor–liquid equilibrium data or vapor pressure data are available for the two components. Only measured normal boiling points are available, together with critical temperatures and pressures that have been estimated from the structure of the components. Vapor pressures can be estimated from this data assuming the behavior follows the Clausius–Clapeyron Equation from the normal boiling point to the critical point:

$$\ln P_i^{SAT} = A_i + \frac{B_i}{T} \qquad (9.67)$$

where P_i^{SAT} is the vapor pressure, T is the absolute temperature and A_i and B_i are constants for each component. The physical property data are summarized in Table 9.8.

Table 9.8 Physical properties of components α and β.

Component	Critical temp (K)	Critical pressure (bar)	Normal boiling point (K)
α	369.8	42.5	231.0
β	425.2	39.0	272.6

Assuming the vapor–liquid equilibrium to be ideal, at what pressure would the distillation column have to operate on the basis of the temperature in the condenser?

6. A saturated liquid mixture of ethane, propane, n-butane, n-pentane and n-hexane given in Table 9.9 is to be separated by distillation such that 95% of the propane is recovered in the distillate and 90% of the butane is recovered in the bottoms. Estimate the distribution of the other components for a column operating at 10 bar. Assume that the K-values can be correlated by

$$K_i = \frac{1}{P}\exp\left[A_i - \frac{B_i}{C+T}\right] \qquad (9.68)$$

where T is the absolute temperature (K), P the pressure (bar) and constants A_i, B_i and C_i are given in the Table 9.9.

Table 9.9 Feed and physical property for constants.

Component	Formula	Feed (kmol·h^{-1})	A_i	B_i	C_i
Ethane	C_2H_6	5	9.0435	1511.4	−17.16
Propane	C_3H_8	25	9.1058	1872.5	−25.16
n-Butane	C_4H_{10}	30	9.0580	2154.9	−34.42
n-Pentane	C_5H_{12}	20	9.2131	2477.1	−39.94
n-Hexane	C_6H_{14}	20	9.2164	2697.6	−49.78

7. The second column in the distillation train of an aromatics plant is required to split toluene and ethylbenzene. The recovery of toluene in the overheads must be 95%, and 90% of the ethylbenzene must be recovered in the bottoms. In addition to toluene and ethylbenzene, the feed also contains benzene and xylene. The feed enters the column under saturated conditions at a temperature of 170°C, with component flowrates given in Table 9.10. Estimate the mass balance around the column using the Fenske Equation. Assume that the K-values can be correlated by Equation 9.68 with constants A_i, B_i and C_i given in Table 9.10.

Table 9.10 Data for mixture of aromatics.

Component	Feed (kmol·h^{-1})	A_i	B_i	C_i
Benzene	1	9.2806	2789.51	−52.36
Toluene	26	9.3935	3096.52	−53.67
Ethylbenzene	6	9.3993	3279.47	−59.95
Xylene	23	9.5188	3366.99	−59.04

8. A distillation column separating 150 kmol·h^{-1} of a four-component mixture is estimated to have a minimum reflux

Table 9.11 Feed characteristics of a four-component mixture.

Component	Mole fraction in the feed	Relative volatility in the feed (relative to component D)
A	0.10	3.9
B	0.35	2.5
C	0.30	1.6
D	0.25	1.0

ratio of 3.5. The composition and the relative volatility of the feed are shown in Table 9.11. The column recovers 95% of Component B to the distillate and 95% of Component C to the bottom product. The feed to the column is a saturated liquid.

a. Calculate the molar flowrate and composition of the distillate and bottoms products for this column. State any assumptions you need to make.
b. Estimate the vapor load of the reboiler to this column if the reflux ratio is 25% greater than the minimum reflux ratio. Constant molar overflow can be assumed in the column and that a total condenser is used.
c. Use the Gilliland correlation to estimate the minimum number of theoretical stages required to carry out this separation.

9. A distillation column uses a partial condenser as shown in Figure 9.19. Assume that the reflux ratio and the overhead product composition and flowrate and the operating pressure are known and that the behavior of the liquid and vapor phases in the column is ideal (i.e. Raoult's Law holds). How can the flowrate and composition of the vapor feed to the condenser and its liquid products be estimated, given the vapor pressure data for the pure components. Set up the equations that need to be solved.

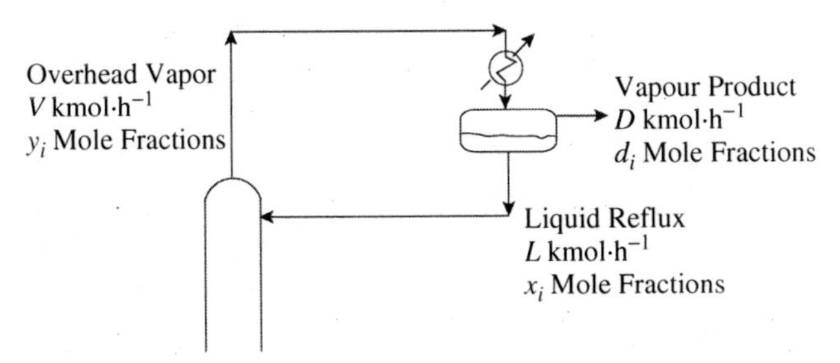

Figure 9.19 Partial condenser for a distillation column.

10. A mixture of ethane, propane, n-butane, n-pentane and n-hexane given in Table 9.12 is to be separated by distillation such that 95% of the propane is recovered in the distillate and 90% of the butane is recovered in the bottoms. The column will operate at 18 bar. Data for the feed and relative volatility are given in Table 9.12.

For the separation, calculate:

a. the distribution of the nonkey components using the Fenske Equation
b. the minimum reflux ratio from the Underwood Equations
c. the actual number of stages at $R/R_{min} = 1.1$ from the Gilliland Correlation.

Table 9.12 Data for five-component system.

Component	Formula	Feed (kmol·h^{-1})	α_{ij}
Ethane	C_2H_6	5	16.0
Propane	C_3H_8	25	7.81
n-Butane	C_4H_{10}	30	3.83
n-Pentane	C_5H_{12}	20	1.94
n-Hexane	C_6H_{14}	20	1.00

11. For the same feed, operating pressure and relative volatility as Exercise 10, the heavy key component is changed to pentane. Now 95% of the propane is recovered in the overheads and 90% of the pentane in the bottoms. Assuming that all lighter than light key components go to the overheads and all the heavier than heavy key go to the bottoms, estimate the distribution of the butane and the minimum reflux ratio using the Underwood Equations.

REFERENCES

1. King C.J (1980) *Separation Processes*, 2nd Edition, McGraw-Hill.
2. Douglas J.M (1988) *Conceptual Design of Chemical Processes*, McGraw-Hill.
3. McCabe W.L and Thiele E.W (1925) Graphical Design of Fractionating Columns, *Ind Eng Chem*, **17**: 605.
4. Sinnott R.K (1996) *Chemical Engineering*, Vol. 6, Butterworth-Heinemann.
5. Fenske M.R (1932) Fractionation of Straight-run Pennsylvania Gasoline, *Ind Eng Chem*, **24**: 482.
6. Treybal R.E (1980) *Mass Transfer Operations*, 3rd Edition, McGraw-Hill.
7. Geankopolis C.J (1993) *Transport Processes and Unit Operations*, 3rd Edition, Prentice Hall.
8. Seader J.D and Henley E.J (1998) *Separation Process Principles*, Wiley, New York.
9. Underwood AJV (1946) Fractional Distillation of Multicomponent Mixtures – Calculation of Minimum Reflux Ratio, *J Inst Petrol*, **32**: 614.
10. Gilliland E.R (1940) Multicomponent Rectification – Estimation of the Number of Theoretical Plates as a Function of the Reflux Ratio, *Ind Eng Chem*, **32**: 1220.
11. Rusche F.A (1999) Gilliland Plot Revisited, *Hydrocarbon Process*, **Feb**: 79.
12. O'Connell H.E (1946) Plate Efficiency of Fractionating Columns and Absorbers, *Trans AIChE*, **42**: 741.
13. Kessler D.P and Wankat P.C (1988) Correlations for Column Parameters, *Chem Eng*, **Sept**: 72.
14. Kister H.Z (1992) *Distillation Design*, McGraw-Hill.
15. Fair J.R (1961) How to Predict Sieve Tray Entrainment and Flooding, *Petrol Chem Eng*, **33**: 45.
16. Liebert T.C (1993) Distillation Feed Preheat – Is It Energy Efficient? *Hydrocarbon Process*, **Oct**: 37.

10 Choice of Separator for Homogeneous Fluid Mixtures II – Other Methods

10.1 ABSORPTION AND STRIPPING

The most common alternative to distillation for the separation of low molar mass materials is absorption. In absorption, a gas mixture is contacted with a liquid solvent that preferentially dissolves one or more components of the gas. Absorption processes often require an extraneous material to be introduced into the process to act as liquid solvent. If it is possible to use one of the materials already in the process, this should be done in preference to introducing an extraneous material. Liquid flowrate, temperature and pressure are important variables to be set.

The vapor–liquid equilibrium for such systems can often be approximated by Henry's Law (see Chapter 4):

$$p_i = H_i x_i \tag{10.1}$$

where p_i = partial pressure of Component i
H_i = Henry's Law constant (determined experimentally)
x_i = mole fraction of Component i in the liquid phase

Assuming ideal gas behavior ($p_i = y_i P$):

$$y_i = \frac{H_i x_i}{P} \tag{10.2}$$

Thus, the K-value is $K_i = H_i/P$, and a straight line would be expected on a plot of y_i against x_i. A mass balance can be carried out around a part of the absorber to obtain the operating line as shown in Figure 10.1:

$$y = \frac{L}{V}x + \left(y_{in} - \frac{L}{V}x_{out}\right) \tag{10.3}$$

where y, x = vapor and liquid mole fractions
y_{in} = mole fraction of the vapor inlet
x_{out} = mole fraction of liquid outlet
L = liquid flowrate
V = vapor flowrate

This is shown in Figure 10.1 to be a straight line with slope L/V. If it is assumed that the gas flowrate is fixed, the solvent flowrate can be varied to obtain the minimum solvent flowrate, as shown in Figure 10.2.

A graphical construction allows evaluation of the number of stages in the absorber analogous to the McCabe–Theile construction in distillation, as shown in Figure 10.3. In Figure 10.3, Stage 1 changes the liquid composition from x_{in} to x_1 and vapor composition from y_2 to y_{out}, and so on.

Rather than using a graphical construction, the approach can be expressed analytically as the Kremser Equation[1–4]. The Kremser Equation assumes that the equilibrium line is straight and intersects the origin of the x–y diagram. The operating line is also assumed to be straight. It provides an analytical expression for the stepping construction shown in Figure 10.3. The derivation of the equation is lengthy and the reader is referred to other sources[2–4]. If concentrations are known, then the theoretical stages N can be calculated from[1–3]:

$$N = \frac{\log\left[\left(\frac{A-1}{A}\right)\left(\frac{y_{in} - Kx_{in}}{y_{out} - Kx_{in}}\right) + \frac{1}{A}\right]}{\log A} \tag{10.4}$$

where $A = L/KV$
= absorption factor
K = vapor–liquid equilibrium K-value

For $A = 1$, the equation takes the form:

$$N = \frac{y_{in} - y_{out}}{y_{out} - Kx_{in}} \tag{10.5}$$

If the number of theoretical stages is known, the concentrations can be calculated from:

$$\frac{y_{in} - y_{out}}{y_{in} - Kx_{in}} = \frac{A^{N+1} - A}{A^{N+1} - 1} \tag{10.6}$$

For multicomponent systems, Equation 10.4 can be written for the limiting component, that is, the component with the highest K_i. Having determined the number of stages, the concentrations of the other components can be determined from Equation 10.6.

A first estimate of the overall stage efficiency can be obtained from the empirical correlation[4]:

$$\log_{10} E_O = -0.773 - 0.415\log_{10} X - 0.0896(\log_{10} X)^2 \tag{10.7}$$

where E_O = overall stage efficiency ($0 < E_O < 1$)
$X = \dfrac{KM\mu_L}{\rho_L}$
M = molar mass (kg·kmol^{-1})
μ_L = liquid (solvent) viscosity (mN·s·m^{-2} = cP)
ρ_L = liquid (solvent) density (kg·m^3)

Chemical Process Design and Integration R. Smith
 ISBNs: 0-471-48680-9 (HB); 0-471-48681-7 (PB)

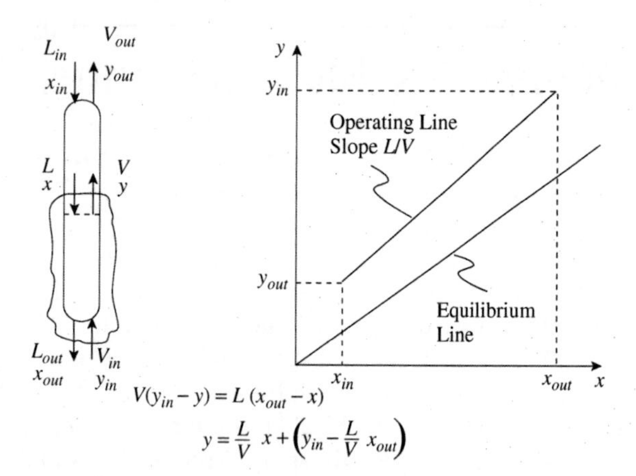

Figure 10.1 Equilibrium and operating lines for an absorber.

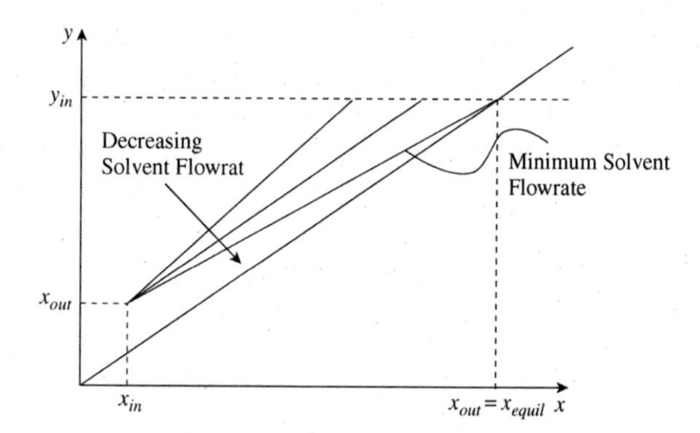

Figure 10.2 Minimum liquid–vapor ratio for absorbers.

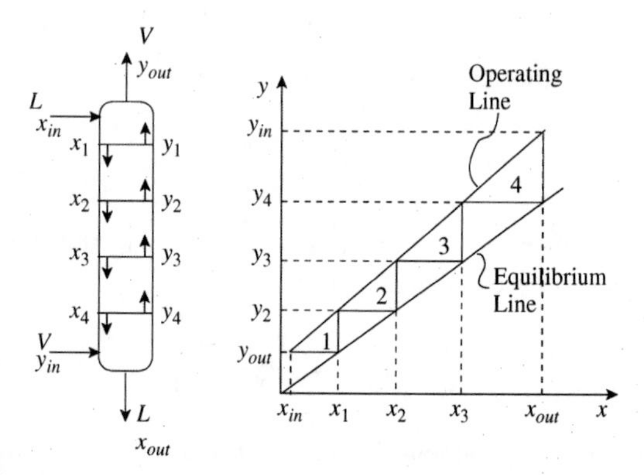

Figure 10.3 Equilibrium countercurrent stage operation for absorbers.

To calculate the overall stage efficiency for absorbers, the liquid viscosity and density need to be specified at average conditions.

As with distillation, the correlation for overall tray efficiency for absorbers, given in Equation 10.7, should only be used to derive a first estimate of the actual number of trays. More elaborate and reliable methods are available, but these require much more information on tray type and geometry and physical properties. If the column is to be packed, then the height of the packing is determined from Equation 9.64. As with distillation, the height equivalent of a theoretical plate (HETP) can vary

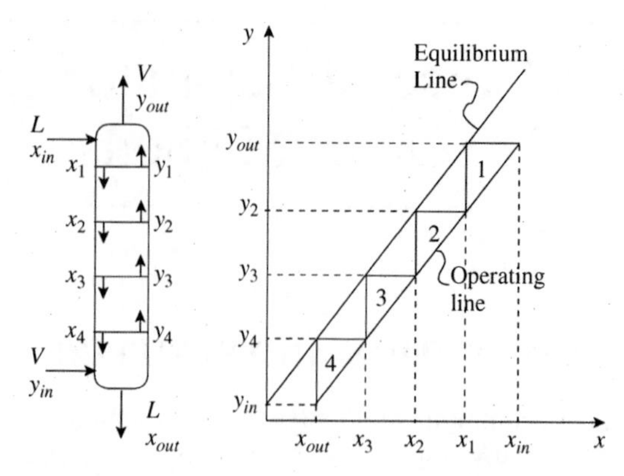

Figure 10.4 Equilibrium countercurrent stage operation for stripping.

significantly according to the packing type, F-factor from Equation 9.65 and the separation. The performance of a given packing can differ significantly between distillation and absorption applications, and reliable data can only be obtained from packing manufacturers.

Stripping is the reverse of absorption and involves the transfer of solute from the liquid to the vapor phase. This is illustrated in Figure 10.4. Note in Figure 10.4 that the operating line is below the equilibrium line. Again, if it is assumed that K_i, L and V are constant, as shown in Figure 10.4, the graphical construction can be expressed by the Kremser Equation[1–4]:

$$N = \frac{\log\left[(1 - A)\left(\dfrac{x_{in} - y_{in}/K}{x_{out} - y_{in}/K}\right) + A\right]}{\log(1/A)} \tag{10.8}$$

For $A = 1$:

$$N = \frac{x_{in} - x_{out}}{x_{out} - y_{in}/K} \tag{10.9}$$

If the number of stages is known, then the concentrations can be calculated from:

$$\frac{x_{in} - x_{out}}{x_{in} - y_{in}/K} = \frac{(1/A)^{N+1} - (1/A)}{(1/A)^{N+1} - 1} \tag{10.10}$$

Equations 10.8 to 10.10 can be written in terms of a stripping factor (S), where $S = 1/A$. For multicomponent systems, Equations 10.8 and 10.9 can be used for the limiting component (i.e. the component with the lowest K_i) and then Equation 10.10 can be used to determine the distribution of the other components.

Equations 10.4 to 10.6 and 10.8 to 10.10 can be written in terms of mole fractions and molar flowrates. Alternatively, mass fractions and mass flowrates can be used instead, as long as a consistent set of units is used.

If absorption is the choice of separation, then some preliminary selection of the major design variables must be made to allow the design to proceed:

1. *Liquid flowrate*. The liquid flowrate is determined by the absorption (L/K_iV) factor for the key separation. The

absorption factor determines how readily Component i will absorb into the liquid phase. When the absorption factor is large, Component i will be absorbed more readily into the liquid phase. The absorption factor must be greater than 1 for a high degree of solute removal; otherwise removal will be limited to a low value by liquid flowrate. Since L/K_iV is increased by increasing the liquid flowrate, the number of plates required to achieve a given separation decreases. However, at high values of L/K_iV, the increase in liquid flowrate brings diminishing returns. This leads to an economic optimum in the range $1.2 < L/K_iV < 2.0$. An absorption factor around 1.4 is often used[5].

2. *Temperature.* Decreasing temperature increases the solubility of the solute. In an absorber, the transfer of solute from gas or vapor to liquid brings about a heating effect. This usually will lead to temperatures increasing down the column. If the component being separated is dilute, the heat of absorption will be small and the temperature rise down the column will also be small. Otherwise, the temperature-rise down the column will be large, which is undesirable since solubility decreases with increasing temperature. To counteract the temperature rise in absorbers, the liquid is sometimes cooled at intermediate points as it passes down the column. The cooling is usually down to temperatures that can be achieved with cooling water, except in special circumstances where refrigeration is used.

If the solvent is volatile, there will be some loss of the gas or vapor. This should be avoided if the solvent is expensive and/or environmentally harmful by using a condenser (refrigerated if necessary) on the vapor leaving the absorber.

3. *Pressure.* High pressure gives greater solubility of solute in the liquid. However, high pressure tends to be expensive to create since this can require a gas compressor. Thus, there is an optimal pressure.

As with distillation, no attempt should be made to carry out any optimization of liquid flowrate, temperature or pressure at this stage in the design.

Having dissolved the solute in the liquid, it is often necessary to then separate the solute from the liquid in a stripping operation so as to recycle the liquid to the absorber. Now, the stripping factor for Component i, (K_iV/L) should be large to concentrate it in the vapor phase and thus be stripped out of the liquid phase. For a stripping column, the stripping factor should be in the range $1.2 < K_iV/L < 2.0$ and is often around 1.4[5]. As with absorbers, there can be a significant change in temperature through the column. This time, however, the liquid decreases in temperature down the column for reasons analogous to those developed for absorbers. Increasing temperature and decreasing pressure will enhance the stripping.

Example 10.1 A hydrocarbon gas stream containing benzene vapor is to have the benzene separated by absorption in a heavy liquid hydrocarbon stream with an average molar mass of 200 kg·kmol^{-1}. The concentration of benzene in the gas stream is 2% by volume and the liquid contains 0.2% benzene by mass. The flowrate of the gas stream is 850 m^3·h^{-1}, its pressure is 1.07 bar and its temperature is 25°C. It can be assumed that the gas and liquid flowrates are constant, the kilogram molecular volume occupies 22.4 m^3 at standard conditions, the molar mass of benzene is 78 kg·kmol^{-1} and that the vapor–liquid equilibrium obeys Raoult's Law. At the temperature of the separation, the saturated liquid vapor pressure of benzene can be taken to be 0.128 bar. If a 95% removal of the benzene is required, estimate the liquid flowrate and the number of theoretical stages.

Solution First, estimate the flowrate of vapor in kmol·h^{-1}:

$$V = 850 \times \frac{1}{22.4\left(\frac{298}{273}\right)\left(\frac{1.013}{1.07}\right)}$$
$$= 36.6 \text{ kmol·h}^{-1}$$

From Raoult's Law:

$$K = \frac{p^{sat}}{P} = \frac{0.128}{1.07} = 0.12$$

Assuming $A = 1.4$:

$$L = 1.4 \times 0.12 \times 36.6 = 6.15 \text{ kmol·h}^{-1} = 6.15 \times 200 \text{ kg·h}^{-1} = 1{,}230 \text{ kg·h}^{-1}$$

$$x_{in} = 0.002 \times \frac{200}{78} = 0.0051$$

$$y_{out} = 0.02(1 - 0.95), \text{ assuming } V \text{ constant}$$
$$= 0.001$$

$$N = \frac{\log\left[\left(\frac{A-1}{A}\right)\left(\frac{y_{in} - Kx_{in}}{y_{out} - Kx_{in}} + \frac{1}{A}\right)\right]}{\log A}$$

$$= \frac{\log\left[\left(\frac{1.4-1}{1.4}\right)\left(\frac{0.02 - 0.12 \times 0.0051}{0.001 - 0.12 \times 0.0051} + \frac{1}{1.4}\right)\right]}{\log 1.4}$$

$$= 8 \text{ theoretical stages}$$

Separation in absorption is sometimes enhanced by adding a component to the liquid that reacts with the solute. The discussion regarding absorption has so far been restricted to physical absorption. In chemical absorption, chemical reactions are used to enhance absorption. Both irreversible and reversible reactions can be used. An example of an irreversible reaction is the removal of SO_2

from gas streams using sodium hydroxide solution:

$$\underset{\text{sodium hydroxide}}{2NaOH} + SO_2 \longrightarrow \underset{\text{sodium sulfite}}{Na_2SO_3} + H_2O$$

$$\underset{\text{sodium sulfite}}{Na_2SO_3} + \tfrac{1}{2}O_2 \longrightarrow \underset{\text{sodium sulfate}}{Na_2SO_4}$$

Another example of an irreversible reaction is the removal of oxides of nitrogen from gas streams using hydrogen peroxide:

$$\underset{\text{nitric oxide}}{2NO} + \underset{\text{hydrogen peroxide}}{3H_2O_2} \longrightarrow \underset{\text{nitric acid}}{2HNO_3} + 2H_2O$$

$$\underset{\text{nitrogen dioxide}}{2NO_2} + \underset{\text{hydrogen peroxide}}{H_2O_2} \longrightarrow \underset{\text{nitric acid}}{2HNO_3}$$

Examples of reversible reactions are the removal of hydrogen sulfide and carbon dioxide from gas streams using a solution of monoethanolamine:

$$\underset{\text{monoethanolamine}}{HOCH_2CH_2NH_2} + H_2S \underset{\text{Regeneration}}{\overset{\text{Absorption}}{\rightleftharpoons}} \underset{\text{monoethanolamine hydrogen sulphide}}{HOCH_2CH_2NH_3HS}$$

$$\underset{\text{monoethanolamine}}{HOCH_2CH_2NH_2} + CO_2 + H_2O \underset{\text{Regeneration}}{\overset{\text{Absorption}}{\rightleftharpoons}} \underset{\text{monoethanolamine hydrogen carbonate}}{HOCH_2CH_2NH_3HCO_3}$$

In this case, the reactions can be reversed at a regeneration stage in a stripping column by the input of heat in a reboiler. If the solvent is to be recovered by stripping the solute from the solvent, the chemical absorption requires more energy than physical absorption. This is because the energy input for chemical absorption must overcome the heat of reaction as well as the heat of solution. However, chemical absorption involves smaller solvent rates than physical absorption.

Unfortunately, the analysis of chemical absorption is far more complex than physical absorption. The vapor–liquid equilibrium behavior cannot be approximated by Henry's Law or any of the methods described in Chapter 4. Also, different chemical compounds in the gas mixture can become involved in competing reactions. This means that simple methods like the Kremser equation no longer apply and complex simulation software is required to model chemical absorption systems such as the absorption of H_2S and CO_2 in monoethanolamine. This is outside the scope of this text.

10.2 LIQUID–LIQUID EXTRACTION

Like gas absorption, liquid–liquid extraction separates a homogeneous mixture by the addition of another phase – in this case, an immiscible liquid. Liquid–liquid extraction carries out separation by contacting a liquid feed with another immiscible liquid. The equipment used for liquid–liquid extraction is the same as that used for the liquid–liquid reactions illustrated in Figure 7.4. The separation occurs as a result of components in the feed distributing themselves differently between the two liquid phases. The liquid with which the feed is contacted is known as the *solvent*. The solvent extracts *solute* from the *feed*. The solvent-rich stream obtained from the separation is known as the *extract* and the residual feed from which the solute has been extracted is known as the *raffinate*.

The distribution of solute between the two phases at equilibrium can be quantified by the K-value or *distribution coefficient*:

$$K_i = \frac{x_{E,i}}{x_{R,i}} = \frac{\gamma_{R,i}}{\gamma_{E,i}} \tag{10.11}$$

where $x_{E,i}, x_{R,i}$ = mole fractions of Component i in the extract and in the raffinate

$\gamma_{E,i}, \gamma_{R,i}$ = activity coefficients of Component i in the extract and in the raffinate

By taking the ratio of the distribution coefficients for the two Components i and j, the *separation factor* can be defined, which is analogous to relative volatility in distillation:

$$\beta_{ij} = \frac{K_i}{K_j} = \frac{x_{E,i}/x_{R,i}}{x_{E,j}/x_{R,j}} = \frac{\gamma_{R,i}/\gamma_{E,i}}{\gamma_{R,j}/\gamma_{E,j}} \tag{10.12}$$

The separation factor (or selectivity) indicates the tendency for Component i to be extracted more readily from the raffinate to the extract than Component j.

When choosing a solvent for an extraction process, there are many issues to consider:

1. *Distribution coefficient.* It is desirable for the distribution coefficient, defined in Equation 10.11, to be large. Large values will mean that less solvent is required for the separation. A useful guide when selecting a solvent is that the solvent should be chemically similar to the solute. In other words, like dissolves like. A polar liquid like water is generally best suited for ionic and polar compounds. Nonpolar compounds like hexane are better for nonpolar compounds. When a solute dissolves in a solvent, some attractions between solvent molecules must be replaced by solute–solvent attractions when a solution forms. If the new attractions are similar to those replaced, very little energy is required to form the solution. This holds true for many systems, but the molecular interactions that contribute to the solubility of one compound in another are many and varied. Hence, it is only a general guide.

2. *Separation factor.* The separation factor, defined by Equation 10.12, measures the tendency of one component to be extracted more readily than another. If the separation needs to separate one component from a feed relative to another, then it is necessary to have a separation factor greater than unity, and preferably as high as possible.
3. *Insolubility of the solvent.* The solubility of the solvent in the raffinate should be as low as possible.
4. *Ease of recovery.* It is always desirable to recover the solvent for reuse. This is often done by distillation. If this is the case, then the solvent should be thermally stable and not form azeotropes with the solute. Also, for the distillation to be straightforward, the relative volatility should be large and the latent heat of vaporization small.
5. *Density difference.* The density difference between the extract and the raffinate should be as large as possible to allow the liquid phases to coalesce more readily.
6. *Interfacial tension.* The larger the interfacial tension between the two liquids, the more readily coalescence will occur. However, on the other hand, the higher the interfacial tension, the more difficult will be the dispersion in the extraction.
7. *Side reactions.* The solvent should be chemically stable and not undergo any side reactions with the components in the feed (including impurities).
8. *Vapor pressure.* The vapor pressure at working conditions should preferably be low if an organic solvent is to be used. High vapor pressure for an organic solvent will lead to the emission of volatile organic compounds (VOCs) from the process, potentially leading to environmental problems. VOCs will be discussed in more depth later when environmental issues are considered.
9. *General properties.* The solvent should be nontoxic for applications such as the manufacture of foodstuffs. Even for the manufacture of general chemicals, the solvent should be preferably nontoxic for safety reasons. Safety also dictates that the solvent should preferably be nonflammable. Low viscosity and high freezing point will also be advantageous.

It will rarely be the case that a solvent can be chosen to satisfy all of the above criteria, and hence some compromise will almost always be necessary. Once the solvent has been chosen, and some information is available on the distribution coefficients, the number of theoretical stages required for the separation needs to be determined. The stage-wise calculation of liquid–liquid extraction has much in common with the stage-wise calculations for distillation, absorbers and strippers. For absorber and stripper design, it has been discussed how straight operating lines can simplify the calculations. In liquid–liquid extraction, the concept of solute-free streams passing countercurrently can be used. Concentrations are then given as the ratio of solute to solvent. These assumptions help in keeping operating lines straight and simplifying the calculations. If the simplifying assumption is made that the distribution coefficients are constant and that the liquid flowrates are also constant on a solute-free basis, the same analysis as that used for stripping in Section 10.1 can be applied. In liquid–liquid extraction, like stripping, solute is transferred from the feed.

If the equilibrium and operating lines are both straight, then the Kremser Equation can be used[6]:

$$N = \frac{\log\left[\left(\frac{\varepsilon - 1}{\varepsilon}\right)\left(\frac{x_F - x_S/K}{x_R - x_S/K}\right) + \frac{1}{\varepsilon}\right]}{\log \varepsilon} \tag{10.13}$$

where N = number of theoretical stages
x_F = mole fraction of solute in the feed based on solute-free feed
x_S = mole fraction of solute in the solvent inlet based on solute-free solvent
x_R = mole fraction of solute in raffinate based on solute-free raffinate
K = slope of the equilibrium line
ε = extraction factor
= KS/F
S = flowrate of solute-free solvent ($kmol{\cdot}s^{-1}$)
F = flowrate of solute-free feed ($kmol{\cdot}s^{-1}$)

For $\varepsilon = 1$:

$$N = \frac{x_F - x_R}{x_R - x_S/K} \tag{10.14}$$

When N is known, the composition can be calculated from:

$$\frac{x_F - x_R}{x_F - x_S/K} = \frac{\varepsilon^{N+1} - \varepsilon}{\varepsilon^{N+1} - 1} \tag{10.15}$$

For multicomponent systems, Equations 10.13 and 10.14 can be used to determine the number of stages for the limiting component (i.e. the component with the lowest K_i). Equation 10.15 can then be applied to determine the compositions of the other components.

Equations 10.13 to 10.15 are written in terms of mole fractions and molar flowrates. However, mass fractions and mass flowrates can also be used, as long as a consistent set of units is used.

Having obtained an estimate of the number of theoretical stages, this must be related to the height of the actual equipment or the number of stages in the actual equipment. The equipment used for liquid–liquid extraction is the same as that described for liquid–liquid reactions illustrated in Figure 7.4. For the mixer–settler arrangement shown in Figure 7.4, these can be combined in multiple stages countercurrently, in which each mixing and settling stage represents a theoretical stage. Much less straightforward is the relationship between the stages, or height of the contactor, in the other arrangements in Figure 7.4 and the number of theoretical stages. Typical HETPs for various designs of contactor are[7]:

Contactor	HETP (m)
Sieve tray	0.5–3.5
Random packing	0.5–2.0
Structured packing	0.2–2.0
Mechanically agitated	0.1–0.3

The relationship between the number of theoretical and actual stages or contactor height depends on many factors, such as geometry, rate of agitation, flowrates of the liquids, physical properties of the liquids, the presence of impurities affecting the surface properties at the interface, and so on. The only reliable way to relate the actual stages to the theoretical stages in liquid–liquid extraction equipment is to scale from the performance of similar equipment carrying out similar separation duties, or to carry out pilot plant experiments.

Like absorption, separation is sometimes enhanced by using a solvent that reacts with the solute. The discussion, so far, regarding liquid–liquid extraction has been restricted to physical extraction. Both irreversible and reversible reactions can be used in chemical extraction. For example, acetic acid can be extracted from water using organic bases that take advantage of the acidity of the acetic acid. These organic bases can be regenerated and recycled. Unfortunately, the analysis of chemical extraction is far more complex than that of physical extraction. The phase equilibrium behavior cannot be approximated by constant distribution coefficients. This means that simple methods like the Kremser Equation no longer apply and complex simulation software is required. This is outside the scope of this text.

Example 10.2 An organic product with a flowrate of 1000 kg·h^{-1} contains a water-soluble impurity with a concentration of 6% wt. A laboratory test indicates that if the product is extracted with an equal mass of water, then 90% of the impurity is extracted. Assume that water and the organic product are immiscible.

a. For the same separation of 90% removal, estimate how much water would be needed if a two-stage countercurrent extraction is used.
b. If an equal mass flowrate of water to feed is maintained for a two-stage countercurrent extraction, estimate the fraction of impurity extracted.

Solution

a. Mass of impurity in feed

$$= 1000 \times 0.06$$
$$= 60 \text{ kg·h}^{-1}$$

Feed flowrate on solute-free basis

$$= 1000 - 60$$
$$= 940 \text{ kg·h}^{-1}$$

$$x_F = \frac{60}{940} = 0.06383$$

Mass of impurity in raffinate

$$= 60 \times 0.1$$
$$= 6 \text{ kg·h}^{-1}$$

Mass of impurity in extract

$$= 60 - 6$$
$$= 54 \text{ kg·h}^{-1}$$

$$x_R = \frac{6}{940} = 6.383 \times 10^{-3}$$

$$x_E = \frac{54}{1000} = 0.054$$

$$K = \frac{x_E}{x_R} = \frac{0.054}{6.383 \times 10^{-3}} = 8.460$$

From Equation 10.15:

$$\frac{x_F - x_R}{x_F - x_{S/K}} = \frac{\varepsilon^{N+1} - \varepsilon}{\varepsilon^{N+1} - 1}$$

$$\frac{0.06383 - 6.383 \times 10^{-3}}{0.06383 - 0/8.460} = \frac{\varepsilon^3 - \varepsilon}{\varepsilon^3 - 1}$$

$$0.9 = \frac{\varepsilon^3 - \varepsilon}{\varepsilon^3 - 1}$$

$$0 = 0.1\varepsilon^3 - \varepsilon + 0.9$$

Solving for ε by trial and error:

$$\varepsilon = 2.541$$

$$S = \frac{\varepsilon F}{K} = \frac{2.541 \times 940}{8.460} = 282.3 \text{ kg·h}^{-1}$$

b.

$$\varepsilon = \frac{8.460 \times 1000}{940} = 10.0$$

From Equation 10.15:

$$\frac{x_F - x_R}{x_F - x_{S/K}} = \frac{\varepsilon^{N+1} - \varepsilon}{\varepsilon^{N+1} - 1}$$

$$\frac{0.06383 - x_R}{0.06383 - 0/8.460} = \frac{9.0^3 - 9.0}{9.0^3 - 1}$$

$$x_R = 7.0 \times 10^{-4}$$

Mass of impurity in raffinate

$$= 940 \times 7 \times 10^{-4}$$
$$= 0.658 \text{ kg·h}^{-1}$$

Fraction of impurity extracted

$$= \frac{60 - 0.658}{60}$$
$$= 0.989$$

Example 10.3 A feed with a flowrate of 1000 kg·h^{-1} contains 30% acetic acid by mass in aqueous solution. The acetic acid (AA) is to be extracted with isopropyl ether to produce a raffinate with 2% by mass on a solvent-free basis. Equilibrium data are given in Table 10.1[1,8].

It can be seen from Table 10.1 that the water and ether have significant mutual solubility and this must be accounted for.

a. Estimate the minimum flowrate of ether for the separation.
b. Estimate the number of theoretical extraction stages if a flowrate of 2500 kg·h^{-1} of ether is used.

Solution

a. The overall flow scheme is shown in Figure 10.5. To maintain a straight operating line, the concentrations must be expressed as ratios of solute to feed solvent (water) and ratios of solute to extraction solvent (isopropyl ether). The data from Table 10.1 are expressed on this basis in Table 10.2.

The equilibrium data from Table 10.2 are plotted in Figure 10.6a. It can be seen that this does not form a straight-line relationship overall. Figure 10.6a shows the operating line of maximum slope that touches the equilibrium line at $x_R = 0.1576$ kg AA/kg Water. The slope of this line is the ratio of feed to extraction flowrates. If the liquids are immiscible, then the flowrate of solute-free feed (F) is equal to the flowrate of the solute-free raffinate (R). Also, the flowrate of solute-free solvent (S) is equal to the flowrate of the solute-free extract (E). Thus, if the liquids are immiscible, then the slope of the operating line is $F/S = R/E$. However, in this case, there is a significant mutual solubility of the water and isopropyl ether that must be accounted for. To simplify the calculations, it can be assumed that the feed stream dissolves the extraction solvent only in the feed stage and that the extraction solvent dissolves a large amount of feed solvent in the feed stage compared with the amount dissolved in the raffinate stage[6]. From this, it can be assumed that the flowrate of feed solvent is R and that the flowrate of extraction solvent is S[6]. Thus, the slope of the operating line in Figure 10.6a is R/S.

In Figure 10.6a, the operating line for minimum solvent flowrate touches the equilibrium line at $x_E = 0.1576$ and $x_R = 0.05166$. Thus:

$$\frac{R}{S} = \frac{0.05166 - 0}{0.1576 - 0.02}$$
$$= 0.3754$$

Table 10.1 Equilibrium data for acetic acid–water–isopropyl ether[1,8]. (Reproduced from Cambell H, 1940, *Trans AIChE*, **36**: 628 by permission of the American Institute of Chemical Engineers).

Mass fraction in water phase			Mass fraction ether phase		
Acetic acid	Water	Isopropyl ether	Acetic acid	Water	Isopropyl ether
0.0069	0.981	0.012	0.0018	0.005	0.993
0.0141	0.971	0.015	0.0037	0.007	0.989
0.0289	0.955	0.016	0.0079	0.008	0.984
0.0642	0.917	0.019	0.0193	0.010	0.971
0.1330	0.844	0.023	0.0482	0.019	0.933
0.2550	0.711	0.034	0.1140	0.039	0.847
0.3670	0.589	0.044	0.2160	0.069	0.715

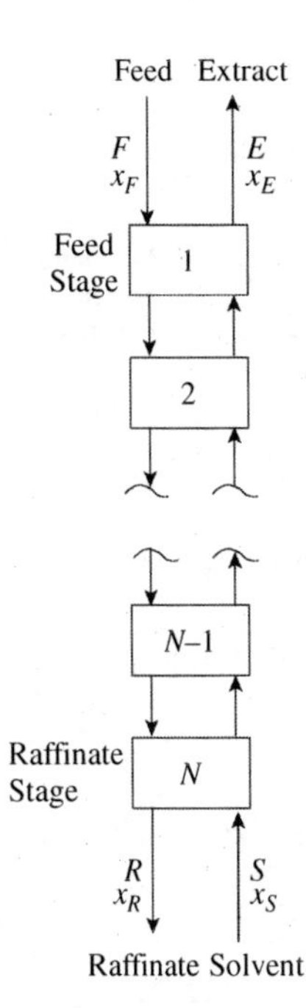

Figure 10.5 Mass balance for countercurrent liquid–liquid extraction.

Table 10.2 Equilibrium data expressed as ratios of feed solvent and extraction solvent.

Mass ratios in water phase		Mass ratios in ether phase	
x_{AA}	x_{Ether}	x_{AA}	x_{Water}
7.034×10^{-3}	0.01223	1.813×10^{-3}	5.035×10^{-3}
0.01452	0.01545	3.741×10^{-3}	7.078×10^{-3}
0.03026	0.01675	8.028×10^{-3}	8.130×10^{-3}
0.07001	0.02072	0.01988	0.01030
0.1576	0.02725	0.05166	0.02036
0.3586	0.04782	0.1346	0.04604
0.6231	0.07470	0.3021	0.09650

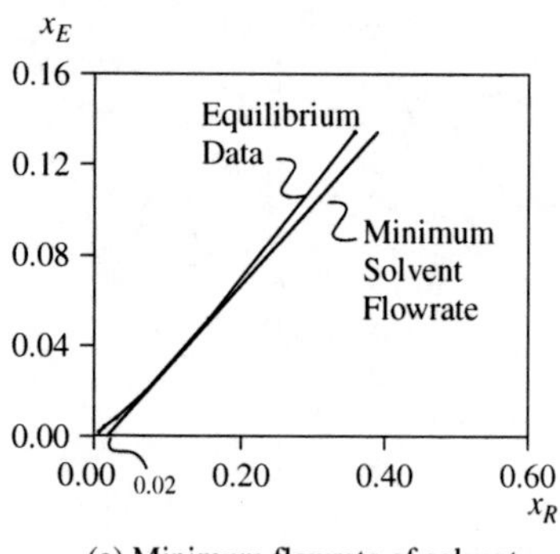

(a) Minimum flowrate of solvent

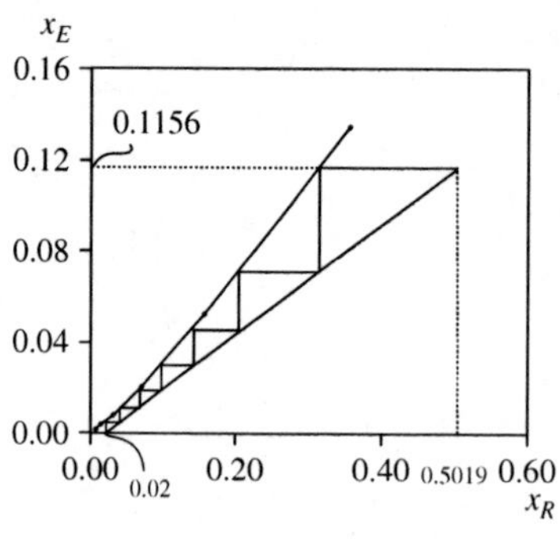

(b) Stepping off reveals the number of theoretical stages.

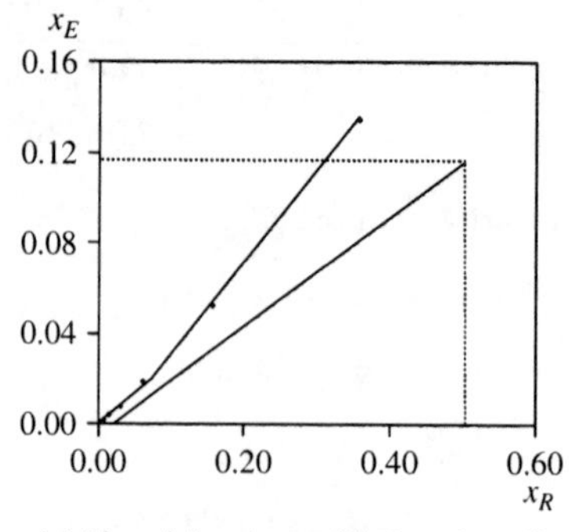

(c) Linearizing the equilibrium data allows the Kremser Equation to be used.

Figure 10.6 Extraction of acetic acid from aqueous solution using isopropyl ether.

Assume initially that:

$$R = F = 1000 \times 0.7 = 700 \text{ kg}\cdot\text{h}^{-1}$$

Thus, from the slope:

$$S = \frac{700}{0.3754}$$
$$= 1864.7 \text{ kg}\cdot\text{h}^{-1}$$

If it is also assumed initially that $S = E$, an overall mass balance can be applied for acetic acid (see Figure 10.5):

$$x_E E = x_F F + x_S S - x_R R \qquad (10.16)$$

$$1864.7 x_E = \frac{0.3}{0.7} \times 700 + 0 \times 1864.7 - 0.02 \times 700$$

$$x_E = 0.1534$$

For the ether phase, from Table 10.2, for $x_E = 0.1534$ kg AA/kg ether, the ether phase contains 0.05170 kg water/kg ether (by interpolation).

Water in extract

$$= 0.05170 \times 1864.7$$
$$= 96.40 \text{ kg h}^{-1}$$

Water in raffinate (R)

$$= 700 - 96.40$$
$$= 603.6 \text{ kg h}^{-1}$$

Thus, the flowrate of solvent can now be reestimated taking into account the mutual solubility[6]:

$$S = \frac{603.6}{0.3754}$$
$$= 1607.9 \text{ kg h}^{-1}$$

For the raffinate, from Table 10.2, at $x_R = 0.02$ kg AA/kg water, the water phase contains 0.01590 kg ether/kg water (by interpolation).

Ether in raffinate

$$= 0.01590 \times 603.6$$
$$= 10.597 \text{ kg}\cdot\text{h}^{-1}$$

Ether in extract (E)

$$= 1607.9 - 10.6$$
$$= 1598.3 \text{ kg}\cdot\text{h}^{-1}$$

Now, return to Equation 10.16 and iterate to coverage:

$$S = 1610.4 \text{ kg}\cdot\text{h}^{-1}$$

b. As before:

$$F = 700 \text{ kg}\cdot\text{h}^{-1}$$
$$x_F = 0.428$$
$$x_R = 0.02$$

Now, the extraction solvent flowrate is fixed:

$$S = 2500 \text{ kg}\cdot\text{h}^{-1}$$

Assume initially, $F = R$ and $S = E$ and perform an overall mass balance on acetic acid from Equation 10.16:

$$x_E = \frac{0.4286 \times 700 + 0 \times 2{,}500 - 0.02 \times 700}{2500}$$
$$= 0.1144$$

For the ether phase, from Table 10.2, at $x_E = 0.1144$ kg AA/kg water the ether phase contains 0.015979 kg ether/kg water.

Water in extract

$$= 0.03979 \times 2500$$
$$= 910.48 \text{ kg}\cdot\text{h}^{-1}$$

Water in raffinate (R)

$$= 700 - 910.48$$
$$= 600.5 \text{ kg}\cdot\text{h}^{-1}$$

For the raffinate, at $x_R = 0.02$ kg AA/kg water, the water phase contains 0.01590 kg ether/kg water.

Ether in raffinate

$$= 0.01590 \times 600.5$$
$$= 10.548 \text{ kg}\cdot\text{h}^{-1}$$

Ether in extract (E)

$$= 2500 - 10.548$$
$$= 2490.5 \text{ kg·h}^{-1}$$

Now, return to Equation 10.16 and iterate to converge:

$$E = 2490.5 \text{ kg·h}^{-1}$$
$$R = 600.0 \text{ kg·h}^{-1}$$
$$x_E = 0.1156$$

Following the assumption that the feed stream only dissolves extraction solvent in the feed stage, then the apparent feed concentration after the mass transfer in the feed stage can be calculated[6]:

$$x'_F R = x_F F + x_E(S - E)$$
$$x'_F\ 600.0 = 0.4286 \times 700 + 0.1156(2500 - 2490.5)$$
$$x'_F = 0.5019$$

In Figure 10.6b, the operating line for this is drawn between

$$x'_F = 0.5019, \quad x_E = 0.1156$$
$$x_R = 0.02 \text{ and} \quad x_S = 0.0$$

However, the Kremser Equation cannot be applied directly. Although the operating line is straight as a result of the assumptions made, the equilibrium line is not straight. Figure 10.6b shows that a graphical stepping off between the operating and equilibrium lines requires around seven equilibrium stages.

The Kremser Equation can still be applied if the equilibrium data can be linearized. From the slope of the equilibrium data at the raffinate stage (low concentrations):

$$x_F = 0.5019$$
$$x_R = 0.02$$
$$x_S = 0.0$$
$$K = 0.26$$
$$\varepsilon = \frac{0.26 \times 2490.5}{600.0}$$
$$= 1.079$$
$$N = \frac{\log\left[\left(\frac{1.079 - 1}{1.079}\right)\left(\frac{0.5019 - 0.0/0.26}{0.02 - 0.0/0.26}\right)\right]}{\log[1.079]}$$
$$= 13.4$$

This is a significant overestimate, caused by the difference between the equilibrium line and operating line being too small for the overall problem. Alternatively, the slope can be averaged over the first six equilibrium points to give $K = 0.36$. This gives $N = 5.5$, which is an underestimate, resulting from the overall difference between the equilibrium and operating lines being too large. A better estimate using the Kremser equation can be obtained by breaking down the overall problem into two linear regions, as shown in Figure 10.6c. The equilibrium data can be linearized as:

$$x_E = 0.28 x_R \qquad \text{for } 0 < x_R < 0.0791$$
$$x_E = 0.4001 x_R - 0.009476 \qquad \text{for } x_R > 0.0791$$

Applying the Kremser Equation for $0 < x_R < 0.0791$:

$$x_F = 0.0791$$
$$x_R = 0.02$$
$$x_S = 0$$
$$K = 0.28$$

Substituting in the Kremser equation gives:

$$N = 2.3$$

In applying the Kremser equation for $x_R > 0.0791$, it must be recognized that the equilibrium line no longer intercepts the origin, but intercepts the x_R axis at 0.02368. The concentrations, x'_F and x'_R, therefore need to be adjusted correspondingly:

$$x'_F = 0.5019 - 0.02368 = 0.4782$$
$$x'_R = 0.0791 - 0.02368 = 0.05542$$
$$x_s = 0.01418 \text{ (from the operating line, } x_E \text{ at } x_R = 0.0791)$$
$$K = 0.4001$$

Substituting in the Kremser Equation gives:

$$N = 4.4$$

This gives a total number of theoretical stages of 6.7, which is in closer agreement with the graphical construction. However, the equilibrium data could have been linearized in a variety of ways and the answer will be sensitive to the way the data is linearized.

10.3 ADSORPTION

Adsorption is a process in which molecules of *adsorbate* become attached to the surface of a solid *adsorbent*. Adsorption processes can be divided into two broad classes:

1. *Physical adsorption,* in which physical bonds form between the adsorbent and the adsorbate.
2. *Chemical adsorption,* in which chemical bonds form between the adsorbent and the adsorbate.

An example of chemical adsorption is the reaction between hydrogen sulfide and ferric oxide:

$$\underset{\text{hydrogen sulphide}}{6H_2S} + \underset{\text{ferric oxide}}{2Fe_2O_3} \longrightarrow \underset{\text{ferric sulfide}}{2Fe_2S_3} + 6H_2O$$

The ferric oxide adsorbent, once it has been transformed chemically, can be regenerated in an oxidation step:

$$\underset{\text{ferric sulfide}}{2Fe_2S_3} + \underset{\text{oxygen}}{3O_2} \longrightarrow \underset{\text{sulphur}}{6S} + \underset{\text{ferric oxide}}{2Fe_2O_3}$$

Table 10.3 Physical and chemical adsorption[9].

	Physical adsorption	Chemical adsorption
Heat of adsorption	Small, same order as heat of vaporization (condensation)	Large, many times greater than the heat of vaporization (condensation)
Rate of adsorption	Controlled by resistance to mass transfer Rapid rate at low temperatures	Controlled by resistance to surface reaction Low rate at low temperatures
Specificity	Low, entire surface availability for physical adsorption	High, chemical adsorption limited to active sites on the surface
Surface coverage	Complete and extendable to multiple molecular layers	Incomplete and limited to a layer, one molecule thick
Activation energy	Low	High, corresponding to a chemical reaction
Quantity adsorbed per unit mass	High	Low

The adsorbent having been regenerated is then available for reuse. By contrast, physical adsorption does not involve chemical bonds between the adsorbent and the adsorbate. Table 10.3 compares physical and chemical adsorption in broad terms[9].

Here, attention will focus on physical adsorption. This is a commonly used method for the separation of gases, but is also used for the removal of small quantities of organic components from liquid streams.

A number of different adsorbents are used for physical adsorption processes. All are highly porous in nature. The main types can be categorized as follows.

1. *Activated carbon.* Activated carbon is a form of carbon that has been processed to develop a solid with high internal porosity. Almost any carbonaceous material can be used to manufacture activated carbon. Coal, petroleum residue, wood or shells of nuts (especially, coconut) can be used. The most common method used for the manufacture of activated carbon starts by forming and heating the solid up to 400 to 500°C. This is followed by controlled oxidation (or *activation*) by heating to a higher temperature (up to 1000°C) in the presence of steam or carbon dioxide to develop the porosity and surface activity. Other methods of preparation include mixing carbonaceous material with an oxidizing agent (e.g. alkali metal hydroxides or carbonates), followed by heating up to 500 to 900°C. Adsorption using activated carbon is the most commonly used method for the separation of organic vapors from gases. It is also used for liquid-phase separations. A common liquid-phase application is for decolorizing or deodorizing aqueous solutions.

2. *Silica gel.* Silica gel is a porous amorphous form of silica (SiO_2) and is manufactured by acid treatment of sodium silicate solution and then dried. The silica gel surface has an affinity for water and organic material. It is primarily used to dehydrate gases and liquids.

3. *Activated aluminas.* Activated alumina is a porous form of aluminum oxide (Al_2O_3) with high surface area, manufactured by heating hydrated aluminum oxide to around 400°C in air. Activated aluminas are mainly used to dry gases and liquids, but can be used to adsorb gases and liquids other than water.

4. *Molecular sieve zeolites.* Zeolites are crystalline aluminosilicates. They differ from the other three major adsorbents in that they are crystalline and the adsorption takes place inside the crystals. This results in a pore structure different from other adsorbents in that the pore sizes are more uniform. Access to the adsorption sites inside the crystalline structure is limited by the pore size, and hence zeolites can be used to absorb small molecules and separate them from larger molecules, as *"molecular sieves"*. Zeolites selectively adsorb or reject molecules on the basis of differences in molecular size, shape and other properties, such as polarity. Applications include a variety of gaseous and liquid separations. Typical applications are the removal of hydrogen sulfide from natural gas, separation of hydrogen from other gases, removal of carbon dioxide from air before cryogenic processing, separation of *p*-xylene from mixed aromatic streams, separation of fructose from sugar mixtures, and so on.

Data used for the design of adsorption processes are normally derived from experimental measurements. The capacity of an adsorbent to adsorb an adsorbate depends on the compound being adsorbed, the type and preparation of the adsorbate, inlet concentration, temperature and pressure. In addition, adsorption can be a competitive process in which different molecules can compete for the adsorption sites. For example, if a mixture of toluene and acetone vapor is being adsorbed from a gas stream onto activated carbon, then toluene will adsorbed preferentially, relative to acetone and will displace the acetone that has already been adsorbed.

The capacity of an adsorbent to adsorb an adsorbate can be represented by adsorption isotherms, as shown in Figure 10.7a, or adsorption isobars, as shown in Figure 10.7b. The general trend can be seen that adsorption increases with decreasing temperature and increases with increasing pressure.

Data for adsorption isotherms can often be correlated by the Freundlich Isotherm Equation. For adsorption from

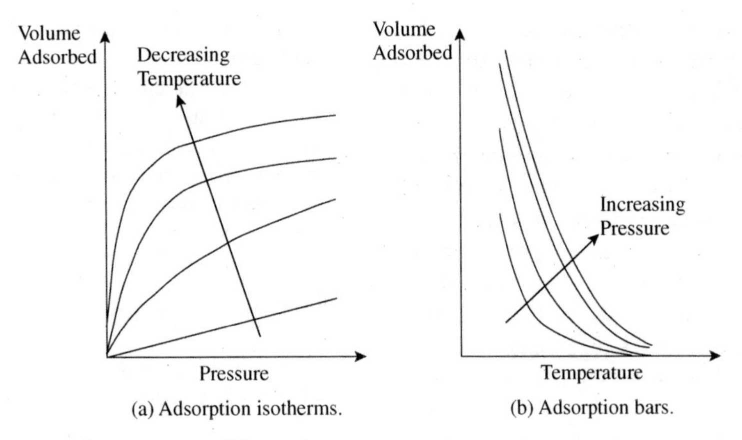

Figure 10.7 Adsorption of gases and vapors on solids.

liquids, this takes the form:

$$w = kC^n \tag{10.17}$$

where w = mass of adsorbate per mass of adsorbent at equilibrium
C = concentration
k, n = constants determined experimentally

Although Equation 10.17 is written in terms of concentration of a specific component, it can also be used if the identity of the solute is unknown. For example, adsorption is used to remove color from liquids. In such cases, the concentration of solute can be measured, for example, by a colorimeter and Equation 10.17 expressed in terms of arbitrary units of color intensity, providing the color scale varies linearly with the concentration of the solute responsible for the color.

Data from the adsorption of gases and vapors are usually correlated in terms of volume adsorbed (at standard conditions of 0°C and 1 atm pressure) and partial pressure. The adsorption can then be represented by:

$$V = k'p^{n'} \tag{10.18}$$

where V = volume of gas or vapor adsorbed at standard conditions ($m^3 \cdot kg^{-1}$)
p = partial pressure (Pa, bar)
k', n' = constants determined experimentally

This means that representing the equilibrium adsorbate mass (or volume) adsorbed versus concentration (or partial pressure) should be a straight line if plotted on logarithmic scales. Other theoretically based correlating equations are available for the adsorption of gases and vapors[9], but all equations have their limitations.

Adsorption can be carried out in fixed-bed arrangements, as illustrated in Figure 10.8a. This is the most common arrangement. However, fluidized beds and traveling beds can also be used, as illustrated in Figure 10.8b and c. Adsorption in a fixed bed is nonuniform and a *front* moves through the bed with time, as illustrated in Figure 10.9. The point at which the concentration in the outlet begins to rise is known as *breakthrough*. Once breakthrough has occurred, or preferably before, the bed online must be taken off-line and regenerated. This can be done by having several beds operating in parallel, one or more online, with one being regenerated. The regeneration options are:

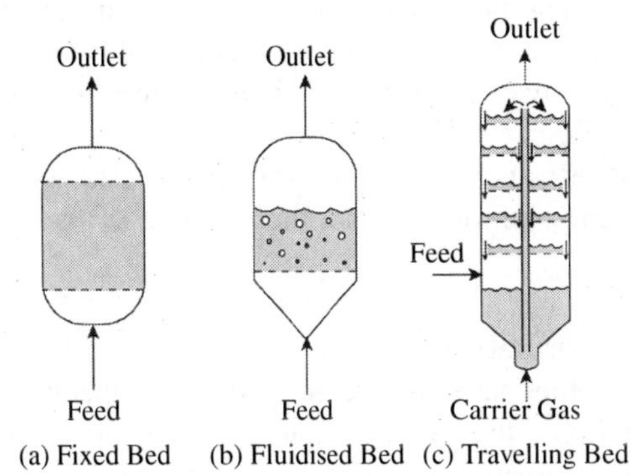

Figure 10.8 Different contacting arrangements for adsorber design.

1. *Steam.* This is the most commonly used method for recovery of organic material from activated carbon. Low-pressure steam is passed through the bed countercurrently. The steam is condensed along with any recovered organic material.

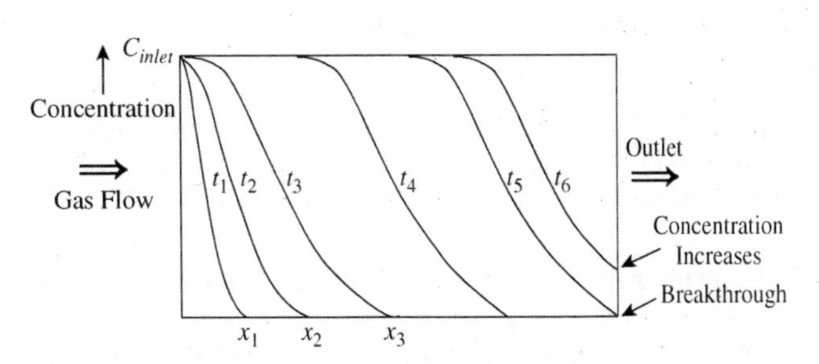

Figure 10.9 Concentration profiles through an adsorption bed exhibit a moving font.

2. *Hot gas*. Hot gas can be used when the regeneration gas is going to be fed to incineration. Air is normally used outside of the flammability range (see process safety later); otherwise nitrogen can be used.
3. *Pressure swing*. The desorption into a gas stream at a pressure lower than that for adsorption can also be used.
4. *Off-site regeneration*. Off-site regeneration is used when regeneration is difficult or infrequent. For example, organic material can polymerize on the adsorbent, making it difficult to regenerate. If activated carbon is being used, the carbon can be regenerated by heat treating in a furnace and activating with steam.

When using fixed beds, there are various arrangements used to switch between adsorption and regeneration. Cycles involving two, three and four beds are used. Clearly, the more beds that are used, the more complex is the cycle of switching between the beds, but the more effective is the overall system.

In actual operation, the adsorption bed is not at equilibrium conditions. Also, there is loss of bed capacity due to:

- heat of adsorption;
- other components in the inlet flow competing for the adsorption sites, for example, moisture in an inlet gas stream competing for adsorption sites and
- for gas adsorption, any moisture in the bed after regeneration will block the adsorption sites.

In practice, two to three times the equilibrium capacity tends to be used.

Example 10.4 A gas mixture with a flowrate of 0.1 $m^3 \cdot s^{-1}$ contains 0.203 $kg \cdot m^{-3}$ of benzene. The temperature is 10°C and the pressure 1 atm (1.013 bar). Benzene needs to be separated to give a gas stream with a benzene concentration of less than 5 $mg \cdot m^{-3}$. It is proposed to achieve this by adsorption using activated carbon in a fixed bed. The activated carbon is to be regenerated using superheated steam. The experimental adsorption isotherms cannot be adequately represented by Freundlich isotherms and, instead, can be correlated at 10°C by the empirical relationship:

$$\ln V = -0.0113(\ln p)^2 + 0.2071 \ln p - 3.0872 \qquad (10.19)$$

where V = volume benzene adsorbed ($m^3 \cdot kg^{-1}$)
p = partial pressure (Pa)

It can be assumed that the gas mixture follows ideal gas behavior and that the kilogram molar mass of a gas occupies 22.4 m^3 at standard conditions of 0°C and 1 atm (1.013 bar)

a. Estimate the mass of benzene that can be adsorbed per kg of activated carbon.
b. Estimate the volume of activated carbon required, assuming a cycle time of 2 hours (minimum cycle time is usually around 1.5 hours). Assume that the actual volume is three times the equilibrium volume and the bulk density of activated carbon is 450 $kg \cdot m^{-3}$.
c. The concentration of benzene in the gas stream must be less than 5 $mg \cdot m^{-3}$ after the bed has been regenerated with steam at 200°C and brought back online at 10°C. What fraction of benzene must be recovered from the bed by the regeneration to achieve this if the bed is assumed to be saturated before regeneration?

Solution

a. Assuming ideal gas behavior, first calculate the partial pressure of benzene at 10°C:

$$y = 0.203 \times \frac{1}{78} \times 22.4 \times \frac{283}{273}$$
$$= 0.0604$$
$$p = yP = 0.0604 \times 1.013 = 0.0612\text{bar}$$

Substitute $p = 6120$ Pa in Equation 10.19:

$$\ln V = -0.0113(\ln 6120)^2 + 0.2071 \ln(6120) - 3.0872$$
$$V = 0.1176 \text{ m}^3 \cdot \text{kg}^{-1}$$

Mass of benzene adsorbed

$$= 0.1176 \times \frac{1}{22.4} \times 78 = 0.410 \text{ kg benzene} \cdot \text{kg carbon}^{-1}$$

b. Carbon required for a 2-hour cycle, assuming equilibrium

$$= 0.1 \times 0.203 \times 2 \times 3{,}600 \times \frac{1}{0.410}$$
$$= 356.5 \text{ kg}$$

Assuming a design factor of 3:

$$= 1069.5 \text{ kg}$$

Volume of bed

$$= \frac{1069.5}{450} = 2.38 \text{ m}^3$$

c. Concentration of benzene when the regenerated bed is brought back online must be less than 5 $mg \cdot m^{-3}$. This results from the partial pressure of any benzene left on the bed after regeneration, until there is a breakthrough from the bed. Thus, for a concentration of 5 $mg \cdot m^{-3}$ at 10°C:

$$y = 5 \times 10^{-6} \times \frac{1}{78} \times 22.4 \times \frac{283}{273} = 1.488 \times 10^{-6}$$
$$p = 1.488 \times 10^{-6} \times 1.013 = 1.507 \times 10^{-6}\text{bar}$$
$$= 0.1507 \text{ Pa}$$

Volume of benzene on the bed at 10°C is given by

$$\ln V = -0.0113(\ln 0.1507)^2 + 0.2071(\ln 0.1507) - 3.0872$$
$$V = 0.0296 \text{ m}^3 \cdot \text{kg}^{-1}$$

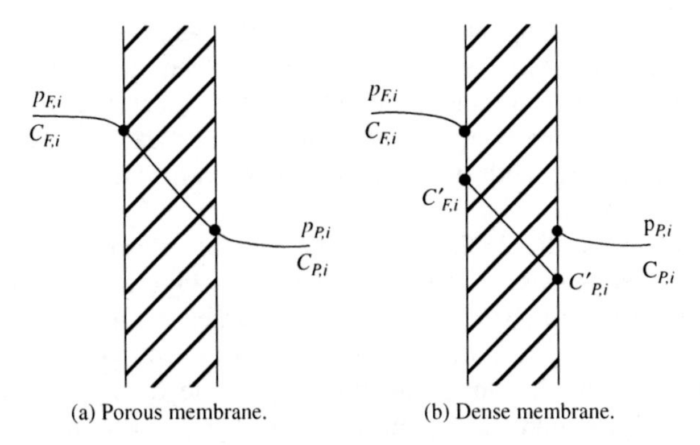

Figure 10.10 Partial pressure and concentration profiles across membranes.

Benzene recovered from the bed during regeneration, assuming saturation before regeneration

$$= \frac{0.1176 - 0.0296}{0.1176} = 0.75$$

During regeneration, 75% of the benzene must be recovered from the bed if the outlet concentration is to be 5 mg·m^{-3}.

10.4 MEMBRANES

Membranes act as a *semipermeable* barrier between two phases to create a separation by controlling the rate of movement of species across the membrane. The separation can involve two gas (vapor) phases, two liquid phases or a vapor and a liquid phase. The feed mixture is separated into a *retentate*, which is the part of the feed that does not pass through the membrane, and a *permeate*, which is that part of the feed that passes through the membrane. The driving force for separation using a membrane is partial pressure in the case of a gas or vapor and concentration in the case of a liquid. Differences in partial pressure and concentration across the membrane are usually created by the imposition of a pressure differential across the membrane. However, driving force for liquid separations can be also created by the use of a solvent on the permeate side of the membrane to create a concentration difference, or an electrical field when the solute is ionic.

To be effective for separation, a membrane must possess high *permeance* and a high *permeance ratio* for the two species being separated. The permeance for a given species diffusing through a membrane of given thickness is analogous to a mass transfer coefficient, that is, the flowrate of that species per unit area of membrane per unit driving force (partial pressure or concentration). The flux (flowrate per unit area) of Component i across a membrane can be written as[4,10–12]:

$$N_i = \overline{P}_{M,i} \times (\text{driving force}) = \frac{P_{M,i}}{\delta_M} \times (\text{driving force}) \tag{10.20}$$

where N_i = flux of Component i
$\overline{P}_{M,i}$ = permeance of Component i
$P_{M,i}$ = permeability of Component i
δ_M = membrane thickness

The flux, and hence the permeance and permeability, can be defined on the basis of volume, mass or molar flowrates. The accurate prediction of permeabilities is generally not possible and experimental values must be used. Permeability generally increases with increasing temperature. Taking a ratio of two permeabilities defines an *ideal separation factor* or *selectivity* α_{ij}, which is defined as:

$$\alpha_{ij} = \frac{P_{M,i}}{P_{M,j}} \tag{10.21}$$

Another important variable that needs to be defined is the cut or fraction of feed permeated θ that is defined as:

$$\theta = \frac{F_P}{F_F} \tag{10.22}$$

where θ = fraction of feed permeated
F_P = volumetric flowrate of permeate
F_F = volumetric flowrate of feed

Membrane materials can be divided into two broad categories:

1. *Microporous*. *Microporous* membranes are characterized by interconnected pores, which are small, but large in comparison to the size of small molecules. If the pores are of the order of size of the molecules for at least some of the components in the feed mixture, the diffusion of those components will be hindered, resulting in a separation. Molecules of size larger than the pores will be prevented from diffusing through the pores by virtue of a *sieving* effect.

For microporous membranes, the partial pressure profiles, in the case of gas (vapor) systems, and concentration profiles are continuous from the bulk feed to the bulk permeate, as illustrated in Figure 10.10a. Resistance to mass transfer by films adjacent to the upstream and downstream membrane interfaces create partial pressure and concentration differences between the bulk concentration and the concentration adjacent to the membrane interface. Permeability for microporous membranes is high but selectivity is low for small molecules.

2. *Dense*. Nonporous *dense* solid membranes are also used. Separation in dense membranes occurs by components in a gas or liquid feed diffusing to the surface of the membrane, dissolving into the membrane material, diffusing through the solid and desorbing at the downstream interface. The permeability depends on both the solubility and the diffusivity of the permeate in the membrane material. Diffusion through the membrane can be slow, but highly selective. Thus, for the separation of small molecules, a high permeability or high separation factor can be achieved,

but not both. This problem is solved by the formation of *asymmetric* membranes involving a thin dense layer, called the *permselective* layer, supported on a much thicker microporous layer of *substrate* that provides support for the dense layer. A microporous substrate can be used on one side of the dense layer or on both sides of the dense layer. Supporting the dense layer on both sides has advantages if the flow through the membrane needs to be reversed for cleaning purposes. The flux rate of a species is controlled by the permeance of the thin permselective layer.

The permeability of dense membranes is low because of the absence of pores, but the permeance of Component i in Equation 10.20 can be high if δ_M is very small, even though the permeability is low. Thickness of the permselective layer is typically in the range 0.1 to 10 μm for gas separations. The porous support is much thicker than this and typically more than 100 μm. When large differences in P_M exist among species, both high permeance and high selectivity can be achieved in asymmetric membranes.

In Figure 10.10a, it can be seen that for porous membranes, the partial pressure and concentration profiles vary continuously from the bulk feed to the bulk permeate. This is not the case with nonporous dense membranes, as illustrated in Figure 10.10b. Partial pressure or concentration of the feed liquid just adjacent to the upstream membrane interface is higher than the partial pressure or concentration at the upstream interface. Also, the partial pressure or concentration is higher just downstream of the membrane interface than in the permeate at the interface. The concentrations at the membrane interface and just adjacent to the membrane interface can be related according to an equilibrium partition coefficient $K_{M,i}$. This can be defined as (see Figure 10.10b):

$$K_{M,i} = \frac{C'_{F,i}}{C_{F,i}} = \frac{C'_{P,i}}{C_{P,i}} \qquad (10.23)$$

Most membranes are manufactured from synthetic polymers The application of such membranes is generally limited to temperatures below 100°C and to the separation of chemically inert species. When operation at high temperatures is required, or the species are not chemically inert, microporous membranes can be made of ceramics, and dense membranes from metals such as palladium.

Four idealized flow patterns can be conceptualized for membranes. These are shown in Figure 10.11. In Figure 10.11a, both the feed and permeate sides of the membrane are well mixed. Figure 10.11b shows a cocurrent flow pattern in which the fluid on the feed or retentate side flows along and parallel to the upstream surface of the membrane. The permeate fluid on the downstream side of the membrane consists of fluid that has just passed through the membrane at that location plus the permeate flowing to that location. The cross-flow case is shown in Figure 10.11c. In this case, there is no flow of permeate fluid along the membrane surface. Finally, Figure 10.11d shows countercurrent flow in which the feed flows along, and parallel to, the upstream of the membrane and the permeate fluid is the fluid that has just passed through the membrane at that location plus the permeate fluid flowing to that location in a countercurrent arrangement.

For these idealized flow patterns, parametric studies show that, in general, for the same operating conditions, countercurrent flow patterns yield the best separation and require the lowest membrane area. The next best performance is given by cross flow, then by cocurrent flow and the well-mixed arrangement shows the poorest performance. In practice, it is not always obvious as to which idealized flow pattern is assumed. The flow pattern not only depends on the geometry of the membrane arrangement but also on the permeation rate and, therefore, cut fraction. The two most common practical arrangements are spiral wound and hollow fiber. In the spiral wound arrangement, flat membrane sheets separated by spacers for the flow of feed and permeate are wound into a spiral and inserted in a pressure vessel. Hollow fiber arrangements, as the name implies, are cylindrical membranes arranged in

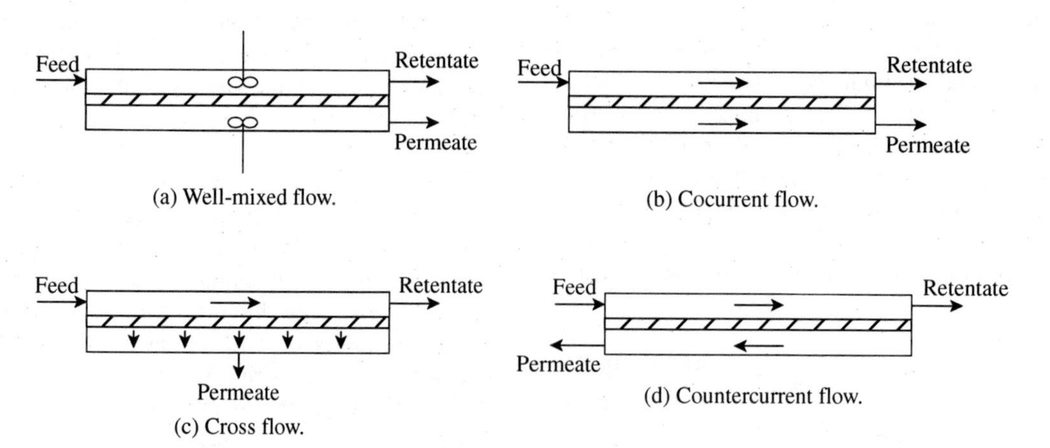

Figure 10.11 Idealized flow patterns in membrane separation.

series in a pressure vessel in an arrangement similar to a shell-and-tube heat exchanger. The permselective layer is on the outside of the fibers. The feed enters the shell-side and permeate passes through the membrane to the center of the hollow fibers. Other arrangements are possible, but less common. For example, arrangements similar to the plate-and-frame filter press illustrated in Figure 8.10a are used for some membrane separations. Membrane separators are usually modular in construction, with many parallel units required for large-scale applications.

With all membrane processes, the condition of the feed has a significant influence on the performance of the unit. This often means that some feed pretreatment is necessary to minimize fouling and the potential to damage the membrane.

Consider the most important membrane separations now.

1. *Gas permeation.* In *gas permeation* applications of membranes, the feed is at high pressure and usually contains some low molar mass species (typically less than 50 kg·kmol^{-1}) to be separated from higher molar mass species. The other side of the membrane is maintained at a low pressure, giving a high-pressure gradient across the membrane, typically in the range of 20 to 40 bar. It is also possible to separate organic vapor from a gas (e.g. air), using a membrane that is more permeable to the organic species than to the gas. Typical applications of gas permeation include:

- separation of hydrogen from methane
- air separation
- removal of CO_2 and H_2S from natural gas
- helium recovery from natural gas
- adjustment of H_2 to CO ratio in synthesis gas
- dehydration of natural gas and air
- removal of organic vapor from air, and so on.

The membrane is usually dense but sometimes microporous. If the external resistances to mass transfer are neglected in Figure 10.10, then $p_{F,i} = p'_{F,i}$ and $p_{P,i} = p'_{P,i}$ and Equation 10.20 can be written in terms of the volumetric flux as:

$$N_i = \frac{P_{M,i}}{\delta_M}(p_{F,i} - p_{P,i}) \qquad (10.24)$$

where
N_i = molar flux of Component i (kmol·m^{-2}·s^{-1})
$P_{M,i}$ = permeability of Component i (kmol·m·s^{-1}·m^{-2}·bar^{-1})
δ_M = membrane thickness (m)
$p_{F,i}$ = partial pressure of Component i in the feed (bar)
$p_{P,i}$ = partial pressure of Component i in the permeate (bar)

Low molar mass gases and strongly polar gases have high permeabilities and are known as *fast gases. Slow gases* have high molar mass and symmetric molecules. Thus, membrane gas separators are effective when the gas to be separated is already at a high pressure and only a partial separation is required. A near perfect separation is generally not achievable. Improved performance, overall, can be achieved by creating membrane networks in series, perhaps with recycles. Figure 10.12 illustrates some common membrane networks.

There are important trade-offs to be considered when designing a gas separation using a membrane. The cost of a membrane separation is dominated by the capital cost of the membrane, which is proportional to its area, the capital cost of any compression equipment required and the operating

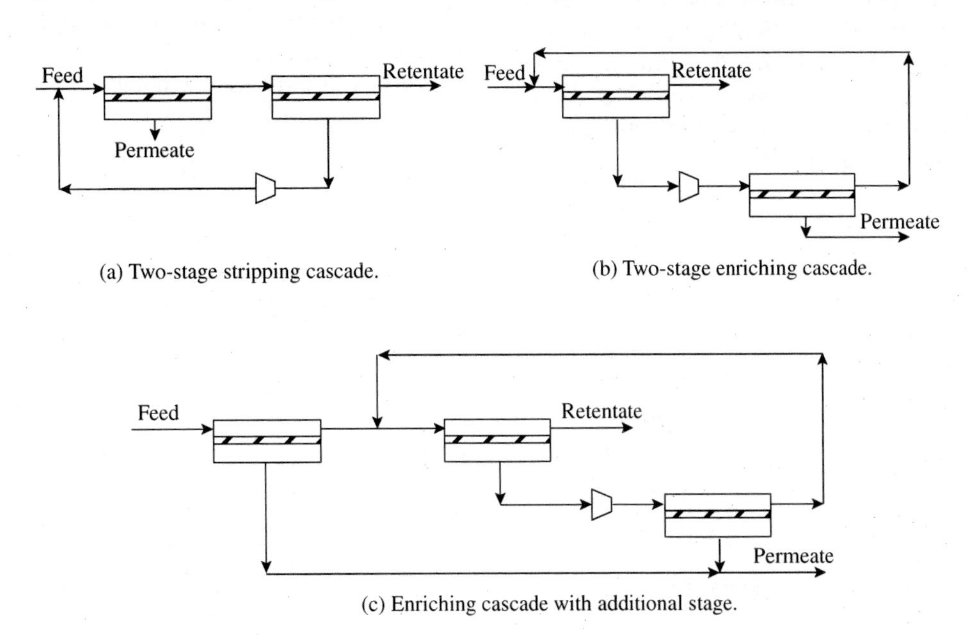

Figure 10.12 Membrane networks.

costs of the compression equipment. If the feed is at a low pressure, then its pressure will need to be increased. The low-pressure permeate might need to be recompressed for further processing. Also, if a network of membranes has been used, then compressors might be needed within the membrane network.

In Equation 10.24, if a certain total molar flowrate of a component through the membrane is specified, then choosing a membrane material with a high permeability will decrease the required membrane area and hence the capital cost. Gas permeability increases with increasing temperature. Thus, increasing the temperature for a given flux will decrease the membrane area. This would suggest that the feed to membrane systems should be heated to reduce the area requirements. Unfortunately, the polymer membranes most often used in practice do not allow high-temperature feeds and feed temperatures are normally restricted to below 100°C. It is common practice to heat the feed to gas membranes to avoid the possibility of condensation, which can damage the membrane. It is also clear from Equation 10.24 that, for a given membrane material, the thinner the membrane and the higher the pressure difference across the membrane, the lower the area requirement for a given component flux. However, the trade-offs are even more complex when the relative flux of components through the membrane and the product purity are considered. Figure 10.13a shows that, as the stage cut increases, the purity of the permeate decreases. In other words, if a component is being recovered as permeate, the more that is recovered, the less pure will be the product that is recovered. This is another important degree of freedom. Figure 10.13b shows the variation of permeate purity with separation factor. As expected, the higher the separation factor, as defined in Equation 10.21, the higher the product purity. Also, in Figure 10.13b, the higher the pressure ratio across the membrane, the higher the permeate purity. However, Figure 10.13b also shows that above a certain separation factor, the product purity is not greatly affected by an increase in the separation factor.

For the designer, given a required separation, there are three major contributions to the total cost to be traded off against each other:

- capital cost of the membrane unit (membrane modules and pressure housings);
- capital and operating costs of compression equipment (for compression of feed, recompression of the permeate or recycling within the membrane network);
- raw material losses.

2. *Reverse osmosis.* In *reverse osmosis*, a solvent permeates through a dense asymmetric membrane that is permeable to the solvent but not to the solute. The solvent is usually water and the solutes are usually dissolved salts. The principle of reverse osmosis is illustrated in Figure 10.14. In Figure 10.14a, a solute dissolved in a solvent in a concentrated form is separated from the same solvent in a dilute form by a dense membrane. Given the difference in concentration across the membrane, a natural process known as *osmosis* occurs, in which the solvent permeates across the membrane to dilute the more concentrated solution. The osmosis continues until equilibrium is established, as illustrated in Figure 10.14b. At equilibrium, the flow of solvent in both directions is equal and a difference in pressure is established between the two sides of the membrane, the *osmotic pressure*. Although a separation has occurred as a result of the presence of the membrane, the osmosis is not useful because the solvent is transferred in the wrong direction, resulting in mixing rather than separation. However, applying a pressure to the concentrated solution, as shown in Figure 10.14c, can reverse the direction of transfer of solvent through the membrane. This causes the solvent to permeate through the membrane from a concentrated solution to the dilute solution. This separation process, known as *reverse osmosis*, can be used to separate a solvent from a solute–solvent mixture.

The flux through the membrane can be written as

$$N_i = \frac{P_{M,i}}{\delta_M}(\Delta P - \Delta\pi) \tag{10.25}$$

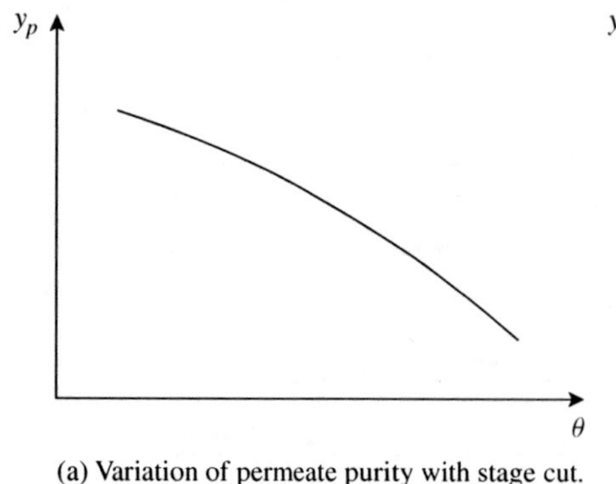

(a) Variation of permeate purity with stage cut.

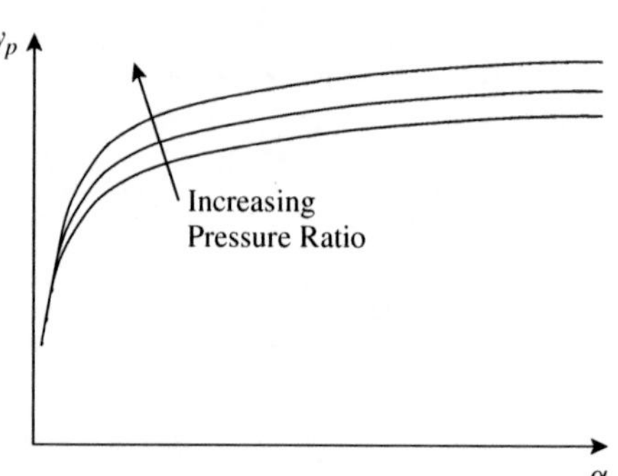

(b) Variation of permeate purity with separation factor.

Figure 10.13 Trade-offs in membrane design for gas separation.

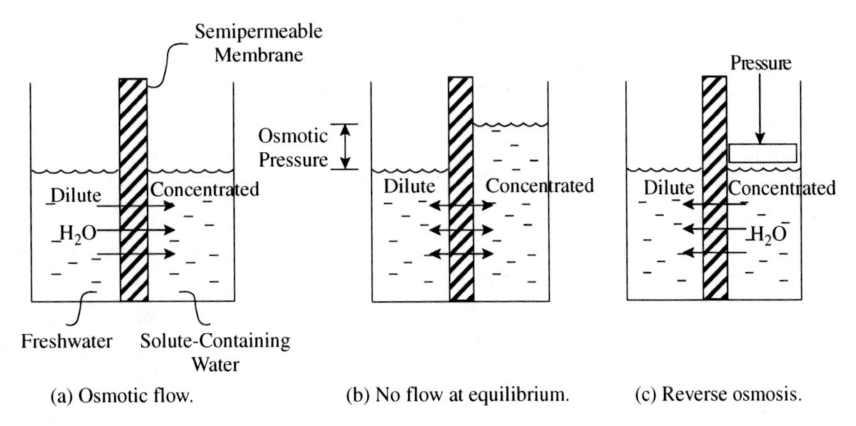

Figure 10.14 Reverse osmosis.

where N_i = solvent (water) flux through the membrane ($kg \cdot m^{-2} \cdot s^{-1}$)
$P_{M,i}$ = solvent membrane permeability ($kg\ solvent \cdot m^{-1} \cdot bar^{-1} \cdot s^{-1}$)
δ_M = membrane thickness (m)
ΔP = pressure difference across the membranes bar)
$\Delta\pi$ = difference between the osmotic pressures of the feed and permeate solutions (bar)

Hence, as the pressure difference is increased, the solvent flow increases. The pressure difference used varies according to the membrane and the application, but is usually in the range 10 to 50 bar but can also be up to 100 bar. The osmotic pressure in Equation 10.25 for dilute solutions can be approximated by the Van't Hoff equation:

$$\pi = iRT\frac{N_S}{V} \tag{10.26}$$

where π = osmotic pressure (bar)
i = number of ions formed if solute molecule dissociates (e.g. for NaCl, $i = 2$, for $BaCl_2$, $i = 3$)
R = universal gas constant (0.083145 $bar \cdot m^3 \cdot K^{-1} \cdot kmol^{-1}$)
T = absolute temperature (K)
N_S = number of moles of solute (kmol)
V = volume of pure solvent (m^3)

Applications of reverse osmosis are normally restricted to below 50°C. Reverse osmosis is now widely applied to the desalination of water to produce potable water. Other applications include:

- dewatering/concentrating foodstuffs
- concentration of blood cells
- treatment of industrial wastewater to remove heavy metal ions
- treatment of liquids from electroplating processes to obtain a metal ion concentrate and a permeate that can be used as rinse water
- separation of sulfates and bisulfates from effluents in the pulp and paper industry
- treatment of wastewater in dying processes
- treatment of wastewater inorganic salts, and so on.

Reverse osmosis is particularly useful when it is necessary to separate ionic material from an aqueous solution. A wide range of ionic species is capable of being removed with an efficiency of 90% or greater in a single stage. Multiple stages can increase the separation.

A phenomenon that is particularly important in the design of reverse osmosis units is that of *concentration polarization*. This occurs on the feed-side (concentrated side) of the reverse osmosis membrane. Because the solute cannot permeate through the membrane, the concentration of the solute in the liquid adjacent to the surface of the membrane is greater than that in the bulk of the fluid. This difference causes mass transfer of solute by diffusion from the membrane surface back to the bulk liquid. The rate of diffusion back into the bulk fluid depends on the mass transfer coefficient for the boundary layer on feed-side. Concentration polarization is the ratio of the solute concentration at the membrane surface to the solute concentration in the bulk stream. Concentration polarization causes the flux of solvent to decrease since the osmotic pressure increases as the boundary layer concentration increases and the overall driving force ($\Delta P - \Delta\pi$) decreases.

When used in practice, the membranes for reverse osmosis must be protected by pretreatment of the feed to reduce membrane fouling and degradation[10–12]. Spiral wound modules should be treated to remove particles down to 20 to 50 μm, while hollow fiber modules require particles down to 5 μm to be removed. If necessary, pH should be adjusted to avoid extremes of pH. Also, oxidizing agents such as free chlorine must be removed. Other pretreatments are also used depending on the application, such as removal of calcium and magnesium ions to prevent scaling of the membrane. Even with elaborate pretreatment, the membrane may still need to be cleaned regularly.

3. *Nano-filtration. Nano-filtration* is a pressure-driven membrane process similar to reverse osmosis, using asymmetric membranes but that are more porous. It can be considered to be a "coarse" reverse osmosis and is used to separate covalent ions and larger univalent ions such as heavy metals. Small univalent ions (e.g. Na^+, K^+ Cl^-) pass through the membrane to a major extent. For example, a nano-filtration membrane might remove 50% NaCl and 90% $CaSO_4$. Nano-filtration membranes use less fine membranes than reverse osmosis and the pressure drop across the membrane is correspondingly lower. Pressure drops are usually in the range of 5 to 20 bar. Also, the fouling rate of nano-filtration membranes tends to be lower than that of reverse osmosis membranes. Typical applications include:

- water softening (removal of calcium and magnesium ions)
- water pretreatment prior to ion exchange or electrodialysis
- removal of heavy metals to allow reuse of water
- concentration of foodstuffs
- desalination of foodstuffs, and so on.

As with reverse osmosis, feed pretreatment can be used to minimize membrane fouling and degradation, and regular cleaning is necessary.

4. *Ultra-filtration. Ultra-filtration* is again a pressure-driven membrane process similar to reverse osmosis, using asymmetric membranes but that are significantly more porous. Particles and large molecules do not pass through the membrane and are recovered as a concentrated solution. Solvent and small solute molecules pass through the membrane and are collected as permeate. Ultra-filtration is used to separate very fine particles (typically in the range 0.001 to 0.02 μm), microorganisms and organic components with molar mass down to 1000 $kg \cdot kmol^{-1}$. The flux through the membrane is described by Equation 10.25. However, the ultra-filtration does not retain species for which the bulk solution osmotic pressure is significant. Virtually all commercial applications of ultra-filtration are flux limited by concentration polarization. A boundary layer forms near the membrane surface, which has a substantially higher solute concentration than the bulk stream. As a result, the flux and the separation correspond with a substantially higher feed concentration than is measured in the bulk stream. The result is that above a threshold pressure difference, the flux is independent of pressure difference. Below the threshold, there is a linear dependence of flux on pressure difference, as implied by Equation 10.25. Pressure drops are usually in the range 1.5 to 10 bar.

Typical applications of ultra-filtration are:

- clarification of fruit juice
- recovery of vaccines and antibiotics from fermentation broths
- oil–water separation
- color removal, and so on.

Temperatures are restricted to below 70°C. Again, feed pretreatment can be used to minimize membrane fouling and degradation, and regular cleaning is necessary.

5. *Microfiltration. Microfiltration* is a pressure-driven membrane filtration process and has already been discussed in Chapter 8 for the separation of heterogeneous mixtures. Microfiltration retains particles down to a size of around 0.05 μm. Salts and large molecules pass through the membrane but particles of the size of bacteria and fat globules are rejected. A pressure difference of 0.5 to 4 bar is used across the membrane. Typical applications include:

- clarification of fruit juice
- removal of bacteria from foodstuffs
- removal of fat from foodstuffs, and so on.

6. *Dialysis. Dialysis* uses a membrane to separate species by virtue of their different diffusion rates in a microporous membrane. Both plate-and-frame and hollow fiber membrane arrangements are used. The feed solution, or *dialysate* containing the solutes to be separated flows on one side of the membrane. A solvent, or *diffusate* stream, flows on the other side of the membrane. Some solvent may also diffuse across the membrane in the opposite direction, which reduces the performance by diluting the dialysate. Dialysis is used to separate species that differ appreciably in size and have reasonably large differences in diffusion rates. The driving force for separation is the concentration gradient across the membrane. Hence, dialysis is characterized by low flux rates when compared with other membrane processes such as reverse osmosis, microfiltration and ultra-filtration that depend on applied pressure. Dialysis is generally used when the solutions on both sides of the membrane are aqueous. Applications include recovery of sodium hydroxide, recovery of acids from metallurgical process liquors, purification of pharmaceuticals and separation of foodstuffs. An important application of dialysis is as an artificial kidney in the biomedical field where it is used for the purification of human blood. Urea, uric acid and other components that have elevated concentrations in the blood diffuse across the membrane to an aqueous dialyzing solution, without removing essential high molar mass materials and blood cells.

7. *Electrodialysis. Electrodialysis* enhances the dialysis process with the aid of an electrical field and ion-selective membranes to separate ionic species from solution. It is used to separate an aqueous electrolyte solution into a concentrated and a dilute solution. Figure 10.15 illustrates

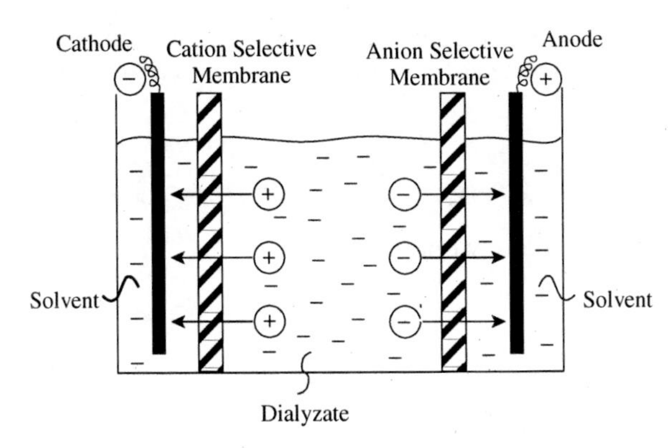

Figure 10.15 Electrodialysis.

the principle. The cation-selective membrane passes only cations (positively charged ions). The anion-selective membrane passes anions (negatively charged ions). The electrodes are chemically neutral. When a direct current charge is applied to the cell, the cations are attracted to the cathode (negatively charged) and anions are attracted to the anode (positively charged). Ions in the feed will pass through the appropriate membranes according to their charge, thus separating the ionic species.

Since electrodialysis is suited only for the removal or concentration of ionic species, it is suited to recovery of metals from solution, recovery of ions from organic compounds, recovery of organic compounds from their salts, and so on.

8. *Pervaporation. Pervaporation* differs from the other membrane processes described so far in that the phase-state on one side of the membrane is different from that on the other side. The term pervaporation is a combination of the words *per*mselective and e*vaporation*. The feed to the membrane module is a mixture (e.g. ethanol–water mixture) at a pressure high enough to maintain it in the liquid phase. The liquid mixture is contacted with a dense membrane. The other side of the membrane is maintained at a pressure at or below the dew point of the permeate, thus maintaining it in the vapor phase. The permeate side is often held under vacuum conditions. Pervaporation is potentially useful when separating mixtures that form azeotropes (e.g. ethanol–water mixture). One of the ways to change the vapor–liquid equilibrium to overcome azeotropic behavior is to place a membrane between the vapor and liquid phases. Temperatures are restricted to below 100°C, and as with other liquid membrane processes, feed pretreatment and membrane cleaning are necessary.

With all membrane processes, there is a potential fouling problem that must be addressed when specifying the unit. With liquid feeds in particular, this usually means pretreating the feed to remove solids, potentially down to very fine particle sizes, as well as other pretreatments. Membranes used for liquid separations often need regular cleaning, perhaps on a daily basis or even more regularly. Cleaning can be carried out by reversal of flow (*back-flushing*) and chemical treatment. The system used for cleaning is usually cleaning-in-place, whereby the membrane is taken off-line and connected to a cleaning circuit. However, membranes are easily damaged – chemically, by aggressive components, mechanically, in cleaning cycles and by excessive temperatures.

Example 10.5 A gaseous purge stream from a process has a mole fraction of hydrogen of 0.7. The balance can be assumed to be methane. It is proposed to recover the hydrogen using a membrane. The flowrate of the purge gas is 0.05 kmol·s^{-1}. The pressure is 20 bar and temperature of 30°C. The permeate will be assumed to be 1 bar. Assume that the gas is well mixed on each side of the membrane and that there is no pressure drop across the membrane surface. The permeance ($P_{M,i}/\delta_M$) of hydrogen and methane for the membrane are given in Table 10.4.

Assume that 1 kmol of gas occupies 22.4 m^3 at standard temperature and pressure (STP). For stage-cut fractions from 0.1 to 0.9, calculate the purity of hydrogen in the permeate, the membrane area and the fractional hydrogen recovery for a single-stage membrane.

Solution If the gas is assumed to be well mixed, then on the high-pressure (feed-side) side of the membrane, the mole fraction is that of the retentate leaving the membrane. Assuming no pressure drop across the membrane and a binary separation, Equation 10.24 can be written for Component A as:

$$\frac{F_P y_{P,A}}{22.4 A_M} = \frac{P_{M,A}}{22.4 \delta_M}(P_F y_{R,A} - P_P y_{P,A}) \tag{10.27}$$

where
F_P = volumetric flowrate of the permeate (m^3 STP·s^{-1})
A_M = membrane area (m^2)
$P_{M,A}$ = permeability of Component A (m^3·m·s^{-1}·m^{-2}·bar^{-1})
δ_M = membrane thickness (m)
P_F = pressure of feed (bar)
P_P = pressure of permeate (bar)
$y_{P,A}$ = mole fraction of Component A in the permeate
$y_{R,A}$ = mole fraction of Component A in the retentate

Similarly, for Component B,

$$\frac{F_P y_{P,B}}{22.4 A_M} = \frac{P_{M,B}}{22.4 \delta_M}(P_F y_{R,B} - P_P y_{P,B}) \tag{10.28}$$

where $y_{P,B}$ = mole fraction of Component B in the permeate
$y_{R,B}$ = mole fraction of Component B in the retentate

Table 10.4 Permeance data for Example 10.5.

	Permeance (m^3 STP·m^{-2}·s^{-1}·bar^{-1})
H_2	7.5×10^{-5}
CH_4	3.0×10^{-7}

For a binary mixture:

$$y_{P,B} = 1 - y_{P,A}$$
$$y_{R,B} = 1 - y_{R,A} \qquad (10.29)$$

Substituting Equation 10.29 in Equation 10.28 gives:

$$\frac{F_P(1 - y_{P,A})}{A_M} = \frac{P_{M,B}}{\delta_M}[P_F(1 - y_{R,A}) - P_P(1 - y_{P,A})] \qquad (10.30)$$

Dividing Equation 10.27 by 10.30:

$$\frac{y_{P,A}}{1 - y_{P,A}} = \frac{\alpha\left[y_{R,A} - \left(\frac{P_P}{P_F}\right)y_{P,A}\right]}{(1 - y_{R,A}) - \left(\frac{P_P}{P_F}\right)[1 - y_{P,A}]} \qquad (10.31)$$

where $\alpha = \dfrac{P_{M,A}}{P_{M,B}}$

The value of $y_{R,A}$ is usually unknown, but it can be eliminated from Equation 10.31 by making an overall balance at STP:

$$F_F = F_R + F_P \qquad (10.32)$$

and an overall balance for Component A at STP:

$$F_F y_{F,A} = F_R y_{R,A} + F_P y_{P,A} \qquad (10.33)$$

Equations 10.32 and 10.33 can be combined to give:

$$y_{R,A} = \frac{y_{F,A} - \theta y_{P,A}}{(1 - \theta)} \qquad (10.34)$$

where $\theta = \dfrac{F_P}{F_F}$

Substituting Equation 10.34 in Equation 10.31 and rearranging gives:

$$0 = a_0 + a_1 y_{P,A} + a_2 y_{P,A}^2 \qquad (10.35)$$

where $a_0 = -\alpha y_{F,A}$

$$a_1 = 1 - (1 - \alpha)(\theta + y_{F,A}) - \frac{P_P}{P_F}(1 - \theta)(1 - \alpha)$$
$$a_2 = \theta(1 - \alpha) + \frac{P_P}{P_F}(1 - \theta)(1 - \alpha)$$

Thus, given α, θ, P_F, P_F and y_F, Equation 10.35 can be solved to give y_P. This can be done numerically or using the general analytical solution to quadratic equations[11,12]:

$$y_{P,A} = \frac{-a_1 + \sqrt{a_1^2 - 4a_2a_0}}{2a_2} \qquad (10.36)$$

Once y_P has been determined, A_M can be determined by substituting Equation 10.34 and $F_p = \theta F_F$ in Equation 10.27 to obtain:

$$A_M = \frac{\theta(1 - \theta)F_F y_{P,A}\delta_M}{P_{M,A}[P_F(y_{F,A} - \theta y_{P,A}) - P_P y_{P,A}(1 - \theta)]} \qquad (10.37)$$

Also, the recovery can be defined as:

$$R = \frac{F_P y_{P,A}}{F_F y_{F,A}} = \theta\frac{y_{P,A}}{y_{F,A}} \qquad (10.38)$$

From the feed data:

$$F_F = 0.05 \times 22.4 = 1.12 \text{ m}^3\cdot\text{s}^{-1}$$

Values of θ can now be substituted in Equations 10.35, 10.37 and 10.38 to obtain the results in Table 10.5.

The values of $y_{P,A}$, A_M and R are plotted against θ in Figure 10.16.

Example 10.5 illustrates the trade-offs for gas permeation. The feed was assumed to be a binary mixture, which simplifies the calculations. For multicomponent mixtures, the same basic equations can be written for each component and solved simultaneously[12]. The approach is basically the same, but numerically more complex.

It was also assumed that the gas on both sides of the membrane was well mixed. Again, this simplifies the

Table 10.5 Results for a range of values of stage cut for Example 10.5.

θ	a_0	a_1	a_2	$y_{P,A}$	A_M	R
0.1	−175.0	211.4	−36.1	0.998	120.8	0.143
0.2	−175.0	235.1	−59.8	0.997	258.7	0.285
0.3	−175.0	258.7	−83.4	0.997	426.9	0.427
0.4	−175.0	282.4	−107.1	0.996	656.2	0.569
0.5	−175.0	306.0	−130.7	0.993	1039.0	0.710
0.6	−175.0	329.7	−154.4	0.987	2007.3	0.846
0.7	−175.0	353.3	−178.0	0.951	7353.6	0.951
0.8	−175.0	377.0	−201.7	0.859	22761.7	0.981
0.9	−175.0	400.6	−225.3	0.772	40852.5	0.993

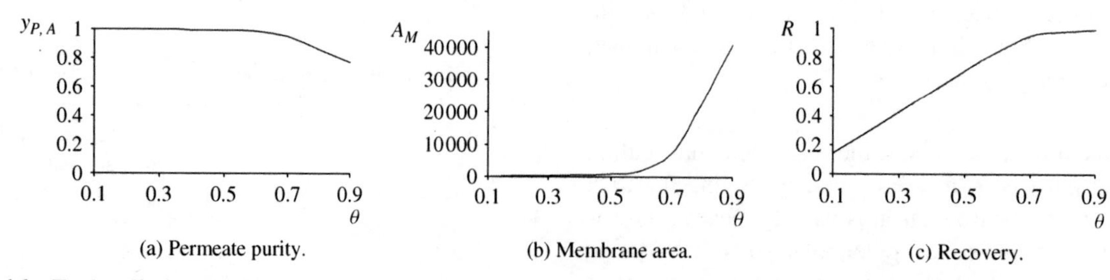

(a) Permeate purity. (b) Membrane area. (c) Recovery.

Figure 10.16 Trade-offs for permeate mole fraction, membrane area and hydrogen recovery for Example 10.5.

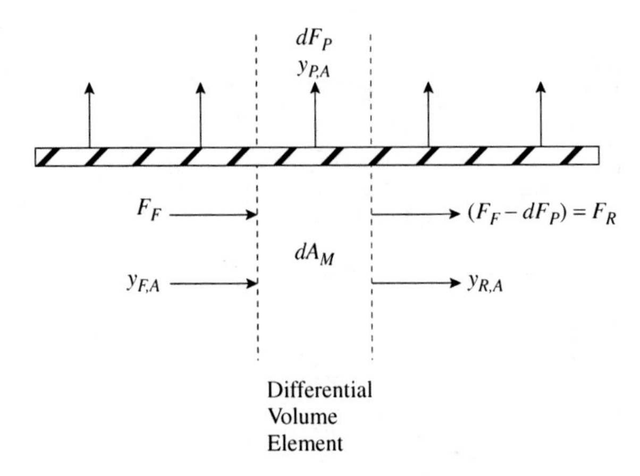

Figure 10.17 Cross-flow model for membranes.

calculations. In practice, cross-flow is more likely to represent the actual flow pattern. This is again more numerically complex as the separation varies along the membrane. The assumption of well-mixed feed and permeate will tend to overestimate the membrane area requirements. To perform the calculations assuming cross-flow, the model illustrated in Figure 10.17 can be used. If the simplifying assumptions are made that the pressure drop across the membrane is fixed, there is no pressure drop along the membrane and that the selectivity is fixed, the equations can be solved either analytically or numerically. A simple numerical approach is to assume that the membrane area is divided into incremental areas, as illustrated in Figure 10.17. Each incremental area can be modeled by the well-mixed model developed in Example 10.5. The calculation starts from the feed-side of the membrane for the first incremental area. A value of the stage cut θ needs to be assumed for the first incremental area. This allows $y_{P,A}$ to be calculated from Equation 10.36 and the incremental area dA_M to be calculated from Equation 10.37. The retentate concentration $y_{R,A}$ can then be calculated from Equation 10.34. The flow through the membrane dF_P can be calculated from the stage cut θ, and hence, F_R from a balance around the incremental area. For the next incremental area across the membrane, the feed flowrate is F_R from the first increment and the feed concentration is the retentate concentration from the first increment, which is $y_{R,A}$. The well-mixed model is then applied to the next incremental area, and so on, across the membrane. To ensure that the numerical approach is accurate, the assumed stage cut must be small enough to allow the well-mixed model for the incremental area to be representative of cross-flow. The procedure is carried on across the membrane until the recovery or product purity meets the required specification. Alternatively, if the size of the membrane module is known, the integration across the membrane is continued until the calculated membrane area equals the specified area. The equations for cross-flow can also be solved analytically[12].

Example 10.6 Reverse osmosis is to be used to separate sodium chloride (NaCl) from water to produce 45 $m^3 \cdot h^{-1}$ of water with a concentration of less than 250 ppm NaCl. The initial concentration of the feed is 5000 ppm. A membrane is available for which the following test results have been reported:

Feed concentration	2000 ppm NaCl
Pressure difference	16 bar
Temperature	25°C
Solute rejection	99%
Flux	10.7×10^{-6} $m^3 \cdot m^{-2} \cdot s^{-1}$

The following assumptions can be made.

- Feed and permeate sides of the membrane are both well mixed.
- Solute rejection is constant for different feeds.
- There is no pressure drop along the membrane surfaces.
- Test conditions are such that the cut fraction of water recovered is low enough for the retentate concentration to be equal to the feed concentration.
- Solute concentrations are low enough for the osmotic pressure to be represented by the Van't Hoff Equation ($R =$ 0.083145 $bar \cdot m^3 \cdot K^{-1} \cdot kmol^{-1}$)
- Density of all solutions is 997 $kg \cdot m^{-3}$ and molar mass of NaCl is 58.5.

For a pressure difference of 40 bar across the membrane, estimate the permeate and retentate concentrations and membrane area for cut fractions of 0.1 to 0.5.

Solution Start by writing an overall balance and a balance for the solute:

$$F_F = F_R + F_P \tag{10.39}$$

$$C_F F_F = C_R F_R + C_P F_P \tag{10.40}$$

where F_F = flowrate of feed ($m^3 \cdot s^{-1}$)
F_R = flowrate of retentate ($m^3 \cdot s^{-1}$)
F_P = flowrate of permeate ($m^3 \cdot s^{-1}$)
C_F = concentration of feed ($kg \cdot m^{-3}$)
C_R = concentration of retentate ($kg \cdot m^{-3}$)
C_P = concentration of permeate ($kg \cdot m^{-3}$)

Combining Equations 10.39 and 10.40 with the definition of the cut fraction, $\theta = F_P/F_F$, after rearranging gives:

$$C_F = C_R(1 - \theta) + C_P\theta \tag{10.41}$$

The solute rejection is defined as the ratio of the concentration difference across the membrane to the bulk concentration on the feed-side of the membrane. If it is assumed that both sides are well – mixed:

$$SR = \frac{C_R - C_P}{C_R} \tag{10.42}$$

where SR = solution rejection(–)

Combining Equations 10.41 and 10.42 gives:

$$C_R = \frac{C_F}{1 - \theta \times SR} \tag{10.43}$$

$$C_P = \frac{C_F(1 - SR)}{1 - \theta \times SR} \tag{10.44}$$

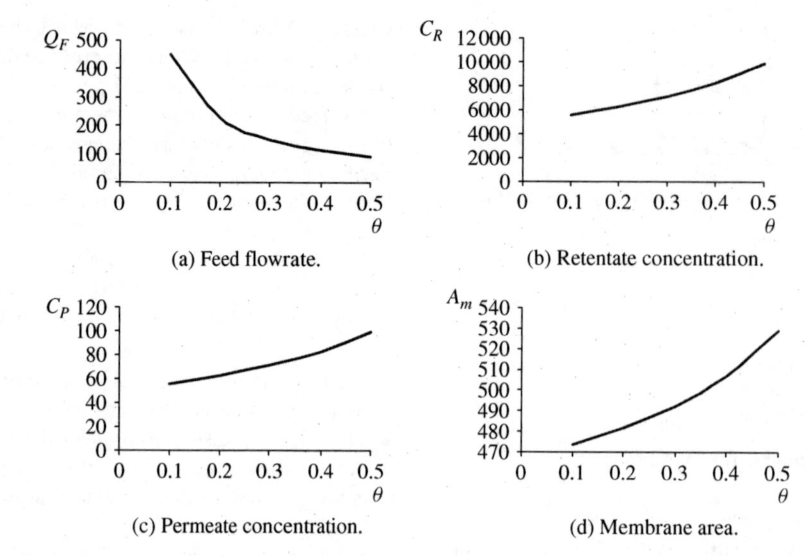

Figure 10.18 Trade-offs for stage cut, retentate concentration, permeate concentration and membrane area for the reverse osmosis separation in example.

First, calculate the osmotic pressure on each side of the membrane under test conditions:

$$C_R = 2000 \text{ ppm}$$
$$= 0.002 \text{ kg solute·kg solution}^{-1}$$
$$= \frac{0.002}{0.998} \text{ kg solute·kg solvent}^{-1}$$
$$\frac{N_R}{V_R} = \frac{0.002}{0.998} \times \frac{1}{58.5} \times 997$$
$$= 0.0342 \text{ kmol solute·m}^3\text{·solvent}^{-1}$$
$$\pi_R = iRT\frac{N_R}{V_R}$$
$$= 2 \times 0.083145 \times 298 \times 0.342$$
$$= 1.69 \text{ bar}$$

Similarly, for the permeate side,

$$\pi_P = 0.0169 \text{ bar}$$
$$\Delta\pi = 1.69 - 0.0169$$
$$= 1.68 \text{ bar}$$
$$\overline{P}_M = \frac{10.7 \times 10^{-6}}{(16 - 1.68)}$$
$$= 7.470 \times 10^{-7} \text{ m}^3\text{·bar}^{-1}\text{·m}^{-2}\text{·s}^{-1}$$

The required permeate flux is

$$F_P = 45 \text{ m}^3\text{·h}^{-1}$$
$$= 0.0125 \text{ m}^3\text{·s}^{-1}$$

For $\theta = 0.1$:

$$C_R = \frac{5000}{1 - 0.1 \times 0.99}$$
$$= 5549 \text{ ppm}$$
$$C_P = \frac{5000(1 - 0.99)}{1 - 0.1 \times 0.99}$$
$$= 55.5 \text{ ppm}$$
$$\pi_R = 4.71 \text{ bar}$$
$$\pi_P = 0.047 \text{ bar}$$
$$A_M = \frac{Q_P}{\overline{P}_M(\Delta P - \Delta\pi)}$$
$$= \frac{0.0125}{7.470 \times 10^{-7}(40 - 4.71 + 0.047)}$$
$$= 473.5 \text{ m}^2$$

Table 10.6 gives the results of these calculations for other values of θ:

Table 10.6 Results for a range of values of stage cut for Example 10.6.

θ	F_F (m^3·h^{-1})	C_R (ppm)	C_P (ppm)	A_M (m^2)
0.1	450	5549	55.5	473.5
0.2	225	6234	62.3	481.5
0.3	150	7112	71.1	492.0
0.4	112.5	8278	82.8	506.8
0.5	90	9901	99.0	528.9

These trends are plotted in Figure 10.18. It can be seen that as θ increases, Q_F decreases and C_R, C_P and A_M all increase.

A number of points need to be noted regarding these calculations:

1. The basic assumption of well-mixed fluid on the feed-side of the membrane does not reflect the flow patterns for the configurations used in practice. The assumption simplifies the calculations and allows the basic trends to be demonstrated.

Cross flow is a better reflection of practical configurations. The well-mixed assumption is only reasonable for low concentrations and low values of θ. As a comparison, seawater desalination involves a feed with a concentration of the order of 35,000 ppm.

2. The membrane test data assumed that the value of θ was effectively zero. In practice, measurements are usually taken at low values of θ and the solute recovery is based on the average of the feed and retentate concentrations.
3. The basic assumption was made that the solute recovery was constant and independent of both the feed concentration and θ. This is only a reasonable assumption for high values of solute recovery where the flux of solute is inherently very low.
4. The volume of the equipment for a given area requirement depends on the chosen membrane configuration. For example, spiral wound membranes have a typical packing density of around 800 $m^2 \cdot m^{-3}$, whereas the packing density for hollow fiber membranes is much higher, at around 6000 $m^2 \cdot m^{-3}$.

10.5 CRYSTALLIZATION

Crystallization involves formation of a solid product from a homogeneous liquid mixture. Often, crystallization is required as the product is in solid form. The reverse process of crystallization is dispersion of a solid in a *solvent*, termed *dissolution*. The dispersed solid that goes into solution is the *solute*. As dissolution proceeds, the concentration of the solute increases. Given enough time at fixed conditions, the solute will eventually dissolve up to a maximum solubility where the rate of dissolution equals the rate of crystallization. Under these conditions, the solution is saturated with solute and is incapable of dissolving further solute under equilibrium conditions. In fact, the distinction between the solute and solvent is arbitrary as either component can be considered to be the solute or solvent. If one of the components is extraneous to the process, this would normally be termed the solvent.

Figure 10.19a shows the solubility of a typical binary system of two Components A and B. The line CED in Figure 10.19a represents the conditions of concentration and temperature that correspond to a saturated solution. If a mixture along the line CE is cooled, then crystals of pure B are formed, leaving a residual solution. This continues along line CE until Point E is reached, which is the *eutectic point*. At the eutectic point, both components crystallize and further separation is not possible. If a mixture along the line DE is cooled, then crystals of pure A are formed, leaving a residual solution. Again, this continues along line DE until Point E is reached, the eutectic point, and further separation is not possible. Point C in Figure 10.19a is the melting point of pure B and Point D is the melting point of pure A. Below the eutectic temperature, a solid mixture of A and B forms. Examples of binary mixtures that exhibit the kind of behavior illustrated in Figure 10.19a are benzene–naphthalene and acetic acid–water. Not all binary systems follow the behavior shown in Figure 10.19a. There are other forms of behavior, some of which resemble vapor–liquid equilibrium behavior.

In Figure 10.19a, no new compound was formed between the solute and solvent. Some solutes can form compounds with their solvents. Such compounds with definite proportions between solutes and solvents are termed *solvates*. If the solvent is water, the compounds formed are termed *hydrates*.

Figure 10.19b shows the equilibrium solubility of various salts in water. Usually, the solubility increases as temperature increases. The solubility of copper sulfate increases significantly with increasing temperature. The solubility of sodium chloride increases with increasing temperature, but

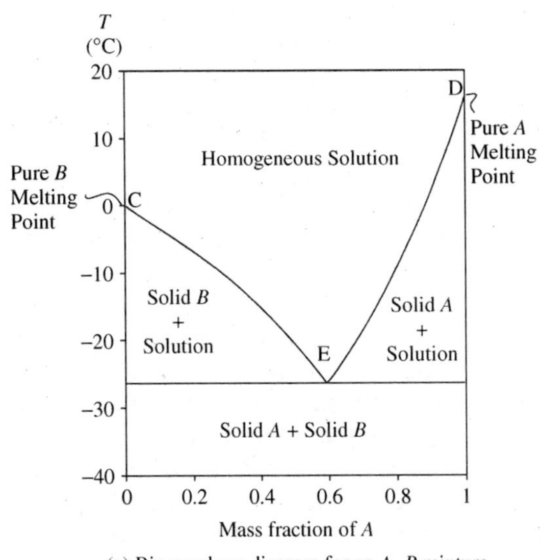

(a) Binary phase diagram for an A–B mixture (A = acetic acid, B = water in this example).

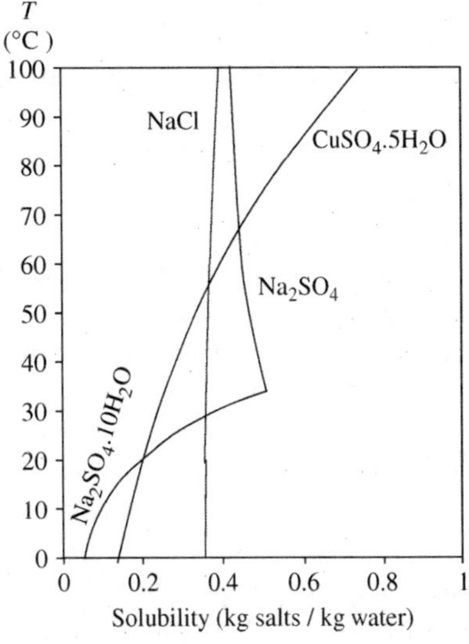

(b) Solubility of various salts in water.

Figure 10.19 Equilibrium solubility of solutes versus temperature.

the effect of temperature on solubility is small. The solubility of sodium sulfate decreases with increasing temperature. Such reverse solubility behavior is unusual.

In general, solubility is mainly a function of temperature, generally increasing with increasing temperature. Pressure has a negligible effect on solubility.

It might be expected that if a solute is dissolved in a solvent at a fixed temperature until the solution achieves saturation and any excess solute is removed and the saturated solution is then cooled, the solute would immediately start to crystallize from solution. However, solutions can often contain more solute than is present at saturation. Such *supersaturated* solutions are thermodynamically metastable and can remain unaltered indefinitely. This is because crystallization first involves *nucleus formation* or *nucleation* and then *crystal growth* around the nucleus. If the solution is free of all solid particles, foreign or of the crystallizing substance, then nucleus formation must first occur before crystal growth starts. *Primary nucleation* occurs in the absence of suspended product crystals. Homogeneous primary nucleation occurs when molecules of solute come together to form clusters in an ordered arrangement in the absence of impurities or foreign particles. The growing clusters become crystals as further solute is transferred from solution. As the solution becomes more supersaturated, more nuclei are formed. This is illustrated in Figure 10.20. The curve *AB* in Figure 10.20 represents the equilibrium solubility curve. Starting at Point *a* in the unsaturated region and cooling the solution without any loss of solvent, the equilibrium solubility curve is crossed horizontally into the *metastable* region. Crystallization will not start until it has been subcooled to Point *c* on the *supersolubility* curve. Crystallization begins at Point *c*, continues to Point *d* in the *labile* region and onwards. The curve *CD*, called the *supersolubility curve*, represents where nucleus formation appears spontaneously and hence, where crystallization can start. The supersolubility curve is more correctly thought of as a region where the nucleation rate increases rapidly, rather than a sharp boundary. Primary nucleation can also occur heterogeneously on solid surfaces such as foreign particles.

Figure 10.20 also shows another way to create supersaturation instead of reducing the temperature. Starting again at Point *a* in the unsaturated region, the temperature is kept constant and the concentration is increased by removing the solvent (for example, by evaporation). The equilibrium solubility curve is now crossed vertically at Point *e* and the metastable region entered. Crystallization is initiated at Point *f*, it continues to Point *g* in the labile region and onwards.

Secondary nucleation requires the presence of crystalline product. Nuclei can be formed through attrition either between crystals or between crystals and solid walls. Such attrition can be created either by agitation or by pumping. The greater the intensity of agitation, the greater

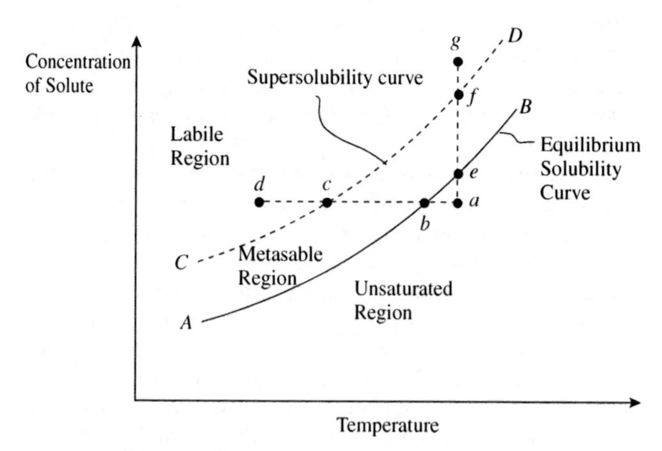

Figure 10.20 Supersaturation in crystallization processes.

the rate of nucleation. Referring back to Figure 10.20, the equilibrium solubility curve *AB* is unaffected by the intensity of agitation. However, the supersolubility curve *CD* moves closer to the equilibrium solubility curve *AB*, as the intensity of agitation becomes greater. Another way to create secondary nucleation is by adding *seed crystals* to start crystal growth in the supersaturated solution. These seeds should be a pure product. As solids build up in the crystallization, the source of new nuclei is often a combination of primary and secondary nucleation, although secondary nucleation is normally the main source of nuclei. Secondary nucleation is likely to vary with position in the crystallization vessel, depending on the geometry and the method of agitation.

Crystal growth can be expressed in a number of ways such as the change in a characteristic dimension or the rate of change of mass of a crystal. The different measures are related through crystal geometry. The growth is measured as increase in length in linear dimension of the crystals. This increase in length is for geometrically corresponding distances on the crystals.

The size of the crystals and the size distribution of the final crystals are both important in determining the quality of the product. Generally, it is desirable to produce large crystals. Large crystals are much easier to wash and filter and therefore allow a purer final product to be achieved. Because of this, when operating a crystallizer, operation in the labile region is generally avoided. Operation in the labile region, where primary nucleation mechanisms will dominate, will produce a large number of fine crystals. The two phenomena of nucleation and crystal growth compete for the crystallizing solute.

In addition to crystal size and size distribution, the shape of the crystal product might also be important. The term *crystal habit* is used to describe the development of faces of the crystal. For example, sodium chloride crystallizes from aqueous solution with cubic faces. On the other hand, if sodium chloride is crystallized from an aqueous solution containing a small amount of urea, the crystals will have octahedral faces. Both crystals differ in habit.

The choice of crystallizer for a given separation will depend on the method used to bring about supersaturation. Batch and continuous crystallizers can be used. Continuous crystallizers are generally preferred, but special circumstances often dictate the use of batch operation, as will be discussed further in Chapter 14. The methods used to bring about supersaturation can be classified as:

1. *Cooling* a solution through indirect heat exchange. This is most effective when the solubility of the solute decreases significantly with temperature. Rapid cooling will cause the crystallization to enter the labile region. Controlled cooling, perhaps with seeding, can be used to keep the process in the metastable region. Care must be taken to prevent fouling of the cooling surfaces by maintaining a low temperature difference between the process and the coolant. Scraped surface heat exchange equipment might be necessary.
2. *Evaporation* of the solvent can be used to generate supersaturation. This can be used if the solute has a weak dependency of solubility on temperature.
3. *Vacuum* can be used to assist evaporation of the solvent and reduce the temperature of the operation. One arrangement is for the hot solution to be introduced into a vacuum, where the solvent evaporates and cools the solution as a result of the evaporation.
4. *Salting* or *knockout* or *drown-out* involves adding an extraneous substance, sometimes call a *nonsolvent*, which induces crystallization. The nonsolvent must be miscible with the solvent and must change the solubility of the solute in the solvent. The nonsolvent will usually have a polarity different from that of the solvent. For example, if the solvent is water, the nonsolvent might be acetone, or if the solvent is ethanol, the nonsolvent might be water. This method has the advantage that fouling of heat exchange surfaces is minimized. On the other hand, an additional (extraneous) component is introduced into the system that must be separated and recycled.
5. *Reaction* can create metastable conditions directly. This is an attractive option when the reaction to produce the desired product and the separation can be carried out simultaneously.
6. *pH Switch* can be used to adjust the solubility of sparingly soluble salts in aqueous solution.

Given these various methods of creating supersaturation, which is preferred?

- Reaction is preferable if the situation permits. It requires low solubility of the solute formed, but can produce tiny crystals if the solubility is too low.
- Cooling crystallization is also preferred. Here, there are options. The mixture can be crystallized directly in *melt crystallization* or a solvent can be used in *solution crystallization*. If crystallization is carried out without an extraneous solvent in melt crystallization, then a high temperature might be needed for the mixture to melt in order to carry out the separation. High temperature might lead to product decomposition. If this is the case, there might be no choice other than to use an extraneous solvent. It might also be the case that a solvent is forced on the design. For example, a prior step, such as reaction, might require a certain solvent, and crystallization must be carried out from this solvent. Otherwise, there might be freedom to choose the solvent. The initial criteria for the choice of solvent will be related to the solubility characteristics of the solute in the solvent. A solvent is preferred that exhibits a high solubility at high temperature, but a low solubility at low temperature. In other words, it should have a steep solubility curve. The logic here is that it is desirable to use as little solvent as possible, hence high solubility at high temperature is required, but it is necessary to recover as much solute as possible, hence low solubility at low temperature is desirable. In the pursuit of a suitable solvent, a pure solvent or a mixed solvent might be used. In addition to the solvent having suitable solubility characteristics, it should preferably have low toxicity, low flammability, low environmental impact, low cost, it should be easily recovered and recycled and have ease of handling, such as suitable viscosity characteristics.
- pH switch is preferred if water can be used as the solvent and if the solute has a solubility in water that is sensitive to changes in pH.
- Evaporative crystallization is not preferred if the product needs to be of high purity. In addition to evaporation concentrating the solute, it also concentrates impurities. Such impurities might form crystals to contaminate the product or might be present in the residual liquid occluded within the solid product.
- Knockout or drown-out is generally not preferred as it involves adding a further extraneous material to the process. If it is to be successful, it requires a steep solubility curve versus the fraction of nonsolvent added.

Although the crystals are likely to be pure, the mass of crystals will retain some liquid when the solid crystals are separated from the residual liquid. If the adhering liquid is dried on the crystals, this will contaminate the product. In practice, the crystals will be separated from the residual liquid by filtration or centrifuging. Large uniform crystals separated from a low-viscosity liquid will retain the smallest proportion of liquid. Nonuniform crystals separated from a viscous liquid will retain a higher proportion of liquid. It is common practice to wash the crystals in the filter or centrifuge. This might be with fresh solvent, or in the case of melt crystallization, with a portion of melted product.

Example 10.7 A solution of sucrose in water is to be separated by crystallization in a continuous operation. The solubility of sucrose in water can be represented by the expression:

$$C^* = 1.524 \times 10^{-4}T^2 + 8.729 \times 10^{-3}T + 1.795 \qquad (10.45)$$

where C^* = solubility at the operating temperature (kg sucrose·kg H_2O^{-1})
T = temperature (°C)

The feed to the crystallizer is saturated at 60°C (C^* = 2.867 kg sucrose·kg H_2O^{-1}). Compare cooling and evaporative crystallization for the separation of sucrose from water.

a. For cooling crystallization, calculate the yield of sucrose crystals as a function of the temperature of the operation for the crystallizer.
b. For evaporative crystallization, calculate the energy requirement as a function of crystal yield.

Solution

a. First, it is necessary to define the yield. Since there is no change in the volume of water, the yield can be defined as:

$$\text{Yield} = \frac{C_{in} - C_{out}}{C_{in}} \times 100(\%) \qquad (10.46)$$

The operating conditions in the crystallizer will be under supersaturated conditions. Calculation of the supersolubility curve is possible, but complex. The crystallizer will be designed to operate under supersaturated conditions in the metastable region. The degree of supersaturation is an important degree of freedom in the crystallizer design. Detailed design is required to define this with any certainty. Therefore, the yield will be defined here assuming the outlet concentration to be saturated. Then assuming a temperature, the outlet concentration can be calculated from Equation 10.45 and the yield from Equation 10.46. The results are presented in Table 10.7.

Table 10.7 shows that the temperature must be decreased to low values to obtain a reasonable yield from the crystallization process. It should also be noted that cooling to 40°C should be possible against cooling water, and perhaps even down to 30°C. Any cooler than this, and refrigeration of some kind is required. This increases the cost of the cooling significantly.

Table 10.7 Yield versus temperature for Example 10.7.

Temperature (°C)	Yield (%)
60	0.0
50	8.9
40	16.7
30	23.5
20	29.7
10	33.8
5	35.7

b. A mass balance on the solvent gives:

$$F_{in} = F_{out} + F_V \qquad (10.47)$$

where F_{in} = inlet flowrate of liquid solvent
F_{out} = outlet flowrate of liquid solvent
F_V = flowrate of vaporized solvent

A mass balance on the solute gives:

$$C_{in}F_{in} = C_{out}F_{out} + m_{out}F_{out} \qquad (10.48)$$

where C_{in} = inlet solute concentration
C_{out} = outlet solute concentration
m_{out} = mass of crystals leaving the crystallizer

$$\begin{aligned}\text{Yield} &= \frac{m_{out}F_{out}}{C_{in}F_{in}} \\ &= \frac{C_{in}F_{in} - C_{out}F_{out}}{C_{in}F_{in}} \\ &= \frac{C_{in}F_{in} - C_{out}(F_{in} - F_V)}{C_{in}F_{in}} \\ &= 1 - \frac{C_{out}}{C_{in}} + \frac{C_{out}}{C_{in}}\frac{F_V}{F_{in}} \end{aligned} \qquad (10.49)$$

Equation 10.49 Indicates that as the rate of evaporation F_V increases, the yield increases. However, the energy input must also increase. If the simplifying assumption is made that the outlet concentration is the saturated equilibrium concentration C^* at 60°C, then:

$$C_{in} = C_{out} = C^*$$

and from Equation 10.49:

$$\text{Yield} = \frac{F_V}{F_{in}}$$

Thus, if 10% of the solvent is vaporized, this will lead to a yield of 10%, and so on. Energy input is required to vaporize the solvent at 60°C. The latent heat of water is 2350 kJ·kg^{-1}. The product of the mass of evaporation and latent heat gives the energy input:

$$\text{Energy input} = 2350F_V$$

10.6 EVAPORATION

Evaporation separates a volatile solvent from a solid. Single-stage evaporators tend to be used only when the capacity needed is small. For larger capacity, it is more usual to employ multistage systems that recover and reuse the latent heat of the vaporized material. Three different arrangements for a three-stage evaporator are illustrated in Figure 10.21.

1. *Forward-feed* operation is shown in Figure 10.21a. The fresh feed is added to the first stage and flows to the next stage in the same direction as the vapor flow. The boiling temperature decreases from stage to stage, and this

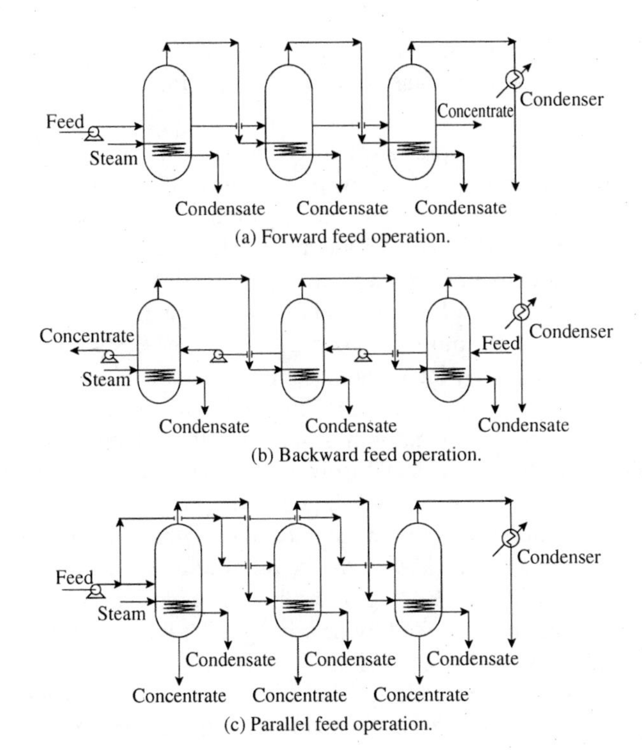

Figure 10.21 Three possible arrangements for a three-stage evaporator.

arrangement is thus used when the concentrated product is subject to decomposition at higher temperatures. It also has the advantage that it is possible to design the system without pumps to transfer the solutions from one stage to the next.

2. *Backward-feed* operation is shown in Figure 10.21b. Here, the fresh feed enters the last and coldest stage and leaves the first stage as concentrated product. This method is used when the concentrated product is highly viscous. The high temperatures in the early stages reduce viscosity and give higher heat transfer coefficients. Because the solutions flow against the pressure gradient between stages, pumps must be used to transfer solutions between stages.
3. *Parallel-feed* operation is illustrated in Figure 10.21c. Fresh feed is added to each stage, and product is withdrawn from each stage. The vapor from each stage is still used to heat the next stage. This arrangement is used mainly when the feed is almost saturated, particularly when solid crystals are the product.

Many other *mixed-feed* arrangements are possible, which combine the individual advantages of each type of arrangement. Figure 10.22 shows a three-stage evaporator in temperature–enthalpy terms, assuming that inlet and outlet solutions are at saturated conditions and that all evaporation and condensation duties are at constant temperature.

The three principal degrees of freedom in the design of stand-alone evaporators are:

1. Temperature levels can be changed by manipulating the operating pressure. Figure 10.22a shows the effect of a decrease in pressure.
2. The temperature difference between stages can be manipulated by changing the heat transfer area. Figure 10.22b shows the effect of a decrease in heat transfer area.
3. The heat flow through the system can be manipulated by changing the number of stages. Figure 10.22c shows the effect of an increase from three to six stages.

Given these degrees of freedom, how can an initialization be made for the design? The most significant degree of

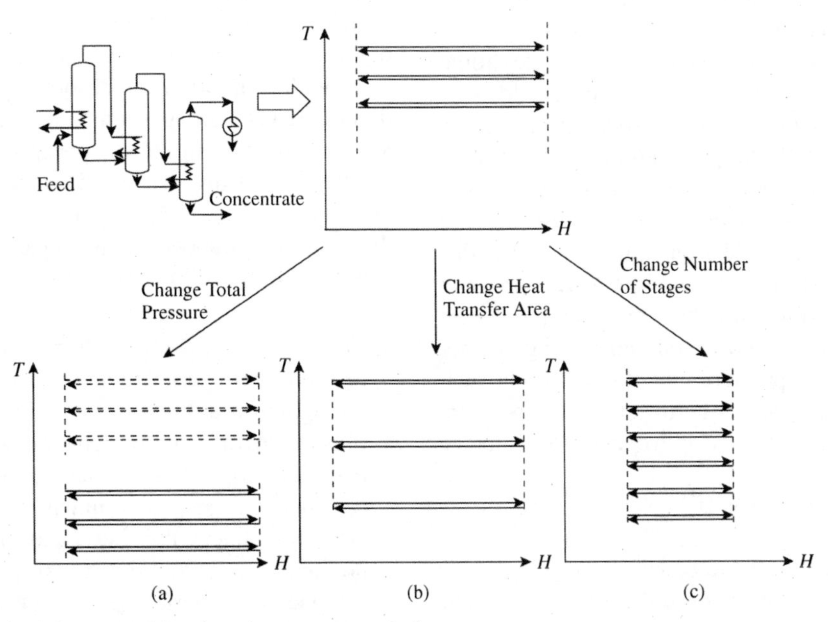

Figure 10.22 The principal degrees-of-freedom in evaporator design.

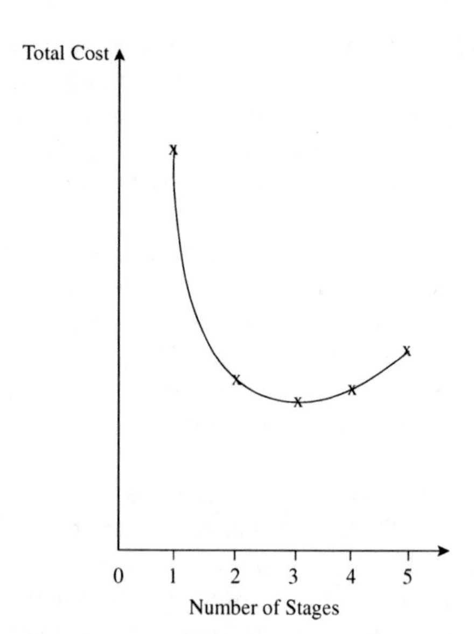

Figure 10.23 Variation of total cost with number of stages indicates that three stages is the optimum number for a stand-alone system in this case.

freedom is the choice of number of stages. If the evaporator is operated using hot and cold utility, as the number of stages is increased, a trade-off might be expected, as shown in Figure 10.23. Here, starting with a single stage, it has a low capital cost but requires a large energy cost. Increasing the stages to two decreases the energy cost in return for a small increase in capital cost, and the total cost decreases. However, as the stages are increased, the increase in capital cost at some point no longer compensates for the corresponding decrease in energy cost, and the total cost increases. Hence, there is an optimal number of stages. However, no attempt should be made to carry out this optimization in the early stages of a design, since the design is almost certain to change significantly when heat integration is considered later.

All that can be done is to make a reasonable initial assessment of the number of stages. Having made a decision for the number of stages, the heat flow through the system is temporarily fixed so that the design can proceed. Generally, the maximum temperature in evaporators is set by product decomposition and fouling. Therefore, the highest-pressure stage is operated at a pressure low enough to be below this maximum temperature. The pressure of the lowest-pressure stage is normally chosen to allow heat rejection to cooling water or air-cooling. If decomposition and fouling are not a problem, then the stage pressures should be chosen such that the highest-pressure stage is below steam temperature and the lowest-pressure stage is above cooling water or air-cooling temperature.

For a given number of stages, if

- all heat transfer coefficients are equal,
- all evaporation and condensation duties are at constant temperature,
- boiling point rise of the evaporating mixture is negligible,
- latent heat is constant through the system,

then minimum capital cost is given when all temperature differences are equal[13]. If evaporator pressure is not limited by the steam temperature but by product decomposition and fouling, then the temperature differences should be spread out equally between the upper practical temperature limit and the cold utility. This is usually a good enough initialization for most purposes, given that the design might change drastically later when heat integration is considered.

Another factor that can be important in the design of evaporators is the condition of the feed. If the feed is cold, then the backward-feed arrangement has the advantage that a smaller amount of liquid must be heated to the higher temperatures of the second and first stages. However, factors such as this should not be allowed to dictate design options at the early stages of flowsheet design because preheating the cold feed by heat integration with the rest of the process might be possible.

If the evaporator design is considered against a background process and heat integration with the background process is possible, then very different designs can emerge. When making an initial choice of separator, a simple, low-capital-cost design of evaporator should be chosen.

10.7 SEPARATION OF HOMOGENEOUS FLUID MIXTURES BY OTHER METHODS – SUMMARY

The most common alternative to distillation for the separation of low molar mass materials is absorption. Liquid flowrate, temperature and pressure are important variables to be set, but no attempt should be made to carry out any optimization in the early stages of a design.

As with distillation and absorption, when evaporators and dryers are chosen, no attempt should be made to carry out any optimization in the early stages of a design.

When choosing a separation technique (azeotropic distillation, absorption, stripping, liquid–liquid extraction, etc.), the use of extraneous mass-separating agents should be avoided for the following reasons:

- The introduction of extraneous material can create new problems in achieving product purity specifications throughout the process.
- It is often difficult to separate and recycle extraneous material with high efficiency. Any material not recycled can become an environmental problem. As will be discussed later, the best way to deal with effluent problems is not to create them in the first place.
- Extraneous material can create additional safety and storage problems.

Occasionally, a component that already exists in the process can be used as the mass separation agent, thus avoiding the introduction of extraneous material. However, clearly in many instances, practical difficulties and excessive cost might force the use of extraneous material.

10.8 EXERCISES

1. A gas stream with a flowrate of 8 $Nm^3 \cdot s^{-1}$ contains 0.1% SO_2 by volume. It is necessary to remove 95% of the SO_2 by absorption in water at 10°C at 1.01 bar before the gas stream is discharged to atmosphere. The Henry's Law constant for SO_2 in water at 10°C is 22.3 bar and gas behavior can be assumed to be ideal. Assuming the concentration of water at the exit of the absorber to be 90% of equilibrium, calculate:
   (a) How much water will be required?
   (b) How many theoretical stages will be required in the absorber?
2. A batch process produces 5 t of aqueous waste containing 25% wt acetic acid. The acetic acid is to be recovered by extraction with isopropyl ether. Equilibrium data are given in Table 10.1.
   (a) If 95% of the acetic acid is to be recovered from the aqueous waste in a single-stage extraction, how much isopropyl ether will be required?
   (b) For the single-stage extraction, how much isopropyl ether will be lost to the aqueous waste?
   (c). If a two-stage countercurrent batch extraction is to be used, sketch the flowsheet.
   (d) For the two-stage extraction, how much solvent is required for a 95% recovery of the acetic acid?
3. Air containing 21% by volume oxygen is to be enriched to provide a gas with higher oxygen content. A membrane is available for the separation, with permeabilities shown in the Table 10.8.

Table 10.8 Permeability data for Exercise 3.

Component	Permeance ($m^3STP \cdot m^{-2} \cdot s^{-1} \cdot bar^{-1}$)
Oxygen	1.52×10^{-5}
Nitrogen	2.28×10^{-6}

The thickness of the permselective layer of the membrane is 0.1 μm. The flowrate of air to the membrane is 0.01 $kmol \cdot s^{-1}$ at a pressure of 4.0 bar. The permeate pressure is 1 bar. Examine the relationship between permeate purity, membrane area and stage cut by varying the stage cut between 0.1 and 0.9. To simplify the calculations, assume the gas to be well mixed on both sides of the membrane.

4. A well-mixed continuous crystallizer is to be used to separate potassium sulfate from an aqueous solution by cooling crystallization. The solubility of potassium sulfate can be represented by the expression:

$$C^* = 0.0666 + 0.0023T - 6 \times 10^{-6}T^2$$

where C^* = solubility at the operating temperature (kg solute per kg solvent)
T = temperature (°C)

The feed to the crystallizer is saturated at 50°C. Estimate the crystal yield at 40, 30, 20, 10 and 5°C.

REFERENCES

1. Treybal RE (1980) *Mass Transfer Operations*, 3rd Edition, McGraw-Hill.
2. King CJ (1980) *Separation Processes*, 2nd Edition, McGraw-Hill.
3. Geankopolis CJ (1993) *Transport Processes and Unit Operations*, 3rd Edition, Prentice Hall.
4. Seader JD and Henley EJ (1998) *Separation Process Principles*, Wiley, New York.
5. Douglas JM (1988) *Conceptual Design of Chemical Processes*, McGraw-Hill, New York.
6. Schweitzer PA (1997) *Handbook of Separation Process Techniques for Chemical Engineers*, 3rd Edition, McGraw-Hill, New York.
7. Humphrey JL and Keller GE (1997) *Separation Process Technology*, McGraw-Hill.
8. Campbell H (1940) Report on the Estimated Costs of Doubling the Production of the Acetic Acid Concentrating Department, *Trans AIChE*, **36**: 628.
9. Hougen OA, Watson KM and Ragatz RA (1959) *Chemical Process Principles Part I Material and Energy Balances*, John Wiley.
10. Winston WS and Sirkar KK (1992) *Membrane Handbook*, Chapman & Hall.
11. Mulder M (1996) *Basic Principles of Membrane Technology*, Kluwer Academic Publishers.
12. Hwang S-T and Kammermeyer K (1975) *Membranes in Separations*, John Wiley Interscience.
13. Smith R and Jones PS (1990) The Optimal Design of Integrated Evaporation Systems, *Heat Recovery Systems and CHP*, **10**: 341.

11 Distillation Sequencing

Consider now the particular case in which a homogeneous multicomponent fluid mixture needs to be separated into a number of products, rather than just two products. As noted previously, distillation is the most common method of separating homogeneous fluid mixtures and in this chapter, the choice of separation will be restricted such that all separations are carried out using distillation only. If this is the case, generally there is a choice of order in which the products are separated that is, the choice of *distillation sequence*.

11.1 DISTILLATION SEQUENCING USING SIMPLE COLUMNS

Consider first the design of distillation systems comprising only simple columns. These simple columns employ:

- one feed split into two products;
- key components adjacent in volatility, or any components that exist in small quantities between the keys will become impurities in the products;
- a reboiler and a condenser.

If there is a three-component mixture to be separated into three relatively pure products and simple columns are employed, then the decision is between two sequences, as illustrated in Figure 11.1. The sequence shown in Figure 11.1a is known as the *direct sequence* in which the lightest component is taken overhead in each column. The *indirect sequence,* as shown in Figure 11.1b, takes the heaviest component as bottom product in each column.

If the distillation columns have both reboiling and condensation supplied by utilities, then the direct sequence in Figure 11.1a often requires less energy than the indirect sequence in Figure 11.1b. This is because the light material (Component A) is only vaporized once in the direct sequence. However, the indirect sequence can be more energy-efficient if the feed to the sequence has a low flowrate of the light material (Component A) and a high flowrate of heavy material (Component C). In this case, vaporizing the light material twice in the indirect sequence is less important than feeding a high flowrate of heavy material to both of the columns in the direct sequence.

For a three-component mixture to be split into three relatively pure products, there are only two alternative sequences. The complexity increases significantly as the number of products increases. Figure 11.2 shows the alternative sequences for a four-product mixture. Table 11.1 shows the relationship between the number of products and the number of possible sequences for simple columns[1].

Thus, there may be many ways in which the separation can be carried out to produce the same products. The problem is that there may be significant differences in the capital and operating costs between different distillation sequences that can produce the same products. In addition, heat integration may have a significant effect on operating costs. Heat integration of distillation will be considered later in Chapter 21.

11.2 PRACTICAL CONSTRAINTS RESTRICTING OPTIONS

Process constraints often reduce the number of options that can be considered. Examples of constraints of this type are:

1. Safety considerations might dictate that a particularly hazardous component be removed from the sequence as early as possible to minimize the inventory of that material.
2. Reactive and heat-sensitive components must be removed early to avoid problems of product degradation.
3. Corrosion problems often dictate that a particularly corrosive component be removed early to minimize the use of expensive materials of construction.
4. If thermal decomposition in the reboilers contaminates the product, then this dictates that finished products cannot be taken from the bottoms of columns.
5. Some compounds tend to polymerize when distilled unless chemicals are added to inhibit polymerization. These polymerization inhibitors tend to be nonvolatile, ending up in the column bottoms. If this is the case, it normally prevents finished products being taken from column bottoms.
6. There might be components in the feed to a distillation sequence that are difficult to condense. Total condensation of these components might require low-temperature condensation using refrigeration and/or high operating pressures. Condensation using both refrigeration and operation at high pressure increases operating costs significantly. Under these circumstances, the light components are normally removed from the top of the first column to minimize the use of refrigeration and high pressures in the sequence as a whole.

Chemical Process Design and Integration R. Smith
 ISBNs: 0-471-48680-9 (HB); 0-471-48681-7 (PB)

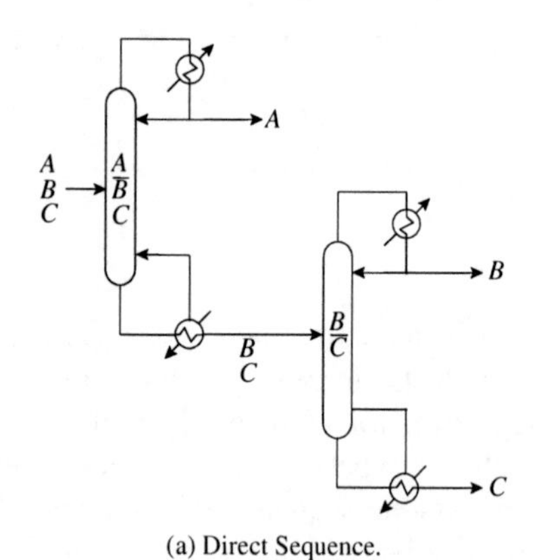

(a) Direct Sequence.

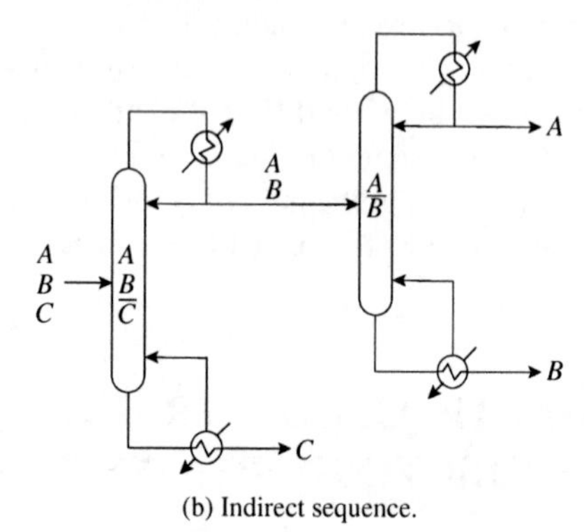

(b) Indirect sequence.

Figure 11.1 The direct and indirect sequences of simple distillation columns for a three-product separation (From Smith R and Linnhoff B, 1998, *Trans IChemE ChERD*, **66**, 195, reproduced by permission of the Institution of Chemical Engineers.)

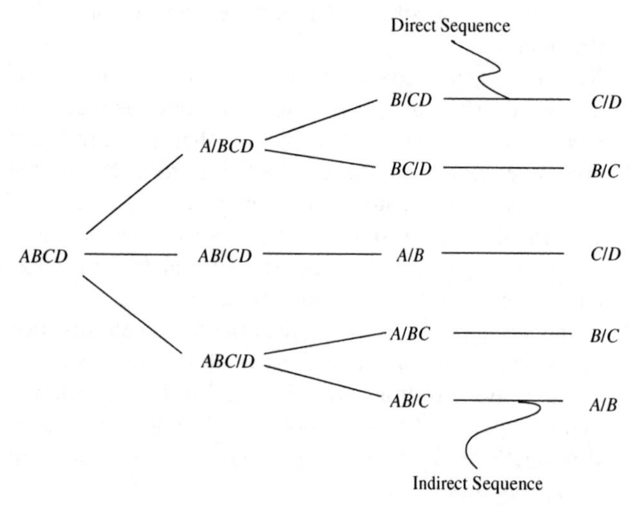

Figure 11.2 Alternative sequences for the separation of a four-product mixture.

11.3 CHOICE OF SEQUENCE FOR SIMPLE NONINTEGRATED DISTILLATION COLUMNS

Heuristics have been proposed for the selection of the sequence for simple nonintegrated distillation columns[1].

Table 11.1 Number of possible distillation sequences using simple columns.

Number of products	Number of possible sequences
2	1
3	2
4	5
5	14
6	42
7	132
8	429

These heuristics are based on observations made in many problems and attempt to generalize the observations. Although many heuristics have been proposed, they can be summarized by the following four[2]:

Heuristic 1. Separations where the relative volatility of the key components is close to unity or that exhibit azeotropic behavior should be performed in the absence of nonkey components. In other words, do the most difficult separation last.

Heuristic 2. Sequences that remove the lightest components alone one by one in column overheads should be favored. In other words, favor the direct sequence.

Heuristic 3. A component composing a large fraction of the feed should be removed first.

Heuristic 4. Favor splits in which the molar flow between top and bottom products in individual columns is as near equal as possible.

In addition to being restricted to simple columns, the observations are based on no heat integration (i.e. all reboilers and condensers are serviced by utilities). Difficulties can arise when the heuristics are in conflict with each other, as the following example illustrates.

Example 11.1 Each component for the mixture of alkanes in Table 11.2 is to be separated into relatively pure products. Table 11.2 shows normal boiling points and relative volatilities to indicate the order of volatility and the relative difficulty of the separations. The relative volatilities have been calculated on the basis of the feed composition to the sequence, assuming a pressure of 6 barg using the Peng–Robinson Equation of State with interaction parameters set to zero (see Chapter 4). Different pressures can, in practice, be used for different columns in the sequence and if a single set of relative volatilities is to be used, the pressure at which the relative volatilities are calculated needs, as much as possible, to be chosen to represent the overall system.

Use the heuristics to identify potentially good sequences that are candidates for further evaluation.

Solution

Heuristic 1. Do D/E split last since this separation has the smallest relative volatility.

Table 11.2 Data for a mixture of alkanes to be separated by distillation.

Component	Flowrate ($kmol \cdot h^{-1}$)	Normal boiling point (K)	Relative volatility	Relative volatility between adjacent components
A. Propane	45.4	231	5.78	1.94
B. *i*-Butane	136.1	261	2.98	1.26
C. *n*-Butane	226.8	273	2.36	1.95
D. *i*-Pentane	181.4	301	1.21	1.21
E. *n*-Pentane	317.5	309	1.00	

Heuristic 2. Favor the direct sequence.

A
B
C
D
E

Heuristic 3. Remove the most plentiful component first.

A
B
C
D
E

Heuristic 4. Favor near-equimolar splits between top and bottom products.

A
B 408.3 $kmol \cdot h^{-1}$
C
D 498.9 $kmol \cdot h^{-1}$
E

All four heuristics are in conflict here. Heuristic 1 suggests doing the D/E split last, whereas Heuristic 3 suggests it should be done first. Heuristic 2 suggests the A/B split first and Heuristic 4 the C/D split first.

Take one of the candidates and accept, say, the A/B split first.

Heuristic 1. Do D/E split last

Heuristic 2. *B*
C
D
E

Heuristic 3. *B*
C
D
E

Heuristic 4. *B* 362.9 $kmol \cdot h^{-1}$
C
D 498.9 $kmol \cdot h^{-1}$
E

Again the heuristics are in conflict. Heuristic 1 again suggests doing the D/E split last, whereas again Heuristic 3 suggests it should be done first. Heuristic 2 suggests the B/C split first and Heuristic 4 the C/D split first.

This process could be continued and possible sequences identified for further consideration. Some possible sequences would be eliminated, narrowing the number down suggested by Table 11.1.

The conflicts that have arisen in this problem have not been helpful in identifying sequences that are candidates for further evaluation. A little more intelligence could be used in the application of the heuristics and they could be ranked in order of importance. However, the rank order might well change from process to process. Although in the above example the heuristics do not give a clear indication of good candidate sequences, in some problems they might. It does seem though that a more general method than the heuristics is needed.

Rather than relying on heuristics that are qualitative, and can be in conflict, a quantitative measure of the relative performance of different sequences would be preferred. A physical measure that can be readily calculated is the vapor flow up the columns. This provides an indication of both capital and operating costs. There is clearly a relationship between the heat duty required for the reboiler and condenser to run the distillation and the vapor rate since the latent heat of vaporization relates these parameters. The heat duty in the reboiler relates directly to the cost of hot utility for the distillation (e.g. cost of steam). The heat duty in the condenser relates directly to the cost of cold utility for the distillation (e.g. cost of cooling water or refrigeration). However, there is also a link between vapor rate and capital cost, since a high vapor rate leads to a large diameter column. High vapor rate also requires large reboilers and condensers. Thus, vapor rate is a good indication of both capital and operating costs for individual columns. Consequently, sequences with lower total vapor load would be preferred to those with a high total vapor load. But how is the total vapor load predicted?

In Chapter 9, it was shown how the Underwood Equations can be used to calculate the minimum reflux ratio. A simple mass balance around the top of the column for constant molar overflow, as shown in Figure 11.3, at minimum reflux gives:

$$V_{min} = D(1 + R_{min}) \qquad (11.1)$$

where V_{min} = minimum vapor load ($kmol \cdot s^{-1}$)
R_{min} = minimum reflux ratio (–)
D = distillate flowrate ($kmol \cdot s^{-1}$)

Equation 11.1 can also be written at finite reflux, Figure 11.3. Defining R_F to be the ratio R/R_{min} (typically $R/R_{min} = 1.1$):

$$V = D(1 + R_F R_{min}) \qquad (11.2)$$

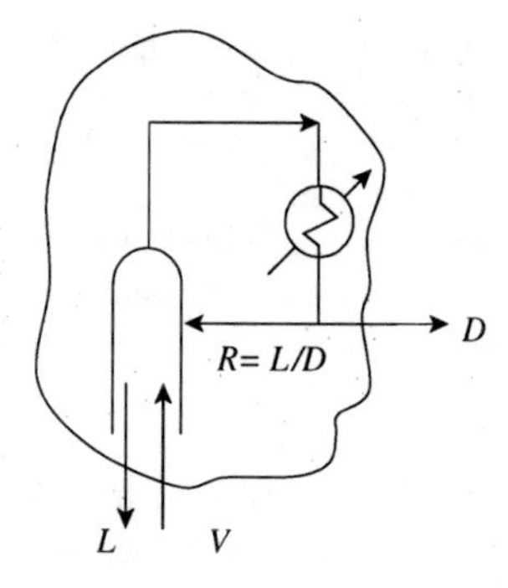

Figure 11.3 Mass balance around top of a distillation column.

If the feed is partially vaporized, the vapor flow below the feed will be lower than the top of the column. For above ambient temperature separations, the cost of operating the distillation will be dominated by the heat load in the reboiler and the vapor flow in the bottom of the column. For below ambient temperature separations, the cost of operating the column will be dominated by the cost of operating the refrigerated condenser and hence the vapor flow in the top of the column. If constant molar overflow is assumed, the vapor flow in the bottom of the column V' is related to the vapor flow in the top of the column by

$$V' = V - F(1 - q) \tag{11.3}$$

where V' = vapor flow below the feed ($\text{kmol}\cdot\text{s}^{-1}$)
V = vapor flow above the feed ($\text{kmol}\cdot\text{s}^{-1}$)
F = feed flowrate ($\text{kmol}\cdot\text{s}^{-1}$)
q = thermal condition of the feed
$= \dfrac{\text{heat required vaporize one mole of feed}}{\text{molar latent heat of vaporization of feed}}$
= 0 for saturated vapor feed, or 1 for saturated liquid

In most cases, for reasons discussed in Chapter 9, a saturated liquid feed is normally preferred.

The calculation is repeated for all columns in the sequence and the vapor loads summed to obtain the overall vapor load for the sequence. Different sequences can then be compared on the basis of total vapor load.

Example 11.2 Using the Underwood Equations, determine the best distillation sequence, in terms of overall vapor load, to separate the mixture of alkanes in Table 11.2 into relatively pure products. The recoveries are to be assumed to be 100%. Assume the ratio of actual to minimum reflux ratio to be 1.1 and all columns are fed with a saturated liquid. Neglect pressure drop across each column. Relative volatilities can be calculated from the Peng–Robinson Equation of State with interaction parameters assumed to be zero (see Chapter 4). Determine the rank order of the distillation sequences on the basis of total vapor load for:

a. All column pressures fixed to 6 barg with relative volatility calculated from the feed to the sequence.
b. All column pressures fixed to 6 barg with relative volatility recalculated for each column based on the feed composition of each column.
c. Pressures allowed to vary through the sequence with relative volatilities recalculated on the basis of the feed composition for each column. Pressures of each column are to be minimized such that either the bubble point of the overhead product is 10°C above the cooling water return temperature of 35°C (i.e. 45°C) or a minimum of atmospheric pressure. This will avoid the necessity for refrigeration.

Solution The results for the three cases are shown in Tables 11.3, 11.4 and 11.5.

Table 11.3 Sequences for the separation of the mixture of alkanes, with pressure fixed at 6 barg and relative volatilities fixed by feed to sequence.

Rank order	Vapor rate ($\text{kmol}\cdot\text{h}^{-1}$)	% Best	Sequence			
1	5849	100.0	AB/CDE	A/B	CD/E	C/D
2	5862	100.2	ABCD/E	AB/CD	A/B	C/D
3	5875	100.5	AB/CDE	A/B	C/DE	D/E
4	5883	100.6	ABC/DE	AB/C	D/E	A/B
5	6009	102.7	ABC/DE	A/BC	D/E	B/C
6	6037	103.2	ABCD/E	ABC/D	AB/C	A/B
7	6040	103.3	ABCD/E	A/BCD	B/CD	C/D
8	6080	104.0	A/BCDE	BCD/E	B/CD	C/D
9	6086	104.1	A/BCDE	BC/DE	B/C	D/E
10	6099	104.3	A/BCDE	B/CDE	CD/E	C/D
11	6126	104.7	A/BCDE	B/CDE	C/DE	D/E
12	6162	105.4	ABCD/E	A/BCD	BC/D	B/C
13	6163	105.4	ABCD/E	ABC/D	A/BC	B/C
14	6202	106.0	A/BCDE	BCD/E	BC/D	B/C

In Case a, the relative volatilities were assumed to be constant, based on the feed. But the relative volatilities will change through the sequence and might change significantly, owing to:

1. changes in the concentration as separation is carried out;
2. changes in column pressures.

It can be seen from Tables 11.3 to 11.5 that for each case there is very little difference between the best few sequences in terms of total vapor load. For this problem, there is not even too much difference between the best and the worst sequences. Also, when the results of Cases a, b and c are compared, it should be noted that the rank order changes according to the calculation of the relative volatilities. The best three sequences for the three cases are illustrated in Figure 11.4. This sensitivity of the rank order to changes in the assumptions is not surprising given the small differences between the various sequences. All of the columns in all sequences are above atmospheric pressure, varying from 0.7 to 14.4 barg.

Example 11.3 The mixture of aromatics in Table 11.6 is to be separated into five products. The xylenes are to be taken

Table 11.4 Sequences for the separation of the mixture of alkanes, with pressure fixed at 6 barg and relative volatility recalculated.

Rank order	Vapor rate (kmol·h^{-1})	% Best	Sequence			
1	5708	100.0	*ABCD/E*	*AB/CD*	*A/B*	*C/D*
2	5752	100.8	*ABCD/E*	*ABC/D*	*AB/C*	*A/B*
3	5940	104.1	*ABCD/E*	*ABC/D*	*A/BC*	*B/C*
4	5971	104.6	*ABCD/E*	*A/BCD*	*B/CD*	*C/D*
5	5988	104.9	*ABCD/E*	*A/BCD*	*BC/D*	*B/C*
6	6064	106.3	*AB/CDE*	*A/B*	*CD/E*	*C/D*
7	6171	108.1	*A/BCDE*	*BCD/E*	*B/CD*	*C/D*
8	6188	108.4	*A/BCDE*	*BCD/E*	*BC/D*	*B/C*
9	6229	109.1	*ABC/DE*	*AB/C*	*D/E*	*A/B*
10	6417	112.4	*ABC/DE*	*A/BC*	*D/E*	*B/C*
11	6436	112.8	*A/BCDE*	*B/CDE*	*CD/E*	*C/D*
12	6474	113.4	*AB/CDE*	*A/B*	*C/DE*	*D/E*
13	6586	115.4	*A/BCDE*	*BC/DE*	*B/C*	*D/E*
14	6846	120.0	*A/BCDE*	*B/CDE*	*C/DE*	*D/E*

as a mixed xylenes product. The C9's in Table 11.6 are to be characterized as C_9H_{12} (1-methylethylbenzene). The recoveries are to be assumed to be 100%. Relative volatilities are to be calculated from the Peng–Robinson Equation of State, assuming that all interaction parameters are zero (see Chapter 4). Pressures

Table 11.5 Sequences for the separation of the mixture of alkanes, with pressure fixed for cooling water in condensers.

Rank order	Vapor rate (kmol·h^{-1})	% Best	Sequence			
1	5178	100.0	*ABC/DE*	*AB/C*	*D/E*	*A/B*
2	5292	102.2	*ABC/DE*	*A/BC*	*D/E*	*B/C*
3	5371	103.7	*AB/CDE*	*A/B*	*C/DE*	*D/E*
4	5405	104.4	*ABCD/E*	*AB/CD*	*A/B*	*C/D*
5	5450	105.3	*AB/CDE*	*A/B*	*CD/E*	*C/D*
6	5483	105.9	*ABCD/E*	*ABC/D*	*AB/C*	*A/B*
7	5544	107.1	*A/BCDE*	*BC/DE*	*B/C*	*D/E*
8	5598	108.1	*ABCD/E*	*ABC/D*	*A/BC*	*B/C*
9	5622	108.6	*ABCD/E*	*A/BCD*	*B/CD*	*C/D*
10	5661	109.3	*ABCD/E*	*A/BCD*	*BC/D*	*B/C*
11	5745	111.0	*A/BCDE*	*BCD/E*	*B/CD*	*C/D*
12	5747	111.0	*A/BCDE*	*B/CDE*	*C/DE*	*D/E*
13	5784	111.7	*A/BCDE*	*BCD/E*	*BC/D*	*B/C*
14	5826	112.5	*A/BCDE*	*B/CDE*	*CD/E*	*C/D*

of each column are to be minimized such that either the bubble point of the overhead product is 10°C above the cooling water return temperature of 35°C (i.e. 45°C) or a minimum of atmospheric pressure. Assume the ratio of actual to minimum reflux to be 1.1 and that all columns are fed with a saturated liquid. Neglect pressure drop across each column. Determine the

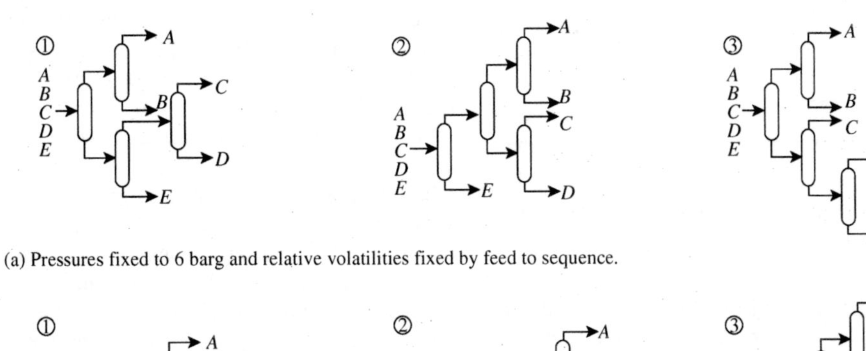

(a) Pressures fixed to 6 barg and relative volatilities fixed by feed to sequence.

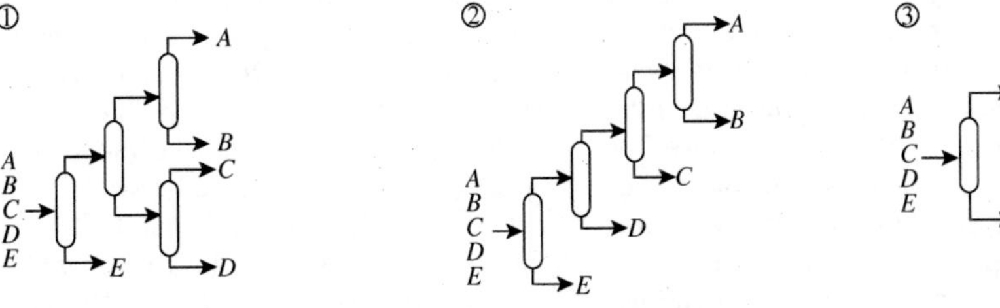

(b) Pressures fixed to 6 barg with relative volatilities recalculated for each column.

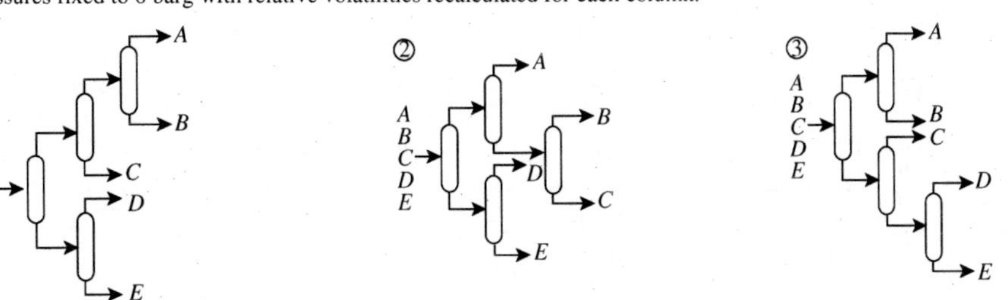

(c) Column pressures fixed for cooling water in condensers.

Figure 11.4 The best sequences in terms of vapor load for the separation of the mixture of alkanes from Example 11.2.

Table 11.6 Data for five-product mixture of aromatics to be separated by distillation.

Component	Flowrate (kmol·h^{-1})
A. Benzene	269
B. Toluene	282
C. Ethyl benzene	57
D. *p*-Xylene	47
m-Xylene	110
o-Xylene	58
E. C9's	42

Table 11.7 Relative volatilities of the feed to the sequence at 1 atm.

Component	Relative volatility	Relative volatility between adjacent components
Benzene	7.577	
		2.34
Toluene	3.245	
		2.07
Ethyl benzene	1.565	
		1.07
p-Xylene	1.467	
		1.04
m-Xylene	1.417	
		1.16
o-Xylene	1.220	
		1.22
C9's	1.000	

rank order of the distillation sequences on the basis of total vapor load calculated from the Underwood Equations.

Solution Although the relative volatilities are recalculated for each column, Table 11.7 shows the relative volatilities of the feed mixture to the sequence at a pressure of 1 atm. This shows clearly that the ethyl benzene/xylenes separation is by far the most difficult with relative volatilities for the xylenes close to unity. The volatilities of the components are such that all separations can be carried out at atmospheric pressure and at the same time allow the use of cooling water in the condensers. Thus, column pressures are fixed to atmospheric pressure with the relative volatilities recalculated for the feed composition at this pressure as the concentration changes through the sequence.

Table 11.8 Sequences for the separation of the mixture of aromatics in Example 11.3.

Rank order	Vapor rate (kmol·h^{-1})	% Best	Sequence			
1	5707	100.0	*ABC/DE*	*A/BC*	*D/E*	*B/C*
2	5982	104.8	*ABC/DE*	*AB/C*	*D/E*	*A/B*
3	6074	106.4	*A/BCDE*	*BC/DE*	*B/C*	*D/E*
4	6441	112.9	*ABCD/E*	*ABC/D*	*A/BC*	*B/C*
5	6511	114.1	*A/BCDE*	*B/CDE*	*C/DE*	*D/E*
6	6537	114.5	*A/BCDE*	*BCD/E*	*BC/D*	*B/C*
7	6588	115.4	*A/BCDE*	*B/CDE*	*CD/E*	*C/D*
8	6649	116.5	*AB/CDE*	*A/B*	*C/DE*	*D/E*
9	6715	117.7	*ABCD/E*	*ABC/D*	*AB/C*	*A/B*
10	6726	117.9	*AB/CDE*	*A/B*	*CD/E*	*C/D*
11	6782	118.8	*ABCD/E*	*A/BCD*	*BC/D*	*B/C*
12	6944	121.7	*A/BCDE*	*BCD/E*	*B/CD*	*C/D*
13	7190	126.0	*ABCD/E*	*A/BCD*	*B/CD*	*C/D*
14	7344	128.7	*ABCD/E*	*AB/CD*	*A/B*	*C/D*

Table 11.8 gives the total vapor flow for different sequences in rank order.

Again it can be noted from Table 11.8 that there is little difference between the best sequences in terms of the overall vapor load. The three sequences with the lowest overall vapor flows are shown in Figure 11.5. At first sight, the structure of the best sequences seems to be surprising. In each case, the most difficult separation (C/D) is carried out in the presence of other components. Heuristic 1 would suggest that such a difficult separation should be carried out in isolation from other components. Further investigation reveals that the relative volatilities for this most difficult separation are sensitive to the presence of other components. The presence of benzene and toluene improves the relative volatility of the C/D separation slightly, making it beneficial to carry out the difficult separation in the presence of nonkey components. This problem illustrates some of the dangers faced when dealing with separations that are very close in relative volatility. Special care should be taken in such problems to make sure that the vapor–liquid equilibrium data are specified to be as accurate as possible. The assumption of zero for the interaction parameters is questionable. In fact, the inclusion of interaction parameters in this case does not change the order of the best few sequences. It does, however, change the absolute values

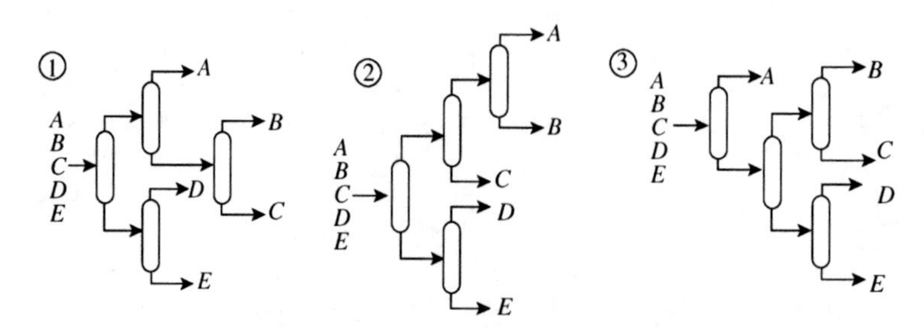

Figure 11.5 The best sequences in terms of vapor load for the separation of the mixture of aromatics from Example 11.3.

of the calculated vapor loads. The other questionable assumption is that of complete recovery for the products. Again, changing the assumed recovery to be less than complete does not change the order of the best few sequences in this case, but does change the value of the total vapor load. Finally, the assumption at this stage is that utilities will be used to satisfy all of the heating and cooling requirements. This leads to all columns operating at atmospheric pressure. Later it will be seen that it is beneficial to operate some of the columns at higher pressures to allow heat recovery to take place between some of the reboilers and condensers in the sequences.

The errors associated with the Underwood Equations were discussed in Chapter 9, which tend to underpredict the minimum reflux ratio. This introduces uncertainty in the way that the calculations were carried out in Examples 11.2 and 11.3. The differences in the total vapor load between different sequences are small and these differences are smaller than the errors associated with the prediction of minimum reflux ratio and minimum vapor load using the Underwood Equations. However, as long as the errors are consistently low for all of the distillation calculations, the vapor load from the Underwood Equations can still be used to screen between options. Nevertheless, the predictions should be used with caution and options not ruled out because of some small difference in the total vapor load.

The use of total vapor flow, even without any calculation errors, is still only a guide and might not give the correct rank order. The calculation could be taken one step further to calculate the energy consumption or even energy costs. In fact, given some computational aids, it is straightforward to size and cost all of the possible sequences using a shortcut sizing calculation, such as the Fenske–Gilliland–Underwood approach discussed in Chapter 9, together with cost correlations.

Whatever the method used to screen possible sequences, it is important not to give exclusive attention to the one sequence that appears to have the lowest vapor load, lowest energy consumption or lowest total cost. There is often little to choose in this respect between the best few sequences, particularly when the number of possible sequences is large. Other considerations such as heat integration, operability, safety, and so on, might also have an important bearing on the final decision. Thus, the screening of sequences should focus on the best few sequences rather than exclusively on the single best sequence.

11.4 DISTILLATION SEQUENCING USING COLUMNS WITH MORE THAN TWO PRODUCTS

When separating a three-product mixture using simple columns, there are only two possible sequences, Figure 11.1. Consider the first characteristic of simple

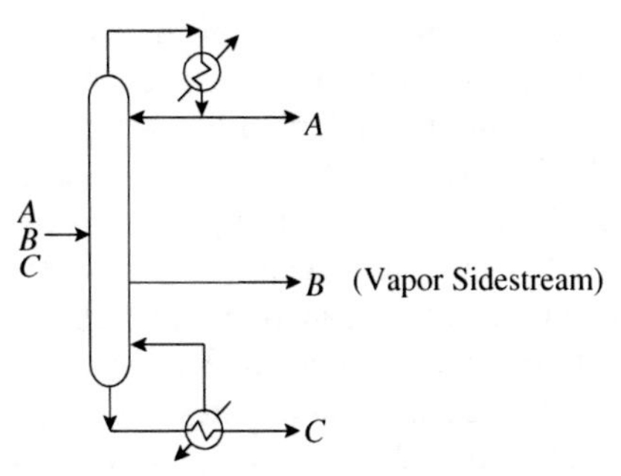

(a) More than 50% middle component and less than 5% heaviest component.

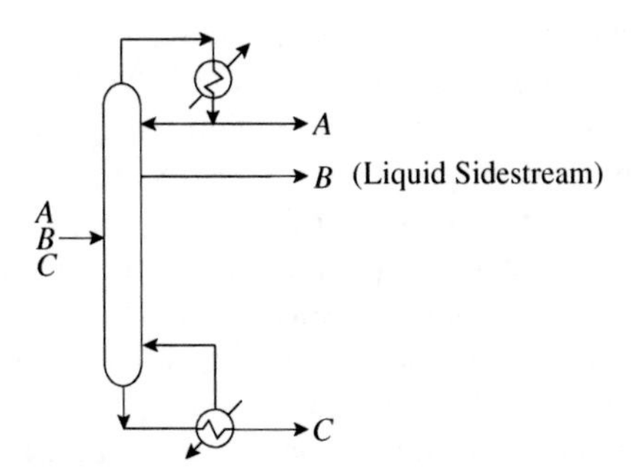

(b) More than 50% middle component and less than 5% lightest component.

Figure 11.6 Distillation columns with three products (From Smith R and Linnhoff B, 1988, *Trans IChemE ChERD*, **66**, 195 reproduced by permission of the Institution of Chemical Engineers.)

columns. A single feed is split into two products. As a first alternative to two simple columns, the possibilities shown in Figure 11.6 can be considered. Here three products are taken from one column. The designs are in fact both feasible and cost-effective when compared to simple arrangements on a stand-alone basis (i.e. reboilers and condensers operating on utilities) for certain ranges of conditions. If the feed is dominated by the middle product (typically more than 50% of the feed) and the heaviest product is present in small quantities (typically less than 5%), then the arrangements shown in Figure 11.6a can be an attractive option[3]. The heavy product must find its way down the column past the sidestream. Unless the heavy product has a small flow and the middle product a high flow, a reasonably pure middle product cannot be achieved. In these circumstances, the sidestream is usually taken as a vapor product to obtain a reasonably pure sidestream. A large relative volatility between the sidestream Product *B* and the bottom Product *C* is also necessary to obtain a high purity sidestream. A practical difficulty with this arrangement should, however, be noted. Whilst it is straightforward to split a liquid flow in a column, it is far less straightforward to split a vapor flow as shown. The flow of vapor up the column must somehow be restricted to allow a split to be made between the vapor that must continue up the column and the vapor sidestream. This can be achieved by the use of a baffle across the column diameter, or vapor tunnels in the tray downcomers. However, taking a vapor sidestream is more problematic than taking a liquid sidestream.

If the feed is dominated by the middle product (typically more than 50%) and the lightest product is present in small quantities (typically less than 5%), then the arrangement shown in Figure 11.6b can be an attractive option[3]. This time the light product must find its way up the column past the sidestream. Again, unless the light product is a small flow and the middle product a high flow, a reasonably pure middle product cannot be achieved. This time the sidestream is taken as a liquid product to obtain a reasonably pure sidestream. A large relative volatility between the sidestream Product B and the overhead Product A is also necessary to obtain a high purity sidestream.

In summary, single-column sidestream arrangements can be attractive when the middle product is in excess and one of the other components is present in only minor quantities. Thus, the sidestream column only applies to special circumstances for the feed composition. More generally applicable arrangements are possible by relaxing the restriction that separations must be between adjacent key components.

Consider a three-product separation as in Figure 11.7a in which the lightest and heaviest components are chosen to be the key separation in the first column. Two further columns are required to produce pure products, Figure 11.7a. This arrangement is known as *distributed distillation* or *sloppy distillation*. The distillation sequence provides parallel flow paths for the separation of a product. At first sight, the arrangement in Figure 11.7a seems to be inefficient in the use of equipment in that it requires three columns instead of two, with the bottoms and overheads of the second and third columns both producing pure B. However, it can be a useful arrangement in some circumstances. In new design, the three columns can, in principle, all be operated at different pressures. Also, the distribution of the middle Product B between the second and third columns is an additional degree of freedom in the design. The additional freedom to vary the pressures and the distribution of the middle product gives significant extra freedom to vary the loads and levels at which the heat is added to or rejected from the distillation. This might mean that the reboilers and condensers can be matched more cost-effectively against utilities, or heat integrated more effectively.

If the second and third columns in Figure 11.7a are operated at the same pressure, then the second and third columns could simply be connected and the middle product taken as a sidestream as shown in Figure 11.7b. The arrangement in Figure 11.7b is known as a *prefractionator* arrangement. Note that the first column in Figure 11.7b, the prefractionator, has a partial condenser to reduce the overall energy consumption.

Comparing the distributed (sloppy) distillation in Figure 11.7a and the prefractionator arrangement in Figure 11.7b with the conventional arrangements in Figure 11.1, the distributed and prefractionator arrangements typically require 20 to 30% less energy than conventional arrangements for the same separation duty. The reason for this difference is rooted in the fact that the distributed distillation and prefractionator arrangements are fundamentally thermodynamically more efficient than a simple arrangement. Consider why this is the case.

Consider the sequence of simple columns shown in Figure 11.8. In the direct sequence shown in Figure 11.8, the composition of Component B in the first column increases below the feed as the more volatile Component A decreases. However, moving further down the column, the composition of Component B decreases again as the composition of the less-volatile Component C increases.

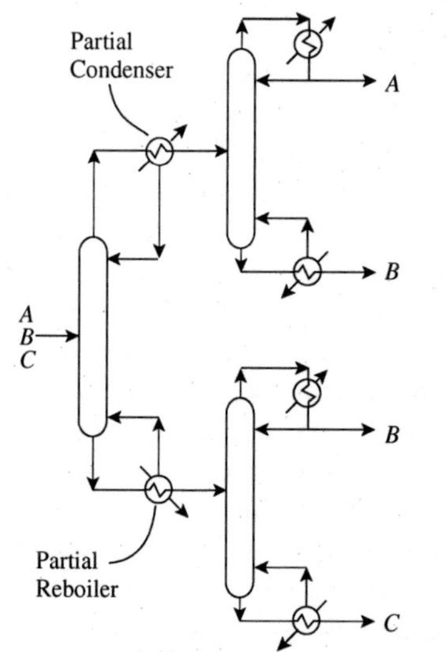

(a) Distributed distillation or sloppy distillation separates nonadjacent key components.

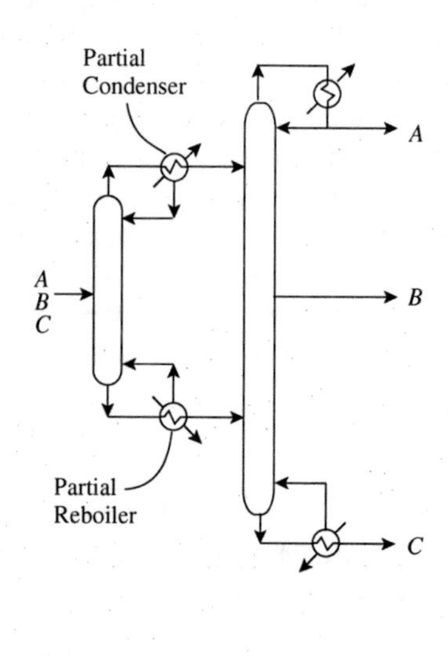

(b) Prefractionator arrangement.

Figure 11.7 Choosing nonadjacent keys leads to the prefractionator arrangement.

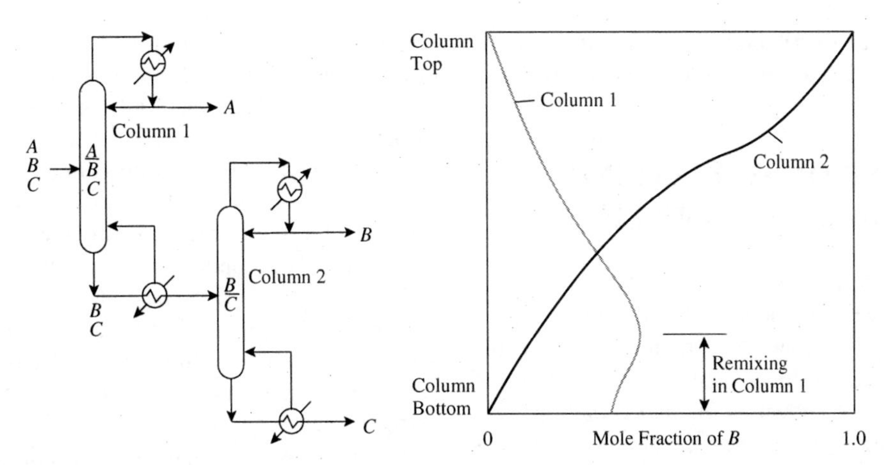

Figure 11.8 Composition profiles for the middle product in the columns of the direct sequence show remixing effects (From Triantafyllou C and Smith R, 1992, *Trans IChemE*, **70A**: 118, reproduced by permission of the Institution of Chemical Engineers.)

Thus the composition of B reaches a peak only to be remixed[4].

Similarly, with the first column in the indirect sequence, the composition of Component B first increases above the feed as the less-volatile Component C decreases. It reaches a maximum only to decrease as the more volatile Component A increases. Again, the composition of Component B reaches a peak only to be remixed.

This remixing that occurs in both sequences of simple distillation columns is a source of inefficiency in the separation. By contrast, consider the prefractionator arrangement shown in Figure 11.9. In the prefractionator, a crude split is performed so that Component B is distributed between the top and bottom of the column. The upper section of the prefractionator separates AB from C, whilst the lower section separates BC from A. Thus, both sections remove only one component from the product of that column section and this is also true for all four sections of the main column. In this way, the remixing effects that are a feature of both simple column sequences are avoided[4].

In addition, another feature of the prefractionator arrangement is important in reducing mixing effects. Losses occur in distillation operations due to mismatches between the composition of the column feed and the composition at the feed stage. In theory, for binary distillation in a simple column, a good match can be found between the feed composition and the feed stage. However, because the changes from stage to stage are finite, an exact match is not always possible. For multicomponent distillation in a simple column, except under extreme circumstances, it is not possible to match the feed composition and the feed stage. In a prefractionator arrangement, because the prefractionator distributes Component B between top and bottom, this allows greater freedom to match the feed composition with one of the trays in the column to reduce mixing losses at the feed tray.

The elimination of mixing losses in the prefractionator arrangement means that it is inherently more efficient than an arrangement using simple columns. The same basic arguments apply to both distributed distillation and prefractionator arrangements, with the additional degree of

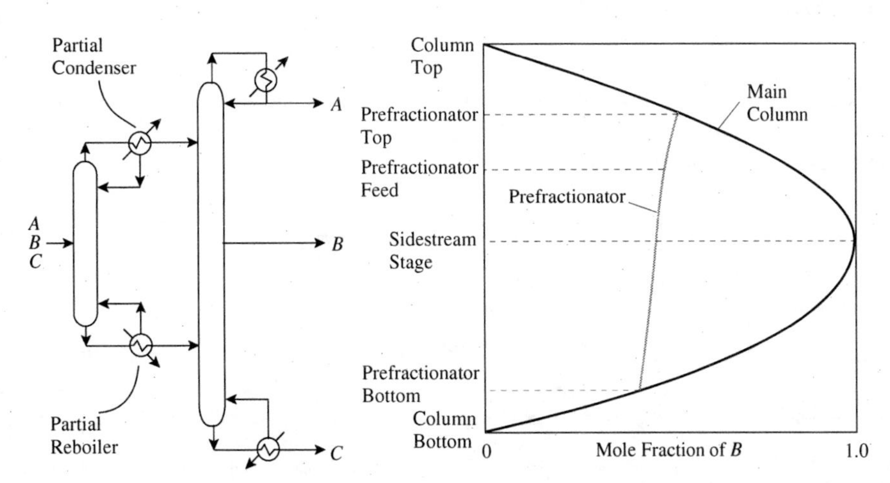

Figure 11.9 Composition profiles for the middle product in the prefractionator arrangement show that there are no remixing effects (From Triantafyllou C and Smith R, 1992, *Trans IChemE*, **70A**: 118, reproduced by permission of the Institution of Chemical Engineers.)

freedom in the case of distributed distillation to vary the pressures of the second and third columns independently.

11.5 DISTILLATION SEQUENCING USING THERMAL COUPLING

The final restriction of simple columns stated earlier was that they should have a reboiler and a condenser. It is possible to use material flows to provide some of the necessary heat transfer by direct contact. This transfer of heat via direct contact is known as *thermal coupling*.

First consider thermal coupling of the simple sequences from Figure 11.1. Figure 11.10b shows a thermally coupled direct sequence. The reboiler of the first column is replaced by a thermal coupling. Liquid from the bottom of the first column is transferred to the second as before, but now the vapor required by the first column is supplied by the second column, instead of a reboiler on the first column. The four column sections are marked as 1, 2, 3 and 4 in Figure 11.10a. In Figure 11.10c, the four column sections from Figure 11.10b are rearranged to form a *side-rectifier* arrangement[5]. There is a practical difficulty in engineering a side-rectifier arrangement. This is the difficulty in taking a vapor sidestream from the main column. As already noted for vapor sidestream columns, whilst it is straightforward to split a liquid flow in a column, it is not straightforward to split a vapor flow as required in Figure 11.10c. This problem can be avoided by constructing the side-rectifier in a single shell with a *partition wall* as shown in Figure 11.10d. The partition wall in Figure 11.10d should be insulated to avoid heat transfer across the wall as different separations are carried out on each side of the wall and the temperatures on each side will differ. Heat transfer across the wall will have an overall detrimental effect on column performance[6].

Similarly, Figure 11.11a shows an indirect sequence and Figure 11.11b the corresponding thermally coupled indirect sequence. The condenser of the first column is replaced by thermal coupling. The four column sections are again marked as 1, 2, 3 and 4 in Figure 11.11b. In Figure 11.11c, the four column sections are rearranged to form a *side-stripper* arrangement[5]. As with the side-rectifier, the side-stripper can be arranged in a single shell with a partition wall, as shown in Figure 11.11d. Again the partition wall should be insulated, otherwise heat transfer across the wall will have an overall detrimental effect on the separation[6].

Both the side-rectifier and side-stripper arrangements have been shown to reduce the energy consumption compared to simple two-column arrangements[3,7]. This results from reduced mixing losses in the first (main) column. As with the first column of the simple sequence, a peak in composition occurs with the middle product. But now, advantage of the peak is taken by transferring material to the side-rectifier or side-stripper.

The side-rectifier and side-stripper arrangements have some important degrees of freedom for optimization. In these arrangements, there are four column sections. For the side-rectifier, the degrees of freedom to be optimized are:

- number of stages in each of the four column sections;
- reflux ratios (ratio of liquid reflux to overhead product flowrates) in the main column and sidestream column;

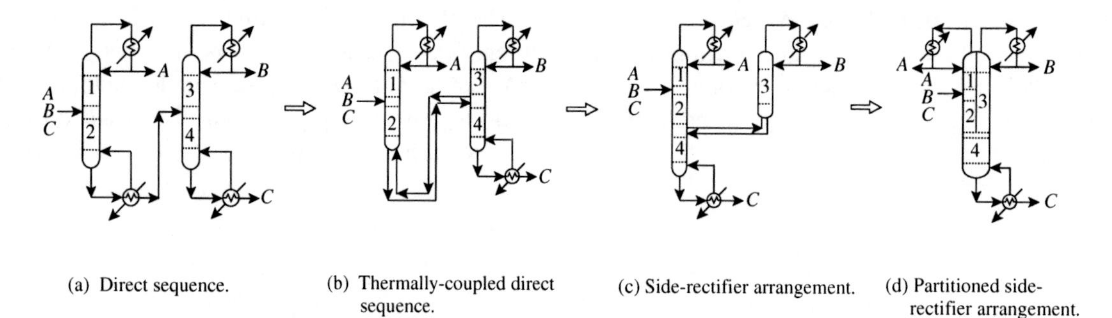

(a) Direct sequence. (b) Thermally-coupled direct sequence. (c) Side-rectifier arrangement. (d) Partitioned side-rectifier arrangement.

Figure 11.10 Thermal coupling of the direct sequence.

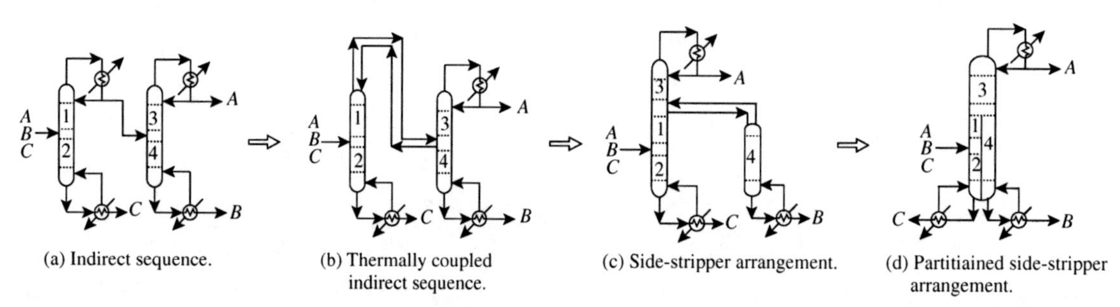

(a) Indirect sequence. (b) Thermally coupled indirect sequence. (c) Side-stripper arrangement. (d) Partitiained side-stripper arrangement.

Figure 11.11 Thermal coupling of the indirect sequence.

- vapor split between the main column and sidestream column;
- feed condition.

For the side-stripper, the degrees of freedom to be optimized are:

- number of stages in each of the four column sections;
- reboil ratios (ratio of stripping vapor to bottom product flowrates) in the main column and sidestream column;
- liquid split between the main column and sidestream column;
- feed condition.

All of these variables must be optimized simultaneously to obtain the best design. Some of the variables are continuous and some are discrete (the number of stages in each column section). Such optimizations are far from straightforward if carried out using detailed simulation. It is therefore convenient to carry out some optimization using shortcut methods before proceeding to detailed simulation where the optimization can be fine-tuned.

A simple model for side-rectifiers suitable for shortcut calculation is shown in Figure 11.12. The side-rectifier can be modeled as two columns in the thermally coupled direct sequence. The first column is a conventional column with a condenser and partial reboiler. The second column is modeled as a sidestream column, with a vapor sidestream one stage below the feed stage[4]. The liquid entering the reboiler and vapor leaving can be calculated from vapor–liquid equilibrium (see Chapter 4). The vapor and liquid streams at the bottom of the first column can then be matched with the feed and sidestream of the second column to allow the calculations for the second column to be carried out.

The side-stripper can be modeled as two columns in the thermally coupled indirect sequence, as shown in Figure 11.13. The first column is modeled as a conventional column with a partial condenser and partial reboiler. The second column is modeled as a sidestream column with a liquid sidestream on stage above the feed stage[4]. The vapor entering the condenser and liquid leaving can be calculated from vapor–liquid equilibrium (see Chapter 4). This allows

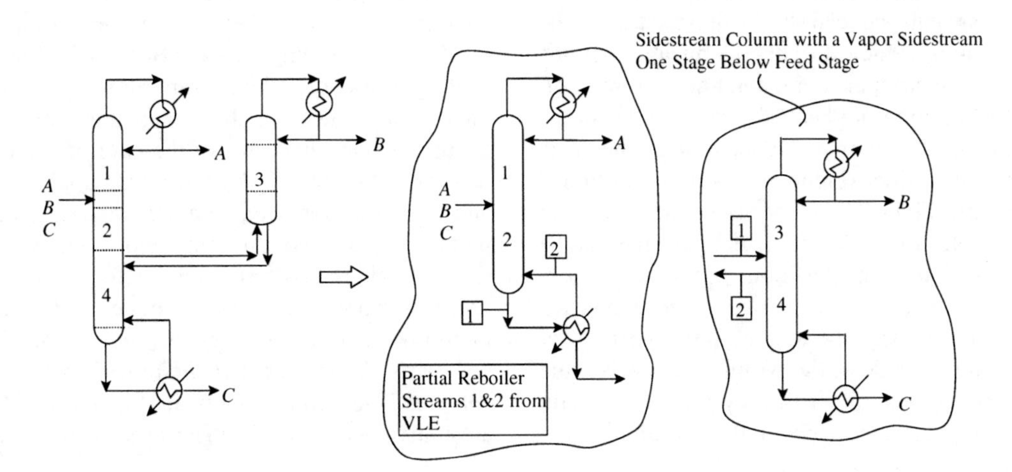

Figure 11.12 A side-rectifier can be modeled as a sequence of two simple columns in the direct sequence.

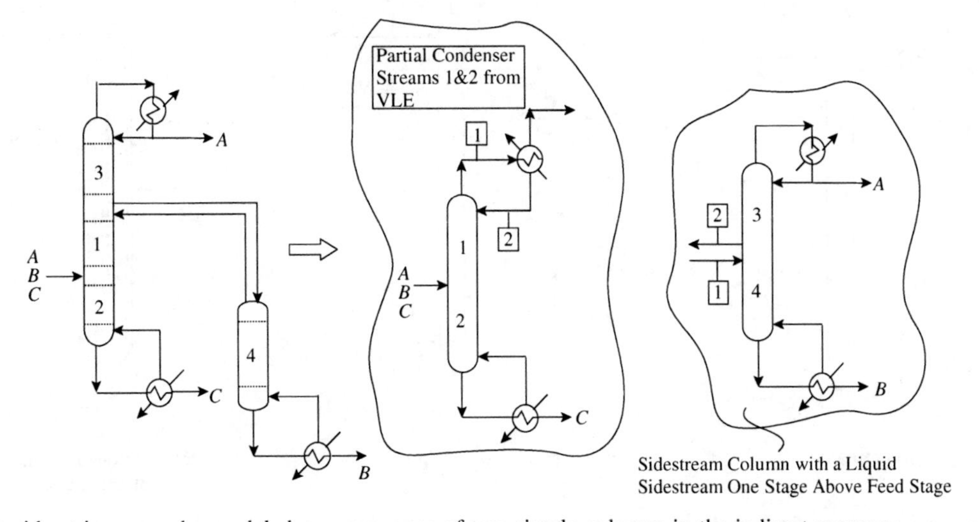

Figure 11.13 A side-stripper can be modeled as a sequence of two simple columns in the indirect sequence.

the two columns to be linked and the calculations for the second column to be carried out.

In both the cases of the side-rectifier and side-stripper, the first column in the two-column model can be modeled using the Fenske–Underwood–Gilliland Equations, as described in Chapter 9. The second column is modeled as a sidestream column that can also be modeled using the Fenske–Underwood–Gilliland Equations[4]. The minimum reflux ratio for a liquid sidestream column one stage above the feed stage can be estimated using the Underwood Equations by combining the sidestream and overhead product as a net overhead product[1]. The minimum reflux ratio for a vapor sidestream column one stage below the feed stage can also be estimated using the Underwood Equations but this time by combining the sidestream and bottoms product as a net bottoms product[1].

The optimization can be carried out using nonlinear optimization techniques such as SQP (see Chapter 3). The nonlinear optimization has the problems of local optima if techniques such as SQP are used for the optimization. Constraints need to be added to the optimization in order that a mass balance can be maintained and the product specifications achieved. The optimization of the side-rectifier and side-stripper in a capital-energy trade-off determines the distribution of plates, the reflux ratios in the main and sidestream columns and condition of the feed. If a partitioned side-rectifier (Figure 11.10d) or partitioned side-stripper (Figure 11.11d) is to be used, then the ratio of the vapor flowrates on each side of the partition can be used to fix the location of the partition across the column. The partition is located such that the ratio of areas on each side of the partition is the same as the optimized ratio of vapor flowrates on each side of the partition. However, the vapor split for the side-rectifier will only follow this ratio if the pressure drop on each side of the partition is the same. Rather than locate the partition in this way, some deterioration in the design performance might be accepted by locating the partition across the diameter of the column (i.e. equal areas on each side of the partition) for mechanical simplicity. The sensitivity of the design performance to the location of the partition should be explored before the location is finalized.

Now consider thermal coupling of the prefractionator arrangement from Figure 11.7b. Figure 11.14a shows a prefractionator arrangement with partial condenser and reboiler on the prefractionator. Figure 11.14b shows the equivalent thermally coupled prefractionator arrangement, sometimes known as the Petlyuk column. To make the two arrangements in Figures 11.14a and 14b equivalent, the thermally coupled prefractionator requires extra plates to substitute for the prefractionator condenser and reboiler[8]. The prefractionator arrangement in Figure 11.14a and the thermally coupled prefractionator (Petlyuk Column) in Figure 11.14b are almost the same in terms of total heating and cooling duties[8]. There are differences between the vapor and liquid flows in the top and bottom sections of the main column of the designs in Figures 11.14a and 11.14b, resulting from the presence of the partial reboiler and condenser in Figure 11.14a. However, although the total heating and cooling duties are almost the same, there are greater differences in the temperatures at which the heat is supplied and rejected. In the case of the prefractionator in Figure 11.14a, the heat load is supplied at two points and two different temperatures and rejected from two points and at two different temperatures. Figure 11.14c shows an alternative configuration for the thermally coupled prefractionator that uses a single shell with a vertical partition dividing the central section of the shell into two parts, known as a *dividing-wall column* or *partition column*. The arrangements in Figures 11.14b and 11.14c are equivalent if there is no heat transfer across the partition.

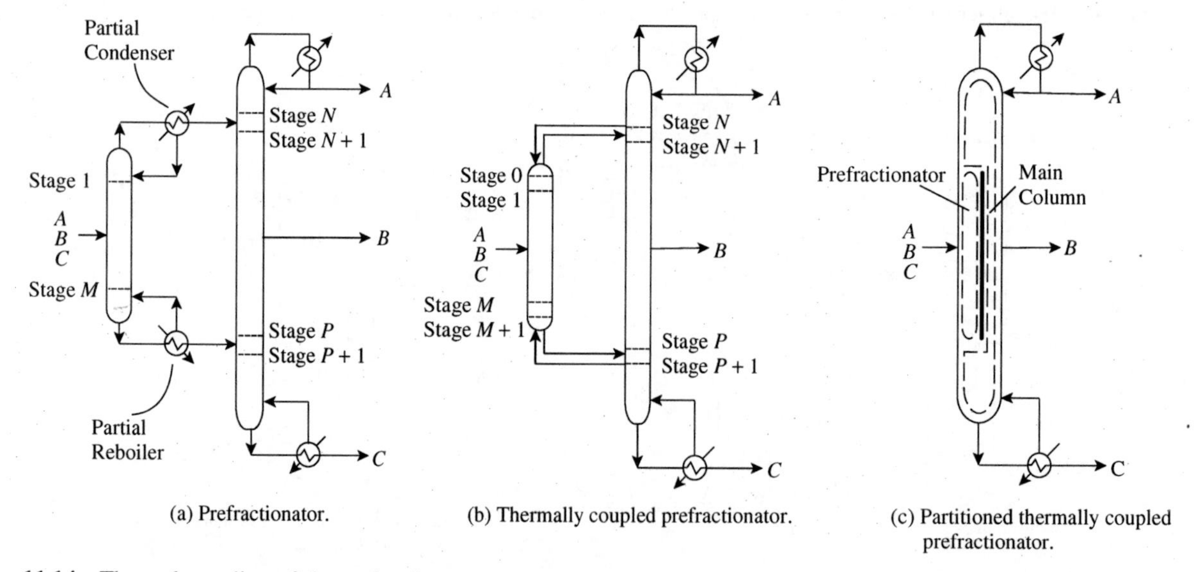

Figure 11.14 Thermal coupling of the prefractionator arrangement.

As with side-rectifiers and side-strippers, the partition wall should be insulated to avoid heat transfer across the wall as different separations are carried out on each side of the wall and the temperatures on each side will differ. Heat transfer across the wall will have an overall detrimental effect on column performance[6].

Partition columns offer a number of advantages over conventional arrangements:

1. Various studies[7,9–16] have compared the thermally coupled arrangement in Figures 11.14b and 11.14c, with a conventional arrangement using simple columns on a stand-alone basis. These studies show that the prefractionator arrangement in Figure 11.14 requires typically 20 to 30% less energy than the best conventional arrangement using simple columns (Figure 11.1). The prefractionator column also requires less energy than the side-rectifier and side-stripper arrangements, for the same separation[7]. The energy saving for the thermally coupled arrangement is the same as that for the prefractionator, with reduced mixing losses as illustrated in Figures 11.8 and 11.9.
2. In addition, the partition column in Figure 11.14c requires typically 20 to 30% less capital cost than a two-column arrangement of simple columns.
3. The partition column has one further advantage over the conventional arrangements in Figure 11.1. In partitioned columns, the material is only reboiled once and its residence time in the high-temperature zones is minimized. This can be important if distilling heat-sensitive materials.

Standard distillation equipment can be used for the fabrication. Either packing or plates can be used, although packing is more commonly used in practice. Also, the control of partitioned columns is straightforward[13,14].

The partition column does have a number of disadvantages relative to simple column arrangements:

1. Even though the arrangement might require less energy than a conventional arrangement, all of the heat must be supplied at the highest temperature and all of the heat rejected at the lowest temperature of the separation. This can be particularly important if the distillation is at low temperature using refrigeration for condensation. In such circumstances, minimizing the amount of condensation at the lowest temperatures can be very important. If differences in the temperature at which the heat is supplied or rejected are particularly important, then distributed distillation or a prefractionator might be better options. These offer the advantage of being able to supply and reject heat at different temperatures. This issue will be addressed again later when dealing with heat integration of distillation.

2. If it is appropriate for the two columns in a simple column arrangement to operate at very different pressures, then this can pose problems for a partition column to replace them, as the partition column must carry out the whole separation at effectively the same pressure (usually the highest pressure). In general, partition columns are not suited to replace sequences of two simple columns that operate at very different pressures.

3. Another disadvantage of partitioned columns might arise from materials of construction. If the two columns of a conventional arrangement require two different materials of construction, one being much more expensive than the other, then any savings in capital cost arising from using a partitioned column will be diminished. This is because the whole partitioned column will have to be fabricated from the more expensive material.

4. The hydraulic design of the partitioned column is such that the pressure must be balanced on either side of the partition. This is usually achieved by designing with the same number of stages on each side of the partition. This constrains the design. Alternatively, a different number of stages can be used on each side of the partition, and different column internals with different pressure drop per stage used to balance the pressure drop. Also, if foaming is more likely to occur on one side of the partition than the other, this can lead to a hydraulic imbalance and the vapor split changing from design conditions.

Although side-stripper arrangements have been routinely used in the petroleum industry and side-rectifiers in air separation, designers have been reluctant to use the fully thermally coupled arrangements in practical applications until recently[12,15,16].

The partitioned thermally coupled prefractionator in Figure 11.14c can be simulated using the arrangement in Figure 11.14b as the basis of the simulation. However, like side-rectifiers and side-strippers, fully thermally coupled columns have some important degrees of freedom for optimization. In the fully thermally coupled column, there are six column sections (above and below the partition, above and below the feed in the prefractionator and above and below the sidestream from the main column side of the partition). The degrees of freedom to be optimized in partitioned columns are:

- number of stages in these six column sections;
- reflux ratio;
- liquid split on each side of the partition flowing down the column;
- vapor split on each side of the partition flowing up the column;
- feed condition.

These basic degrees of freedom can be represented in different ways, but all must be optimized simultaneously to obtain the best design. Again some of the variables

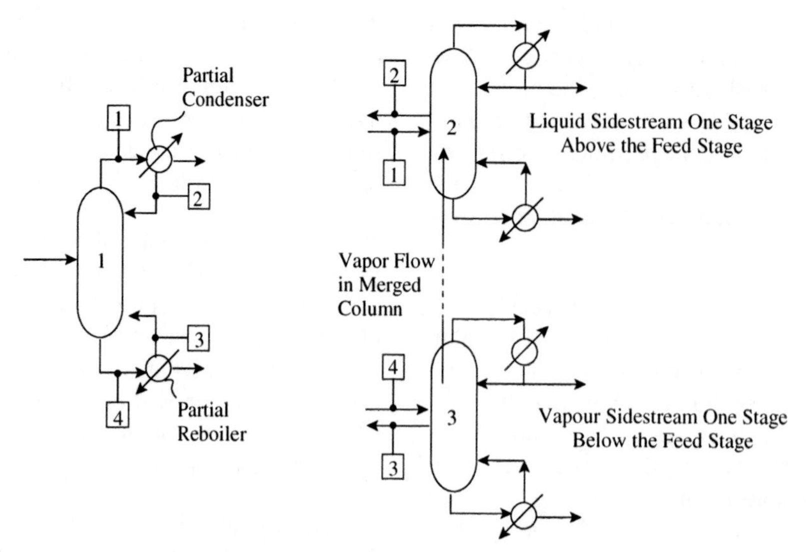

Figure 11.15 Three-column model of partitioned thermally coupled prefractionator.

are continuous and some are discrete (the number of stages in each column section). It is again convenient to carry out some optimization using shortcut methods before proceeding to detailed simulation and fine-tuning of the optimization. Like side-rectifiers and side-strippers, the fully thermally coupled column can be represented as simple columns. This time three simple columns are needed to represent the fully thermally coupled column, as shown in Figure 11.15[4]. The prefractionator is modeled as Column 1 in Figure 11.15, which has a partial condenser and partial reboiler initially. Column 2 in Figure 11.15 is the top section of the main column and is modeled as a liquid sidestream column with a liquid sidestream one stage above the vapor feed. A vapor–liquid equilibrium calculation around the partial condenser in Column 1 allows the vapor entering the condenser and liquid leaving the condenser to be determined. These can then be connected to the sidestream and feed for Column 2. Column 3 in Figure 11.15 is the lower part of the main column and can be modeled as a vapor sidestream column with the sidestream one stage below the liquid feed. A vapor–liquid equilibrium calculation around the partial reboiler in the prefractionator allows the vapor leaving the reboiler and liquid entering the reboiler to be determined and can then be connected with the sidestream and feed to Column 3. The three columns can then be represented by the shortcut calculations described above for side-rectifiers and side-strippers on the basis of the Fenske–Gilliland–Underwood correlations. The optimization can again be carried out using nonlinear optimization techniques such as SQP (see Chapter 3). Constraints need to be added to the optimization in order that it can fulfill the required objectives. As illustrated in Figure 11.15, Columns 2 and 3 together represent the main column, and it is necessary to ensure that the vapor flowrate in Columns 2 and 3 are the same. Also, the bottom product from Column 2 and top product from Column 3 represent the same sidestream from the main column of the thermally coupled configuration. Thus, the bottom product of Column 2 and top product of Column 3 must be constrained to be the same. Additional constraints must also be added to ensure that the mass balance is maintained and product specifications are achieved.

11.6 RETROFIT OF DISTILLATION SEQUENCES

It might be the case that an existing plant needs to be modified, rather than a new design carried out. For example, a retrofit study might require the capacity of the unit to be increased. When such a revamp is carried out, it is essential to try and make as effective use as possible of existing equipment. For example, consider a simple two-column sequence like the ones illustrated in Figure 11.1. If the capacity needs to be increased, then, instead of replacing the two existing columns with new ones, a new column might be added to the two existing ones and reconfigured to the distributed distillation arrangement in Figure 11.7a. The vapor and liquid loads for the increased capacity are now spread between three columns, instead of two, taking advantage of the existing two-column shells. Of course, the design of the existing shells might not be ideally suited to the new role to which they will be put, but at least it will allow existing equipment to carry on being used and avoid unnecessary investment in new equipment. Of the three shells shown in Figure 11.7a, any of the three could be the new shell. The arrangement of new shell and existing shells will depend on how the existing shells can best be adapted to the new role.

If the existing shells differ significantly from what is required in the retrofitted sequence, then, in addition to reconfiguring the distillation sequence, the existing shells can also be modified. For example, the number of theoretical stages can be increased in the columns by changing the design of the column internals. But, generally, the fewer the number of modifications the better.

Rather than retrofit a two-column sequence to the distributed distillation arrangement in Figure 11.7a, it can be retrofitted to the prefractionator arrangement in Figure 11.7b. This time, rather than have two-column shells, as shown in Figure 11.7b, three columns are used with the second and third connected directly by vapor and liquid flows. Thus, there is no reboiler in the second column and no condenser in the third. The liquid flow between the two shells that function as the main column is split to provide the reflux for the lower part of the main column and the middle product. Two-column shells are thus used to mimic the main column of the prefractionator. As with the distributed distillation retrofit arrangement, which of the two existing shells and the new shell are used in which role depends on the details of the design of the existing shells and how they can best be fitted to the new role. As with the distributed distillation retrofit arrangement, the existing shells can also be modified.

The retrofit of more complex distillation sequences can also be considered. Within a larger, more complex sequence, any pair of columns that are together in a sequence can be considered as candidates for the same retrofit modifications as those discussed for two-column retrofit.

11.7 CRUDE OIL DISTILLATION

One particularly important case of distillation sequencing is worthy of special consideration. This is the case of crude oil distillation, which is the fundamental process underlying the petroleum and petrochemicals industry. Crude oil is an extremely complex mixture of hydrocarbons containing small quantities of sulfur, oxygen, nitrogen and metals. A typical crude oil contains many millions of compounds, most of which cannot be identified. Only the lightest compounds for example, methane, ethane, propane, benzene, and so on, can normally be identified.

In the first stage of processing crude oil, it is distilled under conditions slightly above atmospheric pressure (typically 1 barg where the feed enters the column). A range of products are taken from the crude oil distillation, based on their boiling temperatures. Designs are normally thermally coupled. Most configurations follow the thermally coupled indirect sequence as shown in Figure 11.16a. However, rather than build the configuration in Figure 11.16a, the configuration of Figure 11.16b is the one normally constructed. The two arrangements are equivalent and the corresponding column sections are shown in Figure 11.16. Unfortunately, a practical crude oil distillation cannot be operated in quite the way shown in Figure 11.16b. The first problem is that if high-temperature reboiling is used, as would be required for the higher boiling products in the lower part of the column in Figure 11.16b, extremely high-temperature sources of heat would be required. Steam is usually not distributed for process heating at such high temperatures. Also, high temperatures in the reboilers would result in significant fouling of the reboilers from decomposition of the hydrocarbons to form coke. The maximum temperature for using reboiling in petroleum refining is usually of the order of 300°C. Therefore, in practice, some or all of the reboiling is substituted by the direct injection of steam into the distillation. The steam has two functions:

- It provides some of the heat required for the distillation.
- It lowers the partial pressure of the boiling components, making them more volatile.

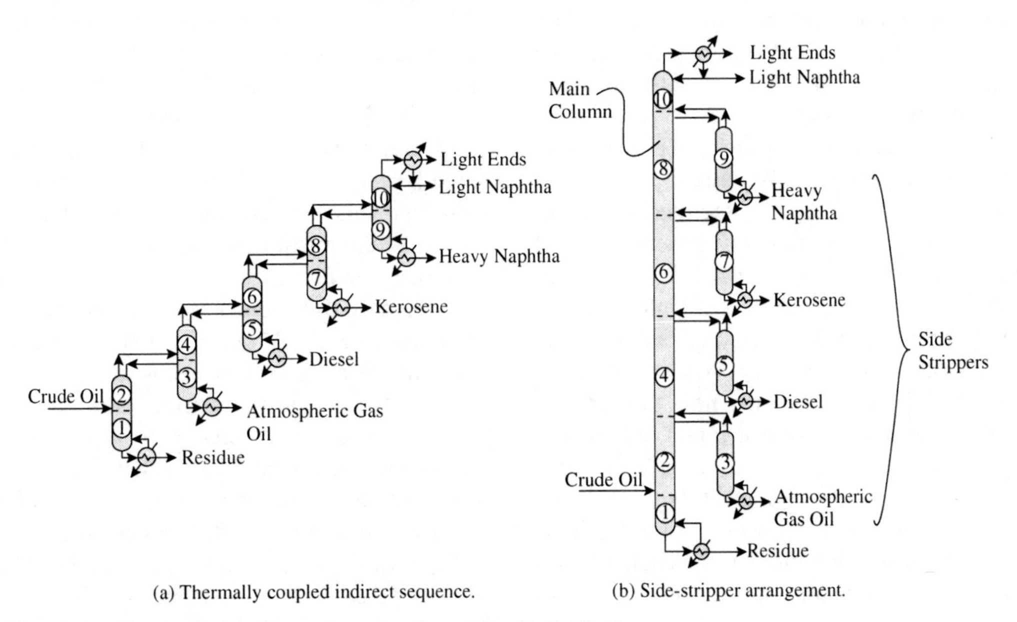

Figure 11.16 The thermally coupled indirect sequence for crude oil distillation.

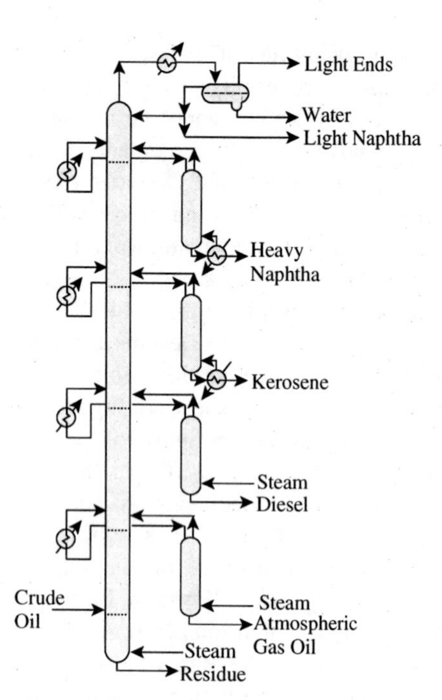

Figure 11.17 Substitute some (or all) of the reboiling with direct steam injection and introduce intermediate condensation.

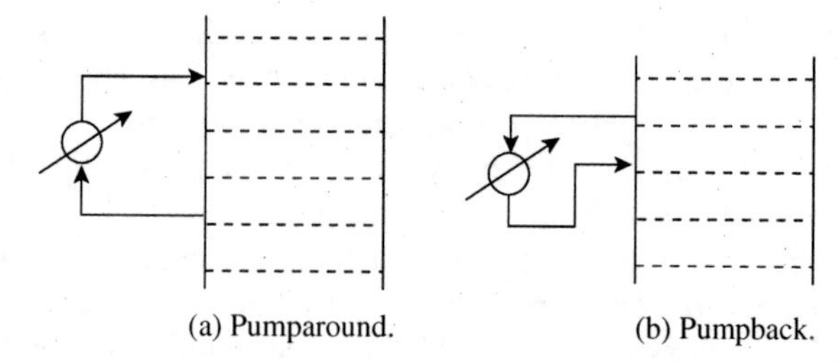

Figure 11.18 Partial condensation can be achieved by subcooling liquid for the column.

Because steam is injected into the operation in various places in Figure 11.17, the steam is condensed overhead and is separated in a decanter from the condensing hydrocarbons and the hydrocarbons that do not condense.

Another problem with the arrangement in Figure 11.16b is that as the vapor rises up the main column, its flowrate increases significantly. In Figure 11.17, heat is being removed from the main column at intermediate points. This corresponds with introducing some condensation of the vapor at the top of intermediate columns in the arrangement shown in Figure 11.16a. It should be noted that the introduction of some condensation between the columns does not necessarily eliminate all thermal coupling between the columns in Figure 11.16a, but can leave partial thermal coupling, rather than full thermal coupling. How much condensation and how much thermal coupling are used at each point are important degrees of freedom to be optimized.

To remove heat from the column at intermediate points as in Figure 11.17, it would be best to condense part of the vapor flowing up the column and return the condensed liquid to the column at saturated conditions. Unfortunately, as already noted for vapor sidestream and side-rectifier designs, it is difficult to take a vapor sidestream from a column. In crude oil distillation, the heat is therefore removed from the column by taking a liquid sidestream, subcooling the liquid and returning it to the column. This provides the condensation by direct contact, but introduces inefficiency into the design. Returning subcooled liquid to the column is inefficient in distillation, as subcooled liquid cannot take part in the distillation until it has returned to saturation conditions.

The heat is removed from the main column in one of two ways. These are shown in Figure 11.18. The first, in Figure 11.18a, is a *pumparound*. Liquid is taken from the column, subcooled and returned to the column at a higher point. Another arrangement shown in Figure 11.18b is a *pumpback*. This takes liquid from the column, subcools it, but this time returns it to a lower point in the column. The problem with the pumpback is that the flowrate that can be withdrawn must be less than the liquid flowing down the column. It is therefore restricted in its capacity to remove heat. The pumparound, on the other hand, is not restricted by the flow of liquid down the column and can recirculate as much liquid as required around a section of the column. By choosing the most appropriate flowrate and temperature for the pumparound, the heat load to be removed can be adjusted to whatever is desired. The trays between the liquid draw and return in a pumparound have more to do with heat transfer than mass transfer. In addition to returning a subcooled liquid to the column, mixing occurs as material is introduced to a higher point in the column.

One final point needs to be made about the arrangement shown in Figure 11.17. The crude oil entering the main column needs to be preheated. This is done initially by heat recovery to a temperature in the range 90 to 150°C, and the crude oil is extracted with water to remove salt. The desalting process mixes with water and then separates into two layers, the salt dissolving in the water. The desalted crude oil is then heated further by heat recovery to a temperature usually around 280°C and then by a furnace (fired heater) to around 400°C before entering the column. Note that this temperature is higher than that previously quoted as the maximum that can be supplied in a reboiler. However, the decomposition depends on both temperature and residence time, and a high temperature can be tolerated in the furnace if it is only for a short residence time.

All of the material that needs to leave as product above the feed point must vaporize as it enters the column. In addition to this, some extra vapor over and above this flowrate must be created that will be condensed and flow back down through the column as reflux. This extra vaporization to create reflux is known as *overflash*.

The majority of atmospheric crude oil distillation columns follow the configuration shown in Figure 11.17, which is basically the partially thermally coupled indirect sequence.

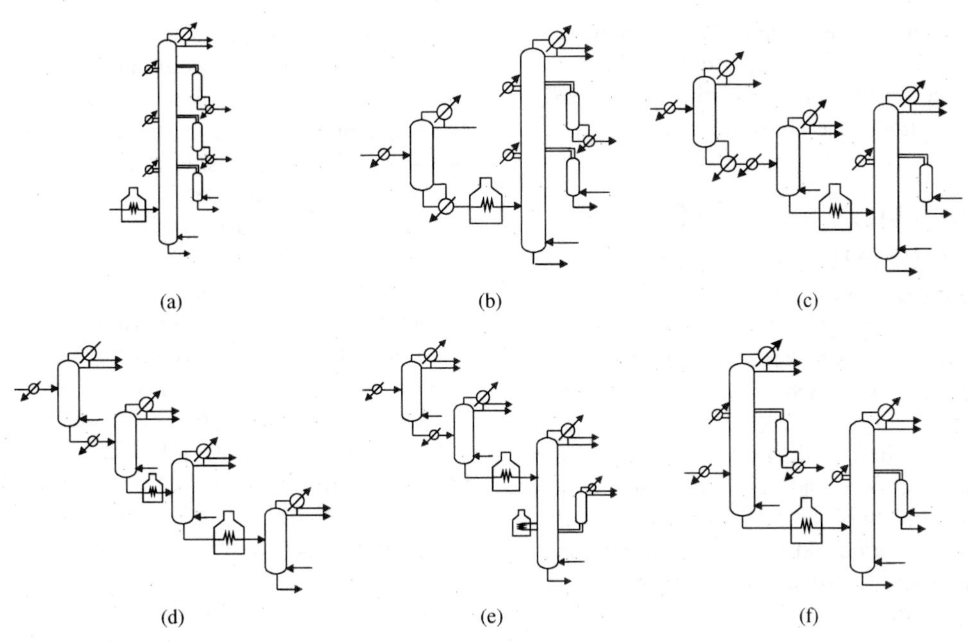

Figure 11.19 Some possible arrangements for crude oil distillation.

However, there are many other possibilities for different column arrangements that have the potential to reduce the energy consumption. Some of these are illustrated in Figure 11.19, but there are many other possibilities.

The distillation of crude oil under conditions slightly above atmospheric pressure is limited by the maximum temperature that can be tolerated by the materials being distilled, otherwise there would be decomposition. Further separation of the bottoms of the column (the *atmospheric residue*) would require higher temperature, and therefore bring about the decomposition of material. However, this leaves a significant amount of valuable material that could still be recovered from the atmospheric residue. Therefore, the residue from the atmospheric crude oil distillation is usually reheated to a temperature around 400°C or slightly higher and fed to a second column, the *vacuum column*, which operates under a high vacuum (very low pressure) to allow further recovery of material from the atmospheric residue, as shown in Figure 11.20. The pressure of the vacuum column is typically 0.006 bar at the top of the

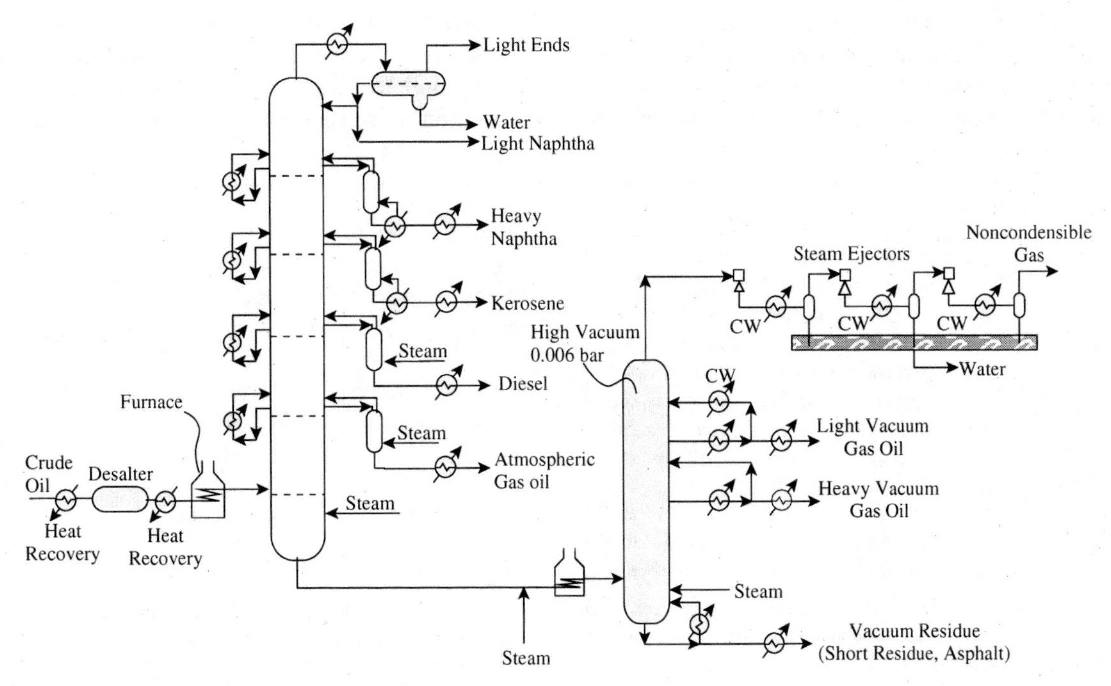

Figure 11.20 A typical complete crude oil distillation system.

column where the vacuum is created. The design of the vacuum column is simpler than the atmospheric column with typically two sidestream products being taken, leaving a heavy *vacuum residue* at the bottom of the column.

11.8 DISTILLATION SEQUENCING USING OPTIMIZATION OF A SUPERSTRUCTURE

Consider now ways in which the best arrangement of a distillation sequence can be determined more systematically. Given the possibilities for changing the sequence of simple columns or the introduction of prefractionators, side-strippers, side-rectifiers and fully thermally coupled arrangements, the problem is complex with many structural options. The problem can be addressed using the optimization of a superstructure. As discussed in Chapter 1, this approach starts by setting up a "grand" flowsheet in which all structural features for an optimal solution are embedded.

The creation of a superstructure for a distillation sequence and its optimization is, in principle, straightforward[17]. Unfortunately, unless care is taken with the problem formulation it becomes a difficult mixed integer nonlinear programming (MINLP) problem[17]. Such problems are to be avoided if at all possible. It is possible to avoid the formulation of an MINLP problem in the case of distillation sequencing by following a philosophy of building the superstructure from distillation *tasks*[18,19].

Figure 11.21a shows the different sequences for the separation of a five-product system. Four simple columns are needed for this separation and there are 14 possible sequences, as shown in Figure 11.21a. If an approach is to be followed that will allow screening and optimizing networks, then 14 possible networks each with 4 columns would have to be considered, involving 56 column sizing and costing calculations. An alternative way to analyze the system is to work in terms of tasks, as illustrated in Figure 11.21b[19]. A task could be to separate component *A* from *BCDE*, or to separate *A* from *BCD* or separate *A* from *BC*, and so on. There are only 20 basic tasks involved in the five-product separation problem. The 14 sequences can be produced by combining these tasks together in different ways, as illustrated in Figure 11.21. Thus, instead of carrying out 56 column sizing and costing calculations, only 20 need to be carried out and then combined in different ways to be able to evaluate all possible sequences of simple columns. In Figure 11.22, the direct indirect sequences are shown for illustration, but all sequences can be generated by the appropriate combination of tasks.

For complex columns, tasks can be combined together to produce *hybrid tasks*, as shown in Figure 11.23[19]. A hybrid task can then be evolved for different complex column arrangements, depending on whether the hybrid involves a direct or indirect sequence combination.

Before a task or hybrid task can be modeled, the material balance and the pressure of the operation must first be specified. To specify the material balance, a matrix of product recoveries can be specified as illustrated in Figure 11.24. This example involves six components and three products. Components are arranged in order of their volatility. It follows that components in a product must all be adjacent, and a component can only distribute between adjacent products. Also, the lightest component of Product $(i + 1)$ must be no lighter than the lightest component in Product i and the heaviest component of Product $(i + 1)$ must be at least as heavy as the heaviest component of Product i. In this way, the material balance for each of the tasks and hybrid tasks can be specified.

However, before sizing and costing calculations can be carried out, the pressure must be also specified. Pressure is a critically important optimization variable, and the appropriate pressure is not known until optimization calculations have been carried out. If the pressure is allowed

Sequences				
1	*A/BCDE*	*B/CDE*	*C/DE*	*D/E*
2	*A/BCDE*	*B/CDE*	*CD/E*	*C/D*
3	*A/BCDE*	*BC/DE*	*B/C*	*D/E*
4	*A/BCDE*	*BCD/E*	*BC/D*	*B/C*
5	*A/BCDE*	*BCD/E*	*BC/D*	*B/C*
6	*AB/CDE*	*C/DE*	*A/B*	*D/E*
7	*AB/CDE*	*CD/E*	*A/B*	*C/D*
8	*ABC/DE*	*A/BC*	*B/C*	*D/E*
9	*ABC/DE*	*AB/C*	*B/C*	*D/E*
10	*ABCD/E*	*A/BCD*	*B/CD*	*C/D*
11	*ABCD/E*	*A/BCD*	*BC/D*	*B/C*
12	*ABCD/E*	*AB/CD*	*A/B*	*C/D*
13	*ABCD/E*	*ABC/D*	*A/BC*	*B/C*
14	*ABCD/E*	*ABC/D*	*AB/C*	*A/B*

(a) 14 sequences of 4 simple columns each can separate a five product system.

Tasks			
A/BCDE	*A/BCDA*	*A/BC*	*A/B*
AB/CDE	*AB/CDA*	*AB/C*	*B/C*
ABC/DE	*ABC/DB*	*B/CD*	*C/D*
ABCD/E	*B/CDEB*	*BC/D*	*D/E*
	BC/DEC	*C/DE*	
	BCD/EC	*CD/E*	

(b) 20 different tasks are used in the 14 sequences.

Figure 11.21 A limited number of tasks are required to perform a given separation.

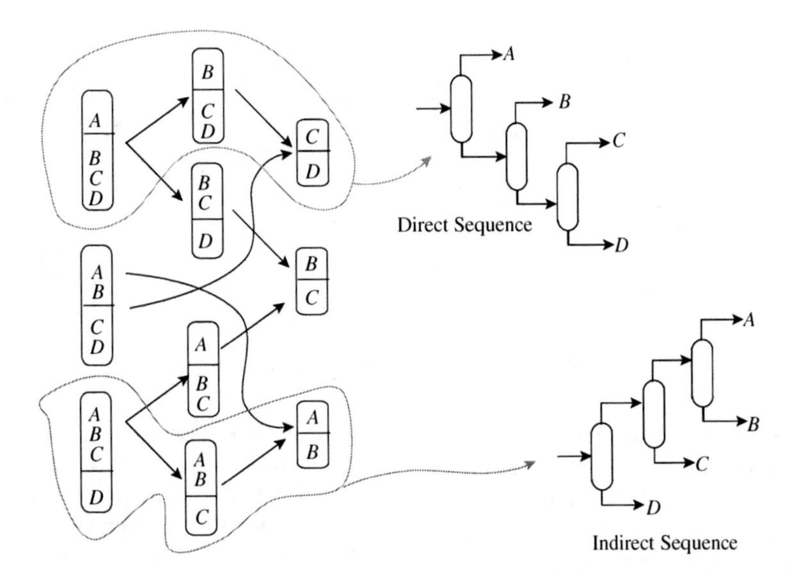

Figure 11.22 Superstructure for distillation sequence for a four-product mixture based on task representation.

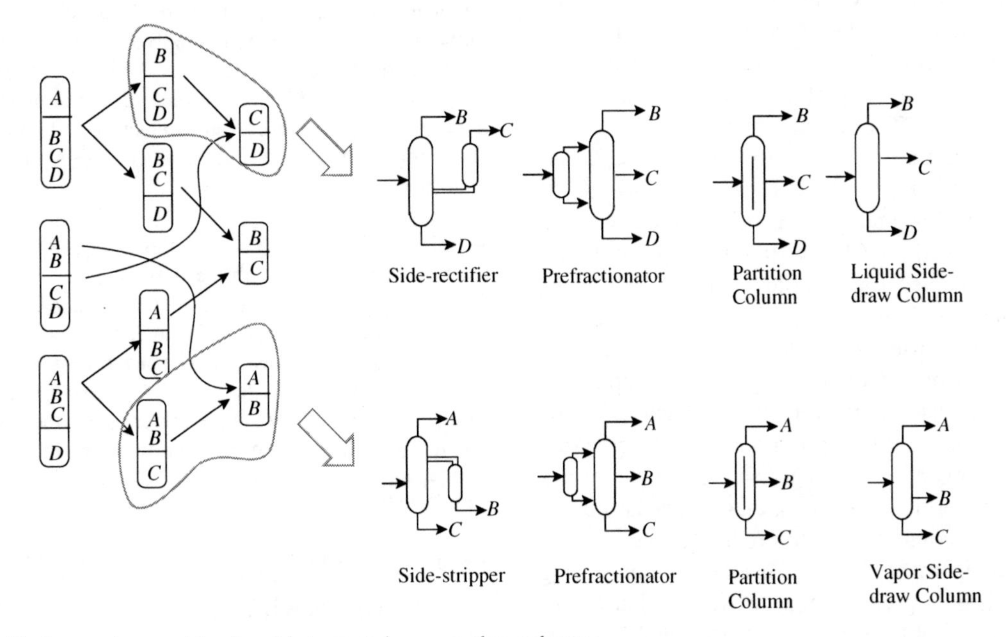

Figure 11.23 Tasks can be combined and interpreted as complex columns.

Component	Recovery to Products		
	Product A	Product B	Product C
1	1.0	0	0
2	0.2	0.80	0
3	0.05	0.95	0
4	0	0.5	0.5
5	0	0	1
6	0	0	1

Figure 11.24 A typical product recovery matrix.

to vary continuously during the optimization, then nonlinear optimization will have to be carried out as the sizing and costing equations are nonlinear. However, it is desirable to avoid nonlinear optimization if at all possible. Instead, the pressure of each task and each hybrid task can be chosen to operate only at discrete pressures. But how can the appropriate discrete pressures be chosen? This can be done either from the condenser or the reboiler of the columns in the task or hybrid task. For example, suppose that a distillation is being carried out in which one of three levels of steam is to be chosen for the reboiler. Three discrete pressures would be chosen such that the temperature of the reboiler (the dew point of the vapor leaving the reboiler) was some reasonable temperature difference below the temperature of the steam at each level. In another case, there might be a low-temperature distillation in which the

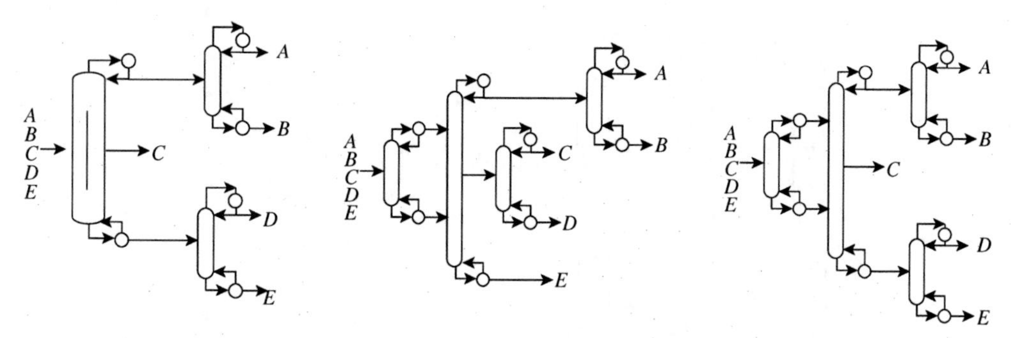

Figure 11.25 Complex column arrangements for the separation of alkanes from Example 11.4.

condensers of the columns could operate against one of six different levels of refrigeration or cooling water, depending on the pressure chosen. In this case, there would be seven discrete operating pressures fixed such that the condenser duty was a reasonable temperature difference above the temperature of the cooling utility.

To carry out the optimization, the overall material balance is first specified via the product recovery matrix. Because of the task representation, it is possible to evaluate each task independently from the performance of the upstream or downstream tasks. The product streams are known prior to the optimization of the superstructure. Each task (simple column) or hybrid task (complex column) can be evaluated by designing and costing using either shortcut calculations (described in Chapter 9 for simple columns and earlier this chapter for complex columns) or using detailed simulation and costing calculations. Although the approach commonly uses shortcut calculations, such as the Fenske–Underwood–Gilliland approach, the approach is not restricted to use shortcut methods. Each task and hybrid task is evaluated at the appropriate discrete pressure levels. Having evaluated each of the columns and complex columns at the appropriate pressures, they can be combined together in feasible combinations to produce sequences and the optimum arrangement chosen. This can be done using mixed integer linear programming (MILP). Constraints are imposed on the objective function to maintain mass balances. Integer constraints can be used to control the complexity of the design by restricting structural options, such as the number of thermally coupled columns allowed in the sequence.

It is important to note that in formulating the problem in this way, it is a linear formulation, guaranteeing (within the bounds of the assumptions made) a global optimum solution. The potential problems associated with nonlinear optimization have been avoided. Even though the models are nonlinear, the problems associated with nonlinear optimization have been avoided. The approach can use either shortcut models or detailed models and the linearity of the optimization is maintained.

Example 11.4 The mixture of alkanes given in Table 11.2 is to be separated into relatively pure products. Using the Underwood Equations, determine sequences of simple and complex columns that minimize the overall vapor load. The recoveries will be assumed to be 100%. Assume the actual to minimum reflux ratio to be 1.1 and all the columns, with the exception of thermal coupling and prefractionator links, are fed with saturated liquid. Neglect pressure drop through columns. Relative volatilities can be calculated from the Peng–Robinson Equation of State with interaction parameters set to zero. Pressures are allowed to vary through the sequence with relative volatilities recalculated on the basis of the feed composition for each column. Pressures of each column are allowed to vary to a minimum such that the bubble point of the overhead product is 10°C above the cooling water return temperature of 35°C (i.e. 45°C) or a minimum of atmospheric pressure.

Solution Figure 11.25 shows three sequences that reduce the vapor load compared with the best sequence of simple columns. The performance of the three sequences is effectively the same, given the assumptions made for the calculations. The total vapor load is around 10% lower than that the best sequence of simple columns from Table 11.5. The differences between networks of simple and complex column in terms of the vapor load vary according to the problem and can be much larger than the result in this example.

11.9 DISTILLATION SEQUENCING – SUMMARY

The best few non-heat-integrated sequences can be identified most simply using the total vapor load as a criterion. If this is not satisfactory, then the alternative sequences can be sized and costed using shortcut techniques.

Complex column arrangements, such as the prefractionator and thermally coupled arrangements, offer large potential savings in energy compared to sequences of simple columns. Partitioned columns (or dividing-wall columns) also offer large potential savings in capital cost. However, caution should be exercised in fixing the design at this stage, as the optimum sequence can change later in the design once heat integration is considered.

Crude oil distillation is carried out in a complex column sequence in which live steam is injected into the separation to provide the heat required and to reduce the partial pressure of the components to be distilled. The design most often used is the equivalent of the partially thermally coupled indirect sequence. However, other design configurations can also be used.

The design of sequences of simple and complex columns can be carried out on the basis of the optimization of a superstructure. The overall separation problem is first decomposed into tasks. Combining tasks together allows sequences of simple distillation columns to be constructed. This can be extended to include complex columns by combining tasks to produce hybrid tasks that can be sized and costed as complex columns. The pressure of each task and each hybrid task can be fixed to operate only at discrete pressures. The discrete pressures are dictated at this stage in the design by the temperatures of either the available heating or cooling utilities.

11.10 EXERCISES

1. Table 11.9 shows the composition of a four-component mixture to be separated by distillation. The K-values for each component at the bubble point temperature of this mixture are given. The liquid- and vapor-phase mixtures of these components may be assumed to behave ideally. Figure 11.26 shows a sequence that has been proposed to separate the mixture shown in Table 11.9 into pure component products. The minimum reflux ratio for each column in this sequence is shown. The feed and all distillation products may be assumed to be liquids at their bubble point temperatures.
   (a) Calculate the flowrates and composition of all streams shown in Figure 11.26. Assume that the recovery of the light and heavy key components in each column is complete.
   (b) Calculate the vapor load (the total flow of vapor produced in all reboilers in the sequence) for this sequence assuming a ratio of actual to minimum reflux ratio of 1:1.

Table 11.9 Data for four-component mixture to be separated.

Component	Mole fraction in feed	K-value	Normal boiling point (°C)
n-Pentane	0.05	3.726	36.3
n-Hexane	0.30	1.5373	69.0
n-Heptane	0.45	0.6571	98.47
n-Octane	0.20	0.284	125.7
Total flow rate ($kmol{\cdot}h^{-1}$)	150		

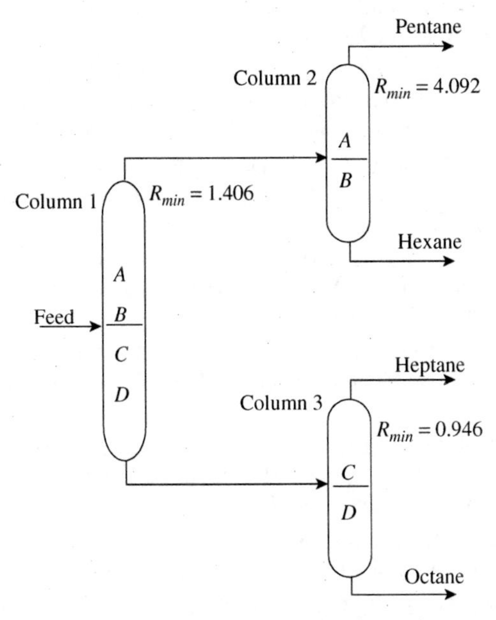

Figure 11.26 Sequence of simple distillation columns.

2. Apply the heuristics for the sequencing of simple distillation columns to the problem in Table 11.9. Is the sequence shown in Figure 11.26 likely to be a good or bad sequence based on the heuristics?

Table 11.10 Heuristics for separating a mixture of components A, B and C using complex distillation columns. Component A is the most volatile and Component C is the least volatile[7].

Complex column	Heuristic for using columns
Sidestream column (sidestream below feed)	$B > 50\%$ of feed, $C < 5\%$ of feed; $\alpha_{BC} >> \alpha_{AB}$
Sidestream column (sidestream above feed)	$B > 50\%$ of feed, $A < 5\%$ of feed; $\alpha_{AB} >> \alpha_{BC}$
Side-rectifier	$B > 30\%$ of feed; less C than A in the feed
Side-stripper	$B > 30\%$ of feed; less A than C in the feed
Prefractionator arrangement	Fraction of B in feed is large; α_{AB} is similar to α_{BC}

3. Table 11.10 presents some heuristics for using complex distillation columns to separate a ternary mixture into its pure component products. On the basis of these heuristics and those for simple columns, suggest two sequences containing complex columns that can be used to separate the mixture described in Table 11.9 into relatively pure products.
4. A mixture of 100 $kmol{\cdot}h^{-1}$ of benzene (B), toluene (T) and xylenes (X) with a feed composition $x_B = 0.3$, $x_T = 0.15$ and $x_x = 0.55$ is to be separated in an indirect sequence of distillation columns. Assume 99% recovery of the light key overheads and heavy key in the bottom of each column.
   (a) Calculate the flowrate and composition of the distillate of the first column.
   (b) The first column has a partial condenser and the composition of the liquid leaving the partial condenser is $x_B = 0.429$, $x_T = 0.524$ and $x_x = 0.047$. Calculate the flowrate of the vapor entering the partial condenser and flowrate of liquid leaving, assuming $R_{min} = 0.97$ and the ratio of actual to minimum reflux is 1.1.
   (c) The pair of columns to be thermally coupled. Estimate the flowrate and composition of the vapor feed to the downstream column and the liquid side-draw from the downstream column.

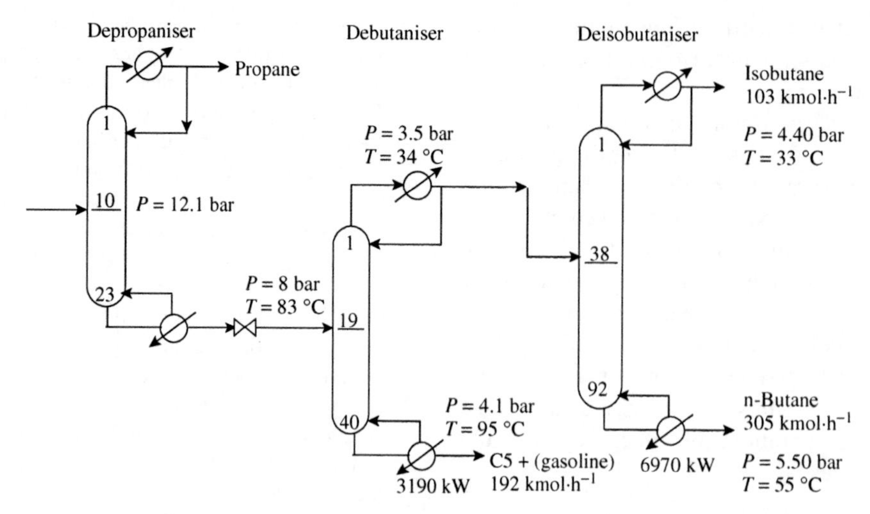

Figure 11.27 Existing arrangement of columns for the separation of light hydrocarbons (From Lestak F and Collins C, 1997, *Chem Engg*, July: 72, reproduced by permission.)

(d) The thermally coupled downstream column is to be modeled by the Underwood Equations. Determine the flowrate and composition of the net overhead product.
(e) Set up the Underwood Equations for the downstream column.

5. Figure 11.27 shows an existing distillation arrangement for the separation of light hydrocarbons[20]. The system is to be retrofitted to reduce its energy consumption by changing to a complex distillation arrangement. As much of the existing equipment as possible is to be retained.
   (a) If two of the three columns are to be transformed into a complex column arrangement, which two columns would be the most appropriate ones to merge?
   (b) Devise a number of complex column arrangements involving merging of two of the existing columns that reuse the existing column shells and are likely to bring a significant reduction in the energy consumption.
6. Develop a network superstructure for the separation of a mixture of five components (*A-B-C-D-E*) into relatively pure products using simple tasks.
7. Vapor flowrate in kmol·h^{-1} for each task for the separation of a four-component mixture are:

A/BCD	100	*B/CD*	90	*A/B*	70
AB/CD	120	*BC/D*	250	*B/C*	100
ABC/D	240	*A/BC*	130	*C/D*	220
		AB/C	140		

Determine the best distillation sequence for minimum total vapor flowrate.

8. Derive a mixed integer model for the problem in Exercise 7 to determine the best sequence for the minimum vapor flowrate.
   (a) Develop task-based network superstructure for the mixture *ABCD*.
   (b) Write logic constraints for the selection of tasks. For example, if Task 4 is selected, then Task 8 must also be selected, but Task 8 can be selected even if Task 4 is not selected.
   (c) Assign a binary variable (y_i) for each task and translate these constraints into mathematical constraints. For example, if Task 4 is selected ($y_4 = 1$), then Task 8 is also selected ($y_8 = 1$), but if Task 4 is not selected ($y_4 = 0$), then Task 8 may or may not be selected ($y_8 = 0$ or 1), therefore $(y_4 - y_8) \leqslant 0$.
   (d) Write an objective function for vapor load in terms of the binary variables.

REFERENCES

1. King CJ (1980) *Separation Processes*, 2nd Edition, McGraw-Hill.
2. Stephanopoulos G, Linnhoff B and Sophos A (1982) Synthesis of Heat Integrated Distillation Sequences, *IChemE Symp Ser*, **74**: 111.
3. Tedder DW and Rudd DF (1978) Parametric Studies in Industrial Distillation, *AIChE J*, **24**: 303.
4. Triantafyllou C and Smith R (1992) The Design and Optimization of Fully Thermally-coupled Distillation Columns, *Trans IChemE*, **70A**: 118.
5. Calberg NA and Westerberg AW (1989) Temperature-Heat Diagrams for Complex Columns 2, Underwood's Method for Side-Strippers and Enrichers, *Ind Eng Chem Res*, **28**(9): 1379.
6. Lestak F, Smith R and Dhole VR (1994) Heat Transfer Across the Wall of Dividing Wall Columns, *Trans IChemE*, **49**: 3127.
7. Glinos K and Malone MF (1988) Optimality Regions for Complex Column Alternatives in Distillation Columns, *Trans IChemE, Chem Eng Res Dev*, **66**: 229.
8. Aichele P (1992) *Sequencing Distillation Operations Being Serviced by Multiple Utilities*, MSc Dissertation, UMIST, UK.
9. Petlyuk FB, Platonov VM and Slavinskii DM (1965) Thermodynamically Optimal Method for Separating Multicomponent Mixtures, *Int Chem Eng*, **5**: 555.
10. Stupin WJ and Lockhart FJ (1972) Thermally-coupled Distillation – A Case History, *Chem Eng Prog*, **68**: 71.
11. Kaibel G (1987) Distillation Columns With Vertical Partitions, *Chem Eng Technol*, **10**: 92.

12. Kaibel G (1988) Distillation Column Arrangements With Low Energy Consumption, *IChemE Symp Ser*, **109**: 43.
13. Mutalib MIA and Smith R (1998) Operation and Control of Dividing Wall Distillation Columns Part I: Degrees of Freedom and Dynamic Simulation, *Trans IChemE*, **76A**: 308.
14. Mutalib MIA, Zeglam AO and Smith R (1998) Operation and Control of Dividing Wall Distillation Columns Part II: Simulation and Pilot Plant Studies Using Temperature Control, *Trans IChemE*, **76A**: 319.
15. Becker H, Godorr S, Kreis and Vaughan J (2001) Partitioned Distillation Columns – Why, When and How, *Chem Eng*, **Jan**: 68.
16. Shultz MA, Stewart DG, Harris JM, Rosenblum SP, Shakur MS and O'Brien DE (2002) Reduce Costs with Dividing-Wall Columns, *Chem Eng Prog*, **May**: 64
17. Eliceche AM and Sargent RWH (1981) Synthesis and Design of Distillation Systems, *IChemE Symp Ser*, **61**(1): 1.
18. Hendry JE and Hughes RR (1972) Generating Separation Process Flowsheets, *Chem Eng Prog*, **68**: 71.
19. Shah PB and Kokossis AC (2002) New Synthesis Framework for the Optimization of Complex Distillation Systems, *AIChE J*, **48**: 527.
20. Lestak F and Collins C (1997) Advanced Distillation Saves Energy and Capital, *Chem Eng*, **July**: 72.

12 Distillation Sequencing for Azeotropic Distillation

12.1 AZEOTROPIC SYSTEMS

In Chapter 11, the distillation sequence for the separation of homogeneous liquid mixtures into a number of products was considered. The mixtures considered in Chapter 11 did not involve the kind of highly nonideal vapor–liquid equilibrium behavior discussed in Chapter 4 that would result in azeotropic behavior or two-liquid phase separations. If a multicomponent mixture needs to be separated, in which some of the components form azeotropes, the mixture is often separated to isolate the azeotropic-forming components and then the azeotropic system (or systems) separated in the absence of other components. Thus, for the original mixture, the separation of the various components could be performed as addressed in Chapter 11, but treating the azeotrope-forming components as if they were a single pseudocomponent. However, caution should be exercised when following such an approach as it will not always lead to the best solution. Complex interactions can occur between components, and the presence of components other than those involved with the azeotrope can sometimes make the azeotropic separation easier. Indeed, if a mass separation agent is used to achieve the separation, it is preferable to use a component that already exists in the process.

In many cases, the azeotropic behavior is between two components, involving binary azeotropes. In other cases, azeotropes can involve more than two components, involving multicomponent azeotropes.

The problem with azeotropic behavior is that, for highly nonideal mixtures, at a specific composition, a constant boiling mixture can form that has the same composition for the vapor and liquid at an intermediate composition away from the required product purity. This depends on the vapor–liquid equilibrium physical properties. Distillation can thus be used to separate to a composition approaching the azeotropic composition, but the azeotropic composition cannot be approached closely in a finite number of distillation stages, and the composition cannot be crossed, even with an infinite number of stages. Thus, if the light and heavy key components form an azeotrope, then something more sophisticated than simple distillation is required. If a mixture that forms an azeotrope is to be separated using distillation, then the vapor–liquid equilibrium behavior must somehow be changed to allow the azeotrope to be crossed. There are three ways in which this can be achieved.

Chemical Process Design and Integration R. Smith
 ISBNs: 0-471-48680-9 (HB); 0-471-48681-7 (PB)

1. *Change in pressure*. The first option to consider when separating a mixture that forms an azeotrope is exploiting change in azeotropic composition with pressure. If the composition of the azeotrope is sensitive to pressure and it is possible to operate the distillation over a range of pressures without any material decomposition occurring, then this property can be used to carry out a separation. A change in azeotropic composition of at least 5% with a change in pressure is usually required[1].
2. *Add an entrainer to the distillation*. A mass separation agent, known as an *entrainer*, can be added to the distillation. The separation becomes possible because the entrainer interacts more strongly with one of the azeotrope-forming components than the other. This can in turn alter in a favorable way the relative volatility between the key components.
3. *Use a membrane*. If a semipermeable membrane is placed between the vapor and liquid phases, it can alter the vapor–liquid equilibrium and allow the separation to be achieved. This technique is known as *pervaporation* and was discussed in Chapter 10.

12.2 CHANGE IN PRESSURE

The first option to consider when separating an azeotropic mixture is exploiting change in azeotropic composition with pressure. Azeotropes are often insensitive to change in pressure. However, if the composition of the azeotrope is sensitive to pressure and it is possible to operate the distillation over a range of pressures without any decomposition of material occurring, then this property can be used to carry out a separation. A change in azeotropic composition of at least 5% with a change in pressure is usually required[1].

Figure 12.1a shows the temperature-composition diagram for a mixture that forms a minimum-boiling azeotrope and is sensitive to change in pressure[1]. This mixture can be separated using two columns operating at different pressures, as shown in Figure 12.1b. Feed with composition f_1 is fed to the high-pressure column. The bottom product from this high-pressure column b_1 is relatively pure B, whereas the overhead product d_1 approaches the high-pressure azeotropic composition. This distillate d_1 is fed to the low-pressure column as f_2. The low-pressure column produces a bottom product b_2 that is relatively pure A and an overhead product d_2 that approaches the high-pressure azeotropic composition. This distillate d_2 is recycled to the high-pressure column.

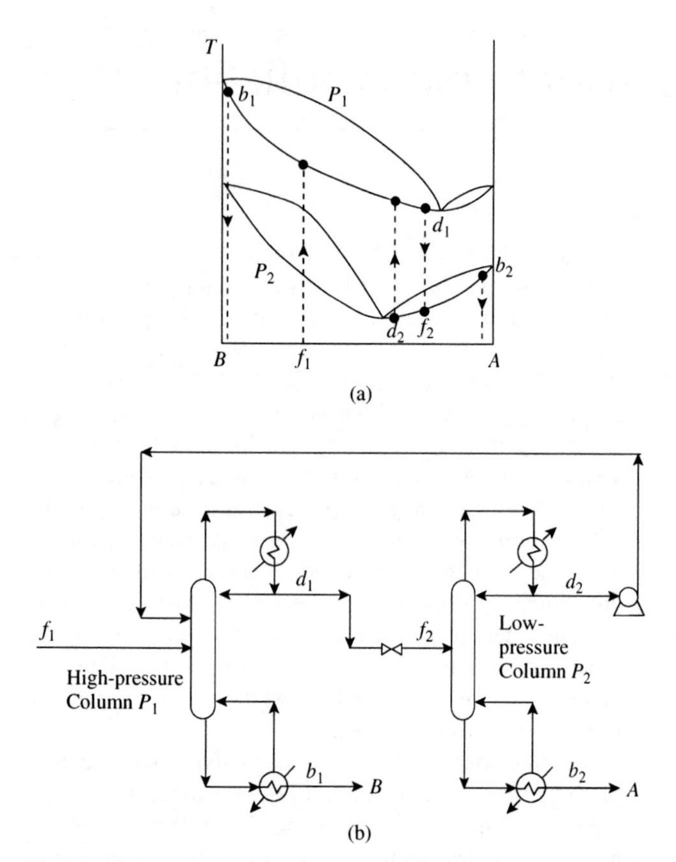

Figure 12.1 Separation of minimum-boiling azeotrope by pressure change.

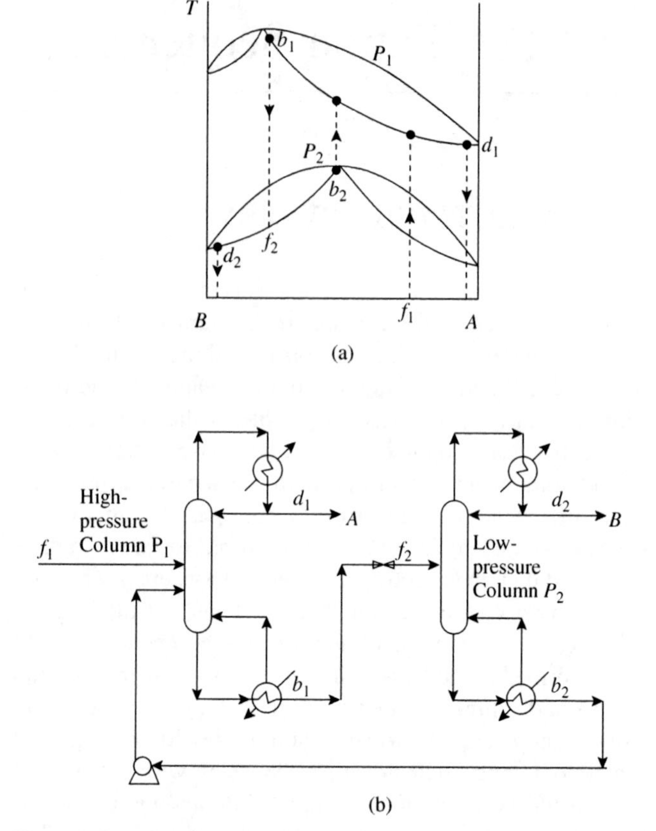

Figure 12.2 Separation of a maximum-boiling azeotrope by pressure change.

Figure 12.2a shows the temperature-composition diagram for a mixture that forms a maximum-boiling azeotrope and is sensitive to change in pressure[1]. Again, this mixture can be separated using two columns operating at different pressures, as shown in Figure 12.2b. Feed with composition f_1 is fed to the high-pressure column. The overhead product from this high-pressure column d_1 is relatively pure A, whereas the bottom product b_1 approaches the high-pressure azeotropic composition. This bottom product b_1 is fed to the low-pressure column as f_2. The low-pressure column produces an overhead product d_2 that is relatively pure B and a bottom product b_2 that approaches the low-pressure azeotropic composition. This bottom product b_2 is recycled to the high-pressure column.

The problem with using a pressure change is that the smaller the change in azeotropic composition, the larger is the recycle in Figures 12.1b and 12.2b. A large recycle means a high energy consumption and high capital cost for the columns, resulting from large feeds to the columns.

If the azeotrope is not sensitive to changes in pressure, then an entrainer can be added to the distillation to alter in a favorable way the relative volatility of the key components. Before the separation of an azeotropic mixture using an entrainer is considered, the representation of azeotropic distillation in ternary diagrams needs to be introduced.

12.3 REPRESENTATION OF AZEOTROPIC DISTILLATION

To gain conceptual understanding of the separation of such systems, they can be represented graphically on triangular diagrams[2]. Figure 12.3 shows a triangular diagram for three components, A, B and C. Different forms of triangular diagrams can be used, such as equilateral triangular and right-triangular[3]. The principles are the same for each form of triangular diagram, and the form of diagram used is only a matter of preference. Here, the right-triangular diagram will be adopted, as illustrated in Figure 12.3. The three edges of the triangle represent A–B, A–C and B–C mixtures. Anywhere within the triangle represents an A–B–C mixture. An example is shown in Figure 12.3 involving a mixture with mole fraction of $A = 0.4$, $B = 0.4$ and $C = 0.2$. The location of the mixture is at the intersection of a horizontal line corresponding with $A = 0.4$ from the axis on the A–B edge of the triangle with a vertical line corresponding with $C = 0.2$ from the axis on the B–C edge of the triangle. The concentration of B is not read from any scale but is obtained by difference and is 0.4. Lines of constant mole fraction of B run parallel with the AC edge of the triangle.

First, consider the representation of a mixer on the triangular diagram[3]. Figure 12.4a shows a mixer for two streams P and R producing a mixed product Q. An overall

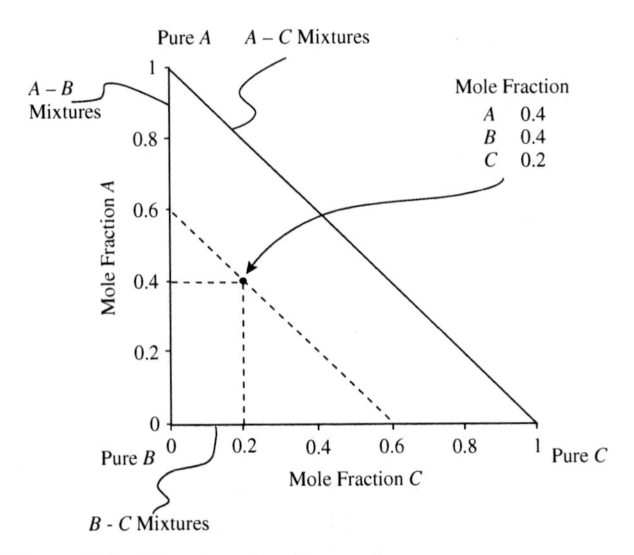

Figure 12.3 The triangular diagram.

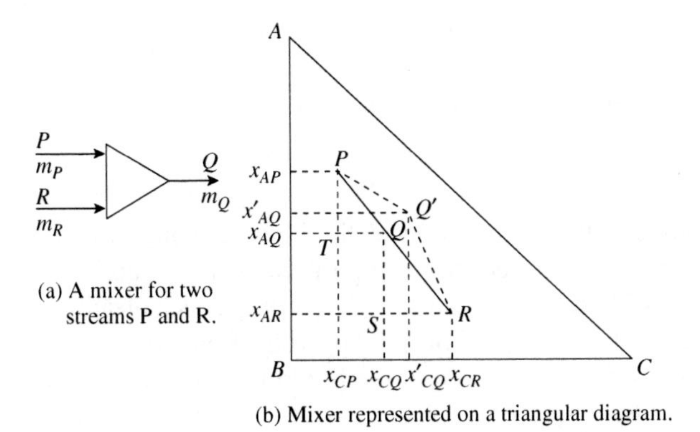

Figure 12.4 The Lever Rule.

mass balance gives:

$$m_Q = m_P + m_R \tag{12.1}$$

A mass balance on Component A gives:

$$m_Q x_{A,Q} = m_P x_{A,P} + m_R x_{A,R} \tag{12.2}$$

A mass balance on Component C gives:

$$m_Q x_{C,Q} = m_P x_{C,P} + m_R x_{C,R} \tag{12.3}$$

Substituting m_Q from Equation 12.1 into Equation 12.2 gives:

$$\frac{m_R}{m_P} = \frac{x_{A,P} - x_{A,Q}}{x_{A,Q} - x_{A,R}} \tag{12.4}$$

Substituting m_Q from Equation 12.1 into Equation 12.3 gives:

$$\frac{m_R}{m_P} = \frac{x_{C,Q} - x_{C,P}}{x_{C,R} - x_{C,Q}} \tag{12.5}$$

Equating 12.4 and 12.5 gives:

$$\frac{x_{A,P} - x_{A,Q}}{x_{A,Q} - x_{A,R}} = \frac{x_{C,Q} - x_{C,P}}{x_{C,R} - x_{C,Q}} \tag{12.6}$$

Equation 12.6 was developed purely from mass balance arguments, without any reference to triangular diagrams. Thus, point Q must be located in a triangular diagram so that Equation 12.6 is satisfied. If Q lies on a straight line between P and R as shown in Figure 12.4b, the triangles PQT and QRS in Figure 12.4b are similar triangles from geometry, and from the properties of similar triangles Equation 12.6 is satisfied. Had Q not been on a straight line between P and R, at say Q' in Figure 12.4b, the corresponding triangles would not have been similar, and Equation 12.6 would not have been satisfied. Thus, the composition of the mixture Q from Figure 12.4a when represented in a triangular diagram must lie on a straight line between the compositions of the two feeds to the mixer. From Figure 12.4b:

$$\frac{x_{AP} - x_{AQ}}{x_{AQ} - x_{AR}} = \frac{PQ}{QR} = \frac{x_{CQ} - x_{CP}}{x_{CR} - x_{CQ}} \tag{12.7}$$

Substituting Equation 12.7 in Equations 12.4 and 12.5gives

$$\frac{m_R}{m_P} = \frac{PQ}{QR} \tag{12.8}$$

Equation 12.8 indicates that the ratio of molar flowrates of two streams being mixed are given by the ratio of the opposite line segments on a straight line, when represented in a triangular diagram[3]. As with the relationship developed in Chapter 9 for a single vapor–liquid equilibrium stage, this is again known as the *Lever Rule* from the analogy with a mechanical level and fulcrum.

Figure 12.5a shows again a mixer and how the mixed stream lies on a straight line between the two streams being mixed at a point, located by the Lever Rule. Figure 12.5b shows a distillation column represented on the triangular diagram. The representation is basically the same, as distillation columns are basically reverse mixers. The feed distillate and bottoms compositions all lie on a straight line with the relative distances again dictated by the Lever Rule.

Figure 12.6a now shows a sequence of two distillation columns represented on a triangular diagram. The separation is assumed to be straightforward with no azeotropes forming in this case. In Figure 12.6a, Component A is the most volatile, and Component C the least volatile. A mixture of A, B and C at Point F is first separated into pure A and a B–C mixture, as represented by the intersection of a line between pure A, F and the B–C edge of the triangle in Figure 12.6a. This B–C mixture is then separated in a second distillation column to produce pure B and C. Again, the relative flowrates are dictated by the Lever Rule. Figure 12.6b shows a slightly more complex arrangement that involves a recycle and a mixer. The feed mixture F is a mixture of A and B. This mixes with pure C to form feed F' that forms the feed to the first column. F' is now therefore a mixture of A, B and C. This is separated in the first column into pure A overhead and bottoms product of B–C. In the

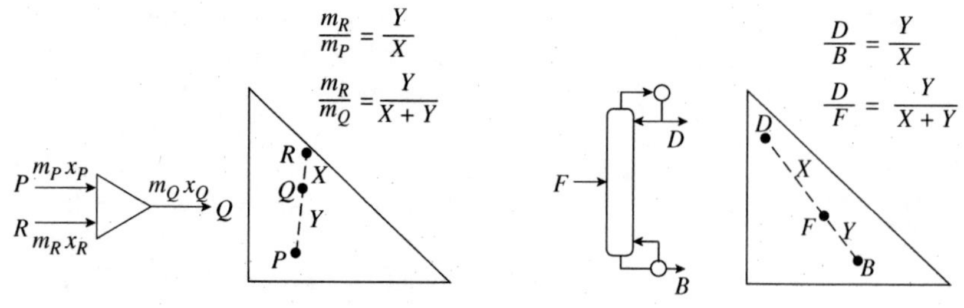

(a) Representing mixers. (b) Distillation column are reverse mixers.

Figure 12.5 Mixers and separators on the triangular diagram.

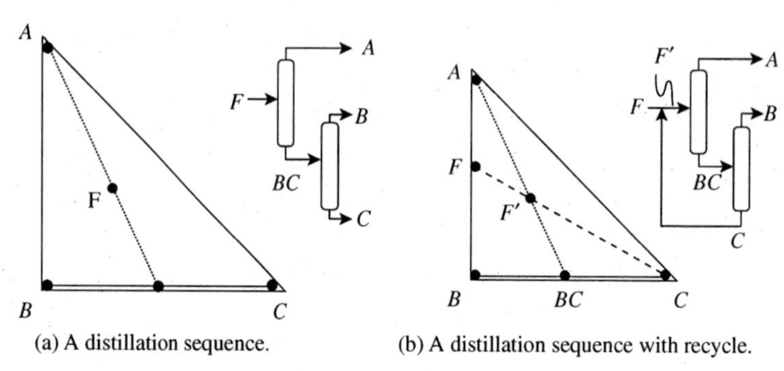

(a) A distillation sequence. (b) A distillation sequence with recycle.

Figure 12.6 Representation of distillation systems on triangular diagrams.

second column, B and C are separated into pure products, with the C being recycled to mix with the original feed F.

12.4 DISTILLATION AT TOTAL REFLUX CONDITIONS

Like conventional distillation, when studying azeotropic distillation, it is helpful to consider the two limiting cases of distillation under total reflux conditions and minimum reflux conditions[2]. Consider distillation at total reflux. Both staged columns and packed columns need to be considered at total reflux. Consider first staged distillation columns at total reflux conditions. The stripping section of a staged distillation column under total reflux conditions is illustrated in Figure 9.14.

In Chapter 9, a relationship was developed for total reflux conditions at any Stage n:

$$y_{i,n} = x_{i,n-1} \tag{9.27}$$

Starting from an assumed bottoms composition, the liquid entering and vapor leaving the reboiler are the same at total reflux conditions. This is the same as the vapor entering and liquid leaving the bottom stage. Knowing the composition of the liquid leaving the bottom stage, the composition of the vapor leaving the bottom stage can be calculated from vapor–liquid equilibrium (see Chapter 4). Then, from Equation 9.27 the composition of the liquid entering the bottom stage is equal to the composition of the vapor leaving the bottom stage at total reflux conditions. Now, knowing the composition of the liquid leaving the second from bottom stage, the composition of the vapor leaving the second from bottom stage can be calculated from vapor–liquid equilibrium. Applying Equation 9.27 then allows the composition of the liquid leaving the third stage from bottom to be calculated, and so on up the column. In stepping up the column, the pressure is usually assumed to be constant. If the liquid composition at each stage is plotted on a triangular diagram, a *distillation line* is obtained[4]. The distillation line is unique for any given ternary mixture and depends only on vapor–liquid equilibrium data, pressure and the composition of the starting point. Repeating this for different assumed bottoms compositions allows us to plot a *distillation line map*, as illustrated in Figure 12.7[5,6]. The distillation line map in Figure 12.7a involves a simple system with no azeotropes. Distillation lines in a distillation line map can never cross each other, otherwise vapor–liquid equilibrium would not be unique for the system. The points on the distillation lines represent the discrete liquid compositions at each stage in a staged column. The *tie lines* between the points have no real physical significance, as the changes in a staged column are discrete from stage to stage. The tie lines between the points simply help clarify the overall picture. Arrows can also be assigned to the tie lines to assist in the interpretation of the diagram. Here, arrows will be assigned in the direction of increasing temperature.

A more complex distillation line map is shown in Figure 12.7b. This involves two binary azeotropes. The closeness of the dots on a distillation line is indicative of the difficulty of separation. As an azeotrope is approached, the

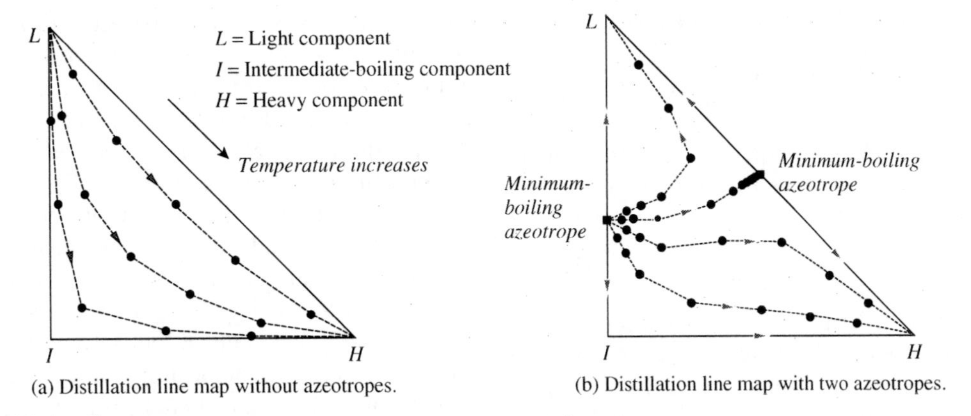

Figure 12.7 Distillation line maps.

points become closer together, indicating a smaller change in composition from stage to stage.

The distillation lines in the distillation line map were in this case developed by carrying out a balance around the bottom of the column, as indicated in Figure 9.13. Equally well, the distillation line could have been developed by drawing an envelope around the top of the column at total reflux, and the calculation developed down the column in the direction of increasing temperature.

The distillation lines in Figure 12.7 show the change in composition through a staged column at total reflux. The analogous principle for the packed column at total reflux could also be developed[7]. To begin, consider the mass balance around the rectifying section of a staged column, this time not at total reflux, as illustrated in Figure 12.8a. To simplify the development of the equations, constant molar overflow will be assumed. A mass balance around the top of the column for Component i gives:

$$Vy_{i,n+1} = Lx_{i,n} + Dx_{i,D} \tag{12.9}$$

Also, by definition of the reflux ratio R:

$$L = RD \tag{12.10}$$

An overall balance around the top of the column gives:

$$V = (R + 1)D \tag{12.11}$$

Combining Equations 12.9 to 12.11 gives:

$$y_{i,n+1} = \frac{Rx_{i,n}}{R+1} + \frac{x_{i,D}}{R+1} \tag{12.12}$$

Then carrying out a balance across Stage n for Component i, as illustrated in Figure 12.8b gives:

$$Lx_{i,n-1} + Vy_{i,n+1} = Lx_{i,n} + Vy_{i,n} \tag{12.13}$$

Rearranging Equation 12.13 gives:

$$\begin{aligned} x_{i,n} - x_{i,n-1} &= \frac{V}{L}(y_{i,n+1} - y_{i,n}) \\ &= \frac{R+1}{R}(y_{i,n+1} - y_{i,n}) \end{aligned} \tag{12.14}$$

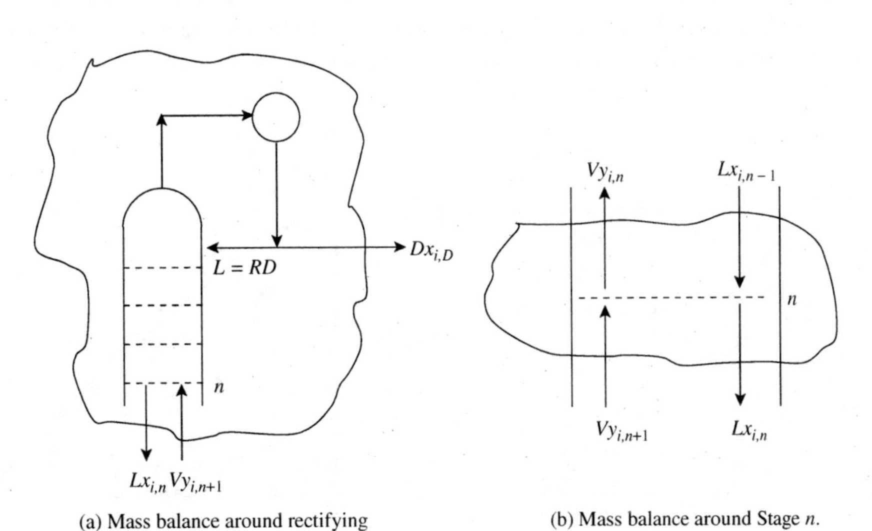

Figure 12.8 Mass balances for rectifying section at finite reflux conditions.

Now combining Equations 12.12 and 12.14 gives:

$$x_{i,n} - x_{i,n-1} = x_{i,n} + \frac{x_{i,D}}{R} - \frac{(R+1)y_{i,n}}{R} \quad (12.15)$$

Equation 12.15 is a difference equation that, for small changes, can be approximated by a differential equation:

$$\frac{dx_i}{dh} = x_i + \frac{x_i}{R} - \frac{(R+1)y_i}{R} \quad (12.16)$$

where h = dimensionless height in a packed column

Equation 12.16 can be considered to represent the change of concentration with height within a packed column, in which change in composition is continuous. At total reflux, $R = \infty$ and Equation 12.16 simplifies to:

$$\frac{dx_i}{dh} = x_i - y_i \quad (12.17)$$

Substituting the vapor–liquid equilibrium K-value:

$$\frac{dx_i}{dh} = x_i(1 - K_i) \quad (12.18)$$

This equation can be integrated at constant pressure from an initial composition $x_{i,0}$ through the dimensionless height h. However, the integration must be carried out numerically. Since K_i is a function of x_i, Equation 12.18 is nonlinear. Also, the equations for the individual components are coupled. Hence, the integration must be carried out numerically, typically using a fourth-order Runge–Kutta integration technique[7].

Equations 12.17 and 12.18 for packed columns at total reflux could also have been developed by considering a mass balance around the bottom of the column. The same result would have been obtained.

Equation 12.18 allows us to predict the concentration profile through a packed column operating at total reflux conditions. The resulting profile is known as a *residue curve*, since they were first developed by considering the batch vaporization of a mixture through time[8,9]. Such batch vaporization is often termed *simple distillation*. An analysis of a simple distillation process results in the same form of expression as Equations 12.17 and 12.18[8]. However, in the case of simple distillation, the expressions feature dimensionless time instead of the dimensionless packing height. The significance of the residue curve in simple distillation is therefore a plot of the residual liquid composition through time as material is vaporized, hence the name residue curve. The arrows assigned to residue curves then follow increasing time as well as temperature. Within the present context, the interpretation of the residue curve as change of composition through height in a packed column at total reflux conditions is more useful. The residue curve is unique for any given ternary mixture and depends only on vapor–liquid equilibrium data, pressure and the composition of the starting point.

Given a number of different starting points, a number of residue curves can be plotted on the same triangular diagram, giving a *residue curve map*[10]. Figure 12.9a shows a residue curve map that does not exhibit any azeotropes. Residue curves in a residue curve map can never cross each other, otherwise vapor–liquid equilibrium would not be unique for the system. Unlike distillation lines, residue curves are continuous. Arrows can again be assigned to help in the interpretation. Here, the arrows will be assigned in the direction of increasing temperature. A slightly more complex residue curve map is shown in Figure 12.9b. This residue curve shows a binary minimum-boiling azeotrope between the light and intermediate-boiling components. An even more complex residue curve map is shown in Figure 12.10. This has binary azeotropes between each of the three components. It also has a ternary minimum-boiling azeotrope involving all three components.

Figure 12.11 superimposes distillation lines and residue curves for the same ternary systems. Figure 12.11a shows the system n-pentane, n-hexane and n-heptane, which is a relatively wide boiling mixture. It can be observed in Figure 12.11a that there are significant differences between the paths of the distillation lines and the residue curves. By

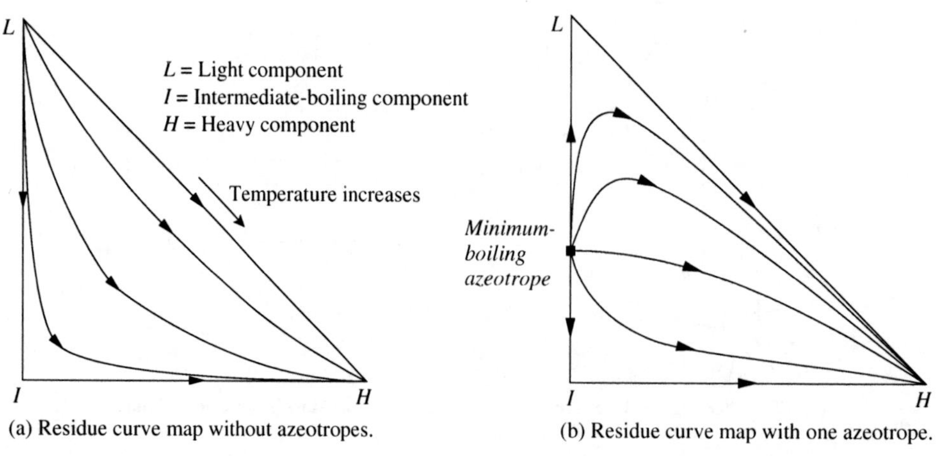

Figure 12.9 Residue curve maps.

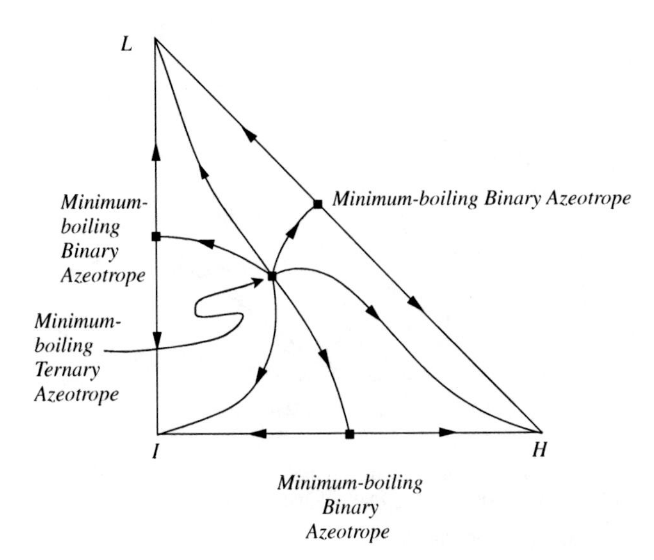

Figure 12.10 A residue curve map with three binary azeotropes and one ternary azeotrope.

contrast, the ternary system ethanol, isopropanol and water shown in Figure 12.11b is a more close-boiling mixture. This time, the distillation lines and residue curves follow each other fairly closely because the difficult separation means that the changes from stage to stage in a staged column become smaller and approach the continuous changes in a packed column. It is important to note that distillation lines and residue curves have the same properties at fixed points (when the distillation lines and residue curves converge to a pure product or an azeotrope).

Both distillation lines and residue curves can be used to make an initial assessment of the feasibility of an azeotropic separation at high reflux rates. Both will give a similar picture. Figure 12.12a shows a residue curve map involving a binary azeotrope between the light and intermediate-boiling components. Also shown in Figure 12.12a are two suggested distillation separations drawn as straight lines. At total reflux, for a packed column to be feasible, both the distillate and bottoms product should be located on the same residue curve, as well as the feed and products being located on the same straight line. The upper separation in Figure 12.12a shows this to be the case and might in principle be expected to be feasible. The lower separation in Figure 12.12a shows a separation in which the products

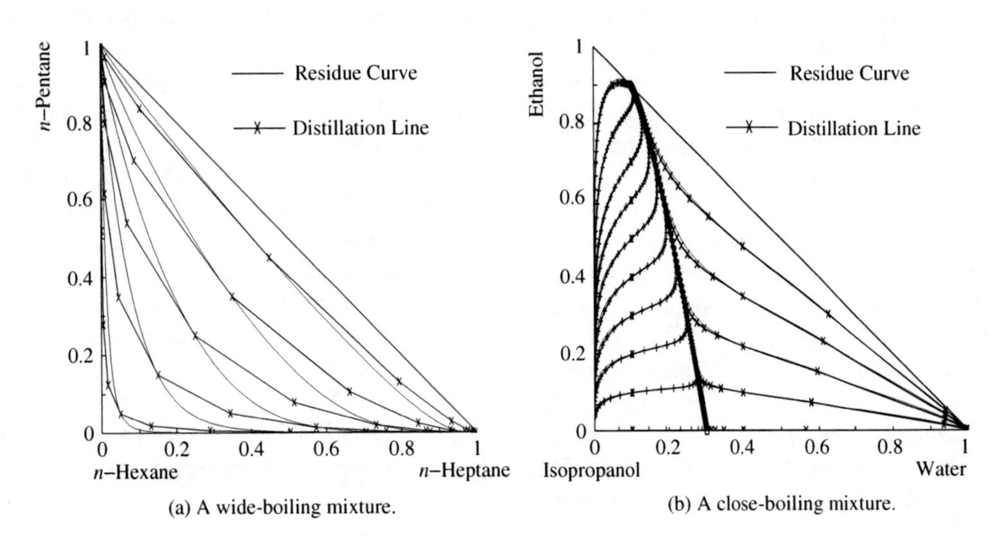

Figure 12.11 Comparison of distillation lines and residue curves.

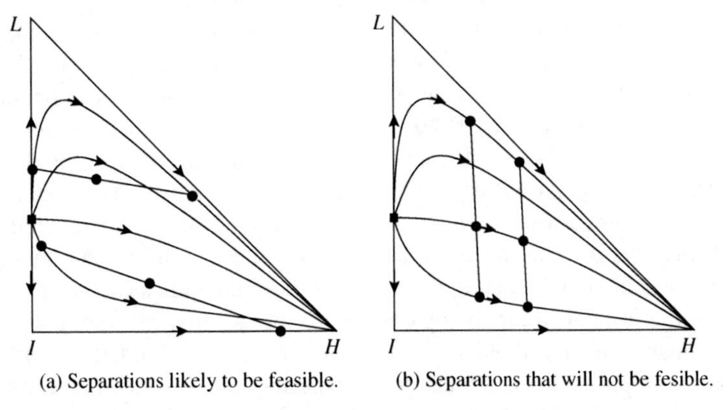

Figure 12.12 Residue curves give an indication of what separations are likely to be feasible.

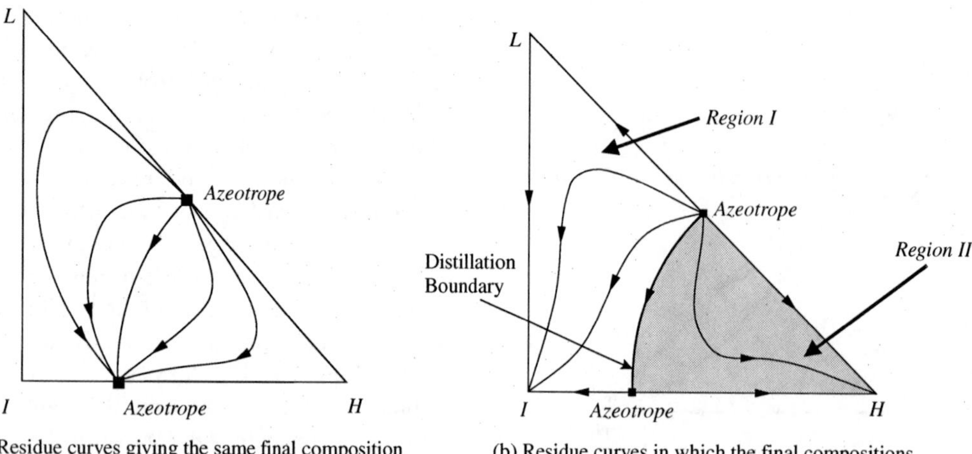

(a) Residue curves giving the same final composition for any intermediate starting point.

(b) Residue curves in which the final compositions are different divide the map into different regions.

Figure 12.13 The residue curve maps can be divided into regions by distillation boundaries.

do not quite lie on the same residue curve. However, total reflux will not be used in the final design and there may be slight changes to the feed and product specifications. All that can be determined from the two separations shown in Figure 12.12a is that they are likely to be feasible. However, there is no guarantee of this. Later, more rigorous assessment of the feasibility of an individual column will be considered. By contrast, two different separations are shown for the same residue curve map in Figure 12.12b. This time, the suggested separations run across the residue curves. In this case, the separations will not be feasible.

Another important feature of residue curve maps (and distillation line maps) is illustrated in Figure 12.13. The residue curve map in Figure 12.13a shows two binary azeotropes. Starting anywhere in the residue curve map will always terminate at the same point, in this case at the binary azeotrope between the intermediate-boiling and heavy components in Figure 12.13a. The residue curve map in Figure 12.13b shows a quite different behavior. Again, there are two binary azeotropes. This time, however, the residue curve map can be divided into two distinct *distillation regions* separated by a *distillation boundary*[10,11]. Starting at any point to the left of the distillation boundary in Figure 12.13b will always terminate at the intermediate-boiling component. Starting anywhere to the right of the distillation boundary will always terminate at the heavy boiling component. The residue curve connecting the two azeotropes divides the residue curve map into two distinct regions. The significance of this is critical. Starting with a composition anywhere in Region *I* in Figure 12.13b, a composition in Region *II* cannot be accessed, and vice versa. This means that distillation problems involving distillation boundaries are likely to be extremely problematic. For more complex systems, the residue curve map can be divided into more than two regions. For example, referring back to Figure 12.10, this residue curve map is divided into three regions. The three distillation boundaries are created by the residue curves connecting the binary azeotropes with the ternary azeotrope.

Distillation line maps also exhibit distillation boundaries. For example, referring back to Figure 12.7b, this is divided into two regions. The boundary is formed by the distillation line connecting the two binary azeotropes. The boundary for a residue curve on the system in Figure 12.7b would start and end at the same point but might follow a slightly different path because the residue curve follows a continuous profile rather than the discrete profile of the distillation line boundary in Figure 12.7b. Figure 12.11b also has two regions. The distillation lines and residue curves are superimposed in Figure 12.11b, and it can be seen that the distillation line and residue curve distillation boundaries follow each other closely in this case.

12.5 DISTILLATION AT MINIMUM REFLUX CONDITIONS

So far, only the limiting case of total reflux condition has been investigated in developing distillation lines and residue curves for staged and packed columns. The other limiting condition of minimum reflux will now be explored[2,12]. Consider the rectifying section of a column as illustrated in Figure 12.14a. This is operating at minimum reflux, as there is a pinch in the column in which the changes in composition are incrementally small. This could either be an incrementally small change from stage to stage in a staged column or an incrementally small change with height in a packed column. A mass balance around the top of the column indicates that the liquid composition leaving the top section of the column, the liquid composition of the distillate and the vapor composition of the vapor entering the top section must all be on a straight line on a triangular diagram, as illustrated in Figure 12.14b.

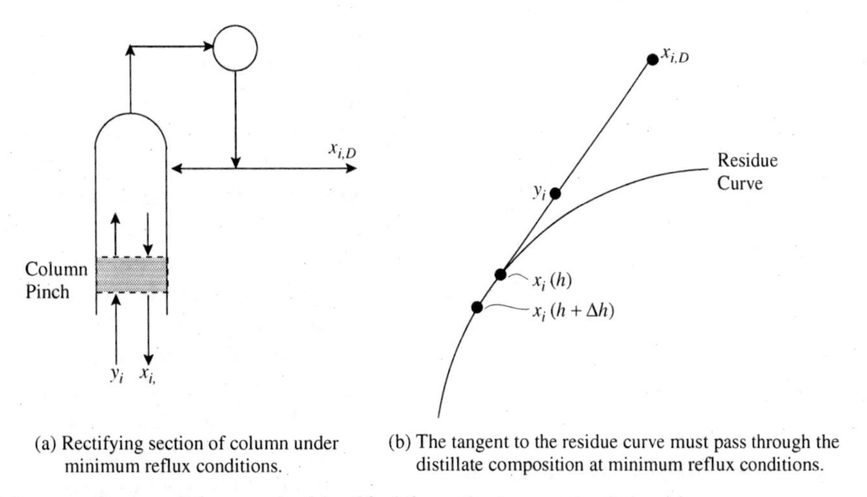

(a) Rectifying section of column under minimum reflux conditions.

(b) The tangent to the residue curve must pass through the distillate composition at minimum reflux conditions.

Figure 12.14 The minimum reflux condition can be identified from the tangent to the residue curve.

Also, at a pinch condition the change in composition is incremental. For an incremental change in composition with height in the residue curve, as illustrated in Figure 12.14b, both incremental points of the pinch must be on the residue curve, and an incremental change defines the line to be a tangent to the residue curve[12]. Thus, it follows that the tangent to the residue curve at a pinch point x_i must pass through the distillate composition $x_{i,D}$. This principle gives a very simple way to follow the path of pinch conditions for any given distillate composition. Figure 12.15 shows a distillate composition D and a series of tangents drawn to the residue curves. The locus of points joining the tangents to the residue curves defines the *pinch point curve*[12]. The same principle can also be applied to the bottom of the column. The same pinch point curve applies to both staged and packed columns, as by definition at a pinch condition the changes are incrementally small.

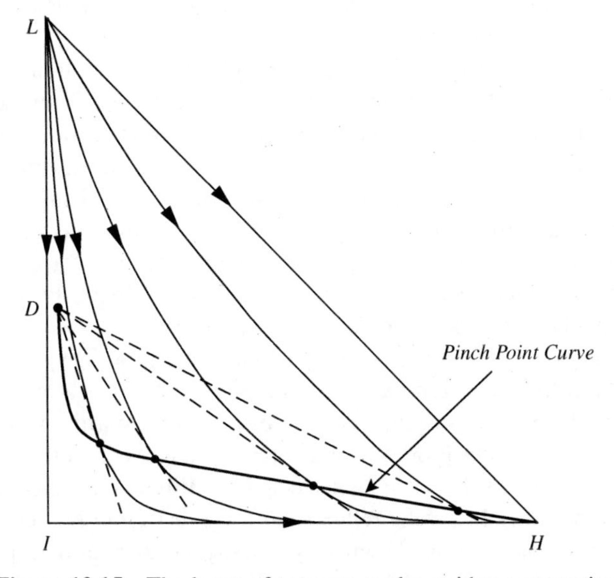

Figure 12.15 The locus of tangents to the residue curves gives the pinch point curve.

12.6 DISTILLATION AT FINITE REFLUX CONDITIONS

Consider again the column rectifying section shown in Figure 12.8a at finite reflux conditions. Equation 12.12 relates the composition of passing vapor and liquid streams at any stage, given the reflux ratio and distillate composition. Specifying the distillate composition from the rectifying section in Figure 12.8a allows the vapor composition leaving the top stage of the distillation column to be calculated from vapor–liquid equilibrium. This would allow the composition of the vapor leaving the first stage and the liquid composition entering the first stage (being the same as the distillate composition) to be determined. The liquid composition leaving the first stage will be at equilibrium with the vapor composition leaving the first stage. Thus, from a vapor–liquid equilibrium calculation, the liquid leaving the first stage can be calculated from the composition of the vapor leaving the first stage. The composition of the vapor entering the first stage can then be calculated from Equation 12.12. This is the vapor leaving the second stage. Thus, the composition of the liquid leaving the second stage can be calculated from vapor–liquid equilibrium. The vapor entering the second stage can then be calculated from Equation 12.12, and so on down the column. In this way, the composition through the rectifying section can be calculated by specifying the reflux ratio and distillate composition, based on the assumption of constant molar overflow[13].

The corresponding mass balance can also be carried out around the stripping section of a column, as illustrated in Figure 12.16. A mass balance around the bottom of the column for Component i gives:

$$Lx_{i,n-1} = Vy_{i,n} + Bx_{i,B} \tag{12.19}$$

Also, by definition of the reboil ratio S:

$$V = SB \tag{12.20}$$

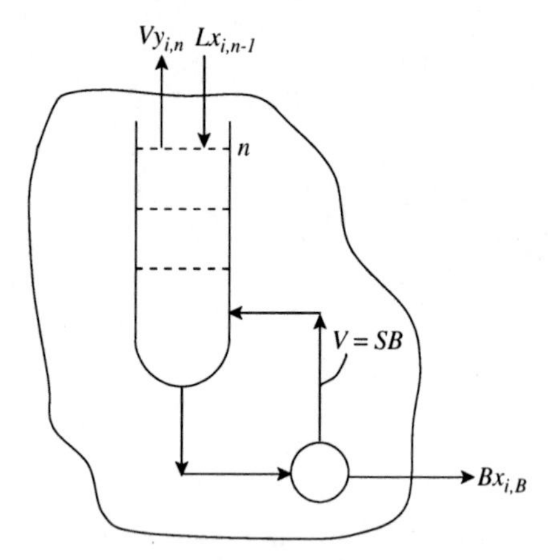

Figure 12.16 Mass balance for the stripping section at finite reflux conditions.

An overall balance around the bottom of the column gives:

$$L = (S + 1)B \tag{12.21}$$

Combining Equations 12.19 to 12.21 gives:

$$x_{i,n-1} = \frac{Sy_{i,n}}{S+1} + \frac{x_{i,B}}{S+1} \tag{12.22}$$

Equation 12.22 relates the compositions of vapor and liquid streams passing each other in the stripping section of a column. Equation 12.22 can be used together with vapor–liquid equilibrium calculations to calculate a composition profile in the stripping section of the column, similar to that of the rectifying section of the column as described above. The calculation is started with an assumed bottoms composition and Equation 12.22 applied repeatedly with vapor–liquid equilibrium calculations working up the column.

Figure 12.17 shows a plot of *section profiles* developed in this way[13]. Starting from an assumed distillate composition D, the calculation works down the column with a given reflux ratio. Working down the column, the changes from stage to stage gradually decrease as the lighter components become depleted. The distillate section profile approaches a pinch condition towards its termination. Also shown in Figure 12.17 is a section profile for the stripping section of a column. Starting at bottoms composition B with a given reboil ratio, the section profile works up the column. Towards the end of the section profile for the stripping section, again the changes from stage to stage become smaller as the section becomes depleted in the heavier components and again approaches a pinch condition. The section profiles starting from D and B in Figure 12.17 intersect each other before reaching a pinch conditions. The intersection would correspond with the feed condition. Figure 12.17 indicates that because the section profiles for the rectifying and stripping sections intersect, these settings would, in principle, lead to a feasible column.

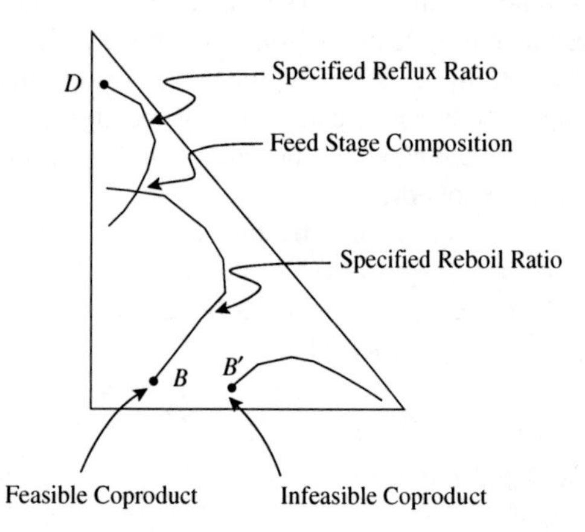

Figure 12.17 Section profiles.

Two issues should be noted. Firstly, the section profiles are actually discrete changes from stage to stage, and strictly speaking the compositions of discrete points corresponding with stages for the rectifying and stripping sections need to intersect for feasibility. If the discrete points do not intersect, then usually some small adjustments to the product compositions, reflux ratio or reboil ratio can be used to fine-tune for feasibility. Secondly, the section profiles were developed by assuming constant molar overflow, which is an approximation.

Also shown in Figure 12.17 is a section profile for a stripping section starting at B'. The stripping section profile starting at B' follows a path different from the profile starting at B and does not intersect the rectifying section profile. This means that the combination of distillate compositions D and B' could not lead to a feasible column design.

The intersection of section profiles as illustrated in Figure 12.17 can be used to test the feasibility of given product compositions and settings for the reflux ratio and reboil ratio. However, the profiles would change as the reflux ratio and reboil ratios change, and trial and error will be required.

It should be noted that although the assumption of constant molar overflow has been used here to develop the section profiles, it is not difficult in principle to develop rigorous profiles. A rigorous profile requires the material and energy balances to be solved simultaneously. Thus, the rigorous profile would step through the column one stage at a time, solving the material and energy balance for each stage, rather than just the material balance. Obviously, the calculations become more complex.

Figure 12.18 shows a product composition for the system chloroform, benzene and acetone with a series of section profiles for the stripping section of a column starting from bottoms composition B. For a defined product composition, the section profiles can be projected for different settings of the reboil ratio (or the reflux ratio for a rectifying section). The reboil ratio is shown to gradually be increased from $S = 1$ to $S = 100$. The profile changes according to the setting of the reboil ratio. Each of the section profiles

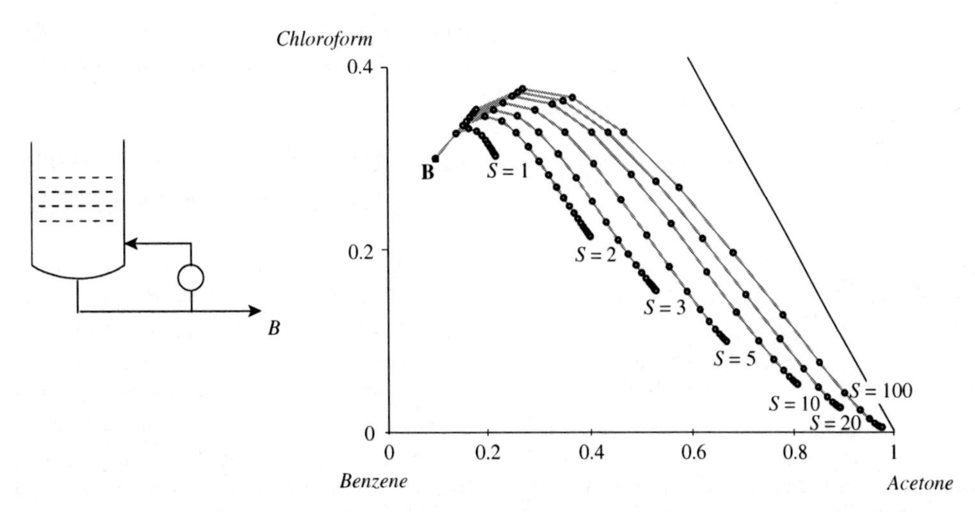

Figure 12.18 For a given bottoms composition, a range of section profiles can be generated by varying the reboil ratio. (From Castillo F, Thong DY-C, Towler GP, 1998, *Ind Eng Chem Res*, **37**: 987 reproduced by permission of the American Chemical Society).

terminates in a pinch condition. This shows that there is a large area of the triangular diagram that can be accessed from a starting composition B by a column stripping section with the appropriate setting of the reboil ratio. However, it would be convenient to determine the region of all possible composition profiles for a given product without having to plot all of the section profiles, as illustrated in Figure 12.18.

The rectifying or stripping section of a column must operate somewhere between total reflux and minimum reflux conditions. The range of feasible operation of a column section can thus be defined for a given product composition. It can be seen in Figure 12.19 that these section profiles are bounded for a stage column by the distillation line and the pinch point curve. As noted previously, the pinch point curve provides a minimum reflux boundary for both staged and packed columns, as by definition at a pinch condition the changes are incrementally small.

Also shown in Figure 12.19 is the residue curve projected from the same product composition. The area enclosed within the residue curve and the pinch point curve thus provides the feasible compositions that can be obtained by a packed column section from a given product composition. For any given product composition, the *operation leaf* of feasible operation for a column section can be defined by plotting the distillation line (or residue curve) and the pinch point curve[13,14]. The column section must operate somewhere between the total and minimum reflux conditions.

For the stripping and rectifying sections to become a feasible column, the two operation leaves must overlap. Figure 12.20 shows the system chloroform, benzene and acetone. The operation leaf for a distillate composition D intersects with the operation leaf for a bottoms composition

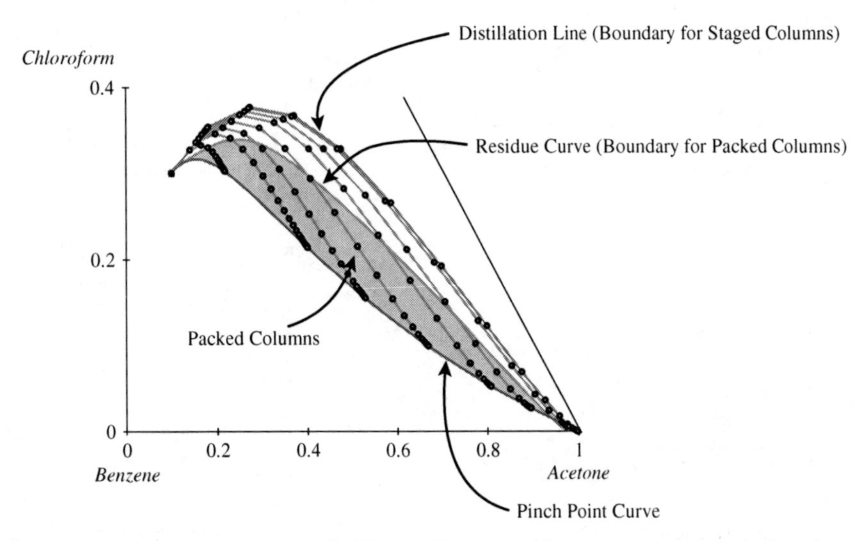

Figure 12.19 The operation leaf is bounded by the distillation line or residue curve and the pinch point curve. (From Castillo F, Thong DY-C, Towler GP, 1998, *Ind Eng Chem Res*, **37**: 987 reproduced by permission of the American Chemical Society).

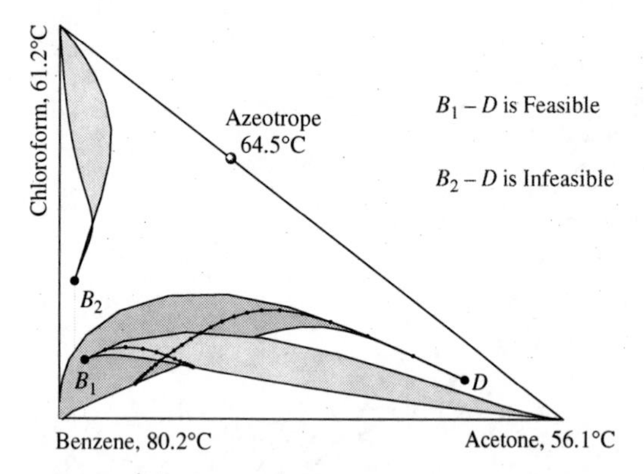

Figure 12.20 Overlap of column section operating leaves is a necessary condition for feasible separation. (From Castillo F, Thong DY-C, Towler GP, 1998, *Ind Eng Chem Res*, **37**: 987 reproduced by permission of the American Chemical Society).

B_1 in Figure 12.20. This means that there is some combination of settings for the reflux ratio and reboil ratio that will allow the section profiles to intersect and become a feasible column design. By contrast, bottoms composition B_2 shows an operation leaf that does not intersect with the operation leaf of distillate D. This means that the two products D and B_2 cannot be produced in the same column, and the design is infeasible. No settings of reboil ratio or reflux ratio can make the combination of B_2 and D a feasible design.

The operation leaves are particularly interesting for cases when separations are not feasible at high reflux ratios but can be feasible at lower reflux ratios. The reason that this kind of behavior can happen will be discussed later.

12.7 DISTILLATION SEQUENCING USING AN ENTRAINER

Consider the separation of a mixture containing mole fraction acetone of 0.7 and mole fraction heptane of 0.3. This needs to be separated to produce pure acetone. Unfortunately, there is an azeotrope between acetone and heptane at a mole fraction of acetone of 0.95. Figure 12.21a shows the mass balance for the separation of this mixture into pure acetone and heptane using benzene as an entrainer. The feed is first mixed with benzene. Figure 12.21a shows three different mass balances corresponding with different amounts of benzene being mixed with the feed, producing feed mixtures at *A*, *B* and *C* in Figure 12.21a. From the Lever Rule, Point *A* in Figure 12.21a mixes the largest amount of benzene and Point *C* the lowest amount of benzene. The first column *C*1 separates the ternary mixture into pure heptane and a mixture of acetone and benzene, with a composition that depends on the amount of benzene mixed with the original feed. A second column *C*2 then separates pure acetone and benzene, with the benzene being recycled. Figure 12.21b shows the distillation sequence that would correspond with this mass balance.

Even though a feasible mass balance has been set up, there is no guarantee that vapor–liquid equilibrium will

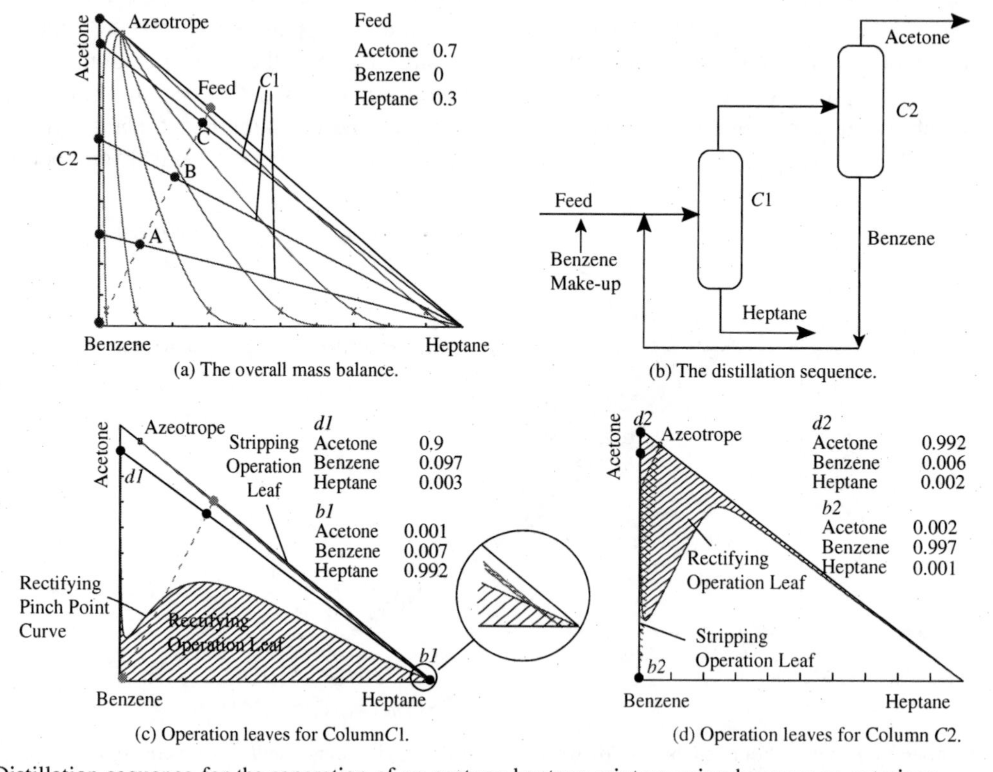

Figure 12.21 Distillation sequence for the separation of an acetone-heptane mixture using benzene as entrainer.

allow the separation. To test whether the columns are feasible, the operation leaves corresponding with the products suggested by the mass balance need to be plotted. Take the benzene flowrate corresponding with Point C in Figure 12.21a. This would correspond with the lowest flowrate of benzene of the three settings. Figure 12.21c shows the operation leaves for Column $C1$. The stripping section leaf is very narrow and follows the acetone–heptane edge of the triangular diagram from b_1. The rectifying leaf starts at d_1 and is long and narrow along the acetone-benzene edge, after which it opens up and converges towards pure heptane, where it intersects with the stripping operation leaf. The rectifying operation leaf is bounded by the rectifying pinch point curve, as shown in Figure 12.21c, and the residue curve from d_1 that follows the acetone–benzene and benzene–heptane edges of the triangular diagram. Because the stripping and rectifying leaves for Column $C1$ intersect, in principle it is therefore a feasible column. One further point should be noted from the rectifying section leaf. From the shape of the leaf, it would be expected that the concentration of benzene in the column would be low at each end of the column and be higher in the middle of the column. This follows from the fact that benzene is an *intermediate-boiling entrainer*, with a volatility between that of acetone and heptane. A detailed simulation of Column $C1$ would show a benzene concentration through the column following these trends. Figure 12.21d shows the operation leaves for the second Column $C2$, which intersect. Thus, the distillation sequence would be expected to be feasible, and optimization of the design can be considered.

In fact, it is possible to carry out the separation of acetone and heptane using benzene as entrainer in a different sequence to that shown in Figure 12.21 by separating the acetone in the first column as an overhead product. The heptane is separated in the second column as the bottom product with the overhead of benzene from the second column being recycled.

In Figure 12.21a, different settings were possible for the mass balance, depending on the benzene recycle. At first sight, it might be expected that the lower the benzene recycle, the better. In other words, Point C in Figure 12.21a will be better than Point A. However, it is not so straightforward. A higher concentration of benzene in Column $C1$ will help the separation. Setting C in Figure 12.21a requires a very high reflux ratio for Column $C1$, in turn requiring a large amount of energy for the reboiler. As the benzene recycle increases, moving towards Point A in Figure 12.21a, the reflux ratio for Column $C1$ decreases considerably. It would be expected that as the benzene recycle is increased, the improvements in the design of Column $C1$ would reach diminishing returns, and thereafter an excessive load on the system would be created as the recycle of benzene increases. Thus, the flowrate of benzene in the recycle is an important optimization parameter that affects the sizes of both Columns $C1$ and $C2$ in Figure 12.21.

Consider a second example involving the separation of a mixture of ethanol and water that forms an azeotrope at around a mole fraction of ethanol of 0.88. It is proposed to use ethylene glycol as entrainer. An overall mass balance for the separation is shown in Figure 12.22a. As with the

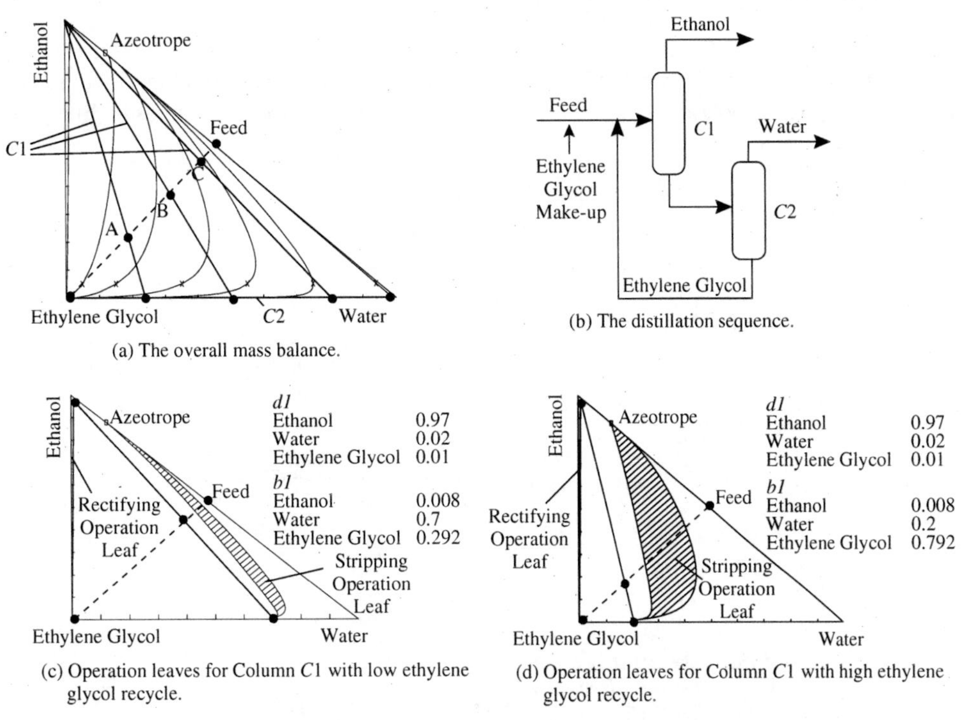

Figure 12.22 Distillation sequence for the separation of an ethanol-water mixture using ethylene glycol as entrainer is infeasible using single-feed column.

previous example in Figure 12.21, the feed is mixed with the entrainer in varying amounts to give Points *A*, *B* and *C*, and the resulting mixture first distilled to produce pure ethanol, Figure 12.22b. The water and ethylene glycol are then separated in the second column, with the ethylene glycol being recycled. Operation leaves for the first column can now be constructed to test the feasibility with a low ethylene glycol recycle, as shown in Figure 12.22c. The operation leaves do not overlap, and therefore the column is not feasible. Figure 12.22d also tests the feasibility of Column *C*1, but this time with a higher flowrate of ethylene glycol. Unfortunately, this is also an infeasible design, as the section leaves do not overlap.

One feature common to both designs in Figures 12.21 and 12.22 is that single-feed columns were used with the entrainer being mixed with the feed. In the case of the ethanol-water-ethylene glycol, the operation leaves for the top and bottom sections of the column do not meet, and there is a gap. In some systems, it is possible to bridge the gap between the operation leaves between the top and bottom sections by creating a middle section in the column. This is achieved by using a two-feed column, as shown in Figure 12.23, in which a heavy entrainer (sometimes termed a *solvent*) is fed to the column at a point above the feed point for the feed mixture[2]. The entrainer must not form any new azeotropes in the system. This arrangement, shown in Figure 12.23, is known as *extractive distillation*. In the first column, the *extraction column*, a heavy entrainer is fed to the column above the feed point for the feed mixture. This entrainer flows down the column and creates a bridge between the top and bottom sections. One of the components is distilled overhead as pure *A*. The other component and the entrainer leave the bottom of the extraction column and are fed to the *entrainer recovery column*, which separates pure *B* from the entrainer, and the entrainer is recycled. The critical part of the design in Figure 12.23 is the two-feed extraction column. The design of the entrainer recovery column is straightforward, as it is a standard design. For the two-feed column, the middle section operation leaf needs to be constructed to see whether it intersects with the operation leaves for the top section (above the entrainer feed) and bottom section (below the feed point for the feed mixture). Figure 12.24 shows the mass balance around the top of the column including the entrainer feed. A *difference point* can be defined to be the composition of the net overhead product, that is, the difference between the distillate product and entrainer feed[15]. Since:

$$\Delta = D - E \tag{12.23}$$

Also, a stripping section mass balance gives:

$$\Delta = F - B \tag{12.24}$$

Thus, the difference point (net overhead product) is given by the intersection of a straight line joining the Feed *F*,

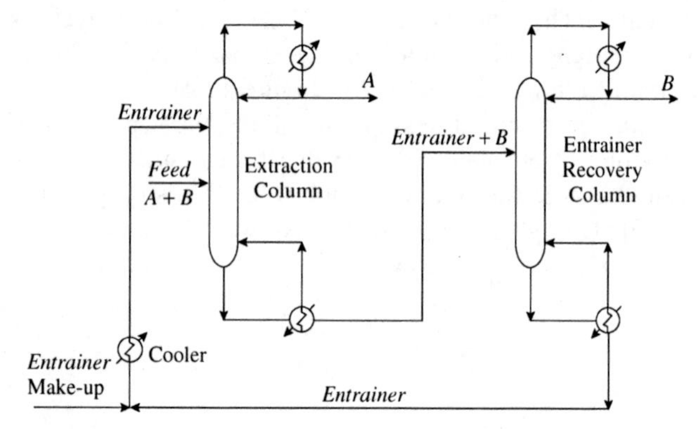

Figure 12.23 Extractive distillation flowsheet.

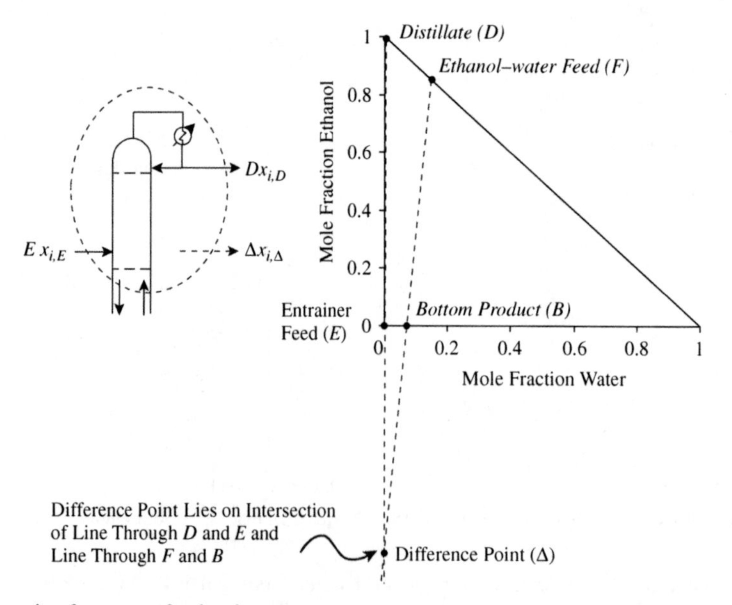

Figure 12.24 The difference point for a two-feed column.

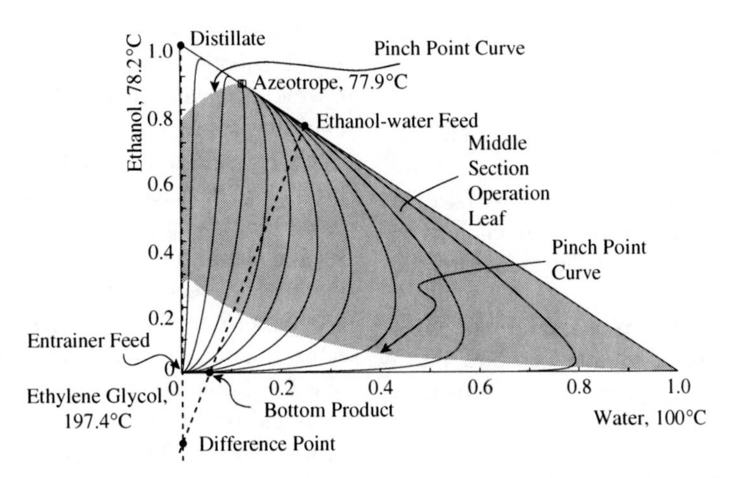

Figure 12.25 The difference point allows the pinch point curves for the middle section operation leaf to be constructed.

Bottom Product B with a straight line joining the Distillate D and Entrainer Feed E, as shown in Figure 12.24. Pinch point curves for the middle section can now be constructed by drawing tangents to the residue curves from the difference point (net overhead product). This is shown in Figure 12.25 for the system ethanol-water-ethylene glycol. The area bounded by the pinch point curves defines the middle section operation leaf.

As long as this middle section operation leaf intersects with those for the top section (above the entrainer feed) and the bottom section (below the feed point for the feed mixture), the column design will be feasible. Note that there will always be a maximum reflux ratio, above which the separation will not be feasible because the profiles in the top and bottom sections will tend to follow residue curves, which cannot intersect. Also, the separation becomes poorer at high reflux ratios as a result of the entrainer being diluted by the reflux of lower boiling components.

The size and shape of the middle section operation leaf depends on the location of the difference point, which in turn depends on the flowrate of entrainer. There will be a minimum flowrate of entrainer for feasible design. Above the minimum flowrate, the actual flowrate of the entrainer is an important degree of freedom for optimization.

Section profiles can also be developed for a two-feed column in a similar way to the section profiles for a single-feed column[16]. The section profiles will be the same as a single-feed column above the entrainer feed (Equation 12.12) and below the feed point for the feed mixture (Equation 12.22). Figure 12.26 shows middle section mass balances. The mass balance can be created either around the top of the column, as shown in Figure 12.26a, or around the bottom of the column, as shown in Figure 12.26b. A mass balance around the top of the column from Figure 12.26a gives:

$$Vy_{i,m+1} + Ex_{i,E} = Lx_{i,m} + Dx_{i,D} \tag{12.25}$$

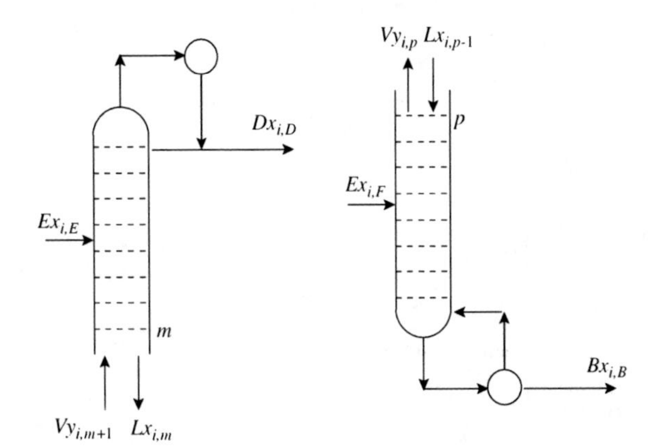

Figure 12.26 Middle section mass balances for a two-feed column.

Since:

$$V = (R + 1)D \tag{12.26}$$

and

$$L = RD + E \tag{12.27}$$

Equation 12.25 can be combined with Equations 12.26 and 12.27 to give:

$$y_{i,m+1} = \frac{(RD + E)x_{i,m} + Dx_{i,D} - Ex_{i,E}}{(R + 1)D} \tag{12.28}$$

Alternatively, a mass balance can be carried out around the bottom of the column from Figure 12.26b to give:

$$Lx_{,p-1} + Fx_{i,F} = Vy_{i,p} + Bx_{i,B} \tag{12.29}$$

Since

$$V = SB \tag{12.30}$$

and

$$\begin{aligned} L &= V + B - F \\ &= SB + B - F \end{aligned} \tag{12.31}$$

Equation 12.29 can be combined with Equations 12.30 and 12.31 to give:

$$x_{i,p-1} = \frac{SBy_{i,p} + Bx_{i,B} - Fx_{i,F}}{SB + B - F} \quad (12.32)$$

Either Equation 12.28 or Equation 12.32 can be used in conjunction with vapor–liquid equilibrium calculations to calculate the section profile for the middle section of the two-feed column.

Figure 12.27 shows the section profiles for the three sections of a two-feed column. The three profiles intersect in Figure 12.27, and the column will, in principle, be feasible.

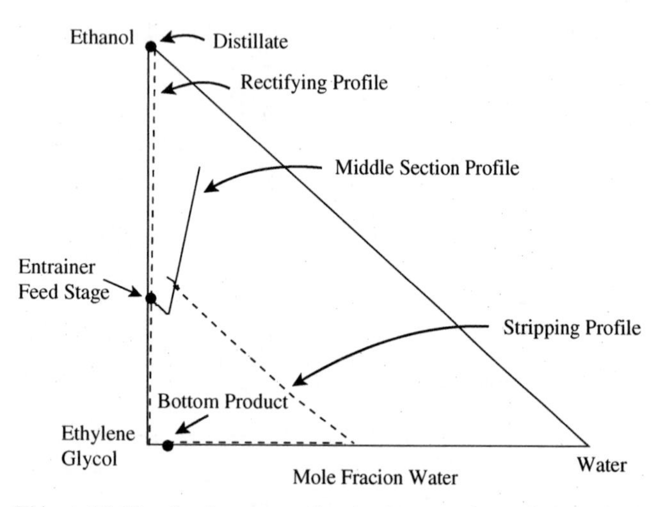

Figure 12.27 Section curves for the three sections of a two-feed column must intersect for a feasible two-feed column design.

Thus, the first step in the design of an extractive two-feed distillation column is to setup a system of intersecting operating leaves for the three sections of the column by choosing appropriate products and entrainer feed flowrate. Then by selecting a reflux ratio, the rectifying section profile can be calculated, as described previously. From the feasible entrainer flowrate, the section profile for the middle section can be calculated using Equation 12.28 to calculate the middle section profile. Calculation of the stripping section profile requires first the assumption of a reboil ratio. If three section profiles intersect, then the design is feasible. Some trial and error may be required for the reflux ratio and reboil ratio to obtain an intersection of the three profiles and feasible design. Figure 12.27 shows the section profiles for a two-feed column successfully intersecting to provide a feasible design. The number of stages in each of the column sections can also be obtained from the section profiles. This can then provide all the information required for a rigorous simulation.

One final point regarding extractive distillation is illustrated in Figure 12.28. The order in which the separation occurs depends on the change in relative volatility between the two components to be separated. Figure 12.28 shows both the residue curves and the equi-volatility curve for the system *A–B–entrainer*. This equi-volatility curve shows where the relative volatility between Components A and B is unity. On either side of the equi-volatility curve, the order of volatility of A and B changes. In Figure 12.28a, if the equi-volatility curve intersects the *A–entrainer* axis, then Component A should be recovered first. However, if the equi-volatility curve intersects the *B–entrainer* axis,

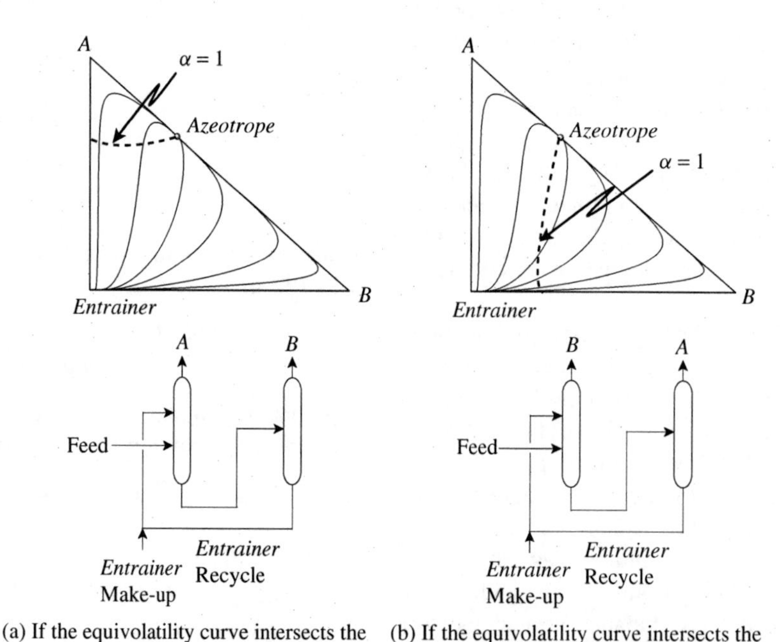

(a) If the equivolatility curve intersects the *Entrainer-A* axis, component *A* is recovered first.

(b) If the equivolatility curve intersects the *Entrainer-B* axis, component *B* is recovered first.

Figure 12.28 The order of separation in extractive distillation.

then Component B should be recovered first, as shown in Figure 12.28b.

All of the systems discussed so far involve homogeneous separation, that is, a single-phase liquid throughout. Consider now distillation systems involving two-liquid phases.

12.8 HETEROGENEOUS AZEOTROPIC DISTILLATION

Figure 12.29 shows a triangular diagram for a ternary system that forms two-liquid phases at certain compositions. From Figure 12.29, a binary I-L system is completely miscible, and a binary I-H system is also completely miscible. However, a binary L-H system shows partial miscibility in a range of concentrations. For L-I-H mixtures, there is a region of immiscibility in which two-liquid phases form. A mixture of composition F in the two-phase region in Figure 12.29 will separate into two phases of composition E and R. Phases in equilibrium with each other are termed *conjugate phases*, and the Line $E - R$ connecting the conjugate phases E and R is termed a *tie line*. Any number of tie lines may be constructed in the two-phase region as shown in Figure 12.29. The ratio of the two phases resulting in the separation of F into E and R is given by the Lever Rule. Thus, the flowrate of E is given by the length of Line F-R in Figure 12.29, and the flowrate of R is given by the length of the Line F-E in Figure 12.29. Figure 12.29 illustrates how to design decanters using a ternary diagram.

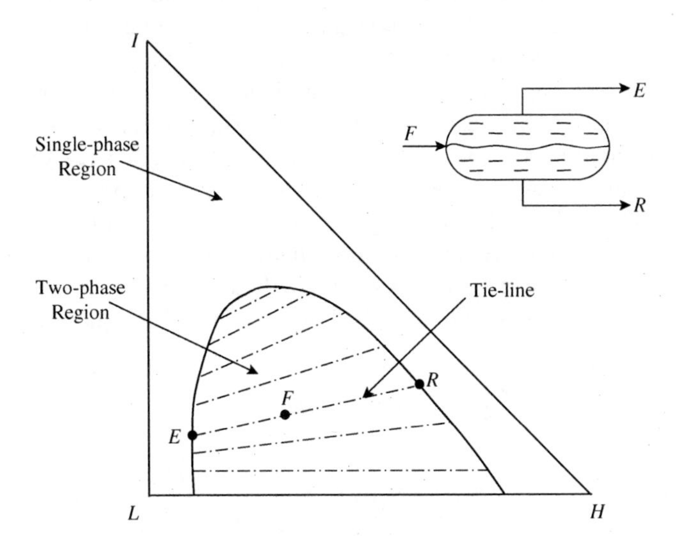

Figure 12.29 Phase splitting in a ternary system.

Now consider the distillation with the decanter. There are two basic arrangements possible when using a decanter in conjunction with distillation. The first is shown in Figure 12.30 where the overhead from the column is condensed to form a two-liquid phase mixture. Part is refluxed, and part goes to a decanter that separates two layers E and R. The basic problem with the arrangement shown in Figure 12.30 is that the reflux to the column at D is a two-phase mixture. It is preferable not to have two-liquid phases inside the column unless this cannot be avoided. Another arrangement is shown in Figure 12.31. This time the overhead vapor is condensed to form two-liquid phases D and R. D is taken as a top product, and R (a single-phase lower layer) is refluxed to the column. The overall separation is between F, D and B to give the column mass balance as shown in Figure 12.31.

One extremely powerful feature of heterogeneous distillation is the ability to cross distillation boundaries. It was noted previously that distillation boundaries divide the compositions into two regions that cannot be accessed from each other. Decanters allow distillation boundaries to be crossed, as illustrated in Figure 12.32. The feed to the decanter at F is on one side of the distillation boundary. This splits in the decanter to two-liquid phases E and R. These two-liquid phases are now on opposite sides of the distillation boundary. Phase splitting in this way is not constrained by a distillation boundary, and exploiting a two-phase separation in this way is an extremely effective way to cross distillation boundaries.

An example is shown in Figure 12.33 in which a feed containing a mole fraction of isopropyl alcohol 0.6 and a mole fraction of water 0.4 needs to be split into relatively pure isopropyl alcohol and water. There is an azeotrope

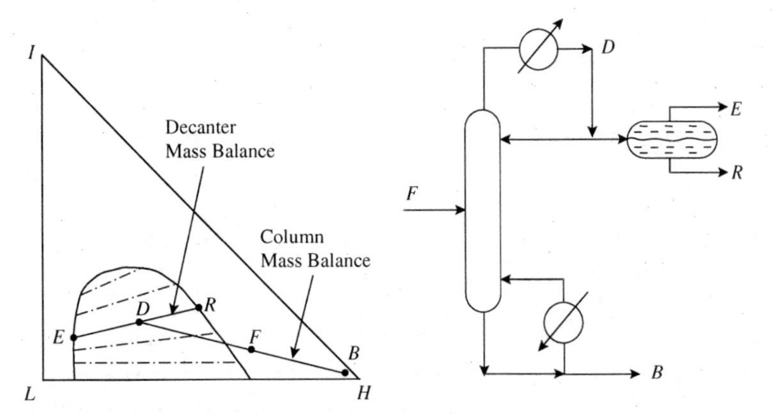

Figure 12.30 Heterogeneous distillation with liquid phases separated after distillation.

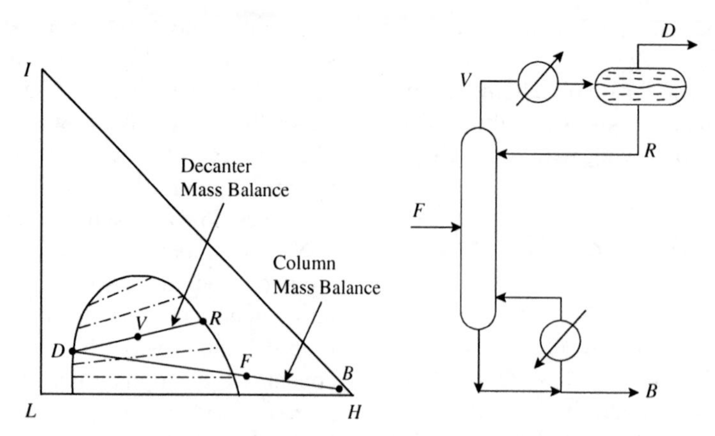

Figure 12.31 Heterogeneous distillation with phases separated before reflux is returned to the column.

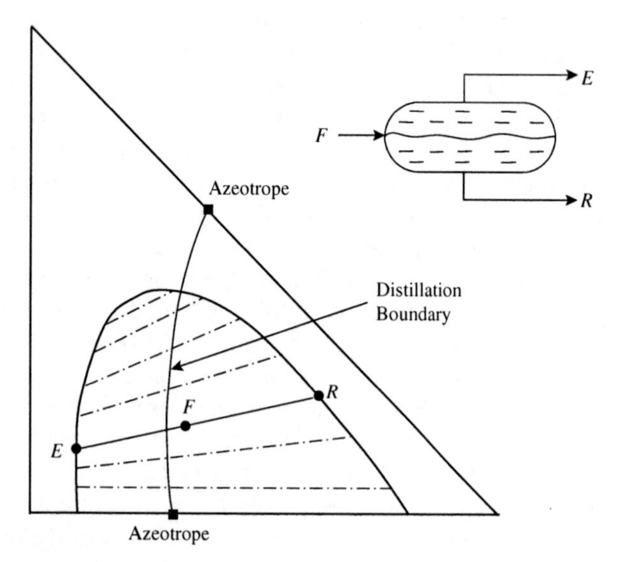

Figure 12.32 Phase splitting can be used to cross distillation boundaries.

between isopropyl alcohol (IPA) and water with a mole fraction isopropyl alcohol around 0.68. It is proposed to use di-isopropyl ether (DIPE) as an entrainer to separate the mixture. The ternary diagram in Figure 12.33a shows that the system IPA-DIPE-water forms a complex behavior. New azeotropes are formed between the IPA and DIPE and between DIPE and water. A ternary azeotrope is also formed. The distillation boundaries shown in Figure 12.33a would make this separation extremely difficult. However, the two-phase region shown in Figure 12.33a allow the distillation boundaries to be crossed. The synthesis of the separation system can be started by placing a decanter at the ternary azeotrope, as shown in Figure 12.33a. This follows the tie line and separates the ternary azeotrope into a DIPE-rich layer and a water-rich layer. Next, a distillation column is placed to produce IPA and the ternary azeotrope, as shown in Figure 12.33b. Finally, the DIPE-rich layer

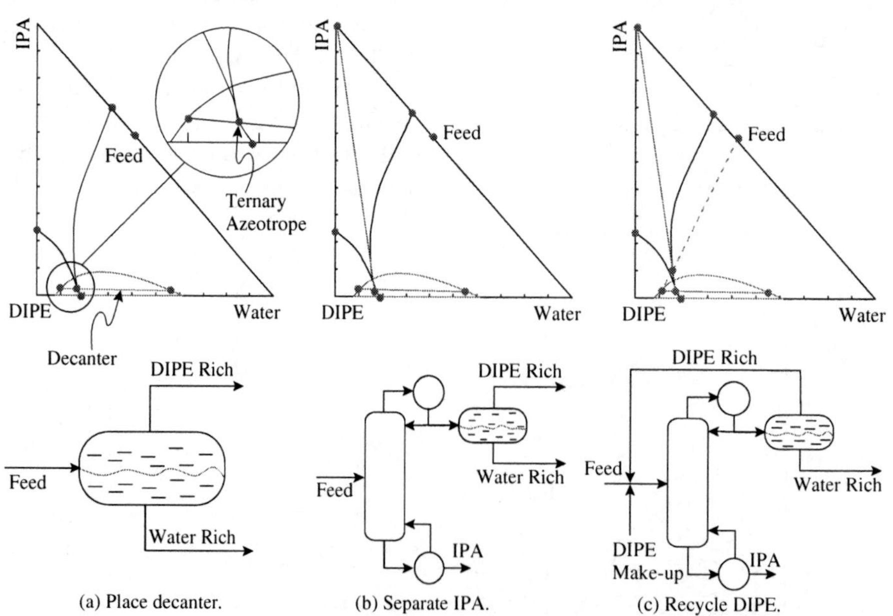

Figure 12.33 Separation of isopropyl alcohol (IPA) and water mixture using di-isopropyl ether (DIPE) as entrainer in heterogeneous azeotropic distillation.

is recycled from the decanter to the feed for the column. This mixes with the incoming feed mixture to provide the feed for the column, as shown in Figure 12.33c. The water product from the decanter in Figure 12.33c is not pure and may require further separation. This judgment is based on the prediction of the design of the decanter from Figure 12.33.

The phase equilibrium in the triangular diagrams in Figure 12.33 was predicted using the NRTL equation (see Chapter 4). The NRTL equation is capable of predicting vapor–liquid equilibrium, liquid–liquid equilibrium and vapor–liquid–liquid equilibrium. However, it is difficult to find a single set of parameters to represent all of these kinds of behavior well. The parameters for the triangular diagrams in Figure 12.33 were correlated from vapor–liquid equilibrium behavior. Figure 12.34b shows the triangular diagram for the same system, but with the two-phase region calculated from the NRTL equation, based on parameters correlated from liquid–liquid equilibrium data. The two-phase region is much larger, and the decanter is capable of producing almost pure water from this prediction.

Even further caution should be exhibited regarding the two-phase region in such diagrams. In Figures 12.33, 12.34a and 12.34b, the phase equilibrium is based on saturated conditions throughout. This is useful in judging the design of the distillation system and where two-liquid phase behavior will occur, as distillation by definition operates under saturated conditions. However, this is not necessarily the case in the decanter. The temperature in the decanter can be fixed, as it is outside of the column. Figure 12.34c shows the two-phase region again on the basis of NRTL parameters correlated from liquid–liquid equilibrium data, but this time plotted at a fixed temperature of 30°C. The two-phase region is slightly larger at 30°C when compared with saturated conditions. Generally, the lower the temperature, the larger will be the two-phase region. Lowering the temperature lowers the mutual solubility of the two-liquid phases. This is an important degree of freedom when designing a decanter. A better separation in the decanter can be obtained by condensing the distillation overheads and subcooling before the two-liquid phase separation.

It is important to understand whether there will be two-liquid phases present in the column. If two-liquid phases form in a large part of the column, it can make the column difficult to operate. The formation of two-liquid phases also affects the hydraulic design and mass transfer in the distillation (and hence stage efficiency). If it is possible to avoid the formation of two-liquid phases inside the column, then such behavior should be avoided. Unfortunately, there will be many instances when two-liquid phases on some plates cannot be avoided. The formation of two-liquid phases can also be sensitive to changes in the reflux ratio.

12.9 ENTRAINER SELECTION

When separating azeotropic mixtures, if possible, changes in the azeotropic composition with pressure should be exploited rather than using an extraneous mass-separating agent, since:

- The introduction of extraneous material can create new problems in achieving product purity specifications throughout the process.
- It is often difficult to separate and recycle extraneous material with high efficiency. Any material not recycled can become an environmental problem. As will be discussed later, the best way of dealing with effluent problems is not to create them in the first place.
- Extraneous material can create additional safety and storage problems.

Occasionally, a component that already exists in the process can be used as the entrainer, thus avoiding the introduction of extraneous materials for azeotropic distillation. However, in many instances practical difficulties and excessive cost might force the use of extraneous material.

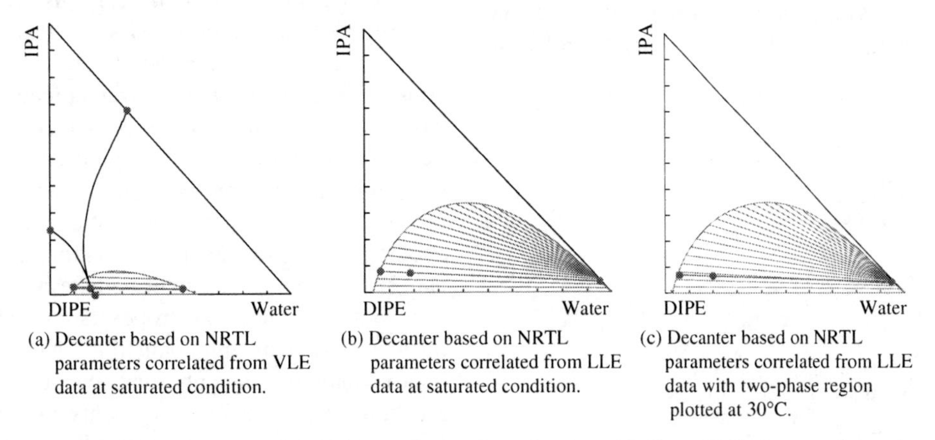

(a) Decanter based on NRTL parameters correlated from VLE data at saturated condition.

(b) Decanter based on NRTL parameters correlated from LLE data at saturated condition.

(c) Decanter based on NRTL parameters correlated from LLE data with two-phase region plotted at 30°C.

Figure 12.34 The two-phase region can be plotted from correlations fitted to VLE or LLE data, and can be plotted at saturated conditions or at a fixed temperature.

Whether a component is used that already exists within the process as an entrainer or an extraneous material is used as entrainer, it is necessary to be able to select between different entrainers. Distillation line maps and residue curve maps are particularly useful for entrainer selection, as the difficulty of separation can be judged from the configuration of the map. For example, it has been noted how distillation boundaries divide distillation line and residue curve maps into different regions and that distillation cannot access one region from another. However, in theory, it is possible to cross a distillation boundary as illustrated in Figure 12.35[16]. The distillation boundary in Figure 12.35 has a marked curvature, and a column can be placed as shown such that the feed is one side of the boundary and the products are on the other side of the boundary. Depending on the shape of the distillation lines or residue curves, the products *D* and *B* can be on the same distillation line or residue curve. In this way, a distillation boundary can be crossed using distillation, rather than relying on a liquid–liquid separation as discussed previously to cross distillation boundaries. Whilst arrangements like those in Figure 12.35 are possible in theory, there are many potential dangers associated with this as follows.[16]

1. The distillation boundary must be curved as shown in Figure 12.35. However, even if there is very significant curvature across the boundary, a column placed like the one in Figure 12.35 is going to be highly constrained in its operation.
2. There is always uncertainty and inaccuracy with vapor–liquid equilibrium data and correlations. Any errors in this data could mean an incorrect prediction of the location and shape of the boundary.
3. All of the discussions so far regarding distillation lines, residue curves and distillation boundaries have assumed equilibrium behavior. Real columns do not work at equilibrium, and stage efficiency must be accounted for. Each component will have its own stage efficiency, which means that each composition will deviate from equilibrium behavior differently. This means that if nonequilibrium behavior is taken into account, the shape of the distillation lines, residue curves and distillation boundaries will change[17]. Thus, the shape of the distillation boundary will be different in a real column when compared with equilibrium predictions. If a system is designed assuming equilibrium, there is no guarantee that it will still work in a real column with nonideal stages. These nonequilibrium effects can, in principle, be included into the analysis, but there is also considerable uncertainty regarding stage efficiency calculations and require a considerable amount of information on the geometry of the column and distillation internals[17].
4. Even with reassurance regarding uncertainties in the vapor–liquid equilibrium data and nonequilibrium effects, there is often a significant difference between the way a column is required to operate in practice compared with its design. The operation of the overall plant can often be different from the design because of a whole range of reasons. If the design is highly constrained and cannot be flexible to accommodate changes in operation, then it might not be able to function.

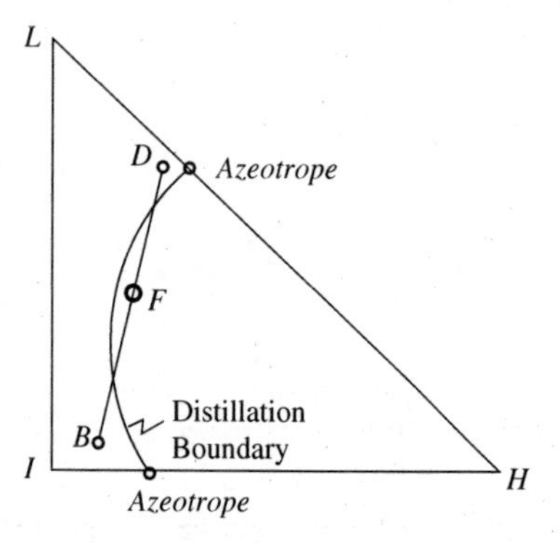

Figure 12.35 Crossing distillation boundaries.

Thus, while it is possible in theory to cross a curved distillation boundary as shown in Figure 12.35, it is generally more straightforward to follow designs that will be feasible over a wide range of reflux ratios and in the presence of uncertainties. Such designs can be readily developed using distillation line and residue curve maps.

When introducing an entrainer, it will need to have a significant effect on the relative volatility between the azeotropic components to be separated, and it must be possible to separate the entrainer relatively easily. One way of making sure the entrainer can be easily separated is to choose a component that will introduce a two-liquid phase separation. Such entrainers typically introduce additional distillation boundaries, but the overall separation can be efficient if the two-liquid separation produces mixtures with compositions in the different distillation regions[10].

When using an entrainer for separation of a homogeneous mixture, it is best to select components that do not introduce any additional azeotropes. The classical method for the separation of homogeneous mixtures separation is extractive distillation, which relies on the effect of a high-boiling entrainer on the relative volatility in the column sections below the entrainer feed. Such columns can work very well, but they exhibit sometimes counterintuitive behavior, in particular, with regard to the detrimental effect of high reflux diluting the entrainer composition. However, in most cases high-boiling entrainers will be the best choice for homogeneous distillation. Another possibility can be to choose an intermediate-boiling entrainer that does not introduce azeotropes, since this will lead to a residue curve map with no distillation boundaries. However, intermediate-boiling entrainers can only be practical for breaking azeotropes of components with large boiling point differences, otherwise an intermediate-boiling entrainer will lead to very difficult, energy-consuming separations of close-boiling mixtures. Finally, using low-boiling entrainers that do not introduce azeotropes is generally not practical, since such components would not tend to accumulate sufficiently in the liquid phase to make a difference on the relative volatility between the components to be separated.

Thus, distillation line and residue curve maps are excellent tools to evaluate feasibility of azeotropic separations, with just one exception, namely, the use of high-boiling entrainers for separation. In such cases, the equi-volatility curves discussed in this chapter are a better way of determining separation feasibility.

12.10 TRADE-OFFS IN AZEOTROPIC DISTILLATION

For a simple distillation column separating a ternary system, once the feed composition has been fixed, three-product component compositions can be specified, with at least one for each product. The remaining compositions will be determined by colinearity in the ternary diagram. For a binary distillation only two product compositions can be specified independently, one in each product. Once the mass balance has been specified, the column pressure, reflux (or reboil ratio) and feed condition must also be specified.

By contrast with nonazeotropic systems, for azeotropic systems there is a maximum reflux ratio above which the separation deteriorates[16]. This is because an increase in reflux ratio results in two competing effects. Firstly, as in nonazeotropic distillation, the relative position of the operating surface relative to the equilibrium surface changes to improve the separation. This is countered by a reduction in the entrainer concentration, owing to dilution by the increased reflux, which results in a reduction in the relative volatility between the azeotropic components, leading to a poorer separation[16].

However, this so far assumes that the feed to the column is fixed. Even if the overall feed to the separation system is fixed, the feed to each column can be changed by changing the amount of entrainer recycled. Such a trade-off has already been seen in Figure 12.21. As the amount of entrainer recycled is increased, this helps the azeotropic separation. This allows the reflux ratio to be decreased. However, as the entrainer recycle increases, it creates an excessive load on the overall system. The amount of entrainer recycled is therefore an important degree of freedom to be optimized.

12.11 MULTICOMPONENT SYSTEMS

All of the discussion in this chapter has so far related to binary or ternary systems. It will most often be the case that systems involving azeotropic behavior will also be multicomponent. The concepts developed here for ternary systems are readily extended to quaternary systems. The difference is that this cannot be represented on a ternary diagram but must be represented in three dimensions as a pyramid. Lines in the ternary diagram become surfaces in the quaternary diagram. It simply becomes more difficult to represent graphically and to interpret. The concepts can be extended beyond quaternary systems but cannot be represented graphically at all, unless three or four components are picked out and represented to the exclusion of the other components.

When dealing with multicomponent systems, one possible approach is to simplify down the problem by lumping components together and representing it in a ternary analysis. Such an approach should be exercised with great caution. Even trace amounts of components can change the analysis very significantly in azeotropic systems. For example, the designer might consider that the system has been represented reasonably well if 99% of the mixture can be accounted for as a ternary mixture, and the influence of the other 1% neglected. However, in some systems, varying the makeup of the 1% can significantly change the shape of operation leaves, and their equivalent in multidimensional space, and render a design infeasible that appears to be feasible on the basis of an analysis of the behavior of 99% of the mixture[18]. Thus, the residue curves, distillation lines, pinch point curves and operation leaves for multicomponent mixtures should be constructed from a full multicomponent calculation, even if the dominant components are to be represented on a ternary diagram.

Once a design has been synthesized, it should be checked carefully with the most detailed simulation possible. Even if the design is confirmed to be feasible by this simulation, the sensitivity of the design should be checked carefully by simulation for:

- errors in the phase equilibrium behavior by perturbing the phase equilibrium data
- changes in the feed composition.

12.12 MEMBRANE SEPARATION

So far, the separation of azeotropic systems has been restricted to the use of pressure shift and the use of entrainers. The third method is to use a membrane to alter the vapor–liquid equilibrium behavior. *Pervaporation* differs from other membrane processes in that the phase-state on one side of the membrane is different from the other side. The feed to the membrane is a liquid mixture at a high-enough pressure to maintain it in the liquid phase. The other side of the membrane is maintained at a pressure at or below the dew point of the permeate, maintaining it in the vapor phase. Dense membranes are used for pervaporation, and selectivity results from chemical affinity (see Chapter 10). Most pervaporation membranes in commercial use are hydrophyllic[19]. This means that they preferentially allow

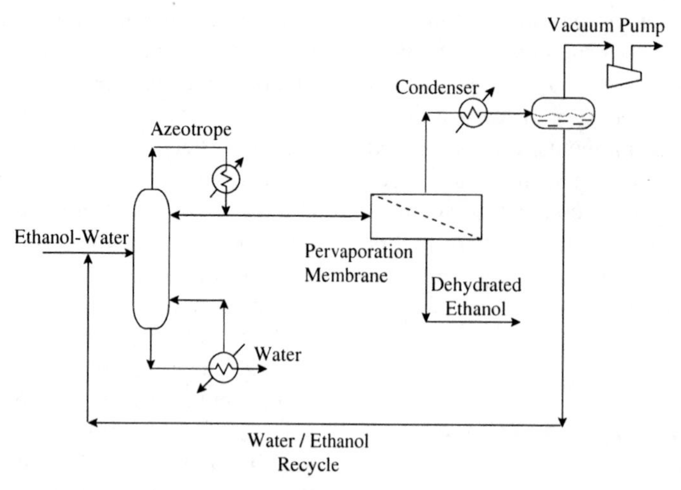

Figure 12.36 Flowsheet for dehydration of ethanol using pervaporation.

water to permeate and are therefore suitable for the dehydration of organics. Typical applications include the dehydration of ethanol–water and isopropanol–water mixtures, both of which form azeotropes[19]. A flowsheet for ethanol dehydration is shown in Figure 12.36. An ethanol–water mixture is fed to a standard distillation column that separates excess water at the bottom from a mixture in the overhead approaching the azeotropic composition. This is then fed to a pervaporation membrane that dehydrates the ethanol past the azeotrope by allowing water to permeate through the membrane. The low-pressure side of the membrane in Figure 12.36 is maintained under vacuum to ensure that the water leaves in the vapor phase. This is condensed and recycled to the distillation column as there is still a significant amount of ethanol that permeates through the membrane and should be recovered.

Figure 12.37 shows a flowsheet for the separation of an azeotropic mixture using a membrane, but this time using vapor permeation. The mixture is first distilled to approach the azeotrope using a distillation column with a partial condenser. The uncondensed vapor is fed to a vapor permeation membrane that preferentially permeates the organic material. The retentate vapor is passed back to the distillation column. In this way, the membrane allows the azeotrope to be crossed.

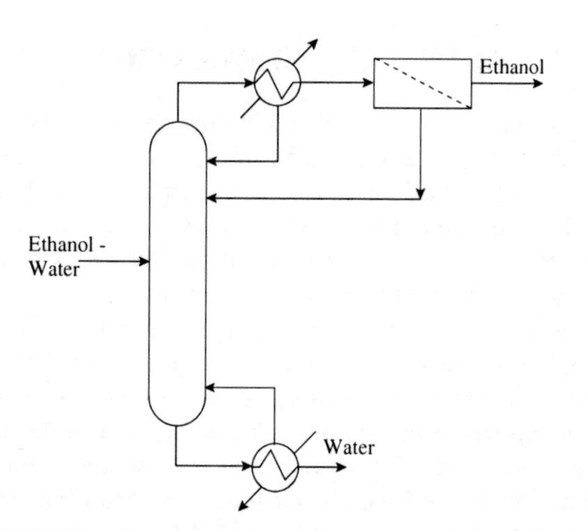

Figure 12.37 Flowsheet for dehydration of ethanol using vapor permeation.

12.13 DISTILLATION SEQUENCING FOR AZEOTROPIC DISTILLATION – SUMMARY

When liquid mixtures exhibit azeotropic behavior, it presents special challenges for distillation sequencing. At the azeotropic composition, the vapor and liquid are both at the same composition for the mixture. The order of volatility of components changes, depending on which side of the azeotrope the composition occurs. There are three ways of overcoming the constraints imposed by an azeotrope.

- Pressure shift
- Use of an entrainer
- Membrane separation.

Pressure shift should always be explored as the first option when separating an azeotropic system. Adding extraneous components to a separation should always be avoided if possible. Unfortunately, most azeotropes are insensitive to change in pressure, and at least a 5% change in composition with pressure is required for a feasible separation using pressure shift[1].

If pressure shift cannot be exploited, then the next option is to add an entrainer to the mixture that interacts differently with the components in the mixture to alter the vapor–liquid equilibrium behavior in a favorable way. When dealing with ternary systems, the mass balance and vapor–liquid equilibrium behavior can be represented on a triangular diagram. The two limiting cases of distillation at total reflux and minimum reflux conditions can be used to understand the system. For staged columns at total reflux conditions, distillation lines can be plotted. Residue curves represent the behavior of packed columns at total reflux. The feasibility of a column section can be represented by the area between the total reflux and pinch point lines, as an operating leaf. If the operating leaves for the rectifying and stripping sections of a column intersect, in principle, the column will be feasible.

Some systems form two-liquid phases for certain compositions and this can be exploited in heterogeneous azeotropic distillation. The use of liquid–liquid separation in a decanter can be extremely effective and can be used to cross distillation boundaries.

When selecting entrainers for homogeneous mixture separation, the entrainer should preferably not introduce any new azeotropes, otherwise it will be difficult to separate the entrainer from the components to be separated. When separating multicomponent mixtures, the first thing to check generally is if there are components in the feed that can facilitate the separation of azeotrope-forming components, because using such components will typically lead to more cost-effective designs than processes in which the azeotropic separations are left to the end of a sequence and extraneous separating agents are chosen. Using components that are not already in the feed will generally require dedicated additional recovery steps.

Membranes can also be used to alter the vapor–liquid equilibrium behavior and allow separation of azeotropes. The liquid mixture is fed to one side of the membrane, and the permeate is held under conditions to maintain it in the vapor phase. Most separations use hydrophyllic membranes that preferentially pass water rather than organic material. Thus, pervaporation is commonly used for the dehydration of organic components.

12.14 EXERCISES

1. An equimolar mixture of ethanol and ethyl acetate is to be separated by distillation into relatively pure products. The mixture forms a minimum-boiling azeotrope, as detailed in Table 12.1. However, the composition of the azeotrope is sensitive to pressure, showing a significant increase in mole fraction of ethanol with increasing pressure, as indicated in Table 12.1. Sketch a flow scheme for the separation of the binary mixture that exploits change in pressure.
2. An equimolar mixture of ethyl acetate and methanol is to be separated by distillation into relatively pure products. Data from Table 12.1 indicate that the mixture forms a minimum-boiling azeotrope. The data in Table 12.1 also show a significant decrease in mole fraction of ethyl acetate for the azeotrope as pressure increases. Sketch a flow scheme for the separation that exploits pressure change.
3. Figure 12.38 shows a distillation sequence, together with its mass balance. Sketch a representation of the mass balance in a triangular diagram.
4. Sketch the distillation line map (residue curve map) for the system ethanol-ethyl acetate-methanol at 1 atm and 5 atm from the data in Table 12.1. Does the system have a distillation boundary? Is the position of the boundary sensitive to pressure?
5. A ternary mixture of mole fraction ethanol of 0.15, ethyl acetate of 0.6 and methanol 0.25 is to be separated into relatively pure products. Sketch a system of distillation columns and mixer arrangements in the triangular diagram to carry out the separation by exploiting the shift in the distillation boundary with pressure. Sketch the flowsheet corresponding with this mass balance.
6. The vapor–liquid equilibrium for a ternary system of Components *A*, *B* and *C* can be represented by:

$$y_A = 0.2x_A$$

$$y_B = 2.0x_B$$

$$y_C = 1 - y_A - y_C$$

Starting from a bottoms composition $x_A = 0.95$, $x_B = 0.04$ and $x_C = 0.01$, calculate the stripping section profile for the

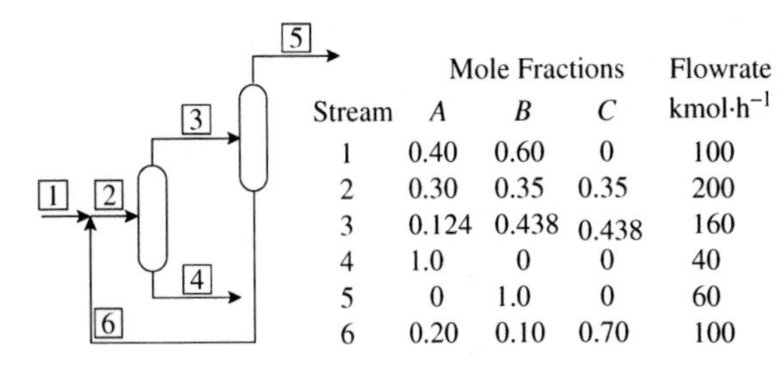

Stream	Mole Fractions *A*	*B*	*C*	Flowrate kmol·h^{-1}
1	0.40	0.60	0	100
2	0.30	0.35	0.35	200
3	0.124	0.438	0.438	160
4	1.0	0	0	40
5	0	1.0	0	60
6	0.20	0.10	0.70	100

Figure 12.38 Sequence representation in a triangular diagram.

Table 12.1 Data for the ethanol-ethyl acetate-methanol system.

Component	Boiling temperature (°C) and Azeotrope composition at 1 atm	Boiling temperature (°C) and a azeotrope composition at 5 atm
Ethanol	78.2	125.6
Ethyl acetate	77.1	135.8
Methanol	64.5	111.8
Ethanol-ethyl acetate azeotrope	72.2°C 0.465 mole fraction ethanol	122.7°C 0.677 mole fraction ethanol
Ethyl acetate-methanol azeotrope	62.3°C 0.291 mole fraction ethyl acetate	110.6°C 0.180 mole fraction ethyl acetate

reboiler and bottom 5 stages of the column for a reboil ratio of 1. Sketch the profile on a ternary diagram.

7. The separation in Figure 12.21 can be carried out in an alternative sequence. Sketch the mass balance for the alternative sequence in a triangular diagram and the resulting flowsheet.

REFERENCES

1. Holland CD, Gallun SE and Lockett MJ (1981) Modeling Azeotropic and Extractive Distillations, *Chem Eng*, **88** (March 23): 185.
2. Doherty MF and Malone MF (2001) *Conceptual Design of Distillation Systems*, McGraw-Hill.
3. Hougen OA, Watson KM and Ragatz RA (1954) *Chemical Process Principles Part I Material and Energy Balances*, 2nd Edition, John Wiley.
4. Zharov VT (1968) Phase Representations and Rectification of Multi-component Solutions, *J Appl Chem USSR*, **41**: 2530.
5. Petlyuk FB and Avetyan VS (1971) Investigation of Three Component Distillation at Infinite Reflux, *Theor Found Chem Eng*, **5**: 499.
6. Petlyuk FB, Kievskii VY and Serafimov LA (1975) Thermodynamic and Topological Analysis of the Phase Diagrams of Polyazeotropic Mixtures: II. Algorithm for Construction of Structural Graphs for Azeotropic Ternary Mixtures, *Russ J Phys Chem*, **49**: 1836.
7. Doherty MF and Perkins JD (1979) The Behaviour of Multicomponent Azeotropic Distillation Processes, *IChemE Symp Ser*, **56**:4.2/21.
8. Doherty MF and Perkins JD (1978) On the Dynamics of Distillation Processes I The Simple Distillation of Multicomponent Non-reacting Homogeneous Liquid Mixtures, *Chem Eng Sci*, **33**: 281.
9. Schreinemakers FAH (1901) Dampfdrueke im System: Wasser Aceton and Phenol, *Z Phys Chem*, **39**: 440.
10. Doherty MF and Perkins JD (1979) On the Dynamics of Distillation Process: III. Topological Structure of Ternary Residue Curve Maps, *Chem Eng Sci*, **34**: 1401.
11. Petlyuk FB, Kievskii VY and Serafimov LA (1975) Thermodynamic and Topological Analysis of the Phase Diagrams of Polyazeotropic Mixtures: I. Definition of Distillation Regions Using a Computer, *Russ J Phys Chem*, **49**: 1834.
12. Wahnschafft OM, Koehler JW, Blass E and Westerberg AW (1992) The Product Composition Regions of Single-Feed Azeotropic Distillation Columns, *Ind Eng Chem Res*, **31**: 2345.
13. Levy SG, Van Dongen DB and Doherty MF (1985) Design and Synthesis of Homogeneous Azeotropic Distillation: 2. Minimum Reflux Calculations for Non-ideal and Azeotropic Columns, *Ind Eng Chem Fund*, **24**: 463.
14. Castillo FJL, Thong DYC and Towler GP (1998) Homogeneous Azeotropic Distillation 1. Design Procedure for Single-Feed Columns at Non-total Reflux, *Ind Eng Chem Res*, **37**: 987.
15. Wahnschafft OM and Westerberg AW (1993) The Product Composition Regions of Azeotropic Distillation Columns: II. Separability in Two-Feed Columns and Entrainer Selection, *Ind Eng Chem Res*, **32**: 1108.
16. Laroche L, Bekiaris N, Andersen HW and Morari M (1992) Homogeneous Azeotropic Distillation: Separability and Flowsheet Synthesis, *Ind Eng Chem Res*, **31**: 2190.
17. Castillo FJL and Towler GP (1998) Influence of Multicomponent Mass Transfer on Homogeneous Azeotropic Distillation, *Chem Eng Sci*, **53**: 963.
18. Thong DYC and Jobson M (2001) Multi-component Azeotropic Distillation 1. Assessing Product Feasibility, *Chem Eng Sci*, **56**: 4369.
19. Wynn N (2001) Pervaporation Comes of Age, *Chem Eng Prog*, **97**(10): 66.

13 Reaction, Separation and Recycle Systems for Continuous Processes

After making some basic decisions regarding the choice of reactor, and the resulting separation system, these two systems need to be brought together. Raw materials need to be brought from storage, purified or treated if necessary, and fed to the reaction system. The effluent from the reactor is passed to the separation system and the product isolated, along with byproducts and unreacted feed material. The product will most likely go forward to product storage. However, the direction of material flow is not just forward and some material, especially unreacted feed material, might be *recycled* to earlier steps in the process. Byproducts and unreacted feed material might also require storage. Completing the structure of the reactor, separation and recycle system allows a material and energy balance for the basic process to be completed. Knowing the size and conditions of the material flows for the basic process allows a first evaluation of the pumping and compression requirements to be evaluated. A first evaluation of the storage requirements can also be carried out.

13.1 THE FUNCTION OF PROCESS RECYCLES

The recycling of material is an essential feature of most chemical processes. Thus, it is necessary to consider the main factors that dictate the recycle structure of a process. Start by restricting consideration to continuous processes.

1. *Reactor conversion.* Earlier in Chapters 5 to 7, the initial choice of reactor type, operating conditions and conversion was discussed. The initial assumption for the conversion varies according to whether there are single reactions or multiple reactions producing byproducts and whether reactions are reversible. Consider the simple reaction:

$$FEED \longrightarrow PRODUCT \qquad (13.1)$$

To achieve complete conversion of *FEED* to *PRODUCT* in the reactor might require an extremely long residence time, which is normally uneconomic. Thus, if there is no byproduct formation, the initial reactor conversion is set to be around 95% as discussed in Chapter 5. The reactor effluent thus contains unreacted *FEED* and *PRODUCT* (Figure 13.1).

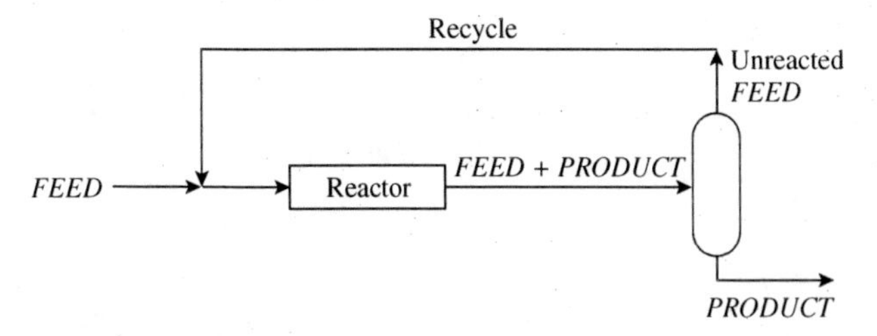

Figure 13.1 Incomplete conversion in the reactor requires a recycle for unconverted feed material.

Because a pure product is required, a separator is needed. The unreacted *FEED* is usually too valuable to be disposed of and is therefore recycled to the reactor inlet via a pump or compressor, Figure 13.1. In addition, disposal of unreacted *FEED*, rather than recycling, could create an environmental problem.

2. *Byproduct formation.* Consider now the case where a byproduct is formed either by the primary reaction such as:

$$FEED \longrightarrow PRODUCT + BYPRODUCT \qquad (13.2)$$

or via a secondary reaction such as:

$$FEED \longrightarrow PRODUCT$$
$$PRODUCT \longrightarrow BYPRODUCT \qquad (13.3)$$

An additional separator is now required (Figure 13.2a). Again the unreacted *FEED* is normally recycled but the *BYPRODUCT* must be removed to maintain the overall material balance. An additional complication now arises with two separators because the separation sequence can be changed (Figure 13.2b).

Also, instead of using two separators, a purge can be used (Figure 13.2c). Using a purge saves the cost of a separator but incurs raw material losses, and possibly waste treatment and disposal costs. This might be worthwhile if the *FEED-BYPRODUCT* separation is expensive. To use a purge, the *FEED* and *BYPRODUCT* must be adjacent to each other in the order of separation (e.g. adjacent to each other in order of volatility if distillation is used as the means of separation). Care should be taken to ensure that the resulting increase in concentration of *BYPRODUCT* in the reactor does not have an adverse effect

Chemical Process Design and Integration R. Smith
 ISBNs: 0-471-48680-9 (HB); 0-471-48681-7 (PB)

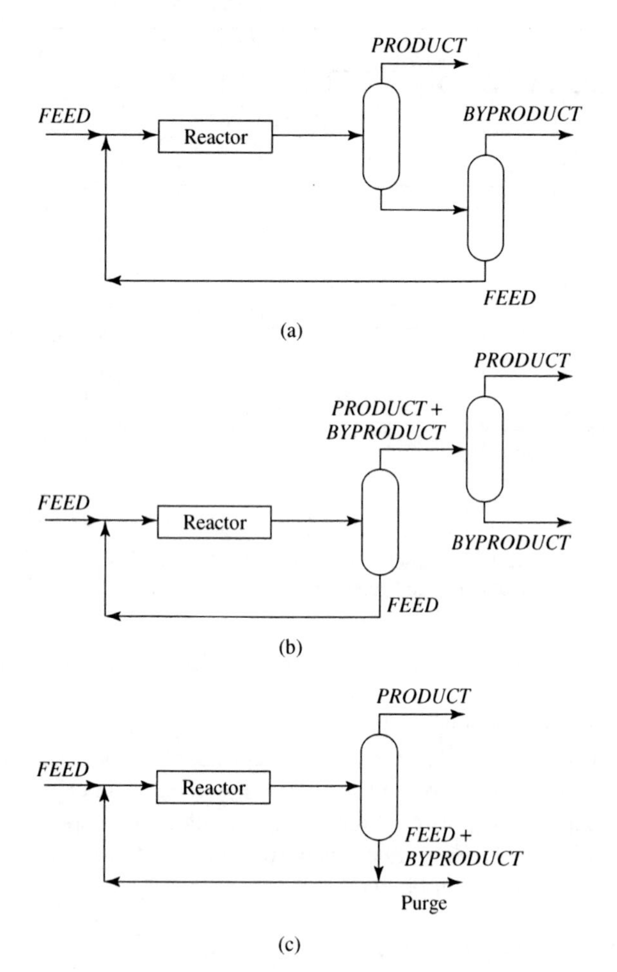

Figure 13.2 If a byproduct is formed in the reactor, then different recycle structures are possible.

on reactor performance. Too much *BYPRODUCT* might, for example, cause deterioration in the performance of the reactor catalyst.

Clearly, the separation configurations shown in Figure 13.2 change between different processes if the properties on which the separation is based change the order of separation. For example, if distillation is to be used for the separation and the order of volatility between the components changes, then the order of the separation will also change from that shown in Figure 13.2.

3. *Recycling byproducts for improved selectivity.* In systems of multiple reactions, byproducts are sometimes formed in secondary reactions that are reversible, such as:

$$\begin{aligned} &FEED \longrightarrow PRODUCT \\ &FEED \rightleftharpoons BYPRODUCT \end{aligned} \quad (13.4)$$

The three recycle structures shown in Figure 13.2 can also be used with this case. However, because the *BYPRODUCT* is now being formed by a secondary reaction that is reversible, its formation at source can be inhibited

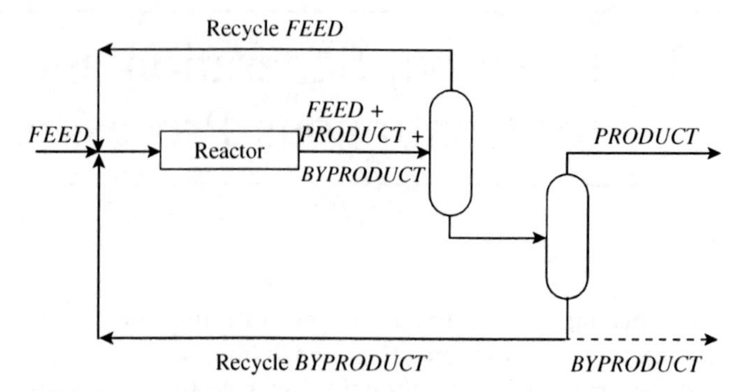

Figure 13.3 If a byproduct is formed via a reversible secondary reaction then recycling the byproduct can inhibit its formation at source.

by recycling *BYPRODUCT*, as shown in Figure 13.3. In Figure 13.3, the *BYPRODUCT* formation might be inhibited to the extent that it is effectively stopped, or it may be only reduced. If the formation is only reduced, then the net *BYPRODUCT* formation must be removed, as shown in Figure 13.3. Again, the separation configuration will change between different processes as the physical properties on which the separation is based change the order of separation.

4. *Recycling byproducts or contaminants that damage the reactor.* When recycling unconverted feed material, it is possible that some byproducts or contaminants, such as products of corrosion, can poison the catalyst in the reactor. Even trace quantities can sometimes be damaging to the catalyst. It is clearly desirable to remove such damaging components from the recycle in arrangements similar to Figure 13.2.

5. *Feed impurities.* So far it has been assumed that the feed material is pure. An impurity in the feed affects the recycle structure and opens up further options. The first option in Figure 13.4a shows the impurity being separated before entering the process. If the impurity has an adverse effect on the reaction or poisons the catalyst, this is the obvious solution. However, if the impurity does not have a significant effect on the reaction, then it could perhaps be passed through the reactor and be separated as shown in Figure 13.4b. Alternatively, the separation sequence could be changed as shown in Figure 13.4c[1].

The fourth option shown in Figure 13.4d uses a purge[1]. As with its use to separate byproducts, the purge saves the cost of a separation, but incurs raw material losses. This might be worthwhile if the *FEED-IMPURITY* separation is expensive. To use a purge, the *FEED* and *IMPURITY* must be adjacent to each other in the order of separation (e.g. adjacent in the order of volatility if a single-stage flash or distillation is used for the separation). Care should be taken to ensure that the resulting increase in concentration of *IMPURITY* in the reactor does not have an adverse effect on reactor performance.

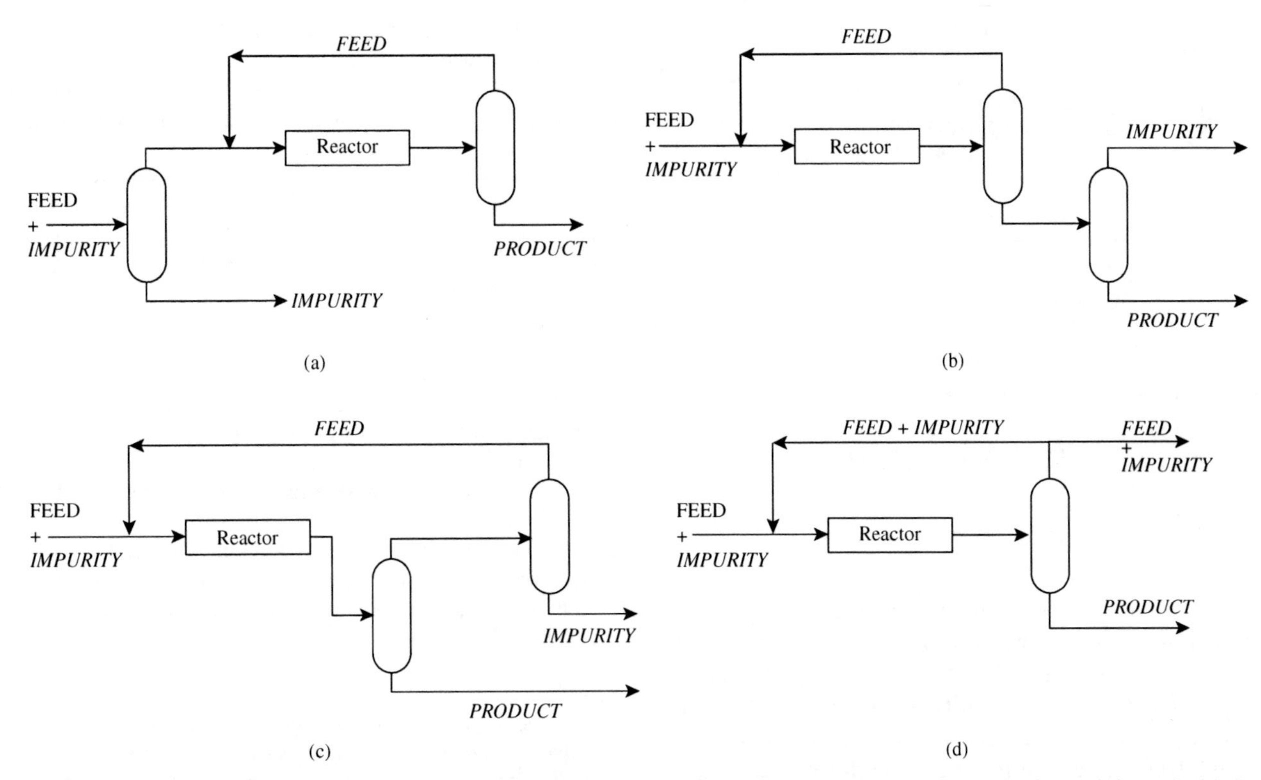

Figure 13.4 Introduction of an impurity in the feed creates further options for recycle structures. (Reproduced from Smith R and Linnhoff B, 1988, *Trans IChemE, CHERD*, **66**:195 by permission of the Institution of Chemical Engineers.)

Again, the separation configuration will change between different processes as the properties on which the separation is based change the order of separation, when compared with Figure 13.4.

6. *Reactor diluents and solvents*. As pointed out in Chapter 6, an inert diluent such as steam is sometimes needed in the reactor to lower the partial pressure of reactants in the vapor phase. Diluents are normally recycled. An example is shown in Figure 13.5. The actual configuration used depends on the order of separation.

Many liquid-phase reactions are carried out in a solvent. If this is the case, then the solvent, in the first instance, should be assumed to be separated and recycled in arrangements similar to that in Figure 13.5. In some cases, the solvent will be contaminated with byproducts of reaction that will need to be separated and disposed of (e.g. by thermal oxidation). In this case, it might be cheaper to dispose of the entire solvent, rather than separate and recycle the solvent. However, for the efficient use of materials, every effort should be made to recycle solvents, as illustrated in Figure 13.5.

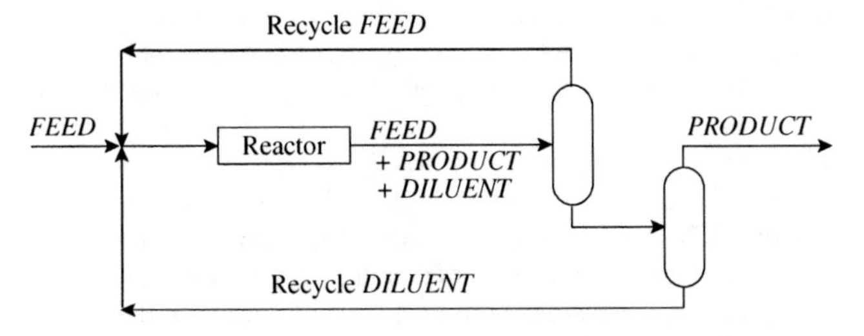

Figure 13.5 Diluents and solvents are normally recycled.

7. *Reactor heat carrier*. As pointed out in Chapter 7, if adiabatic operation is not possible and it is not possible to control temperature by indirect heat transfer, then an inert material can be introduced to the reactor to increase its heat capacity flowrate (i.e. product of mass flowrate and specific heat capacity). This will reduce temperature rise for exothermic reactions or reduce temperature decrease for endothermic reactions. The introduction of an extraneous component as a heat carrier effects the recycle structure of the flowsheet. Figure 13.6a shows an example of the recycle structure for just such a process.

Where possible, introducing extraneous materials into the process should be avoided, and material already present in the process should be used. Figure 13.6b illustrates the use of the product as the heat carrier. This simplifies the recycle structure of the flowsheet and removes the need for one of the separators (Figure 13.6b). The use of the product as heat carrier is obviously restricted to situations where the product does not undergo secondary reactions to unwanted byproducts. Unconverted feed that is recycled also acts as a heat carrier. Thus, rather than relying on recycled product to limit the temperature rise (fall), simply opt for a low conversion, a high recycle of feed and a resulting small temperature change in the reactor.

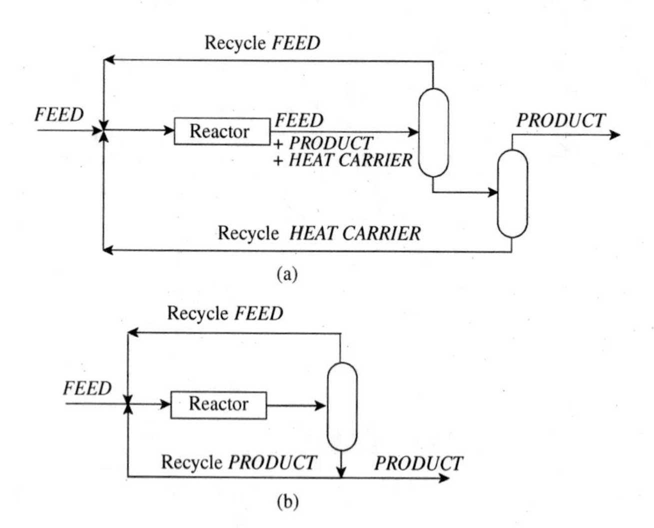

Figure 13.6 Heat carriers are normally recycled.

Other options are possible. If the process produces a byproduct of reaction, then this can be recycled, provided it does not have an adverse effect on the reactor performance. If the feed enters with an impurity, then the impurity could also be recycled as a heat carrier, provided it too does not have an adverse effect on the reactor performance.

Whether an extraneous component, product, feed, byproduct or feed impurity is used as heat carrier, as before, the actual configuration of the separation configuration will change between different processes as the properties on which the separation is based change the order of separation.

Having considered the main factors that determine the need for recycles, care should be taken if a flowsheet requires multiple recycles. It is counterproductive to separate two components adjacent in the order of separation that are to be recycled to the reactor, since they would only be re-mixed at some point before entering the reactor. The designer should always be on guard to avoid unnecessary separation and unnecessary mixing.

Example 13.1 Monochlorodecane (MCD) is to be produced from decane (DEC) and chlorine via the reaction[2,3]:

$$\underset{\text{DEC}}{C_{10}H_{22}} + \underset{\text{chlorine}}{Cl_2} \longrightarrow \underset{\text{MCD}}{C_{10}H_{21}Cl} + \underset{\text{hydrogen chloride}}{HCl}$$

A side reaction occurs in which dichlorodecane (DCD) is produced:

$$\underset{\text{MCD}}{C_{10}H_{21}Cl} + \underset{\text{chlorine}}{Cl_2} \longrightarrow \underset{\text{DCD}}{C_{10}H_{20}Cl_2} + \underset{\text{hydrogen chloride}}{HCl}$$

The byproduct DCD is not required. Hydrogen chloride can be sold to a neighboring plant. Assume at this stage that all separations can be carried out by distillation. The normal boiling points are given in the Table 13.1.

Table 13.1 Data for the materials involved with the production of monochlorodecane.

Material	Molar mass ($kg{\cdot}kmol^{-1}$)	Normal boiling point (K)	Value ($\${\cdot}kg^{-1}$)
Hydrogen chloride	36	188	0.35
Chlorine	71	239	0.21
Decane	142	447	0.27
Monochlorodecane	176	488	0.45
Dichlorodecane	211	514	0

a. Determine alternative recycle structures for the process by assuming different levels of conversion of raw materials and different excesses of reactants.
b. Which structure is most effective in suppressing the side reaction?
c. What is the minimum selectivity of decane that must be achieved for profitable operation? The values of the materials involved, together with their molar mass, are given in the Table 13.1.

Solution

a. Four possible arrangements can be considered:

(i) *Complete conversion of both feeds.* Figure 13.7a shows the most desirable arrangement; complete conversion of the decane and chlorine in the reactor. The absence of reactants in the reactor effluent means that no recycles are needed.

Although the flowsheet shown in Figure 13.7a is very attractive, it is not practical. This would require careful control of the stoichiometric ratio of decane to chlorine, taking into account both the requirements of the primary and byproduct reactions. Even if it were possible to balance out the reactants exactly, a small upset in process conditions would create an excess of either decane or chlorine and these would then appear as components in the reactor effluent. If these components appear in the reactor effluent of the flowsheet in Figure 13.7a, there are no separators to deal with their presence and no means of recycling unconverted raw materials.

Also, although there are no selectivity data for the reaction, the selectivity losses would be expected to increase with increasing conversion. Complete conversion would tend to produce high byproduct formation and poor selectivity. Finally, the reactor volume required to give complete conversion would be extremely large.

(ii) *Incomplete conversion of both feeds.* If complete conversion is not practical, consider incomplete conversion. This is shown in Figure 13.7b. In this case, all components are present in the reactor effluent, and one additional separator and a recycle are required. Thus the complexity of the flowsheet is significantly increased, compared with that for complete conversion.

Note that no attempt has been made to separate the chlorine and decane since they are remixed after recycling to the reactor.

(iii) *Excess chlorine.* Use of excess chlorine in the reactor can force the decane to effectively complete conversion (Figure 13.7c). Now there is effectively no decane in the reactor effluent and again three separators and a recycle are required.

In practice, there is likely to be a trace of decane in the reactor effluent. However, this should not be a problem since it can either

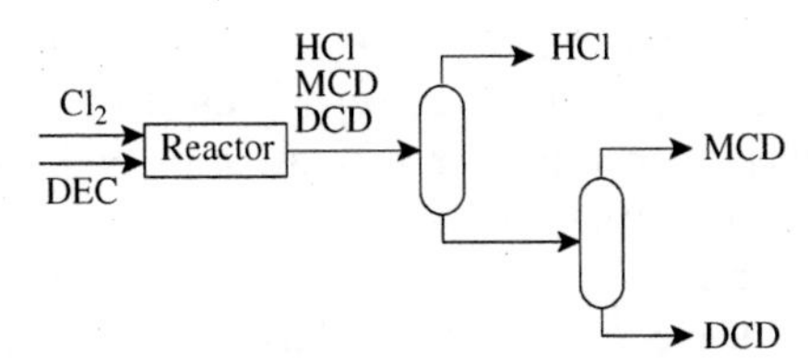

(a) Assume complete conversion of both feeds.

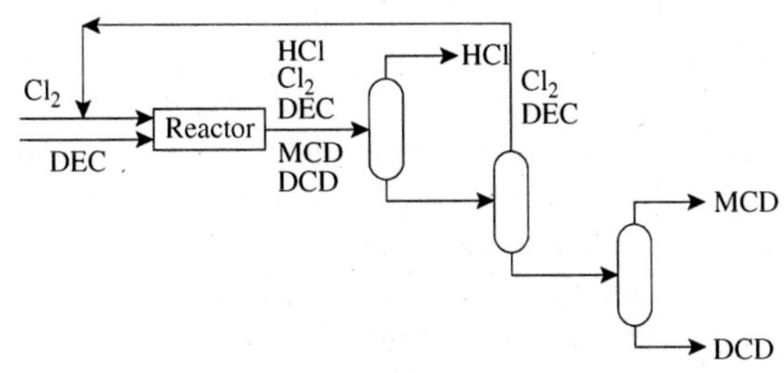

(b) Assume neither feed is completely converted.

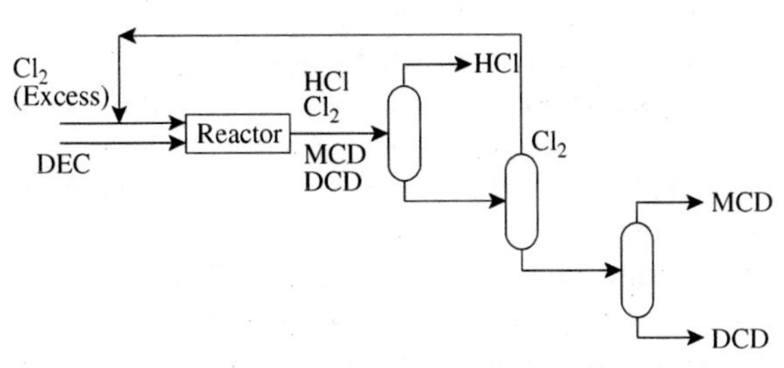

(c) Assume chlorine is fed in excess.

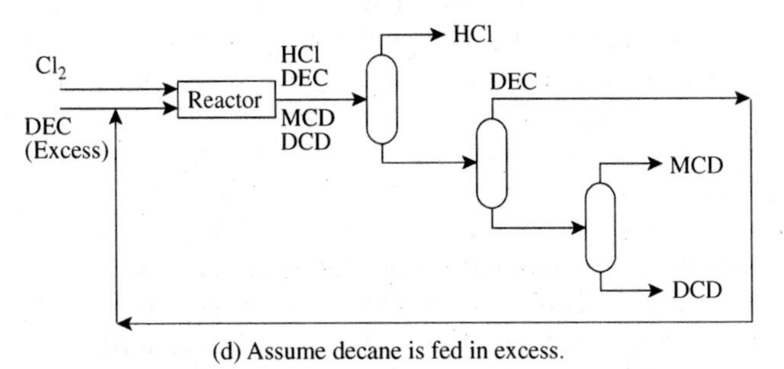

(d) Assume decane is fed in excess.

Figure 13.7 Different recycle structures for the production of monochlorodecane.

be recycled with the unreacted chlorine or it can leave with the product monochlorodecane (providing it can still meet product specifications).

At this stage, how large the excess of chlorine should be for Figure 13.7c to be feasible cannot be specified. Experimental data on the reaction chemistry would be required in order to establish this. However, the size of the excess does not change the basic structure.

(iv) *Excess decane.* Use of excess decane in the reactor forces the chlorine to effectively complete conversion (Figure 13.7d). Now there is effectively no chlorine in the reactor effluent. Again, three separators are required and a recycle of unconverted raw material.

In practice, there is likely to be a trace of chlorine in the reactor effluent. This can be recycled to the reactor with the unreacted decane or allowed to leave with the hydrogen chloride byproduct (providing this meets with the byproduct specification).

Again at this stage, it cannot be determined exactly how large an excess of decane would be required in order to make Figure 13.7d feasible. This would have to be established from experimental data, but the size of the excess does not change the basic flowsheet structure.

b. An arrangement is to be chosen to inhibit the side reaction, that is, to give high selectivity. The side reaction is suppressed by starving the reactor of either monochlorodecane or chlorine. Since the reactor is designed to produce monochlorodecane, the former option is not sensible. However, it is practical to use an excess of decane.

The last of the four flowsheet options generated above, which features excess decane in the reactor, is therefore preferred as shown in Figure 13.7d.

c. The selectivity (S) is defined by

$$S = \frac{\text{(MCD produced in the reactor)}}{\text{(DEC consumed in the reactor)}} \times \text{stoichiometric factor}$$

In this case, the stoichiometric factor is one. This is a measure of the MCD obtained from the DEC consumed. To assess the selectivity losses, the MCD produced in the primary reaction is split into that fraction that will become final product and that which will become the byproduct. Thus the reaction stoichiometry is:

$$C_{10}H_{22} + Cl_2 \longrightarrow SC_{10}H_{21}Cl + (1 - S)C_{10}H_{21}Cl + HCl$$

and, for the byproduct reaction:

$$(1 - S)C_{10}H_{21}Cl + (1 - S)Cl_2 \longrightarrow (1 - S)C_{10}H_{20}Cl_2 + (1 - S)HCl$$

Adding the two reactions gives overall:

$$C_{10}H_{22} + (2 - S)Cl_2 \longrightarrow SC_{10}H_{21}Cl + (1 - S)C_{10}H_{20}Cl_2 + (2 - S)HCl$$

Considering raw materials costs only, the economic potential, EP, of the process is defined as:

$$\begin{aligned} EP &= \text{Value of Products} - \text{Raw Materials Costs} \\ &= [176 \times S \times 0.45 + 36 \times (2 - S) \times 0.35] \\ &\quad - [142 \times 1 \times 0.27 + 71 \times (2 - S) \times 0.21] \\ &= 79.2S - 2.31(2 - S) - 38.34 \ (\$\cdot\text{kmol}^{-1}\text{decane reacted}) \end{aligned}$$

The minimum selectivity that can be tolerated is given when the economic potential is just zero:

$$0 = 79.2S - 2.31(2 - S) - 38.34$$

$$S = 0.53$$

In other words, the process must convert at least 53% of the decane that reacts to monochlorodecane rather than to dichlorodecane for the process to be economic. This figure assumes that the hydrogen chloride is sold to a neighboring process. If this is not the case, there is no value associated with the hydrogen chloride. Assuming there are no treatment and disposal costs for the now waste hydrogen chloride, the minimum economic potential is given by

$$0 = [176 \times S \times 0.45] - [142 \times 1 \times 0.27 + 71 \times (2 - S) \times 0.21]$$

$$= 79.2S - 14.91(2 - S) - 38.4$$

$$S = 0.72$$

Now the process must convert at least 72% of the decane to monochlorodecane.

If the hydrogen chloride cannot be sold, it must be disposed of somehow. Alternatively, it could be converted back to chlorine via the reaction:

$$2HCl + \tfrac{1}{2}O_2 \rightleftharpoons Cl_2 + H_2O$$

and then recycled to the MCD reactor. Now the overall stoichiometry changes since the $(2 - S)$ moles of HCl that were being produced as byproduct are now being recycled to substitute fresh chlorine feed:

$$(2 - S)HCl + \tfrac{1}{4}(2 - S)O_2 \longrightarrow \tfrac{1}{2}(2 - S)Cl_2 + \tfrac{1}{2}(2 - S)H_2O$$

Thus the overall reaction now becomes

$$C_{10}H_{22} + \tfrac{1}{2}(2 - S)Cl_2 + \tfrac{1}{4}(2 - S)O_2 \longrightarrow$$

$$SC_{10}H_{21}Cl + (1 - S)C_{10}H_{20}Cl_2 + \tfrac{1}{2}(2 - S)H_2O$$

The economic potential is now given by

$$0 = [176 \times S \times 0.45] - [142 \times 1 \times 0.27 + 71 \times \tfrac{1}{2}(2 - S) \times 0.21]$$

$$= 79.2S - 7.455(2 - S) - 38.34$$

$$S = 0.61$$

The minimum selectivity that can now be tolerated becomes 61%.

13.2 RECYCLES WITH PURGES

As discussed in the previous section, a purge can be used to avoid the cost of separating a component from a recycle. Purges can, in principle, be used either with liquid or vapor (gas) recycles. However, purges are most often used to remove low-boiling components from vapor (gas) recycles.

A common situation is encountered when the effluent from a chemical reactor contains components with large relative volatilities. As discussed in Chapter 9, a partial condensation of the mixture from the vapor phase followed by a simple phase split can often produce an effective separation. Cooling below cooling water temperature is not preferred, otherwise refrigeration is required. A dew-point calculation (see Chapter 4) at the system pressure reveals whether partial condensation above cooling water temperatures is effective. If partial condensation does not occur, even down to cooling water temperature, increasing the reactor pressure or using refrigeration (or both) can be considered to accomplish a phase split. Increasing the pressure of the reactor needs careful evaluation as far as the implications for reactor design are concerned. However, by its very nature, a single-stage separation does not produce pure products, hence further separation of both liquid and vapor streams is often required.

In Chapter 9, it was concluded that if a component is required to leave in the vapor phase, its K-value should be large (typically greater than 10)[4]. If a component is required to leave in the liquid phase, its K-value should be small (typically less than 0.1)[4]. Ideally, the K-value for the light key component in the phase separation should be greater than 10 and, at the same time, the K-value for the heavy key less than 0.1. Having such circumstances leads to a good separation in a single stage. However, use of phase separators might still be effective in the flowsheet if the K-values for the key components are not so favorable. Under such circumstances, a more crude separation must be accepted.

Phase separation in this way is most effective if the light key component is significantly above its critical temperature. If a component is above its critical temperature, it will not condense. However, any condensed liquid will still contain a small amount of the component above its critical temperature as it "dissolves" in the liquid phase. This means that a component above its critical temperature is bound to have an extremely high K-value. Many processes, particularly in the petroleum and petrochemical industries, produce a reactor effluent that consists of a mixture of low-boiling components such as hydrogen and methane that are above their critical temperature, together with much less-volatile organic components. In such circumstances, simple phase splits can give a very effective separation.

If the vapor from the phase split is either predominantly product or predominantly byproduct, then it can be removed from the process. If the vapor contains predominantly unconverted feed material, it is normally recycled to the reactor. If the vapor stream consists of a mixture of unconverted feed material, products and byproducts, then some separation of the vapor may be needed. The vapor from the phase split will be difficult to condense if the feed to the phase split has been cooled to cooling water temperature. If separation of the vapor is needed in such circumstances, one of the following methods can be used:

1. *Refrigerated condensation*. Separation by condensation relies on differences in volatility between the condensing components. Refrigeration, or a combination of high pressure and refrigeration, is needed. If a single-stage separation using refrigerated condensation does not give an adequate separation, the process can be repeated in a *refluxed condenser* or *dephlegmator*. The vapor flows up through the condenser and the condensed liquid flows down over the heat exchange surface countercurrently. Mass is exchanged between the upward flowing vapor and the downward flowing liquid. A separation of typically up to eight theoretical stages can be achieved in such a device.

2. *Low-temperature/high-pressure distillation.* Rather than use a low-temperature single-stage condensation or refluxed condenser, a conventional distillation can be used. To carry out the separation under these circumstances will require a low-temperature condenser for the column, or operation at high pressure, or a combination of both.

3. *Absorption.* Absorption was discussed in Chapter 10. If possible, a component that already exists in the flowsheet should be used as a solvent. Introducing an extraneous component into the flowsheet introduces additional complexity and the possibility of increased environmental and safety problems later in the design.

4. *Adsorption.* Adsorption involves the transfer of a component onto a solid surface (Chapter 10). The adsorbent will need to be regenerated by a gas or vapor when the bed approaches saturation. As discussed in Chapter 10, a vapor or gas can be used for regeneration. However, this can introduce extraneous material into the process, with the regeneration stream needing further separation. Thus, for such applications, regeneration by a change in pressure (pressure swing adsorption) would normally be preferred.

5. *Membrane separation.* Membranes, as discussed in Chapter 10, separate gases by means of a pressure gradient across a membrane, typically 40 bar or greater. Some gases permeate through the membrane faster than others and concentrate on the low-pressure side. Low molar mass gases and strongly polar gases have high permeabilities and are known as *fast gases. Slow gases* have higher molar mass and symmetrical molecules. Thus, membrane gas separators are effective when the gas to be separated is already at a high pressure and only a partial separation is required.

In situations where a large vapor (gas) flow having a dew point below cooling water temperature is to be recycled back to the reactor, it is often expensive to carry out such separations on the recycle. This is especially true when relatively small amounts of material need to be separated from the recycle. Rather than carry out a separation on the recycle vapor (gas) stream, a purge from the recycle stream can allow such material to be removed without the need to carry out a separation. Although the purge removes the need for a separator, it incurs raw material losses. Not only can these material losses be expensive, but they can also create environmental problems. However, another option is to use a combination of a purge with a separator on the purge.

As an example, consider ammonia synthesis. In an ammonia synthesis loop, hydrogen and nitrogen are reacted to ammonia. The reactor effluent is partially condensed to separate ammonia as a liquid. Unreacted gaseous hydrogen and nitrogen are recycled to the reactor. A purge on the recycle prevents the buildup of argon and methane that enter the system as feed impurities. Although the purge can be burnt as fuel, considerable quantities of hydrogen are lost and therefore recovery of this hydrogen is usually economic. For such hydrogen recovery systems, adsorption, a membrane or cryogenic condensation could be used. For hydrogen recovery from ammonia purge gas, a membrane is usually the most economic option. The membrane allows fast gases, such as hydrogen to be separated from slow gases such as methane. The driving force for the permeation of the fast gas (and hence the separation of the fast gas from the other slower components) is the difference in partial pressure from one side of the membrane to the other. Hence, for recovery of hydrogen, the product stream must be at a substantially lower pressure than the feed stream.

If the liquid from the phase split requires separation, then this can normally be accomplished by distillation, except under special circumstances. A distillation sequence is most often required with products and byproducts removed from the process and unreacted feed material recycled. In some situations, byproducts might be recycled for reasons discussed in the previous section.

The following example illustrates the quantitative relationships involving the use of a purge on a gas recycle stream.

Example 13.2 Benzene is to be produced from toluene according to the reaction[5]:

$$\underset{\text{toluene}}{C_6H_5CH_3} + \underset{\text{hydrogen}}{H_2} \longrightarrow \underset{\text{benzene}}{C_6H_6} + \underset{\text{methane}}{CH_4}$$

The reaction is carried out in the gas phase and normally operates at around 700°C and 40 bar. Some of the benzene formed undergoes a series of secondary reactions. These are characterized here by the single secondary reversible reaction to an unwanted byproduct, diphenyl, according to the reaction:

$$\underset{\text{benzene}}{2C_6H_6} \rightleftharpoons \underset{\text{diphenyl}}{C_{12}H_{10}} + \underset{\text{hydrogen}}{H_2}$$

Laboratory studies indicate that a hydrogen/toluene ratio of 5 at the reactor inlet is required to prevent excessive coke formation in the reactor. Even with a large excess of hydrogen, the toluene cannot be forced to complete conversion. The laboratory studies indicate that the selectivity (i.e. fraction of toluene reacted that is converted to benzene) is related to the conversion (i.e. fraction of toluene fed that is reacted) according to[5]:

$$S = 1 - \frac{0.0036}{(1 - X)^{1.544}}$$

where S = selectivity
X = conversion

The reactor effluent is thus likely to contain hydrogen, methane, benzene, toluene and diphenyl. Because of the large differences in volatility of these components, it seems likely that partial condensation will allow the effluent to be split into a vapor stream containing predominantly hydrogen and methane, and a liquid stream containing predominantly benzene, toluene and diphenyl.

The hydrogen in the vapor stream is a reactant and hence should be recycled to the reactor inlet (Figure 13.8). The methane enters the process as a feed impurity and is also a byproduct from the primary reaction and must be removed from the process. The hydrogen–methane separation is likely to be expensive but the methane can be removed from the process by means of a purge (Figure 13.8).

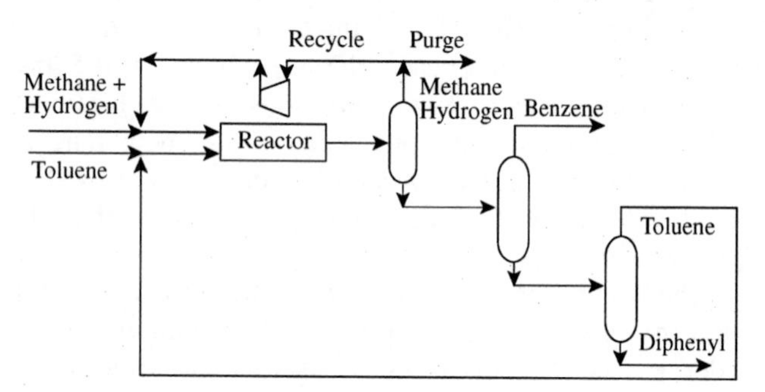

Figure 13.8 A flowsheet for the production of benzene uses purge to remove the methane that enters as a feed impurity and is also formed as a byproduct.

The liquid stream can readily be separated into relatively pure components by distillation, the benzene taken off as product, diphenyl as an unwanted byproduct and the toluene recycled. It is possible to recycle the diphenyl to improve selectivity, but it will be assumed that is not done here. The hydrogen feed contains methane as an impurity at a mole fraction of 0.05. The production rate of benzene required is 265 kmol·h^{-1}. Assume initially that a phase split can separate the reactor effluent into a vapor stream containing only hydrogen and methane, and a liquid containing only benzene, toluene and diphenyl, and that it can be separated to produce essentially pure products. For a conversion in the reactor of 0.75,

a. Determine the relation between the fraction of vapor from the phase split sent to purge (α) and the fraction of methane in the recycle and purge (y), as shown in Figure 13.9.

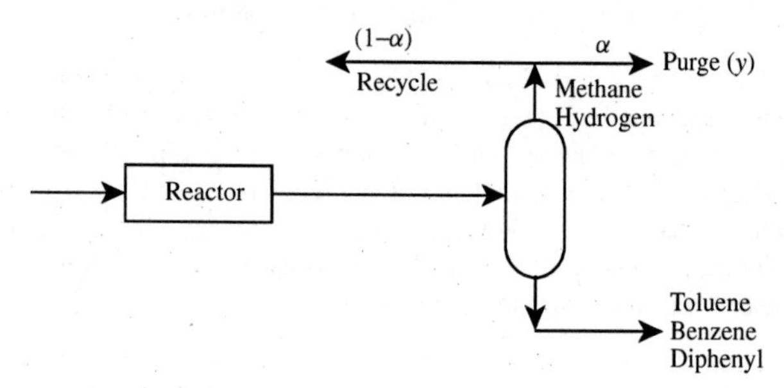

Figure 13.9 Purge fraction for the recycle.

b. Given the assumptions, estimate the composition of the reactor effluent for fraction of methane in the recycle and purge of 0.4.

Solution

a. Following Douglas[5], the benzene produced by the primary reaction is split into two parts, one part forming the final product and the other reacting to byproduct:

$$\begin{array}{ccc} C_6H_5CH_3 + H_2 & \longrightarrow & C_6H_6 + C_6H_6 + CH_4 \\ \dfrac{P_B}{S} + \dfrac{P_B}{S} & & P_B + P_B\left(\dfrac{1}{S}-1\right) + \dfrac{P_B}{S} \end{array}$$

$$\begin{array}{ccc} 2C_6H_6 & \rightleftharpoons & C_{12}H_{10} + H_2 \\ P_B\left(\dfrac{1}{S}-1\right) & & \dfrac{P_B}{2}\left(\dfrac{1}{S}-1\right) + \dfrac{P_B}{2}\left(\dfrac{1}{S}-1\right) \end{array}$$

For $X = 0.75$,

$$S = 1 - \frac{0.0036}{(1-0.75)^{1.544}} = 0.9694$$

Toluene Balance

$$\text{Fresh toluene feed} = \frac{P_B}{S}$$

$$\text{Toluene recycle} = R_T$$

$$\text{Toluene entering the reactor} = \frac{P_B}{S} + R_T$$

$$\text{Toluene in reactor effluent} = \left(\frac{P_B}{S} + R_T\right)(1-X) = R_T$$

For $P_B = 265$ kmol·h^{-1}, $X = 0.75$ and $S = 0.9694$,

$$R_T = 91.12 \text{ kmol·h}^{-1}$$

$$\text{Toluene entering the reactor} = \frac{265}{0.9694} + 91.12$$

$$= 364.5 \text{ kmol·h}^{-1}$$

Hydrogen Balance

$$\text{Hydrogen entering the reactor} = 5 \times 364.5$$

$$= 1823 \text{ kmol·h}^{-1}$$

$$\text{Net hydrogen consumed in reaction} = \frac{P_B}{S} - \frac{P_B}{2}\left(\frac{1}{S}-1\right)$$

$$= \frac{P_B}{S}\left(1 - \frac{1-S}{2}\right)$$

$$= 269.2 \text{ kmol·h}^{-1}$$

$$\text{Hydrogen in reactor effluent} = 1823 - 269.2$$

$$= 1554 \text{ kmol·h}^{-1}$$

$$\text{Hydrogen lost in purge} = 1554\alpha$$

$$\text{Hydrogen feed to the process} = 1554\alpha + 269.2$$

Methane Balance

$$\text{Methane feed to process as impurity} = (1554\alpha + 269.2)\frac{0.05}{0.95}$$

$$\text{Methane produced by reactor} = \frac{P_B}{S}$$

$$\text{Methane in purge} = \frac{P_B}{S} + (1554\alpha + 269.2)\frac{0.05}{0.95}$$

$$= 81.79\alpha + 287.5$$

$$\text{Total flowrate of purge} = 1554\alpha + 81.79\alpha + 287.5$$

$$= 1636\alpha + 287.5$$

Fraction of methane in the purge (and recycle):

$$y = \frac{81.79\alpha + 287.5}{1636\alpha + 287.5} \tag{13.5}$$

Figure 13.10 shows a plot of Equation 13.5. As the purge fraction α is increased, the flowrate of purge increases but the concentration of methane in the purge and recycle decreases. This variation (along with reactor conversion) is an important degree of freedom in the optimization of reaction and separation systems as will be discussed later.

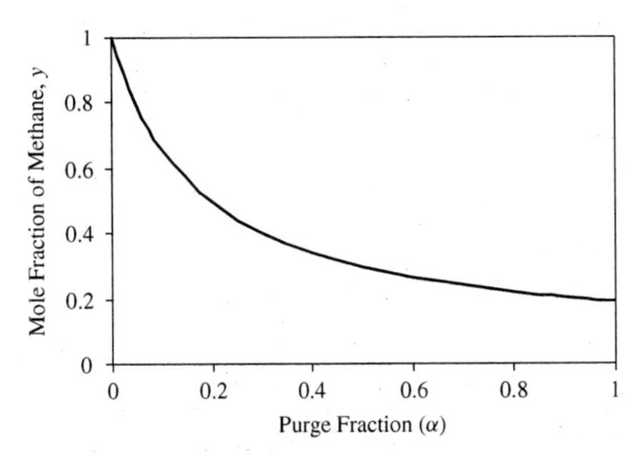

Figure 13.10 Variation of vapor composition with purge fraction.

b. **Methane Balance**

Mole fraction of methane in vapor from phase separator = 0.4

$$\text{Methane in reactor effluent} = \frac{0.4}{0.6} \times 1554$$

$$= 1036 \text{ kmol·h}^{-1}$$

Diphenyl Balance

$$\text{Diphenyl in reactor effluent} = \frac{P_B}{2}\left(\frac{1}{S} - 1\right)$$

$$= 4 \text{ kmol·h}^{-1}$$

The estimated composition of the reactor effluent are given in Table 13.2:

Table 13.2 Composition of reactor effluent for Example 13.2.

Component	Flowrate (kmol·h^{-1})
Hydrogen	1554
Methane	1036
Benzene	265
Toluene	91
Diphenyl	4

This calculation assumes that all separations in the phase split are sharp.

The above example illustrates some important principles for the design of recycles with purges:

1. There must be an overall mass balance in which the mass of the component to be separated that enters with the feed, or is formed in the reactor, must equal the mass of that component leaving with the purge gas, plus that which leaves with the liquid. If the mass that leaves with the liquid is extremely small relative to the purge, then effectively all of this mass must leave with the purge.
2. The concentration of the recycle can be controlled by varying the purge fraction. Decreasing the purge fraction causes the concentration of the component being purged to increase and vice versa.
3. A given mass of material can be purged with a low flowrate and high concentration, leading to a low loss of useful material in the purge. Alternatively, a given mass of material can be purged with a high flowrate and low concentration, leading to a high loss of useful material in the purge.
4. Purging with a low flowrate and high concentration leads to a relatively high recycle flowrate, and hence high costs for the recycle. On the other hand, purging with a high flowrate and low concentration leads to a relatively low recycle flowrate, and hence low costs for the recycle.
5. There are important costs to be traded-off. These will be considered later in the chapter.

13.3 PUMPING AND COMPRESSION

Feed entering the reaction and separation system needs to be pumped in the case of a liquid and compressed in the case of a gas. Liquids and gases need to be transferred between operations, some involving recycle from the separation system to the reaction system. The cost of a pressure increase for liquids is usually small relative to that for gases. Pumps usually have low capital and operating costs relative to other plant items. On the other hand, to increase the pressure of material in the gas phase requires a compressor, which tend to have a high capital cost and large power requirements, leading to high operating costs.

When recycling material to the reactor for whatever reason, the pressure drop through the reactor, separator (if there is one), the heat transfer equipment upstream and downstream of the reactor, control valves, and so on must be overcome. This means the pressure of any material to be recycled must be increased. Again, for the case of a liquid recycle, the cost of this pressure increase is usually small. On the other hand, to increase the pressure of material in the gas phase for recycle requires a compressor and is expensive.

When synthesizing the reaction, separation and recycle configuration, gas recycles should be avoided if it is feasible to use a liquid recycle without excessive cost. In some cases, this might be possible by increasing the pressure of the whole reaction, separation and recycle system. However, there will be many consequences from an increase in the system pressure that might overwhelm the benefits from avoiding a gas recycle. The capital cost of the reactor and separators will be higher to handle the higher pressure, the separation might be more difficult, hazards will be increased, and so on. It is often not possible to avoid gas recycles without using very high pressures or very low levels of refrigeration, in which case the expense of a recycle compressor must be accepted.

For gas-phase reactions, the pressure drop through the reactor is usually less than 10% of the inlet pressure[6]. The pressure drop through trickle-bed reactors is usually less than 1 bar. A value of 0.5 bar is often a reasonable first estimate for packed and trickle-bed reactors, although pressure drops can be higher. The pressure drop through fluidized-bed reactors is usually between 0.02 and 0.1 bar.

Heat exchangers for liquids normally have a pressure drop in the range 0.35 to 0.7 bar (see Chapter 15). For gases, heat exchangers have a pressure drop typically between 1 bar for high-pressure gases (10 bar and above), down to 0.01 bar for gases under vacuum conditions (see Chapter 15).

Pressure drop in the transmission pipes is a combination of pressure losses in the pipes and pipe fittings[7]. Pipe fittings include bends, isolation valves, control valves, orifice plates, expansions, reductions, and so on. If the fluid is assumed to be incompressible and the change in kinetic energy from inlet to outlet is neglected, then:

$$\frac{\Delta P}{\rho g} = \sum h_L + \Delta z \qquad (13.6)$$

where $\dfrac{\Delta P}{\rho g}$ = loss in pressure head (m)

ΔP = pressure drop ($N{\cdot}m^{-2}$)

ρ = fluid density ($kg{\cdot}m^{-3}$)

g = gravitational constant (9.81 $m{\cdot}s^{-2}$)

$\sum h_L$ = sum of head losses for straight pipes and pipe fittings (m)

Δz = change in elevation (m)

For straight pipes[7]:

$$h_L = 2c_f \frac{L}{d_I}\frac{v^2}{g} \qquad (13.7)$$

where c_f = Fanning friction factor

L = pipe length (m)

d_I = internal diameter of pipe (m)

v = mean velocity in the pipe ($m{\cdot}s^{-1}$)

The Fanning friction factor is given by[8]:

$$c_f = \frac{16}{Re} \quad Re < 2000 \qquad (13.8)$$

$$c_f = 0.046Re^{-0.2} \quad 2000 < Re < 20{,}000 \qquad (13.9)$$

$$c_f = 0.079Re^{-0.25} \quad Re > 20{,}000 \qquad (13.10)$$

where $Re = \dfrac{\rho d_I v}{\mu}$

μ = fluid viscosity ($N{\cdot}s{\cdot}m^{-2}$)

It should be noted that Equations 13.9 and 13.10 apply to smooth pipes, whereas the pipes used for transmission of fluids usually have some surface roughness, which increases the friction factor. However, for short fluid transmission pipes, the overall pressure drop is usually dominated by the pressure drop in the pipe fittings (valves, bends, etc). Thus, for short transmission pipes, there is little point in calculating the straight pipe pressure drop accurately. If the transmission pipe is long (>100 m) and straight, then the Fanning friction factor can be correlated as[7]:

$$\frac{1}{\sqrt{c_f}} = -1.77\ln\left[0.27\frac{\varepsilon}{d_I} + \frac{1.25}{Re\sqrt{c_f}}\right] \qquad (13.11)$$

where ε = surface roughness (m)

Table 13.3 gives some typical values of surface roughness[7].

The head loss in the pipe fittings can be correlated as[7]:

$$h_L = c_L \frac{v^2}{2g} \qquad (13.12)$$

where c_L = loss coefficient (–)

For laminar flow,

$$c_L = f(Re, \text{geometry of fitting})$$

For turbulent flow,

$$c_L = f(\text{geometry of fitting})$$

Table 13.4 gives some typical values of the loss coefficient for various fittings[9]. It should be noted that values for loss coefficient will vary for the same fitting, but from different manufacturers, as a result of differences in geometry. Table 13.5 gives head losses for sudden contractions, sudden expansions and orifice plates. Note that the relationship for orifice plates in Table 13.5 relates to the overall pressure drop and not the pressure drop between the pressure tappings used to determine the flowrate.

In preliminary design, the fluid transmission lines can be designed on the basis of an assumed fluid velocity. For non-viscous liquids (μ < 10 $mN{\cdot}s{\cdot}m^{-2}$ = cP), use a pipe velocity of 1 to 2 $m{\cdot}s^{-1}$.

For viscous fluids, the velocity may be constrained by the allowable pressure drop or shear degradation of the fluid (e.g. large molecules breaking down into smaller molecules from high shear rates). Typical values are given in Table 13.6.

Table 13.3 Surface roughness of pipes.

Pipe	ε (mm)
Drawn tubing	0.0015
Commercial steel	0.046
Cast iron	0.26
Concrete	0.3–3.0

Table 13.4 Loss coefficients for various pipe fittings.

Pipe fitting	Laminar flow c_L	Turbulent flow c_L	Correction for partial closure of valve in turbulent flow (α = fraction open)
Bend (standard)	$\frac{840}{Re}$ (Re <1100)	0.8	–
Globe valve (composite seat)	$\frac{100}{Re^{0.33}}$ (Re <5000)	6.0	$\frac{c_L}{\alpha^{0.5}}$
Globe valve (plug disc)	$\frac{70}{Re^{0.26}}$ (Re <2700)	9.0	$\frac{c_L}{\alpha^{1.84}}$
Gate valve	$\frac{1200}{Re}$ (Re <6000)	0.2	$\frac{c_L}{\alpha^{3.7}}$
Plug cock	–	0.05	–
Butterfly valve	–	0.24	–
Check valve	$\frac{1200}{Re^{0.86}}$ (Re <1700)	2.0	–

Table 13.5 Head losses in sudden contractions, sudden expansions and orifice plates.

Fitting	h_L
Sudden contraction	$\left[0.5\left(1-\frac{A_2}{A_1}\right)\right]\left(\frac{v_2^2}{2g}\right)$
Sudden expansion	$\left[1-\frac{A_1}{A_2}\right]^2\left(\frac{v_1^2}{2g}\right)$
Orifice plate	$\frac{1}{c_D^2}\left[1-\frac{A_O}{A}\right]\left[\left(\frac{A}{A_O}\right)^2-1\right]\left(\frac{v^2}{2g}\right)$

Where v_1, v_2 = upstream and downstream velocities (m·s^{-1})
A_1, A_2 = upstream and downstream pipe areas (m^2)
g = acceleration due to gravity (9.81 m·s^{-2})
A, A_O = areas of pipe and orifice (m^2)
v = velocity in the pipe (m·s^{-1})
c_D = 0.62 for preliminary design

Table 13.6 Typical fluid velocities for viscous liquids.

Viscosity (mN·s·m^{-2} = cP)	Velocity (m·s^{-1})
50	0.5–1.0
100	0.3–0.6
1000	0.1–0.3

For gases and vapors, typical fluid velocities are in the range of 15 to 30 m·s^{-1}. The fluid velocity must take account of standard pipe sizes. Table 13.7 gives details of a number of commonly used pipe sizes.

If the piping layout is known, then the above correlations can be used to estimate the pressure drop through the pipes. This might be the case, for example, in a retrofit situation. In preliminary design, it might be necessary to make some allowance for a cost of pumping and compression without knowledge of the piping layout. If this is the case, then it is not difficult to make a first estimate of the distances required for transportation of the fluid. What is uncertain is the pipe fittings that will be involved. Some general guidelines are therefore required in order to make a first estimate of the pressure drop to take account of the pipe fittings:

- Vessels will often have isolation valves (but this varies between different sectors of the industry).
- Equipment that needs to be taken out of service for maintenance will normally have an isolation valve on each side. This will include pumps, compressors and control valves.
- Pumps will normally have a check valve to prevent reversal of flow.
- Flow control will usually be based on the measurement of pressure drop across an orifice plate.
- A line going between vessels or between a vessel and a pipe junction will normally have at least three bends.

Table 13.7 Commonly used standard pipe sizes.

Pipe size (mm)	Outside diameter (mm)	Inside diameter (mm)
15	21.34	15.80
20	26.67	20.93
25	33.40	26.64
32	41.16	35.05
40	48.26	40.89
50	60.32	52.50
65	73.03	62.71
80	88.90	77.93
100	114.3	102.3
150	168.3	154.1
200	219.1	202.7
250	273.1	254.5
300	323.9	304.8

For example, suppose a liquid is being pumped from one vessel into another vessel using a pump and under the action of flow control using an orifice plate to measure the flowrate. The head losses involved will be

- a sudden contraction from the feed vessel into the discharge line;
- an isolation valve for the vessel;
- two isolation valves and a check valve for the pump;
- an orifice plate for flow measurement;
- a control valve;
- two isolation valves for the control valve;
- typically three pipe bends from changes in direction of the pipes;
- an isolation valve for the discharge vessel;
- a sudden expansion for the fluid entering the discharge vessel.

Example 13.3 Water is to be pumped between two vessels separated by an estimated distance of 30 m under the action of flow control. An increase in elevation of 5 m is also estimated. The flowrate of water is 100 m^3 per hour, its viscosity is 0.8 mN·s·m^{-2} (equal to centipoise) and its density is 993 kg·m^{-3}. Estimate the pressure drop required to be produced by the pump.

Solution First determine the pipe diameter from an assumed velocity of say 2 m·s^{-1}. The area of the pipe (A) is given by:

$$A = 100 \times \frac{1}{3{,}600} \times \frac{1}{2}$$

$$= 0.01389 \text{ m}^2$$

The internal diameter (d_I) is given by

$$d_I = \sqrt{\frac{4A}{\pi}}$$

$$= \sqrt{\frac{4 \times 0 \cdot 01389}{\pi}}$$

$$= 0.133 \text{ m}$$

This needs to be rounded to the next largest standard pipe diameter of internal diameter 0.154 m.

The actual fluid velocity is therefore

$$v = 100 \times \frac{1}{3600} \times \frac{4}{\pi \times 0.154^2}$$

$$= 1.49 \text{ m·s}^{-1}$$

Reynolds number for the straight pipes is

$$Re = \frac{\rho d_I v}{\mu}$$

$$= \frac{993 \times 0.154 \times 1.49}{0.8 \times 10^{-3}}$$

$$= 2.85 \times 10^5$$

Head loss in the straight pipe sections:

$$h_L = 2c_f \frac{L}{d_I} \frac{v^2}{g}$$

$$= 2 \times \frac{0.079}{Re^{0.25}} \times \frac{30}{0.154} \times \frac{1.49^2}{9.81}$$

$$= 0.30 \text{ m}$$

For the isolation valves, take gate valves fully open, one for each vessel, two for the pump and two for the control valve:

$$h_L = 6 \times c_L \frac{v^2}{2g}$$

$$= 6 \times 0.2 \times \frac{1.49^2}{2 \times 9.81}$$

$$= 0.14 \text{ m}$$

Assume a check valve for the pump:

$$h_L = c_L \frac{v^2}{2g}$$

$$= 2 \times \frac{1.49^2}{2 \times 9.81}$$

$$= 0.23$$

To estimate the control valve, take a globe valve to be half open:

$$h_L = \frac{c_L}{\alpha^{1 \cdot 84}} \frac{v^2}{2g}$$

$$= \frac{9}{0.5^{1 \cdot 84}} \frac{1.49^2}{2 \times 9.81}$$

$$= 3.65 \text{ m}$$

Assume 3 bends:

$$h_L = 3 \times 0.8 \times \frac{1.49^2}{2 \times 9.81}$$

$$= 0.27 \text{ m}$$

Assume an orifice plate to measure the flowrate with diameter ratio of 0.4 and discharge coefficient of 0.62:

$$h_L = \frac{1}{c_D^2}\left[1 - \frac{A_O}{A}\right]\left[\left(\frac{A}{A_O}\right)^2 - 1\right]\frac{v^2}{2g}$$

$$= \frac{1}{0.62^2}[1 - 0.4^2]\left[\left(\frac{1}{0.4}\right)^4 - 1\right] \times \frac{1.49^2}{2 \times 9.81}$$

$$= 9.41 \text{ m}$$

The entrance loss from the feed vessel is given by

$$h_L = 0.5\left[1 - \frac{A_2}{A_1}\right]\frac{v_2^2}{2g}$$

$$= 0.5(1-0)\frac{1.49^2}{2 \times 9.81}$$

$$= 0.06 \text{ m}$$

The exit loss into the receiving vessel is given by

$$h_L = \left[1 - \frac{A_1}{A_2}\right]^2 \frac{v_1^2}{2g}$$

$$= [1-0]^2 \frac{1.49^2}{2 \times 9.81}$$

$$= 0.11 \text{ m}$$

$$\Delta P = \rho g\left[\sum h_L + \Delta z\right]$$

$$= 993 \times 9.81[(0.30 + 0.14 + 0.23 + 3.65 + 0.27$$

$$+ 9.41 + 0.06 + 0.11) + 5]$$

$$= 1.87 \times 10^5 \text{N·m}^{-2}$$

1. *Pumping.* The two main types of pump used in the process industries are *centrifugal* and *positive displacement.* In a centrifugal pump, the liquid enters near the center of a rotating impeller and is thrown outwards by centrifugal action. The resulting increase in kinetic energy is converted to pressure energy as the liquid leaves the pump. Centrifugal pumps deliver a volume that is dependent on the discharge pressure and the energy added. On the other hand, in a positive displacement pump, a volume of liquid is trapped in a chamber, which is decreased in volume and increased in pressure before discharge. These can be reciprocating (e.g. piston and cylinder) or rotational (e.g. two interlocking rotating gears). Positive displacement pumps therefore deliver a definite quantity for each stroke or partial rotation of the device. Most industrial applications favor the use of centrifugal pumps as they can handle a wide range of fluids and a wide range of pumping conditions at low cost relative to positive displacement devices. Positive displacement pumps are used when the liquid has a high viscosity, low flowrate, or a combination of the two. The power required for a given pumping duty can be calculated from:

$$W = \frac{F\Delta P}{\eta} \tag{13.13}$$

where W = power required for pumping ($\text{N·m·s}^{-1} = \text{J·s}^{-1} = \text{W}$)
F = volumetric flowrate ($\text{m}^3\text{·s}^{-1}$)
ΔP = pressure drop across pump (N·m^{-2})
η = pump efficiency (–)

The efficiency of a pump is a function of both its design and its capacity. The efficiency is a strong function of capacity and might be as high as 90% for a large pump and as low as 30% for a small one. For centrifugal pumps, a first estimate of the pump efficiency can be obtained from

$$\eta = -0.01(\ln F)^2 + 0.15\ln(F) + 0.3 \quad 1 < F < 1000 \tag{13.14}$$

where F = volumetric flowrate ($\text{m}^3\text{·h}^{-1}$)
η = fractional pump efficiency (–)

It should be noted that the efficiency suggested by Equation 13.14 is only approximate. The efficiency for a given capacity will depend not only on the pump design but also on the liquid being pumped. The performance of pumps is normally quoted on the basis of the pumping of water and this needs to be corrected for the actual fluid. However, Equation 13.14 is good enough for a first estimate for centrifugal pumps on a low-viscosity duty.

Example 13.4 A centrifugal pump is required to deliver 100 m^3 of water per hour with an increase in pressure of 5 bar. If the driver for the pump is to be an electric motor with an efficiency of 90%, electricity costs \$0.06 KW·h^{-1}, operating for 8300 hours per year, estimate the annual cost of power.

Solution Pump efficiency:

$$\eta = -0.01(\ln F)^2 + 0.15\ln F + 0.3$$

$$= -0.01(\ln 100)^2 + 0.15\ln(100) + 0.3$$

$$= 0.78$$

$$W = \frac{F\Delta P}{\eta} \times \frac{1}{0.9}$$

$$= \frac{100}{3600} \times 5 \times 10^5 \times \frac{1}{0.78} \times \frac{1}{0.90}$$

$$= 19{,}800 \text{ W}$$

$$\text{Annual Cost} = \frac{19{,}800}{10^3} \times 0.06 \times 8300$$

$$= \$9900 \text{ y}^{-1}$$

2. *Compression.* As noted previously, pumps are cheap to buy and cheap to operate relative to gas compressors. Gas compressors can generally be classified as:

a. *Positive displacement compressors* in which the machine traps a volume of gas in a chamber and is decreased in volume with a resulting increase in pressure.
b. *Dynamic compressors* or *turbo-compressors* in which energy is transferred to the gas by dynamic means from a rotating impeller or blades. The kinetic energy of the gas is increased, which is then converted to pressure energy. A dynamic compressor with a low-pressure drop is normally termed a *fan* or *blower.*
c. *Ejectors* in which the kinetic energy of a high-velocity *working fluid* or *motive fluid* (steam or a gas) entrains and compresses a second fluid stream. The device has no moving parts. They are inefficient devices and normal

usage is for vacuum service where small quantities of gas are handled. They will not be considered further here.

The maximum compression ratio (ratio of outlet to inlet pressure) for compressors depends on the design of the machine, the properties of the lubricating oil used in the machine, the ratio of heat capacities of the gas ($C_P/C_V = \gamma$), other properties of the gas (e.g. tendency to polymerize when heated), and the inlet temperature. The most common types of compressor used for gas compression in the process industries are:

a. *Reciprocating*. Reciprocating compressors (a piston moving backwards and forwards in a cylinder, see Appendix B.1) can be used over a wide range of pressures and flowrates. Diatomic gases with $\gamma = 1.4$ can have a pressure ratio per cylinder of up to 4 and hydrocarbon gases with $\gamma = 1.2$ up to 9. Maximum discharge pressures can be in excess of 20 bar for a single stage or in excess of 5000 bar for multiple stages. Suction flowrates of up to 1 $m^3{\cdot}s^{-1}$ for a single stage or up to 2 $m^3{\cdot}s^{-1}$ for multiple stages are possible. A significant disadvantage of reciprocating compressors is that the compressed gas is not delivered continuously. The resulting pulsations in flow and pressure can lead to vibrations (and, in extreme cases, mechanical failure) and poor compressor efficiency (to overcome high-pressure peaks). *Surge drums* and *acoustic filters* are required to dampen the pulsations.

b. *Positive displacement rotary*. Rotary machines, as their name implies, involve one or two rotating shafts to create chambers with decreasing size from inlet to outlet to increase the pressure. There are four broad classes of positive displacement rotary compressor:

- *Screw* compressors use two counterrotating screw-like shafts (Figure 13.11a). Lubricating oil is required in some designs to lubricate the rotors, seal the gaps between the rotors and reduce the temperature rise of the gas during compression. These have the disadvantage of contaminating the gas with lubricating oil. On the other hand, oil-free machines do not feature mixing of the gas with lubricating oil. In these designs, contact between the rotors is prevented by timing gears outside the working chamber. However, they are more expensive than oil-injected machines. Diatomic gases with $\gamma = 1.4$ can have a pressure ratio per casing of up to 4.5 and hydrocarbon gases with $\gamma = 1.2$ can have a pressure ratio per casing of up to 10. Maximum discharge pressures are up to 30 bar. Suction flowrates of up to 15 $m^3{\cdot}s^{-1}$ and greater are possible.
- *Rotary piston* or *rotary lobe* or *roots* compressors use two counterrotating matching lobe-shaped rotors (Figure 13.11b). Each revolution of the lobes delivers four pulses of gas. Maximum discharge pressures in a single stage are up to 2.5 bar. Suction flowrates of up to 3 $m^3{\cdot}s^{-1}$ are possible.
- *Sliding vane* compressors use a single rotating shaft with an eccentrically located rotor in the compressor casing with sliding vanes in the rotor (Figure 13.11c). Maximum pressure ratio for each casing is restricted to around 3.5. Maximum discharge pressures are up to 10 bar. Suction flowrates of up to 3 $m^3{\cdot}s^{-1}$ are possible.
- *Liquid ring* compressors use a single rotating shaft with an eccentrically located rotor in the compressor casing with static vanes (Figure 13.11d). A flow of low-viscosity liquid (usually water) through the casing draws the gas into the cells between the vanes where it is compressed by the movement of the rotor. A settling drum after the compressor separates the liquid from the gas and the liquid is usually recycled. Maximum discharge pressures can be greater than 5 bar. Flowrates of up to 3 $m^3{\cdot}s^{-1}$ are possible. Liquid ring compressors are most commonly used for vacuum service.

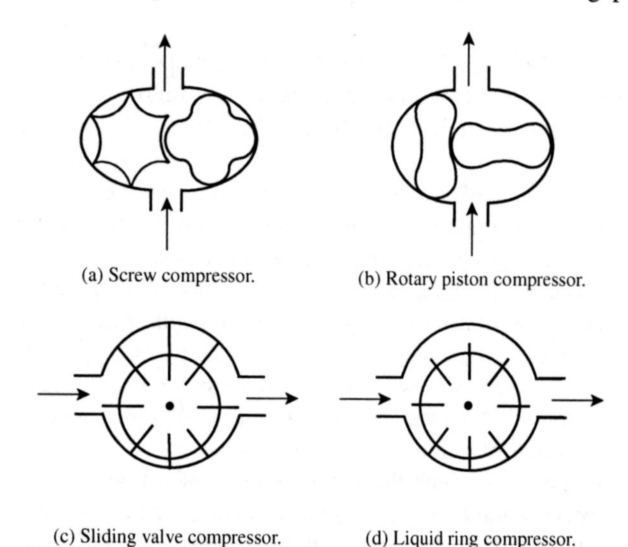

Figure 13.11 Positive displacement rotary compressors.

c. *Centrifugal*. Centrifugal compressors increase gas pressure by accelerating it radially outwards through an *impeller* or *wheel*. Centrifugal compressors might use a single impeller, but pressure ratios are limited to around 2 to 2.5. For higher-pressure ratios, multiple impellers are used. The increased kinetic energy is then converted to pressure energy as the gas leaves the impeller. Centrifugal compressors are used for flowrates of up to 90 $m^3{\cdot}s^{-1}$ at low pressure (typically up to 2 barg discharge) or 1 $m^3{\cdot}s^{-1}$ at high pressure for single stage machines. Multistage machines can be used for flowrates up to 140 $m^3{\cdot}s^{-1}$. Discharge pressures for a single stage machine are up to 130 bar at low flowrates or 3 bar at high flowrates. Multistage machines have discharge pressures up to 700 bar. A significant disadvantage of centrifugal compressors is their limited turndown capacity. It is frequently necessary to operate a compressor at flows below the design capacity. If the flow is reduced far enough, the compressor enters an unstable region known as the *surge region*. Figure 13.12 shows the typical behavior

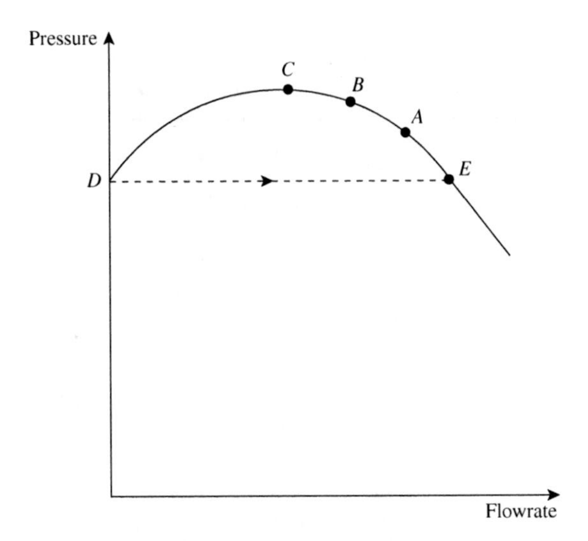

Figure 13.12 Surging in centrifugal compressors.

of pressure versus flowrate for a centrifugal compressor. If the compressor is operating at Point *A* in Figure 13.12 and the flow reduced to Point *B* by restricting the outlet flow, the increased pressure drop caused by the restriction is met by the compressor. Reducing the flow beyond the peak Point *C* results in a discharge mains pressure higher than the compressor pressure. The flow then reverses with the operating point moving to zero flow at Point *D*. Once the discharge main clears, the operating point moves to Point *E*. The excessive flow then moves the operation back to beyond Point *C*. The cycle is then repeated. The surge limit depends on the speed of the compressor, decreasing with decreasing speed. Unstable operations can start at 50 to 70% of the rated flow.

d. *Axial.* Axial compressors increase gas pressure by accelerating it in the direction of the flow using blades mounted on a rotating shaft that rotate between stationary blades mounted on the compressor casing. Axial compressors are used for very high flowrates and moderate pressure drops. Discharge pressures can be up to 10 bar. Flowrates of up to 150 $m^3 \cdot s^{-1}$ are possible. As with centrifugal compressors, axial compressors can also suffer from surge.

For low-pressure difference and large flows, the capital cost of a reciprocating compressor can be twice that of a centrifugal machine with the same flowrate. If the pressure difference is high and the flowrate low, then the costs tend to be more equal. Also, centrifugal machines tend to be more reliable than reciprocating machines. Standby machines might need to be installed if reciprocating machines are used. The capital cost of axial compressors tends to be of the same order as centrifugal machines. At flowrates typically less than 30 $m^3 \cdot s^{-1}$, the capital cost of axial compressors tends to be higher than centrifugal machines. Rotary screw compressors tend to be cheaper than both reciprocating and centrifugal machines in their operating range. Sliding vane compressors tend to be cheaper than reciprocating machines in their operating range.

For high demand compression duties, the overall compression can be broken down into stages and different types of compressor used in the different casings. For example, a large flowrate with a large pressure differential might use an axial compressor followed by a centrifugal compressor.

For reciprocating compressors, the power required can be estimated from (see Appendix B):

$$W = \left(\frac{\gamma}{\gamma - 1}\right)\frac{P_{in}F_{in}}{\eta_{IS}}\left[1 - \left(\frac{P_{out}}{P_{in}}\right)^{\frac{\gamma-1}{\gamma}}\right] \tag{B.15}$$

where W = power required for compression ($N \cdot m \cdot s^{-1} = J \cdot s^{-1} = W$)
P_{in}, P_{out} = inlet and outlet pressures ($N \cdot m^{-2}$)
F_{IN} = inlet volumetric flowrate ($m^3 \cdot s^{-1}$)
γ = ratio of heat capacities C_P/C_V (–)
η_{IS} = isentropic efficiency (–)

Table 13.8 gives the ratio of heat capacities C_P/C_V for a number of common gases.

Alternatively, the value of C_P/C_V can be estimated from C_P data from the relationship (see Appendix B):

$$\gamma = \frac{C_P}{C_V} = \frac{C_P}{C_P - R} \tag{13.15}$$

where R = universal gas constant

If an equation of state is to be used, then:

$$\gamma = \frac{C_P}{C_V} = \frac{C_P}{C_P - (C_P - C_V)} \tag{13.16}$$

where $C_P - C_V = -T\left(\frac{\partial P}{\partial T}\right)_V \left(\frac{\partial V}{\partial P}\right)_T$

The value of $(\partial P/\partial T)_V$ and $(\partial V/\partial P)_T$ can be determined from an equation of state (e.g. Peng–Robinson).

The isentropic efficiency (see Appendix B) is a function of the machine design and pressure ratio (P_{out}/P_{in}). A first estimate of the isentropic efficiency of a reciprocating compressor can be obtained from:

$$\eta_{IS} = 0.1091(\ln r)^3 - 0.5247(\ln r)^2 + 0.8577\ln r + 0.3727 \quad 1.1 < r < 5 \tag{13.17}$$

where r = pressure ratio P_{out}/P_{in}

For an ideal gas compression, the corresponding temperature rise is given by (see Appendix B):

$$\frac{T_{out}}{T_{in}} = \left(\frac{P_{out}}{P_{in}}\right)^{\frac{\gamma-1}{\gamma}} \tag{B.21}$$

where T_{in}, T_{out} = inlet and outlet temperatures (K)

Equation B.21 assumes the compression to be adiabatic ideal gas compression. In practice, the compression will be neither perfectly adiabatic nor ideal. To allow for this,

Table 13.8 Heat capacity ratio for various gases.

	Typical value		Values at 10°C	Values at 25°C	Values at 50°C
Monatomic gases	1.67	He	1.67	1.67	1.67
		Ar	1.66	1.66	1.66
Diatomic gases	1.40	H_2	1.42	1.42	1.41
		N_2	1.40	1.40	1.40
		CO	1.40	1.40	1.40
		Air	1.40	1.40	1.40
		NO	1.39	1.38	1.38
		O_2	1.40	1.40	1.39
		Cl_2	1.32	1.32	1.32
		HCl	1.39	1.39	1.39
Polyatomic gases	1.30	CH_4	1.31	1.30	1.29
		NH_3	1.30	1.30	1.29
		H_2O	1.33	1.33	1.33
		C_2H_4	1.24	1.23	1.22
		C_2H_6	1.19	1.19	1.17
		H_2S	1.32	1.32	1.31
		CO_2	1.28	1.28	1.27
		C_3H_6	1.16	1.15	1.14
		C_3H_8	1.13	1.12	1.12

the gas compression can be assumed to follow a *polytropic* compression (see Appendix B). A polytropic compression is neither adiabatic nor isothermal, but specific to the physical properties of the gas and the design of the compressor. If the real compression is assumed to follow a polytropic compression:

$$T_{out} = T_{in}\left(\frac{P_{out}}{P_{in}}\right)^{\frac{n-1}{n}} \qquad \text{(B.26)}$$

where n = polytropic coefficient

The polytropic coefficient n must therefore be determined experimentally, and can be estimated from (see Appendix B):

$$n = \frac{\ln\left(\dfrac{P_{out}}{P_{in}}\right)}{\ln\left[\dfrac{\eta_{IS}\left(\dfrac{P_{out}}{P_{in}}\right)}{\eta_{IS} - 1 + \left(\dfrac{P_{out}}{P_{in}}\right)^{\frac{\gamma-1}{\gamma}}}\right]} \qquad \text{(B.28)}$$

Equation B.28 is useful to estimate the polytropic coefficient n if the inlet and outlet pressures are known, along with an estimate of the isentropic efficiency. Knowing the polytropic coefficient allows the outlet temperature for a real gas compression to be estimated from Equation B.26.

For centrifugal and axial compressors, the compression path is usually based on polytropic compression. The power required for a polytropic compression can be expressed as (see Appendix B):

$$W = \frac{n}{n-1}\,\frac{P_{in}F_{in}}{\eta_P}\left[1 - \left(\frac{P_{out}}{P_{in}}\right)^{\frac{n-1}{n}}\right] \qquad \text{(B.35)}$$

where η_P = polytropic efficiency (ratio of polytropic power to actual power)

The polytropic efficiency is a function of the physical properties of the gas and the machine design. Generally, the higher the feed flowrate, the higher the polytropic efficiency of the compressor. A first estimate of the polytropic efficiency of a centrifugal compressor can be obtained from:

$$\eta_P = 0.017 \ln F + 0.7 \qquad \text{(13.18)}$$

where F = inlet flowrate of gas ($m^3 \cdot s^{-1}$)

It should be emphasized that the actual efficiency depends on the design of the machine and the physical properties of the gas.

Axial flow compressors have polytropic efficiencies up to 10% higher than centrifugal compressors for the same pressure ratio. A reasonable estimate is for the efficiency to be 5% higher than that of the corresponding centrifugal machine (i.e. multiply the estimate from Equation 13.18 by 1.05).

The polytropic coefficient can be estimated from the heat capacity ratio and an estimate of the polytropic efficiency from the relationship (see Appendix B):

$$n = \frac{\gamma\eta_P}{\gamma\eta_P - \gamma + 1} \qquad \text{(B.38)}$$

The temperature rise accompanying gas compression might be unacceptably high because of the properties of the gas (e.g. decomposition, polymerization etc.), materials of construction of the compressor or the properties of the lubricating oil used in the machine. The temperature must be below the flash point of the lubricating oil (i.e. the temperature at which it gives off enough vapor to form an ignitable mixture). If this is the case, the overall compression can be broken down into a number of stages with intermediate cooling. Also, intermediate cooling will reduce the volume of gas between stages and reduce the power for compression of the next stage. On the other hand, the intercoolers will have a pressure drop that will increase the power requirements, but this effect is usually small compared with the reduction in power from gas cooling. The power for the staged compression of a gas is given by (see Appendix B):

$$W = \frac{\gamma}{\gamma - 1}\frac{P_{in}F_{in}N}{\eta_{IS}}[1 - r^{\frac{\gamma-1}{\gamma}}] \qquad \text{(B.49)}$$

where N = number of compression stages
r = compression ratio

$$= \sqrt[N]{\frac{P_{out}}{P_{in}}} \qquad \text{(B.47)}$$

Equations B.49 and B.47 minimize the overall compression power for an N-stage compression. However, the basis is for adiabatic ideal gas compression and therefore not strictly valid for real gas compression (see Appendix B). It is also assumed that intermediate cooling is back to initial conditions, which might not be the case with real intercoolers. The power for the corresponding expression for a polytropic compression is given by:

$$W = \frac{n}{n - 1}\frac{P_{in}F_{in}N}{\eta_P}[1 - r^{\frac{n-1}{n}}] \qquad \text{(B.50)}$$

Staged compression should not be confused with compressor stages. A centrifugal compressor will often have multiple impellers (or wheels) mounted on the shaft to form a multistage machine without cooling. Staged compression is where the overall compression is broken down into intermediate stages with intercooling.

Whilst compression ratios for staged compression of 7 or greater can be used, the maximum per stage is normally taken to be around 4. If the maximum temperature is known, then the maximum pressure ratio can be calculated.

Example 13.5 A recycle gas stream containing 88% hydrogen and 12% methane is to be increased in pressure from 81 bar to 98 bar. The inlet temperature is 40°C and the flowrate is 170,000 $Nm^3 \cdot h^{-1}$ (Nm^3 = normal m^3). Estimate the power requirements for a centrifugal compressor for this duty.

Solution

$$\text{Pressure ratio} = \frac{98}{81}$$

$$= 1.21$$

Thus, compression in a single stage will be acceptable.

Suction volume of gas at design conditions of 40°C and 81 bar

$$= 170{,}000 \times \frac{313}{273} \times \frac{1.013}{81}$$

$$= 2{,}438 \text{ m}^3\cdot\text{h}^{-1}$$

$$= 0.677 \text{ m}^3\cdot\text{s}^{-1}$$

The ratio of heat capacities for the mixture can be taken as a weighted mean of the values in Table 13.8.

$$\gamma = 0.88 \times 1.40 + 0.12 \times 1.30$$

$$= 1.39$$

From Equation 13.18, the polytropic efficiency is given by:

$$\eta_P = 0.017 \ln F + 0.7$$

$$= 0.017 \ln 0.677 + 0.7$$

$$= 0.69$$

The polytropic coefficient can then be calculated from Equation B.38:

$$n = \frac{\gamma\eta_P}{\gamma\eta_P - \gamma + 1}$$

$$= \frac{1.39 \times 0.69}{1.39 \times 0.69 - 1.39 + 1}$$

$$= 1.69$$

Now calculate the power from Equation B.35:

$$W = \frac{n}{n - 1}\frac{P_{in}F_{in}}{\eta_P}\left[1 - \left(\frac{P_{out}}{P_{in}}\right)^{\frac{n-1}{n}}\right]$$

$$= \frac{1.69}{1.69 - 1}\frac{81 \times 10^5 \times 0.677}{0.69}\left[1 - \left(\frac{98}{81}\right)^{\frac{1.69-1}{1.69}}\right]$$

$$= -1.57 \times 10^6 \text{ W}$$

This calculation assumed the gas to be ideal. For comparison, the calculation can be based on the Peng–Robinson Equation of State (see Chapter 4). A number of commercial physical property software packages allow the prediction of gas density and γ for a mixture of hydrogen and methane using the Peng–Robinson Equation of State. Using this, the gas density at normal conditions is 0.1651 $kg\cdot m^{-3}$. At 40°C and 81 bar, the density is 11.2101 $kg\cdot m^{-3}$. Thus, suction volume of gas

$$= 170{,}000 \times \frac{0.1651}{11.2101}$$

$$= 2504 \text{ m}^3\cdot\text{h}^{-1}$$

$$= 0.695 \text{ m}^3\cdot\text{s}^{-1}$$

At suction conditions:

$$\gamma = 1.38$$

This is close to the estimated value. Greater accuracy again could be obtained from γ at the average compression conditions. From Equation 13.18:

$$\eta_P = 0.69$$

From Equation B.38:

$$n = 1.66$$

From Equation B.35:

$$W = -1.61 \times 10^6 \text{ W}$$

This is a relatively small error given the high pressure of the compression.

13.4 SIMULATION OF RECYCLES

Having created process recycles, for what might be a variety of reasons, an evaluation of the design requires the material and energy balance to be evaluated with greater accuracy. In turn this will allow a preliminary sizing of equipment and an economic evaluation of the design. Example 13.2 presented a problem in which the recycle was solved for a simple recycle system by making some approximations. Such an approach is limited in its application. More complex problems demand a more sophisticated approach. Computer simulation packages are normally used to evaluate the material and energy balance once the recycle structure has been established.

To understand how such computer packages function, consider the simple flowsheet in Figure 13.13a. This involves an isomerization of Component A to Component B. The mixture of A and B from the reactor is separated into relatively pure A, which is recycled, and relatively pure B, which is the product. No byproducts are formed and the reactor performance can be characterized by its conversion. The performance of the separator is to be characterized by the recovery of A to the recycle stream (r_A) and recovery of B to the product (r_B).

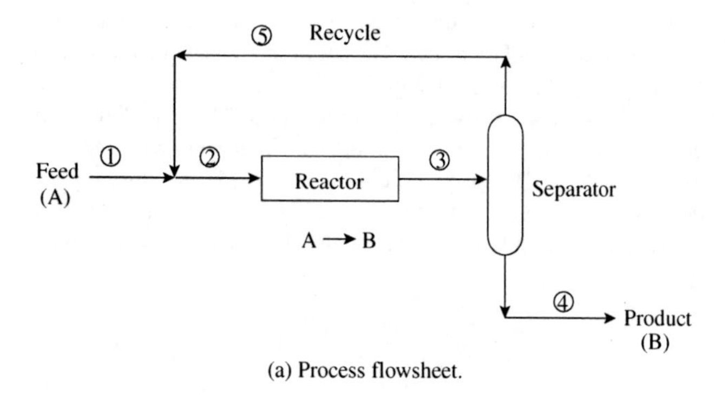

(a) Process flowsheet.

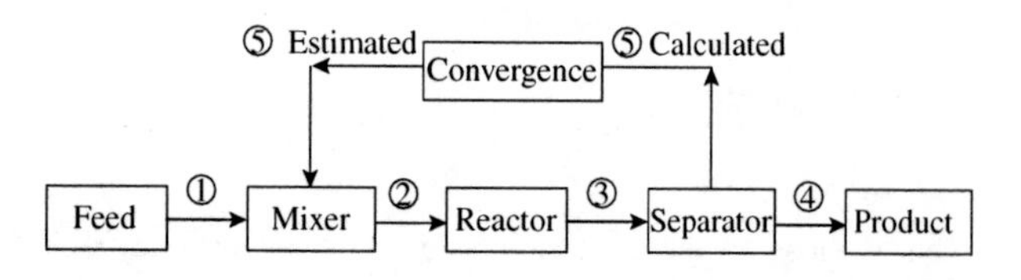

(b) Block structure of sequential modular calculation.

Figure 13.13 A simple process with recycle.

In this case, only the material balance will be solved in order to keep the problem simple. If the material balance is to be solved, then a series of material balance equations can be written for the flowsheet in Figure 13.13a:

Mixer

$$m_{A,2} = m_{A,1} + m_{A,5} \tag{13.19}$$

$$m_{B,2} = m_{B,1} + m_{B,5} \tag{13.20}$$

Reactor

$$m_{A,3} = m_{A,2}(1 - X) \tag{13.21}$$

$$m_{B,3} = m_{B,2} + Xm_{A,2} \tag{13.22}$$

Separator

$$m_{A,4} = m_{A,3}(1 - r_A) \tag{13.23}$$

$$m_{A,5} = r_A m_{A,3} \tag{13.24}$$

$$m_{B,4} = r_B m_{B,3} \tag{13.25}$$

$$m_{B,5} = m_{B,3}(1 - r_B) \tag{13.26}$$

where $m_{i,j}$ = molar flowrate of Component i in Stream j
X = reactor conversion
r_i = fractional recovery of Component i

Equations 13.19 to 13.26 form a set of 8 equations and 13 variables (m_A and m_B for each stream, X, r_A, r_B). Specifying the feed stream $m_{A,1}$ and m_{B1}, and X, r_A and r_B allows the set of equations to be solved. There are two basic approaches that could be adopted.

1. *Equation-oriented.* The *equation-oriented* or *equation-based* approach solves the set of equations simultaneously. If the problem involves n design variables, with p equations (equality constraints) and q inequality constraints, the problem becomes one of:

$$\begin{aligned} &\text{solve } h_i(x_1, x_2, \ldots\ldots, x_n) = 0 \ (i = 1, p) \\ &\text{subject to } g_i(x_1, x_2, \ldots\ldots, x_n) \geq 0 \ (i = 1, q) \end{aligned} \tag{13.27}$$

Whilst this approach seems straightforward for the simple mass balance above, for more complex recycle systems with energy balance equations and phase equilibrium equations, it is not straightforward. Equations describing the flowsheet connectivity are combined with equations describing the various unit operations in the flowsheet and, if possible, the physical property correlations into one large equation set[10]. The solution of the set of equations can be performed by a general-purpose nonlinear equation solver. Because of the difficulties of including the physical property equations, these are often formulated as distinct procedures and kept separate from equations describing the flowsheet connectivity and unit operations[10].

For example, in the simple problem above, if the values are set to:

$$\begin{aligned} m_{A,1} &= 100 \text{ kmol} \\ m_{B,1} &= 0 \text{ kmol} \\ X &= 0.7 \\ r_A &= 0.95 \\ r_B &= 0.95 \end{aligned} \tag{13.28}$$

Then the equations can be solved simultaneously to give:

$$\begin{aligned} m_{A,5} &= 39.8601 \text{ kmol} \\ m_{B,5} &= 5.1527 \text{ kmol} \\ m_{A,4} &= 2.0979 \text{ kmol} \\ m_{B,4} &= 97.9021 \text{ kmol} \end{aligned}$$

2. *Sequential modular.* In the *sequential modular* approach, the process equations are grouped within *unit operation blocks*. Each unit operation block contains the equations that relate the outlet stream and the performance variables for the block to the inlet stream variables and specified parameters. Each unit operation block is then solved one at a time in sequence[10]. The output calculated from each block becomes the feed to the next block, and so on. Figure 13.13b shows the block structure for the flowsheet in Figure 13.13a. The direction of information flow usually follows that of the material flow. First the feed stream must be created. This then goes to a mixer where the fresh feed is mixed with the recycle stream. Here a problem is encountered, as the flowrate and composition of the recycle are unknown. The sequential modular solution technique is to *tear* one of the streams in the recycle loop. In Figure 13.13b, the recycle stream itself has been torn. In general, tearing the recycle stream itself is only one option for tearing a stream in a loop. It is often best to tear a stream for which a good initial estimate can be provided. Tearing determines those streams or information flows that must be torn to render the system (or subsystem) to be acyclic. A *recycle convergence unit* or *solver* is inserted in the tear stream (Figure 13.13b). To start the calculation of the material balance in Figure 13.13b, values for the component molar flowrates for the recycle stream (tear stream) must be estimated. This allows the material balance in the reactor and separator to be solved. In turn, this allows the molar flowrates for the recycle stream to be calculated. The calculated and estimated values can then be compared to test whether errors are within a specified tolerance. It is usual to specify a scaled error in the form:

$$-Tolerance \leq \frac{G(x) - x}{x} \leq Tolerance \tag{13.29}$$

where x = estimate of the variable
$G(x)$ = resulting calculated value of the variable

If a material balance is to be solved, then the convergence variables can be taken to be the component molar flowrates. When a material and energy balance is to be solved, the additional convergence variables are usually taken to be pressure and enthalpy.

Care needs to be taken if some components are present in trace quantities. If an estimated concentration is 0.5 ppm and the calculated value is 1 ppm, the scaled error is 100%. This is much too large an error for most variables and yet the absolute error might be acceptable for a trace component. In other situations, it might be necessary to define trace components with a high precision. A trace component threshold can be set, below which the convergence criterion is ignored.

It is unlikely that the estimated values for the recycle stream will be within tolerance for the initial estimate. If the convergence criteria are not met, then the convergence block needs to update the value of the recycle stream. The simplest approach to this is *direct substitution* or *repeated substitution*[10]. In this approach, the sequence is calculated from an initial estimate. The calculated value then becomes the value for the next iteration. This is repeated until all convergence criteria are met. For example, in the example from Figure 13.13, assuming the values in Equation 13.28, the initial estimate could be:

$$\begin{aligned} m_{A,5} &= 50 \text{ kmol} \\ m_{B,5} &= 5 \text{ kmol} \end{aligned}$$

Table 13.9 follows the iterations until convergence is achieved.

Figure 13.14a shows a schematic representation of the direct substitution strategy. The problem with this approach is that convergence might require many iterations and some problems might fail to converge to the required tolerance. Rather than use direct substitution, the convergence unit can *accelerate* the convergence. Figure 13.14b illustrates one such method. The direct substitution iterations are linearized. A straight-line equation can be written for the two iterations as:

$$G(x) = ax + b \tag{13.30}$$

where a = slope of the line $= \dfrac{G(x_k) - G(x_{k-1})}{x_k - x_{k-1}}$
$G(x_k), G(x_{k-1})$ = calculated values of variables for iterations k and $k-1$
x_k, x_{k-1} = estimated values of variables for iterations k and $k-1$

For Iteration k, Equation 13.30 can be written to define the intercept:

$$b = G(x_k) - ax_k \tag{13.31}$$

Substituting Equation 13.31 into Equation 13.30 gives:

$$G(x_{k+1}) = ax_{k+1} + [G(x_k) - ax_k] \tag{13.32}$$

Table 13.9 Solution of material balance by direct substitution.

Iteration	Assumed		Calculated		Scaled residual	
	$m_{A,5}$ (kmol)	$m_{B,5}$ (kmol)	$m_{A,5}$ (kmol)	$m_{B,5}$ (kmol)	$m_{A,5}$	$m_{B,5}$
1	50	5	42.7500	5.5000	−0.1450	0.1000
2	42.7500	5.500	40.6838	5.2713	−0.0483	−0.0416
3	40.6838	5.2713	40.0949	5.1875	−0.0145	−0.0159
4	40.0949	5.1875	39.9270	5.1627	−0.0042	−0.0048
5	39.9270	5.1627	39.8792	5.1556	−0.0012	−0.0014
6	39.8792	5.1556	39.8656	5.1536	−0.0003	−0.0004
7	39.8656	5.1536	39.8617	5.1530	−0.0001	−0.0001
8	39.8617	5.1530	39.8606	5.1528	0.0000	0.0000

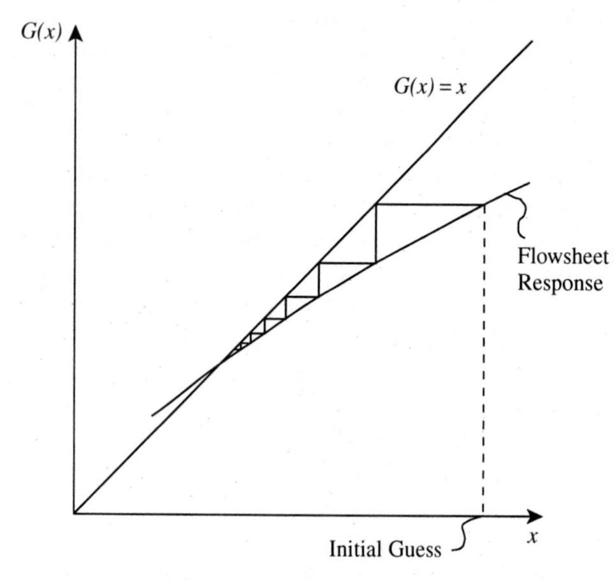

(a) Direct substitution.

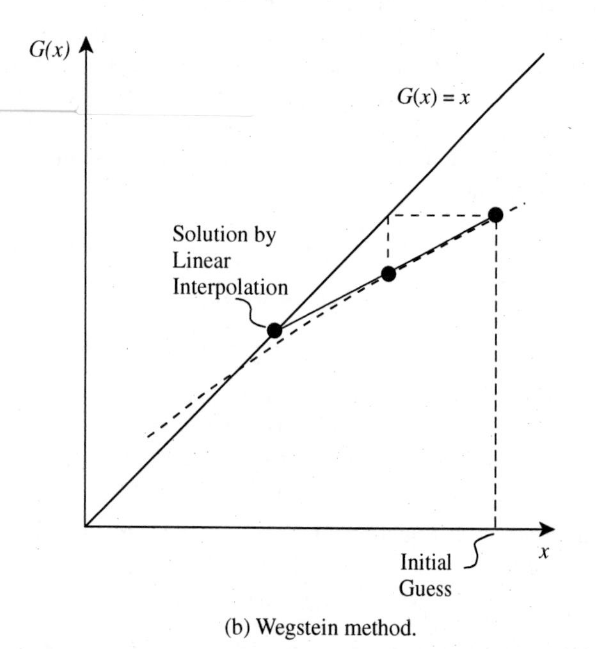

(b) Wegstein method.

Figure 13.14 Convergence of recycle loops using the sequential modular approach.

The intersection is required for Equation 13.32 with the equation:

$$G(x_{k+1}) = x_{k+1} \tag{13.33}$$

Substituting Equation 13.33 into Equation 13.32 gives:

$$x_{k+1} = ax_{k+1} + [G(x_k) - ax_k] \tag{13.34}$$

Rearranging Equation 13.34 gives:

$$x_{k+1} = \left(\frac{a}{a-1}\right)x_k - \left(\frac{1}{a-1}\right)G(x_k) \tag{13.35}$$

Substituting $q = a/(a-1)$ gives:

$$x_{k+1} = qx_k + (1-q)G(x_k) = x_k + (1-q)[G(x_k) - x_k] \tag{13.36}$$

Thus, Equation 13.36 can be used to accelerate the convergence and is known as the Wegstein method[11]. If $q = 0$ in Equation 13.36, the method becomes direct substitution. If $q < 0$, acceleration of the solution occurs. Bounds are normally set for the value of q to prevent unstable behavior.

Returning to the example from Figure 13.13, the solution by direct substitution is followed in Table 13.9. If the Wegstein Method is applied after the first two iterations:

$$a = \frac{42.7500 - 40.6838}{50 - 42.7500}$$

$$= 0.2850$$

$$q = -0.3986$$

Substituting in Equation 13.36 gives:

$$x_{k+1} = 42.7500 + (1 + 0.3986)[40.6838 - 42.7500]$$

$$= 39.8602 \text{ kmol}$$

Compared with direct substitution, this approaches the solution much more rapidly. The procedure is repeated until the convergence criteria are met.

When dealing with more complex flowsheets than the one in Figure 13.13, the order in which the calculations

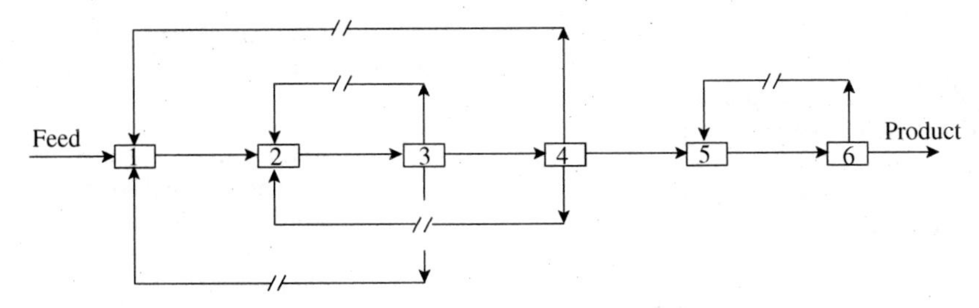

(a) Initial calculation sequence showing 5 tear streams.

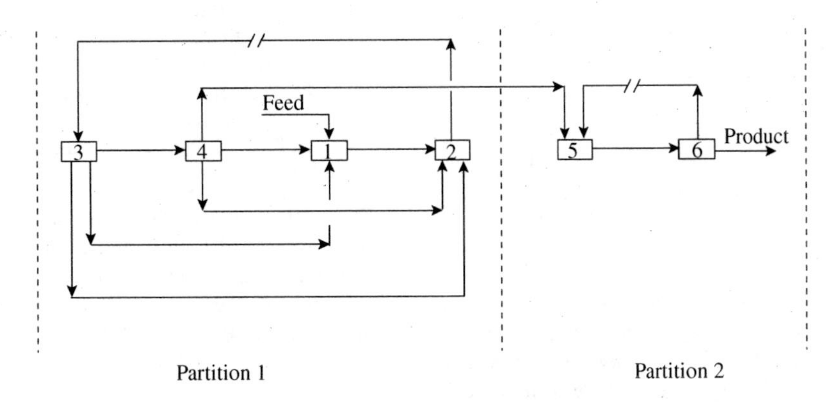

(b) Re-ordered calculation sequence showing 2 tear streams.

Figure 13.15 Partitioning and tearing a complex flowsheet.

take place in the sequential modular approach is important. The first consideration is to tear streams for which a good initial estimate can be provided. Thereafter the choice of tear streams should be to reduce the complexity of the calculation of the solution. Consider the flowsheet in Figure 13.15a. At first sight there appears to be five tear streams. Figure 13.15b shows a reordered calculation sequence. This calculation order requires tearing of only two streams rather than five. This will greatly simplify the calculations. Also in Figure 13.15b, the calculation sequence has been *partitioned* into two sets of blocks. Partitioning identifies the sets of blocks that must be solved together. In Figure 13.15b, there is little point solving the second partition until the first partition has been solved. If there are multiple tear streams in a partition, then the tear streams can either be converged sequentially or simultaneously. Various algorithms are available for the systematic *partitioning and tearing* of flowsheets[10].

The equation-oriented approach and the sequential modular approach each have their relative advantages and disadvantages. The sequential modular approach is intuitive and easy to understand. It allows the designer to interact with the solution as it develops and errors tend to be more straightforward to understand than with the equation-oriented approach. However, large problems may be difficult to converge with the sequential modular approach. On the other hand, the equation-oriented approach can make it difficult to diagnose errors. It is generally not as robust as the sequential modular approach and generally requires a good initialization to solve. One major advantage of the equation-oriented approach is the ability to formulate the problem as an optimization problem, as design problems almost invariably involve some optimization. Thus, the formulation in Equation 13.27 can be readily transformed to the corresponding optimization problem and the material and energy balance solved simultaneously with the optimization problem:

$$\begin{aligned} &\text{minimize} f(x_1, x_2, \ldots\ldots, x_n) \\ &\text{subject to } h_i(x_1, x_2, \ldots\ldots, x_n) = 0 (i = 1, p) \\ &g_i(x_1, x_2, \ldots\ldots, x_n) \geq 0 (i = 1, q) \end{aligned} \quad (13.37)$$

Of course, the two approaches can be combined and the sequential modular approach used to provide an initialization for the equation-oriented approach.

Example 13.6

a. Given the estimate of the reactor effluent from Example 13.2 for fraction of methane in the recycle and purge of 0.4, calculate the actual separation in the phase separator assuming the temperature to be 40°C. Phase equilibrium for this mixture can be represented by the Peng–Robinson Equation of State with binary interaction parameters assumed to be zero. Many computer simulation programs are available commercially to carry out such calculations.
b. Repeat the calculation from Example 13.2 with actual phase equilibrium data in the phase separation instead of assuming a sharp split.

Solution

a. If such a phase separation is carried out assuming the feed in Table 13.2, the results are given in Table 13.10:

Table 13.10 Phase separation calculated using the Peng–Robinson Equation of State.

Component	Reactor effluent flowrate ($kmol\cdot h^{-1}$)	Vapor flowrate from phase separator ($kmol\cdot h^{-1}$)	Liquid flowrate from phase separator ($kmol\cdot h^{-1}$)
Hydrogen	1554	1550	4
Methane	1036	1020	16
Benzene	265	17.8	247.2
Toluene	91	2.3	88.7
Diphenyl	4	0	4

The phase separation at 40°C gives a good separation of the hydrogen and methane into the vapor phase and benzene, toluene and diphenyl into the liquid phase. Under these conditions, the hydrogen and methane are above their critical temperatures and are effectively non-condensibles. However, some hydrogen and methane dissolve in the liquid phase. Also, some aromatics are carried with the vapor. An important consequence of this is that the flowsheet in Figure 13.8 would need to be modified to separate the hydrogen and methane carried forward with the liquid from the phase separation. This will require another distillation column to separate the hydrogen and the methane from the aromatics before separating the aromatics.

The reader might like to check that as the temperature of the phase separation is increased or its pressure decreased, the separation between the hydrogen, methane and the other components becomes worse.

b. Assuming the phase split operates at 40 bar and 40°C, a rigorous solution of the phase equilibrium using the Peng–Robinson Equation of State and the recycle equations using flowsheet simulation software gives a composition of the reactor effluent is given in Table 13.11:

Table 13.11 Composition of the reactor effluent and phase separation calculated using the Peng–Robinson Equation of State and solving the recycle for Example 13.3.

Component	Reactor effluent flowrate ($kmol\cdot h^{-1}$)	Vapor flowrate from phase separator ($kmol\cdot h^{-1}$)	Liquid flowrate from phase separator ($kmol\cdot h^{-1}$)
Hydrogen	1536	1532	4
Methane	1053	1036	17
Benzene	283	18	265
Toluene	93	3	90
Diphenyl	4	0	4

Comparing this solution with that based on a sharp phase separation in Example 13.2, the errors are surprisingly small. However, studying at the K-values in the phase separator given in Table 13.12, it is not so surprising.

Table 13.12 K-values for the phase separation based on the Peng–Robinson Equation of State.

Component	K_i
Hydrogen	54
Methane	8.9
Benzene	0.010
Toluene	0.0037
Diphenyl	1.2×10^{-5}

The temperature of the phase split is well above the critical temperatures of both hydrogen and methane, leading to large K-values. On the other hand, the K-values of the benzene, toluene and diphenyl are very low, hence the assumption of a sharp split in Example 13.2 was a good one in this case.

13.5 THE PROCESS YIELD

Having considered the feed, reaction, separation and recycling of material, the streams entering and leaving the process can be established. Figure 13.16 illustrates typical input and output streams. Feed streams enter the process and product, byproduct and purge streams leave after the separation and recycle system has been established.

Raw materials costs dominate the operating costs of most processes (see Chapter 2). Also, if raw materials are not used efficiently, this creates waste that becomes an environmental problem. It is therefore important to have a measure of the efficiency of raw materials usage. The process yield is defined as:

$$\text{Process yield} = \frac{(\text{desired product produced})}{(\text{reactant fed to the process})} \times \text{stoichiometric factor} \qquad (13.38)$$

where the stoichiometric factor is the stoichiometric moles of reactant required per mole of product. When more than one reactant is used (or more than one desired product produced) Equation 13.38 can be applied to each reactant (or product).

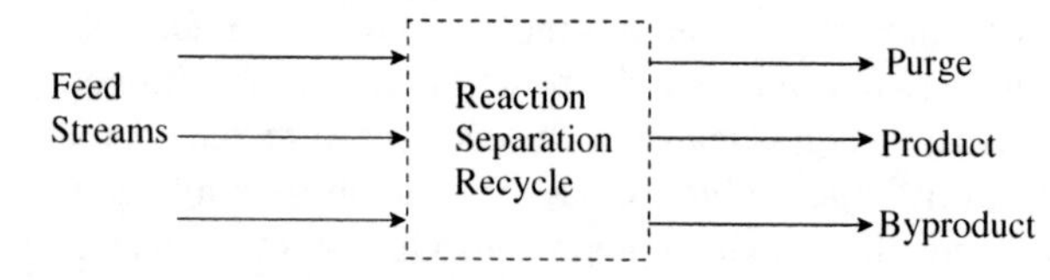

Figure 13.16 The overall process material balance for the process yield.

In broad terms, there are two sources of yield loss in the process:

- losses in the reactor due to byproduct formation (selectivity losses) or unconverted feed material if recycling is not possible
- losses from the separation and recycle system

Addressing the streams entering and leaving the process in Figure 13.16, there are material losses in the byproducts and purges that should be reduced if possible. Thus, before proceeding further, a number of questions should be considered:

1. Can byproduct formation be avoided or reduced by recycling? This is often possible when the byproduct is formed by secondary reversible reactions.
2. If a byproduct is formed by reaction involving feed impurities, can this be avoided or reduced by purification of the feed?
3. Can the byproduct be subjected to further reaction and its value upgraded? For example, most organic chlorination reactions produce hydrogen chloride as a byproduct. If this cannot be sold it must be disposed of, an alternative as discussed in Example 13.1 is to convert the hydrogen chloride back to chlorine via the reaction:

$$2HCl + \tfrac{1}{2}O_2 \rightleftharpoons Cl_2 + H_2O$$

   The chlorine can then be recycled.
4. Can the loss of useful material in the purge streams be avoided or reduced by feed purification?
5. Can the loss of useful material in the purge be avoided or reduced by additional separation on the purge? The roles of refrigerated condensation, low-temperature distillation, absorption, adsorption and membranes in this respect have already been discussed.
6. Can the useful material lost in the purge streams be reduced by additional reaction to useful products? If the purge stream contains significant quantities of reactants, then placing a reactor and additional separation on the purge can sometimes be justified. This technique is used in some designs of ethylene oxide processes.

Example 13.7 Calculate the process yield of benzene from toluene and benzene from hydrogen for the approximate phase split in Example 13.2.

Solution

$$\text{Benzene yield} = \frac{\text{(benzene produced)}}{\text{(toluene fed to the process)}} \times \text{stoichiometric factor}$$

Stoichiometric factor = stoichiometric moles of toluene required per mole of benzene produced

$$= 1$$

$$\text{Benzene yield from toluene} = \frac{(P_B)}{(P_B/S)} \times 1$$

$$= S = 0.97$$

In this case, because there are no raw materials losses in the separation and recycle system, the only yield loss is in the reactor and the process yield equals the reactor selectivity.

$$\text{Benzene yield from hydrogen} = \frac{\text{(benzene produced)}}{\text{(hydrogen fed to the process)}} \times \text{stoichiometric factor}$$

Stoichiometric factor = stoichiometric moles of hydrogen required per mole of benzene produced

$$= 1$$

For $y = 0.4$, $\alpha = 0.3013$:

$$\text{Benzene yield from hydrogen} = \frac{P_B}{(1554\alpha + 269.2)} \times 1$$

$$= \frac{265}{1554 \times 0.3013 + 269.2)}$$

$$= 0.36$$

13.6 OPTIMIZATION OF REACTOR CONVERSION

Once the structure of the recycle and separation has been established, some important degrees of freedom can be optimized that can have a very significant effect on the overall process economics. Start by considering the optimization of reactor conversion.

If the reactor conversion is changed so as to optimize its value, then not only is the reactor affected in size and performance but also the separation system, since it now has a different separation task. The size of the recycle will also change. If the recycle requires a compressor, then the capital and operating costs of the recycle compressor will change. In addition, the heating and cooling duties associated with the reactor and the separation and recycle system change.

As the reactor conversion increases, the reactor volume increases and hence reactor capital cost increases. At the same time, the amount of unconverted feed needing to be separated decreases and hence the cost of recycling unconverted feed decreases.

Consider a simple process in which *FEED* is reacted to *PRODUCT* via the reaction in Equation 13.1. The flowsheet synthesis is started at the reactor. The effluent from the reactor contains both *PRODUCT* and unreacted *FEED* that must be separated. Unreacted *FEED* is recycled to the reactor via a pump if the recycle is liquid, or a compressor if the recycle is vapor.

Optimization of the system can be carried out by minimizing cost or maximizing economic potential (*EP*), as discussed in Chapter 2. Costs for the process to carry out the reaction in Equation 13.1 are illustrated in Figure 13.17 decomposed according to the layers of the onion model[1]. In Figure 13.17, the annualized reactor cost (capital only) increases since high conversion requires a large volume and hence high capital cost. The annualized separation and recycle cost (capital only in this case) decreases with increasing reactor conversion, since the amount of unreacted *FEED* to separate and recycle decreases. If the recycle had required a compressor, the capital and operating costs of the compressor would have been included in the separation and recycle cost. The cost of the heat exchanger network and utilities is a combination of annualized energy cost and annualized capital cost of all exchangers, heaters and coolers. Later, it will be explained how to estimate the energy and capital cost of the heat exchanger network without having to carry out its detailed design. Figure 13.17 shows the cost of the heat exchanger network and utilities decreases with increasing conversion, since the separation duty is decreased and also the heating and cooling in the recycle. Combining the reactor, separation and recycle and heat exchanger network costs into a total annual cost (energy and capital) reveals that there is an optimum reactor conversion. From Figure 13.17, for this example, heat integration and the cost of the heat exchanger network and utilities has a significant influence on the optimum conversion. In other cases, the relative importance of the component costs will be different.

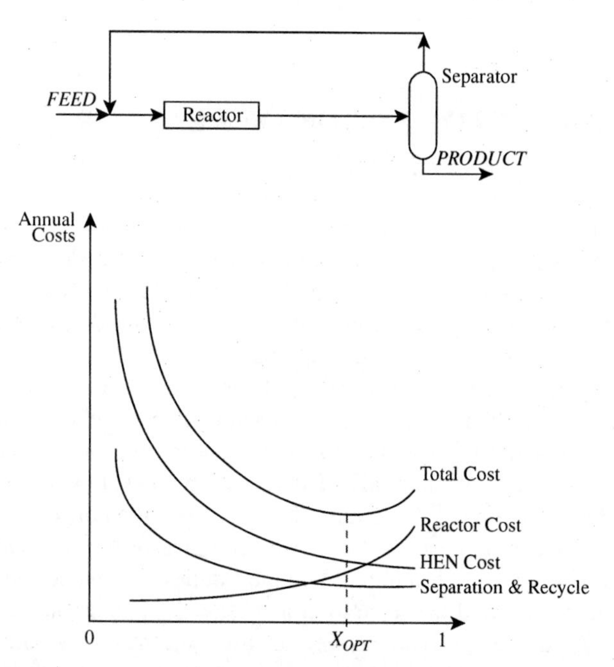

Figure 13.17 Overall cost trade-offs for a simple process as a function of reactor conversion. (Reproduced from Smith R and Linnhoff B, 1988, *Trans IChemE, CHERD*, **66**: 195 by permission of the Institution of Chemical Engineers.)

If the cost of the heat exchanger network changes, perhaps through a change in energy cost, then the optimum reactor conversion will change. This change will likely dictate a different optimum reactor conversion and hence different separator design and process flowrates. However, such sensitivity is easily explored by changing the component costs in Figure 13.17.

In Figure 13.17, the only cost forcing the optimum conversion back from high values is that of the reactor. Hence, for such simple reaction systems, a high optimum conversion would be expected. This was the reason in Chapter 5 that an initial value of reactor conversion of 0.95 of the maximum conversion was chosen for simple reaction systems.

In Figure 13.17, the curves are limited by a maximum reactor conversion of 1.0. If the reaction had been reversible, then a similar picture would have been obtained. However, instead of being limited by a reactor conversion of 1.0, the curves would have been limited by the equilibrium conversion (see Chapter 6).

Consider the example of a process that involves the multiple reactions in Equation 13.3. Because there is a mixture of *FEED, PRODUCT* and *BYPRODUCT* in the reactor effluent, an additional separator is required. The economic trade-offs now become more complex and a new cost must be added to the trade-offs. This is a raw materials efficiency cost due to byproduct formation. If the *PRODUCT* formation is kept constant, despite varying levels of *BYPRODUCT* formation, then the cost can be defined to be[11,12]:

$$\begin{aligned}\text{Cost due to } \mathit{BYPRODUCT} \text{ formation} \\ = \text{cost of } \mathit{FEED} \text{ lost to } \mathit{BYPRODUCT} \\ - \text{value of } \mathit{BYPRODUCT}\end{aligned} \tag{13.39}$$

The value of *PRODUCT* formation and the raw materials cost of *FEED* that reacts to *PRODUCT* are constant. Alternatively, if the byproduct has no value, the cost of disposal should be included as:

$$\begin{aligned}\text{Cost due to } \mathit{BYPRODUCT} \text{ formation} \\ = \text{cost of } \mathit{FEED} \text{ lost to } \mathit{BYPRODUCT} \\ + \text{cost of disposal of } \mathit{BYPRODUCT}\end{aligned} \tag{13.40}$$

By considering only those raw materials that undergo reaction to undesired byproduct, only the raw materials costs that are in principle avoidable are considered. Those raw materials costs that are inevitable, that is, the stoichiometric requirements for *FEED* that converts into the

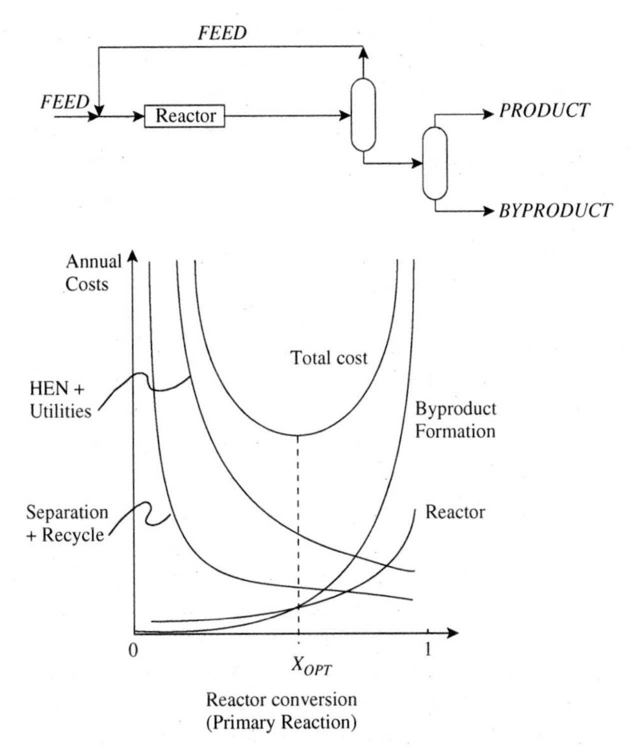

Figure 13.18 Cost trade-offs for a process with byproduct formation from secondary reactions. (Reproduced from Smith R and Linnhoff B, 1988, *Trans IChemE, CHERD*, **66**: 195 by permission of the Institution of Chemical Engineers.)

desired *PRODUCT* are not included. Raw materials costs that are in principle avoidable are distinguished from those that are inevitable from the stoichiometric requirements of the reaction[12,13].

Figure 13.18 shows typical cost trade-offs for this case. At high conversion, the raw materials cost due to byproduct formation is dominant. This is because the reaction to the undesired *BYPRODUCT* is series in nature, which results in the selectivity becoming very low at high conversions. In Chapter 5, the initial setting for reactor conversion was set to be 0.5 of the maximum conversion for such reaction systems. Figure 13.18 shows clearly why a high setting for reactor conversion would be inappropriate. The byproduct formation cost forces the optimum to lower values of conversion. Again, if the primary reaction had been reversible, then a similar picture would have obtained. However, instead of being limited by a reactor conversion of 1.0, the curves would have been limited by the equilibrium conversion (see Chapter 6).

Also, if there are two separators' the order of separation can change. The trade-offs for these two alternative flowsheets will be different. The choice between different separation sequences can be made using the methods described in Chapters 11 and 12. However, as the reactor conversion changes, the most appropriate sequence can also change. In other words, different separation system structures can become appropriate for different reactor conversions.

13.7 OPTIMIZATION OF PROCESSES INVOLVING A PURGE

If an impurity entering with the feed, or a byproduct of reaction, needs to be removed via a purge, the concentration of impurity in the recycle can be varied as a degree of freedom. If the impurity is allowed to build up to a high concentration, then this reduces the loss of valuable raw materials in the purge. However, this decrease in raw materials cost is offset by an increase in the cost of recycling the additional impurity, together with increased capital cost of equipment in the recycle. Changes in the recycle concentration again make changes throughout the flowsheet.

As with the case of byproduct losses, another cost needs to be added to the trade-offs when there is a purge. This is a raw materials efficiency cost due to purge losses. If the *PRODUCT* formation is constant, this cost can be defined to be[12,13]:

$$\text{Cost of Purge Losses} = \text{Cost of } \textit{FEED} \text{ Lost to Purge} - \text{Value of Purge} \qquad (13.41)$$

The purge by its nature is mixture and usually only has value in terms of its fuel value. Alternatively, if the purge must be disposed of by effluent treatment:

$$\text{Cost of Purge Losses} = \text{Cost of } \textit{FEED} \text{ Lost to Purge} + \text{Cost of Disposal of Purge} \qquad (13.42)$$

Again, as with the byproduct case, those raw materials costs that are in principle avoidable (i.e. the purge losses) are distinguished from those that are inevitable (i.e. the stoichiometric requirements for *FEED* entering the process that converts to the desired *PRODUCT*). Consider the trade-offs for the reaction in Equation 13.1, but now with *IMPURITY* entering with the *FEED*.

Now there are two variables in the optimization. These are the reactor conversion (as before) but now also the concentration of *IMPURITY* in the recycle. For each setting of the *IMPURITY* concentration in the recycle, a set of trade-offs can be produced analogous to those shown in Figures 13.17 and 13.18. Figure 13.19 shows the trade-offs for the feed impurity case and a purge with fixed concentration of impurity in the recycle[12,13].

As the concentration of impurity in the recycle is varied, each component cost shows a family of curves when plotted against reactor conversion. Reactor cost (capital only) increases as before with increasing conversion (Figure 13.20a). Separation and recycle costs decrease as before (Figure 13.20b). Figure 13.20c shows the cost of the heat exchanger network and utilities to again decrease with increasing conversion. Each of the component costs in Figure 13.20 increase with increasing concentration of impurity in the recycle.

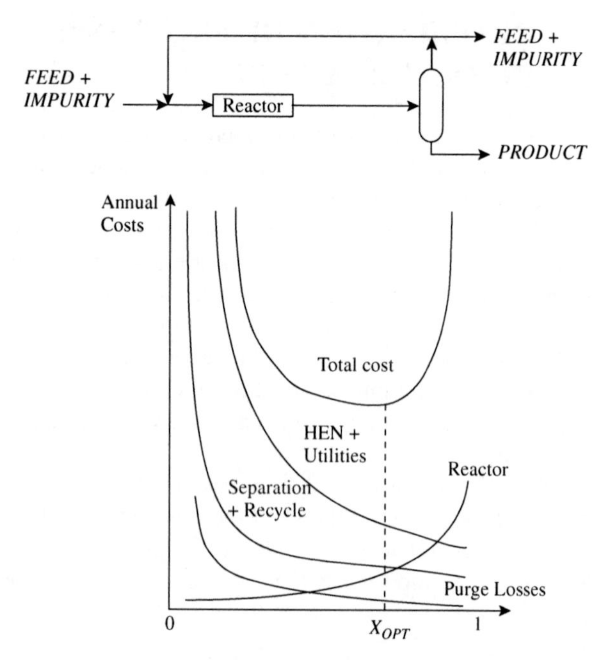

Figure 13.19 Cost trade-offs for a process with a purge for a fixed concentration of impurity in the recycle.

Figure 13.21 shows the corresponding component costs for the purge losses. Figure 13.21a shows component costs that are typical when the value of the raw materials lost in the purge is high relative to the fuel value of the purge. Figure 13.21b shows component costs that are typical when the value of the raw materials lost in the purge is low relative to the fuel value of the purge. If the process produces a byproduct to be purged and the purge has a value, then the component cost of the purge can show more complex behavior, as shown in Figure 13.21c. This shows a complex pattern that depends on the relative costs of raw materials and fuel value of the purge[12,13].

Figure 13.22 shows the component costs combined to give a total cost that varies with both reactor conversion and recycle concentration of impurity. Each setting of the recycle impurity concentration shows a cost profile with an optimum reactor conversion. As the recycle impurity concentration is increased, the total cost initially decreases but then increases for larger values of recycle concentration. The optimum conditions in Figure 13.22 are in the region of $y = 0.6$ and $X = 0.5$, although the optimum is quite flat.

An alternative way to view the trade-offs shown in Figure 13.23 is as a contour diagram. The contours in Figure 13.23a are lines of constant total cost. The objective of the optimization is to find the lowest point. A simple strategy would fix the first variable then optimize the second variable, and then fix the second variable and optimize the first, and so on in a univariate search, as discussed in Chapter 3. This is illustrated in Figure 13.23b where reactor conversion (X) is first fixed and impurity concentration (y) optimized. Impurity concentration is then fixed and conversion optimized. In this case, after two searches the solution is close to the optimum (Figure 13.23c). Whether this is adequate depends on how flat the solution space is in the region of the optimum. Whether such a strategy will find the actual optimum depends on the shape of the solution space and the initialization for the optimization, as discussed in Chapter 3. Other strategies for the optimization have been discussed in Chapter 3.

Obviously, the use of purges is not restricted to dealing with impurities. Purges can also be used to deal with byproducts. As with the optimization of reactor conversion, changes in the recycle concentration of impurity might change the most appropriate separation sequence.

13.8 HYBRID REACTION AND SEPARATION

So far, it has been assumed that the reaction and separation would be carried out consecutively and connected with recycle streams if appropriate. Consider a liquid-phase

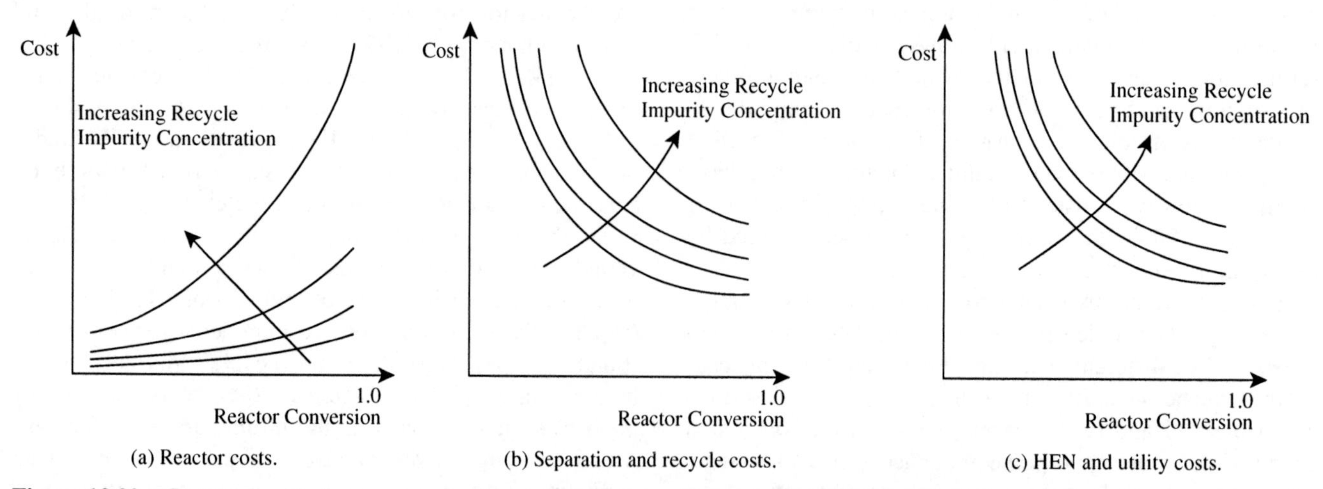

Figure 13.20 Cost trade-offs for processes with a purge as recycle inert concentration changes. (Reproduced from Smith R and Linnhoff B, 1988, *Trans IChemE, CHERD*, **66**: 195 by permission of the Institution of Chemical Engineers.)

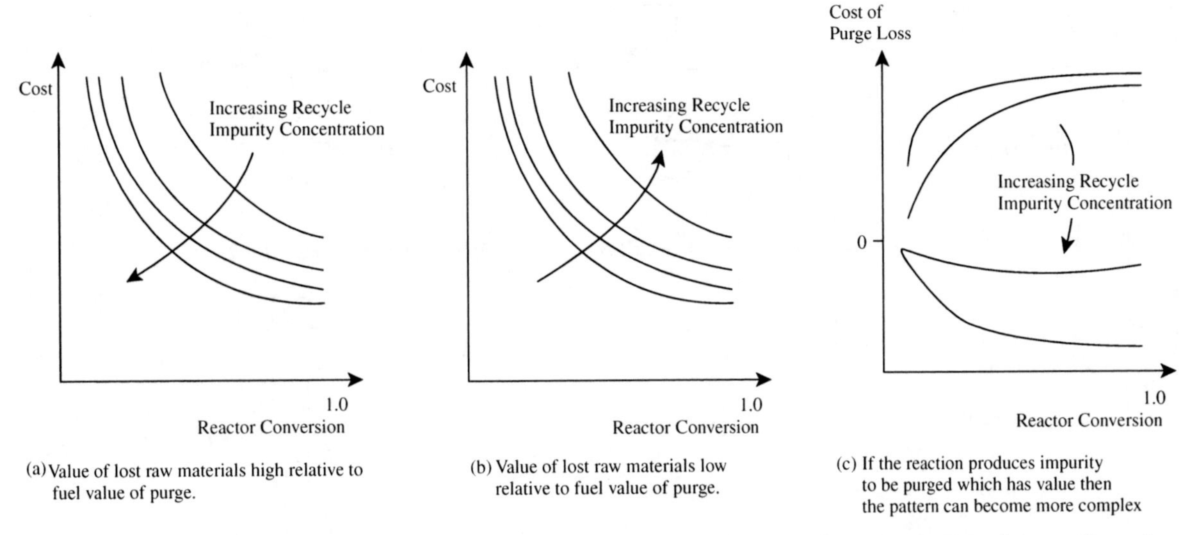

Figure 13.21 Cost trade-offs for purge losses as reactor conversion and recycle impurity concentration change. (Reproduced from Smith R and Linnhoff B, 1988, *Trans IChemE, CHERD*, **66**: 195 by permission of the Institution of Chemical Engineers.)

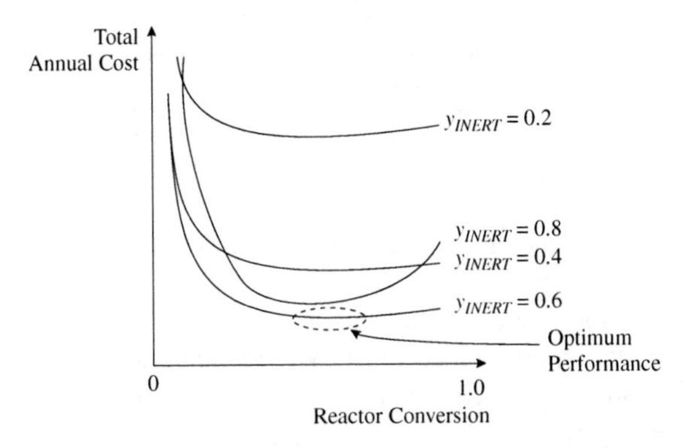

Figure 13.22 Putting all the costs together allows the optimum conversion and recycle inert concentration to be determined. (Reproduced from Smith R and Linnhoff B, 1988, *Trans IChemE, CHERD*, **66**: 195 by permission of the Institution of Chemical Engineers.)

exothermic equilibrium reaction, such as:

$$FEED1 + FEED2 \rightleftharpoons PRODUCT + BYPRODUCT \quad (13.43)$$

If the reaction is carried out in a reactor, such as that shown in Figure 13.24a, in which the product is allowed to vaporize and leave the reactor, then the equilibrium is shifted to higher conversion, as discussed in Chapter 6. However, the vaporization would be expected to be such that not just the product would vaporize, but also perhaps feed material and byproduct. A distillation rectifying section could be added to the reactor as shown in Figure 13.24b to carry out rectification and produce a pure *PRODUCT* from the top of the distillation. Of course, for this to be possible, the relative volatilities of the various components must be appropriate both in their order and their magnitude. Also, the temperature at which the reaction and distillation take place need to be similar. The arrangement shown in Figure 13.24b, in addition to having advantages for the reactor, is also energy efficient in that the vapor supplied to the distillation comes from the heat of reaction. This idea is carried a step further by adding a distillation stripping section, as shown in Figure 13.24c, to separate the pure *BYPRODUCT* from the liquid leaving the reactor and return the feed material to the reactor. Again, the

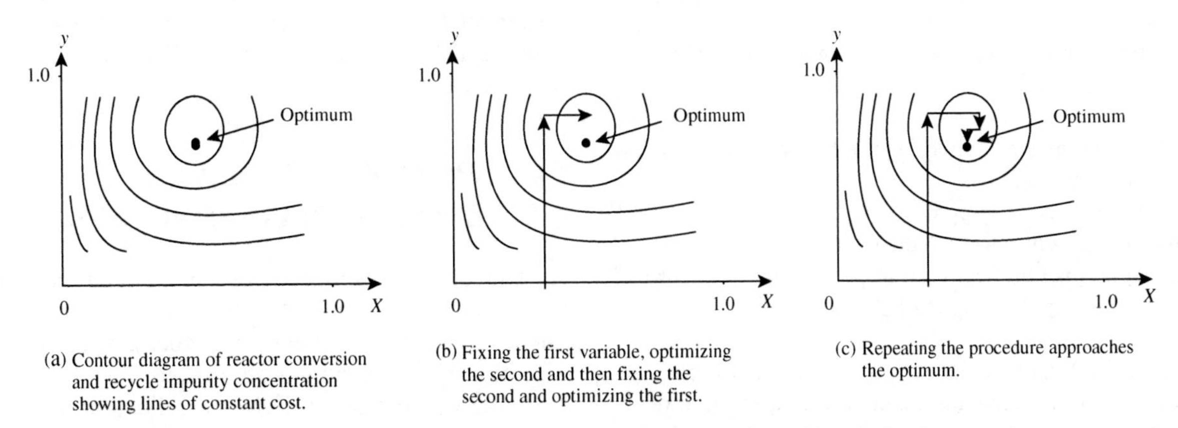

Figure 13.23 Optimization of reactor conversion and recycle impurity concentration using a univariate search.

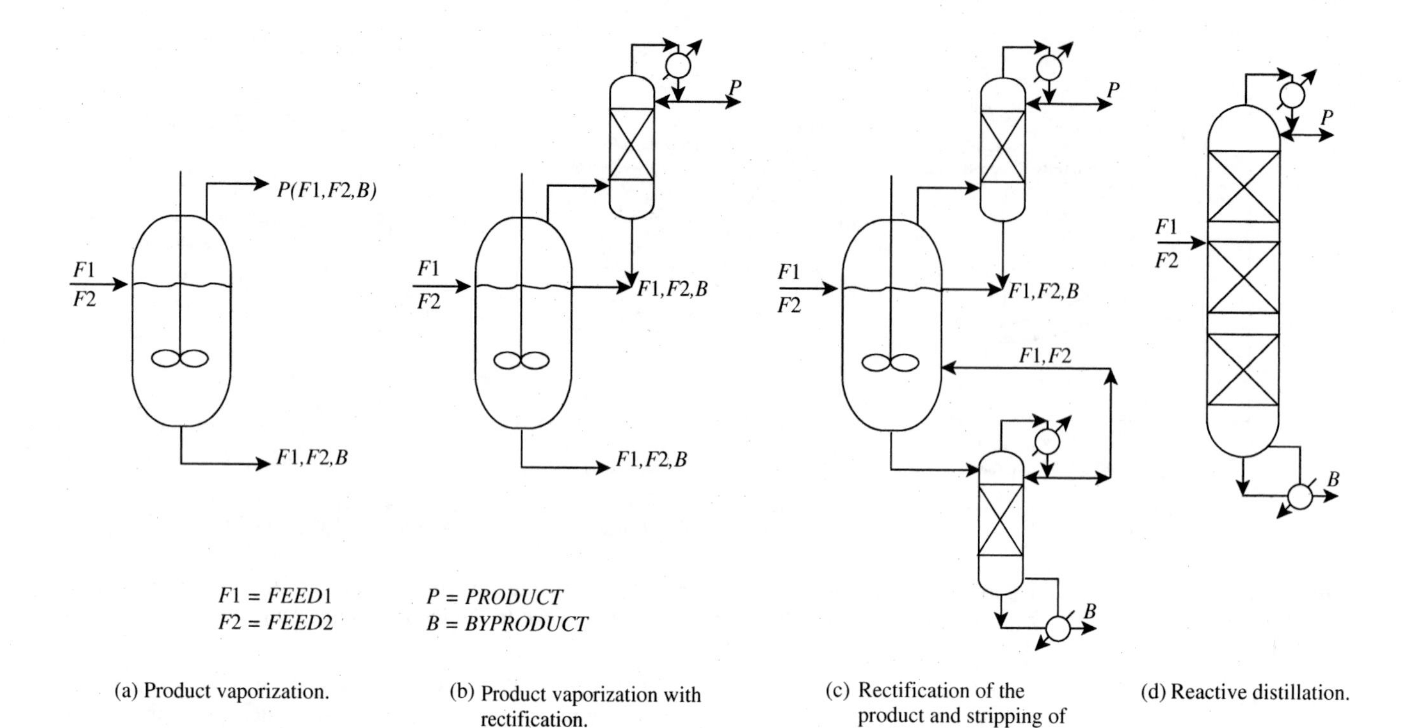

Figure 13.24 Hybrid reaction and distillation – reactive distillation.

relative volatilities of the components must be appropriate both in their order and their magnitude to allow this to be achieved. Finally, the whole system is fitted into a single shell and the result is a *reactive distillation*, as shown in Figure 13.24d.

It may also be useful to carry out such hybrid reaction and separation when byproducts are formed from competing reactions, such as:

$$\begin{aligned} FEED1 + FEED2 &\longrightarrow PRODUCT \\ PRODUCT &\longrightarrow BYPRODUCT \end{aligned} \quad (13.44)$$

It is desirable to remove product from the reaction as soon as it is formed to prevent it from reacting further to byproduct. An arrangement such as that shown in Figure 13.24d would allow this to happen if the relative volatilities have the appropriate order and magnitude.

Thus, there are a number of potential advantages for reactive distillation:

- reaction equilibrium can be shifted to higher (or even complete) conversion
- side reactions can be suppressed and selectivity increased
- capital investment can be reduced
- exothermic reaction heat used to provide the heat for the separation and reduce operating costs

In some fortunate circumstances, azeotropes can be eliminated from the separation that would need to be dealt with if reaction and separation are carried out consecutively.

The disadvantages of reactive distillation are:

- favorable relative volatilities are needed
- distillation conditions must give adequate reaction rate
- thorough research, testing and even pilot plant trials are required

Figure 13.25 shows another example of hybrid reaction and separation. This shows an esterification reaction in which water is removed between the reaction stages using pervaporation[14]. The esterification uses a heterogeneous catalyst and the process in Figure 13.25 runs in four stages. Each stage includes a reactor where the components are brought close to equilibrium and then the mixture flows through a pervaporation stage where the water generated in the reaction step is removed. This shifts the equilibrium conversion favorably. In the next reaction step, equilibrium is reestablished and again the reaction water is removed, and so on.

13.9 FEED, PRODUCT AND INTERMEDIATE STORAGE

Most processes require storage for the feed and product. Storage of feed is required if the delivery of the feed is in batches (e.g. barge, rail car, road truck). Even if the feed is being delivered continuously via pipelines for gases and liquids or conveyors in the case of solids, there will be no guarantee that feed will be free from interruptions in supply.

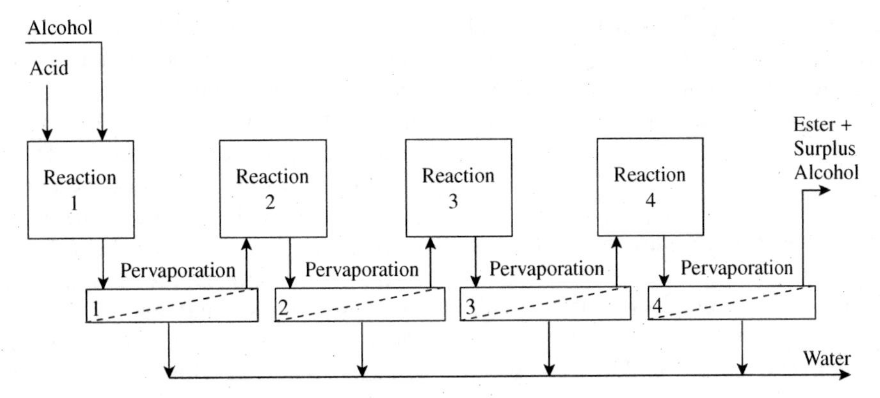

Figure 13.25 Hybrid reaction and pervaporation. (From Wynn N, 2001, *Chem Eng Progr*, Oct: 66, reproduced by permission.)

For example, the upstream plant providing the continuous feed will need to be shut down for various reasons, and there might be unexpected failures of the delivery, for example, as a result of breakdowns.

Whilst solids and liquids are straightforward to store, gases are difficult. Relatively small quantities of gas can be stored in the gaseous state at ambient temperature in pressurized vessels. Larger quantities of gas storage require the gas to be liquefied. This can be achieved by decreasing the temperature using refrigeration, or increasing the pressure, or a combination of both. High-pressure storage has a high capital cost, as it requires thick-walled vessels. Low-temperature storage also has a high capital cost, as it requires capital investment in refrigeration equipment. Low-temperature storage also has a significant operating cost for power to run the refrigeration. The most appropriate method of storage for gases depends on a number of factors and involves safety, as well as capital and operating cost considerations.

If the feed is delivered in batches and used continuously, there is a fluctuating amount of storage, known as the *active stock*. For example, suppose a plant operating with a liquid feed is at steady state. The liquid feed tank after a delivery might be perhaps 80% full. As the plant operates at steady state, the liquid level falls continuously to say 20% when the next delivery arrives and the level returns to 80% as a result of the delivery. The amount of liquid between the 20% and 80% levels is the active stock and 20% is *inactive*. Storage tanks for liquids should not be designed to operate with less than 10% inactive stack at the minimum, as this would create difficulties in operation. On the other hand, tanks for liquids should not be designed to operate more than 90% full at the maximum. An empty space above the liquid (known as *ullage*) is required when the tank is full to allow for safety and expansion. If the flowrate of feed material is m_{FEED} tons per year and the maximum active stock is m_{STOCK}, the number of deliveries per year will be the ratio m_{FEED}/m_{STOCK}.

The capital cost of the storage equipment will be approximately proportional to the storage capacity. This involves the capital cost of storage tanks in the case of gases and liquids, and silos in the case of solids, capital cost of materials handling equipment (e.g. pumps, conveyors, etc.) and capital cost of refrigeration equipment. In addition to the capital cost of the equipment, there is also the working capital associated with the value of the material being stored. The greater the value of the material in storage the greater the disincentive to store large quantities of material.

The feed storage

- provides a supply of feed to the plant between feed deliveries;
- compensates for interruptions in feed delivery due to unforeseen circumstances (e.g. breakdown in the plant manufacturing the feed material);
- allows short-term increases in production if the market for the product is favorable;
- compensates for interruptions in feed delivery due to holiday periods;
- compensates for seasonal variations in feed supply;
- allows feed to be bought under favorable market conditions when it is cheaper and stored for later use;
- dampens out variations in the feed properties.

The amount of feed storage will depend on:

- the frequency of deliveries;
- the size of deliveries;
- the reliability of deliveries;
- the capacity of the plant;
- the phase of the feed (gas, liquid or solid);
- the hazardous nature of the feed material (the inventory of hazardous feed should be kept to a minimum);
- the capital and operating costs (e.g. refrigeration system for the storage of liquefied gas) associated with the feed storage equipment;
- the working capital locked up in the stored feed;
- the economic benefit to be gained from being able to take advantage of market fluctuations in the purchase cost of the raw materials.

The product must also be stored, for similar reasons to those for feed storage. The product delivery will often not be continuous. Also, product will often be delivered to different customers. If the product is being delivered via pipeline in the case of a liquid or gas, or continuous conveyer in the case of a solid, then product storage can be minimized.

The product storage

- balances the difference between the rates of production and dispatch;
- maintains product delivery during plant shutdown for maintenance;
- maintains product delivery during unforeseen plant shutdown;
- compensates for peak and seasonal demands;
- holds materials when holidays prevent dispatch;
- allows material to be held up for later sale if short-term market conditions are unfavorable, leading to a short-term decrease in the sales price;
- allows variations in product quality to be dampened out.

The amount of product storage will depend on:

- the frequency of product dispatches;
- the size of dispatches;
- the reliability of the dispatches;
- the capacity of the plant;
- the phase of the product (gas, liquid or solid);
- the hazardous nature of the product (the inventory of hazardous product should be kept to a minimum);
- the capital and operating costs (e.g. refrigeration system for the storage of liquefied gas) associated with product storage equipment;
- the working capital locked up in the stored product;
- the economic benefit to be gained from being able to take advantage of market fluctuations in the sales price of the product.

In addition to the storage of feed and product, chemical intermediates are also often stored within the process. *Intermediate storage* for chemical intermediates is required particularly when the process requires a number of transformation steps between the feed and the product. It creates flexibility in the operation of the plant. For example, consider a process involving a complex reaction system, followed by a complex separation system. The start-up and control of the two sections can be simplified if they can be decoupled. This can be achieved by introducing intermediate storage between the reaction and separation sections. For start-up, the reaction section can be started up independently of the separation section. When starting up, the reaction section produces an intermediate chemical that is accumulated in the intermediate storage. When the reaction section is producing material of a suitable quality to be fed to the separation section, then the separation section can be started up by feeding from intermediate storage. Off-specification material can be kept separate for reworking at a later time, or disposal. The intermediate storage allows the two sections to be operated independently of each other. This is not only important for start-up and shutdown, but allows one of the sections to be operated even if the other breaks down for a short period. The intermediate storage also decouples the control of the two sections.

The intermediate storage between the reaction and separation system can also help dampen out variations in composition, temperature and flowrate between the two sections (for gases and non-viscous liquids, but not solids). Variations in the outlet properties from the storage are reduced compared with variations in the inlet properties.

The greater the amount of intermediate storage the greater the flexibility created in the operation of the process and the simpler will be the control. However, like feed and product storage there are significant costs associated with intermediate storage, involving capital, working and operating costs. In addition, intermediate storage will bring additional safety problems if the material being stored is of a hazardous nature.

In summary, the amount of feed, product and intermediate storage will depend on capital, working and operating costs, together with operability, control and safety considerations.

13.10 REACTION, SEPARATION AND RECYCLE SYSTEMS FOR CONTINUOUS PROCESSES – SUMMARY

The use of excess reactants, diluents or heat carriers in the reactor design has a significant effect on the flowsheet recycle structure. Sometimes the recycling of unwanted byproduct to the reactor can inhibit its formation at source. If this can be achieved, it improves the overall usage of raw materials and reduces effluent disposal problems. However, the recycling results in an increase of some costs.

When a mixture in a reactor effluent contains components with a wide range of volatilities, then a partial condensation from the vapor phase followed by a simple phase split can often produce a good separation. If the vapor from such a phase split is difficult to condense, then further separation needs to be carried out in a vapor separation process such as a membrane. The liquid from the phase split can be sent to a liquid separation unit such as distillation.

The process yield is an important measure of both raw materials efficiency and environmental impact.

Interactions between the reactor and the rest of the process are extremely important. Reactor conversion is the most significant optimization variable, since it tends to influence most operations through the process. When a purge is used to remove impurities from the process, the concentration in the recycle is another important optimization variable, again influencing operations throughout the process.

As the reactor conversion and recycle impurity changes in the optimization, the most appropriate separation sequence can also change. In other words, different separation system structures become appropriate for different settings of the reactor conversion and concentration of impurity in the recycle.

Feed, product and intermediate storage can be very significant cost components.

13.11 EXERCISES

1. Ethylene is to be converted by catalytic air oxidation to ethylene oxide. The air and ethylene are mixed in the ratio 10:1 by volume. This mixture is combined with a recycle stream and the two streams are fed to the reactor. Of the ethylene entering the reactor, 40% is converted to ethylene oxide, 20% is converted to carbon dioxide and water, and the rest does not react. The exit gases from the reactor are treated to remove substantially all of the ethylene oxide and water, and the residue recycled. Purging of the recycle is required to avoid accumulation of carbon dioxide and hence maintain a constant feed to the reactor. Calculate the ratio of purge to recycle if not more than 8% of the ethylene fed is lost in the purge. What will be the composition of the corresponding reactor feed gas?
2. Benzene is to be produced by the hydrodealkylation of toluene according to the reaction:

$$\underset{\text{toluene}}{C_6H_5CH_3} + \underset{\text{hydrogen}}{H_2} \longrightarrow \underset{\text{benzene}}{C_6H_6} + \underset{\text{methane}}{CH_4}$$

Some of the benzene formed undergoes secondary reactions in series to unwanted byproducts that can be characterized by the reaction to diphenyl, according to the reaction:

$$\underset{\text{benzene}}{2C_6H_6} \rightleftharpoons \underset{\text{diphenyl}}{C_{12}H_{10}} + \underset{\text{hydrogen}}{H_2}$$

Laboratory studies have established that the selectivity (i.e. fraction of toluene reacted that is converted to benzene) is related to the conversion (i.e. fraction of toluene fed that is reacted) according to:

$$S = 1 - \frac{0.0036}{(1 - X)^{1.544}}$$

where S = selectivity
X = conversion

The hydrogen feed to the plant contains methane as an impurity at a mole fraction of 0.05. In the first instance, it can be assumed that the reactor effluent will contain hydrogen, methane, benzene, toluene and diphenyl, and that a simple phase split will produce vapor stream containing all of the hydrogen and methane and a liquid stream containing all of the aromatics. The hydrogen and methane will be recycled to the reactor with a purge to prevent the buildup of methane. The liquid stream containing the aromatics will be separated into pure products and the toluene recycled. The values of the feeds and products are given in Table 13.13.

The purge stream containing hydrogen and methane and the byproduct diphenyl will be burned in a furnace and can be attributed with their fuel value given in Table 13.14

Table 13.13 Values of feeds and products for Exercise 2.

	Molar mass ($kg{\cdot}kmol^{-1}$)	Value ($\${\cdot}kg^{-1}$)
Hydrogen	2	1.06
Toluene	92	0.21
Benzene	78	0.34

Table 13.14 Fuel value of waste streams for Exercise 2.

	Molar mass ($kg{\cdot}kmol^{-1}$)	Fuel value ($\${\cdot}kg^{-1}$)
Hydrogen	2	0.53
Methane	16	0.22
Diphenyl	154	0.17

a. For a production rate of benzene of 300 $kmol{\cdot}h^{-1}$ and mole fraction of hydrogen in the purge of 0.35, determine the flowrate of hydrogen feed and the flowrate of the purge stream as a function of the selectivity S.
b. Determine the range of reactor conversions over which the plant is profitable.

3. Toluene is to be pumped between two vessels using a centrifugal pump with a flowrate of 30 $t{\cdot}h^{-1}$. The pipe diameter is 80 mm (internal diameter 77.93 mm). The pipeline is 35 m long, with 4 isolation valves (plug cock), a check valve and 5 bends. The discharge tank is 3 m in elevation above the feed tank. The density of toluene is 778 $kg{\cdot}m^{-3}$ and viscosity of 0.251×10^{-3} $N{\cdot}s{\cdot}m^{-2}$.
a. Estimate the pressure drop through the pipeline.
b. Estimate the power consumed by the centrifugal pump.
4. Toluene is to be pumped between two vessels under flow control at a rate of 30 $t{\cdot}h^{-1}$. The piping arrangement has yet to be laid out in detail but the approximate distance between the two tanks in plan is 50 m. The density of the toluene is 778 $kg{\cdot}m^{-3}$ and viscosity 0.251×10^{-3} $N{\cdot}s{\cdot}m^{-2}$.
a. Estimate the pressure drop through the pipeline and the power required if the pumping is accomplished using a centrifugal pump and the tanks are at the same elevation.
b. How high would the feed tank have to be elevated if the flow was to be accomplished by gravity?
5. A two-stage compression is from ambient pressure (1.013 bar) to a final pressure of 10 bar. What should be the pressures across each compression stage for:
a. no pressure drop between stages;
b. a pressure drop of 0.2 bar between stages in the intercooler and associated piping.
6. A compressor is required to compress natural gas with a flowrate of 100,000 $m^3{\cdot}d^{-1}$ (measured at 15°C and 1.013 bar) from 1.013 bar to 10 bar. The inlet temperature of the gas can be assumed to be 20°C and $\gamma = 1.3$. A two-stage reciprocating compressor is to be used with intercooling to 40°C. Estimate the power requirements in kW for:
a. no pressure drop in the intercooler;
b. a pressure drop of 0.3 bar in the intercooler;
c. a pressure drop of 0.3 bar in the intercooler and intercooling to 30°C.

7. A centrifugal compressor is to be used in place of the reciprocating compressor in Exercise 6. Estimate the power requirements. If multistage compression is to be used, assume no pressure drop in the intercooler with intercooling to 40°C.
8. A process is to be simulated involving a simple reaction, separation and recycle system, as illustrated in Figure 13.13. The reaction is:

$$A \longrightarrow B$$

   The feed contains 95 kmol of A and 5 kmol of B. The reactor operates at a conversion of 70%. The performance of the separator gives a recovery of A in the recycle stream of 95% and a recovery of B in the product stream of 98%. In a spreadsheet:
   a. Solve the material balance using a sequential modular approach by tearing the recycle stream and using direct (repeated) substitution to a scaled residual convergence of 0.0001 for both components.
   b. Solve the material balance by accelerating the solution using the Wegstein Method.
   c. Check the solutions by solving the equations simultaneously.
9. From the material balance in Exercise 8, calculate the process yield for the manufacture of B.
10. Consider a process in which each delivery of raw material provides a 10 day supply and is stored in a tank. Delivery from the supplier takes between 5 and 15 days. The minimum inventory in the tank is to be 20 days supply. For what period should the storage tank be sized?

REFERENCES

1. Smith R and Linnhoff B (1988) The Design of Separators in the Context of Overall Processes, *Trans IChemE ChERD*, **66**: 195.
2. Powers GJ (1972) Heuristics Synthesis in Process Development, *Chem Eng Prog*, **68**: 88.
3. Rudd DF, Powers GJ and Siirola JJ (1973) *Process Synthesis*, Prentice-Hall, New Jersey.
4. Douglas JM (1988) *Conceptual Design of Chemical Processes*, McGraw Hill.
5. Douglas JM (1985) A Hierarchical Decision Procedure for Process Synthesis, *AIChE J*, **31**: 353.
6. Rase HF (1990) *Fixed-bed Reactor Design and Diagnostics – Gas Phase Reactions*, Butterworth.
7. Coulson JM and Richardson JF (1999) *Chemical Engineering, Volume 1 Fluid Flow, Heat Transfer and Mass Transfer*, 6th Edition, Butterworth Heinemann.
8. Hewitt GF, Shires GL and Bott TR (1994) *Process Heat Transfer*, CRC Press Inc.
9. Perry RH (1997) *Chemical Engineers Handbook*, 7th Edition, McGraw Hill.
10. Biegler LT, Grossmann IE and Westerberg AW (1997) *Systematic Methods of Chemical Process Design*, Prentice Hall.
11. Wegstein JH (1958) Accelerating Convergence of Iterative Processes, *Commun Assoc Comp Mach*, **1**: 9.
12. Smith R and Omidkhah Nasrin M (1993) Trade-offs and Interactions in Reaction and Separation Systems, Part I: Reactions With No Selectivity Losses, *Trans IChemE ChERD*, **A5**: 467.
13. Smith R and Omidkhah Nasrin M (1993) Trade-offs and Interactions in Reaction and Separation Systems, Part II: Reactions With Selectivity Losses, *Trans IChemE ChERD*, **A5**: 474.
14. Wynn N (2001) Pervaporation Comes of Age, *Chem Eng Prog*, **97**: 66.

14 Reaction, Separation and Recycle Systems for Batch Processes

14.1 BATCH PROCESSES

As pointed out in Chapter 1, in a batch process the main steps operate discontinuously. This means that temperature, concentration, mass and other properties vary with time. Also as pointed out in Chapter 1, most batch processes are made up of a series of batch and semicontinuous steps. A semicontinuous step operates continuously with periodic start-ups and shutdowns.

Many batch processes are designed on the basis of a scale-up from the laboratory, particularly for the manufacture of specialty chemicals. If this is the case, the process development will produce a *recipe* for the manufacturing process. The recipe is not unlike a recipe used in cookery. It is a step-by-step procedure that resembles the laboratory procedure, but scaled to the quantities required for manufacturing. It provides information on the quantities of material to be used in any step in the manufacturing, the conditions of temperature, pressure, and so on at any time, and the times over which the various steps take place. The recipe can be thought of as the equivalent of the material and energy balance in a continuous process. However, care should be taken to avoid taking artificial constraints from the laboratory to the manufacturing process (i.e. those constraints imposed by the laboratory procedures that do not apply to industrial plant).

As noted in Chapter 1, the priorities in batch processes are often quite different from those in large-scale continuous processes. Particularly when manufacturing specialty chemicals, the shortest time possible to get a new product to market is often the biggest priority (accepting that the product must meet the specifications and regulations demanded and the process must meet the required safety and environmental standards). This is particularly true if the product is protected by patent. The period over which the product is protected by patent must be exploited to its full. This means that product development, testing, pilot plant work, process design and construction should be *fast tracked* and carried out as much as possible in parallel.

Before a system of batch reaction and separation processes is considered, the main operations that will be used in batch processes need to be reviewed, but with the emphasis on how they will differ from the corresponding operations in continuous processes.

Chemical Process Design and Integration R. Smith
 ISBNs: 0-471-48680-9 (HB); 0-471-48681-7 (PB)

14.2 BATCH REACTORS

As with continuous processes, the heart of a batch chemical process is its reactor. Idealized reactor models were considered in Chapter 5. In an ideal-batch reactor, all fluid elements have the same residence time. There is thus an analogy between ideal-batch reactors and plug-flow reactors. There are four major factors that effect batch reactor performance:

- Contacting pattern
- Operating conditions
- Agitation for agitated vessel reactors
- Solvent selection.

1. *Contacting pattern.* Most batch reactors are agitated vessels with a standard configuration. However, this is only one way that such reactions can be accomplished. If the reaction is multiphase, then any of the contacting patterns in Figures 7.3 and 7.4 might be considered to enhance the mass transfer relative to that which can be achieved using an agitated vessel. For example, a recirculation system could be used in which material was taken from the reactor and returned to the reactor via a pump in the case of a liquid, or a compressor in the case of a gas. If such possibilities are considered, then there is no reason why the kind of equipment used for continuous processes could not be applied, with recirculation of the materials used to create the batch mode. Indeed, the reactor could simply be operated in a semicontinuous mode. If the reaction is rapid, and the temperature can be controlled, whether single-phase or multiphase, then a simple static mixer arrangement operated in semicontinuous mode might be the best solution.

Figure 14.1 shows various modes of operation for agitated vessel batch and semibatch reactors. In Figure 14.1a, the feeds are loaded into the reactor at the beginning of the batch, the reaction then proceeds for a specified time, after which the products are removed. By contrast, Figure 14.1b shows semibatch operation in which one of the feeds is initially charged to the reactor and the other feed is charged progressively. This mode of operation was discussed in Chapter 5. The semibatch operation in Figure 14.1b is one in which both the composition and the volume in the reactor change with time. However, batch reactors offer additional degrees of freedom. Figure 14.1c shows a semibatch operation in which the feeds are charged to the reactor as the

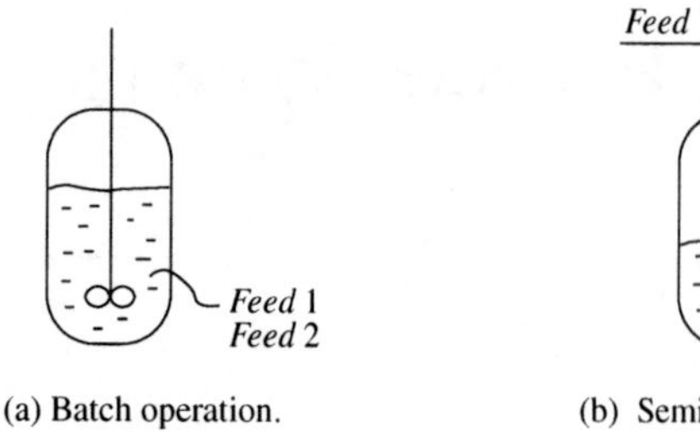

(a) Batch operation.

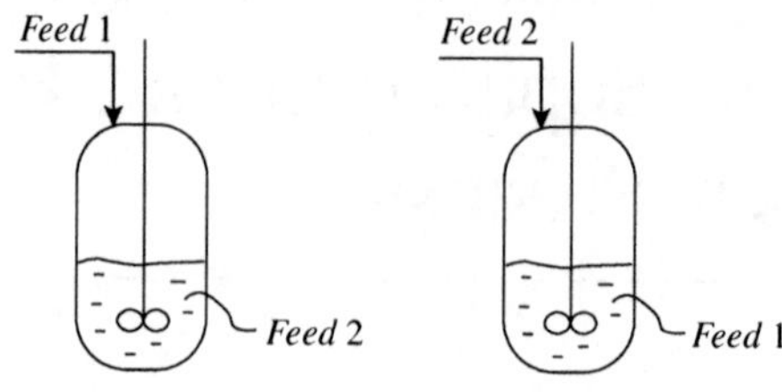

(b) Semibatch operation in which both composition and volume change.

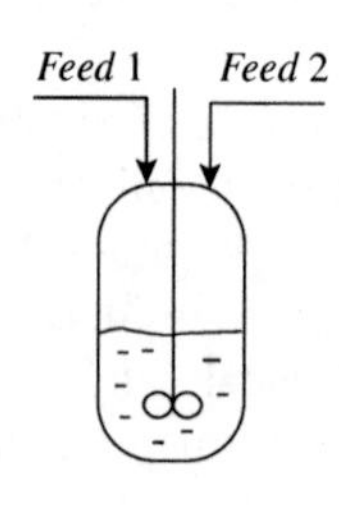

(c) Semibatch operation in which composition is maintained constant, but volume changes.

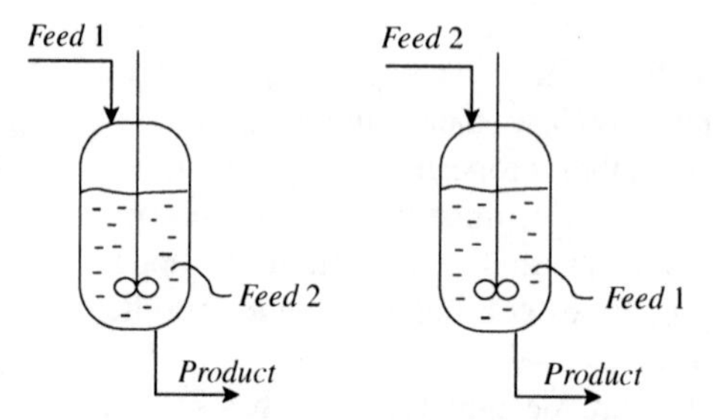

(d) Semibatch operation in which composition changes, but volume is maintained constant.

Figure 14.1 Various modes of operation for batch and semibatch reactors.

batch progresses. In this way, composition can in principle be maintained constant, but volume changes. Finally, Figure 14.1d shows semibatch operation in which the product is withdrawn before the end of a batch reaction. In the arrangements shown in 14.1d, the composition changes but the volume can in principle be maintained constant. In Chapter 5, some simple heuristics were developed to help decide the contacting pattern if some knowledge of the reaction kinetics is available. For example, for the parallel reaction:

$$\begin{aligned} &FEED1 + FEED2 \longrightarrow PRODUCT \\ &\qquad r_1 = k_1 C_{FEED1}^{a_1} C_{FEED2}^{b_1} \\ &FEED1 + FEED2 \longrightarrow BYPRODUCT \\ &\qquad r_2 = k_2 C_{FEED1}^{a_2} C_{FEED2}^{b_2} \end{aligned} \tag{14.1}$$

Taking the ratio of the reaction rates:

$$\frac{r_2}{r_1} = \frac{k_2}{k_1} C_{FEED2}^{a_2 - a_1} C_{FEED2}^{b_2 - b_1} \tag{14.2}$$

In order to minimize the formation of the unwanted *BYPRODUCT*, the ratio of the reaction rates in Equation 14.2 should be minimized. Thus[1,2]:

- $a_2 > a_1$ and $b_2 > b_1$. The concentration of both feeds should be minimized and each added progressively as the reaction proceeds. Predilution of the feeds might be considered.
- $a_2 > a_1$ and $b_2 < b_1$. The concentration of *FEED*1 should be minimized by charging *FEED*2 at the beginning of the batch and adding *FEED*1 progressively as the reaction proceeds. Predilution of *FEED*1 might be considered.
- $a_2 < a_1$ and $b_2 > b_1$. The concentration of *FEED*2 should be minimized by charging *FEED*1 at the beginning of the batch and adding *FEED*2 progressively as the reaction proceeds. Predilution of *FEED*2 might be considered.
- $a_2 < a_1$ and $b_2 < b_1$. The concentration of *FEED*1 and *FEED*2 should be maximized by rapid addition and mixing.

Whilst heuristics like these are helpful, they have also severe limitations:

- Knowledge of the reaction chemistry and the kinetics is needed.
- They only are helpful when dealing with simple reaction systems.
- They are only qualitative.

Even if it is decided to use semibatch operation in the above example in which *FEED*2 is charged to the reactor initially and then *FEED*1 as the reaction progresses, it is not known how fast the feed should be added to obtain the optimum performance. Feed addition rate and product takeoff rate are degrees of freedom that need to be optimized.

If a model is available for the reaction chemistry and kinetics, then a *temporal superstructure* can be developed to represent a batch reactor in the time dimension with a series of reactor compartments that connect to each other sequentially in the time dimension[3]. This temporal superstructure network, representing a batch reactor, is created

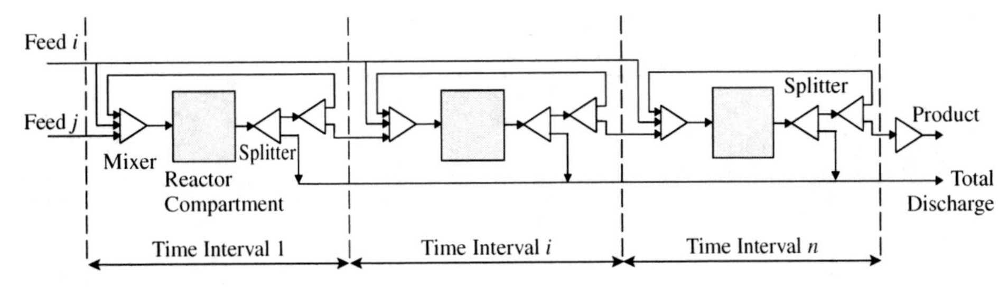

Figure 14.2 A temporal superstructure for a well-mixed batch reactor.

along the batch cycle time, rather than a superstructure with physical connections. An example of a temporal superstructure for a batch reactor is shown in Figure 14.2. Figure 14.2 illustrates the superstructure for a single-phase batch or semibatch reactor involving two feeds. In this network, the batch cycle time is divided into a number of time intervals. The reactor compartment in each time interval connects to the reactor compartments in the time intervals that are immediately before and after. The compartment in each time interval allows feed material to be added, creating semibatch or multistep operation. Also, Figure 14.2 allows for product takeoff at intermediate points in the batch. Given that intermediate feed addition or product takeoff is likely to be semicontinuous, the time intervals need to be short enough to approximate such semicontinuous operation. The greater the number of the time intervals, the closer the model approaches the batch reactor modeled.

Structural changes of the temporal superstructure can generate different operating modes (batch or semibatch). The reactants that are fed at the beginning of a batch are treated as feed streams at the start of the network, while the intermediate batch feeds, semicontinuous feeds and product takeoffs are represented as sidestreams feeding to, or withdrawing from, the network at different time intervals[3].

A simulation model needs to be developed for each reactor compartment within each time interval. An ideal-batch reactor has neither inflow nor outflow of reactants or products while the reaction is carried out. Assuming the reaction mixture is perfectly mixed within each reactor compartment, there is no variation in the rate of reaction throughout the reactor volume. The design equation for a batch reactor in differential form is from Chapter 5:

$$\frac{dN_i}{dt} = \frac{d[N_{i0}(1 - X_i)]}{dt} = -N_{i0}\frac{dX_i}{dt} = r_i V \qquad (5.39)$$

where t = batch time
N_{i0} = initial moles of Component i
X_i = conversion of Component i after time t
r_i = reaction rate for Component i
V = volume of reaction mixture

Thus within each time interval, the batch reactor can be modeled using Equation 5.39. This differential form of the design equation reflects the fundamental dynamics of a batch reactor. Equation 5.39 needs to be integrated within each time interval (e.g. using a fourth order Runge–Kutta integration technique).

Rather than use Equation 5.39 to model the batch reactor, an alternative approach can be developed on the basis of the analogy between ideal-batch and continuous plug-flow reactors[3]. Like an ideal-batch reactor, the residence time in a plug-flow reactor is the same for all fluid elements. This means that a differential element within a plug-flow reactor can be taken to be perfectly mixed as it travels along the reactor, but will not exchange any fluid with the elements in front or behind. In this way, it may be considered to behave as a differential batch reactor. Thus, the difference between ideal-batch and plug-flow reactors is that composition changes take place temporally in the first and spatially in the second (see Chapter 5). Correspondingly, a plug-flow reactor with sidestreams that feed at different locations along the reactor has the same performance as an ideal semibatch reactor, given that the fresh feeds in both cases have been arranged to have the same residence times. Also, as discussed in Chapter 5, plug-flow operation can be approached by using a number of mixed-flow reactors in series. The total volume of the plug-flow reactor is modeled by dividing it into a number of mixed-flow reactors with equal volumes. The greater the number of mixed-flow reactors in series, the closer is the approach to plug-flow operation. Sidestreams can be added to the mixed-flow reactors in series to model a plug-flow reactor with sidestreams. Thus, a batch reactor can also be modeled as a series of mixed-flow reactors. Each time interval comprises a mixed-flow reactor to approach the performance of a batch reactor within this time interval. The greater the number of time intervals, the closer the model will approach batch and semibatch behavior. Thus, if the time interval is short, then a mixed-flow model can approximate a batch reactor within each time interval. From Chapter 5:

$$\frac{N_{i,out}}{\Delta t} = \frac{N_{i,in}}{\Delta t} + r_i V \qquad (14.3)$$

where $N_{i,in}$ = inlet moles of Component i
$N_{i,out}$ = outlet moles of Component i
Δt = time interval

For the first time interval, the input to the reactor compartment are the reactants charged to the vessel

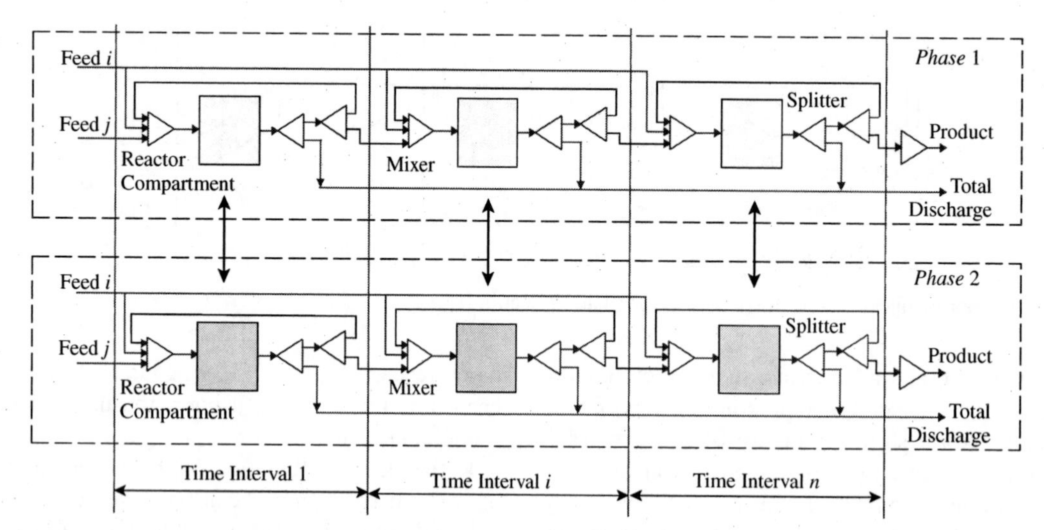

Figure 14.3 A temporal superstructure for a multiphase well-mixed batch reactor.

initially. The output from the reactor compartment in the final time interval is the product generated. Equation 4.3 is applied to each time interval and requires the reaction rate to be evaluated at the outlet conditions. Unfortunately, the outlet conditions are unknown. A simple approach would be to calculate the reaction rate from the inlet conditions, rather than the outlet conditions. However, this is often too crude, leading to poor numerical reliability. Instead, Equation 14.3 can be solved simultaneously with the equation(s) describing the reaction rate.

Thus, the design equations for a batch reactor for the optimization of a temporal superstructure can be based on differential or algebraic equations.

Figure 14.3 shows a temporal superstructure for a multiphase batch reactor[3]. As with the continuous steady state reactors discussed in Chapter 7, mass transfer is only allowed between adjacent reactor compartments.

In Figures 14.2 and 14.3, a single reactor compartment is used within each time interval for each phase. Perfect mixing is assumed within each reactor compartment. Unfortunately, in practice, the mixing will be imperfect and this can cause deterioration in the performance relative to perfect mixing. If an agitated vessel reactor is used, different types of agitators, baffles, feed locations and other reactor vessel configurations create different flow patterns. To reflect this, a network of perfectly mixed compartments can be used to replace the single compartment within each time interval[3]. The reactor is divided into a number of homogeneous regions, but different in intensity of mixing (regions with different values of rate of energy dissipation). Each mixing compartment is assumed to be represented by a mixed-flow reactor with input and output streams that feed to, or are fed by, other mixing compartments. Thus, the interconnected mixing compartments and their connections constitute a mixing compartment network, which can be used to model the mixing pattern inside the reactor vessel. Figure 14.4 shows some examples of mixing compartment networks. The mixing compartment network can be used to simulate more accurately the reactor performance within each time interval and increase the reliability of the optimization calculations. Even further, the structure of the mixing compartment network within each time interval can be optimized[3].

However, it should be noted that there are many practical issues that need to be considered when choosing mixing equipment and mixing patterns, in addition to those for maximizing yield, selectivity or conversion[4]. This is especially the case when dealing with multiphase reactions[4].

Thus, the design of a batch reactor can be based on the optimization of a temporal superstructure. Given a simulation model with a mathematical formulation, the next step is to determine the optimal values for the control variables of a batch reaction system.

2. *Operating conditions*. Optimization variables such as batch cycle time and total amount of reactants have fixed values for a given batch reactor system. However, variables such as temperature, pressure, feed addition rates and product takeoff rates are dynamic variables that change through the batch cycle time. The values of these variables form a profile for each variable across the batch cycle time.

If the rate of feed addition, rate of product takeoff, temperature and pressure are known in each time interval, a simulation of the reactor can be carried out in that time interval. The problem is that the conditions will change from one time interval to subsequent time intervals. The *profile* of the dynamic variables (feed addition, product takeoff, temperature and pressure) need to be known through time. In the approach described in Chapter 3 for profile optimization[5], a shape can be imposed for a given variable through time and the dynamic variables optimized in conjunction with the temporal superstructure. One profile for each dynamic variable is assigned to the

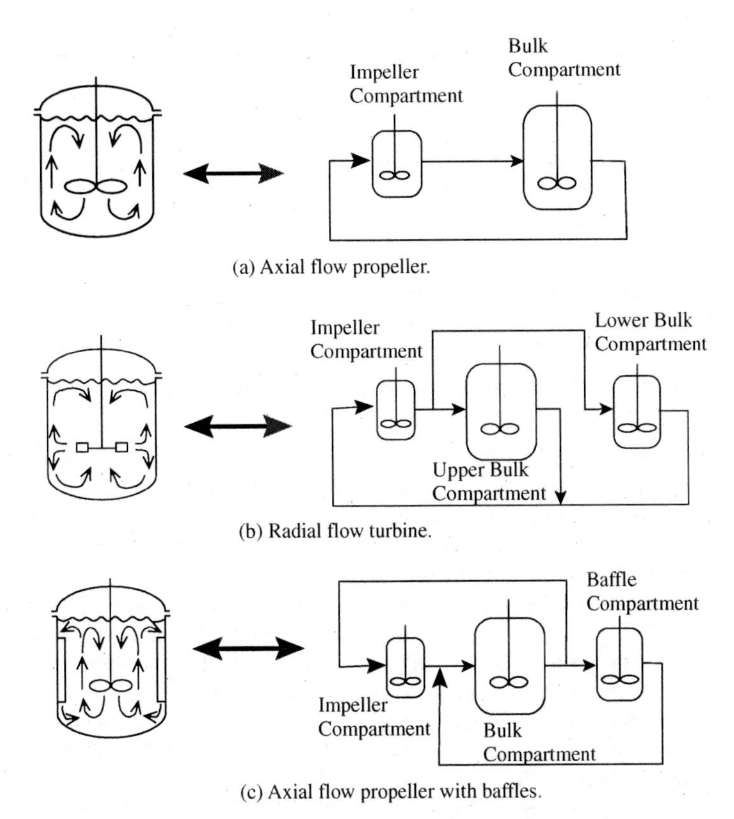

Figure 14.4 Some examples of mixing compartment networks to represent agitated vessels. (Reproduced from Zhang J and Smith R, 2004, *Chem Eng Sci*, **59**:459, by permission of Elsevier Ltd).

batch reactor system. Thereafter, the performance of the batch reactor under these assigned feed, product takeoff, temperature and pressure profiles can be evaluated against the corresponding objective function. The optimization imposes different shapes with different initial and final values, location and values of any maximum or minimum. The optimization requires structural decisions (e.g. should the feeds be added at the beginning of the batch or added progressively in the appropriate time interval?). Feed addition rate, product takeoff rate, temperature and pressure are treated as continuous variables, along with the batch reaction time. The objective function needs to reflect the performance of the batch reactor. Maximization of the yield, selectivity or profit, minimization of batch cycle time or total cost could all be possible objectives. The optimization is a mixed integer nonlinear optimization. However, it can be tackled successfully using stochastic optimization[3,5].

The resulting optimal profiles for the operating conditions will need to be evaluated in terms of their practicality from the point of view of control and safety. If a complex profile offers only a marginal benefit relative to a fixed value, then simplicity (and possibly safety) will dictate a fixed value to be maintained. But an optimized profile might offer a significant increase in the performance, in which the complex control problem will be worth addressing.

Example 14.1 Consider again the chlorination reaction in Example 7.3. This was examined as a continuous process. Now assume it is carried out in batch or semibatch mode. The same reactor model will be used as in Example 7.3. The liquid feed of butanoic acid is 13.3 kmol. The butanoic acid and chlorine addition rates and the temperature profile need to be optimized simultaneously through the batch, and the batch time optimized. The reaction takes place isobarically at 10 bar. The upper and lower temperature bounds are 50°C and 150°C respectively. Assume the reactor vessel to be perfectly mixed and assume that the batch operation can be modeled as a series of mixed-flow reactors. The objective is to maximize the fractional yield of α-monochlorobutanoic acid with respect to butanoic acid. Specialized software is required to perform the calculations, in this case using simulated annealing[3].

Solution From the kinetic expressions given in Example 7.3, it can be concluded that $r_2/r_1 = k_3 C_{LCl2}$. In order to increase the selectivity of the reactions, the C_{LCl2} should be minimized. A semibatch reactor would thus be preferred, in which *BA* is charged at the beginning of the batch and Cl_2 fed continuously during the reaction. In any case, charging all of the chlorine at the beginning of the reaction will be impractical.

The most straightforward way to operate such a process is to maintain a constant chlorine addition rate and a constant temperature. However, both the constant value of the chlorine addition rate and the fixed temperature should be optimized. The temperature of the reaction system is allowed to vary within the set temperature range, but kept constant throughout a batch cycle. The batch time is divided into twenty time

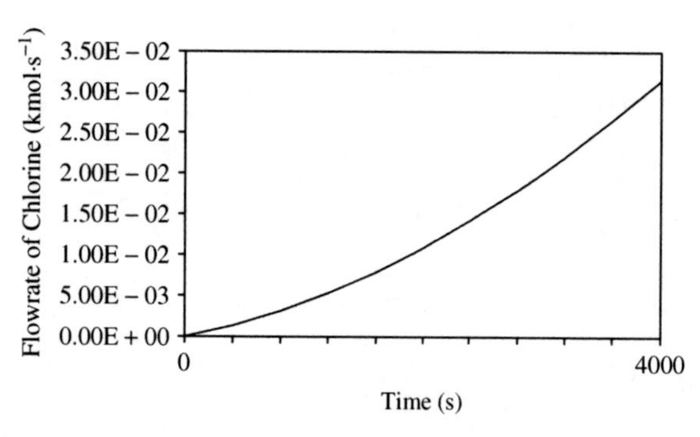

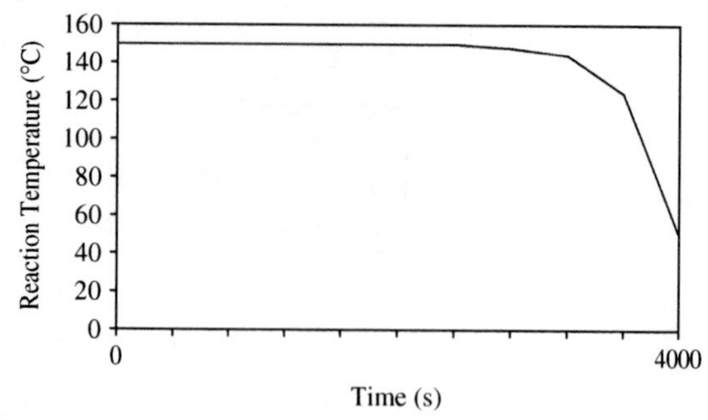

Figure 14.5 One possible combination of chlorine feed trajectory and temperature profile.

intervals, although this needs to be examined to ensure that a greater number will not give a significantly different answer. The optimum yield under such conditions is 85.8%. In terms of batch operation, this strategy has the benefit of simplicity, with a batch cycle time at 0.98 h. The optimization favors high temperature, with the reactor operating isothermally at the upper bound of 150°C.

The optimization is now constrained to be at a fixed (optimized) chlorine addition rate, but the temperature profile optimized. Profile optimization is used for the temperature, as discussed in Chapter 3. The batch cycle time required is 1.42 h. The resulting fractional yield of *MBA* from *BA* now reaches 92.7%.

Conversely, the optimization is now constrained to be at a fixed (optimized) temperature, but the chlorine addition profile optimized. Both the feed addition profile and the total chlorine feed are optimized. The optimum temperature reaches its upper bound of 150°C. Chlorine addition is 75.0 kmol and the batch cycle is 1.35 h. The resulting fractional yield of *MBA* from *BA* now reaches 97.4%.

The final option is to allow both the chlorine addition profile and temperature profile to be varied through the batch. The optimization shows a further improvement of the objective to 99.8%. It requires 1.35 h of batch cycle time and 75.0 kmol of chlorine. The optimized profiles for reaction temperature and feed addition rate of chlorine are shown in Figure 14.5.

Table 14.1 presents the yields of the *MBA* from different operations. It shows that batch operation for the production of *MBA* is favorable when compared with standard design configurations in continuous operation. An advanced control system is needed to apply the results in practice, but the incentives to succeed are significant. Although the optimized temperature and feed profiles give a better yield, operation with constant feed and temperature might be preferred for control and safety reasons. A high conversion of chlorine is preferred. However, the model predicts that the mass transfer becomes very slow at a high conversion of chlorine. Therefore, a large amount of chlorine is used in order to maintain a high mass transfer rate. The unreacted chlorine can be recovered by scrubbing using dilute hydrochloric acid or water to remove the hydrogen chlorine produced. Unlike the continuous process, recycle of the chlorine is difficult in batch operation.

Finally, the reactor vessel has been assumed to be perfectly mixed. Imperfect mixing and a flow pattern created by different types of agitators, baffles, feed locations and other reactor vessel configurations will cause the performance to be below that indicated by perfect mixing.

Table 14.1 Fractional yields of *MBA* from *BA* under different reactors and operating modes.

Reactor and operation mode	Yield of *MBA* from *BA* (%)
Semibatch with optimized constant addition rate of chlorine and optimized constant temperature	85.8
Semibatch with optimized constant addition rate of chlorine and optimized temperature profile	92.7
Semibatch with optimized addition rate profile of chlorine under optimized constant temperature	97.4
Semibatch with optimized profiles of addition rate of chlorine and temperature	99.8

3. *Agitation.* If an agitated vessel is being used for the batch reactor, then various types of agitator are available. Processing duties are central to the selection of the agitator. The duties of the agitator are:

- Mixing
- Solids suspension (suspending sinking solids or incorporating floating solids)
- Dispersion of two liquid phases
- Gas dispersion
- Heat transfer.

Selection of the most appropriate design of agitator is a specialized subject outside of the scope of this text (see, for example, Harnby, Nienow and Edwards[5]). However, it is important to note that the selectivity of the system can be effected by agitation when competing reactions are involved that produce byproducts. For example, consider the reaction system:

$$\begin{aligned} &FEED1 + FEED2 \longrightarrow PRODUCT \\ &PRODUCT + FEED2 \longrightarrow BYPRODUCT \end{aligned} \qquad (14.4)$$

Where mixing is imperfect, then regions of excess *FEED*2 produce excessive *BYPRODUCT*. Regions low in *FEED*2 give lower rates for both the *PRODUCT* and

BYPRODUCT. This means that imperfect mixing favors the production of the *BYPRODUCT*.

Often the reaction kinetics is not known when trying to design a batch reactor. Indeed, the reaction chemistry is often not known. In such circumstances, scale-up from small-scale experiments to production scale is used. Two broad scales of mixing can be distinguished. Mixing through the bulk flow is termed *macromixing*. Mixing at the molecular level through diffusion is termed *micromixing*. The rate of micromixing depends on the operating conditions, viscosity and the energy dissipation rate. If the energy dissipation rate from the agitator in the reaction zone is held constant, the product distribution for the same operating conditions should be independent of the scale. If the reaction is very slow, then the reaction will proceed throughout the whole reactor volume. On the other hand, if the reaction is very fast, the reaction will proceed in only a small fraction of the whole reactor volume. The fraction within which the reaction takes place is scale-dependent and reaction zones spread out to different extents at different scales. This introduces complexities in the scale-up of reaction systems.

4. *Solvent selection*. Solvents are very common in batch reaction systems. The solvent has a number of purposes[2]:

- mobilize solid reactants or solid products with high melting points;
- allows reaction and mass transfer;
- modify reaction pathway;
- act as a heat sink;
- temperature control where the solvent evaporates from the reactor is condensed and returned to the reactor;
- allows material transfer to other units.

There are many issues to consider in solvent selection[2]:

- a large working range between the melting and boiling points is desirable;
- viscosity should allow good mixing;
- interfacial tension should be considered for liquid–liquid reactions;
- the solvent should be easily recovered for recycling (i.e. no azeotropes if distillation is to be used, low latent heat, etc.);
- low boiling point can create environmental problems through the release of volatile organic compounds (VOC's) that create environmental problems;
- the solvent should be easily disposed of if waste is formed (e.g. chlorinated solvents should be avoided);
- effect on reactions (solvent polarity, bond type, effect on reaction rate, activation energy, equilibrium, changes in reaction mechanism).

14.3 BATCH SEPARATION PROCESSES

1. *Batch distillation*. Batch distillation has a number of advantages when compared with continuous distillation:

- The same equipment can be used to process many different feeds and produce different products.
- There is flexibility to meet different product specifications.
- One distillation column can separate a multicomponent mixture into relatively pure products.

The disadvantages of batch distillation are:

- High purity products require the careful control of the column because of its dynamic state.
- The mixture is exposed to a high temperature for extended periods.
- Energy requirements are generally higher.

The simplest batch distillation would involve charging a feed to a batch *distillation pot*, as shown in Figure 14.6, and carrying out a vaporization of the material from the pot. The vapor formed is removed from the system. Since this vapor is richer in the more volatile components than the liquid, the liquid remaining in the pot becomes steadily leaner in these components. The result of this vaporization is that the composition of the product changes progressively through time. The vapor leaving the pot at any time is assumed to be in equilibrium with the liquid in the pot, but since the vapor is richer in the more volatile components, the composition of the liquid and vapor are not constant. To show how the compositions change with time, consider vaporization of an initial charge to the distillation pot in Figure 14.6.

Let B be the total number of moles and $x_{i,B}$ the liquid mole fraction of Component i of liquid in the batch pot at time t. If a small amount of liquid dB with a vapor mole fraction $x_{i,D}$ is vaporized, a material balance on Component i gives:

$$\begin{aligned} x_{i,D}\,dB &= d(Bx_{i,B}) \\ &= B\,dx_{i,B} + x_{i,B}\,dB \end{aligned} \quad (14.5)$$

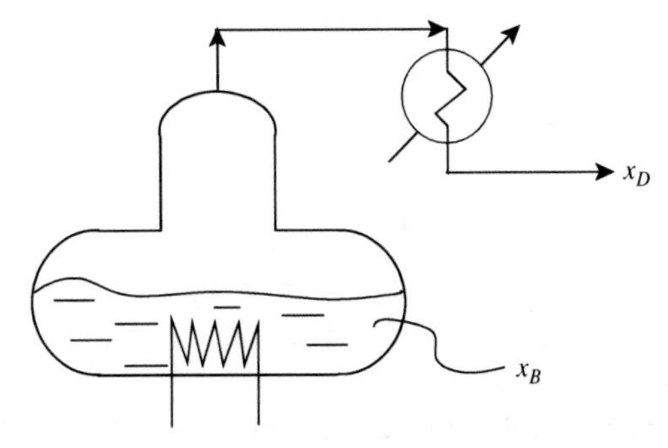

Figure 14.6 Simple distillation from a batch pot.

Rearranging Equation 14.5 and integrating gives:

$$\int_F^B \frac{dB}{B} = \int_{x_{i,F}}^{x_i} \frac{dx_{i,B}}{x_{i,D} - x_{i,B}} \tag{14.6}$$

or

$$\ln\frac{B}{F} = \int_{x_{i,F}}^{x_i} \frac{dx_{i,B}}{x_{i,D} - x_{i,B}} \tag{14.7}$$

where B = total moles in the batch pot at time t
F = total moles in the batch pot at the beginning of the operation
$x_{i,B}$ = mole fraction of Component i in the liquid in the batch pot at time t
$x_{i,F}$ = mole fraction of Component i in the liquid in the batch at the beginning of the operation
$x_{i,D}$ = mole fraction of Component i in the vaporized material leaving the batch pot at time t

Equation 14.7 is known as the Rayleigh Equation and describes the material balance around the distillation pot.

Batch distillation from a pot will not provide a good separation unless the relative volatility is very high. In most cases, a rectifying column with reflux is added to the pot, as shown in Figure 14.7. The operation of a batch distillation can be analyzed for binary systems at a given instant in time using a McCabe–Thiele diagram. The operating line is the same as that for the rectifying section of a continuous distillation (see Chapter 9).

$$y_{i,n+1} = \frac{R}{R+1}x_{i,n} + \frac{1}{R+1}x_{i,D} \tag{9.13}$$

This is shown in Figure 14.6, where the slope of the operating line is given by $R/(R+1)$. The feed is charged to the distillation pot at the beginning of the batch and is subjected to continuous vaporization. This vapor would then flow upward through trays or packing to the condenser and reflux would be returned, as with continuous distillation. However, unlike continuous distillation, the overhead product will change with time. The first material to be distilled will be the more volatile components. As the vaporization proceeds and product is withdrawn overhead, the product will become gradually richer in the less-volatile components. Thus, batch distillation allows different fractions to be taken from the same feed. The batch distillation strategy depends on both the feed mixture and the products required from the distillation.

Figure 14.7 illustrates what happens if the reflux ratio is fixed for a binary separation. Because the reflux ratio is fixed, the slope of the operating line is fixed, but its position varies through time. For a fixed number of stages, the distillate and the bottoms gradually become more concentrated in the heavier component. This means

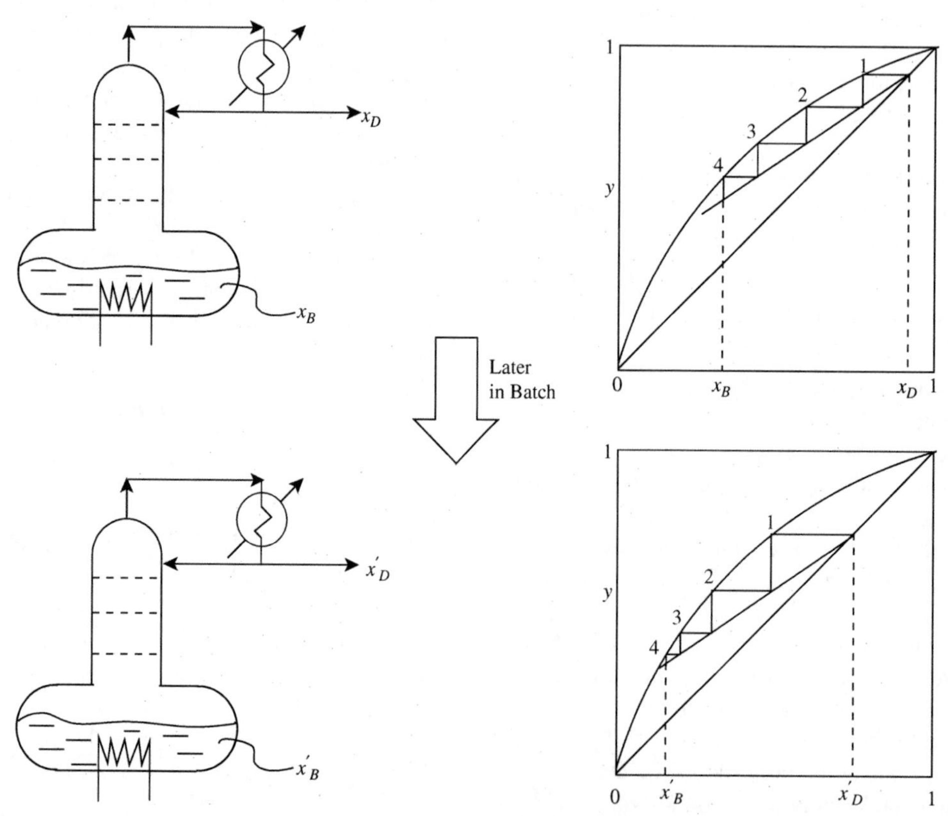

Figure 14.7 If reflux ratio is fixed in batch distillation, the overhead product purity decreases. (Reproduced from Smith R and Jobson B, 2000, *Distillation in Encyclopedia of Separation Science*, by permission of Academic Press Ltd).

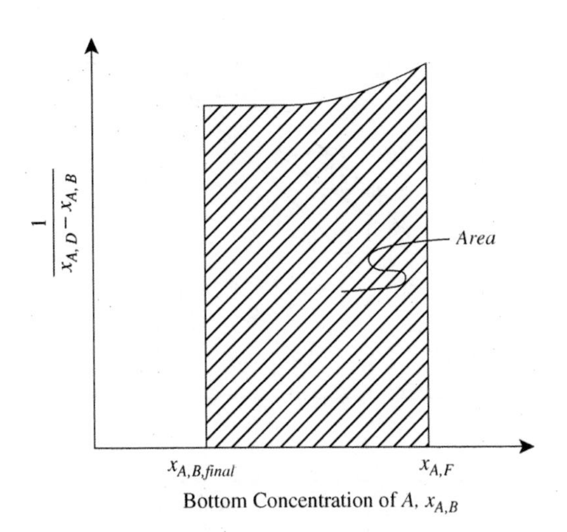

Figure 14.8 Integration of the Rayleigh Equation for constant reflux ratio.

that the product quality deteriorates with time and blending is required to meet the appropriate purity specification. The Rayleigh Equation (Equation 14.7) can be integrated graphically, as illustrated in Figure 14.8 for the separation of an AB mixture. The initial charge to the distillation pot is F moles with a mole fraction of $x_{A,F}$. The McCabe–Thiele construction can be used to determine the distillate mole fraction $x_{A,D}$ for various bottoms moles fractions $x_{A,B}$ for a fixed reflux ratio (fixed slope of operating line) and a given number of stages (4 stages in the case of Figure 14.7). Thus, in Figure 14.8, $1/(x_{A,D} - x_{A,B})$ is plotted versus the bottoms mole fraction from $x_{A,F}$ to the final bottoms mole fraction $x_{A,B,final}$. Integrating to find the area under the graph (e.g. using Simpson's Rule), the amount of liquid left in the pot at the end of the batch is given by:

$$B_{final} = F \exp(-Area) \quad (14.8)$$

where B_{final} = total moles in the batch pot at the end of the batch

The amount of distillate is by mass balance:

$$D_{final} = F[1 - \exp(-Area)] \quad (14.9)$$

The amount of A in the final product

$$= x_{A,F}F - x_{A,B,final}B_{final}$$

Thus the average overhead composition of $\overline{x}_{A,D}$ is given by:

$$\begin{aligned} \overline{x}_{A,D} &= \frac{x_{A,F}F - x_{A,B,final}B_{final}}{F - B} \\ &= \frac{x_F - x_{A,B,final}(B_{final}/F)}{1 - (B_{final}/F)} \end{aligned} \quad (14.10)$$

The total amount of material vaporized during the batch

$$= (F - B_{final})(R + 1) \quad (14.11)$$

Thus the total heat requirement for the batch is given by:

$$Q = F\Delta H_{VAP}(1 - B_{final}/F)(R + 1) \quad (14.12)$$

where Q = heat requirement
ΔH_{VAP} = latent heat of vaporization

Alternatively, by careful control of the reflux ratio, it is possible to hold the composition of the distillate constant for a time until the required reflux ratio becomes intolerably large, as illustrated in Figure 14.9. A mass balance gives:

$$D_{final}x_{A,D} = Fx_{A,F} - B_{final}x_{A,B,final} \quad (14.13)$$

Substituting $B_{final} = F - D_{final}$ in Equation 14.13 and rearranging gives:

$$D_{final} = \frac{F(x_{A,F} - x_{A,B,final})}{(x_{A,D} - x_{A,B,final})} \quad (14.14)$$

Substituting $D_{final} = F - B_{final}$ in Equation 14.13 and rearranging gives:

$$B_{final} = \frac{F(x_{A,D} - x_{A,F})}{(x_{A,D} - x_{A,B,final})} \quad (14.15)$$

In this case, the energy requirement is a variable as reflux ratio varies. For a differential time interval:

$$\begin{aligned} dQ &= \Delta H_{VAP}\, dD(R + 1) \\ &= \Delta H_{VAP}\, dB(R + 1) \end{aligned} \quad (14.16)$$

Substituting Equation 14.15 into Equation 14.16:

$$dQ = \Delta H_{VAP} B\left(\frac{dx_{A,B}}{x_{A,D} - x_{A,B}}\right)(R + 1) \quad (14.17)$$

From Equation 14.15:

$$B = \frac{F(x_{A,D} - x_{A,F})}{x_{A,D} - x_{A,B}} \quad (14.18)$$

Substituting Equation 14.18 into Equation 14.17 and integrating gives:

$$Q = F\Delta H_{VAP}(x_{A,D} - x_{A,F})\int_{x_{A,F}}^{x_{A,D}} \frac{R + 1}{(x_{A,D} - x_{A,B})^2}\, dx_{A,B} \quad (14.19)$$

Thus, from Figure 14.9, the McCabe–Thiele diagram can be used to determine the reflux ratio for a given $x_{A,B}$. This allows Equation 14.19 to be integrated graphically, as shown in Figure 14.10.

Operation at constant reflux ratio is better than operation with constant distillate composition for high-yield batch separations. However, operation with constant distillate composition might be necessary if high product purity is required. In fact, it is not necessary to operate in one of these two special cases of constant reflux ratio or constant distillate composition. Given the appropriate control scheme, the reflux ratio can be varied through the batch

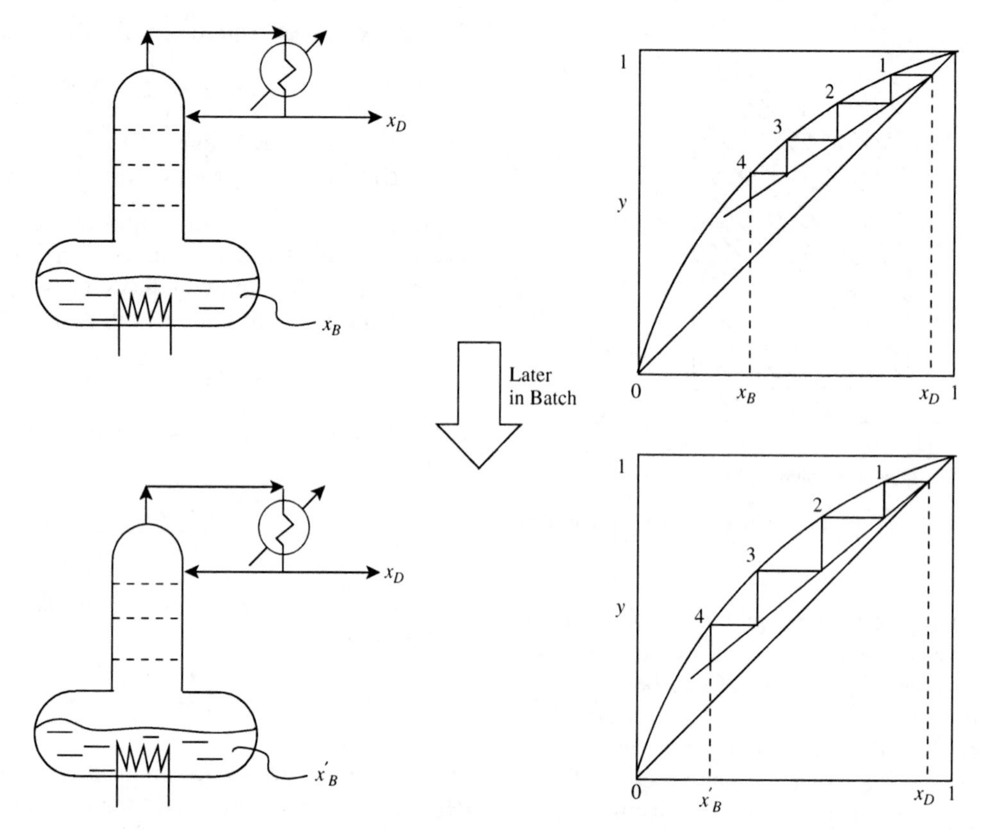

Figure 14.9 Reflux ratio can be varied in batch distillation to maintain overhead product purity. (Reproduced from Smith R and Jobson B, 2000, *Distillation in Encyclopedia of Separation Science*, by permission of Academic Press Ltd).

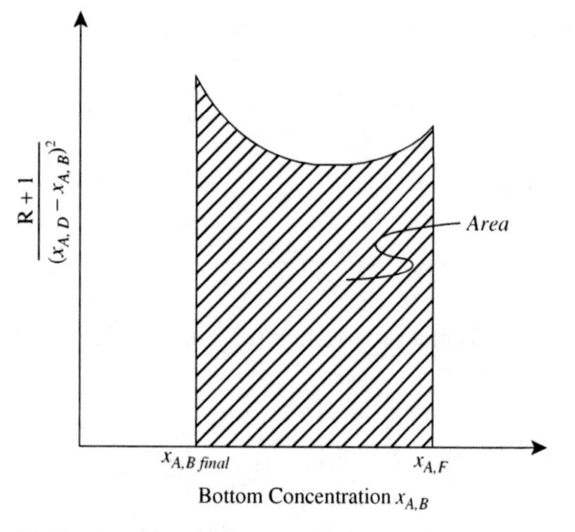

Figure 14.10 Integration of the equation for constant distillate composition allows the energy requirement to be determined.

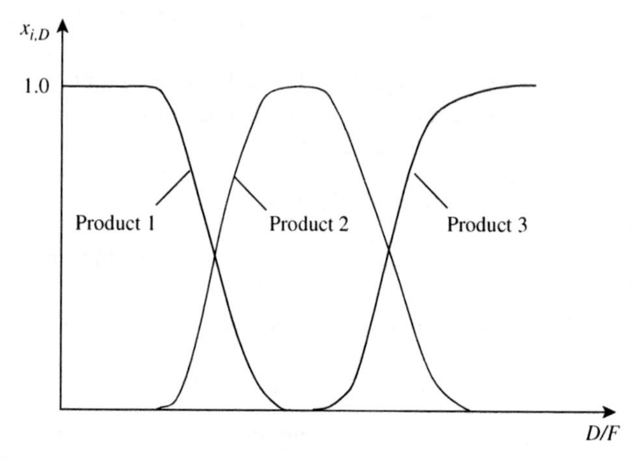

Figure 14.11 Batch distillation of a multiproduct mixture.

with an objective different from that of maintaining constant distillate composition. As with other batch processes, profile optimization can be used to optimize for a variety of objective functions.

The calculation procedures can be extended to multicomponent systems using shortcut or rigorous simulation of the column as the batch progresses[6].

Whilst most batch distillation operations involve the purification of a single-product overhead, it can also be applied to the separation of multiproduct systems[6]. In principle, a number of products need to be purified from the same initial charge to the batch distillation. Different product receivers are then used to collect the separate products. Figure 14.11 shows a profile of distillate composition versus the ratio of distillate to feed flowrate for a variety of products distilled overhead. Initially, Product 1 is distilled overhead. Once Product 1 has largely been distilled overhead, its concentration begins to fall in the distillate and the composition of Product 2 begins to rise.

Then Product 2 is distilled overhead for a period, after which its concentration begins to fall as the concentration of Product 3 increases in the distillate. The distillation is finished by collecting Product 3 overhead.

When evaluating a batch distillation, either through simulation or from scale-up of a pilot plant, one thing that needs special attention is hold-up in the column. A large hold-up in the column reduces the sharpness of separation and will require increased reflux or increased number of stages. If hold-up in the column is not accounted for, products might be off specification[6].

2. *Batch crystallization*. Crystallization is extremely common in the production of fine and specialty chemicals. Many chemical products are in the form of solid crystals. Also, crystallization has the advantage that it can produce a product with a high purity and can be more effective than distillation from the separation of heat-sensitive materials. Crystallization has already been discussed in Chapter 10 and has two main steps. Firstly the solute to be crystallized is dissolved in a suitable solvent, unless it is already dissolved, for example, solute dissolved in a solvent from a previous a reaction step. Secondly, the solid is then deposited in the form of crystals from the solution by cooling, evaporation and so on.

Two of the main objectives in crystallization are to maximize the average crystal size and to minimize the coefficient of variation of crystal size. As pointed out in Chapter 10, large crystals are easier to filter and wash in order to produce a high purity product.

Table 14.2 contrasts continuous versus batch crystallization.

From Table 14.2, it can be concluded that both batch and continuous crystallization have their relative advantages and disadvantages. The biggest single advantage of batch crystallization is its flexibility. Batch crystallization operations are most often carried out in an agitated vessel. However, the vessel might be fitted with baffles or a *draft tube*. A draft tube is a vertical cylinder placed inside the crystallizer with a diameter typically 70% of the vessel diameter. The agitator induces a vertical flow through the inside of the draft tube and circulation vertically in the opposite direction on the outside of the draft tube. This can help maintain good circulation rates for the slurry.

Table 14.2 Batch versus continuous crystallization.

Batch	Continuous
Flexible	Not flexible
Low capital investment	High capital investment
Small process development requirements	Large process development requirements
Poor reproducibility	Good reproducibility

Cooling crystallization (see Chapter 10) is the most common method of achieving supersaturation. A solvent is preferred that exhibits a high solubility at high temperature, but a low solubility at low temperature. The initial temperature is reduced gradually to the final temperature. Figure 14.12 illustrates a batch cooling crystallization. Starting at Point *A* in the unsaturated region, this can be cooled and the metastable region entered. A cooling profile can then be followed within the metastable region to the final temperature. As pointed out in Chapter 10, it is desirable to stay out of the labile region, otherwise too many fine crystals are created. In cooling from the initial to the final temperature, different cooling profiles can be followed, as illustrated in Figure 14.13. Figure 14.13a shows *natural cooling*, whereby there is a large temperature decrease initially, but as the temperature difference between the

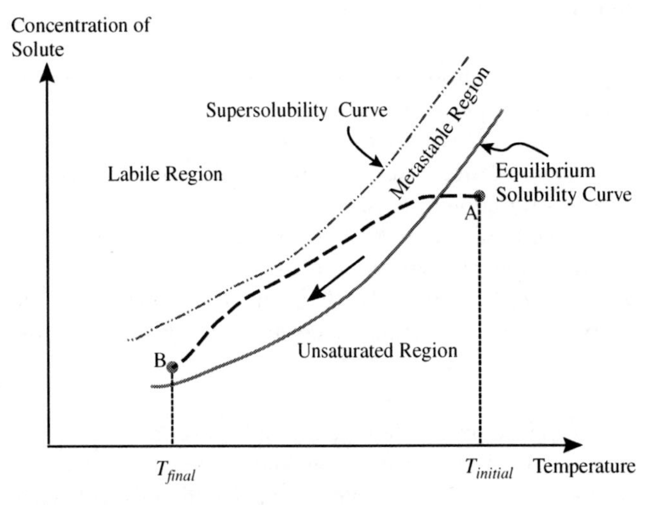

Figure 14.12 Batch cooling crystallization.

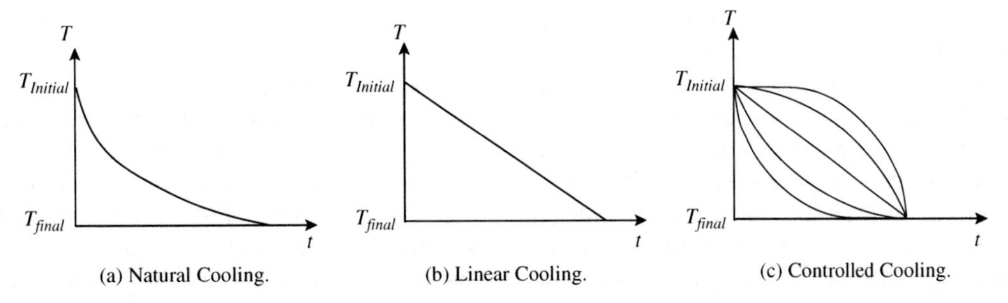

Figure 14.13 Types of batch cooling crystallization.

crystallization and the ambient temperature decreases, the rate of cooling decreases. Natural cooling is uncontrolled and tends to lead to poor crystal quality. Rather than rely on natural cooling, a coolant such as cooling water could be exploited. Figure 14.13b illustrates *linear cooling* in which the coolant flowrate is used to maintain a constant rate of cooling. This tends to lead to better crystal quality than natural cooling. Finally, Figure 14.13c illustrates controlled cooling in which the flowrate of the coolant is varied in order to follow a specified profile from the initial to the final temperature. The cooling profile is a degree of freedom for optimization and can be optimized using profile optimization[5].

Although cooling crystallization is the most common method of inducing supersaturation in batch crystallization processes, other methods can be used, as discussed in Chapter 10. For example, evaporation can be used, in which case the profile of the rate of evaporation through the batch can also be optimized[7]. Indeed, the profiles of both temperature and rate of evaporation can be controlled simultaneously to obtain greater control over the level of supersaturation as the batch proceeds[7]. However, it should be noted that there is often reluctance to use evaporation in the production of fine, specialty and pharmaceutical products, as evaporation can concentrate any impurities and increase the level of contamination of the final product.

The agitator is a key component of crystallizer design and is required to

- generate appropriate levels of secondary nucleation throughout the vessel;
- prevent local excessive levels of supersaturation;
- provide adequate temperature control throughout the vessel for cooling crystallization;
- provide adequate rates of heat transfer for evaporation if evaporative crystallization is used;
- keep crystals from settling on the bottom of the crystallizer to maintain crystal growth and a low coefficient variation of crystal size.

Increasing the power input and fluid shear from agitation, increases both crystal breakage and secondary nucleation. As the speed of an agitator is increased, this generally reduces the size of the metastable region. Scale-up of crystallizers from laboratory size to full-scale operation is far from straightforward. If a crystallization operation is scaled up for a constant power input per unit volume of slurry based on geometric similarity, the maximum shear rate at the agitator increases, but the average shear rate decreases. This can result in a greater coefficient variation of crystal size from scale-up. Secondary nucleation is likely to vary significantly with location in the crystallizer. Changes in the agitation and vessel geometry can change the local rates of energy dissipation and result in significant changes in total secondary nucleation.

Table 14.3 Optimization variables for batch and semibatch cooling crystallization.

Batch	Semibatch
1. Temperature profile	1. Solute/feed flowrate
2. Number of seeds	2. Temperature profile
3. Size of seeds	3. Number of seeds
4. Agitation	4. Seed addition profile
5. Initial level of supersaturation	5. Size of seeds
6. Batch time	6. Agitation
	7. Initial level of supersaturation
	8. Batch Time

As pointed out in Chapter 10, another way to create secondary nucleation is to add seed crystals to start the crystals growing in the supersaturated solution. These seeds should be pure product.

Just as a reactor can be operated in batch or semibatch mode, a crystallizer can also be operated in batch or semibatch mode. Table 14.3 contrasts the optimization variables for batch and semibatch operation of cooling crystallization.

As mentioned previously, scale-up of crystallization processes from the laboratory is far from straightforward. Various parameters need to be maintained to be as close to those used in the laboratory as possible in order to reproduce the results from the laboratory. For scale-up, supersolubility, agitation (and its effect on secondary nucleation throughout the vessel), fraction of solids in the slurry, seed number and sizes, contact time between growing crystals and liquid all need to be maintained.

3. *Batch filtration.* Batch filtration involves the separation of suspended solids from a slurry of associated liquid. The required product could be either the solid particles or the liquid filtrate. In batch filtration, the filter medium presents an initial resistance to the fluid flow that will change as particles are deposited. The driving forces used in batch filtration are[2]:

- gravity
- vacuum
- pressure
- centrifugal.

If a constant pressure difference is maintained across a filter medium and filter cake, then for the filtration of volume of liquid F, filtration time is proportional to the square of F.

Filter aids can be added to the slurry to reduce the filter cake resistance. These are materials that have high porosity. Their application is normally restricted to cases where the filtrate is valuable and the solid cake is a waste. In cases where the solid is valuable, the filter aid should be readily separable from the filter cake. Sometimes the filter aid is precoated onto the filter medium. If the solid filter cake is the product, then the cake is normally washed to remove dissolved solutes and solvents present in the original feed. Liquid may be retained within the cake within clusters of particles or at points of contact between particles. In an ideal situation, perfect washing would require only one volume of wash liquid equivalent to the void in the solid bed to wash away unwanted solute and solvent. In practice, perfect washing is not achieved and the actual washing efficiency depends on mixing and mass transfer. After washing the filter cake, the product would normally be dried to remove any wash liquid remaining within the solid.

Experimental results from the laboratory or pilot plant may often be scaled up by a factor of 100 times or more. However, to reduce errors in scale-up, a similar filter, the same slurry mixture, the same filter aid and approximately the same pressure drop should be used.

14.4 GANTT CHARTS

Now consider the complete batch process. Figure 14.14 shows a simple process. Feed material is withdrawn from storage using a pump. The feed material is preheated in a heat exchanger before being fed to a batch reactor. Once the reactor is full, further heating takes place inside the reactor using steam to the reactor jacket, before the reaction proceeds. During the later stages of the reaction, cooling water is applied to the reactor jacket. Once the reaction is complete, the reactor product is withdrawn using a pump. The reactor product is passed to a batch distillation that produces a finished product in the overhead and a residue left in the distillation. The product and residue are sent to storage.

The process is also shown in Figure 14.14 as a *Gantt* or *time event chart*[8,9]. The first two steps, pumping for reactor filling and feed preheat are both semicontinuous. The heating inside the reactor, the reaction itself and the cooling using the reactor jacket are all batch. The pumping to empty the reactor and charge to the batch distillation is again semicontinuous. The distillation step is batch. It can be seen from the Gantt chart in Figure 14.14 that there is very poor utilization of equipment. There are considerable periods over which the equipment is standing idle, sometimes termed *dead time*. The batch *cycle time* is the time interval between successive batches of product being produced.

High utilization of equipment is one of the goals of batch process design. This can be achieved by *overlapping* batches. Overlapping means that more than one batch, at different stages, resides in the process at any time, as shown in Figure 14.15. This allows the batch cycle time, that is, the time interval between producing successive batches of product, to be decreased considerably. The step with the longest time limits the cycle time. Alternatively, if more than one step is carried out in the same equipment, the cycle time is limited by the longest series of steps in the same equipment. The batch cycle time must be at least as

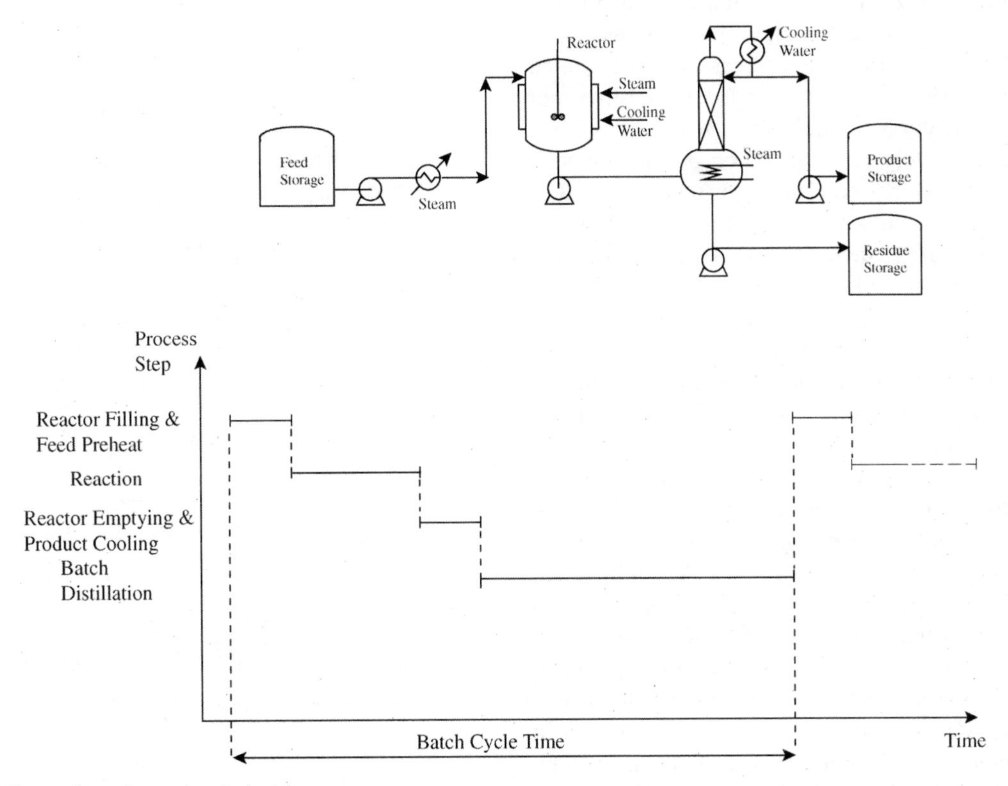

Figure 14.14 Gantt chart for a simple batch process.

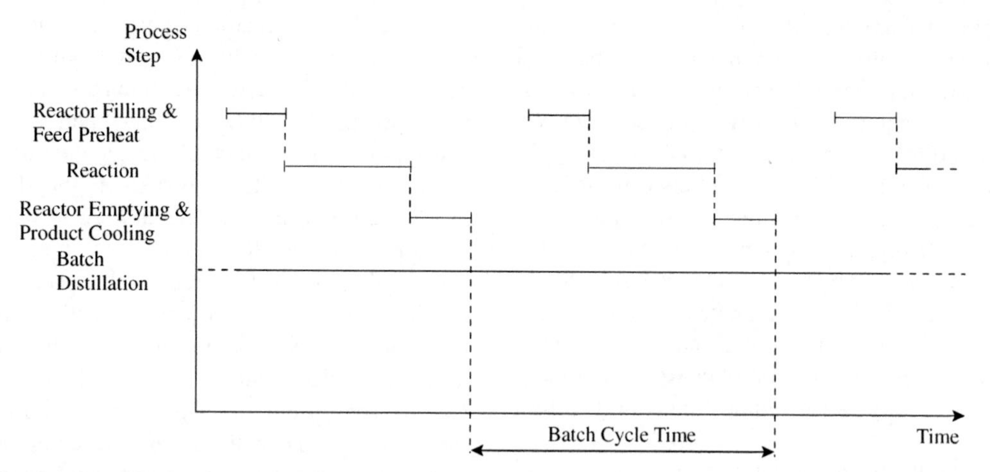

Figure 14.15 Overlapping batches allows the batch cycle time to be decreased.

long as the longest step. The rest of the equipment other than the limiting step is then idle for some fraction of the batch cycle.

14.5 PRODUCTION SCHEDULES FOR SINGLE PRODUCTS

Batch processes can be dedicated to the production of a single product or can produce multiple products. Start by considering the simplest case in which the process produces only a single product. Consider the process shown in Figure 14.16 involving three steps (Step A, Step B and Step C) in which Step A takes 10 h, Step B takes 5 h and Step C also takes 5 h. Figure 14.16a shows a *sequential* production schedule. Subsequent batches are only started once the previous batch has been completely finished. For this sequential production schedule, the cycle time is 20 h. This clearly leads to very poor utilization of equipment. It has already been noted that *overlapping* batches can reduce the cycle time. This is illustrated in Figure 14.16b, where subsequent batches are started as soon as the appropriate equipment becomes available. Cycle time in Figure 14.16b decreases to 10 h for overlapping batches (the length of the longest step). If a specified volume of production needs to be achieved over a given period of time, then the equipment in the process that uses overlapping batches in Figure 14.16b can in principle be half the size of the equipment for sequential production in Figure 14.16a.

Even with overlapping batches in Figure 14.16b, Steps B and C are under utilized. Step A is fully utilized and this is the limiting step. Figure 14.16c shows a design in which there are two items of equipment operating Step A, but in parallel. This allows both Step B and Step C to be carried out with complete utilization. If the sizes of the equipment are compared to the sequential production schedule, then each of the two Steps $A1$ and $A2$ in Figure 14.16c can in principle be one-quarter the size of the equipment for Step A for sequential production in Figure 14.16a. The size of

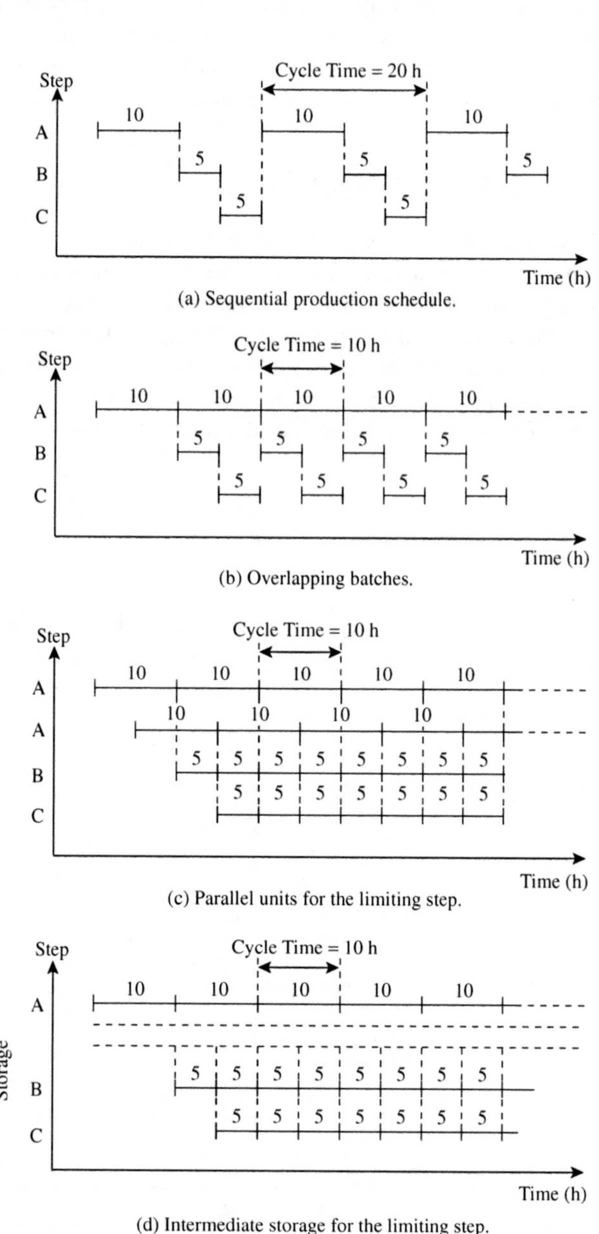

Figure 14.16 Production schedules for a three-step process.

the equipment for Steps B and C in Figure 14.16c will also be one-quarter the size of those in the sequential production schedule in Figure 14.16a.

The final option shown in Figure 14.16d is to use *intermediate storage* for the limiting step. Material from Step A is sent to storage, from which Step B draws its feed. Material is still passed directly from Step B to Step C. Now all three steps are fully utilized. For the same rate of production over a period of time, the size of Step A can in principle be half that relative to the sequential production in Figure 14.16a and the sizes of Steps B and C can in principle be one-quarter those for sequential production. However, this is at the cost of introducing intermediate storage.

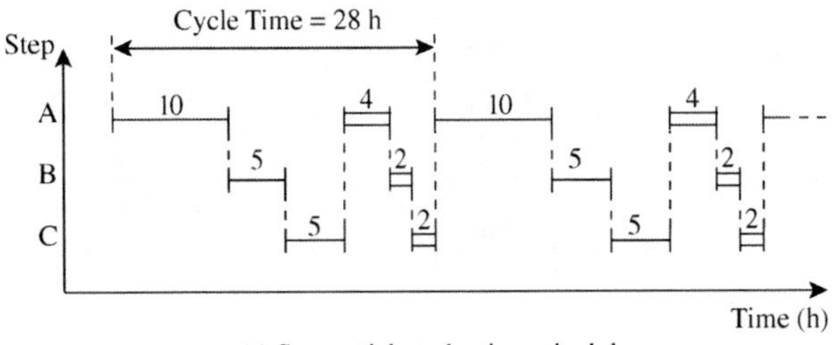

(a) Sequential production schedule.

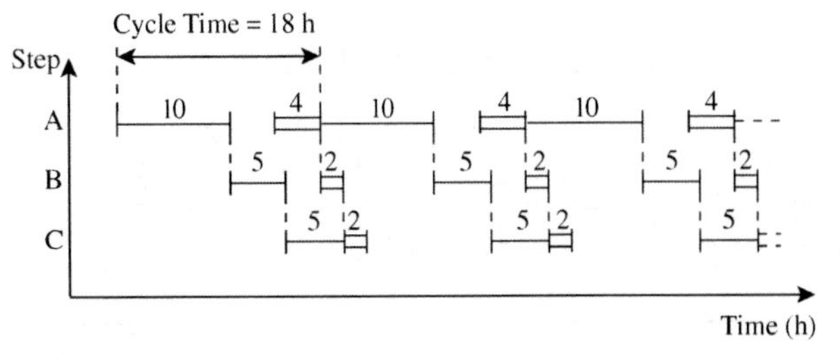

(b) Overlapping batches.

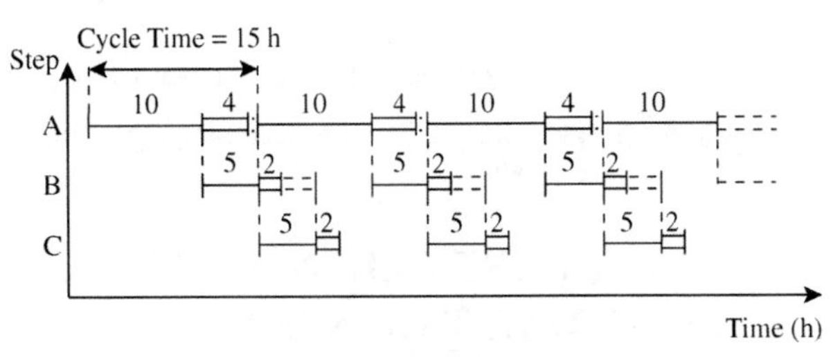

(c) Equipment hold-up.

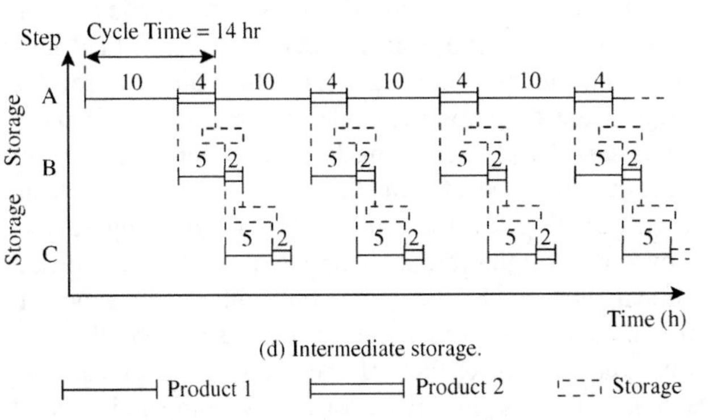

(d) Intermediate storage.

Product 1 Product 2 Storage

Figure 14.17 Production schedule for two products with a three-step process.

14.6 PRODUCTION SCHEDULES FOR MULTIPLE PRODUCTS

So far plants have been considered involving a single product. However, batch processes often produce multiple products in the same equipment. Here two broad types of process can be distinguished. In *flowshop* or *multiproduct* plants, all products produced require all steps in the process and follow the same sequence of operations. In *jobshop* or *multipurpose* processes, not all products require all steps and/or might follow a different sequence of steps[9].

Figure 14.17 shows a process that produces two products, Product 1 and 2, in a flowshop process. Figure 14.17a shows a production cycle involving a sequential production schedule. Production alternates between Product 1 and Product 2. The cycle time to produce a batch each of Product 1 and 2 is 28 h.

The first thing that can be considered in order to reduce the cycle time and increase equipment utilization is to overlap the batches as shown in Figure 14.17b. This reduces the cycle time to 18 h.

All of the schedules considered so far involved transferring material from one step to another, from a step to storage or from storage to a step without any time delay. This is known as *zero-wait transfer*. An alternative is to exploit the equipment in which a production step has taken place to provide *hold-up*. In this situation, material is held in the equipment until it is required by the production schedule. A schedule using equipment hold-up is shown in Figure 14.17c. This reduces the cycle time to 15 h.

Finally, Figure 14.17d shows the use of intermediate storage. The use of storage is only necessary for Product 2. Use of intermediate storage in this way reduces the cycle time to 14 h.

Consider now another problem involving the production of two products (Product 1 and 2) each involving two steps (Step A and B) in a flowshop plant. Figure 14.18a shows the production cycle for three batches each of Product 1 and Product 2. It can be seen from Figure 14.18a that the batches have been overlapped to increase equipment utilization. In order to produce three products each of Product 1 and Product 2, the schedule in Figure 14.18a involves single-product *campaigns*. Three batches of Product 1 and three batches of Product 2 follow directly from each other. For this production schedule, the cycle time is 47 h. The total time required to produce a given number of batches, in this case three batches of each Product 1 and Product 2, is known as the *makespan*. From Figure 14.18a, for single-product *campaigns* the makespan is 53 h.

An alternative production schedule can be suggested by following a mixed-product campaign, as illustrated in Figure 14.18b. Alternating between batches of Product 1 and Product 2 in Figure 14.14b allows the cycle time to be reduced to 45 h and the makespan to be reduced to 51 h.

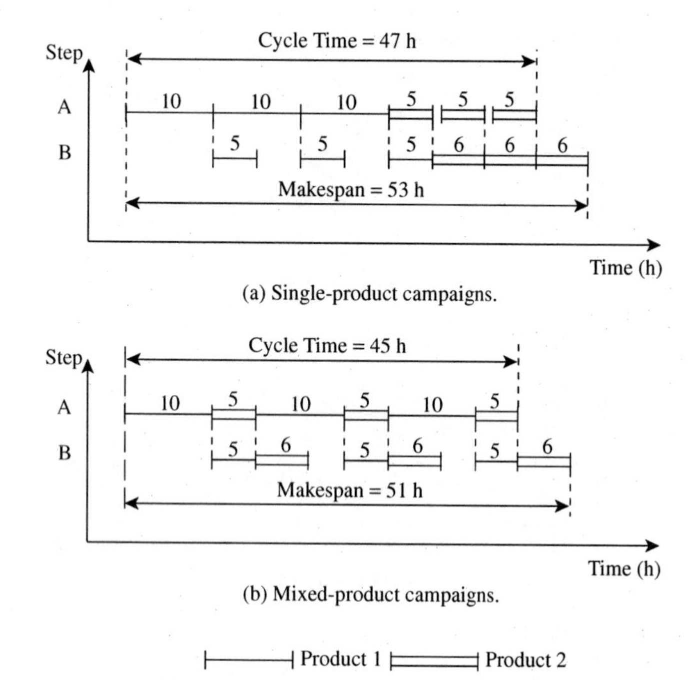

Figure 14.18 Single versus mixed-product campaigns for three batches each of two products.

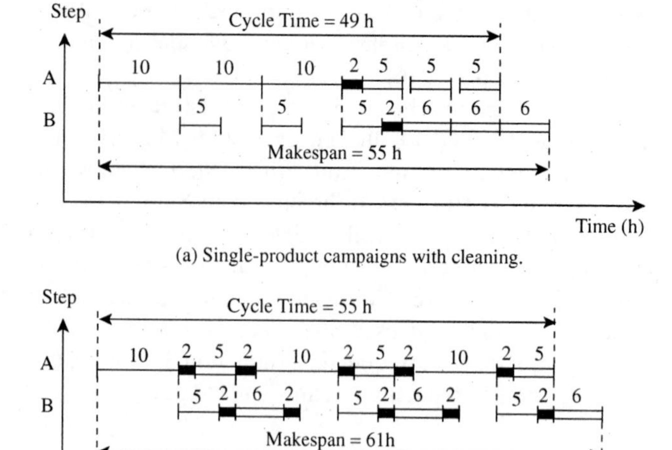

Figure 14.19 Cleaning between product changes extends the cycle times.

14.7 EQUIPMENT CLEANING AND MATERIAL TRANSFER

So far, in the discussion of production schedules, some practical issues have been neglected. Two practical issues can be encountered in production scheduling that can have a significant effect on the cycle time and the makespan[9].

Consider first the changeover between two different products. It is usual for the equipment to be cleaned when changing from one product to another. Figure 14.19 shows the single-product campaigns and mixed-product campaigns from Figure 14.18, but with cleaning between product changes. The cleaning increases both the cycle time and the makespan. If the single-product campaign without cleaning in Figure 14.18a is compared with the single-product campaign with cleaning in Figure 14.19a, then the cycle time increases from 47 to 49 h and the makespan from 53 to 55 h. However, when the mixed-product campaign in Figure 14.18a is compared with the mixed-product campaign with cleaning in Figure 14.19b, it can be seen that there is a more significant increase in both the cycle time and the makespan. Cleaning increases the cycle time from 45 to 55 h and the makespan from 51 to 61 h.

Without cleaning the mixed-product campaign would have been considered to be more efficient than the single-product campaigns. Once cleaning is introduced, the mixed-product campaigns are seen to be less efficient than single-product campaigns.

Whether cleaning introduces such a decrease in the overall equipment utilization as that presented in Figures 14.18 and 14.19 depends on the problem. However, it is something that must be taken into account when planning production schedules. Another issue to be addressed later is that when products are changed in a batch production system, and equipment needs to be cleaned, this can produce a significant amount of waste from the process that can create a significant environmental problem. This will be discussed again later under cleaner production.

Another important issue that has been neglected so far in the production schedules is that of transfer times between different steps. Figure 14.20 shows again the

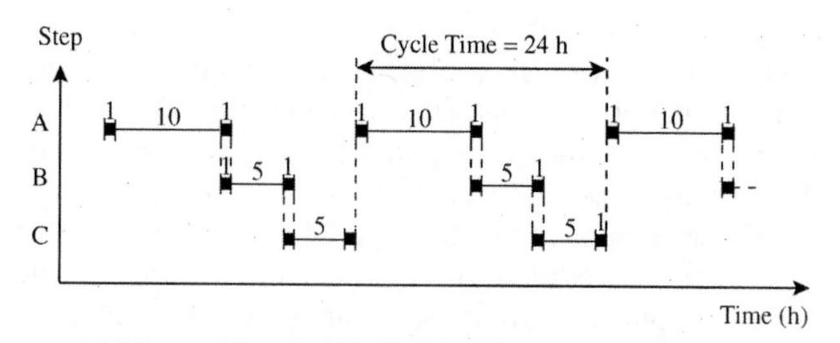

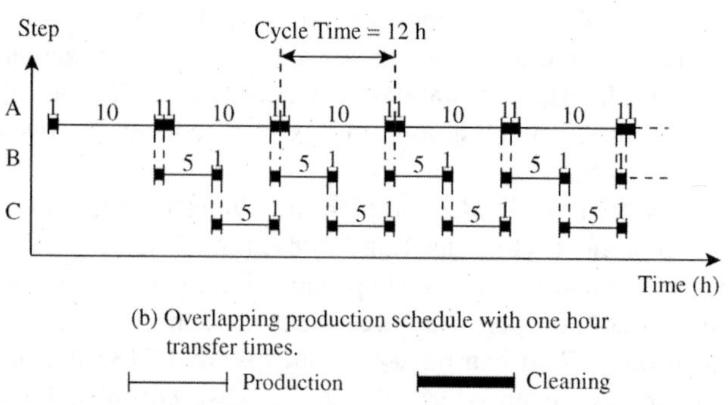

Figure 14.20 Transfer times extend the cycle times.

production schedule from Figure 14.16, but this time introducing an allowance of 1 h to transfer material between storage and the production steps, from one production step to another and from the outlet of a production step to storage. If material is being transferred from one step to another step, then the emptying of one step and filling the next step can be carried out simultaneously, hence the transfer from Step *A* to Step *B* and from Step *B* to Step *C* overlaps. The cycle time for sequential production as shown in Figure 14.20a increases from 20 h to 24 h for one-hour transfer times. Figure 14.20b shows overlapping production with 1 h transfer times. In this case, the cycle time increases from 10 h to 12 h.

14.8 SYNTHESIS OF REACTION AND SEPARATION SYSTEMS FOR BATCH PROCESSES

Now consider how to synthesize the reaction and separation system for a batch process. Start by assuming the process to be continuous and then, if choosing to use batch operation, the continuous steps are replaced by batch steps[10]. It is simpler to start with continuous process operation because the time dependency of batch operation adds additional constraints over and above those for continuous operation.

However, there is one very significant difference between batch and continuous processes as far as the synthesis of reaction and separation systems is concerned. Continuous processes involve connections in space between processing steps. Batch processes also have connections in space between processing steps. However in addition, batch processes have connections in time between processing steps. In batch processes, the connections in space can sometimes be substituted by connections in time. Consider the sequential production schedule in Figure 14.16a. Suppose that Step *B* and Step *C* could be carried out in the same equipment as Step *A*. This would mean that only one piece of equipment would be needed rather than three, but the production schedule would look the same as shown in Figure 14.16a. This would serve to reduce the capital cost of the equipment. It would also give advantages in terms of material transfer. Thus, the transfer times between the steps in Figure 14.20 would be eliminated. Another advantage in terms of cleaning is that there is less equipment to clean and less waste resulting from cleaning. Of course, the option to overlap batches would no longer be available if steps were merged. An example of how steps can be merged is a reaction followed by cooling crystallization. In principle, both steps can be carried out in the same equipment.

Before steps are merged into a single piece of equipment, it must be ensured that the equipment is suitable for multiple purposes in terms of its function, size, materials of construction and pressure rating. Also, it is clear that merging will affect the production schedule and the schedule needs to be considered when merging.

Finally, recycling of materials is difficult in batch processes because the connection in time cannot usually be made between the steps involved in the recycling. This is because different steps take place in different time periods. However, time can be bridged through the use of intermediate storage for the recycle.

The approach is illustrated by the following example.

Example 14.2 Butadiene sulfone (or 3-sulfolene) is an intermediate used for the production of solvents. It can be produced from butadiene and sulfur dioxide according to the reaction[10,11]:

$$CH_2{=}CHCH{=}CH_2 + SO_2 \rightleftharpoons \underset{\text{(ring: } CH{=}CH,\ CH_2\ CH_2,\ SO_2)}{C_4H_6SO_2}$$

CH=CH
\ /
CH₂ CH₂ bonded to SO₂ (ring)

butadiene — sulfur dioxide — butadiene sulfone

This is an exothermic, reversible, homogeneous reaction that takes place in a single liquid phase. The liquid butadiene feed contains 0.5% *n*-butane as an impurity. The sulfur dioxide is essentially pure. The mole ratio of sulfur dioxide to butadiene must be kept above 1 to prevent unwanted polymerization reactions. A value of 1.2 will be assumed. The temperature in the process must be kept above 65°C to prevent crystallization of the butadiene sulfone but below 100°C to prevent its decomposition. The product must contain less than 0.5 wt% butadiene and less than 0.3 wt% sulfur dioxide.

The normal boiling points of the materials are given in the Table 14.4.

Table 14.4 Normal boiling points for the components.

Material	Normal boiling point (°C)
Sulfur dioxide	−10
Butadiene	−4
n-Butane	−1
Butadiene sulfone	151

Synthesize a continuous reaction, separation and recycle system for the process, bearing in mind that the process will later become batch.

Solution The reversible nature of the reaction means that neither of the feed materials can be forced to complete conversion. The reactor design in Figure 14.21a shows that the reactor product contains a mixture of both feed and product materials together with the n-butane impurity. These must be separated, but how?

If the relative boiling points of the components in the reactor product are considered, there is a wide range of volatilities. The sulfur dioxide, butadiene and n-butane are all

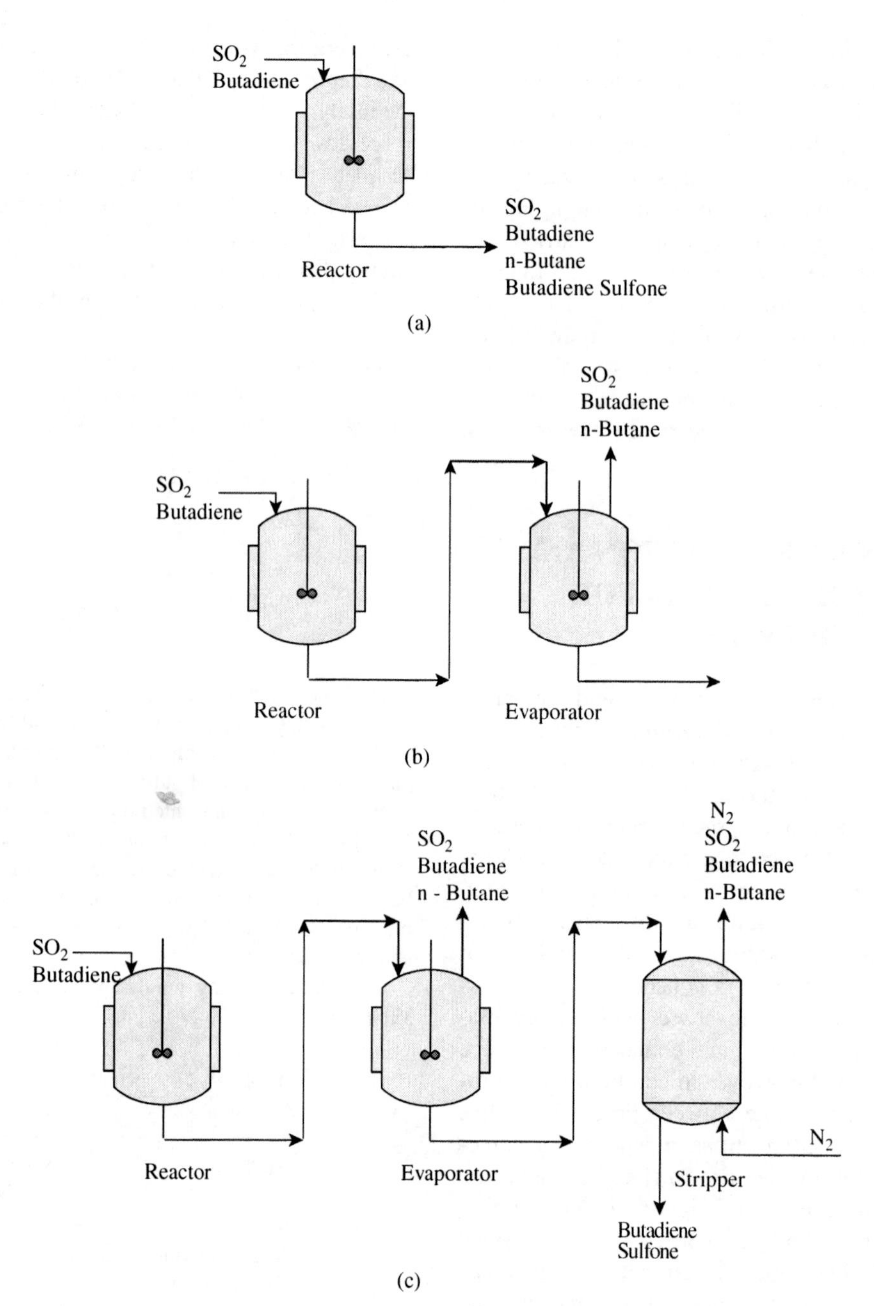

Figure 14.21 Reaction and separation system for the production of butadiene sulfone.

low boilers and the butadiene sulfone is a much higher boiling material by comparison. Given that the reaction takes place in the liquid phase, a partial vaporization might well give a good separation between the butadiene sulfone and the other components (Figure 14.21b).

A vapor–liquid equilibrium calculation shows that a good separation is obtained but the required product purity of butadiene <0.5 wt% and sulfur dioxide <0.3 wt% is not obtained. Further separation of the liquid is needed. Distillation of the liquid is difficult because of the narrow temperature limits between which the distillation must operate. However, the liquid can be stripped using nitrogen (Figure 14.21c).

The type of equipment illustrated in Figure 14.21 is more typical of batch operation than continuous, even though continuous operation is being contemplated at the moment. For example, the evaporator is a stirred tank with a heating jacket. In continuous plant, a more elaborate design with tubular heating of some type would probably have been used, perhaps with multiple stages.

Now consider recycling unconverted feed material to the reactor. Figure 14.22a shows recycles for unconverted feed material. The recycle from the evaporator to the reactor has been made possible by pressurizing the evaporator with the evaporator feed pump. Had this not been done, the vapor recycle would have required a compressor. The stripper works at a lower pressure to allow the unconverted material to be stripped. Thus, the recycle from the stripper requires a compressor. It is then condensed and fed back to the reactor.

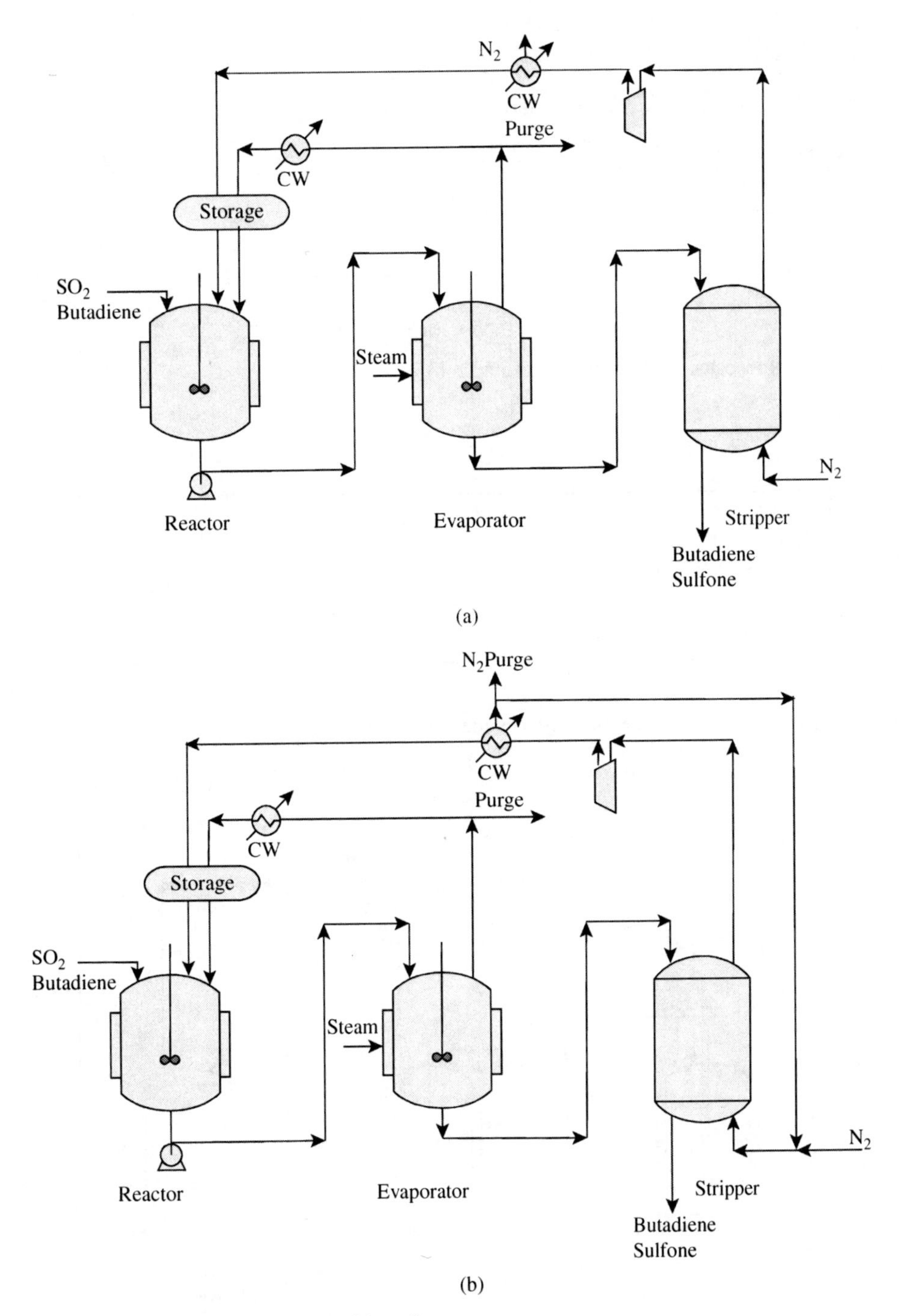

Figure 14.22 The recycle system for the production of butadiene.

Another problem is that most of the n-butane impurity that enters with the feed enters the vapor phase from the evaporator. Thus the n-butane builds up in the recycle unless a purge is provided (Figure 14.22a). Finally, the possibility of a nitrogen recycle should be considered to minimize the use of fresh nitrogen (Figure 14.22b).

Example 14.3 Convert the continuous process from Example 14.2 into a batch process. Preliminary sizing of the equipment indicates that the duration of the processing steps are given in the Table 14.5[10]:

Table 14.5 Duration of processing steps.

Processing step	Duration (h)
Reaction	2.1
Evaporation	0.45
Stripping	0.65
Vessel filling	0.25
Vessel emptying	0.25

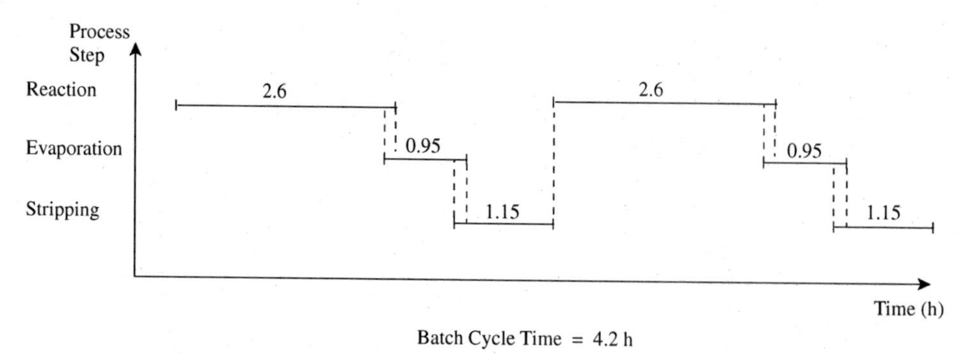

Figure 14.23 Gantt chart for a repeated batch cycle for Example 14.2.

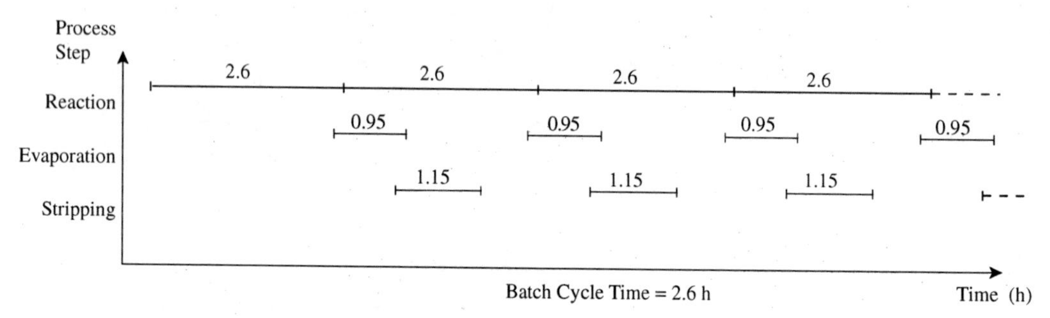

Figure 14.24 Overlapping batches for Example 14.2 reduces the batch cycle time.

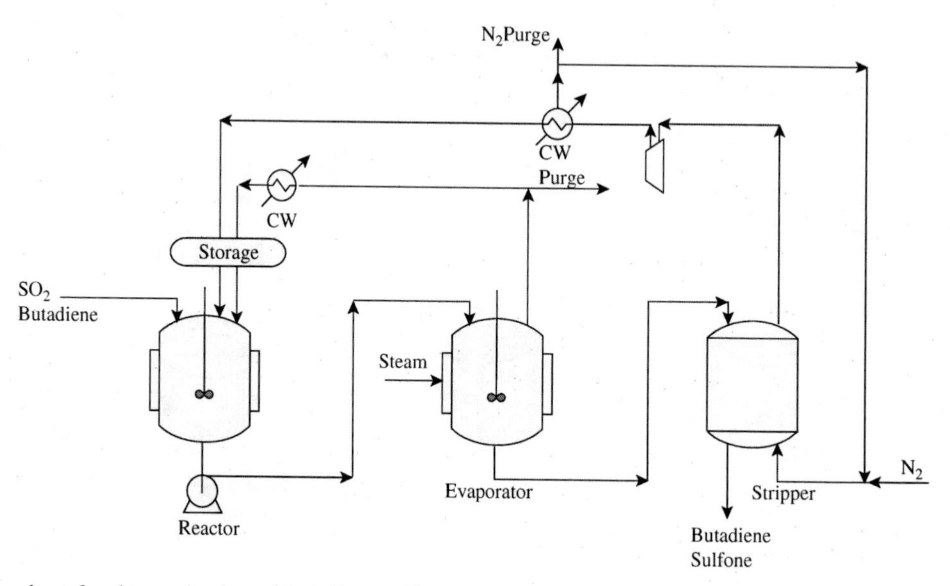

Figure 14.25 Flowsheet for the production of butadiene sulfone in a batch process.

Solution Having synthesized the continuous flowsheet shown in Figure 14.22b, now convert this into batch operation. The reactor now becomes batch, requiring the reaction to be completed before the separation can take place. Figure 14.23 shows the production schedule for a sequential batch schedule. Note in Figure 14.23 that there is a small overlap between the process steps. This is to allow for the fact that emptying of one step and filling the following step take place during the same time period.

The Gantt chart shown in Figure 14.23 indicates that individual items of equipment have a poor utilization. To improve the equipment utilization, overlapping batches are shown in Figure 14.24. Clearly, it is not possible to recycle directly from the separators to the reactor since the reactor is fed at a time different from that at which the separation is carried out. A storage tank is needed to hold the recycle material. This material is then used to provide part of the feed for the next batch. The final flowsheet for batch operation is shown in Figure 14.25.

In Figure 14.24, the reactor limits the batch cycle time, that is, it has no dead time. On the other hand, the evaporator and stripper both have significant dead time. Figure 14.26 shows the schedule for an arrangement with two reactors operating in parallel. With parallel operation, the reaction operations can

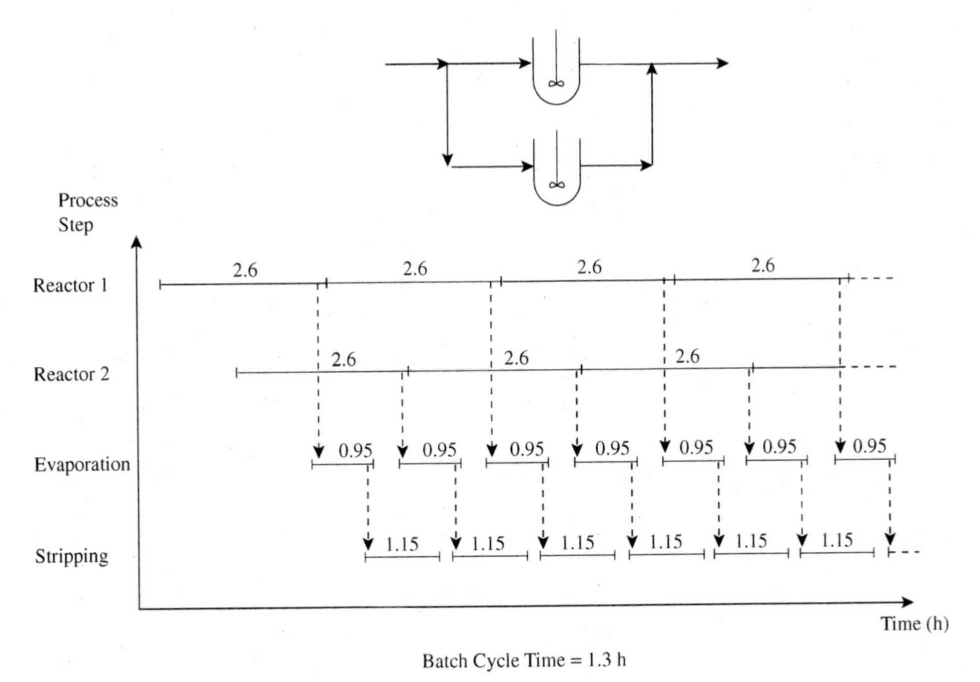

Figure 14.26 Placing units in parallel for the limiting step reduces the batch cycle time.

overlap, allowing the evaporation and stripping operations to be carried out more frequently. This improves the overall utilization of equipment, and in principle allows the size of equipment to be reduced.

The batch cycle time has been reduced from 2.6 to 1.3 h. This means that a greater number of batches can be processed and hence, if there are two reactors each with the original capacity, the process capacity has increased. However, the increase in capacity has been achieved at the expense of an increased capital cost for the second reactor. An economic assessment is required before it can judge whether the trade-off is justified.

Perhaps the additional capacity might not be needed. If it is not needed then the size of the reactors, evaporator and stripper can be reduced. Keeping the original process capacity with parallel operation of the reactors would mean a trade-off between the increased capital cost of two (smaller) reactors versus reduced capital cost of the evaporator and stripper. An economic comparison would be required to judge whether this would be beneficial.

Another option to improve utilization of equipment is, instead of adding a reactor in parallel, installing intermediate storage. Figure 14.27 shows a production schedule with intermediate storage between the reactor and evaporator and between the reactor and stripper. The evaporation step is no longer constrained to start on completion of the reaction step and start the stripping step on completion of the evaporation step. The individual steps can be decoupled via the intermediate storage. This maintains the original batch cycle time of 2.6 h but allows, as shown in Figure 14.27, the elimination of dead time in the evaporation and stripping steps. Now more evaporation and stripping steps can be carried out and the size of the evaporator and stripper reduced accordingly. This time the capital cost of intermediate storage is traded off against reduced capital cost of the evaporator and stripper. In Figure 14.27, the intermediate storage between the reactor and evaporator has a significant effect on equipment utilization. The intermediate storage between the evaporator and stripper has a less pronounced effect and would be more difficult to justify economically.

Finally, merging of operations into the same equipment could be considered to replace connections through space by those through time. For example, it might be possible to carry out the reaction and evaporation in the same equipment. Overlapping of the reaction and evaporation would no longer be possible, but there would be savings in capital cost.

14.9 OPTIMIZATION OF BATCH PROCESSES

In general, utilization can be improved by:

- merging more than one operation into a single piece of equipment (e.g. feed preheating and reaction in the same vessel), providing these operations are not limiting the cycle time;
- overlapping batches, that is, more than one batch at different processing stages resides in the plant at any given time;
- introducing parallel operations to the steps, which limit the batch cycle time;
- introducing multiple operations in series to the steps that limit the batch cycle time;
- increasing the size of equipment in the steps that limit the batch cycle time to reduce the dead time for those steps that are not limiting;
- decreasing the size of equipment for those steps that are not limiting to increase the time required for those steps that are not limiting and hence reduce the dead time for the nonlimiting steps;
- introducing intermediate storage between batch steps.

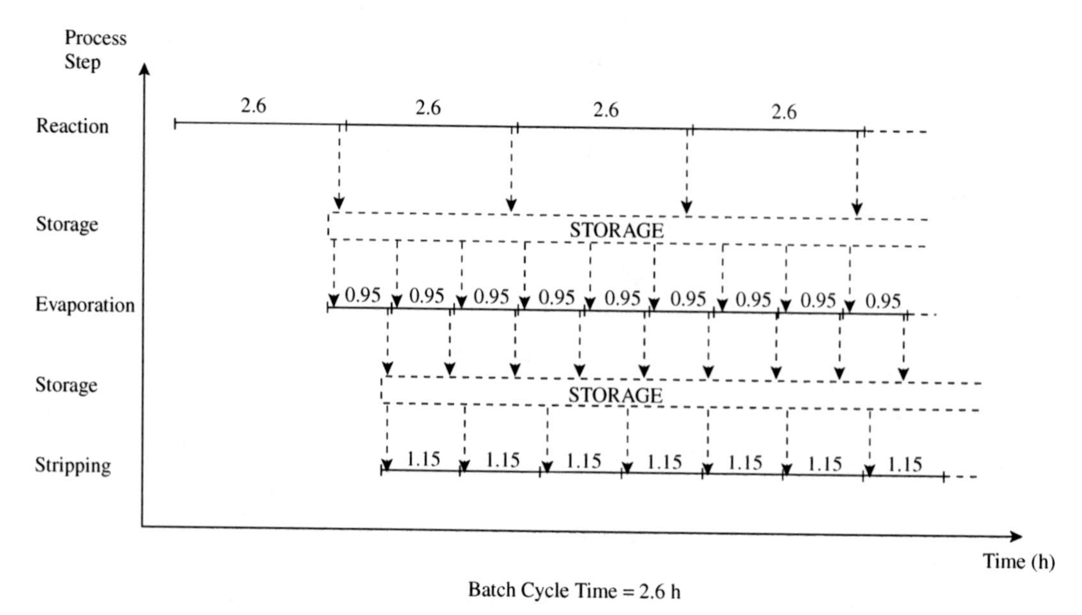

Figure 14.27 Gantt chart for Example 14.2 with intermediate storage.

Whether parallel operations, larger or smaller items of equipment and intermediate storage should be used can only be judged based on economic trade-offs. However, this is still not the complete picture as far as the batch process trade-offs are concerned. So far the batch size has not been varied. Batch size can be varied as a function of cycle time. Overall, the variables are:

- batch size
- batch cycle time
- number of batch units in parallel
- number of batch units in series
- size of equipment
- intermediate storage.

In addition to these variables, which result from the batch nature of the process, there are still the variables considered earlier for continuous processes:

- reactor conversion
- recycle inert concentration.

All of these variables must be varied in order to minimize the total cost or maximize the economic potential (see Chapter 2). This is a complex optimization problem involving both continuous variables (e.g. batch size) and integer variables (e.g. number of units in parallel) and is outside the scope of the present text[9].

However, the designer should not loose sight of the fact that, whilst schedule planning and optimization provides the basis for the design of batch processes, production schedules are often disrupted once production has commenced. For example, what happens if a key item of equipment breaks down and needs to be repaired? What happens if delivery of some of the raw materials is delayed? In a multiproduct plant, what happens if there is a rush order for one of the products, or cancellation of an order? These and many other possible disruptions mean that a carefully planned and optimized production schedule would need to be completely changed. Thus, the design must make allowance as much as possible for such disruptions by building flexibility into the design.

14.10 STORAGE IN BATCH PROCESSES

The storage requirements associated with continuous processes were discussed in Chapter 13. The minimum storage volume in batch processes is equal to the delivered batch size for feed materials, or the manufactured quantity per batch for the product.

Because batch processes are in a dynamic state, it can be difficult to maintain the required product specification throughout the batch. Thus, storage can help to dampen out variations in product quality for fluid products. Variations in the outlet properties from the storage are reduced compared with variations in the inlet properties for fluid products. As long as the mean quality is within specifications, all of the product can be sold.

Batch manufacture often takes place in the form of campaigns. Each campaign manufactures a quantity of a given product that is stored until dispatched. Product campaigns might involve a single batch or a number of batches. The production might then be switched to a different product and another campaign carried out for that product. Changing product leads to lost production time, creates waste through equipment cleaning and decontamination and might result in some off-specification product as a result of the changeover. Thus, operation of the plant is favored by large batches and long production campaigns, but this increases storage costs (capital, working and operating).

As with storage for continuous processes, storage tanks for liquids should not be designed to operate with less that 10% inventory at the minimum or more than 90% full at the maximum. As with storage for continuous processes, the amount of feed product and intermediate storage will depend on capital, working and operating costs, together with operability, control and safety considerations.

14.11 REACTION AND SEPARATION SYSTEMS FOR BATCH PROCESSES – SUMMARY

Many batch processes are designed on the basis of a scale-up from the laboratory, particularly for the manufacture of specialty chemicals. Particularly when manufacturing specialty chemicals, the shortest time possible to get a new product to market is often the biggest priority (accepting that the product must meet the specifications and regulations demanded and the process must meet the required safety and environmental standards). This is particularly true if the product is protected by patent.

For batch reactors the four major factors that effect batch reactor performance are:

- Contacting pattern
- Operating conditions
- Agitation
- Solvent selection.

Batch distillation has a number of advantages when compared with continuous distillation:

- The same equipment can be used to process many different feeds and produce different products.
- There is flexibility to meet different product specifications.
- One distillation column can separate a multicomponent mixture into relatively pure products.

Crystallization is extremely common in the production of fine and specialty chemicals. Many chemical products are in the form of solid crystals. Also, crystallization has the advantage that it can produce a product with a high purity and can be more effective than distillation from the separation of heat-sensitive materials. Batch crystallization is:

- Flexible
- Incurs low capital investment
- Features small process development requirements
- But can give poor reproducibility.

Batch filtration involves the separation of suspended solids from a slurry of associated liquid. The required product could be either the solid particles or the liquid filtrate. In batch filtration, the filter medium presents an initial resistance to the fluid flow that will change as particles are deposited. The driving forces used in batch filtration are:

- gravity
- vacuum
- pressure
- centrifugal.

Production schedules for batch processes can be sequential, overlapping, parallel, use intermediate storage, or use a combination of these. Such schedules can be analyzed using Gantt charts. Batch processes often produce multiple products in the same equipment and can be distinguished as flowshop or multiproduct plants. Equipment cleaning and material transfer policy have a significant effect on the production schedule.

Synthesis of the reaction and separation system for a batch process can be carried out by assuming the process to be continuous and then, replacing the continuous steps by batch steps. When synthesizing a batch process, connections through both space and time can be exploited.

14.12 EXERCISES

1. Benzyl acetate is to be manufactured in a batch reactor from the reaction between benzyl chloride and sodium acetate in a solution of xylene in the presence of triethylamine as catalyst according to the reaction:

$$C_6H_5CH_2Cl + CH_3COONa \longrightarrow CH_3COOC_6H_5CH_2 + NaCl$$

or

$$A + B \longrightarrow C + D$$

The kinetic model for the reaction is given by

$$-r_A = k_A C_A$$

where r_A = rate of reaction of benzyl chloride
k_A = reaction rate constant
= 0.01306 h^{-1}
C_A = molar concentration of benzyl chloride

Calculate the residence time required for a conversion of 40% and 60%.

2. Ethyl acetate is to be manufactured in a batch reactor from the reaction between ethanol and acetic acid in the liquid phase according to the reaction:

$$\underset{\text{aceticacid}}{CH_3COOH} + \underset{\text{ethanol}}{C_2H_5OH} \rightleftharpoons \underset{\text{ethylacetate}}{CH_3COOC_2H_5} + \underset{\text{water}}{H_2O}$$

or

$$A + B \rightleftharpoons C + D$$

The reaction rate expression is of the form

$$-r_A = k_A C_A^2 - k'_A C_C$$

where k_A = 0.002688 $m^3 \cdot kmol^{-1} \cdot min^{-1}$
k'_A = 0.004644 min^{-1}

Molar masses and densities for the components are given in Table 14.6.

Table 14.6 Molar masses for the components.

Component	Molar mass ($kg{\cdot}kmol^{-1}$)	Density at 60°C ($kg{\cdot}m^{-3}$)
Acetic acid	60.05	1018
Ethanol	46.07	754
Ethyl acetate	88.11	847
Water	18.02	980

The reactants are equimolar with product concentrations of initially zero.

a. Calculate the residence time required for a conversion of 50%.
b. The operating schedule for the reactor requires it to be shut down between batch reactions for 1 h. Calculate the volume of the reactor required to produce $10\ t \cdot d^{-1}$ based on a conversion of 50% and operation of 24 h per day.

3. A solvent *A* is to be recovered from a residue *B* by batch distillation. The feed mixture contains a mole fraction of *A* of 60%. The relative volatility between *A* and *B* is 3.5. The distillation is required to recover 90% of *A*. Calculate composition of the recovered *A*:
   a. If only a batch distillation pot is used (i.e. no distillation trays and no reflux).
   b. If a rectification section is added to the batch distillation pot with four ideal stages using a constant reflux ratio of 3.
4. A batch process consists of the four steps given in Table 14.7. For repeated batch cycles of the same process, calculate the cycle time for:
   a. a sequential production schedule;
   b. overlapping batches.

Table 14.7 Batch steps for Exercise 4.

Step	Duration (h)
A	0.75
B	3
C	0.75
D	6

5. A batch process consists of three steps given in Table 14.8. For repeated cycles of the same process, calculate the cycle time for:
   a. a sequential production schedule
   b. overlapping batches
   c. two parallel units for the limiting step
   d. intermediate storage for the limiting step.

Table 14.8 Batch steps for Exercise 5.

Step	Duration (h)
A	12
B	5
C	5

6. A batch process manufactures Product 1 and Product 2 in the same process. The manufacture of both products involves three steps with durations given in Table 14.9. Calculate the cycle time and makespan for one batch each of Product 1 and Product 2 with no delay between the two batches for:
   a. a sequential production schedule with zero-wait transfer;
   b. overlapping batches with zero-wait transfer;
   c. overlapping batches with hold-up;
   d. overlapping batches with intermediate storage.

Table 14.9 Batch steps for Exercise 6.

Step	Product 1 (h)	Product 2 (h)
A	12	4
B	5	3
C	5	3

7. The following process is proposed for the production of Product C. For the scheduling of the production campaign, overlapping is allowed and a zero-wait transfer is applied for storage policy. The process is represented in Figure 14.28.

Stage I

- Liquid raw materials *A* and *B* are reacted in a batch reactor for 6 h.
- 1 kg of *A* and 1 kg of *B* are mixed to produce *C*.

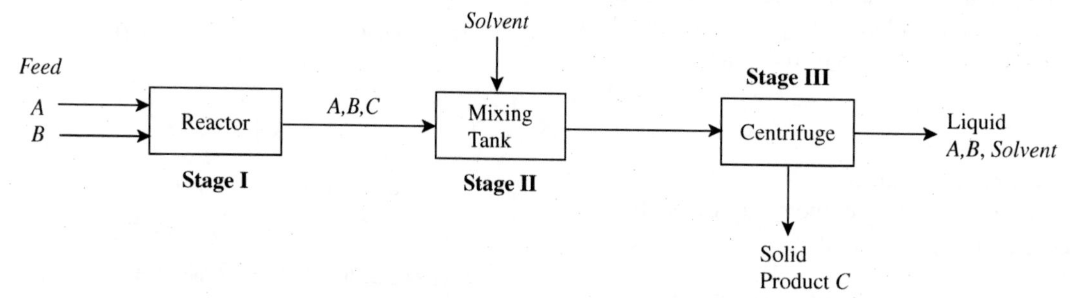

Figure 14.28 A simple batch process involving three steps.

- A reactor is operated with 70% of yield by mass.
- The mixture density is 800 kg·m^{-3}.

Stage II

- The liquid mixture of A, B and C is fed into a tank and mixed with a solvent for 3 h. The product C is converted into solids.
- 1 kg of solvent is added and mixed with A, B and C.
- The mixture density is 950 kg·m^{-3}.

Stage III

- The product C is separated by centrifuge for 4 h.
- 90% recovery by mass for product C is obtained.
- The mixture density is 900 kg·m^{-3}.

(a) Calculate the cycle time for the production of Product C when only one unit is used for each stage. Use a Gantt chart to show at least two production batches.

(b) The current production campaign must produce 100,000 kg·y^{-1} of product C. The plant is operated for 7200 h·y^{-1}. What is the size of the reactor (Stage I) required? (Use the cycle time calculated in Part a).

(c) The addition of parallel units can increase the efficiency of equipment utilization. When two reactors are operated for Stage I, one tank for Stage II, and two centrifuges for Stage III, how is the cycling time changed? Use a Gantt chart to show at least four production batches.

8. The plant from Exercise 7 is modified to produce Products D and E according to the schedule in Table 14.10.

Table 14.10 Modified schedule.

	Product	Stage I	Stage II	Stage III
Processing times (h)	D	8	3	5
	E	4	2	4

a. When no intermediate storage is applied as the storage policy, calculate the cycle time for the production of D and E in a sequence *DEDEDE*...... Use a Gantt chart to show at least two cycles of production (i.e. *DEDE*).

b. When unlimited intermediate storage is applied for storage policy, calculate the cycle time for the production of D and E in a sequence *DEEDEEDEE*...... Use a Gantt chart to show at least two cycles of production (i.e. *DEEDEE*).

REFERENCES

1. Smith R and Petela EA (1992) Waste Minimisation in the Process Industries: Part 2 - Reactors, *Chem Eng*, **12**, 509–510:12.
2. Sharratt PN (1997) *Handbook of Batch Process Design*, Blackie Academic and Professional.
3. Zhang J and Smith R (2004) Design and Optimisation of Batch and Semi-batch Reactors, *Chem Eng Sci*, **59**: 459.
4. Harnby N, Nienow AW and Edwards MF (1997) *Mixing in the Process Industries*, Butterworth-Heineman.
5. Choong KL and Smith R (2004) Optimisation of Batch Cooling Crystallization, *Chem Eng Sci*, **59**: 313.
6. Diwekar UM (1995) *Batch Distillation – Simulation, Optimal Design and Control*, Taylor & Francis.
7. Choong KL and Smith R (2004) Novel Strategies for Optimization of Batch, Semi-batch and Heating/Cooling Evaporative Crystallization, *Chem Eng Sci*, **59**: 329.
8. Mah RSH (1990) *Chemical Process Structures and Information Flows*, Butterworth.
9. Biegler LT, Grossmann IE and Westerberg AW (1997) *Systematic Methods of Chemical Process Design*, Prentice Hall.
10. Myriantheos CM (1986) *Flexibility Targets for Batch Process Design*, M.S. Thesis, University of Massachusetts, Amherst.
11. McKetta JJ (1977) *Encyclopedia of Chemical Processing and Design*, Vol. 5, Marcel Dekker Inc, New York.

15 Heat Exchanger Networks I – Heat Transfer Equipment

Many types of heat transfer equipment are used in the process industries. By far, the most commonly used type is the shell-and-tube heat exchanger. Two of the more common types of shell-and-tube heat exchanger are illustrated in Figure 15.1. The first, shown in Figure 15.1a, shows the fluid flowing through the tubes in a *single pass*. Figure 15.1b shows a design in which the fluid flows in *two passes* through the tube-side. The two-pass arrangement can be constructed using "U" tubes to turn the tube-side fluid, rather than relying on the heat exchanger head to reverse the flow, as shown in Figure 15.1b. The fluid flowing in the shell is made to flow repeatedly across the outside of the tubes through the use of baffles. Figure 15.2a shows an ideal shell-side flow pattern using *segmented* baffles with combinations of ideal cross-flow and ideal axial flow. Cross-flow gives both higher rates of heat transfer and higher pressure drops than axial flow. In practice, the flow pattern is not ideal, as illustrated in Figure 15.2a, but leakage occurs through the tube-to-baffle clearance, as illustrated in Figure 15.2b. Also, bypassing occurs between the tube bundle and the shell and is a function of the shell-to-baffle clearance, as illustrated in Figure 15.2b. Sealing devices are usually included in the design to minimize bypassing. Both leakage and bypassing act to reduce the rate of heat transfer on the shell-side. Baffle designs other than segmented baffles are available, giving different flow patterns on the shell-side. There are many other shell-and-tube designs than those illustrated in Figure 15.1[1–4].

It should be noted that in Figure 15.1a, the hot fluid flows vertically down. This is normal practice since a hot liquid will become more dense as it is cooled, and therefore less

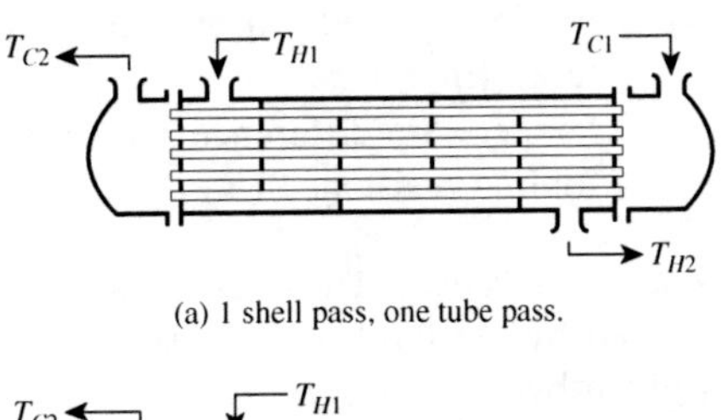

(a) 1 shell pass, one tube pass.

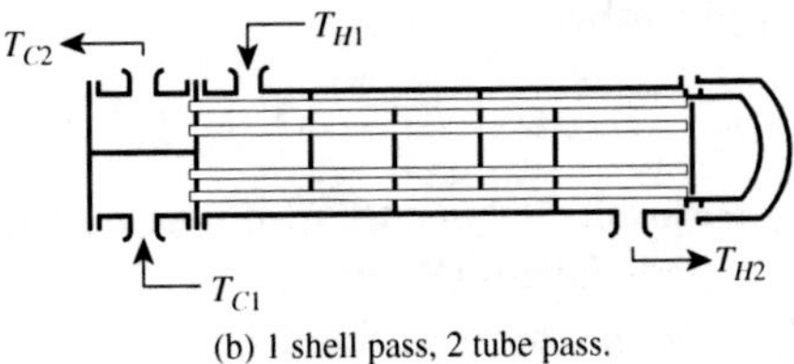

(b) 1 shell pass, 2 tube pass.

Figure 15.1 Shell-and-tube heat exchangers.

Ideal Cross Flow

Ideal Axial Flow

(a) Ideal shell-side flow.

Baffle Leakage

Bypassing

(b) Nonideal shell-side flow.

Figure 15.2 Shell-side flow patterns.

buoyant, and would tend to naturally flow downwards as a result of the buoyancy forces. Also, if some condensation of the vapor were occurring, this would also tend to flow naturally downwards. Similarly, in Figure 15.1b, the cold fluid on the tube-side of the heat exchanger flows upwards. This is because a cold liquid being heated up would become less dense and therefore more buoyant, and would tend to naturally flow upwards as a result of the buoyancy forces. Alternatively, if a liquid were being partially vaporized, any vapor would tend to naturally flow upwards.

Consider first the resistance to heat transfer across the wall of the tubes.

15.1 OVERALL HEAT TRANSFER COEFFICIENTS

Figure 15.3 illustrates the resistance to heat transfer across the wall of the tube. There are five resistances to heat transfer. Each can be characterized by a *heat transfer coefficient*.

Chemical Process Design and Integration R. Smith
 ISBNs: 0-471-48680-9 (HB); 0-471-48681-7 (PB)

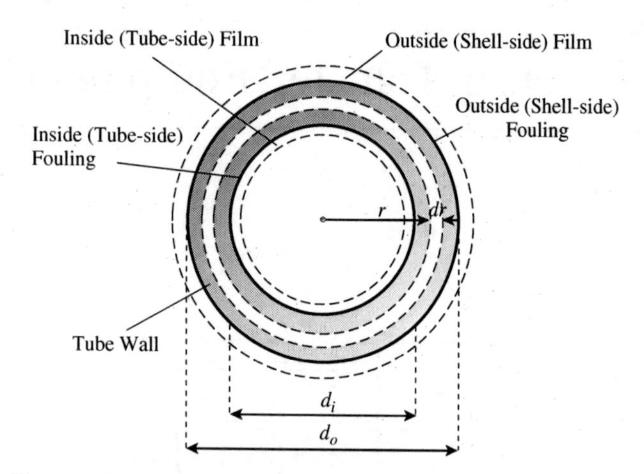

Figure 15.3 Resistance to heat transfer across the tube.

1. *Shell-side film coefficient.* The heat transfer through the resistance created by the fluid on the outside (shell-side) of the tubes is given by:

$$Q = h_S A_O \Delta T_S \tag{15.1}$$

where Q = heat transferred per unit time ($J \cdot s^{-1} = W$)
h_S = film heat transfer coefficient on the outside (shell-side) of the tubes ($W \cdot m^{-2} \cdot K^{-1}$)
A_O = heat transfer area outside (shell-side) of the tubes (m^2)
ΔT_S = temperature difference across the outside (shell-side) film (K)

2. *Shell-side fouling coefficient.* Heat transfer is usually impeded by surface deposits on the heat transfer surface (*fouling*). The material deposited as fouling usually has a low thermal conductivity. Fouling is time dependent and depends on the fluid velocity, temperature and many other factors. Fouling is difficult to predict and allowances are usually based on experience. Design is based on an assumed value to be expected after a reasonable period of time before the exchanger is cleaned.

The heat transfer through the resistance created by the outside (shell-side) fouling is quantified by a *fouling coefficient* given by:

$$Q = h_{SF} A_O \Delta T_{SF} \tag{15.2}$$

where h_{SF} = outside (shell-side) fouling coefficient ($W \cdot m^{-2} \cdot K^{-1}$)
ΔT_{SF} = temperature difference across the outside (shell-side) fouling resistance (K)

3. *Tube wall coefficient.* Heat transfer across the tube wall is described by the Fourier Equation[1]:

$$Q = -kA \frac{dT}{dr} \tag{15.3}$$

where k = thermal conductivity of the tube wall material ($W \cdot m^{-1} \cdot K^{-1}$)
r = radial distance (m)
A = heat transfer area at radial distance r(m)

Consider an incremental thickness of tube wall dr with radius r as illustrated in Figure 15.3.

$$A = 2\pi r L \tag{15.4}$$

where L = tube length (m)

Substituting Equation 15.4 into Equation 15.3 and integrating:

$$-\frac{Q}{2\pi k L} \int_{r_I}^{r_O} \frac{dr}{r} = \int_{T_I}^{T_O} dT \tag{15.5}$$

where r_O = outside tube radius (m)
r_I = inside tube radius (m)
T_O = outside surface temperature of the tube (°C)
T_I = inside surface temperature of the tube (°C)

Integrating Equation 15.5 gives:

$$-\frac{Q}{2\pi k L} \ln\left(\frac{r_O}{r_I}\right) = T_O - T_I \tag{15.6}$$

Thus:

$$Q = \frac{2\pi k L}{\ln\left(\dfrac{d_O}{d_I}\right)} \Delta T_W \tag{15.7}$$

where d_O, d_I = outside and inside tube diameters (m)
ΔT_W = temperature difference across the wall (K)

4. *Tube-side fouling coefficient.* The heat transfer through the resistance created by the inside (tube-side) fouling is given by:

$$Q = h_{TF} A_I \Delta T_{TF} \tag{15.8}$$

where h_{TF} = inside (tube-side) fouling coefficient ($W \cdot m^{-2} \cdot K^{-1}$)
A_I = inside (tube-side) heat transfer area of tubes (m^2)
ΔT_{TF} = temperature difference across the tube-side fouling resistance (K)

5. *Tube-side film coefficient.* The heat transfer through the resistance created by the fluid on the inside (tube-side) of the tubes is given by:

$$Q = h_T A_I \Delta T_T \tag{15.9}$$

where h_T = inside (tube-side) film heat transfer coefficient ($W \cdot m^{-2} \cdot K^{-1}$)
ΔT_T = temperature difference across the inside (tube-side) film (K)

The five resistances can be added. If ΔT represents the temperature difference between the bulk temperature of the fluid on the outside and inside of the tubes, then the temperature differences across the individual resistances can be added to give:

$$\Delta T = \Delta T_S + \Delta T_{SF} + \Delta T_W + \Delta T_{TF} + \Delta T_T$$
$$= \frac{Q}{h_S A_O} + \frac{Q}{h_{SF} A_O} + \frac{Q}{2\pi k L}\ln\left(\frac{d_O}{d_I}\right) + \frac{Q}{h_{TF} A_I} + \frac{Q}{h_T A_I} \quad (15.10)$$

Rearranging Equation 15.10 gives:

$$\Delta T = \frac{Q}{A_O}\left[\frac{1}{h_S} + \frac{1}{h_{SF}} + \frac{d_O}{2k}\ln\left(\frac{d_O}{d_I}\right) + \frac{d_O}{d_I}\cdot\frac{1}{h_{TF}} + \frac{d_O}{d_I}\cdot\frac{1}{h_T}\right] \quad (15.11)$$

If the overall heat transfer is written as:

$$\Delta T = \frac{Q}{A_O}\frac{1}{U} \quad (15.12)$$

where U = overall heat transfer coefficient based on the outside area of the tube ($W \cdot m^{-2} \cdot K^{-1}$), then comparing Equations 15.11 and 15.12:

$$\frac{1}{U} = \frac{1}{h_S} + \frac{1}{h_{SF}} + \frac{d_O}{2k}\ln\left(\frac{d_O}{d_I}\right) + \frac{d_O}{d_I}\cdot\frac{1}{h_{TF}} + \frac{d_O}{d_I}\cdot\frac{1}{h_T} \quad (15.13)$$

Table 15.1 lists typical values for the film transfer coefficients[1–4].

Table 15.2 gives typical values of fouling coefficients[1–4].

Fouling is often quoted as a *fouling factor*. This is simply the reciprocal of the fouling coefficient.

Table 15.1 Typical values for film transfer coefficients.

	h_S or h_T ($W \cdot m^{-2} \cdot K^{-1}$)
No change of state	
Water	2000–6000
Gases	10–500
Organic liquid (low viscosity)	1000–3000
Organic liquid (high viscosity)	100–1000
Condensing	
Steam	5000–15,000
Organic (low viscosity)	1000–2500
Organic (high viscosity)	500–1000
Ammonia	3000–6000
Evaporation	
Water	2000–10,000
Organic (low viscosity)	500–2000
Organic (high viscosity)	100–500
Ammonia	1000–2500

Table 15.2 Typical values of fouling coefficients.

	h_{SF} or h_{TF} ($W \cdot m^{-2} \cdot K^{-1}$)
Water	
Distilled	11,000
Boiler feedwater	6000–11,000
Town water	2000–5000
Well water	1000–3000
Clear river	2000–6000
Good-quality cooling water	3000–6000
Poor-quality cooling water	1000–2000
Sea	6000–11,000
Boiler blowdown	3000
Steam	
Good quality	20,000
Contaminated	5000–11,000
Liquids	
Aqueous salt solutions	3000–6000
Organic (low viscosity)	3000–11,000
Organic (high viscosity)	1000–3000
Machinery oil	6000
Fuel oils	1000
Tars	500–1000
Vegetable oils	2000
Gases	
Air	2000–4000
Organic vapor	5000–11,000

As pointed out previously, fouling is dependent on time, fluid velocity, temperature and other factors. Table 15.3 illustrates this with typical fouling coefficients for crude oil as a function of fluid velocity and temperature.

It is also interesting to consider the heat transfer coefficient of the tube wall relative to the film and fouling coefficients. Table 15.4 tabulates the coefficients for a variety of materials for some standard tube sizes. It should be noted that the sizes in Table 15.4 are not universally standard.

From Table 15.4, the heat transfer coefficient for the tube wall in many cases is so high that its contribution to the overall heat transfer coefficient can be neglected.

15.2 HEAT TRANSFER COEFFICIENTS AND PRESSURE DROPS FOR SHELL-AND-TUBE HEAT EXCHANGERS

Calculation of the overall heat transfer coefficient from Equation 15.13 requires knowledge of the film transfer coefficients. Although Table 15.1 presents typical values,

Table 15.3 Fouling coefficient for crude oil as a function of temperature and fluid velocity ($W \cdot m^{-2} \cdot K^{-1}$)[1].

Below 90°C		90 to 150°C		150 to 260°C		Over 260°C	
Less than 1 $m \cdot s^{-1}$	Greater than 1 $m \cdot s^{-1}$	Less than 1 $m \cdot s^{-1}$	Greater than 1 $m \cdot s^{-1}$	Less than 1 $m \cdot s^{-1}$	Greater than 1 $m \cdot s^{-1}$	Less than 1 $m \cdot s^{-1}$	Greater than 1 $m \cdot s^{-1}$
2000	3000	2000	3000	1000	3000	1000	2000

Table 15.4 Tube wall coefficients based on the outer diameter for a variety of metals at 100°C.

Metal	$k(W \cdot m^{-1} \cdot K^{-1})$	$\frac{2k}{d_O \ln d_O/d_I}(W \cdot m^{-2} \cdot K^{-1})$			
		$d_O = 20$ mm		$d_O = 25$ mm	
		$d_I = 16.8$ mm	$d_I = 16$ mm	$d_I = 21$ mm	$d_I = 19.8$ mm
Copper	378	216,800	169,400	173,400	129,700
Aluminum	206	118,200	92,300	94,500	70,700
Nickel	61	35,000	27,300	28,000	20,900
Cupro-nickel	45	25,800	20,200	20,600	15,400
Steel	45	25,800	20,200	20,600	15,400
Monel	30	17,200	13,400	13,800	10,300
Stainless steel	16	9200	7200	7300	5500
Titanium	16	9200	7200	7300	5500

the actual values depend on the velocities (flowrates) of the fluids, fluid physical properties and exchanger geometry. A method is needed that will allow an estimate of the film transfer coefficients for both the tube-side and the shell-side. In addition to heat transfer coefficients, it is also necessary to be able to estimate the pressure drops.

At the conceptual stage for heat exchanger network synthesis, the calculation of heat transfer coefficients and pressure drops should depend as little as possible on the detailed geometry. However, some assumptions must be made regarding the geometry.

1. *Tube diameter*. Standard sizes are used, but standards are not universal. Common sizes are $d_O = 20$ mm with $d_I = 16$ mm and $d_O = 25$ mm with $d_I = 19.8$ mm.

2. *Tube length*. Standard tube lengths are preferred, but again, standards are not universal. However, in principle, any length of tube can be used. The working length of the tube is slightly shorter than its end-to-end length from the length taken up by the tube sheets on which it is mounted. An allowance of 0.05 m for two tube sheets is a reasonable assumption in the preliminary design stage. The choice of tube length is a degree of freedom at the discretion of the designer. The same heat transfer area can be made available either with a small number of long tubes in a small diameter but long shell or a large number of short tubes in a large diameter but short shell. The ratio of tube length to shell diameter is usually in the range 5 to 10.

3. *Tube pitch*. Tube *pitch* (p_T) is the center-to-center distance between adjacent tubes and is usually $1.25d_O$.

4. *Tube configuration*. Tubes can be arranged in either a *square* or *triangular configuration*, as illustrated in Figure 15.4. A square configuration is used for fouling

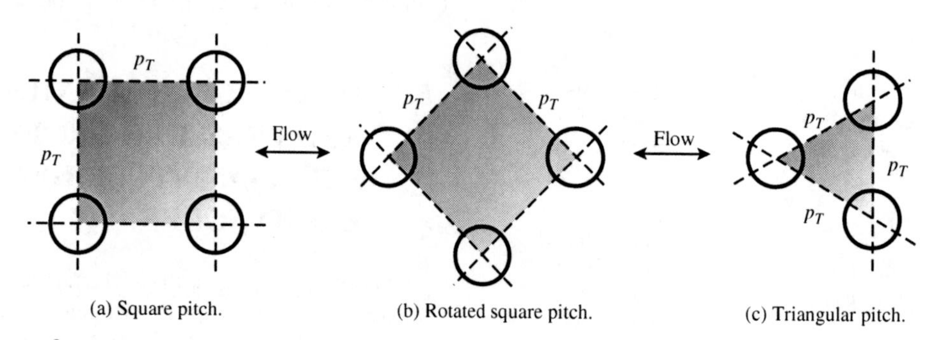

Figure 15.4 Tube configurations.

fluids to provide *lanes* between the rows for mechanical cleaning. Cleaning lanes should be continuous through the entire tube bundle. A triangular configuration is restricted to nonfouling fluids because it is more difficult to clean mechanically. However, for a given tube pitch, a triangular configuration allows a greater number of tubes in a given shell diameter. A square configuration will normally be a conservative assumption in conceptual design. For a square pitch, each tube is contained in an area of p_T^2. For a triangular pitch, each triangular pitch with sides p_T having an area of $0.5p_T^2 \sin 60^\circ$ contains half a tube. Thus, a single tube in a triangular pitch is contained in an area of $p_T^2 \sin 60^\circ = 0.866 p_T^2$. This means that for the same pitch, a triangular configuration can allow a greater number of tubes than a square configuration in the same size of shell. The assumption of a square configuration will be conservative.

5. *Baffle cut*. Baffles are used to direct the fluid stream across the tubes. The *baffle cut* is the height of the segment removed to form the baffle as a fraction of the baffle disc diameter. Baffle cuts from 0.15 to 0.45 are used. In conceptual design, a value of 0.25 can be assumed.

Simple models are developed in Appendix C in which heat transfer coefficient and pressure drop are both related to velocity[5]. It is then possible to derive a correlation between the heat transfer coefficient, pressure drop and the surface area by using velocity as the bridge between the two[5].

For the tube-side:

$$h_T = K_{hT} v_T^{0.8} \tag{15.14}$$

$$\Delta P_T = K_{PT} \frac{A\, d_I^2}{4F_I\, d_O} v_T^{2.8} + 1.25 N_{PT} \rho v_T^2 \tag{15.15}$$

$$\Delta P_T = K_{PT1} A h_T^{3.5} + K_{PT2} h_T^{2.5} \tag{15.16}$$

where

$$K_{hT} = C \left(\frac{k}{d_I}\right) Pr^{1/3} \left(\frac{d_I \rho}{\mu}\right)^{0.8} \tag{15.17}$$

$$K_{PT} = 0.092 \rho^{0.8} \mu^{0.2} d_I^{-1.2} \tag{15.18}$$

$$K_{PTI} = \frac{0.023 \rho^{0.8} \mu^{0.2} d_I^{0.8}}{F_I\, d_O} \left[\frac{1}{K_{hT}}\right]^{3.5} \tag{15.19}$$

$$K_{PT2} = 1.25 N_{TP} \rho \left[\frac{1}{K_{hT}}\right]^{2.5} \tag{15.20}$$

h_T = tube-side film transfer coefficient
v_T = fluid velocity for the tube-side
ΔP_T = tube-side pressure drop
N_{PT} = number of tube passes
L = tube length
ρ = fluid density
A = heat transfer area
C = constant (0.021 for gases, 0.023 for nonviscous liquids, 0.027 for viscous liquids)
k = fluid thermal conductivity
d_I = inside(tube-side)diameter
Pr = Prandtl Number $= \dfrac{C_P \mu}{k}$
C_P = fluid heat capacity
μ = fluid viscosity at the bulk fluid temperature
F_I = inside(tube-side)volumetric flowrate
d_O = outside(shell-side)diameter

For the shell-side:

$$h_S = K_{hS} v_S^{0.64} \tag{15.21}$$

$$\Delta P_S = 2K_{PS1} v_S^{1.83} + K_{PS2} A v_S^{2.83} - K_{PS3} v_S^{1.83} + K_{PS4} A v_S^3 \tag{15.22}$$

$$\Delta P_S = K_{S1} h_S^{2.86} + K_{S2} A h_S^{4.42} + K_{S3} A h_S^{4.69} \tag{15.23}$$

where

$$K_{hS} = \frac{0.24 F_{hn} F_{hw} F_{hb} F_{hL} \rho^{0.64} C_P^{\frac{1}{3}} k^{\frac{2}{3}}}{\mu^{0.307} d_O^{0.36}} \tag{15.24}$$

$$K_{PS1} = \frac{1.298 F_{Pb} D_S (1 - B_C) \rho^{0.83} \mu^{0.17}}{p_T\, d_O^{0.17}} \tag{15.25}$$

$$K_{PS2} = \frac{0.5261 F_{Pb} F_{PL} p_C (1 - 2B_C)(p_T - d_O) \rho^{0.83} \mu^{0.17}}{d_O^{1.17} F_O} \tag{15.26}$$

$$K_{PS3} = \frac{2.596 F_{Pb} F_{PL} (1 - 2B_C) D_S \rho^{0.83} \mu^{0.17}}{p_T\, d_O^{0.17}} \tag{15.27}$$

$$K_{PS4} = \frac{0.2026 F_{PL} p_C p_T (p_T - d_O) \rho}{d_O F_O} \left(\frac{2}{D_S} + \frac{0.6 B_C}{p_T}\right) \tag{15.28}$$

$$K_{S1} = \frac{2K_{PS1} - K_{PS3}}{K_{hS}^{2.86}} \tag{15.29}$$

$$K_{S2} = \frac{K_{PS2}}{K_{hS}^{4.42}} \tag{15.30}$$

$$K_{S3} = \frac{K_{PS4}}{K_{hS}^{4.69}} \tag{15.31}$$

where
h_S = shell-side film transfer coefficient
v_S = fluid velocity for the shell-side

$$= \frac{F_O D_S}{p_T}(p_T - d_O) L_B \tag{15.32}$$

F_O = outside (tube-side) volumetric flowrate
B_C = baffle cut
D_S = shell diameter
p_C = pitch configuration factor ($p_C = 1$ for square pitch, $p_C = 0.866$ for triangular pitch)

p_T = tube pitch (i.e. center-to-center distance between adjacent tubes)
L_B = distance between baffles
F_{hn} = correction factor to allow for the effect of the number of tube rows crossed (see Appendix C)
F_{hw} = the window correction factor (see Appendix C)
F_{hb} = the bypass stream correction factor (see Appendix C)
F_{hL} = the leakage correction factor (see Appendix C)
F_{Pb} = bypass correction factor for pressure drop to allow for flow between the tube bundle and the shell wall and is a function of the shell to bundle clearance (see Appendix C)
F_{PL} = leakage correction factor to allow for leakage through the tube-to-baffle clearance and the baffle-to-shell clearance (see Appendix C)

The shell-side correction factors $F_{hn}, F_{hw}, F_{hb}, F_{hL}, F_{Pb}$ and F_{PL} are adjustable parameters that characterize the flow pattern on the shell-side. The values depend on the design of the heat exchanger, the fluids being heated and cooled and the fouled condition on the shell-side. For preliminary design in the clean condition, reasonable values are $F_{hn} = F_{hw} = 1$, $F_{hb} = F_{hL} = 0.8$, $F_{Pb} = 0.8$ and $F_{PL} = 0.5$. As the shell-side fouls, bypassing and leakage will decrease, and the factors will tend towards 1 (but not necessarily reach 1). Choosing the most appropriate values is at the discretion of the designer.

Before these equations can be applied, a number of issues need to be considered:

1. *Shell diameter.* The equations for the shell-side feature the shell diameter, which is unknown at the conceptual design stage. This can be eliminated by considering the heat transfer area.

$$A = N_T \pi d_O L \tag{15.33}$$

where A = heat transfer area based on the tube outside surface (m^2)
N_T = total number of tubes (–)
d_O = tube outside diameter (m)
L = tube length (m)

For a square pitch, each tube is contained in an area of p_T^2. The number of tubes can then be approximated as:

$$N_T = \frac{\frac{\pi}{4} D_S^2}{p_T^2} \tag{15.34}$$

For a triangular pitch, each tube is contained in an area of $0.866 p_T^2$. The number of tubes can then be generalized as:

$$N_T = \frac{\frac{\pi}{4} D_S^2}{p_C p_T^2} \tag{15.35}$$

where p_C = pitch configuration factor
= 1 for square pitch
= 0.866 for triangular pitch

In practice, the number of tubes in a given shell diameter will be less than that predicted by Equation 15.35 because of the mechanical features such as pass partition plates, not accounted for. Combining Equations 15.33 and 15.35 gives:

$$A = \frac{\pi^2 d_O D_S^2 L}{4 p_C p_T^2} \tag{15.36}$$

Equation 15.36 can be rearranged to give:

$$D_S = \left(\frac{4 p_C p_T^2 A}{\pi^2 d_O L} \right)^{\frac{1}{2}} \tag{15.37}$$

Equation 15.37 can be used to approximate the shell diameter if the tube length L is specified. This might be preferred, as standard tube lengths will tend to be used. Alternatively, the tube length to shell diameter (L/D_S) can be specified. If this is specified, then Equation 15.36 can be rearranged to give:

$$D_S = \left(\frac{4 p_C p_T^2 A}{\pi^2 d_O (L/D_S)} \right)^{\frac{1}{3}} \tag{15.38}$$

The ratio of tube length to shell diameter is usually in the range 5 to 10.

2. *Fluid velocity.* Equations 15.14, 15.15, 15.21 and 15.22 require knowledge of the fluid velocity. Liquid velocity on the tube-side is usually of the order of 1 to 3 $m{\cdot}s^{-1}$. On the shell-side, liquid velocity is usually of the order of 0.5 to 2 $m{\cdot}s^{-1}$. Gas velocity on both the tube-side and the shell-side typically vary in the range 5 to 70 $m{\cdot}s^{-1}$. The higher the pressure, the lower the gas velocity.

High velocities will result in high heat transfer coefficients and will tend to reduce fouling. However, high velocities also result in a high pressure drop. Also, if solids are present in the fluids, then high velocities can result in erosion. High velocities on the shell-side for liquids (typically greater than 3 $m{\cdot}s^{-1}$) can also induce vibration within the heat exchanger that can cause long-term damage to the heat exchanger.

3. *Pressure drop.* Rather than specify a fluid velocity, it is often preferred to specify the pressure drop across the heat exchanger. In retrofit situations, where a new heat exchanger is to be installed in an existing plant, the allowable pressure drop is often highly constrained. This is because in a retrofit situation, it is often desirable to avoid

the installation of new pumps. In the absence of a specific retrofit constraint for pressure drop, the value for liquids is normally in the range 0.35 to 0.7 bar. For low-viscosity liquids (less than 1 $mN{\cdot}s{\cdot}m^{-2}$), the pressure drop is likely to be less than 0.35 bar. For gases, the maximum allowable pressure drop varies typically between 1 bar for high-pressure gases (10 bar and above) down to 0.01 bar for gases under vacuum conditions. For gases, the cost associated with the pressure drop is more sensitive to aspects of the design outside of the immediate considerations for the heat exchanger design than is the case for liquids.

4. *Fouling*. Fouling is the accumulation of undesired material at the heat transfer surfaces. It is a transient process that begins with a clean heat transfer surface and continues until the point where the surface becomes fouled to the point where the heat exchanger can no longer be used effectively. At this point, the heat exchanger must be taken out of service and cleaned. Cleaning can be accomplished by mechanical or chemical processes. Mechanical cleaning involves the use of high-pressure water jets and so on. Chemical cleaning exploits cleaning fluids that react with and/or dissolve surface deposits. This usually involves creating a cleaning circuit involving the heat exchanger and a pump in which the cleaning fluid is recirculated through the heat exchanger at a relatively high velocity.

In some instances (e.g. cooling water circuits), chemicals can be added to the heat transfer fluids to inhibit fouling. The chemicals added to inhibit fouling and the means of cleaning depend on the nature of the fouling. Fouling can be classified as:

- *particulate*, in which solid particles suspended in the process stream are carried to the heat transfer surface and accumulate;
- *scaling*, in which solid material is precipitated from solution on the heat transfer surface through inverse solubility (e.g. calcium carbonate deposit from hardness in water);
- *crystallization*, in which solid material is crystallized from solution as a result of the change in temperature and/or the presence of nucleation sites at the surface;
- *freezing*, in which material is decreased in temperature locally to below its freezing point;
- *chemical reaction*, which results from reactions at the heat transfer surface (e.g. polymerization and cracking reactions);
- *corrosion*, in which the heat transfer surface is exposed to a corrosive fluid that reacts to produce byproducts of corrosion on the surface;
- *biological fouling* or *biofouling*, in which a layer of microorganisms grows on the heat transfer surface producing *slime*.

The rate at which fouling occurs depends on many factors, but the two most important are temperature and fluid velocity. Fouling can be removed from the surface by high surface shear stress created by high fluid velocity. Indeed, for certain types of fouling, it has been demonstrated that for a given surface temperature, there is a threshold velocity above which fouling will not occur.

As pointed out in Section 15.1, fouling adds additional resistances to heat transfer on both the tube-side and the shell-side. If one of the fluids in the heat exchanger is significantly more fouling than the other, the fouling fluid is normally placed on the tube-side. If this fouling fluid placed on the tube-side has a particularly high tendency to foul (e.g. crude oil), then designing for a low pressure drop can be a false economy. As the fluid fouls the tube-side, not only will the overall heat transfer coefficient decrease significantly but the pressure drop will also increase significantly as a direct result of the fouling. If, for this highly fouling fluid, the exchanger had been designed for an initial high pressure drop on the tube-side, the resulting high velocity would not only have increased the initial rate of heat transfer, but also decreased the rate of deterioration of the rate of heat transfer and the rate of increase of pressure drop. An initial design for a tube-side pressure drop of 0.7 bar for a highly fouling fluid (e.g. crude oil) might result in a fouled pressure drop of say 4 to 5 bar after a year of operation. Had the initial design been for, say, 2.5 bar, this could have led to a lower pressure drop after a year of operation and an overall better performance.

In addition to the use of antifouling chemical agents to mitigate the effect of fouling on the tube-side, *twisted tubes* can be used rather than plain tubes. These have surface irregularities. Plain tubes can also be fitted on the inside with *tube inserts*. Twisted tubes and tube inserts promote additional turbulence and pressure drop and reduce the surface temperature of the tube to mitigate fouling. Tube inserts will be dealt with in more detail later.

The effect of fouling on the shell-side is more complex. On the shell-side, the heat transfer and pressure drop are also affected by the fouling changing the flow pattern. Bypassing and leakage reduces both the shell-side heat transfer coefficient and pressure drop. Fouling will tend to block the clearances, increasing the amount of cross-flow. This will have the effect of increasing the shell-side heat transfer coefficient and pressure drop. This increase will be countered by film resistances of the heat transfer surfaces. Initially, the fouling might even increase the heat transfer coefficient by promoting better cross-flow, only to decrease it later as the film resistance increases.

The segmented baffles normally used in shell-and-tube heat exchangers are not the best arrangement for fouling fluids placed on the shell-side, as stagnant zones are created by the flow pattern. Rather than use segmented baffles, helical flow baffles can be used, which induce a spiral flow pattern along the exchanger on the shell-side, eliminating the stagnant zones.

5. *Standard dimensions.* Applying the equations described here will result in certain specifications for heat transfer area, tube length and shell diameter. In practice, standard tube lengths, shell diameters and tube layouts are used. The preliminary design would then need to be adjusted to meet standard size and layout specifications.

15.3 TEMPERATURE DIFFERENCES IN SHELL-AND-TUBE HEAT EXCHANGERS

Consider the heat exchange process in Figure 15.5a. The heat transfer arrangement shows a *concentric pipe* heat exchanger. The flow of the fluids in such a device is truly countercurrent. To understand the heat transfer, assume that it is a steady state process in which all of the fluid properties and overall heat transfer coefficient U are constant. It is also assumed that there is no phase change and that there is no heat loss from the system. The heat transfer process is shown in Figure 15.5b in a plot of heat transfer versus temperature. The slope of the ΔT line in Figure 15.5b is given by:

$$\frac{d(\Delta T)}{dQ} = \frac{\Delta T_H - \Delta T_C}{Q} \tag{15.39}$$

Applying Equation 15.39 across the differential element in Figure 15.5b:

$$dQ = U \cdot dA \cdot \Delta T \tag{15.40}$$

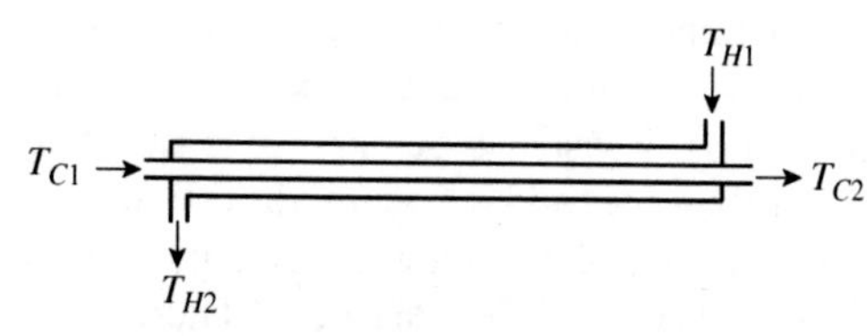

(a) Countercurrent heat exchange.

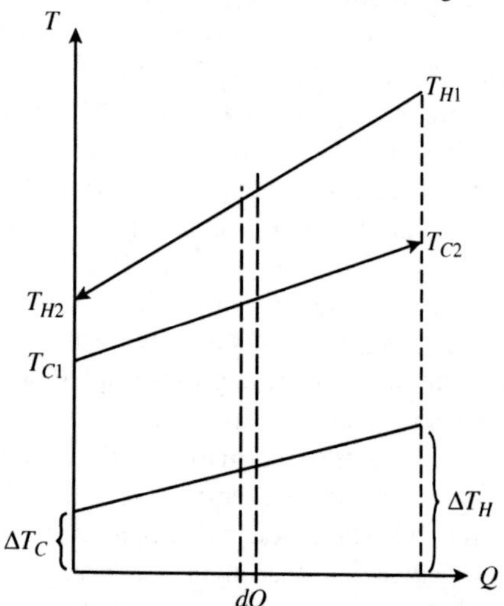

(b) Heat transfer versus temperature.

Figure 15.5 Heat exchange temperature difference.

Combining Equations 15.39 and 15.40 gives:

$$\frac{\Delta T_H - \Delta T_C}{Q} = \frac{d(\Delta T)}{U \cdot dA \cdot \Delta T} \tag{15.41}$$

Rearranging and integrating gives:

$$\int_o^A dA = \frac{Q}{\Delta T_H - \Delta T_C} \cdot \frac{1}{U} \int_{\Delta T_C}^{\Delta T_H} \frac{d(\Delta T)}{\Delta T} \tag{15.42}$$

Thus:

$$A = \frac{Q}{\Delta T_H - \Delta T_C} \cdot \frac{1}{U} \ln \frac{\Delta T_H}{\Delta T_C} \tag{15.43}$$

$$Q = UA \left[\frac{\Delta T_H - \Delta T_C}{\ln \dfrac{\Delta T_H}{\Delta T_C}} \right] \tag{15.44}$$

$$= UA\Delta T_{LM} \tag{15.45}$$

where ΔT_{LM} = logarithmic mean temperature difference

Note that this result would have been obtained if

a. the fluid positions inside and outside of the tube in Figure 15.5b had been reversed;
b. the direction of one of the fluids had been reversed giving cocurrent flow;
c. either fluid had been isothermal.

If both fluids had been isothermal, then the logarithmic and average temperature differences would be equal and $\Delta T_H = \Delta T_C = \Delta T$. It should also be noted that:

$$\frac{\Delta T_H - \Delta T_C}{\ln \dfrac{\Delta T_H}{\Delta T_C}} = \frac{\Delta T_C - \Delta T_H}{\ln \dfrac{\Delta T_C}{\Delta T_H}} \tag{15.46}$$

Although the result in Equation 15.45 applies to both countercurrent and cocurrent flow, in practice, cocurrent flow is almost never used as, given fixed fluid inlet and outlet temperatures, the logarithmic mean temperature difference for countercurrent flow is always larger. This in turn leads to smaller surface area requirements. Also, as shown in Figure 15.6a for countercurrent flow, the final temperature of the hot fluid can be lower than the final temperature of the cold fluid (sometimes known as *temperature cross*), whereas in Figure 15.6b, it is clear that there can never be a temperature cross.

For a given heat duty and overall heat transfer coefficient, the 1–1 design (1 shell pass – 1 tube pass), as illustrated in Figure 15.7a, offers the lowest requirement for surface area for shell-and-tube heat exchangers. Many flow arrangements other than the 1–1 design exist, the most common of which is the 1–2 design (1 shell pass – 2 tube passes), as illustrated in Figure 15.7b. Because the flow arrangement involves part countercurrent and part cocurrent flow, the effective temperature difference for heat exchange is

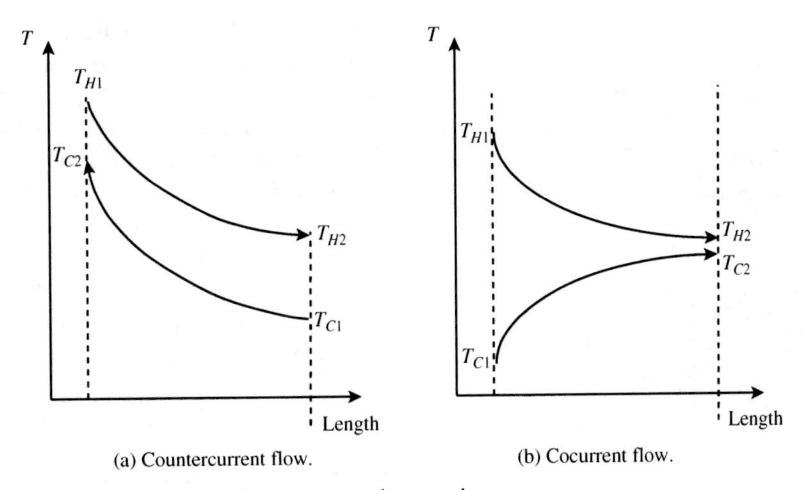

Figure 15.6 Fluid temperatures can never cross in a cocurrent heat exchanger.

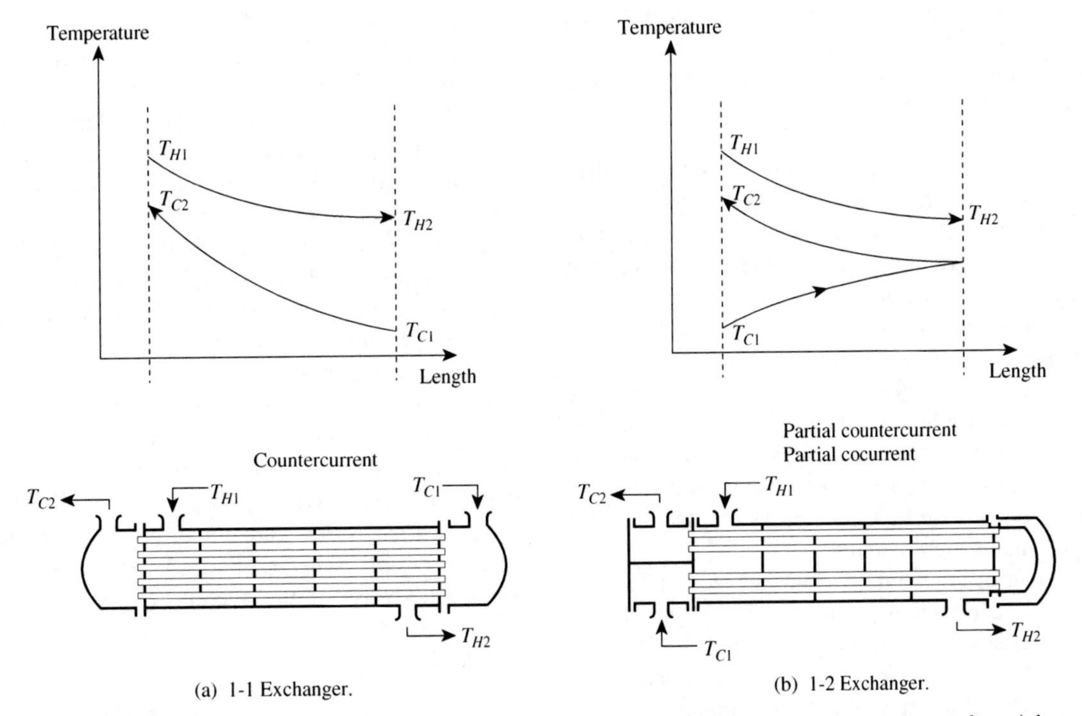

Figure 15.7 1-1 shells approach pure countercurrent flow, whereas 1-2 shells exhibit partial countercurrent and partial cocurrent flow.

reduced compared to a pure countercurrent device. This is accounted for, in design, by the introduction of the F_T factor into the basic heat exchanger design equation[6]:

$$Q = UA\Delta T_{LM} F_T \quad \text{where} \quad 0 < F_T < 1 \qquad (15.47)$$

Thus, for a given exchanger duty and overall heat transfer coefficient, the 1–2 design needs a larger area than the 1–1 design. However, the 1–2 design offers many practical advantages. These include, in particular, allowance for thermal expansion, easy mechanical cleaning and good heat transfer coefficients on the tube-side (due to higher velocity).

The F_T correction factor is usually correlated in terms of two dimensionless ratios, the ratio of the two heat capacity flowrates (R) and the thermal effectiveness of the exchanger (P)[6]:

$$F_T = f(R, P) \qquad (15.48)$$

where

$$R = CP_C/CP_H = (T_{H1} - T_{H2})/(T_{C2} - T_{C1}) \qquad (15.49)$$

and

$$P = (T_{C2} - T_{C1})/(T_{H1} - T_{C1}) \qquad (15.50)$$

Note therefore that F_T depends only on the inlet and outlet temperatures of the streams in a 1–2 heat exchanger. An expression can be developed for the F_T in a 1–2 heat exchanger. The proof is lengthy and can be found elsewhere[1,6].

For $R \neq 1$:

$$F_T = \frac{\sqrt{R^2+1}\ln\left[\dfrac{(1-P)}{(1-RP)}\right]}{(R-1)\ln\left[\dfrac{(2-P(R+1-\sqrt{R^2+1}))}{(2-P(R+1+\sqrt{R^2+1}))}\right]} \quad (15.51)$$

For $R = 1$:

$$F_T = \frac{\left[\dfrac{\sqrt{2}P}{(1-P)}\right]}{\ln\left[\dfrac{(2-P(2-\sqrt{2}))}{(2-P(2+\sqrt{2}))}\right]} \quad (15.52)$$

The shape of this function is illustrated in Figure 15.8. It can be seen that the slope of the F_T curve for a given R becomes very steep and approaches an asymptote as the thermal effectiveness P increases.

Three basic situations can be encountered when using 1–2 exchangers (Figure 15.8):

1. The final temperature of the hot stream is higher than the final temperature of the cold, as illustrated in Figure 15.8a. This is a so-called *temperature approach.* This situation is straightforward to design for, since it can always be accommodated in a single 1–2 shell.
2. The final temperature of the hot stream is slightly lower than the final temperature of the cold stream, as illustrated in Figure 15.8b. This is a so-called *temperature cross*. This situation is usually straightforward to design for, provided the temperature cross is small, because again it can probably be accommodated in a single shell. However, the decrease in F_T increases the heat transfer area requirements significantly.
3. As the amount of temperature cross increases, however, problems are encountered as illustrated in Figure 15.8c. The F_T decreases significantly, causing a dramatic increase in the heat transfer area requirements. Local reversal of heat flow may also be encountered, which is wasteful in heat transfer area. The design may even become infeasible. Thus, for a given R, the design of the heat exchanger becomes less and less efficient as the asymptote for the F_T curve is approached.

The maximum temperature cross that can be tolerated is often set by rules of thumb, for example, $F_T > 0.75$[1]. It is important to avoid low values of F_T because:

1. Low values of F_T indicate inefficient use of the heat transfer area.
2. Any violation of the simplifying assumptions used in the approach tends to have a particularly significant effect in areas of the F_T chart where slopes are particularly steep.
3. Any uncertainties or inaccuracies in design data also have a more significant effect when slopes are steep.

Consequently, to be confident in a design, those parts of the F_T chart where slopes are steep should be avoided, irrespective of $F_T > 0.75$[7]. A simple method to achieve this is based upon the fact that for any value of R there is a maximum asymptotic value for P, say P_{max}, which is given as F_T tends to $-\infty$, and is given by[7]:

$$P_{max} = \frac{2}{R+1+\sqrt{R^2+1}} \quad (15.53)$$

Equation 15.53 is derived in Appendix D. Practical designs will be limited to some fraction of P_{max}, that is[7]:

$$P = X_P P_{max},\ 0 < X_P < 1 \quad (15.54)$$

where X_P is a constant defined by the designer.

A line of constant X_P is compared with a line of constant F_T in Figure 15.9[8]. It can be seen that the line of constant X_P avoids the regions of steep slope.

Situations are often encountered where the design is infeasible in a single 1–2 shell, because the F_T is too low or the F_T slope too large. If this happens, either different types of shell or multiple shell arrangements must be considered[1–3]. Here, consideration will be restricted to multiple shell arrangements of the 1–2 type. By using two 1–2 shells in series (Figure 15.10), the temperature cross in each individual shell is reduced below that for a single 1–2 shell for the same duty. The profiles shown in Figure 15.10 could in principle be achieved either by two 1–2 shells in series or by a single 2–4 shell.

For a number of 1–2 shells in series, a transformation can be developed based on the fact that for N_{SHELLS} in series, each shell pass has the same value of F_T, which also equals the F_T across all N_{SHELLS} passes[6]. Also, all values of P of each shell pass (P_{1-2}) are equal, but not equal to the value of P across all N_{SHELLS} (P_{N-2N}). Of course, R is constant across all shells and the overall design.

For $R \neq 1$[6]:

$$P_{N-2N} = \frac{1-\left[\dfrac{1-P_{1-2}R}{1-P_{1-2}}\right]^{N_{SHELLS}}}{R-\left[\dfrac{1-P_{1-2}R}{1-P_{1-2}}\right]^{N_{SHELLS}}} \quad (15.55)$$

For $R = 1$[6]:

$$P_{N-2N} = \frac{P_{1-2}N_{SHELLS}}{P_{1-2}N_{SHELLS} - P_{1-2} + 1} \quad (15.56)$$

Thus, given an overall value of P_{N-2N} for the duty involving N_{SHELLS}, Equations 15.55 and 15.56 allow the value of P_{1-2} to be calculated for each shell. To do this, first define a variable Z:

$$Z = \left[\frac{1-P_{1-2}R}{1-P_{1-2}}\right]^{N_{SHELLS}} \quad (15.57)$$

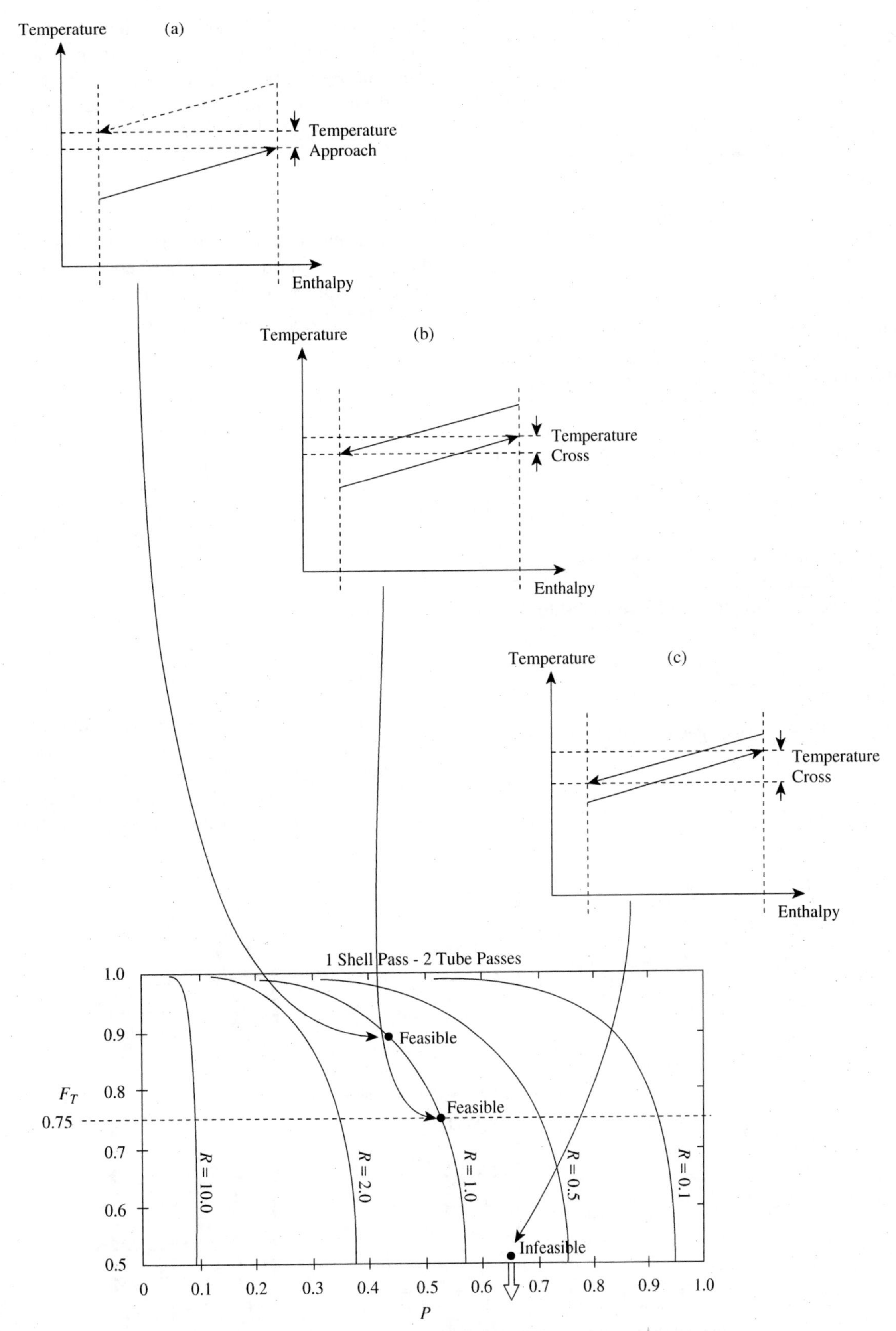

Figure 15.8 Designs with a temperature approach or small temperature cross can be accommodated in a single 1-2 shell, whereas designs with a large temperature cross become infeasible. (From Ahmad S, Linnhoff B and Smith R, 1988, *Trans ASME J Heat Transfer*, **110**: 304, reproduced by permission of the American Society of Mechanical Engineers).

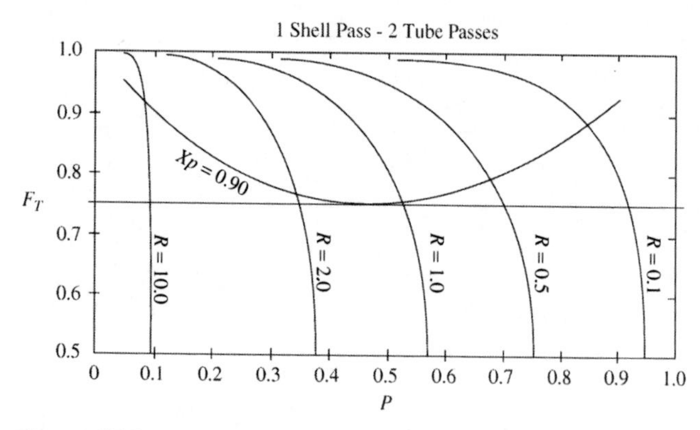

Figure 15.9 The X_P parameter avoids steep slopes on the F_T curves, whereas minimum F_T does not. (From Ahmad S, Linnhoff B and Smith R, 1990, *Computers and Chem Eng*,**7**: 751, reproduced by permission of Elsevier Ltd).

Substituting Z into Equation 15.55 and rearranging gives:

$$Z = \frac{1 - P_{N-2N} R}{1 - P_{N-2N}} \tag{15.58}$$

Thus, starting with the overall P_{N-2N} for the duty, Z is first calculated from Equation 15.58. Then P_{1-2} is calculated for each shell by inverting Equation 15.57 and substituting the value of Z:

$$P_{1-2} = \frac{Z^{1/N_{SHELLS}} - 1}{Z^{1/N_{SHELLS}} - R} \quad for R \neq 1 \tag{15.59}$$

For $R = 1$, a simple rearrangement of Equation 15.56 gives:

$$P_{1-2} = \frac{P_{N-2N}}{P_{N-2N} - N_{SHELLS} P_{N-2N} + N_{SHELLS}} \tag{15.60}$$

Traditionally, the designer would approach a design for an individual unit by trial and error. Starting by assuming one shell, the F_T can be evaluated. If the F_T is not acceptable, then the number of shells in series is progressively increased until a satisfactory value of F_T is obtained for each shell. For a number of 1–2 shells in series:

$$F_{1-2} < F_{2-4} < F_{3-6} < F_{4-8} \ldots\ldots\ldots < 1 \tag{15.61}$$

Adopting the design criterion given by Equation 15.54 as the basis, any need for trial and error can be eliminated since an explicit expression for the number of shells for a given unit is derived in Appendix E[7].

$R \neq 1$:

$$N_{SHELLS} = \frac{\ln\left(\dfrac{1 - RP}{1 - P}\right)}{\ln W} \tag{15.62}$$

where

$$W = \frac{R + 1 + \sqrt{R^2 + 1} - 2RX_P}{R + 1 + \sqrt{R^2 + 1} - 2X_P} \tag{15.63}$$

$R = 1$:

$$N_{SHELLS} = \frac{\left(\dfrac{P}{1 - P}\right)\left(1 + \dfrac{\sqrt{2}}{2} - X_P\right)}{X_P} \tag{15.64}$$

X_P is chosen to satisfy the minimum allowable F_T (for example, for $F_{Tmin} > 0.75$, $X_P = 0.9$ is used). Once the real (noninteger) number of shells is calculated from Equation 15.62 or 15.64, this is rounded up to the next largest number to obtain the number of shells. Generally,

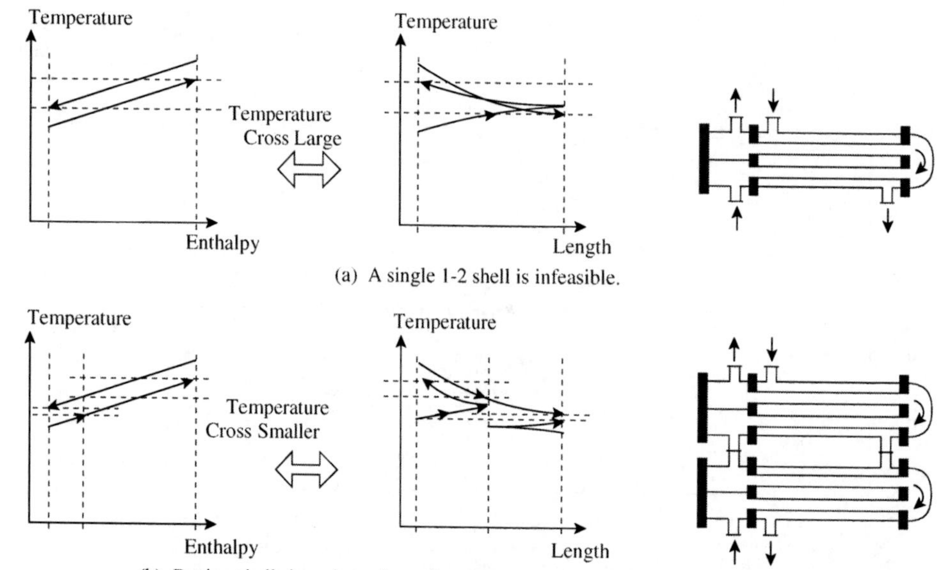

Figure 15.10 A large overall temperature cross requires shells in series to reduce the cross in individual exchangers. (From Ahmad S, Linnhoff B and Smith R, 1988, *Trans ASME J Heat Transfer*, **110**: 304, reproduced by permission of the American Society of Mechanical Engineers).

the smaller the number of shells for a given overall duty, the cheaper will be the design. The higher the value of X_P chosen, the larger will be the number of shells, but the safer the design. Thus, a compromise is required. A value of $X_P = 0.9$ is reasonable for conceptual design.

In addition to the F_T limiting the temperature cross in each shell and dividing the overall duty into a number of shells, there is a maximum physical size that can be fabricated in a single shell. For shell-and-tube heat exchangers with removable tube bundles, the maximum size of shell is around 1000 m^2. However, a significantly lower figure might well be preferred for maintenance and cleaning purposes. Fixed bundle heat exchangers can be much larger, typically up to 4500 m^2.

Example 15.1 A hot stream is to be cooled from 300 to 100°C by exchange with a cold stream being heated from 60 to 200°C in a single unit. 1–2 shell-and-tube heat exchangers are to be used subject to $X_P = 0.9$. The duty for the exchanger is 3.5 MW and the overall heat transfer coefficient is estimated to be 100 W·m^{-2}·K^{-1}. Calculate:

a. number of shells required
b. P_{1-2} for each shell
c. F_T for the shells in series
d. heat transfer area

Solution

a.
$$R = \frac{T_{H1} - T_{H2}}{T_{C2} - T_{C1}} = \frac{300 - 100}{200 - 60} = 1.4286$$

$$P_{N-2N} = \frac{T_{C2} - T_{C1}}{T_{H1} - T_{C1}} = \frac{200 - 60}{300 - 60} = 0.5833$$

$$W = \frac{R + 1 + \sqrt{R^2 + 1} - 2RX_P}{R + 1 + \sqrt{R^2 + 1} - 2X_P} = 0.6748$$

$$N_{SHELLS} = \frac{\ln\left[\dfrac{(1 - RP_{N-2N})}{(1 - P_{N-2N})}\right]}{\ln W} = 2.33$$

Thus, the unit requires three shells.

b.
$$Z = \frac{1 - P_{N-2N}R}{1 - P_{N-2N}} = \frac{1 - 0.5833 \times 1.4286}{1 - 0.5833} = 0.4$$

$$P_{1-2} = \frac{Z^{1/N_{SHELLS}} - 1}{Z^{1/N_{SHELLS}} - R} = \frac{(0.4)^{\frac{1}{3}} - 1}{(0.4)^{\frac{1}{3}} - 1.4286} = 0.3805$$

c.
$$F_T = \frac{\sqrt{R^2 + 1}\ln\left[\dfrac{(1 - P)}{(1 - RP)}\right]}{(R - 1)\ln\left\{\dfrac{[2 - P(R + 1 - \sqrt{R^2 + 1})]}{[2 - P(R + 1 + \sqrt{R^2 + 1})]}\right\}}$$

Substituting $R = 1.4286$ and $P = 0.3805$:

$$F_T = 0.86$$

d.
$$A = \frac{Q}{U\Delta T_{LM}F_T} = \frac{3.5 \times 10^6}{100 \times 65.48 \times 0.86} = 619 \text{ m}^2$$

15.4 ALLOCATION OF FLUIDS IN SHELL-AND-TUBE HEAT EXCHANGERS

In a new design, the allocation of streams to the tube-side or shell-side will be unknown at this stage. The issues to be considered for the allocation include:

a. *Materials of construction.* If expensive materials of construction are required for one of the fluids because of its corrosive nature or high temperature, then this fluid should normally be allocated to the tube-side. This can reduce the cost of expensive materials of construction.

b. *Fouling.* The fluid that has the greatest tendency to foul the heat transfer surfaces should be allocated to the tube-side. This will give better control over the design fluid velocity, and the higher allowable velocity in the tubes will reduce fouling. Stagnant zones and zones with low velocity can occur on the shell-side, leading to accelerated fouling. Also, the tubes are easier to clean than the shell-side.

c. *Operating pressures.* The stream with the higher pressure should be allocated to the tube-side. The smaller the diameter of a tube, the thinner the wall needed to contain the same pressure. The tubes are therefore more effective at containing high pressure than the shell.

d. *Pressure drop.* For the same pressure drop, higher heat transfer coefficients will be obtained on the tube-side than the shell-side. The fluid with the lowest allowable pressure drop should normally be allocated to the tube-side.

e. *Viscosity.* Generally, a higher heat transfer coefficient will be obtained by allocating the more viscous material to the shell-side, provided the flow is turbulent. The critical Reynolds number for turbulent flow on the shell-side is in the region of 200.

f. *Stream flowrates.* Allocating the fluid with the lower flowrate to the shell-side will normally allow a higher heat transfer coefficient to be obtained for that fluid and give the most economic design.

g. *Fluid temperatures.* Placing the hotter fluid in the tubes will reduce the shell surface temperature, and hence the need for lagging to reduce heat loss and might be desirable for safety reasons.

These general guidelines can contradict each other. If this is the case, some compromise must be made.

Example 15.2 A crude oil stream is to be preheated by recovering heat from a kerosene product in a shell-and-tube heat exchanger. The flowrates, temperatures and physical properties (at the mean temperatures) are given in Table 15.5.

Table 15.5 Data for a heat recovery problem.

	Kerosene	Crude oil
Flowrate ($m^3 \cdot s^{-1}$)	0.037	0.103
Initial temperature (°C)	200	35
Final temperature (°C)	95	75
Density ($kg \cdot m^{-3}$)	730	830
Heat capacity ($J \cdot kg^{-1} \cdot K^{-1}$)	2470	2050
Viscosity ($N \cdot s \cdot m^{-2}$)	4.0×10^{-4}	3.6×10^{-3}
Thermal conductivity ($W \cdot m^{-1} \cdot K^{-1}$)	0.132	0.133

The exchanger type to be used is 1 shell-pass, 2 tube-passes. Because crude oil will have a greater tendency to foul than kerosene, crude oil will be allocated to the tube-side. Assume the fouling coefficient for kerosene on the shell-side is 5000 $W \cdot m^{-2} \cdot K^{-1}$ and for the crude oil on the tube-side is 2000 $W \cdot m^{-2} \cdot K^{-1}$.

Steel tubes will be used and the following assumptions can be made regarding the heat exchanger geometry:

$$d_O = 20 \text{ mm}$$

$$d_I = 16 \text{ mm}$$

$$p_T = 1.25 d_O$$

$$p_C = 1 \text{(square pitch)}$$

$$B_C = 0.25$$

$$L/D_S = 7$$

a. Calculate the number of shells required and the F_T based on $X_P = 0.9$.

b. Estimate the overall heat transfer coefficient, heat transfer area and pressure drops for the tube-side and shell-side assuming a fluid velocity of 1 $m \cdot s^{-1}$ on both the tube-side and shell-side. Assume the flow on the shell-side is characterized by the clean condition, that is, $F_{hn} = F_{hw} = 1$, $F_{hb} = F_{hL} = 0.8$, $F_{Pb} = 0.8$ and $F_{PL} = 0.5$.

c. Estimate the overall heat transfer coefficient and heat transfer area for a tube-side pressure drop of 0.3 bar and a shell-side pressure drop of 0.6 bar. Again, assume the flow on the shell-side is characterized by the clean condition.

d. Repeat the calculation in Part *c* above but assuming the flow is characterized by the fouled condition. Assume $F_{hn} = F_{hw} = F_{hb} = F_{hL} = F_{Pb} = F_{PL} = 1$.

Solution

a. The calculations can be conveniently carried out in a spreadsheet.

$$R = \frac{T_{H1} - T_{H2}}{T_{C2} - T_{C1}} = \frac{200 - 95}{75 - 35} = 2.625$$

$$P = \frac{T_{C2} - T_{C1}}{T_{H1} - T_{C1}} = \frac{75 - 35}{200 - 35} = 0.2424$$

$$W = \frac{R + 1 + \sqrt{R^2 + 1} - 2RX_P}{R + 1 + \sqrt{R^2 + 1} - 2X_P}$$

$$= 0.3688$$

$$N_{SHELLS} = \frac{\ln\left[\dfrac{1 - RP}{1 - P}\right]}{\ln W}$$

$$= 0.74$$

Thus, the unit requires 1 shell.

$$F_T = \frac{\sqrt{R^2 + 1} \ln\left[\dfrac{(1 - P)}{(1 - RP)}\right]}{(R - 1) \ln\left\{\dfrac{[2 - P(R + 1 - \sqrt{R^2 + 1})]}{[2 - P(R + 1 + \sqrt{R^2 + 1})]}\right\}}$$

Substituting $R = 2.625$ and $P = 0.2424$:

$$F_T = 0.90$$

b. $$K_{hT} = C\left[\frac{k}{d_I}\right] Pr^{\frac{1}{3}} \left[\frac{d_I \rho}{\mu}\right]^{0.8}$$

$$= 0.023 \left[\frac{0.133}{0.016}\right]\left[\frac{2050 \times 3.6 \times 10^{-3}}{0.133}\right]^{\frac{1}{3}}$$

$$\times \left[\frac{0.016 \times 830}{3.6 \times 10^{-3}}\right]^{0.8}$$

$$= 520.4$$

For an assumed tube-side velocity of 1 $m \cdot s^{-1}$:

$$h_T = K_{hT} v_T^{0.8}$$

$$= 520.4 \times 1^{0.8}$$

$$= 520.4 \text{ W} \cdot \text{m}^{-2} \cdot \text{K}^{-1}$$

$$K_{hS} = \frac{0.24 F_{hn} F_{hw} F_{hb} F_{hL} \rho^{0.64} C_P^{\frac{1}{3}} k^{\frac{2}{3}}}{\mu^{0.307} d_O^{0.36}}$$

$$= \frac{0.24 \times 1 \times 1 \times 0.8 \times 0.8 \times (730)^{0.64} \times (2470)^{\frac{1}{3}} \times (0.132)^{2/3}}{(4 \times 10^{-4})^{0.307} \times (0.02)^{0.36}}$$

$$= 1653$$

For an assumed shell-side velocity of 1 m·s^{-1}:

$$h_S = K_{hS} v_S^{0.64}$$
$$= 1653 \times 1^{0.64}$$
$$= 1653 \text{ W·m}^{-2}\text{·K}^{-1}$$

Taking the wall coefficient from Table 15.4, the overall heat transfer coefficient can be estimated from:

$$\frac{1}{U} = \frac{1}{h_S} + \frac{1}{h_{SF}} + \frac{1}{h_W} + \frac{d_o}{d_I}\frac{1}{h_{TF}} + \frac{d_o}{d_I}\frac{1}{h_T}$$
$$= \frac{1}{1653} + \frac{1}{5000} + \frac{1}{20{,}200}$$
$$+ \frac{0.02}{0.016}\left(\frac{1}{2000} + \frac{1}{520.4}\right)$$
$$U = 257.6 \text{ W·m}^{-2}\text{·K}^{-1}$$

$$\Delta T_{LM} = \frac{(200 - 75) - (95 - 35)}{\ln\left[\dfrac{200 - 75}{95 - 35}\right]}$$
$$= 88.6°\text{C}$$

$$A = \frac{Q}{U \Delta T_{LM} F_T}$$
$$= \frac{0.037 \times 730 \times 2470 \times (200 - 95)}{257.4 \times 88.6 \times 0.9}$$
$$= 341 \text{ m}^2$$

For the tube-side pressure drop:

$$K_{PT} = 0.092\rho^{0.8}\mu^{0.2} d_I^{-1.2}$$
$$= 0.092 \times (830)^{0.8} \times (3.6 \times 10^{-3})^{0.2} \times (0.016)^{-1.2}$$
$$= 923.4$$

$$\Delta P_T = K_{PT}\frac{Ad_I^2}{4F_I d_O} v_T^{2.8} + 1.25 N_{PT}\rho v_T^2$$
$$= \frac{923.4 \times 341 \times (0.016)^2 \times 1^{2.8}}{4 \times 0.103 \times 0.02} + 1.25 \times 2 \times 830 \times 1^2$$
$$= 11{,}900 \text{ N·m}^{-2}$$

For the shell-side pressure drop, first estimate the shell diameter:

$$D_S = \left(\frac{4 p_C p_T^2 A}{\pi^2 d_O (L/D_S)}\right)^{\frac{1}{3}}$$
$$= \left(\frac{4 \times 1 \times 0.025^2 \times 341}{\pi^2 \times 0.02 \times 7}\right)^{\frac{1}{3}}$$
$$= 0.85 \text{ m}$$

$$K_{PSI} = \frac{1.298 F_{pb} D_S (1 - B_C)\rho^{0.83}\mu^{0.17}}{p_T d_O^{0.17}}$$
$$= \frac{1.298 \times 0.8 \times 0.85 \times (1 - 0.25) \times (730)^{0.83} \times (4 \times 10^{-4})^{0.17}}{0.025 \times (0.02)^{0.17}}$$
$$= 3246$$

$$K_{PS2} = \frac{0.5261 F_{Pb} F_{PL} p_C (1 - 2B_C)(p_T - d_O)\rho^{0.83}\mu^{0.17}}{d_O^{1.17} F_I}$$
$$= \frac{0.5261 \times 0.8 \times 0.5 \times 1 \times (1 - 2 \times 0.25) \times (0.025 - 0.02) \times (730)^{0.83} \times (4 \times 10^{-4})^{0.17}}{(0.02)^{1.17} \times 0.037}$$
$$= 87.01$$

$$K_{PS3} = \frac{2.596 F_{Pb} F_{PL} (1 - 2B_c) D_S \rho^{0.83}\mu^{0.17}}{p_T d_O^{0.17}}$$
$$= \frac{2.596 \times 0.8 \times 0.5 \times (1 - 2 \times 0.25) \times 0.85 \times (730)^{0.83} \times (4 \times 10^{-4})^{0.17}}{0.025 \times (0.02)^{0.17}}$$
$$= 2164$$

$$K_{PS4} = \frac{0.2026 F_{PL} p_C p_T (p_T - d_o)\rho}{d_O F_O}\left[\frac{2}{D_S} + \frac{0.6 B_C}{p_T}\right]$$
$$= \frac{0.2026 \times 0.5 \times 1 \times 0.025(0.025 - 0.02) \times 730}{0.02 \times 0.037} \times \left[\frac{2}{0.85} + \frac{0.6 \times 0.25}{0.025}\right]$$
$$= 104.3$$

$$\Delta P_S = 2K_{PS1} v_S^{1.83} + K_{PS2} A v_S^{2.83} - K_{PS3} v_S^{1.83} + K_{PS4} A v_S^3$$
$$= 2 \times 3246 \times 1^{1.83} + 87.01 \times 341 \times 1^{2.83} - 2164 \times 1^{1.83} + 104.3 \times 341 \times 1^3$$
$$= 69{,}600 \text{ N·m}^{-2}$$

The shell-side pressure drop is high. The fluid velocity on the shell-side could be decreased and the calculation repeated until an acceptable pressure drop is obtained. However, the approach allows the pressure drop to be specified directly.

c. The tube-side and shell-side pressure drops are now fixed to be:

$$\Delta P_T = 30{,}000 \text{ N·m}^{-2}$$
$$30{,}000 = K_{PTI} A h_T^{3.5} + K_{PT2} h_T^{2.5} \qquad (15.65)$$
$$\Delta P_S = 60{,}000 \text{ N·m}^{-2}$$
$$60{,}000 = K_{S1} h_s^{2.86} + K_{S2} A h_s^{4.42} + K_{S3} A h_s^{4.69} \qquad (15.66)$$

Thus, h_T and h_S can be varied by trial and error to satisfy these two equations. At each new value of h_T and h_S, the U and A need to be calculated for fixed Q, ΔT_{LM} and F_T. Also, for each iteration, the new shell diameter needs to be calculated in order to update K_{PS1}, K_{PS3}, K_{PS4}, K_{S1} and K_{S3}.

For flow characterized on the shell-side by the clean condition:

$$F_{hn} = F_{hw} = 1$$
$$F_{hb} = F_{hL} = 0.8$$
$$F_{Pb} = 0.8, F_{PL} = 0.5$$

The calculation is iterative and can be conveniently carried out using a solver in a spreadsheet to satisfy the Equations 15.65 and 15.66 simultaneously. To reach the two variables simultaneously, the objective can be set up such that the difference between the left- and right-hand sides of Equations 15.65 and 15.66 are given the values, say *Objective* 1 and *Objective* 2 respectively. Then the spreadsheet solver is used to search for:

$$(\textit{Objective}\ 1)^2 + (\textit{Objective}\ 2)^2 = 0 \text{ (within a tolerance)}$$

The result is:

$$\Delta P_T = 30{,}000 \text{ N·m}^{-2}$$
$$h_T = 721 \text{ W·m}^{-2}\text{·K}^{-1}$$
$$\Delta P_S = 60{,}000 \text{ N·m}^{-2}$$
$$h_S = 1662 \text{ W·m}^{-2}\text{·K}^{-1}$$
$$U = 311.5 \text{ W·m}^{-2}\text{·K}^{-1}$$
$$A = 282 \text{ m}^2$$
$$D_S = 0.8 \text{ m}$$

Also, given the specified pressure drop, the corresponding fluid velocities can be calculated from:

$$v_T = \left(\frac{h_T}{K_{ht}}\right)^{\frac{1}{0.8}}$$
$$= 1.5 \text{ m·s}^{-1}$$
$$v_S = \left(\frac{h_S}{K_{hS}}\right)^{\frac{1}{0.64}}$$
$$= 1.0 \text{ m·s}^{-1}$$

d. Fouling will tend to reduce *bypass* and leakage on the shell-side. As an extreme case, all shell-side factors can be assumed to be 1.0 to illustrate the trend that fouling will have. As in Part C above, h_T and h_S are varied simultaneously to satisfy the pressure drop specifications. The result is:

$$\Delta P_T = 30{,}000 \text{ N·m}^{-2}$$
$$h_T = 731 \text{ W·m}^{-2}\text{·K}^{-1}$$
$$\Delta P_S = 60{,}000 \text{ N·m}^{-2}$$
$$h_S = 2226 \text{ W·m}^{-2}\text{·K}^{-1}$$
$$U = 329.7 \text{ W·m}^{-2}\text{·K}^{-1}$$
$$A = 267 \text{ m}^2$$
$$D_S = 0.78 \text{ m}$$

The effect of the fouling on the shell-side flow is to increase the cross-flow and increase the overall heat transfer coefficient for a fixed pressure drop (assuming the same fouling coefficients in both cases).

Given the uncertainties associated with the calculations, especially those on the shell-side, a sensible design basis for the heat transfer area specification would be the shell-side flow characterized by the clean condition. Of course, the fouling coefficients for the shell-side and tube-side should be included to account for the surface fouling resistance.

This approach provides a preliminary specification for the heat exchanger. The actual heat exchanger will be restricted to standard tube lengths, tube layout and shell size. The preliminary design would then be adjusted up to meet standard size and layout specifications.

15.5 EXTENDED SURFACE TUBES

In situations in which the film transfer coefficient on the outside of the tubes in a heat exchanger is much lower than the inside, the outside becomes the controlling coefficient. Considering Equation 15.13 for the overall heat transfer coefficient, if the outside film is controlling, then no matter what is done to increase the inside coefficient, it will have little effect on the overall heat transfer coefficient. This will happen when heating or cooling viscous liquids and gases. In such circumstances, a more compact unit will result if a greater surface area is presented to the controlling fluid by the use of extended-surface tubes. Extended surfaces increase the rate of heat transfer per unit length of tube and the resulting exchanger can be much smaller and cheaper than the corresponding plain-tube exchanger. The external tube surface can be extended in one of two general ways:

- *Integrally formed tubes*, in which the extended surface is produced by cold-forming fins onto the surface of the tube by extrusion of the parent tube.
- *Nonintegrally formed fins*, in which the surface is extended by attaching pieces of metal in the form of longitudinal or transverse strips, wire or spines by welding, brazing, grooving and peening, or shrink fitting the extended surface to the tube. The method of fixing the extended surface to the parent tube might create a resistance to heat transfer.

By far the most common design of extended surface tube uses "high" transverse fins, as illustrated in Figure 15.11a. High transverse fins can increase the surface area by a factor of up to 16 relative to plain tubes. Such tubes are commonly used in heat exchangers for cooling duties where hot fluid is passed through the inside of horizontal tubes with ambient air from a fan flowing vertically across the outside. These *air-cooled heat exchangers* or *fin-fan exchangers* are illustrated in Figures 15.11b and 15.11c. The ambient air can be drawn across the tubes in an *induced draft* arrangement, as shown in Figure 15.11b, or driven across the tubes in a *forced draft* arrangement, as shown in Figure 15.11c. The enhancement in the heat transfer depends on the dimensions of the tube, the dimensions of the fins, the number of fins per unit length, bundle arrangement (square or triangular pitch), tube and fin materials of construction, as well as the fluid velocities, physical properties, temperatures and fouling characteristics. Enhancements in the outside film transfer coefficients for such exchangers are of the order 50 to 70% when compared with plain tubes. Typical outside film transfer coefficients would usually be of the order of 60 W·m^2·°C. Air-cooled heat exchangers of the type in Figure 15.11 are used extensively for cooling utility, particularly when cooling water is scarce.

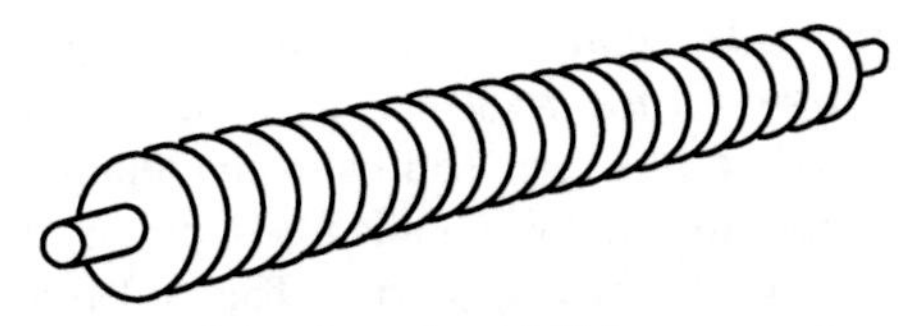

(a) Extended surface tube with high transverse fins.

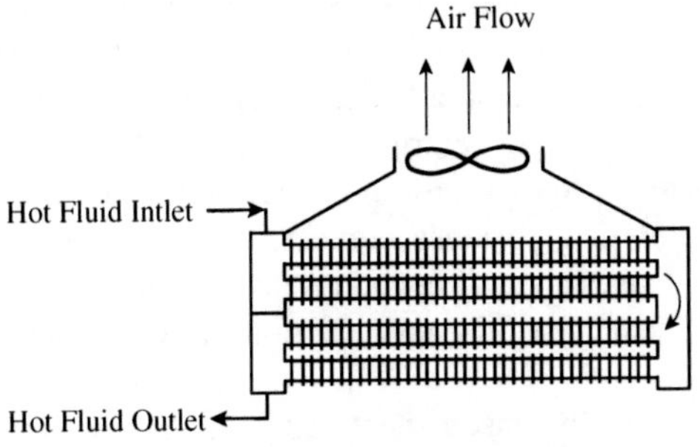

(b) Induced draft air-cooled heat exchanger.

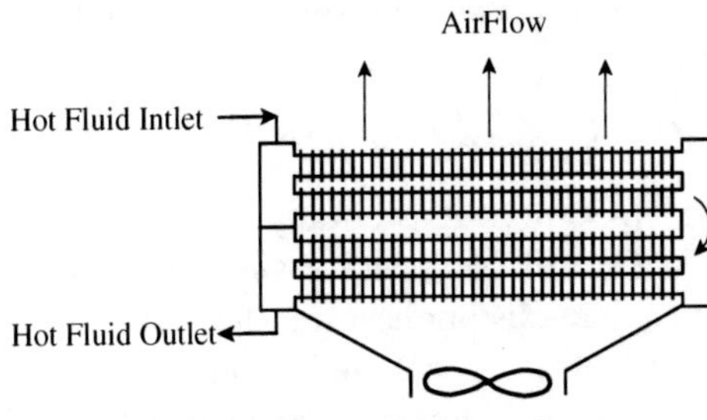

(c) Forced draft air-cooled heat exchanger.

Figure 15.11 Extended surface heat transfer.

Extended surfaces created from integrally formed "low" transverse fins can be used in conventional shell-and-tube heat exchangers to enhance the outside film transfer coefficient. Low transverse fins can increase the surface area by a factor of around 2.5 relative to plain tubes.

Longitudinal fins can also be used, but their application is restricted to small heat exchangers in the form of a concentric pipe heat exchanger, similar to the schematic in Figure 15.5a. In this arrangement, the inner tube would be the extended surface tube with the fins in the annular space to enhance the heat transfer. Longitudinal fins can increase the surface area by a factor of 14 to 20 relative to plain tubes.

15.6 RETROFIT OF HEAT EXCHANGERS

In retrofit situations, existing exchangers might be subjected to changes in flowrate. For example, a debottlenecking project might require an increase in throughput. If this is the case, then the change in the heat transfer coefficients can be approximated by:

For the tube-side coefficient

$$\frac{h_{T2}}{h_{T1}} = \left(\frac{F_{I2}}{F_{I1}}\right)^{0.8} \tag{15.67}$$

where h_{T1}, h_{T2} = tube-side film transfer coefficients for the initial and new flowrates

F_{I1}, F_{I2} = tube-side volumetric flowrates for the initial and new flowrates

For the shell-side coefficient

$$\frac{h_{S2}}{h_{S1}} = \left(\frac{F_{O2}}{F_{O1}}\right)^{0.64} \tag{15.68}$$

where h_{S1}, h_{S2} = shell-side film transfer coefficients for the initial and new flowrates

F_{O1}, F_{O2} = shell-side volumetric flowrates for the initial and new flowrates

The change in pressure drop can be approximated by:

For the tube-side pressure drop

$$\frac{\Delta P_{T2}}{\Delta P_{T1}} = \left(\frac{F_{I2}}{F_{I1}}\right)^{1.9} \tag{15.69}$$

where $\Delta P_{T1}, \Delta P_{T2}$ = tube-side pressure drop for the initial and new flowrates

For the shell-side pressure drop

$$\frac{\Delta P_{S2}}{\Delta P_{S1}} = \left(\frac{F_{O2}}{F_{O1}}\right)^{2.9} \tag{15.70}$$

where $\Delta P_{S1}, \Delta P_{S2}$ = shell-side pressure drop for the initial and new flowrates.

Although an increase in flowrate can result in increased film transfer coefficients, an increase in flowrate is also usually accompanied by an increase in heat transfer duty. In turn, this might lead to a need for increased heat transfer area.

For example, an increase in the process throughput might result in an increase in the heat duty, and hence an increase in the heat transfer area, if the temperatures are unchanged. More generally, additional heat transfer area might be required as a result of increased heat duty, operation under reduced temperature differences, operation under reduced heat transfer coefficients or increased fouling.

If shell-and-tube heat exchangers are being used and additional heat transfer area is required, it might be possible to install a new tube bundle into existing shells if the additional area requirement is small. This might be possible if the tube pitch can be decreased or the pitch configuration changed from a square to a triangular pitch. If a significant amount of additional area is required, then the existing area can be supplemented by adding a new shell (or more than one shell if there is a large area requirement). New heat exchanger shells can be added to an existing match in one of two ways:

a. *Series.* When new exchanger shells are added in series to an existing match, then the full flowrate goes through the existing exchangers of the match. The pressure drop and heat transfer coefficients of the existing exchangers of the match will only change significantly if the changes lead to significant changes in the operating conditions of the existing match (e.g. increase in flowrate in a debottlenecking study). The addition of new exchangers in series to the existing match will lead to an increase in the overall pressure drop across the match. This might be important if the pump (or compressor) is close to its maximum capacity.

b. *Parallel.* If new exchanger shells are added in parallel to an existing match, then the flowrate through the existing exchangers is decreased. This will decrease the flowrate and pressure drop through the existing exchangers of the match. However, it will also decrease the heat transfer coefficient, decreasing its performance. The addition of new exchangers will therefore leave the pressure difference largely unchanged. The pressure drop across the match will be the largest between that of the existing exchangers under the new conditions and the new exchangers installed in parallel.

Increasing the heat transfer area, either by replacing tube bundles or by adding new shells, is likely to be expensive. Rather than install additional heat transfer area to cater for the new operational requirements, the overall heat transfer coefficient of the heat exchanger can be increased. It might be possible to modify the tube-side flow pattern to increase the number of tube passes, thereby increasing the tube-side film transfer coefficient. On the shell-side, it might be possible to modify the baffle arrangement to increase the shell-side film transfer coefficient, for example, by decreasing the baffle spacing. Heat transfer enhancement can also be considered. Changing the heat transfer surface from a plain surface to a finned, ribbed or nonuniform surface can increase the rate of heat transfer of the tube surfaces. This can be applied to the inside or outside surface of the tube, or both, but requires the tubes to be changed. Alternatively, devices can be inserted into the inside of existing plain tubes to enhance the inside heat transfer coefficient, at the expense of increased pressure drop. Insertion devices include[8,9]:

- *Twisted tapes.* These consist of a thin strip of twisted metal with the same width as the tube inside diameter that is slid into the tube. The flow is caused to spiral along the tube length.
- *Static mixers.* Static mixers are manufactured from thin metal strips with the same width as the tube internal diameter. In contrast with twisted tapes, the helix is not continuous. The flow is caused to twist through 180° and then twist in the reverse direction through 180°, and so on. The continuous splitting, reorientation and recombination of the fluid cause the enhancement as it flows through the insert.
- *Coiled wire.* The coil is manufactured by tightly wrapping wire onto a rod such that the outside diameter of the coil is slightly larger than the inside diameter of the tube. This ensures that when it is fitted into the tube there is no movement when in service. The coil functions as nonintegral tube roughness.
- *Mesh inserts.* Mesh inserts are manufactured from a matrix of thin wire filaments. The matrix consists of small loops offset in a helical arrangement and supported by two thicker central wires twisted together. The diameter of the wire mesh is slightly larger than the tube diameter to give close contact with the tube wall. The mesh causes a swirling and random mixing from the flow and is usually restricted to laminar flow applications.

If the new increased heat duty and new log mean temperature difference are fixed, then the additional surface area above the existing area using a plain tube surface is given by[9]:

$$Q = U(A_{existing} + \Delta A)\Delta T_{LM} F_T \tag{15.71}$$

where
Q = heat transfer duty
U = overall heat transfer coefficient
$A_{existing}$ = existing heat transfer area
ΔA = additional area requirement
ΔT_{LM} = logarithmic mean temperature difference
F_T = logarithmic mean temperature difference correction factor

If the same duty is to be serviced by enhanced heat transfer without additional area:

$$Q = U_e A_{existing} \Delta T_{LM} F_T \tag{15.72}$$

where U_e = enhanced overall heat transfer coefficient

Thus, for heat transfer enhancement to solve the problem without installation of new area:

$$U_e A_{existing} \geq U(A_{existing} + \Delta A) \tag{15.73}$$

Rearranging this equation gives:

$$\frac{U_e}{U} \geq \frac{A_{existing} + \Delta A}{A_{existing}} \tag{15.74}$$

The overall heat transfer coefficient is given by:

$$\frac{1}{U} = \frac{1}{h_O} + \frac{1}{h_W} + \frac{d_O}{d_I}\frac{1}{h_I} \tag{15.75}$$

where
$$\frac{1}{h_O} = \frac{1}{h_S} + \frac{1}{h_{SF}}$$
$$\frac{1}{h_I} = \frac{1}{h_T} + \frac{1}{h_{TF}}$$
U = overall heat transfer coefficient
h_O, h_I = heat transfer coefficients for the outside and inside of plain tubes including the fouling resistance

h_W = heat transfer coefficients for the tube wall
d_O, d_I = outside and inside tube diameters
h_S, h_T = film heat transfer coefficients for the shell-side and tube-side of plain tubes
h_{SF}, h_{TF} = fouling coefficients for the shell-side and tube-side of plain tubes

If heat transfer enhancement is only for the inside of the tubes using inserts, then:

$$\frac{1}{U_e} = \frac{1}{h_O} + \frac{1}{h_W} + \frac{d_O}{d_I}\frac{1}{h_{Ie}} \tag{15.76}$$

where $\frac{1}{h_{Ie}} = \frac{1}{h_{Te}} + \frac{1}{h_{TFe}}$

U_e = enhanced overall heat transfer coefficient
h_{Ie} = enhanced film transfer coefficient for the inside of tubes including the fouling resistance
h_{Te} = enhanced film heat transfer coefficients for the tube-side
h_{TFe} = enhanced fouling resistance for the tube-side.

Dividing Equation 15.75 by 15.76 gives:

$$\frac{U_e}{U} = \frac{h_{Ie}(d_O h_O h_W + d_O h_O h_I + d_I h_W h_I)}{h_I(d_O h_O h_W + d_O h_O h_{Ie} + d_I h_W h_{Ie})} \tag{15.77}$$

Combining Equations 15.74 and 15.77 gives:

$$\frac{h_{Ie}}{h_I} \geq \frac{d_O h_O h_W(A_{existing} + \Delta A)}{d_O h_O h_W A_{existing} - \Delta A h_I(d_O h_O + d_I h_W)} \tag{15.78}$$

Equation 15.78 gives the criterion for heat transfer enhancement to cater to the new duty without increasing the heat transfer area. If it is assumed that the resistance to heat transfer across the tube wall is negligible (i.e. h_W goes to infinity) and the difference between the inside and outside diameters is negligible (i.e. $d_O = d_I$), then Equation 15.78 simplifies to[9]:

$$\frac{h_{Ie}}{h_I} \geq \frac{h_O(A_{existing} + \Delta A)}{h_O A_{existing} - h_I \Delta A} \tag{15.79}$$

The enhancement ratio (h_{Ie}/h_I) for tube inserts can be correlated as a function of Reynolds number based on the internal diameter of the tube $Re = \rho V d_i/\mu$[10]. Different tube inserts can be compared on the basis of plots of enhancement ratio (h_{Te}/h_T) versus Reynolds number.

The major disadvantage in using heat transfer enhancement is that it increases the pressure drop. In retrofit, this can be important, as the pumps driving the flow might be limited in their capacity to meet the required increase in pressure drop.

Example 15.3 Suppose the data in Table 15.5 now relate to the performance of an existing heat exchanger. The details of the exchanger geometry (which uses steel tubes) are:

$$d_O = 20 \text{ mm}$$
$$d_I = 16 \text{ mm}$$
$$p_T = 25 \text{ mm (square pitch)}$$
$$B_C = 0.25$$
$$L = 5.95 \text{(after subtracting tube sheet thickness)}$$
$$D_S = 0.84 \text{ m (internal diameter)}$$
$$N_T = 824$$
$$N_{TP} = 2$$

This exchanger is required to operate under new conditions that call for an increase in flowrate of 30% for both fluids. The temperatures at the inlet of the exchanger are unchanged, but the temperatures at the exit must be maintained at their current values. The pumps feeding the exchanger have spare capacity and therefore some increase in the pressure drop of both fluids is acceptable.

a. Estimate the increase in the overall heat transfer coefficient and the resulting outlet temperatures for the new conditions.
b. Can the increase in heat duty be accommodated by keeping the existing tube bundle and using tube inserts? A number of inserts are available. Their thermal performance is given in Table 15.6 and friction factor performance given in Table 15.7 as a function of the Reynolds number based on the internal tube diameter, d_I[10]. The value of H in Tables 15.6 and 15.7 relates to the axial distance between a 180° twist in the device[10].

Table 15.6 Thermal performance of tube insert devices.

Type of insert	Enhancement ratio h_{Ie}/h_I
Coiled wire H/d_I 5.5	4.191 $Re^{-0.1}$
Coiled wire H/d_I 2.8	5.309 $Re^{-0.1}$
Coiled wire H/d_I 1.12	7.313 $Re^{-0.1}$
Twisted tape H/d_I 8.5	4.687 $Re^{-0.1203}$
Twisted tape H/d_I 3.6	15.88 $Re^{-0.2206}$

Table 15.7 Friction factor performance of tube insert devices.

Type of insert	Enhancement ratio c_{fe}/c_f
Coiled wire Hd_I 5.5	4.948 $Re^{-0.0953}$
Coiled wire Hd_I 2.8	5.465 $Re^{-0.0345}$
Coiled wire Hd_I 1.12	5.685 $Re^{0.0432}$
Twisted tape Hd_I 8.5	4.765 $Re^{-0.1166}$
Twisted tape Hd_I 3.6	12.65 $Re^{-0.1787}$

Solution

a. Existing heat duty

$$= 0.037 \times 730 \times 2470 \times (200 - 95)$$
$$= 7.01 \times 10^6 \text{ W}$$

From Example 15.2, $\Delta T_{LM} = 88.6°C$. Since the F_T is based on temperatures only and the temperatures are the same as in Example 15.2, the existing $F_T = 0.90$ as calculated in Example 15.2.

$$A = \pi\, d_O L N_T$$
$$= \pi \times 0.02 \times 5.95 \times 824$$
$$= 308 \text{ m}^2$$
$$U = \frac{Q}{A \Delta T_{LM} F_T}$$
$$= \frac{7.01 \times 10^6}{308 \times 88.6 \times 0.90}$$
$$= 285 \text{ W·m}^{-2}\text{·K}^{-1}$$

While it is possible to calculate the existing overall heat transfer coefficient from the operating data, it is not possible to calculate the individual film transfer coefficients. The individual film transfer coefficients can be combined in any number of ways to add up to an overall value of 285 W·m^{-2}·K^{-1}. However, the film transfer coefficients can be estimated from the correlations in Appendix C. Given that the tube-side correlations are much more reliable than the shell-side correlations, the best way to determine the individual coefficients is to calculate the coefficient for the tube-side and allocate the shell-side coefficient to add up to $U = 285$ W·m^{-2}·K^{-1}. Thus, to calculate the tube-side film transfer coefficient, K_{hT} must first be determined.

$$K_{hT} = C\left[\frac{k}{d_I}\right] Pr^{\frac{1}{3}} \left[\frac{d_I \rho}{\mu}\right]^{0.8}$$
$$= 0.023\left[\frac{0.133}{0.016}\right]\left[\frac{2050 \times 3.6 \times 10^{-3}}{0.133}\right]^{\frac{1}{3}}\left[\frac{0.016 \times 830}{3.6 \times 10^{-3}}\right]^{0.8}$$
$$= 520.4$$
$$v_T = \frac{F_I}{\dfrac{\pi\, d_I^2}{4}\dfrac{N_T}{N_{TP}}}$$
$$= \frac{0.103}{\dfrac{\pi \times 0.016^2}{4}\dfrac{824}{2}}$$
$$= 1.24 \text{ m·s}^{-1}$$
$$h_T = K_{hT} V_T^{0.8}$$
$$= 520.4 \times 1.24^{0.8}$$
$$= 618 \text{ W·m}^{-2}\text{·K}^{-1}$$

Assume fouling coefficients:

$$h_{TF} = 2000 \text{ W·m}^{-2}\text{·K}^{-1}$$
$$h_{SF} = 5000 \text{ W·m}^{-2}\text{·K}^{-1}$$
$$\frac{1}{U} = \frac{1}{h_S} + \frac{1}{h_{SF}} + \frac{1}{h_W} + \frac{d_O}{d_I}\frac{1}{h_{TF}} + \frac{d_O}{d_I}\frac{1}{h_T}$$
$$\frac{1}{285} = \frac{1}{h_S} + \frac{1}{5000} + \frac{1}{20{,}200} + \frac{0.02}{0.016}\left(\frac{1}{2000} + \frac{1}{618}\right)$$
$$h_S = 1635 \text{ W·m}^{-2}\text{·K}^{-1}$$

The change in the tube-side and shell-side film transfer coefficients as a result of a 30% increase in both flowrates can be estimated from Equations 15.67 and 15.68:

$$\frac{h_{T2}}{h_{T1}} = \left(\frac{F_{I2}}{F_{I1}}\right)^{0.8}$$
$$h_{T2} = 618 \times 1.3^{0.8}$$
$$= 762 \text{ W·m}^{-2}\text{·K}^{-1}$$
$$\frac{h_{S2}}{h_{S1}} = \left(\frac{F_{O2}}{F_{O1}}\right)^{0.64}$$
$$h_{S2} = 1635 \times 1.3^{0.64}$$
$$= 1934 \text{ W·m}^{-2}\text{·K}^{-1}$$

If the fouling coefficients are assumed to be unchanged, the new overall heat transfer coefficient is given by:

$$\frac{1}{U} = \frac{1}{1934} + \frac{1}{5000} + \frac{1}{20{,}200} + \frac{0.02}{0.016}\left(\frac{1}{2000} + \frac{1}{762}\right)$$
$$U = 330 \text{ W·m}^{-2}\text{·K}^{-1}$$

At the increased flowrate condition, the outlet temperatures and heat duty for the existing area are unknown.

$$Q = 0.037 \times 1.3 \times 730 \times 2470(200 - T_{H2})$$
$$= 0.103 \times 1.3 \times 830 \times 2050(T_{C2} - 35)$$
$$A = \frac{Q}{U \Delta T_{LM} F_T}$$

Thus, the outlet temperature of the hot or cold stream can be assumed. This specifies the heat duty and hence the outlet temperature of the other stream. In this case, assume the outlet temperature of the hot stream and calculate the final temperature of the cold stream from:

$$T_{C2} = T_{C1} + \frac{Q}{F_I \rho C_P}$$

This allows ΔT_{LM} and F_T to be calculated. Given the value of U, this allows the area to be calculated. The outlet temperature of the hot stream can then be varied by trial and error until the calculated area equals the actual area of 308 m^2.

$$\text{Try } T_{H2} = 100°\text{C}$$
$$Q = 8.67 \times 10^6 \text{ W}$$
$$T_{C2} = 73.1°\text{C}$$
$$\Delta T_{LM} = 92.5°\text{C}$$
$$F_T = 0.92$$
$$A = 309 \text{ m}^2$$

This is close; an exact value is given by further trial and error to give:

$$T_{H2} = 100.2°\text{C}$$
$$Q = 8.66 \times 10^6 \text{ W}$$
$$T_{C2} = 73°\text{C}$$

$$\Delta T_{LM} = 92.7°C$$

$$F_T = 0.92$$

$$A = 308 \text{ m}^2$$

b. Maintaining the existing temperatures, the new duty is given by:

$$Q = 0.037 \times 1.3 \times 730 \times 2470(200–95)$$

$$= 9.11 \times 10^6 \text{ W}$$

$$A = \frac{Q}{U \cdot \Delta T_{LM} \cdot F_T}$$

$$= \frac{9.11 \times 10^6}{330 \times 88.6 \times 0.90}$$

$$= 346 \text{ m}^2$$

$$\Delta A = 346–308$$

$$= 38 \text{ m}^2$$

Calculate the outside and inside coefficients including the fouling:

$$\frac{1}{h_O} = \frac{1}{h_S} + \frac{1}{h_{SF}}$$

$$= \frac{1}{1934} + \frac{1}{5000}$$

$$h_O = 1395 \text{ W·m}^{-2}\text{·K}^{-1}$$

$$\frac{1}{h_I} = \frac{1}{h_T} + \frac{1}{h_{TF}}$$

$$= \frac{1}{762} + \frac{1}{2000}$$

$$h_I = 552 \text{ W·m}^{-2}\text{·K}^{-1}$$

The heat transfer enhancement ratio can now be calculated from Equation 15.78:

$$\frac{h_{Ie}}{h_I} = \frac{d_O h_O h_W (A_{existing} + \Delta A)}{d_O h_O h_W A_{existing} - \Delta A h_I (d_O h_O + d_I h_W)}$$

$$= \frac{0.02 \times 1395 \times 20{,}200(308 + 38)}{0.02 \times 1395 \times 20{,}200 \times 308 - 38 \times 552 \;(0.02 \times 1395 + 0.016 \times 20{,}200)}$$

$$= 1.17$$

For the purpose of comparison, now neglect the wall resistance and difference between the inside and outside tube diameters. From Equation 15.79:

$$\frac{h_{Ie}}{h_I} = \frac{h_O(A_{existing} + \Delta A)}{h_O A_{existing} - h_I \Delta A}$$

$$= \frac{1395(308 + 38)}{1395 \times 308 - 552 \times 38}$$

$$= 1.18$$

For most applications, Equation 15.79 is accurate enough.

The enhancement ratio for a given device can be correlated as a function of the Reynolds Number[9]:

$$Re = \frac{\rho d_I v_T}{\mu}$$

$$= \frac{830 \times 0.016 \times 1.24 \times 1.3}{3.6 \times 10^{-3}}$$

$$= 5946$$

Table 15.8 shows the enhancement ratios for various tube inserts.

Table 15.8 Enhancement ratios for heat transfer of various tube inserts.

Type of insert	Enhancement ratio
Coiled wire H/d_I 5.5	$4.191\ Re^{-0.1} = 1.76$
Coiled wire H/d_I 2.8	$5.309\ Re^{-0.1} = 2.23$
Coiled wire H/d_I 1.12	$7.313\ Re^{-0.1} = 3.07$
Twisted tape H/d_I 8.5	$4.687\ Re^{-0.1203} = 1.65$
Twisted tape H/d_I 3.6	$15.88\ Re^{-0.2206} = 2.33$

Thus, all of the tube inserts will potentially provide the required enhancement without the need for extra heat transfer area. However, it would be preferred to have the enhancement with minimum increase in pressure drop. Table 15.9 shows the increase in friction factor for the various inserts.

Table 15.9 Enhancement ratios for friction factor of various tube inserts.

Type of insert	Friction factor ratio
Coiled wire H/d_I 5.5	$4.948\ Re^{-0.0953} = 2.16$
Coiled wire H/d_I 2.8	$5.465\ Re^{-0.0345} = 4.05$
Coiled wire H/d_I 1.12	$5.685\ Re^{0.0432} = 8.28$
Twisted tape H/d_I 8.5	$4.765\ Re^{-0.1166} = 1.73$
Twisted tape H/d_I 3.6	$12.654\ Re^{-0.1787} = 2.68$

On the basis of heat transfer enhancement and pressure drop considerations, Twisted Tape H/d_I 8.5 would be chosen. However, a detailed examination of capital cost might cause this to be revised.

15.7 CONDENSERS

The construction of shell-and-tube condensers is very similar to shell-and-tube heat exchangers for duties that do not involve a change of phase. Condensers can be horizontally or vertically oriented with the condensation on the tube-side or the shell-side. The magnitude of the condensing film coefficient for a given quantity of vapor condensation on a given surface is significantly different depending on the orientation of the condenser. The condensation normally takes place on the shell-side of horizontal exchangers and the tube-side of vertical exchangers. Horizontal shell-side condensation is normally preferred, as the condensing film transfer coefficients are higher. Condensation on the tube-side of horizontal condensers is normally restricted to the use of condensing steam as a heating medium.

Condensation can take place by one of two mechanisms:

a. *film-wise condensation*, in which the condensing vapor wets the surface of the tube forming a continuous film
b. *drop-wise condensation*, in which droplets of condensation do not wet the surface and after growing, fall from the tube to expose fresh condensing surface without forming a continuous film.

Although drop-wise condensation can produce much higher condensing film transfer coefficients, it is unpredictable, and the design is carried out on the basis of film-wise condensation.

The basic equations describing film-wise condensation were developed by Nusselt[11]. The derivation of the equations has been given by Kern[1] and others. A number of assumptions are made in the derivation:

- the liquid film flows smoothly and steadily by gravitational forces;
- the liquid film is in laminar flow;
- no noncondensable gases are present in the vapor phase;
- no vapor shear force acts on the liquid–vapor interface;
- momentum terms are negligible;
- temperature distribution in the condensate film is linear;
- the only heat transferred across the liquid film is the latent heat of condensation released at the liquid–vapor interface (transfer of sensible heat in the liquid film is negligible);
- temperature of the liquid–vapor interface is equal to the saturation temperature;
- physical properties of the liquid film are constant.

For condensation outside of a horizontal tube[1]:

$$h_C = 0.725\left(\frac{k_L^3 \rho_L^2 \Delta H_{VAP}\, g}{d_O \mu_L \Delta T}\right)^{\frac{1}{4}} \qquad (15.80)$$

where h_C = condensing film coefficient ($W \cdot m^{-2} \cdot K^{-1}$)
k_L = thermal conductivity of the liquid ($W \cdot m^{-1} \cdot K^{-1}$)
ρ_L = density of the liquid ($kg \cdot m^{-3}$)
ΔH_{VAP} = latent heat ($J \cdot kg^{-1}$)
g = gravitational constant ($9.81\ m \cdot s^{-2}$)
d_O = outside diameter of tube (m)
μ_L = viscosity of the liquid ($N \cdot s \cdot m^{-2}$ or $kg \cdot m^{-1} \cdot s^{-1}$)
ΔT = temperature difference across the condensate film (K)

The analysis was later modified to include some of the factors neglected by Nusselt[12,13]. One of these is the effect of buoyancy forces acting on the liquid film. This results in the ρ_L^2 term in Equation 15.80 being replaced by $\rho_L(\rho_L - \rho_V)$. Such buoyancy forces are usually only important close to the critical point. In most cases, the two most important factors that cause a significant deviation from Equation 15.80 are the presence of vapor shear forces and noncondensable gases in the vapor. Vapor shear forces act to increase the heat transfer coefficient, whereas noncondensable gases act to decrease it.

Since the ΔT across the film is unknown, it is best eliminated from Equation 15.80. By definition of the condensing film coefficient:

$$m \Delta H_{VAP} = h_C \pi\, d_O L N_T \Delta T \qquad (15.81)$$

where m = flowrate of condensate ($kg \cdot s^{-1}$)
L = tube length (m)
N_T = number of tubes (–)

Substituting Equation 15.81 into Equation 15.80 and rearranging gives:

$$h_C = 0.954 k_L \left(\frac{\rho_L^2 L g N_T}{\mu_L m}\right)^{\frac{1}{3}} \qquad (15.82)$$

For the shell-side of a horizontal tube bundle, dripping of condensate over successive rows acts to decrease the condensing coefficient. This can be accounted for by multiplying the condensing coefficient for a single tube by an empirical correction involving the number of tubes in a vertical row. However, in a tube bundle, the number of tubes in the vertical rows varies according to the position in the bundle. A simple empirical correction is[14]:

$$h_C = 0.954 k_L \left(\frac{\rho_L^2 L g N_T}{\mu_L m}\right)^{\frac{1}{3}} N_R^{-\frac{1}{6}} \qquad (15.83)$$

N_R = number of tubes in a vertical row(–)
$\approx 0.78 \dfrac{D_S}{p_T}$
D_S = shell diameter(m)
p_T = vertical tube pitch(m)

For condensation inside horizontal tubes, the Nusselt Equation can be applied with a correction for the reduction in condensing coefficient caused by the accumulation of condensation. The correction usually applied is 0.8. No correction for the number of tubes is required. Thus, for condensation inside horizontal tubes:

$$h_C = 0.763 k_L \left(\frac{\rho_L^2 L g N_T}{\mu_L m}\right)^{\frac{1}{3}} \qquad (15.84)$$

The Nusselt Equations apply to laminar flow of the condensing film. For horizontal condensation the equations are applicable for:

$$Re = \frac{2m}{\mu_L L N_T} < 2000 \qquad (15.85)$$

For condensation on a vertical surface, the Nusselt Equation takes the form[1]:

$$h_C = 0.943\left(\frac{k_L^3 \rho_L^2 \Delta H_{VAP} g}{L \mu_L \Delta T}\right)^{\frac{1}{4}} \qquad (15.86)$$

where L = length of the condensing surface (m)
ΔT = temperature difference across the condensate film (K)

Combining Equation 15.81 for condensation on the outside of a tube with Equation 15.86 gives:

$$h_C = 1.35 k_L \left(\frac{\rho_L^2 d_O g N_T}{\mu_L m}\right)^{\frac{1}{3}} \qquad (15.87)$$

For condensation on the inside of the vertical tubes:

$$m \Delta H_{VAP} = h_C \pi\, d_I L N_T \Delta T \qquad (15.88)$$

Combining Equations 15.86 and 15.88 gives the Nusselt Equation for condensation inside vertical tubes:

$$h_C = 1.35 k_L \left(\frac{\rho_L^2 d_I g N_T}{\mu_L m}\right)^{\frac{1}{3}} \qquad (15.89)$$

For vertical condensation, the Nusselt Equations are valid for a laminar film according to:

$$Re = \frac{4m}{\pi \mu_L d_I N_T} < 2000 \qquad (15.90)$$

In the above equations, the film thickness and hence the condensing coefficient varies across the surface. The correlations give an average coefficient applicable to the entire surface. The condensing coefficients are independent of shell-side geometry (e.g. baffle cut, distance etc.). The Nusselt Equations give reasonably good agreement with experimental data for laminar flow of the condensate film in the absence of vapor shear forces and noncondensable gases in the vapor. In the absence of noncondensable gases, the Nusselt Equations will tend to give a conservative prediction of the condensing coefficient. Vapor shear and turbulence in the film can lead to considerably higher values than those predicted by the Nusselt Equations.

For a simple total condenser involving isothermal condensation:

$$Q = m \Delta H_{VAP} = U A \Delta T_{LM} \qquad (15.91)$$

where U is defined by Equation 15.13. Note that if the condensing fluid is pure and therefore isothermal, no F_T correction factor is required if multipass exchangers are used, that is, $F_T = 1$.

If the heat exchange involves desuperheating as well as condensation, then the exchanger can be divided into zones with linear temperature–enthalpy profiles in each zone. Figure 15.12a illustrates desuperheating and condensation on the shell-side of a horizontal condenser. The total heat transfer area is the sum of the values for each zone:

$$A = A_{DS} + A_{CN}$$
$$= \frac{Q_{DS}}{U_{DS} \Delta T_{LM,DS}} + \frac{Q_{CN}}{U_{CN} \Delta T_{LM,CN}} \qquad (15.92)$$

where A = total heat transfer area
A_{DS} = heat transfer area for the desuperheating zone
A_{CN} = heat transfer area for the condensing zone
Q_{DS} = heat transfer duty for the desuperheating zone
Q_{CN} = heat transfer duty for the condensing zone
U_{DS} = overall heat transfer coefficient for the desuperheating zone
U_{CN} = overall heat transfer coefficient for the condensing zone
$\Delta T_{LM,DS}$ = logarithmic mean temperature difference for the desuperheating zone
$\Delta T_{LM,CN}$ = logarithmic mean temperature difference for the condensing zone

To calculate the condensing heat transfer coefficient requires the length of the condensing zone L to be specified. Thus, a value of L must be estimated before the calculation can be made. For the value of L to be correct, it must comply with:

$$L = \frac{A_{CN}}{A} \times tube\ length \qquad (15.93)$$

The value of L is then varied until there is agreement with Equation 15.93.

It might also be necessary to subcool the condensate. As with desuperheating, if subcooling is required, the heat exchanger can be divided into zones. Figure 15.12b illustrates subcooling on the shell-side of a vertical condenser. The subcooling arrangement in Figure 15.12b is achieved by using a loop seal to create a partially submerged tube bundle[1]. For subcooling, the heat transfer area is given by:

$$A = A_{CN} + A_{SC}$$
$$= \frac{Q_{CN}}{U_{CN} \Delta T_{LM,CN}} + \frac{Q_{SC}}{U_{SC} \Delta T_{LM,SC}} \qquad (15.94)$$

where A_{SC} = heat transfer area for the subcooling zone
Q_{SC} = heat transfer duty for the subcooling zone

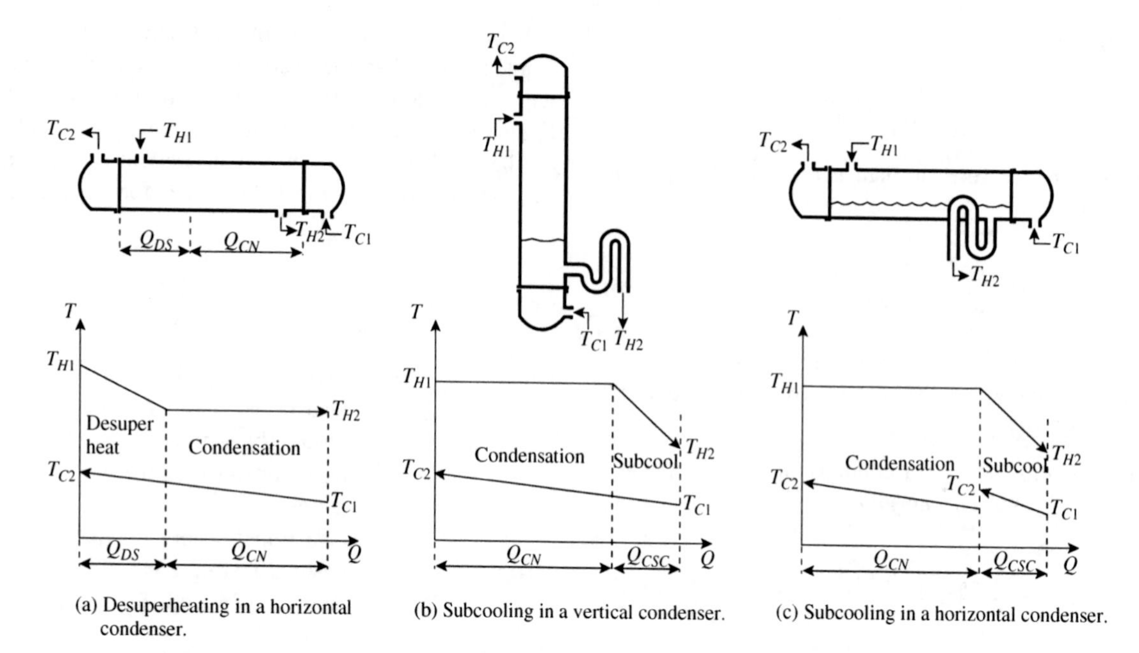

Figure 15.12 Condensation with desuperheating and subcooling.

U_{SC} = overall heat transfer coefficient for the subcooling zone

$\Delta T_{LM,SC}$ = logarithmic mean temperature difference for the subcooling zone

To calculate the condensing heat transfer coefficient again requires the length of the condensing zone L to be specified. Thus, a value of L must be estimated and adjusted until it complies with Equation 15.93.

Figure 15.12c illustrates subcooling on the shell-side of a horizontal condenser. The subcooling arrangement in Figure 15.12c is again achieved by using a loop seal to create a partially submerged tube bundle[1]. Rather than use a loop seal, a dam baffle can be used to partially submerge the bundle[1]. Figure 15.12c shows the zones this time represented in parallel, rather than the series arrangements in Figures 15.12a and 15.12b. Calculation of the condensing heat transfer coefficient for a horizontal exchanger requires the number of tubes in the condensing zone $N_{T,CN}$ to be specified. Thus, a value of $N_{T,CN}$ must be estimated before the calculation can be made. For the value of $N_{T,CN}$ to be correct, it must comply with:

$$N_{T,CN} = \frac{A_{CN}}{A} \times number\ of\ tubes \qquad (15.95)$$

The value of $N_{T,CN}$ is then varied until there is agreement with Equation 15.95.

For multicomponent condensation, the condensation will not be isothermal, leading to a nonlinear temperature–enthalpy profile for the condensation. If this is the case, then the exchanger can be divided into a number of zones with the temperature–enthalpy profiles linearized in each zone. Each zone is then modeled separately and zones summed to obtain the overall area requirement[1].

Particular care needs to be adopted if a vapor to be condensed has noncondensable gases present. Here the vapor diffuses through the gas to the cold surface where it condenses. But as the condensation proceeds, the concentration of the noncondensable gas increases, which increases the diffusional resistance and decreases the condensing coefficient. To take this into account requires complex models, which is outside the scope of this text.

Pressure drop during condensation results essentially from the vapor flow. As condensation proceeds, the vapor flowrate decreases. The equations described previously for pressure drop in shell-and-tube heat exchangers are only applicable under constant flow conditions. Again the exchanger can be divided into zones. However, in preliminary design, a reasonable estimate of the pressure drop can usually be obtained by basing the calculation on the mean of the inlet and outlet vapor flowrates.

Example 15.4 A flowrate of 0.1 $kmol{\cdot}s^{-1}$ of essentially pure acetone vapor from the overhead of a distillation column is to be condensed without any condensate subcooling. The condensation is to take place on the shell-side of a horizontal shell-and-tube heat exchanger against cooling water flowing in two passes on the tube-side. The operating pressure of the condenser is 1.52 bar. At this pressure, the acetone condenses at 67°C. The cooling water can be assumed to be at 25°C and to be returned to the cooling tower at 35°C. The condenser can be assumed to be steel with tubes with 20-mm outside diameter and 2-mm wall thickness. The tube pitch can be assumed to be $1.25d_O$ and a square configuration. The ratio of tube length to shell diameter can be assumed to be 5. The physical property data are given in Table 15.10. The properties of acetone are at 67°C. Although the average film temperature will be lower than this, the value of $k(\rho^2/\mu)^{1/3}$ tends not to be very sensitive to temperature. The molar mass of acetone can be assumed to be 58 $kg{\cdot}kmol^{-1}$. Assume the fouling coefficients to be 11,000 $W \cdot m^{-2} \cdot K^{-1}$ and 5000 $W \cdot m^{-2} \cdot K^{-1}$ for the shell-side and tube-side respectively. For an allowable tube-side velocity of 2 $m \cdot s^{-1}$, estimate the heat transfer area.

Table 15.10 Physical property data for acetone and water.

Property	Acetone (67°C)	Water (30°C)
Density ($kg \cdot m^{-3}$)	736	996
Heat capacity ($J \cdot kg^{-1} \cdot K^{-1}$)	2320	4180
Viscosity ($N \cdot s \cdot m^{-2}$)	0.213×10^{-3}	0.797×10^{-3}
Thermal conductivity ($W \cdot m^{-1} \cdot K^{-1}$)	0.137	0.618
Heat of vaporization ($J \cdot kg^{-1}$)	494,000	–

Solution

$$\text{Flowrate of acetone} = 0.1 \times 58 = 5.8 \text{ kg} \cdot \text{s}^{-1}$$

$$\text{Duty on condenser} = 5.8 \times 494{,}000 = 2.865 \times 10^6 \text{ W}$$

$$\text{Flowrate of cooling water} = \frac{2.865 \times 10^6}{4180(35-25)} = 68.54 \text{ kg} \cdot \text{s}^{-1} = 0.0688 \text{ m}^3 \cdot \text{s}^{-1}$$

To determine the condensing film coefficient using Equation 15.83 requires the number of tubes to be known.

$$N_T = \frac{\pi D_S^2}{4 p_C p_T^2}$$

$$\text{where } D_S = \left(\frac{4 p_C p_T^2 A}{\pi^2 d_O (L/D_S)} \right)^{\frac{1}{3}}$$

Thus, the solution must be iterative as the heat transfer area A is unknown. An initial estimate of A (say 100 m^2) is required, giving:

$$D_S = 0.633 \text{ m}$$

$$N_T = 503$$

$$h_C = 0.954 k_L \left(\frac{\rho_L^2 L g N_T}{\mu_L m} \right)^{\frac{1}{3}} N_R^{-\frac{1}{6}}$$

$$= 0.954 k_L \left(\frac{\rho_L^2 (L/D_S) D_S g N_T}{\mu_L m} \right)^{\frac{1}{3}} \left(0.78 \frac{D_S}{p_T} \right)^{-\frac{1}{6}}$$

$$= 0.954 \times 0.137 \left(\frac{736^2 \times 5 \times 0.633 \times 9.81 \times 503}{2.13 \times 10^{-4} \times 5.8} \right)^{\frac{1}{3}} \times \left(0.78 \frac{0.633}{0.02 \times 1.25} \right)^{-\frac{1}{6}}$$

$$= 1510 \text{ W} \cdot \text{m}^{-2} \cdot \text{K}^{-1}$$

The tube-side heat transfer coefficient is given by:

$$K_{hT} = C \left[\frac{k}{d_I} \right] Pr^{\frac{1}{3}} \left[\frac{d_I \rho}{\mu} \right]^{0.8}$$

$$= 0.023 \left[\frac{0.618}{0.016} \right] \left[\frac{4180 \times 7.97 \times 10^{-4}}{0.618} \right]^{\frac{1}{3}} \times \left[\frac{0.016 \times 996}{7.97 \times 10^{-4}} \right]^{0.8}$$

$$= 4298$$

For a tube-side velocity of 2 $m \cdot s^{-1}$:

$$h_T = K_{hT} V_T^{0.8} = 4298 \times 2^{0.8} = 7483 \text{ W} \cdot \text{m}^{-2} \cdot \text{K}^{-1}$$

Now the overall heat transfer coefficient can be estimated.

$$\frac{1}{U} = \frac{1}{h_C} + \frac{1}{h_{SF}} + \frac{d_O}{2k} \ln \frac{d_O}{d_I} + \frac{d_O}{d_I} \frac{1}{h_{TF}} + \frac{d_O}{d_I} \frac{1}{h_T}$$

$$= \frac{1}{1510} + \frac{1}{11{,}000} + \frac{1}{20{,}200} + \frac{0.02}{0.016} \left[\frac{1}{5000} + \frac{1}{7483} \right]$$

$$U = 820 \text{ W} \cdot \text{m}^{-2} \cdot \text{K}^{-1}$$

$$\Delta T_{LM} = \frac{(67-35)+(67-25)}{\ln \left[\frac{67-35}{67-25} \right]} = 36.8 \text{ K}$$

Now the duty can be calculated from:

$$Q = U A \Delta T_{LM} = 820 \times 100 \times 36.8 = 3.01 \times 10^6 \text{ W}$$

This does not agree with the specified duty of 2.865×10^6 W. To make the heat duty balanced requires the heat transfer area A to be adjusted by trial and error. At each value of A, the number of tubes and h_C must be calculated. This can be readily done using a spreadsheet solver. The result is:

$$Q = 2.865 \times 10^6 \text{ W}$$

$$h_C = 1491 \text{ W} \cdot \text{m}^{-2} \cdot \text{K}^{-1}$$

$$h_T = 7483 \text{ W} \cdot \text{m}^{-2} \cdot \text{K}^{-1}$$

$$U = 814 \text{ W} \cdot \text{m}^{-2} \cdot \text{K}^{-1}$$

$$A = 96 \text{ m}^2$$

$$D_S = 0.623 \text{ m}$$

$$N_T = 488$$

Rather than specify the tube-side velocity, the tube-side pressure drop could have been specified (e.g. $\Delta P_T = 30{,}000$ $N \cdot m^{-2}$). Had this been the case, then the calculation would have required Equation 15.16 to be solved simultaneously with the above equations by varying A and h_T simultaneously, similar to the solution of Example 15.1c.

15.8 REBOILERS AND VAPORIZERS

Reboilers are required for distillation columns to vaporize a fraction of the bottom product, as discussed in Chapter 9. It may also be the case that a liquid needs to be vaporized for other purposes, for example, a liquid feed needs to be vaporized before entering a reactor. The discussion here will focus on reboiling a distillation column, but the same principles apply to other types of vaporizers.

Three common designs of the reboiler are illustrated in Figure 15.13. The first shown in Figure 15.13a is a *kettle reboiler*. Vaporization takes place on the outside of tubes immersed in a pool of liquid. The bottom product is taken from an overflow from the liquid pool and there is no recirculation between the reboiler and the column. In some designs, the tube bundle can be installed in the base of the column as an internal reboiler. The kettle reboiler incorporates a volume above the liquid pool and tube bundle for vapor and liquid disengagement. The shell diameter is typically 40% greater than the bundle diameter to allow for this. The second type of reboiler shown in Figure 15.13b, the *horizontal thermosyphon*, also features vaporization on the outside of the tubes. However, in this case, there is recirculation around the base of the column. A mixture of vapor and liquid leaves the reboiler and enters the base of the column where it separates. The third type of reboiler, the *vertical thermosyphon*, is illustrated in Figure 15.13c. Again, there is a recirculation around the base of the column, but this time the vaporization takes place inside the tubes. Boiling on the tube-side will normally be carried out in a 1–1 exchanger.

The three reboilers in Figure 15.13 are shown under *natural circulation*. The flow of liquid from the column to the reboiler is created by the difference in hydrostatic head between the column of liquid feeding the reboiler and the vapor–liquid mixture created by the reboiler.

The amount of liquid vaporized in the reboiler should not be more than 80%, otherwise this will tend to lead to excessive fouling of the reboiler. For kettle reboilers, there is no recirculation. But for thermosyphon reboilers, a *recirculation ratio* can be defined as:

$$\text{Recirculation ratio} = \frac{\text{flowrate of liquid at reboiler outlet}}{\text{flowrate of vapor at reboiler outlet}}$$

This usually lies between 0.25 and 6. The greater the value of recirculation ratio, the less fouling there is in the reboiler. Lower values tend to be used in horizontal thermosyphons and higher values (greater than 4) used in vertical thermosyphons. The recirculation ratio is a degree of freedom at the discretion of the designer. This should be fixed later when the detailed design is carried out.

The kettle reboiler has the advantage that it is equivalent to a theoretical stage for the distillation but is relatively expensive due to the extra volume required for vapor disengagement. Also, the liquid has a high residence time in the heating zone, which can be a problem if the material is prone to thermal decomposition. If the reboiler must operate at a high pressure, then the large diameter of the kettle shell is a disadvantage. A large diameter cylindrical shell requires a thicker wall to withstand a given pressure than a small diameter shell. The horizontal and vertical thermosyphons both have the disadvantage of not providing a theoretical stage for the distillation. But the thermosyphon reboilers are less prone to fouling than kettle reboilers and have a lower residence time in the heating zone. Thermosyphon reboilers require additional height inside the column shell than kettle reboilers to allow for vapor disengagement as the vapor–liquid mixture enters the column. Vertical thermosyphon reboilers require the column to be at a higher elevation than kettle and horizontal thermosyphons. Horizontal thermosyphons tend to be preferred to vertical ones if the heat transfer area is large, as horizontal arrangements are easier to maintain. Although thermosyphon reboilers can be used under vacuum conditions, care must be exercised, as the effect of pressure on the boiling point of the fluid entering the reboiler must be considered. When reboiling multicomponent systems, the vaporization can take place over a range of temperature. The forced flow in a thermosyphon can give a higher mean temperature difference than a kettle for the same percentage of vaporization, as the kettle boiling temperature is uniform. Generally the heat flux (heat transferred per unit area) and heat transfer coefficients are in the order

$$\text{kettle} < \text{horizontal thermosyphon} < \text{vertical thermosyphon}$$

Given these arguments, it is not surprising that the most common design of reboiler is the vertical thermosyphon.

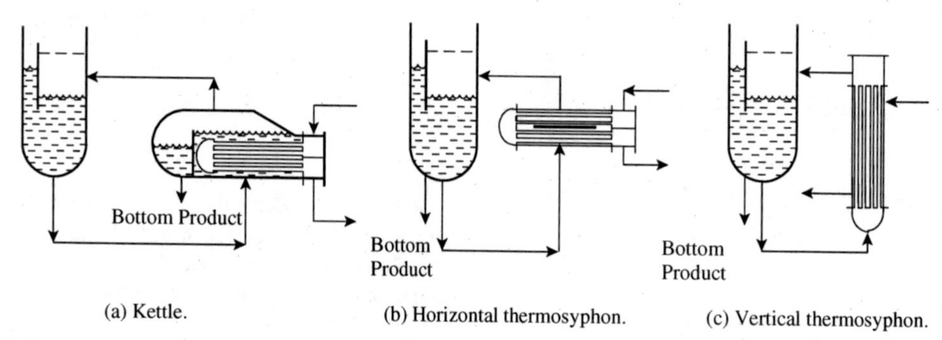

Figure 15.13 Reboiler designs.

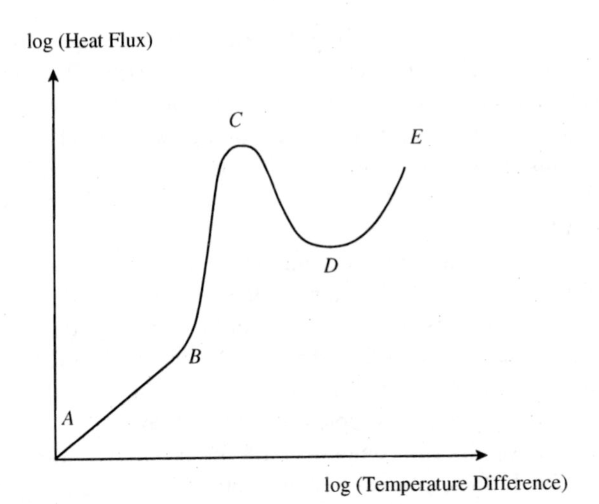

Figure 15.14 Heat transfer characteristics of boiling.

The basic characteristics of the boiling process are illustrated in Figure 15.14. This shows a plot of heat flux versus the temperature difference between the vaporizing surface and the bulk liquid plotted on logarithmic axes. Initially, between Points A and B in Figure 15.14, heat transfer is by natural convection. Superheated liquid rises to the liquid surface where evaporation takes place. As the temperature difference increases beyond Point B in Figure 15.14, nucleate boiling occurs in which vapor bubbles are formed at the heating surface and released from the surface. Nucleate boiling, as its name implies, depends on the presence of nuclei. In boiling, the nuclei are preexisting inclusions of noncondensable gas or vapor in cavities on the heat transfer surface. It thus depends on the character of the heat transfer surface.

Point C in Figure 15.14 is termed the *critical heat flux* or *maximum boiling flux* or *peak boiling flux* as bubbles coalesce on the surface creating a vapor blanket. Critical heat flux occurs because insufficient liquid is able to reach the heat transfer surface due to the rate at which vapor is leaving. Beyond Point D, the surface is dry and entirely blanketed by vapor and heat is transferred by conduction and radiation.

Reboilers are designed to operate below the peak flux, as beyond it either the heat flux would be lower, or much higher temperate difference would be required. Design is normally restricted to have a heat flux less than 70% of the critical flux. Preliminary design of kettle and horizontal thermosyphon reboilers can be based on *pool boiling*. In pool boiling, the heating surface is surrounded by a large body of fluid in which the fluid motion is only induced by natural convection currents and the motion of bubbles. A simple correlation for nucleate boiling that can be used for the preliminary design of kettle and horizontal thermosyphon reboilers is that due to Palen[15]:

$$h_{NB} = 0.182 P_C^{0.67} q^{0.7} \left(\frac{P}{P_C}\right)^{0.17} \quad (15.96)$$

where h_{NB} = nucleate boiling coefficient ($W \cdot m^{-2} \cdot K^{-1}$)
P_C = liquid critical pressure (bar)
P = operating pressure (bar)
q = heat flux ($W \cdot m^{-2}$)
$= h_{NB}(T_w - T_{SAT})$
T_W = wall temperature of the heating surface (°C)
T_{SAT} = saturation temperature of the boiling liquid (°C).

This equation applies to vaporization of single components, but can be used for close boiling mixtures without too much error. Coefficients for wide boiling mixtures will be overestimated.

Mostinski[16] gives a correlation for the estimation of the critical heat flux for single tubes:

$$q_{C1} = 3.67 \times 10^4 P_C \left(\frac{P}{P_C}\right)^{0.35} \left[1 - \frac{P}{P_C}\right]^{0.9} \quad (15.97)$$

where q_{C1} = critical heat flux for single tube ($W \cdot m^{-2}$)

This can be corrected for tube bundles by[15]:

$$q_C = q_{C1} \phi_B \quad (15.98)$$

where q_C = critical heat flux for the tube bundle ($W \cdot m^{-2} \cdot K^{-1}$)
$\phi_B = \dfrac{3.1 \pi D_B L}{A}$
D_B = tube bundle diameter (m)
L = tube length (m)
A = heat transfer area (m^2)

The design of vertical thermosyphon reboilers requires iterative calculations in which the exchanger needs to be divided into zones. The energy and pressure balances need to be performed simultaneously. Frank and Prickett[17] performed a range of detailed simulations and presented the results graphically. This can be used as the basis of preliminary design.

The graphical data can be correlated as:

For *Aqueous Solutions*

$$q = 384.52 \Delta T + 130.07 \Delta T^2 - 2.4204 \Delta T^3 \quad (15.99)$$

where q = heat flux ($W \cdot m^{-2}$)
ΔT = mean temperature difference between the heat transfer surface and the bulk fluid (K)

For *Organic Liquids*

$$q = -100.98 + 1705.9 \Delta T + 26.37 \Delta T^2 - 0.288 \Delta T^3 - 5902.8 T_R \Delta T + 6031.3 T_R^2 \Delta T \quad (15.100)$$

where T_R = reduced temperature of fluid
$= T/T_C$
T = temperature (K)
T_C = critical temperature (K)

For the simulations, the heating was assumed to be supplied using saturated steam on the shell-side with a combined condensation and fouling coefficient of 5700 $W \cdot m^{-2} \cdot K^{-1}$. The fouling coefficient on the tube-side was assumed to be 5700 $W \cdot m^{-2} \cdot K^{-1}$.

If these values are appropriate, then the overall heat transfer coefficient can be calculated from:

$$U = \frac{q}{\Delta T} \tag{15.101}$$

The correlation should be used with caution outside the range $0.6 < T_R < 0.8$ and should not be used below a pressure of 0.3 bar. When dealing with a clean, nondegrading material, the process fouling coefficient should be increased to around 11,000 $W \cdot m^{-2} \cdot K^{-1}$, but should be reduced to 1400 to 1900 $W \cdot m^{-2} \cdot K^{-1}$ for material that has a tendency to polymerize[17]. If a shell-side coefficient of process fouling coefficient different from 5700 $W \cdot m^{-2} \cdot K^{-1}$ is required, the corrected overall heat transfer coefficient can be calculated from[17]:

$$\frac{1}{U'} = \frac{1}{U} - \frac{1}{5700} + \frac{1}{h_S} - \frac{1}{5700} + \frac{1}{h_{TF}} \tag{15.102}$$

where U' = corrected overall heat transfer coefficient ($W \cdot m^{-2} \cdot K^{-1}$)

h_S = required shell-side coefficient including fouling ($W \cdot m^{-2} \cdot K^{-1}$)

h_{TF} = required process fouling coefficient ($W \cdot m^{-2} \cdot K^{-1}$)

The mean temperature difference should be less than 35 to 55°C. This will avoid excessive fouling and excessive vaporization per pass (i.e. low recirculation ratio), leading to poor heat transfer in the upper parts of the tubes as heat transfer to a liquid annulus is replaced by heat transfer to a mist.

Great caution should be exercised regarding correlations for boiling: they are notoriously unreliable. Unlike other heat transfer phenomena, the goal of reliable prediction of boiling rates has proved to be elusive. Many correlations other than those given here are available in the literature. Their predictions of boiling coefficients for the same duty can differ by an order of magnitude. Even the predictions from detailed simulations should be treated with great caution.

Rather than use natural circulation, as in the designs in Figure 15.13, the liquid can be fed to the reboiler by a pump using *forced circulation*. Vaporization in forced convection reboilers is usually limited to be less than 1 to 5%. In some cases, it might be desirable to suppress boiling inside the heat exchanger completely by installing a control valve at the exchanger outlet to increase the pressure in the exchanger and suppress boiling. The liquid leaving the exchanger will then partially vaporize as the liquid pressure is decreased across the valve. Suppression of boiling inside the heat exchanger might be desirable if it leads to excessive fouling. In preliminary design, forced convection reboilers can be based on the assumption that heat transfer is by forced convection only. This will give a conservative design if some boiling is allowed in the exchanger and will overestimate the heat transfer area. High tube velocities of the order of 3 to 5 $m \cdot s^{-1}$ or higher are used to reduce fouling.

If forced-convective boiling is to be carried out, the boiling can take place either on the shell-side or the tube-side. The designs for shell-side boiling are essentially the same as any 1–1 or 1–2 exchanger. Boiling on the tube-side will normally be carried out in a 1–1 exchanger.

Forced convection suppresses nucleate boiling but introduces a significant component of forced convection heat transfer. In fact, in most practical situations, including natural circulation, both forced convection and nucleate boiling are important. In forced-convective boiling, the boiling heat transfer coefficient can be estimated by a combination of convective and nucleate boiling heat transfer with the nucleate boiling component reduced by a suppression factor. There is significant uncertainty associated with the prediction of such components. Because the amount of vapor changes through the exchanger, the exchanger needs to be divided into zones and the correlations applied in each zone.

Natural circulation will lead to cheaper designs than forced circulation, but forced circulation can lead to lower fouling than the corresponding natural circulation designs.

Finally, if vaporization of a liquid is required for applications other than distillation, then the same principles and methods can be applied. The one distinctive difference is that with all designs, with the exception of kettle designs, a vapor–liquid separation device will be required at the vaporizer outlet, as illustrated in Figure 15.15.

Example 15.5 A reboiler is required to supply 0.1 $kmol \cdot s^{-1}$ of vapor to a distillation column. The column bottom product is almost pure butane. The column operates with a pressure at the bottom of the column of 19.25 bar. At this pressure, the butane vaporizes at a temperature of 112°C. The vaporization can be assumed to be essentially isothermal and is to be carried out using steam with a condensing temperature of 140°C. The heat of vaporization for butane is 233,000 $J \cdot kg^{-1}$, its critical pressure 38 bar, critical temperature 425.2 K and molar mass 58 $kg \cdot kmol^{-1}$. Steel tubes with 30 mm outside diameter, 2 mm wall thickness and length 3.95 m are to be used. The thermal conductivity of the tube wall can be taken to be 45 $W \cdot m^{-1} \cdot K^{-1}$. The film coefficient (including fouling) for the condensing steam can be assumed to be 5700 $W \cdot m^{-2} \cdot K^{-1}$. Estimate the heat transfer area for

a. kettle reboiler
b. vertical thermosyphon reboiler

Solution

a. Heat load $= 0.1 \times 58 \times 233{,}000$

$$= 1.351 \times 10^6 \text{ W}$$

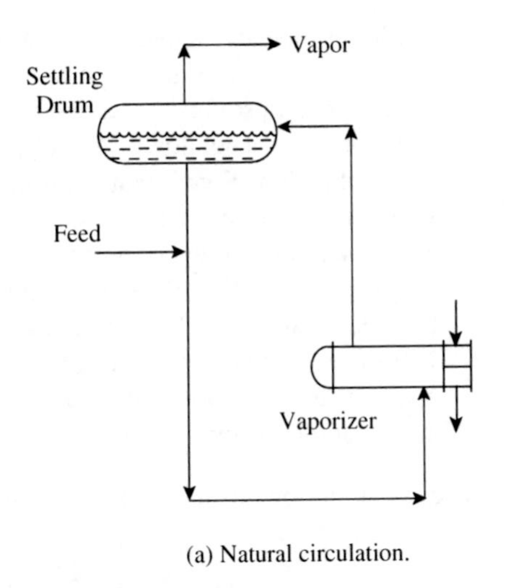

(a) Natural circulation.

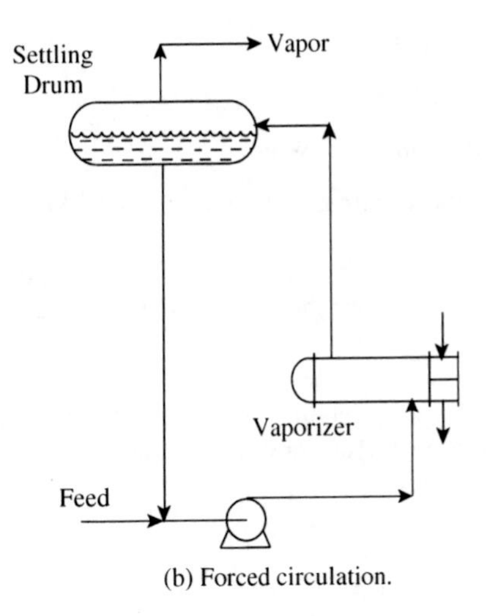

(b) Forced circulation.

Figure 15.15 Process vaporizer arrangements.

The boiling film coefficient for a kettle reboiler can be estimated from the correlation for pool boiling. Equation 15.96 gives one such method due to Palen[15]. However, the correlation requires the heat flux to be known, and therefore the heat transfer area to be known. Hence the calculation will need to be iterative. An initial estimate of the overall heat transfer coefficient of 2000 W·m^{-2}·K^{-1} gives:

$$A = \frac{Q}{U\Delta T} = \frac{1.351 \times 10^6}{2000 \times (140 - 112)} = 24.1 \text{ m}^2$$

$$q = \frac{Q}{A} = \frac{1.351 \times 10^6}{24.1} = 56{,}100 \text{ W·m}^{-2}$$

Now the boiling film coefficient can be calculated:

$$h_{NB} = 0.182 P_C^{0.67} q^{0.7} \left(\frac{P}{P_C}\right)^{0.17}$$

$$= 0.182 \times 38^{0.67} \times 56{,}100^{0.7} \left(\frac{19.25}{38}\right)^{0.17}$$

$$= 3914 \text{ W·m}^{-2}\text{·K}^{-1}$$

The overall heat transfer coefficient can be calculated from:

$$\frac{1}{U} = \frac{1}{h_{NB}} + \frac{1}{h_{SF}} + \frac{d_O}{2k}\ln\frac{d_O}{d_I} + \frac{d_O}{d_I}\frac{1}{h_{TF}} + \frac{d_O}{d_I}\frac{1}{h_T}$$

$$= \frac{1}{3914} + \frac{1}{5700} + \frac{0.03}{2 \times 45}\ln\frac{0.03}{0.026} + \frac{0.03}{0.026}\left(\frac{1}{5700}\right)$$

$$U = 1468 \text{ W·m}^{-2}\text{·K}^{-1}$$

$$Q = UA\Delta T = 1468 \times 24.1 \times 28 = 0.991 \times 10^6 \text{ W}$$

The calculated heat duty does not agree with the specified duty of 1.351×10^6 W. To make the duty balance requires the heat transfer area to be adjusted by trial and error. At each value of A, a new heat flux and boiling film coefficient is calculated. This allows the new overall heat transfer coefficient to be calculated, and so on, until the calculated heat duty agrees with a specified duty. This can be readily done using a spreadsheet solver. The result is:

$$Q = 1.351 \times 10^6 \text{ W}$$

$$h_{NB} = 2883 \text{ W·m}^{-2}\text{·K}^{-1}$$

$$U = 1295 \text{ W·m}^{-2}\text{·K}^{-1}$$

$$A = 37.3 \text{ m}^2$$

$$q = 36{,}250 \text{ W·m}^{-2}$$

This heat flux must be checked to see if it is below the critical heat flux. The critical heat flux for a single tube is given by the Mostinski Equation (Equation 15.97):

$$q_{C1} = 3.67 \times 10^4 P_C \left(\frac{P}{P_C}\right)^{0.35}\left[1 - \frac{P}{P_C}\right]^{0.9}$$

$$= 5.82 \times 10^5 \text{ W·m}^{-2}$$

To correct this for the bundle requires the bundle diameter to be calculated from Equation 15.37:

$$D_B = \left(\frac{4 p_C p_T^2 A}{\pi\, d_O L}\right)^{\frac{1}{2}}$$

Assume a pitch of $1.25d_O$ and square configuration with a tube length of 3.95 m:

$$D_B = \left(\frac{4 \times 1 \times (0.03 \times 1.25)^2 \times 37.3}{\pi \times 0.03 \times 3.95}\right)^{\frac{1}{2}} = 0.75 \text{ m}$$

$$\phi_B = \frac{3.1\pi D_B L}{A} = \frac{3.1 \times \pi \times 0.75 \times 3.95}{37.3} = 0.77$$

$$q_C = q_{C1}\phi_B$$
$$= 5.82 \times 10^5 \times 0.77$$
$$= 4.48 \times 10^5 \text{ W·m}^{-2}$$

The flux is well below the maximum predicted by the Mostinski Equation.

It should be noted that the shell diameter will be around 40% greater than that of the tube bundle to allow for vapor disengagement.

b. For a vertical thermosyphon reboiler, the heat flux can be approximated using Equation 15.100 for organic liquids.

$$T_R = \frac{T}{T_C}$$
$$= \frac{112 + 273.15}{425.2}$$
$$= 0.91$$

This is outside the range over which the original data was correlated and hence the results should be treated with caution.

$$q = -100.98 + 1705.9 \times 28 + 26.37 \times 28^2 - 0.288 \times 28^3$$
$$-5902.8 \times 0.91 \times 28 + 6031.3 \times 0.91^2 \times 28$$
$$= 51{,}460 \text{ W·m}^{-2}$$

$$U = \frac{q}{\Delta T}$$
$$= \frac{51{,}460}{28}$$
$$= 1838 \text{ W·m}^{-2}\text{·K}^{-1}$$

This does not need to be corrected using Equation 15.102 as the steam condensing film coefficient and process fouling coefficient agree with the assumptions on which the correlation is based.

$$A = \frac{Q}{q}$$
$$= \frac{1.351 \times 10^6}{51{,}460}$$
$$= 26.3 \text{ m}^2$$

On the basis of these calculations, the vertical thermosyphon would appear to be the best option. However, the correlations to predict boiling have an extremely low reliability. Predicted minor variations in heat transfer area should not be used to choose options, as the predictions for boiling are so unreliable.

15.9 OTHER TYPES OF HEAT EXCHANGE EQUIPMENT

While the shell-and-tube heat exchanger is the most common type of heat exchanger in the process industries, it does have some significant limitations:

a. The flow is not truly countercurrent. Even the 1–1 exchanger is not truly countercurrent as the flow on the shell-side is partially cross-flow. This means that the shell-and-tube exchanger is mostly limited to transfer heat with a minimum temperature difference of 10°C. Designs can in some cases achieve a smaller temperature difference (perhaps down to 5°C), but care is needed in the application. Some heat transfer duties demand very small temperature differences.
b. The *area density* (heat transfer area per unit of volume of exchanger) is relatively low. A conventional shell-and-tube heat exchanger has an area density of the order of 100 m^2·m^{-3}. Other designs of the heat exchanger can achieve area densities of 300 m^2·m^{-3} to 700 m^2·m^{-3} and in some designs greater than 1000 m^2·m^{-3}.

The most important alternatives to shell-and-tube designs are:

1. *Gasketed plate heat exchanger.* After the shell-and-tube heat exchanger, probably the most commonly used alternative is the *gasketed plate heat exchanger.* This consists of a series of parallel plates with gaskets between the plates to provide a fluid seal. The plates are corrugated both to increase the turbulence (and hence the film transfer coefficients) and to give mechanical rigidity. The enhancement of heat transfer coefficients from the corrugations is a particular advantage when heating and cooling more viscous materials. The turbulence promoted by the corrugations also helps to reduce fouling relative to plain surfaces. The plates are held together and compressed in a frame by the use of lateral bolts. The basic arrangement is illustrated in Figure 15.16. Each corrugated plate is provided with four ports. The gasket arrangement around the four ports forces the hot and cold fluids to flow down alternate channels. This provides countercurrent flow, allowing temperature differences between hot and cold fluids down to around 1°C. Most applications for gasketed plate heat exchangers are for liquid–liquid duties, but can also be applied to condensing and evaporating duties. The limitations of the gasket seals mean that applications are normally restricted to be between −30 and 200°C with pressures up to 20 bar.

A gasketed plate heat exchanger is usually significantly cheaper than a shell-and-tube heat exchanger for the same duty. This is especially the case if the shell-and-tube heat exchanger must be fabricated in more expensive materials such as stainless steel.

Another advantage in retrofit is that an existing frame can often accommodate additional plates if a higher capacity is required. For the same flowrate, an increase in the number of plates decreases the flowrate through the channels and therefore the heat transfer coefficients. However, if the flowrate increases, the number of plates can be increased to accommodate a higher duty (at the expense of increased pressure drop).

The flow arrangement shown in Figure 15.16 involves a single pass for each fluid. More complex flow arrangements can also be used.

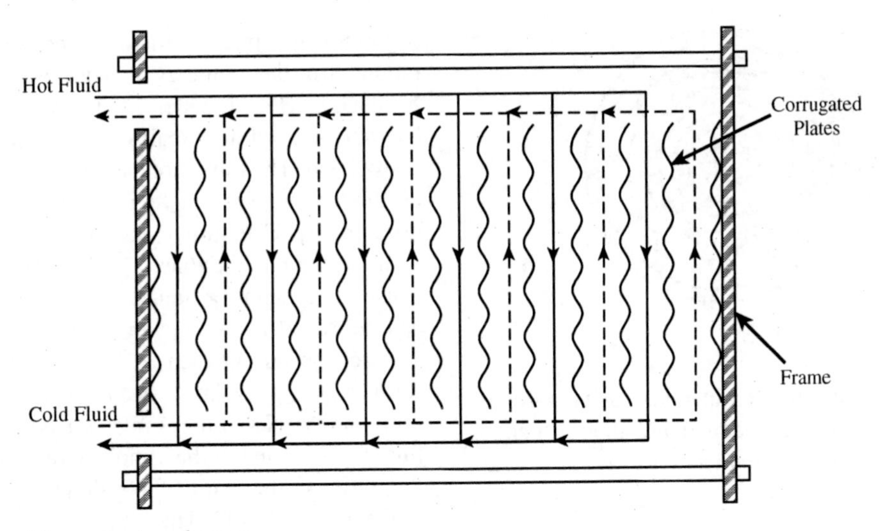

Figure 15.16 Plate-and-frame heat exchanger.

2. *Welded plate heat exchangers*. The limitations of the gaskets in gasketed plate heat exchangers can be overcome by welding the plates together. This eliminates both the gaskets and the frame from the design. Elimination of the gaskets extends the range of application to a wider range of temperatures and pressures. The highest pressures for welded plate heat exchangers can be achieved by mounting the plates within a shell. Welded plate heat exchangers can achieve an area density of up to 300 $m^2 \cdot m^{-3}$. The operating temperature range varies between −200 and 900°C. Pressures up to 300 bar can be accommodated.

The costs for this form of exchanger are higher than those for gasketed plate heat exchangers. An important limitation is that they can only be cleaned chemically and not mechanically.

3. *Plate-fin heat exchangers*. Another type of plate heat exchanger is the *plate-fin heat exchanger*. This is illustrated in Figure 15.17. The plate-fin heat exchanger consists of a series of flat plates, between which is a matrix formed from corrugated metal that provides a large extended heat transfer area. The components are bonded together either by brazing or diffusion bonding. Many different surface geometries are available to promote heat transfer. The space between the plates and the surface geometry chosen for each fluid are important degrees of freedom in the design of such units. The area density of such units is typically in the range 850 to 1500 $m^2 \cdot m^{-3}$. The operating range for such units depends both on the bonding technique used and the material of construction. Aluminum-brazed plate-fin heat exchangers are used for cryogenic applications, but can also be used up to temperatures of around 100°C. Stainless steel plate-fin heat exchangers are able to operate up to 650°C and titanium units up to 550°C. Aluminum-brazed units can operate up to 100 bar, stainless steel units up to 50 bar and higher. Higher pressures require diffusion-bonded units. Plate-fin heat exchangers not only have the advantages of high area density and high heat transfer coefficients but also have a number of other advantages that make them overwhelmingly attractive for certain applications. Temperature differences of 1°C or less can be tolerated in such units. Also, the units can be designed to handle multiple streams through the use of complex header arrangements. This allows, in effect, for a heat exchanger network to be accommodated within a single unit.

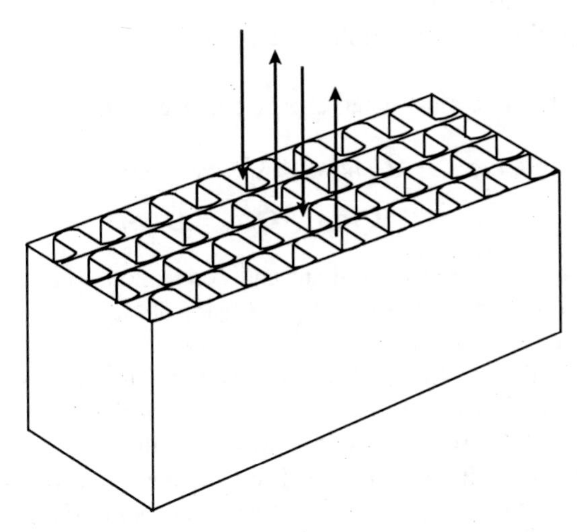

Figure 15.17 Plate-fin heat exchanger.

4. *Spiral heat exchangers*. *Spiral heat exchangers* can be thought of as plate heat exchangers in which the plates are formed into a spiral, as illustrated in Figure 15.18. The channels are closed by gasketed end-plates. The hot fluid enters at the center of the unit and flows from the inside outwards. The cold fluid enters at the periphery and flows towards the center in a countercurrent flow arrangement. The gap between the plates can be adjusted to suit the application. Operating temperatures up to 400°C and operating pressures up to 20 bar are possible. The units have a low tendency to foul, but are easily cleaned by removing the end-plates. Again, true countercurrent flow

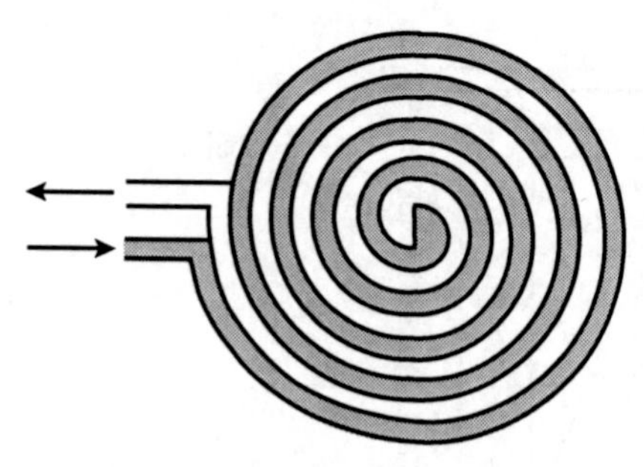

Figure 15.18 Spiral heat exchanger.

is possible and therefore lower temperature driving forces can be tolerated. Spiral heat exchangers are suited to small heat transfer duties that have a tendency to foul.

15.10 FIRED HEATERS

In some situations, process heat needs to be supplied:

a. with a high heat duty (e.g. a very large reboiler)
b. at a high temperature (e.g. at a temperature above which heat can be supplied by steam)
c. with a high heat flux (e.g. heat of reaction in situations where a short residence time in the reactor is required)

In such cases, radiant heat transfer is used from the combustion of fuel in a *fired heater* or *furnace*. Sometimes the function is to purely provide heat; sometimes the fired heater is also a reactor and provides heat of reaction. The special case of steam generation in a fired heater (a steam boiler) will be dealt with in Chapter 23. Fired heater designs vary according to the function, heating duty, type of fuel and the method of introducing combustion air. However, process furnaces have a number of features in common. A simple design is illustrated in Figure 15.19. The chamber where combustion takes place, the *radiant section*

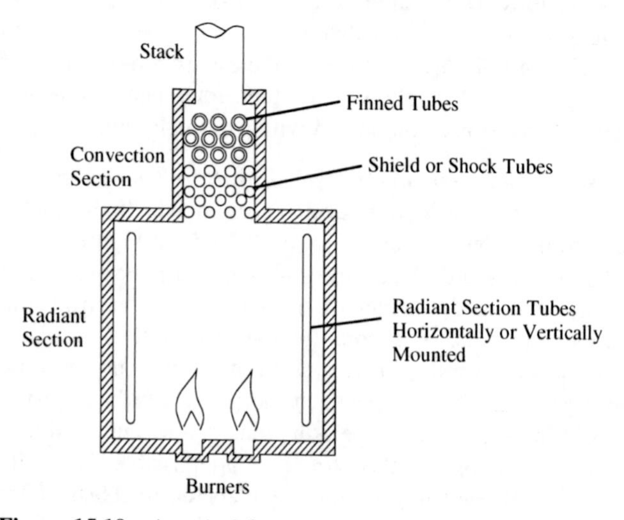

Figure 15.19 A typical furnace arrangement.

or *firebox*, is refractory lined and heat is transferred to tubes mounted in the chamber through which passes the fluid to be heated. Heat transfer in the radiant section is mainly by radiation, with a small contribution (less than 10%) by convection. The tubes are usually mounted around the walls of the radiant section and can be mounted vertically or horizontally. The burners can be mounted in the base or the walls of the radiant zone. The fuel for process-fired heaters is usually gaseous or liquid. The shape of the radiant section can be cylindrical or rectangular.

After the flue gas leaves the combustion chamber, most furnace designs extract further heat from the flue gas in horizontal banks of tubes in a *convection section*, before the flue gas is vented to the atmosphere. The temperature of the flue gases at the exit of the radiant section is usually in the range 700 to 900°C. The first few rows of tubes at the exit of the radiant section are plain tubes, known as *shock tubes* or *shield tubes*. These tubes need to be robust enough to be able to withstand high temperatures and receive significant radiant heat from the radiant section. Heat transfer to the shock tubes is both by radiation and by convection. After the shock tubes, the hot flue gases flow across banks of tubes that usually have extended surfaces to increase the rate of heat transfer to the flue gas. The heat transferred in the radiant section will usually be between 50 and 70% of the total heat transferred.

Heat input to the fired heater is from three sources:

a. net heat of combustion (Q_{COMB})
b. sensible heat of combustion air (Q_{AIR})
c. sensible heat of the fuel, together with heat from atomizing steam if it is used for heavy fuel oil combustion (Q_{FUEL})

The heat output from the fired heater is to four sinks:

a. heat transferred to the radiant section tubes ($Q_{RADIANT}$)
b. heat transferred in the convection section (Q_{CONV})
c. casing losses (Q_{CASING})
d. sensible heat of exit flue gas (Q_{LOSS})

Overall, there must be an energy balance such that:

$$Q_{COMB} = Q_{RADIANT} + Q_{CONV} + Q_{CASING} + Q_{LOSS} - Q_{AIR} - Q_{FUEL} \qquad (15.103)$$

The split between the radiant and convection section heat varies according to the design. Casing losses are usually between 1 and 3% of the heat release from combustion. The heat loss from the stack is constrained by the desire to avoid any condensation of water vapor in the convection section. If there is any sulfur present in the fuel, then the condensate will be corrosive. The temperature at which the flue gas starts to condense is the acid dew point. For sulfur-bearing fuels, the temperature of the flue gas is normally kept above 150 to 160°C. For combustion of sulfur-free gaseous fuels, the temperature can be decreased to below 100°C.

Three methods are used to produce airflow through the fired heater:

a. *Natural Draft*. In natural draft, the pressure in the furnace is maintained slightly below atmospheric by a flow of air created from the difference in density between the hot flue gas in the stack and the ambient air. The stack must therefore be high enough to provide adequate draft.
b. *Forced Draft*. Forced draft uses a fan to create the airflow by blowing air into the furnace. The furnace is slightly above atmospheric pressure. The stack height is now only required to provide adequate gas dispersion.
c. *Induced Draft*. Induced draft uses a fan between the furnace and the stack, leading to a pressure in the furnace slightly below atmospheric. Again, the stack height is only required to provide adequate gas dispersion.

Natural draft is a low capital cost option, but the draft requirements tend to lead to high stack temperature with a low efficiency for the furnace. Forced draft and induced draft are higher capital cost options than natural draft but tend to lead to higher furnace efficiencies as the stack temperature can be lowered significantly.

The preliminary specification of fired heaters is largely based on heat duty and furnace efficiency. A simple model can be developed on the basis of the concept of the *theoretical flame temperature* or *adiabatic combustion temperature*. Theoretical flame temperature is the temperature attained when a fuel is burnt in air or oxygen without loss or gain of heat. A combustion process is the reaction of carbon, hydrogen, sulfur and nitrogen in the fuel with oxygen to produce carbon dioxide, carbon monoxide, water, sulfur dioxide, sulfur trioxide and oxides of nitrogen. Depending on the fuel being solid, liquid or gaseous, the carbon, hydrogen, sulfur and nitrogen in the fuel can be in their elemental forms or as compounds. Nitrogen in combustion air or elemental nitrogen in a gaseous fuel can react with oxygen to form oxides of nitrogen at high temperatures. However, the formation of oxides of nitrogen in this way can be neglected when calculating the theoretical flame temperature, as only small quantities are formed. Also, in a well-designed and operated combustion device, there should be no carbon monoxide formed, but all carbon should be oxidized to carbon dioxide. Any sulfur present in the fuel can be assumed to react to form sulfur dioxide. In practice, some formation of sulfur trioxide will occur, but measurements on combustion systems indicate that sulfur trioxide is only formed in small quantities and less than what would be predicted by thermodynamic equilibrium with the sulfur dioxide. Thus, it can be assumed that all sulfur in the fuel reacts to sulfur dioxide.

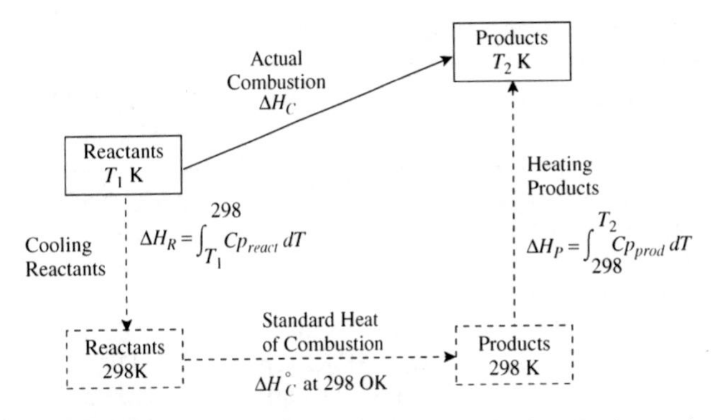

Figure 15.20 The enthalpy change from initial to final state in a combustion process is independent of path.

To calculate the heat release from combustion and the temperature of the products of combustion, the thermodynamic path shown in Figure 15.20 can be followed[18]. The actual combustion process goes from reactants at temperature T_1 to products at temperature T_2. However, it is more convenient to follow the alternative path from reactants at temperature T_1 that are initially cooled (or heated) to standard temperature of 298 K. The combustion reactions are then carried out at a constant temperature of 298 K. Standard heats of combustion are available for this. The products of combustion are then heated from 298 K to the final temperature of T_2. The actual heat of combustion is given by[18]:

$$\Delta H_{COMB} = \Delta H_R + \Delta H_{COMB}^O + \Delta H_P \qquad (15.104)$$

where ΔH_{COMB} = heat of combustion ($J \cdot kmol^{-1}$)
ΔH_R = heat to bring reactants from their initial temperature to standard temperature ($J \cdot kmol^{-1}$)
$= \int_{T_1}^{298} Cp_{react}\, dT$
ΔH_{COMB}^O = standard heat of combustion at 298 K ($J \cdot kmol^{-1}$)
ΔH_P = heat to bring products from standard temperature to the final temperature ($J \cdot kmol^{-1}$)
$= \int_{298}^{T_2} Cp_{prod}\, dT$
Cp_{react} = heat capacity of reactants ($J \cdot kmol^{-1} \cdot K^{-1}$)
Cp_{prod} = heat capacity of products ($J \cdot kmol^{-1} \cdot K^{-1}$)

For adiabatic combustion, $\Delta H_{COMB} = 0$ and Equation 15.89 becomes:

$$\Delta H_P = -\Delta H_R - \Delta H_{COMB}^O \qquad (15.105)$$

Standard heats of combustion are widely available. Table 15.11 provides a list of some combustion reactions.

When using data for standard heats of combustion, care should be taken regarding the initial state of the reactants and the final state of the products. If these do not correspond with the conditions for the actual combustion, then errors can arise. Note in Table 15.11, the final state of water was

Table 15.11 Standard heats of combustion.

Reaction	ΔH^O_{COMB} at 298 K ($MJ \cdot kmol^{-1}$)
$C(sol) + O_2 \rightarrow CO_2$	−393.5
$S(sol) + O_2 \rightarrow SO_2$	−297.1
$CO + \frac{1}{2}O_2 \rightarrow CO_2$	−283.0
$H_2 + \frac{1}{2}O_2 \rightarrow H_2O(vap)$	−241.8
$CH_4(vap) + 2O_2 \rightarrow CO_2 + 2H_2O(vap)$	−802.8
$C_2H_6(vap) + 3\frac{1}{2}O_2 \rightarrow 2CO_2 + 3H_2O(vap)$	−1428.7
$C_3H_8(vap) + 5O_2 \rightarrow 3CO_2 + 4H_2O(vap)$	−2043.2
$nC_4H_{10}(vap) + 6\frac{1}{2}O_2 \rightarrow 4CO_2 + 5H_2O(vap)$	−2657.7
$iC_4H_{10}(vap) + 6\frac{1}{2}O_2 \rightarrow 4CO_2 + 5H_2O(vap)$	−2649.1

taken in all cases to be vapor rather than liquid. This relates to the condition of the combustion products as they would normally be in practice. Combustion data for fuels are given in terms of different conditions for the water that is formed. When the water vapor remains in the vapor phase, the heat of combustion is known as the *net calorific value* or *lower heating value*. When the water condenses, it is known as the *gross calorific value* or *higher heating value*. Data can obviously be converted from one to the other from knowledge of how many molecules of water are formed and the heat of vaporization of water. When dealing with solid fuels, extra care needs to be taken as data can be quoted "as fired" or on a "dry ash-free basis".

Data for some typical gaseous and liquid fuels are given in Tables 15.12 and 15.13.

One further problem needs to be considered before a calculation can be attempted. The heat balance as illustrated in Figure 15.20 involves extremely large temperature changes for the combustion process. This means that the heat capacity will vary significantly, especially for the products of combustion. The variation in heat capacity can be expressed as a polynomial in temperature, as discussed in Chapter 6. However, with combustion processes, care should be taken that the data have been correlated over a wide range of temperature. Thus, for example[18]:

$$C_P = \alpha_O + \alpha_1 T + \alpha_2 T^2 + \alpha_3 T^3 \tag{15.106}$$

where C_P = heat capacity ($J \cdot kmol^{-1} \cdot K^{-1}$ or $kJ \cdot kmol^{-1} \cdot K^{-1}$)
$\alpha_0, \alpha_1, \alpha_2, \alpha_3$ = constants
T = absolute temperature (K)

Thus,

$$\Delta H = \int_{T_1}^{T_2} C_P \, dT$$
$$= \int_{T_1}^{T_2} (\alpha_0 + \alpha_1 T + \alpha_2 T^2 + \alpha_3 T^3) \, dT$$
$$= \left[\alpha_0 T + \frac{\alpha_1 T^2}{2} + \frac{\alpha_2 T^3}{3} + \frac{\alpha_3 T^4}{4}\right]_{T_1}^{T_2} \tag{15.107}$$

where ΔH = enthalpy change from T_1 to T_2 ($J \cdot kmol^{-1}$, $kJ \cdot kmol^{-1}$)

Equation 15.92 and Table 15.14 can be used to calculate the enthalpy changes, for example, in a spreadsheet.

Table 15.14 Heat capacity constants.

Component	C_P ($kJ \cdot kmol^{-1} \cdot K^{-1}$)			
	α_0	$\alpha_1 \times 10^2$	$\alpha_2 \times 10^5$	$\alpha_3 \times 10^9$
O_2	25.4767	1.5202	−0.7155	1.3117
N_2	28.9015	−0.1571	0.8081	−2.8726
H_2O	32.2384	0.1923	1.0555	−3.5952
CO_2	22.2570	5.9808	−3.5010	7.4693
SO_2	25.7781	5.7945	−3.8112	8.6122
CO	28.3111	0.1675	0.5372	−2.2219
H_2	29.1066	−0.1916	0.4004	−0.8704
CH_4	19.8873	5.0242	1.2686	−11.0113
C_2H_6	6.8998	17.2664	−6.4058	7.2850
C_3H_8	−4.0444	30.4757	−15.7214	31.7359
n-C_4H_{10}	3.9565	37.1495	−18.3382	35.0016
i-C_4H_{10}	−7.9131	41.6000	−23.0065	49.9067

Table 15.13 Data for some typical liquid fuels.

Fuel	Composition (% by mass)				Calorific value ($MJ \cdot kg^{-1}$)	
	C	H	S	O+N +Ash	Gross	Net
Light fuel oil	85.6	11.7	2.5	0.2	43.5	41.1
Medium fuel oil	85.6	11.5	2.6	0.3	43.1	40.8
Heavy fuel oil	85.4	11.4	2.8	0.4	42.9	40.5

Table 15.12 Data for some typical natural gases.

Gas	Composition (% by volume)								Calorific value ($MJ \cdot m^{-3}$)	
	CO_2	N_2	CH_4	C_2H_6	C_3H_8	$C_{10}H_{14}$	C_5H_{12}	C_5H_{14}	Gross	Net
North Sea	0.2	1.5	94.4	3.0	0.5	0.2	0.1	0.1	38.62	34.82
Groningen	0.9	15.0	81.8	2.7	0.4	0.1	0.1	–	33.28	30.00
Algeria	0.2	5.5	83.8	7.1	2.1	0.9	0.4	–	39.1	

Table 15.15 Mean heat capacity data.

T(°C)	$\overline{C_P}$ (kJ·kmol^{-1}·K^{-1})											
	O_2	N_2	H_2O	CO_2	SO_2	CO	H_2	CH_4	C_2H_6	C_3H_8	n-C_4H_{10}	i-C_4H_{10}
25	29.41	29.07	33.65	37.17	39.89	29.12	28.87	35.69	52.86	73.65	99.30	96.95
100	29.82	29.18	33.94	38.65	41.24	29.40	28.88	37.76	57.87	81.65	109.20	107.56
200	30.33	29.34	34.36	40.47	42.87	29.64	28.92	40.51	64.22	91.59	121.56	120.68
300	30.81	29.54	34.82	42.12	44.33	29.89	28.98	43.25	70.20	100.75	132.99	132.70
400	31.26	29.76	35.31	43.61	45.61	30.16	29.05	45.97	75.84	109.19	143.55	143.69
500	31.67	30.00	35.83	44.96	46.74	30.44	29.14	48.64	81.13	116.95	153.29	153.73
600	32.05	30.26	36.38	46.17	47.74	30.73	29.25	51.26	86.08	124.07	162.27	162.88
700	32.41	30.53	36.94	47.25	48.60	31.02	29.37	53.80	90.72	130.61	170.53	171.23
800	32.73	30.80	37.52	48.23	49.34	31.31	29.51	56.24	95.05	136.61	178.14	178.84
900	33.04	31.09	38.10	49.10	49.99	31.59	29.65	58.58	99.08	142.12	185.14	185.80
1000	33.32	31.36	38.68	49.88	50.54	31.88	29.80	60.79	102.82	147.19	191.59	192.18
1100	33.58	31.64	39.27	50.58	51.02	32.15	29.97	62.86	106.28	151.86	197.53	198.05
1200	33.82	31.90	39.84	51.21	51.43	32.40	30.14	64.77	109.48	156.19	203.03	203.49
1300	34.04	32.15	40.39	51.78	51.79	32.64	30.32	66.51	112.42	160.22	208.12	208.58
1400	34.25	32.39	40.93	52.31	52.12	32.86	30.50	68.05	115.12	163.99	212.88	213.38
1500	34.44	32.60	41.44	52.80	52.42	33.06	30.69	69.39	117.58	167.57	217.34	217.98
1600	34.63	32.78	41.93	53.27	52.70	33.23	30.87	70.50	119.83	170.98	221.56	222.44
1700	34.81	32.93	42.37	53.73	52.99	33.37	31.06	71.37	121.86	174.28	225.59	226.84
1800	34.97	33.05	42.78	54.18	53.29	33.48	31.25	71.98	123.69	177.53	229.49	231.27
1900	35.14	33.13	43.13	54.65	53.62	33.55	31.44	72.32	125.33	180.76	233.31	235.78
2000	35.30	33.16	43.44	55.14	53.99	33.58	31.62	72.37	126.79	184.02	237.09	240.47
2100	35.46	33.14	43.69	55.66	54.41	33.60	31.80	72.11	128.08	187.36	240.91	245.39
2200	35.62	33.07	43.87	56.22	54.89	33.51	31.98	71.52	129.22	190.84	244.79	250.63

Alternatively, the heat capacity data can be used to derive mean heat capacities. The mean heat capacity can be defined as[18]:

$$\overline{C_P} = \frac{\int_{T_1}^{T_2} C_P \, dT}{T_2 - T_1} \tag{15.108}$$

where $\overline{C_P}$ = mean heat capacity between temperatures T_1 and T_2 (J·kmol^{-1}·K^{-1} or kJ·kmol^{-1}·K^{-1})

Table 15.15 presents mean heat capacity data between 25°C and a given temperature for a range of temperatures.

Example 15.6 A gas, which can be considered to be pure methane, is to be used as fuel in a furnace. Both the fuel gas and combustion air are both at 25°C. Calculate the theoretical flame temperature if the methane is burnt in:

a. its stoichiometric ratio in dry air
b. 15% excess dry air
c. 15% excess air with a relative humidity of 60%

Solution

a. Stoichiometric air requirements are defined by

$$\underset{1\text{ kmol}}{CH_4} + \underset{2\text{ kmol}}{2O_2} \longrightarrow \underset{1\text{ kmol}}{CO_2} + \underset{2\text{ kmol}}{2H_2O}$$

2 kmol oxygen are required per kmol of methane burnt. If air is assumed to be 21% of oxygen and nitrogen is assumed inert, then combustion products are:

$$O_2 = 2 \text{ kmol}$$

$$N_2 = \frac{2 \times 0.79}{0.21} = 7.52 \text{ kmol}$$

$$H_2O = 2 \text{ kmol}$$

$$CO_2 = 1 \text{ kmol}$$

Since the fuel and combustion air are at the standard temperature of 25°C, $\Delta H_R = 0$. To calculate ΔH_p, start by estimating the theoretical flame temperature to be 2000°C. From:

$$\overline{C_{P\,O_2}} = 35.30 \text{ kJ·kmol}^{-1}\text{·K}^{-1}$$

$$\overline{C_{P\,N_2}} = 33.16 \text{ kJ·kmol}^{-1}\text{·K}^{-1}$$

$$\overline{C_{P\,H_2O}} = 43.44 \text{ kJ·kmol}^{-1}\text{·K}^{-1}$$

$$\overline{C_{P\,CO_2}} = 55.14 \text{ kJ·kmol}^{-1}\text{·K}^{-1}$$

$$\Delta H_P = (7.52 \times 33.16 + 2 \times 43.44 + 1 \times 55.14)(T - 25)$$

$$= 391.38(T - 25)$$

From Table 15.11, $\Delta H^o_{COMB} = 802.8 \times 10^3$ kJ·kmol^{-1}. Thus, from an energy balance:

$$\Delta H_P = -\Delta H_R - \Delta H^o_{COMB}$$

$$391.38(T_{TFT} - 25) = 0 - (-802.8 \times 10^3)$$

$$T_{TFT} = 2076°\text{C}$$

This agrees well with the initial estimate. Had there been significant disagreement, then revised mean heat capacities would need to be taken and the calculation repeated.

b. If 15% excess air is used, then combustion products are:

$$O_2 = 2 \times 0.15 = 0.3 \text{ kmol}$$

$$N_2 = \frac{2 \times 0.79}{0.21} \times 1.15 = 8.65 \text{ kmol}$$

$$H_2O = 2 \text{ kmol}$$

$$CO_2 = 1 \text{ kmol}$$

Again, estimate the theoretical flame temperature to be 2000°C. The mean heat capacities are as before and the heat balance is now:

$$\Delta H_P = (0.3 \times 35.3 + 8.65 \times 33.16 + 2 \times 43.44 + 1 \times 55.14)(T_{TFT} - 25)$$

$$= 439.44(T_{TFT} - 25)$$

$$\Delta H_P = \Delta H_R - \Delta H^o_{COMB}$$

$$439.44(T_{TFT} - 25) = 0 - (-802.8 \times 10^3)$$

$$T_{TFT} = 1852°\text{C}$$

c. So far the combustion air has been assumed to be dry. If the combustion air is humid, then the water vapor will act as another inert in the combustion. The relative humidity of air is the percentage relative to saturation.

Saturated vapor pressure of water at 25°C (from steam tables)

$$= 0.03166 \text{ bar}$$

For 60% relative humidity, the partial pressure of water vapor is given by:

$$p_{H_2O} = 0.03166 \times 0.6$$

$$= 0.0190 \text{ bar}$$

Thus, the mole fraction of water vapor in the combustion air for a pressure of 1 atm (1.013 bar) is given by:

$$y_{H_2O} = \frac{p_{H_2O}}{P}$$

$$= \frac{0.0190}{1.013}$$

$$= 0.0188$$

The combustion air for 15% excess

$$= (2 + 7.52)1.15$$

$$= 10.95 \text{ kmol}$$

Water from combustion air

$$= 10.95 \times \frac{y_{H_2O}}{1 - y_{H_2O}}$$

$$= 10.95 \times \frac{0.0188}{1 - 0.0188}$$

$$= 0.21 \text{ kmol}$$

The total water vapor in the combustion products

$$= 2 + 0.21$$

$$= 2.21 \text{ kmol}$$

Again, estimate the theoretical flame temperature to be 2000°C. The mean heat capacities are as before and the heat balance is:

$$\Delta H_P = (0.3 \times 35.3 + 8.65 \times 33.16 + 2.21 \times 43.44 + 1 \times 55.14) \times (T_{TFT} - 25)$$

$$= 448.57(T_{TFT} - 25)$$

$$448.57(T_{TFT} - 25) = 0 - (-802.8 \times 10^3)$$

$$T_{TFT} = 1815°\text{C}$$

Again, iterate with revised heat capacities for greater accuracy.

It can be seen that excess air and humidity in the combustion air both act to reduce the theoretical flame temperature. However, the excess air has the more significant effect. In some combustion processes, steam is injected into the combustion process to decrease the flame temperature to decrease NO_x formation. This will be discussed later in Chapter 25.

In these calculations, the fuel and combustion air were both at the standard temperature of 25°C. If the temperature of either had been below 25°C, then ΔH_R would have acted to decrease the theoretical flame temperature. If either had been above 25°C, the effect would have been to increase the theoretical flame temperature. One energy conservation technique sometimes used in furnace design is to use waste heat to preheat the combustion air. This has the effect of increasing the theoretical flame temperature, and as will be seen later, increases the fuel efficiency.

It should be emphasized that the theoretical and real flame temperatures will be significantly different. The real flame temperature will be lower than the theoretical flame temperature because, in practice, heat is lost from the flame (mainly due to radiation). Also, part of the heat released provides heat for a variety of endothermic dissociation reactions, which occur at high temperatures, such as:

$$CO_2 \rightleftharpoons CO + O$$

$$H_2O \rightleftharpoons H_2 + O$$

$$H_2O \rightleftharpoons H + OH$$

However, as the temperature of the flue gas decreases, as heat is extracted, the dissociation reactions reverse and the heat is released. Thus, although theoretical flame temperature does not reflect the true flame temperature, it does provide a convenient reference to indicate how much heat is actually released by combustion as the flue gas is cooled. Figure 15.21 shows the flue gas starting from the theoretical flame temperature. This is cooled

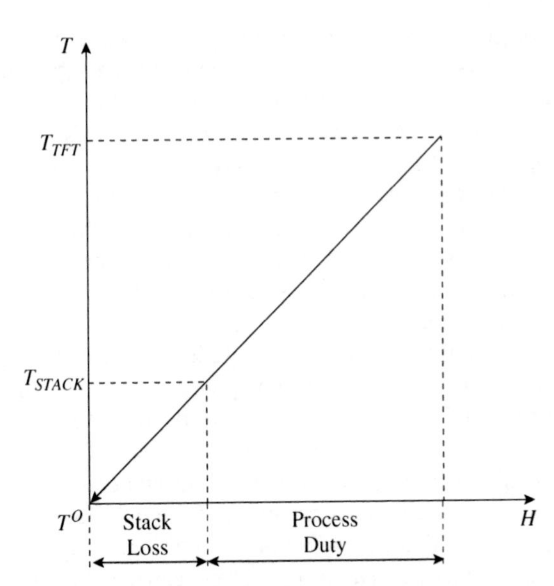

Figure 15.21 The flue gas profile for a fired heater.

in the radiant and convection sections to provide the process duty. The temperature at the exit of the convection section is the stack temperature. The cooling from the stack temperature to the ambient temperature is the *stack loss*. The temperatures in the radiant section (above 700 to 900°C) are thus not accurately represented by the profile based on the theoretical flame temperature. The temperature differences between the sources of radiant heat and the furnace tubes in the radiant zone of the furnace cannot be represented accurately by simple models, and requires a detailed simulation model. Such a model must account for the burner design, the absorption and emission of radiant heat for the surfaces in the radiant zone of the particular geometry, and so on. Fortunately, this does not usually present a problem in using the simple model represented in Figure 15.21 for preliminary design, since temperature differences in the radiant zone are very large anyway. Also, as the flue gas is cooled and passes through the convection section of the furnace, the temperatures are more representative of what they would be in practice. However, it should be emphasized again that the simple model in Figure 15.21 does allow the correct heat duty to be represented. Furnace efficiency can be defined as:

$$\text{Furnace efficiency} = \frac{\text{Heat to process}}{\text{Heat released by fuel}} \qquad (15.109)$$

The profile shown in Figure 15.21 represents the furnace efficiency, if the casing heat losses are neglected. Making this assumption, the process duty plus the stack loss represents the heat released by the fuel.

Example 15.6 For the three cases in Example 15.5, determine the furnace efficiency for an assumed stack temperature of 100°C.

Solution

a. For dry, stoichiometric air $T_{TFT} = 2076°C$:

$$\text{Furnace efficiency} = \frac{\text{Heat to process}}{\text{Heat released by fuel}}$$
$$= \frac{391.38(2076 - 100)}{802.8 \times 10^3}$$
$$= 0.963$$

b. For dry air in 15% excess, $T_{TFT} = 1852°C$:

$$\text{Furnace efficiency} = \frac{439.44(1852 - 100)}{802.8 \times 10^3}$$
$$= 0.96$$

c. For 15% excess air with 60% relative humidity $T_{TFT} = 1815°C$:

$$\text{Furnace efficiency} = \frac{448.57(1815 - 100)}{802.8 \times 10^3}$$
$$= 0.958$$

For a given stack temperature, the higher the theoretical flame temperature, the higher the furnace efficiency. However there is a minimum excess air required to ensure that the combustion is itself efficient.

All combustion processes work with an excess of air or oxygen to ensure complete combustion of the fuel. Excess air typically ranges between 5 and 25% depending on the fuel, burner design and furnace design. Typical excess air is 10 to 15% for gaseous fuels, 15 to 20% for liquid fuels and 20 to 25% for solid fuels. Natural draft furnaces normally work with a higher excess air than forced draft designs. The excess oxygen in the flue gas should normally be around 3%. However, in large furnaces, the excess oxygen might be significantly less (e.g. 2% for fuel oil, 1% for natural gas).

As excess air is reduced, theoretical flame temperature increases. This has the effect of reducing the stack loss and increasing the thermal efficiency of the furnace for a given process heating duty. Alternatively, if the combustion air is preheated (e.g. by heat recovery), then again the theoretical flame temperature increases, reducing the stack loss.

Although, higher flame temperatures reduce the fuel consumption for a given process heating duty, there is one significant disadvantage. Higher flame temperatures increase the formation of oxides of nitrogen, which are environmentally harmful.

Obviously, the lower the stack temperature, the higher the furnace efficiency. As already noted, it is desirable to avoid condensation in the convection section and the stack. If there is any sulfur in the fuel, the condensate will be corrosive. There is thus a practical minimum to which a flue gas can be cooled without condensation causing corrosion in the stack, known as the *acid dew point*. If there is any sulfur in the fuel, the stack temperature is normally kept above 150 to 160°C. Natural gas can normally be cooled to a significantly lower temperature (100°C or less), because it is usually treated to remove sulfur down to less than 1 ppm.

In preliminary design, the heat duty and furnace efficiency are the prime considerations. However, if the tube area needs to be specified, a preliminary estimate can be obtained from an assumed flux. In the radiant section, this usually lies in the range of 45,000 $W \cdot m^{-2}$ to 65,000 $W \cdot m^{-2}$ of tube surface, with a value of around 55,000 $W \cdot m^{-2}$ most often used. The heat flux is particularly important if a reaction is being carried out in the furnace tubes. Overall heat transfer coefficients in the convection section are in the range 20 to 50 $W \cdot m^{-2} \cdot K^{-1}$.

15.11 HEAT TRANSFER EQUIPMENT – SUMMARY

Of the many types of heat transfer equipment used in the process industries, the shell-and-tube heat exchanger is by far the most common. A number of different flow arrangements are possible with shell-and-tube heat exchangers, but the one shell pass–one tube pass and the one shell pass–two tube pass arrangements are the most common.

Resistance to heat transfer across the tube wall for shell-and-tube heat exchangers is made up of five individual resistances to heat transfer:

- shell-side film coefficient
- shell-side fouling coefficient
- tube wall coefficient
- tube-side fouling coefficient
- tube-side film coefficient

These are combined to give an overall heat transfer coefficient in which one of the individual coefficients might be controlling.

At the conceptual stage for the design of shell-and-tube heat exchangers, the calculation of heat transfer coefficients and pressure drops should depend as little as possible on the detailed geometry. Simple models are available in which both heat transfer coefficients and pressure drops can be related to the fluid velocity. It is then possible to derive a correlation between the heat transfer coefficient, pressure drop and heat transfer surface area by using the velocity as a bridge between the two. The 1–1 heat exchanger is designed as a countercurrent device. The 1–2 heat exchanger has an element of cocurrent flow as well as the countercurrent flow. This must be corrected by the introduction of a correction to the logarithmic mean temperature difference.

In retrofit situations, existing heat exchangers might be subjected to changes in flowrate, heat transfer duty, temperature differences or fouling characteristics. Heat transfer coefficients and pressure drops can be approximated from their original values from the change in flowrate. Increased duties can be accommodated by changing the tube bundle to incorporate a greater heat transfer area in the same shell, or by the use of tube insert enhancement devices.

Shell-and-tube heat exchangers are also used extensively for condensing duties. Condensers can be horizontally or vertically mounted with the condensation on the tube-side or the shell-side. Condensation normally takes place on the shell-side of horizontal exchangers and the tube-side of vertical exchangers.

Reboilers are required on distillation columns to vaporize a fraction of the bottom product. Three common designs of reboiler are used: the kettle reboiler, the horizontal and vertical thermosyphon reboilers. The preliminary design of kettle and horizontal thermosyphon reboilers can be based on correlations for pool boiling. The design of vertical thermosyphon reboilers requires the hydraulic and thermal design to be carried out simultaneously. Preliminary design of such units can be based on correlations derived from a large number of detailed designs. Great care must be exercised in the preliminary design of reboilers as the predictions of the correlations are extremely unreliable.

While the shell-and-tube heat exchanger is the most commonly used in the process industries, it has the disadvantages that the flow is not truly countercurrent, which limits the minimum temperature difference that can be accommodated, and the area density is relatively low. Commonly used alternatives for shell-and-tube heat exchangers are:

- Gasketed plate heat exchangers
- Welded plate heat exchangers
- Plate-fin heat exchangers
- Spiral heat exchangers

Some heat transfer operations demand a high heat duty, high temperature and/or high heat flux. In such cases, radiant heat transfer is used from the combustion of fuel in a fired heater. Fired heater designs vary according to the function, heating duty, type of fuel and method of introducing combustion air. In conceptual design, the heat duty in the various sections of the furnace is more important than the heat transfer area, as the cost in preliminary design is based on the heat duty. The theoretical flame temperature provides a simple model to allow the heat duty to be determined.

15.12 EXERCISES

1. A distillation operation separating a low-viscosity hydrocarbon mixture requires three shell-and-tube heat exchangers. The liquid feed is to be preheated to saturated liquid by heat recovery from another low-viscosity hydrocarbon stream. The reboiler is to be a vertical thermosyphon using steam heating.

The condenser is to be a horizontal exchanger with the condensing stream on the shell-side and the cooling serviced by cooling water. What would be the expected rank order of the three heat exchangers in terms of their overall heat transfer coefficient?

2. For the three heat exchangers from Exercise 1, make a first estimate of the order of magnitude of the overall heat coefficients from tabulated values of film transfer coefficients and fouling coefficients. Neglect the resistance from the tube walls.
3. Under what circumstances might a thermosyphon reboiler be orientated vertically or horizontally?
4. Under what circumstances might a distillation condenser be orientated vertically or horizontally.
5. The demethanizer distillation column of an ethylene process works at extremely low temperatures. The feed is cooled with extremely small temperature differences of the order of 1°C to minimize the refrigeration costs associated with the cooling. What type of heat exchanger would you expect to be used for this duty?
6. Gasketed plate heat exchangers are commonly used in food processing. What advantages does the design offer in such applications?
7. A hot stream is to be cooled from 210 to 80°C by heating a cold stream from 60 to 150°C with a duty of 1.7 MW. A 1–1 shell-and-tube heat exchanger is to be used and the overall heat transfer coefficient has been estimated to be 120 $W \cdot m^{-2} \cdot K^{-1}$. Calculate the heat transfer area of the unit.
8. Instead of using a 1–1 design in Example 7, a 1–2 design is to be used subject to $X_P = 0.9$. Assume that the overall heat transfer coefficient is unchanged. (In practice, it would be expected to increase). Calculate
   a. the number of shells required
   b. P_{1-2} for each shell
   c. F_T for the shells in series
   d. the heat transfer area
9. Liquid n-butanol at 115° is to be cooled to 45°C against cooling water between 25 and 35°C. The flowrate of n-butanol is 10 $kg \cdot s^{-1}$. A 1–1 shell-and-tube heat exchanger is to be used. The cooling water has a greater fouling tendency than the n-butanol and will have a higher flowrate; hence the cooling water will be allocated to the tube-side and the n-butanol to the shell-side. Physical property data for the fluids are given in Table 15.16.

Table 15.16 Physical property data for n-butanol and water.

	n-Butanol	Water
Density ($kg \cdot m^{-3}$)	712	993
Heat capacity ($J \cdot kg^{-1} \cdot K^{-1}$)	3200	4190
Viscosity ($N \cdot s \cdot m^{-2}$)	4.0×10^{-4}	8.0×10^{-4}
Thermal conductivity ($W \cdot m^{-1} \cdot K^{-1}$)	0.127	0.616

Assume that the fouling coefficient for n-butanol is 10,000 $W \cdot m^{-2} \cdot K^{-1}$ and that for cooling water is 3,000 $W \cdot m^{-2} \cdot K^{-1}$. Steel tubes are to be used and the following assumptions can be made regarding the heat exchanger geometry:

$$d_O = 20 \text{ mm}$$

$$d_I = 16 \text{ mm}$$

$$p_T = 1.25\, d_O$$

$$p_C = 0.866 \text{(triangular pitch)}$$

$$B_C = 0.25$$

$$L/D_S = 7$$

Estimate the overall heat transfer coefficient, heat transfer area and pressure drops for the tube-side:
   a. for a fluid velocity of 1 $m \cdot s^{-1}$ on both tube-side and the shell-side
   b. for a tube-side and shell-side pressure drop of 0.6 bar.

Assume the shell-side is characterized by:

$$F_{hn} = F_{hw} = 1,\ F_{hb} = F_{hL} = 0.8,\ F_{Pb} = 0.8 \text{ and } F_{PL} = 0.5$$

10. Table 5.17 gives estimates of the film transfer coefficients of an existing shell-and-tube heat exchanger, assuming the tube-wall resistance to be negligible.

Table 15.17 Estimated film transfer coefficients for an existing heat exchanger.

	Tube-side	Shell-side
Film transfer coefficient ($W \cdot m^{-2} \cdot K^{-1}$)	500	750
Fouling coefficient ($W \cdot m^{-2} \cdot K^{-1}$)	1000	2000

The flowrate on the tube-side is to be increased by 20% with the shell-side flowrate fixed.
   a. Estimate the change in the tube-side film transfer coefficient.
   b. The tube-side pressure drop before the increase is 0.7 bar. Estimate the change in the tube-side pressure drop.
   c. The data for an existing heat exchanger are given in Table 15.18.

Table 15.18 Data for existing heat exchanger.

	Tube-side	Shell-side
Flowrate ($kg \cdot s^{-1}$)	7.7	11.0
Initial temperature (°C)	90	30
Final temperature (°C)	40	50
Heat capacity ($J \cdot kg^{-1} \cdot K^{-1}$)	2400	4180

Estimate the change in the overall heat transfer coefficient and the new outlet temperatures for a 20% increase in the tube-side flow.

11. A condenser is required to condense a flowrate of 7 $kg \cdot s^{-1}$ of isopropanol. The condensation takes place isothermally at 83°C without subcooling of the condensate. The cooling is provided by cooling water between 25 and 35°C. The condenser can be assumed to be steel with 20 mm tubes

with 2 mm wall thickness. The tube pitch can be assumed to be $1.25d_O$ with a square configuration. The length to shell diameter can be assumed to be 5. The physical property data are given in Table 15.19.

Table 15.19 Physical property data for isopropanol and water.

Property	Isopropanol (83°C)	Water (30°C)
Density ($kg \cdot m^{-3}$)	732	996
Liquid heat capacity ($J \cdot kg^{-1} \cdot K^{-1}$)	3370	4180
Viscosity ($N \cdot s \cdot m^{-2}$)	0.502×10^{-3}	0.797×10^{-3}
Thermal conductivity ($W \cdot m^{-1} \cdot K^{-1}$)	0.131	0.618
Heat of vaporization ($J \cdot kg^{-1}$)	678,000	–

Assume the fouling coefficients to be 10,000 $W \cdot m^{-2} \cdot K^{-1}$ and 5000 $W \cdot m^{-2} \cdot K^{-1}$ for isopropanol and cooling water respectively. For an assumed cooling water velocity of 1 $m \cdot s^{-1}$, estimate the heat transfer area for:

a. a horizontal condenser with shell-side condensation.
b. a vertical condenser with tube-side condensation.

12. For the case of vertical tube-side condensation from Exercise 11, the condensate is to be subcooled to 45°C. By dividing the condenser into two zones for condensation and subcooling, estimate the heat transfer area.
13. A reboiler of a distillation column is required to supply 10 $kg \cdot s^{-1}$ of toluene vapor. The column operating pressure at the bottom of the column is 1.6 bar. At this pressure, the toluene vaporizes at 127°C and can be assumed to be isothermal. Steam at 160°C is to be used for the vaporization. The latent heat of vaporization of toluene is 344,000 $J \cdot kg^{-1}$, the critical pressure is 40.5 bar and critical temperature is 594 K. Steel tubes with 30 mm outside diameter, 2 mm wall thickness and length 3.95 m are to be used. The film coefficient (including fouling) for the condensing steam can be assumed to be 5700 $W \cdot m^{-2} \cdot K^{-1}$. Estimate the heat transfer area for:
    a. kettle reboiler.
    b. vertical thermosyphon reboiler.
14. The purge gas from a petrochemical process is at 25°C and contains a mole fraction of methane of 0.6, the balance being hydrogen. This purge gas is to be burnt in a furnace to provide heat to a process with a cold stream pinch temperature of 150°C ($\Delta T_{min} = 50$°C). Ambient temperature is 10°C.
    a. Calculate the theoretical flame temperature if 15% excess air is used in the combustion. Standard heats of combustion are given in Table 15.11 and mean molar heat capacities in Table 15.15.
    b. Calculate the furnace efficiency.
    c. Suggest ways in which the furnace efficiency could be improved.

REFERENCES

1. Kern DQ (1950) *Process Heat Transfer*, McGraw-Hill.
2. Hewitt GF (1992) *Handbook of Heat Exchangers Design*, Begell House Inc.
3. Hewitt GF, Shires GL and Bott TR (1994) *Process Heat Transfer*, CRC Press Inc.
4. Sinnott RK (1996) *Chemical Engineering, Volume 6 Chemical Engineering Design*, Butterworth Heinemann.
5. Nie X-R (1998) *Optimisation Strategies for Heat Exchanger Network Design Considering Pressure Drop Aspects*, PhD Thesis, UMIST, UK.
6. Bowman RA, Mueller AC and Nagle WM (1940) Mean Temperature Differences in Design, *Trans ASME*, **62**: 283.
7. Ahmad S, Linnhoff B and Smith R (1988) Design of Multipass Heat Exchangers: an Alternative Approach, *Trans ASME J Heat Transfer*, **110**: 304.
8. Ahmad S, Linnhoff B and Smith R (1990) Cost Optimum Heat Exchanger Networks II Targets and Design for Detailed Capital Cost Models, *Comp Chem Eng*, **7**: 751.
9. Zhu XX, Zanfir M and Klemes J (2000) Heat Transfer Enhancement for Heat Exchanger Network Retrofit, *Heat Transfer Eng*, **21**: 7.
10. Polley GT, Reyes Athie CM and Gough M (1992) Use of Heat Transfer Enhancement in Process Integration, *Heat Recovery Syst CHP*, **12**: 191.
11. Nusselt W (1916) Die Oberflachenkondensation des Wasserdampfes, *Z Ver Deut Ing*, **60**: 541, 569.
12. Rosenow WM (1956) Heat Transfer and Temperature Distribution in Laminar Film Condensation, *Trans ASME*, **78**: 1645.
13. Rosenow WM, Webber JH and Ling AT (1956) Effect of Velocity on Laminar and Turbulent Film Condensation, *Trans ASME*, **78**: 1645.
14. Mueller AC in Hewitt GF (1992) *Handbook of Heat Exchanger Design*, Begell House Inc.
15. Palen JW in Hewitt GF (1992) *Handbook of Heat Exchanger Design*, Begell House Inc.
16. Mostinski IL (1963) Calculation of Boiling Heat Transfer Coefficients, Based on the Law of Corresponding States, *Br Chem Eng*, **8**: 580.
17. Frank O and Prickett RD (1973) Design of Vertical Thermosyphon Reboilers, *Chem Eng*, **3**: 107.
18. Hougen OA, Watson KM and Ragatz RA (1954) *Chemical Process Principles Part I Material and Energy Balances*, 2nd Edition, John Wiley.

16 Heat Exchanger Networks II – Energy Targets

The design philosophy started at the heart of the onion with the reactor and moved out to the next layer, the separation and recycle system (Figure 1.7). Acceptance of the major processing steps (reactors, separators and recycles) in the inner two layers of the onion fixes the material and energy balance. Thus, the heating and cooling duties for the next layer of the onion, the heat recovery system, are known. However, completing the design of the heat exchanger network is not necessary in order to assess the completed design. *Targets* can be set for the heat exchanger network to assess the performance of the complete process design without actually having to carry out the network design. These targets allow both energy and capital cost for the heat exchanger network to be assessed. Moreover, the targets allow the designer to suggest process changes for the reactor and separation and recycle systems to improve the targets for energy and capital cost of the heat exchanger network.

Using targets for the heat exchanger network, rather than designs, allows many design options for the overall process to be screened quickly and conveniently. Screening many design options by completed designs is usually simply not practical in terms of the time and effort required. First consider the details of how to set energy targets. Capital cost targets will be considered in the next chapter. In later chapters, energy targets will be used to suggest design improvements to the reaction, separation and recycle systems.

16.1 COMPOSITE CURVES

The analysis of the heat exchanger network first identifies sources of heat (termed *hot streams*) and sinks (termed *cold streams*) from the material and energy balance. Consider first a very simple problem with just one hot stream (heat source) and one cold stream (heat sink). The initial temperature (termed *supply temperature*), final temperature (termed *target temperature*) and enthalpy change of both streams are given in Table 16.1.

Steam is available at 180°C and cooling water at 20°C. Clearly, it is possible to heat the cold stream using steam and cool the hot stream, in Table 16.1, using cooling water. However, this would incur excessive energy cost. It is also incompatible with the goals of sustainable industrial activity, which call for use of the minimum energy consumption. Instead, it is preferable to try to recover the heat between process streams, if this is possible. The scope for heat recovery can be determined by plotting both streams from Table 16.1 on temperature-enthalpy axes. For feasible heat exchange between the two streams, the hot stream must at all points be hotter than the cold stream. Figure 16.1a shows the temperature-enthalpy plot for this problem with a minimum temperature difference (ΔT_{min}) of 10°C. The region of overlap between the two streams in Figure 16.1a determines the amount of heat recovery possible (for $\Delta T_{min} = 10$°C). For this problem, the heat recovery (Q_{REC}) is 11 MW. The part of the cold stream that extends beyond the start of the hot stream in Figure 16.1a cannot be heated by recovery and requires steam. This is the minimum hot utility or *energy target* (Q_{Hmin}), which for this problem is 3 MW. The part of the hot stream that extends beyond the start of the cold stream in Figure 16.1a cannot be cooled by heat recovery and requires cooling water. This is the minimum cold utility (Q_{Cmin}), which for this problem is 1 MW. Also shown at the bottom of Figure 16.1a is the arrangement of heat exchangers corresponding with the temperature-enthalpy plot.

Table 16.1 Two-stream heat recovery problem.

Stream	Type	Supply temperature T_S (°C)	Target temperature T_T (°C)	ΔH (MW)
1	Cold	40	110	14
2	Hot	160	40	−12

The temperatures or enthalpy change for the streams (and hence their slope) cannot be changed, but the relative position of the two streams can be changed by moving them horizontally relative to each other. This is possible since the reference enthalpy for the hot stream can be changed independently from the reference enthalpy for the cold stream. Figure 16.1b shows the same two streams moved to a different relative position such that ΔT_{min} is now 20°C. The amount of overlap between the streams is reduced (and hence heat recovery is reduced) to 10 MW. A greater amount of the cold stream now extends beyond the start of the hot stream, and hence the amount of steam is increased to 4 MW. Also, more of the hot stream extends beyond the start of the cold stream, increasing the cooling water demand to 2 MW. Thus, the approach of plotting a hot and a cold stream on the same temperature-enthalpy axes can determine hot and cold utility for a given value of ΔT_{min}.

Chemical Process Design and Integration R. Smith
 ISBNs: 0-471-48680-9 (HB); 0-471-48681-7 (PB)

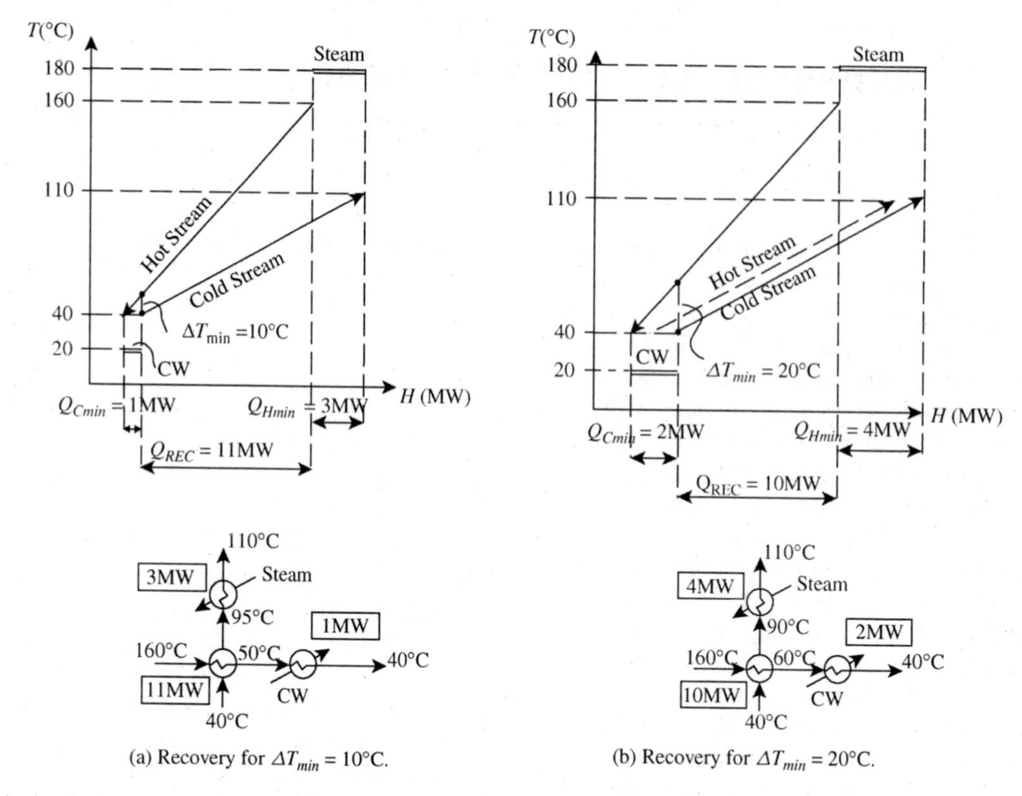

Figure 16.1 A simple heat recovery problem with one hot stream and one cold stream.

The importance of ΔT_{min} is that it sets the relative location of the hot and cold streams in this two-stream problem, and therefore the amount of heat recovery. Setting the value of ΔT_{min} or Q_{Hmin} or Q_{Cmin} sets the relative location and the amount of heat recovery.

Consider now the extension of this approach to several hot streams and cold streams. Figure 16.2 shows a simple flowsheet. Flowrates, temperatures and heat duties for each stream are shown. Two of the streams in Figure 16.2 are sources of heat (hot streams) and two are sinks for heat (cold streams). Assuming that the heat capacities are constant, the data for the hot and cold streams can be extracted as given in Table 16.2. Note that the heat capacities (CP) are total heat capacities, being the product of mass flowrate and specific heat capacity ($CP = m \cdot C_P$). Had the heat capacities varied significantly, the nonlinear temperature-enthalpy behavior could have been represented by a series of linear *segments* (see Chapter 19).

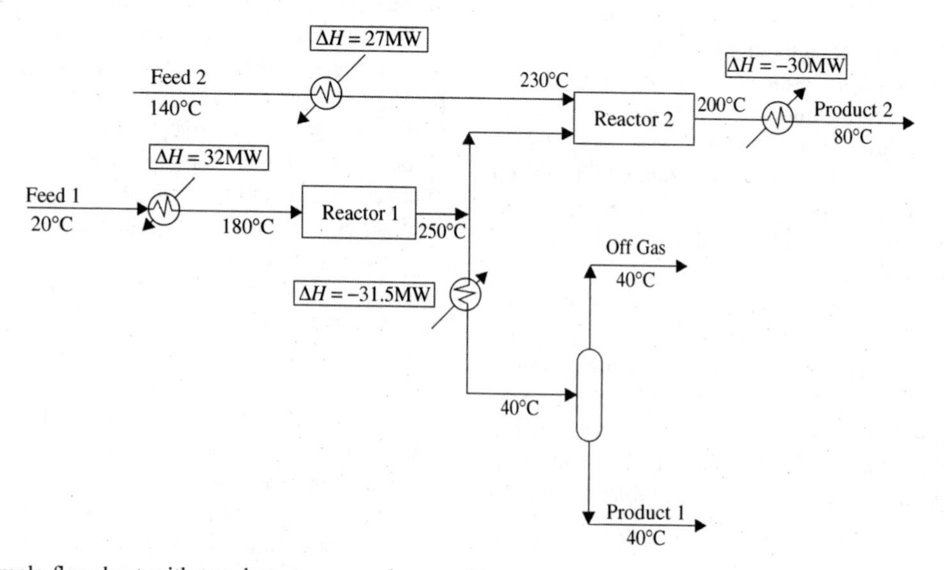

Figure 16.2 A simple flowsheet with two hot streams and two cold streams.

Table 16.2 Heat exchange stream data for the flowsheet Figure 16.2.

Stream	Type	Supply temperature T_S (°C)	Target temperature T_T(°C)	ΔH (MW)	Heat capacity flowrate CP(MW·K^{-1})
1. Reactor 1 feed	Cold	20	180	32.0	0.2
2. Reactor 1 product	Hot	250	40	−31.5	0.15
3. Reactor 2 feed	Cold	140	230	27.0	0.3
4. Reactor 2 product	Hot	200	80	−30.0	0.25

Instead of dealing with individual streams as given in Table 16.1, an overview of the process is needed. Figure 16.3a shows the two hot streams individually on temperature-enthalpy axes. How these hot streams behave overall can be quantified by combining them in the given temperature ranges[1,2,3]. The temperature ranges in question are defined by changes in the overall rate of change of enthalpy with temperature. If heat capacities are constant, then changes will occur only when streams start or finish. Thus, in Figure 16.3, the temperature axis is divided into ranges defined by the supply and target temperatures of the streams.

Within each temperature range, the streams are combined to produce a composite hot stream. This composite hot stream has a CP in any temperature range that is the sum of the individual streams. Also, in any temperature range, the enthalpy change of the composite stream is the sum of the enthalpy changes of the individual streams. Figure 16.3b shows the *composite curve* of the hot streams[1,2,3]. The composite hot stream is a single stream that is equivalent to the individual hot streams in terms of temperature and enthalpy. Similarly, the composite curve of the cold streams for the problem can be produced, as shown in Figure 16.4. Again, the composite cold stream is a single stream that is equivalent to the individual cold streams in terms of temperature and enthalpy.

The composite hot and cold curves can now be plotted on the same axes, as in Figure 16.5. Plotting the composite hot and cold curves is analogous to plotting the single hot and cold streams in Figure 16.1. The composite curves in Figure 16.5a are set to have a minimum temperature difference (ΔT_{min}) of 10°C. Where the curves overlap in Figure 16.5a, heat can be recovered vertically from the hot streams that comprise the hot composite curve into the cold streams that comprise the cold composite curve. The way in which the composite curves are constructed (i.e. monotonically decreasing hot composite curve and monotonically increasing cold composite curve) allows maximum overlap between the curves and hence maximum heat recovery. Maximizing the energy recovery thereby minimizes the external requirements for heating and cooling duties and minimizes the energy consumption. In this problem, for $\Delta T_{min} = 10$°C, the maximum heat recovery (Q_{REC}) is 51.5 MW.

Where the cold composite curve extends beyond the start of the hot composite curve in Figure 16.5a, heat recovery is not possible, and the cold composite must be supplied with external hot utility such as steam. This represents the target for hot utility (Q_{Hmin}). For this problem, with $\Delta T_{min} =$ 10°C, $Q_{Hmin} = 7.5$ MW. Where the hot composite curve

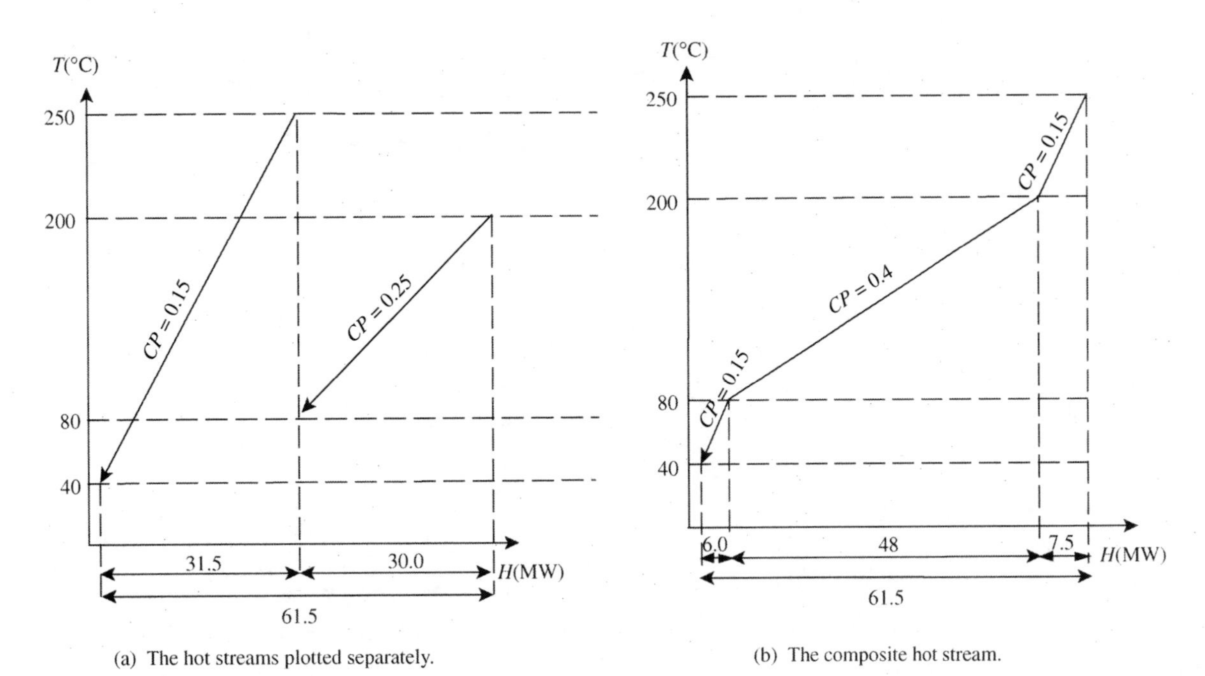

(a) The hot streams plotted separately.

(b) The composite hot stream.

Figure 16.3 The hot streams can be combined to obtain a composite stream.

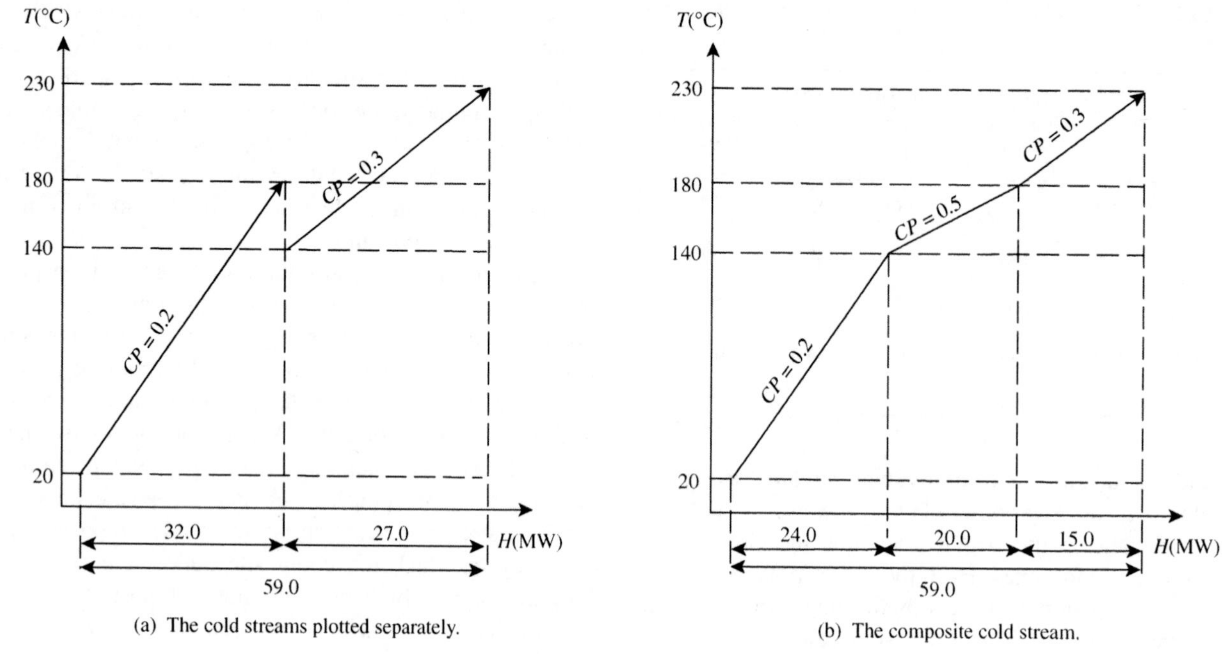

(a) The cold streams plotted separately.

(b) The composite cold stream.

Figure 16.4 The cold streams can be combined to obtain a composite cold stream.

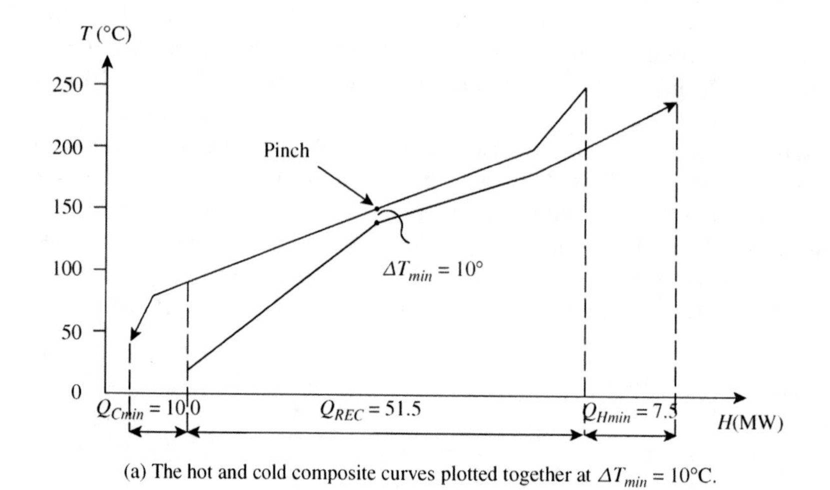

(a) The hot and cold composite curves plotted together at $\Delta T_{min} = 10°C$.

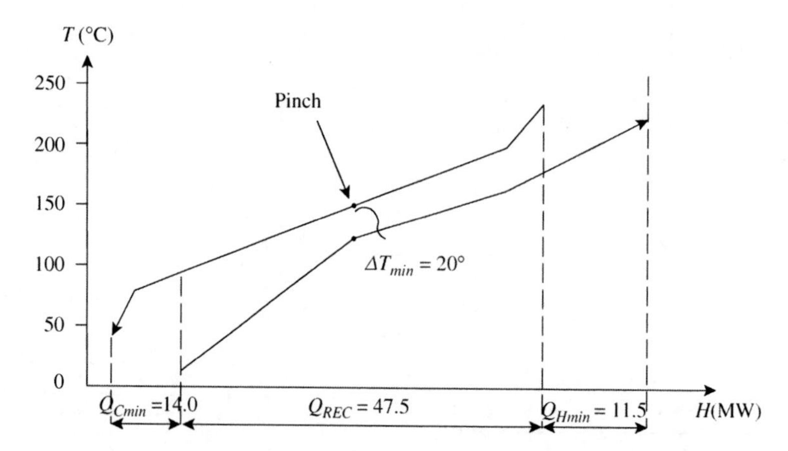

(b) Increasing ΔT_{min} from 10°C to 20°C increases the hot and cold utility targets.

Figure 16.5 Plotting the hot and cold composite curves together allows the targets for hot and cold utility to be obtained.

extends beyond the start of the cold composite curve in Figure 16.5a, heat recovery is again not possible and the hot composite curve must be supplied with external cold utility such as cooling water. This represents the target for cold utility (Q_{Cmin}). For this problem, setting $\Delta T_{min} = 10°C$ gives $Q_{Cmin} = 10.0$ MW.

Specifying the hot utility or cold utility heat duty or ΔT_{min} fixes the relative position of the two curves. As with the simple problem in Figure 16.1, the relative position of the two curves is a degree of freedom[4]. Again, the relative position of the two curves can be changed by moving them horizontally relative to each other. Clearly, to consider heat recovery from hot streams into cold streams, the hot composite curve must be in a position such that it is always above the cold composite curve for feasible heat transfer. Thereafter, the relative position of the curves can be chosen. Figure 16.5b shows the curves with $\Delta T_{min} = 20°C$. The hot and cold utility targets are now increased to 11.5 MW and 14 MW respectively.

Figure 16.6 illustrates what happens to the cost of the system as the relative position of the composite curves is changed over a range of values of ΔT_{min}. When the curves just touch, there is no driving force for heat transfer at one point in the process, which would require infinite heat transfer area and hence infinite capital cost. As the energy target (and hence ΔT_{min} between the curves) is increased, the capital cost decreases. This results from increased temperature differences throughout the process, decreasing the heat transfer area. On the other hand, the energy cost increases as ΔT_{min} increases. There is a trade-off between energy and capital cost and an economic amount of energy recovery. Later, it will be shown how this trade-off can be carried out using energy and capital cost targets.

However, care should be taken not to ignore practical constraints when setting ΔT_{min}. To achieve a small ΔT_{min} in a design requires heat exchangers that exhibit pure countercurrent flow. With shell-and-tube heat exchangers

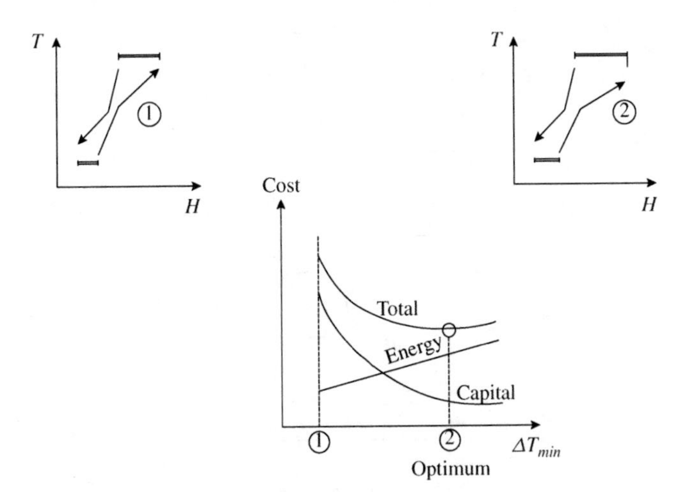

Figure 16.6 The correct setting for ΔT_{min} is fixed economic trade-offs.

this is not possible, even if single-shell pass and single-tube pass designs are used, because the shell-side stream takes periodic cross-flow. Consequently, operating with a ΔT_{min} less than 10°C should be avoided, unless under special circumstances[5]. A smaller value of 5°C or less can be achieved with plate heat exchangers, and the value can go as low as 1 to 2°C with plate-fin designs[5]. It should be noted, however, that such constraints only apply to the exchangers that operate around the point of closest approach between the composite curves. Additional constraints apply if vaporization or condensation is occurring at the point of closest approach (see Chapter 15).

16.2 THE HEAT RECOVERY PINCH

As discussed above, the correct setting for the composite curves is determined by an economic trade-off between energy and capital, corresponding to an economic minimum temperature difference between the curves, ΔT_{min}. Accepting for the moment that the correct economic ΔT_{min} is known, this fixes the relative position of the composite curves and hence the energy target. The value of ΔT_{min} and its location between the composite curves have important implications for design, if the energy target is to be achieved in the design of a heat exchanger network. The ΔT_{min} is normally observed at only one point between the hot and the cold composite curves, called the *heat recovery pinch*[3,6–8]. The pinch point has a special significance.

The trade-off between energy and capital in the composite curves suggests that, "on average", individual exchangers should have a temperature difference no smaller than ΔT_{min}. A good initialization in heat exchanger network design is to assume that no individual heat exchanger has a temperature difference smaller than the ΔT_{min} between the composite curves.

With this rule in mind, divide the process at the pinch as shown in Figure 16.7a. Above the pinch (in temperature terms) the process is in heat balance with the minimum hot utility, Q_{Hmin}. Heat is received from hot utility and no heat is rejected. The process acts as a heat sink. Below the pinch (in temperature terms), the process is in heat balance with the minimum cold utility, Q_{Cmin}. No heat is received but heat is rejected to cold utility. The process acts as a heat source.

Consider now the possibility of transferring heat between these two systems. Figure 16.7b shows that it is possible to transfer heat from hot streams above the pinch to cold streams below it. The pinch temperature for hot streams for the problem is 150°C and for cold streams 140°C. Transfer of heat from above the pinch to below, as shown in Figure 16.7b, means transfer of heat from hot streams with a temperature of 150°C or greater into cold streams with a temperature of 140°C or less. This is clearly possible. By contrast, Figure 16.7c shows that heat transfer from hot streams below the pinch to cold streams

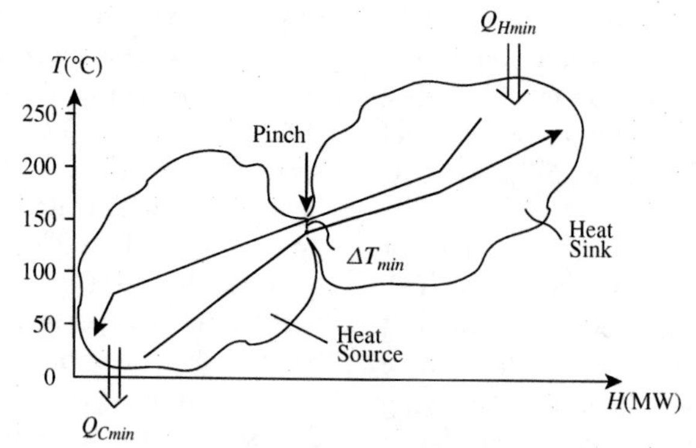

(a) The pinch divides the process into a heat source and a heat sink.

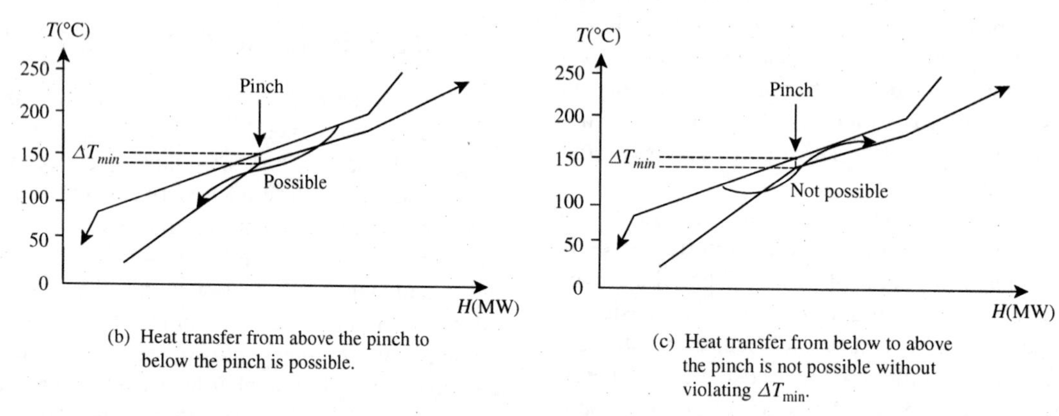

(b) Heat transfer from above the pinch to below the pinch is possible.

(c) Heat transfer from below to above the pinch is not possible without violating ΔT_{min}.

Figure 16.7 The composite curves set the energy target and the location of the pinch.

above is not possible. Such transfer requires heat being transferred from hot streams with a temperature of 150°C or less into cold streams with a temperature of 140°C or greater. This is clearly not possible (without violating the ΔT_{min} constraint).

If an amount of heat *XP* is transferred from the system above the pinch to the system below the pinch, as in Figure 16.8a, this will create a deficit of heat *XP* above the pinch and an additional surplus of heat *XP* below the pinch. The only way this can be corrected is by importing an extra *XP* amount of heat from hot utility and exporting an extra *XP* amount of heat to cold utility[3,4].

Analogous effects are caused by the inappropriate use of utilities. Utilities are appropriate if they are necessary to satisfy the enthalpy imbalance in that part of the process. Above the pinch, hot utility (in this case, steam) is needed to satisfy the enthalpy imbalance. Figure 16.8b illustrates what happens if inappropriate use of utilities is made. If cooling to cold utility *XP* is used to cool hot streams above the pinch, this creates an enthalpy imbalance in the system above the pinch. To satisfy the enthalpy imbalance above the pinch, an import of $(Q_{Hmin} + XP)$ heat from hot utility is required. Overall, $(Q_{Cmin} + XP)$ of cold utility is used[3,4].

Another inappropriate use of utilities involves heating of some of the cold streams below the pinch by hot utility (steam in this case). Below the pinch, cold utility is needed to satisfy the enthalpy imbalance. Figure 16.8c illustrates

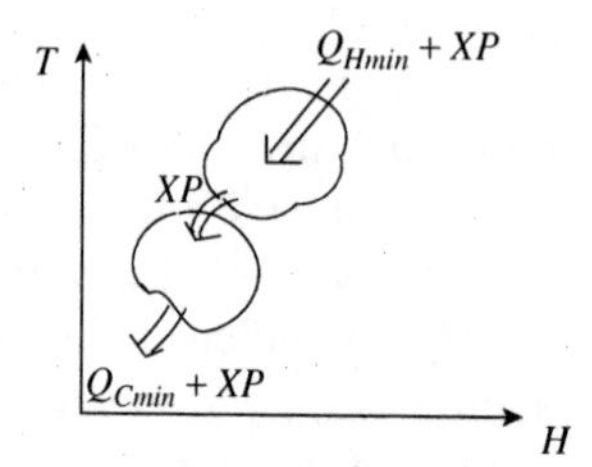

(a) Process–process heat transfer across the pinch.

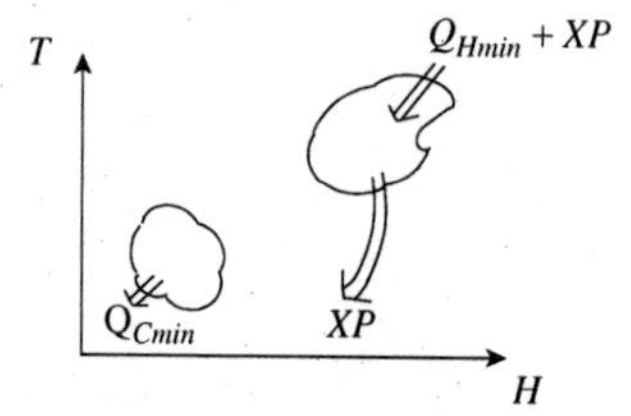

(b) Cold utility above the pinch.

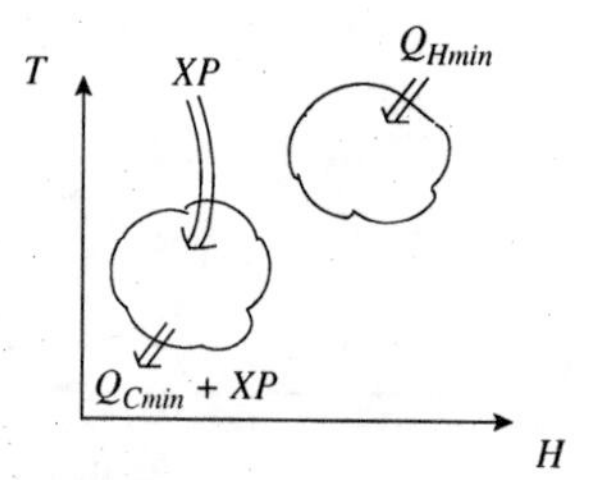

(c) Hot utility below the pinch.

Figure 16.8 Three forms of cross pinch heat transfer.

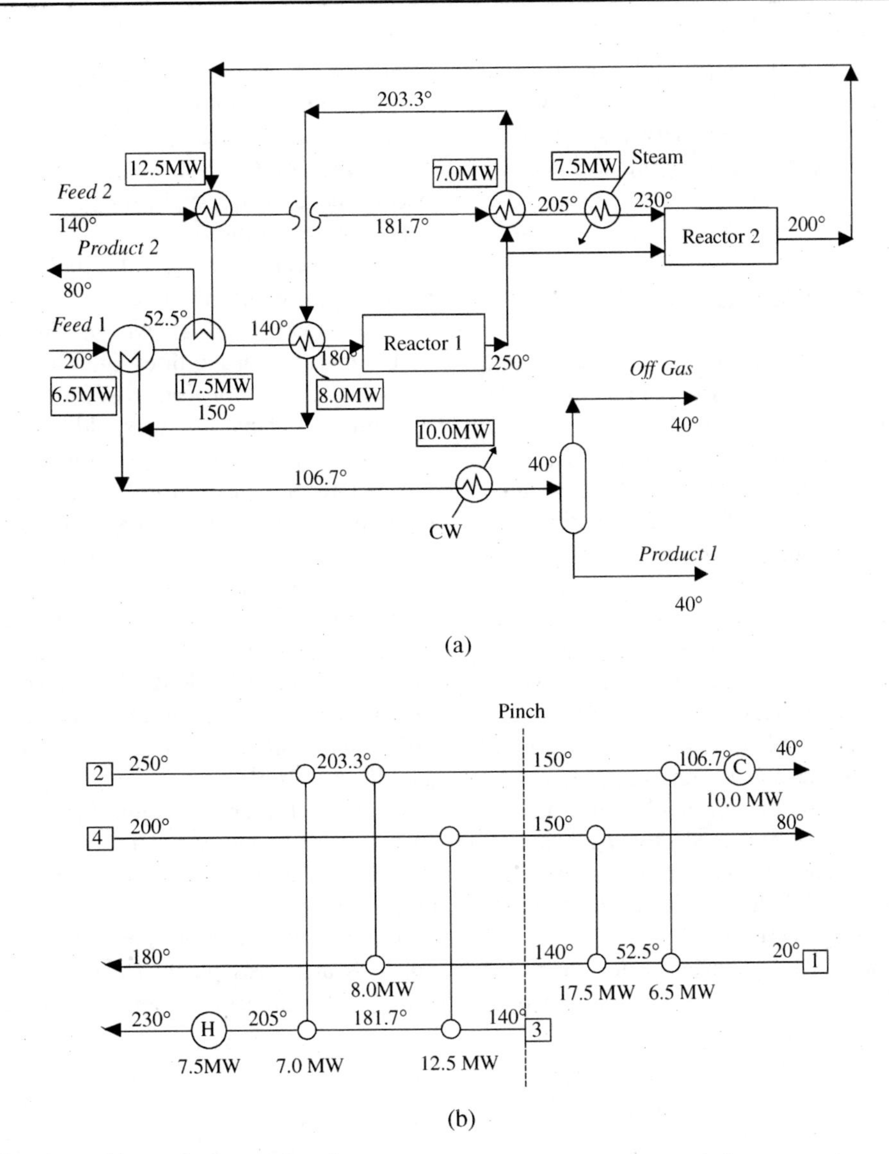

Figure 16.9 A design that achieves the energy target.

what happens if an amount of heat XP from hot utility is used below the pinch. Q_{Hmin} must still be supplied above the pinch to satisfy the enthalpy imbalance above the pinch. Overall, $(Q_{Hmin} + XP)$ of steam is used and $(Q_{Hmin} + XP)$ of cooling water[3,4].

In other words, to achieve the energy target set by the composite curves, the designer must not transfer heat across the pinch by[3,4]:

a. Process-to-process heat transfer
b. Inappropriate use of utilities

These rules are both necessary and sufficient to ensure that the target is achieved, providing that the initialization rule is adhered to that no individual heat exchanger should have a temperature difference smaller than ΔT_{min}.

Figure 16.9a shows a design corresponding to the flowsheet in Figure 16.2, which achieves the target of $Q_{Hmin} = 7.5$ MW and $Q_{Cmin} = 10$ MW for $\Delta T_{min} = 10°C$.

Figure 16.9b shows an alternative representation of the flowsheet in Figure 16.9a, known as the *grid diagram*[9]. The grid diagram shows only heat transfer operations. Hot streams are at the top running left to right. Cold streams are at the bottom running right to left. A heat exchange match is represented by a vertical line joining two circles on the two streams being matched. An exchanger using hot utility is represented by a circle with an "H". An exchanger using cold utility is represented by a circle with a "C". The importance of the grid diagram is clear in Figure 16.9b, since the pinch, and how it divides the process into two parts, is easily accommodated. Dividing the process into two parts on a conventional diagram such as shown in Figure 16.9a is both difficult and extremely cumbersome.

Details of how the design in Figure 16.9 was developed are explained in Chapter 18. For now, simply take note that the targets set by the composite curves are achievable in design, providing that the pinch is recognized and there is no transfer of heat across the pinch, either process-to-process or through inappropriate use of utilities. However, the insight of the pinch is needed to analyze some of the important decisions still to be made before the network design is addressed.

16.3 THRESHOLD PROBLEMS

Not all problems have a pinch to divide the process into two parts[4]. Consider the composite curves in Figure 16.10a. At this setting, both steam and cooling water are required. As the composite curves are moved closer together, both the steam and cooling water requirements decrease until the setting shown in Figure 16.10b. At this setting, the composite curves are in alignment at the hot end, indicating that there is no longer a demand for hot utility. Moving the curves closer together as shown in Figure 16.10c, decreases the cold utility demand at the cold end but opens up a demand for cold utility at the hot end corresponding with the decrease at the cold end. In other words, as the curves are moved closer together, beyond the setting in Figure 16.10b, the utility demand is constant. The setting shown in Figure 16.10b marks a threshold, and problems that exhibit this feature are known as *threshold problems*[4]. In some threshold problems, the hot utility requirement disappears as in Figure 16.10. In others, the cold utility disappears as shown in Figure 16.11.

Considering the capital-energy trade-off for threshold problems, there are two possible outcomes as shown in Figure 16.12. Below the threshold ΔT_{min}, energy costs are constant, since utility demand is constant. Figure 16.12a shows a situation where the optimum occurs at the threshold ΔT_{min}. Figure 16.12b shows a situation where the optimum occurs above the threshold ΔT_{min}. The flat profile of energy costs below the threshold ΔT_{min} means that the optimum can never occur below the threshold value. It can only be at or above the threshold value.

In a situation, as shown in Figure 16.12a, with the optimum ΔT_{min} at the threshold, there is no pinch. On the other hand, in a situation as shown in Figure 16.12b with the optimum above the threshold value, there is a demand for both utilities and there is a pinch.

It is interesting to note that threshold problems are quite common in practice and although they do not have a process pinch, pinches are introduced into the design when multiple utilities are added. Figure 16.13a shows composite curves similar to the composite curves from Figure 16.10 but with two levels of cold utility used instead of one. In this case, the second cold utility is steam generation. The introduction of this second utility causes a pinch. This is known as a *utility pinch* since it is caused by the introduction of an additional utility[4].

Figure 16.13b shows composite curves similar to those from Figure 16.11, but with two levels of steam used. Again, the introduction of a second steam level causes a utility pinch.

In design, the same rules must be obeyed around a utility pinch as a process pinch. Heat should not be transferred across it by process-to-process transfer and there should

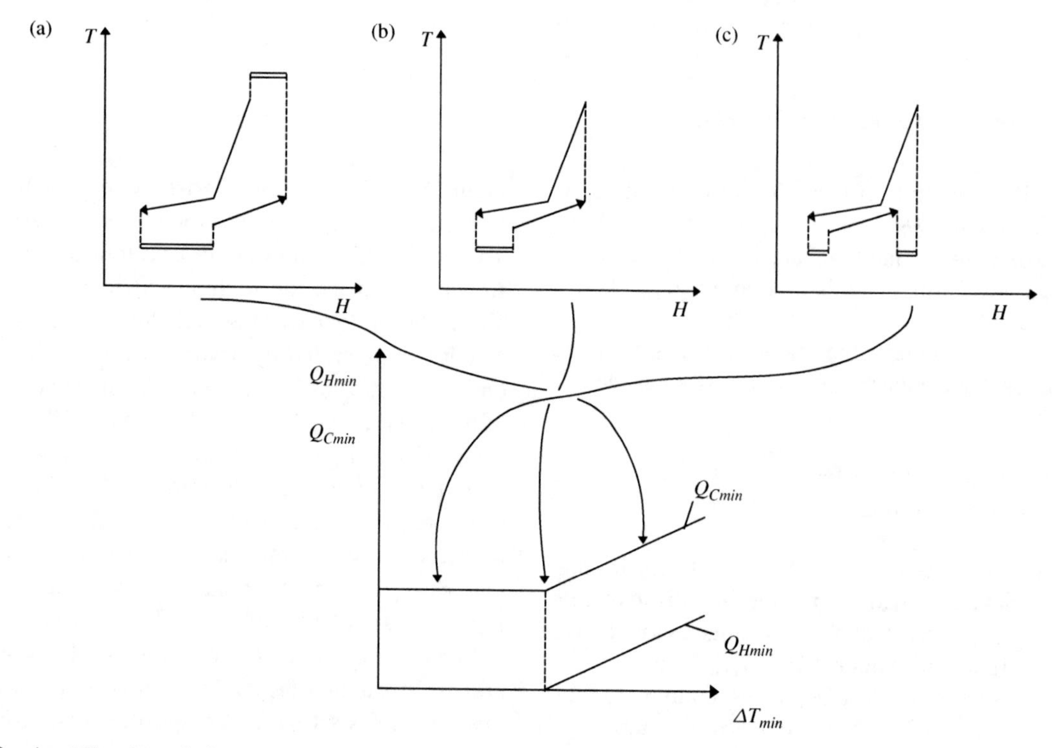

Figure 16.10 As ΔT_{min} is varied, some problems require only cold utility below a threshold value.

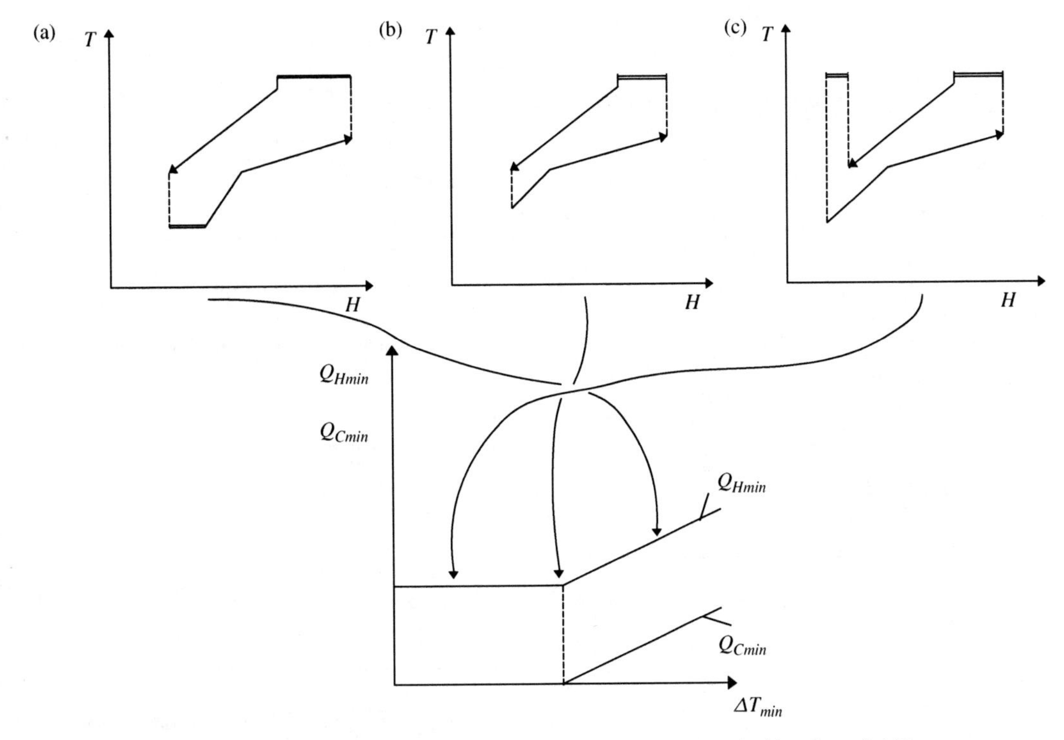

Figure 16.11 In some threshold problems, only hot utility is required below the threshold value of ΔT_{min}.

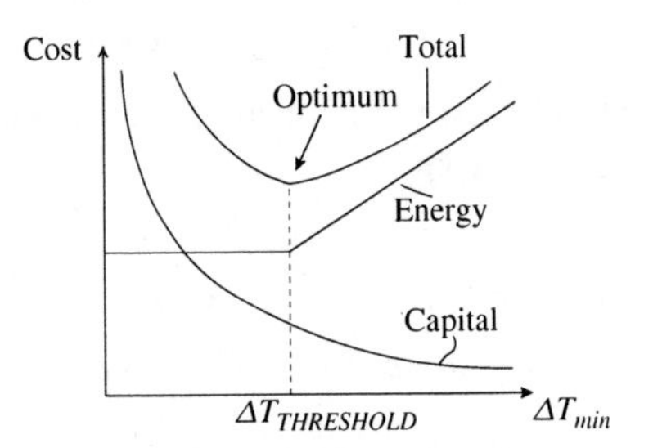

(a) The capital - energy trade-off can lead to an optimum at threshold ΔT_{min}

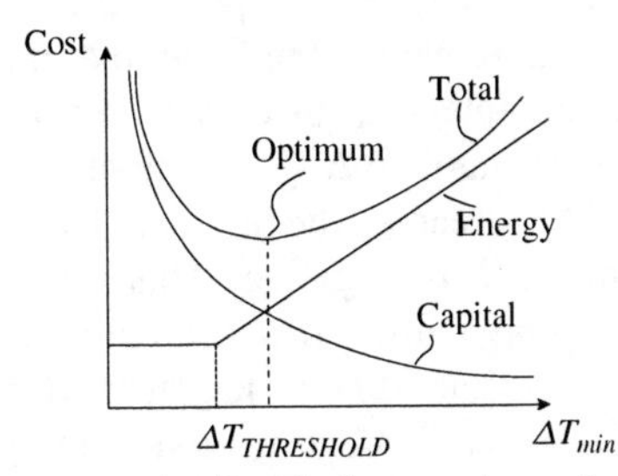

(b) The capital - energy trade-off can lead to an optimum above $\Delta T_{THRESHOLD}$

Figure 16.12 The optimum setting of the capital/energy trade off for threshold problems.

be no inappropriate use of utilities. In Figure 16.13a, this means that the only utility to be used above the utility pinch is steam generation and only cooling water below. In Figure 16.13b, this means that the only utility to be used above the utility pinch is high-pressure steam and only low-pressure steam below.

16.4 THE PROBLEM TABLE ALGORITHM

Although composite curves can be used to set energy targets, they are inconvenient since they are based on a graphical construction. A method of calculating energy targets directly without the necessity of graphical construction can be developed[1,9]. The process is first divided into temperature intervals in the same way as was done for construction of the composite curves. Figure 16.14a shows that it is not possible to recover all of the heat in each temperature interval since temperature driving forces are not feasible throughout the interval. Some heat recovery is possible, but all of the heat cannot be recovered. The amount that can be recovered depends on the relative slopes of the two curves in the temperature interval. This problem can be overcome if, purely for the purposes of construction, the hot composite is shifted to be $\Delta T_{min}/2$ colder than it is in practice and that the cold composite is shifted to be $\Delta T_{min}/2$ hotter than it is in practice as shown in Figure 16.14b. The *shifted composite curves* now touch at the pinch. Carrying out a heat balance between the shifted composite curves within a *shifted temperature interval* shows that heat transfer is feasible throughout each shifted temperature interval, since hot streams in practice are actually $\Delta T_{min}/2$ hotter and cold streams $\Delta T_{min}/2$ colder. Within each shifted interval, the hot streams are in reality hotter than the cold streams by ΔT_{min}.

It is important to note that shifting the curves vertically does not alter the horizontal overlap between the curves. It therefore does not alter the amount by which the

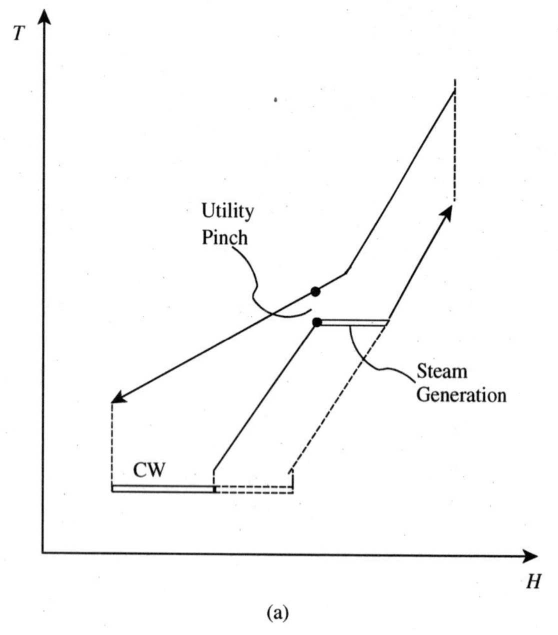

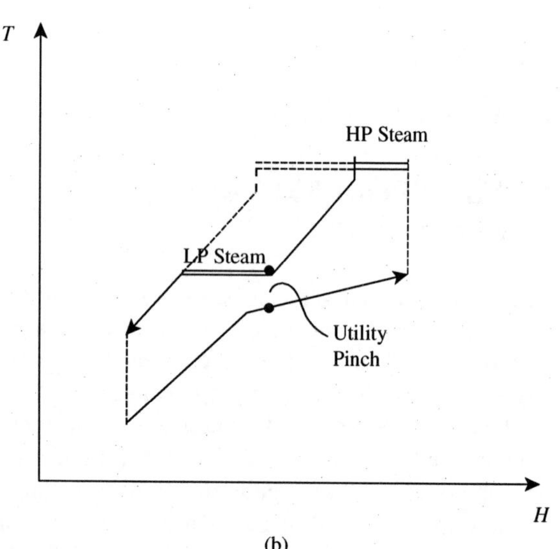

Figure 16.13 Threshold problems are turned into pinched problem when additional utilities are added.

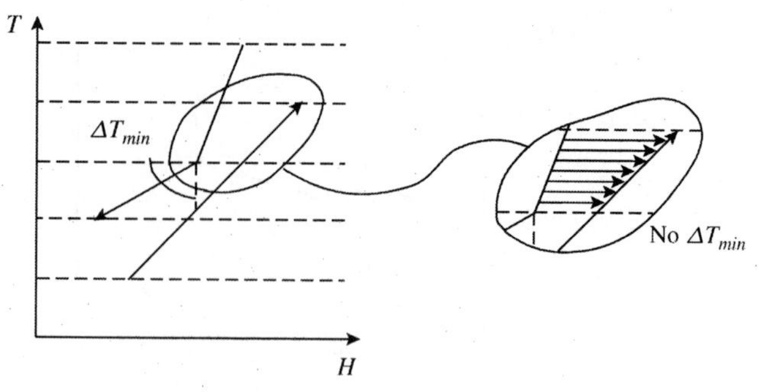

(a) Driving forces not feasible within each interval.

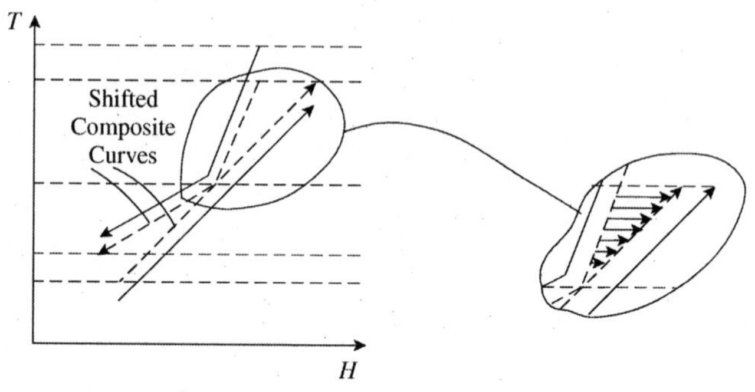

(b) Heat transfer within temperature intervals now feasible.

Figure 16.14 Shifting the composite curves in temperature allows complete heat recovery within temperature intervals.

cold composite curve extends beyond the start of the hot composite curve at the hot end of the problem. Also, it does not alter the amount by which the hot composite curve extends beyond the start of the cold composite curve at the cold end. The shift simply removes the problem of ensuring temperature feasibility within temperature intervals.

This shifting technique can be used to develop a strategy to calculate the energy targets without having to construct composite curves[1,9]:

1. Set up shifted temperature intervals from the stream supply and target temperatures by subtracting $\Delta T_{min}/2$ from the hot streams and adding $\Delta T_{min}/2$ to the cold streams (as in Figure 16.14b).
2. In each shifted temperature interval, calculate a simple energy balance from:

$$\Delta H_i = \left[\sum_{Cold\ streams} CP_C - \sum_{Hot\ streams} CP_H \right] \Delta T_i \quad (16.1)$$

   where ΔH_i is the heat balance for shifted temperature interval i and ΔT_i is the temperature difference across it. If the cold streams dominate the hot streams in a temperature interval, then the interval has a net deficit of heat, and ΔH is positive. If hot streams dominate cold streams, the interval has a net surplus of heat, and ΔH is negative. This is consistent with standard thermodynamic convention, for example, for an exothermic reaction, ΔH is negative. If no hot utility is used, this is equivalent to constructing the shifted composite curves shown in Figure 16.15a.
3. The overlap in the shifted curves as shown in Figure 16.15a means that heat transfer is infeasible. At some point, this overlap is a maximum. This maximum overlap is added as hot utility to correct the overlap. The shifted curves now touch at the pinch as shown in Figure 16.15b. Since the shifted curves just touch, the actual curves are separated by ΔT_{min} at this point, Figure 16.15b.

This basic approach can be developed into a formal algorithm known as the *problem table algorithm*[9]. The

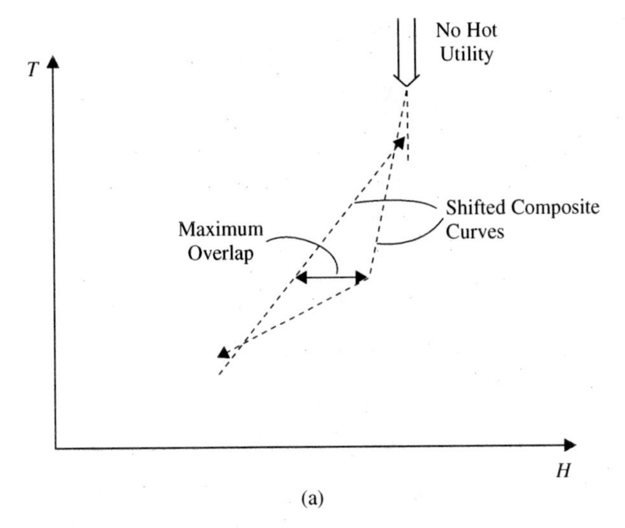

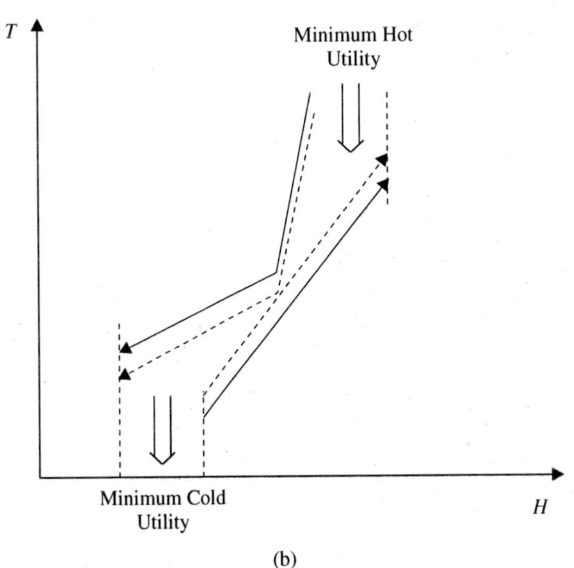

Figure 16.15 The utility target can be determined from the maximum overlap between the shifted composite curves.

algorithm will be explained using the data from Figure 16.2 given in Table 16.2 for $\Delta T_{min} = 10°C$.

First determine the shifted temperature intervals (T^*) from actual supply and target temperatures. Hot streams are shifted down in temperature by $\Delta T_{min}/2$ and cold streams up by $\Delta T_{min}/2$ as detailed in Table 16.3.

The stream population is shown in Figure 16.16 with a vertical temperature scale. The interval temperatures

Table 16.3 Shifted temperatures for the data from Table 16.2.

Stream	Type	T_S	T_T	T_S^*	T_T^*
1	Cold	20	180	25	185
2	Hot	250	40	245	35
3	Cold	140	230	145	235
4	Hot	200	80	195	75

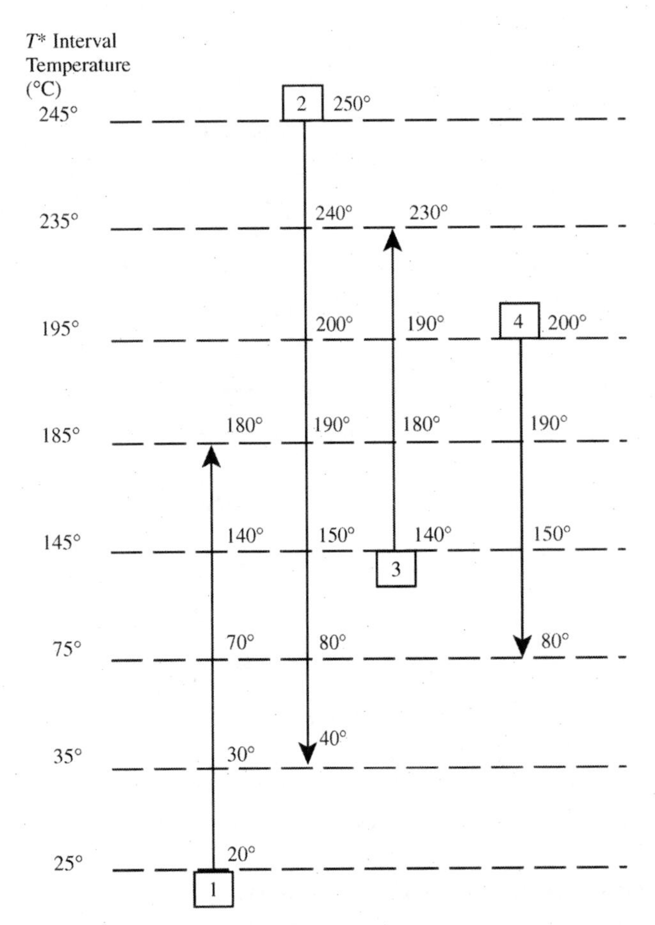

Figure 16.16 The stream population for the data from Figure 16.2.

Interval Temperature	Stream Population	$\Delta T_{INTERVAL}$ (°C)	$\Sigma CP_C - \Sigma CP_H$ $MW \cdot K^{-1}$	$\Delta H_{INTERVAL}$ (MW)	Surplus/ Deficit
245°–235°		10	−0.15	−1.5	Surplus
235°–195°		40	0.15	6.0	Deficit
195°–185°		10	−0.1	−1.0	Surplus
185°–145°		40	0.1	4.0	Deficit
145°–75°		70	−0.2	−14.0	Surplus
75°–35°		40	0.05	2.0	Deficit
35°–25°		10	0.2	2.0	Deficit

Figure 16.17 The temperature interval heat balances.

shown in Figure 16.16 are set to $\Delta T_{min}/2$ below hot stream temperatures and $\Delta T_{min}/2$ above cold stream temperatures.

Next, a heat balance is carried out within each shifted temperature interval according to Equation 16.1. The result is given in Figure 16.17, in which some of the shifted intervals are seen to have a surplus of heat and some have a

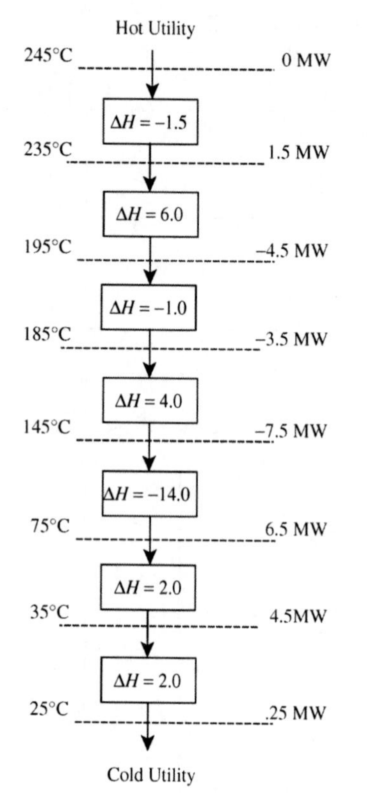

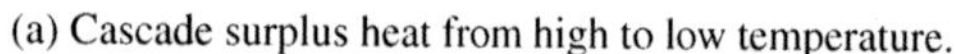
(a) Cascade surplus heat from high to low temperature.

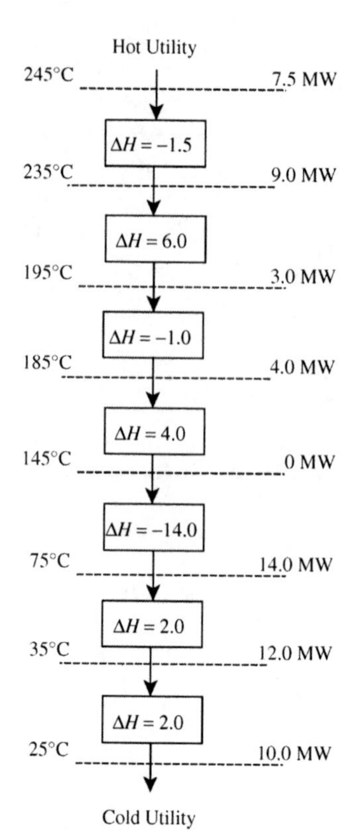

(b) Add heat from hot utility to make all heat flows zero or positive.

Figure 16.18 The problem table cascade.

deficit. The heat balance within each shifted interval allows maximum heat recovery within each interval. However, recovery must also be allowed between intervals.

Now, *cascade* any surplus heat down the temperature scale from interval to interval. This is possible because any excess heat available from the hot streams in an interval is hot enough to supply a deficit in the cold streams in the next interval down. Figure 16.18 shows the cascade for the problem. First, assume no heat is supplied to the first interval from hot utility, Figure 16.18a. The first interval has a surplus of 1.5 MW, which is cascaded to the next interval. This second interval has a deficit of 6 MW, which leaves the heat cascaded from this interval to be −4.5 MW. In the third interval, the process has a surplus of 1 MW, which leaves −3.5 MW, to be cascaded to the next interval, and so on.

Some of the heat flows in Figure 16.18a are negative, which is infeasible. Heat cannot be transferred up the temperature scale. To make the cascade feasible, sufficient heat must be added from hot utility to make the heat flows to be at least zero. The smallest amount of heat needed from hot utility is the largest negative heat flow from Figure 16.18a, that is 7.5 MW. In Figure 16.18b, 7.5 MW is added from hot utility to the first interval. This does not change the heat balance within each interval, but increases all of the heat flows between intervals by 7.5 MW, giving one heat flow of just zero at an interval temperature of 145°C.

More than 7.5 MW could be added from hot utility to the first interval, but the objective is to find the minimum hot and cold utility, hence only the minimum is added. Thus from Figure 16.18b, $Q_{Hmin} = 7.5$ MW and $Q_{Cmin} = 10$ MW. This corresponds with the values obtained from the composite curves in Figure 16.5. One further important piece of information can be deduced from the cascade in Figure 16.18b. The point where the heat flow goes to zero at $T^* = 145$°C corresponds to the pinch. Thus, the actual hot and cold stream pinch temperatures are 150°C and 140°C respectively. Again, this agrees with the result from the composite curves in Figure 16.5.

The initial setting for the heat cascade in Figure 16.18a corresponds to the setting for the shifted composite curves in Figure 16.15a where there is an overlap. The setting of the heat cascade for zero or positive heat flows in Figure 16.18b corresponds to the shifted composite curve setting in Figure 16.15b.

The composite curves are useful in providing conceptual understanding of the process but the problem table algorithm is a more convenient calculation tool.

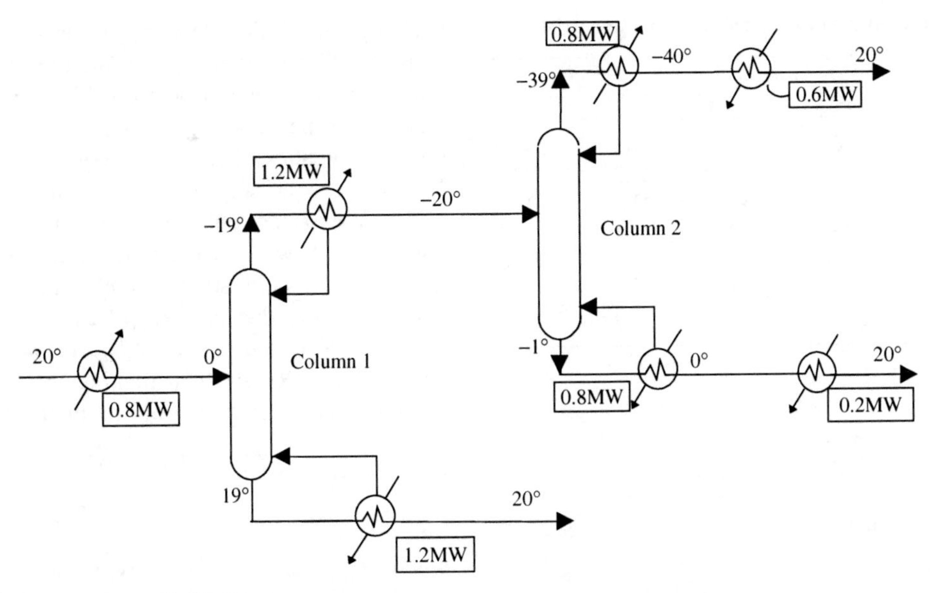

Figure 16.19 A low-temperature distillation process.

Example 16.1 The flowsheet for a low-temperature distillation process is shown in Figure 16.19. Calculate the minimum hot and cold utility requirements and the location of the pinch assuming $\Delta T_{min} = 5°C$.

Solution First extract the stream data from the flowsheet. This is given in Table 16.4 below.

Next, calculate the shifted interval temperatures. Hot stream temperatures are shifted down by 2.5°C and cold stream temperatures shifted up by 2.5°C, as shown in Table 16.5.

Now carry out a heat balance within each shifted temperature interval as shown in Figure 16.20.

Finally, the heat cascade is shown in Figure 16.21. Figure 16.21a shows the cascade with zero hot utility. This leads to negative heat flows, the largest of which is −1.84 MW. Adding 1.84 MW from hot utility as shown in Figure 16.21b gives $Q_{Hmin} = 1.84$ MW, $Q_{Cmin} = 1.84$ MW, hot stream pinch temperature = −19°C and cold stream pinch temperature = −24°C.

Table 16.4 Stream data for low-temperature distillation process.

Stream	Type	Supply temperature T_S(°C)	Target temperature T_T(°C)	ΔH (MW)	CP (MW·K^{-1})
1. Feed to column 1	Hot	20	0	−0.8	0.04
2. Column 1 condenser	Hot	−19	−20	−1.2	1.2
3. Column 2 condenser	Hot	−39	−40	−0.8	0.8
4. Column 1 reboiler	Cold	19	20	1.2	1.2
5. Column 2 reboiler	Cold	−1	0	0.8	0.8
6. Column 2 bottoms	Cold	0	20	0.2	0.01
7. Column 2 overheads	Cold	−40	20	0.6	0.01

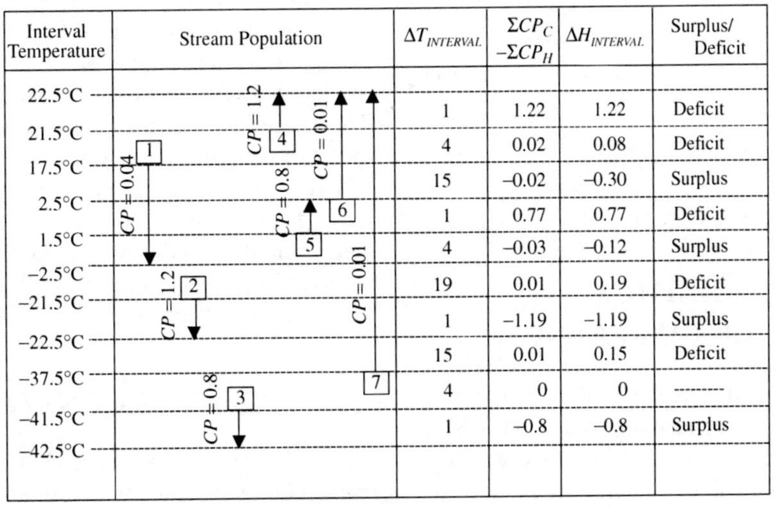

Interval Temperature	Stream Population	$\Delta T_{INTERVAL}$	$\Sigma CP_C - \Sigma CP_H$	$\Delta H_{INTERVAL}$	Surplus/ Deficit
22.5°C – 21.5°C		1	1.22	1.22	Deficit
21.5°C – 17.5°C		4	0.02	0.08	Deficit
17.5°C – 2.5°C		15	−0.02	−0.30	Surplus
2.5°C – 1.5°C		1	0.77	0.77	Deficit
1.5°C – −2.5°C		4	−0.03	−0.12	Surplus
−2.5°C – −21.5°C		19	0.01	0.19	Deficit
−21.5°C – −22.5°C		1	−1.19	−1.19	Surplus
−22.5°C – −37.5°C		15	0.01	0.15	Deficit
−37.5°C – −41.5°C		4	0	0	--------
−41.5°C – −42.5°C		1	−0.8	−0.8	Surplus

Stream population: 1 ($CP = 0.04$), 2 ($CP = 1.2$), 3 ($CP = 0.8$), 4 ($CP = 1.2$), 5 ($CP = 0.8$), 6 ($CP = 0.01$), 7 ($CP = 0.01$)

Figure 16.20 Temperature interval heat balances for Example 16.1.

Table 16.5 Shifted temperatures for the data in Table 16.4.

Stream	Type	T_S	T_T	T_S^*	T_T^*
1	Hot	20	0	17.5	−2.5
2	Hot	−19	−20	−21.5	−22.5
3	Hot	−39	−40	−41.5	−42.5
4	Cold	19	20	21.5	22.5
5	Cold	−1	0	1.5	2.5
6	Cold	0	20	2.5	22.5
7	Cold	−40	20	−37.5	22.5

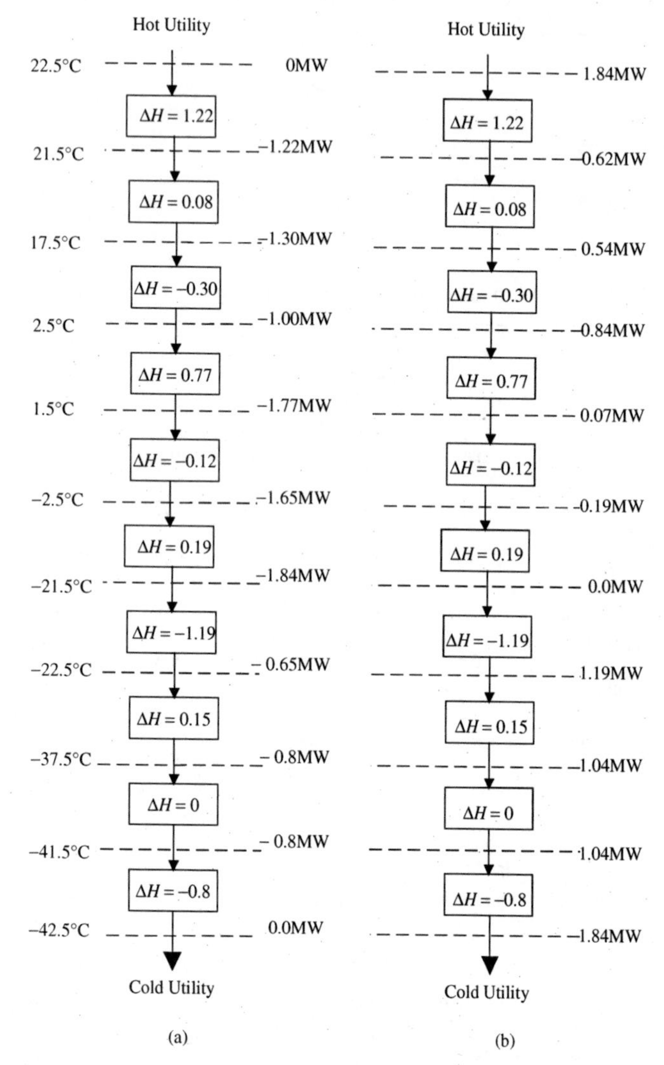

Figure 16.21 The problem table cascade for Example 16.1.

16.5 NONGLOBAL MINIMUM TEMPERATURE DIFFERENCES

So far, it has been assumed that the minimum temperature difference for a heat exchanger network applies globally between all streams in the network. However, there are occasions when nonglobal minimum temperature differences might be required. For example, suppose a heat exchanger network is servicing some streams that are liquid and some that are gaseous. For the liquid–liquid heat transfer matches, a value of perhaps $\Delta T_{min} = 10°C$ is appropriate. But for the gas–gas matches, a larger temperature minimum temperature difference is required, say $\Delta T_{min} = 20°C$. How can this be accommodated in the targeting?

When carrying out the problem table algorithm, the temperatures were shifted according to $\Delta T_{min}/2$ being added to the cold streams and subtracted from the hot streams. This value of $\Delta T_{min}/2$ can be considered to be a *contribution* to the overall ΔT_{min} between the hot and the cold streams. Rather than making the ΔT_{min} contribution equal for all streams, it could be made stream-specific:

$$T_{H,i}^* = T_{H,i} - \Delta T_{min,cont,i}$$

$$T_{C,j}^* = T_{C,j} + \Delta T_{min,cont,j}$$

where $T_{H,i}^*$, $T_{H,i}$ are the shifted and actual temperatures for Hot Stream i, $T_{C,j}^*$, $T_{C,j}$ are the shifted and actual temperatures for Cold Stream j, and $\Delta T_{min,cont,i}$ and $\Delta T_{min,cont,j}$ are the contributions to ΔT_{min} for Hot Stream i and Cold Stream j. Thus, for the above example, if the ΔT_{min} contribution for liquid streams is taken to be 5°C and for gas streams 10°C, then a liquid–liquid match would lead to $\Delta T_{min} = 10°C$, a gas–gas match would lead to $\Delta T_{min} = 20°C$ and a liquid–gas match would lead to $\Delta T_{min} = 15°C$[4]. To include this in the problem table algorithm is straightforward. All that needs to be done is that the appropriate ΔT_{min} contribution is to be allocated to each stream and then that ΔT_{min} contribution is subtracted in the case of hot streams and added in the case of cold streams. This would lead to different interval temperatures compared with a global minimum temperature difference. The remainder of the problem table algorithm would be the same. Once the interval temperatures based on ΔT_{min} contributions have been established, the interval heat balances can be performed and the cascade set up in the same way as for a global ΔT_{min}.

From the point of view of the composite curves, the location of the pinch and the ΔT_{min} at the pinch would depend on which kind of streams were located in the region of the point of closest approach between the composite curves. If only liquid streams were present around the point of closest approach of the composite curves, then in the above example, $\Delta T_{min} = 10°C$ will apply. If there were only gas streams in the region around the point of closest approach, then in the above example, $\Delta T_{min} = 20°C$ would apply. If there was a mixture of liquid and gas streams at the point of closest approach, then $\Delta T_{min} = 15°C$ would apply.

16.6 PROCESS CONSTRAINTS

So far it has been assumed that any hot stream could, in principle, be matched with any cold stream providing there is feasible temperature difference between them. Often, practical constraints prevent this. For example, it might be the case that if two streams are matched in a heat exchanger and a leak develops, such that the two streams come into contact, this might produce an unacceptably hazardous situation. If this were the case, then no doubt a constraint would be imposed to prevent the two streams being matched. Another reason for a constraint might be that two streams are expected to be geographically very distant from each other, leading to unacceptably long pipe runs. Potential control and start-up problems might also call for constraints. There are many reasons why constraints might be imposed.

One common reason for imposing constraints results from *areas of integrity*[10]. A process is often normally designed to have logically identifiable sections or areas. An example might be the "reaction area" and "separation area" of the process. These areas might need to be kept separate for reasons such as start-up, shutdown, operational flexibility, safety, and so on. The areas are often made operationally independent by use of intermediate storage between the areas. Such independent areas are generally described as areas of integrity and impose constraints on the ability to transfer heat. Clearly, to maintain operational independence, two areas cannot be dependent on each other for heating and cooling by recovery.

The question now is, given that there are often constraints to deal with, how to evaluate the effect of these constraints on the system performance? The problem table algorithm cannot be used directly if constraints are imposed. However, often the effect of constraints on the energy performance can be evaluated using the problem table algorithm, together with a little common sense. The following example illustrates how[10].

Example 16.2 A process is to be divided into two operationally independent areas of integrity, Area A and Area B. The stream data for the two areas are given in Table 16.6[10].

Calculate the penalty in utility consumption to maintain the two areas of integrity for $\Delta T_{min} = 20°C$.

Solution To identify the penalty, first calculate the utility consumption of the two areas separate from each other as shown in Figure 16.22a. Next, combine all of the streams from both areas and again calculate the utility consumption, Figure 16.22b. Figure 16.23a shows the problem table cascade for Area A, the cascade for Area B is shown in Figure 16.23b, and that for Areas A and B combined in Figure 16.23c.

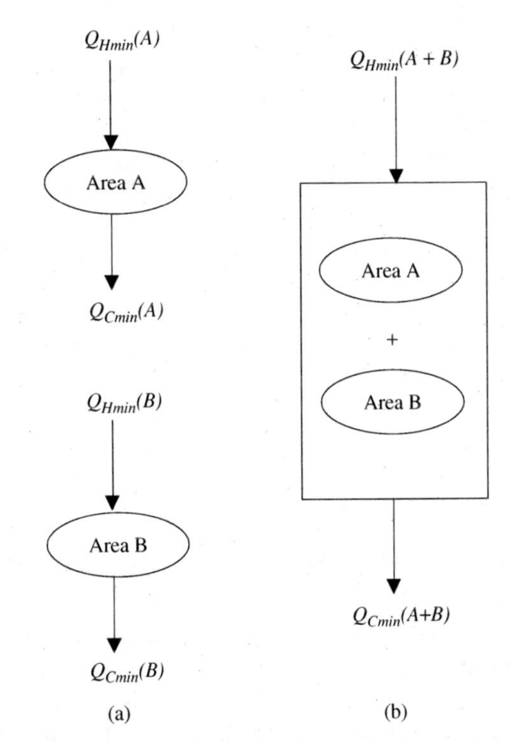

Figure 16.22 The areas of integrity can be targeted separately and then the combination targeted.

With Areas A and B separate, the total hot utility consumption is $(1400 + 0) = 1400$ kW and the total cold utility consumption is $(0 + 1350) = 1350$ kW. With Areas A and B combined, the total utility consumption is 950 kW of hot utility and 900 kW of cold utility.

The penalty for maintaining the areas of integrity is thus $(1400 - 950) = 450$ kW of hot utility and $(1350 - 900) = 450$ kW of cold utility.

Having quantified the penalty as a result of imposing constraints, the designer can exercise judgment as to whether it is acceptable or too expensive. If it is too expensive, there is a choice between two options:

a. Reject the constraints and operate the process as a single system.
b. Find a way to overcome the constraint. This is often possible by using a heat transfer fluid. The simplest option is via the existing utility system. For example, rather than have a direct match between two streams,

Table 16.6 Stream data for heat recovery between two areas of integrity.

Area A				Area B			
Stream	T_S (°C)	T_T (°C)	CP (kW·K^{-1})	Stream	T_S (°C)	T_T (°C)	CP (kW·K^{-1})
1	190	110	2.5	3	140	50	20.0
2	90	170	20.0	4	30	120	5.0

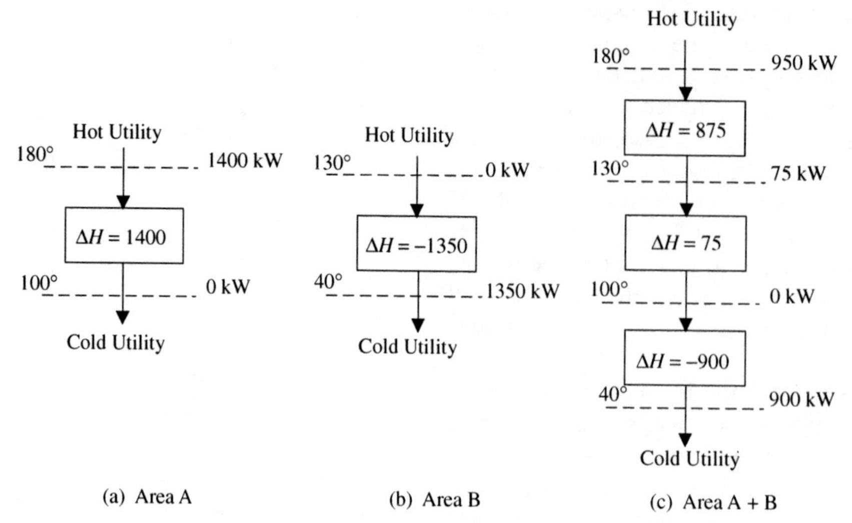

Figure 16.23 Problem table cascade for the separate and combined areas of integrity.

it might be possible for the heat source to generate steam to be fed into the steam mains and the heat sink to use steam from the same mains. The utility system then acts as a buffer between the heat sources and sinks. Another possibility might be to use a heat transfer fluid such as hot oil that circulates between the two streams being matched. To maintain operational independence, a standby heater and cooler supplied by utilities can be provided in the hot oil circuit, so that if either the heat source or sink is not operational, utilities could substitute heat recovery for short periods.

Many constraints can be evaluated by scoping out the problem with different boundaries. In Example 16.2, the sets of streams that were constrained to be separate were collected to be within each boundary for targeting. Comparing the targets for the streams within each boundary with that for all the streams put together allows the penalty of the constraint to be evaluated. The approach is more widely applicable than just areas of integrity. Whenever a stream, or set of streams, is to be maintained separate from any other set of streams, the same approach can be used. However, this approach of scoping out the problem with different boundaries has limitations in the evaluation of constraints. More complex constraints require linear programming to obtain the energy target[11,12].

16.7 UTILITY SELECTION

After maximizing heat recovery in the heat exchanger network, those heating and cooling duties not serviced by heat recovery must be provided by external utilities. The most common hot utility is steam. It is usually available at several levels. High-temperature heating duties require furnace flue gas or a hot oil circuit. Cold utilities might be refrigeration, cooling water, air-cooling, furnace air preheating, boiler feedwater preheating or even steam generation if heat needs to be rejected at high temperatures.

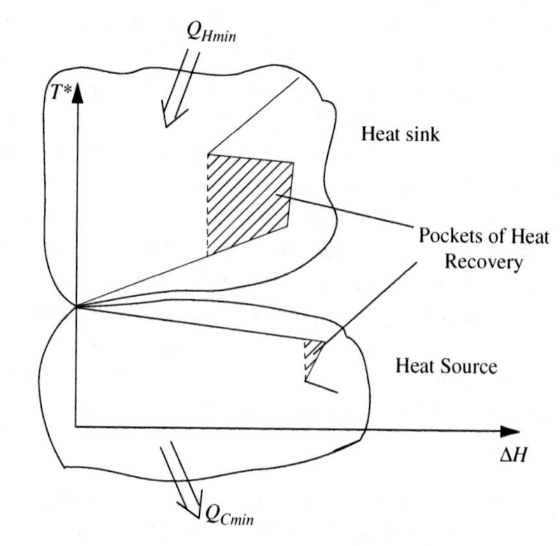

Figure 16.24 The grand composite curve shows the utility requirements both in enthalpy and temperature terms.

Although the composite curves can be used to set energy targets, they are not a suitable tool for the selection of utilities. The *grand composite curve* is a more appropriate tool for understanding the interface between the process and the utility system[4,13,14]. It will also be shown in later chapters to be a useful tool to study the interaction between heat-integrated reactors and separators and the rest of the process.

The grand composite curve is obtained by plotting the problem table cascade. A typical grand composite curve is shown in Figure 16.24. It shows the heat flow through the process against temperature. It should be noted that the temperature plotted here is *shifted temperature* (T^*) and not actual temperature. Hot streams are represented $\Delta T_{min}/2$ colder and cold streams $\Delta T_{min}/2$ hotter than they

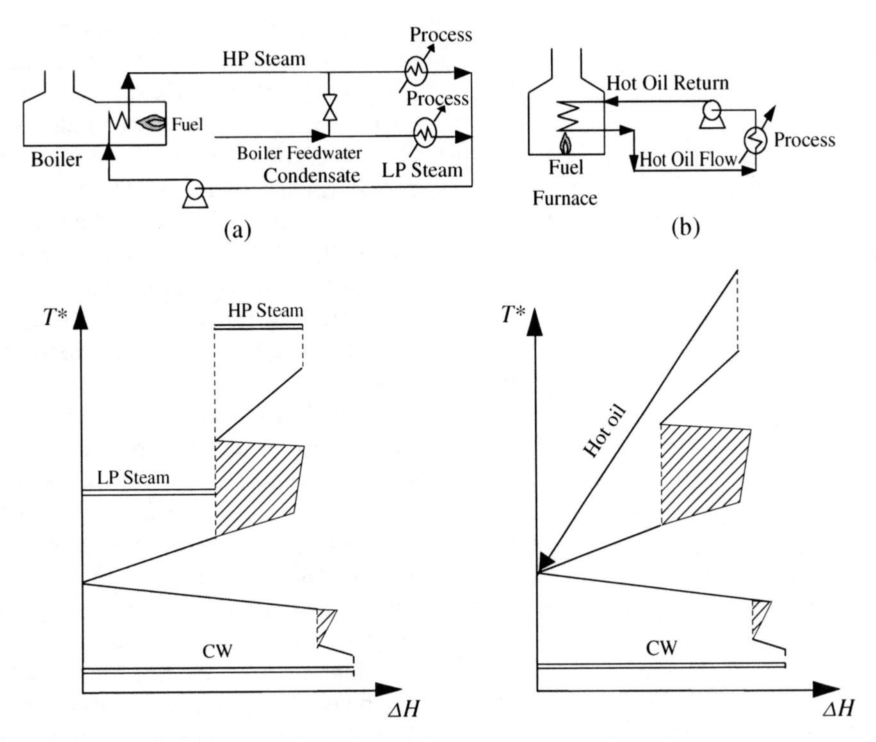

Figure 16.25 The grand composite curve allows alternative utility mixes to be evaluated.

are in practice. Thus, an allowance for ΔT_{min} is built into the construction.

The point of zero heat flow in the grand composite curve in Figure 16.24 is the heat recovery pinch. The open "jaws" at the top and bottom are Q_{Hmin} and Q_{Cmin} respectively. Thus, the heat sink above the pinch and heat source below the pinch can be identified as shown in Figure 16.24. The shaded areas in Figure 16.24, known as *pockets*, represent areas of additional process-to-process heat transfer. Remember that the profile of the grand composite curve represents residual heating and cooling demands after recovering heat within the shifted temperature intervals in the problem table algorithm. In the pockets in Figure 16.24, a local surplus of heat in the process is used at temperature differences in excess of ΔT_{min} to satisfy a local deficit. This reflects the cascading of excess heat from high-temperature intervals to lower temperature intervals in the problem table algorithm.

Figure 16.25a shows the same grand composite curve with two levels of steam used as hot utility. Figure 16.25b shows again the same grand composite curve but with hot oil used as hot utility.

Example 16.3 The problem table cascade for the process in Figure 16.2 is given in Figure 16.18. Using the grand composite curve:

a. For two levels of steam at saturation conditions and temperatures of 240°C and 180°C, determine the loads on the two steam levels that maximizes the use of the lower pressure steam.

b. Instead of using steam, a hot oil circuit is to be used with a supply temperature of 280°C and $C_P = 2.1\ \text{kJ}\cdot\text{kg}^{-1}\cdot\text{K}^{-1}$. Calculate the minimum flowrate of hot oil.

Solution

a. For $\Delta T_{min} = 10$°C, the two steam levels are plotted on the grand composite curve at temperatures of 235°C and 175°C. Figure 16.26a shows the loads that maximize the use of the lower pressure steam. Calculate the load on the low-pressure steam by interpolation of the cascade heat flows. At $T^* =$ 175°C:

$$\text{Load of 180°C steam} = \frac{175 - 145}{185 - 145} \times 4$$

$$= 3\ \text{MW}$$

$$\text{Load of 240°C steam} = 7.5 - 3$$

$$= 4.5\ \text{MW}$$

b. Figure 16.26b shows the grand composite curve with hot oil providing the hot utility requirements. If the minimum flowrate is required, then this corresponds to the steepest slope and minimum return temperature. For this problem, the minimum return temperature for the hot oil is pinch temperature ($T^* =$ 145°C, $T = 150$°C for hot streams). Thus,

$$\text{Minimum flowrate} = 7.5 \times 10^3 \times \frac{1}{2.1} \times \frac{1}{(280 - 150)}$$

$$= 27.5\ \text{kg}\cdot\text{s}^{-1}$$

In other problems, the shape of the grand composite curve away from the pinch could have limited the flowrate.

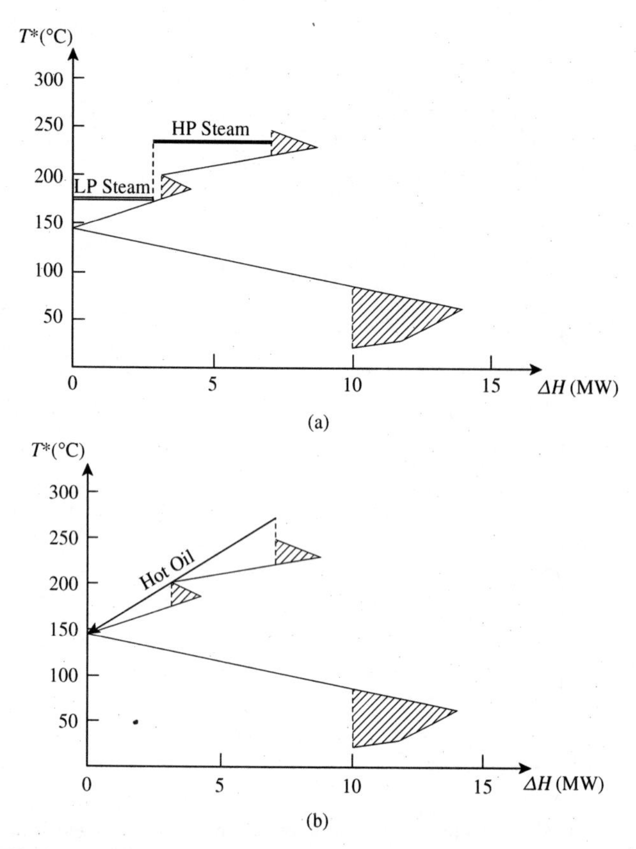

Figure 16.26 Alternative utility mixes for the process in Figure 16.2.

16.8 FURNACES

When hot utility needs to be at a high temperature and/or provide high heat fluxes, radiant heat transfer is used from combustion of fuel in a furnace. Furnace designs vary according to the function of the furnace, heating duty, type of fuel and the method of introducing combustion air (see Chapter 15). Sometimes the function is to purely provide heat, sometimes the furnace is also a reactor and provides heat of reaction. However, process furnaces have a number of features in common. In the chamber in which combustion takes place, heat is transferred mainly by radiation to tubes around the walls of the chamber, through which passes the fluid to be heated. After the flue gas leaves the combustion chamber, most furnace designs extract further heat from the flue gas in a convection section before the flue gas is vented to atmosphere.

Figure 16.27 shows a grand composite curve with a flue gas matched against it to provide hot utility[15]. The flue gas starts at its theoretical flame temperature (see Chapter 15) shifted for ΔT_{min} on the grand composite curve and presents a sloping profile because it is giving up sensible heat. Theoretical flame temperature is the temperature attained when a fuel is burnt in air or oxygen without loss or gain of heat, as explained in Chapter 15.

In Figure 16.27, the flue gas is cooled to pinch temperature before being released to atmosphere. The heat released from the flue gas between pinch temperature and ambient is the stack loss. Thus in Figure 16.27, for a given grand composite curve and theoretical flame temperature, the heat from fuel and stack loss can be determined.

All combustion processes work with an excess of air or oxygen to ensure complete combustion of the fuel. Excess air typically ranges between 5 and 20% depending on the fuel, burner design and furnace design. As excess air is reduced, theoretical flame temperature increases as shown in Figure 16.28. This has the effect of reducing the stack loss and increasing the thermal efficiency of the furnace for a given process heating duty. Alternatively, if

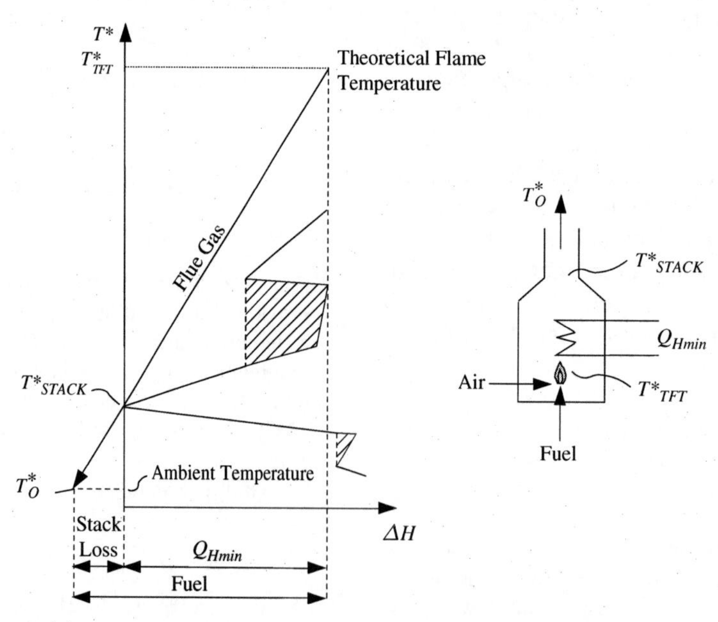

Figure 16.27 Simple furnace model.

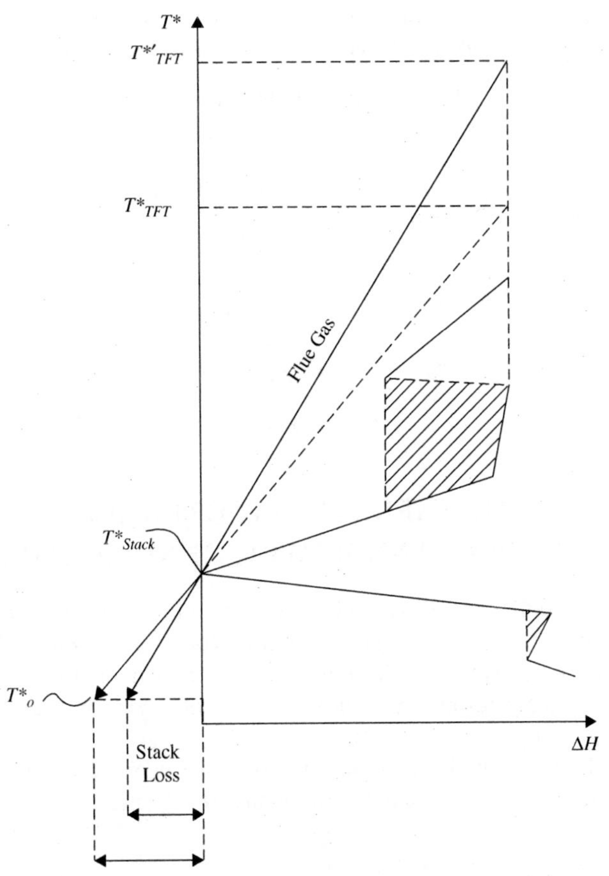

Figure 16.28 Increasing the theoretical flame temperature by reducing excess air or preheating combustion air reduces the stack loss.

the combustion air is preheated (say by heat recovery), then again the theoretical flame temperature increases as shown in Figure 16.28, reducing the stack loss.

Although, the higher flame temperatures in Figure 16.28 reduce the fuel consumption for a given process heating duty, there is one significant disadvantage. Higher flame temperatures increase the formation of oxides of nitrogen, which are environmentally harmful. This point will be returned to in Chapter 25.

In Figures 16.27 and 16.28, the flue gas is capable of being cooled to pinch temperature before being released to atmosphere. This is not always the case. Figure 16.29a shows a situation in which the flue gas is released to atmosphere above pinch temperature for practical reasons. There is a practical minimum, the acid dew point, to which a flue gas can be cooled without condensation causing corrosion in the stack (see Chapter 15). The minimum stack temperature in Figure 16.29a is fixed by acid dew point. Another case is shown in Figure 16.29b where the process away from the pinch limits the slope of the flue gas line and hence the stack loss.

Example 16.4 The process in Figure 16.2 is to have its hot utility supplied by a furnace. The theoretical flame temperature for combustion is 1800°C and the acid dew point for the flue gas is 160°C. Ambient temperature is 10°C. Assume $\Delta T_{min} = 10°C$ for process-to-process heat transfer but $\Delta T_{min} = 30°C$ for flue gas to process heat transfer. A high value for ΔT_{min} for flue gas to process heat transfer has been chosen because of poor heat-transfer coefficients in the convection bank of the furnace. Calculate the fuel required, stack loss and furnace efficiency.

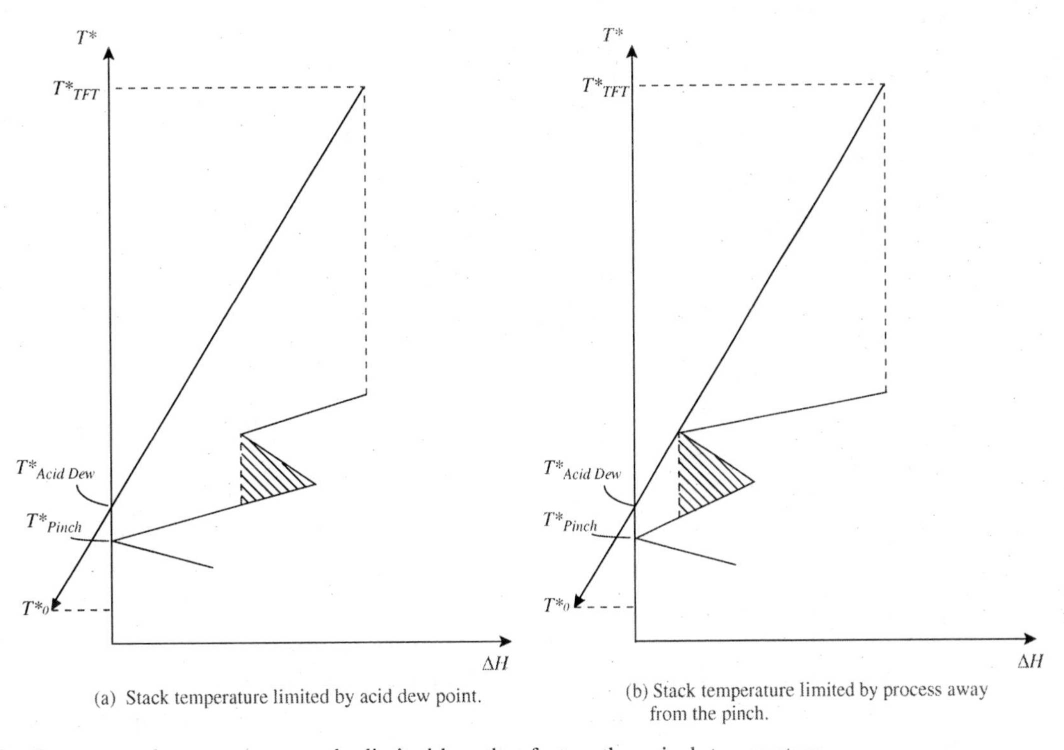

(a) Stack temperature limited by acid dew point.

(b) Stack temperature limited by process away from the pinch.

Figure 16.29 Furnace stack temperature can be limited by other factors than pinch temperature.

Solution The first problem is that a different value of ΔT_{min} is required for different matches. The problem table algorithm is easily adapted to accommodate this. This is achieved by assigning ΔT_{min} contributions to streams. If the process streams are assigned a contribution of 5°C and flue gas a contribution of 25°C, then a process/process match has a ΔT_{min} of $(5 + 5) = 10$°C and a process/flue gas match has a ΔT_{min} of $(5 + 25) = 30$°C. When setting up the interval temperatures in the problem table algorithm, the interval boundaries are now set at hot stream temperatures minus their ΔT_{min} contribution, rather than half the global ΔT_{min}. Similarly, boundaries are now set on the basis of cold stream temperatures plus their ΔT_{min} contribution.

Figure 16.30 shows the grand composite curve plotted from the problem table cascade in Figure 16.18b. The starting point for the flue gas is an actual temperature of 1800°C, which corresponds to a shifted temperature of $(1800 - 25) = 1775$°C on the grand composite curve. The flue gas profile is not restricted above the pinch and can be cooled to pinch temperature corresponding with a shifted temperature of 145°C before venting to atmosphere. The actual stack temperature is thus $(145 + 25) = 170$°C. This is just above the acid dew point of 160°C. Now calculate the fuel consumption.

$$Q_{Hmin} = 7.5 \text{ MW}$$

$$CP_{FLUE\ GAS} = \frac{7.5}{1775 - 145} = 0.0046 \text{ MW·K}^{-1}$$

The fuel consumption is now calculated by taking the flue gas from theoretical flame temperature to ambient temperature:

$$\text{Fuel required} = 0.0046(1800 - 10) = 8.23 \text{ MW}$$

$$\text{Stack loss} = 0.0046(170 - 10) = 0.74 \text{ MW}$$

$$\text{Furnace efficiency} = \frac{Q_{H\,min}}{Fuel\ Required} \times 100 = \frac{7.5}{8.23} \times 100 = 91\%$$

16.9 COGENERATION (COMBINED HEAT AND POWER GENERATION)

More complex utility options are encountered when combined heat and power generation (or *cogeneration*) is exploited. Here the heat rejected by a heat engine such as a steam turbine, gas turbine or diesel engine is used as hot utility.

Fundamentally, there are two possible ways to integrate a heat engine exhaust[14]. In Figure 16.31, the process is

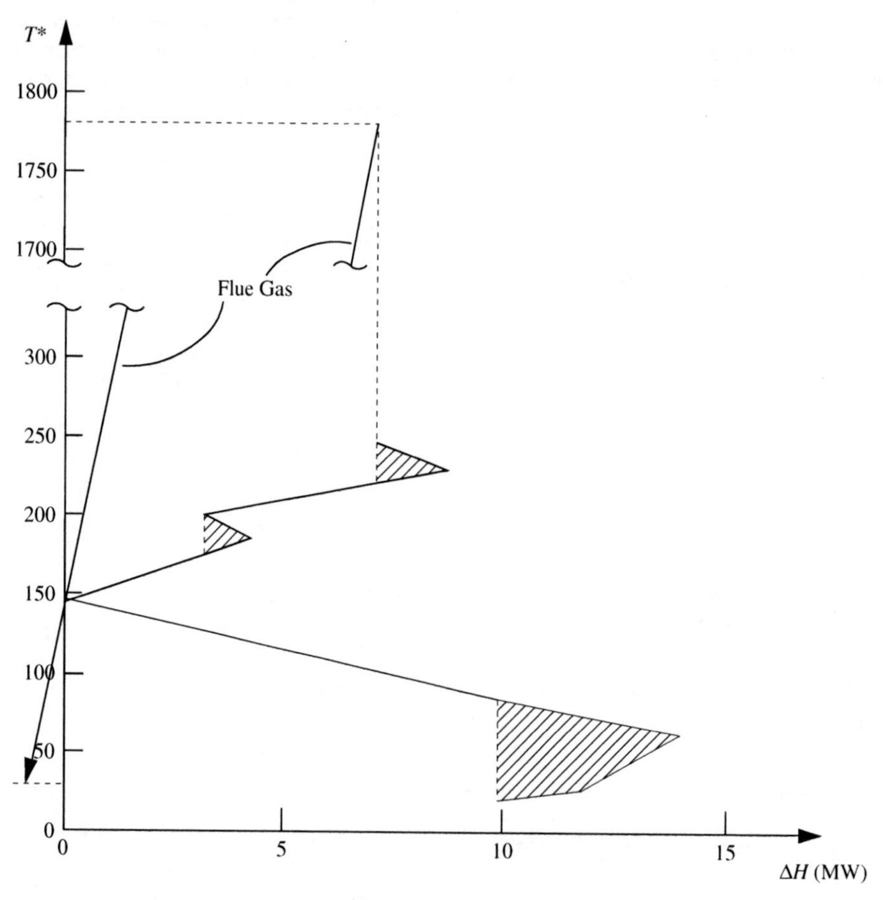

Figure 16.30 Flue gas matched against the grand composite curve of the process in Figure 16.2.

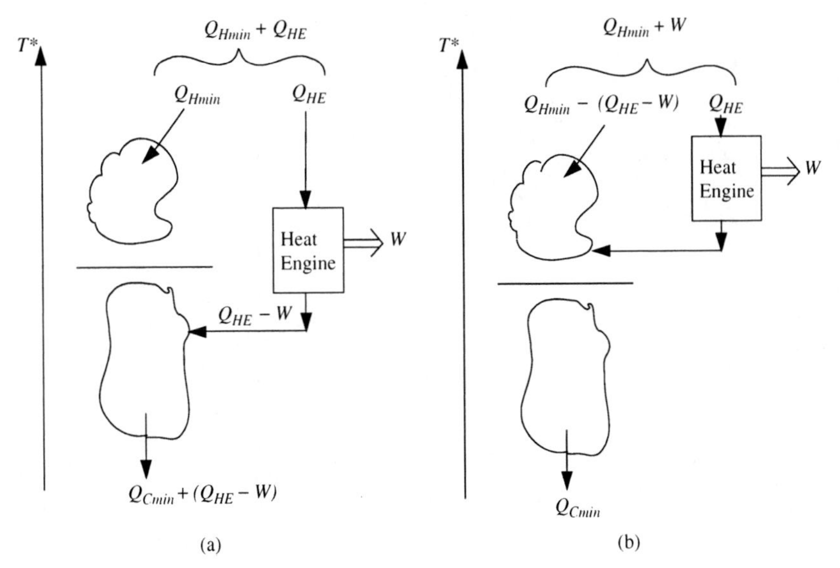

Figure 16.31 Heat engine exhaust can be integrated either across or not across the pinch.

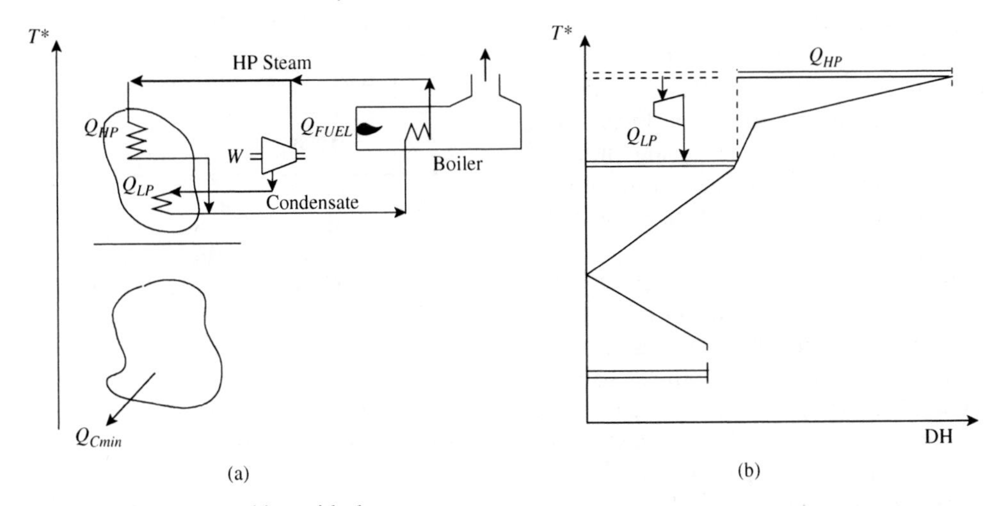

Figure 16.32 Integration of a steam turbine with the process.

represented as a heat sink and heat source separated by the pinch. Integration of the heat engine across the pinch as shown in Figure 16.31a is counterproductive. The process still requires Q_{Hmin} and the heat engine performs no better than operated stand-alone. There is no saving by integrating a heat engine across the pinch[14].

Figure 16.31b shows the heat engine integrated above the pinch. In rejecting heat above the pinch, it is rejecting heat into the part of the process, which is overall a heat sink. In so doing, it is exploiting the temperature difference that exists between the utility source and the process sink, producing power at high efficiency. The net effect in Figure 16.31b is the import of extra energy W from heat sources to produce W power. Owing to the process and heat engine acting as one system, apparently conversion of heat to power at 100% efficiency is achieved[14].

Now consider the two most commonly used heat engines (steam and gas turbines) in more detail to see whether they achieve this in practice. To make a quantitative assessment of any combined heat and power scheme, the grand composite curve should be used and the heat engine exhaust treated like any other utility.

Figure 16.32 shows a steam turbine integrated with the process above the pinch. A steam turbine is conceptually like a centrifugal compressor, but working in reverse. Steam is expanded from high to low pressure in the machine, producing power. In Figure 16.32 heat Q_{HP} is taken into the process from high-pressure steam. The balance of the hot utility demand Q_{LP} is taken from the steam turbine exhaust. In Figure 16.32a, heat Q_{FUEL} is taken into the boiler from fuel. An overall energy balance gives:

$$Q_{FUEL} = Q_{HP} + Q_{LP} + W + Q_{LOSS} \tag{16.2}$$

The process requires $(Q_{HP} + Q_{LP})$ to satisfy its enthalpy imbalance above the pinch. If there were no losses from

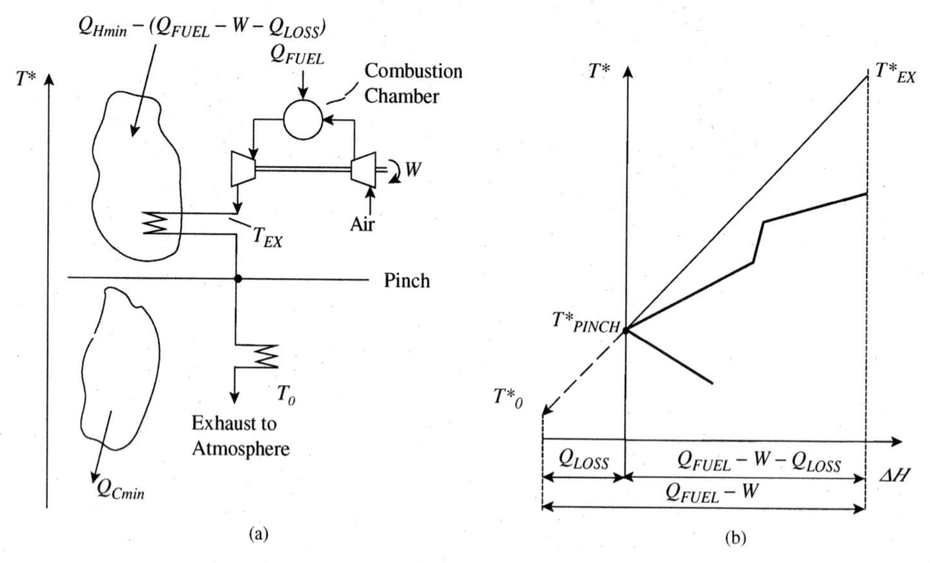

Figure 16.33 A gas turbine exhaust matched against the process (same as a flue gas).

the boiler, then fuel W would be converted to power W at 100% efficiency. However, the boiler losses Q_{LOSS} reduce this to below 100% conversion. In practice, in addition to the boiler losses, there can also be significant losses from the steam distribution system. Figure 16.32b shows how the grand composite curve can be used to size steam turbine cycles[14]. Steam turbines will be dealt with in more detail in Chapter 23.

Figure 16.33 shows a schematic of a simple gas turbine. The machine is essentially a rotary compressor mounted on the same shaft as a turbine. Air enters the compressor where it is compressed before entering a combustion chamber. Here the combustion of fuel increases its temperature. The mixture of air and combustion gases is expanded in the turbine. The input of energy to the combustion chamber allows enough power to be developed in the turbine to both drive the compressor and provide useful power. The performance of the machine is specified in terms of the power output, airflow rate through the machine, efficiency of conversion of heat to power and the temperature of the exhaust. Gas turbines are normally used only for relatively large-scale applications, and will be dealt with in more detail in Chapter 23.

Figure 16.33 shows a gas turbine matched against the grand composite curve[14]. As with the steam turbine, if there was no stack loss to atmosphere (i.e. if Q_{LOSS} was zero), then W heat would be turned into W power. The stack losses in Figure 16.33 reduce the efficiency of conversion of heat to power. The overall efficiency of conversion of heat to power depends on the turbine exhaust profile, the pinch temperature and the shape of the process grand composite.

Example 16.5 The stream data for a heat recovery problem are given in Table 16.7 below.

Table 16.7 Stream data for Example 16.5.

Stream No.	Stream Type	T_S (°C)	T_T (°C)	Heat capacity flowrate ($MW \cdot K^{-1}$)
1	Hot	450	50	0.25
2	Hot	50	40	1.5
3	Cold	30	400	0.22
4	Cold	30	400	0.05
5	Cold	120	121	22.0

A problem table analysis for $\Delta T_{min} = 20°C$ results in the heat cascade given in Table 16.8.

Table 16.8 Problem table cascade for Example 16.5.

T^* (°C)	Cascade heat flow (MW)
440	21.9
410	29.4
131	23.82
130	1.8
40	0
30	15

The process also has a requirement for 7 MW of power. Two alternative cogeneration schemes are to be compared economically.

a. A steam turbine with its exhaust saturated at 150°C used for process heating is one of the options to be considered. Superheated steam is generated in the central boiler house at 41 bar with a temperature of 300°C. This superheated steam can be expanded in a single-stage turbine with an isentropic efficiency of 85%. Calculate the maximum generation of power possible by matching the exhaust steam against the process.

b. A second possible scheme uses a gas turbine with a flowrate of air of 97 kg·s^{-1}, which has an exhaust temperature of 400°C. Calculate the power generation if the turbine has an efficiency of 30%. Ambient temperature is 10°C.

c. The cost of heat from fuel for the gas turbine is \$4.5 GW^{-1}. The cost of imported electricity is \$19.2 GW^{-1}. Electricity can be exported with a value of \$14.4 GW^{-1}. The cost of fuel for steam generation is \$3.2 GW^{-1}. The overall efficiency of steam generation and distribution is 80%. Which scheme is most cost-effective, the steam turbine or the gas turbine?

Solution

a. This is shown in Figure 16.34a. The steam condensing interval temperature is 140°C.

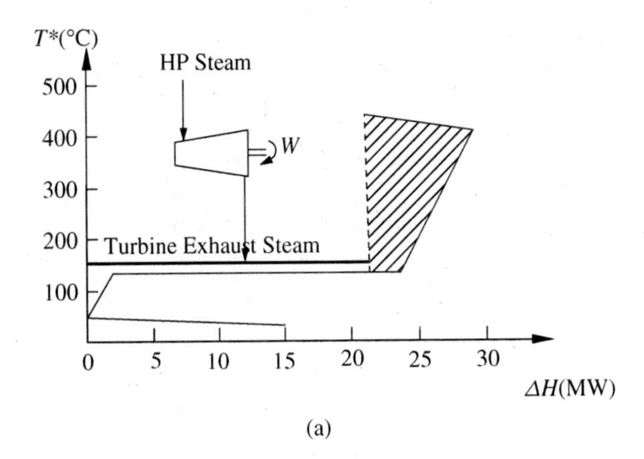

(a)

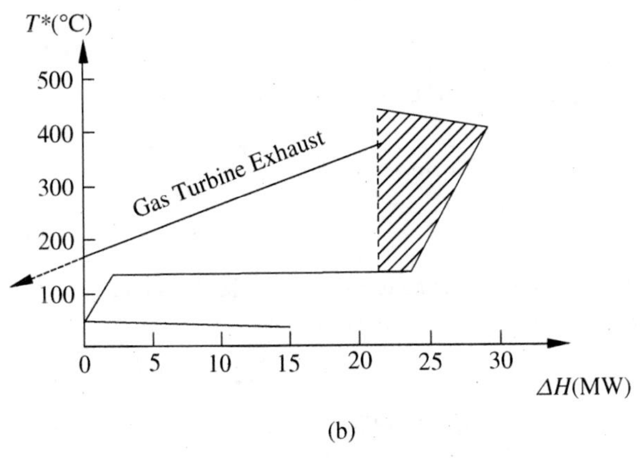

(b)

Figure 16.34 Alternative combined heat and power schemes for Example 16.5.

Heat flow required from the turbine exhaust

$$= 21.9 \text{ MW}$$

From steam tables, inlet conditions at $T_1 = 300°\text{C}$ and $P_1 =$ 41 bar are:

$$H_1 = 2959 \text{ kJ·kg}^{-1}$$

$$S_1 = 6.349 \text{ kJ·kg}^{-1}\text{·K}^{-1}$$

Turbine outlet conditions for isentropic expansion to 150°C from steam tables are:

$$P_2 = 4.77 \text{ bar}$$

$$S_2 = 6.349 \text{ kJ·kg}^{-1}\text{·K}^{-1}$$

The wetness fraction (X) can be calculated from

$$S_2 = XS_L + (1 - X)S_V$$

where S_L and S_V are the saturated liquid and vapor entropies. Taking saturated liquid and vapor entropies from steam tables at 150°C and 4.77 bar:

$$6.349 = 1.842X + 6.838(1 - X)$$

$$X = 0.098$$

The turbine outlet enthalpy for an isentropic expansion can now be calculated from:

$$H_2 = XH_L + (1 - X)H_V$$

where H_L and H_V are the saturated liquid and vapor enthalpies. Taking saturated liquid and vapor enthalpies from steam tables at 150°C and 4.77 bar:

$$H_2 = 0.098 \times 632 + (1 - 0.098)2747$$

$$= 2540 \text{ kJ·kg}^{-1}$$

For a single-stage expansion with isentropic efficiency of 85%:

$$H_2' = H_1 - \eta_{IS}(H_1 - H_2)$$

$$= 2959 - 0.85(2959 - 2540)$$

$$= 2603 \text{ kJ·kg}^{-1}$$

The actual wetness fraction (X') can be calculated from:

$$H_2' = X'H_L + (1 - X')H_V$$

where H_L and H_V are the saturated liquid and vapor enthalpies.

$$H_2' = 2603 = 632X' + 2747(1 - X')$$

$$X' = 0.068$$

Assume in this case that the saturated steam and condensate are separated after the turbine and only the saturated steam used for process heating.

$$\text{Steam flow to process} = \frac{21.9 \times 10^3}{2747 - 632}$$

$$= 10.35 \text{ kg·s}^{-1}$$

$$\text{Steam flow through turbine} = \frac{10.35}{(1 - 0.068)}$$

$$= 11.11 \text{ kg·s}^{-1}$$

$$\text{Power generated } W = 11.11(2959 - 2603) \times 10^{-3}$$

$$= 3.96 \text{ MW}$$

b. The exhaust from the gas turbine is primarily air with a small amount of combustion gases. Hence, the CP of the exhaust can be approximated to be that of the airflow. Assuming C_P for air = 1.03 kJ·kg^{-1}·K^{-1}.

$$CP_{EX} = 97 \times 1.03$$
$$= 100 \text{ kW·K}^{-1}$$

The gas turbine option is shown in Figure 16.34b.

$$Q_{EX} = CP_{EX}(T_{EX} - T_0)$$
$$= 0.1 \times (400 - 10)$$
$$= 39 \text{ MW}$$
$$Q_{FUEL} = \frac{Q_{EX}}{(1 - \eta_{GT})}$$
$$= \frac{39}{(1 - 0.3)}$$
$$= 55.71 \text{ MW}$$
$$W = Q_{FUEL} - Q_{EX}$$
$$= 16.71 \text{ MW}$$

c. Steam turbine economics:

$$\text{Cost of fuel} = (21.9 + 3.96) \times \frac{3.2 \times 10^{-3}}{0.8}$$
$$= \$0.10 \text{ s}^{-1}$$
$$\text{Cost of imported electricity} = (7 - 3.96) \times 19.2 \times 10^{-3}$$
$$= \$0.06 \text{ s}^{-1}$$
$$\text{Net cost} = \$0.16 \text{ s}^{-1}$$

Gas turbine economics:

$$\text{Cost of fuel} = 55.71 \times 4.5 \times 10^{-3}$$
$$= \$0.25 \text{ s}^{-1}$$
$$\text{Electricity credit} = (16.71 - 7) \times 14.4 \times 10^{-3}$$
$$= \$0.14 \text{ s}^{-1}$$
$$\text{Net cost} = \$0.11 \text{ s}^{-1}$$

Hence the gas turbine is the most profitable in terms of energy costs. However, this is only part of the story since the capital cost of a gas turbine installation is likely to be significantly higher than that of a steam turbine installation.

Example 16.6 The problem table cascade for a process is given in Table 16.9 below for $\Delta T_{min} = 10°$C.

It is proposed to provide process cooling by steam generation from boiler feedwater with a temperature of 100°C.

Table 16.9 Problem table cascade.

Interval temperature (°C)	Heat flow (MW)
495	3.6
455	9.2
415	10.8
305	4.2
285	0
215	16.8
195	17.6
185	16.6
125	16.6
95	21.1
85	18.1

a. Determine how much steam can be generated at a saturation temperature of 230°C.
b. Determine how much steam can be generated with a saturation temperature of 230°C and superheated to the maximum temperature possible against the process.
c. Calculate how much power can be generated from the superheated steam from Part *b*, assuming a single-stage condensing steam turbine is to be used with an isentropic efficiency of 85%. Cooling water is available at 20°C and is to be returned to the cooling tower at 30°C.

Solution

a. Heat available for steam generation at 235°C interval temperature

$$= 12.0 \text{ MW}$$

From steam tables, the latent heat of water at a saturated temperature of 230°C is 1812 kJ·kg^{-1}.

$$\text{Steam production} = \frac{12.0 \times 10^3}{1812}$$
$$= 6.62 \text{ kg·s}^{-1}$$

Taking the heat capacity of water to be 4.3 kJ·kg^{-1}·K^{-1}, heat duty on boiler feedwater preheating

$$= 6.62 \times 4.3 \times 10^{-3}(230 - 100)$$
$$= 3.70 \text{ MW}$$

The profile of steam generation is shown against the grand composite curve in Figure 16.35a. The process can support both boiler feedwater preheat and steam generation.

b. Maximum superheat temperature

$$= 285°\text{C interval}$$
$$= 280°\text{C actual}$$

The profile is shown against the grand composite curve in Figure 16.35b.

From steam tables, enthalpy of superheated steam at 280°C and 28 bar

$$= 2947 \text{ kJ·kg}^{-1}$$

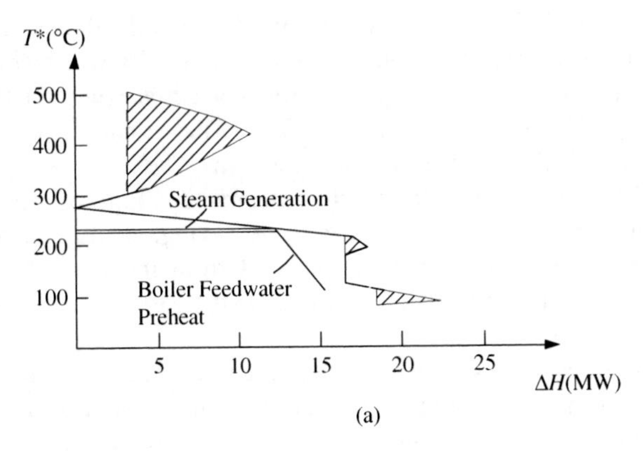

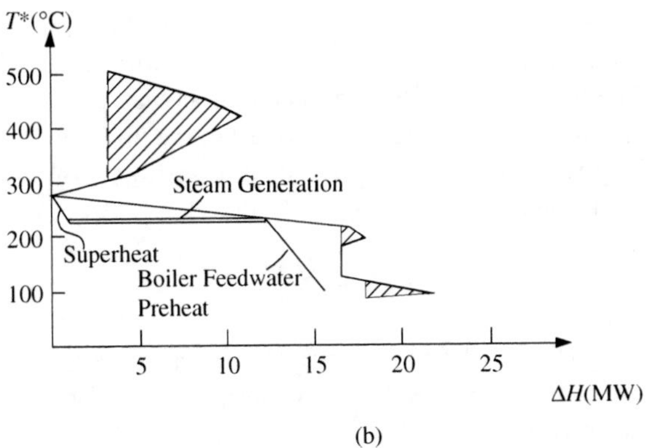

Figure 16.35 Alternative cold utilities for Example 16.6.

and enthalpy of saturated water at 230°C and 28 bar

$$= 991 \text{ kJ·kg}^{-1}$$

$$\text{Steam production} = \frac{12.0 \times 10^3}{(2947 - 991)}$$

$$= 6.13 \text{ kg·s}^{-1}$$

c. In a condensing turbine, the exhaust from the turbine is condensed under vacuum against cooling water. The lower the condensing temperature, the greater the power generation. The lowest condensing temperature for this problem is cooling water temperature plus ΔT_{min}, that is, $30 + 10 = 40$°C. From steam tables, inlet conditions at $T_1 = 280$°C and $P_1 =$ 28 bar are:

$$H_1 = 2947 \text{ kJ·kg}^{-1}$$

$$S_1 = 6.488 \text{ kJ·kg}^{-1}\text{·K}^{-1}$$

Turbine outlet conditions for isentropic expansion to 40°C from steam tables are:

$$P_2 = 0.074 \text{ bar}$$

For $S_2 = 6.488$ kJ·kg^{-1}·K^{-1}, the wetness fraction (X) and outlet enthalpy H_2 can be calculated as shown in Example 14.4.

$$X = 0.23$$

$$H_2 = 2020 \text{ kJ·kg}^{-1}$$

For a single-stage expansion with isentropic efficiency of 85%:

$$H_2' = 2947 - 0.85(2947 - 2020)$$

$$= 2159 \text{ kJ·kg}^{-1}$$

The power generation (W) is given by:

$$W = 6.13(2947 - 2159) \times 10^{-3}$$

$$= 4.8 \text{ MW}$$

The wetness fraction for the real expansion is given by:

$$H_2' = 2159 = X'H_L + (1 - X')H_V$$

$$= 167.5X' + 2574(1 - X')$$

$$X' = 0.17$$

This wetness fraction is possibly too high, since high levels of wetness can cause damage to the turbine. To allow a lower wetness fraction of say $X = 0.15$, the outlet pressure of the turbine must be raised to 0.2 bar, corresponding to a condensing temperature of 60°C. However, in so doing, the power generation decreases to 4.2 MW.

16.10 INTEGRATION OF HEAT PUMPS

A heat pump is a device that takes in low-temperature heat and upgrades it to a higher temperature to provide process heat. A schematic of a simple vapor compression heat pump is shown in Figure 16.36. In Figure 16.36, the heat pump absorbs heat at a low temperature in the evaporator, consumes power when the working fluid is compressed and rejects heat at a higher temperature in the condenser. The condensed working fluid is expanded and partially vaporizes. The cycle then repeats. The working fluid is usually a pure component, which means that the evaporation and condensation take place isothermally. When considering the integration of a heat pump with the process, there are appropriate and inappropriate ways to integrate heat pumps.

There are two fundamental ways in which a heat pump can be integrated with the process; across and not across the pinch[14]. Integration not across (above) the pinch is illustrated in Figure 16.37a. This arrangement imports W

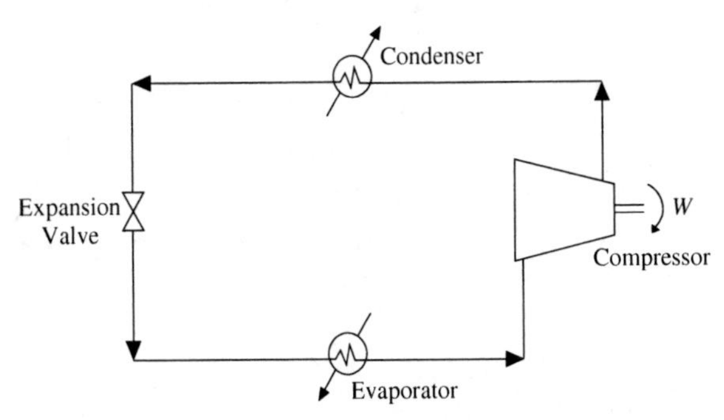

Figure 16.36 Schematic of a simple vapor compression heat pump.

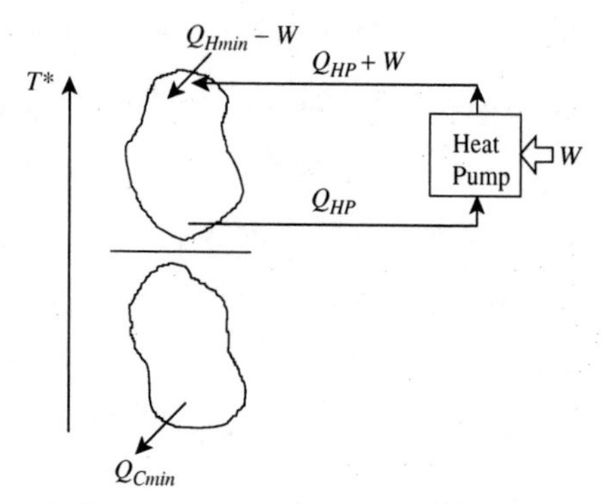

(a) Integration of a heat pump abovethe pinch.

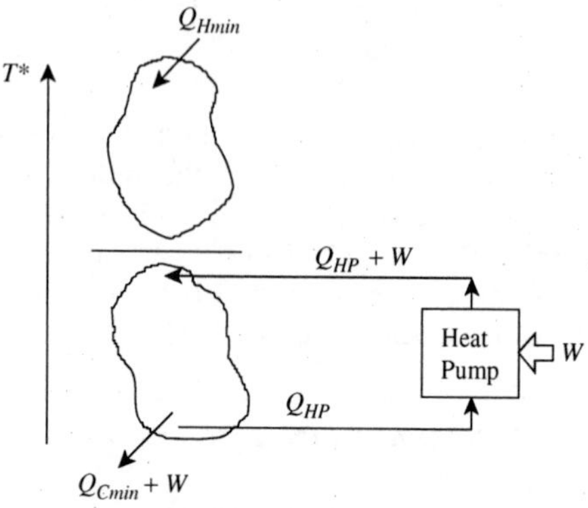

(b) Integration of a heat pump below the pinch.

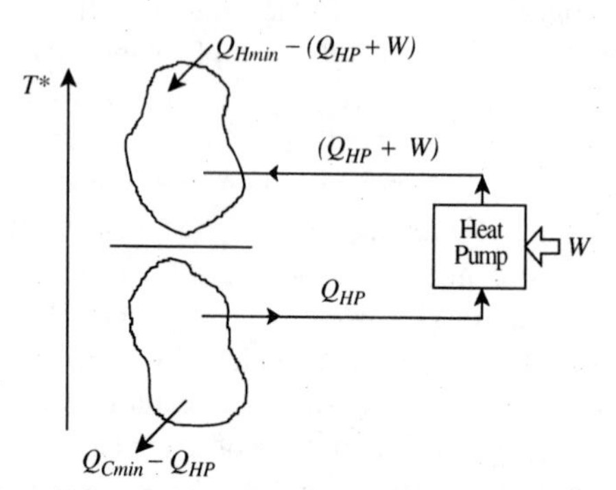

(c) Integration of a heat pump across the pinch.

Figure 16.37 Integration of heat pump with the process.

power and saves W hot utility. In other words, the system converts power into heat, which is not normally worthwhile economically. Another integration not across (below) the pinch is shown in Figure 16.37b. The result is worse economically. Power is turned into waste heat[14].

Integration across the pinch is illustrated in Figure 16.37c. This arrangement brings a genuine saving. It also makes overall sense since heat is pumped from the part of the process that is overall a heat source to the part that is overall a heat sink.

Figure 16.38 shows a heat pump appropriately integrated against a process. Figure 16.38a shows the overall balance. Figure 16.38b illustrates how the grand composite curve can be used to size the heat pump. How the heat pump performs determines its coefficient of performance. The coefficient of performance for a heat pump can generally be defined as the useful energy delivered to the process divided by the power expended to produce this useful energy. From Figure 16.38a:

$$COP_{HP} = \frac{Q_{HP} + W}{W} \tag{16.3}$$

where COP_{HP} is the heat pump coefficient of performance, Q_{HP} the heat absorbed at low temperature and W the power consumed.

For any given type of heat pump, a higher COP_{HP} leads to better economics. Having a better COP and hence better economics means working across a small temperature lift with the heat pump. The smaller the temperature lift, the better the COP_{HP}. For most applications, a temperature lift greater than 25°C is rarely economic. Attractive heat pump application normally requires a lift much less than 25°C.

Using the grand composite curve, the loads and temperatures of the cooling and heating duties and hence the COP_{HP} of integrated heat pumps can be readily assessed.

Thus, the appropriate placement of heat pumps is that they should be placed across the pinch[14]. Note that the principle needs careful interpretation if there are utility pinches. In such circumstances, heat pump placement above the process pinch or below it can be economic, providing

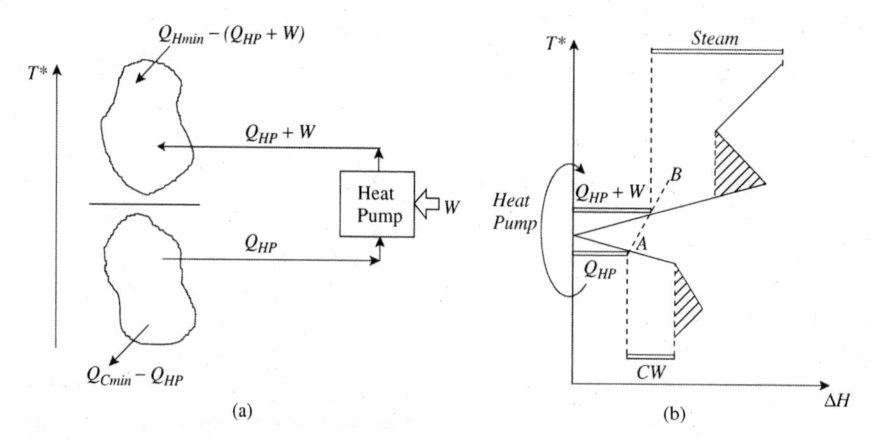

Figure 16.38 The grand composite curve allows heat pump cycles to be sized.

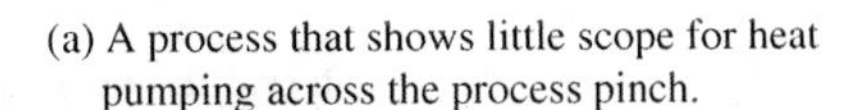

(a) A process that shows little scope for heat pumping across the process pinch.

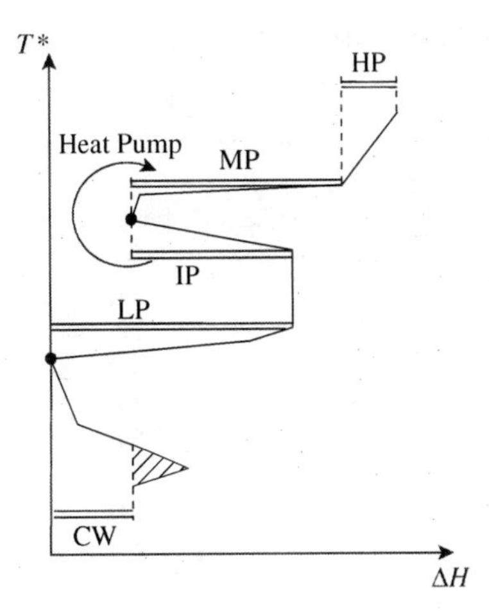

(b) Heat pump placed across a utility pinch.

Figure 16.39 Heat pumping can be applied across utility pinches as well as the process pinch.

the heat pump is placed across a utility pinch. Figure 16.39a shows a process that does not show a significant potential for heat pumping across the process pinch. The heat source just below the process pinch is small. The heat sink just above the process pinch is also small. Significant heat pumping across the process pinch can only be accomplished across a large temperature difference in this case, which will result in a poor COP_{HP}. However, Figure 16.39b shows another possible heat pump arrangement. A heat source in the pocket of the grand composite curve is pumped through a relatively small temperature difference to replace medium-pressure (MP) steam. To compensate for taking heat from within the pocket of the grand composite curve, additional low-pressure (LP) steam is used for process heating to maintain the heat balance. Thus, the heat pump results in a saving in MP steam for a sacrifice of extra LP steam usage and the power required for the heat pump. This might be economic if there is a large cost difference between MP and LP steam. Although the heat pump does not operate across the process pinch, it does not violate the principle of the appropriate placement of heat pumps. The heat pump in Figure 16.39b operates across a utility pinch.

16.11 HEAT EXCHANGER NETWORK ENERGY TARGETS – SUMMARY

The energy targets for the process can be set without having to design the heat exchanger network and utility system. These energy targets can be calculated directly from the material and energy balance. Thus, energy costs can be established without design for the outer layers of the process onion. Using the grand composite curve, different utility scenarios can be screened quickly and conveniently, including cogeneration and heat pumps.

16.12 EXERCISES

1. A heat recovery problem consists of two streams given in Table 16.10:

Table 16.10 Stream data for Exercise 1.

Stream	Type	Supply temperature (°C)	Target temperature (°C)	Enthalpy change (MW)
1	Hot	100	40	12
2	Cold	10	150	7

Steam is available at 180°C and cooling water at 20°C.

   a. For a minimum permissible temperature difference (ΔT_{min}) of 10°C, calculate the minimum hot and cold utility requirements.
   b. What are the hot and cold stream pinch temperatures?
   c. If the steam is supplied as the exhaust from a steam turbine, is the heat engine appropriately placed?
   d. If the (ΔT_{min}) is increased to 20°C, what will happen to the utility requirements?

2. The stream data for a process are given in Table 16.11.

Table 16.11 Stream data for Exercise 2.

Stream No.	Stream Type	T_S (°C)	T_T (°C)	Heat duty (MW)
1	Hot	160	40	3.6
2	Hot	50	50	5.0
3	Hot	140	110	1.5
4	Cold	160	160	5.0
5	Cold	60	150	1.8

   a. Sketch the composite curves for $\Delta T_{min} = 10$°C.
   b. From the composite curves, determine the target for hot and cold utility for ΔT_{min} 10°C.

3. The problem table cascade for a process is given in the Table 16.12 for $\Delta T_{min} = 10°C$.

Table 16.12 Problem table cascade for Exercise 3.

Interval temperature (°C)	Cascade heat flow (MW)
360	9.0
310	7.6
230	8.0
210	4.8
190	5.8
190	0
170	1.0
150	7.6
60	3.0

The minimum permissible temperature differences for various options are given in Table 16.13.

Table 16.13 ΔT_{min} for different matches.

Type of match	ΔT_{min} (°C)
Process/process	10
Process/steam	10
Process/flue gas (furnace or gas turbine)	50

a. Calculate the amount of fuel required to satisfy the heating requirements for a furnace used to supply hot utility for a theoretical flame temperature of 1800°C and acid dew point of 150°C.

b. If the furnace from Part *a* is used in conjunction with saturated steam at 250°C for heating, such that the heat duty for the steam is maximized, what is the heat duty on the steam and the furnace heat duty?

4. The stream data for a process involving an exothermic chemical reaction are given in the Table 16.14.

Table 16.14 Stream data for Exercise 4.

Stream		Enthalpy Change (kW)	T_S (°C)	T_T (°C)
No.	Type			
1	Hot	−7000	377	375
2	Hot	−3600	376	180
3	Hot	−2400	180	70
4	Cold	2400	60	160
5	Cold	200	20	130
6	Cold	200	160	260

A problem table analysis of the data indicates that it is a threshold problem requiring only cold utility. The threshold value of ΔT_{min} is 117°C, corresponding with a cold utility duty of 10,100 kW. It is proposed to use steam generation as cold utility for which $\Delta T_{min} = 10°C$.

a. Assume that saturated boiler feedwater is available and that the steam generated is saturated, in order to calculate how much steam can be generated by the process at a pressure of 41 bar. The temperature of saturated steam at this pressure is 252°C and the latent heat is 1706 $kJ \cdot kg^{-1}$.

b. If the steam is superheated to a temperature of 350°C, calculate how much steam can be generated at 41 bar. Assume the heat capacity of steam is 4.0 $kJ \cdot kg^{-1} \cdot K^{-1}$.

c. What would happen if the steam in Part *b* was generated from boiler feedwater at 100°C with a heat capacity of 4.2 $kJ \cdot kg^{-1} \cdot K^{-1}$. How would the steam generation be calculated under these circumstances?

5. The problem table cascade for a process is given in Table 16.15 for $\Delta T_{min} = 10°C$.

Table 16.15 Problem table cascade for Exercise 5.

Interval temperature (°C)	Heat flow (MW)
495	3.6
455	9.2
415	10.8
305	4.2
285	0
215	16.8
195	17.6
185	16.6
125	16.6
95	21.1
85	18.1

It is proposed to provide process cooling by steam generation from boiler feedwater with a temperature of 80°C.

a. Determine how much saturated steam can be generated at a temperature of 230°C. The latent heat of steam under these conditions is 1812 $kJ \cdot kg^{-1}$. The heat capacity of water can be taken as 4.2 $kJ \cdot kg^{-1} \cdot K^{-1}$.

b. Determine how much steam can be generated with a saturation temperature of 230°C and superheated to the maximum temperature possible against the process. The heat capacity of steam can be taken as 3.45 $kJ \cdot kg^{-1} \cdot K^{-1}$.

c. Calculate how much power can be generated from the superheated steam from Part *b*, assuming the exhaust steam is saturated at a pressure of 4 bar.

6. The problem table cascade for a process is given in Table 16.16 for $\Delta T_{min} = 20°C$.

Table 16.16 Problem table cascade for Exercise 6.

Interval temperature (°C)	Heat flow (MW)
440	21.9
410	29.4
130	23.82
130	1.8
100	0
95	15

The process also has a requirement for 7 MW of power. Three alternative utility schemes are to be compared economically:

a. A steam turbine with its exhaust saturated at 150°C used for process heating is one of the options to be considered. Steam is raised in the central boiler house at 41 bar with an enthalpy of 3137 kJ·kg^{-1}. The enthalpy of saturated steam at 150°C is 2747 kJ·kg^{-1} and its latent heat is 2115 kJ·kg^{-1}. Calculate the maximum power generation possible by matching the exhaust steam against the process.
b. A second possible scheme uses a gas turbine with an exhaust temperature of 400°C and heat capacity flowrate of 0.1 MW·K^{-1}. Calculate the power generation if the turbine converts heat to power with an efficiency of 30%. Ambient temperature is 10°C.
c. A third possible scheme integrates a heat pump with the process. The power required by the heat pump is given by

$$W = \frac{Q_H}{0.6}\frac{T_H - T_C}{T_H}$$

where Q_H is the heat rejected by the heat pump at T_H (K). Heat is absorbed into the heat pump at T_C (K). Calculate the power required by the heat pump.
d. The cost utilities are given in the Table 16.17.

Table 16.17 Utility costs for Exercise 6.

Utility	Cost ($·MW^{-1})
Fuel for gas turbine	0.0042
Fuel for steam generation	0.0030
Imported power	0.018
Credit for exported power	0.014
Cooling water	0.00018

The overall efficiency of steam generation and distribution is 60%. Which scheme is most cost-effective, the steam turbine, the gas turbine or the heat pump?

7. The Table 16.18 represents a problem table cascade (ΔT_{min} = 20°C).

Table 16.18 Problem table cascade for Exercise 7.

Interval temperature (°C)	Heat flow (kW)
160	1000
150	0
130	1100
110	1400
100	900
80	1300
40	1400
10	1800
-10	1900
-30	2200

The following utilities are available:

(i) MP steam at 200°C
(ii) LP steam at 107°C raised from boiler feed water at 60°C
(iii) Cooling water (20 to 40°C)
(iv) Refrigeration at 0°C
(v) Refrigeration at −40°C

For matches between process and refrigeration, ΔT_{min} = 10°C. Draw the process grand composite curve and set the targets for the utilities. Below the pinch use of higher temperature, cold utilities should be maximized. For boiler feedwater, the specific heat capacity is 4.2 kJ·kg·K^{-1} and the latent heat of vaporization is 2238 kJ·kg^{-1}.

REFERENCES

1. Hohman EC (1971) *Optimum Networks of Heat Exchange*, PhD Thesis, University of Southern California.
2. Huang F and Elshout RV (1976) Optimizing the Heat Recover of Crude Units, *Chem Eng Prog*, **72**: 68.
3. Linnhoff B, Mason DR and Wardle I (1979) Understanding Heat Exchanger Networks, *Comp Chem Eng*, **3**: 295.
4. Linnhoff B, Townsend DW, Boland D, Hewitt GF, Thomas BEA, Guy AR and Marsland RH (1982) *A User Guide on Process Integration for the Efficient Use of Energy*, IChemE, UK.
5. Polley GT (1993) Heat Exchanger Design and Process Integration, *Chem Eng*, **8**: 16.
6. Umeda T, Itoh J and Shiroko K (1978) Heat Exchange System Synthesis, *Chem Eng Prog*, **74**: 70.
7. Umeda T, Harada T and Shiroko K (1979) A Thermodynamic Approach to the Synthesis of Heat Integration Systems in Chemical Processes, *Comp Chem Eng*, **3**: 273.
8. Umeda T, Niida K and Shiroko K (1979) A Thermodynamic Approach to Heat Integration in Distillation Systems, *AIChE J*, **25**: 423.
9. Linnhoff B and Flower JR (1978) Synthesis of Heat Exchanger Networks, *AIChE J*, **24**: 633.
10. Ahmad S and Hui DCW (1991) Heat Recovery Between Areas of Integrity, *Comp Chem Eng*, **15**: 809.
11. Cerda J, Westerberg AW, Mason D and Linnhoff B (1983) Minimum Utility Usage in Heat Exchanger Network Synthesis – A Transportation Problem, *Chem Eng Sci*, **38**: 373.
12. Papoulias SA and Grossmann IE (1983) A Structural Optimization Approach in Process Synthesis – II Heat Recovery Networks, *Comp Chem Eng*, **7**: 707.
13. Itoh J, Shiroko K and Umeda T (1982) Extensive Application of the T-Q Diagram to Heat Integrated System Synthesis, *International Conference on Proceedings Systems Engineering (PSE-82)*, Kyoto, 92.
14. Townsend DW and Linnhoff B (1983) Heat and Power Networks in Process Design, *AIChE J*, **29**: 742.
15. Linnhoff B and de Leur J (1988) Appropriate Placement of Furnaces in the Integrated Process, *IChemE Symp Ser*, **109**: 259.

17 Heat Exchanger Networks III – Capital and Total Cost Targets

In addition to being able to predict the energy costs of the heat exchanger network directly from the material and energy balance, it would be useful to be able to calculate the capital cost of the network.

The principal components that contribute to the capital cost of the heat exchanger network are:

- number of units (matches between hot and cold streams)
- heat exchange area
- number of shells
- materials of construction
- heat exchanger type
- pressure rating.

Now consider each of these components in turn and explore whether they can be estimated from the material and energy balance without having to perform heat exchanger network design.

17.1 NUMBER OF HEAT EXCHANGE UNITS

To understand the minimum number of matches or units in a heat exchanger network, some basic results of *graph theory* can be used[1,2]. A *graph* is any collection of points in which some pairs of points are connected by lines. Figures 17.1a and 17.1b give two examples of graphs. Note that the lines such as *BG* and *CE* and *CF* in Figure 17.1a are not supposed to cross, that is, the diagram should be drawn in three dimensions. This is true for the other lines in Figure 17.1 that appear to cross.

In this context, the points correspond to process and utility streams, and the lines to heat exchange matches between the heat sources and heat sinks.

A *path* is a sequence of distinct lines that are connected to each other. For example, in Figure 17.1a *AECGD* is a path. A graph forms a single *component* (sometimes called a *separate system*) if any two points are joined by a path. Thus, Figure 17.1b has two components (or two separate systems), and Figure 17.1a has only one.

A *loop* is a path that begins and ends at the same point, like *CGDHC* in Figure 17.1a. If two loops have a line in common, they can be linked to form a third loop by deleting the common line. In Figure 17.1a, for example, *BGCEB* and *CGDHC* can be linked to give *BGDHCEB*. In this case, this last loop is said to be *dependent* on the other two.

From graph theory, the main result needed in the present context is that the number of independent loops for a graph is given by[1]:

$$N_{UNITS} = S + L - C \tag{17.1}$$

where N_{UNITS} = number of matches or units (lines in graph theory)
S = number of streams including utilities (points in graph theory)
L = number of independent loops
C = number of components

In general, the final network design should be achieved in the minimum number of units to keep down the capital cost (although this is not the only consideration to keep down the capital cost). To minimize the number of units in Equation 17.1, L should be zero and C should be a maximum. Assuming L to be zero in the final design is a reasonable assumption. However, what should be assumed about C? Consider the network in Figure 17.1b that has two components. For there to be two components, the heat duties for streams A and B must exactly balance the duties for streams E and F. Also, the heat duties for streams C and D must exactly balance the duties for streams G and H. Such balances are likely to be unusual and not easy to predict. The safest assumption for C thus appears to be that there will be one component only, that is, $C = 1$. This leads to an important special case when the network has a single component and is loop-free. In this case[1,2]:

$$N_{UNITS} = S - 1 \tag{17.2}$$

Equation 17.2 put in words states that the minimum number of units required is one less than the number of streams (including utility streams).

This is a useful result since, if the network is assumed to be loop-free and has a single component, the minimum number of units can be predicted simply by knowing the number of streams. If the problem does not have a pinch, then Equation 17.2 predicts the minimum number of units. If the problem has a pinch, then Equation 17.2 is applied on each side of the pinch separately[2]:

$$N_{UNITS} = [S_{ABOVE\ PINCH} - 1] + [S_{BELOW\ PINCH} - 1] \tag{17.3}$$

Chemical Process Design and Integration R. Smith

ISBNs: 0-471-48680-9 (HB); 0-471-48681-7 (PB)

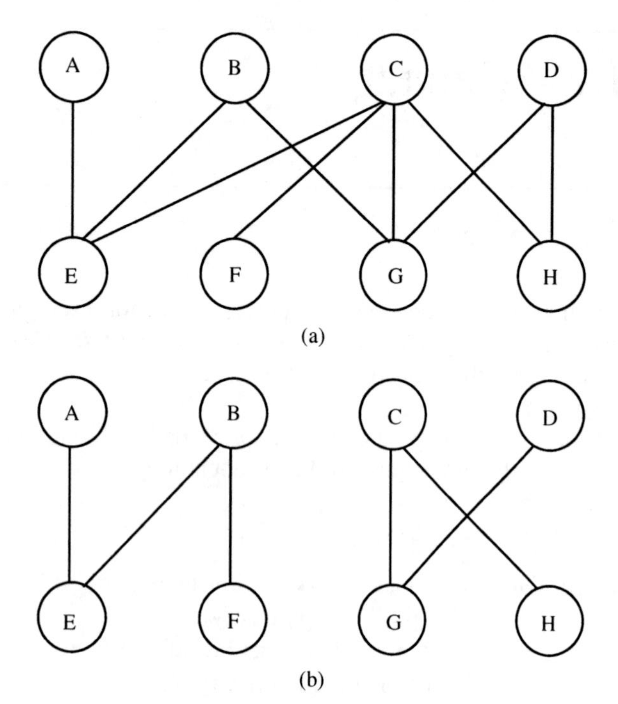

Figure 17.1 Two alternative graphs. (From Linnhoff B, Mason D and Wardle I, 1979, *Comp and Chem Engg*, **3**: 279, reproduced by permission of Elsevier Ltd.)

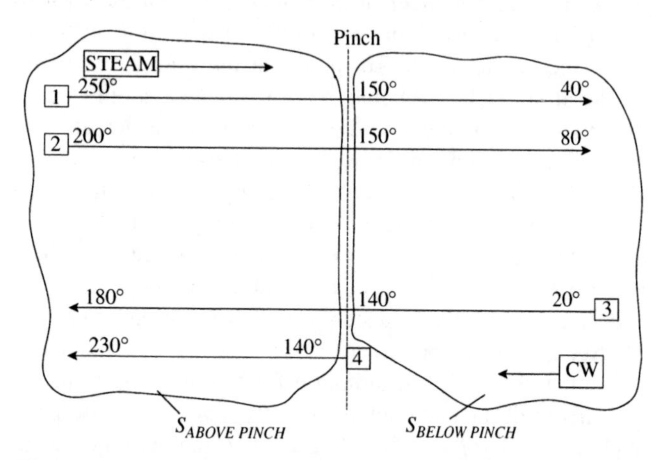

Figure 17.2 To target the number of units for pinched problems the streams above and below the pinch must be counted separately with the appropriate utilities included.

Example 17.1 For the process in Figure 17.2, calculate the minimum number of units given that the pinch is at 150°C for the hot streams and 140°C for the cold streams.

Solution Figure 17.2 shows the stream grid with the pinch in place dividing the process into two parts. Above the pinch, there are five streams, including the steam. Below the pinch, there are four streams, including the cooling water. Applying Equation 17.3:

$$N_{UNITS} = (5 - 1) + (4 - 1)$$
$$= 7$$

Looking back at the design presented for this problem in Figure 16.9, it does in fact use the minimum number of units of 7. In the next chapter, design for the minimum number of units will be addressed.

17.2 HEAT EXCHANGE AREA TARGETS

In addition to giving the necessary information to predict energy targets, the composite curves also contain the necessary information to predict network heat transfer area. To calculate the network area from the composite curves, utility streams must be included with the process streams in the composite curves to obtain the *balanced composite curves*[3], going through the same procedure as illustrated in Figures 16.3 and 16.4 but including the utility streams. The resulting balanced composite curves should have no residual demand for utilities. The balanced composite curves are divided into vertical *enthalpy intervals* as shown in Figure 17.3. Assume initially that the overall heat transfer coefficient U is constant throughout the process. Assuming true countercurrent heat transfer, the area requirement for enthalpy interval k for this *vertical heat transfer* is given by[1,3]:

$$A_{NETWORKk} = \frac{\Delta H_k}{U\,\Delta T_{LMk}} \tag{17.4}$$

where $A_{NETWORKk}$ = heat exchange area for vertical heat transfer required by interval k

ΔH_k = enthalpy change over interval k

ΔT_{LMk} = log mean temperature difference for interval k

U = overall heat transfer coefficient

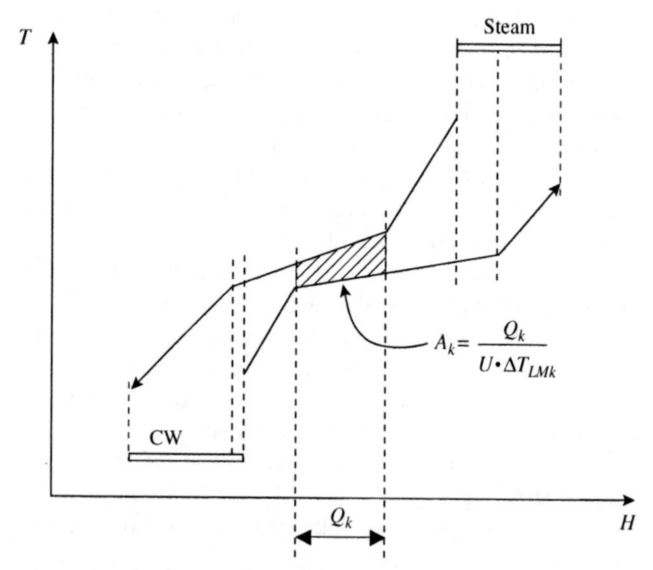

Figure 17.3 To determine the network area the balanced composite curves are divided into enthalpy intervals.

To obtain the network area, Equation 17.4 is applied to all enthalpy intervals[1,3]:

$$A_{NETWORK} = \frac{1}{U} \sum_{k}^{INTERVALS\ K} \frac{\Delta H_k}{\Delta T_{LMk}} \quad (17.5)$$

where $A_{NETWORK}$ = heat exchange area for vertical heat transfer for the whole network

K = total number of enthalpy intervals

The problem with Equation 17.5 is that the overall heat transfer coefficient is not constant throughout the process. Is there some way to extend this model to deal with the individual heat transfer coefficients?

The effect of individual stream film transfer coefficients can be included using the following expression, which is derived in Appendix F[3,4]:

$$A_{NETWORK} = \sum_{k}^{INTERVALS\ K} \frac{1}{\Delta T_{LMk}} \left[\sum_{i}^{HOT\ STREAMS\ I} \frac{q_{i,k}}{h_i} + \sum_{j}^{COLD\ STREAMS\ J} \frac{q_{j,k}}{h_j} \right] \quad (17.6)$$

where $q_{i,k}$ = stream duty on hot stream i in enthalpy interval k

$q_{j,k}$ = stream duty on cold stream j in enthalpy interval k

h_i, h_j = film transfer coefficients for hot stream i and cold stream j (including wall and fouling resistances)

I = total number of hot streams in enthalpy interval k

J = total number of cold streams in enthalpy interval k

K = total number of enthalpy intervals

This simple formula allows the network area to be targeted, on the basis of a vertical heat exchange model if film transfer coefficients vary from stream to stream. However, if there are large variations in film transfer coefficients, Equation 17.6 does not predict the true minimum network area. If film transfer coefficients vary significantly, then deliberate nonvertical matching is required to achieve minimum area. Consider Figure 17.4a. Hot stream A with a low heat transfer coefficient is matched against cold stream C with a high coefficient. Hot stream B with a high heat transfer coefficient is matched with cold stream D with a low coefficient. In both matches, the temperature difference is taken to be the vertical separation between the curves. This arrangement requires 1616 m^2 area overall.

By contrast, Figure 17.4b shows a different arrangement. Hot stream A with a low heat transfer coefficient is matched with cold stream D, which also has a low coefficient but uses temperature differences greater than vertical separation. Hot stream B is matched with cold stream C, both with high heat transfer coefficients, but with temperature differences less than vertical. This arrangement requires 1250 m^2 area overall, less than the vertical arrangement.

If film transfer coefficients vary significantly from stream to stream, the true minimum area must be predicted using linear programming[5,6]. However, Equation 17.6 is still a

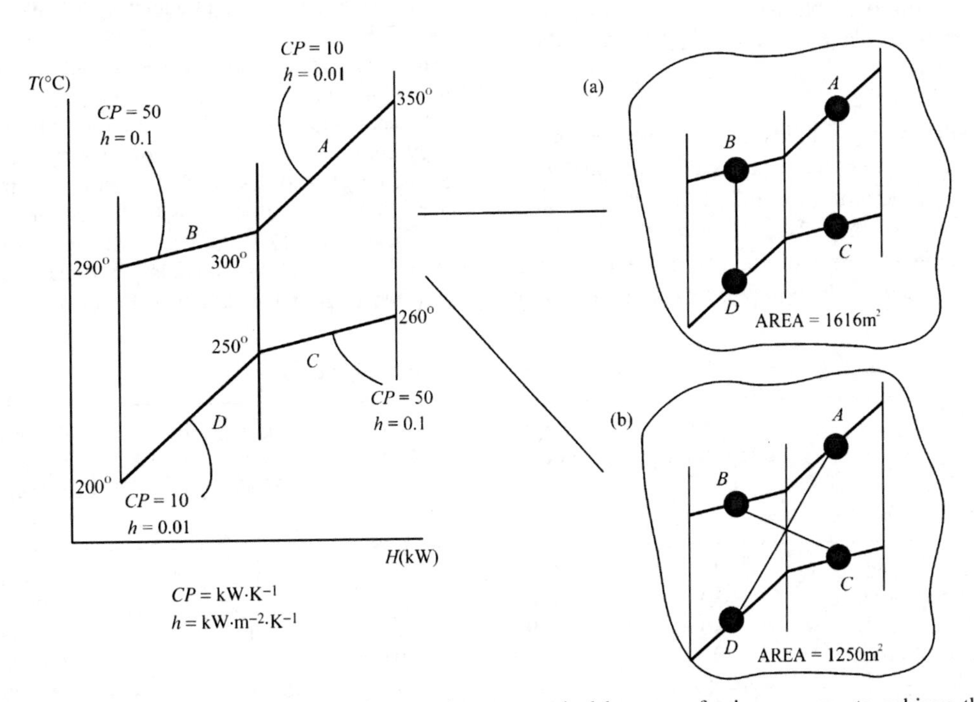

Figure 17.4 If film transfer coefficients differ significantly then nonvertical heat transfer is necessary to achieve the minimum area. (From Linnhoff B and Ahman S, 1990, *Comp and Chem Engg*, **7**: 729 with permission of Elsevier Science Ltd.)

useful basis to calculate the network area for the purposes of capital cost estimation, for the following reasons.

1. Providing film coefficients vary by less than one order of magnitude, then Equation 17.6 has been found to predict network area to within 10% of the actual minimum[6].
2. Network designs tend *not* to approach the true minimum in practice, since a minimum area design is usually too complex to be practical. Putting the argument the other way around, starting with the complex design required to achieve minimum area, then a significant reduction in complexity usually only requires a small penalty in area.
3. The area target being predicted here is used for predesign optimization of the capital–energy trade-off and the evaluation of alternative flowsheet options, such as different reaction and separation configurations. Thus, the area prediction is used in conjunction with capital cost data for heat exchangers that often have a considerable degree of uncertainty. The capital cost predictions obtained later from Equation 17.6 are likely to be more reliable than the capital cost predictions for the major items of equipment, such as reactors and distillation columns.

One significant problem remains; where to get the film transfer coefficients h_i and h_j from. There are the following three possibilities.

1. Tabulated experience values (see Chapter 15).
2. By assuming a reasonable fluid velocity, together with fluid physical properties, heat transfer correlations can be used (see Chapter 15).
3. If the pressure drop available for the stream is known, the expressions from Chapter 15, derived in Appendix C can be used.

The detailed allocation of fluids to tube-side or shell-side can only be made later in the heat exchanger network design. Also, the area targeting formula does not recognize fluids to be allocated to the tube-side or shell-side. Area targeting only recognizes the individual film heat transfer coefficients. All that can be done in network area targeting is to make a preliminary estimate of the film heat transfer coefficient on the basis of an initial assessment as to whether the fluid is likely to be suited to tube-side or shell-side allocation in the final design. Thus, in addition to the approximations inherent within the area targeting formula, there is uncertainty regarding the preliminary assessment of the film heat transfer coefficients.

However, one other issue that needs to be included in the assessment often helps mitigate the uncertainties in the assessment of the film heat transfer coefficient. A fouling allowance needs to be added to the film transfer coefficient according to Equation 15.13.

Example 17.2 For the process in Figure 16.2, calculate the target for network heat transfer area for $\Delta T_{min} = 10°C$. Steam at 240°C and condensing to 239°C is to be used for hot utility. Cooling water at 20°C and returning to the cooling tower at 30°C is to be used for cold utility. Table 17.1 presents the stream data, together with utility data and stream heat transfer coefficients.

Calculate the heat exchange area target for the network.

Solution First, the balanced composite curves must be constructed using the complete set of data from Table 17.1. Figure 17.5 shows the balanced composite curves. Note that the steam has been incorporated within the construction of the hot composite curve to maintain the monotonic nature of composite curves. The same is true of the cooling water in the cold composite curve. Figure 17.5 also shows the curves divided into enthalpy intervals where there is a change of slope either on the hot composite curve or on the cold composite curve.

Figure 17.6 now shows the stream population for each enthalpy interval together with the hot and cold stream temperatures. Now set up a table to compute Equation 17.6. This is shown below in Table 17.2.

Thus, the network area target for this problem for $\Delta T_{min} = 10°C$ is 7410 m^2.

The network design in Figure 16.9 already achieves minimum energy consumption, and it is now possible to judge how close the area target is to design if the area for the individual units in Figure 16.9 is calculated. Using the same heat transfer coefficients as given in Table 17.1, the design in Figure 16.9 requires some 8341 m^2, which is 13% above target. Remember that no attempt was made to steer the design in Figure 16.9 towards minimum

Table 17.1 Complete stream and utility data for the process from Figure 16.2.

Stream	Supply temperature T_S (°C)	Target temperature T_T (°C)	ΔH (MW)	Heat capacity flowrate, CP ($MW \cdot K^{-1}$)	Film heat transfer coefficient, h ($MW \cdot m^{-2} \cdot K^{-1}$)
1. Reactor 1 feed	20	180	32.0	0.2	0.0006
2. Reactor 1 product	250	40	−31.5	0.15	0.0010
3. Reactor 2 feed	140	230	27.0	0.3	0.0008
4. Reactor 2 product	200	80	−30.0	0.25	0.0008
5. Steam	240	239	7.5	7.5	0.0030
6. Cooling water	20	30	10.0	1.0	0.0010

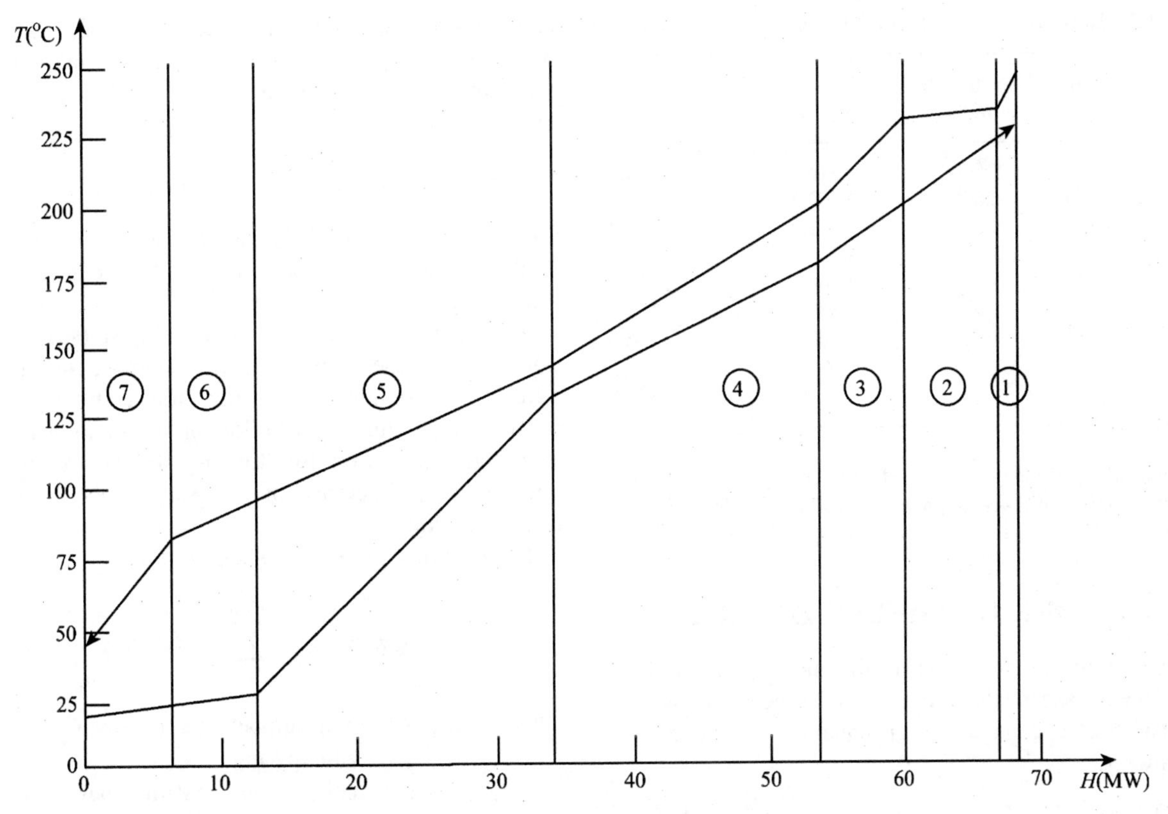

Figure 17.5 The enthalpy intervals for the balanced composite curves of Example 17.2.

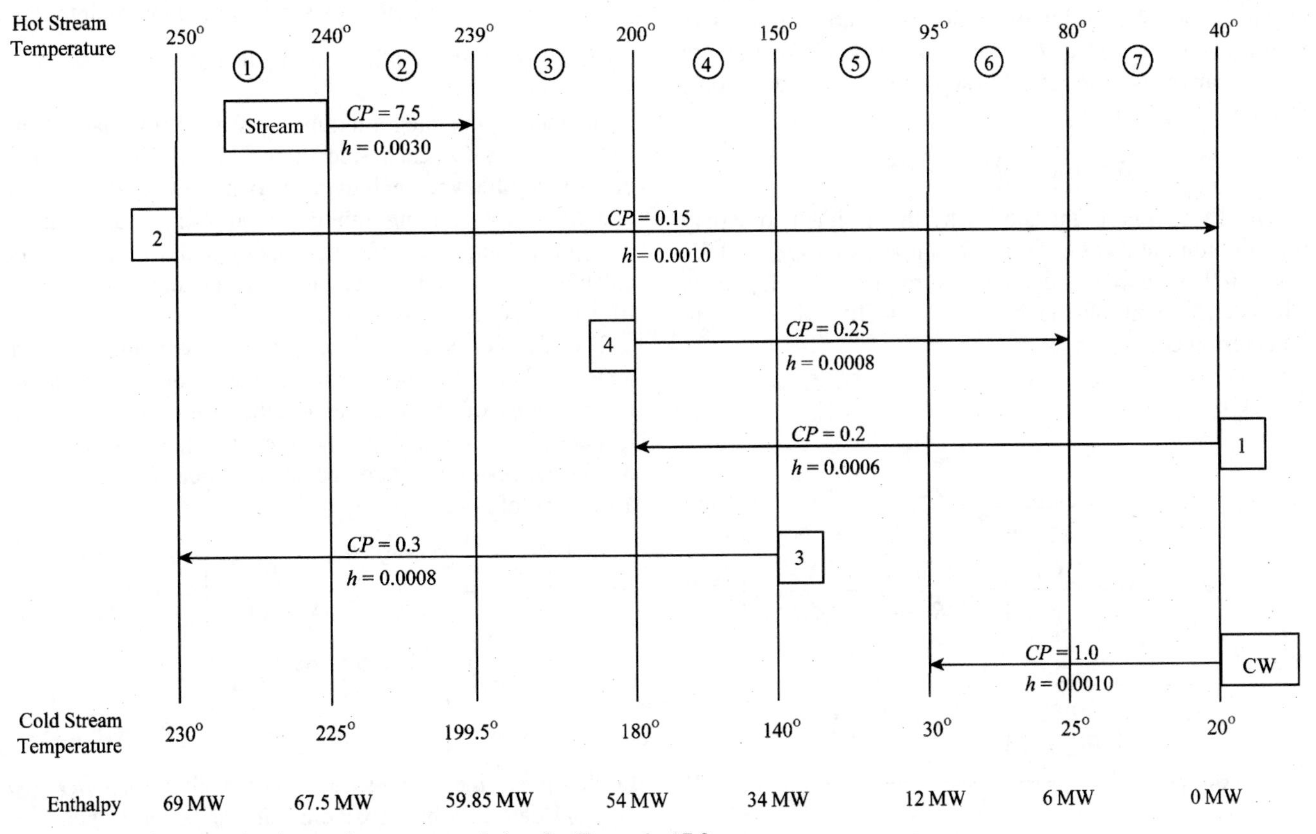

Figure 17.6 The enthalpy interval stream population for Example 17.2.

Table 17.2 Network area target for the process from Figure 17.2.

Enthalpy interval	ΔT_{LMk}	Hot streams $\Sigma(q_i/h_i)_k$	Cold streams $\Sigma(q_j/h_j)_k$	A_k
1	17.38	1500	1875	194.2
2	25.30	2650	9562.5	482.7
3	28.65	5850	7312.5	459.4
4	14.43	23,125	28,333.3	3566.1
5	29.38	25,437.5	36,666.7	2113.8
6	59.86	6937.5	6666.7	227.3
7	34.60	6000	6666.7	366.1
			ΣA_k	7409.6

area. Instead, the design achieved the energy target in the minimum number of units that tends to lead to simple designs.

17.3 NUMBER-OF-SHELLS TARGET

The shell-and-tube exchanger is the most common type of exchanger used in the chemical and process industries. The basic heat exchanger design equation was developed in Chapter 15:

$$Q = UA\Delta T_{LM} F_T \quad \text{where } 0 < F_T < 1 \qquad (15.47)$$

In Chapter 15, the F_T correction factor was correlated in terms of two dimensionless ratios, the ratio of the two heat capacity flowrates (R) and the thermal effectiveness of the exchanger (P). Practical designs were limited to some fraction of P_{max}, that is[7]:

$$P = X_P P_{max}, \quad 0 < X_P < 1 \qquad (15.54)$$

where X_P is a constant defined by the designer to satisfy the minimum allowable F_T (for example, for $F_{Tmin} > 0.75$, $X_P = 0.9$ is used). Equations were also developed in Chapter 15 to allow the number of shells for a unit to be determined:

$R \neq 1$[7]:

$$N_{SHELLS} = \frac{\ln\left(\frac{1 - RP}{1 - P}\right)}{\ln W} \qquad (15.62)$$

$$\text{where } W = \frac{R + 1 + \sqrt{R^2 + 1} - 2RX_P}{R + 1 + \sqrt{R^2 + 1} - 2X_P} \qquad (15.63)$$

$R = 1$[7]:

$$N_{SHELLS} = \frac{\left(\frac{P}{1 - P}\right)\left(1 + \frac{\sqrt{2}}{2} - X_P\right)}{X_P} \qquad (15.64)$$

If exchangers are countercurrent devices, then the number of units equals the number of shells, providing individual shells do not exceed some practical upper size limit. If, however, equipment is used that is not completely countercurrent, as with the 1–2 shell and tube heat exchanger, then:

$$N_{SHELLS} \geq N_{UNITS} \qquad (17.7)$$

Since the number of shells can have a significant influence on the capital cost, it would be useful to be able to predict it as a target ahead of design.

A simple algorithm can be developed (see Appendix G) to target the minimum total number of shells (as a real, i.e. noninteger, number) for a stream-set based on the temperature distribution of the composite curves. The algorithm starts by dividing the composite curves into enthalpy intervals in the same way as the area target algorithm.

The resulting number of shells is[8]:

$$N_{SHELLS} = \sum_{k}^{INTERVALS\ K} N_k(S_k - 1) \qquad (17.8)$$

where N_{SHELLS} = total number of shells over K enthalpy intervals
N_k = real (or fractional) number of shells resulting from the temperatures of enthalpy interval k
S_k = number of streams in enthalpy interval k

N_k is given by the application of Equations 15.62 to 15.64 to interval k.

In practice, the integer number of shells is evaluated from Equation 17.8 for each side of the pinch. This maintains consistency between achieving maximum energy recovery and the corresponding minimum number of units target N_{UNITS}. In summary, the number-of-shells target can be calculated from the basic stream data and an assumed value of X_P (or equivalently, F_{Tmin}).

The F_T correction factor for each enthalpy interval depends both on the assumed value of X_P and the temperatures of the interval on the composite curves. It is possible to modify the simple area target formula to obtain the resulting increased overall area, $A_{NETWORK}$, for a network of 1–2 exchangers[8].

$$A_{NETWORK,1-2} = \sum_{k}^{INTERVALS\ K} \frac{1}{\Delta T_{LMk} F_{Tk}} \left[\sum_{i}^{HOT\ STREAMS\ I} \frac{q_{i,k}}{h_i} + \sum_{j}^{COLD\ STREAMS\ J} \frac{q_{j,k}}{h_j} \right] \qquad (17.9)$$

Furthermore, the average area per shell ($A_{NETWORK,1-2}/N_{SHELL}$) can also be considered at the targeting stage. If this is greater than the maximum allowable area per shell, $A_{SHELL,max}$, then the shells target needs to be increased to the next largest integer above $A_{NETWORK,1-2}/A_{SHELL,max}$. Again, this can be applied each side of the pinch.

17.4 CAPITAL COST TARGETS

To predict the capital cost of a network, it must first be assumed that a single heat exchanger with surface area A can be costed according to a simple relationship such as:

$$\textit{Installed Capital Cost of Exchanger} = a + bA^c \quad (17.10)$$

where a, b, c = cost law constants that vary according to materials of construction, pressure rating and type of exchanger.

When cost targeting, the distribution of the targeted area between network exchangers is unknown. Thus, to cost a network using Equation 17.10, some area distribution must be assumed, the simplest being that all exchangers have the same area:

$$\textit{Network Capital Cost} = N[a + b(A_{NETWORK}/N)^c] \quad (17.11)$$

where N = number of units or shells, whichever is appropriate.

At first sight, the assumption of equal area exchangers used in Equation 17.11 might seem crude. However, from the point of view of predicting capital cost, the assumption turns out to be a remarkably good one[6]. At the targeting stage, no given distribution can be judged consistently better than another, since the network is not yet known. The implications of these inaccuracies, together with others, will be discussed later.

If the problem is dominated by equipment with a single specification (i.e. a single material of construction, equipment type and pressure rating), then the capital cost target can be calculated from Equation 17.11 with the appropriate cost coefficients. However, if there is a mix of specifications such as different streams requiring different materials of construction, then the approach must be modified.

Equation 17.11 uses a single cost function in conjunction with the targets for the number of units (or shells) and network area. Differences in cost can be accounted for by either introducing new cost functions or adjusting the heat exchange area to reflect the cost differences[9]. This can be done by weighting the stream heat transfer coefficients in the calculation of network area with a factor ϕ to account for these differences in cost. If, for example, a corrosive stream requires more expensive materials of construction than the other streams, it has a greater contribution to the capital cost than a similar noncorrosive stream. This can be accounted for by artificially decreasing its heat transfer coefficient to increase the contribution the stream makes to the network area. This fictitious area when turned into a capital cost using the cost function for the noncorrosive materials returns a higher capital cost, reflecting the increased cost resulting from special materials.

Heat exchanger cost data can usually be manipulated such that the fixed costs, represented by the coefficient a in Equation 17.10, do not vary with exchanger specification[9]. If this is done, then Equation 17.6 can be modified, as derived in Appendix H, to[9]:

$$A^*_{NETWORK} = \sum_k^{INTERVALS\ K} \frac{1}{\Delta T_{LMk}} \left[\sum_i^{HOT\ STREAMS\ I} \frac{q_{i,k}}{\phi_i h_i} + \sum_j^{COLD\ STREAMS\ J} \frac{q_{j,k}}{\phi_j h_j} \right] \quad (17.12)$$

where

$$\phi_j = \left(\frac{b_1}{b_2}\right)^{\frac{1}{c_1}} \left(\frac{A_{NETWORK}}{N}\right)^{1-\frac{c_2}{c_1}} \quad (17.13)$$

where ϕ_i = cost-weighting factor for hot stream i
ϕ_j = cost-weighting factor for cold stream j
a_1, b_1, c_1 = cost law coefficients for the reference cost law
a_2, b_2, c_2 = cost law coefficients for the special cost law
N = number of units or shells, whichever is applicable

Heat exchanger cost laws can often be adjusted with little loss of accuracy such that the coefficient c is constant for different specifications, that is, $c_1 = c_2 = c$. In this case, Equation 17.13 simplifies to[9]:

$$\phi = \left(\frac{b_1}{b_2}\right)^{\frac{1}{c}} \quad (17.14)$$

Thus, to calculate the capital cost target for a network comprising mixed exchanger specifications, the procedure is as follows.

1. Choose a reference cost law for the heat exchangers. Greatest accuracy is obtained if the category of streams that make the largest contribution to capital cost is chosen as reference[9].
2. Calculate ϕ-factors for those streams that require a specification different from that of the reference, using Equations 17.13 or 17.14. If Equation 17.13 is to be used, then the actual network area $A_{NETWORK}$ must first be calculated using either Equation 17.6 or 17.9 and N_{UNITS} or N_{SHELLS}, whichever is appropriate.
3. Calculate the weighted network area $A^*_{NETWORK}$ from Equation 17.12. When the weighted h-values (ϕh) vary appreciably, say by more than one order of magnitude, an improved estimate of $A^*_{NETWORK}$ can be evaluated by linear programming[5,6].

4. Calculate the capital cost target for the mixed specification heat exchanger network from Equation 17.11 using the cost law coefficients for the reference specification.

Example 17.3 For the process in Figure 16.2, the stream and utility data are given in Table 17.1. Pure countercurrent (1-1) shell and tube heat exchangers are to be used. For $\Delta T_{min} = 10°C$:

a. calculate the capital cost target if all individual heat exchangers can be costed by the relationship:

$$Heat\ Exchanger\ Capital\ Cost = 40{,}000 + 500A\ (\$)$$

where A is the heat transfer area in m^2.

b. calculate the capital cost target if cold stream 3 from Table 17.1 required a more expensive material. Individual heat exchangers made entirely from this more expensive material can be costed by the relationship:

$$Heat\ Exchanger\ Capital\ Cost\ (special) = 40{,}000 + 1{,}100A(\$)$$

where A is the heat transfer area in m^2.

Solution

a. The capital cost target of the network can be calculated from Equation 17.11. To apply this equation requires the target for both the number of units (N_{UNITS}) and the heat exchange area ($A_{NETWORK}$). In Example 17.1, $N_{UNITS} = 7$ was calculated, and in Example 17.2 $A_{NETWORK} = 7410$ m^2. Thus:

$$Network\ Capital\ Cost = 7[40{,}000 + 500(7410/7)^1] = 3.99 \times 10^6\ \$$$

b. To calculate the capital cost target of the network with mixed materials of construction, a reference material is first chosen. In principle, either of the materials can be chosen as reference. However, greater accuracy is obtained if the reference is taken to be that category of streams that makes the largest contribution to capital cost. In this case, the reference should be taken to be the cheaper material of construction. Now calculate ϕ-factors for those streams that require a specification different from the reference. In this problem, it only applies to stream 3. Since the c constant is the same for both cost laws, Equation 17.14 can be used.

$$\phi_3 = (500/1100)^{1/1} = 1/2.2$$

$$\phi_3 h_3 = 0.0008/2.2$$

Now recalculate the network area target substituting $\phi_3 h_3$ for h_3 in Figure 17.6. Table 17.2 is revised to the values shown in Table 17.3.

Thus, the weighted network area $A^*_{NETWORK}$ is 9547 m^2. Now calculate the network capital cost for mixed materials of construction by using $A^*_{NETWORK}$ in conjunction with the cost coefficients for the reference material in Equation 17.11.

Table 17.3 Area target per enthalpy interval.

Enthalpy interval	ΔT_{LMk}	Hot streams $\Sigma(q_i/h_i)_k$	Cold streams $\Sigma(q_j/h_j)_k$	A_k
1	17.38	1500	4125.0	323.6
2	25.30	2650	21,037.5	936.1
3	28.65	5850	16,087.5	765.6
4	14.43	23,125	46,333.4	4814.5
5	29.38	25,437.5	36,666.7	2113.8
6	59.86	6937.5	6666.7	227.3
7	34.60	6000	6666.7	366.1
			ΣA_k	9546.9

$$Network\ Capital\ Cost\ (mixed\ materials) = 7[40{,}000 + 500(9547/7)^1] = 5.05 \times 10^6\ \$$$

Consider now how accurate the capital cost targets are likely to be. It was discussed earlier how the basic area targeting equation (Equations 17.6 or 17.9) represents a true minimum network area if all heat transfer coefficients are equal but slightly above the true minimum if there are significant differences in heat transfer coefficients. Providing heat transfer coefficients vary by less than one order of magnitude, Equations 17.6 and 17.9 predict an area that is usually within 10% of the minimum. However, this does not turn into a 10% error in capital cost of the final design since practical designs are almost invariably slightly above the minimum. There are also the following two errors inherent in the approach to capital cost targets.

- Total heat transfer area is assumed to be divided equally between exchangers. This tends to overestimate the capital cost.
- The area target is usually slightly less than the area observed in design.

These small positive and negative errors partially cancel each other. The result is that capital cost targets predicted by the methods described in this chapter are often within 5% of the final design, providing heat transfer coefficients vary by less than one order of magnitude. If heat transfer coefficients vary by more than one order of magnitude, then a more sophisticated approach can sometimes be justified[6].

If the network comprises mixed exchanger specification, then an additional degree of uncertainty is introduced into the capital cost target. Applying the ϕ-factor approach to a single exchanger, where both streams require the same specification, there is no error. In practice, there can be different specifications on two streams being matched, and $\phi_H \neq \phi_C$ for specifications involving shell-and-tube heat exchangers with different materials of construction and pressure rating. In principle, this does not present a problem since the exchanger can be designed for different materials of construction or pressure rating on the shell-side and the tube-side of a heat exchanger. If, for example, there is a mix of streams, some requiring carbon steel and some stainless steel, then some of the matches involve a corrosive stream on one side of the exchanger and a noncorrosive stream on the other. The capital cost of such exchangers will lie somewhere between the cost based on the sole use of either material. This is what the capital cost target predicts. Thus, introducing mixed specifications for materials of construction and pressure rating does not significantly decrease the accuracy of the capital cost predictions.

By contrast, the same is not true of mixed exchanger types. For networks comprising different exchanger types (e.g. shell-and-tube, plate-and-frame, spiral, etc.), it is not possible to mix types in a single unit. Although a cost-weighting factor may be applied to one stream in targeting, this assumes that different exchanger types can be mixed. In practice, such a match is forced to be a special-type exchanger. Thus, there may be some discrepancy between cost targets and design cost when dealing with mixed exchanger types.

Overall, the accuracy of the capital cost targets is usually good enough for the purposes for which they are used:

- Screening design options from the material and energy balance. For example, changes in reactor design or separation system design can be screened effectively without performing repeated network design.
- Different utility options such as furnaces, gas turbines and different steam levels can be assessed more easily and with greater confidence, knowing the capital cost implications for the heat exchanger network.
- Preliminary process optimization is greatly simplified.
- The design of the heat exchanger network is greatly simplified if the design is initialized with an optimized value for ΔT_{min}.

17.5 TOTAL COST TARGETS

Increasing the chosen value of process energy consumption also increases all temperature differences available for heat recovery and hence decreases the necessary heat exchanger surface area, Figure 16.6. The network area can be distributed over the targeted number of units or shells to obtain a capital cost using Equation 17.11. This capital cost can be annualized as detailed in Chapter 2. The annualized capital cost can be traded off against the annual utility cost as shown in Figure 16.6. The total cost shows a minimum at the optimum energy consumption.

Example 17.4 For the process in Figure 16.2, determine the value of ΔT_{min} and the total cost of the heat exchanger network at the optimum setting of the capital–energy trade-off. The stream and utility data are given in Table 17.1. The utility costs are:

$$\text{Steam Cost} = 120{,}000\ (\$\cdot\text{MW}^{-1}\cdot\text{y}^{-1})$$

$$\text{Cooling Water Cost} = 10{,}000\ (\$\cdot\text{MW}^{-1}\cdot\text{y}^{-1})$$

The heat exchangers to be used are single-tube and shell-pass. The installed capital cost is given by:

$$\textit{Heat Exchanger Capital Cost} = 40{,}000 + 500A\,(\$)$$

where A is the heat transfer area in m^2. The capital cost is to be paid back over five years at 10% interest.

Solution From Equation 2.7 from Chapter 2:

Annualized Heat Exchanger Capital Cost

$$= \textit{Capital Cost} \times \frac{i(1+i)^n}{(1+i)^n - 1}$$

where i = fractional interest rate per year
n = number of years

Annualized Heat Exchanger Capital Cost

$$= [40{,}000 + 500A] \times \frac{0.1(1+0.1)^5}{(1+0.1)^5 - 1}$$

$$= [40{,}000 + 500A]0.2638$$

$$= 10{,}552 + 131.9A$$

Annualized Network Capital Cost

$$= N_{UNITS}\left[10,552 + \frac{131.9A_{NETWORK}}{N_{UNITS}}\right]$$

Now scan a range of values of ΔT_{min} and calculate the targets for energy, number of units and network area and combine these into a total cost. The results are given in Table 17.4.

The data from Table 17.4 are presented graphically in Figure 17.7. The optimum ΔT_{min} is at 10°C, confirming the initial value used for this problem in Chapter 16. The total annualized cost at the optimum setting of the capital–energy trade-off is $2.05 \times 10^6\ \$\cdot\text{y}^{-1}$.

For more complex examples, total cost profiles return step changes shown in Figure 17.8 (due to changes in N_{UNITS} and N_{SHELLS}). These step changes are easily located, prior to design, through simple software. Most importantly, experience has shown that predicted overall costs are typically accurate within 5% or better[6].

17.6 HEAT EXCHANGER NETWORK AND UTILITIES CAPITAL AND TOTAL COSTS – SUMMARY

There are parts of the flowsheet synthesis problem that can be predicted without having to study actual designs. These are the layers of the process onion relating to the

Table 17.4 Variation of annualized costs with ΔT_{min}.

ΔT_{min}	Q_{Hmin} (MW)	Annual hot Utility Cost (10^6 \$·y^{-1})	Q_{Cmin} (MW)	Annual cold utility cost (10^6 \$·y^{-1})	$A_{NETWORK}$ (m^2)	N_{UNITS}	Annualized capital cost (10^6 \$·y^{-1})	Annualized total cost (10^6 \$·y^{-1})
2	4.3	0.516	6.8	0.068	15,519	7	2.121	2.705
4	5.1	0.612	7.6	0.076	11,677	7	1.614	2.302
6	5.9	0.708	8.4	0.084	9645	7	1.346	2.138
8	6.7	0.804	9.2	0.092	8336	7	1.173	2.069
10	7.5	0.900	10.0	0.100	7410	7	1.051	2.051
12	8.3	0.996	10.8	0.108	6716	7	0.960	2.064
14	9.1	1.092	11.6	0.116	6174	7	0.888	2.096

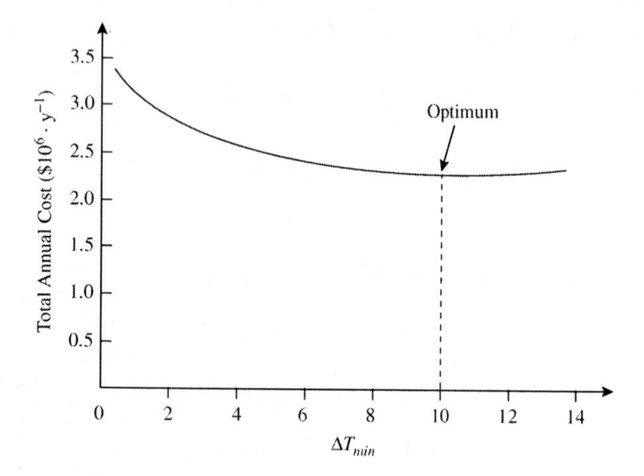

Figure 17.7 Optimization of the capital–energy trade-off for Example 17.5.

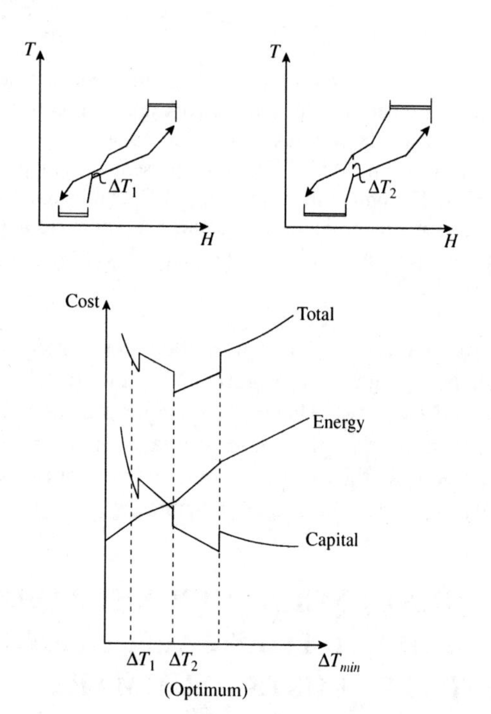

Figure 17.8 Energy and capital cost targets can be combined to optimize prior to design (From Smith R and Linnhoff B, 1988, *ChERD*,**66**: 195 reproduced by permission of the Institution of Chemical Engineers.).

heat exchanger network and utilities. For these parts of the process design, targets can be set for energy costs and capital costs directly from the material and energy balance without having to resort to heat exchanger network design for evaluation.

Once a design is known for the reaction and separation systems, the overall total cost is the total cost of all reactors and separators (evaluated explicitly) plus the total cost target for heat exchanger network.

17.7 EXERCISES

1. The problem table cascade for a process is given in Table 17.5 for $\Delta T_{min} = 10°C$. The process hot utility requirement is to be provided by a steam turbine exhaust. The exhaust steam is saturated. The performance of the turbine can be described by the equation:

$$W = 0.6 \times Q_H \frac{T_H - T_C}{T_H}$$

where W = power generated (MJ)
Q_H = heat input from high-pressure steam (MJ)
T_H = temperature of input steam (K)
T_C = temperature of exhaust steam (K)

High-pressure steam is available at 400°C. Table 17.6 below presents the relevant cost data.

a. Determine the optimum exhaust temperature for $\Delta T_{min} = 10°C$. The variation of network area (with the exhaust steam

Table 17.5 Cascade heat flow.

Interval temperature (°C)	Heat flow (MW)
435	8.160
395	12.160
116	9.928
115	1.120
35	0.480
25	0.000

Table 17.6 Cost data for Exercise 1.

Power cost	=	50 $\$\cdot MWh^{-1}$
High-pressure steam cost	=	15 $\$\cdot MWh^{-1}$
Process availability	=	8000 $h\cdot y^{-1}$
Heat exchanger cost	=	$350A$ ($)
(where A is exchanger area in m^2)		
Plant lifetime	=	5 years
Interest rate	=	8%

Table 17.7 Variation of heat exchanger network area with steam turbine exhaust temperature.

Steam turbine exhaust temperature (°C)	Network area (m^2)
120	3345
130	2415
140	2078
150	1914
160	1818

included) is given in Table 17.7 for a range of exhaust temperatures.

b. The optimization of the steam turbine exhaust temperature used a fix value of ΔT_{min}. Is this correct? If not, why, and what would have been a better way for the calculation to have been done?

2. Figure 17.9 shows a heat exchanger network designed with the minimum number of units and to satisfy the energy target at $\Delta T_{min} = 20°C$. On the basis of the following utilities and cost data, it has a total annual cost of 14.835×10^6 ($\$\cdot y^{-1}$).

$$\text{Cost of saturated steam } (200°C) = 70\ \$\cdot kW^{-1}\cdot y^{-1}$$

$$\text{Cost of cooling water } (10°C \text{ to } 40°C) = 7.5\ \$\cdot kW^{-1}\cdot y^{-1}$$

$$\text{Heat exchanger installed cost } (\$) = 15{,}000 + 700A$$

where A = heat exchanger area (m^2)

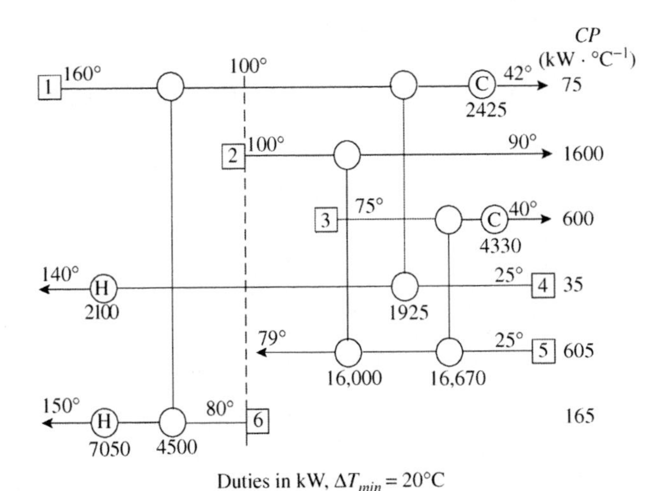

Figure 17.9 A heat exchanger network.

Table 17.8 Targets for Exercise 2.

ΔT_{min} (°C)	Hot utility (kW)	Area (m^2)	Units
10	7150	86,730	9
20	9150	75,920	8
30	14,945	62,721	9
40	21,345	56,486	9
45	24,545	54,796	9
50	27,745	53,703	9

Assume plant lifetime is 5 years and that all streams (process and utilities) have a heat transfer coefficient of 0.05 $kW\cdot m^{-2}\cdot K^{-1}$. Rather than accepting this network design, a designer proposes to examine the energy–capital trade-off for the stream data by evaluating the targets in Table 17.8.

$$\text{Cold utility (kW)} = \text{hot utility (kW)} - 2395 \text{ (kW)}$$

(a) From the targets and cost data given, construct a table to show the targets for energy cost, annualized capital cost and total annual cost for each of the values of ΔT_{min} in Table 17.8. Plot the total annual cost target against ΔT_{min}. What value of ΔT_{min} can be suggested for a network design featuring optimum total cost?

(b) Using the optimum value of ΔT_{min}, design a network with the minimum number of units and minimum energy (the pinch corresponds with the start of Hot Stream 2). Ignore ΔT_{min} violations in utility exchangers.

(c) Calculate the total annual cost of your network design. Locate this and the network in Figure 17.9 against the total annual cost profile from Part *a*.

(d) Why is the structure of your network design necessarily different from that in Figure 17.9? Could this have been known before the design? What can you conclude about ΔT_{min} initialization for obtaining optimum cost networks?

REFERENCES

1. Hohman EC (1971) *Optimum Networks for Heat Exchange*, PhD Thesis, University of Southern California.
2. Linnhoff B, Mason DR and Wardle I (1979) Understanding Heat Exchanger Networks, *Comp Chem Eng*, **3**: 295.
3. Townsend DW and Linnhoff B (1984) *Surface Area Targets for Heat Exchanger Networks*, IChemE Annual Research Meeting, Bath, UK.
4. Linnhoff B and Ahmad S (1990) Cost Optimum Heat Exchanger Networks – 1. Minimum Energy and Capital Using Simple Models for Capital Cost, *Comp Chem Eng*, **14**: 729.
5. Saboo AK, Morari M and Colberg RD (1986) RESHEX - An Interactive Software Package for the Synthesis and Analysis of Resilient Heat Exchanger Networks - II Discussion of Area Targeting and Network Synthesis Algorithms, *Comp Chem Eng*, **10**: 591.
6. Ahmad S, Linnhoff B and Smith R (1990) Cost Optimum Heat Exchanger Networks - 2. Targets and Design for Detailed Capital Cost Models, *Comp Chem Eng*, **14**: 751.

7. Ahmad S, Linnhoff B and Smith R (1988) Design of Multipass Heat Exchangers: an Alternative Approach, *Trans ASME J Heat Transfer*, **110**: 304.

8. Ahmad S and Smith R (1989) Targets and Design for Minimum Number of Shells in Heat Exchanger Networks, *Trans IChemE ChERD*, **67**: 481.

9. Hall SG, Ahmad S and Smith R (1990) Capital Cost Targets for Heat Exchanger Networks Comprising Mixed Materials of Construction, Pressure Ratings and Exchanger Types, *Comp Chem Eng*, **14**: 319.

18 Heat Exchanger Networks IV – Network Design

Having explored the targets for the heat exchanger network, it now remains to develop the design of the heat exchanger network.

18.1 THE PINCH DESIGN METHOD

The capital-energy trade-off in the heat exchanger networks was discussed in Chapter 16. Varying ΔT_{min}, as shown in Figure 16.6, changed the relative position of the process composite curves. As ΔT_{min} is changed from a small to a large value, the capital cost decreases but the energy cost increases. When the two costs are combined to obtain a total cost, the optimum point in the capital-energy trade-off is identified, corresponding with an optimum value of ΔT_{min} (Figure 16.6). As pointed out in Chapter 16, the trade-off between energy and capital suggests that individual exchangers should have a temperature difference no smaller than the ΔT_{min} between the composite curves.

It was suggested in Chapter 16 that a good initialization would be to assume that no individual exchanger should have a temperature difference smaller than ΔT_{min}. Having made this assumption, two rules were deduced in Chapter 16. If the energy target set by the composite curves (or the problem table algorithm) is to be achieved, the design must not transfer heat across the pinch by:

- Process-to-process heat transfer
- Inappropriate use of utilities

These rules are necessary for the design to achieve the energy target, given that no individual exchanger should have a temperature difference smaller than ΔT_{min}. To comply with these two rules, the process should be divided at the pinch. As pointed out in Chapter 16, this is most clearly done by representing the stream data in the grid diagram. Figure 18.1 shows the stream data from Table 16.2 in grid form with the pinch marked. Above the pinch, steam can be used (up to Q_{Hmin}), and below the pinch cooling water can be used (up to Q_{Cmin}). But what strategy should be adopted for the design? A number of simple criteria can be developed to help[1].

1. *Start at the pinch.* The pinch is the most constrained region of the problem. At the pinch, ΔT_{min} exists between all hot and cold streams. As a result, the number of feasible matches in this region is severely restricted. Quite often there are essential matches to be made. If such matches are not made, the result will be either use of temperature differences smaller than ΔT_{min} or excessive use of utilities resulting from heat transfer across the pinch. If the design was started away from the pinch at the hot end or cold end of the problem, then initial matches are likely to need follow-up matches that violate the pinch or the ΔT_{min} criterion as the pinch is approached. Putting the argument the other way around, if the design is started at the pinch, then initial decisions are made in the most constrained part of the problem. This is much less likely to lead to difficulties later.

2. *The CP inequality for individual matches.* Figure 18.2a shows the temperature profiles for an individual exchanger at the pinch, above the pinch[1,2]. Moving away from the pinch, temperature differences must increase. Figure 18.2a shows a match between a hot stream and a cold stream that has a *CP* smaller than the hot stream. At the pinch, the match starts with a temperature difference equal to ΔT_{min}. The relative slopes of the temperature–enthalpy profiles of the two streams mean that the temperature differences become smaller moving away from the pinch, which is infeasible. On the other hand, Figure 18.2b shows a match involving the same hot stream but with a cold stream that has a larger *CP*. The relative slopes of the temperature–enthalpy profiles now cause the temperature differences to become larger moving away from the pinch, which is feasible. Thus, starting with ΔT_{min} at the pinch, for temperature differences to increase moving away from the pinch[1,2]:

$$CP_H \leq CP_C \text{ (above pinch)} \tag{18.1}$$

Figure 18.3 shows the situation below the pinch at the pinch. If a cold stream is matched with a hot stream with smaller *CP*, as shown in Figure 18.3a (i.e. a steeper slope), then the temperature differences become smaller (which is infeasible). If the same cold stream is matched with a hot stream with a larger *CP* (i.e. a less steep slope), as shown in Figure 18.3b, then temperature differences become larger, which is feasible. Thus, starting with ΔT_{min} at the pinch, for temperature differences to increase moving away from the pinch[1,2]:

$$CP_H \geq CP_C \text{ (below pinch)} \tag{18.2}$$

Note that the *CP* inequalities given by Equations 18.1 and 18.2 only apply at the pinch and when both ends of the match are at pinch conditions.

3. *The CP-table.* Identification of the essential matches in the region of the pinch is clarified by use of the *CP-table*[1,2].

Chemical Process Design and Integration R. Smith

ISBNs: 0-471-48680-9 (HB); 0-471-48681-7 (PB)

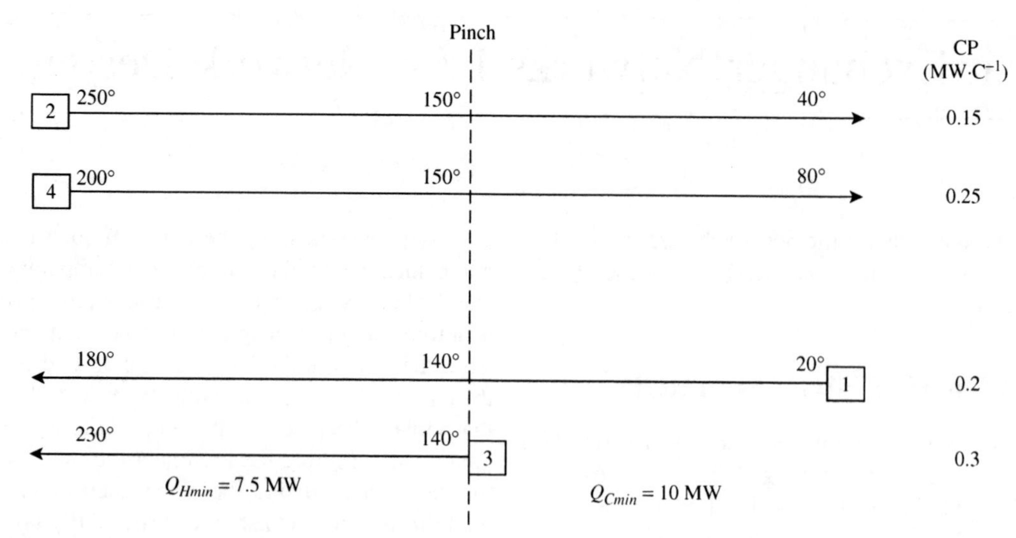

Figure 18.1 The grid diagram for the data from the Table 16.2.

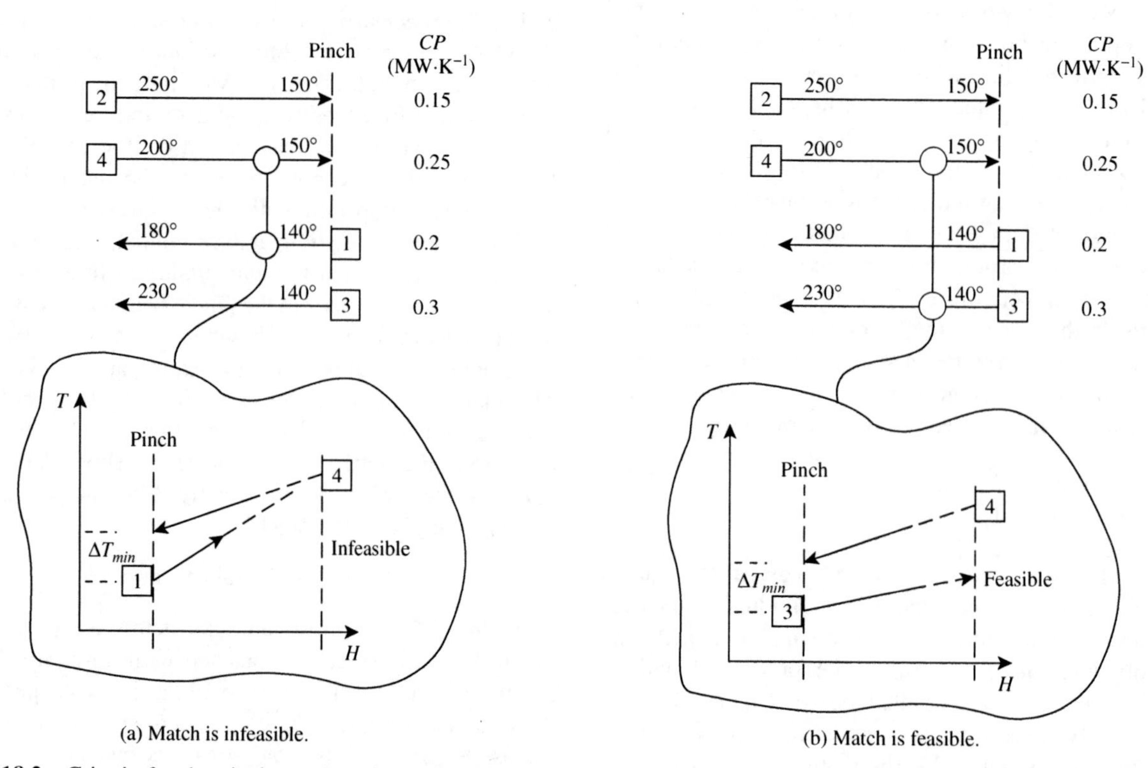

Figure 18.2 Criteria for the pinch matches above the pinch.

In the CP-table, the *CP* values of the hot and cold streams for the streams at the pinch are listed in descending order.

Figure 18.4a shows the grid diagram with CP-table for the design above the pinch. Cold utility must not be used above the pinch, which means that hot streams must be cooled to pinch temperature by heat recovery. Hot utility can be used, if necessary, on the cold streams above the pinch. Thus, it is essential to match hot streams above the pinch with a cold partner. In addition, if the hot stream is at pinch conditions, the cold stream it is to be matched with must also be at pinch conditions, otherwise the ΔT_{min} constraint will be violated. Figure 18.4a shows a feasible design arrangement above the pinch that does not use temperature differences smaller than ΔT_{min}. Note again that the *CP* inequality only applies when a match is made between two streams that are both at the pinch. Away from the pinch, temperature differences increase, and it is no longer essential to obey the *CP* inequalities.

Figure 18.4b shows the grid diagram with CP-table for the design below the pinch. Hot utility must not be used below the pinch, which means that cold streams must be heated to pinch temperature by heat recovery. Cold utility

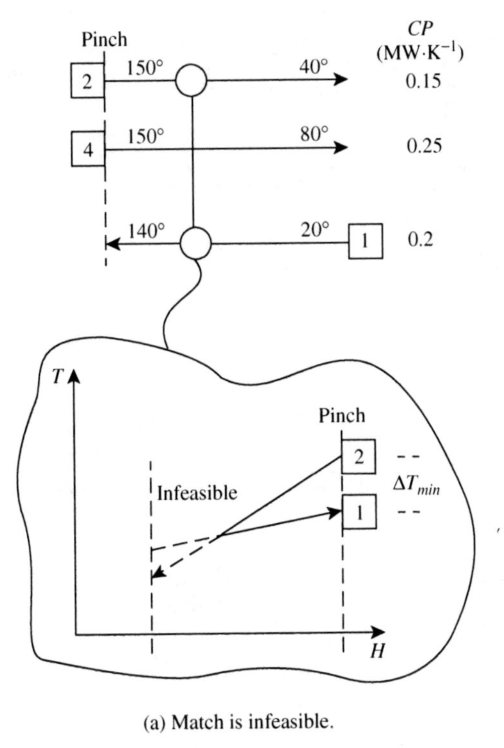

(a) Match is infeasible.

(b) Match is feasible.

Figure 18.3 Criteria for pinch matches below the pinch.

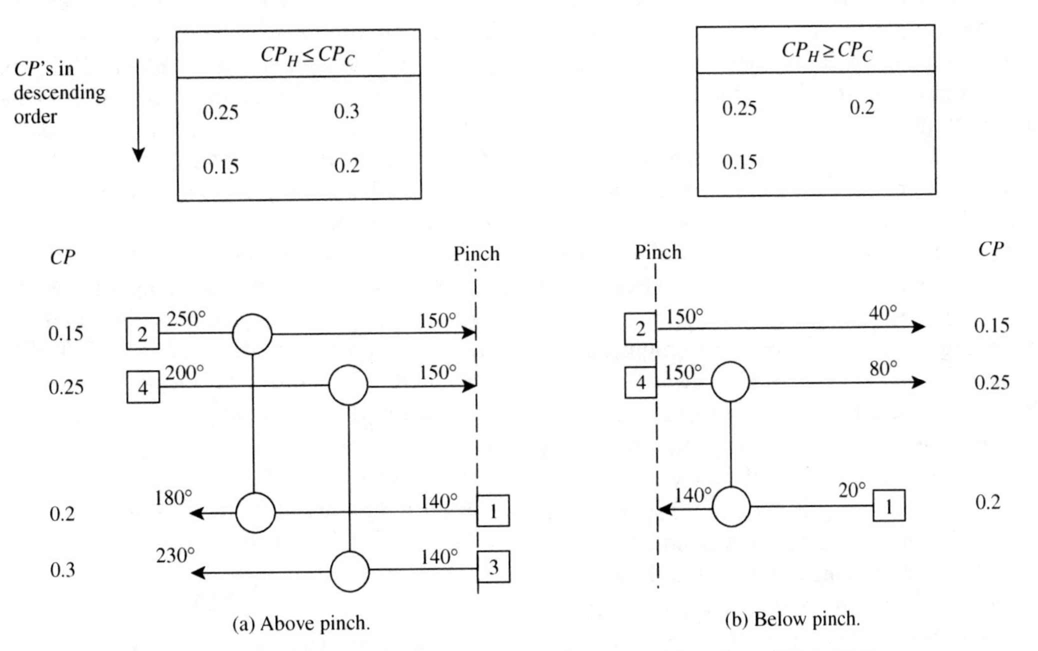

(a) Above pinch.

(b) Below pinch.

Figure 18.4 The CP table for the designs above and below the pinch for the problem from Table 16.2.

can be used, if necessary, on the hot streams below the pinch. Thus, it is essential to match cold streams below the pinch with a hot partner. In addition, if the cold stream is at pinch conditions, the hot stream it is to be matched with must also be at pinch conditions, otherwise the ΔT_{min} constraint will be violated. Figure 18.4b shows a design arrangement below the pinch that does not use temperature differences smaller than ΔT_{min}.

Having decided that some essential matches need to be made around the pinch, the next question is how big should the matches be?

4. *The "tick-off" heuristic*. Once the matches around the pinch have been chosen to satisfy the criteria for minimum energy, the design should be continued in such a manner as to keep capital costs to a minimum. One important criterion

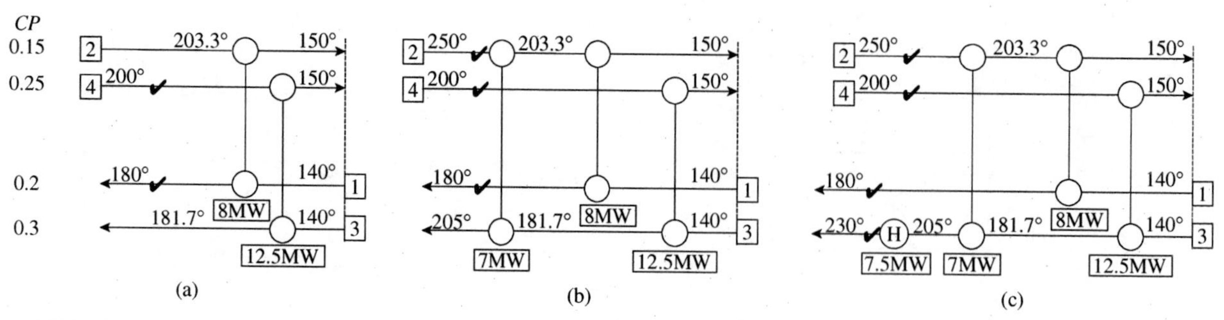

Figure 18.5 Sizing the units above the pinch using the tick-off heuristic.

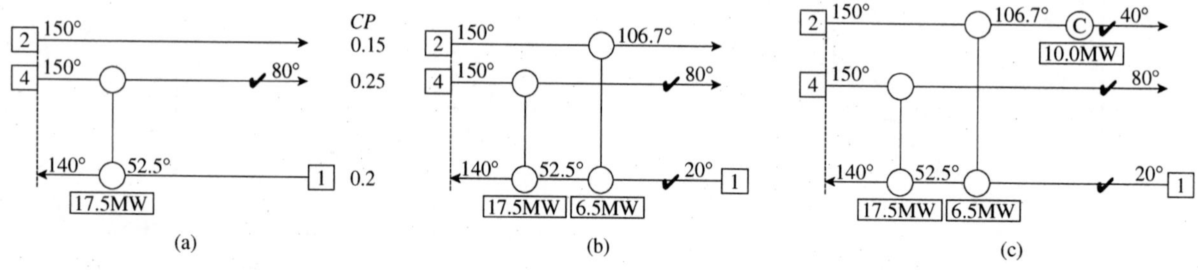

Figure 18.6 Sizing the units below the pinch using the tick-off heuristic.

in the capital cost is the number of units (there are others, of course, which shall be addressed later). Keeping the number of units to a minimum can be achieved using the *tick-off heuristic*. To tick off a stream, individual units are made as large as possible, that is, the smaller of the two heat duties on the streams being matched.

Figure 18.5a shows the matches around the pinch from Figure 18.4a with their duties maximized to tick off streams. It should be emphasized that the tick-off heuristic is only a heuristic and can occasionally penalize the design. Methods will be developed later, which allow such penalties to be identified as the design proceeds.

The design in Figure 18.5a can now be completed by satisfying the heating and cooling duties away from the pinch. Cooling water must not be used above the pinch. Therefore, if there are hot streams above the pinch for which the pinch matches do not satisfy the duties, additional process-to-process heat recovery is required. Figure 18.5b shows an additional match to satisfy the residual cooling of the hot streams above the pinch. Again, the duty on the unit is maximized. Finally, above the pinch, the residual heating duty on the cold streams must be satisfied. Since there are no hot streams left above the pinch, hot utility must be used as shown in Figure 18.5c.

Turning now to the cold end design, Figure 18.6a shows the pinch design with the streams ticked off. If there are any cold streams below the pinch for which the pinch matches do not satisfy the duties, then additional process-to-process heat recovery is required, since hot utility must not be used. Figure 18.6b shows an additional match to satisfy the residual heating of the cold streams below the pinch. Again, the duty on the unit is maximized. Finally, below the pinch, the residual cooling duty on the hot streams must be satisfied. Since there are no cold streams left below the pinch, cold utility must be used (Figure 18.6c).

The final design shown in Figure 18.7 amalgamates the hot end design from Figure 18.5c and cold end design from Figure 18.6c. The duty on hot utility of 7.5 MW agrees with Q_{Hmin} and the duty on cold utility of 10.0 MW agrees with Q_{Cmin} predicted by the composite curves and the problem table algorithm.

Note one further point from Figure 18.7. The number of units is 7 in total (including the heater and cooler). Referring back to Example 17.1, the target for the minimum number of units was predicted to be 7. It therefore appears that there was something in the procedure that naturally steered the design to achieve the target for the minimum number of units.

It is in fact the tick-off heuristic that steered the design toward the minimum number of units[1,2]. The target for the minimum number of units was given by Equation 17.2:

$$N_{UNITS} = S - 1 \tag{17.2}$$

Before any matches are placed, the target indicates that the number of units required is equal to the number of streams (including utilities) minus one. The tick-off heuristic satisfied the heat duty on one stream every time one of the units was used. The stream that has been ticked off is no longer part of the remaining design problem. The tick-off heuristic ensures that having placed a unit (and used up one of the available units), a stream is removed from the problem. Thus, Equation 17.2 is satisfied if every match satisfies the heat duty on a stream or a utility.

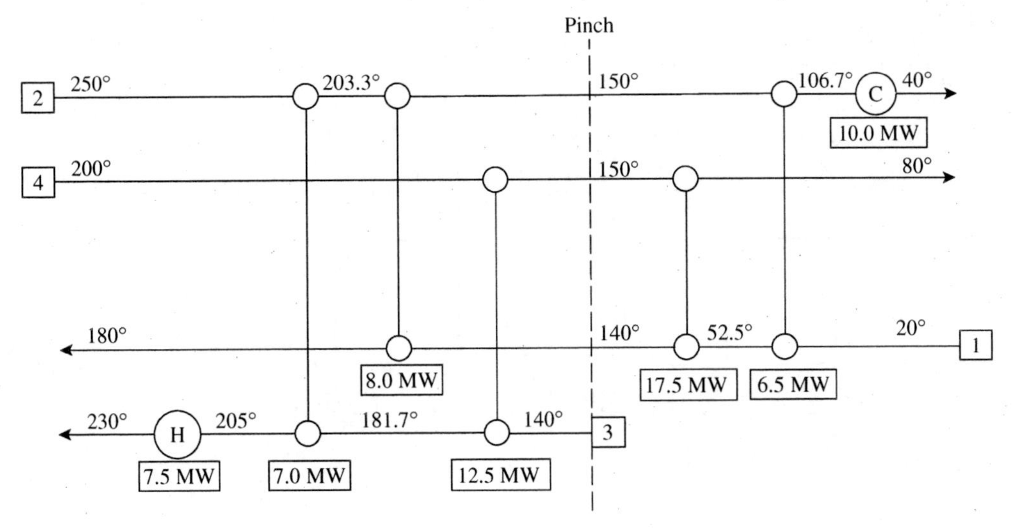

Figure 18.7 The completed design for the data from Table 16.2.

This design procedure is known as the *pinch design method* and can be summarized in five steps[1].

- Divide the problem at the pinch into separate problems.
- The design for the separate problems is started at the pinch, moving away.
- Temperature feasibility requires constraints on the *CP* values to be satisfied for matches between the streams at the pinch.
- The loads on individual units are determined using the tick-off heuristic to minimize the number of units. Occasionally, the heuristic causes problems.
- Away from the pinch, there is usually more freedom in the choice of matches. In this case, the designer can discriminate on the basis of operability, plant layout and so on.

Example 18.1 The process stream data for a heat recovery network problem are given in Table 18.1.

Table 18.1 Stream data for Example 18.1.

Stream		Supply temperature	Target temperature	Heat capacity flowrate
No.	Type	(°C)	(°C)	($MW \cdot K^{-1}$)
1	Hot	400	60	0.3
2	Hot	210	40	0.5
3	Cold	20	160	0.4
4	Cold	100	300	0.6

A problem table analysis on this data reveals that the minimum hot utility requirement for the process is 15 MW and the minimum cold utility requirement is 26 MW for a minimum allowable temperature difference of 20°C. The analysis also reveals that the pinch is located at a temperature of 120°C for hot streams and 100°C for cold streams. Design a heat exchanger network for maximum energy recovery featuring the minimum number of units.

Solution Figure 18.8a shows the hot end design with the CP-table. Above the pinch, adjacent to the pinch, $CP_H \leq CP_C$. The duty on the units has been maximized according to the tick-off heuristic.

Figure 18.8b shows the cold end design with the CP-table. Below the pinch, adjacent to the pinch, $CP_H \geq CP_C$. Again the duty on units has been maximized according to the tick-off heuristic.

The completed design is shown in Figure 18.8c. The minimum number of units for this problem is given by

$$N_{UNITS} = (S - 1)_{ABOVE\ PINCH} + (S - 1)_{BELOW\ PINCH}$$
$$= (5 - 1) + (4 - 1)$$
$$= 7$$

The design in Figure 18.8 is seen to achieve the minimum number of units target.

The pinch design method, as discussed so far, has assumed the same ΔT_{min} applied between all stream matches. In Chapter 16, it was discussed how the basic targeting methods for the composite curves and the problem table algorithm can be modified to allow stream-specific values of ΔT_{min}. The example was quoted in which liquid streams were required to have a ΔT_{min} contribution of 5°C and gas streams a ΔT_{min} contribution of 10°C. For liquid–liquid matches, this would lead to a $\Delta T_{min} = 10$°C. For gas–gas matches, this would lead to a $\Delta T_{min} = 20$°C. For liquid–gas matches, it will lead to a $\Delta T_{min} = 15$°C [2]. Modifying the problem table and the composite curves to account for these stream-specific values of ΔT_{min} is straightforward. But how is the pinch design method modified to take account of such ΔT_{min} contributions? Figure 18.9 illustrates the approach. Suppose the interval pinch temperature from the problem table is 120°C. This would correspond with hot stream pinch temperatures of 125°C and 130°C for hot streams with ΔT_{min} contributions of 5°C and 10°C respectively. For

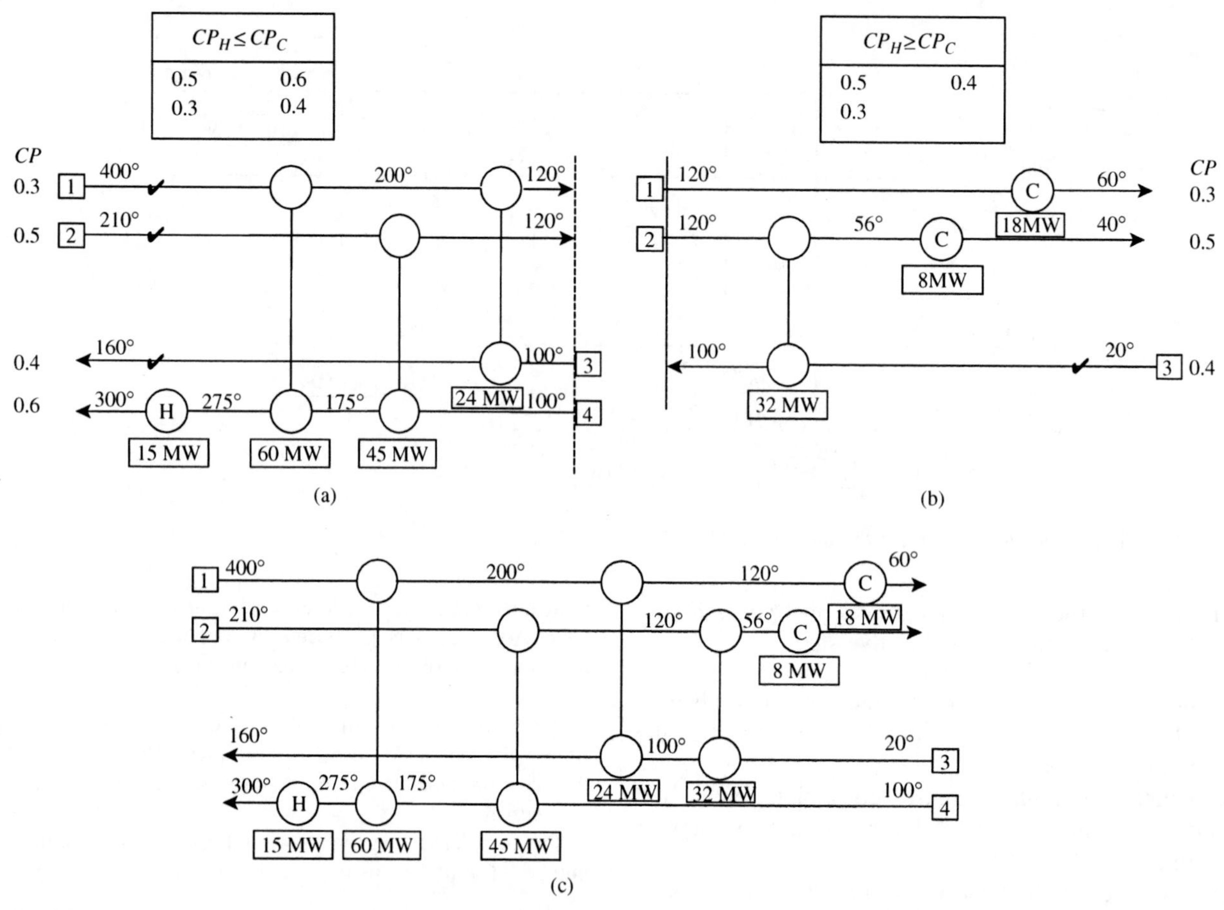

Figure 18.8 Maximum energy recovery design for Example 18.1.

an interval pinch temperature of 120°C, the corresponding cold stream pinch temperatures would be 115°C and 110°C for cold streams with ΔT_{min} contributions of 5°C and 10°C respectively. The stream grid is set up as shown in Figure 18.9 with the pinch matches being made and given their appropriate value of ΔT_{min} depending on the streams being matched. The constraints regarding the *CP* inequalities apply in this case, just as the case with a global value of ΔT_{min}.

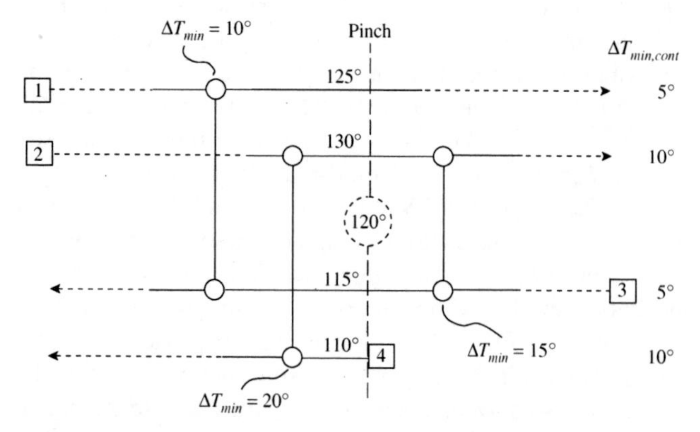

Figure 18.9 The design grid for a problem stream-specific with ΔT_{min} contributions.

18.2 DESIGN FOR THRESHOLD PROBLEMS

In Section 16.3, it was discussed that some problems, known as threshold problems, do not have a pinch. They need either hot utility or cold utility but not both. How should the approach be modified to deal with the design of threshold problems?

The philosophy in the pinch design method was to start the design where it was most constrained. If the design is pinched, the problem is most constrained at the pinch. If there is no pinch, where is the design most constrained? Figure 18.10a shows one typical threshold problem that requires no hot utility, just cold utility. The most constrained part of this problem is the no-utility end[2]. This is where temperature differences are smallest and there may be constraints, as shown in Figure 18.10b, where the target temperatures on some of the hot streams can only be satisfied by specific matches. Also, if individual matches are required to have a temperature difference no smaller than the threshold ΔT_{min}, the *CP* inequalities described in the pinch design method must be applied. For the most part, problems similar to those in Figure 18.10a are treated as one half of a pinched problem.

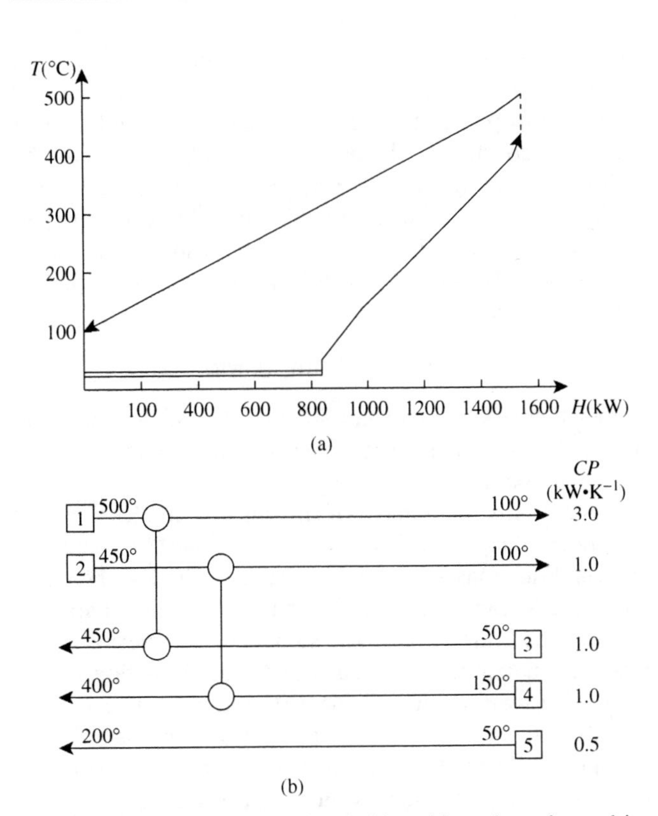

Figure 18.10 Even though threshold problems have large driving forces, there are still often essential matches to be made, especially at the no-utility end.

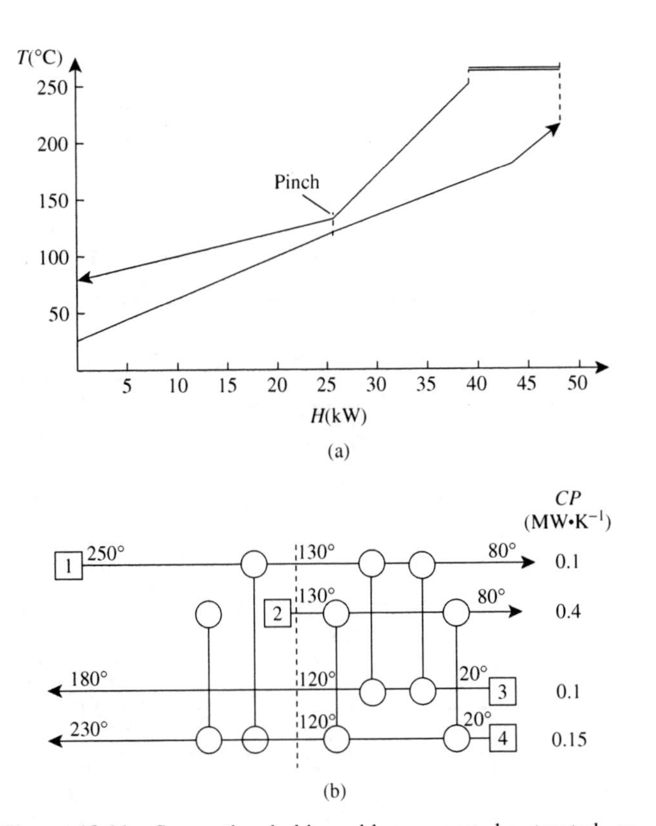

Figure 18.11 Some threshold problems must be treated as pinched problems requiring essential matches at both the no utility end and the pinch.

Figure 18.11 shows another threshold problem that requires only hot utility. This problem is different in characteristic from the one in Figure 18.10. Now the minimum temperature difference is in the middle of the problem causing a pseudo-pinch. The best strategy to deal with this type of threshold problem is to treat it as a pinched problem. In Figure 18.11, the problem is divided into two parts at the pseudo-pinch, and the pinch design method followed. The only complication in applying the pinch design method for such problems is that one half of the problem (the cold end in Figure 18.11) will not feature the flexibility offered by matching against utility.

18.3 STREAM SPLITTING

The pinch design method developed earlier followed several rules and guidelines to allow design for minimum utility (or maximum energy recovery) in the minimum number of units. Occasionally, it appears not to be possible to create the appropriate matches because one or other of the design criteria cannot be satisfied.

Consider Figure 18.12a that shows the above-pinch part of a design. Cold utility must not be used above the pinch, which means that all hot streams must be cooled to pinch temperature by heat recovery. There are three hot streams

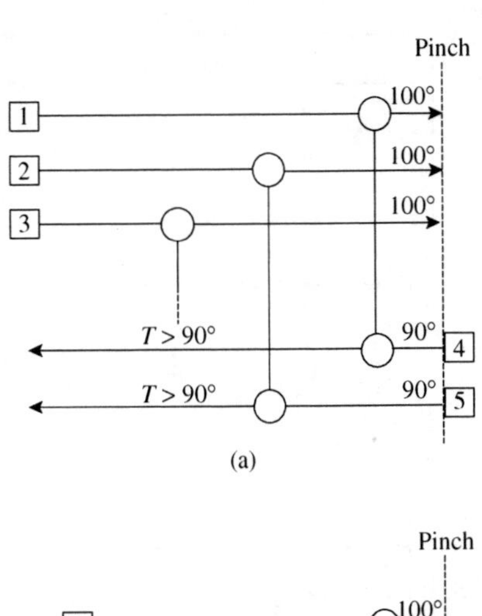

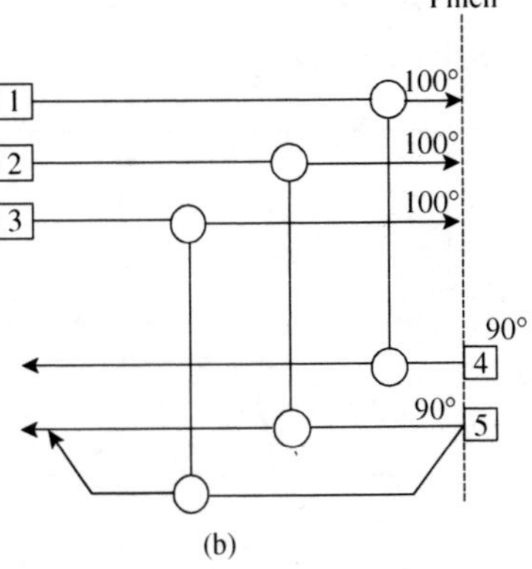

Figure 18.12 If the number of hot streams at the pinch, above the pinch, is greater than the number of cold streams, then stream splitting of the cold streams is required.

and two cold streams in Figure 18.12a. Thus, regardless of the CP values of the streams, one of the hot streams cannot be cooled to pinch temperature without some violation of the ΔT_{min} constraint. The problem can only be resolved by splitting a cold stream into two parallel branches as shown in Figure 18.12b. Now each hot stream has a cold partner with which to match, capable of cooling it to pinch temperature. Thus, in addition to the CP inequality criteria introduced earlier, there is a stream number criterion above the pinch such that[1,2]:

$$S_H \leq S_C \text{ (above pinch)} \tag{18.3}$$

where S_H = number of hot streams at the pinch (including branches)

S_C = number of cold streams at the pinch (including branches)

If there had been a greater number of cold streams than hot streams in the design above the pinch, this would not have created a problem, since hot utility can be used above the pinch.

By contrast, now consider part of a design below the pinch, as shown in Figure 18.13a. Here hot utility must not be used, which means that all cold streams must be heated to pinch temperature by heat recovery. There are now three cold streams and two hot streams in Figure 18.13a. Again, regardless of CP values, one of the cold streams cannot be heated to pinch temperature without some violation of the ΔT_{min} constraint. The problem can only be resolved by splitting a hot stream into two parallel branches, as shown in Figure 18.13b. Now each cold stream has a hot partner with which to match and capable of heating it to pinch temperature. Thus there is a stream number criterion below the pinch, such that[1,2]:

$$S_H \geq S_C \text{ (below pinch)} \tag{18.4}$$

Had there been more hot streams than cold below the pinch, this would not have created a problem since cold utility can be used below the pinch.

It is not only the number of streams that creates the need to split streams at the pinch. Sometimes the CP inequality criteria, Equations 18.1 and 18.2, cannot be met at the pinch without a stream split. Consider the above-pinch part of a problem in Figure 18.14a. The number of hot streams is less than the number of cold streams, and hence Equation 18.3 is satisfied. However, the CP inequality, Equation 18.1, must be satisfied. Neither of the two cold streams has a large enough CP. The hot stream can be made smaller by splitting it into two parallel branches (Figure 18.14b).

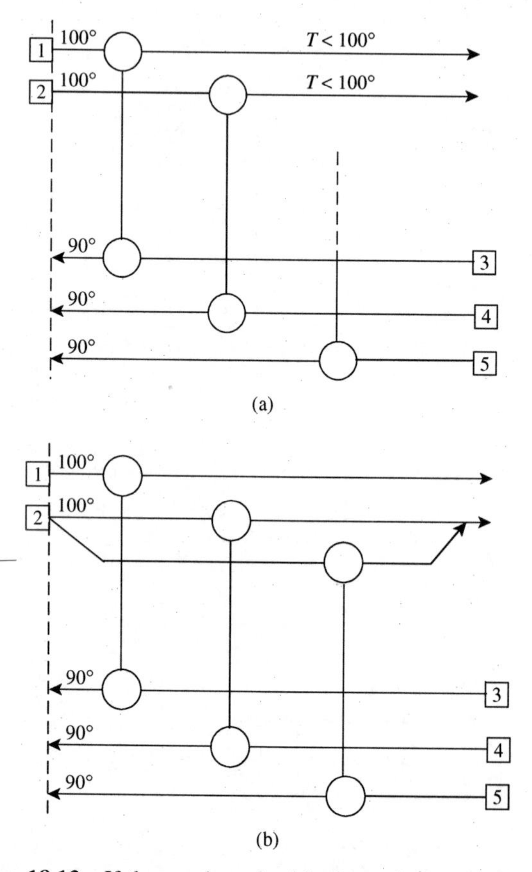

Figure 18.13 If the number of cold streams below the pinch, at the pinch, is greater than the pinch number of hot streams, then stream splitting of the hot steam is required.

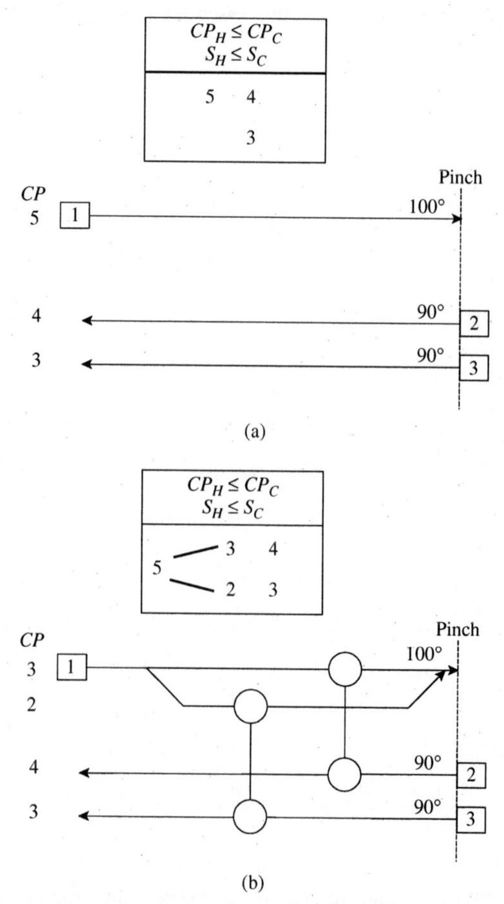

Figure 18.14 The CP in equality rules can necessitate stream splitting above the pinch.

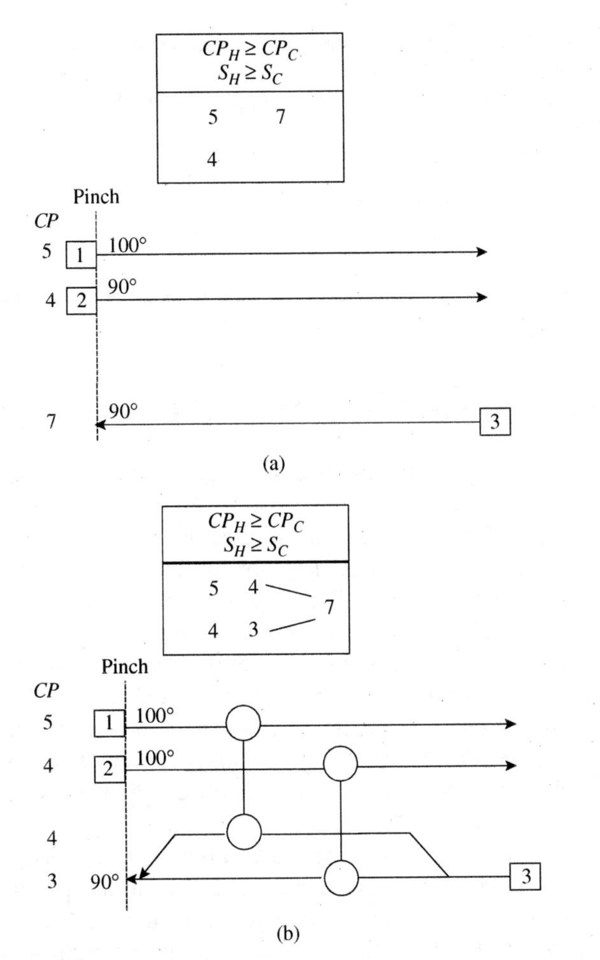

Figure 18.15 The CP in equality rules can necessitate stream splitting below the pinch.

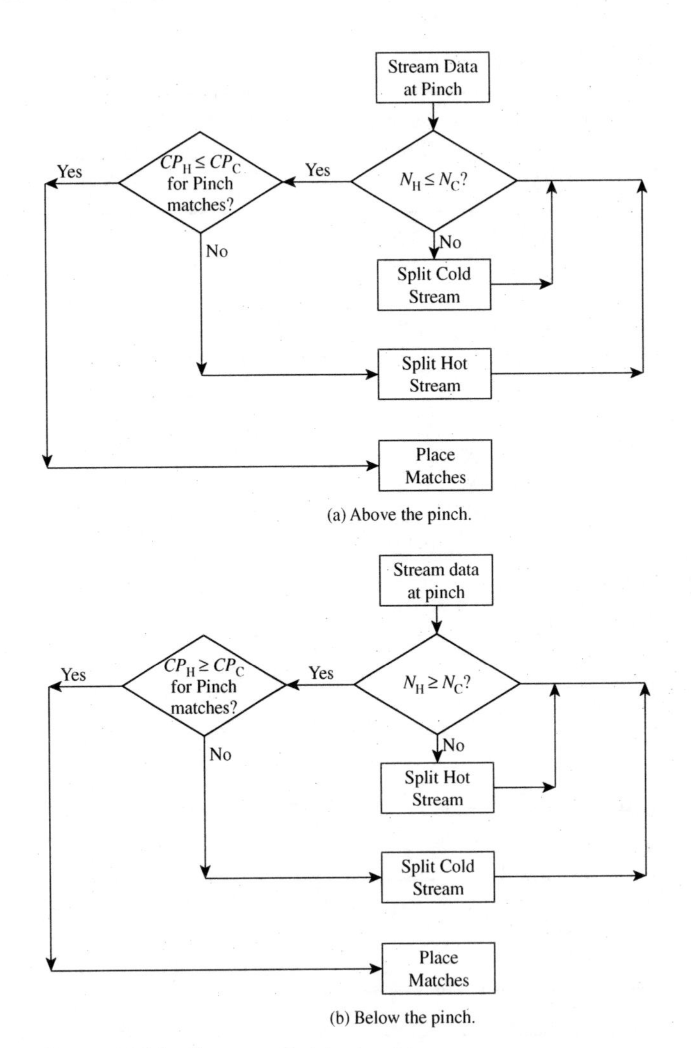

Figure 18.16 Stream splitting algorithms.

Figure 18.15a shows the below-pinch part of a problem. The number of hot streams is greater than the number of cold streams, and hence Equation 18.4 is satisfied. However, neither of the two hot streams has a large enough *CP* to satisfy the *CP* inequality, Equation 18.2. The cold stream can be made smaller by splitting it into two parallel branches (Figure 18.14b).

Clearly, in designs different from those in Figures 18.14 and 18.15, when streams are split to satisfy the *CP* inequality, this might create a problem with the number of streams at the pinch such that Equations 18.3 and 18.4 are no longer satisfied. This would then require further stream splits to satisfy the stream number criterion. Figures 18.16a and 18.16b present algorithms for the overall approach[1,2].

One further important point needs to be made regarding stream splitting. In Figure 18.14, the hot stream is split into two branches with *CP* values of 3 and 2 to satisfy the *CP* inequality criteria. However, a different split could have been chosen. For example, the split could have been into branch *CP* values of 4 and 1, or 2.5 and 2.5, or 2 and 3 (or any setting between 4 and 1, and 2 and 3). Each of these would also have satisfied the *CP* inequalities. Thus, there is a degree of freedom in the design to choose the branch flowrates. By fixing the heat duties on the two units in Figure 18.14b and changing the branch flowrates, the temperature differences across each unit are changed. The best choice can only be made by sizing and costing the various units in the completed network for different branch flowrates. This is an important degree of freedom when the network is optimized. Similar arguments could be made regarding the cold end design in Figure 18.15b.

Example 18.2 A problem table analysis for part of a high-temperature process reveals that for $\Delta T_{min} = 20°C$ the process requires 9.2 MW of hot utility, 6.4 MW of cold utility and the pinch is located at 520°C for hot streams and 500°C for cold streams. The process stream data are given in Table 18.2. Design a heat exchanger network for maximum energy recovery that features the minimum number of units.

Solution Figure 18.17a shows the stream grid with the CP-tables for the above- and below-pinch designs. Following the algorithms in Figure 18.16, a hot stream must be split above the pinch to satisfy the *CP* inequality, as shown in Figure 18.17b.

Table 18.2 Stream data.

Stream No.	Type	Supply Temp. (°C)	Target Temp. (°C)	Heat Capacity Flowrate ($MW{\cdot}K^{-1}$)
1	Hot	720	320	0.045
2	Hot	520	220	0.04
3	Cold	300	900	0.043
4	Cold	200	550	0.02

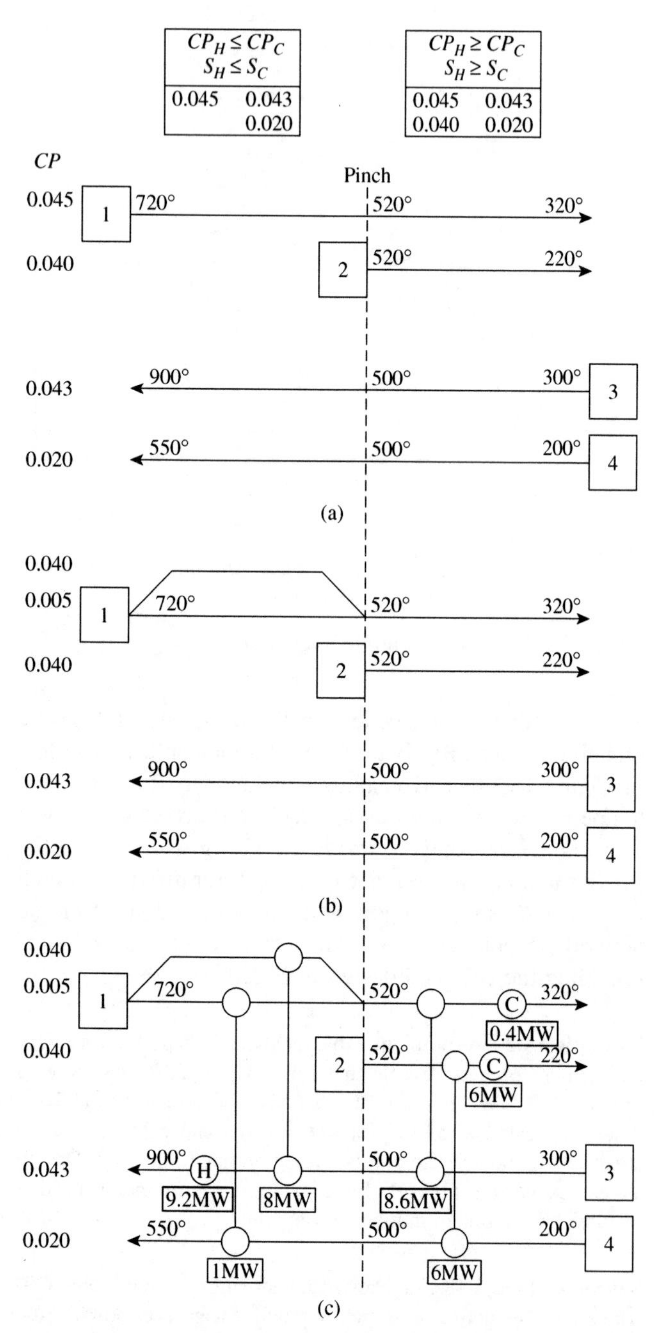

Figure 18.17 Maximum energy recovery design for Example 18.2.

Thereafter the design is straightforward, and the final design is shown in Figure 18.17c.

The target for the minimum number of units is given by:

$$\begin{aligned} N_{UNITS} &= (S-1)_{ABOVE\ PINCH} + (S-1)_{BELOW\ PINCH} \\ &= (4-1) + (5-1) \\ &= 7 \end{aligned}$$

The design in Figure 18.17c is seen to achieve the minimum units target.

18.4 DESIGN FOR MULTIPLE PINCHES

In Chapter 16, it was discussed as to how the use of multiple utilities could give rise to multiple pinches. For example, the design for the process in Figure 16.2 could have used either a single level of hot utility or two steam levels (Figure 16.26a). The targeting indicated that instead of using 7.5 MW of high-pressure steam at 240°C, 3 MW of this could be substituted with low-pressure steam at 180°C. Where the low-pressure steam touches the grand composite curve in Figure 16.26a results in a utility pinch. Figure 18.18a shows the grid diagram when two steam levels are used with the utility pinch dividing the process into three parts.

Following the pinch rules, heat should not be transferred across either the process pinch or the utility pinch by process-to-process heat exchange. Also, there must be no inappropriate use of utilities. This means that above the utility pinch shown in Figure 18.18a, high-pressure steam should be used and no low-pressure steam or cooling water. Between the utility pinch, and the process pinch low-pressure steam should be used and no high-pressure steam or cooling water. Below the process pinch in Figure 18.18 only cooling water should be used. The appropriate utility streams have been included with the process streams in Figure 18.18.

The network can now be designed using the pinch design method[1,2]. The philosophy of the pinch design method is to start at the pinch, and move away. At the pinch, the rules for the *CP* inequality and the number of streams must be obeyed. Above the utility pinch in Figure 18.18, and below the process pinch in Figure 18.18 there is clearly no problem in applying this philosophy. However, between the two pinches there is a problem, since designing away from both pinches could lead to a clash.

More careful examination of Figure 18.18a reveals that, between the two pinches, one is more constrained than the other. Below the utility pinch, $CP_H \geq CP_C$ is required and low-pressure steam is available as a hot stream with an extremely large *CP*. In fact, if steam is assumed to condense or vaporize isothermally, it will have a *CP* that is infinite. Thus, following the philosophy of starting the design in the most constrained region, the design between the pinches in Figure 18.18a should be started at the most constrained pinch, which is the process pinch.

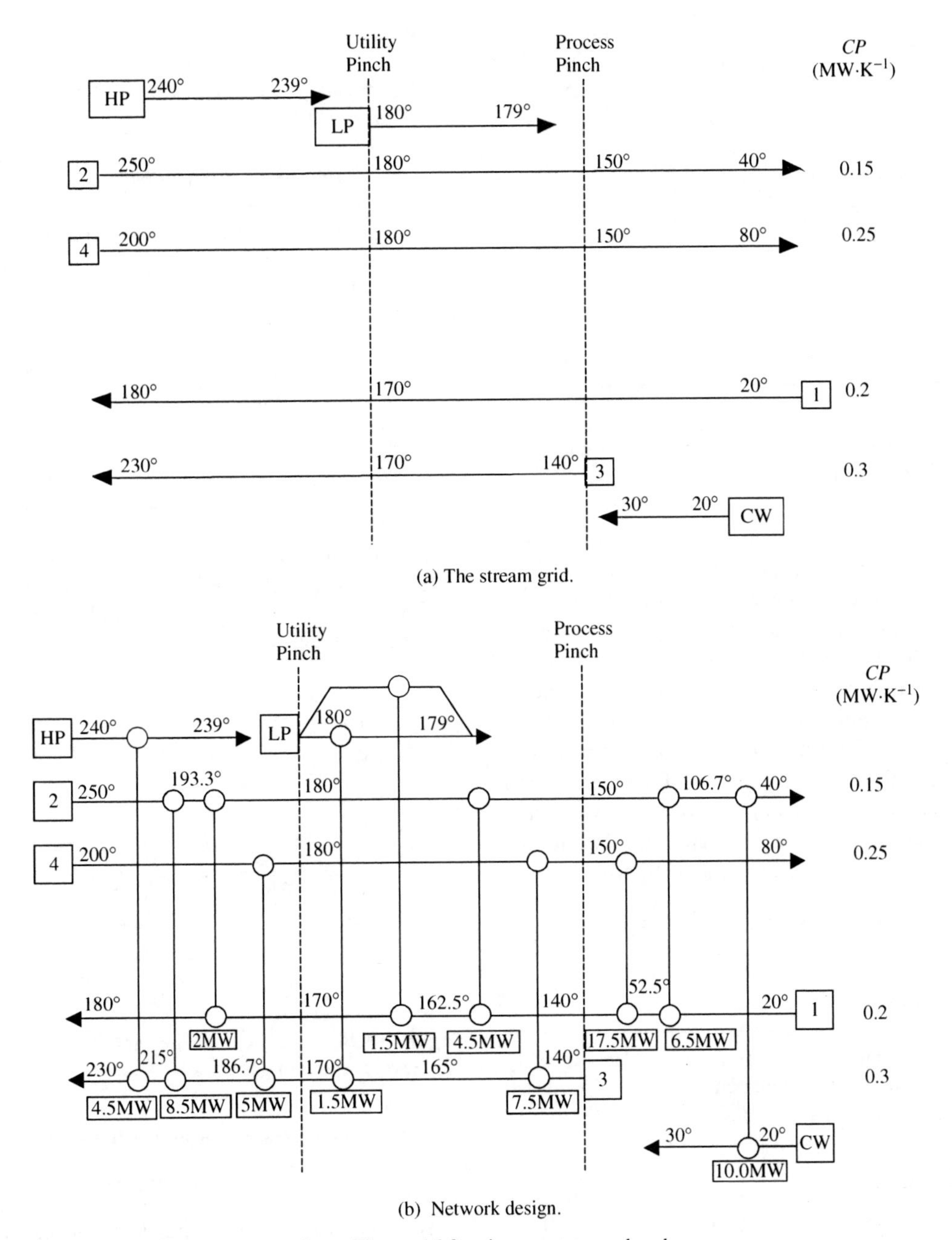

Figure 18.18 Network design for the process from Figure 16.2 using two steam levels.

Following this approach, the design is straightforward and the final design is shown in Figure 18.18b. It achieves the target set in Example 16.3 and in the minimum number of units. Remember that, in this case, to calculate the minimum number of units, the stream count must be performed separately in the three parts of the problem. Note that the stream split on the low-pressure steam in Figure 18.18b is not strictly necessary, but is made for practical reasons. Without the stream split, steam would have to partially condense in one unit and the steam-condensate mixture transferred to the next unit. The stream split allows two conventional steam heaters on low-pressure steam to be used. It is clear from Figure 18.18b that the use of two steam levels has increased the complexity of the design considerably. However, the complexity of the design can be reduced later when the structure is subjected to optimization. The optimization can remove units that are uneconomic.

It is rare for there to be two process pinches in a problem. Multiple pinches usually arise from the introduction of additional utilities causing utility pinches. However, cases such as that shown in Figure 18.19 are not uncommon, where there is strictly speaking, only one pinch (one place where ΔT_{min} occurs), but there is a near-pinch. This near-pinch is a point in the process where the temperature difference becomes small enough to be effectively another pinch, even

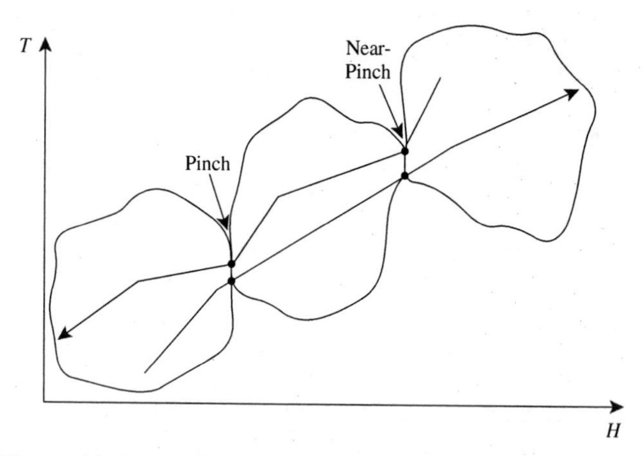

Figure 18.19 A near-pinch might require the design to be treated as if it had multiple pinches.

though the temperature difference is slightly larger than ΔT_{min}. Because the region around the near-pinch will be almost as constrained as the pinch, the best strategy is often to treat the near-pinch as if it was another pinch and divide the problem into three parts as shown in Figure 18.19. The initial design would therefore avoid heat transfer across the near-pinch and the process pinch and use hot utility only above the near-pinch and cold utility only below the pinch. The designer can then exploit the small amount of freedom around the near-pinch. Some heat transfer across the near-pinch is possible without causing an energy penalty. Exploiting this freedom allows the design to be simplified slightly.

Example 18.3 The stream data for a process are given in Table 18.3:

Table 18.3 Stream data.

Stream No.	Stream Type	Supply temperature (°C)	Target temperature (°C)	Heat capacity flowrate ($MW \cdot K^{-1}$)
1	Hot	635	155	0.044
2	Cold	10	615	0.023
3	Cold	85	250	0.020
4	Cold	250	615	0.020

It has been decided to integrate a gas turbine (GT) exhaust with the process. The exhaust temperature of the GT is 400°C with $CP = 0.05\ MW \cdot K^{-1}$. Ambient temperature is 10°C.

a. Calculate the problem table cascade for $\Delta T_{min} = 20°C$.
b. Saturated steam is to be generated by the process at a high-pressure level of 250°C and low-pressure level of 140°C, each from saturated boiler feedwater. The generation of the higher-pressure steam is to be maximized. How much steam can be generated at the two levels assuming boiler feedwater and final steam condition are both saturated?
c. Design a network for maximum energy recovery for $\Delta T_{min} =$ 20°C that generates steam at these two levels.
d. What is the residual cooling demand?

Solution

a. The problem table cascade is shown in Table 18.4 for $\Delta T_{min} = 20°C$.

Table 18.4 Problem table cascade.

Interval temperature (°C)	Cascade heat flow (MW)
625	0
390	0.235
260	6.865
145	12.73
95	13.08
20	15.105
0	16.105

b. For high-pressure steam, $T^* = 260°C$, for low-pressure steam, $T^* = 150°C$. Figure 18.20 shows the grand composite curve plotted from the problem table cascade. The two levels of steam generation are shown.

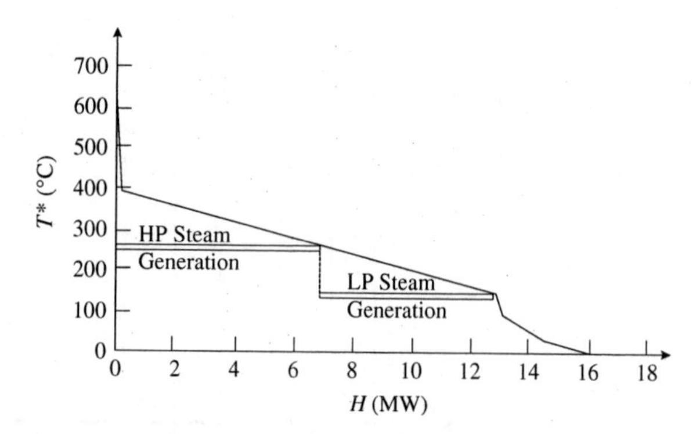

Figure 18.20 Grand composite curve for Example 18.3 showing two levels of steam generation.

Duty on high-pressure steam generation

$$= 6.865 \text{ MW}$$

By interpolation from the problem table cascade, heat flow at $T^* = 150°C$

$$= 6.865 + \frac{(260 - 150)}{(260 - 145)} \times (12.73 - 6.865)$$

$$= 12.475 \text{ MW}$$

Duty on low-pressure steam generation

$$= 12.475 - 6.865$$

$$= 5.61 \text{ MW}$$

c. The use of two levels of steam generation in Figure 18.20 creates two utility pinches. Thus, the stream grid needs to be divided into three parts. Figure 18.21 shows the final design, which achieves the targets set for both high-pressure (HP) and low-pressure (LP) steam generation. In Figure 18.21 above the HP pinch, the CP of Stream 1 and the GT stream are too large

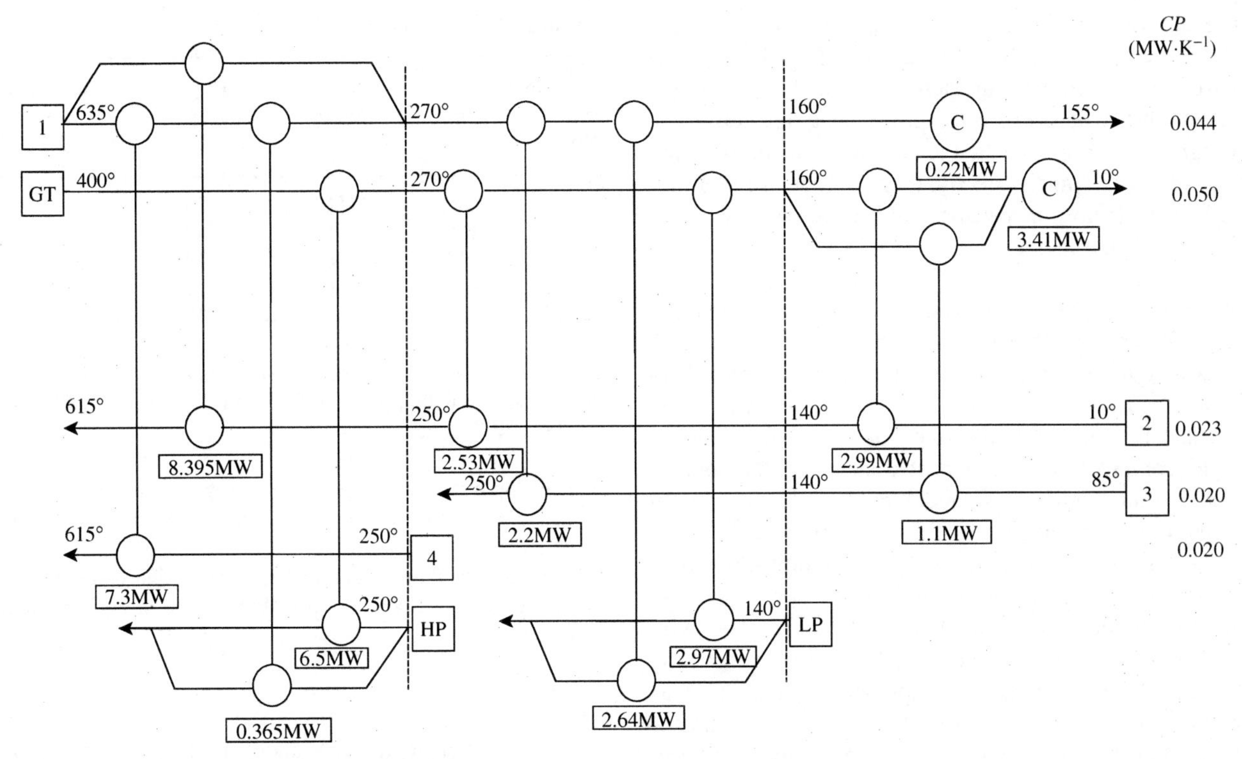

Figure 18.21 Network design for Example 18.3.

to match directly against Streams 3 and 4. This is overcome in Figure 18.21 by splitting Stream 1 and exploiting the infinite *CP* of the HP steam generation. Between the pinches, the design is started below the HP pinch and developed toward the LP pinch, where the infinite *CP* of the LP steam generation can be exploited to satisfy the *CP* inequalities. Below the LP pinch, although the *CP* inequalities can be satisfied by direct matches, the heat duty on Stream 1 is small compared with the other streams. If Stream 1 did not exist below the LP pinch, then this would call for the GT stream to be split. This has been done in Figure 18.21 because of the small duty on Stream 1. Although the steam generation for both high-pressure and low-pressure steam are shown with steam splits in Figure 18.21, in practice these units would be individual steam generators, each fed with boiler feedwater. Also, the stream split for the GT exhaust could be accommodated by splitting the gas turbine flow into two ducts, or by placing two sets of tubes in the same GT exhaust located in parallel. The design in Figure 18.21 has significant scope for simplification, but at the penalty of reduced energy efficiency. Such trade-offs are at the discretion of the designer.

d. There is a cooling demand of 0.22 MW on Stream 1 that needs to be satisfied by cold utility and cooling demand of 3.41 MW required by the GT exhaust. The cooling of the GT exhaust is satisfied by simply venting it to atmosphere after heat recovery has been completed.

18.5 REMAINING PROBLEM ANALYSIS

The considerations addressed so far in network design have been restricted to those of energy performance and number of units. In addition, the problems have all been straightforward to design for maximum energy recovery in the minimum number of units by ticking-off streams. Not all problems are so straightforward. Also, heat transfer area, number of shells when using 1–2 shells, capital cost and so on should be considered when placing matches. Here, a more sophisticated approach is needed[3].

When a match is placed, the duty needs to be chosen with some quantitative assessment of the match in the context of the whole network, without having to complete the network. This can be done by exploiting the powers of targeting using a technique known as *remaining problem analysis*.

Consider first design for minimum energy in a more complex problem than has so far been addressed. If a problem table analysis is performed on the stream data, Q_{Hmin} and Q_{Cmin} can be calculated. When the network is designed and a match placed, it would be useful to assess whether there will be any energy penalty caused by some feature of the match without having to complete the design. Whether there will be a penalty can be determined by performing a problem table analysis on the *remaining problem*. The problem table analysis is simply repeated on the stream data, leaving out those parts of the hot and cold stream satisfied by the match. One of the two results would then occur:

1. The algorithm may calculate Q_{Hmin} and Q_{Cmin} to be unchanged. In this case, the designer knows that the match will not penalize the design in terms of increased utility usage.

2. The algorithm may calculate an increase in Q_{Hmin} and Q_{Cmin}. This means that the match is transferring heat across the pinch or that there is some feature of the design that will cause cross-pinch heat transfer if the design was completed. If the match is not transferring heat across the pinch directly, then the increase in utility will result from the match being too big as a result of the tick-off heuristic.

The remaining problem analysis technique can be applied to any feature of the network that can be targeted, such as a minimum area. In Chapter 17, the approach to targeting for heat transfer area (Equation 17.6) was based on vertical heat transfer from the hot composite curve to the cold composite curve. If heat transfer coefficients do not vary significantly, this model predicts the minimum area requirements adequately for most purposes[3]. Thus, if heat transfer coefficients do not vary significantly, then the matches created in the design should come as close as possible to the conditions that would correspond with vertical transfer between the composite curves. Remaining problem analysis can be used to approach the area target, as closely as a practical design permits, using a minimum (or near-minimum) number of units. Suppose a match is placed, then its area requirement can be calculated. A remaining problem analysis can be carried out by calculating the area target for the stream data, leaving out those parts of the data satisfied by the match. The area of the match is now added to the area target for the remaining problem. Subtraction of the original area target for the whole-stream data $A_{NETWORK}$ gives the area penalty incurred.

If heat transfer coefficients vary significantly, then the vertical heat transfer model adopted in Equation 17.6 predicts a network area that is higher than the true minimum, as illustrated in Figure 17.4. Under these circumstances, a careful pattern of nonvertical matching is required to approach the minimum network area. However, the remaining problem analysis approach can still be used to steer the design toward a minimum area under these circumstances. When heat transfer coefficients vary significantly, the minimum network area can be predicted using linear programming.[4,5] The remaining problem analysis approach can then be applied using these more sophisticated area targeting methods. Under such circumstances, the design is likely to be difficult to steer toward the minimum area, and an automated design method based on the optimization of a superstructure can be used, as will be discussed later.

Targets for number of shells, capital cost and total cost also can be set. Thus, remaining problem analysis can be used on these design parameters also.

Example 18.4 The stream data for a process are given in Table 18.5 below:

Table 18.5 Stream data.

Stream No.	Stream Type	Supply temp. (°C)	Target temp. (°C)	Heat capacity flowrate ($MW \cdot K^{-1}$)
1	Hot	150	50	0.2
2	Hot	170	40	0.1
3	Cold	50	120	0.3
4	Cold	80	110	0.5

Steam is available condensing between 180 and 179°C and cooling water between 20 and 30°C. All film transfer coefficients are 200 $W \cdot m^{-2} \cdot K^{-1}$. For $\Delta T_{min} = 10°C$, the minimum hot and cold utility duties are 7 MW and 4 MW respectively, corresponding with a pinch at 90°C on the hot streams and 80°C on the cold streams.

a. Develop a maximum energy recovery design above the pinch that comes close to the area target in the minimum number of units.
b. Develop a maximum energy recovery design below the pinch that comes as close as possible to the minimum number of units.

Solution

a. The area target for the above-pinch problem shown in Figure 18.22 is 8859 m^2. If the design is started at the pinch

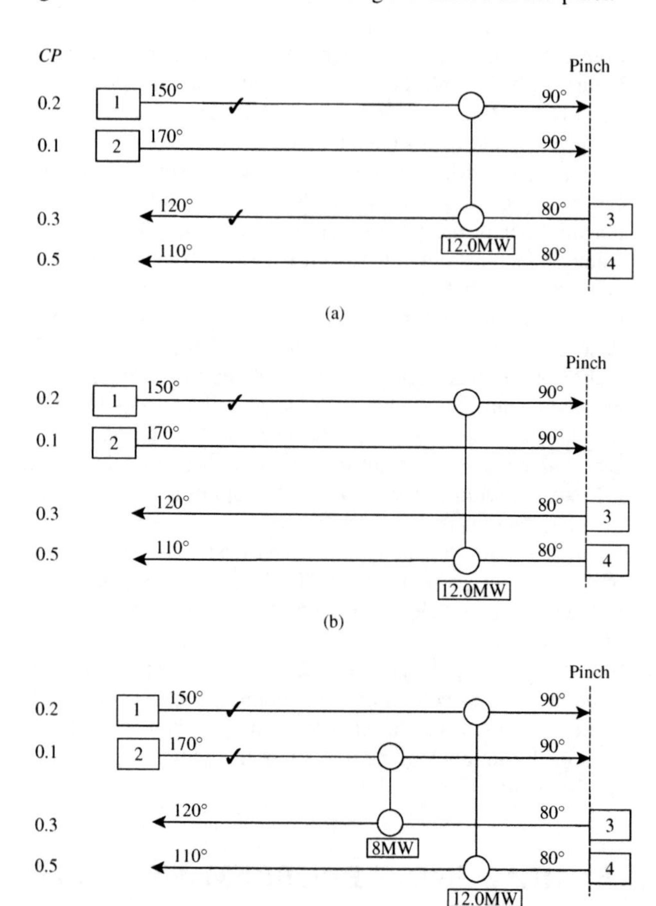

Figure 18.22 Above the pinch design for Example 18.4.

with Stream 1, then Figure 18.22a shows a feasible match that obeys the *CP* inequality. Maximizing its duty to 12 MW allows two streams to be ticked off simultaneously. This results from a coincidence in the stream data, the duties for Streams 1 and 3 being equal above the pinch. The area of the match is 6592 m^2 and the target for the remaining problem above the pinch is 3419 m^2, giving a total of 10,011 m^2. Thus, the match in Figure 18.22a causes the overall target to be exceeded by 1152 m^2 (13%). This does not seem to be a good match.

Figure 18.22b shows an alternative match for Stream 1 that also obeys the *CP* inequality. The tick-off heuristic also fixes its duty to be 12 MW. The area for this match is 5087 m^2, and the target for the remaining problem above the pinch is 3788 m^2, giving a total of 8,875 m^2. Thus, the match in Figure 18.22b causes the overall target to be exceeded by 16 m^2 (0.2%). This seems to be a better match and therefore is accepted.

Placing the next match above the pinch as shown in Figure 18.22c also allows the *CP* inequality to be obeyed. The area for both matches in Figure 18.22c is 7856 m^2 and the target for the remaining problem is 1020 m^2, giving a total of 8876 m^2. Accepting both matches causes the overall area target to be exceeded by 17 m^2 (0.2%). This seems to be reasonable, and both matches are accepted. No further process-to-process matches are possible, and it remains to place hot utility.

b. The cold utility target for the problem shown in Figure 18.23 is 4 MW. If the design is started at the pinch with Stream 3, then Stream 3 must be split to satisfy the CP inequality (Figure. 18.23a). Matching one of the branches against Stream 1 and ticking off Stream 1 results in a duty of 8 MW. This is a case in which the tick-off heuristic has caused problems. The match is infeasible, because the temperature difference between the streams at the cold end of the match is infeasible. Its duty must be reduced to 6 MW to be feasible without either stream being ticked off (Figure 18.23b).

Figure 18.23c shows an additional match placed on the other branch for Stream 3 with its duty maximized to 3 MW to tick off Stream 3. No further process-to-process matches are possible, and it remains to place cold utility.

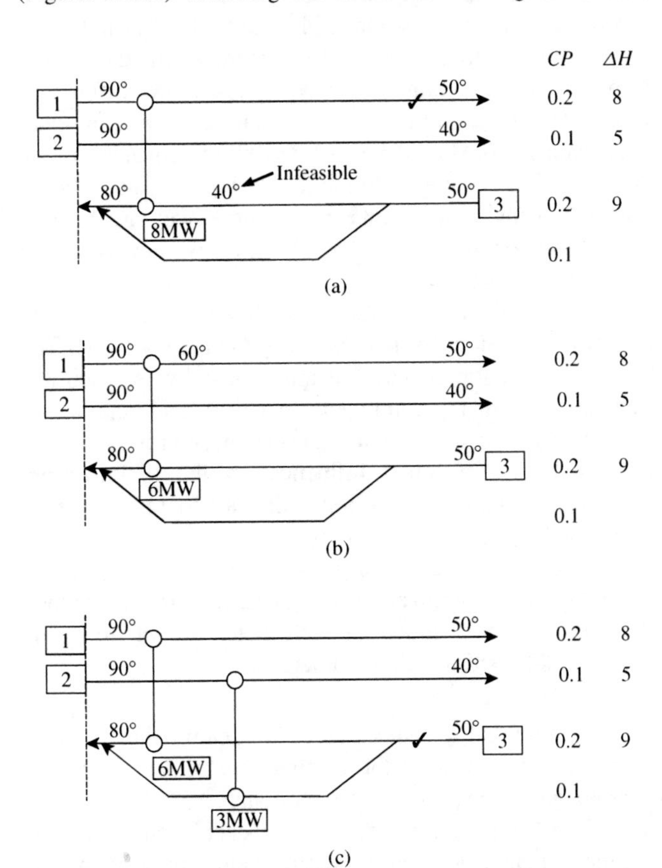

Figure 18.23 Below the pinch design for Example 18.4.

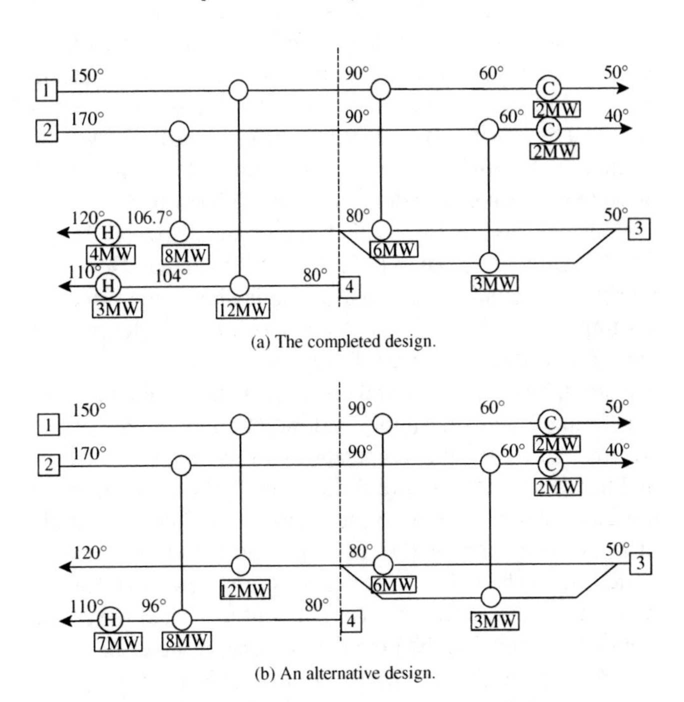

Figure 18.24 Alternative designs for Example18.4.

Figure 18.24a shows the complete design, achieving maximum energy recovery in one more unit than the target minimum, due to the inability to tick off streams below the pinch. If the match in Figure 18.22a had been accepted and the design completed, then the design in Figure 18.24b would have been obtained. This achieves the target for the minimum number of units of 7 (at the expense of excessive area). This results from the coincidence of data mentioned earlier in Figure 18.22a, which allowed two streams to be ticked off simultaneously. The result is that the design above the pinch uses one fewer unit than target, owing to the formation of two components above the pinch (see Section 17.1). The design below the pinch uses one more than target, and the net result is that the overall design achieves the target for the minimum number of units.

18.6 NETWORK OPTIMIZATION

The pinch design method creates a network structure based on the assumption that no heat exchanger should have a temperature difference smaller than ΔT_{min}. Having now created a structure for the heat exchanger network, the structure can now be subjected to continuous optimization. The constraint that no exchanger should have a temperature difference smaller than ΔT_{min} can now be relaxed. The continuous optimization of heat exchanger networks is based on the redistribution of the exchanger duties. Some exchangers should perhaps be larger, some smaller and some perhaps removed from the design altogether. Exchangers are removed from the design if the optimization sets their duty to zero.

Given a network structure, it is possible to identify loops and paths for it, as in Section 17.1. Within the context of optimization, only those paths that connect two different utilities need to be considered. This could be a path from steam to cooling water or a path from high-pressure steam used as hot utility to low-pressure steam also used as hot utility. These paths between two different utilities will be termed to be *utility paths*. Loops and utility paths both provide degrees of freedom in the optimization[1,2].

Consider Figure 18.25a that shows the network design from Figure 18.7, but with a loop highlighted. Heat can be shifted around loops. Figure 18.25a shows the effect of shifting heat duty U around the loop. In this loop, heat duty U is simply moved from Unit E to Unit B. The change in heat duties around the loop maintains the network heat balance and the supply and target temperatures of the streams. However, the temperatures around the loop change and hence the temperature differences of the exchangers in the loop change in addition to their duties. The magnitude of U could be changed to different values and the network costed at each value to find the optimum setting for U. If the optimum setting for U turns out to be 6.5 MW (the original duty on E), then the duty on one of the exchangers is zero and should be removed from the design.

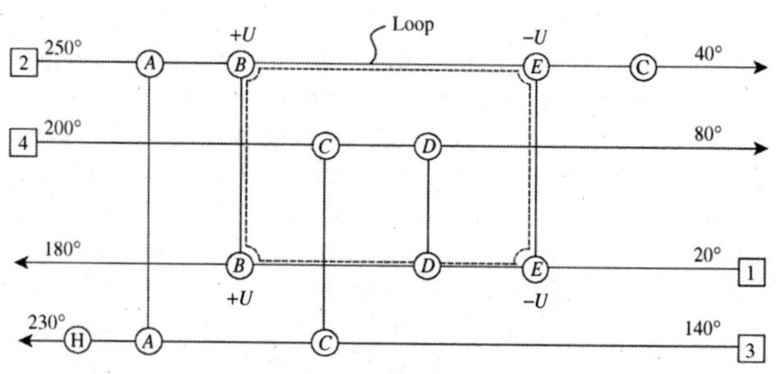

(a) Heat duties can be changed within a loop without changing the utility consumption.

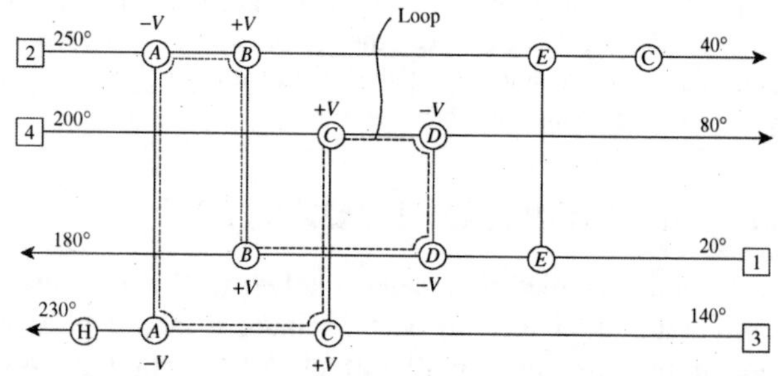

(b) Another loop allowing heat duties to be changed without changing the utility consumption.

Figure 18.25 The loops that can be exploited for the optimization of the design for Figure 16.7.

Figure 18.25b shows the network with another loop marked. Figure 18.25b shows the effect of shifting heat duty V around the loop. Again the heat balance is maintained, but the temperatures as well as the duties around the loop change. As before, the value of V can be optimized by costing the network at different settings of V. If V is optimized to 7.0 MW (the original duty on A), then the duty on A becomes zero and this unit is removed from the design. Note again that once optimization is started, there is no longer a constraint to maintain temperature differences to be larger than ΔT_{min}.

Figure 18.26a shows the network with a utility path highlighted. Heat duty can be shifted along utility paths in a similar way to that for loops. Figure 18.26a shows the effect of shifting heat duty W along the path. This time the heat balance changes because the loads imported from hot utility and exported to cold utility both change by W. The supply and target temperatures are maintained. If W is optimized to 7.0 MW, this will result in Unit A being removed from the design. Different values of W can be taken and the network sized and costed at each value to find the optimal setting for W. Figure 18.26b shows other utility paths that can be exploited for optimization.

In fact, the optimization of the network requires that U, V, W, X, Y and Z in Figures 18.25 and 18.26 must be optimized simultaneously. Furthermore, stream splits may exist in the design, and variations of their branch flowrates can be superimposed on the exploitation of loops and paths in the optimization. During this optimization, the design is no longer constrained to have temperature differences larger than ΔT_{min} (although very small values in individual heat exchangers should be avoided for practical reasons). Also, pinches no longer divide the design into independent thermodynamic regions, and there is no longer any concern about cross-pinch heat transfer. The objective now is simply to minimize cost.

Thus, loops, utility paths and stream splits offer the degrees of freedom for manipulating the network cost. The objective function in new design is usually to minimize total cost, that is, combined operating and annualized capital cost. The annualization period chosen for the capital cost will have a direct influence on the optimization. A longer annualization period will lead to more energy-efficient designs.

In practice, rather than manipulate loops and paths explicitly, the optimization is normally formulated such that the individual duties on each match are varied in the multivariable optimization, subject to:

- the total enthalpy change on each stream being within a specified tolerance of the original stream data,
- nonnegative heat duty for each match,
- positive temperature difference for each exchanger to be greater than a practical minimum value for a given type of heat exchanger,

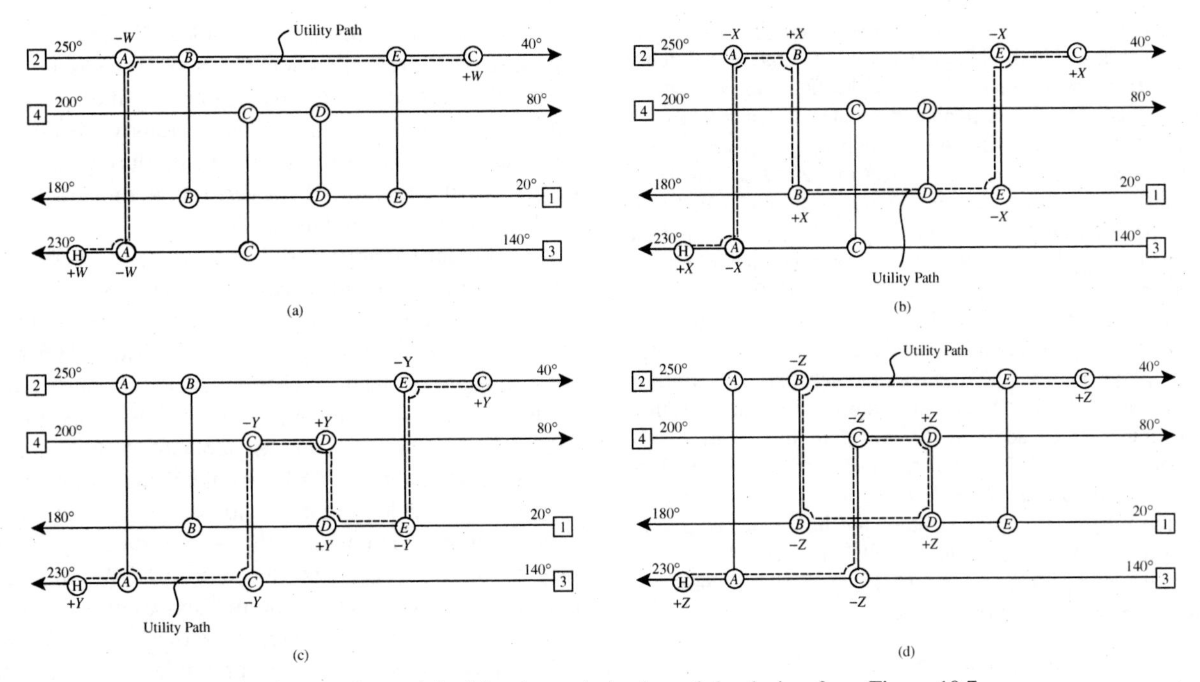

Figure 18.26 The utility paths that can be exploited for the optimization of the design from Figure 18.7.

- for stream splits, branch flowrates must be positive and above a practical minimum flowrate.

In a network, some of the duties on the matches will not be able to be varied because they are not in a loop or a utility path. This simplifies the optimization. The problem is one of multivariable nonlinear continuous optimization[6].

If the network is optimized at fixed energy consumption, then only loops and stream splits are exploited. When energy consumption is allowed to vary, utility paths must also be included. As the network energy consumption increases, the overall capital cost tends to decrease and vice versa.

Example 18.5 Evolve the heat exchanger network in Figure 18.7 to simplify its structure.

a. Remove the smallest heat recovery unit from the network by exploiting the degree of freedom in a loop.
b. Recalculate the network temperatures and identify any violations of the $\Delta T_{min} = 10°C$ constraint.
c. Restore the original $\Delta T_{min} = 10°C$ throughout the network by exploiting a utility path.

Solution

a. Figure 18.27a shows the network from Figure 18.7 with a loop highlighted involving the unit with the smallest heat duty (6.5 MW). A heat duty of 6.5 MW has been shifted around the loop to adjust the smallest unit to a duty of zero. This will change the temperatures around the loop.
b. Heat balances around the units for the new heat duties allow the new temperatures in the network to be calculated, as shown in Figure 18.27b. Also highlighted in Figure 18.27b, is a unit with an infeasible temperature difference. Not only is it less than

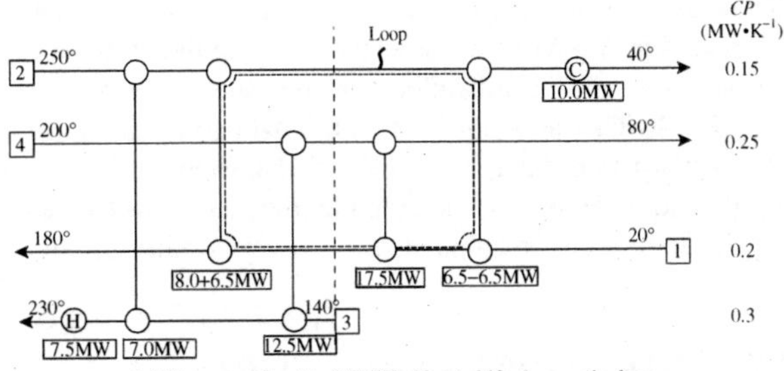

(a) The network with 6.5MW of heat shifted around a loop.

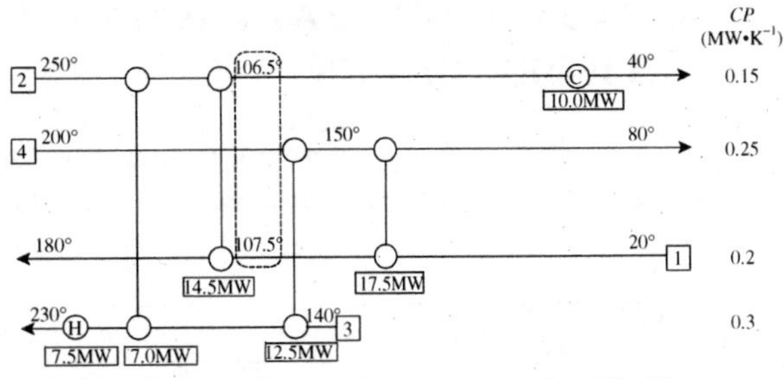

(b) Calculating intermediate network temperatures reveals an infeasible temperature difference.

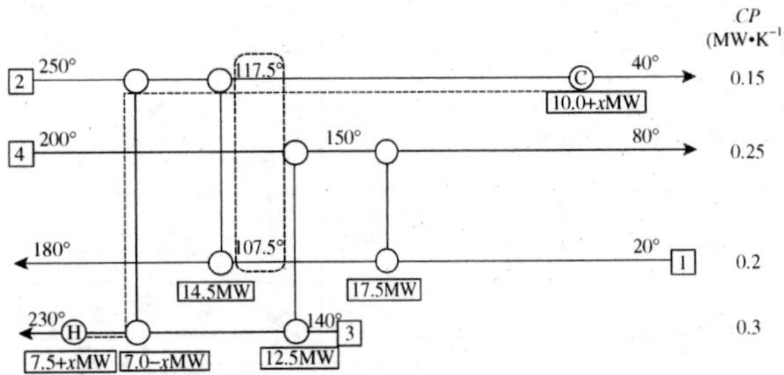

(c) Increasing the heat flow along a utility path allows the feasible temperature difference to be restored.

Figure 18.27 Evolution of a network to remove a unit.

the original ΔT_{min} of 10°C, it is actually negative. Removing a unit in this way will always create a temperature difference smaller than the original ΔT_{min}, if the hot and cold utility remain fixed. There is a minimum number of units to satisfy the problem to hot utility consumption of Q_{Hmin} and a cold utility consumption of Q_{Cmin}, subject to the ΔT_{min} constraint. If a unit is then removed, something must be violated. By its nature, shifting heat load around a loop does not change the energy consumption, but does change internal temperatures.

c. Given the infeasible temperature difference in Figure 18.27b, this can be corrected by exploiting a utility path to change the temperatures in the network at the expense of increased energy consumption. Figure 18.27c shows the network with a utility path highlighted. The utility path allows one of the infeasible temperatures to be adjusted (Stream 2 in this case). If the original ΔT_{min} of 10°C is to be restored, then the intermediate temperature of Stream 2 needs to be adjusted to 117.5°C as shown in Figure 18.27c. The unknown is how much additional heat duty (x MW) needs to be shifted along the utility path to restore the temperature to 117.5°C. This can be determined by a simple heat balance around the cooler.

$$10.0 + x = 0.15\ (117.5 - 40)$$

$$x = 1.6 \text{ MW}$$

Thus the hot and cold utility consumption both need to be increased by 1.6 MW to restore the ΔT_{min} to the original 10°C. In fact there is no justification to restore the ΔT_{min} back to the original 10°C. The amount of additional energy shifted along the utility path is a degree of freedom that should be set by cost optimization. However, the example illustrates how the degrees of freedom can be manipulated in network optimization.

18.7 THE SUPERSTRUCTURE APPROACH TO HEAT EXCHANGER NETWORK DESIGN

The approach to heat exchanger network design discussed so far was based on the creation of an irreducible structure. No redundant features were included. However, after the structure was created, when the network was optimized, some of the features might be removed by the optimization. The scope for the optimization to remove features is a consequence of the assumptions made during the creation of the initial structure. However, no attempt was made to deliberately include redundant features.

An alternative approach is to create a superstructure that deliberately includes redundant features and then subject this to optimization. Redundant features are then removed by the optimization. Floudas, Ciric and Grossman[7] showed how a heat exchanger network superstructure could be set up with all structural features included. Figure 18.28a shows such a superstructure for part of a heat exchanger network problem involving two hot streams, two cold streams and steam. All possibilities have been included within this superstructure. The basic idea is then to optimize the superstructure in order to remove the unnecessary features and minimize the cost, possibly leading to the design as shown in Figure 18.28b. While this looks simple in principle, the optimization required is a mixed integer nonlinear programming problem (MINLP, see Chapter 3)[8]. This is a difficult optimization problem with all of the issues associated with local optima.

One of the ways to avoid this problem is to simplify the superstructure to remove some of the structural options in Figure 18.28a[9]. This is done in Figure 18.29. This structure is created by splitting each hot stream into a number of branches equal to the number of cold streams and splitting each cold stream into a number of branches equal to the number of hot streams. In this way, a structure is created that allows each hot stream to be matched with each cold stream[9].

One significant advantage of the simplified superstructure in Figure 18.29 is that each exchanger can be modeled by a linear equation if the supply and target temperatures of the streams are fixed. The area for each heat exchanger is

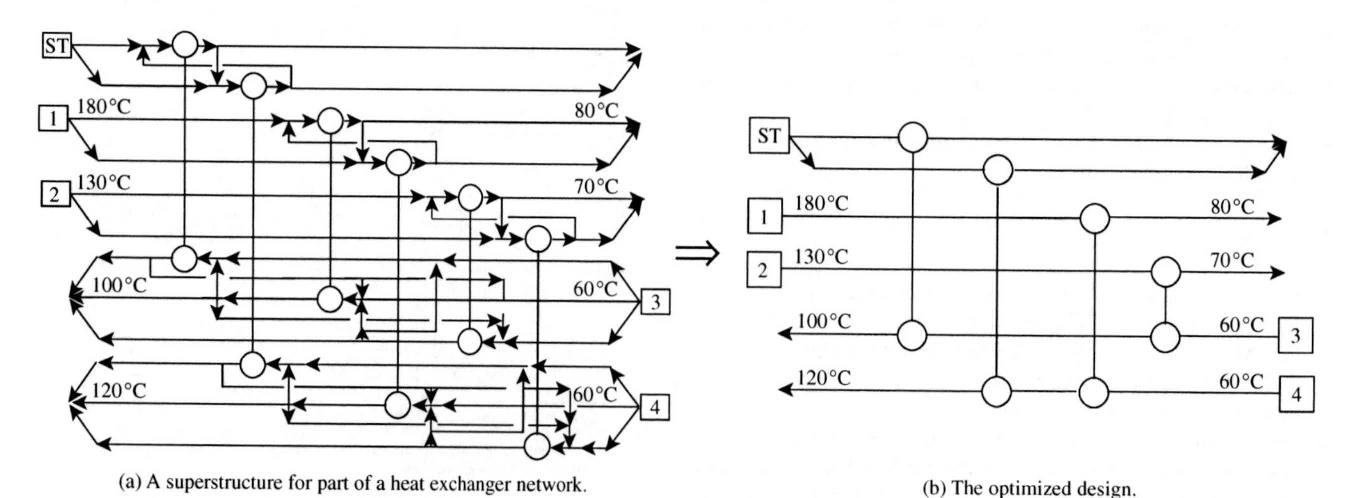

(a) A superstructure for part of a heat exchanger network.

(b) The optimized design.

Figure 18.28 Heat exchanger network design from the optimization of a superstructure.

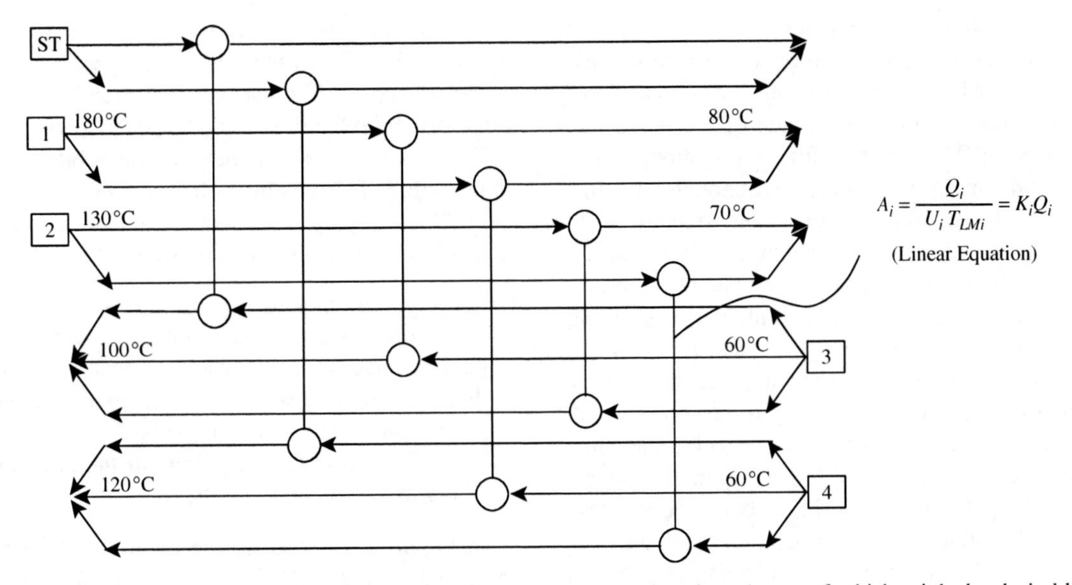

Figure 18.29 Starting with a simpler superstructure removes many structural options (some of which might be desirable, but makes the optimization linear.

given by:

$$A = \frac{Q}{U \Delta T_{LM}} \tag{18.5}$$

where A = heat transfer area
Q = heat exchanger duty
U = overall heat transfer coefficient
ΔT_{LM} = logarithmic mean temperature difference

If U is assumed to be constant, despite changing conditions during the optimization, then because ΔT_{LM} is fixed for each match in Figure 18.29, Equation 18.5 becomes a linear equation in Q. If the capital cost of the heat exchangers is taken to be a linear function, then the optimization simplifies to be mixed integer linear programming (MILP) rather than MINLP. The problem with this is that the initial superstructure shown in Figure 18.29 features parallel configurations, as shown in 18.30a. If the linear optimization needs to retain all three matches on the stream, as shown in Figure 18.30a, then this cannot evolve to all of the series configurations, 1×2 parallel configurations, series-parallel configurations and parallel-series configurations as shown in Figure 18.30b.

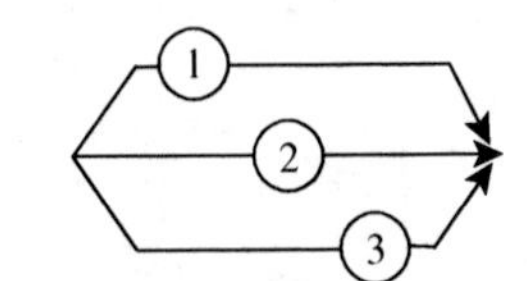

(a) Possible parallel feature of the network.

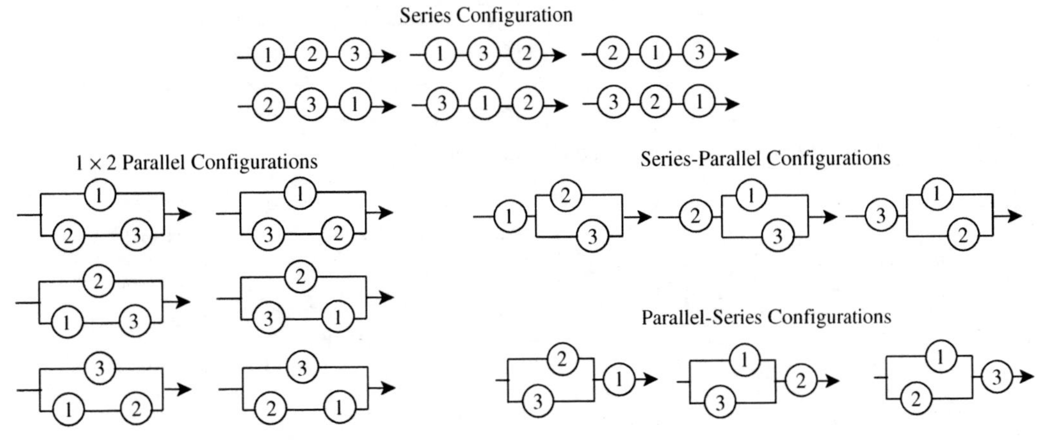

(b) Possible evolutions of the parallel configuration.

Figure 18.30 The simplified superstructure cannot be evolved to all structural options.

Another issue is how to divide the overall problem in order to set up the initial superstructure. The problem could be divided at the pinch, as done in the pinch design method, and a superstructure set up on each side of the pinch like the one in Figure 18.29. However, for large complex problems, this would not be comprehensive enough in terms of the number of structural options. The overall problem could therefore be divided into enthalpy intervals as shown in Figure 18.31 and a simplified superstructure created within each enthalpy interval[10]. Rather than dividing the composite curves into enthalpy intervals, as shown in Figure 18.31, with a superstructure within each enthalpy interval, enthalpy intervals can be merged into *blocks*[11]. A superstructure is then created within each block rather than each enthalpy interval. This simplifies the complete superstructure for the whole network. This network superstructure could then be subjected to optimization using MILP.

Note that the ΔT_{min} for the problem must be fixed in order to remain an MILP problem. Fixing ΔT_{min} fixes the composite curves and the temperatures across each enthalpy interval or block. Unfortunately, this would not necessarily lead to the best network, as the initial superstructure was already simplified with many structural options missing. But this can be allowed for by first carrying out the optimization on the basis of the simplified superstructure in Figure 18.29. If then the optimized superstructure featured structural options like the one in Figure 18.32a, then additional structural features could be added to give the structure in Figure 18.32b, which then can be subjected to MINLP optimization. This has the potential then to optimize to any parallel, series-parallel or parallel-series arrangement. This breaks the overall optimization down into two steps.

1. A simplified superstructure is first optimized using MILP.
2. If the solution from the simplified superstructure features parallel arrangements, such as that in Figure 18.32a, then additional features are added to the network, as shown in Figure 18.32b. This is then subjected to MINLP optimization to allow all the structural options in Figure 18.30 to be accessed.

This approach is one way to avoid the difficulties inherent with applying MINLP to a large complex superstructure. Another way might be to apply stochastic optimization[12]. However, this has the disadvantage of being computationally very demanding for a reasonable size of heat exchanger network.

The major advantage of the superstructure approach to heat exchanger network design is that, in principle, it is capable

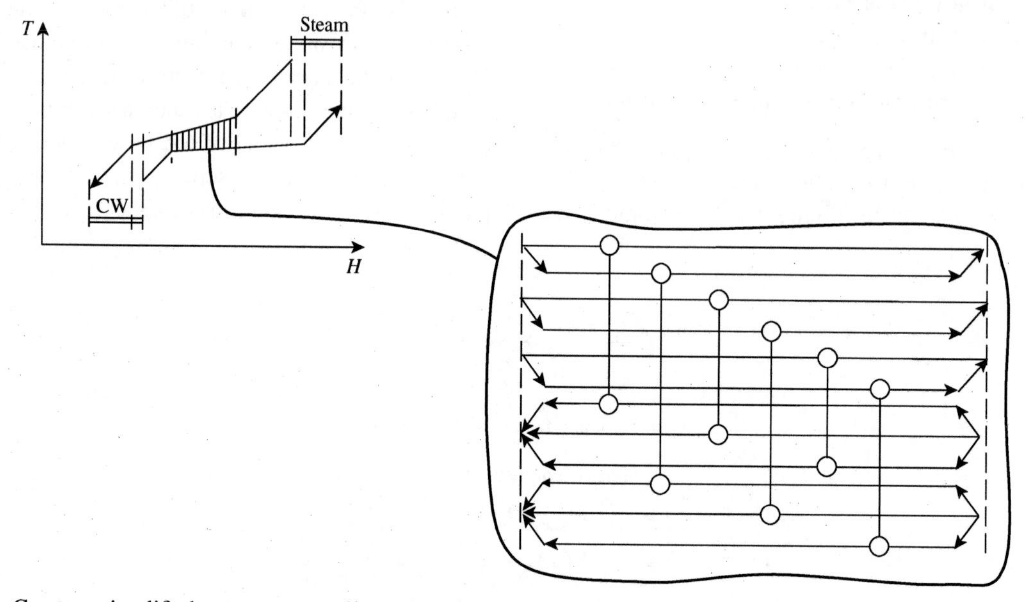

Figure 18.31 Create a simplified superstructure for each enthalpy interval.

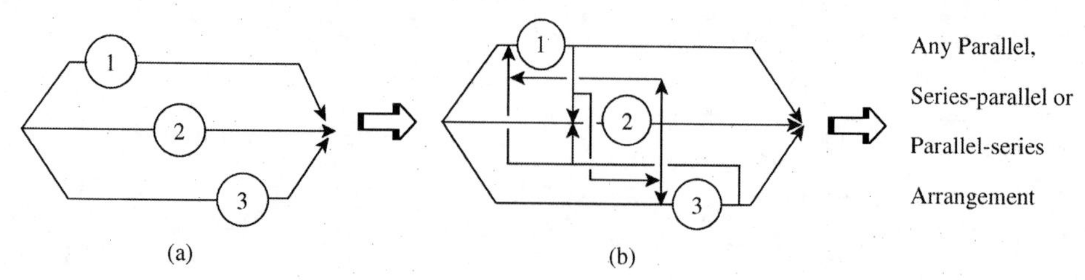

Figure 18.32 If the solution from the optimization of a simplified superstructure features parallel branches, then add additional structural features and re-optimize.

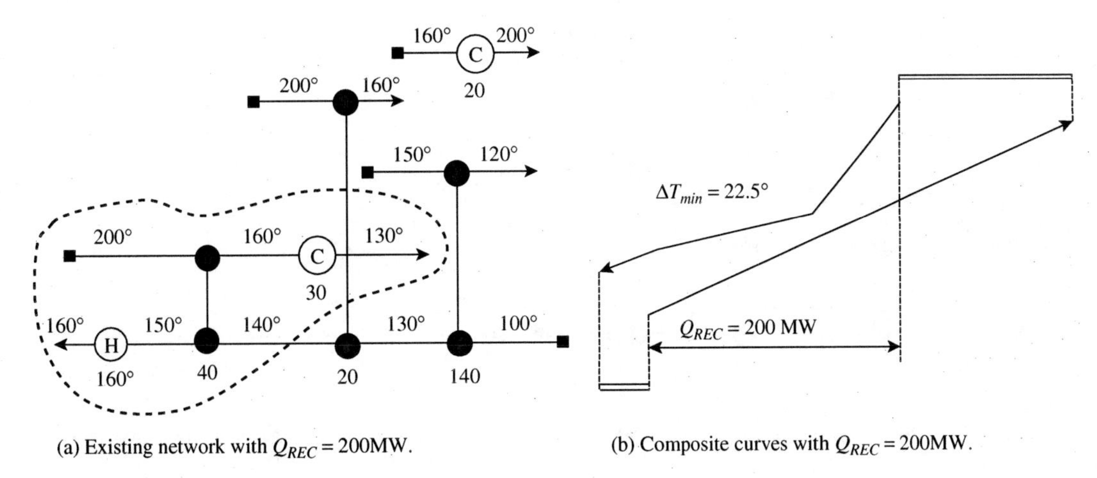

(a) Existing network with $Q_{REC} = 200$MW. (b) Composite curves with $Q_{REC} = 200$MW.

Figure 18.33 An existing heat exchanger network. (From Asante NDK and Zhu XX, 1997, *Trans IChemE*, **75A**: 349, reproduced by permission of the Institution of Chemical Engineers.)

of designing large networks with complex constraints, mixed materials of construction, equipment types and so on. One major disadvantage of the approach is that the optimization is a difficult MINLP. Another significant disadvantage is that, because a computer carries out the optimization, the designer is removed from the decision-making.

18.8 RETROFIT OF HEAT EXCHANGER NETWORKS

So far, all considerations regarding the targeting and design of heat exchanger networks have related to new, or grassroot, design. It is often necessary to retrofit an existing heat exchanger network. The need to retrofit might arise from a desire to reduce the utility consumption of the existing network, need to increase the throughput, modification of the feed to the process or a modification to the product specification. All of these objectives might require heat duties within the network to be changed.

One approach to retrofit would be to try and evolve the network toward an ideal grassroot design. Following this approach, stream data would be targeted using the composite curves or the problem table to determine the location of the pinch for an assumed ΔT_{min}. Knowing the location of the pinch, the existing units in the network could then be located relative to the pinch. Any cross-pinch heat transfer could then be eliminated by disconnecting the units that are transferring heat across the pinch. These could be process-to-process, or the inappropriate use of utilities (e.g. use of steam below the pinch). The network could then be reconnected, with as many features of the existing network retained as possible[13]. This would no doubt improve the energy performance of an existing heat exchanger network, but it has a number of fundamental problems:

- Which ΔT_{min} should be used? As the value of ΔT_{min} changes, the location of the pinch changes and therefore the assessment of which unit to transfer heat across the pinch also changes.
- Existing equipment is only reused in an ad hoc way.
- The retrofit is likely to involve a large number of modifications to the existing network.
- Constraints associated with the existing network are not readily included.

The problem with this approach is that it is attempting to change the network to a grassroot design rather than accepting the features that already exist. A better approach is to evolve the network from the existing structure in order to identify only the most critical, and therefore cost-effective, changes to the network structure[14].

Suppose that it is desired to reduce the energy consumption of the existing heat exchanger network shown in Figure 18.33a. Figure 18.33b shows the composite curves for the basic stream data. The composite curves have been adjusted such that the overlap between the composite curves corresponds with the actual existing heat recovery duty of 200 MW. For the composite curves, this corresponds with a ΔT_{min} of 22.5°C. On the other hand, it can be seen that the existing heat exchanger network has minimum temperature differences of 20°C. First consider how the energy consumption of the existing network might be decreased without changing the network structure. As discussed previously, this is only possible by the exploitation of a utility path. The existing network in Figure 18.33a has only one degree of freedom that can be exploited for network optimization. This is highlighted within the bubble superimposed on the network. It involves a utility path between the heater and the cooler. The matches in the network outside of this bubble are all constrained by the heat duties on individual streams. The only way, therefore, that the utility consumption of the existing network can be decreased is by shifting heat load along the path between the heater and the cooler.

Figure 18.34a shows the network evolved to reduce the utility consumption to the point where the temperature

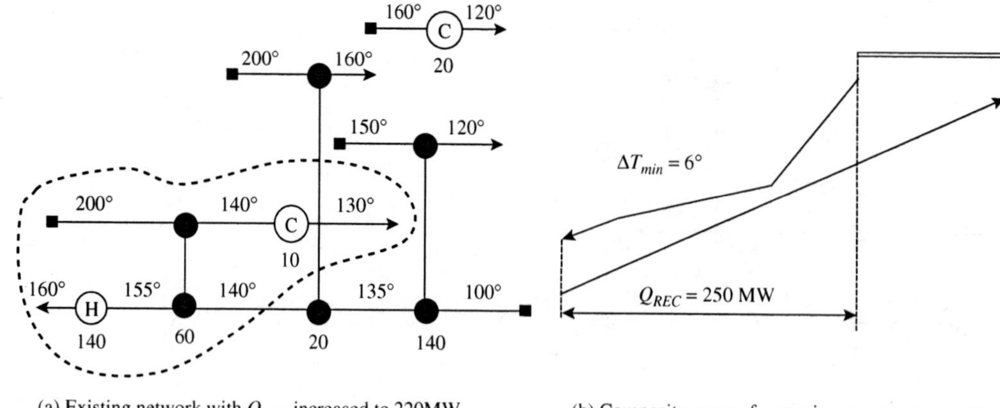

(a) Existing network with Q_{REC} increased to 220MW. (b) Composite curves for maximum energy recovery indicates $Q_{REC} = 250$MW.

Figure 18.34 Maximizing the energy recovery of the existing heat exchanger network even down to $\Delta T = 0°C$ gives an energy performance worse than the energy target. (From Asante NDK and Zhu XX, 1997, *Trans IChemE*, **75A**: 349, reproduced by permission of the Institution of Chemical Engineers.)

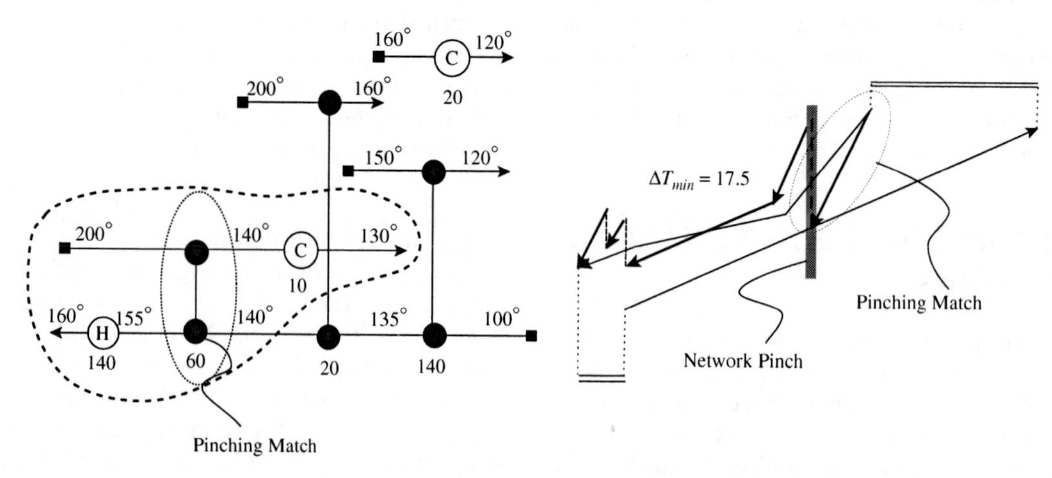

(a) Maximum heat recovery condition with $Q_{REC} = 220$ MW. (b) Composite curves with $Q_{REC} = 220$ MW.

Figure 18.35 The network pinch limits the energy recovery in the existing heat exchanger network. (From Asante NDK and Zhu XX, 1997, *Trans IChemE*, **75A**: 349, reproduced by permission of the Institution of Chemical Engineers.)

difference in the existing network is now 0°C[14]. This is not a limit that could be achieved in practice, but is taken here for the sake of illustration. For a minimum temperature difference of 0°C in the existing network, the energy recovery must increase to 220 MW. Compare the composite curves for increased heat recovery. These are shown in Figure 18.34b for their maximum overlap, which is 250 MW. At this setting, the composite curves still have a ΔT_{min} of 6°C. This reveals that the maximum heat recovery within the existing heat exchanger network structure is different from that theoretically possible from the composite curves. The difference is caused by the fact that the existing heat exchanger network structure is not appropriate for maximum energy recovery. How can the existing network structure be modified to improve its performance?

The existing heat exchanger network is shown again in Figure 18.35 with its recovery increased to an absolute maximum. Figure 18.35a highlights what is limiting the existing heat exchanger network in terms of its heat recovery. One of the existing units features minimum temperature difference (0°C in this case for the sake of illustration). The composite curves are also shown in Figure 18.35b, but set to the same overall heat recovery load as that featuring the existing heat exchanger network at its limit (220 MW). Superimposed on the composite curves are the temperature profiles for the hot streams in each of the existing units. One of the matches limits the heat recovery, in this case the one featuring a temperature difference of 0°C. This match that limits the heat recovery is known as the *pinching match*[14]. The point at which this occurs is known as the *network pinch*[14]. The network pinch limits the heat recovery for the existing heat exchanger network.

In practice, if the network pinch is being identified in a design study, then a practical minimum temperature difference of say 10°C or 20°C would be taken rather than the 0°C used here for the sake of illustration. The principle is exactly the same. One or more matches in the existing network feature ΔT_{min} when the network is pinched.

Consider now how the network pinch might be overcome. There are four ways in which the network pinch can be overcome and the performance of the existing heat exchanger network improved beyond that for the pinched condition[14].

a. *Resequencing*. Figure 18.36 illustrates how *resequencing* can be used to overcome the network pinch. Resequencing moves the unit to a new location in the network, but between the same streams as the original match. Figure 18.36 shows a cold stream being heated by two hot streams. One of the hot stream profiles indicates that it is a pinching match and features a minimum temperature difference of 0°C (again for the sake of illustration). In Figure 18.36, the pinching match is adjacent to another hot stream that has a finite temperature difference for the heat exchange. If the position of the two hot streams is swapped by a simple resequence, as shown in Figure 18.36, then the pinching match no longer limits. This means that there is now new scope to reduce the energy consumption of the network by exploiting a utility path, as shown in Figure 18.36. If the utility path is exploited to its limit, then a new network pinch is created, but now at a lower utility consumption for the network.

b. *Repiping*. *Repiping* is very similar to resequencing. Like resequencing, repiping moves the unit to a new location in the network. However, in repiping, the unit can be moved to a location involving streams other than in the original location, rather than be restricted to operate between the same streams. Repiping is a more general case than resequencing, but might not be practical for a variety of reasons, for example, materials of construction being unsuitable for other streams. The basic principle of repiping is the same as that for resequencing, but a distinction needs to be made for practical reasons.

c. *Inserting a new match*. Figure 18.37 shows how the network pinch can be overcome by inserting a new match. Again the principle is illustrated by two hot streams providing heat to a cold stream. One of the matches is pinched. If a new match is inserted such that the heat duty on the hot stream adjacent to the pinching match is decreased and replaced by the new match, then the position of the pinching match can be changed such that it is no longer pinching. This introduces scope to exploit the utility path to reduce the utility consumption of the network, until it is again pinched. The network is now pinched again, but at a lower utility consumption.

d. *Introduce additional stream splitting*. The fourth way to overcome the network pinch is by introducing additional stream splitting to the existing network, as illustrated in Figure 18.38. In this case, it can be seen that two matches are pinched simultaneously. By introducing a stream split, the cold stream profiles in the two pinched units are now such that one of the pinching matches is no longer pinched. This means that there is scope to exploit a utility path and reduce the energy consumption. This is being carried out to its limit in Figure 18.38 such that the network is again pinched, but at a lower utility consumption.

Again, it should be emphasized that in practice a finite practical ΔT_{min} should be used rather than 0°C. However, any assumption of ΔT_{min} to identify, and then overcome, the network pinch does not guarantee

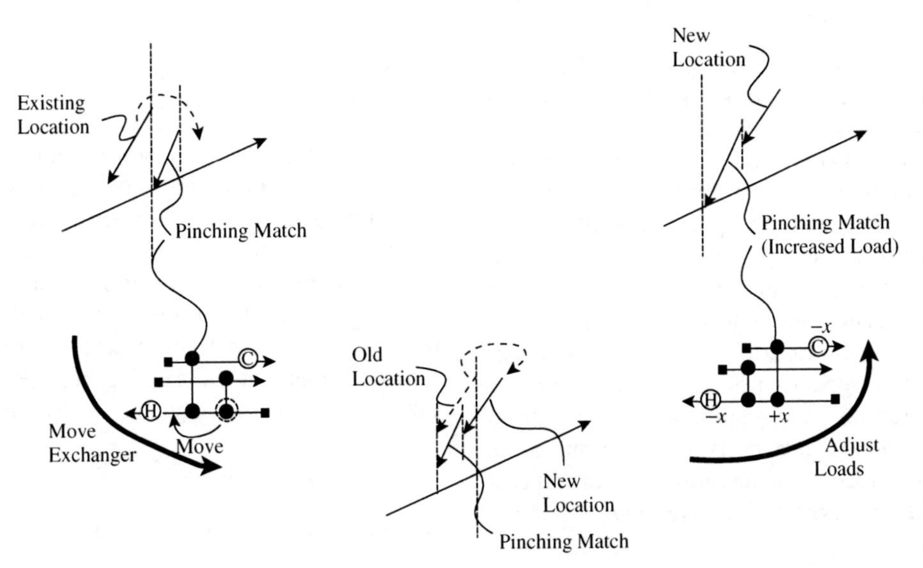

Figure 18.36 Resequencing a match can be used to overcome the network pinch. (From Asante NDK and Zhu XX, 1997, *Trans IChemE*, **75A**: 349, reproduced by permission of the Institution of Chemical Engineers.)

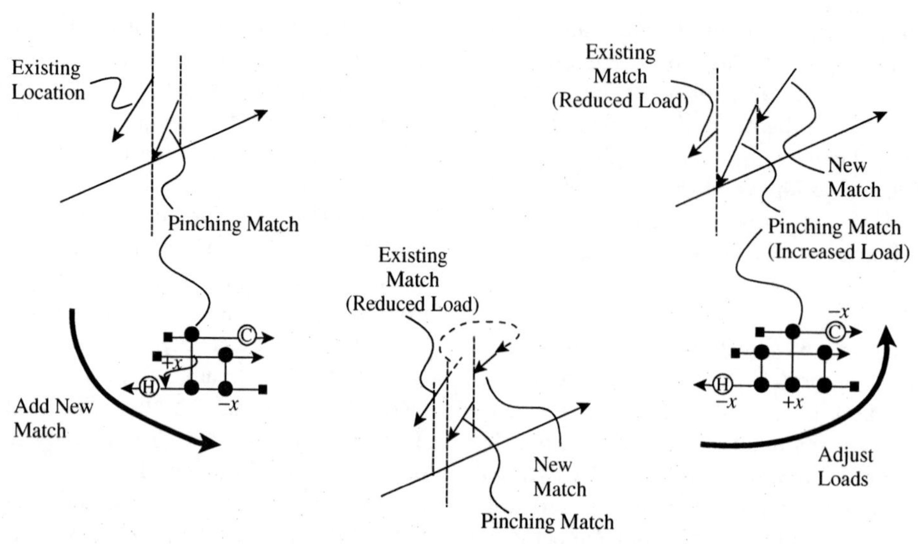

Figure 18.37 Adding a new heat exchanger can be used to overcome the heat network pinch. (From Asante NDK and Zhu XX, 1997, *Trans IChemE*, **75A**: 349, reproduced by permission of the Institution of Chemical Engineers.)

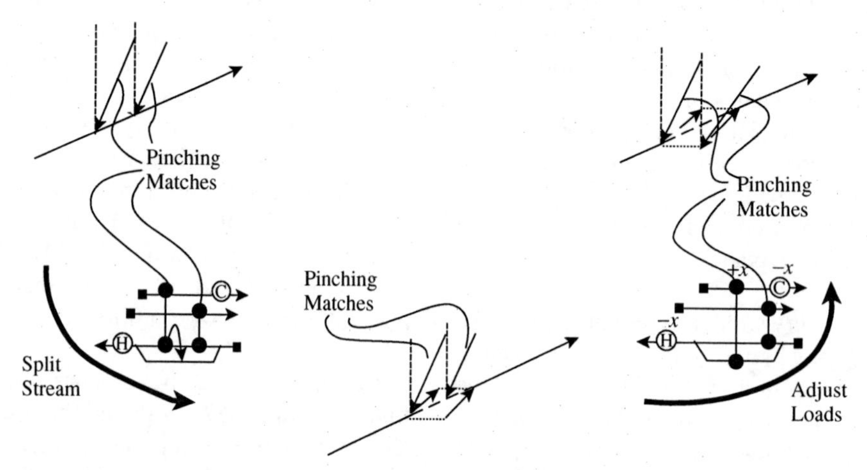

Figure 18.38 Changing the stream splitting arrangement can be used to overcome the network pinch. (From Asante NDK and Zhu XX, 1997, *Trans IChemE*, **75A**: 349, reproduced by permission of the Institution of Chemical Engineers.)

an optimum retrofit. The modified network should be subjected to cost optimization in order to obtain the correct setting for the capital-energy trade-off. The optimization of a heat exchanger network with a fixed structure is a Nonlinear Programming (NLP) optimization, as discussed previously. When dealing with the optimization of existing heat exchanger networks, it needs to be recognized that there is existing heat transfer area in place, but new area (or enhanced heat transfer) needs to be installed in some parts of the network to improve the network performance. The existing heat transfer area has zero capital cost and only the new heat transfer area and pipework modifications need to be included in the capital costs for the optimization. Care is required when specifying the form of the capital cost correlation in retrofit. Designers often like to use a cost per unit area, such as:

$$Capital\ Cost = bA \qquad (18.6)$$

where A = heat transfer area
b = cost coefficient

If the capital cost of new heat transfer area is expressed in the form of Equation 18.6, then this will lead to poor retrofit projects. The problem with Equation 18.6 is that the optimization is likely to spread the new heat transfer area in the network in many locations, without incurring a cost penalty associated with the many modifications that would result. To ensure that new heat transfer area is not spread around throughout the existing heat exchanger network, a capital cost correlation should be used that is of the form:

$$Capital\ Cost = a + bA^c \qquad (18.7)$$

where a, b, c = cost coefficients

In Equation 18.7, the coefficient a is a threshold cost that is incurred even if a small amount of heat transfer area

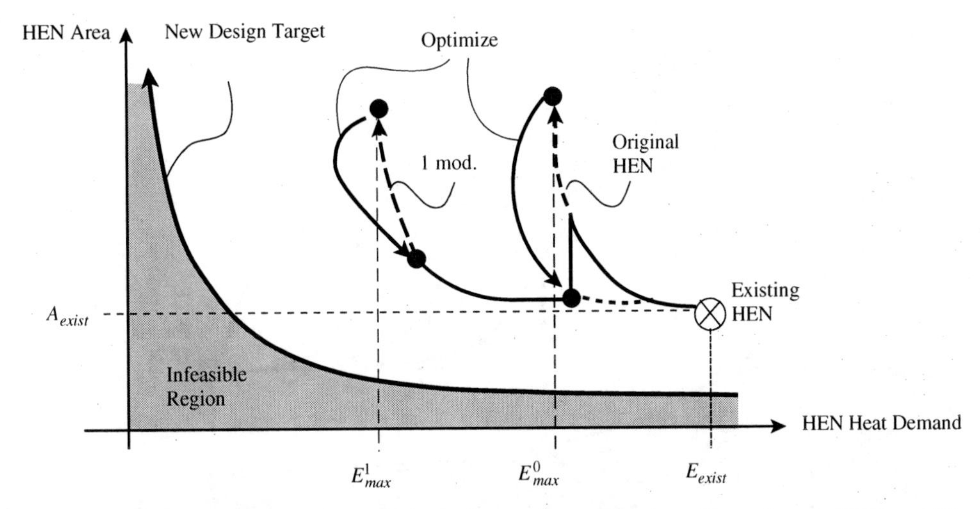

Figure 18.39 The heat exchanger network can be optimized after each modification. (From Asante NDK and Zhu XX, 1997, *Trans IChemE*, **75A**: 349, reproduced by permission of the Institution of Chemical Engineers.)

is installed. If a large threshold cost is used in the cost correlation, then this would lead an optimizer to attempt to concentrate any new heat transfer area required in as few locations as possible in the network. This, in turn, will lead to fewer modifications in the retrofit project to accommodate the new heat transfer area requirements.

It should be noted that a resequence or repipe does not involve zero capital cost, even though no new heat exchanger equipment might be purchased. The pipework modifications for a resequence or repipe might be very expensive. Also, equipment might need to be relocated. Methods for capital cost estimation for retrofit were discussed in Chapter 2.

Figure 18.39 shows the capital-energy trade-off on a plot of heat exchanger network area versus energy consumption. The new (grassroot) design target marks the infeasible region in Figure 18.39. Targeting methods for energy were discussed in Chapter 16 and targeting methods for network area were discussed in Chapter 17. The existing network performance cannot be better than the target, but it can be worse, as illustrated in Figure 18.39. If the existing network structure is retained, then there is a limit beyond which the network cannot be improved in terms of its energy performance. This is when the existing network structure is pinched at a practical minimum temperature difference for the heat exchangers. In order to approach this condition, it will require an additional heat transfer area. The more the energy consumption is decreased, the additional heat transfer area required will increase per unit energy reduction, as the network pinch is approached. For the existing heat exchanger network structure, the capital-energy trade-off can be optimized, as shown in Figure 18.39. Alternatively, the structure of the network could be modified, in which case, the limit for heat recovery for the modified network is now at a lower energy consumption, Figure 18.39. Again, the network has been modified on the basis of an assumed practical minimum temperature difference that is not necessarily providing the correct setting for the capital-energy trade-off. Therefore, the modified network can be optimized to provide the optimized network for one structural modification. This procedure could be repeated for a second and third modification, and so on. In practice, retrofits are usually only economic for a small number of network modifications.

This approach to heat exchanger network retrofit based on the concept of the network pinch can be automated[15]. A superstructure can be created for network resequencing (or repiping) and the best resequence (or repipe) identified by optimizing the superstructure of resequences (or repipes) for an assumed practical ΔT_{min}[15]. This can be formulated as an MILP optimization if the network is optimized for minimum energy consumption with a fixed ΔT_{min}. Inclusion of area calculations to estimate heat exchanger capital costs would make the optimization nonlinear. Similarly, a superstructure can be created for positioning new matches or stream split modifications in the existing network and the superstructure optimized for minimum energy cost for an assumed practical ΔT_{min}. The problem can again be formulated as an MILP problem[15].

After each suggested modification has been identified, the network can be subjected to a detailed capital-energy trade-off requiring NLP optimization. Optimization of the capital-energy trade-off is illustrated in Figure 18.40. Structural modifications are first explored using MILP and then the correct setting of the capital-energy trade-off corrected using NLP. By decomposing the problem in this way, what is overall an MINLP problem, is carried out by MILP followed by NLP, which is a more robust approach to the optimization.

This approach to heat exchanger network retrofit allows modifications to be introduced one at a time. In this way, the designer has control over the complexity of the network retrofit. At each stage, a suggested modification can be

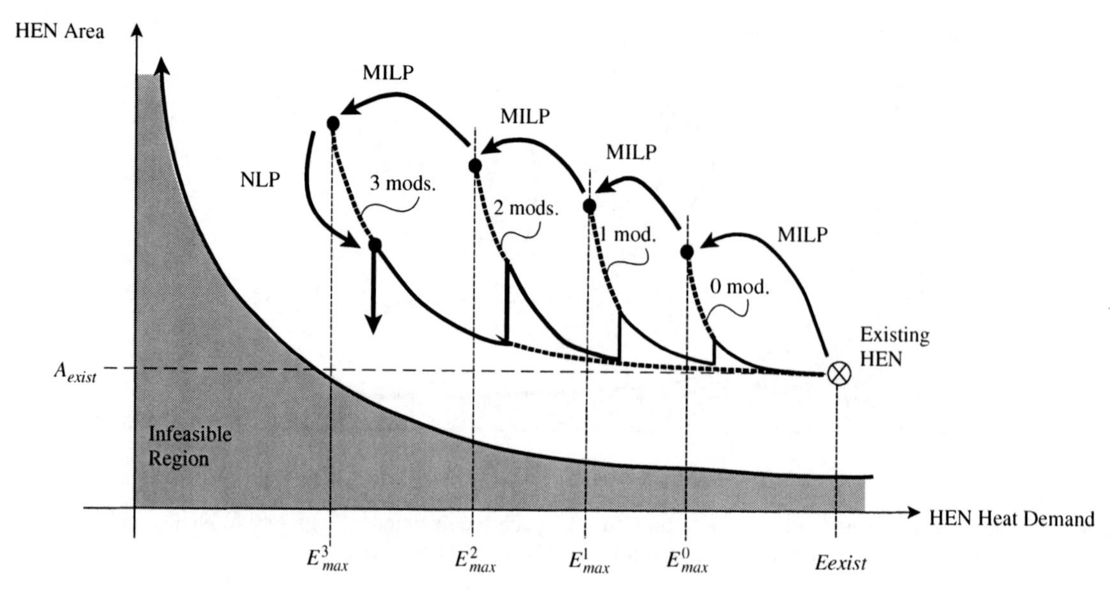

Figure 18.40 The retrofit can be proceed in a stepwise approach. (From Asante NDK and Zhu XX, 1997, *Trans IChemE*, **75A**: 349, reproduced by permission of the Institution of Chemical Engineers.)

evaluated in terms of its practicality and true costs. If the best structural option identified is considered to be impractical (e.g. the plant is too congested), then another modification can be suggested instead. For example, the optimization could be carried out to identify the best resequence, only to find that this is impractical. If this is the case, then the second best resequence could be taken, or the option of a new match chosen and so on. The designer has control and can assess each modification as is suggested by the optimization. Having identified an acceptable number of modifications, the modifications are subjected to a detailed capital-energy trade-off to assess their true economic performance.

The approach leads to simple and practical retrofit designs and has the major advantage that it allows the designer to assess modifications one at a time and to keep control over the complexity of the retrofit. Its disadvantage is that different combinations of modifications can be taken and there is no guarantee that this will lead to an optimum network retrofit. However, it is almost impossible to say that any retrofit is optimal or nonoptimal. The features of each retrofit are unique and it is difficult to formulate all the constraints for a retrofit in order to guarantee the very best retrofit.

18.9 ADDITION OF NEW HEAT TRANSFER AREA IN RETROFIT

Retrofit of heat exchanger networks requires resequencing and repiping of existing heat exchangers, installation of new heat exchange matches and changes to stream splitting arrangements. However, existing matches that have not been moved might require additional heat transfer area as a result of changes in the operation. This might result from increased heat duty, operation under reduced temperature differences or operation under reduced heat transfer coefficients. Methods for the retrofit of heat exchangers have been discussed in Chapter 15. If shell-and-tube heat exchangers are being used and additional heat transfer area is required, it might be possible to install a new tube bundle into existing shells if the additional area requirement is small. If this is not feasible, then the existing area can be supplemented by adding a new shell (or more than one shell if there is a large area requirement) in:

a. *Series*. The addition of new exchangers in series to the existing match will lead to an increase in the overall pressure drop across the match. This might be important if the pump (or compressor) is close to its maximum capacity.

b. *Parallel*. The addition of new exchangers in parallel will leave the pressure difference largely unchanged. The pressure drop across the match will be the largest between that of the existing exchangers under the new conditions and new exchangers installed in parallel.

Also, as discussed in Chapter 15, rather than install additional heat transfer area to cater for the new operational requirements, heat transfer enhancement can be considered. Changes to the number of tube passes or the baffle arrangement might allow the heat transfer coefficient to be enhanced. Alternatively, tube inserts could be used. This was discussed in Chapter 15. The major disadvantage in using heat transfer enhancement is that it increases the pressure drop. In retrofit this can be important, as the pumps driving the flow might be limited in their capacity to meet the required increase in pressure drop.

Rather than shell-and-tube heat exchangers, the network might involve plate heat exchangers. Adding additional area to plate heat exchangers is generally more straightforward than shell-and-tube exchangers. Plate heat exchangers can often have their area increased by adding additional plates to the existing frame. If an additional frame needs to be added to provide the additional area, then the new frame can be added in series or parallel with the existing frame. Series arrangements increase the pressure drop. Parallel arrangements decrease the flowrate through each frame but decrease the heat transfer coefficient in the existing frame. The pressure drop across the match will be the largest between that of the existing exchangers under the new conditions and new exchangers installed in parallel.

18.10 HEAT EXCHANGER NETWORK DESIGN – SUMMARY

A good initialization for heat exchanger network design is to assume that no individual exchanger should have a temperature difference smaller than ΔT_{min}. Having decided that no exchanger should have a temperature difference smaller than ΔT_{min} two rules were deduced in Chapter 16. If the energy target set by the composite curves (or the problem table algorithm) is to be achieved, there must be no transfer heat across the pinch by:

- Process-to-process heat transfer
- Inappropriate use of utilities.

These rules are both necessary and sufficient for the design to achieve the energy target given that no individual exchanger should have a temperature difference smaller than ΔT_{min}.

The design of heat exchanger networks can be summarized in five steps.

1. Divide the problem at the pinch into separate problems.
2. The design for the separate problems is started at the pinch, moving away.
3. Temperature feasibility requires constraints on the *CP*'s to be satisfied for matches between the streams at the pinch.
4. The loads on individual units are determined using the tick-off heuristic to minimize the number of units. Occasionally, the heuristic causes problems.
5. Away from the pinch, there is usually more freedom in the choice of matches. In this case, the designer can discriminate on the basis of operability, plant layout and so on.

If the number of hot streams at the pinch above the pinch is greater than the number of cold streams, then the cold streams must be split to satisfy the ΔT_{min} constraint. If the number of cold streams at the pinch below the pinch is greater than the number of hot streams, then the hot streams must be split to satisfy the ΔT_{min} constraints. If the *CP* inequalities for all streams at the pinch cannot be satisfied, this also can necessitate stream splitting.

If the problem involves more than one pinch, then between pinches the design should be started from the most constrained pinch.

Remaining problem analysis can be used to make a quantitative assessment of matches in the context of the whole network without having to complete the network.

Once the initial network structure has been defined, then loops, utility paths and stream splits offer the degrees of freedom for manipulating network cost in multivariable continuous optimization. When the design is optimized, any constraint that temperature differences should be larger than ΔT_{min} or that there should not be heat transfer across the pinch no longer applies. The objective is simply to design for minimum total cost.

For more complex network designs, especially those involving many constraints, mixed equipment specifications and so on, design methods based on the optimization of a superstructure can be used.

Network retrofit can be performed on the basis of the concept of the network pinch. Resequencing and repiping of heat exchangers, the introduction of new exchangers and additional stream splitting can all be used to overcome the network pinch. The identification of structural modifications to the network can be done on the basis of minimizing the energy consumption for a ΔT_{min} using MILP. Once structural modifications have been identified, these should be subjected to a detailed capital-energy trade-off using NLP optimization. New heat transfer area to existing matches can be added through the addition of new heat exchanger shells in series or parallel or through heat transfer enhancement to existing exchangers.

18.11 EXERCISES

1. The stream data for a process are given in the Table 18.6:

Table 18.6 Stream data for Exercise 1.

Stream No.	Stream Type	Supply temperature (°C)	Target temperature (°C)	Heat capacity flowrate ($MW \cdot K^{-1}$)
1	Hot	200	100	0.4
2	Hot	200	100	0.2
3	Hot	150	60	1.2
4	Cold	50	140	1.1
5	Cold	80	120	2.4

a. For a ΔT_{min} of 10°C, the pinch is at 90°C for hot streams and 80°C for cold streams. Design a heat exchanger network for maximum energy recovery that features the minimum number of units.
b. By relaxing the constraint of $\Delta T_{min} = 10$°C, shift heat load through a utility path such that the network uses only hot utility and no cold utility at all.

2. The process stream data for a heat recovery network problem are given in Table 18.7

Table 18.7 Stream data for Exercise 2.

Stream	T_S (°C)	T_T (°C)	CP (MW·K^{-1})
1. Hot	300	80	0.15
2. Hot	200	40	0.225
3. Cold	40	180	0.2
4. Cold	140	280	0.3

a. Determine the energy targets for hot and cold utility for $\Delta T_{min} = 20$°C.
b. Design a heat exchanger network to achieve the energy targets in the minimum number of units.
c. Identify the scope to reduce the number of units by manipulating loops and utility paths by sacrificing energy consumption.
d. Design a network to realize this scope and restore $\Delta T_{min} =$ 20°C.

3. The heat recovery stream data for a process are given in Table 18.8.

Table 18.8 Stream data for Exercise 3.

Stream	T_S (°C)	T_T (°C)	CP (kW·K^{-1})
1. Hot	180	40	200
2. Hot	150	40	400
3. Cold	60	180	300
4. Cold	30	130	220

A problem table analysis for $\Delta T_{min} = 10$°C results in the heat recovery cascade given in Table 18.9.

Table 18.9 Heat recovery cascade for Exercise 3.

Interval temperature (°C)	Heat flow (MW)
185	6.0
175	3.0
145	0
135	3.0
65	8.6
35	20.0

a. Design a heat recovery network in the minimum number of units. The number of units is below what might have been expected from the target for the number of units; why?
b. Low-pressure saturated steam can be generated from saturated boiler feedwater at an interval temperature of 115°C. Determine how much low-pressure steam can be generated and design a network to achieve this duty.
c. The network design from Part *b* results in two steam generators. Evolve the network to remove one of these by suffering a penalty in steam generation, but maintain $\Delta T_{min} =$ 10°C.

4. The process stream data for a heat recover problem are given in Table 18.10.

Table 18.10 Stream data for Exercise 4.

Stream	T_S (°C)	T_T (°C)	CP (MW·K^{-1})
1. Cold	18	123	0.0933
2. Cold	118	193	0.1961
3. Cold	189	286	0.1796
4. Hot	159	77	0.2285
5. Hot	267	80	0.0204
6. Hot	343	90	0.0538

A problem table analysis reveals that $Q_{Hmin} = 13.95$ MW, $Q_{Cmin} = 8.18$ MW, hot stream pinch temperature is 159°C and cold stream pinch temperature is 149°C for $\Delta T_{min} = 10$°C.

a. Design a heat recovery network with the minimum number of units. (Hint: below the pinch it may be necessary to split a stream away from the pinch to achieve the minimum number of units).
b. Evolve the design to eliminate stream splits below the pinch, while maintaining the same number of units, by allowing a temperature difference slightly smaller than 10°C.

5. The data for a heat recovery problem are given in the Table 18.11.

Table 18.11 Stream data for Exercise 5.

Stream		T_S	T_T	Heat capacity
		(°C)	(°C)	flowrate
No.	Type			(MW·K^{-1})
1	Hot	120	65	0.5
2	Hot	80	50	0.3
3	Hot	135	110	0.29
4	Hot	220	95	0.02
5	Hot	135	105	0.26
6	Cold	65	90	0.15
7	Cold	75	200	0.14
8	Cold	30	210	0.1
9	Cold	60	140	0.05
Steam		250	–	–
Cooling water		15	–	–

The problem table cascade is given in Table 18.12 for $\Delta T_{min} =$ 20°C.

Given this data:

a. Set out the stream grid.
b. Design a maximum energy recovery network.

Table 18.12 Problem table cascade for Exercise 5.

Interval temperature (°C)	Heat flow (MW)
220	20.95
210	19.95
150	6.75
125	0
110	4.2
100	12
95	13.7
85	14.5
75	16.5
70	18.25
55	28.75
40	31.75

Table 18.13 Stream data for Exercise 6.

Stream No.	Stream Type	T_s (°C)	T_T (°C)	Stream heat duty (MW)
1	Hot	150	50	−20
2	Hot	170	40	−13
3	Cold	50	120	21
4	Cold	80	110	15

6. The stream data for a process are given in the Table 18.13. Steam is available between 180 and 179°C and cooling water between 20 and 40°C. For $\Delta T_{min} = 10$°C, the minimum hot and cold utility duties are 7 MW and 4 MW respectively. The pinch is at 90°C on the hot streams and 80°C on the cold streams.
   a. Calculate the target for the minimum number of units for maximum energy recovery.
   b. Develop two alternative maximum energy recovery designs, keeping units to a minimum.
   c. Explain why the design below the pinch cannot achieve the target for the minimum number of units.
   d. How many degrees of freedom are available for network optimization?
7. The stream data for a process are given in the Table 18.14:

Table 18.14 Stream data for Exercise 7.

Stream No.	Stream Type	T_s (°C)	T_T (°C)	CP (kW·K^{-1})
1	Hot	170	88	23
2	Hot	278	90	2
3	Hot	354	100	5
4	Cold	30	135	9
5	Cold	130	205	20
6	Cold	200	298	18

Table 18.15 Heat flow cascade for Exercise 7.

Interval temperature (°C)	Heat flow (kW)
349	1528
303	1758
273	1368
210	675
205	520
165	0
140	250
135	255
95	1095
85	1255
83	1283
35	851

A problem table heat cascade for $\Delta T_{min} = 10$°C is given the Table 18.15. Hot utility is to be provided by a hot oil circuit with a supply temperature of 400°C. Cooling water is available at 20°C.
   a. Calculate the minimum flowrate of hot oil if C_P for the hot oil is 2.1 kJ·kg^{-1}·K^{-1}, assuming ΔT_{min} process-to-process heat recovery is 10°C and process to hot oil to be 20°C.
   b. Design a heat exchanger network for maximum energy recovery in the minimum number of units ensuring $\Delta T_{min} = 10$°C for all process-to-process heat exchangers throughout the network.
   c. Suggest an alternative network design below the pinch to eliminate stream splits by accepting a violation of ΔT_{min}.
   d. What tools could have been used to develop the design below the pinch more systematically?
8. The stream data for a process are given in Table 18.16

Table 18.16 Stream data for Exercise 8.

Stream No.	Stream Type	T_S (°C)	T_T (°C)	Heat duty (MW)
1	Hot	150	30	7.2
2	Hot	40	40	10
3	Hot	130	100	3
4	Cold	150	150	10
5	Cold	50	140	3.6

   a. Sketch the composite curves for $\Delta T_{min} = 10$°C.
   b. Determine the target for hot and cold utility for $\Delta T_{min} =$ 10°C.
   c. Design a maximum energy recovery network in the minimum number of units for $\Delta T_{min} = 10$°C.
   d. Can the number of units be reduced by evolution of the network?
9. The stream data for a process are given in Table 18.17. $\Delta T_{THRESHOLD}$ for the problem is 50°C and ΔT_{min} is 20°C. A problem table analysis on this data produces the cascade given in Table 18.18 for $\Delta T_{min} =$ 20°C.

Table 18.17 Stream data for exercise 9.

Stream No.	Type	T_S (°C)	T_T (°C)	Heat capacity flowrate (MW·K^{-1})
1	Hot	500	100	4
2	Cold	50	450	1
3	Cold	60	400	1
4	Cold	40	420	0.75

Table 18.18 Heat flow cascade for exercise 9.

Interval temperature (°C)	Heat flow (MW)
490	0
460	120
430	210
410	255
90	655
70	600
60	582
50	575

a. Design a heat exchanger network for this problem that achieves maximum energy recovery in the minimum number of units.
b. Determine how much steam at a condensing temperature of 180°C can be generated by this process
c. Sketch the composite curves for the process showing maximum steam generation at 180°C.
d. Design a network that achieves maximum energy recovery in the minimum number of units and which generates the maximum possible steam at 180°C.

REFERENCES

1. Linnhoff B and Hindmarsh E (1983) The Pinch Design Method of Heat Exchanger Networks, *Chem Eng Sci*, **38**: 745.
2. Linnhoff B, Townsend DW and Boland D, Hewitt GF, Thomas BEA, Guy AR and Marsland RH (1982) *A User Guide on Process Integration for the Efficient Use of Energy*, IChemE, Rugby, UK.
3. Linnhoff B and Ahmad S (1990) Cost Optimum Heat Exchanger Networks: I. Minimum Energy and Capital Using Simple Models for Capital Cost, *Comp Chem Eng*, **14**: 729.
4. Saboo AK, Morari M and Colberg RD (1986) RESHEX: An Interactive Software Package for the Synthesis and Analysis of Resilient Heat Exchanger Networks: II. Discussion of area Targeting and Network Synthesis Algorithms, *Comp Chem Eng*, **10**: 591.
5. Ahmad S, Linnhoff B and Smith R (1990) Cost Optimum Heat Exchanger Networks: II. Targets and Design for Detailed Capital Cost Models, *Comp Chem Eng*, **14**: 751.
6. Gunderson T and Naess L (1988) The Synthesis of Cost Optimal Heat Exchanger Networks: An Industrial Review of the State of the Art, *Comp Chem Eng*, **12**: 503.
7. Floudas CA, Ciric AR and Grossmann IE (1986) Automatic Synthesis of Optimum Heat Exchanger Network Configurations, *AIChE J*, **32**: 276.
8. Grossmann IE (1990) Mixed-Integer Non-Linear Programming Techniques for the Synthesis of Engineering Systems, *Res Eng Des*, **1**: 205.
9. Grossmann IE and Sargent RWH (1978) Optimum Design of Heat Exchanger Networks, *Comp Chem Eng*, **2**: 1.
10. Yee TF, Grossmann IE and Kravanja Z (1990) Simultaneous Optimization Models for Heat Integration – I. Area and Energy Targeting of Modeling of Multi-stream Exchangers, *Comp Chem Eng*, **14**: 1151.
11. Zhu XX, O'Neill BK, Roach JR and Wood RM (1995) New Method for Heat Exchanger Network Synthesis Using Area Targeting Procedures, *Comp Chem Eng*, **19**: 197.
12. Dolan WB, Cummings PT and Le Van MD (1990) Algorithm Efficiency of Simulated Annealing for Heat Exchanger Network Design, *Comp Chem Eng*, **14**: 1039.
13. Tjoe TN and Linnhoff B (1986) Using Pinch Technology for Process Retrofit, *Chem Eng*, **April**: 47.
14. Asante NDK and Zhu XX (1997) An Automated and Interactive Approach for Heat Exchanger Network Retrofit, *Trans IChemE*, **75**: 349.
15. Zhu XX and Asante NDK (1999) Diagnosis and Optimization Approach for Heat Exchanger Network Retrofit, *AIChE J*, **45**: 1488.

19 Heat Exchanger Networks V – Stream Data

The heat exchanger network targeting and design methods presented in Chapters 16, 17 and 18 maximize heat recovery for a given set of process conditions. However, before any analysis can be performed, the material and energy balance needs to be represented as a set of hot and cold streams. This is often not straightforward. Firstly, the process conditions are often not fixed rigidly. Process conditions such as pressures, temperatures and flowrates might have the freedom to be changed within certain limits. The flexibility to change the processing conditions, where this is possible, can be exploited to improve the heat recovery further. Also, even if the process conditions are fixed for a given process flowsheet and material and energy balance, it is still not straightforward to interpret the flowsheet as a set of hot and cold streams.

Consider first the issue of changing the process conditions and how those changes might be directed to improve the heat recovery.

19.1 PROCESS CHANGES FOR HEAT INTEGRATION

Consider the composite curves in Figure 19.1a. Any process change that[1,2]:

- increases the total hot stream heat duty above the pinch;
- decreases the total cold stream heat duty above the pinch;
- decreases the total hot stream heat duty below the pinch;
- increases the total cold stream heat duty below the pinch

will bring about a decrease in utility requirements. This is known as the *plus–minus principle*[3,4]. These simple guidelines provide a reference for appropriate design changes to improve the targets. The changes apply throughout the process to reactors, recycle flowrates, distillation columns, and so on.

If a process change, such as a change in distillation column pressure, allows shifting a hot stream from below the pinch to above, it has the effect of increasing the overall hot stream duty above the pinch and therefore decreasing hot utility. Simultaneously, it decreases the overall hot stream duty below the pinch and decreases the cold utility. Shifting a cold stream from above the pinch to below decreases the overall cold stream duty above the pinch, decreasing the hot utility, and increases the overall cold stream duty below the pinch, reducing the cold utility. Thus, one way to implement the plus–minus principle, as illustrated in Figure 19.1b, is[5]:

- shifting hot streams from below to above the pinch, or
- shifting cold streams from above to below the pinch.

Another way to relate these principles is to remember that heat integration will always benefit by keeping hot streams hot and keeping cold streams cold[5].

19.2 THE TRADE-OFFS BETWEEN PROCESS CHANGES, UTILITY SELECTION, ENERGY COST AND CAPITAL COST

Although the plus–minus principle is the basic reference in guiding process changes to reduce utility costs, it takes no account of capital costs. Process changes to reduce utility consumption will normally bring about a reduction in temperature driving forces as indicated in Figure 19.1. Thus, the capital-energy trade-off (and hence ΔT_{min}) should be readjusted after process changes.

Having to readjust the capital-energy trade-off after every process change would be a real problem if it were not for the existence of total cost targeting procedures discussed in Chapter 17.

In addition, the decrease in driving forces in Figure 19.1 caused by the process changes also affects the potential for using multiple utilities. For example, as the driving forces above the pinch become smaller, the potential to switch duty from high-pressure to low-pressure steam, as discussed in Section 16.7, decreases. Process changes are competing with better choice of utility levels, heat engines and heat pumps for available spare driving forces. Each time either a process change or a different choice of utilities is suggested, the capital-energy trade-off should be readjusted. If multiple utilities are used, the optimization of the capital-energy trade-off is not straightforward, since each pinch (process and utility) can have its own value of ΔT_{min}. The optimization thus becomes multidimensional.

Chemical Process Design and Integration R. Smith

ISBNs: 0-471-48680-9 (HB); 0-471-48681-7 (PB)

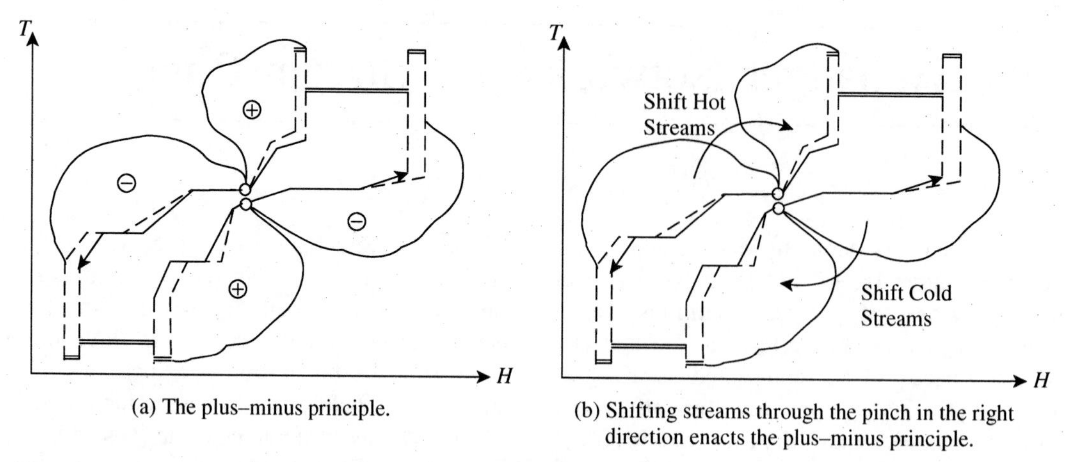

Figure 19.1 The plus–minus principle guides process changes to reduce utility consumption. (From Smith R and Linnhoff B, 1988, *Trans IChemE ChERD*, **66**: 195 reproduced by permission of the Institution of Chemical Engineers.)

19.3 DATA EXTRACTION

Having discussed the way in which changes to the basic stream data can improve targets, an even more fundamental question now needs to be addressed. Before any heat integration analysis can be carried out, the basic stream data needs to be extracted from the material and energy balance. In some cases, the representation of the stream data from the material and energy balance is straightforward. However, there are a number of pitfalls that can lead to errors and missed opportunities. Missed opportunities can arise through extracting too many constraints.

Consider now the basic principles of data extraction for heat integration.

1. *Stream identification.* Figure 19.2a shows part of a flowsheet in which a feed stream is heated from 10 to 70°C before being filtered. After the filter, it is heated from 70°C to 135°C and then from 135°C to 200°C before it is fed to a distillation column. A fundamental question for

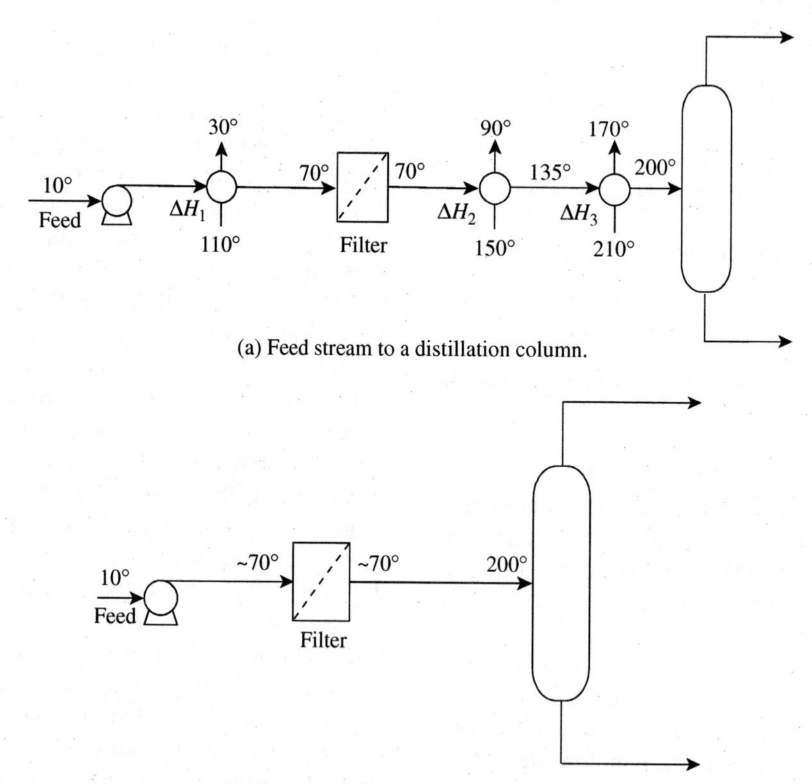

Figure 19.2 Stream identification.

representation of stream data for heat integration is "how many streams are there in this part of the flowsheet?" It could be assumed that there is a stream from 10°C to 70°C, another from 70°C to 135°C and a third from 135°C to 200°C. These are three cold streams. There are also three hot streams currently preheating the cold streams. One stream is cooled from 110°C to 30°C, another from 150°C to 90°C and a third from 210°C to 170°C. If the data are extracted in this way, then there would be some very neat matches between hot and cold streams: the ones that already exist. Extracting the data in this way does not seem to open up any opportunities for improvement. Fundamental questions need to be asked about what exactly is essential as far as the stream data are concerned? Heating the outlet of the filter from 70°C to 135°C is part of a previously suggested solution. It is not a constraint that the process heating should stop at 135°C. The feed stream needs to be heated to 200°C, but the midpoint at 135°C is not a constraint. The feed is heated from 10°C to 70°C before entering a filter. It may be that the temperature at which the filtration can take place has some flexibility and does not need to be rigidly at 70°C. Thus, for this problem, it would seem to be appropriate to extract the feed stream to the distillation as two streams, one from 10°C to 70°C and a second from 70°C to 200°C. The operation of the filter at 70°C is kept flexible, as shown in Figure 19.2b.

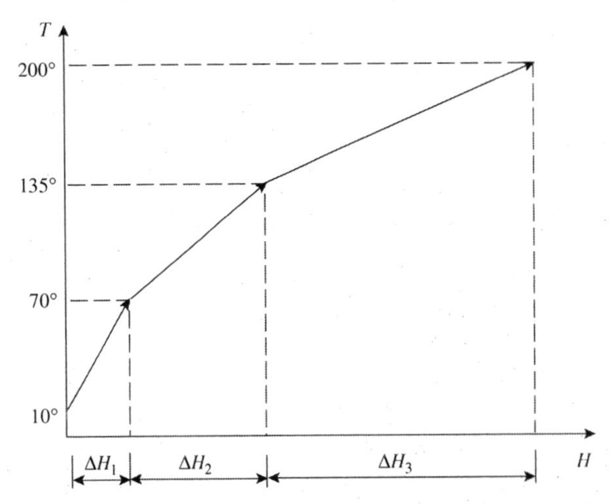

Figure 19.3 Estimates of temperature-enthalpy profiles from existing exchanger heat duties and temperature.

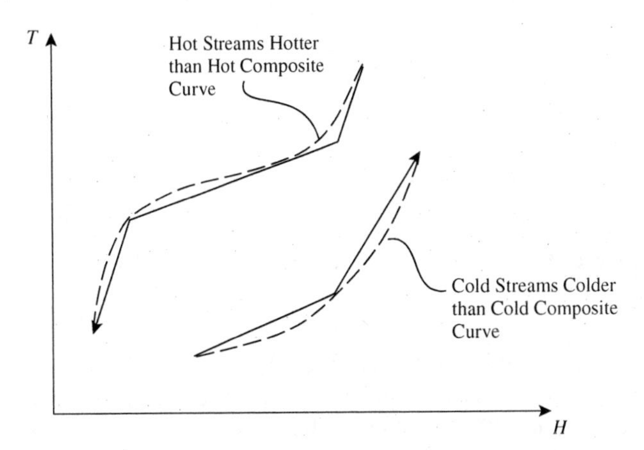

Figure 19.4 Linearization of nonlinear temperature-enthalpy profiles.

2. *Temperature-enthalpy profiles*. Targeting of heat exchanger networks has assumed that the heat capacities of the streams are constant, leading to straight lines when plotted on a temperature-enthalpy plot. However, heat capacities are often not constant and this must somehow be represented, otherwise serious errors might occur. Consider again the feed preheat for the distillation shown in Figure 19.2a. If the physical properties are known for the stream, then the temperature-enthalpy profile of the feed stream can be obtained from physical property correlations. This is one possibility. Another possibility is to represent the temperature-enthalpy profiles approximately, as illustrated in Figure 19.3. The known temperatures and heat duties for the existing heat exchangers can be plotted on a temperature-enthalpy profile as shown in Figure 19.3. This shows a nonlinear temperature-enthalpy profile taken directly from the flowsheet but represented as three linear *segments*. Such an approximation is good enough for many purposes. It is also a particularly convenient approach to adopt when dealing with retrofit of existing flowsheets.

Suppose that the actual behavior of temperature versus enthalpy is known and is highly nonlinear, as shown in Figure 19.4. How can the nonlinear data be linearized so that the construction of composite curves and the problem table algorithm can be performed? Figure 19.4 shows the nonlinear streams being represented by a series of linear segments. The linearization of the hot streams should be carried out on the underside (low-temperature side) of the curve and the linearization of the cold streams on the upper side (high-temperature side) of the curve. This is a conservative way to represent the nonlinear data, as the linearizations will come closer than the actual curves.

There is a great temptation to use a large number of linear segments to represent nonlinear stream data. This is rarely necessary. Most nonlinear stream data can be represented reasonably well by two or three linear segments, as illustrated in Figure 19.4.

3. *Mixing*. Figure 19.5 illustrates a mixing junction in which a stream at 100°C is mixed with a stream at 50°C to produce a combined stream with a temperature of 70°C. Great caution must be exercised with such mixing points, as the mixing acts as a heat transfer unit. The mixing transfers heat directly rather than indirectly through a heat exchanger. This is illustrated in Figure 19.5. The data for a problem table analysis could be taken as that for the mixer in Figure 19.6a where one stream is cooled from

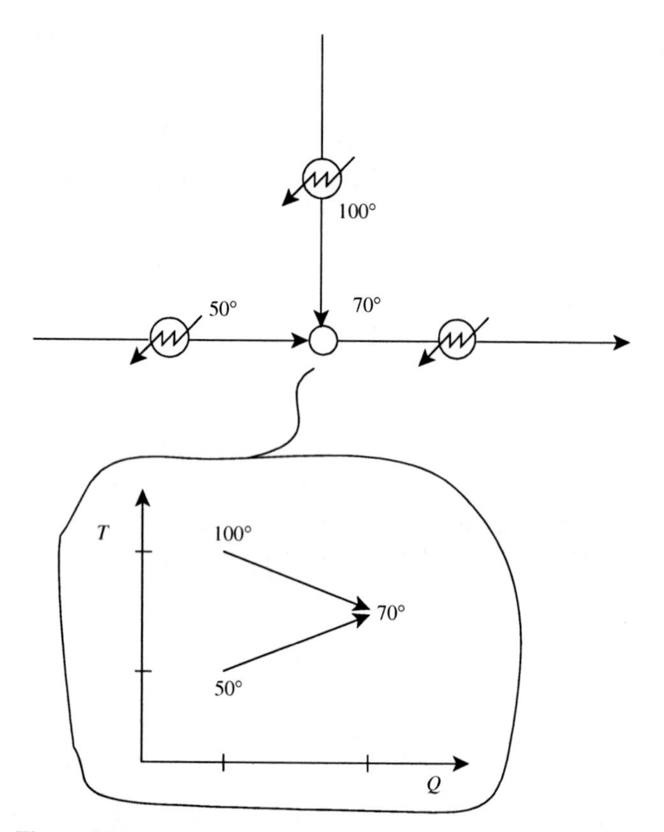

Figure 19.5 Nonisothermal mixing carries out direct contact heat transfer.

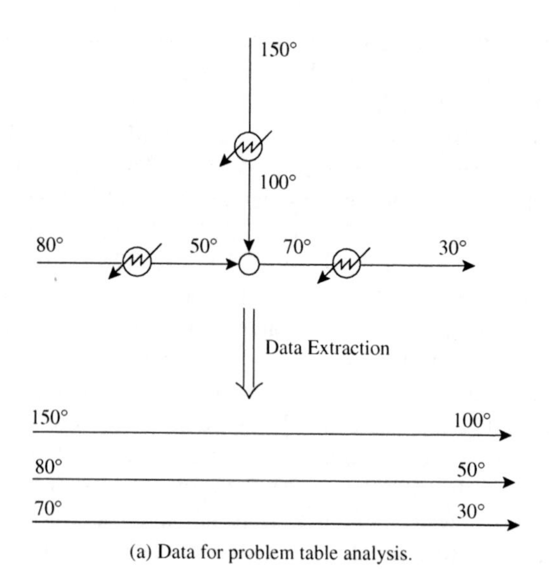

(a) Data for problem table analysis.

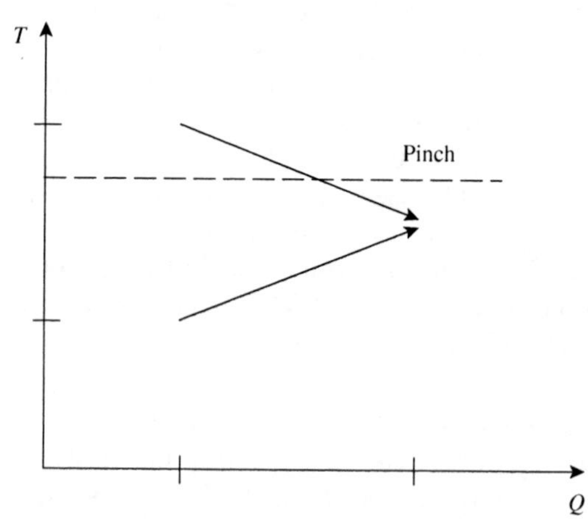

(b) Direct contact heat transfer might transfer heat across the pinch.

Figure 19.6 Nonisothermal mixing degrades temperature driving forces and might transfer heat across the pinch.

150°C to 100°C, another is cooled from 80°C to 50°C and both of these combined to produce a stream with 70°C. This 70°C stream is then cooled to 30°C. If the stream data are extracted as illustrated in Figure 19.6a, then the heat transfer that takes place in the mixing junction is embedded within the stream data as being inevitable, and this might lead to lost opportunities. It might be, for example in Figure 19.6b, that the pinch temperature is somewhere between 100°C and 50°C. If this is the case, then some cross-pinch heat transfer will be embedded within the stream data. This will be a lost opportunity to improve the energy performance of the system. The cross-pinch heat transfer that occurs in the mixing junction can never be corrected. Mixing streams nonisothermally therefore carries out heat transfer that should be avoided. To avoid this, streams need to be mixed isothermally. Figure 19.7 illustrates how the data should be extracted for this mixing junction. For there to be no prejudice in terms of heat transfer, it should be assumed that the streams are mixed at 30°C in the first instance. This means extracting the streams as two separation streams, one from 150°C to 30°C and another from 80°C to 30°C. This assumes that the mixing takes place at 30°C, for which there can be no heat transfer that takes place as a result of the mixing. Even if the mixing junction does not transfer heat across the pinch, it does degrade driving forces for the overall heat integration. In the example in Figure 19.6, high temperature heat at 100°C is degraded to 70°C. The overall heat exchange can never benefit by such degradation.

4. *Utilities*. In general, utilities should not be extracted from an existing flowsheet and included in the heat integration. For example, suppose a high-temperature heating duty could be carried out either with high-pressure steam or with hot oil generated in a furnace. The cold stream that the utility is heating needs to be included. However, the high-pressure steam or hot oil should not be included. The targeting should be carried out, the grand composite curves constructed and then the designer should make the decision as to the best choice of hot utility. Most uses of utilities, either hot or cold, fall into this category. However, there are other cases that are not so straightforward. Suppose steam is being used in a distillation, but injected as live steam directly into the distillation, in order to reduce the partial pressure of the vaporizing components. This is common in refinery distillation, as discussed in Chapter 11. In this case,

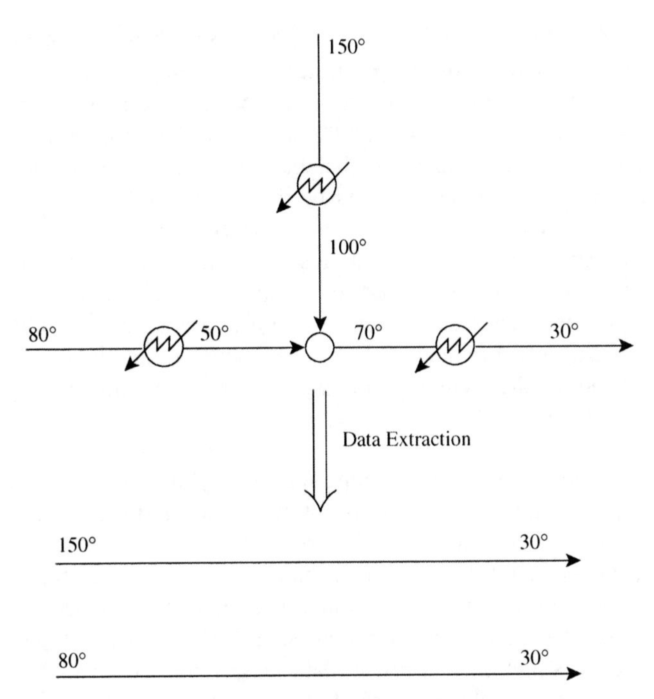

Figure 19.7 Streams should be assumed to mix isothermally for energy targeting.

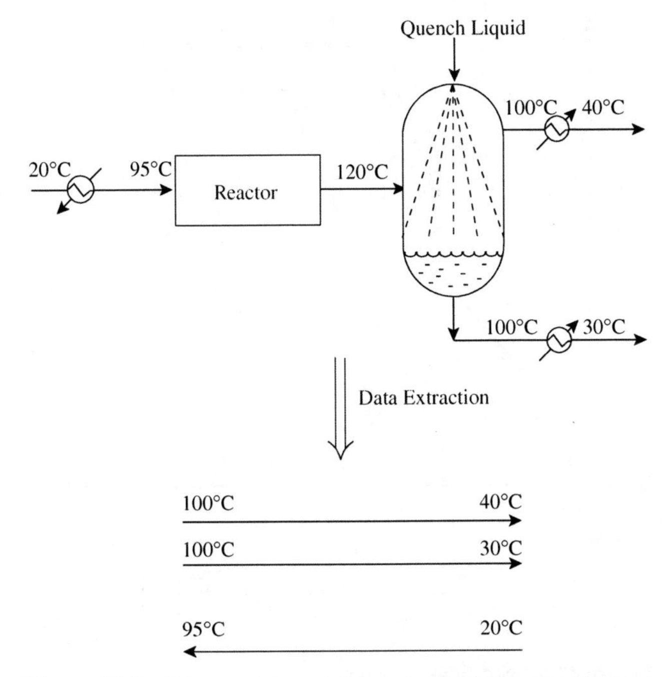

Figure 19.8 Data must be extracted at effective temperatures.

although the steam is generated in a boiler that is part of the utility system, the steam, when it enters the distillation, becomes part of the process. The distillation will not work without the injection of live steam. In this case, the steam that is injected into the process should be treated as a process stream, rather than a utility. It could then be represented within the problem as a cold stream of boiler feed water being preheated, vaporized and superheated to the conditions required for the operation.

5. *Effective temperatures.* When extracting stream data to represent the heat sources and heat sinks for the heat exchanger network problem, care must be exercised so as to represent the availability of heat at its effective temperature. For example, consider the part of the process represented in Figure 19.8. The feed stream to a reactor is preheated from 20°C to 95°C before entering the reactor. The effluent from the reactor is at 120°C and enters a quench that cools the reactor effluent from 120°C to 100°C. The vapor leaving the quench is at 100°C and needs to be cooled to 40°C. The quenched liquid also leaves at 100°C but needs to be cooled to 30°C. How should the data be extracted?

The fundamental question regarding representation of the part of the flowsheet in Figure 19.8 as heat sources and heat sinks is at what temperature the heat becomes available for heat recovery opportunities. Even though the reactor effluent is at 120°C, it is not available at this temperature because the reactor effluent needs to be quenched. The heat only becomes available at 100°C as a vapor stream that needs to be condensed and cooled to 40°C and as a quench liquid at 100°C that needs to be cooled to 30°C.

6. *Soft constraints.* In Figure 19.2b, there was a situation in which there was some flexibility to change the temperature at which a filtration takes place. These are termed *soft constraints*. There is not complete freedom to choose the conditions under which the operation takes place, but there is some flexibility to change the conditions. Another example of a soft constraint is product storage temperature. There is sometimes flexibility to choose the temperature at which material is stored. How should such soft constraints be directed to benefit the overall heat integration problem?

When extracting soft constraints, the plus–minus principle needs to be invoked, such that the flexibility in choosing conditions is directed to reduce the utility consumption, as indicated in Figure 19.1a. Consider the example in Figure 19.9, in which a product stream needs to be cooled before entering the storage tank. There is some flexibility to choose the storage temperature, but what temperature should be chosen? Composite curves are shown in Figure 19.9, and these indicate that the hot stream pinch temperature is 65°C. The product cooling is a hot stream, and it is only of use down to a temperature of 65°C. Thus, 65°C would seem to be a sensible storage temperature, as far as the heat integration is concerned.

7. *Enforced matches.* There are often some features of the heat exchanger network that the designer might wish to accept as fixed. This is often the case in retrofit. For example, a match between a hot and a cold stream might exist that is considered to be already appropriate or too expensive to change. If this is the case, then the part of the hot and cold stream involved in the enforced match should be left out of the analysis and added back in at the

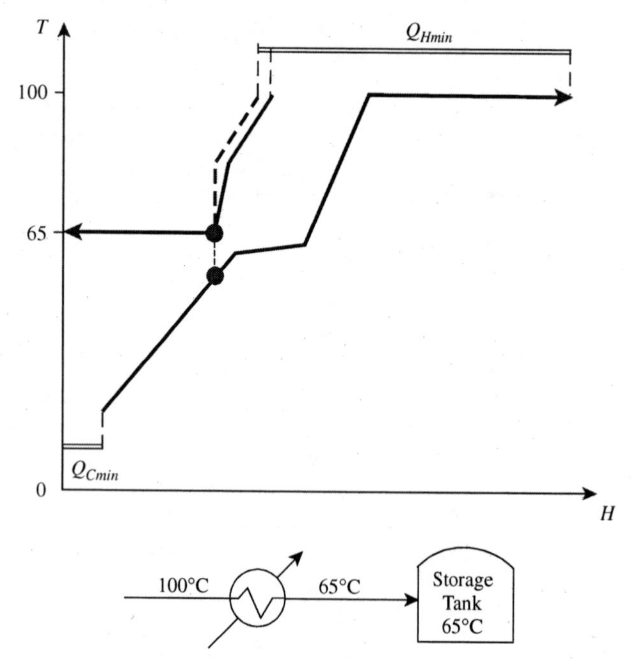

Figure 19.9 Soft temperatures.

end of the analysis. Another typical case encountered in retrofit situations is where hot and cold streams with small heat duties are being serviced by utilities. Even though the temperatures might make the streams tempting to add into the analysis, if their heat duties are small, they will not make a significant difference, and changing them is unlikely to be economic. Again, such streams can be accepted as enforced matches and left out of the analysis. Only streams with large heat duties are likely to have a significant influence on the outcome.

8. *Heat losses*. For the majority of cases, heat losses to the environment from hot surfaces will be small compared to other heat duties and can be neglected. Occasionally, it might be necessary to accept a significant heat loss and to account for it in the energy balance. If the heat loss is from a cold stream, then it should be included with the process duty, as the heat loss must be serviced either from heat recovery or hot utility. If the loss is from a hot stream, the heat loss can be accounted for by splitting the stream from which the heat loss is occurring into two components: the process duty and the heat loss. The heat loss should then be left out of the analysis if the heat loss from the hot stream is to be accepted.

9. *Data accuracy*. It is important to try and work with a set of data that is consistent when designing the heat exchanger network. Data inconsistencies are more likely to occur in retrofit situations than in new design. If there are inconsistencies in the data, it should be adjusted to make the energy balance consistent. Also, especially in retrofit situations, accurate data might not be available across the whole problem. If this is the case, then it is better to adjust the data to make it self-consistent and carry out a preliminary analysis. The most accurate data will be required where the problem is most constrained, in the region of the heat recovery pinch. Until a preliminary analysis is performed, the location of the heat recovery pinch is unknown. Having obtained insights into where data errors are likely to be a problem, then better data can be obtained for only those parts of the problem that are sensitive to data errors. Care should be taken to avoid spending considerable effort refining data that will have no significant influence on the analysis.

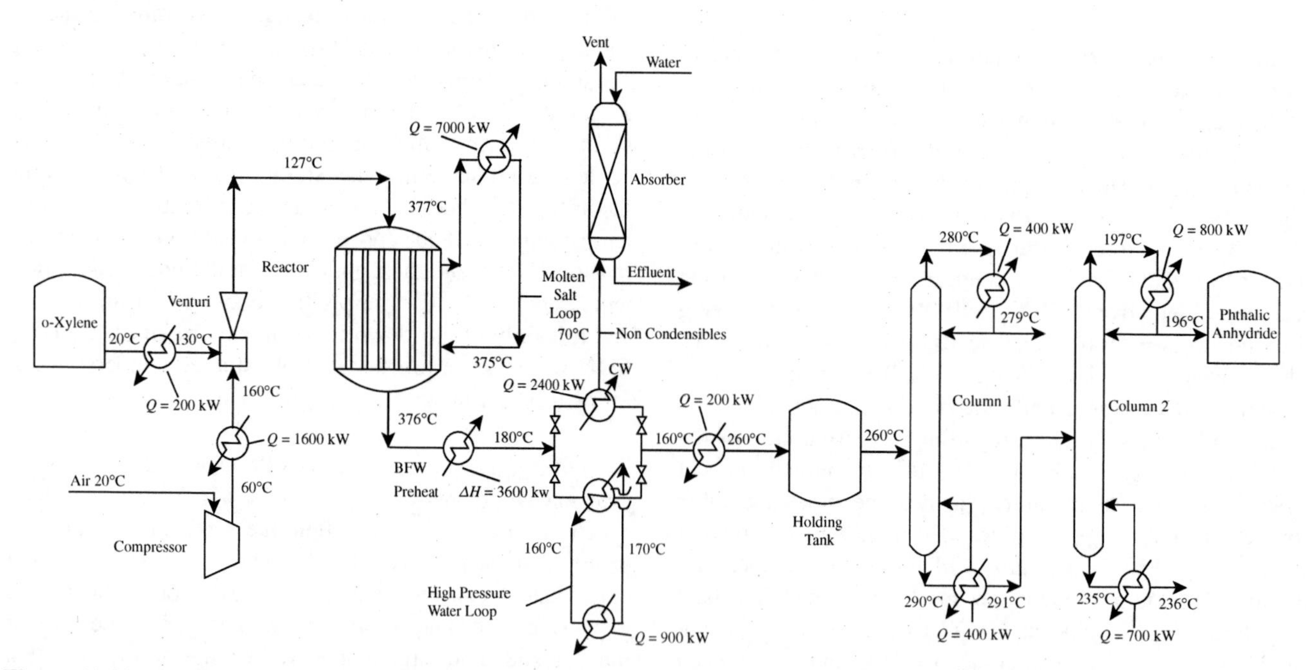

Figure 19.10 Outline of phthalic anhydride flowsheet.

Example 19.1 Phthalic anhydride is an important intermediate for the plastics industry. Manufacture is by the controlled oxidation of *o*-xylene or naphthalene. The most common route uses *o*-xylene via the reaction:

$$\underset{o\text{-xylene}}{C_8H_{10}} + 3O_2 \longrightarrow \underset{\text{phthalic anhydride}}{C_8H_4O_3} + \underset{\text{water}}{3H_2O}$$

A side reaction occurs in parallel:

$$\underset{o\text{-xylene}}{C_8H_{10}} + \frac{5}{2}O_2 \longrightarrow \underset{\text{carbon dioxide}}{8CO_2} + \underset{\text{water}}{5H_2O}$$

The reaction uses a fixed-bed vanadium pentoxide–titanium dioxide catalyst that gives good selectivity for phthalic anhydride, providing temperature is controlled within relatively narrow limits. The reaction is carried out in the vapor phase with reactor temperatures typically in the range 380 to 400°C.

The reaction is exothermic, and multitubular reactors are employed with direct cooling of the reactor via a heat transfer medium. A number of heat transfer media have been proposed to carry out the reactor cooling such as hot oil circuits, water, sulfur, mercury, and so on. However, the favored heat transfer medium is usually a molten heat transfer salt, which is a eutectic mixture sodium–potassium nitrate–nitrite.

Figure 19.10 shows a flowsheet for the manufacture of phthalic anhydride by the oxidation of *o*-xylene. Air and *o*-xylene are heated and mixed in a venturi, where the *o*-xylene vaporizes. The reaction mixture enters a tubular catalytic reactor. The heat of reaction is removed from the reactor by recirculation of molten salt. The temperature control in the reactor would be difficult to maintain by methods other than molten salt.

The gaseous reactor product is cooled first by boiler feedwater before entering a cooling water condenser. The cooling duty provided by the boiler feedwater has been fixed to avoid condensation. The phthalic anhydride in fact forms a solid on the tube walls in the cooling water condenser and is cooled to 70°C. Periodically, the on-line condenser is taken off-line and the phthalic anhydride melted off the surfaces by recirculation of high-pressure hot water. Two condensers are used in parallel, one on-line performing the condensation duty and one off-line recovering the phthalic anhydride. The heat duties shown in Figure 19.10 are time-averaged values. The noncondensible gases contain small quantities of byproducts and traces of phthalic anhydride and are scrubbed before being vented to atmosphere.

The crude phthalic anhydride is heated and held at 260°C to allow some byproduct reactions to go to completion. Purification is by continuous distillation in two columns. In the first column, maleic anhydride and benzoic and toluic acids are removed overhead. In the second column, pure phthalic anhydride is removed overhead. High-boiling residues are removed from the bottom of the second column. The reboilers of both distillation columns are serviced by a fired heater via a hot oil circuit.

There are two existing steam mains. These are high-pressure steam at 41 bar superheated to 270°C and medium-pressure steam at 10 bar superheated at 180°C. Boiler feedwater is available at 80°C and cooling water at 25°C to be heated to 30°C.

a. Extract the data from the flowsheet
b. Plot the composite curves and the grand composite curve

Solution

a. From the flowsheet in Figure 19.10, the stream data for the heat recovery problem are presented in Table 19.1. A number of points should be noted about the data extraction from the flowsheet:
   (1) The reactor is highly exothermic, and the data have been extracted as the molten salt being a hot stream. The basis of this is that it is assumed that the molten salt circuit is an essential feature of the reactor design. Thereafter, there is freedom within reason to choose how the molten salt is cooled.
   (2) The product sublimation and melting are both carried out on a noncontinuous basis. Thus, time-averaged values have been taken.
   (3) The product sublimation and product melting imply a linear change in enthalpy over a relatively large change in temperature. However, changes of phase normally take place with a relatively small change in temperature. Thus, the product sublimation might involve desuperheating over a relatively large range of temperature, change of

Table 19.1 Stream data for the process in Figure 19.1.

Stream			T_s (°C)	T_T (°C)	ΔH (kW)
No.	Name	Type			
1	Reactor cooling	Hot	377	375	−7000
2	Reactor product cooling	Hot	376	180	−3600
3	Product sublimation	Hot	180	70	−2400
4	Column 1 condenser	Hot	280	279	−400
5	Column 2 condenser	Hot	197	196	−800
6	Air feed	Cold	60	160	1600
7	o-xylene feed	Cold	20	130	200
8	Product melting	Cold	70	160	900
9	Holding tank feed	Cold	160	260	200
10	Column 1 reboiler	Cold	290	291	400
11	Column 2 reboiler	Cold	235	236	700

phase over a relatively small change in temperature and subcooling over a relatively large range in temperature. Product melting might involve heating to melting point over a relatively large range of temperature, followed by melting over a relatively small change in temperature. Thus, representation of the product sublimation and product melting as a linear change in enthalpy seems to be inappropriate. To overcome this, these two streams could be broken down into linear segments to represent this nonlinear temperature-enthalpy behavior. Here, for the sake of simplicity, the streams will be assumed to have a linear temperature-enthalpy behavior.

(4) The air starts at 20°C, but it is heated to 60°C in the compressor by the increase in pressure. If the compressor is an essential feature of the process, then the heating between 20 and 60°C is serviced by the compressor and should not be included in the heat recovery problem.

(5) The air and *o*-xylene are mixed at unequal temperature in the venturi, where the *o*-xylene vaporizes. Mixing at unequal temperatures provides heat transfer by direct contact and might in principle be direct contact heat transfer across the pinch, the location of which is as yet unknown. Thus, accepting the direct contact heat transfer might lead to unnecessarily high energy targets if the mixing causes heat transfer across the pinch. The problem is avoided in targeting by mixing streams, where possible, at the same temperature, thus avoiding any direct contact heat transfer. Of course, once the targets have

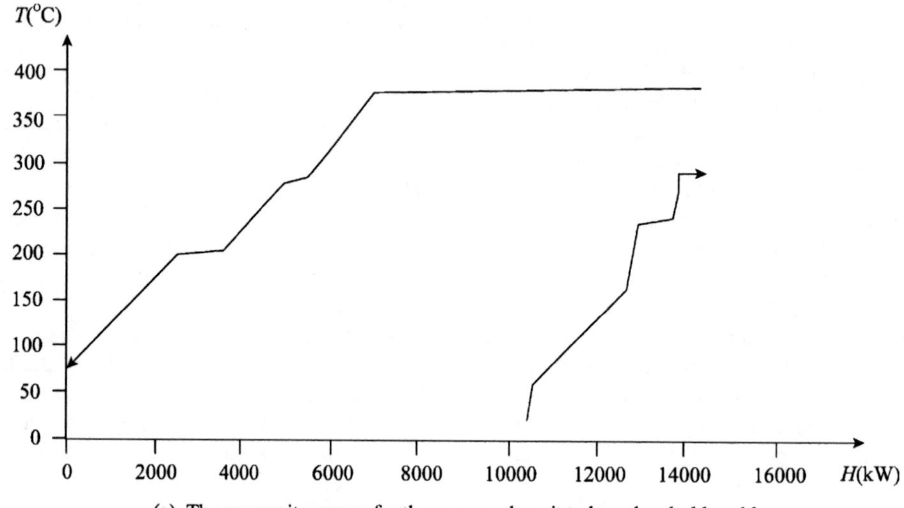

(a) The composite curves for the process show it to be a threshold problem.

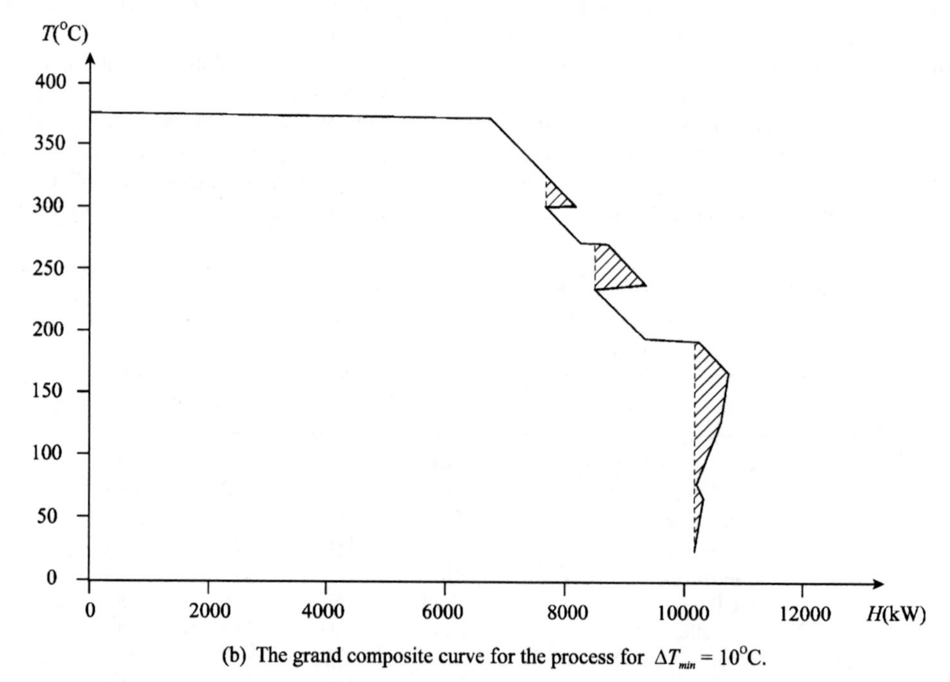

(b) The grand composite curve for the process for $\Delta T_{min} = 10°C$.

Figure 19.11 The composite curves and grand composite curve for the phthalic anhydride process.

been established and the location of the pinch known, streams can then be mixed at unequal temperatures in the design away from the pinch in the knowledge that there is no cross-pinch heat transfer. In this case, the process conditions will be accepted, initially at least, because of the vaporization occurring in the mixing.

b. Figure 19.11a shows the composite curves for the process. The problem is clearly threshold in nature, requiring only cooling, with a threshold ΔT_{min} of 86°C. Figure 19.11b shows the grand composite curves for $\Delta T_{min} = 10$°C. The reason ΔT_{min} has been taken to be 10°C and not 86°C is that the cooling will be supplied by the introduction of steam generation, which will turn the threshold problem into a pinched problem, as discussed in Chapter 16, for which the value of ΔT_{min} is 10°C for this problem.

19.4 HEAT EXCHANGER NETWORK STREAM DATA – SUMMARY

The basic reference in guiding process changes to reduce utility costs is the plus–minus principle. However, process changes so identified prompt changes in the capital-energy trade-off and utility selection. Using the total cost targeting techniques described in Chapter 17, it is possible to effectively screen a wide range of options using relatively simple computation.

For data extraction from a flowsheet:

- Only essential constraints are included in the stream data.
- Stream data should be linearized on the safe side.
- Mixing should take place isothermally.
- Utilities should not be extracted with the stream data.
- Data should be extracted at effective temperatures.
- Soft constraints should exploit the plus–minus principle to improve the targets.
- Heat losses from a cold stream can be accounted for by the inclusion of a fictitious cold stream.
- Heat losses from a hot stream can be accounted for by splitting the hot stream to represent the loss.
- Accurate data are not always required throughout the whole problem.
- Enforced matches can be accounted for by leaving parts of the stream data out of the problem.

19.5 EXERCISES

1. The process flowsheet for a cellulose acetate fibers process is shown in Figure 19.12. Solvent is removed from the fibers in a dryer by recirculating air. The air is cooled before it enters an absorber where the solvent is absorbed in water. The solvent–water mixture is separated in a distillation column and

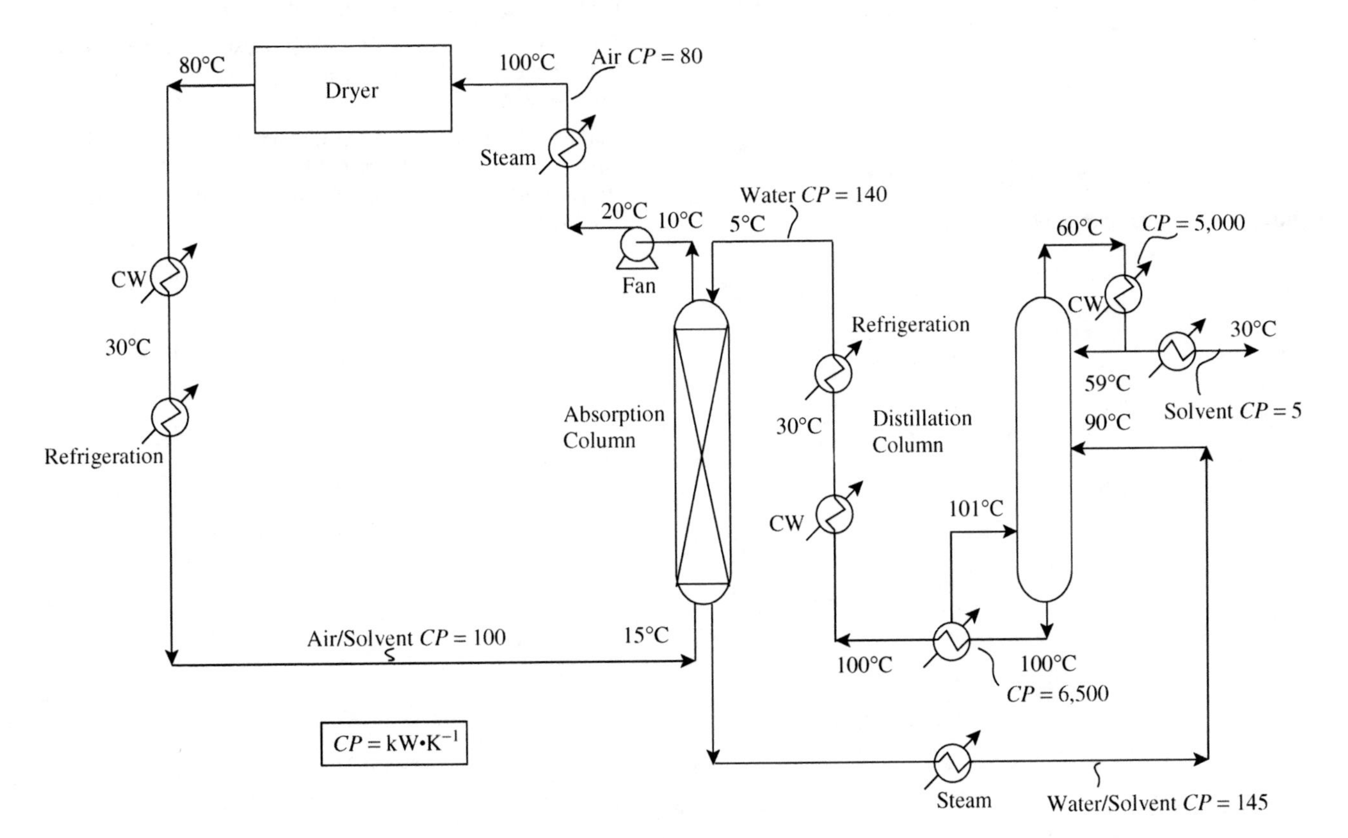

Figure 19.12 Flowsheet of a process for the manufacture of cellulose acetate fiber.

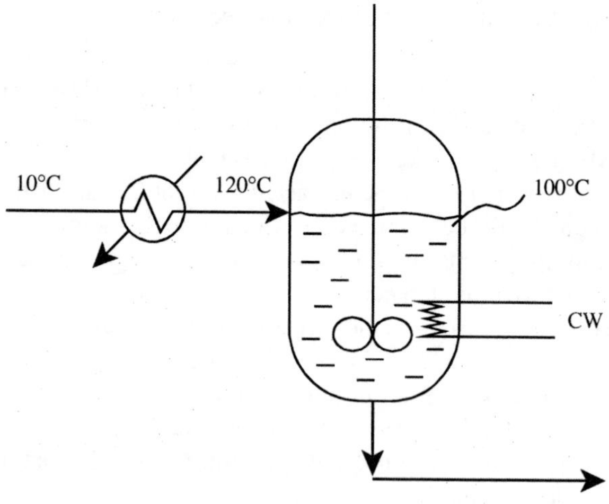

(a) Reactor feed pre-heated to 120°C.

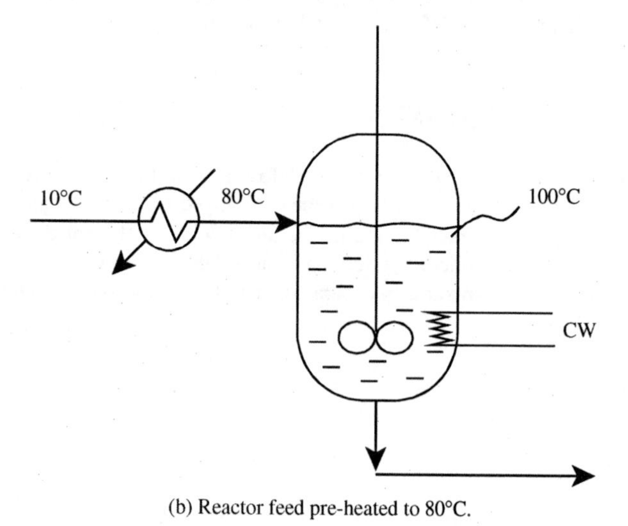

(b) Reactor feed pre-heated to 80°C.

Figure 19.13 Data extraction for a reactor feed.

the water recycled. The process is serviced by saturated steam at 150°C, cooling water at 20°C and refrigerant at −5°C. The temperature rise of both the cooling water and refrigerant can be neglected. Extract the stream data from the flowsheet and present them as hot and cold streams with supply and target temperatures and heat capacity flowrates. Sketch the composite curves for the process for $\Delta T_{min} = 10$°C.

2. Figure 19.13 shows two situations where a feed stream is heated before entering an agitated reactor vessel. Heat of reaction is removed from the reactor by cooling water and the temperature of the reactor controlled to be 100°C. It is considered essential to remove the heat of reaction by cooling water for safety reasons. In the first case (Figure 19.13a), the feed is preheated to a higher temperature than the reactor. In the second case (Figure 19.13b), the feed is preheated to a lower temperature than the reactor. In each case, how should the data for the feed stream be extracted?

REFERENCES

1. Umeda T, Harada T and Shiroko K (1979) A Thermodynamic Approach to Synthesis of Heat Integration Systems in Chemical Processes, *Comp Chem Eng*, **3**: 273.
2. Umeda T, Nidda K and Shiroko K (1979) A Thermodynamic Approach to Heat Integration in Distillation Systems, *AIChE J*, **25**: 423.
3. Linnhoff B and Vredeveld DR (1984) Pinch Technology Has Come of Age, *Chem Eng Prog*, **July**: 40.
4. Linnhoff B and Parker SJ (1984) Heat Exchanger Network with Process Modifications, *IChemE Annual Research Meeting*, Bath, April.
5. Linnhoff B, Townsend DW, Boland D, Hewitt GF, Thomas BEA, Guy AR and Marsland RHA (1982) *User Guide on Process Integration for the Efficient Use of Energy*, IChemE, Rugby, UK.

20 Heat Integration of Reactors

20.1 THE HEAT INTEGRATION CHARACTERISTICS OF REACTORS

The heat integration characteristics of reactors depend both on the decisions that have been made for the removal or addition of heat and the reactor mixing characteristics. In the first instance, adiabatic operation should be considered since this gives the simplest design.

1. *Adiabatic operation.* If adiabatic operation leads to an acceptable temperature rise for exothermic reactors or an acceptable decrease for endothermic reactors, then this is the option that would normally be chosen. If this is the case, then the feed stream to the reactor requires heating and the effluent stream requires cooling. The heat integration characteristics are thus a cold stream (the reactor feed) if the feed needs to be increased in temperature or vaporized, and a hot stream (the reactor effluent) if the product needs to be decreased in temperature or condensed. The heat of reaction appears as increased temperature of the effluent stream in the case of exothermic reaction or decreased temperature in the case of endothermic reaction.

2. *Heat carriers.* If adiabatic operation produces an unacceptable rise or fall in temperature, then the option discussed in Chapters 7 and 13 is to introduce a heat carrier. The operation is still adiabatic, but an inert material is introduced with the reactor feed as a heat carrier. The heat integration characteristics are as before. The reactor feed is a cold stream and the reactor effluent a hot stream. The heat carrier serves to increase the heat capacity flowrate of both streams.

3. *Cold shot.* Injection of cold fresh feed for exothermic reactions or preheated feed for endothermic reactions to intermediate points in the reactor can be used to control the temperature in the reactor. Again, the heat integration characteristics are similar to adiabatic operation. The feed is a cold stream if it needs to be increased in temperature or vaporized and the product a hot stream if it needs to be decreased in temperature or condensed. If heat is provided to the cold shot or hot shot streams, these are additional cold streams.

Chemical Process Design and Integration R. Smith
 ISBNs: 0-471-48680-9 (HB); 0-471-48681-7 (PB)

4. *Indirect heat transfer with the reactor.* Although indirect heat transfer with the reactor tends to bring about the most complex reactor design options, it is often preferable to the use of a heat carrier. A heat carrier creates complications elsewhere in the flowsheet. A number of options for indirect heat transfer were discussed earlier in Chapter 7.

The first distinction to be drawn, as far as heat transfer is concerned, is between the plug-flow and mixed-flow reactor. In the plug-flow reactor shown in Figure 20.1, the heat transfer can take place over a range of temperatures. The shape of the profile depends on the following.

- Inlet feed concentration
- Inlet temperature
- Inlet pressure and pressure drop (gas-phase reactions)
- Conversion
- Byproduct formation
- Heat of reaction
- Rate of cooling/heating
- Presence of catalyst diluents or changes in catalyst through the reactor

Figure 20.1a shows two possible thermal profiles for exothermic plug-flow reactors. If the rate of heat removal is low and/or the heat of reaction if high, then the temperature of the reacting stream will increase along the length of the reactor. If the rate of heat removal is high and/or the heat of reaction is low, then the temperature will decrease. Under conditions between the two profiles shown in Figure 20.1a, a maximum can occur in the temperature at an intermediate point between the reactor inlet and exit.

Figure 20.1b shows two possible thermal profiles for endothermic plug-flow reactors. This time, the temperature decreases for low rates of heat addition and/or high heat of reaction. The temperature increases for the reverse conditions. Under conditions between the profiles shown in Figure 20.1b, a minimum can occur in the temperature profile at an intermediate point between the inlet and exit.

The thermal profile through the reactor will, in most circumstances, be carefully optimized to maximize selectivity, extend catalyst life, and so on. Because of this, direct heat integration with other process streams is almost never carried out. The heat transfer to or from the reactor is instead usually carried out by a heat transfer intermediate. For example, in exothermic reactions, cooling might occur by boiling water to generate steam, which, in turn, can be used to heat cold streams elsewhere in the process or across the site.

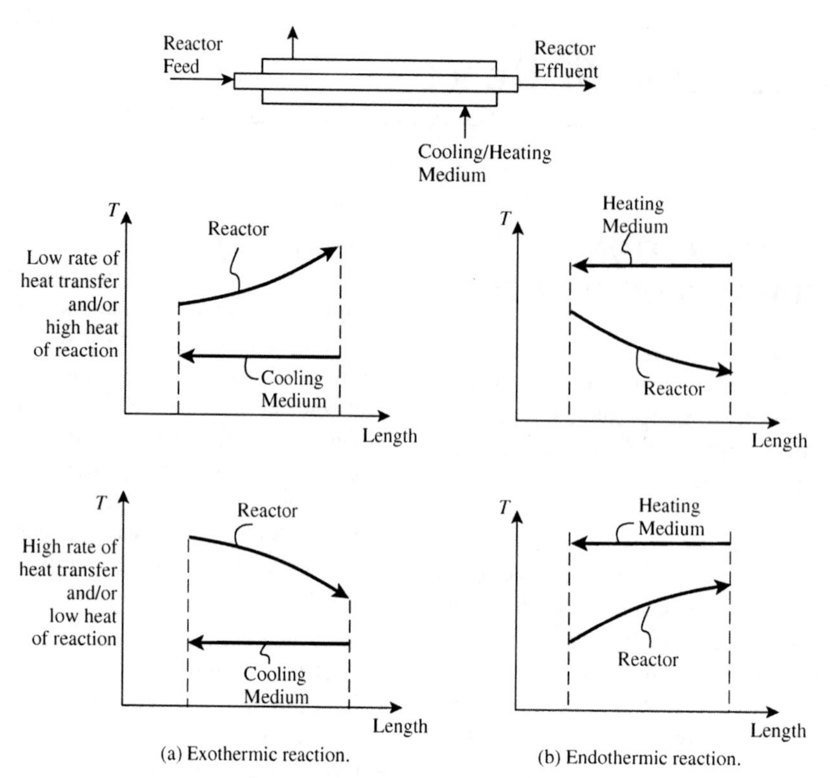

Figure 20.1 The heat transfer characteristics of plug-flow reactors.

By contrast, if the reactor is mixed-flow, then the reactor is isothermal. This behavior is typical of stirred tanks used for liquid-phase reactions or fluidized-bed reactors used for gas-phase reactions. The mixing causes the temperature in the reactor to be effectively uniform.

For indirect heat transfer, the heat integration characteristics of the reactor can be broken down into the following three cases.

a. If the reactor can be matched directly with other process streams (which is unlikely), then the reactor profile should be included in the heat integration problem. This would be a hot stream in the case of an exothermic reaction or a cold stream in the case of an endothermic reaction.
b. If a heat transfer intermediate is to be used and the cooling/heating medium is fixed, then the cooling/heating medium should be included and not the reactor profile itself. Once the cooling medium leaves an exothermic reactor, it is a hot stream requiring cooling before being returned to the reactor. Similarly, once the heating medium leaves an endothermic reactor, it is a cold stream requiring heating before being returned to the reactor.
c. If a heat transfer intermediate is to be used but the temperature of the cooling/heating medium is not fixed, then both the reactor profile and the cooling/heating medium should be included. The temperature of the heating/cooling medium can then be varied within the content of the overall heat integration problem to improve the targets, as described in Chapter 19.

In addition to the indirect cooling/heating within the reactor, the reactor feed is an additional cold stream, if it needs to be increased in temperature or vaporized and the reactor product an additional hot stream, if it needs to be decreased in temperature or condensed.

For the ideal-batch reactor, the temperature can be assumed to be uniform throughout the reactor at any instant in time. Figure 20.2a shows typical variations in temperature with time for an exothermic reaction in a batch reactor. A family of curves illustrates the effect of increasing the rate of heat removal and/or decreasing heat of reaction. Each individual curve assumes the rate of heat transfer to the cooling medium to be constant for that curve throughout the batch cycle. Figure 20.2b shows typical curves for endothermic reactions. Again, each individual curve in Figure 20.2b assumes the rate of heat addition from the heating medium to be constant throughout the batch process.

Fixing the rate of heat transfer in a batch reactor is often not the best way to control the reaction. The heating or cooling characteristics can be varied with time to suit the characteristics of the reaction (see Chapter 14). Because of the complexity of batch operation and the fact that operation is usually small scale, it is rare for any attempt to be made to recover heat from a batch reactor or supply heat by recovery. Instead, utilities are normally used.

The heat duty on the heating/cooling medium is given by

$$Q_{REACT} = -(\Delta H_{STREAMS} + \Delta H_{REACT}) \qquad (20.1)$$

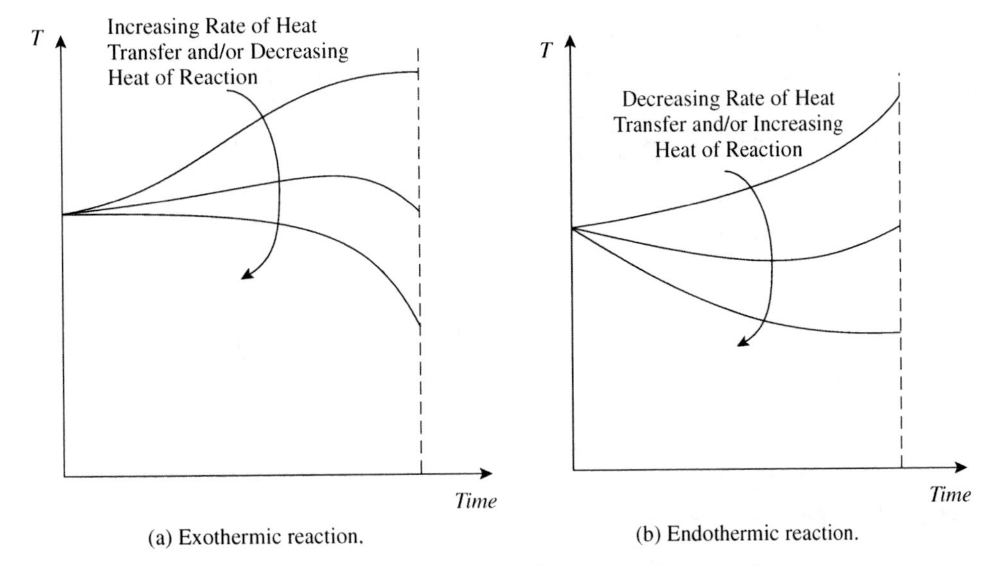

Figure 20.2 The heat transfer characteristics of batch reactors for a fixed rate of heat transfer.

where

Q_{REACT} = reactor heating or cooling required

$\Delta H_{STREAMS}$ = enthalpy change between feed and product streams

ΔH_{REACT} = reaction enthalpy (negative in the case of exothermic reactions)

5. *Quench.* As discussed in Chapter 7, the reactor effluent may need to be cooled rapidly (quenched). This can be by indirect heat transfer using conventional heat transfer equipment or by direct heat transfer by mixing with another fluid.

If indirect heat transfer is used with a large temperature difference to promote high rates of cooling, then the cooling fluid (e.g. boiling water) is fixed by process requirements. In this case, the heat of reaction is not available at the temperature of the reactor effluent. Rather, the heat of reaction becomes available at the temperature of the quench fluid. Thus, the feed stream to the reactor is a cold stream, the quench fluid is a hot stream, and the reactor effluent after the quench is also a hot stream. This was discussed under data extraction in Chapter 19.

The reactor effluent might require cooling by direct heat transfer because the reaction needs to be stopped quickly, or a conventional heat exchanger would foul, or the reactor products are too hot or corrosive to pass to a conventional heat exchanger. The reactor product is mixed with a liquid that can be recycled, cooled product, or an inert material such as water. The liquid vaporizes partially or totally and cools the reactor effluent. Here, the reactor feed is a cold stream, and the vapor and any liquid from the quench are hot streams.

Now consider the placement of the reactor in terms of the overall heat integration problem.

20.2 APPROPRIATE PLACEMENT OF REACTORS

In Chapter 16, it was seen how the pinch takes on fundamental significance in improving heat integration. Now consider the consequences of placing reactors in different locations relative to the pinch.

Figure 20.3 shows the background process represented simply as a heat sink and heat source divided by the pinch. Figure 20.3a shows the process with an exothermic reactor integrated above the pinch. The minimum hot utility can be reduced by the heat released by reaction.

By comparison, Figure 20.3b shows an exothermic reactor integrated below the pinch. Although heat is being recovered, it is being recovered into part of the process, which is a heat source. The hot utility requirement cannot be reduced, since the process above the pinch needs at least Q_{Hmin} to satisfy its enthalpy imbalance.

There is no benefit by integrating an exothermic reactor below the pinch. The appropriate placement for exothermic reactors is above the pinch[1].

Figure 20.4a shows an endothermic reactor integrated above the pinch. The endothermic reactor removes Q_{REACT} from the process above the pinch. The process above the pinch needs at least Q_{Hmin} to satisfy its enthalpy imbalance. Thus, an extra Q_{REACT} must be imported from hot utility to compensate. There is no benefit by integrating an endothermic reactor above the pinch. Locally, it might seem that a benefit is being derived by running the reaction by recovery. However, additional hot utility must be imported elsewhere to compensate.

By contrast, Figure 20.4b shows an endothermic reactor integrated below the pinch. The reactor imports Q_{REACT} from part of the process that needs to reject heat anyway. Thus, integration of the reactor serves to reduce the

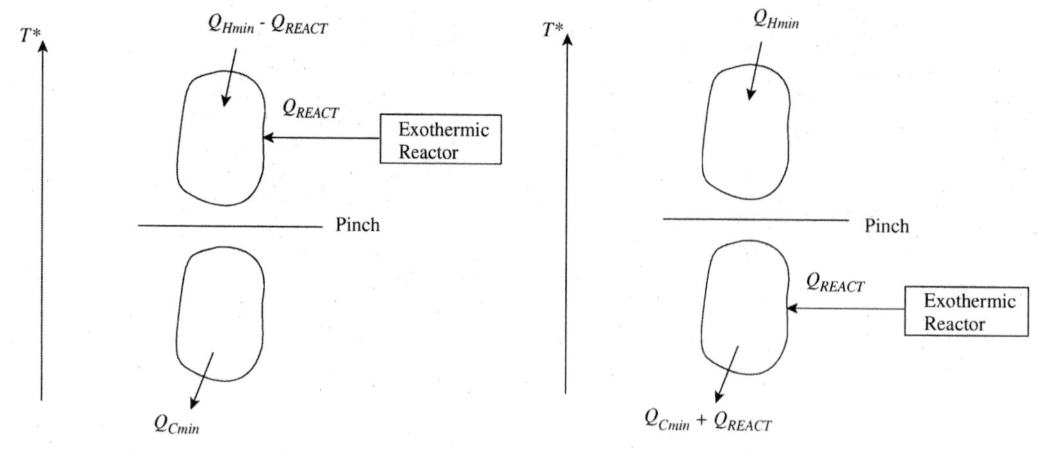

Figure 20.3 Appropriate placement of an exothermic reactor.

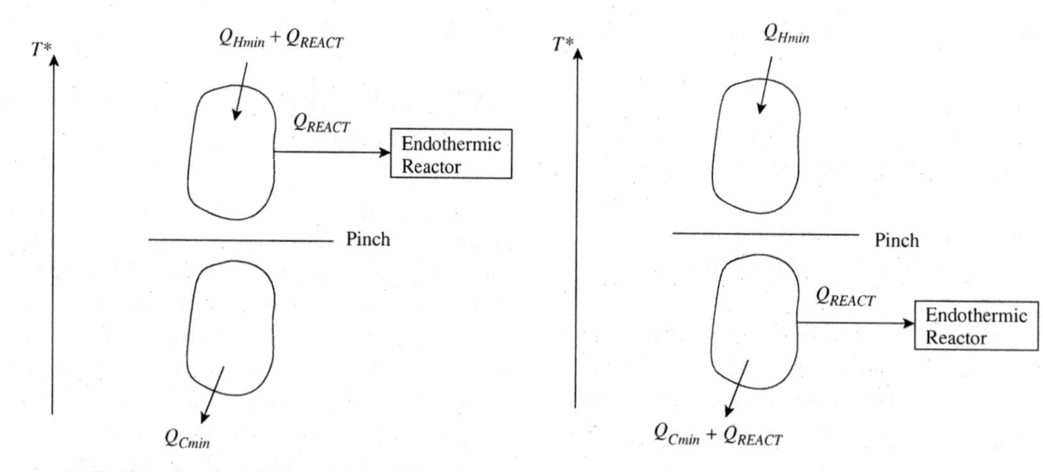

Figure 20.4 Appropriate placement of an endothermic reactor.

cold utility consumption by Q_{REACT}. There is an overall reduction in hot utility because, without integration, the process and reactor would require ($Q_{Hmin} + Q_{REACT}$) from the utility.

There is no benefit by integrating an endothermic reactor above the pinch. The appropriate placement for endothermic reactors is below the pinch[1].

20.3 USE OF THE GRAND COMPOSITE CURVE FOR HEAT INTEGRATION OF REACTORS

The above appropriate placement arguments assume that the process has the capacity to accept or give up the reactor heat duties at the given reactor temperature. A quantitative tool is needed to assess the capacity of the background process. For this purpose, the grand composite curve is used and the reactor profile treated as if it was a utility, as explained in Chapter 16.

The problem with representing a reactor profile is that, unlike utility profiles, the reactor profile might involve several streams. The reactor profile involves not only streams such as those for indirect heat transfer shown in Figure 20.1, but also the reactor feed and effluent streams that can be an important feature of the reactor heating and cooling characteristics. The various streams associated with the reactor can be combined to form a grand composite curve for the reactor. This can then be matched against the grand composite curve for the rest of the process. The following example illustrates the approach.

Example 20.1 Consider again the process for the manufacture of phthalic anhydride discussed in Example 19.1. The data was extracted from the flowsheet in Figure 19.10 and listed in Table 19.1. The composite curves and grand composite curve are shown in Figure 19.11.

a. Examine the placement of the reactor relative to the rest of the process.
b. Determine the utility requirements of the process.

Solution

a. The stream data used to construct the grand composite curve in Figure 20.5a include those associated with the reactor and those for the rest of the process. If the placement of the reactor relative to the rest of the process is to be examined, those streams associated with the reactor need to be separated from the rest of the process. Figure 20.5b shows the grand composite curves for the two parts of the process. Figure 20.5b is based on Streams 1, 2, 6 and 7 from Table 19.1 and Figure 20.5c is based on Streams 3, 4, 5, 8, 9, 10 and 11.

   In Figure 20.5d, the grand composite curves for the reactor and that for the rest of the process are superimposed. To obtain maximum overlap, one of the curves must be taken as a mirror image. It can be seen in Figure 20.5d that the reactor is appropriately placed relative to the rest of the process. Had the reactor not been appropriately placed, it would have been extremely unlikely that the reactor would have been changed to make it so. Rather, to obtain appropriate placement of the reactor, the rest of the process would more likely have been changed.

b. Figure 20.6 shows the grand composite curve for all the streams with a steam generation profile matched against it. The process cooling demand is satisfied by the generation of high-pressure (41 bar) steam from boiler feedwater, which is superheated to 270°C. High-pressure steam generation is preferable to low-pressure generation. There is apparently no need for cooling water.

   A greater amount of steam would be generated if the non-condensible vent was treated using catalytic thermal oxidation (see Chapter 25) rather than absorption. The exotherm from catalytic thermal oxidation would create an extra hot stream for steam generation.

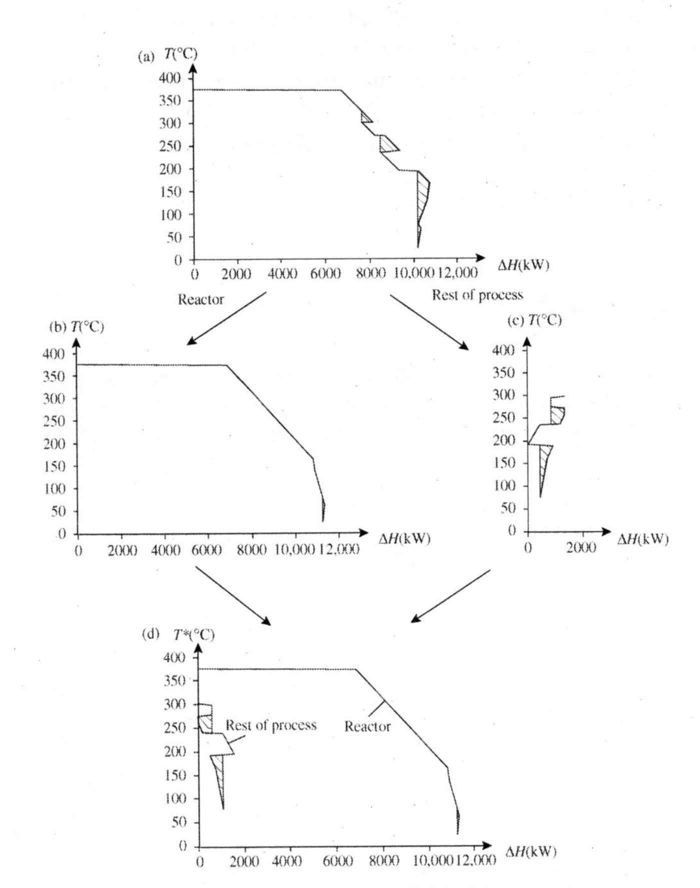

Figure 20.5 The problem can be divided into two parts, one associated with the reactor and the other with the rest of the process ($\Delta T_{min} = 10°C$) and then superimposed.

20.4 EVOLVING REACTOR DESIGN TO IMPROVE HEAT INTEGRATION

If the reactor proves to be inappropriately placed, then the process changes might make it possible to correct this. One option would be to change the reactor conditions to bring this about. Most often, however, the reactor conditions will probably have been optimized for selectivity, catalyst performance, and so on, which, taken together with safety, materials-of-construction constraints, control, and so on, makes it unlikely that the reactor conditions would be changed to improve heat integration. Rather, to obtain appropriate placement of the reactor, the rest of the process would most likely be changed.

If changes to the reactor design are possible, then the simple criteria introduced in Chapter 19 can be used to direct those changes. Heat integration will always benefit by making hot streams hotter and cold streams colder. This applies whether the heat integration is carried out directly between process streams or through an intermediate such as steam. For example, consider the exothermic reactions in Figure 20.1a. Allowing the reactor to work at higher temperature improves the heat integration potential if this does not interfere with selectivity or catalyst life or introduce safety and control problems, and so on. However, if the reactor must work with a fixed intermediate cooling fluid, such as steam generation, then the only benefit will be a reduced heat transfer area in the reactor. The steam becomes a hot stream available for heat integration after leaving the reactor. If the pressure of steam generation can be increased, then there may be energy or heat transfer area benefits when it is integrated with the rest of the process.

Care should be taken when preheating reactor feeds within the reactor using the heat of reaction. This is achieved in practice simply by passing the cold feeds directly to the reactor and allowing them to be preheated by mixing with hot materials within the reactor. However, if the exothermic reactor is appropriately placed above the pinch and the feeds start below the pinch, then the preheating within the reactor is cross-pinch heat transfer. In this case, feeds should be preheated by recovery using streams below the pinch before being fed to the reactor. This increases the heat generated within the reactor, and heat integration will benefit from the increased heat available for recovery from the reactor.

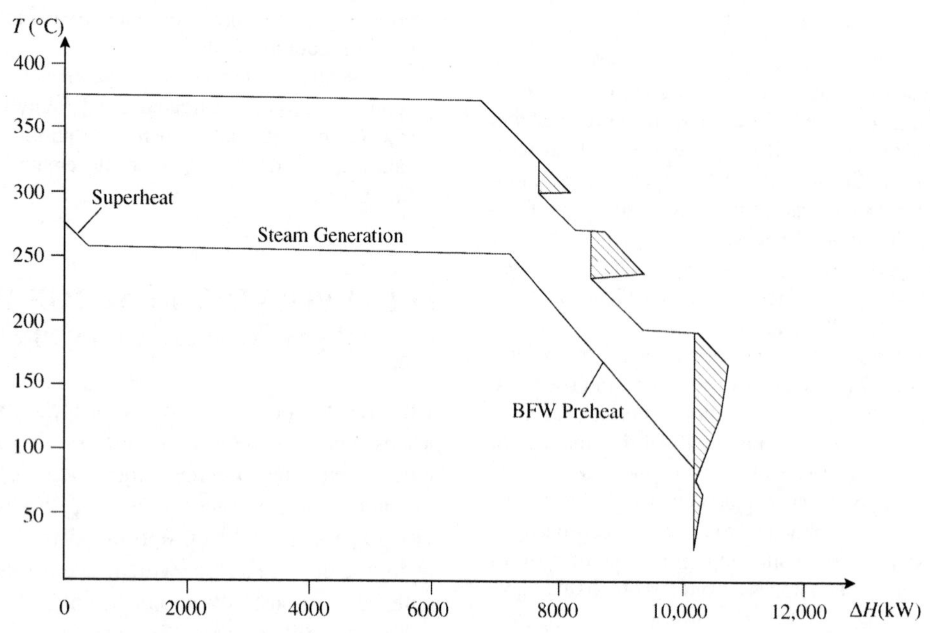

Figure 20.6 The grand composite curve for the whole process apparently requires only high-pressure steam generation from boiler feedwater.

20.5 HEAT INTEGRATION OF REACTORS – SUMMARY

The appropriate placement of reactors, as far as heat integration is concerned, is that exothermic reactors should be integrated above the pinch and endothermic reactors below the pinch. Care should be taken when reactor feeds are preheated by heat of reaction within the reactor for exothermic reactions. This can constitute cross-pinch heat transfer. The feeds should be preheated to pinch temperature by heat recovery before being fed to the reactor.

Appropriate placement can be assessed quantitatively using the grand composite curve. The streams associated with the reactor can be represented as a grand composite curve for the reactor and then matched against the grand composite curve for the rest of the process.

If the reactor is not appropriately placed, then it is more likely that the rest of the process would be changed to bring about appropriate placement rather than changing the reactor. If changes to the reactor design are possible, then the simple criterion of making hot streams hotter and cold streams colder can be used to bring about beneficial changes.

REFERENCE

1. Glavic P, Kravanja Z and Homsak M (1988) Heat Integration of Reactors: I. Criteria for the Placement of Reactors into Process Flowsheet, *Chem Eng Sci*, **43**: 593.

21 Heat Integration of Distillation Columns

21.1 THE HEAT INTEGRATION CHARACTERISTICS OF DISTILLATION

The dominant heating and cooling duties associated with a distillation column are the reboiler and condenser duties. In general, however, there will be other duties associated with heating and cooling of feed and product streams. These sensible heat duties usually will be small in comparison with the latent heat changes in reboilers and condensers.

Both the reboiling and condensing processes normally take place over a range of temperature. Practical considerations, however, usually dictate that the heat to the reboiler must be supplied at a temperature above the dew point of the vapor leaving the reboiler and that the heat removed in the condenser must be removed at a temperature lower than the bubble point of the liquid. Hence, in preliminary design at least, both reboiling and condensing can be assumed to take place at constant temperatures.

21.2 THE APPROPRIATE PLACEMENT OF DISTILLATION

Consider now the consequences of placing simple distillation columns (i.e. one feed, two products, one reboiler and one condenser) in different locations relative to the heat recovery pinch. The separator takes heat Q_{REB} into the reboiler at temperature T_{REB} and rejects heat Q_{COND} at a lower temperature T_{COND}. There are two possible ways in which the column can be heat integrated with the rest of the process. The reboiler and condenser can be integrated either across, or not across, the heat recovery pinch.

1. *Distillation across the pinch.* This arrangement is shown in Figure 21.1a. The background process (which does not include the reboiler and condenser) is represented simply as a heat sink and heat source divided by the pinch. Heat Q_{REB} is taken into the reboiler above pinch temperature and heat Q_{COND} rejected from the condenser below pinch temperature. Because the process sink above the pinch requires at least Q_{Hmin} to satisfy its enthalpy balance, the Q_{REB} removed by the reboiler must be compensated for by introducing an extra Q_{REB} from hot utility. Below the pinch, the process needs to reject Q_{Cmin} anyway, and an extra heat load Q_{COND} from the condenser has been introduced.

By heat integrating the distillation column with the process and by considering only the reboiler, it might be concluded that energy has been saved. The reboiler has its heat requirements provided by heat recovery. However, the overall situation is that heat is being transferred across the heat recovery pinch through the distillation column and that the consumption of hot and cold utility in the process must increase correspondingly. There are fundamentally no savings available from the integration of a separator across the pinch[1,2].

2. *Distillation not across the pinch.* Here the situation is somewhat different. Figure 21.1b shows a distillation column entirely above the pinch. The distillation column takes heat Q_{REB} from the process and returns Q_{COND} at a temperature above the pinch. The hot utility consumption changes by $(Q_{REB} - Q_{COND})$. The cold

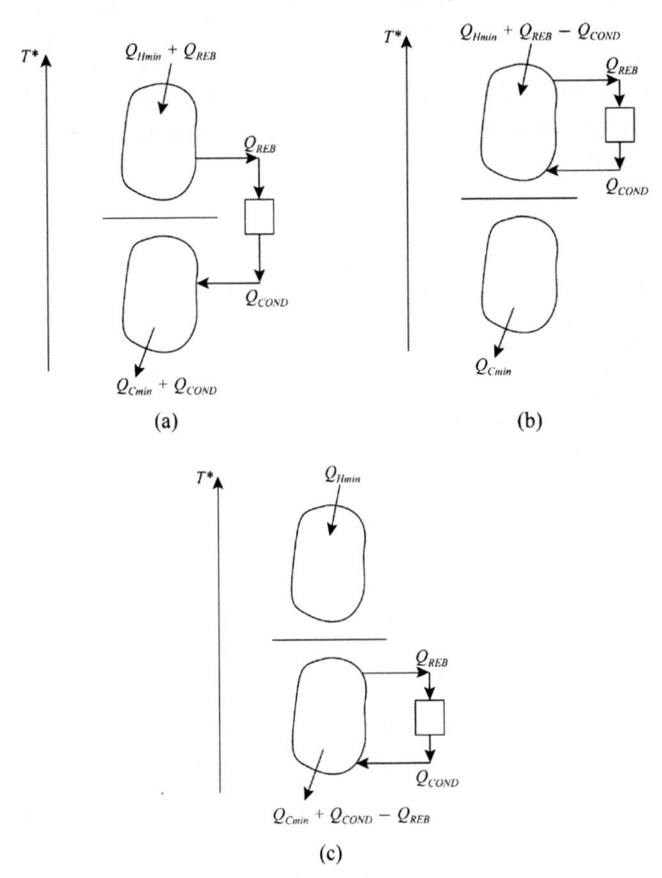

Figure 21.1 The Appropriate placement of distillation columns. (From Smith R and Linnhoff B, 1988. *Trans IChemE ChERD*, **66**: 195, reproduced by permission of the Institution of Chemical Engineers.)

Chemical Process Design and Integration R. Smith
 ISBNs: 0-471-48680-9 (HB); 0-471-48681-7 (PB)

utility consumption is unchanged. Usually, Q_{REB} and Q_{COND} have a similar magnitude. If $Q_{REB} \approx Q_{COND}$, then the hot utility consumption is Q_{Hmin}, and there is no additional hot utility required to run the column. It takes a "free ride" from the process. Heat integration below the pinch is illustrated in Figure 21.1c. Now the hot utility is unchanged, but the cold utility consumption changes by $(Q_{COND} - Q_{REB})$. Again, given that Q_{REB} and Q_{COND} usually have similar magnitudes, the result is similar to heat integration above the pinch.

All these arguments can be summarized by a simple statement: the appropriate placement for separators is not across the pinch[1,2]. Although the principle was originally stated with regard to distillation columns, it clearly applies to any separator that takes in heat at higher temperature and rejects heat at lower temperature.

If both the reboiler and condenser are integrated with the process, this can make the column difficult to start up and control. However, when the integration is considered more closely, it becomes clear that both the reboiler and condenser do not need to be integrated. Above the pinch, the reboiler can be serviced directly from the hot utility with the condenser integrated above the pinch. In this case, the overall utility consumption will be the same as that shown in Figure 21.1b. Below the pinch, the condenser can be serviced directly by cold utility with the reboiler integrated below the pinch. Now the overall utility consumption will be the same as that shown in Figure 21.1c.

21.3 USE OF THE GRAND COMPOSITE CURVE FOR HEAT INTEGRATION OF DISTILLATION

The appropriate placement principle can only be applied if the process has the capacity to provide or accept the required heat duties. A quantitative tool is needed to assess the source and sink capacities of any given background process. For this purpose, the grand composite curve can be used. Given that the dominant heating and cooling duties associated with the distillation column are the reboiler and condenser duties, a convenient representation of the column is therefore a simple "box" representing the reboiler and condenser loads[2]. This "box" can be matched with the grand composite representing the remainder of the process. The grand composite curve would include all heating and cooling duties for the process, including those associated with separator feed and product heating and cooling, but excluding reboiler and condenser loads.

Consider now a few examples of the use of this simple representation. A grand composite curve is shown in Figure 21.2a. The distillation column reboiler and condenser duties are shown separately and are matched against it. The reboiler and condenser duties are on opposite sides of the heat recovery pinch and the column does not fit. In Figure 21.2b, although the reboiler and condenser duties are both above the pinch, the heat duties prevent a fit. Part of the duties can be accommodated, and if heat integrated,

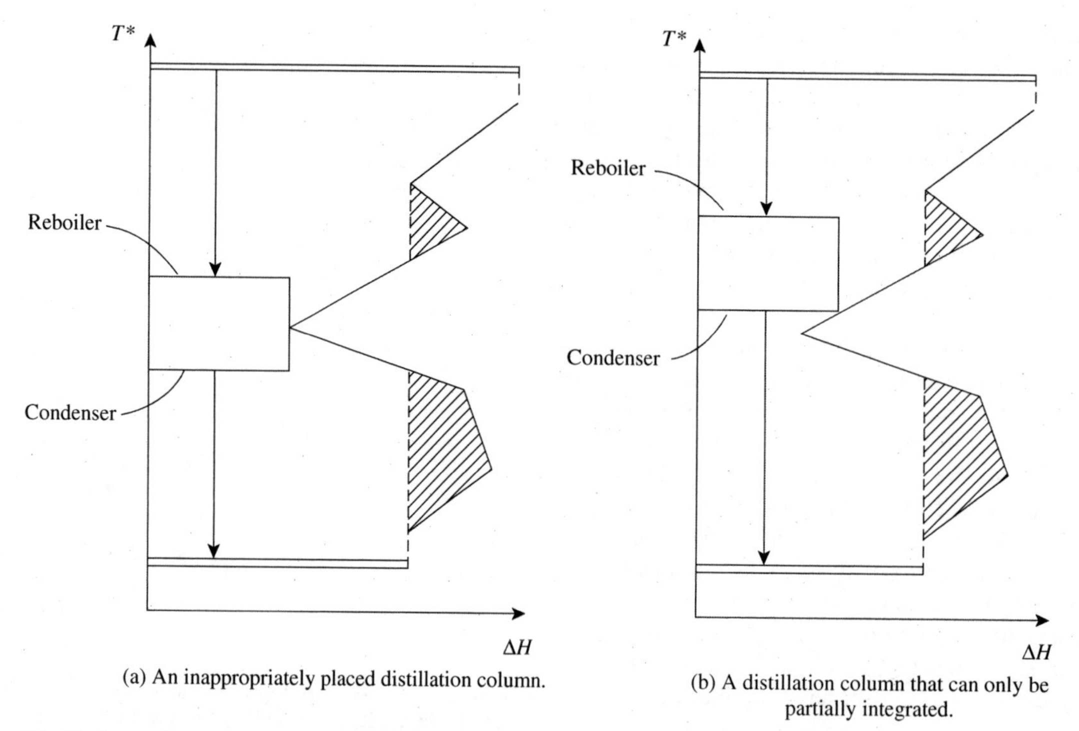

(a) An inappropriately placed distillation column.

(b) A distillation column that can only be partially integrated.

Figure 21.2 Distillation columns that do not fit against the grand composite curve. (From Smith R and Linnhoff B, 1988, *Trans IChemE ChERD*, **66**: 195, reproduced by permission of the Institution of Chemical Engineers.)

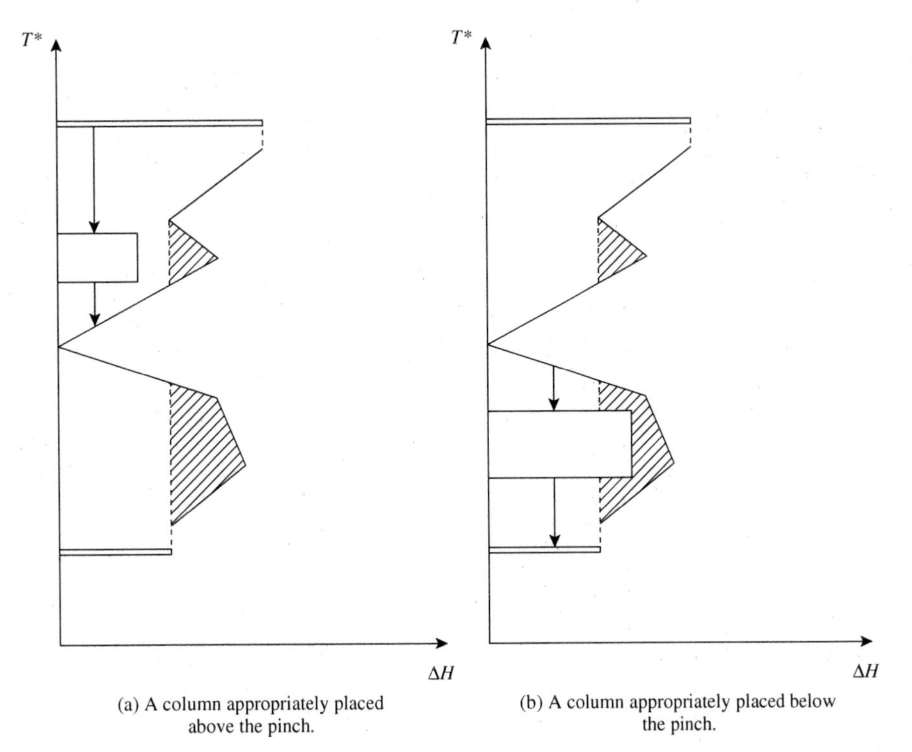

(a) A column appropriately placed above the pinch.

(b) A column appropriately placed below the pinch.

Figure 21.3 Distillation columns that fit against the grand composite curve. (From Smith R and Linnhoff B, 1988, *Trans IChemE ChERD*, **66**: 195, reproduced by permission of the Institution of Chemical Engineers.)

that would be a saving, but less than the full reboiler and condenser duties.

The distillation columns shown in Figure 21.3 both fit. Figure 21.3a shows a case in which the reboiler duty can be supplied by hot utility. The condenser duty must be integrated with the rest of the process. Another example is shown in Figure 21.3b. This distillation column also fits. The reboiler duty must be supplied by integration with the process. Part of the condenser duty in Figure 21.3b must also be integrated, while the remainder of the condenser duty can be rejected to cold utility.

21.4 EVOLVING THE DESIGN OF SIMPLE DISTILLATION COLUMNS TO IMPROVE HEAT INTEGRATION

If an inappropriately placed distillation column is shifted above the heat recovery pinch by changing its pressure, the condensing stream, which is a hot stream, is shifted from below to above the pinch. The reboiling stream, which is a cold stream, stays above the pinch. If the inappropriately placed distillation column is shifted below the pinch, then the reboiling stream, which is a cold stream, is shifted from above to below the pinch. The condensing stream stays below the pinch. Thus appropriate placement is a particular case of shifting streams across the pinch, which in turn is a particular case of the plus–minus principle (see Chapter 19).

If a distillation column is inappropriately placed across the pinch, it may be possible to change its pressure to achieve appropriate placement. Of course, as the pressure is changed, the shape of the "box" also changes, since not only do the reboiler and condenser temperatures change but also the difference between them. The relative volatility will also be affected, generally decreasing with increasing pressure. Thus, both the height and the width of the box will change as the pressure changes. Changes in pressure also affect the heating and cooling duties for column feed and products. These streams normally would be included in the background process. Hence, the shape of the grand composite curve will also change to some extent as the column pressure changes. However, as pointed out previously, it is likely that these effects will not be significant in most processes, since the sensible heat loads involved will usually be small by comparison with the latent heat changes in condensers and reboilers.

If the distillation column will not fit either above or below the pinch, then other design options can be considered. One possibility is *double-effect distillation* as shown in Figure 21.4a[3]. The column feed is split and fed to two separate parallel columns. The classical application of double-effect distillation is to choose the relative pressures of the columns such that the heat from the condenser of the high-pressure column can be used to provide the reboiler heat to the low-pressure column. In isolation, the scheme would save energy, approximately halving the energy consumption by using the same energy twice in different temperature ranges. The energy reduction will

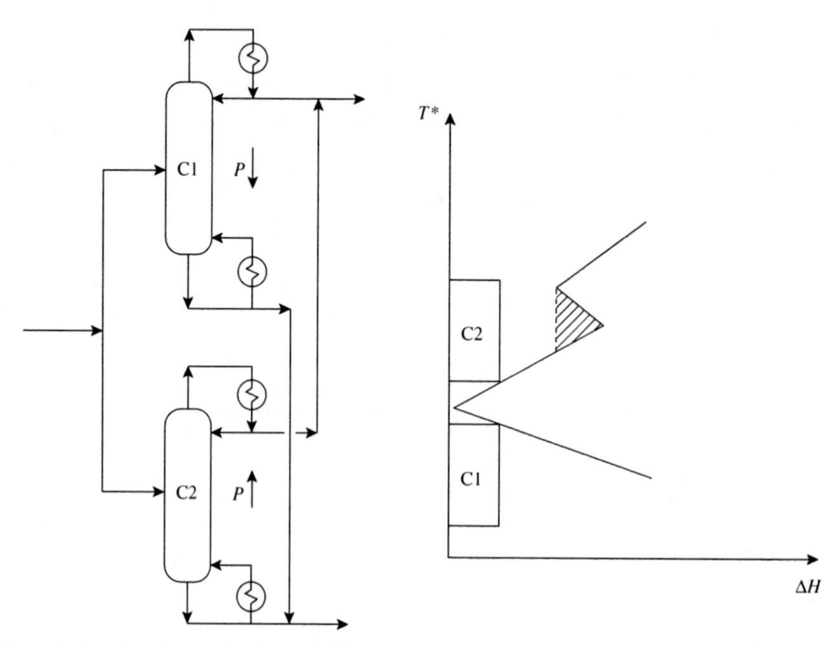

Figure 21.4 Double-effect distillation. (From Smith R and Linnhoff B, 1988, *Trans IChemE ChERD*, **66**: 195, reproduced by permission of the Institution of Chemical Engineers.)

be at the expense of increased capital cost. However, used on a stand-alone basis in this way, in reducing the heat load on the system, the temperature difference over the distillation system increases. If an attempt is made to heat integrate this double-effect distillation with the rest of the process, the increased temperature difference across the system might create problems. The increased temperature difference might prevent integration above the pinch (perhaps because the required high temperature in the reboiler creates fouling) or below the pinch (perhaps because the required low temperature in the condenser would require expensive refrigeration).

However, there is no fundamental reason why these two columns must be linked together thermally. Figure 21.4b shows two columns that are not linked thermally and, as a result, each can be individually appropriately placed. Obviously, the capital cost of such a scheme will be higher than that of a single column, but it may be justified by favorable energy savings.

Another design option that can be considered if a column will not fit into the grand composite curve is the use of an intermediate condenser, as illustrated in Figure 21.5. The shape of the "box" is now altered, because the intermediate condenser changes the heat flow through the column with some of the heat being rejected at a higher temperature in the intermediate condenser. The particular design shown in Figure 21.5, the match with the grand composite curve would require that at least part of the heat rejected from the intermediate condenser should be passed to the process. An analogous approach can be used to evaluate the possibilities for the use of intermediate reboilers. For intermediate reboilers, part of the reboiler heat is supplied at an intermediate point in the column,

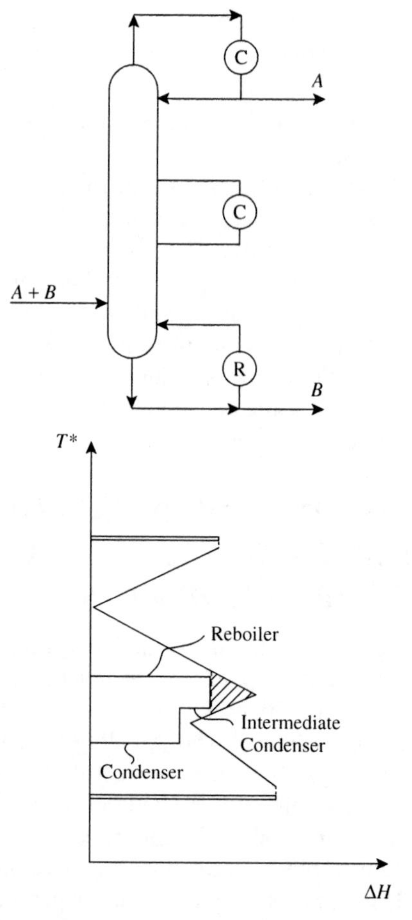

Figure 21.5 Distillation of column with intermediate condenser. The profile can be designed to fit the background process. (From Smith R and Linnhoff B, 1988, *Trans IChemE ChERD*, **66**: 195, reproduced by permission of the Institution of Chemical Engineers.)

at a temperature lower than the reboiler temperature. Flower and Jackson[4], Kayihan[5] and Dhole and Linnhoff[6] have presented procedures for the location of intermediate reboilers and condensers.

21.5 HEAT PUMPING IN DISTILLATION

Various heat-pumping schemes have been proposed as a means of saving energy in distillation. Of these schemes, use of the column overhead vapor as the heat pumping fluid is usually the most economically attractive. Such schemes are known as *vapor recompression*, as illustrated in Figure 21.6.

For heat pumping to be economic on a stand-alone basis, it must operate across a small temperature difference, which for distillation means close boiling mixtures. In addition, the use of the scheme is only going to make sense if the column is constrained to operate either on a stand-alone basis or at a pressure that would mean it would be across the pinch. Otherwise, heat integration with the process might be a much better option. Vapor recompression schemes for distillation therefore only make sense for the distillation of close boiling mixtures in constrained situations[3].

21.6 CAPITAL COST CONSIDERATIONS

The design changes suggested so far for distillation columns have been motivated by the incentive to reduce energy costs by more effective integration between the distillation column and the rest of the process. There are, however, capital cost implications when the distillation design and the heat integration scheme are changed. These implications fall into two broad categories: changes in distillation capital cost and changes in heat exchanger network capital cost. Obviously, these capital cost changes should be considered together, along with the energy cost changes, in order to achieve an optimum trade-off between capital and energy costs.

1. *Distillation capital costs.* The classical optimization in distillation, as discussed in Chapter 9, is to trade-off capital cost of the column against energy cost for the distillation by changing the reflux ratio. In Chapter 9, this was discussed from the situation of distillation columns operating on utilities and not integrated with the rest of the process. Experience gained with this traditional optimization has led to rules of thumb for the selection of reflux ratios. Typically, the optimum ratio of actual to minimum reflux ratio is usually around 1.1. Practical considerations often prevent a ratio of less than 1.1 being used, as discussed in Chapter 9.

If the column is inappropriately placed with the process, then an increase in reflux ratio causes a corresponding overall increase in energy, and the trade-off rules apply. If, however, the column is appropriately placed, then the reflux ratio can often be increased without changing the overall energy consumption as shown in Figure 21.7. Increasing the heat flow through the column decreases the requirement for distillation stages but increases the vapor rate. In designs initialized by traditional rules of thumb, this would have the effect of decreasing the capital cost of the column. However, the corresponding decrease in heat flow through

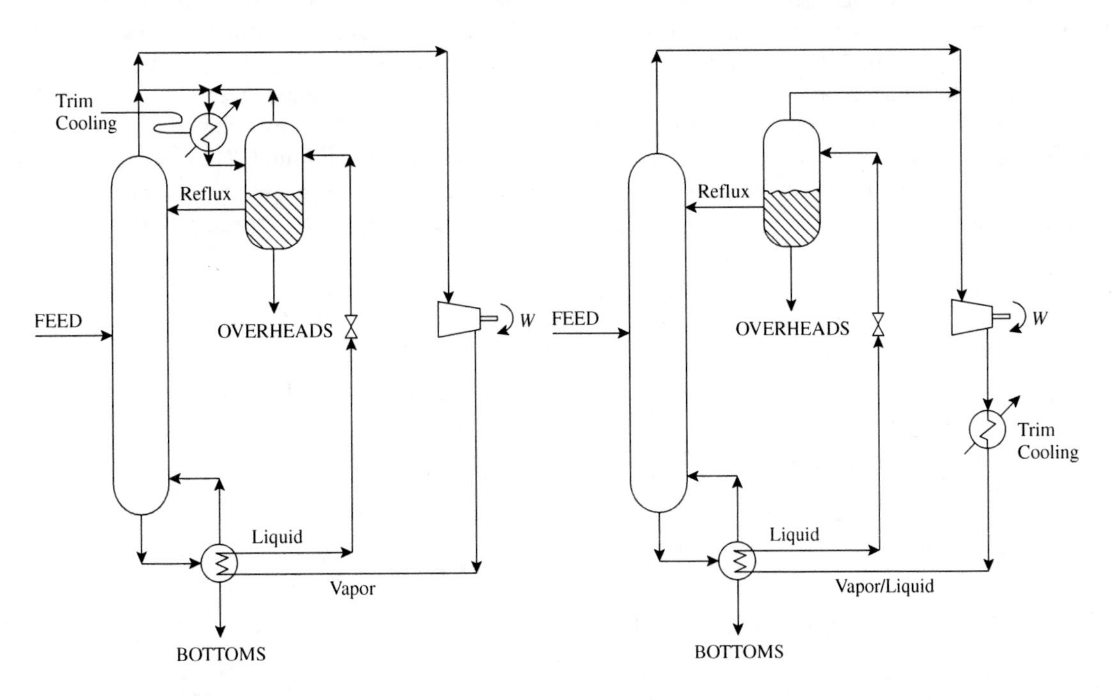

(a) A typical vapor recompression scheme for above ambient temperature distillation.

(b) A typical vapor recompression scheme for below ambient temperature distillation.

Figure 21.6 Heat pumping in distillation. Vapor recompression schemes. (From Smith R. and Linnhoff B, 1988, *Trans IChemE ChERD*, **66**: 195, reproduced by permission of the Institution of Chemical Engineers.)

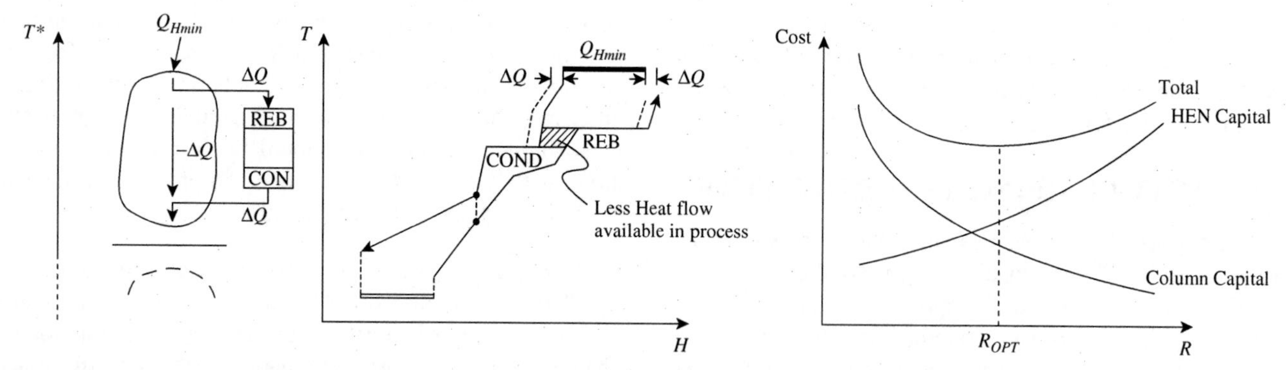

Figure 21.7 The capital/capital trade-off for an appropriately integrated distillation column. (From Smith R and Linnhoff B, 1988, *Trans IChemE ChERD*, **66**: 195, reproduced by permission of the Institution of Chemical Engineers.)

the process will have the effect of decreasing temperature differences and increasing the capital cost of the heat exchanger network, as shown in Figure 21.7. Thus, the trade-off for an appropriately integrated distillation column becomes one between the capital cost of the column and the capital cost of the heat exchanger network[3], Figure 21.7.

Consequently, the optimum reflux ratio for an appropriately integrated distillation column will be problem specific and is likely to be quite different from that of a stand-alone column operated from utilities.

2. *Heat exchanger network capital costs.* It is easy for the designer to become carried away with the elegance of "packing boxes" into space around the grand composite curve. However, the full implications of integration are only clear when the corresponding composite curves of the process with the distillation column are considered. Temperature differences become smaller throughout the process as a result of the integration. This means that the capital-energy trade-off should be readjusted, and a larger ΔT_{min} might be required. The optimization of the capital-energy trade-off might undo part of the savings achieved by appropriate integration.

Unfortunately, the overall design problem is even more complex in practice. Large temperature differences in the process (i.e. space in the grand composite curve) could equally well be exploited to allow the use of moderate temperature utilities or the integration of heat engines, heat pumps, and so on, in preference to integration of distillation columns. There is thus a three-way trade-off between distillation design and integration, utility selection and the capital-energy trade-off (ΔT_{min} optimization).

21.7 HEAT INTEGRATION CHARACTERISTICS OF DISTILLATION SEQUENCES

The problem of distillation sequencing was discussed in Chapter 11, where the distillation columns in the sequence were operated on a stand-alone basis using utilities for the reboilers and condensers. Following the approach in Chapter 11, the best few nonintegrated distillation sequences would be found. These sequences would then be heat integrated as discussed above. Figure 21.8 shows how heat integration can be applied within a two-column direct

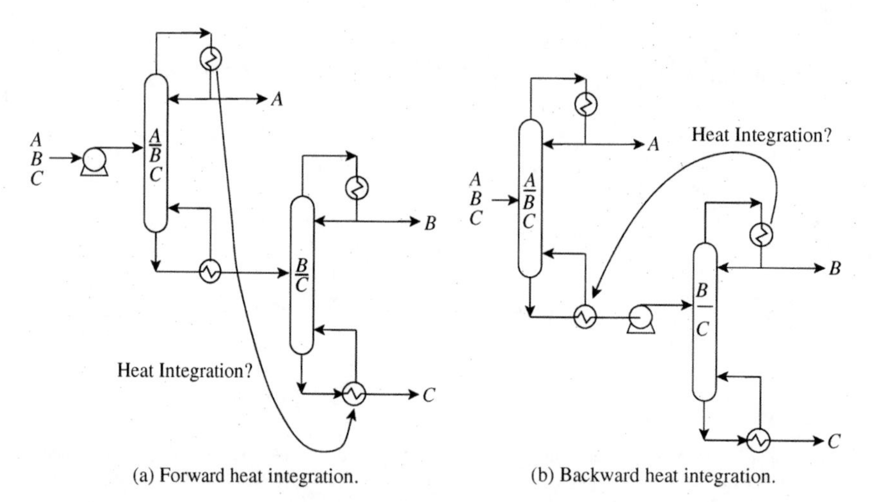

Figure 21.8 Heat integration of a sequence of two simple distillation columns. (From Smith R and Linnhoff B, 1988, *Trans IChemE, CHERD*, **66**: 195, reproduced by permission of the Institution of Chemical Engineers.)

distillation sequence for the separation of three products. In Figure 21.8a, the first distillation column has been increased in pressure such that the condenser of the first column can provide the heat for the reboiler of the second column, sometimes known as *forward integration*. In Figure 21.8b, the pressure of the second column has been increased such that the condenser of the second column can provide the heat for the reboiler of the first, sometimes known as *backward integration*. Both schemes will bring about a significant reduction in the energy requirement.

This approach to the overall problem breaks down the design procedure into two steps of first determining the best nonintegrated sequence and then heat integrating. This assumes that the two problems of distillation sequencing and heat integration can be decoupled. Studies have been carried out where sequences of simple distillation columns were first developed without heat integration and all reboilers and condensers serviced by utilities[7,8]. These sequences were then heat integrated. Comparison of the nonintegrated and integrated sequences showed that the configurations that achieved the greatest energy saving by integration were already amongst those with the lowest energy requirement prior to integration[7,8]. The result is illustrated in Figure 21.9. The best few nonintegrated sequences tend to turn out to be the best few integrated sequences in terms of total cost (capital and operating). It must be emphasized that this decoupling depends on the absence of significant constraints limiting the heat integration potential within the distillation sequence. If there are significant constraints, for example, limitations on the pressure of some of the columns due to reboiler fouling that limit the heat integration potential, then the result might not apply.

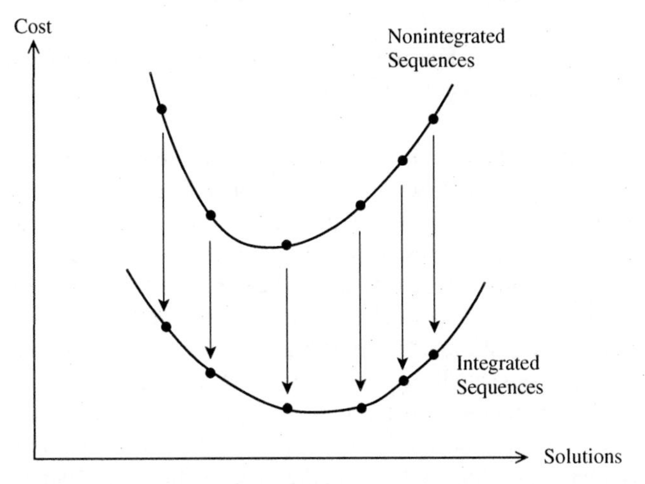

Figure 21.9 For unconstrained sequences of simple columns, the best few nonintegrated sequences tend to turn into the best few integrated sequences. (From Smith R and Linnhoff B, 1988, *Trans IChemE ChERD*, **66**:195 reproduced by permission of the Institution of Chemical Engineers.)

It is important to emphasize that, when considering distillation sequences, the focus should not be exclusively on the single sequence with the lowest overall cost. Rather, because there is often little difference in cost between the best few sequences, and because of the uncertainties in the calculations and the fact that other factors need to be considered in a more detailed evaluation, the best few should be evaluated in more detail rather than just one.

If the design problem in the absence of significant constraints can be decoupled in this way, there must be some mechanism behind this. Take two different sequences for the separation of a four-component mixture, Figure 21.10[3]. Summing the feed flowrates of the key components (see Chapter 9) to each column in the sequence, the total flowrate is the same in both cases, Figure 21.10. However, the flow of nonkey components is different, Figure 21.10.

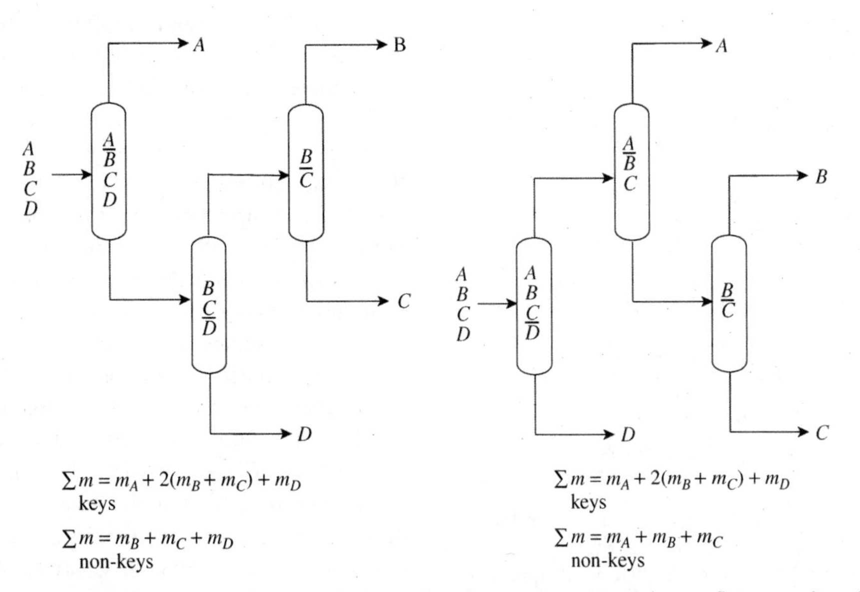

Figure 21.10 The mass flowrate of key components are the same for all sequences, but the mass flowrate of nonkey components varies. (From Smith R and Linnhoff B, 1988, *Trans IChemE ChERD*, **66**:195 reproduced by permission of the Institution of Chemical Engineers.)

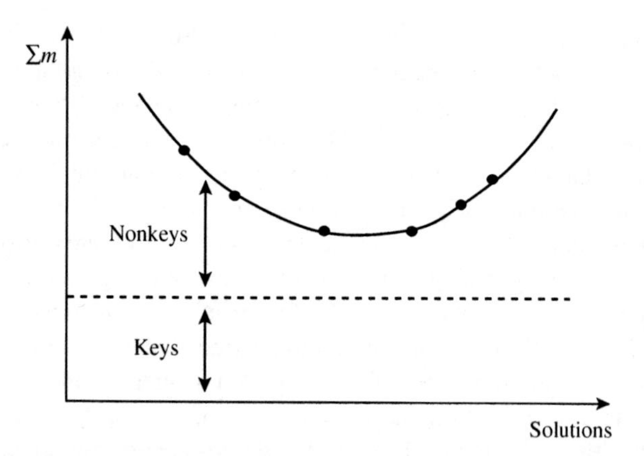

Figure 21.11 The overall flowrate decomposed into key and nonkey components. (From Smith R and Linnhoff B, 1988, *Trans IChemE ChERD*, **66**:195 reproduced by permission of the Institution of Chemical Engineers.)

In general, the flow of key components is constant and independent of the sequence while the flow of nonkeys varies according to the choice of sequence, as illustrated in Figure 21.11[3].

It thus appears that the flowrate of the nonkeys may account for the differences between sequences. Essentially, nonkey components have two effects on a separation. They cause[8]:

1. an unnecessary load on the separation, leading to higher heat loads and vapor rates;
2. a widening of the temperature differences across columns since, light nonkey components cause a decrease in condenser temperature and heavy nonkey components cause an increase in the reboiler temperature.

These effects can be expressed quantitatively in temperature-heat profiles as illustrated in Figure 21.12. A high flowrate of nonkey components leads simultaneously to higher loads and more extreme levels, Figure 21.12.

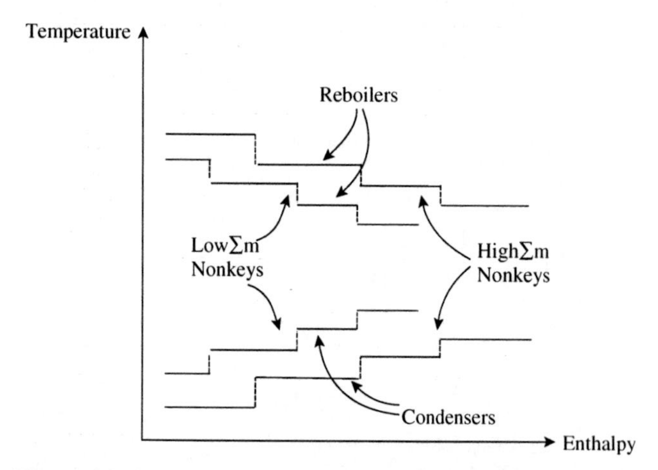

Figure 21.12 Temperature-heat profiles for sequences of simple columns with no constraints having a low flowrate of nonkey components tend to be favorable. (From Smith R and Linnhoff B, 1988, *Trans IChemE ChERD*, **66**:195 reproduced by permission of the Institution of Chemical Engineers.)

Whether heat integration is restricted to the separation system or allowed with the rest of the process, integration always benefits from colder reboiler streams and hotter condenser streams. This point has been dealt with in more general terms in Chapter 19. In addition, when column pressures are allowed to vary, columns with smaller temperature differences are easier to integrate, since smaller changes in pressure are required to achieve suitable integration. However, is there any conflict with capital cost? A column sequence that handles a large amount of heat must have a high capital cost for two reasons:

1. Large heat loads to be transferred result in large reboilers and condensers.
2. Large heat loads will cause high vapor rates and these require large column diameters.

It is thus unlikely that a distillation sequence will have a small capital cost and a large operating cost or vice versa. A nonintegrated sequence with a large heat load has high vapor rates, large heat transfer areas and large column diameters. Even if the heat requirements of this sequence are substantially reduced by integration, large heat transfer areas and large diameters must still be provided. Any savings in utility costs achieved by heat integration must compensate for these high capital costs before this sequence can become competitive with a nonintegrated sequence starting with a smaller heat load. Thus, capital cost considerations reinforce the argument that the best few nonintegrated sequences with the lowest heat load are those with the lowest total cost.

It is interesting to reflect on the heuristics for sequencing of distillation columns in Chapter 11. Heuristics 2, 3 and 4 from Section 11.1 tend to minimize the flowrate of nonkey components. Heuristic 1 relates to special circumstances when there is a particularly difficult separation[8].

The mechanism by which nonkey components affect a given separation is more complex in practice than the broad arguments presented here. There are complex interrelationships between the volatility of the key and nonkey components, and so on. Also, it is often the case that the distillations system has constraints to prevent certain heat integration opportunities. Such constraints will often present themselves as constraints over which the pressure of the distillation columns will operate. For example, it is often the case that the maximum pressure of a distillation column is restricted to avoid decomposition of material in the reboiler. This is especially the case when reboiling high molar mass material. Distillation of high molar mass material is often constrained to operate under vacuum conditions. Clearly, if the pressure of the distillation column is constrained, then this restricts the heat integration opportunities. Another factor that can create problems is that, as the pressure of each column in the sequence is varied, this has implications for the downstream column in terms of the feed condition for the downstream column. If two columns are at the same pressure, then a saturated liquid leaving one column will be saturated as it enters the second column. However, if the pressure of the first column becomes higher than the second as a result of a pressure change, then the saturated liquid from the first column will become a subcooled feed to the second. If the pressure of the first column becomes lower than the second as a result of a pressure change, then the saturated liquid from the first column will become a partially vaporized or superheated feed to the second. The feed condition can have a significant influence on the column design.

Constraints might be applied for the sake of reducing the capital costs (e.g. to avoid long pipe runs). In addition, constraints might be applied to avoid complex heat integration arrangements for the sake of operability and control (e.g. to have heat recovery to a reboiler from a single source of heat, rather than two or three sources of heat).

In addition to these issues regarding constraints for simple columns, there is also the issue of the introduction of complex columns into the sequence. Figure 21.13a illustrates the thermal characteristics of a direct sequence of two simple columns. Once the two columns are thermally coupled, as illustrated in Figure 21.13b, the overall heat load is reduced. However, all of the heat must be supplied at the highest temperature for the system. Thus there is a trade-off in which the load is reduced, but the levels required to supply the heat become more extreme. The corresponding case for the indirect sequence is shown in Figure 21.14. As the indirect sequence is thermally coupled, the heat load is reduced, but now all of the heat must be rejected at the lowest temperature. Thus, there is a benefit of reduced load but a disadvantage of heat rejection at more extreme levels. The same problem occurs with

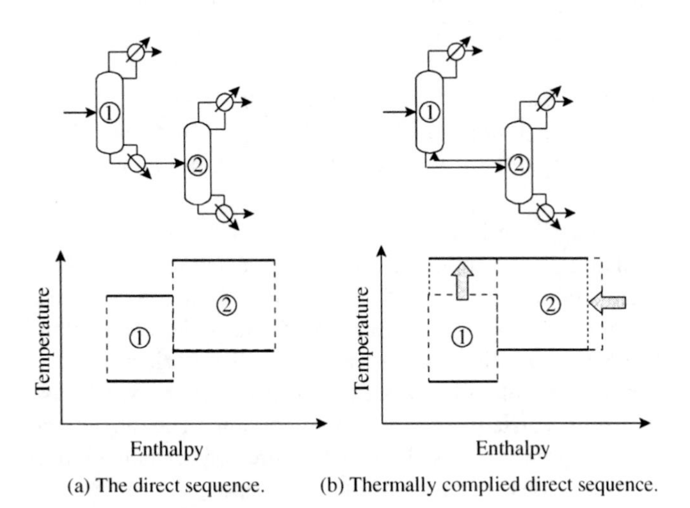

(a) The direct sequence. (b) Thermally complied direct sequence.

Figure 21.13 Thermally coupling the direct sequence changes both the loads and levels.

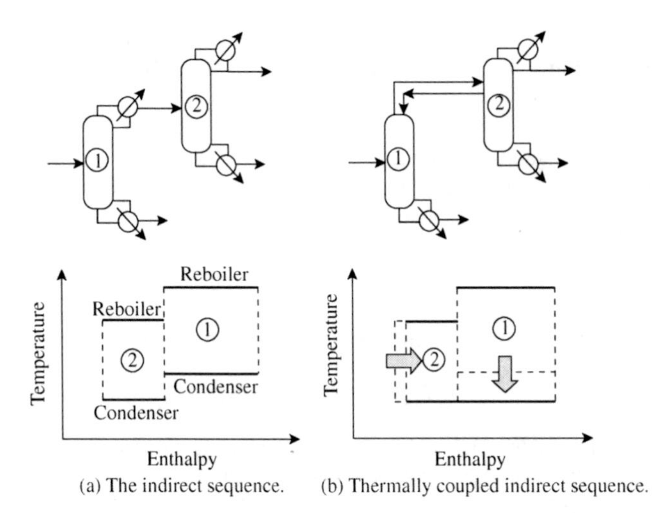

(a) The indirect sequence. (b) Thermally coupled indirect sequence.

Figure 21.14 Thermally coupling the indirect sequence changes both the loads and levels.

thermally coupled prefractionators. However, this time both heat supply and rejection must be carried out at the most extreme temperatures.

All of these arguments demand a more sophisticated approach to the heat integration of distillation sequences, such that the sequence and heat integration (including complex columns) are explored simultaneously.

Example 21.1 Two distillation columns have been sequenced to be in the direct sequence (see Figure 21.8). Opportunities for heat integration between the two columns are to be explored. The operating pressures of the two columns need to be chosen to allow heat recovery. Data for Column 1 and Column 2 at various pressures are given in Tables 21.1 and 21.2.

Medium-pressure (MP) steam is available for reboiler heating at 200°C. Cooling water is available for condensation, to be returned

Table 21.1 Data for Column 1.

P (bar)	T_{COND} (°C)	T_{REB} (°C)	Q_{COND} (kW)	Q_{REB} (kW)
1	90	120	3000	3000
2	130	160	3600	3600
3	140	170	4000	4000
4	160	190	4300	4300

Table 21.2 Data for Column 2.

P (bar)	T_{COND} (°C)	T_{REB} (°C)	Q_{COND} (kW)	Q_{REB} (kW)
1	110	130	5500	5500
2	130	153	6000	6000
3	150	175	6300	6300
4	163	190	6500	6500
5	170	200	6600	6600

to the cooling tower at 30°C. Assume a minimum permissible temperature difference for heat transfer of 10°C. Determine the minimum utility requirements for:

a. both columns operating at 1 bar,
b. forward heat integration,
c. backward heat integration.

Solution

a. Heat integration between condenser and reboiler is feasible if

$$T_{COND} \geq T_{REB} + \Delta T_{min}$$

From Tables 21.1 and 21.2, when both columns operate at 1 bar, no heat integration is possible. Thus,

$$Q_{Hmin} = 3000 + 5500$$
$$= 8500 \text{ kW}$$
$$Q_{Cmin} = 3000 + 5500$$
$$= 8500 \text{ kW}$$

b. Consider forward heat integration by increasing the pressure of Column 1. Because heat duties will increase with increasing pressure, low operating pressures are preferred. Thus, Column 2 is kept at 1 bar with a reboiler temperature of 130°C. This means that the minimum condensing temperature of Column 1 must be 140°C, which corresponds with a pressure of 3 bar. From Tables 21.1 and 21.2:

$$Q_{Hmin} = 4000 + (5500 - 4000)$$
$$= 5500 \text{ kW}$$
$$Q_{Cmin} = 0 + 5500$$
$$= 5500 \text{ kW}$$

c. For backward heat integration, the appropriate operating pressures are 1 bar for Column 1 and 2 bar for Column 2. Thus, from Tables 21.1 and 21.2:

$$Q_{Hmin} = 0 + 6000$$
$$= 6000 \text{ kW}$$
$$Q_{Cmin} = 3000 + (6000 - 3000)$$
$$= 6000 \text{ kW}$$

Thus, to minimize utility costs, forward heat integration should be used with Column 1 at 3 bar and Column 2 at 1 bar.

21.8 HEAT-INTEGRATED DISTILLATION SEQUENCES BASED ON THE OPTIMIZATION OF A SUPERSTRUCTURE

In order to explore distillation sequencing and heat integration simultaneously, a more automated approach to the problem is required. First, the variation in reboiler duty and temperature and condenser duty and temperature with pressure needs to be modeled. If the basic separation task is fixed, then a model can be developed for the separation task and its pressure varied. This will provide the relationship between the duties, temperatures and pressures. The variation of the heat duties and temperatures with pressure will be nonlinear. However, the nonlinear behavior can be approximated using piecewise linearization. Such techniques were discussed in Chapter 3 for modeling nonlinear behavior and turning a nonlinear model into a linear model for use in optimization. This establishes quantitative information on the variation of heat duties and temperatures with pressure, as required for exploring heat integration. All the separation tasks for the sequence need to be modeled in this way. In Chapter 11, an approach was developed for the sequencing of nonintegrated distillation columns based on the optimization of a superstructure of the distillation tasks[9]. In principle, this could be extended to include also the heat integration opportunities. Binary variables can be introduced into the optimization to represent the existence of possible heat integration matches. Each possible match between a condenser and a reboiler, a condenser and a cold steam and a reboiler and a hot stream would require a binary variable to represent its existence. Binary variables would also be needed to represent the relative temperatures of condensers and reboilers (i.e. whether heat transfer would be feasible). This introduces a significant number of new binary variables into the optimization and makes it very difficult to solve for large problems. Thus, a different representation for the superstructure for the heat-integrated columns is required[10].

The minimum number of simple columns required to separate NC end products is $(NC - 1)$. An equipment-based superstructure can be developed, rather than a task-based superstructure. As an example, Figure 21.15a shows sequences of three columns for the separation of a four-product mixture. The tasks that can be carried out by columns can be grouped according to the key components.

- A/B separation – tasks *A/BCD, A/BC* and A/B
- B/C separation – tasks *AB/CD, B/CD, AB/C* and B/C
- C/D separation – tasks *ABC/D, BC/D* and C/D

An equipment-based superstructure can be developed containing the minimum number of columns. This is illustrated in Figure 21.15b where three Columns X, Y and Z can carry out one of a group of tasks. The connections between the columns depend on the task selection. The synthesis problem becomes one of determining which task is to be carried out in which column. The advantage of the equipment-based superstructure for heat integration considerations is that heat-integrated options are related to the number of columns and not to the number of separation tasks. Thus, in Figure 21.15b, only the condenser/reboiler

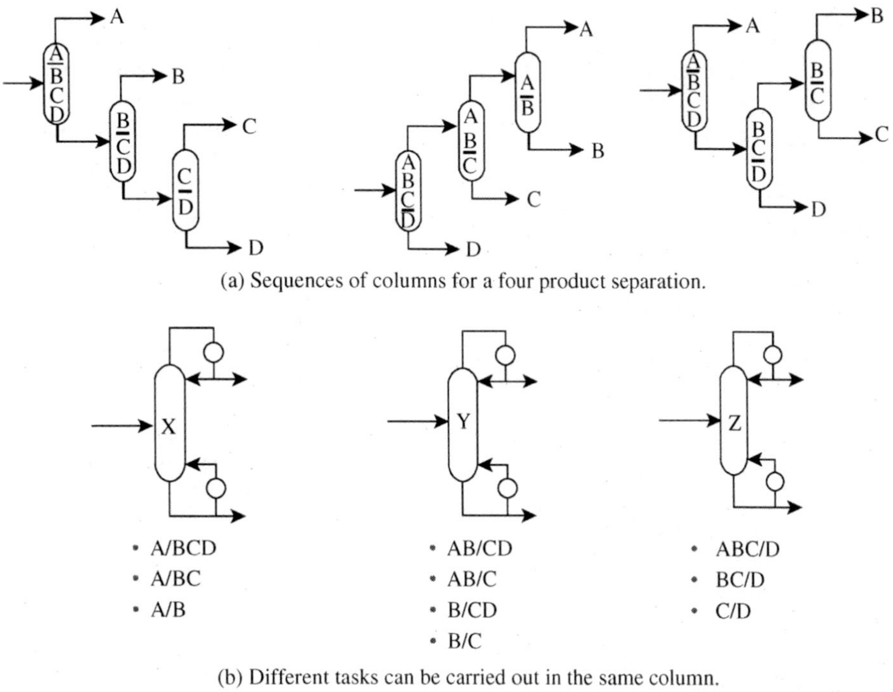

(a) Sequences of columns for a four product separation.

- A/BCD
- A/BC
- A/B

- AB/CD
- AB/C
- B/CD
- B/C

- ABC/D
- BC/D
- C/D

(b) Different tasks can be carried out in the same column.

Figure 21.15 An equipment-based superstructure.

connections between Columns X, Y and Z need to be considered for heat integration, rather than each separation task. Binary variables can be used to represent the matches between condensers and reboilers, relative temperatures, and so on. Using an equipment-based superstructure dramatically reduces the number of variables. Binary variables are associated with assigning a Task i to Column j and also to indicate whether a given heat integration match is active[11].

If nonlinear capital costs considerations are left out of the formulation, then the problem can be formulated as an MILP problem (see Chapter 3). Once nonlinear costs are included, the problem becomes an MINLP, with all of the problems associated with nonlinear optimizations. As with many such problems, if the required formulation turns out to be an MINLP, then an MILP can be used to provide a good initialization for the MINLP, simplifying its solution.

21.9 HEAT INTEGRATION OF DISTILLATION COLUMNS – SUMMARY

The appropriate placement of distillation columns when heat integrated is not across the heat recovery pinch. The grand composite curve can be used as a quantitative tool to assess integration opportunities. The scope for integrating conventional distillation columns into an overall process is often limited. Practical constraints often prevent integration of columns with the rest of the process. If the column cannot be integrated with the rest of the process, or if the potential for integration is limited by the heat flows in the background process, then attention must be turned back to the distillation operation itself and complex arrangements considered.

Once the distillation is integrated, the driving forces between the composite curves become smaller. This in turn means that the capital-energy trade-off for the heat exchanger network should be readjusted accordingly.

When heat integrating nonintegrated simple distillation sequences in the absence of significant constraints, the best few nonintegrated distillation sequences are usually amongst the best few integrated sequences. This results from excessive flowrates of nonkey components increasing reboiler and condenser duties and, at the same time, widening the temperature differences across the sequence as a whole. Larger heat duties and more extreme temperatures tend to reduce the heat integration opportunities. However, this tends to happen only in the absence of significant constraints.

The use of complex columns (side-strippers, side-rectifiers and thermally coupled prefractionators) reduces the overall heat duties for the separation at the expense of more extreme temperatures for reboiling and condensing. Heat integration benefits from smaller duties, but more extreme temperatures make the heat integration more difficult).

Thus, the introduction of constraints and complex columns demands a simultaneous solution of the sequencing and heat recovery problems. This can be carried out on the basis of the optimization of a superstructure.

21.10 EXERCISES

1. Table 21.3 represents a problem table cascade ($\Delta T_{min} = 20°C$).

Table 21.3 Heat flow cascade for Exercise 1.

Interval temperature (°C)	Heat flow (kW)
160	1000
150	0
130	1100
110	1400
100	900
80	1300
40	1400
10	1800
−10	1900
−30	2200

The utilities available are given in Table 21.4. Both low pressure (LP) and medium pressure (MP) steam are available.

Table 21.4 Utility data for Exercise 1.

Utility	Cost
Cooling water (20 to 40°C)	Cost = 0.3 × [\$·kw^{-1}·y^{-1}]
Refrigeration at 0°C and – 40°C	Cost of electricity = 7.5 × [\$·kW^{-1}·y^{-1}]
LP steam at 107°C raised from boiler feed water at 60°C	Credit for LP steam = 1.8 × [\$·kW^{-1}·y^{-1}]
MP steam at 200°C	Cost = 2.25 × [\$·kW^{-1}·y^{-1}]

a. By drawing the process grand composite curve, set refrigeration and other utility loads below the pinch around the full process steam raising capacity. Calculate the duties of all these. Data for boiler feed water are:

$$\text{Specific heat capacity} = 4.2\ (\text{kJ}\cdot\text{kg}^{-1}\cdot\text{K}^{-1})$$

$$\text{Latent heat of vaporization} = 2238\ (\text{kJ}\cdot\text{kg}^{-1})$$

b. The power required by the refrigeration levels is given by

$$W = \frac{Q_C}{0.6}\left(\frac{T_H - T_C}{T_C}\right)$$

where W = power required for the refrigeration cycle
Q_C = the cooling duty
T_C = temperature at which heat is taken into the refrigeration cycle (K)
T_H = temperature at which heat is rejected from the refrigeration cycle (K)

Assume heat rejection from refrigeration is to cooling water. Calculate the total energy cost of the utilities.

c. A distillation column presently using MP steam and cooling water is to be integrated into this process. If the reboiler temperature is 120°C, the condenser is 90°C and each has a duty of 1100 kW, redraw the process grand composite including the integrated distillation column. Show how the above utilities would now be best placed below the pinch. Determine whether the integration of the column has led to an overall energy cost saving.

2. A problem table analysis for a given process produces the heat flow cascade in Table 21.5 for $\Delta T_{min} = 10°C$.

Table 21.5 Problem table cascade for Exercise 2.

Interval temperature (°C)	Cascade heat flow (MW)
295	18.3
285	19.8
185	4.8
145	0
85	10.8
45	12.0
35	14.3

a. A distillation column, which separates a mixture of toluene and diphenyl into relatively pure products, is to be integrated with the process. The operating pressure of the column has been fixed initially to atmospheric (1.013 bara). At this pressure, the toluene condenses overhead at a constant temperature of 111°C and diphenyl is reboiled at a constant temperature of 255°C. What would be the consequence of integrating the distillation column with the process at a pressure of 1.013 bara?

b. Can you suggest a more appropriate operating pressure for the distillation column if it is to be integrated with the process? The reboiler and condenser loads are both 4.0 MW and can be assumed not to change significantly with pressure. The vapor pressures of toluene and diphenyl can be represented by:

$$\ln P_i = A_i - \frac{B_i}{T + C_i}$$

where P_i is the vapor pressure (bar), T is the absolute temperature (K) and A_i, B_i and C_i are constants, which are given in Table 21.6.

Table 21.6 Vapor pressure constants.

Component	A_i	B_i	C_i
Toluene	9.3935	3096.52	−53.67
Diphenyl	10.0630	4602.23	−70.42

Vacuum operation should be avoided and the reboiler temperature kept as low as possible to minimize fouling.

c. Sketch the shape of the grand composite curve after the distillation column has been integrated.

3. In Example 21.1, consider introducing a double-effect column to replace Column 2 for the backward heat integration arrangement. Assume an equal split of flowrate into the double-effect column. Is there any benefit from the arrangement?
4. A direct sequence of two distillation columns produces three products A, B and C. The feed condition and operating pressures are to be chosen to maximize heat recovery opportunities. To simplify the calculations, assume that condenser duties do not change when changing from saturated liquid to saturated vapor feed. This will not be true in practice, but simplifies the exercise. Assume also that the reboiler duty for saturated liquid feed is the sum of the reboiler duty for saturated vapor feed plus the heat duty to vaporize the feed. Data for the two columns are given in Tables 21.7 and 21.8.

Table 21.7 Exercise 4 data for Column 1.

P (bar)	T_{COND} (°C)	T_{REB} (°C)	Saturated liquid feed		Saturated vapor feed	
			Q_{COND} (kW)	Q_{REB} (kW)	T_{FEED} (°C)	Q_{FEED} (kW)
1	90	130	3000	3000	110	2000
2	110	152	4000	4000	130	1800
3	130	173	5000	5000	150	1600
4	150	195	6000	6000	170	1500

Table 21.8 Exercise 4 data for Column 2.

P (bar)	T_{COND} (°C)	T_{REB} (°C)	Saturated liquid feed		Saturated vapor feed	
			Q_{COND} (kW)	Q_{REB} (kW)	T_{FEED} (°C)	Q_{FEED} (kW)
1	120	140	3000	3000	130	1500
2	140	165	4000	4000	150	1300
2.5	150	178	4500	4500	160	1200

Cooling water is available with a return temperature of 30°C and a cost of \$4.5 $kW^{-1} \cdot y^{-1}$. Low-pressure steam is available at a temperature of 140°C and a cost of \$90 $kW^{-1} \cdot y^{-1}$. Medium-pressure steam is available at a temperature of 200°C and a cost of \$135 $kW^{-1} \cdot y^{-1}$. The minimum temperature difference allowed is 10°C.

a. List the possible heat integration opportunities (include steam generation opportunities and feed heating opportunities).
b. Calculate the minimum cost for backward heat integration by optimizing the column pressures.
c. Calculate the minimum cost for backward heat integration by optimizing the column pressures, but disallow heat recovery between condensers and reboilers. Keep both feeds to be saturated liquids.
d. If the feeds to both columns are saturated vapor, calculate the minimum utility cost, but disallow heat recovery between condensers and reboilers. Use the pressures from Part b above.
e. Repeat Part b, but keeping the column pressures to 1 bar.

5. Consider the use of a side-rectifier or side-stripper for the separation of a three-product mixture. Assume that thermally coupled columns operate at the same pressure. Also, assume the feed to be saturated liquid. Data for the operation of the two arrangements are given in Tables 21.9 and 21.10.

Table 21.9 Side-rectifier data for Exercise 5.

P (bar)	T_{COND1} (°C)	T_{REB1} (°C)	Q_{COND1} (kW)	Q_{REB1} (kW)	T_{COND2} (°C)	Q_{COND2} (kW)
1	90	140	2000	4500	120	2500
2	110	165	2500	5500	140	3000
2.5	120	178	3000	6500	150	3500

Table 21.10 Side-stripper data for Exercise 5.

P (bar)	T_{COND1} (°C)	T_{REB1} (°C)	Q_{COND1} (kW)	Q_{REB1} (kW)	T_{REB2} (°C)	Q_{REB2} (kW)
1	90	140	5000	3000	120	2500
2	110	165	6000	3500	140	2500
2.5	120	178	7000	4000	150	3000

Using the utility data from Exercise 4:

a. When both columns operate at 1 bar, which of the two complex column arrangements will have lower utility costs?
b. Compare the results with the direct sequence of heat-integrated simple columns operating at 1 bar.
c. Optimize column pressure in both complex column arrangements to minimize utility costs.

REFERENCES

1. Umeda T, Niida K and Shiroko K (1979) A Thermodynamic Approach to Heat Integration in Distillation Systems, *AIChE J*, **25**: 423.
2. Linnhoff B, Dunford H and Smith R (1983) Heat Integration of Distillation Columns into Overall Processes, *Chem Eng Sci*, **38**: 1175.
3. Smith R and Linnhoff B (1988) The Design of Separators in the Context of Overall Processes, *Trans IChemE ChERD*, **66**: 195.
4. Flower JR and Jackson MA (1964) Energy Requirements in the Separation of Mixture by Distillation, *Trans IChemE*, **42**: T249.
5. Kayihan F (1980) Optimum Distribution of Heat Load in Distillation Columns Using Intermediate Condensers and Reboilers, *AIChE Symp Ser*, **192**(76): 1.
6. Dhole VR and Linnhoff B (1993) Distillation Column Targets, *Comp Chem Eng*, **17**: 549.
7. Freshwater DC and Ziogou E (1976) Reducing Energy Requirements in Unit Operations, *Chem Eng J*, **11**: 215.

8. Stephanopoulos G, Linnhoff B and Sophos A (1982) Synthesis of Heat Integrated Distillation Sequences, *IChemE Symp Ser*, **74**: 111.
9. Shah PB and Kokossis AC (2002) New Synthesis Framework for the Optimization of Complex Distillation Systems, *AIChE J*, **48**: 527.
10. Smith EMB and Pantelides CC (1995) Design of Reaction/Separation Networks Using Detailed Models, *Comp Chem Eng*, **19**: S83.
11. Samanta A (2001) *Modelling and Optimisation for Synthesis of Heat-Integrated Distillation Sequences in the Context of Overall Processes*, PhD Thesis, UMIST, UK.

22 Heat Integration of Evaporators and Dryers

22.1 THE HEAT INTEGRATION CHARACTERISTICS OF EVAPORATORS

Evaporation processes usually separate a single component (typically water) from a nonvolatile material, as discussed in Chapter 10. As such, it is good enough in most cases to assume that the vaporization and condensation processes take place at constant temperatures.

As with distillation, the dominant heating and cooling duties associated with an evaporator are the vaporization and condensation duties. As with distillation, there will be other duties associated with the evaporator for heating or cooling of feed, product and condensate streams. These sensible heat duties will usually be small in comparison with the latent heat changes.

Figure 22.1a shows a single-stage evaporator represented on both actual and shifted temperature scales. Note that in shifted temperature scale, the evaporation and condensation duties are shown at different temperatures even though they are at the same actual temperature. Figure 22.1b shows a similar plot for a three-stage evaporator.

Like distillation, evaporation can be represented as a box. This again assumes that any heating or cooling required by the feed and concentrate will be included with the other process streams in the grand composite curve. However, with evaporation, the temperature difference across the box can be manipulated by varying the heat transfer area. Increasing the heat transfer area between stages allows a smaller temperature difference between stages and hence a smaller overall temperature difference, and vice versa.

22.2 APPROPRIATE PLACEMENT OF EVAPORATORS

The concept of the appropriate placement of distillation columns was developed in the preceding chapter. The principle also applies to evaporators. The heat integration characteristics of distillation columns and evaporators are very similar. Thus, evaporator placement should not be across the pinch[1].

Chemical Process Design and Integration R. Smith
© 2005 John Wiley & Sons, Ltd ISBNs: 0-471-48680-9 (HB); 0-471-48681-7 (PB)

22.3 EVOLVING EVAPORATOR DESIGN TO IMPROVE HEAT INTEGRATION

The thermodynamic profile of an evaporator can also be manipulated. The approach is similar to that used for distillation columns. The degrees of freedom are obviously different[1,2].

Consider the three-stage evaporator against a background process as shown in Figure 22.2a. At the chosen pressure, the evaporator will not fit against the grand composite curve. The most obvious possibility is to first try an increase in pressure to allow appropriate placement above the pinch, Figure 22.2b.

Now suppose that the required increase in pressure in Figure 22.2b would cause unacceptably high levels of decomposition and fouling in the evaporator as a result of the increase in temperature. The possibility of increasing the number of stages from three to, say, six could now be considered in order to allow a fit to the grand composite curve above the pinch, Figure 22.3a. The evaporator fits, but there still might be a problem of product degradation because of high temperatures. However, it is not necessary for all evaporator stages to be linked thermally with each other. Instead, Figure 22.3b shows a six-stage system with three effects appropriately placed above the pinch and three below. This could be either a conventional six-stage system in which the first three and last three stages are not linked thermally or, alternatively, two parallel three-stage systems, analogous to double effecting in distillation.

Yet another design option is shown in Figure 22.3c in which the heat flow (and hence mass flow) is changed between stages in the evaporator. Figure 22.3c shows an arrangement in which part of the vapor from the second stage is used for process heating rather than evaporation in the third stage. This means that more evaporation is taking place in the first two stages than the third and subsequent stages. It should be noted that even if the heat flow through the multistage evaporator is constant, the rate of evaporation will decrease because the latent heat increases as the pressure decreases.

22.4 THE HEAT INTEGRATION CHARACTERISTICS OF DRYERS

The heat input to dryers, as discussed in Chapter 10, is to a gas and, as such, takes place over a range of temperatures.

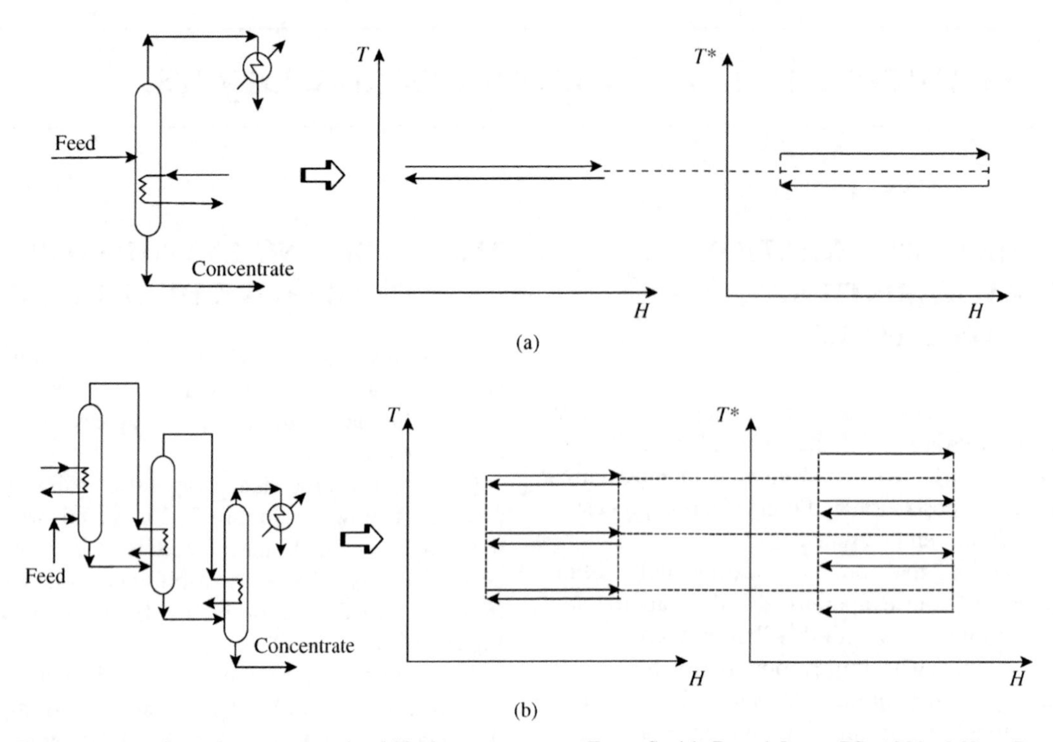

Figure 22.1 The representation of evaporators in shifted temperatures. (From Smith R and Jones PS, 1990, *J Heat Recovery Systems & CHP*, **10**: 34 reproduced by permission of Elsevier Ltd.)

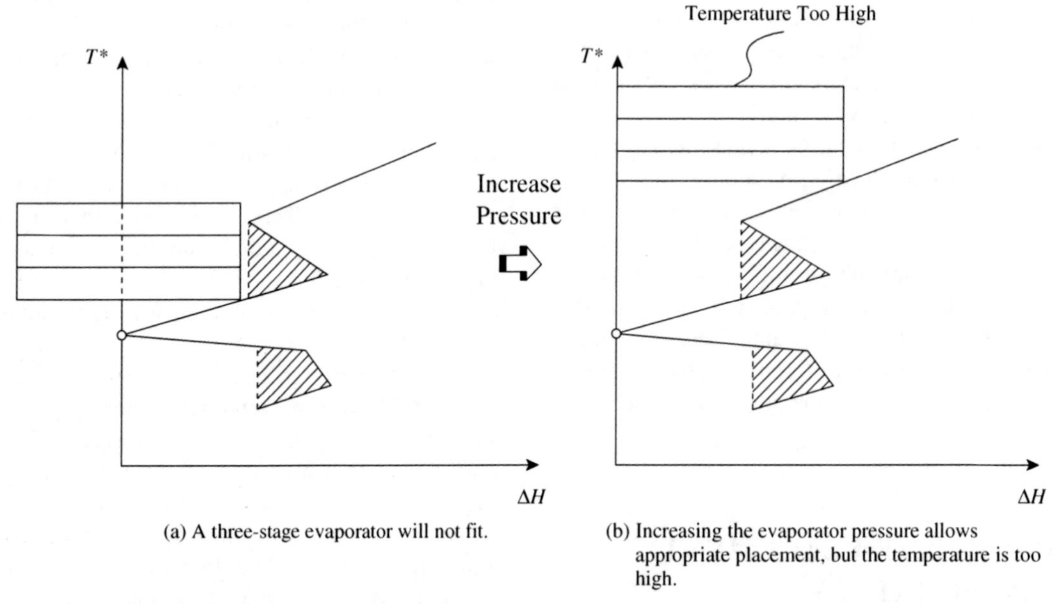

Figure 22.2 Integration of a three-stage evaporator.

Moreover, the gas is heated to a temperature higher than the boiling point of the liquid to be evaporated. The exhaust gases from the dryer will be at a lower temperature than the inlet, but again, the heat available in the exhaust will be available over a range of temperatures. The thermal characteristics of dryers tend to be design specific and quite different in nature from both distillation and evaporation. Figure 22.4 illustrates the heat integration characteristics of a typical dryer.

22.5 EVOLVING DRYER DESIGN TO IMPROVE HEAT INTEGRATION

It was noted that dryers are quite different in character from both distillation and evaporation. However, heat is still taken in at a high temperature to be rejected in the dryer exhaust. The appropriate placement principle as applied to distillation columns and evaporators also applies to

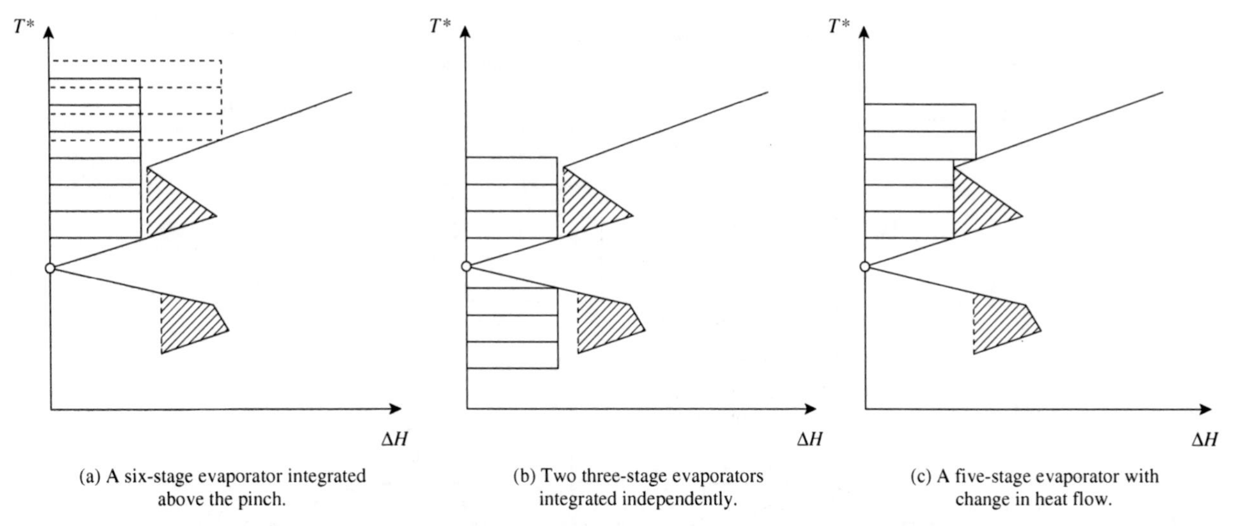

Figure 22.3 Evaporator design with the help of the grand composite curve. (From Smith R and Linnhoff B, 1988, *Trans IChemE ChERD*, **66**, 195, reproduced by permission of the Institution of Chemical Engineers.)

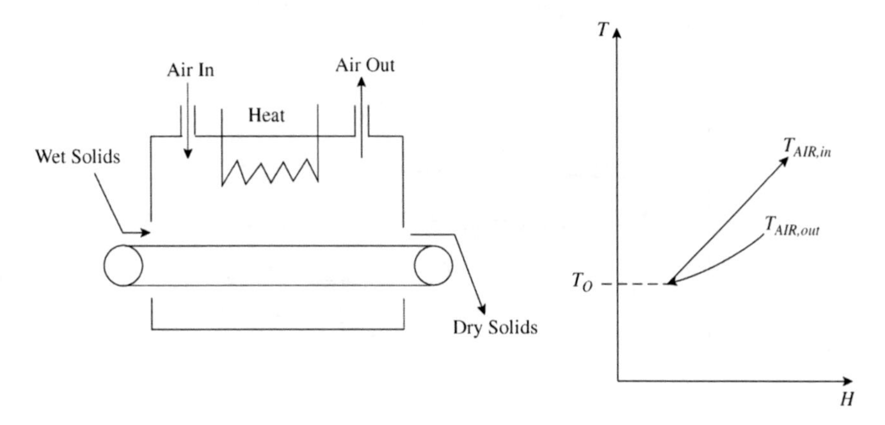

Figure 22.4 The heat integration characteristics of dryers.

dryers. The plus–minus principle from Chapter 19 provides a general tool that can be used to understand the integration of dryers in the overall process context. If the designer has the freedom to manipulate drying temperature and gas flowrates, then these can be changed in accordance with the plus–minus principle in order to reduce overall utility costs.

Figure 22.5 shows a plant for the production of animal feed from spent grains in a whisky distillery. The plant has two feeds, one of low, and one of high concentration solids. Water is removed from the low concentration feed by an evaporator, followed by a rotary dryer. Water is removed from the high concentration feed by a centrifuge, followed by two stages of drying in rotary dryers. As is usual with this type of plant, the evaporators and dryers have been designed on a stand-alone basis without consideration of the process context. Optimization of the evaporator on a stand-alone basis has indicated that heat pumping using a mechanical vapor recompression system would be economic.

Figure 22.5 shows the grand composite curve for this process and the location of the evaporator heat pump. The heating duty for the first dryer has been omitted from the grand composite since the required temperature is too high to allow integration with the rest of the process. The heat pump can be seen to be appropriately placed across the pinch. However, the cold side, below the pinch, encroaches into a pocket in the grand composite curve. If the design of the heat pump is changed so as not to encroach into the pocket, the result shown in Figure 22.6 is obtained. The resulting steam consumption is virtually unchanged, but energy costs will be lower. This results from the reduced load on the heat pump leading to a reduction in electricity demand.

22.6 HEAT INTEGRATION OF EVAPORATORS AND DRYERS – SUMMARY

Like distillation, the appropriate placement of evaporators and dryers is that they should be above the pinch, below the pinch, but not across the pinch. The grand composite curve can be used to assess appropriate placement quantitatively.

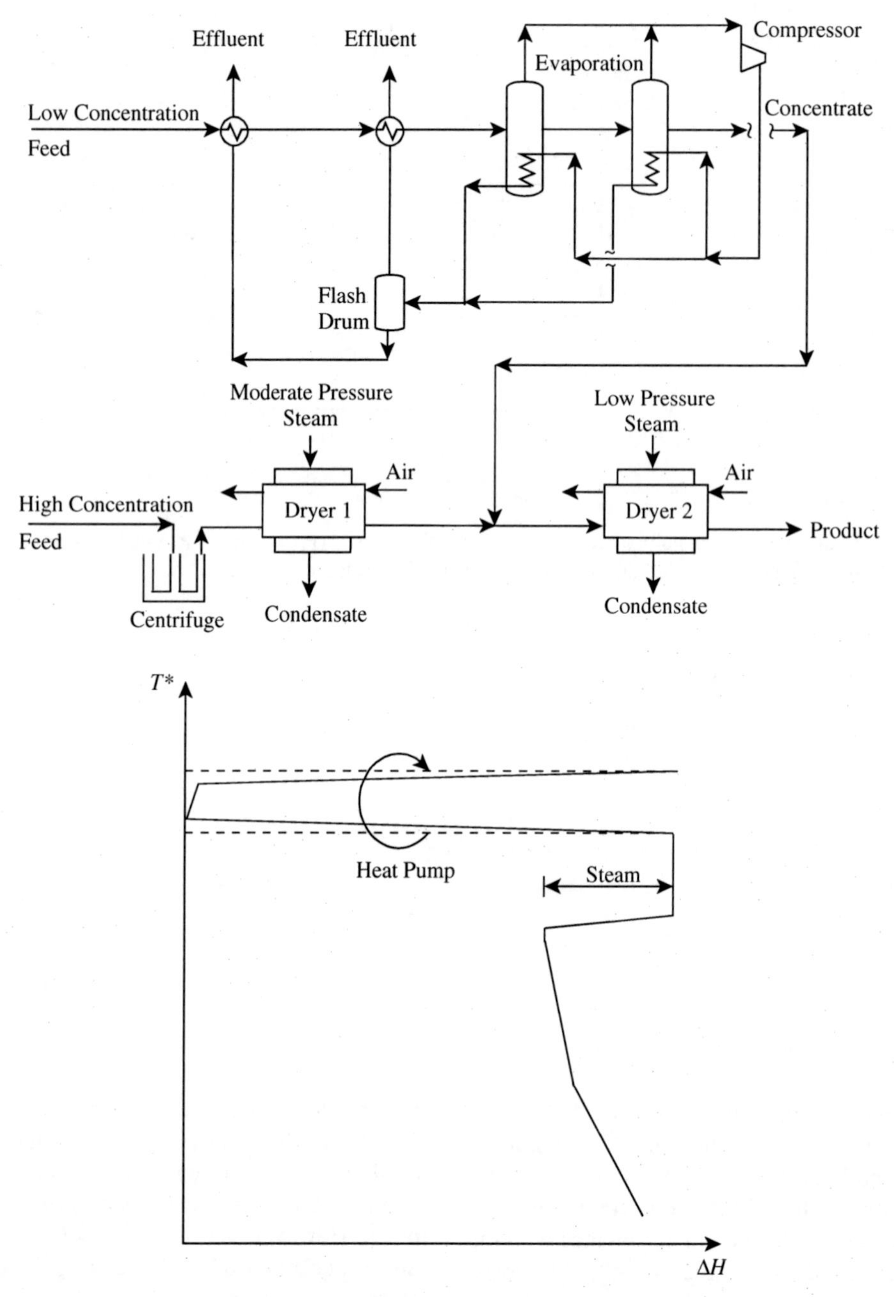

Figure 22.5 A plant for the production of animal feed. The heat pump encroaches into a "pocket" in the grand composite curve. (From Smith R and Linnhoff B, 1988, *Trans IChemE ChERD*, **66**: 195, reproduced by permission of the Institution of Chemical Engineers.)

Also like distillation, the thermal profile of evaporators can be manipulated by changing the pressure. However, the degrees of freedom in evaporator design open up more options.

Dryers are different in characteristic from distillation columns and evaporators in that the heat is added and rejected over a large range of temperature. Changes to drier design can be directed by the plus-minus principle.

22.7 EXERCISES

1. Table 22.1 presents the data for a process. An evaporation process is to be integrated with the process. The evaporator is required to evaporate 1.77 kg·s^{-1} water. The latent heat of vaporization of the water can be assumed to be 2260 kJ·kg^{-1}. For an assumed $\Delta T_{min} = 10$°C, suggest an outline evaporator configuration that will allow heat integration of the evaporator with the background process.

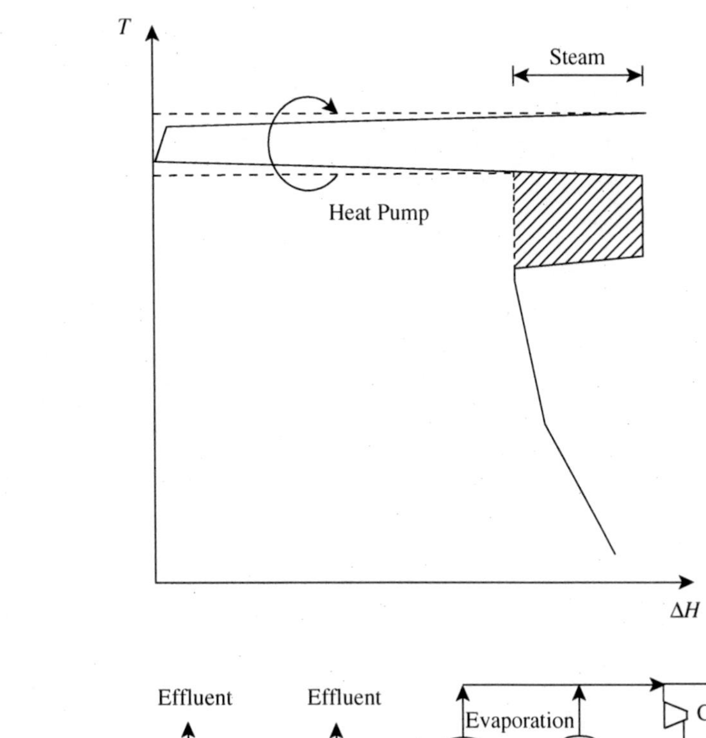

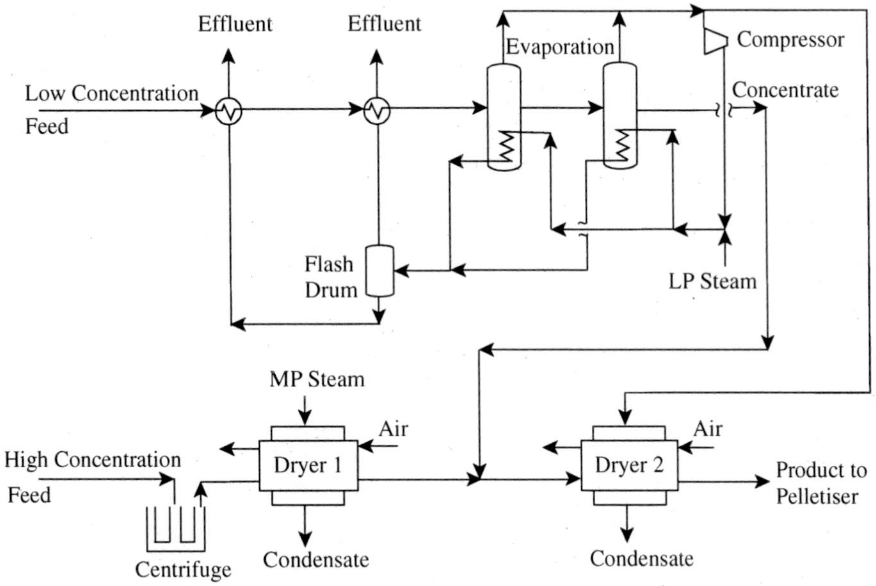

Figure 22.6 A simple modification reduces the load on the heat pump, saving electricity. (From Smith R and Linnhoff B, 1988, *Trans IChemE ChERD*, **66**: 195, reproduced by permission of the Institution of Chemical Engineers.)

Table 22.1 Stream data for background process.

Stream	T_S (°C)	T_T (°C)	ΔH(kW·K^{-1})
Hot	196	65	−10,480
Hot	176	56	−12,000
Hot	216	196	−2800
Hot	110	85	−2000
Cold	90	180	11,700
Cold	40	120	12,000

REFERENCES

1. Smith R and Linnhoff B (1988) The Design of Separators in the Context of Overall Processes, *Trans IChemE ChERD*, **66**: 195.
2. Smith R and Jones PS (1990) The Optimal Design of Integrated Evaporation Systems, *J Heat Recovery Syst CHP*, **10**: 341.

23 Steam Systems and Cogeneration

Most processes operate in the context of an existing site in which a number of processes are linked to the same utility system. The utility systems of most sites have evolved over a period of many years without fundamental questions being addressed as to the design and operation of the utility system. The picture is complicated by individual production processes on a site belonging to different business areas, each assessing investment proposals independently from one another and each planning for the future in terms of their own business. Yet, the efficiency of the site infrastructure and the required investment is of strategic importance and must be considered across the site as a whole, even if this crosses the boundaries of different business areas.

There are a number of reasons site steam and power systems need to be studied in process design:

1. A grassroot utility system may need to be designed. However, this is rare, and most situations would involve modification of an existing system.
2. Modifications might be required to an existing utility system as a result of changes in the demand for steam and power on a site. This might result from a new process starting up, a process closing down, changes in process capacity or basic modifications to the process technology.
3. An old utility system might need a revamp to replace old equipment.
4. Modifications to the configuration might be required to an existing utility system to reduce its operating costs.
5. Changes to the operation of the utility system might allow reduced operating costs.
6. The choice of utility for a new process on the site requires an understanding of the costs and constraints associated with the utility system that is servicing it.
7. Before energy conservation projects are implemented in existing processes on a site, the true value of those savings needs to be established. This is usually not possible without studying the utility system.
8. The true energy costs associated with a production expansion require the true cost implications in the utility system to be established, even if there is no capital investment required in the utility system.

Figure 23.1 shows a typical site utility system. Various processes operate on the site and are connected to a common utility system. In Figure 23.1, a very high-pressure steam is generated in utility stream boilers. This is expanded in steam turbines to provide steam at high, medium and low pressure. The final exhaust steam from the steam turbines is condensed against cooling water. The steam turbine generates power. It may be that this power generation needs to be supplemented by the import of power from an outside power station. It might also be the case that excess power is generated on the site and exported.

In Figure 23.1, three levels of steam are distributed around the site, and the various processes are connected to the steam mains. Process *A* in Figure 23.1 has a local fired heater. It imports low-pressure steam and exports high- and medium-pressure steam via the site distribution system. Process *B* imports both high- and low-pressure steam. Process *C* imports high- and medium-pressure steam and exports low-pressure steam via the low-pressure steam main. Finally, Process *A* and Process *B* require the rejection of waste heat to the ambient, and this is achieved using a cooling water circuit. Some sites use other methods for cooling. These will be discussed in Chapter 24.

This is a complex system to analyze. Steam is available at different pressures and temperatures. Some processes use steam, while others generate steam. There are interactions between the processes and the utility system via steam use and generation. There are also interactions between the processes on the site through the steam mains. Some processes export waste heat into the steam system, while other processes use this waste heat by drawing from the steam system. This is heat recovery between processes on the site using steam as an intermediate for heat transfer. In Figure 23.1, there is *cogeneration* (combined heat and power generation) from steam turbines. Strictly, cogeneration is the production of useful heat and power from the same heat source. Generation of power might be through a steam turbine or gas turbine coupled to an electric generator for the production of electricity in a *turbogenerator*, or a direct machine drive (e.g. a steam turbine driving a process compressor directly). However, the byproduct heat must be useful to class the power generation as cogeneration. Cogeneration is the most efficient way to produce heat and power and lowers energy costs. It also lowers the overall emissions of combustion gases, although this can only be judged properly by including the emissions created by external power generation resulting from any imported power or the savings in emissions from external power generation resulting from the exported power.

Chemical Process Design and Integration R. Smith
 ISBNs: 0-471-48680-9 (HB); 0-471-48681-7 (PB)

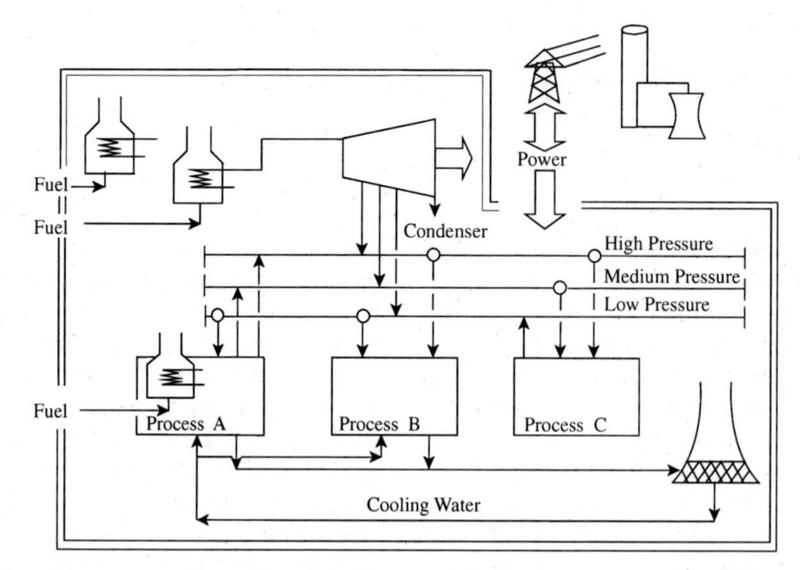

Figure 23.1 A typical site utility system. (From Klemes J, Dhole VR, Raissi K, Perry SJ and Puigjaner L, 1997, *Applied Thermal Engineering*, **17**: 993, reproduced by permission of Elsevier Ltd.)

Steam is used extensively on most sites for:

- indirect heating in steam heaters;
- steam *tracing* for pipes and storage tanks (i.e. steam pipes around the exterior) to keep the materials inside from solidifying;
- direct heating of water through live steam injection;
- creation of vacuum in steam ejectors;
- mass and heat exchange by live steam injection in distillation;
- reduction of partial pressure in gas-phase reactors;
- combustion processes to atomize fuel oil;
- injection into combustion processes to lower NO_x emissions through reduction in the flame temperature (see Chapter 25);
- injection into flares to assist the combustion (see Chapter 25);
- power generation in steam turbines.

It is no accident that steam is used extensively for process heating as it provides a number of useful features that include:

- energy can be generated at one point and distributed;
- it is a convenient form of transferring energy around;
- it has a wide range of operating temperatures;
- it has a high heat content through the latent heat;
- the temperature is easy to control through control of the pressure;
- it can be used to generate power in steam turbines and generate vacuum in steam ejectors;
- it does not require expensive materials of construction;
- it is nontoxic and losses are easily replaced.

The major components of the steam system are:

- boiler feedwater treatment
- steam boilers
- steam turbines
- gas turbines
- steam distribution
- steam users
- condensate collection and recovery.

Consider now each of these components in turn.

23.1 BOILER FEEDWATER TREATMENT

Figure 23.2 shows a schematic representation of a boiler feedwater treatment system. Raw water from a reservoir, river, lake, borehole or a seawater desalination plant is fed to the steam system. However, it needs to be treated before it can be used for steam generation. The treatment required depends both on the quality of the raw water and the requirements of the utility system. The principal problems with raw water are[1,2]:

- suspended solids
- dissolved solids
- dissolved salts
- dissolved gases (particularly, oxygen and carbon dioxide).

Thus, the raw water entering the system in Figure 23.2 may need to be first filtered to remove suspended solids. If required, this would commonly be carried out using sand filtration. Dissolved salts then need to be removed, principally calcium and magnesium ions, that would otherwise cause fouling in the steam boiler. The quality of feedwater required depends on the boiler operating pressure and boiler design. Sodium zeolite *softening* is the simplest process and is used to remove scale-forming ions, such as calcium and magnesium. The softened water produced

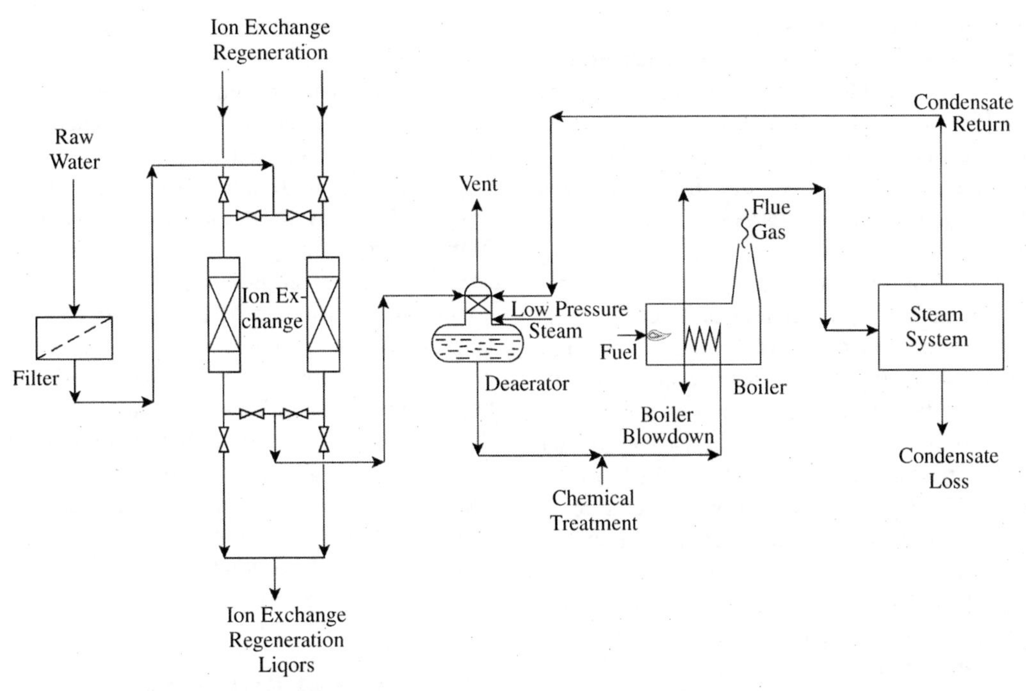

Figure 23.2 Boiler feedwater treatment. (From Smith R and Patela EA, 1992, *The Chemical Engineer*, No 523: 16, reproduced by permission of the Institution of Chemical Engineers.).

can be used for low- to medium-pressure boilers. In this process, the water is passed through a bed of sodium zeolite, and the calcium and magnesium ions are removed according to[1,2]:

$$\begin{bmatrix} Ca \\ Mg \end{bmatrix} \cdot \begin{bmatrix} SO_4 \\ 2Cl \\ 2HCO_3 \end{bmatrix} + Na_2 \cdot Z \longrightarrow Z \cdot \begin{bmatrix} Ca \\ Mg \end{bmatrix} + \begin{bmatrix} Na_2SO_4 \\ 2NaCl \\ 2NaHCO_3 \end{bmatrix}$$

where Z refers to zeolite. The resin is regenerated by treatment with sodium chloride solution[1,2]:

$$Z \cdot \begin{bmatrix} Ca \\ Mg \end{bmatrix} + 2NaCl \longrightarrow Na_2 \cdot Z + \begin{bmatrix} Ca \\ Mg \end{bmatrix} \cdot Cl_2$$

The softening process removes the sparingly soluble calcium and magnesium ions and replaces them with less objectionable sodium ions. Unfortunately, softening alone is not sufficient for most high-pressure boiler feedwaters. Also, the processes on a site often have a requirement for *deionized water*. *Deionization*, in addition to removing hardness, removes other dissolved solids, such as sodium, silica, alkalinity and mineral ions such as chloride, sulfate and nitrate. Deionization is essentially the removal of all inorganic salts. In this process, a *strong acid cation resin* converts dissolved salts into their corresponding acids, and a *strong base anion resin* in the hydroxide form removes these acids. The cation ion-exchange resin exchanges hydrogen for the raw-water ions according to[1,2]:

$$\begin{bmatrix} Ca \\ Mg \\ 2Na \end{bmatrix} \cdot \begin{bmatrix} 2HCO_3 \\ SO_4 \\ 2Cl \\ 2NO_3 \end{bmatrix} + 2Z \cdot H \rightleftharpoons 2Z \cdot \begin{bmatrix} Ca \\ Mg \\ 2Na \end{bmatrix} + \begin{bmatrix} 2H_2CO_3 \\ H_2SO_4 \\ 2HCl \\ 2HNO_3 \end{bmatrix}$$

To complete the deionization process, water from the cation unit is passed through a strong base anion exchange resin in the hydroxide form. The resin exchanges hydrogen ions for both highly ionized mineral ions and the more weakly ionized carbonic and silicic acids according to[1,2]:

$$\begin{bmatrix} H_2SO_4 \\ 2HCl \\ 2H_2SiO_3 \\ 2H_2CO_3 \\ 2HNO_3 \end{bmatrix} + 2Z \cdot OH \rightleftharpoons 2Z \cdot \begin{bmatrix} SO_4 \\ 2Cl \\ 2HSiO_3 \\ 2HCO_3 \\ 2NO_3 \end{bmatrix} + 2H_2O$$

As the ion-exchange beds approach exhaustion, they need to be taken offline and regenerated. The cation resins are regenerated with an acid solution, which returns the exchange sites to the hydrogen form. Sulfuric acid is normally used for this. The anion exchange resin is regenerated using sodium hydroxide solution, which returns the exchange sites to the hydroxyl form.

Rather than use ion exchange, water can be deionized using membrane processes. Both nano-filtration and reverse osmosis can be used, but only reverse osmosis is capable of producing high-quality boiler feedwater (see Chapter 10). Nano-filtration is only capable of separating covalent ions and larger univalent ions such as heavy metals.

Having removed the suspended solids and dissolved salts, the water then needs to have the dissolved gases removed, principally, oxygen and carbon dioxide, which would otherwise cause corrosion in the steam boiler. The usual method to achieve this is deaeration, which removes dissolved gases by raising the water temperature[1,2].

The water to be deaerated enters at the top of the deaerator via a spray system. This water is contacted with steam injected at the bottom of the deaerator. Some form of packing or plates is normally used to assist the contact between the boiler feedwater and steam. Boiler feedwater is heated to within a few degrees of saturation temperature of the steam. Most of the noncondensable gases (principally, oxygen and free carbon dioxide) are released into the steam. Deaerated water falls to a storage tank below, where a steam blanket protects it from recontamination. The deaeration steam flows up through the deaerator and most of it is condensed to become part of the deaerated water. A small portion of the steam, which contains the noncondensable gases released from water, is vented to atmosphere. There is a minimum temperature difference between the deaeration steam and the final boiler feedwater temperature to make the deaerator function effectively. A temperature difference of at least 10°C is required. Some designs of deaerator include a vent condenser that sprays in a small portion of the boiler feedwater feed to a spray system prior to venting the steam and noncondensable gases in order to minimize the vented steam, so reducing the visible plume from the deaerator and improving energy efficiency.

Even after deaeration, there is some residual oxygen that needs to be removed by chemical treatment[1,2]. After the deaerator, oxygen scavengers (e.g. hydroquinone) are added to react with the residual oxygen that would otherwise cause corrosion. Also, the boiler feedwater treatment does not remove all of the solids in the raw water, and the deposition of solids in the boiler is another problem. Phosphates can be added to precipitate any calcium and magnesium away from the heat transfer surfaces of the boiler. *Chelants* (weak organic acids) can also be added. These have the ability to complex with many cations (calcium, magnesium and heavy metals, under boiler conditions). They accomplish this by locking the metals into a soluble organic ring structure. Polymer dispersant can also be added to keep any precipitate dispersed.

The deaerated treated boiler feedwater then enters the boiler. Evaporation takes place in the boiler and the steam generated is fed to the steam system. Solids not removed by the boiler feedwater treatment build up in the boiler, along with products of corrosion. These are removed from the boiler by taking a *blowdown* (purge) from the boiler. The steam from the boiler goes to the steam system to carry out various duties. The steam is ultimately condensed somewhere in the steam system. Some condensate is returned to the deaerator and some lost from the system to effluent. Sometimes, the returned condensate is subjected to deionization in a *condensate-polishing* step to remove any traces of dissolved solids that might have been picked up. Condensate return as high as 90% or greater is possible, but return rates are often significantly lower than this. Higher levels of condensate return than 90% might not be able to be justified because of the capital cost of the pipework required for condensate return or the possibility of some condensate being contaminated. This constant loss of condensate from the steam system means that there must be a constant make up with freshwater. To prevent corrosion in the condensate system, amines are also added to the boiler feedwater. These are volatile and condense with steam, adjusting the pH on condensation to prevent corrosion in the condensate system.

23.2 STEAM BOILERS

There are many types of steam boilers, depending on the steam pressure, steam output and fuel type[3]. Pressures are normally of the following order.

- 100 barg, used for power generation
- 40 barg is the normal maximum pressure for distribution
- 10–40 barg are the conventional distribution pressure levels
- 2–5 barg are typically the lowest pressures used for distribution.

Figure 23.3 illustrates a *fire-tube* (or *shell*) boiler. A large cylindrical shell houses a pool of boiling water that is vaporized by a hot flue gas flowing through the inside of tubes. The steam pressure is contained by the large cylindrical shell, which imposes certain mechanical limitations. A large diameter shell requires a thicker wall to contain the same pressure as a small-diameter shell. The economic limit for such designs is around 20 barg (but they are available at higher pressures). Fire-tube boilers are normally used for small heating duties of low-pressure steam. Fuels are normally restricted to be gas or light fuel oil. Higher duty boilers use a *water-tube* arrangement, as shown in Figure 23.4. Water-tube boilers vaporize steam through the recirculation of water and steam between a steam drum and a water drum connected by the boiler tubes. Steam from the steam drum is then superheated before entering the steam system. Water-tube boilers are suitable for high pressures and high steam output. Fuels can be gas, fuel oil or solid fuel. In the case of solid fuels, the boiler must have some mechanism for the removal of ash produced by combustion from the base of the radiant chamber. Many other designs of boiler are available.

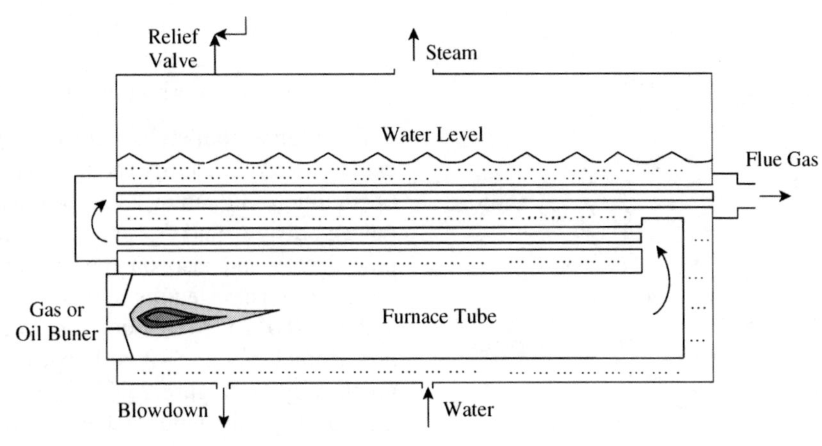

Figure 23.3 Fire-tube (or shell) boiler.

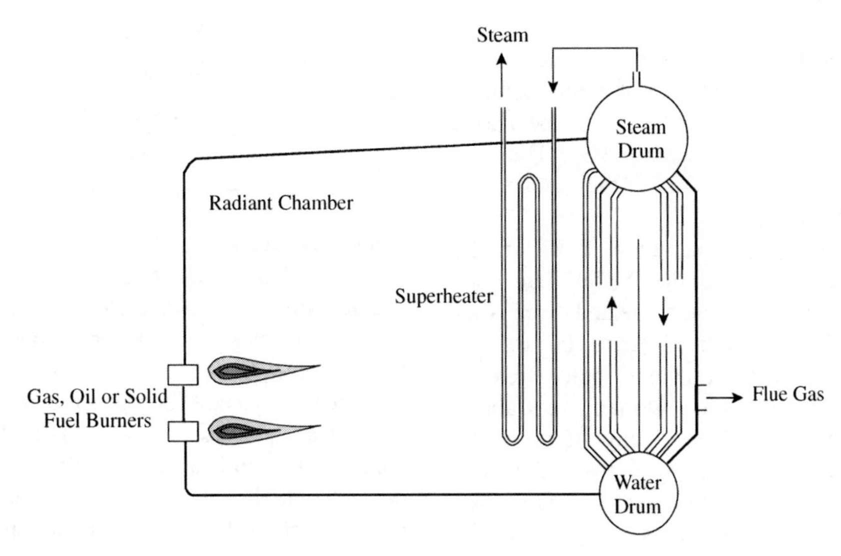

Figure 23.4 Water-tube boiler.

As with furnaces, discussed in Chapter 15, the draft in boilers can be natural, forced draft or induced draft. Forced and induced draft arrangements are the most common. For large boilers requiring a significant amount of equipment in the flue to treat the exhaust gases to remove oxides of sulfur and/or oxides of nitrogen, as well as equipment for heat recovery, a combination of forced and induced draft might need to be used.

As discussed under boiler feedwater treatment, boiler blowdown is required to prevent the build up of solids in the boiler that would otherwise cause fouling and corrosion in the boiler. Carry over of solids from the boiler to the steam system via tiny water droplets should also be avoided. Total dissolved solids (TDS) and silica (SiO_2), as measured by the conductivity of water, are both important to be controlled in the boiler[3]. Dissolved solids carried over from the boiler will be a problem to all components of the steam system. Silica is a particular problem because of its damaging effect on steam turbines, particularly the low-pressure section of steam turbines where some condensation can occur. Blowdown rates are usually quoted as the ratio of flowrate of blowdown to evaporation rate. The blowdown rate required depends on the boiler feedwater treatment system and the pressure of the boiler. Typical blowdown rates are:

- small, low-pressure boilers: 5 to 10%
- large, high-pressure boilers: 2 to 5%
- very large boilers with very pure feed: less than 2%.

Economizers can be used to preheat the incoming boiler feedwater[3]. This is a heat exchange between the incoming boiler feedwater and the hot flue gases before they are vented to atmosphere. This increases the energy efficiency of the boiler. Another option to increase energy efficiency is to use an *air preheater*, in which there is heat recovery from the hot flue gas to the incoming cold combustion air[3]. This has the disadvantage that the combustion temperature is increased with a resulting increase in the formation of oxides of nitrogen (NO_x). The *acid dew point* limits the minimum temperature for flue gases containing oxides

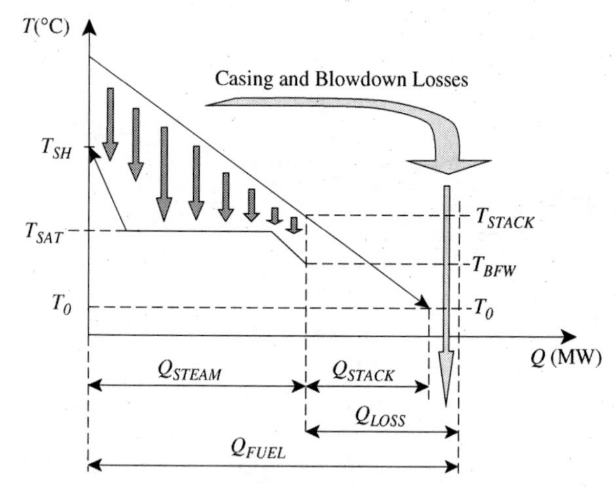

Figure 23.5 Model of steam boiler.

of sulfur (SO_x). If condensation occurs with flue gases containing SO_x, this will cause corrosion of metal surfaces. Metal heat exchange services should normally be above 150 to 160°C in flue gas containing any SO_x. The acid dew point depends on the sulfur content of the fuel and the amount of excess air used.

Figure 23.5 shows the steam generation profile in the boiler as it is matched against the boiler flue gas. Figure 23.5 shows the steam being preheated from subcooled boiler feedwater to superheated steam. In turn, the flue gas is cooled to the stack temperature before release to the atmosphere. The flue gases are normally kept above their dew point, particularly if there is any sulfur in the fuel. The heat loss in the stack Q_{STACK} is the major inefficiency in the use of the fuel. In addition to the stack loss, there are also heat losses from the boiler casing and boiler blowdown. If the total losses from the stack, casing and blowdown are Q_{LOSS}, the efficiency of the boiler can be defined as:

$$\eta_{BOILER} = \frac{Q_{STEAM}}{Q_{FUEL}} = \frac{Q_{STEAM}}{Q_{STEAM} + Q_{LOSS}} \tag{23.1}$$

where η_{BOILER} = boiler efficiency

Q_{FUEL} = heat required from fuel in the boiler

Q_{STEAM} = heat input to boiler feedwater preheating, vaporization and steam superheating

Q_{LOSS} = heat losses from the stack, casing and boiler blowdown

The efficiency of boilers changes with the boiler load. The variation of boiler performance with load can be approximated by[4]:

$$\eta_{BOILER} = \frac{1}{a + b\dfrac{m_{max}}{m_{STEAM}}} \tag{23.2}$$

where m_{max} = maximum flowrate of steam from the boiler

m_{STEAM} = mass flowrate of steam generated in the boiler

a, b = correlating parameters

For a given boiler, the correlating parameters a and b can be determined by fitting data points from part-load operation to Equation 23.2. It must be emphasized that the value of a and b depend on the design of boiler in new design and also on the age and maintenance of the boiler in existing equipment. For example, if $a = 1.1256$ and $b = 0.0126$, from Equation 23.2, this would lead to a boiler efficiency of 88% at maximum load[4]. This is shown graphically in Figure 23.6. Large boilers in the process industries would tend to have a maximum efficiency up to 93%.

The fuel required for the boiler is given by[4]:

$$\begin{aligned} Q_{FUEL} &= \frac{Q_{STEAM}}{\eta_{BOILER}} \\ &= \frac{\Delta H_{STEAM} \cdot m_{STEAM}}{\eta_{BOILER}} \\ &= \Delta H_{STEAM}[a \cdot m_{STEAM} + b \cdot m_{max}] \end{aligned} \tag{23.3}$$

where ΔH_{STEAM} = enthalpy difference between generated steam and boiler feedwater

Equation 23.3 is adequate for most modeling and optimization purposes. It also provides a linear relationship between Q_{FUEL} and m_{STEAM}, with ΔH_{STEAM}, m_{max}, a and b, all being constant for a given boiler design. This has significant advantages over more complex expressions when applied to optimization of utility systems, as will be discussed later. However, the shape of the curve in Figure 23.6 from Equation 23.2 shows a monotonic increase of efficiency with load. Some boiler designs will show a maximum in the efficiency below 100% capacity. This can be captured in modeling by introducing a more complex expression than Equation 23.2. Unfortunately, this will introduce nonlinearity into the model. Rather than introduce a more complex model, Equation 23.2 can be fitted to different ranges of m_{STEAM}/m_{max}. This allows a linear model to be retained.

As noted previously in Chapter 2, care should be exercised when considering boiler efficiency as to whether quoted

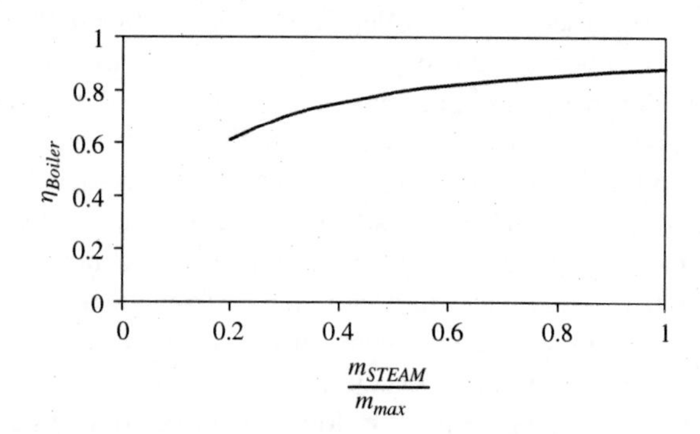

Figure 23.6 Steam boiler performance at part-load.

figures are based on the gross or net calorific value of the fuel. Boiler efficiency is often quoted on the basis of *gross calorific value* of the fuel, which includes the latent heat of the water vapor from combustion. This latent heat is rarely recovered through condensation of the water vapor in the combustion gases. The *net calorific value* of the fuel assumes that the latent heat of water vapor is not recovered and is therefore the most relevant value. Yet, boiler efficiency is still often quoted on the basis of gross calorific value.

Example 23.1 A small package fire-tube boiler has makeup water that contains 500 ppm dissolved solids. The steam system operates with 50% condensate return. Estimate the blowdown rate. Assume that the maximum limit for the TDS is 4500 ppm. Assume that there are no solids in the evaporation or the condensate return.

Solution Referring to Figure 23.7:

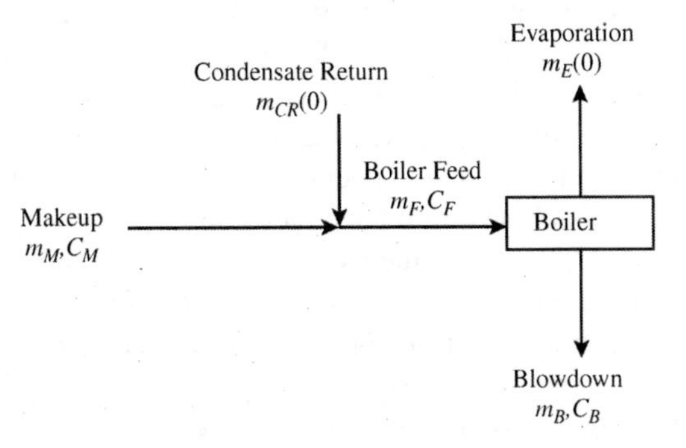

Figure 23.7 Boiler blowdown material balance.

$$m_M C_M = m_F C_F = m_B C_B$$

where m_M, C_M = flowrate and concentration of solids in makeup water
m_F, C_F = flowrate and concentration of solids in the boiler feed
m_B, C_B = flowrate and concentration of solids in the blowdown

Also:

$$m_F = m_M + m_{CR}$$

Combining these equations gives:

$$(m_F - m_{CR})C_M = m_B C_B$$

For a 50% condensate return rate:

$$(m_F - 0.5\ m_E)C_M = m_B C_B$$

$$(m_E + m_B - 0.5\ m_E)C_M = m_B C_B$$

Rearranging:

$$\frac{m_B}{m_E} = \frac{0.5C_M}{C_B - C_M}$$

$$= \frac{0.5 \times 500}{4500 - 500}$$

$$= 0.0625$$

$$= 6.25\%$$

23.3 STEAM TURBINES

A steam turbine converts the energy of steam into power[5]. Steam turbine sizes vary from 0.75 kW up to 100 MW and larger. To maximize the steam turbine efficiency, the steam is expanded in a number of stages. Turbine stages are known as *impulse* or *reaction* stages, although most combine elements of the two[5].

An *impulse turbine* is rather like a water wheel, and a single stage from an impulse turbine is illustrated in Figure 23.8a. Impulse nozzles orient the steam to flow in well-formed high-speed jets. Moving *blades* or *buckets* absorb the kinetic energy of the jet and convert it into mechanical work. As the steam strikes the moving blades, it suffers a change in direction and, therefore, momentum. This gives rise to an impulse on the blades. In a pure impulse turbine, no pressure drop occurs in the moving blades (except that caused by friction). The pressure drop occurs across the stationary blades only.

The principle of a *reaction turbine* is somewhat different. A single turbine stage using reaction blades is illustrated in Figure 23.8b. Steam passing through the fixed blades undergoes a small decrease in pressure and its velocity increases. It then enters the row of moving blades and, just as in the impulse turbine, suffers a change in direction and, therefore, momentum. This gives rise to an impulse on the blades. However, during its passage through the blades, the steam undergoes a further drop in pressure. The resulting increase in velocity gives rise to a reactive force in the direction opposite to that of the added velocity. This is the same reaction principle used to propel rockets and to propel airplanes using jet engines. The gross propelling force in a steam turbine is a combination of impulse and reaction. Thus, in reality, the blades in a so-called reaction turbine are partly impulse and partly reaction. Figure 23.8 illustrates a single stage reaction turbine. Additional stages are created by additional rows of fixed and moving blades.

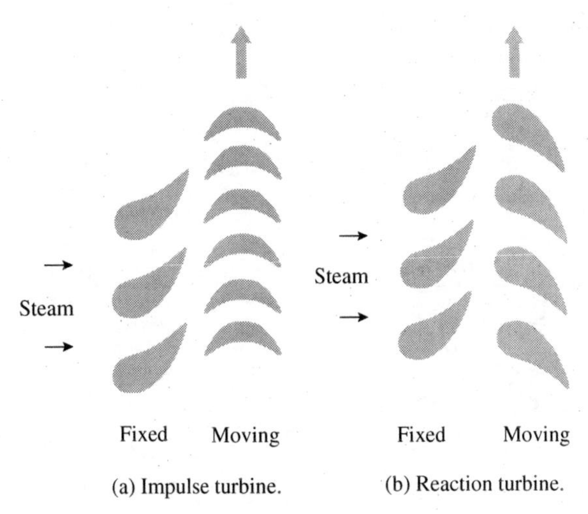

Figure 23.8 Steam turbine types.

The blades are mounted on a rotating shaft. The rotating shaft is supported within a *casing*. The casing supports the bearings for the rotating shaft, the stationary blades and the steam inlet nozzles.

Impulse machines are used for small turbines, and the high-pressure section of large turbines use impulse blades. The low-pressure sections of large turbines tend to use reaction blades.

Figure 23.9 shows the basic ways in which a steam turbine can be configured. Steam turbines can be divided into two basic classes:

- *back-pressure* turbines, which exhaust steam at pressures higher than atmospheric pressure
- *condensing* turbines, which exhaust steam at pressures less than atmospheric pressure.

The exhaust pressure of a steam turbine is fixed by the operating pressure of the downstream equipment. Figure 23.9a shows a back-pressure turbine operating between a high-pressure and low-pressure steam mains. The pressure of the low-pressure steam mains will be controlled elsewhere (see later).

Figure 23.9b shows a condensing turbine. Three types of condensers are used in practice as follows.

1. *Direct contact*, in which cooling water is sprayed directly into the exhaust steam.
2. *Surface condensers*, in which the cooling water and exhaust steam remain separate (see Chapter 15). Rather than using cooling water, boiler feedwater can be preheated to recover the waste heat.
3. *Air coolers* again use surface condensation but reject the heat directly to ambient air (see Chapter 15).

Of the three types, surface condensers using cooling water are by far the most common. When a volume filled with steam is condensed, the resulting condensate occupies a smaller volume and a vacuum is created. Any noncondensable gases remaining after the condensation are removed by steam ejectors or liquid ring pumps. The pressure at the exhaust of the condensing turbine is maintained at a reasonable temperature above cooling water temperature (e.g. 40°C at 0.07 bara to 50°C at 0.12 bara, depending on the temperature of the cooling water). The higher the pressure difference across the steam turbine, the more the energy that can be converted from the turbine inlet steam into power. For a given temperature difference between the condensing steam and the cooling medium, the lower the temperature of cooling, the more the power that can be generated for a given flowrate of steam with fixed inlet conditions.

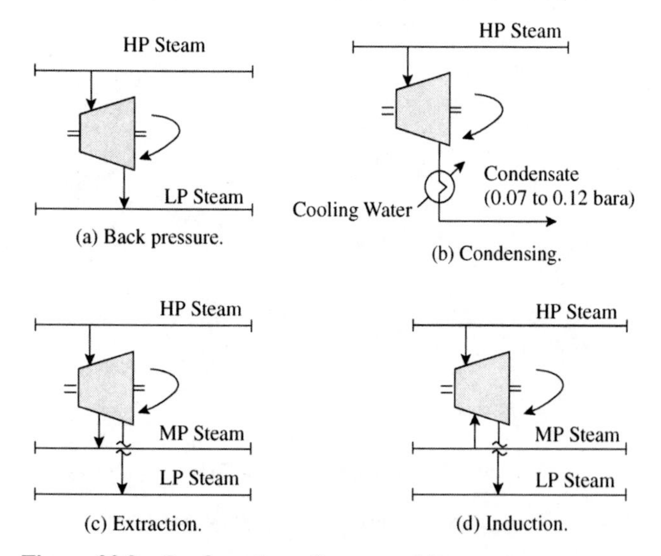

Figure 23.9 Configuration of steam turbines.

Both back-pressure and condensing turbines can be categorized further by the steam that flows through the machine. Figure 23.9c shows an *extraction* machine. Extraction machines bleed off part of the main steam flow at one or more points. Figure 23.9c shows a single extraction with both the extraction and the exhaust being fed to steam mains. The exhaust could have been taken to vacuum conditions and condensed, as in Figure 23.9b. The extraction machine might have partitions with valves inside the machine to control the extracted steam pressure.

Figure 23.9d shows an *induction* turbine. Induction turbines work like extraction machines, except in reverse. Steam at a higher pressure than the exhaust is injected into the turbine to increase the flow part way through the machine and to increase the power production. In a situation like the one shown in Figure 23.9d, an excess of medium-pressure (MP) steam generation over and above that for process heating is used to produce power and exhaust into a low-pressure steam, where there is a demand for the low-pressure steam for process heating.

Any given machine will have minimum and maximum allowable steam flows. In the case of extraction and induction machines, there will be minimum and maximum flows allowable in each turbine section. These minimum and maximum flows are determined by the physical characteristics of individual turbines and specified by the turbine manufacturer.

The expansion process in a steam turbine transforms a certain amount of the energy of inlet steam to power. The total power from the expansion is further split into useful power, delivered to the shaft and energy losses. Losses, occurring in steam turbines are mechanical friction losses, casing heat losses and kinetic energy losses (differences between the inlet and outlet kinetic energy). Steam turbine performance is affected by a number of factors. The most significant among them are:

- turbine size in terms of maximum power load
- pressure drop across the turbine
- operating load.

Generally, the efficiency of steam turbines decreases with decreasing load. The overall turbine efficiency can be represented by two components: the *isentropic efficiency* and the *mechanical efficiency*. The mechanical efficiency reflects the efficiency with which the energy that is extracted from steam is transformed into useful power and accounts for mechanical frictional losses, heat losses, and so on. The mechanical efficiency is high (typically 0.95 to 0.99)[6]. However, the mechanical efficiency does not reflect the efficiency with which energy is extracted from steam. This is characterized by the isentropic efficiency introduced in Figure 2.1 and Equation 2.3, defined as:

$$\eta_{IS} = \frac{H_{in} - H_{out}}{H_{in} - H_{IS}} = \frac{\Delta H_{REAL}}{\Delta H_{IS}} \tag{23.4}$$

where
η_{IS} = turbine isentropic efficiency
H_{in} = specific enthalpy of the inlet steam
H_{out} = specific enthalpy of the outlet steam
H_{IS} = enthalpy of steam at the outlet pressure having the same entropy as the inlet steam
$\Delta H_{REAL} = H_{in} - H_{out}$
$\Delta H_{IS} = H_{in} - H_{IS}$

In addition to the mechanical efficiency being much higher than the isentropic efficiency, in most cases the mechanical efficiency does not change significantly with load. By contrast, the isentropic efficiency does change significantly with load. The overall steam turbine efficiency can be defined as:

$$\eta_{ST} = \eta_{IS} \cdot \eta_{MECH} \tag{23.5}$$

where
η_{ST} = overall steam turbine efficiency
η_{MECH} = mechanical efficiency

Figure 23.10a illustrates the relationship between turbine efficiency and mass flow through the turbine. Because $\eta_{MECH} >> \eta_{IS}$, the major contribution to the nonlinear trend of the overall efficiency with part-load is from the isentropic efficiency η_{IS}. A turbine model needs to capture this behavior. It should also be noted that there will be an efficiency associated with an electricity generating set coupled to the steam turbine (typically 95 to 98%).

The relationship between shaft power and mass flow through a heat engine is sometimes called the Willans' Line. A typical Willans' Line for a steam turbine is illustrated in Figure 23.10b. For many machines this is almost a straight line.

The power–steam flow relationship in Figure 23.10b can be represented over a reasonable range by a linear relationship, as shown in Figure 23.10c[7–9]. The straight-line relationship is given by[7–9]:

$$W = n{\cdot}m - W_{INT} \tag{23.6}$$

where
W = shaft power produced by the turbine
m = mass flowrate of the steam through the turbine
n = slope of the linear Willans' Line
W_{INT} = intercept of the linear Willans' Line

If a single straight-line relationship, as shown in Figure 23.10c, is not adequate, then a series of linear segments can be used, as shown in Figure 23.10d. Each linear segment in Figure 23.10d is represented by an equation of the form of Equation 23.6, each with its own slope and intercept. It is worth preserving the linearity of the model rather than introducing a nonlinear model, as the linear model has advantages for use in optimization, as will be discussed later.

For any load of a given steam turbine with fixed inlet pressure, back-pressure and inlet temperature, the isentropic

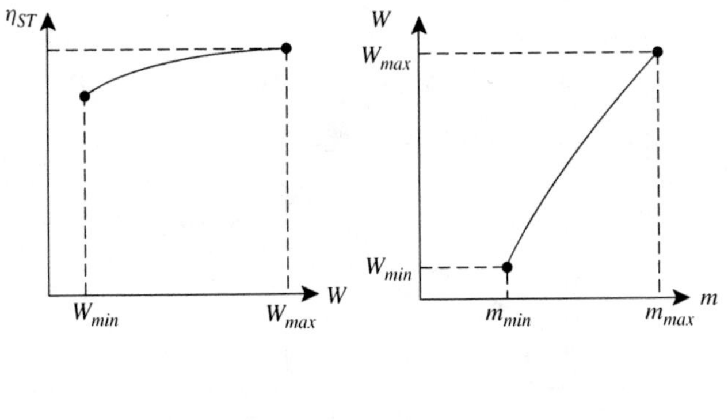

(a) Efficiency versus power output.

(b) Power output versus steam flowrate.

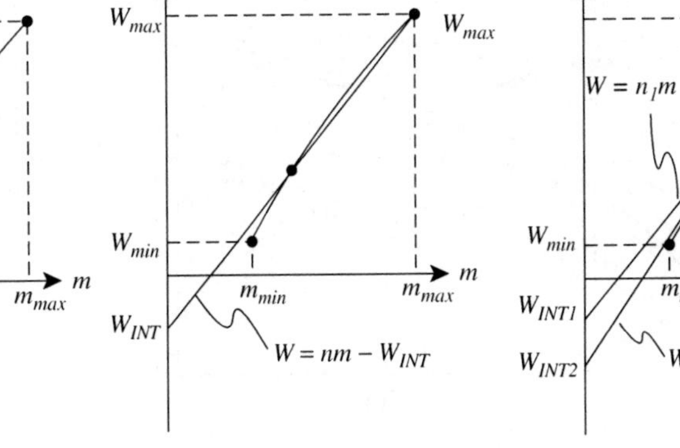

(c) A straight line can be used to represent turbine performance.

(d) Straight line segments can be used to improve the representation.

Figure 23.10 Steam turbine performance versus load.

enthalpy change remains constant. As a result of changing isentropic efficiency, the actual enthalpy drop across the turbine changes with load. However, the change cannot exceed the isentropic enthalpy change. Thus[7–9]:

$$W = \Delta H_{REAL} \cdot m - W_{LOSS} \qquad (23.7)$$

where $\Delta H_{REAL} = H_{in} - H_{out}$
W_{LOSS} = energy loss from the turbine

Equation 23.7 is based on the actual change in steam enthalpy across the turbine. Although both Equations 23.6 and 23.7 have the same form, their coefficients have completely different meanings. Comparing Equations 23.6 and 23.7, it becomes apparent that the slope of the linear Willans' Line (Equation 23.6) is related to the isentropic enthalpy change and turbine isentropic efficiency[9].

The intercept of the line can be assumed to be proportional to the maximum power[7–9]:

$$W_{INT} = L \cdot W_{max} \qquad (23.8)$$

where W_{max} = maximum power from the turbine
L = intercept ratio

The proportionality coefficient L, termed the *intercept ratio*, depends on a number of factors such as the turbine size, the turbine design, the application (electrical or direct drive), and so on. It must be determined from performance data and usually lies in the range of 0.05 and 2[9].

In order to derive the part-load equations, consideration should be given to how the full-load performance depends on the steam pressure drop. Peterson and Mann[10] present curves of the overall efficiency of industrial steam turbines at maximum load. The form of such curves is illustrated in Figure 23.11a. Each curve represents data for many steam turbines at full-load. Performance curves of this form can be used to derive the steam turbine model. Another interpretation of the efficiency curves is to represent them as plots of the isentropic power versus the actual shaft power. In this form, the shape of the plots tends to be linear (Figure 23.11b)[9].

The isentropic power from a steam turbine can be related to the maximum shaft power through the overall turbine efficiency:

$$W_{IS,max} = \Delta H_{IS} \cdot m_{max} = \frac{W_{max}}{\eta_{ST,max}} \qquad (23.9)$$

where $W_{IS,max}$ = isentropic power for maximum flowrate through the turbine
m_{max} = maximum flowrate through the turbine
$\eta_{ST,max}$ = overall steam turbine efficiency for maximum flowrate

The plots in Figure 23.11b can be described by[9]:

$$W_{IS,max} = a + b \cdot W_{max} \qquad (23.10)$$

The coefficients a and b in Equation 23.10 can be correlated as functions of the saturation temperature difference across the turbine[9]. In fact, the coefficients in Equation 23.10 are related to the pressure drop across the turbine. However, in the model, the pressure drop is replaced by its equivalent saturation temperature difference. Use of temperature difference allows easier interface to utility calculations with process heating and cooling demands. Thus:

$$a = a_0 + a_1 \cdot \Delta T_{SAT} \qquad (23.11)$$

$$b = a_2 + a_3 \cdot \Delta T_{SAT} \qquad (23.12)$$

where a_0, a_1, a_2, a_3 = correlating coefficients

Table 23.1 presents coefficients correlated from the data of Peterson and Mann[11]. It should be emphasized that the coefficients may not necessarily be appropriate to predict the performance of all steam turbines. Therefore, it is recommended that coefficients be regressed for a particular

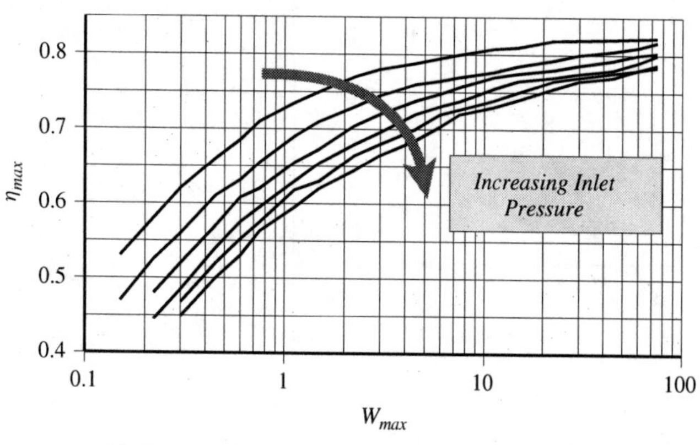

(a) Overall efficiency of steam turbines at maximum load as a function of size.

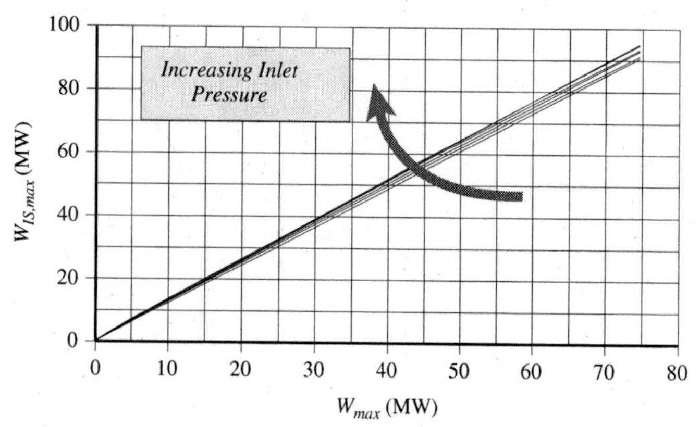

(b) The isentropic power can be related to the actual power at maximum load by a linear function.

Figure 23.11 Modeling steam turbines at maximum load. (From Varbanov PS, Doyle S and Smith R, 2004, *Trans IChemE*, **82A**: 561, reproduced by permission of the Institution of Chemical Engineers.)

Table 23.1 Correlating parameters for steam turbines[9].

	Back-pressure turbines		Condensing turbines	
	$W_{max} < 2000$ kW	$W_{max} > 2000$ kW	$W_{max} < 2000$ kW	$W_{max} > 2000$ kW
a (kW)	$1.08\Delta T_{SAT}$	$4.23\Delta T_{SAT}$	$0.662\Delta T_{SAT}$	$-463 + 3.53\Delta T_{SAT}$
b (-)	$1.097 + 0.00172\Delta T_{SAT}$	$1.155 + 0.000538\Delta T_{SAT}$	$1.191 + 0.000759\Delta T_{SAT}$	$1.220 + 0.000148\Delta T_{SAT}$

type of steam turbine from equipment manufacturer's performance data or, for an existing turbine, actual operating data. The data required to determine the coefficients are:

- W (MW) – the shaft power
- $T_{SAT,in}$ (°C) – the inlet steam saturation temperature
- T_{in} (°C) – the actual inlet steam temperature
- $T_{SAT,out}$ (°C) – the outlet steam saturation temperature
- m (t·h^{-1}) – the mass flow of the expanding steam.

The linear Willans' Line can be written for the maximum load and the intercept substituted by the product of the intercept ratio and W_{max}. The resulting equation can be solved for W_{max}:

$$W_{max} = n \cdot m_{max} - W_{INT} \tag{23.13}$$

$$W_{max} = n \cdot m_{max} - L \cdot W_{max} \tag{23.14}$$

$$W_{max} = \frac{n \cdot m_{max}}{L+1} \tag{23.15}$$

Equation 23.15 can be substituted into Equation 23.10 to give:

$$\Delta H_{IS} \cdot m_{max} = a + b \cdot \frac{n \cdot m_{max}}{L+1} \tag{23.16}$$

Equation 23.16 can be rearranged to give the slope of the Willans' Line:

$$n = \frac{L+1}{b} \cdot \left(\Delta H_{IS} - \frac{a}{m_{max}}\right) \tag{23.17}$$

Substituting Equations 23.15 and 23.17 into Equation 23.13 allows the Willans' Line intercept to be determined[9]:

$$W_{INT} = \frac{L}{b} \cdot (\Delta H_{IS} \cdot m_{max} - a) \tag{23.18}$$

Equations 23.6, 23.17 and 23.18 provide a complete model for steam turbines.

When fitting new data to the model, it is often acceptable to assume the coefficient a to be zero. This simplifies the equations for the model. More importantly, it also simplifies data regression to fit manufacturers' data or operating data to the model. After substituting a to be zero in Equations 23.17 and 23.18, they take the following form[9]:

$$n = \frac{L+1}{b} \cdot \Delta H_{IS} \tag{23.19}$$

$$W_{INT} = \frac{L}{b} \cdot \Delta H_{IS} \cdot m_{max} \tag{23.20}$$

Using this simplified version of the model, it is necessary to regress only two coefficients, b and L.

The model can be used to predict the variation in turbine efficiency with load:

$$\eta_{IS} = \frac{\Delta H}{\Delta H_{IS}} = \frac{W}{m \cdot \eta_{MECH} \cdot \Delta H_{IS}} = \frac{n \cdot m - W_{INT}}{m \cdot \eta_{MECH} \cdot \Delta H_{IS}} \tag{23.21}$$

For back-pressure turbines, the turbine exhausts to a steam main. Therefore, in order to enable estimation of steam main conditions, it is important to predict the condition of the exhausted steam also. With a value of the mechanical efficiency, the enthalpy of the exhaust steam can be calculated from an energy balance[9]:

$$H_{out} = H_{in} - \frac{W}{\eta_{MECH} \cdot m} \tag{23.22}$$

This approach to steam turbine modeling can be extended to complex steam turbines using the decomposition illustrated in Figure 23.12[10]. A complex turbine can be decomposed into its corresponding basic components involving simple turbine units, splitters and mixers[10]. Figure 23.12a shows how an extraction machine can be modeled by a series of simple turbine units with the appropriate connections. Figure 23.12b illustrates how an induction machine can be modeled by simple turbine units with the appropriate connections. Any minor differences between the extraction machine and the decomposed model can be compensated for later in the design by small adjustments in the specifications for the machine in terms of its efficiency or steam flow through the turbine.

Example 23.2 A process heating duty of 25 MW is to be supplied by the exhaust steam of a back-pressure turbine. High-pressure steam at 100 barg with a temperature of 485°C is to be expanded to 20 barg for process heating. The heating duty of the 20 barg steam can be assumed to be the sum of the superheat

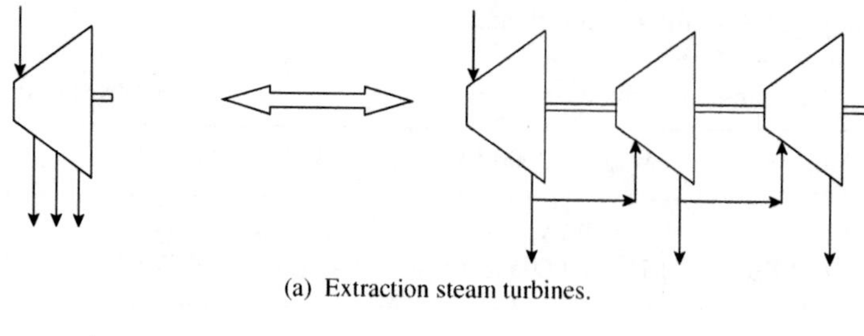

(a) Extraction steam turbines.

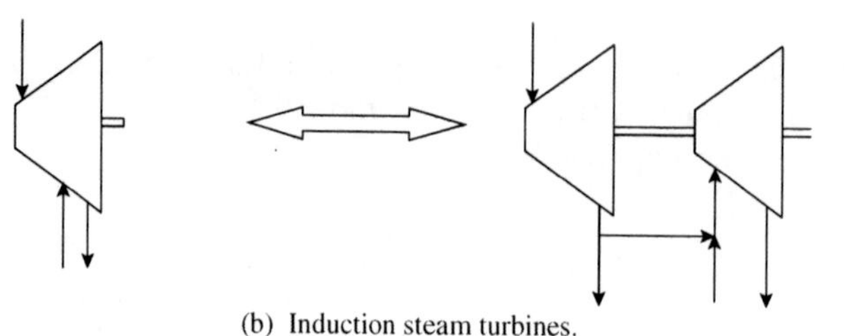

(b) Induction steam turbines.

Figure 23.12 Modeling complex steam turbines.

and latent heat. Assume the intercept ratio L of the turbine to be 0.05 and the mechanical efficiency to be 97%. Estimate the power production and steam flowrate for a fully loaded back-pressure turbine.

Solution The conditions for the inlet steam are fixed, but the conditions of the outlet steam will depend on the performance of the turbine. First, estimate the steam flowrate from the process heating duty. A good approximation is that the sum of the heat content of the superheat and latent heat is constant from inlet to outlet. At the expander inlet, the heat content of the superheat is higher than that of the outlet, but the latent heat is lower in the inlet than in the outlet. The two trends tend to cancel each other out, with the total heat content of superheat and latent heat being approximately constant across the turbine. From steam properties, the enthalpy of the superheated steam H_{SUP}, enthalpy of the saturated steam H_{SAT} and enthalpy of the saturated condensate H_L at 100 barg are:

$$H_{SUP} = 3335 \text{ kJ·kg}^{-1} \text{at } 485^{\circ}$$

$$H_{SAT} = 2726 \text{ kJ·kg}^{-1}$$

$$H_L = 1412 \text{ kJ·kg}^{-1}$$

Assuming the heat content of the superheat and latent heat of steam to be constant, the flowrate of steam can be estimated to be:

$$\begin{aligned} m &= \frac{25 \times 10^3}{3335 - 1412} \\ &= 13.0 \text{ kg·s}^{-1} \end{aligned}$$

Now, the parameters for the Willans' Line Equations need to be calculated. From steam properties:

$$\begin{aligned} T_{SAT} &= 312^{\circ}\text{C at 100 barg} \\ T_{SAT} &= 215^{\circ}\text{C at 20 barg} \\ \Delta T_{SAT} &= 97^{\circ}\text{C} \end{aligned}$$

Thus, from Table 23.1 (assuming a power output greater than 2000 kW):

$$\begin{aligned} a &= 4.23\Delta T_{SAT} \\ &= 4.23 \times 97 \\ &= 410.3 \text{ kW} \\ b &= 1.155 + 0.000538\Delta T_{SAT} \\ &= 1.155 + 0.00538 \times 97 \\ &= 1.207 \end{aligned}$$

Next, determine the isentropic enthalpy change ΔH_{IS}. From steam properties, the entropy of the inlet steam is:

$$S_{SUP} = 6.5432 \text{ kJ·kg}^{-1}\text{·K}^{-1}$$

From steam properties, the enthalpy of steam at 20 barg with this entropy is:

$$H_{SUP} = 2912 \text{ kJ·kg}^{-1}$$

Thus:

$$\begin{aligned} \Delta H_{IS} &= 3335 - 2912 \\ &= 423 \text{ kJ·kg}^{-1} \end{aligned}$$

From Equation 23.17:

$$\begin{aligned} n &= \frac{L+1}{b}\left(\Delta H_{IS} - \frac{a}{m_{max}}\right) \\ &= \frac{0.05+1}{1.207}\left(423 - \frac{410.3}{13.0}\right) \\ &= 340.5 \text{ kJ·kg}^{-1} \end{aligned}$$

From Equation 23.18:

$$\begin{aligned} W_{INT} &= \frac{L}{b}(\Delta H_{IS} m_{max} - a) \\ &= \frac{0.05}{1.207}(423 \times 13.0 - 410.3) \\ &= 210.8 \text{ kW} \end{aligned}$$

From Equation 23.6,

$$W = n \cdot m - W_{INT}$$
$$= 340.5 \times 13.0 - 210.8$$
$$= 4216 \text{ kW}$$

This is consistent with the starting assumption that correlations for power greater than 2 MW were to be used. The outlet steam enthalpy is given by an energy balance from Equation 23.22:

$$H_{out} = H_{in} - \frac{W}{\eta_{MECH} \cdot m}$$
$$= 3335 - \frac{4216}{0.97 \times 13.0}$$
$$= 3001 \text{ kJ}\cdot\text{kg}^{-1}$$

Now revise the steam flow. From steam properties at 20 barg, $H_L = 920 \text{ kJ}\cdot\text{kg}^{-1}$:

$$m = \frac{25 \times 10^3}{3001 - 920}$$
$$= 12.0 \text{ kg}\cdot\text{s}^{-1}$$

Compare this with the original estimate of $13.0 \text{ kg}\cdot\text{s}^{-1}$. The calculation is now repeated with the revised steam flow:

$$n = 338.2 \text{ kW}\cdot\text{kg}^{-1}$$
$$W_{INT} = 193.3 \text{ kW}$$
$$W = 3865 \text{ kW}$$
$$H_{out} = 3003 \text{ kJ}\cdot\text{kg}^{-1}$$
$$m = 12.0 \text{ kg}\cdot\text{s}^{-1}$$

Further iteration will not bring any significant change. The turbine efficiency can now be calculated:

$$\eta_{IS} = \frac{\Delta H_{REAL}}{\Delta H_{IS}} = \frac{3335 - 3003}{423}$$
$$= 0.78$$
$$\eta_{ST} = \frac{W}{m \cdot \Delta H_{IS}} = \frac{3865}{12.0 \times 423}$$
$$= 0.76$$

23.4 GAS TURBINES

In its simplest form, the gas turbine consists of a *compressor* and a *turbine* (or an *expander*). In a *single-shaft* gas turbine, these are mechanically connected and are rotating at the same speed, as shown in Figure 23.13a. Air from ambient is compressed and raised in temperature as a result of the compression. A portion of the compressed air enters a combustion chamber where fuel is fired, and the temperature of the gases is raised further to a temperature in excess of 1500°C. Some of the air from the compressor provides cooling for the combustor walls. The hot, compressed mixture of air and combustion gases then flows to the inlet of the turbine. In the turbine, the gas

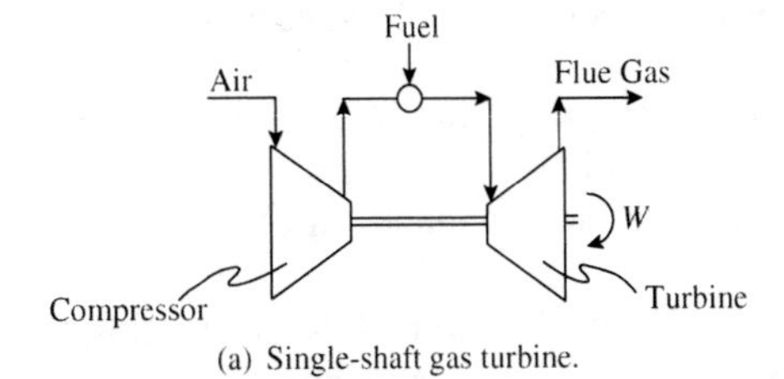

(a) Single-shaft gas turbine.

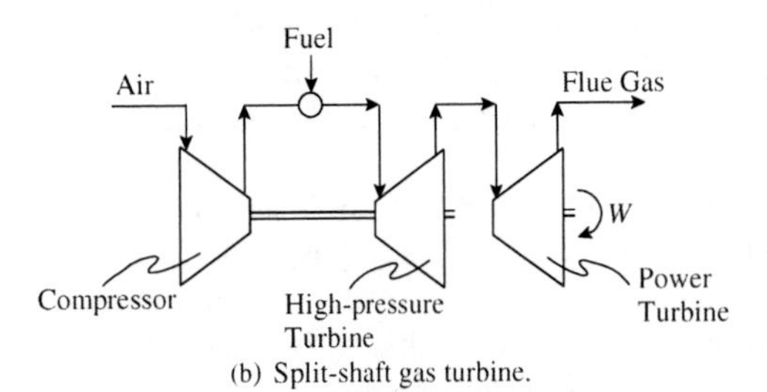

(b) Split-shaft gas turbine.

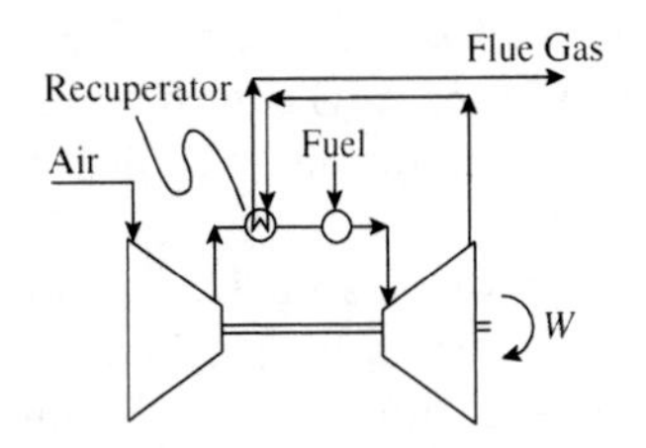

(c) Single-shaft gas turbine with recuperation.

Figure 23.13 Basic gas turbine configurations.

is expanded to develop power to drive the compressor. Around two-thirds of the power produced by the turbine is needed to drive the compressor. However, by virtue of the energy imported in the combustion process, excess power is produced. The higher the expander inlet temperature, the better is the performance of the machine. Temperatures in the turbine are limited by turbine blade materials. Many different machine configurations are possible with gas turbines. Figure 23.13b shows a *split-shaft* arrangement. This is mechanically more complex. The first turbine provides the power necessary to drive the compressor. The second turbine provides the power for the external load. More complex gas turbine configurations divide the compression and expansion into two stages with separate mechanical linkages, one between the low pressure compressor and low pressure expander, and another between the high pressure compressor and high pressure expander in a *twin-spool* design. Many other machine configurations are available. The expanded gas can be discharged directly to atmosphere, as shown in Figures 23.13a and b, or can be used to preheat the air at the exit of the compressor before entering the combustion chamber in a *regenerator*, as illustrated in Figure 23.13c. This increases the efficiency of the machine, but also increases NO_x emissions. NO_x emissions will be discussed in more detail in Chapter 25. However, the increased pressure drop on both the compressed air and turbine exhaust sides of the recuperator creates some inefficiency in the cycle.

The basic characteristics of gas turbines are:

- sizes are restricted to standard frame sizes ranging from 250 kW to 250 MW
- *micro-turbines* are available in the size range 25 kW to 500 kW
- electrical efficiency is 20 to 40% (increasing to 50% for designs with cooled turbine blades)
- exhaust temperatures in the range 450 to 600°C
- fuels need to be at a high pressure
- fuels must be free of particulates and sulfur.

The most common fuels used are gas (natural gas, methane, propane, synthesis gas) and light fuel oils. Contaminants such as ash, alkalis (sodium and potassium) and sulphur result in deposits, which degrade performance and cause corrosion in the hot section of the turbine. Total alkalis and total sulphur in the fuel should both be typically less than 10 ppm. Gas turbines can be equipped with dual firing to allow the machine to be switched between fuels.

Gas turbines can be classified as *industrial* or *frame* machines and *aero-derivative* machines that are lighter weight units derived from aircraft engines. Table 23.2 compares the characteristics of the two broad classes of gas turbine machines.

Aero-derivative gas turbines are typically used for offshore applications where weight and efficiency are a premium, to drive compressors for natural gas pipelines, and stand-alone power generation applications for peak periods of high power demand. For stand-alone applications, gas turbine efficiency becomes a critical issue. However, if heat is to be recovered from the gas turbine exhaust, the efficiency becomes less important as the waste heat is utilized.

Gas turbines require a start-up device, which is usually an electric motor. The power of the start-up device can be up to 15% of the gas turbine power, depending on the size and design of the machine.

The gas turbine performance is a function of a number of important parameters.

1. *Expander inlet temperature*. The power produced by the turbine (expander) is proportional to the absolute temperature of the inlet gases. An increase in the expander inlet temperature increases the power output and efficiency of the machine. The maximum temperature is constrained by the turbine blade materials. Cooling systems for the turbine blades allow higher temperatures. Expander inlet temperatures are in the range 800 to 1300°C (increasing to 1500°C with cooled turbine blades).

2. *Pressure ratio*. As the design pressure ratio across the compressor increases, the power output initially increases to a maximum and then starts to decrease. The optimum pressure ratio increases with increasing expander inlet temperature. Pressure ratios for industrial machines are typically in the range 10 to 15 but can be higher. For aero-derivative machines, pressure ratios are typically 20 to 30.

3. *Ambient conditions*. The performance of the machine is normally specified at International Standards Organization (ISO) conditions of 15°C, 1.013 bar and 60% relative humidity. The power consumed by the compressor is proportional to the absolute temperature of the inlet air. Thus, a decrease in the ambient temperature increases both the power output and the efficiency for a given machine, and vice versa. For example, at 40°C, the power output might typically drop to around 90% of the ISO-rated power. In hot climates, inlet air cooling (e.g. by using a spray of water into the inlet air) can be used to increase performance. Decreasing relative humidity causes the power output and efficiency to decrease. Increasing altitude causes the power output to decrease.

4. *Working load*. The efficiency drops off quickly as the load decreases. The more the load is decreased from 100% rated capacity, the steeper is the decline in the machine performance. The decline in performance depends on the machine and the control system. For example, at 70% of full load, the ratio of the part-load efficiency to the efficiency at full-load might vary typically between 0.95 and 0.85, depending on the machine and the control system. Twin-spool machines have a better part-load performance than single-shaft machines. Part-load performance depends on the size of the machine.

5. *Back-pressure*. Before the gases are vented to atmosphere, they might go through a device that has a pressure drop. This could be a heat recovery steam generator

Table 23.2 Gas turbine types.

Industrial	Aero-derivative
• Lower-pressure ratio (typically, up to 15) • Robust • Lower efficiency • Lower capital cost per unit power output • Sizes up to 250 MW	• Higher-pressure ratio (typically, up to 30) • Lightweight • Higher efficiency • Higher capital cost per unit power output (require advanced materials) • Sizes limited typically up to 50 MW

(HRSG), a furnace or a NO_x treatment unit. The back-pressure created by such a device decreases the power generation. Even if there is no device, changes in elevation changing the ambient pressure also change the machine performance.

The combustion within the gas turbine creates emissions. The principal concern is usually NO_x. NO_x emissions will be dealt with in detail in Chapter 25, but it is worth reviewing the problems associated with gas turbines at this point. NO_x emissions from gas turbines can be dealt with in one of three ways as follows.

1. *Staged combustion.* In a standard combustion arrangement in a gas turbine, the fuel and air are mixed in the combustor. This leads to a high local peak flame temperature in which the nitrogen in the combustion air reacts with oxygen to produce NO_x. *Staged* or *pre-mixed* combustion alleviates this problem by premixing the fuel and air in a substoichiometric mixture. This first stage of combustion involves a fuel rich zone. Additional air is then mixed and the combustion completed. Such staged combustion lowers the peak flame temperature and the NO_x formation. NO_x emissions can be reduced to 10 ppm for gaseous fuels and 25 ppm for liquid fuels.

2. *Steam injection.* Steam can be injected into the combustion zone as an inert material with the purpose of reducing the peak flame temperature and thereby reducing the NO_x formation. NO_x emissions can be reduced by typically 60% by steam injection. An obvious drawback is that the injected steam is lost to atmosphere. A side effect of the steam injection is that it increases the power output due to the higher mass flowrate through the turbine. Indeed, steam injection over and above that required for NO_x suppression can be used to increase power production during times of peak power demand.

3. *Treatment of the exhaust gases.* If staged combustion or steam injection cannot satisfy the regulatory requirements for NO_x emissions, then the gas turbine exhaust must be treated to remove the NO_x. Alternatively, if an existing gas turbine installation requires the NO_x emissions to be decreased, then, again, the exhaust gases must be treated. The usual method to treat NO_x emissions in gas turbine exhausts is by *selective catalytic reduction*. This will be dealt with in more detail in Chapter 25, but involves the injection of ammonia upstream of a catalyst to chemically reduce the NO_x to nitrogen.

To model a gas turbine, it is convenient to define the fuel-to-air ratio and steam-to-air ratio:

$$f = \frac{m_{FUEL}}{m_{AIR}} \qquad s = \frac{m_{STEAM}}{m_{AIR}}$$

A mass balance gives:

$$\begin{aligned} m_{EX} &= m_{AIR} + m_{STEAM} + m_{FUEL} \\ &= \frac{f + s + 1}{f} m_{FUEL} \end{aligned} \tag{23.23}$$

where m_{AIR}, m_{STEAM}, m_{FUEL}, m_{EX}

= mass flowrates of the inlet air, steam injected (if any), fuel and exhaust, respectively

An energy balance across the gas turbine gives[4]:

$$\begin{aligned} & m_{AIR} H_{AIR} + m_{STEAM} H_{STEAM} + m_{FUEL} H_{FUEL} \\ & + m_{FUEL} \Delta H_{COMB} - m_{EX} H_{EX} \\ & - W - W_{LOSS} = 0 \end{aligned} \tag{23.24}$$

where H_{AIR}, H_{STEAM}, H_{FUEL}, H_{EX}
= specific enthalpy of the inlet air, injected steam (if any), fuel and exhaust, respectively
ΔH_{COMB} = net heat of combustion of the fuel
W = power production
W_{LOSS} = power lost due to mechanical inefficiencies

Combining Equations 23.23 and 23.24 and rearranging gives[4,9]:

$$W = n \cdot m_{FUEL} - W_{LOSS} \tag{23.25}$$

where

$$\begin{aligned} n = & \frac{H_{AIR}}{f} + \frac{s}{f} \cdot H_{STEAM} + H_{FUEL} + \Delta H_{COMB} \\ & - \frac{f + s + 1}{f} \cdot H_{EX} \end{aligned}$$

Equation 23.25 has the same basic form as the linear Willans' Line Equation used to model steam turbines. The basic assumption behind the use of Equation 23.25 is that the gas turbine would need to have a control system that would maintain a fixed fuel-to-air ratio and steam-to-air ratio at part-load.

The performance of gas turbines as a function of size can be correlated at ISO conditions (ambient conditions of 15°C, 1.013 bar and 60% relative humidity) by an equation of the form[4,9]:

$$\Delta H_{COMB,max} = \frac{W_{max}}{\eta_{GT,max}} = a + bW_{max} \tag{23.26}$$

where $\Delta H_{COMB,max}$ = maximum fuel heat from combustion
W_{max} = maximum turbine power
$\eta_{GT,max}$ = gas turbine efficiency at maximum load
a, b = correlating parameters

The performance of the turbine will depend on the turbine manufacturer. For example, for GE Turbines in the size range 26.1 MW $\leq W_{max} \leq$ 255.6 MW[9]:

$$a = 21.992(\text{MW}) \tag{23.27}$$

$$b = 2.6683$$

The gas turbine performance at part-load follows the same basic form as that for steam turbines illustrated in Figure 23.10, with the mass flowrate defined in terms of the mass flowrate of fuel. Thus, the part-load gas turbine performance can be modeled as[4,9]:

$$W = n \cdot m_{FUEL} - W_{INT} \tag{23.28}$$

As with steam turbines, it is convenient to characterize the intercept W_{INT} in terms of an intercept ratio as Equation 23.8. At maximum power:

$$W_{max} = \frac{n \cdot m_{FUEL,max}}{L + 1} \tag{23.29}$$

From Equation 23.26:

$$m_{FUEL,max} \Delta H_{COMB} = a + bW_{max} \tag{23.30}$$

Combining Equation 23.29 and 23.30 gives[4,9]:

$$n = \frac{L + 1}{b}\left(\Delta H_{COMB} - \frac{a}{m_{FUEL,max}}\right) \tag{23.31}$$

Combining Equation 23.28 at maximum power with Equation 23.30 and 22.31 gives[5,11]:

$$W_{INT} = \frac{L}{b}(m_{FUEL,max} \Delta H_{COMB} - a) \tag{23.32}$$

The variation in performance of the gas turbine with ambient temperature can be correlated as[9]:

$$W_{max} = a + bT_0 \tag{23.33}$$

where T_0 = ambient temperature (°C)
a, b = correlating parameters for power

The gas turbine heat rate and its variation with temperature can be defined by[9]:

$$HR = \frac{\Delta H_{COMB,max}}{W_{max}} = a + bT_0 \tag{22.34}$$

where HR = heat rate
a, b = correlating parameters for heat rate

The correlating parameters for variation power and heat rate with ambient temperature are specific to a particular gas turbine model. The parameters in the model can be determined from detailed simulation of the gas turbine or by fitting operating data from existing gas turbines, under different operating conditions.

Figure 23.14 shows an integrated gas turbine in which the exhaust from the gas turbine is used to generate steam in a heat recovery steam generator (HRSG) before being vented to atmosphere. It is possible to fire fuel after the gas turbine to increase the temperature of the gas turbine exhaust before entering the HRSG, as gas turbine exhaust is still rich in oxygen (typically, 15% O_2). The steam from the HRSG might be used directly for process heating or be expanded in a steam turbine system to generate additional power. The steam turbine exhaust can either be back-pressure or condensing. Figure 23.15 shows a representation of a HRSG on a temperature–enthalpy diagram. Figure 23.15a shows a single-pressure HRSG. The steam profile includes boiler feedwater preheat, latent heat and superheat. Maximizing the heat recovery from the gas turbine exhaust into steam generation will be limited by a pinch between the gas turbine exhaust and the steam generation profile. Figure 23.15b shows a *dual pressure* HRSG. This includes boiler feedwater preheat and superheat for both the low-pressure (LP) and high-pressure (HP) levels. The dual pressure HRSG allows a better match between the steam profile and the gas turbine exhaust. This means that a greater amount of heat recovery is possible when compared

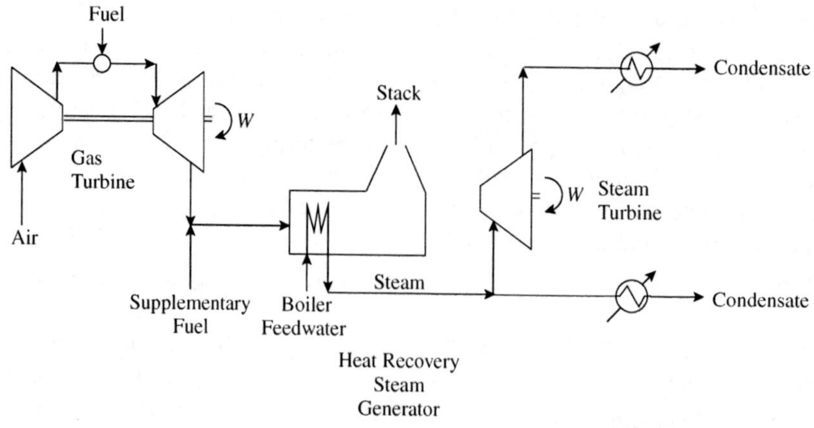

Figure 23.14 Gas turbine with heat recovery steam generator (HRSG).

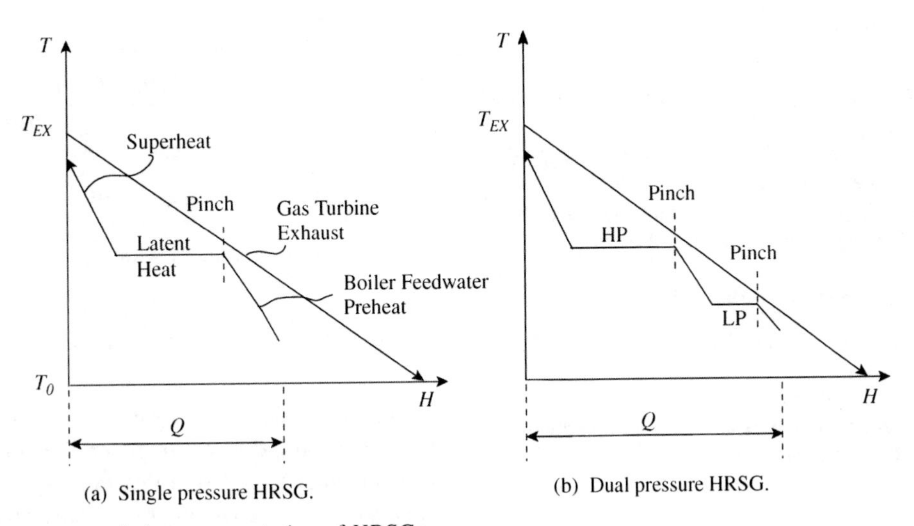

Figure 23.15 Temperature–enthalpy representation of HRSGs.

with a single pressure HRSG. In the case of the dual pressure HRSG, there are degrees of freedom for how much LP and HP steam are generated, leading to two pinches that limit the heat recovery potential. The largest gas turbines sometimes use a triple pressure HRSG, although these are normally reserved for power station applications.

Figure 23.15 shows the potential to generate steam from a gas turbine exhaust. The potential to generate steam can be increased by introducing firing of fuel after the gas turbine. There are three firing modes for gas turbines as follows.

1. *Unfired HRSG*. An unfired HRSG uses the sensible heat in the gas turbine exhaust to raise steam.

2. *Supplementary fired HRSG*. Supplementary (or auxiliary) firing raises the temperature by firing fuel to use a portion of the oxygen in the exhaust. Supplementary firing uses convective heat transfer, and temperatures are limited to a maximum of around 850°C by ducting materials.

3. *Fully fired HRSG*. Fully fired HRSGs make full utilization of the excess oxygen to raise the maximum amount of steam in the HRSG. Full firing means reducing the excess oxygen to a minimum of around the 3% normally demanded by all combustion processes to ensure efficient combustion. However, this means that radiant heat transfer will result from full firing. Essentially, fully firing means that the gas turbine exhaust is used as preheated combustion air to a steam boiler.

Figure 23.16 shows supplementary firing on a temperature versus enthalpy diagram. The supplementary firing increases the temperature of the exhaust, thereby increasing the amount of available heat and the potential for steam generation. Supplementary firing allows greater amounts of steam to be generated with a high efficiency. Higher steam superheat temperatures are possible. Also, importantly, supplementary firing introduces more flexibility to the steam system. As already discussed, gas turbines are inflexible in terms of their part load performance. Supplementary firing allows the amount of steam generated to be varied, without changing the load on the gas turbine. Figure 23.16 shows a situation in which the supplementary firing allows a closer match between the gas turbine exhaust and the steam generation profile when compared with the unfired case. In some cases, this can mean that the additional heat to steam generation from supplementary firing is greater than the heat input from fuel from the supplementary firing.

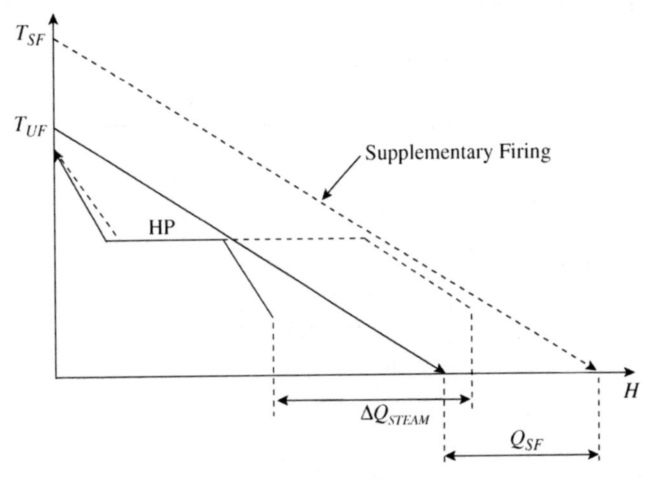

Figure 23.16 Supplementary fired HRSG.

Example 23.3 Data are given for a gas turbine firing natural gas at ISO conditions:

$$W = 26{,}070 \text{ kW}$$

$$HR = 12{,}650 \text{ kJ·kWh}^{-1}$$

$$m_{EX} = 446{,}000 \text{ kg·h}^{-1}$$

$$T_{EX} = 488°\text{C}$$

Using these data calculate:

a. the fuel consumption ΔH_{COMB} and efficiency η_{GT} at standard conditions.
b. how much heat can be recovered from the exhaust if $T_{STACK} = 100°C$ and if C_P of the exhaust is 1.1 kJ·kg^{-1}·K^{-1}?
c. W_{max}, HR, η_{GT} and Q_{COMB} at an ambient temperature of 0°C? The change in performance can be estimated from

$$W_{max}\,(\text{kW}) = 29{,}036 - 189.4T_0$$

$$HR(\text{kJ·kWh}^{-1}) = 12{,}274 + 31.32T_0$$

where T_0 = ambient temperature (°C)

d. estimate the heat available in the exhaust if it is supplementary fired to 850°C for ISO conditions.

Solution

a. From Equation 23.34:

$$\Delta H_{COMB} = W \cdot HR$$
$$= \frac{26{,}070\ (kW) \times 12{,}650\ (\text{kJ·kWh}^{-1})}{3600\ (\text{kJ·kWh}^{-1})}$$
$$= 91{,}607 \text{ kW}$$
$$\eta_{GT} = \frac{1}{HR}$$
$$= \frac{3600\ (\text{kJ·kWh}^{-1})}{12{,}650\ (\text{kJ·kWh}^{-1})}$$
$$= 0.28$$

b.

$$Q_{EX} = mC_P(T_{EX} - T_{STACK})$$
$$= \frac{446{,}000 \times 1.1 \times (488 - 100)}{3600}$$
$$= 52{,}876 \text{ kW}$$
$$= 52.9 \text{ MW}$$

c. At 0°C:

$$W_{max} = 29{,}036 - 189.4T_0$$
$$= 29{,}036 - 189.4 \times 0$$
$$= 29{,}036 \text{ kW}$$
$$HR = 12{,}274 + 31.32T_0$$
$$= 12{,}274 \text{ kJ·kWh}^{-1}$$
$$\eta_{GT} = \frac{3600}{12{,}274}$$
$$= 0.29$$
$$Q_{COMB} = W \cdot HR$$
$$= \frac{29{,}036 \times 12{,}274}{3600}$$
$$= 98{,}997 \text{ kW}$$

d.

$$Q_{SF} = mC_P(T_{SF} - T_{EXHAUST})$$
$$= \frac{446{,}000 \times 1.1 \times (850 - 488)}{3600}$$
$$= 49{,}333 \text{ kW}$$
$$= 49.3 \text{ MW}$$

Total heat in the exhaust with supplementary firing

$$= 52.9 + 49.3$$
$$= 102.2 \text{ MW}$$

The emphasis on gas turbine integration so far has been to utilize the exhaust for generation of steam in unfired, supplementary fired or fully fired systems. However, the exhaust can be used in other ways. In principle, it can be used directly for process heating, as discussed in Chapter 16. This would require a direct match between process fluids and the hot gas turbine exhaust gases. Safety issues might, however, prevent this. If the process fluid is flammable and a leak develops, then flammable material will leak into a duct of hot, oxygen-rich gas with all the potential for fire and explosion. The gas turbine exhaust can also be used for drying. Drying, as discussed in Chapter 8, usually involves the removal of water by evaporation into hot air. In some applications, a gas turbine can be ducted directly into a dryer to provide the hot air. However, care should be taken to ensure that the small amounts of combustion products (including NO_x) do not contaminate the process. Finally, the hot exhaust gases from a gas turbine can be used to supply preheated combustion air (albeit slightly depleted in oxygen) to furnaces. Examples of the application of this type of arrangement are in ethylene production with the furnace reactors and fired heaters in petroleum refinery applications.

23.5 STEAM SYSTEM CONFIGURATION

Figure 23.17 shows a schematic of a typical steam distribution system. Raw water is treated before entering the boilers that fire fuel to generate high-pressure (HP) steam. The HP steam is let down through back-pressure turbines to the medium-pressure (MP) and low-pressure (LP) steam mains. In Figure 23.17, a condensing turbine operates from the HP steam mains to generate additional power. Operating a condensing turbine from the highest-pressure inlet maximizes the power production. The general policy on steam usage for heating is that LP steam should be used in preference to HP steam. Using LP steam for steam heating:

- allows power generation in steam turbines from the high-pressure steam;
- provides a higher latent heat in the steam for the steam heater;
- leads to lower capital cost heat transfer equipment due to the lower pressures.

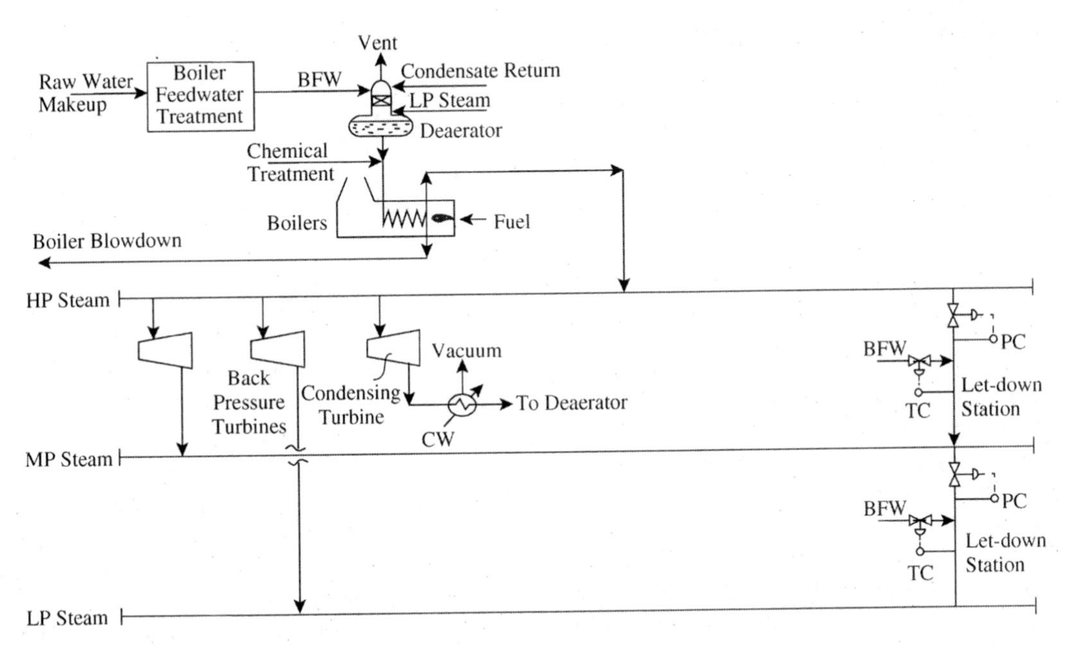

Figure 23.17 Steam distribution system.

Also shown in Figure 23.17 are *let-down stations* between the steam mains to control the mains pressures via a pressure control system. The let-down stations in Figure 23.17 also have *de-superheaters*. When steam is let down from a high to a low pressure under adiabatic conditions, the amount of superheat increases. De-superheating is achieved by the injection of boiler feedwater under temperature control, which evaporates and reduces the superheat. There are two important factors determining the desirable amount of superheat in the steam mains.

a. Steam heating is most efficiently carried out using the latent heat via condensation, rather than having to de-superheat before condensing the steam. Thus, the design of steam heaters benefits from having no superheat in the steam. However, having no superheat in the steam mains is undesirable, as this would lead to excessive condensation in the steam mains. It is desirable to have at least 10 to 20°C superheat in steam mains to avoid excessive condensation in the mains.
b. In addition to using steam for steam heating, it is also used for power generation by expansion through steam turbines. Expansion in steam turbines reduces the superheat in the steam as it is reduced in pressure. If there is not enough superheat in the inlet steam, then condensation can take place in the steam turbine. While some condensation in the machine is acceptable, excessive condensation can be damaging to the machine. Also, if the steam turbine is exhausting to a steam main, then it is desirable to have some degree of superheat in the outlet to maintain some superheat in the outlet low-pressure steam main.

To determine the appropriate amount of superheat in the steam mains, it is best to start with the lowest-pressure steam main that would be used for steam heating and not for power generation. This superheat needs to be set to 10 to 20°C, as discussed above. The degree of superheat in the next highest-pressure main, before expansion through steam turbines to the lowest-pressure steam main, needs to be determined by simulation of the expansion in the steam turbines to maintain the required 10 to 20°C in the lowest-pressure main. In turn, the degree of superheat in the next highest-pressure main, before expansion through steam turbines to the next-to-lowest-pressure steam main, needs to be determined to maintain the required superheat in the next-to-lowest-pressure steam main. In this way, working from the low- to high-pressure mains, the required amount of superheat can be determined for each steam main.

As noted previously, the normal maximum pressure for distribution is around 40 bar. Heat losses in the distribution system might be typically 10% of the fuel fired in the boilers, but can be higher.

Steam heaters are normally provided with a *steam trap* at the outlet of the steam heater that has the function of allowing condensate to pass but closes to prevent steam from passing[3]. Many designs of steam trap are available. Figure 23.18 shows a typical arrangement that uses an expansion bellows. Operation depends on there being a significant difference between the temperature of the condensate and the steam, which can be achieved through heat loss from the pipework upstream of the steam trap. When condensate enters the trap, the expansion bellows contracts and allows the condensate to pass. As soon as steam starts to flow, the bellows expands to close the valve. Steam traps should also preferably allow trapped air to pass

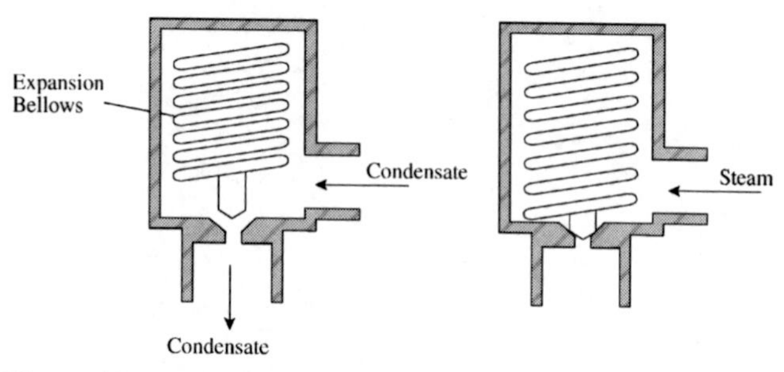

Figure 23.18 A steam trap.

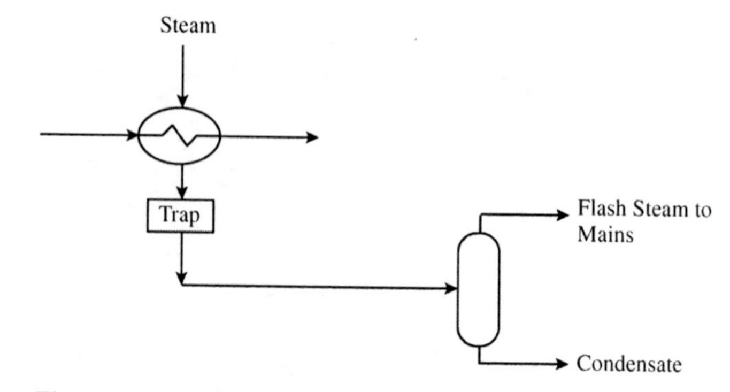

Figure 23.19 Flash steam recovery.

as air reduces the heat transfer coefficient of the condensing steam. Many other designs are available.

For large heat transfer duties, it is good practice to recover the steam that is flashed as the condensate reduces in pressure. Such an arrangement is shown in Figure 23.19. Steam enters the steam heater and condensate (in practice, with some steam) passes through the trap. Flashing occurs before the mixture enters a settling drum that allows the flash steam to be separated from the condensate. The flash steam would then be fed to a steam main at the appropriate pressure.

Figure 23.20 illustrates the features of a typical steam system. It is usual to have at least three levels of steam. On larger sites, steam may also be generated at a very high pressure, which will only be used for power generation in the boiler house. Steam would then be distributed typically at three pressures around the site. Back-pressure turbines let steam down from the high-pressure mains to the lower-pressure mains to generate power. Supplementary power may be generated using condensing turbines. Let-down stations control the steam mains pressures. The system in Figure 23.20 shows flash steam recovery into the medium-pressure and low-pressure mains. Also, as shown in Figure 23.20, the boiler blowdown is flashed and flash steam recovered before being used to preheat incoming boiler feedwater and being sent to the effluent. Whether flash steam recovery is economic is a matter of economy of scale.

23.6 STEAM AND POWER BALANCES

Steam and power balances provide the link between the process utility requirements and the utility supply. The

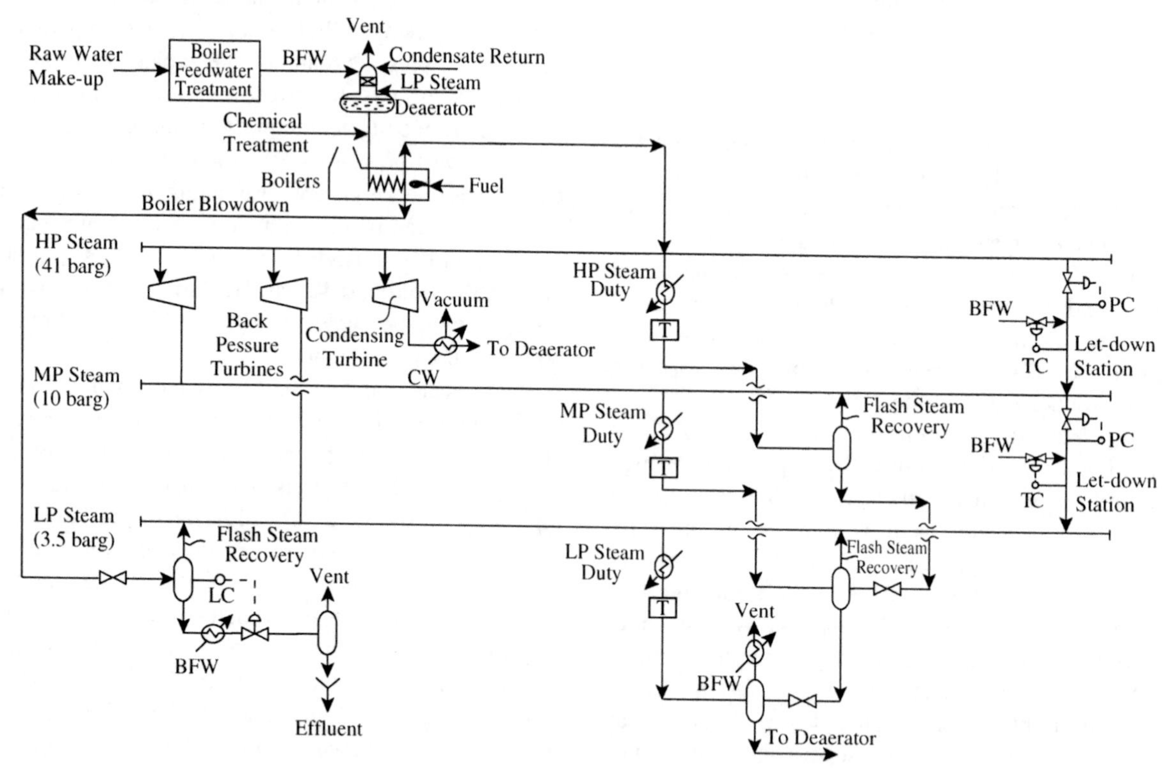

Figure 23.20 Features of a typical steam system.

steam and power balance determines the basis for:

- boiler sizes
- fuel consumption in the boilers
- steam turbine flows
- import power requirements (or export potential)
- let-down station and deaerator capacities
- steam flows in various parts of the system
- boiler feedwater requirements.

To illustrate steam balances, consider a simple example that involves two steam mains with conditions given in Figure 23.21. There are no steam turbines in this simple example, and steam is let down from the high- to low-pressure mains using a let-down station, which does not have de-superheating. Figure 23.21 shows the process steam usage and the process steam generation from high-temperature waste heat recovery. Treated water enters the deaerator at 25°C. The site has a 70% condensate return (based on the rate of steam generation) at a temperature of 80°C, as shown in Figure 23.21. Figure 23.21 shows the various flows around the system that are unknown. The flow from the utility boilers, the let-down flow, deaerator steam, treated water and condensate return flowrates are all unknown. For this particular problem, the steam balance can be closed if the flow of deaeration steam is known. Assume that the flow to deaerator is 5 t·h^{-1}. Also assume that the blowdown rate is 5%. Having assumed the deaerator flow to be 5 t·h^{-1} fixes the let-down flow to be 15 t·h^{-1} and the flow from the utility boilers to be 25 t·h^{-1}, as illustrated in Figure 23.22. Assuming a 5% blowdown rate for the process and utility steam boilers now allows the boiler feedwater flow to be calculated as 42 t·h^{-1}. For a condensate return rate of 70%, the flowrate of condensate return to the deaerator is 28.0 t·h^{-1}. Assuming 5% of the deaeration steam is vented (0.3 t·h^{-1})

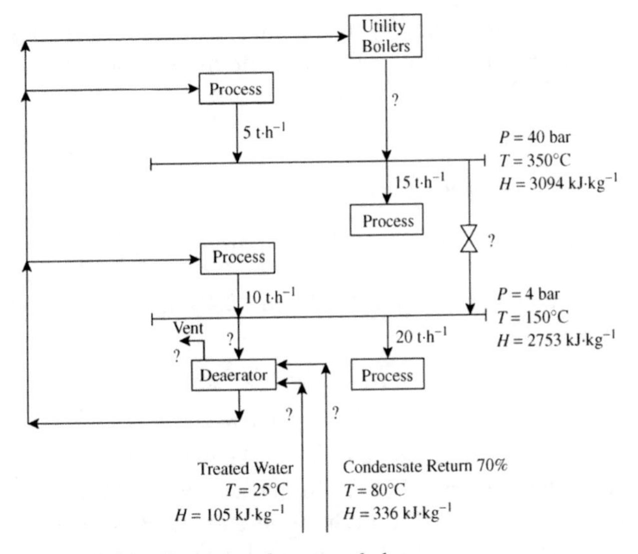

Figure 23.21 Example of a steam balance.

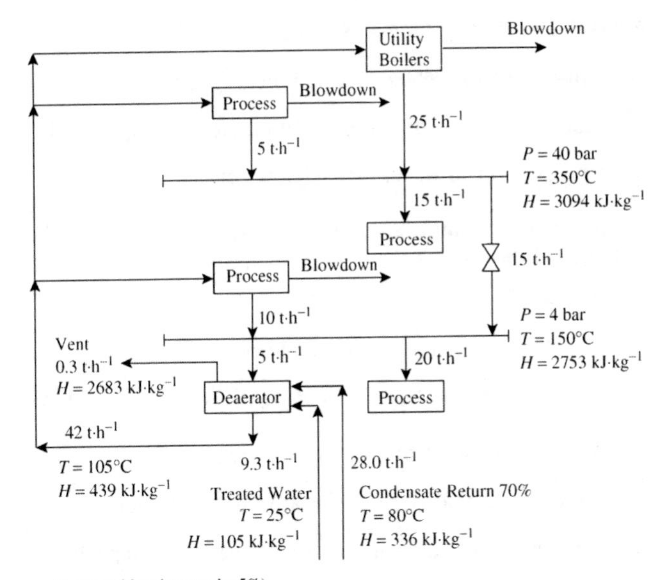

Figure 23.22 Estimation of the deaeration steam allows the missing flows to be determined.

allows the treated water make up to be calculated as 9.3 t·h^{-1}.

Thus, in this example, assumption of the deaeration steam allows the steam balance to be closed. However, this is based on an assumed deaerator flow. The actual flow to the deaerator can be calculated from a heat balance around the deaerator. Figure 23.23 shows the flows into and out of the deaerator. If the boiler feedwater flow and condensate flows are known, together with an assumed value of the vent steam, then the flowrate of deaeration steam can be calculated from an energy balance.

A material balance around the deaerator gives:

$$m_{TW} = m_{BFW} - m_{CR} - m_{STEAM}(1 - \alpha) \qquad (23.35)$$

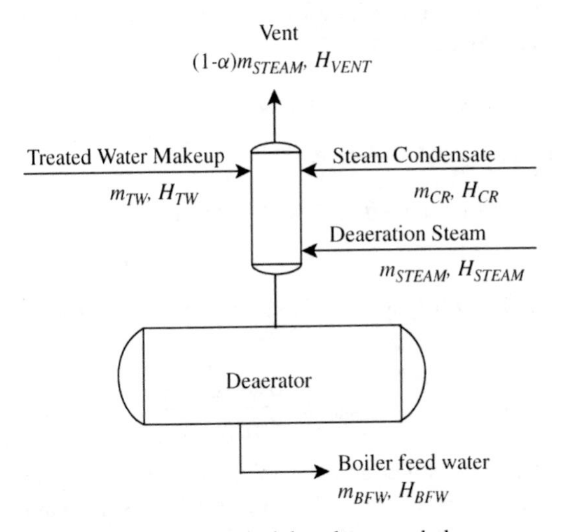

Figure 23.23 Deaerator material and energy balances.

where m_{TW}, m_{BFW}, m_{CR}, m_{STEAM}

= flowrates of treated water, boiler feedwater, condensate return and deaeration steam, respectively

α = proportion of deaeration steam vented

An energy balance around the deaerator assuming the enthalpy of the vented steam to be saturated at the deaerator pressure gives:

$$m_{TW} \cdot H_{TW} + m_{CR} \cdot H_{CR} + m_{STEAM} \cdot H_{STEAM} = m_{BFW} \cdot H_{BFW} + \alpha \cdot m_{STEAM} \cdot H_{VENT} \quad (23.36)$$

where H_{TW}, H_{BFW}, H_{CR}, H_{STEAM}, H_{VENT}

= specific enthalpy of the treated water, boiler feedwater, condensate return deaeration steam and vented steam, respectively

Combining Equations 23.35 and 23.36 and rearranging gives:

$$m_{STEAM} = \frac{m_{BFW}(H_{BFW} - H_{TW}) - m_{CR}(H_{CR} - H_{TW})}{(H_{STEAM} - H_{TW}) - \alpha(H_{VENT} - H_{TW})} \quad (23.37)$$

Substituting the values for the example and assuming $\alpha = 0.05$ gives:

$$m_{STEAM} = \frac{42(439 - 105) - 28.0(336 - 105)}{(2753 - 105) - 0.05(2683 - 105)} = 3.0 \text{ t·h}^{-1}$$

A new steam balance can now be calculated, the energy balance around the deaerator revised and the procedure repeated until convergence is achieved. A converged steam balance is shown in Figure 23.24.

In other problems, the let-down stations might use de-superheating. Figure 23.25 illustrates a de-superheater. A material balance gives:

$$m_{STEAM,out} = m_{STEAM,in} + m_{TW} \quad (23.38)$$

An energy balance gives:

$$m_{STEAM,out} \cdot H_{STEAM,out} = m_{STEAM,in} \cdot H_{STEAM,in} + m_{TW} \cdot H_{TW} \quad (23.39)$$

Combining Equations 23.38 and 23.39 and rearranging gives:

$$m_{STEAM,out} = m_{STEAM,in} \frac{H_{STEAM,in} - H_{TW}}{H_{STEAM,out} - H_{TW}} \quad (23.40)$$

Steam flow and conditions of the inlet steam are known. Outlet conditions are usually fixed by the downstream main conditions, and hence Equation 23.40 can be solved.

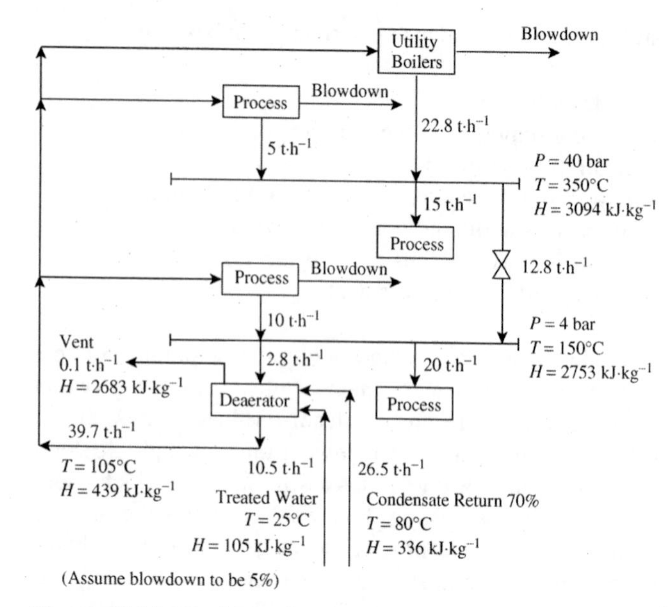

Figure 23.24 Revising the deaerator steam flow allows the steam flows to be updated.

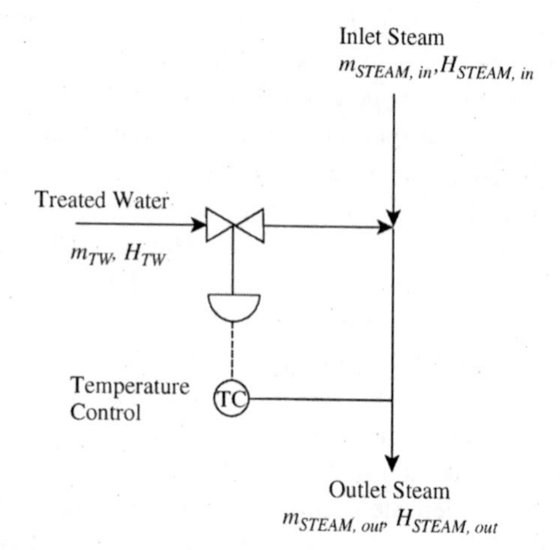

Figure 23.25 A de-superheater.

Flash steam recovery might also feature. Condensate or blowdown is fed to the flash drum, as illustrated in Figure 23.26. A material balance gives:

$$m_{FS} = m_{COND,in} - m_{COND,out} \quad (23.41)$$

An energy balance gives:

$$m_{FS} \cdot H_{FS} = m_{COND,in} \cdot H_{COND,in} - m_{COND,out} \cdot H_{COND,out} \quad (23.42)$$

Combining Equations 22.41 and 22.42 and rearranging gives:

$$m_{FS} = m_{COND,in} \frac{H_{COND,in} - H_{COND,out}}{H_{FS} - H_{COND,out}} \quad (23.43)$$

The flash drum pressure will be known as this will be set by the pressure of the main into which the flash steam will be recovered. The flash steam and condensate outlet enthalpies are set by saturation conditions.

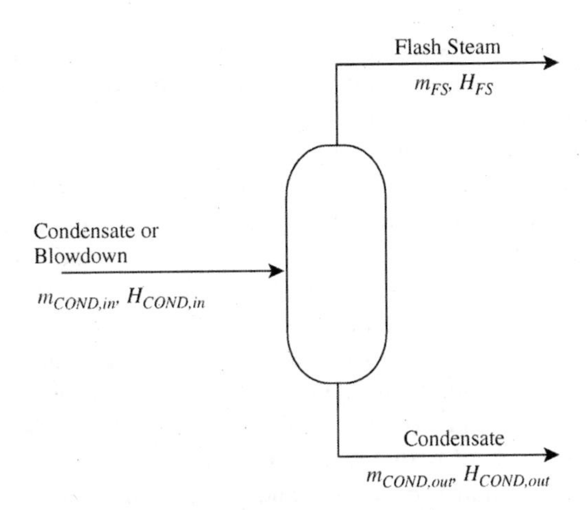

Figure 23.26 Flash steam recovery.

23.7 SITE COMPOSITE CURVES

A method of thermodynamic analysis for site steam systems will next be developed to allow the thermal loads and levels on a complete site to be studied. For this, a temperature–enthalpy picture for the whole site is needed, analogous to the grand composite curve for an individual process, as developed in Chapter 16. There are two ways in which such curves can be developed. The first relates to a new design situation.

A new design situation would start from the grand composite curves of each of the processes on the site and would combine them together to obtain a picture of the overall site utility system[12]. This is illustrated in Figure 23.27, where two processes have their heat sink and heat source profiles from their grand composite curves combined to obtain a *site hot composite curve* and a *site cold composite curve*, using the procedure developed for composite curves in Chapter 16. Wherever there is an overlap in temperature between streams, the heat loads within the temperature intervals are combined together. In Figure 23.27, the pockets of additional heat recovery in the grand composite curve have been left out of the site analysis. This assumes that this part of the heat recovery will take place within the processes. The remaining heat sink profiles and heat source profiles are combined to produce the site composite curves[12].

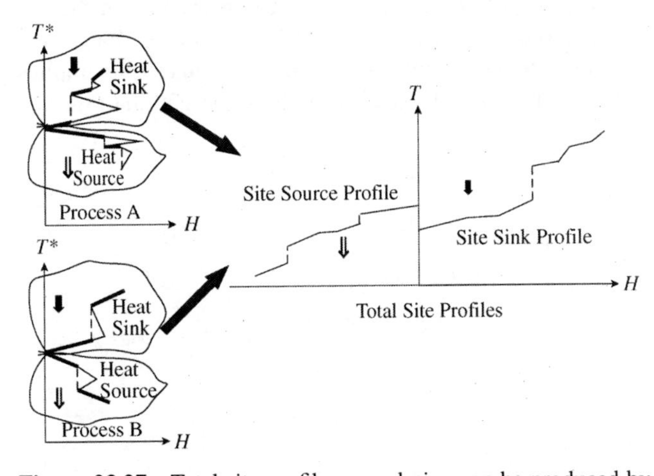

Figure 23.27 Total site profiles new design can be produced by combining the grand opposite curves for the individual processes. (From Klemes J, Dhole VR, Raissi K, Perry SJ and Puigjaner L, 1997, *Applied Thermal Engineering*, **17**: 993, reproduced by permission of Elsevier Ltd.)

While it is usually justified to leave out the pockets of additional heat recovery in a process and accept the in-process heat recovery, some processes demand that information from within the pockets should be included. Figure 23.28 shows a grand composite curve typical of processes involving highly exothermic chemical reactions. The grand composite curve shows only a cooling profile and no heating requirement, as a result of the reaction exothermic heat. There is a large pocket of additional heat recovery, which has not been isolated in Figure 23.28. The temperatures within the pocket are such that high-pressure steam can be generated within the pocket. If high-pressure steam is generated within the pocket as indicated in Figure 23.28, then this disturbs the energy balance within the pocket. To compensate for this, low-pressure steam can be used to provide heating within the pocket, where previously heat recovery would have satisfied the heating requirements within the pocket. In practice, exploiting the pocket in the way indicated in Figure 23.28, using a combination of high-pressure and low-pressure steam, is likely to be economic if there is a significant difference in value between the high- and low-pressure steam. Thus, the extraction of the data from the grand composite curves in such scenarios should include part of the profiles within the pocket, as shown in Figure 23.28. But when should such opportunities be exploited? The situation arises when the temperature difference across the

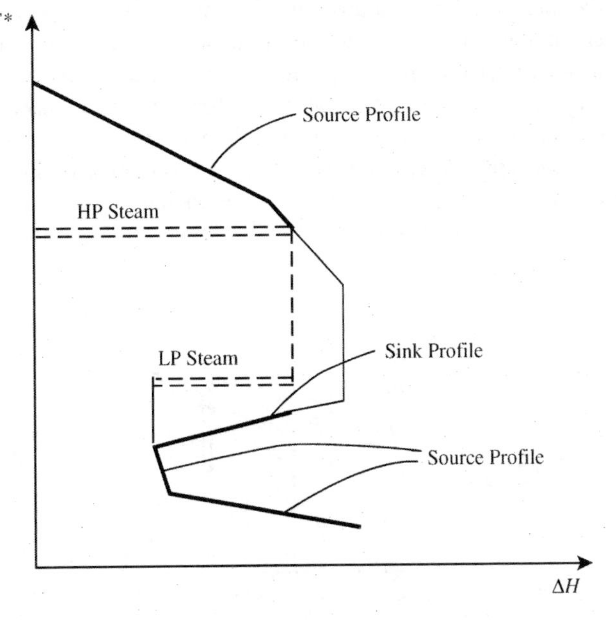

Figure 23.28 Sometimes, the details of the grand composite curve pockets should be included in the site profiles.

pocket spans two steam levels. If the pocket does not span two steam levels, then the pocket should be isolated, as shown in Figure 23.27. Also, even if the pocket spans a large temperature range, the heat duties within the pocket must be large enough to justify the complication of introducing an extra steam level into the design of that process.

One further point needs to be noted regarding the construction of the site composite curves. The temperatures are shifted over and above the shift included in the construction of the grand composite curve. The original hot and cold streams are shifted by $\Delta T_{min}/2$ to produce the grand composite curve. Site composite curves are shifted by an additional $\Delta T_{min}/2$ to give a total ΔT shift of ΔT_{min}, as illustrated in Figure 23.29[13]. If different values of ΔT_{min} apply to different processes, then each set of process data is given its individual shift in ΔT_{min} before the steams are combined in the construction of the site composite curves. The concept of individual shifts for ΔT_{min} was discussed in Chapter 16.

The other way to construct the site composite curves relates more to a retrofit situation. In a retrofit situation, the plants and their heat recovery systems are already in place. The heat recovery might already be maximized and in agreement with the target set by the composite curves and the problem table algorithm. If this is the case, then the grand composite curve will give an accurate reflection of the process utility demand. However, in retrofit, the heat recovery might not be at its maximum. Thus in retrofit situations, the grand composite curve does not necessarily give an accurate picture of utility demands, as it assumes maximum heat recovery. If the existing amount of heat recovery is assumed to be fixed (whether maximized or not), the site profiles can be constructed from the individual process duties within each of the utility heat exchangers on the site. The temperature–enthalpy profiles of the process streams within each of the utility heat exchangers are used to construct the site composite curves in exactly the same way as discussed in Chapter 16 for process composite curves. This way of constructing the site composite curves thus accepts the existing heat recovery system within each process, good or bad. Again, the temperature needs to be shifted for the site composite curves. Starting from the individual process stream data, this needs to have a shift of ΔT_{min}, as indicated in Figure 23.29. One additional advantage of this approach relative to that based on grand composite curves is that fewer data are required for the construction. Construction of the grand composite curve for a process requires data for all heat sources and sinks to be collected, whereas the alternate approach requires only data from the utility heat exchangers.

Following these procedures allows composite curves for the total site to be developed that give a picture of the heating and cooling requirements of the total site, both in terms of enthalpy and temperature[12]. Suppose that the steam mains pressures and temperatures are fixed. Figure 23.30 illustrates how targets can be set for the site utility system in terms of steam generation and steam usage. The site hot composite curve in Figure 23.30 shows the overall site cooling requirements in both temperature and enthalpy terms. In Figure 23.30, the cooling has been satisfied by a combination of steam generation and cooling water. The targets are set for site cooling starting with the highest-temperature cooling utility, in this case, medium-pressure (MP) steam generation. This is first maximized. The second highest-temperature cooling utility is low-pressure (LP) steam generation. This is now maximized, with the residual cooling being satisfied by cooling water. The site cold composite curve in Figure 23.30 shows the heating requirements of the site in temperature and enthalpy terms. To set the targets for steam usage, the lowest-temperature heating utility is first maximized. In this case, the lowest-temperature heating utility is LP steam usage. Having maximized the LP steam usage, the next lowest-heating utility is maximized, in this case, MP steam. The residual high-temperature heating is taken up by high-pressure (HP) steam.

The steam profiles in Figure 23.30 touch the site profiles. This does not imply heat transfer with $\Delta T = 0$. The temperature difference built into the construction of the site composite curves ensures feasible heat transfer. Each time a steam profile touches a site composite curve, this implies ΔT_{min}. In practice, the streams in the construction of the site

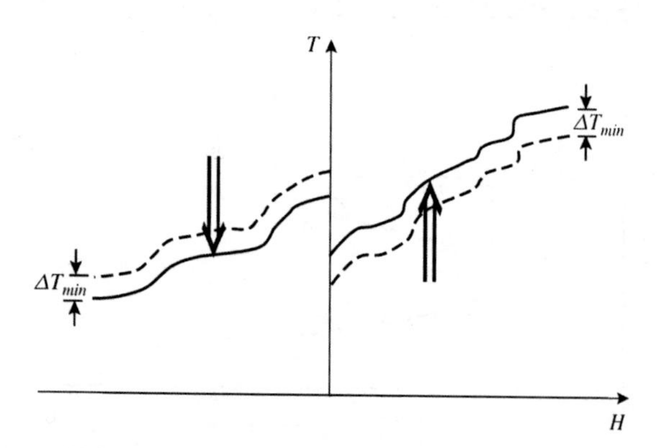

Figure 23.29 Site profiles are required to have a total shift of ΔT_{min} from the original temperature.

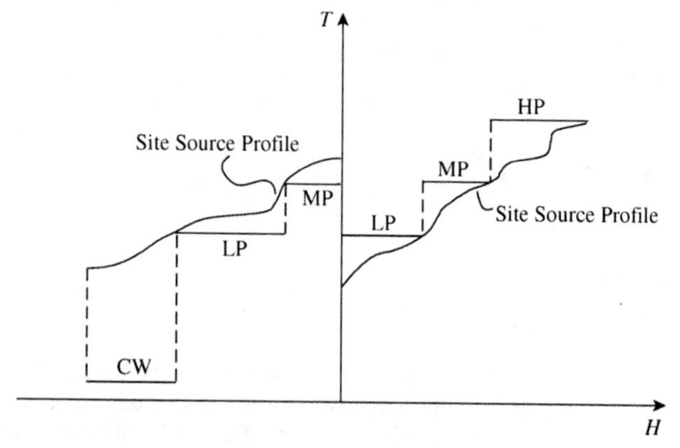

Figure 23.30 Site profiles allow targets for steam usage and steam generation for the whole site to be developed.

hot composite curve are ΔT_{min} hotter than plotted, and the streams in the site cold composite curve are ΔT_{min} colder than plotted. The steam profiles are plotted at their actual temperatures.

The steam profiles plotted in Figure 23.30 show only the latent heat part of the steam profiles. Steam generation will most likely involve boiler feedwater preheating to saturation conditions, followed by vaporization and superheating. Steam usage will involve some de-superheating, followed by condensation of the steam and, possibly, some subcooling of the condensate. These additional features are readily included in the site steam profiles but have been left out of Figure 23.30 for the sake of clarity. If boiler feedwater preheat and superheat need to be included for steam generation, then the steam profiles for the different steam levels might overlap in the boiler feedwater preheat and superheat parts of the profiles. In this case, a composite must be created of the steam generation profiles for the individual levels, in a similar way to the construction of the composite curves, but without any temperature shift. Also, if steam de-superheating and condensate subcooling need to be included in the steam usage, then a composite of the steam usage profiles needs to be constructed. These composite steam profiles would then be matched against the site composite curves. The steam profiles in Figure 23.30 have been simplified to include only the latent heat for the clarity of explanation.

Figure 23.31 shows the site composite curves but now, relating to a retrofit situation. This time, the temperatures have been fixed by the existing steam main pressures, and the loads on the various steam mains have been fixed to their existing duties[12]. There is now a mismatch between the site composite curves and the steam profiles. In the example in Figure 23.31, the steam generation matched against the site hot composite curve is on target. However, the steam usage is poorly matched against the site cold composite curve. In Figure 23.31, the LP steam duty should be higher, the MP steam duty higher and the HP duty lower. The penalty for such a mismatch is lost opportunity for cogeneration in steam turbines. A greater use of the lower-pressure steam for heating allows more steam to be expanded from the HP level through to the lower levels, and hence more power to be generated in steam turbines.

While Figures 22.30 and 22.31 are useful in being able to set targets for steam generation and steam usage and identify missed opportunities in retrofit, there is another feature of the steam system that has not yet been addressed. This is the recovery of heat between processes through the steam system. Figure 23.32a shows the steam generation and steam usage for a site. In this example, in Figure 23.32a, HP steam, MP steam and LP steam are both generated and used on the site. Steam that is generated by waste heat does not need to be generated from the utility system from burning fuel in steam boilers. The steam generated by processes is fed

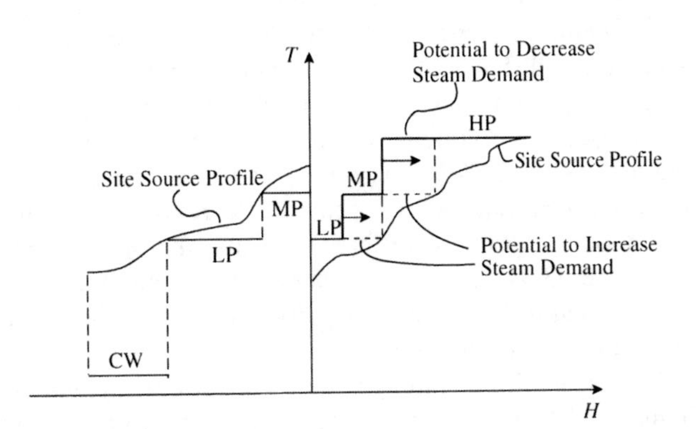

Figure 23.31 In a retrofit situation, plotting the actual steam duties allows lost potential to be identified.

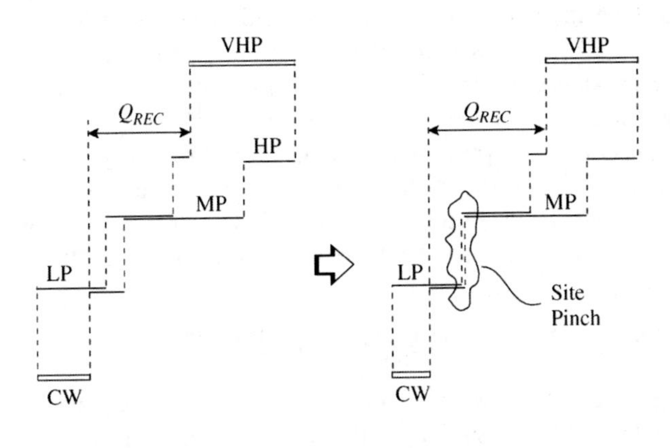

Figure 23.33 Targeting heat recovery for the site through the steam system.

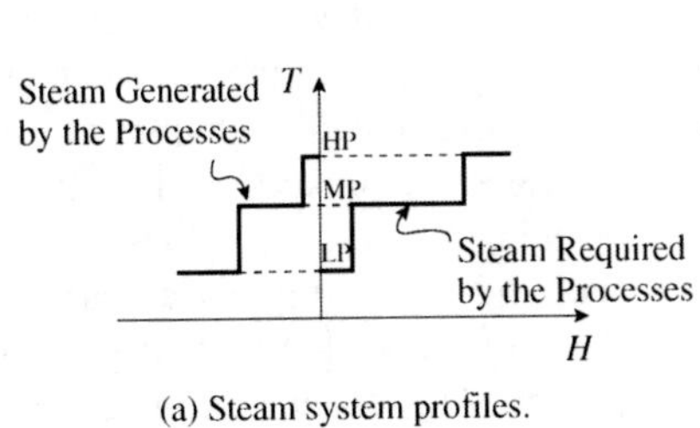

Figure 23.32 Site heat recovery through the steam system.

into the steam mains, which is subsequently used by other processes on the site, as illustrated in Figure 23.32b. This heat recovery between processes through the steam system needs to be included in targeting for the site. Figure 23.33a illustrates how heat recovery for the site can be targeted by overlapping the steam profiles. Figure 23.33a shows the region of overlap between the steam profiles to be a measure of the heat recovery across the site through the steam system[13,14]. Where heat recovery does not satisfy the site heating demand, residual heating needs to be satisfied by the very-high-pressure (VHP) steam generated in the utility boilers. Where the steam generation profile is not required for heating in Figure 23.33a, the residual heat must be rejected to the ambient, for example, to cooling water. If heating and cooling were the only considerations, it would not be sensible to generate LP steam against the process and to condense this against cooling water, as implied by Figure 23.33a. However, there is a power generation potential that needs to be considered later, and it is therefore left in place. The amount of overlap between the steam profiles is a degree of freedom available to the designer. Increasing the heat recovery between the profiles decreases the steam that must be generated in the utility boilers and decreases the site heat rejection. In other words, increasing the heat recovery decreases the heat flow through the system from utility steam generation through site cooling, and vice versa. If the overlap between the site steam profiles is maximized, as shown in Figure 23.33b, this minimizes the steam generation in the utility boilers and the site heat rejection. The limit is set by the *site pinch* between the steam profiles[13,14].

In setting the appropriate amount of heat recovery across the site through the steam system, various trade-offs need to be considered. Before this can be done, the cogeneration implications of the site heat recovery need to be quantified.

23.8 COGENERATION TARGETS

Figure 23.34 shows site hot and cold composite curves, together with the steam profiles, matched together in order to maximize the heat recovery across the site. This means that in this case, the site is pinched. Figure 23.34 shows all of the heat flows involving the site utility system. Steam generation by the processes involves heat flow from the site hot composite curve into steam generation. Steam usage involves heat flow from steam into the site cold composite curve. After recovering heat between process steam generation and process steam usage, the balance of the heating demand in Figure 23.34 is satisfied by fuel fired in the utility boilers to generate VHP steam that satisfies the balance of the site heating demand. The shaded area in Figure 23.34 is a region where high-pressure steam is expanded to low-pressure steam in the utility system. This expansion from high to low pressure, if carried out in steam turbines, will generate power. Below the site pinch, the site needs to reject heat to the environment. This is ultimately rejected to

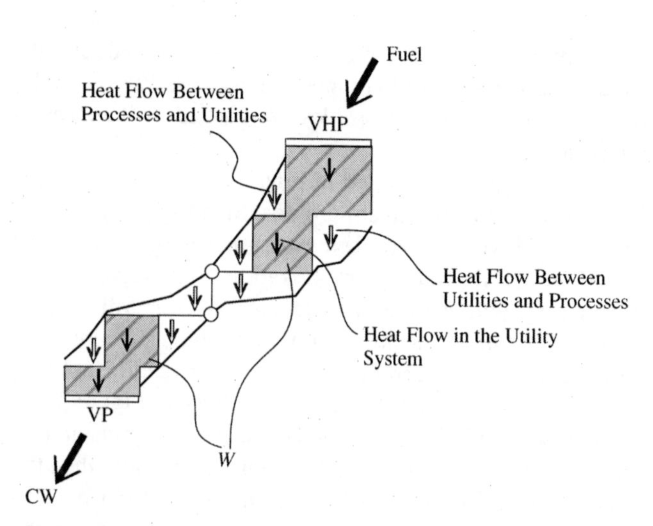

Figure 23.34 The heat flows involving the site utility system. (From Klemes J, Dhole VR, Raissi K, Perry SJ and Puigjaner L, 1997, *Applied Thermal Engineering*, **17**: 993, reproduced by permission of Elsevier Ltd.)

cooling water or air cooling. However, to maximize the cogeneration potential, the low-pressure steam generated is expanded to vacuum pressure (VP) steam, which is in turn condensed against cooling water. Below the site pinch, the shaded area also represents the potential to expand steam in steam turbines to generate power[13].

From the previous discussion of steam turbines, the larger the flow through a steam turbine, the greater the amount of power that will be generated. Also, the larger the pressure difference (and hence the larger the saturation temperature difference) across a steam turbine, the greater the power generation potential. Strictly speaking, the power generation is proportional to the Carnot Factor for a heat engine $(T_H - T_C)/T_H$ (where T_H and T_C are the heat input and heat rejection temperatures in K) rather than the saturation temperature difference. Thus, the shaded area in Figure 23.34 can be considered to be approximately proportional to the amount of power that can be generated by steam turbines in the utility system[13].

The heat flows through the site utility system in Figure 23.34 relate to the setting for maximum heat recovery in a site that is pinched. As pointed out previously, this is only one possible setting that happens to minimize the heat supply from the utility boilers and the heat rejection from the site. It is not necessarily the optimum setting. Consider the steam profiles shown in Figure 23.35 that have been set such that the heat recovery is not maximized. This means that a greater amount of fuel is being fired in the utility boilers, compared with a site in which the heat recovery is being maximized. However, in addition to firing extra fuel and generating extra VHP steam in the utility boilers, the larger area between the steam profiles means a larger potential for cogeneration of power. The site steam profiles corresponding with a setting in which the heat recovery has not been maximized can be decomposed into two parts, as

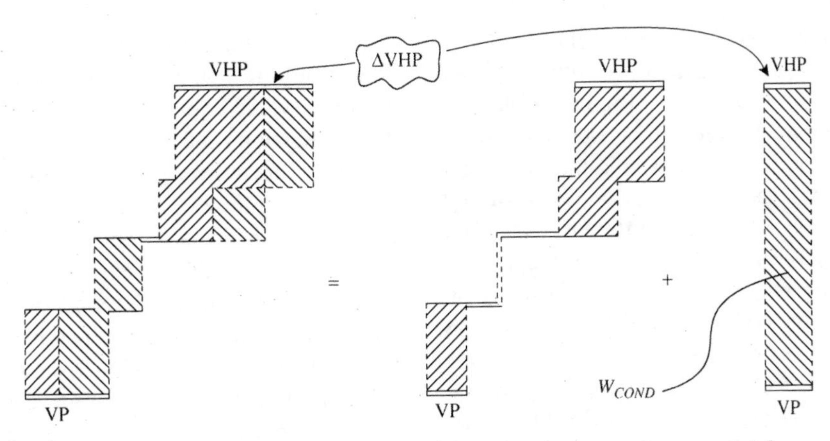

Figure 23.35 Increasing the heat flow through the utility potential for system increases the potential for power generation.

illustrated in Figure 23.35. The two parts correspond with steam profiles corresponding with a site that is pinched and a residual area that expands all the way from VHP to VP. This area between VHP and VP corresponds with condensing power generation. Thus, the setting of the steam profiles with the site pinched presents a cogeneration potential that is all true cogeneration. As more heat is allowed to flow through the utility system, and as the site heat recovery is correspondingly decreased, the additional heat flow corresponds with condensing power generation[13].

It could be argued that a site should not use condensing power generation in the way implied in Figure 23.35. Centralized stand-alone power stations use condensing power generation but with complex and extremely efficient cycles. The simple cycle used on a processing site cannot achieve the same efficiency of power generation as a stand-alone centralized power station using complex power generation cycles. However, it can still be sensible for a site to utilize condensing power generation. This is because the ultimate goal is not to maximize efficiency but to minimize cost. For example, it may be that the site produces waste gases that can be fired to generate steam, and hence power, at no additional cost. Also, it might be cheaper to generate the power in a condensing cycle on the site during peak periods when tariffs for imported power become particularly high. It might also be desirable to generate power through a condensing power generation to ensure utility security in situations where the external supply is subject to interruption. In other words, even though a centralized power station using complex power generation cycles can generate power more efficiently than a process site condensing cycle, it is often desirable to have some element of condensing power generation on the site. The cost of imported power must be balanced against the fuel and other costs associated with power generation, together with considerations of operability and system security to obtain the most appropriate balance between cogeneration and power import (or power export). Depending on the circumstances, this could either be at the extreme case of a pinched site (corresponding with Figure 23.34), or the curves set apart to the extent that all of the power requirements of the site are generated on site by a combination of true cogeneration and condensing power generation (corresponding with Figure 23.35).

How can this trade-off be carried out quantitatively? Can targets be set for cogeneration for the site? Before

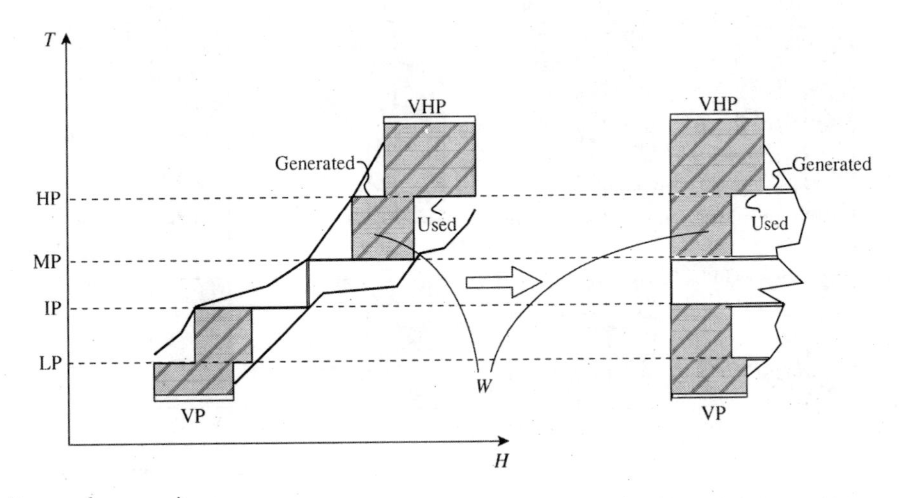

Figure 23.36 The site grand composite curve.

addressing how this can be done, one further graphical tool needs to be introduced. This is illustrated in Figure 23.36. The *site grand composite curve* is simply a plot of the horizontal separation between the site composite curves[13]. Both the site composite curves and the site grand composite curve carry the same information. It is simply represented graphically in a slightly different way. The site grand composite curve is a more convenient graphical representation of the site utility system when focusing on the cogeneration. The site grand composite curve shows the expansion zones in the steam turbines. These can now be quantified in terms of their power generation potential by applying the steam turbine model developed earlier. It can be assumed that there is a single turbine in each of the expansion zones. In practice, size limitations might dictate that there are multiple turbines operating in parallel in each expansion zone. Operability considerations to give the flexibility to operate between different operating cases might also dictate that there should be multiple turbines operating in parallel. However, at this stage, it is desirable to know the maximum power that can be generated. Generally, the larger the turbine, the more efficient it is and hence the greater the power generation for a given steam flow. Thus, to set the targets for generation of power from the site grand composite curve, it should be assumed that there is a single turbine in each of the expansion zones[13].

Having considered how to set targets for the cogeneration, how can these targets be realized in a steam turbine network? Figure 23.37 shows a site grand composite curve with various expansion zones[7,13]. Figure 23.37a shows one possible network design. Above the site pinch, two separate steam turbines have been fitted into the expansion zone, each with a single section, one expanding from VHP to HP, and the other from VHP to MP. The heat flows in Figure 23.37a can be translated into mass flows and hence power generation potential, as described previously. Below the site pinch in Figure 23.37a, the expansion zone has been exploited with a single-section turbine between LP1 and VP (condensation). There is another small expansion zone below the pinch in Figure 23.37a that is shown with a throttle valve expanding between LP2 and condensation. This expansion zone is considered to be too small to be worth exploiting for power generation, and some lost potential is accepted. Figure 23.37b shows an alternative way to exploit the expansion potential above the site pinch. In this case, a single-section turbine expands all of the steam between VHP and HP. Then, a second expands from HP to MP. The design below the site pinch has been kept the same in Figure 23.37a. A third option is shown in Figure 23.37c. This shows an extraction steam turbine expanding the steam from VHP. The extraction level is taken to HP and the exhaust to MP. The steam flows through the sections of the extraction machine are dictated by the heat flow requirements of the HP and MP levels. Again, the design below the site pinch has been kept the same. Finally, Figure 23.37d shows a fourth option. In this case, a single turbine has been fitted between VHP and MP and sized according to the heat requirements for the MP. The expansion zone between VHP and HP has not been exploited for power generation, but simply expanded through a let-down station. In this case, it has been decided that it is not worthwhile to try and exploit the full potential for cogeneration, and instead, some missed potential for cogeneration has been accepted for the sake of capital cost and design simplicity. But which is the best option in Figure 23.37? For those designs that attempt to exploit the full cogeneration potential (Figures 23.37a, b and c), there is little to choose in terms of the power generation potential. The choice would have to be made on the grounds of capital cost, complexity of design, operability and the ability to handle different operating cases. As pointed out previously, steam systems, especially on large sites, are almost never at steady state. Different operating cases need to be identified and the appropriate site grand composite curves constructed. The best fit to the network design across these different operating cases needs to be chosen[7].

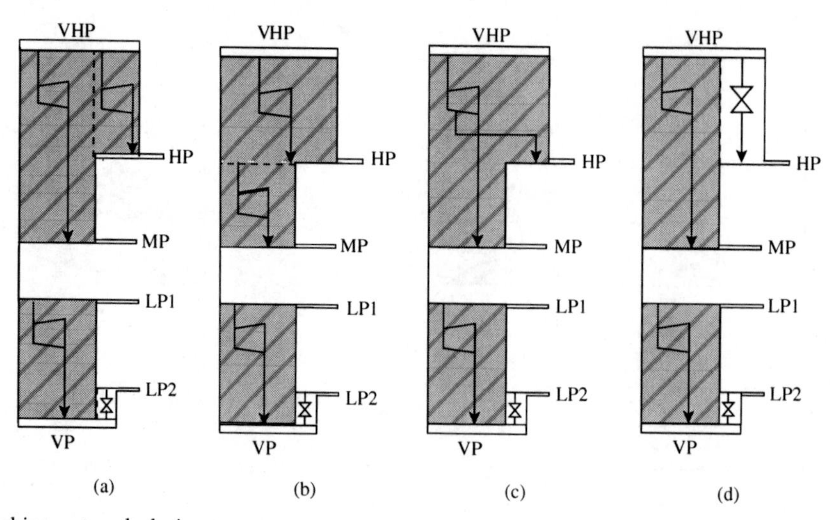

Figure 23.37 Steam turbine network design.

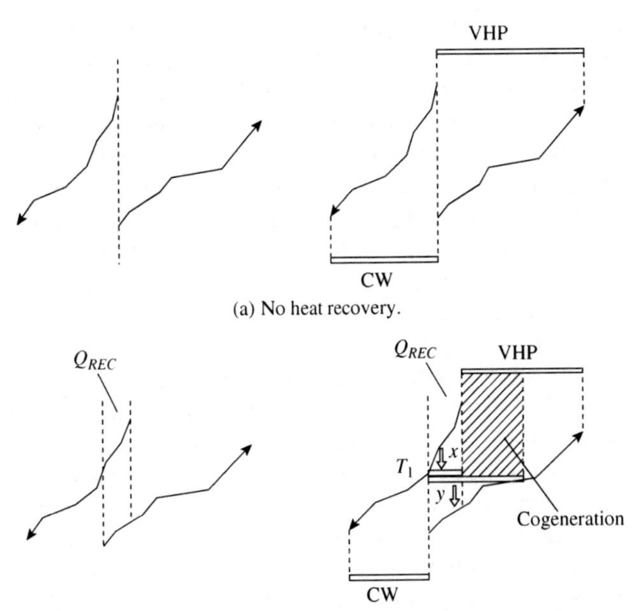

(b) Introducing heat recovery requires an intermediate steam mains.

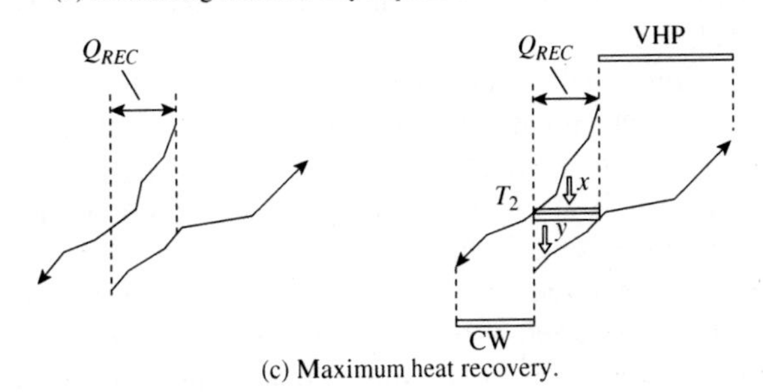

(c) Maximum heat recovery.

Figure 23.38 The effect of steam levels on heat recovery.

23.9 OPTIMIZATION OF STEAM LEVELS

It is important to understand the influence of steam levels on the energy performance of the site utility system. The steam levels affect both the level of heat recovery through the steam system and the cogeneration potential via the steam turbine network[13]. If a system of steam mains already exists, then there is often little opportunity to make significant changes in the pressures of the steam mains. The scope to increase the pressure of an existing steam main and the associated users of that steam level is usually restricted by mechanical limitations preventing a significant increase in the steam pressure. While mechanical limitations do not prevent a decrease in the steam pressure, the scope to decrease the pressure of an existing steam mains is usually restricted by the capacity of the pipework and valves. Decreasing a steam main pressure increases the volumetric flowrate, and it is likely that the capacity of the existing main and the associated users of that steam main cannot cope with a significant decrease in the steam main pressure.

However, it is important to understand where there is scope to improve an existing system, even if changes cannot be justified economically. Steam systems change over many years, and the demands on them change correspondingly. If the existing steam system is understood properly, then long-term infrastructure investment can steadily improve the system, even if short-term investment cannot be justified.

Consider the site composite curves shown in Figure 23.38a. These have been set such that there is no heat recovery through the steam system. All of the heating requirements of the site utility system are being satisfied by VHP steam from utility steam boilers. Cooling water is satisfying all of the cooling requirements of the site utility system. Figure 23.38b shows the introduction of some heat recovery through the steam system resulting from an overlap between the site composite curves. In order to realize this potential for heat recovery, an intermediate level of steam must be introduced, as shown in Figure 23.38b. There is a specific temperature (T_1 in Figure 23.38b) that will allow this level of heat recovery. The intermediate steam main at temperature T_1 allows some steam to be generated at temperature T_1 and also some steam to be used for process heating at that temperature. This not only introduces the potential for heat recovery but also introduces the potential for cogeneration through expansion of steam through a steam turbine between VHP and the intermediate steam main. The amount of heat recovery can be varied by changing the temperature of the intermediate steam main T_1. Changing the heat recovery (and temperature T_1) causes changes in both the amount of VHP steam required from the utility steam boilers and the cogeneration potential in steam turbines. Note that the steam generation (x) and steam usage (y) in Figure 23.38b are unequal. Figure 23.38c shows a situation where heat recovery through the steam system has been maximized. This corresponds with the heat duty on steam generation and steam usage for the intermediate main being equal (T_2 in Figure 23.38c)[13].

If the temperature is increased or decreased from T_2 in Figure 23.38c, then the supply of VHP from the utility boilers will increase. Note that the point of maximum heat recovery in Figure 23.38c also shows no cogeneration potential. Overall, Figure 23.38 shows that for a single intermediate steam main, the temperature of the intermediate main dictates the amount of heat recovery across the site through the steam system and the cogeneration potential. Balancing the steam generation and steam usage for that intermediate main not only minimizes the steam generation in the utility boilers but also removes any cogeneration potential[13].

So far, the objective in setting the steam levels has been to maximize heat recovery through the steam system. Figure 23.38 shows the potential of a single intermediate steam main to reduce steam generation in the utility steam boilers. As additional intermediate steam mains are introduced, as shown in Figure 23.39, increased heat recovery is possible. Moving from a single intermediate steam main, shown in Figure 23.39a, to two intermediate steam mains, shown in Figure 23.39b, allows the heat recovery potential to be increased, decreasing the VHP steam generation in the utility steam boilers correspondingly. In Figures 23.39a and b, the intermediate steam mains have been chosen to be such that the steam generation and steam usage for the intermediate mains are just balanced. For a single intermediate steam main in Figure 23.39a, there is a single degree of freedom to be optimized to maximize the heat recovery. In the case of two intermediate steam mains, as shown in Figure 23.39b, there are two degrees of freedom to be optimized to achieve maximum energy recovery such that the steam generation and steam usage for each intermediate main are balanced. Increasing from two intermediate steam mains to three intermediate steam mains in Figure 23.39c shows a further increase in the heat recovery potential by varying the temperatures of the three intermediate mains until each main has its generation and usage balanced[13].

Whilst minimizing the fuel fired in utility boilers is good from the point of view of reducing fuel costs in utility steam boilers and saving the corresponding emissions (locally), it is unlikely to be the optimum point for heat recovery through the steam system. A cost trade-off is required in order to minimize the overall energy cost by trading off the fuel required for steam generation versus the credit from cogeneration. Thus, in Figure 23.40, involving a single intermediate steam main, a small increase in the temperature of the intermediate main from Figure 23.40a to Figure 23.40b brings about a small increase in the amount of fuel fired in the utility steam boilers, but significantly increases the cogeneration potential. At some setting of the intermediate steam main, the increase in fuel costs will balance the credit from additional cogeneration at an optimum setting for the intermediate main[13].

It should also be noted that in addition to the degree of freedom to vary the level of the steam main in Figure 23.40,

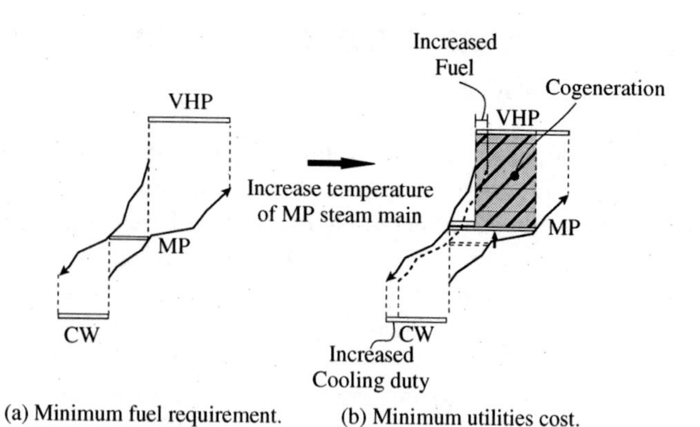

Figure 23.40 The optimum setting of site heat recovery involves a trade-off involving fuel and power costs.

there is an additional degree of freedom to vary the overall degree of heat recovery. This would introduce condensing power generation from the utility steam, as discussed previously. This is an additional degree of freedom in the optimization. Thus, in practice, there are two degrees of freedom to be optimized for a single intermediate steam main, three degrees of freedom for two intermediate steam mains, and so on.

Figure 23.41 illustrates the overall trade-offs as a function of the number of intermediate steam mains for the case of maximized energy recovery[13]. Two curves are shown in Figure 23.41. The upper curve corresponds with minimum fuel requirement and the lower corresponds with minimum utility cost. Both correspond with no condensing power generation from utility steam. Once the number of intermediate mains has been chosen, then the settings for those mains are chosen, either for the upper curve to correspond with the minimum fuel for the utility steam boilers or for the lower curve for minimum utility cost (fuel and power). As the number of intermediate steam mains is increased, the fuel required and the utility cost both decrease. However, the decrease in cost diminishes as the number of steam mains increases.[13]

If a new steam main is added to an existing system, then there are two ways in which the system can benefit[13].

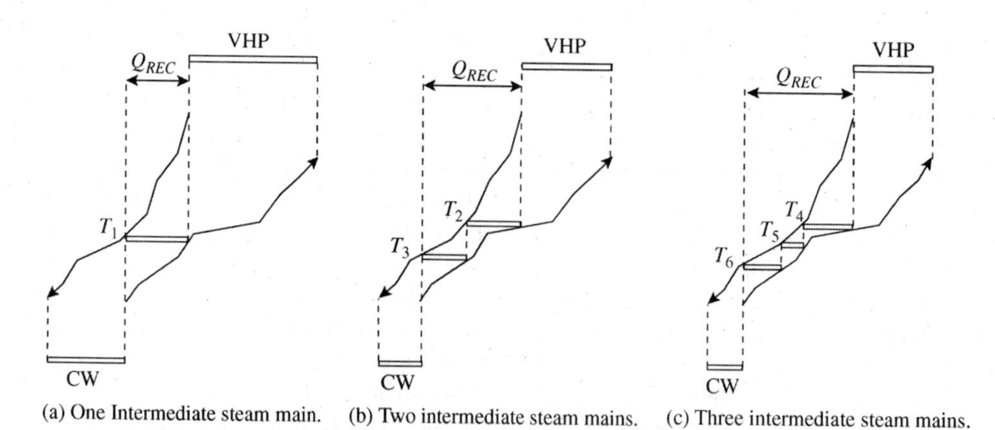

Figure 23.39 Increasing the number of intermediate steam mains increases the heat recovery potential for the site.

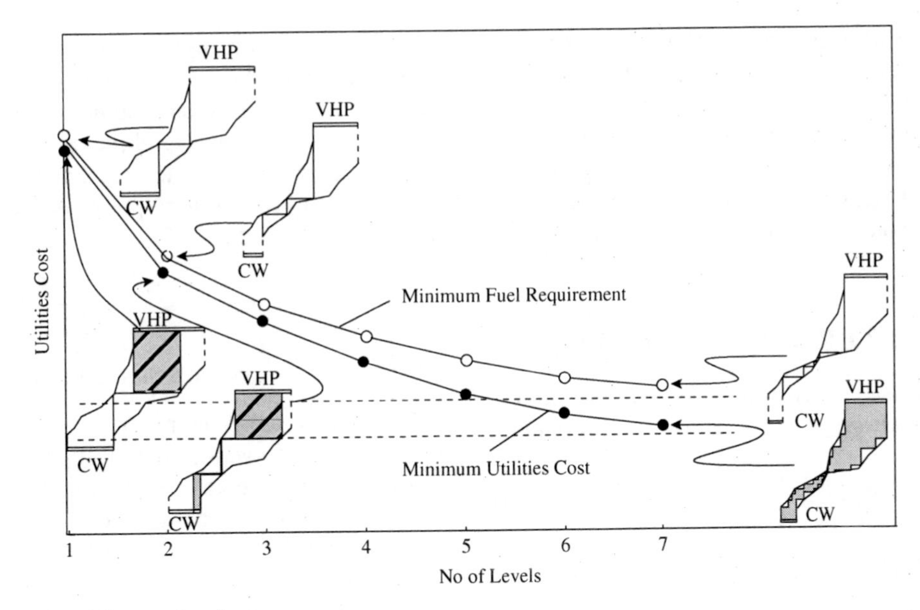

Figure 23.41 The number of steam levels.

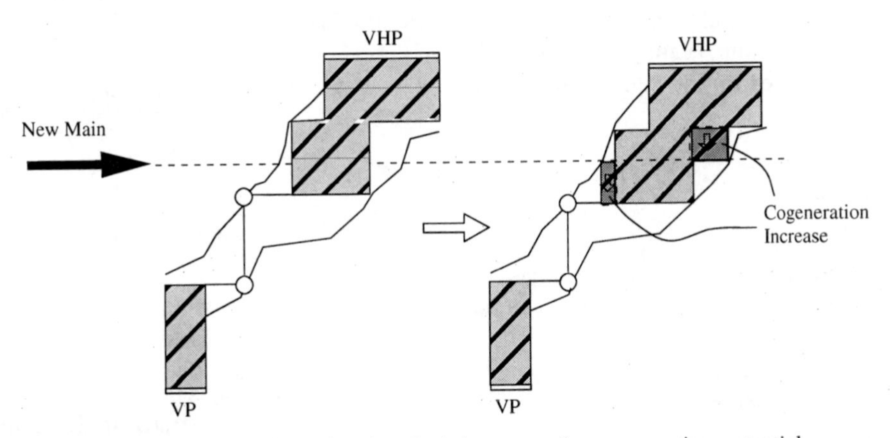

Figure 23.42 Adding a new steam main away from the site pinch increases the cogeneration potential.

Figure 23.42 shows an existing system that is set for maximum energy recovery, in which a new main is introduced away from the site pinch. The new main allows an increase in the steam generation and a change from steam use at a higher pressure to the new pressure. This has the effect of increasing the cogeneration potential, as illustrated in Figure 23.42[13]. On the other hand, Figure 23.43 shows a system for maximum energy recovery in which a new steam main is introduced at the site pinch. A new potential is created to increase steam generation with the new main,

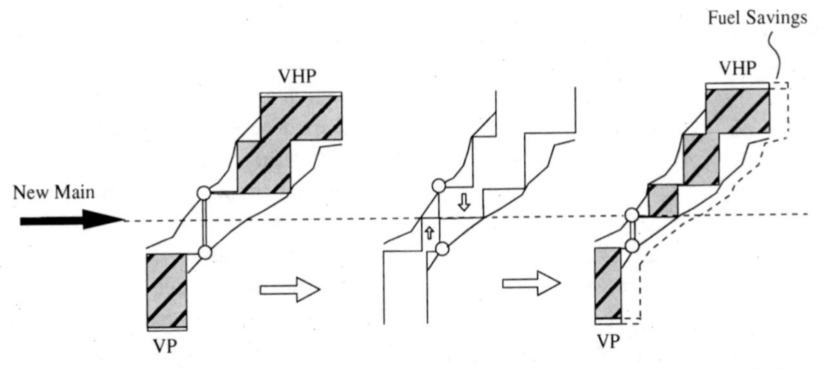

Figure 23.43 Adding a new steam main at the site pinch (for a pinched site) increases the potential for heat recovery through the steam system.

and the ability to use steam at the lower level corresponding with the new main. This allows increased heat recovery, as illustrated in Figure 23.43[13].

Whilst adding a new steam main across an existing site is likely to be prohibitively expensive, a local (as opposed to sitewide) steam main between a few critical processes might allow much of the potential improvement to be realized at much lower capital cost.

23.10 SITE POWER-TO-HEAT RATIO

The most appropriate cogeneration system for a site depends to a large extent on the site power-to-heat ratio, defined as[15,16]:

$$R_{SITE} = \frac{W_{SITE}}{Q_{SITE}} \tag{23.44}$$

where R_{SITE} = site power-to-heat ratio
W_{SITE} = power demand of the site
Q_{SITE} = process heating demand for the site

Figure 23.44 shows a site grand composite curve to illustrate the definition of the site heating demand. The steam demand for the processes on the site is defined by (see Figure 23.44)[17]:

$$Q_{SITE} = \sum_{FIRED\ HEATERS} Q_P + \sum_{STEAM\ MAINS} Q_P \tag{23.45}$$

where Q_p = individual process heating duties

The efficiency with which the power is generated can be defined as:

$$\eta_{POWER} = \frac{W_{GEN}}{Q_{FUEL}} \tag{23.46}$$

where η_{POWER} = power generation efficiency
W_{GEN} = site power generation (not necessarily equal to the site power demand)
Q_{FUEL} = site fuel demand

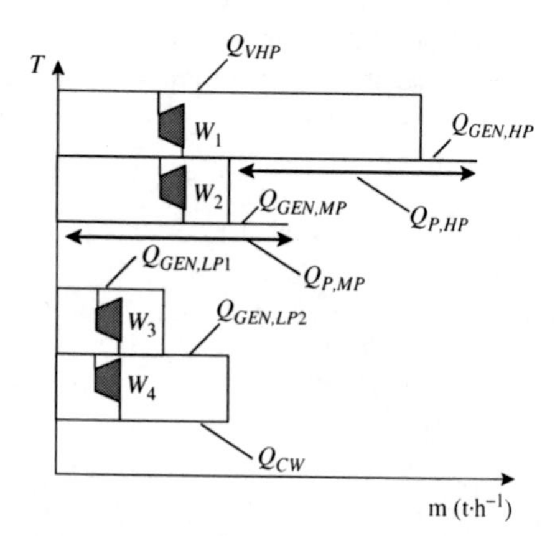

Figure 23.44 The definition of cogeneration efficiency and site power-to-heat ratio. (From Varbanov P, Perry S, Makwana Y, Zhu XX and Smith, 2004, *Trans IChemE*, **82A**: 784, reproduced by permission of the Institution of Chemical Engineers.)

However, a more useful measure of the utility system performance is the cogeneration efficiency. Of the fuel fired in the utility system, some of this energy produces power, some provides useful process heat and some is lost. The cogeneration efficiency recognizes the amount of fuel consumed to produce both power and useful process heat, and can be defined as[15,16]:

$$\eta_{COGEN} = \frac{W_{GEN} + Q_{SITE}}{Q_{FUEL}} \tag{23.47}$$

where η_{COGEN} = cogeneration efficiency

While this basic definition of cogeneration efficiency seems straightforward, complications are created by the process steam generated from waste heat recovery that can be used for power generation or process heating and that does not require any fuel to be fired in the utility system. The heat supply can be defined as the sum of the heat from fuel (both in the utility boilers and fired heaters) and steam generation from the waste heat recovery (see Figure 23.44)[17]:

$$Q_{SUPPLY} = \sum_{FIRED\ HEATERS} Q_{FUEL} + \sum_{BOILERS} Q_{FUEL} + \sum_{STEAM\ MAINS} Q_{GEN} \tag{23.48}$$

$$\text{where} \sum_{BOILERS} Q_{FUEL} = \sum_{BOILERS} \frac{(H_{STEAM} - H_{BFW})m_{STEAM}}{\eta_{BOILER}} \tag{23.49}$$

H_{STEAM}, H_{BFW} = specific enthalpies of the steam generated in the utility steam boiler and boiler feedwater respectively
m_{STEAM} = flowrate of steam generated in the utility steam boiler
η_{BOILER} = boiler efficiency (including stack, casing and blowdown losses)

The cogeneration efficiency is therefore more correctly defined as:

$$\eta_{COGEN} = \frac{W_{GEN} + Q_{SITE}}{Q_{SUPPLY}} \tag{23.50}$$

Having established the basic definitions, a plot can be developed between cogeneration efficiency with site power-to-heat ratio. It will be assumed initially that power is not to be exported from the site. Figure 23.45 shows the variation of η_{COGEN} with R_{SITE}. The development of the curve assumes that the site heating demand is fixed. The power demand gradually increases, increasing the site power-to-heat ratio. Start with a zero demand for power, $R_{SITE} = 0$, as shown in Figure 23.45. At this point, there is no attempt to generate power by expansion in steam turbines. Instead, all of the steam is expanded

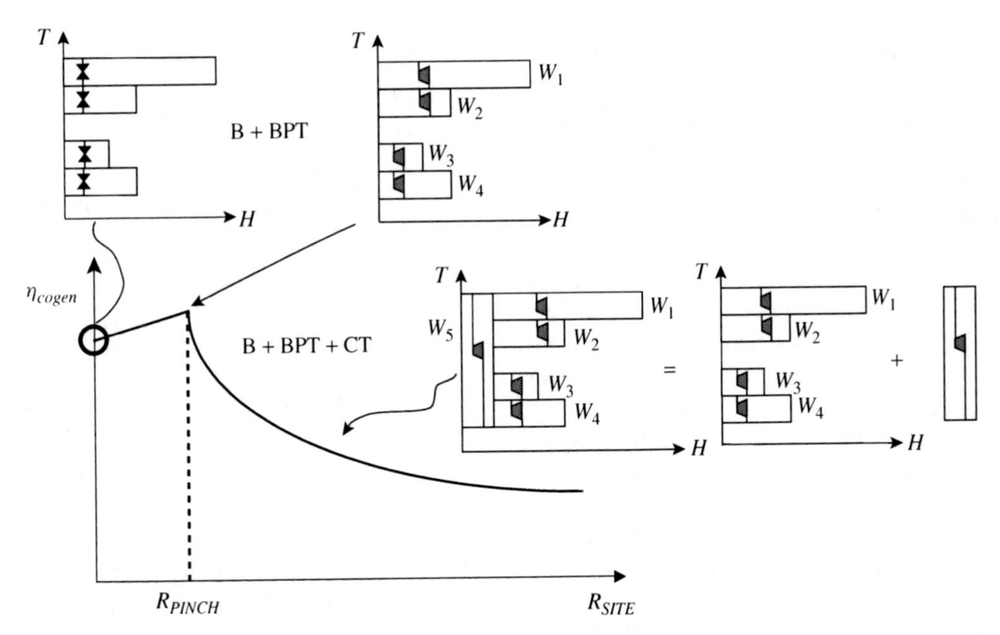

Figure 23.45 The site power-to-heat curve for complete on-site power generation in steam turbines. (From Varbanov P, Perry S, Makwana Y, Zhu XX and Smith, 2004, *Trans IChemE*, **82A**: 784, reproduced by permission of the Institution of Chemical Engineers.)

through let-down stations. As R_{SITE} increases from zero, the steam is expanded through steam turbines to meet the power demand, with the remainder being expanded in let-down stations. As R_{SITE} increases further in Figure 23.45, eventually all of the potential to expand steam through the steam turbines has been realized. This corresponds with the site utility system being pinched at R_{PINCH}. As additional power is generated above R_{PINCH} in Figure 23.45, the additional power generation can be through the use of steam turbines, or combinations of gas turbines and steam turbines. If power generation is restricted to steam turbines, this requires a gradual increase in the amount of condensing power generation from utility steam. This curve asymptotically approaches the efficiency for a stand-alone steam generation cycle.

If cogeneration efficiency (rather than cost) is used to dictate the import and export policy for electricity from the site, then the regions where electricity should be imported and exported are illustrated in Figure 23.46[17]. At low values of R_{SITE} below R_{PINCH}, the site heat demand means that the site has the potential to cogenerate more power than it requires in the steam turbines. Given that power generation in this situation is more efficient than stand-alone power generation, it would be beneficial from the point of view of the efficiency to realize the full potential for cogeneration between $R_{SITE} = 0$ and $R_{SITE} = R_{PINCH}$, and to export the electricity not required on the site. At high values of R_{SITE}, as shown in Figure 23.46, η_{COGEN} decreases and is likely to fall below that for centralized (stand-alone) power generation. Now, from the point of view of efficiency, it would be sensible to stop power generation on the site when η_{COGEN} falls below the efficiency of centralized power generation and to import the balance of power required by the site. Of course, these arguments are based on thermodynamic efficiency, rather than cost, and a full economic analysis is likely to change these thresholds. For example, the availability of large amounts of cheap fuel might make power export attractive over a wide range of values of R_{SITE}, despite poor cogeneration efficiency. It should also be recalled from Chapter 2 that power costs (both import and export) are usually subject to significant tariff variations according to the season of the year (winter versus summer), the time of day (night versus day) and the time of the week (weekend versus weekday). Even though transient economic factors might distort the picture, it is still important to understand the fundamentals of the trade-offs.

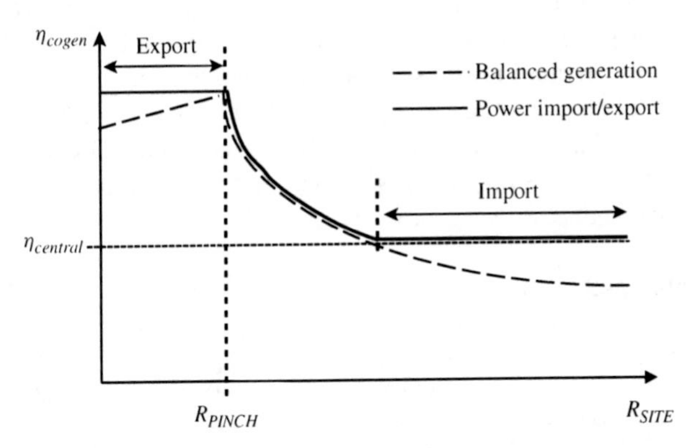

Figure 23.46 The site power-to-heat curve for steam turbines and export to power. (From Varbanov P, Perry S, Makwana Y, Zhu XX and Smith, 2004, *Trans IChemE*, **82A**: 784, reproduced by permission of the Institution of Chemical Engineers.)

Rather than following a path of using steam turbines only, a combination of gas turbines and steam turbines can be used at values above R_{PINCH}. The use of gas turbines will lead to cogeneration efficiencies higher than that for steam turbines used exclusively[15].

Consider an example involving a site where there are four steam mains with their steam usage and generation detailed in Table 23.3. The utility system can use a gas turbine with a power output of 25 MW, if appropriate. The gas turbine is integrated with a HRSG[17].

The calculation of the R-curve assumes the site heating demands from the steam mains to be fixed and the power demand to gradually increase. Cogeneration targets are calculated assuming a single steam turbine in each steam expansion zone. Figure 23.47 shows four curves. The first two curves (B + BPT + CT representing boiler + back-pressure turbine + condensing turbine) assume that the gas turbine and HRSG do not operate and that all of the power generation is through the back-pressure and condensing steam turbines. One curve assumes the simple case of balanced power generation. The other curve assumes a flexible generation strategy for the same system configuration, with export or import of power. The high-power generation efficiency at low values of R_{SITE} below R_{PINCH} means that power should be exported. Centralized power generation is assumed to have an overall efficiency of 30% (including transmission and distribution losses).

The other two curves in Figure 23.47 assume a 25 MW gas turbine (GT) running at full capacity with HRSG in addition to steam turbines[17]. Both curves assume excess power generation at low values of R_{SITE} is exported.

Table 23.3 Data for an R-curve analysis[17].

Steam main	Pressure (bar)	Q_{GEN} (MW)	Q_{USE} (MW)
VHP	120	0	0
HP	50	5	15
MP	14	10	20
LP	3	70	40
VP	0.85	0	0

Other data

Boiler Efficiency = 0.92
Condensate return = 50%
T_{BFW} = 85°C
Gas turbine power = 25 MW
Ambient temperature = 15°C
ΔT_{min} for HRSG = 30°C
T_{STACK} for HRSG = 140°C

Steam turbine performance parameters

$$a = 0.5 + 0.008\Delta T_{SAT}$$
$$b = 1.18 + 0.03\Delta T_{SAT}$$
$$L = 0.15$$

Fuel cost = 0.015 $\$\cdot kWh^{-1}$
Import power cost = 0.050 $\$\cdot kWh^{-1}$
Export power price = 0.040 $\$\cdot kWh^{-1}$

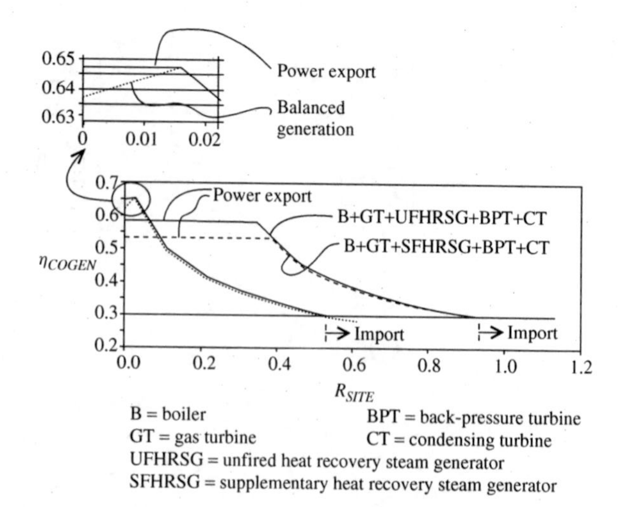

Figure 23.47 An example of site power-to-heat curves. (From Varbanov P, Perry S, Makwana Y, Zhu XX and Smith, 2004, *Trans IChemE*, **82A**: 784, reproduced by permission of the Institution of Chemical Engineers.)

The first of these curves (B + GT + UFHRSG + CT) represents a combination of boiler, gas turbine, unfired HRSG and condensing turbine. The other curve (B + GT + SFHRSG + CT) represents a combination of boiler, gas turbine, supplementary fired HRSG and condensing turbine. It can be seen that substantial benefits can be realized by importing or exporting power, depending on the site demand.

While Figure 23.47 shows typical R-curves, all cases will differ to some extent, depending on the site utility configuration, equipment specifications, site heating demand and the parameters assumed to be fixed and those varied in the analysis.

R-curve analysis can provide insights into efficiency improvements in utility systems. However, it has inherent limitations. Being a purely thermodynamic technique, it does not account for costs. For instance, it does not account for different fuels with different prices or, perhaps, low-efficiency boilers burning cheaper fuels and more efficient boilers burning expensive fuels. In this context, the most efficient operation of the utility system does not always mean the lowest cost operation. Consequently, there is also a need for an economic tool to understand the potential for cost saving in utility systems. However, R-curve analysis provides targets for the cogeneration potential of a site at different values of power requirement and is therefore useful to set an overall framework that can be further optimized by considering economics.

23.11 OPTIMIZING STEAM SYSTEMS

Large and complex utility systems often have significant scope for optimization, even without changing the utility configuration. Consider now the optimization of a fixed utility configuration. First, the degrees of freedom that can be optimized in utility systems need to be identified.

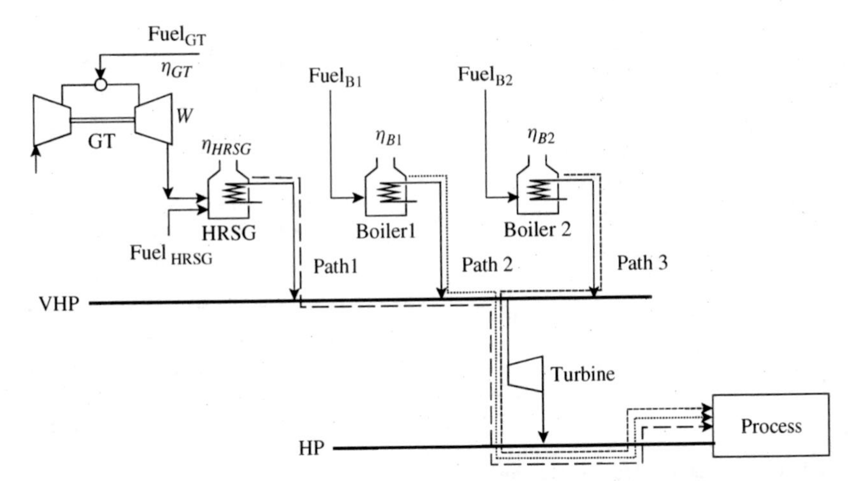

Figure 23.48 Multiple steam generation devices provide degrees of freedom for optimization. (From Varbanov PS, Doyle S and Smith R, 2004, *Trans IChemE*, **82A**: 561, reproduced by permission of the Institution of Chemical Engineers.)

1. *Multiple steam generation devices*. Consider the production of HP steam in Figure 23.48. The HP steam needs to be expanded through a path from the VHP mains. In turn, the VHP steam can be produced by Boiler 1, Boiler 2, or the gas turbine HRSG. If all of these steam generation devices are identical in terms of performance and operating cost, then there is nothing to choose between generating the steam in any of the three steam generation devices. However, this is not likely to be the case. In this example, the two boilers might have different fuels, with different fuel costs and different efficiencies, and the gas turbine (perhaps, with supplementary firing) will have completely different characteristics from the steam boilers. Thus, there are degrees of freedom created by multiple steam generation devices with different costs of fuels, different boiler efficiencies and different power generation potential. Individual steam boilers and HRSGs will have minimum and maximum flows.

2. *Multiple steam turbines*. The HP steam required by the processes in Figure 23.49 can be generated by expanding through either Turbine $T1$ or $T2$, or indeed, through the let-down station, or a combination. Expanding the steam through the turbines is usually more economic than through the let-down station as there is a power credit resulting from expansion through the steam turbines. Moreover, Turbines $T1$ and $T2$ are unlikely to have the same efficiencies for the expansion between VHP and HP steam. Thus, multiple expansion paths create degrees of freedom to choose the most appropriate path through which the steam is expanded.

If a steam turbine generates electricity, then the flow through the turbine can be varied within the minimum and maximum flows allowed by the machine. If a steam turbine is connected directly to a drive (e.g. a back-pressure turbine driving a large pump), then there is likely to be no flexibility to change the flowrate through the steam turbine as this is fixed by the power requirements of the process machine.

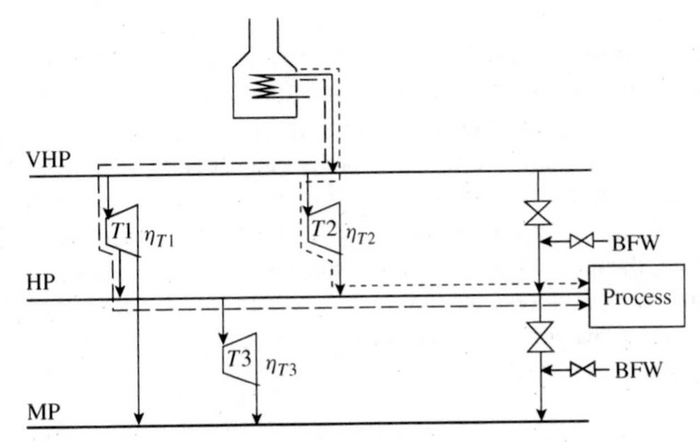

Figure 23.49 Steam turbine networks and let-down stations provide degrees of freedom for optimization. (From Varbanov PS, Doyle S and Smith R, 2004, *Trans IChemE*, **82A**: 561, reproduced by permission of the Institution of Chemical Engineers.)

Process machines can, in some cases, be equipped with both a steam turbine and an electric motor, allowing the drive to be switched between the two according to steam demands in the utility system and operating costs. It should be remembered that operating costs might vary significantly as a result of different electricity tariffs, according to the time of day, the day of the week and the season of the year.

Extraction steam turbines will have minimum and maximum flows in each section of the turbine. This leads to minimum and maximum flows at each extraction level.

3. *Let-down stations*. If there is a requirement for steam to be expanded, then flow through the steam turbines should normally be maximized in order to maximize the power generation. However, a steam balance must be maintained, and let-down stations are still important degrees of freedom in the optimization, as shown in Figure 23.50. Steam expanding through a let-down station might be able to bypass a flow constraint in a steam turbine somewhere in the system and increase the power generation indirectly. Also, if the let-down has de-superheating, this increases the steam flow after

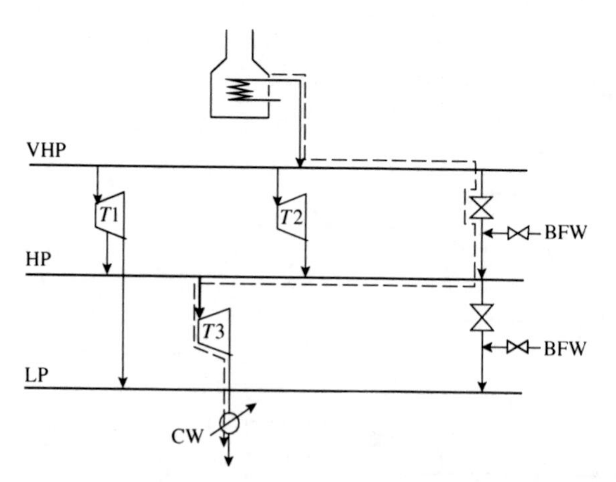

Figure 23.50 Let-down stations provide degrees of freedom to bypass constraints in the utility system. (From Varbanov PS, Doyle S and Smith R, 2004, *Trans IChemE*, **82A**: 561, reproduced by permission of the Institution of Chemical Engineers.)

expansion. This can replace steam generation in the high-pressure boilers and at the same time increase the steam flows in the lower-pressure parts of the system.

Thus, large let-down flows are usually seen to be a missed cogeneration opportunity but might provide a degree of freedom to bypass bottlenecks in the steam system. Also, flow through let-down stations allows the level of superheat in the lower-pressure mains to be controlled independently of the flow through the steam turbines.

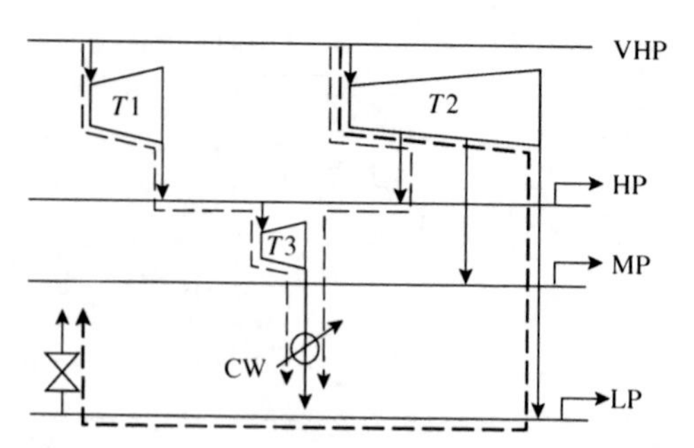

Figure 23.51 Condensing turbines and vents provide additional degrees of freedom for optimization. (From Varbanov PS, Doyle S and Smith R, 2004, *Trans IChemE*, **82A**: 561, reproduced by permission of the Institution of Chemical Engineers.)

4. *Condensing turbines*. Condensing turbines provide extra degrees of freedom for power generation, as shown in Figure 23.51. Although the heat from the exhaust of a condensing turbine is not required, it provides flexibility in power generation to reduce the cost of imported power or to increase the credit from exporting surplus power.

5. *Vents*. Figure 23.51 also shows a vent on the LP steam main. At first sight, it might not seem sensible to vent steam directly to ambient. This steam is being generated at the VHP level, for which fuel costs are incurred. However, in expanding from the VHP main down to the LP main through steam turbines, power is generated. It might

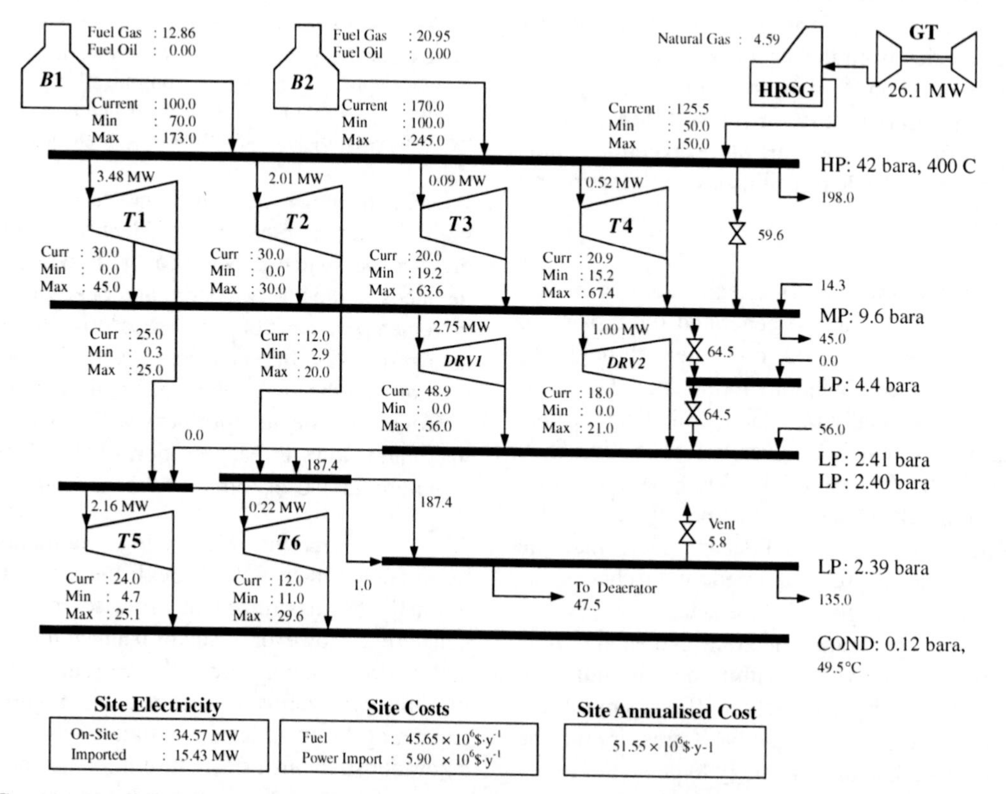

Figure 23.52 Current operating conditions for a site utility system (flows in $t{\cdot}h^{-1}$). (From Varbanov PS, Doyle S and Smith R, 2004, *Trans IChemE*, **82A**: 561, reproduced by permission of the Institution of Chemical Engineers.)

therefore be the case that for a high differential between fuel and power costs, it is economic to vent steam. However, it should always be better to expand the steam through a condensing turbine if there is one, and there is spare capacity, rather than vent steam. Also, in some cases steam venting might not be allowed.

Figure 23.52 shows an example of an existing site utility system[9]. Steam is generated at high pressure (HP), which is distributed around the site, along with steam at medium pressure (MP) and low pressure (LP). Steam is expanded through a network of steam turbines from the HP main to the lower-pressure mains. Let-down stations are used to control the mains pressures. Steam is used for process heating from the HP, MP and LP mains. Turbines $T1$ to $T6$ generate electricity. Turbines $T5$ and $T6$ are condensing turbines. Turbines *DRV1* and *DRV2* are driver turbines connected directly to process machines.

To optimize the system in Figure 23.52 requires a model to be developed for the whole system that accounts for:

- the cost of fuel per unit of steam generation in the boilers
- the cost of fuel per unit of power and steam generation in the gas turbine/HRSG
- the power generation characteristics of the gas turbine
- the power generation characteristics of the steam turbines
- the cost of imported power
- the credit for exported power
- costs for running the steam and power generation, other than fuel costs
- flowrate constraints throughout the system
- cooling water costs.

Figure 23.52 shows the maximum, minimum and current flows in each part of the system. Ancillary data for the utility system in Figure 23.52 are given in Table 23.4[9]. Data for the fuels used by the utility system in Figure 23.52 are given in Table 23.5[9]. Data for the steam turbines are given in Table 23.6.

If the conditions in the steam mains (temperature and pressure) are fixed, and the steam boilers, steam turbines, gas turbines and costs are modeled as linear functions, then the optimization can be carried out with a linear program. In practice, even if the pressures of the steam mains are fixed by pressure control, the temperature (and enthalpy) will vary as the flow through the steam turbines and let-down stations vary. If this is taken into account, then the model becomes a nonlinear optimization. However, the effect of changing the temperature in the steam mains does not introduce too much complexity to the optimization, and an iterative linear programming scheme can usually be used to optimize the system. The steam system model is first simulated, and the pressures and temperatures of the steam mains fixed. The model is then optimized using a linear program with the temperatures of the mains fixed. Resimulation of the model then allows the temperatures of the steam mains to be determined. These temperatures are then fixed, the model reoptimized using a linear program and resimulated, and so on, until convergence is achieved.

For large sites, the conditions in a long steam main will not be uniform throughout the main because the steam main will not be well mixed and can also have a significant pressure drop along the main. Thus, the conditions in the same main can be different in different parts of a large site. If this turns out to be an important issue, then additional steam mains must be created in the model to reflect the different conditions in different geographic locations.

Figure 23.53 shows the conditions of the optimized system. Table 23.7 compares the cost of the base case conditions in Figure 23.52 with those for the optimized system shown in Figure 23.53[9].

Table 23.4 Ancillary data for the site utility system in Figure 23.52[9].

Ambient temperature	25°C
Boiler feedwater temperature	80°C
Cooling water temperature	25°C
Site power demand	50 MW
Minimum power import	10 MW
Maximum power import	50 MW
Minimum power export	0 MW
Maximum power export	20 MW
Power cost (import)	0.045 $\$\cdot kWh^{-1}$
Power value (export)	0.06 $\$\cdot kWh^{-1}$
Cooling water cost	0.0075 $\$\cdot kWh^{-1}$

Table 23.5 Fuel costs for the utility system in Figure 23.52[9].

Fuel	Net heating value ($kJ\cdot kg^{-1}$)	Price	
		($\$\cdot kg^{-1}$)	($\$\cdot kWh^{-1}$)
Fuel oil	40,245	0.071	0.0063
Fuel gas	32,502	0.103	0.0115
Natural gas	46,151	0.160	0.0125

Table 23.6 Steam turbine data for the utility system in Figure 23.52[9].

Turbine: stage	a	b	L
$T1$: HP–MP	0	1.96	0.228
$T1$: MP–LP	0	3.15	0.010
$T2$: HP–MP	0	1.825	2.802
$T2$: MP–LP	0	3.152	0.193
$T3$	0	1.43	0.429
$T4$	0	1.47	0.289
$T5$	0	1.46	0.229
$T6$	0	1.045	0.588
DRV1	0	1.5	0.100
DRV2	0	1.53	0.040

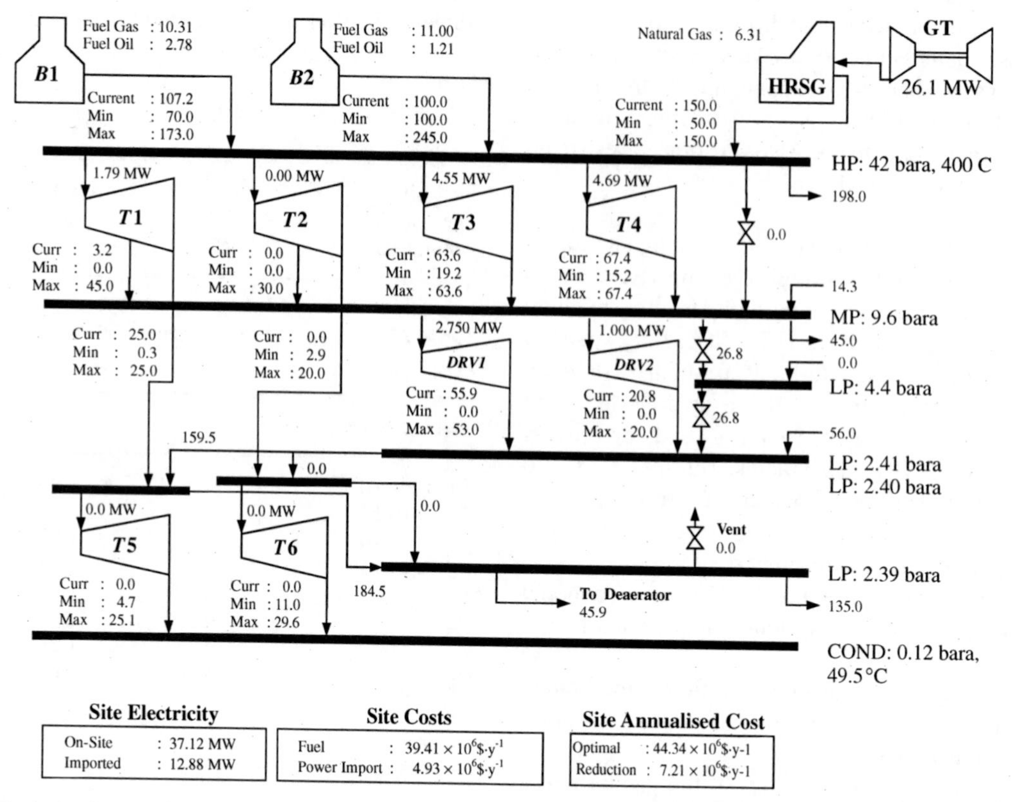

Figure 23.53 Optimized operating conditions for a site utility system (flows in $t \cdot h^{-1}$). (From Varbanov PS, Doyle S and Smith R, 2004, *Trans IChemE*, **82A**: 561, reproduced by permission of the Institution of Chemical Engineers.)

Table 23.7 The performance of the optimized utility system in Figure 23.53[9].

Cost	Base case ($\$ \cdot y^{-1}$)	Optimized system ($\$ \cdot y^{-1}$)
Import power	5.90×10^6	4.93×10^6
Fuel	45.65×10^6	39.41×10^6
Total operating cost	51.55×10^6	44.34×10^6

Thus, in this case, the optimization is capable of reducing the total operating cost by some 14%. The scope for cost reduction through optimization will be case specific but is most often around 5% or less.

Finally, all of the discussions so far have assumed that there is one set of operating conditions that need to be optimized. This is almost never the case. To begin with, shut down and maintenance of process and utility plant dictate that there will be different steam flows and constraints during these periods, giving rise to different operating cases. To add to this complication, the cost of import power and the credit for export power will, in most cases, vary according to the time of day, time of week and season of the year. These tariffs are normally negotiated with the power supplier and form a fixed pattern, but often a complex one. Each of these different tariffs also presents a different operating case for optimization. Thus, many different optimizations are required to reflect the picture for each case throughout the year. The operational set up of the utility system should then be changed for each case to reflect the new conditions. On-line optimization is helpful in this respect.

If an optimization across the whole year is required, perhaps for a design modification, then the different operating cases need to be weighted according to the duration of each case and combined to give an annual picture[18].

23.12 STEAM COSTS

Steam costs were considered in brief in Chapter 2. The philosophy of costing steam was on the basis of the fuel cost to raise the highest-pressure steam and the value of power generated in expanding to lower levels subtracted from this fuel cost. A simple isentropic efficiency model for the steam turbines was used to calculate the value of power generated as a result of the expansion through steam turbines. This represents an idealized picture of steam costs. It does not take account of existing equipment, equipment performance and the constraints associated with both existing equipment and existing steam networks. For an existing steam system, the true cost of steam can be established using the modeling and optimization techniques discussed in this chapter. Two distinct cases need to be considered.

1. *Steam cost for fixed steam heat loads.* In the first case, the steam heat loads for the processes on a site are fixed, and the objective is to calculate the cost of the steam for the allocation of costs to the processes and businesses on a site. To establish the true steam costs for an existing steam system requires that a model for the existing system be first developed. The model should reflect as much as possible the performance of the existing equipment and the constraints in the existing equipment and steam network. The model should also include the existing steam heating demands for the various processes. This model can then be optimized, as described in the previous section. The cost to generate the steam in the utility boilers at the highest pressure can be calculated from the optimized model. This will principally be the cost of fuel but can also include other costs, such as water, water treatment, labor, power for the boiler feedwater pumps and cost of deaeration steam. Having calculated the cost of the highest-pressure steam, the model will also provide the amount of power generated by expansion of the highest-pressure steam to the next highest pressure. The value of the power generated by the expansion can then be subtracted from the cost of generating the highest-pressure steam in the utility boilers to give the cost of the next-to-highest pressure steam. The calculation is repeated for next pressure level lower, and so on. Thus, the cost of each steam level is the cost of the next highest level minus the value of the power generated from the expansion from the next highest level. If there are utility boilers generating steam into the lower pressure steam mains, then the cost of operating these boilers must also be added to the cost of steam at that level.

In extreme cases, this approach to costing steam might lead to steam at low pressure having a negative cost. This can happen if fuel is particularly cheap and power is expensive. This is not a fundamental problem as it reflects the true economics. The site utility system is benefiting from the availability of a heat sink for the low-pressure steam. However, it should be emphasized that if the low-pressure steam has a negative cost this does not necessarily mean that it is economic to increase use of low-pressure steam significantly. As will be discussed later, the cost of steam is likely to change as its consumption changes.

It should also be noted that variation in electricity tariffs that might change through the time of day, day of the week and time of the year will change the optimization and, therefore, the steam costs. An average can be taken according to the relative duration of the tariffs.

2. *Steam cost for changed steam heat loads.* In the second case, the steam costs need to be determined when heat loads are changed. This might be a project for reduction in energy demand (e.g. retrofit of a heat exchanger network for increased heat recovery). Alternatively, a project might involve an increase in heat demand as a result of commissioning of new plant or expansion of an existing plant. The starting case is again a model for the steam system, again optimizing for the existing steam heating loads. It might be suspected that the steam cost calculated from the model for the existing steam loads can be used to provide steam costs for larger or smaller steam loads for a given steam main. However, this is not the case. The optimum settings for the steam system change once the loads for the steam mains change. Constraints on the existing equipment will also be encountered and all of this needs to be accounted for.

Figure 23.54a shows an existing site utility system[17]. Suppose that it is possible to reduce the HP steam demand for the process on a site. This could be done, for example, by improving the heat recovery within the process. But how much is such a steam saving actually worth? The HP steam is being generated by expanding from VHP to HP through Steam Turbines $T4$ and $T6$. The saving in steam from the HP mains means that this amount of steam does not now need to be expanded through the steam turbines from the VHP level. In turn, this means that there is a surplus of steam at the VHP level. The first obvious action to take is to reduce the steam generation in the utility boilers and accept a saving in fuel costs as a result. Unfortunately, the saving in fuel costs would also be accompanied by an

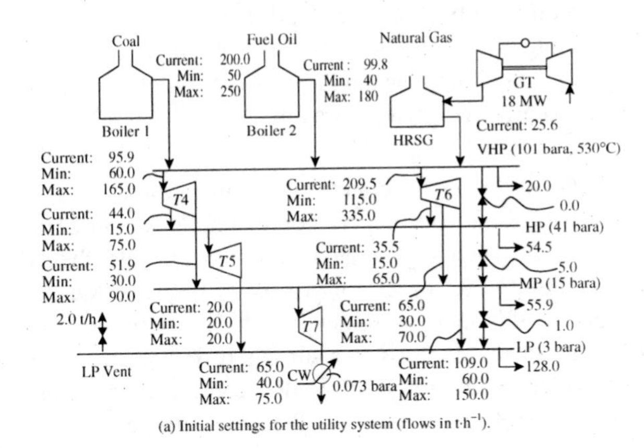

(a) Initial settings for the utility system (flows in t·h^{-1}).

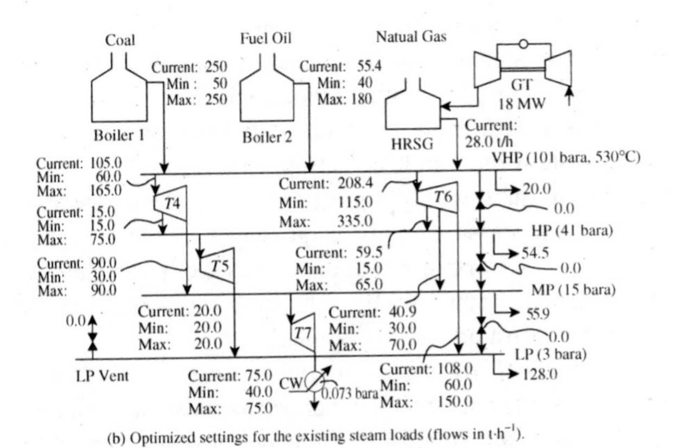

(b) Optimized settings for the existing steam loads (flows in t·h^{-1}).

Figure 23.54 An existing utility system with existing steam loads. (From Varbanov P, Perry S, Makwana Y, Zhu XX and Smith, 2004, *Trans IChemE*, **82A**: 784, reproduced by permission of the Institution of Chemical Engineers.)

increase in power costs. This results from the reduction in the flow of steam through the steam turbines, which, in turn, reduces the cogeneration, and additional power would have to be imported to compensate for this. It is therefore not so straightforward to determine exactly what the cost benefits associated with a steam saving would be. Another way to deal with the surplus of steam at the VHP level created by a steam saving would be to pass the heat through an alternative path, to, say, the condensing turbine. This would allow additional power to be generated, with the resulting cost benefit. In a complex utility system, the heat can flow through the utility system through many paths. The flow through different paths will have different cost implications.

In assessing the true cost benefits associated with a steam saving, the steam and power balance for the site utility system must be considered, together with the costs of fuel and power (or power credit in an export situation). In general, a surplus of steam resulting from a steam saving in a process demand can be exploited by

- saving fuel in the utility boilers;
- generating extra power by passing the steam to a condensing turbine;
- generating extra power by passing the surplus steam to a vent through back-pressure turbines;
- expanding the steam to a lower level than previously used by switching a steam heater from a higher-pressure steam to a lower-pressure steam.

If steam is being generated, rather than used, then the same basic arguments apply. For example, suppose that HP steam was being generated by a process into the HP main in Figure 23.54a[17]. The project to improve the heat recovery in this process might lead to an increase in the HP generation. This leads to a surplus of HP steam, which in turn leads to a surplus of VHP steam. Then the same arguments apply as to what is the most efficient way to exploit the surplus of VHP steam.

The only way to reconcile the true cost implications of a reduction in steam demand created by an energy reduction project is to use the optimization techniques described in the previous section. An optimization model of the existing utility system must first be set up. Starting with the steam load on the main with the most expensive steam (generally the highest pressure), this is gradually reduced and the utility system reoptimized at each setting of the steam load. The steam load can only be reduced to the point where the flowrate constraints are not violated.

The concept of steam marginal cost is used as an indicator in the analysis. It is defined as the change in utility system operating cost for unit change in steam demand *for a given steam main* (change in steam main balance)[17]:

$$MC_{STEAM} = \frac{\Delta Cost}{\Delta m_{STEAM}} \qquad (23.51)$$

where MC_{STEAM} = marginal cost of steam
$\Delta Cost$ = change in cost
Δm_{STEAM} = change in steam flowrate

It is important to emphasize that the change in the operating cost is taken between the optimum operation before the steam demand change and the optimum operation after the steam demand change for the current step. Obviously, the result is context-specific and, for different operating conditions, each steam level will have a different marginal steam cost. Even further, the marginal steam cost can change for a given level, depending on how much steam is saved from the steam main. In general, the steam balance of a main may be changed by:

- increasing/decreasing process steam demand;
- increasing/decreasing process steam generation;
- switching a process demand from one steam main to another;
- changing the utility system, for example, shutdown of a boiler, steam turbine, and so on.

Steam demand for a given main can increase, as a result of an increase in rate of production, or decrease in order to improve the energy efficiency of the site or from a change in production or the site. The approach can be used to deal with any context.

The first step is to optimize the operation of the utility system for the initial steam demands[17]. This allows the true steam prices to be obtained in the subsequent steps. Figure 23.54a shows the initial settings for a utility system, and Figure 23.54b shows the optimized settings. Next, the steam main with the highest steam marginal cost for the current steam demand is identified by calculating the marginal cost for each level. The capacity for decrease in demand of the selected main is then determined by gradually decreasing the demand for that steam. At each step in the reduction, the whole utility system is optimized. If a constraint on decreased usage (or increased generation) is reached, or if the marginal cost for the main changes significantly, the stepwise decrease in the steam demand (increase in steam generation) is terminated. The procedure is then repeated for the steam main with the highest steam marginal cost for the new conditions until no further decrease in steam usage (or increased generation) is possible. The procedure is summarized in Figure 23.55[17].

The steam turbines in Figure 23.54 are all turbogenerators and none are direct drives. Steam turbines for electricity generation allow their flows to be changed smoothly between their minimum and maximum limits. This results in a continuous decrease of the site operating cost. For steam saving potential, this means a monotonic profile of the steam marginal costs. The site data are given in Table 23.8 and the fuels used in Table 23.9[17].

Figure 23.56 shows a plot of steam marginal cost versus the steam savings[17]. It features six segments; three for HP

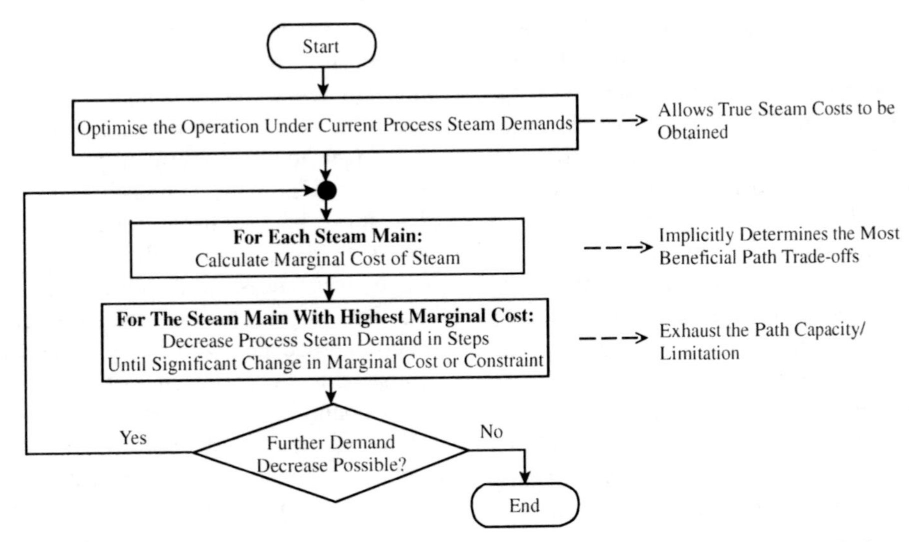

Figure 23.55 Procedure to obtain the marginal costs of steam for a site utility system. (From Varbanov P, Perry S, Makwana Y, Zhu XX and Smith, 2004, *Trans IChemE*, **82**: 784, reproduced by permission of the Institution of Chemical Engineers.)

Table 23.8 Ancillary data for the system in Figure 23.54[17].

Ambient temperature	25°C
Boiler feedwater temperature	110°C
Cooling water temperature	25°C
Site power demand	68 MW
Minimum power import	0 MW
Maximum power import	50 MW
Minimum power export	0 MW
Maximum power export	50 MW
Power value	0.05 $\$\cdot kWh^{-1}$
Cooling water cost	0.005 $\$\cdot kWh^{-1}$

Table 23.9 Fuel costs for the system in Figure 23.54[17].

Fuel	Net heating value ($kJ\cdot kg^{-1}$)	Price	
		($\$\cdot kg^{-1}$)	($\$\cdot kWh^{-1}$)
Coal	28,000	0.065	0.0084
Fuel oil	40,000	0.12	0.0108
Natural gas	52,000	0.22	0.0152

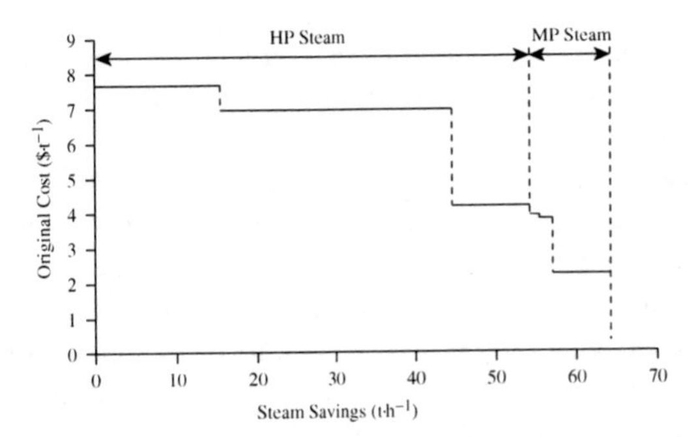

Figure 23.56 Site marginal steam costs for the case study.

steam savings and three MP segments. Saving LP steam has no value. Note that each segment in the steam savings involves slight marginal cost variations, resulting from the nonlinearity of the steam turbine performance.

As described in the methodology, the curve starts from an optimized operation for the initial process steam demands (Figure 23.54b). At this initial point, the HP extraction of Turbine $T4$ is set to its minimum of 15 $t\cdot h^{-1}$. The first HP segment saves steam by reducing the HP extraction of Turbine $T6$. Part of this saving is realized by reduction in steam generation in Boiler 2 and the remainder by increasing the LP exhaust of Turbine $T6$ and the LP vent. This segment is terminated by Boiler 2 reaching its minimum capacity. After this point, any boiler steam savings are realized in Boiler 1. The second HP segment features a further gradual decrease in the $T6$ HP extraction and is terminated at its minimum of 15 $t\cdot h^{-1}$. At this point, both direct paths for saving HP steam are already exhausted after the first two segments. The HP outlet flows of Turbines $T4$ and $T6$ are at their minimum limits of 15 $t\cdot h^{-1}$ and cannot be reduced further. However, because eventual let down of HP steam to the MP main increases the enthalpy of the MP steam, the power generated by the Condensing Turbine $T7$ is always larger than without this let down. This in turn makes it more profitable to save further HP steam, instead of moving to MP steam savings. Finally, after the HP steam demand is reduced to zero, there are three MP steam saving segments. The first one saves steam through the MP extraction of Turbine $T6$ and is quite short. The second MP segment increases the LP exhaust of Turbine $T6$ until the maximum of the exhaust for $T6$ is reached. The third MP segment reflects the reduction of the MP exhaust of Turbine $T4$ until the minimum for $T4$ inlet is reached. Any further saving of MP or LP steam cannot be exploited for load reduction in the boilers, and is vented.

In this case study, the marginal cost of the LP steam is zero. This can be explained from the relatively high efficiency of power generation through the path from the VHP main to Turbine $T6$ to the LP main and the price structure of the site energy resources. In other problems, saving LP steam might have a positive value, or it can be negative. If the marginal value of LP steam is negative, it means that it is counterproductive to save LP steam.

It should also be emphasized that variation in electricity and fuel tariffs will change the picture. Thus, graphs equivalent to Figure 23.56 could be produced for peak versus off-peak electricity costs, and so on. Therefore, to obtain a value of steam savings appropriate for a design study, an average needs to be taken across the different operating scenarios. The average can be taken by weighting the marginal costs for different tariff scenarios according to their relative duration.

The analysis in Figure 23.56 provides the basis for the economics of steam saving on an existing site. Thus, for the example in Figure 23.54, the marginal steam costs in Figure 23.56 indicate that there would be a strong economic incentive to save HP steam, and the utility system allows for a total HP steam saving of around 55 $t \cdot h^{-1}$. This does not mean that it is possible to save around 55 $t \cdot h^{-1}$ in the processes on the site. Figure 23.6 simply shows the economic incentive and the constraints on the utility system as far as the steam system savings are concerned. The task for the designer now is to consider those processes using HP steam and to determine whether economic energy saving projects can be identified. Figure 23.56 shows a limited scope to save MP steam with quite a low marginal cost. In this example, there is no incentive at all to save LP steam by retrofitting those processes using LP steam.

The arguments so far have focused on retrofitting existing processes on a site to reduce the steam consumption or increase the steam generation. On the other hand, it might be the case that a new process is to be added to the site, creating increased demands on the steam system. The same kind of analysis can be performed by increasing the consumption of steam, rather than decreasing it, and determining the marginal cost of steam for the new process. Again, steps are likely to be encountered in the steam marginal cost created by the constraints in the existing utility system. This might then call for the utility system to be modified to overcome the constraints in order to reduce the costs associated with provision of steam to the process[17].

Figure 23.56 shows that it is in general incorrect to attribute a single value for steam at a given level when the load can change (increase or decrease). In general, the value of steam at a given level depends on how much is being consumed, as well as fuel cost, power cost, equipment performance, equipment constraints, and so on. The fact that the price of steam at a given level can vary contrasts with the common practice of attributing a single marginal cost for each level of steam, irrespective of load. Change in steam load might not only arise from an energy saving project or a new process being commissioned but also might arise from the day-to-day variations in steam load as a result of changes in the pattern of production for the existing processes. Using the approach here allows the true cost of steam to be established for any pattern of steam loads. Details of the existing utility system are included in the cost analysis and all of the costs and constraints associated with that system. This can give a very different picture as to the costs associated with changing steam load at any level, compared to an approach that does not take such detail into account.

Finally, the profile of the marginal steam costs, such as the one illustrated in Figure 23.56, provides valuable insights into the potential for retrofit of the utility system itself. The steps in the marginal steam cost are created by constraints in the utility system (e.g. maximum flow in an existing steam turbine). The processes on the site might have the potential to save more steam at higher marginal steam cost than indicated by the length of the steps in the marginal steam costs (e.g. by retrofit of their heat exchanger networks). If this is the case, the constraints imposed by the utility system at those higher marginal steam costs can be relieved by retrofit of the features of the utility system creating the constraint (e.g. by installing a new steam turbine). The utility system modification might then allow increased steam saving with higher marginal steam cost. However, it should be emphasized that the economic justification for the capital expenditure on the modifications to the utility system comes from a combination of improved process and utility system performance, in which the retrofit of the processes and the utility system are considered together.

23.13 CHOICE OF DRIVER

Drivers are required to run various process and utility machines:

- electricity generators
- process gas compressors
- air compressors
- refrigeration compressors
- fans
- pumps
- mixers
- crushing and grinding equipment
- mechanical conveying equipment.

The main types of driver used in the process industries are:

- electric motor
- steam turbine

- turboexpander (analogous to a steam turbine, but expanding a process gas or vapor from high to low pressure)
- gas turbine
- gas engine (reciprocating engine firing gas)
- diesel engine (reciprocating engine firing diesel oil).

Gas turbines produce a hot exhaust gas with a temperature that depends on the particular machine but high enough to generate high-pressure steam. Gas engines and diesel engines produce a hot exhaust gas at around 450°C that can also be used to generate steam. In the case of gas engines and diesel engines, the cylinders need to be cooled by jackets supplied with cooling water. This provides a source of heat up to 95°C. Thus, in the case of gas engines and diesel engines, around half of the waste heat becomes available only at a low temperature. This is appropriate, for example, if large amounts of hot water are needed but inappropriate if large amounts of steam are needed.

The first major decision to be made is whether to generate power and distribute this for use in electric motors or to place direct drives for the other driver types directly onto the duty. A direct drive (e.g. steam turbine driving a large pump) can often be the cheapest solution relative to a large steam turbine producing power, distribution of the power and use of the power in an electric motor to provide the driver. On the other hand, a large single (and efficient) generator can service many electric motor drives. Direct drives are inflexible as far as the utility system is concerned. For example, if the steam turbine is driving a large pump, then the flow through the steam turbine needs to be maintained to keep the pump running, even if the heat produced at the exhaust of the steam turbine might not be required. Electric generators are more flexible as far as the utility system is concerned and can service many drives with electric motors. Also, there are often many small drivers required on a site that are most economically serviced by electric motors, rather than direct drives. The best solution is usually a combination of electric generators and direct drives.

There are many issues to be considered for the selection of the most appropriate combination of drives:

- economic analysis
- fit to the existing infrastructure
- process requirements
- safety issues in the case of power failure
- space limitations.

For example, suppose a new steam turbine is to be installed. The economics of the new steam turbine need to be analyzed in terms of capital and operating costs. However, the issues that need to be addressed are:

- boiler capacity
- fuel system capacity
- steam mains capacity
- water treatment capacity
- cooling tower capacity
- available space, and so on.

The process requirements include consideration of:

- power range
- rotational speed
- speed variability
- operational duration without overhaul
- size and weight
- reliability.

Many issues need to be considered when choosing the most appropriate combination of drivers. Choosing the best combination is a trade-off between capital costs, operating costs, system flexibility and reliability.

23.14 STEAM SYSTEMS AND COGENERATION – SUMMARY

Most process heating is provided from the distribution of steam around sites at various pressure levels. Normally, two or three steam mains will distribute steam at different pressures around the site. The steam system is not only important from the point of view of process heating but is also used to generate a significant amount of power on the site.

Boiler feedwater treatment is required for the raw water to remove suspended solids, dissolved solids, dissolved salts and dissolved gases (particularly, oxygen and carbon dioxide). The raw water entering the site is commonly filtered and dissolved salts removed using softening or ion exchange. The principal problems are calcium and magnesium ions. The water is then stripped to remove the dissolved gases and treated chemically before being used for steam generation.

There are many types of steam boilers, depending on the steam pressure, steam output and fuel type. Blowdown is required to remove the dissolved solids not removed in the boiler feedwater treatment. The efficiency of the boiler depends on its load.

Steam turbines are used to convert part of the energy of the steam into power and can be configured in different ways. Steam turbines can be divided into two basic classes: back-pressure turbines and condensing turbines. The efficiency of the turbine and its power output depend on the flowrate of steam to the turbine. The performance characteristics can be modeled by a simple linear relationship over a reasonable range of operation.

Gas turbines consist of a compressor and a turbine mechanically connected to each other. Release of energy to the compressed gases after the compressor, from the combustion of fuel, creates a net production of power. Gas turbines are available in a wide range of sizes but are restricted to standard frame sizes. The hot exhaust gas is useful for raising steam and the temperature of the exhaust gas can be increased by using supplementary firing. The performance of gas turbines can be modeled by a simple correlation involving the mass flowrate of fuel.

The steam system configuration normally generates most of the steam in boilers at the highest pressure, feeding to a HP steam main. The highest-pressure steam main is normally used for power generation rather than process heating. Steam is expanded from high to lower levels by either steam turbines or the expansion valves. Steam expanded across let-down valves can be de-superheated. It is also good practice to recover the flash steam from large flowrates of HP steam condensate.

Site composite curves can be used to represent the site heating and cooling requirements thermodynamically. This allows the analysis of thermal loads and levels on site. Using the models for steam turbines and gas turbines allows cogeneration targets for the site to be established. Steam levels can be optimized to minimize fuel consumption or maximize cogeneration. A cost trade-off needs to be carried out in order to establish the optimum trade-off between fuel requirements and cogeneration.

The site power-to-heat ratio is very important in determining the most appropriate cogeneration system for the site.

Complex steam systems usually feature many important degrees of freedom to be optimized. Multiple steam generation devices, multiple steam turbines, let-down stations, condensing turbines and vents all provide important degrees of freedom for optimization. To establish the steam costs for retrofit of site processes requires an optimization model to be developed. This allows the steam loads for process heating to be gradually decreased and the steam system reoptimized at each setting. The result in cost savings establishes the true cost of steam for retrofit projects aiming to reduce steam consumption on the site.

Drivers are required for many different types of process machine on a site. Power can be generated and distributed to drive electric motors, or direct drives can be used. Direct drives are inflexible as far as the utility system operation is concerned. A combination of direct drives and electric motors is usually the best solution for the site as a whole.

23.15 EXERCISES

1. A boiler with a capacity of 100,000 $kg \cdot h^{-1}$ is required to produce steam of 40 bar and 350°C. The feedwater from the deaerator is at 100°C and contains 100 ppm dissolved solids. Maximum dissolved solids allowed in the boiler is 2000 ppm. The steam system operates with 60% condensate return. Enthalpy data for steam are given in Table 23.10. Calculate:
   a. the blowdown rate

Table 23.10 Enthalpy data for Exercise 1.

	Enthalpy at 40 bar ($kJ \cdot kg^{-1}$)
Steam at 350°C	3095
Saturated steam	2800
Saturated condensate	1087
Water at 100°C	422

   b. the energy consumption for a boiler efficiency of 88%.
2. A steam turbine operates with inlet steam conditions of 40 barg and 420°C and can be assumed to operate with an isentropic efficiency of 80% and a mechanical efficiency of 95%. Calculate the power production for a steam flowrate of 10 $kg \cdot s^{-1}$ and the heat available per kg in the exhaust steam (i.e. superheat plus latent heat) for outlet conditions of:
   a. 20 barg
   b. 10 barg
   c. 5 barg.

   Determine steam properties from steam tables. What do you conclude about the heat available in the exhaust steam as the outlet pressure varies?
3. A steam turbine is operating between inlet steam of 40 barg and 420°C and outlet steam of 5 barg. Using the Willans' Line Model with parameters from Table 23.1 for large turbines and intercept ratio of 0.05, calculate the power production for a turbine at full load with a flowrate of steam of 10 $kg \cdot s^{-1}$.
4. The inlet steam to an extraction turbine is 30 bara and 400°C. The extraction steam is at 10 bara and the exhaust is condensed at 0.12 bara. The steam turbine is fully loaded with a throttle flow of 40 $kg \cdot s^{-1}$ to the inlet, 15 $kg \cdot s^{-1}$ going to extraction and the remaining 25 $kg \cdot s^{-1}$ going to condensation. Using the Willans' Line Model with parameters for large back-pressure and condensing turbines from Table 23.1 and an intercept ratio of 0.05, calculate the power production.
5. A site steam system needs to supply steam for process heating purposes of 120 MW at 200°C and 80 MW at 150°C. Steam is generated in the site boilers at 40 barg and 420°C and expanded through steam turbines to the appropriate pressures. Saturation temperatures for the steam should be 20°C above the temperature at which the heating is required. Assuming fully loaded turbines, use the Willans' Line Model with parameters for large turbines taken from Table 23.1 and an intercept ratio of 0.05 to calculate the cogeneration target for the site.
6. The problem table cascade for a process is given in Table 23.11 below for $\Delta T_{min} = 30$°C. This process is to be integrated with a steam turbine in a cogeneration scheme. The inlet steam is at 41 barg and 400°C.
   a. The rate of power generation can be determined from the Willans' Line Model with parameters from Table 23.1 and intercept ratio of 0.05, assuming a mechanical efficiency of 95%. Calculate the power generation for a fully loaded turbine if the inlet steam flow to the turbine is fixed to satisfy process heating requirements.
   b. Calculate the heat duty of the fuel consumed in the generation of steam if the theoretical flame temperature in the boiler is 1800°C and the ambient temperature is 15°C.

Table 23.11 Problem table cascade for Exercise 6.

Interval temperature (°C)	Heat flow (MW)
400	15
400	20
170	20
115	21
115	11.5
100	12
100	0
80	1
80	17.5
50	18
50	5

c. The fuel consumption can be reduced by air preheat to the boiler by recovery of waste heat from the process. Calculate the fuel consumption and power generation if air preheat is maximized. Assume that the heat capacity flowrates of combustion air and flue gas are equal. Acid dew point for the flue gas is 150°C.

7. The following data are given for a gas turbine.

Power generation	W	15 MW
Power efficiency	η_{GT}	32.5%
Exhaust flow	m_{EX}	58.32 $kg \cdot s^{-1}$
Exhaust temperature	T_{EX}	488°C
Exhaust heat capacity	$C_{P,EX}$	1.1 $kJ \cdot kg^{-1} \cdot K^{-1}$
Stack temperature	T_{STACK}	100°C

a. Calculate the fuel consumption Q_{FUEL}.

b. Assuming 72% of the heat available in the exhaust can be recovered for steam generation, how much steam (Q_{STEAM}) can be generated from an unfired HRSG in MW? What is the cogeneration efficiency η_{COGEN} for the system of gas turbine and HRSG?

c. With supplementary firing ($T_{SF} = 800$°C), calculate the fuel consumption Q_{SF} for the supplementary firing, available heat from the exhaust Q_{EX} and steam generation for the same HRSG with an 83% efficiency.

d. Calculate the power generation efficiency η_{POWER} and cogeneration efficiency η_{COGEN} for the system of gas turbine and HRSG with supplementary firing. Does supplementary firing improve η_{POWER} or η_{COGEN}?

8. A gas turbine has the following performance data.

Electricity output	13.5 MW
Heat rate	10,810 $kJ \cdot kWh^{-1}$
Exhaust flow	179,800 $kg \cdot h^{-1}$
Exhaust temperature	480°C

The exhaust is to be used to generate steam in an unfired HRSG with a minimum stack temperature of 140°C. The specific heat capacity of the exhaust is 1.1 $kJ \cdot kg^{-1} \cdot K^{-1}$. Enthalpy data for steam are given in Table 23.12. Calculate the following.

Table 23.12 Enthalpy data for Exercise 8.

	Enthalpy at 10 bara ($kJ \cdot kg^{-1}$)
Steam at 200°C	2827
Saturated steam	2776
Saturated condensate	762
Water at 100°C	420

a. fuel requirement in MW

b. power generation efficiency of the turbine

c. amount of steam at 10 bara and 200°C that can be produced from boiler feedwater at 100°C in an unfired HRSG. Assume a $\Delta T_{min} = 40$°C for the heat recovery steam generator.

9. Complete the steam balance provided in Figure 23.57.

a. Calculate extraction flows through $T1$ and $T2$.

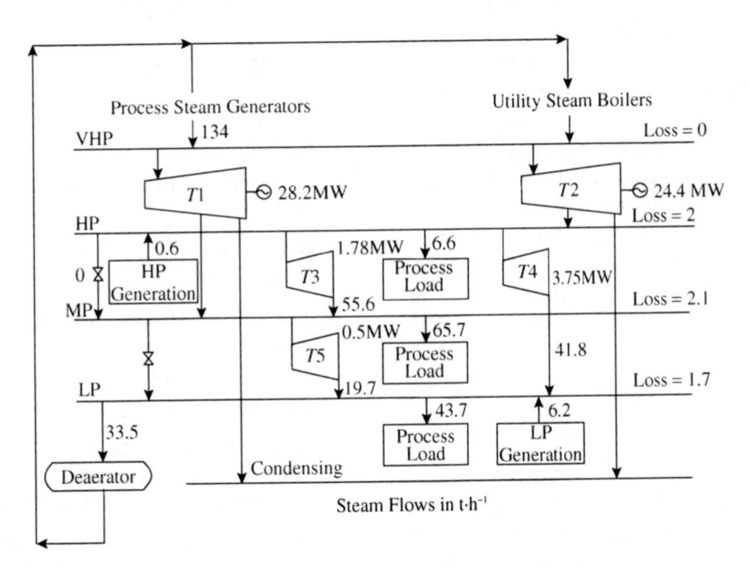

Figure 23.57 Steam balance for a site.

b. Calculate condensing flows of $T1$ and $T2$ from a power balance around each turbine, assuming the power generation is fixed and that the mechanical efficiency to be 97%. Assume steam turbines to be modeled by the Willans' Line Model with parameters from Table 23.1 and an intercept ratio of 0.05. Steam properties for the inlet to steam turbines can be taken to be those of the steam mains, which are given in Table 23.13. Isentropic enthalpy changes between the steam mains in Figure 23.57 are given in Table 23.14.

Table 23.13 Steam properties for the steam mains in Figure 23.57.

	Pressure (bar)	Superheat temperature (°C)	Enthalpy ($kJ{\cdot}kg^{-1}$)	Saturation temperature (°C)
VHP	100	500	3375	311
HP	40	400	3216	250
MP	15	300	3039	198
LP	4.5	200	2858	148
Condensation	0.12		2591	50

Table 23.14 Isentropic enthalpy changes in Figure 23.57.

Inlet	Outlet	Isentropic enthalpy change ($kJ{\cdot}kg^{-1}$)
VHP	HP	272
VHP	MP	508
HP	MP	257
HP	LP	507
HP	Condensation	1049
MP	LP	268
MP	Condensation	824

c. Determine the steam production required from the utility boilers to meet the steam demand.

d. Calculate the isentropic and overall efficiency for Turbine $T4$ (load = 3.75 MW), assuming the inlet conditions are given in Table 23.14 and the mechanical efficiency is 95%.

e. Turbine $T3$ is to be shut down and the driver switched to an electric motor. It can be assumed that the inlet conditions to the turbines do not change as a result of any changes to the steam balance. The steam flows through Turbines $T4$ and $T5$ are to be maintained at their current values. However, the flows through Turbines $T1$ and $T2$ can be changed to rebalance the steam system. How much steam is needed from the utility boilers if this is done? What are the cost savings from the change? Assume that the conditions of the steam in each main are fixed to the values in Table 23.13.

Boiler efficiency	= 0.92
Fuel cost	= 5.69 $\${\cdot}GJ^{-1}$
Power cost	= 55 $\${\cdot}MWh^{-1}$

10. A company is considering implementing a cogeneration scheme. At the moment, the company is supplied with 50,000 MWh of electricity from the grid each year and the company produces steam of 80,000 MWh per year from a gas-fired boiler. It is proposed to replace this arrangement with a gas turbine. With the cogeneration scheme, the waste heat from the gas turbine can be used for the steam generation to satisfy the steam demand.
Assume the power generation efficiency of gas turbine = 0.35, average electricity tariff = 0.063 $\${\cdot}kW^{-1}$, gas price = 0.015 $\${\cdot}kWh^{-1}$, boiler efficiency = 0.9, the installed cost of the gas turbine is 1000 $\${\cdot}kW^{-1}$ and the company operates for 8000 hours annually.
    a. Calculate the operating costs for both the current operating system and the proposed cogeneration scheme.
    b. Determine the savings and payback time that can be achieved by implementing the cogeneration scheme.
    c. What issues should be considered in recovering the exhaust heat from a gas turbine?

11. It is proposed to locate a thermal oxidizer adjacent to a chemical plant, with the purpose of supplying steam generated from the flue gas of the thermal oxidizer. The problem table cascades for the two processes are given in the Table 23.15.

Table 23.15 Data for Exercise 11.

Thermal oxidizer		Chemical plant	
T^* (°C)	Q (MW)	T^* (°C)	Q (MW)
605	0	245	20
255	14	125	14
155	18	45	0

Note that T^* represents shifted temperatures for the process streams. Allow $\Delta T_{min}/2 = 5°C$ for the steam side.
    a. Plot the total site composite curves.
    b. What are the minimum heat requirements for the total site? What steam mains would need to be installed to achieve this, assuming only latent heat to be given from the steam? Give the saturation temperature of the steam.
    c. Plot the steam profiles. Is there a site pinch and if so, at what temperature?

REFERENCES

1. Betz (1991) *Handbook of Industrial Water Conditioning*, 9th Edition.
2. Kemmer FN (1988) *The Nalco Water Handbook*, 2nd Edition, McGraw-Hill.
3. Dryden IGC (1982) *The Efficient Use of Energy*, Butterworth Scientific.
4. Shang Z (2000) *Analysis and Optimisation of Total Site Utility Systems*, PhD Thesis, UMIST, UK.
5. Elliot TC (1989) *Standard Handbook of Powerplant Engineering*, McGraw-Hill.
6. Siddhartha M and Rajkumar N (1999) Performance Enhancement in Coal Fired Thermal Power Plants Part II: Steam Turbines, *Int J Energy Res*, **23**: 489.
7. Mavromatis SP (1996) *Conceptual Design and Operation of Industrial Steam Turbine Networks*, PhD Thesis, UMIST, UK.

8. Mavromatis SP and Kokossis AC (1998) Conceptual Optimisation of Utility Networks for Operational Variations – I Targets and Level Optimisation, *Chem Eng Sci*, **53**: 1585.

9. Varbanov PS, Doyle S and Smith R (2004) Modelling and Optimisation of Utility Systems, *Trans IChemE*, **82A**: 561.

10. Chou CC and Shih YS (1987) A Thermodynamic Approach to the Design and Synthesis of Utility Plant, *Ind Eng Chem Res*, **26**: 1100.

11. Peterson JF and Mann WL (1985) Steam System Design: How it Evolves, *Chem Eng*, **92**(21): 62.

12. Dhole VR and Linnhoff B (1992) *Total Site Targets for Fuel, Cogeneration, Emissions and Cooling, ESCAPE – II Conference*, Toulouse, France.

13. Raissi K (1994) *Total Site Integration*, PhD Thesis, UMIST, UK.

14. Klemes J, Dhole VR, Raissi K, Perry SJ and Puigjaner L (1997) Targeting and Design Methodology for Reduction of Fuel, Power and CO_2 on Total Sites, *J Applied Thermal Eng*, **17**: 993.

15. Kenney WF (1984) *Energy Conservation in Process Industries*, Academic Press.

16. Kimura H and Zhu XX (2000) R-Curve Concept and Its Application for Industrial Energy Management, *Ind Eng Chem Res*, **39**: 2315.

17. Varbanov PS, Perry SJ, Makwana Y, Zu XX and Smith R (2004) Top Level Analysis of Utility Systems, *Trans IChemE*, **82A**: 784.

18. Iyer RR and Grossmann IE (1998) Synthesis and Operational Planning of Utility Systems for Multiperiod Operation, *Comp Chem Eng*, **22**: 979.

24 Cooling and Refrigeration Systems

24.1 COOLING SYSTEMS

Most industrial processes require an external heat sink for heat removal and temperature control. Waste heat rejection should be minimized as much as possible. Priority should be given to recover waste heat in the first instance within the process. If waste heat is available at a high enough temperature and is not required for process heating within the process itself, then opportunities should be explored to pass the heat to other useful heat sinks on the site. Heat can be passed between processes on the site through the steam system, as discussed in Chapter 23. Other heat transfer fluids, such as hot oil, can be used to transfer heat around at high temperatures. Below steam temperature, heat can be passed around a site using hot water. Heat can also sometimes be rejected usefully into the utility system, for example, for boiler feed water preheating or combustion air preheating. Once the opportunity for heat recovery and rejection to useful external heat sinks has been exhausted, then heat must be rejected to the environment.

The most direct way to reject heat above ambient temperature to the environment is by the use of air-cooled heat exchangers, as discussed in Chapter 15[1]. These coolers exploit a flow of ambient air across the outside of tubes through which process fluids are flowing that require cooling. Such air coolers are very common in some industries, particularly when the plant is located in a region where water is scarce.

Another way to reject heat to the environment is through the use of water in once-through cooling systems. For example, water can be taken from a river, canal, lake or the sea, used for cooling, and then returned to its source. Heat rejection to the environment in this way does have environmental consequences and can have an impact on ecosystems. For heat rejection to rivers, canals and lakes, any change of temperature in surface and groundwater resulting from waste heat rejection affects the chemical, biochemical and hydrological properties of the water and potentially has an impact on the overall ecosystem. Once-through cooling systems require large amounts of water to be used on a single-use basis.

When the use of fresh cold water is limited, or environmental regulations limit the heat rejection to rivers, canals, lakes and the sea, then air cooling or recirculating cooling water systems must be used. Recirculating cooling water systems are by far the most common method used for heat rejection to the environment.

If heat rejection is required at temperatures below that which can normally be achieved using cooling water or air cooling, then refrigeration is required. For most applications, refrigeration is required to provide cooling below ambient temperature. A refrigeration system is a heat pump with the purpose of providing low-temperature cooling. This means that heat must be rejected at a higher temperature, to ambient via an external cooling utility (e.g. cooling water), a heat sink within the process or to another refrigeration system.

Cooling systems require integration with the processes on the site, with other cooling systems and also the hot utility systems. Start by considering the above-ambient temperature cooling, using the most common method for this; recirculating cooling water systems.

24.2 RECIRCULATING COOLING WATER SYSTEMS

Figure 24.1 illustrates the basic features of a recirculating cooling water system[1]. Cooling water from the cooling tower is pumped to heat exchangers where waste heat needs to be rejected from the process to the environment. The cooling water is in turn heated in the heat exchanger network and returned to the cooling tower. The hot water returned to the cooling tower flows down over packing and is contacted countercurrently or in cross-flow with air. The packing should provide a large interfacial area for heat and mass transfer between the air and the water. The air is humidified and heated, and rises through the packing. The water is cooled mainly by evaporation as it flows down through the packing. The evaporated water leaving the top of the cooling tower reflects the cooling duty that is being performed. Water is also lost through *drift*. Drift is droplets of water entrained in the air leaving the top of the tower. Drift has the same composition as the recirculating water and is different from evaporation. Drift should be minimized because it wastes water and can also cause staining of buildings, and so on, that are some distance from the cooling tower. *Drift loss* or *windage loss* is around 0.1 to 0.3% of the water circulation rate. *Blowdown*, as shown in Figure 24.1, is necessary to prevent the buildup of contamination in the recirculation. *Makeup water* is required to compensate for the loss of water from evaporation and drift and the blowdown. The makeup water contains solids that build up in the recirculation as a result

Chemical Process Design and Integration R. Smith
 ISBNs: 0-471-48680-9 (HB); 0-471-48681-7 (PB)

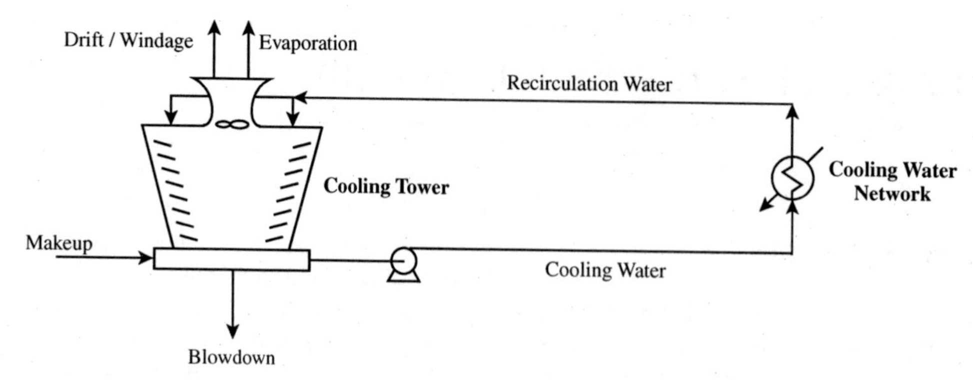

Figure 24.1 Cooling water systems.

of the evaporation. The blowdown purges these, along with products of corrosion and microbiological growth. Both corrosion and microbiological growth need to be inhibited by chemical dosing of the recirculation system. In Figure 24.1, the blowdown is shown to be taken from the cooling tower *basin*. It can be taken as cold blowdown, as shown in Figure 24.1. Alternatively, it can be taken from the hot recirculation water before it is returned to the cooling tower as *hot blowdown*. Taking hot blowdown might be helpful in increasing the heat rejection from the cooling system but might not be acceptable environmentally from the point of view of the resulting increase in effluent temperature.

Many different designs of cooling tower are available. These can be broken down into two broad classes as follows.

1. *Natural draft*. Natural draft cooling towers consist of an empty shell, usually constructed in concrete. The upper, empty portion of the shell merely serves to increase the draft. The lower portion is fitted with the packing. The draft is created by the difference in density between the warm humid air within the tower and the denser ambient air.

2. *Mechanical draft cooling towers*. Mechanical draft cooling towers use fans to move the air through the cooling tower. In a *forced draft* design, fans push the air into the bottom of the tower. *Induced draft* cooling towers have a fan at the top of the cooling tower to draw air through the tower. The tower height for mechanical draft towers does not need to extend much beyond the depth of the packing. Mechanical draft cooling towers for large duties often comprise a series of rectangular *cells* constructed together, but operating in parallel, each with its own fan.

The type of packing used can be as simple as *splash bars* but is more likely to be packing similar in form to that used in absorption and distillation towers. The temperature limitation of the packing needs careful attention. Plastic packing has severe temperature limitations, as far as the cooling water return temperature is concerned. If the temperature is too high, the plastic packing will deform and this will result in a deterioration of cooling tower performance. Polyvinylchloride is limited to a maximum temperature of around 50°C. Other types of plastic packing can withstand temperatures up to around 70°C.

Chemicals are added to the circulation system to prevent fouling. *Dispersants* are added to prevent deposit of solids, *corrosion inhibitors* to prevent corrosion and *biocides* to inhibit biological growth.

Figure 24.2 gives the basis of a cooling water system model. For the cooling tower:

$$T_0 = f(F_2, T_2, T_{WBT}) \tag{24.1}$$

$$F_0 = f(F_2, T_2, T_{WBT}) \tag{24.2}$$

$$F_E = f(F_2, T_2, T_{WBT}) \tag{24.3}$$

where
T_0 = outlet temperature of the cooling tower
T_1, T_2 = inlet temperature and outlet temperature for the heat exchanger network (T_2 = inlet temperature to the cooling tower)
T_{WBT} = wet bulk temperature of the inlet air (the equilibrium temperature attained by water that is vaporizing into the inlet air and is always less than the dry bulk temperature of the air into which vaporization is taking place)
F_0 = flowrate of water from the cooling tower
F_1, F_2 = flowrate of water to and from the heat exchanger network
F_E = evaporation rate

The difference between T_0 and T_{WBT} measures the degree of unsaturation of the inlet air. If the air is initially saturated with water vapor, then neither vaporization of the liquid nor depression of the wet bulb temperature occurs. A simple cooling tower model that can be used in conceptual design is presented elsewhere[2].

There are some general trends that can be observed for the design of cooling towers in terms of the temperature and flowrate of the inlet cooling water to the tower. Increasing

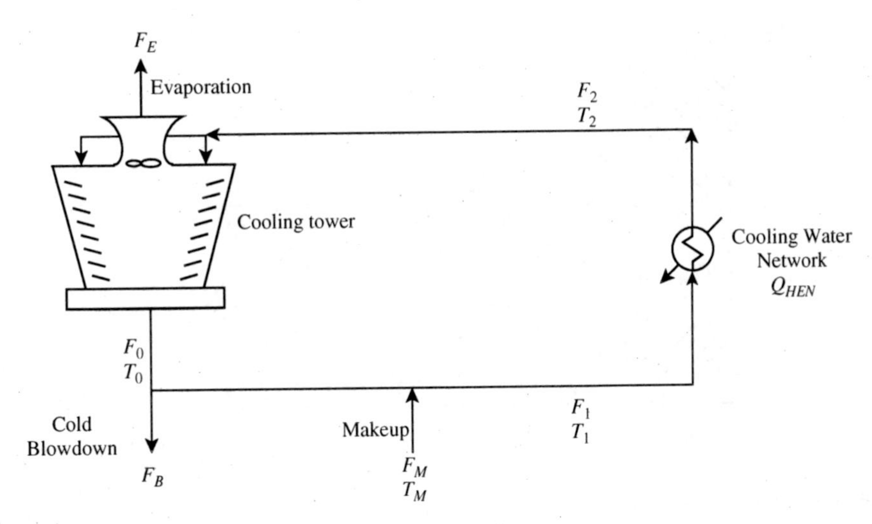

Figure 24.2 Cooling water system model.

the temperature of the inlet to a given cooling tower design for a fixed flowrate increases the performance of the cooling tower and allows more heat to be removed[2]. On the other hand, if the flowrate to the inlet of a given cooling tower is decreased for a fixed inlet temperature, then the performance of the cooling tower improves, allowing more heat to be removed[2]. Thus, the performance of a cooling tower is maximized by maximizing the inlet temperature to the cooling tower, and minimizing the inlet flowrate[2]. There will be constraints on the maximum return temperature determined by the maximum temperature allowable for the cooling tower packing, and fouling and corrosion issues, as discussed above.

Referring back to Figure 24.2, the cooling tower makeup and blowdown form a mass balance according to:

$$F_1 = F_0 - F_B + F_M \tag{24.4}$$

where F_1 = flowrate of water to the heat exchanger network
F_B = flowrate of cooling tower blowdown
F_M = flowrate of makeup water

If the heat capacity of the water is assumed to be constant, then an energy balance can also be written:

$$F_1T_1 = (F_0 - F_B)T_0 + F_MT_M \tag{24.5}$$

where T_1 = temperature to the heat exchanger network (after addition of makeup)
T_0 = temperature from the cooling tower (before addition of makeup)
T_M = temperature of the makeup water

The heat load on the heat exchanger network is given by

$$Q_{HEN} = F_2C_P(T_2 - T_1) \tag{24.6}$$

where Q_{HEN} = heat exchanger network cooling duty
C_P = heat capacity of the water

As evaporated water is pure, solids are left behind in the recirculating water, making it more concentrated than the makeup water. The blowdown purges the solids from the system. Note that the blowdown has the same chemical composition as the recirculated water. *Cycles of concentration* is a comparison of the dissolved solids in the blowdown compared with that in the makeup water. For example, at three cycles of concentration, the blowdown has three times the solids concentration as the makeup water. For calculation purposes, blowdown is defined to be all nonevaporative water losses (drift, leaks and intentional blowdown). In principle, any soluble component in the makeup and blowdown can be used to define the concentration for the cycles, for example, chloride and sulfate being soluble at high concentrations can be used. The cycles of concentration are thus defined to be:

$$CC = \frac{C_B}{C_M} \tag{24.7}$$

where CC = cycles of concentration
C_B = concentration in blowdown
C_M = concentration in makeup

A mass balance for the solids entering with the makeup (assuming zero solids in the evaporation) gives:

$$F_MC_M = F_BC_B \tag{24.8}$$

Combining Equation 24.7 with Equation 24.8 gives:

$$CC = \frac{F_M}{F_B} \tag{24.9}$$

given that:

$$F_M = F_B + F_E \tag{24.10}$$

Combining Equations 24.9 and 24.10 gives:

$$F_B = \frac{F_E}{CC - 1} \tag{24.11}$$

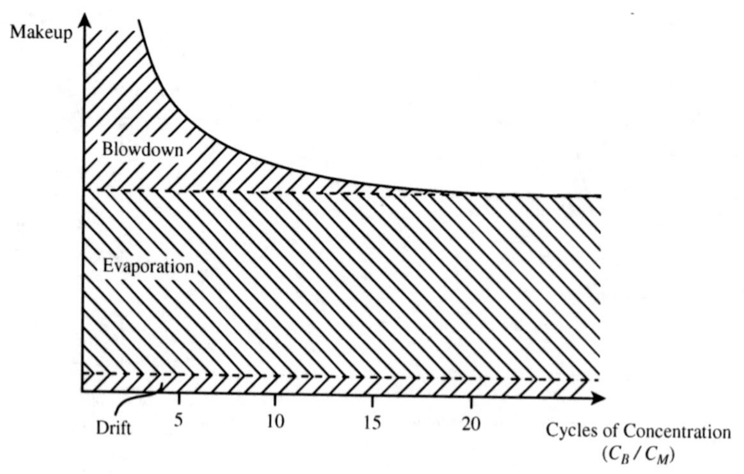

Figure 24.3 Relationship between makeup water and blowdown for cooling towers.

$$F_M = \frac{F_E CC}{CC - 1} \tag{24.12}$$

Figure 24.3 shows the relationship between the makeup, blowdown, evaporation and drift versus the cycles of concentration. For a given design of cooling tower, fixed heat duty and fixed conditions, the evaporation and drift will be constant as the cycles of concentration increase. However, as the cycles of concentration increase, the blowdown decreases and hence the makeup water decreases.

The consequences of an increase in the cycles of concentration are that as the level of dissolved solids increases, corrosion and deposition tendencies also increase. The result is that, although increasing the cycles of concentration decreases the water requirements of the cooling system, the required amount of chemical dosing also increases.

24.3 TARGETING MINIMUM COOLING WATER FLOWRATE

As noted previously, the performance of a cooling tower is maximized by maximizing the inlet temperature to the cooling tower and minimizing the inlet flowrate[2]. It is therefore useful to be able to predict the minimum cooling water flowrate, taking into account the heat transfer limitations of the individual cooling duties.

Most cooling water networks involve the use of cooling water directly from the cooling tower in each heat exchanger. This leads to a *parallel* arrangement of the coolers. Figure 24.4a shows a cooling duty for a hot process stream. The cooling is being supplied by cooling water with a temperature corresponding to temperature T_1 in Figure 24.2. In Figure 24.4a, the flow of cooling water has been minimized until the temperature difference has been minimized. If the cooling duties are configured in parallel, as shown in Figure 24.4b, then minimizing the flowrate in each individual exchanger will minimize the overall flowrate and maximize the return temperature. The minimum flowrate in Figure 24.4 is based on a parallel configuration. Figure 24.5 shows that cooling water can, in some circumstances, be reused and heat exchangers used in series rather than parallel. A series arrangement, if it is acceptable, increases the cooling water return temperature and decreases the recirculation flowrate, increasing the performance of the cooling tower.

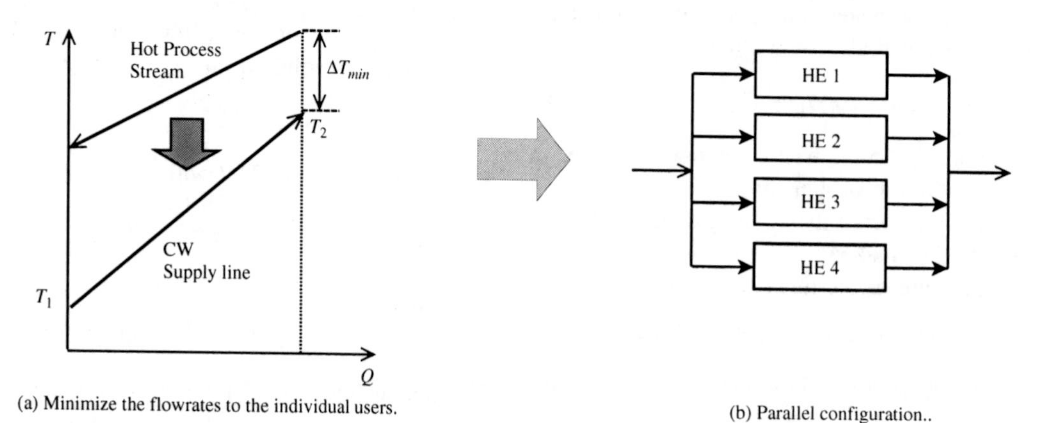

Figure 24.4 Minimizing cooling water flow in a parallel arrangement. (From Kim J-K and Smith R, 2001, *Chem Eng Sci*, **56**: 3641, reproduced by permission of Elsevier Ltd.)

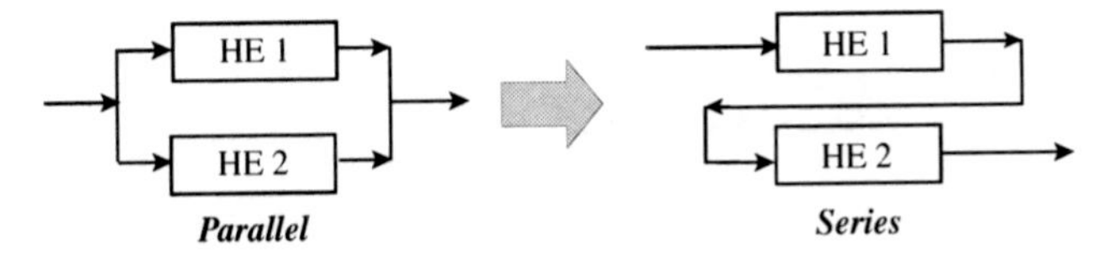

Figure 24.5 Parallel and series arrangements for cooling water coolers. (From Kim J-K and Smith R, 2001, *Chem Eng Sci*, **56**: 3641, reproduced by permission of Elsevier Ltd.)

On the other hand, moving from a parallel to a series arrangement, as shown in Figure 24.5, has the disadvantage that the temperature differences in the heat exchangers will decrease, possibly decreasing their performance. However, an individual heat exchanger might require an increase in flowrate, which can increase the heat transfer coefficient. Also, moving from a parallel to series arrangement will increase the overall pressure drop. For the parallel arrangement, the pressure drop for the system is the largest individual heat exchanger pressure drop. For the series arrangement, the pressure drop is the sum of the pressure drops of the heat exchangers in series. Thus, the series arrangement decreases the flowrate but increases the pressure drop. Generally, the centrifugal pumps used to circulate the cooling water can produce a higher pressure drop (*head*) for a reduced flowrate. In summary, moving from parallel to series arrangements for cooling water networks

- increases the efficiency of the cooling tower
- decreases the temperature differences in the cooling water heat exchangers
- increases the pressure drop through the cooling water network.

Consider now how to systematically determine the minimum overall flowrate required by a cooling water system, such that the resulting high cooling tower inlet temperature and low flowrate enhance the performance of a cooling tower. The parallel configuration shown in Figure 24.4b maximizes the flowrate of cooling water around the system and minimizes the return temperature, as the return to the cooling tower from each individual heat exchanger is limited by the minimum temperature difference in that heat exchanger. Rather than utilize only parallel arrangements, it might be possible in some circumstances to utilize series arrangements, as shown in Figure 24.5. Reusing the cooling water, as shown in Figure 24.5, lowers the overall flowrate as the same cooling water is used twice and increases the overall return temperature to the inlet of the cooling tower. Thus, for a given heat load on the cooling tower, using series arrangements for cooling rather than parallel arrangements can increase the performance of the cooling tower and decrease the cooling tower size. In retrofit situations in which the duty on the cooling tower has increased, the performance of the existing cooling tower can be improved by

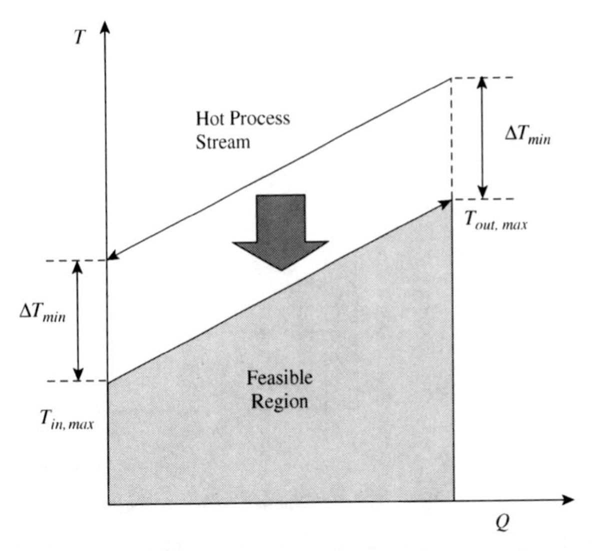

Figure 24.6 Representation of a heat exchanger using cooling water. (From Kim J-K and Smith R, 2001, *Chem Eng Sci*, **56**: 3641, reproduced by permission of Elsevier Ltd.)

adopting a series arrangement for the use of cooling water, rather than a parallel arrangement.

Figure 24.6 shows a representation of a heat exchanger using cooling water in which both the inlet and outlet temperatures of the cooling water have been maximized. This corresponds with ΔT_{min} at the inlet and the outlet of the cooler. The cooling water profile with the inlet and outlet temperatures maximized to give ΔT_{min} throughout the exchanger is known as the *limiting cooling water profile*. This will be used to provide a boundary between feasible and infeasible temperatures in the design of the cooling water network. As long as cooling water is fed to a heat exchanger at a temperature below its limiting cooling water profile, it will be considered to be feasible. Anything above this limiting cooling water profile will be considered to be infeasible. Consider the four cooling operations given in Table 24.1. ΔT_{min} has been assumed to be 10°C.

Figure 24.7a shows the four limiting cooling water profiles from Table 24.1 plotted together on the same axes. Rather than view these profiles individually, an overall picture is required. To achieve this, a composite of the individual cooling water profiles can be developed, as shown in Figure 24.7b. This is analogous to the construction of the composite curves for heat recovery

Table 24.1 Limiting cooling water data for four operations.

Heat exchanger	$T_{CW,in}$ (°C)	$T_{CW,out}$ (°C)	CP (kW·K^{-1})	Q (kW)
1	20	40	20	400
2	30	40	100	1000
3	30	75	40	1800
4	55	75	10	200

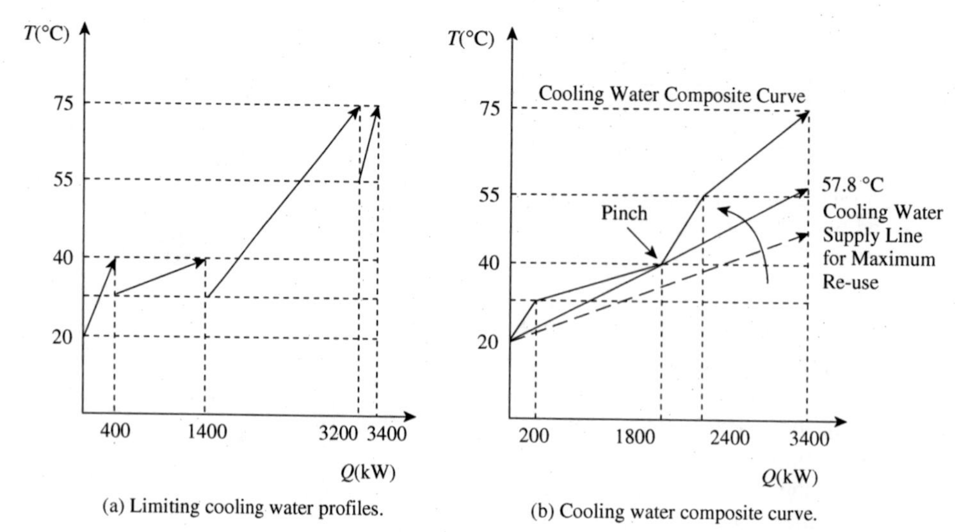

Figure 24.7 The limiting cooling water composite curve. (From Kim J-K and Smith R, 2001, *Chem Eng Sci*, **56**: 3641, reproduced by permission of Elsevier Ltd.)

developed in Chapter 16. The diagram is divided into temperature intervals. Within each temperature interval, the heat duty for the individual streams is combined together to produce the *cooling water composite curve*. This profile represents a single stream equivalent of the four separate streams. To determine the minimum flowrate of cooling water, a *cooling water supply line* is matched against the cooling water composite curve, as shown in Figure 24.7b. The cooling water supply line starts at the cooling water temperature supplied to the heat exchanger network (cooling tower exit temperature after the addition of makeup). The straight line from the supply temperature (outlet temperature from the cooling tower) of 20°C represents the cooling water supply line. Maximizing the slope of this cooling water supply line minimizes the flowrate. The maximum slope is dictated by the *pinch* for the system, which in this example is at 40°C. A steeper line than the one shown in Figure 24.7b would cross the cooling water composite curve, and temperatures would be infeasible at some point in the network. The slope of the cooling water supply line in Figure 24.7b shows that the minimum cooling water flowrate corresponds with a ΔT_{min} at the supply (20°C) and at the pinch (40°C). Elsewhere, the temperature differences for heat transfer will be above the minimum. In this case, the cooling water supply corresponds with a target CP of 90 $kW{\cdot}K^{-1}$, flowrate of 21.5 $kg{\cdot}s^{-1}$ and cooling water return temperature of 57.8°C[2]. It should be noted that practical constraints might limit the cooling water return temperature to be below that which is apparently possible from Figure 24.7. Such constraints might arise from corrosion considerations in the heat exchangers and pipework, temperature limits for the packing materials in the cooling tower or fouling from the cooling water. How such constraints can be included will be addressed in the next section.

The flowrate targets discussed here have been directed to minimizing the flowrate. In addition to temperature constraints, pressure drop constraints might also constrain the flowrate by constraining whether heat exchangers can be placed in series or not. Inclusion of pressure drop constraints is much more complex and outside the scope of the current text[3].

24.4 DESIGN OF COOLING WATER NETWORKS

To achieve the minimum cooling water flowrate target in design requires that the problem is decomposed into design regions, as illustrated in Figure 24.8. This shows

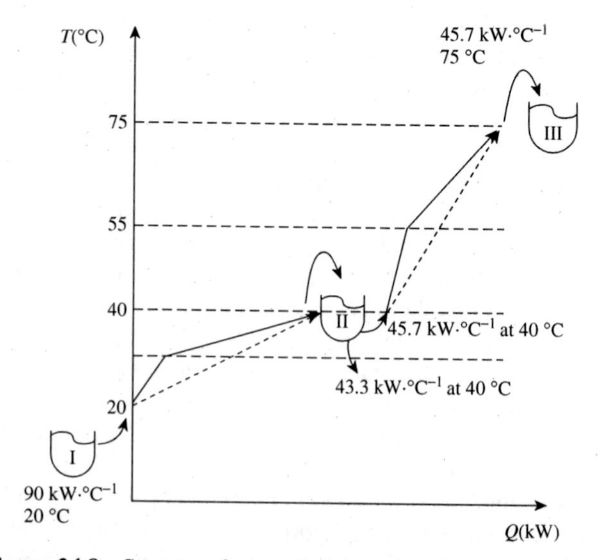

Figure 24.8 Strategy for minimizing the flowrate of cooling water. (From Kim J-K and Smith R, 2001, *Chem Eng Sci*, **56**: 3641, reproduced by permission of Elsevier Ltd.)

the cooling water composite curve from Figure 24.7 decomposed into two regions, above and below the pinch. Below the pinch from 20 to 40°C, the system requires the full flowrate of 21.5 $kg \cdot s^{-1}$. However, above the pinch, a lower flowrate will solve the problem. Figure 24.8 shows a steeper line drawn against the part of the problem above the pinch. Of the CP of 90 $kW \cdot K^{-1}$ from the cooling tower, after use up to the pinch condition of 40°C, only a CP of 45.7 $kW \cdot K^{-1}$ is used, and the balance ($CP =$ 44.3 $kW \cdot K^{-1}$) is returned directly to the cooling tower. In this way, each part of the problem only uses the minimum amount of cooling water. However, it should again be noted that in minimizing the amount of cooling water, the temperature differences in the coolers are also minimized, leading to trade-offs between the designs of the cooling tower system and the heat exchangers.

This strategy of using the cooling water can then be translated into a cooling water design grid, as shown in Figure 24.9. Three *cooling water mains* are conceptualized, one at the supply temperature of 20°C, a second at the pinch temperature of 40°C and a third at the maximum temperature allowable in the system of 75°C. The flowrate of cooling water required in each of the mains is shown at the top of the main. The flowrate to be returned to the cooling tower from each of the mains is shown at the bottom of the main. The cooling streams are superimposed onto the cooling water mains at their appropriate temperatures. In Figure 24.9, Stream 1, which starts at 20°C and terminates

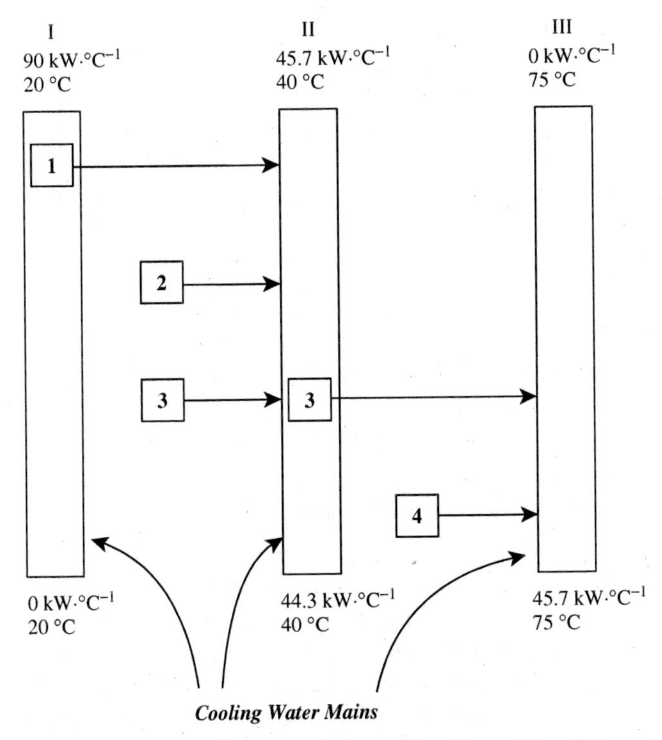

Figure 24.9 Cooling water design grid. (From Kim J-K and Smith R, 2001, *Chem Eng Sci*, **56**: 3641, reproduced by permission of Elsevier Ltd.)

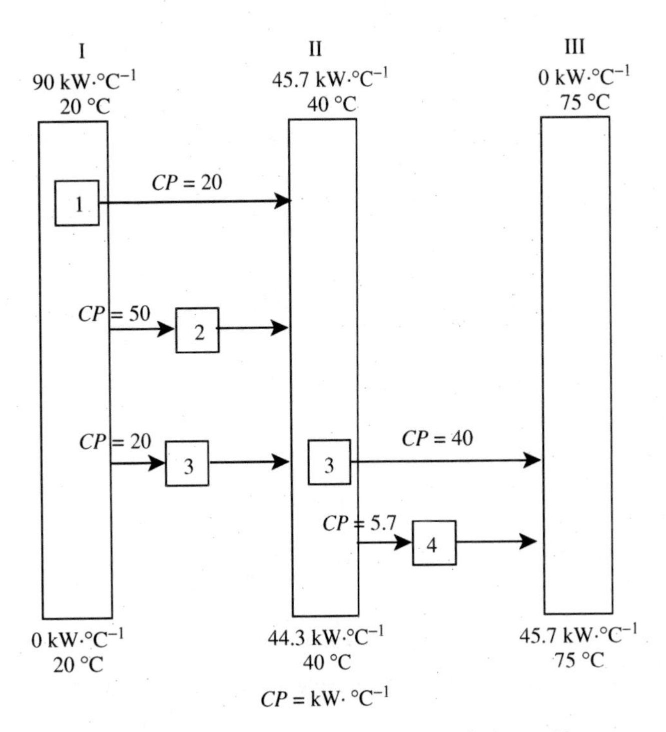

Figure 24.10 The completed cooling water design grid.

at 40°C, is shown between the 20°C and 40°C cooling water mains. Stream 2 starts at 30°C, between the first two cooling water mains. Stream 3 starts at 30°C, again between the first two cooling water mains. However, in this case, Stream 3 terminates at the maximum temperature of 75°C. Initially, therefore, it is broken into two parts, each within the appropriate design region. Finally, Stream 4 starts at 55°C above the pinch water mains and finishes at the 70°C cooling water main. The streams are then connected to the appropriate cooling water main, to satisfy the individual cooling requirement, as shown in Figure 24.10. This provides an initial design for the cooling water network to meet the target requirements.

However, the design has not yet been completed, as there is a fundamental difficulty with the arrangement shown in Figure 24.10. Stream 3 requires a change in flowrate at the pinch cooling water main (40°C). This would mean that the cooling duty corresponding with Stream 3 would have to be broken down into two heat exchangers, each being supplied with different flowrates. The change in flowrate for Stream 3 is in fact easily removed. Consider Figure 24.11 that shows the temperature versus heat duty for an operation with a change in flowrate, as in the case of Stream 3 in Figure 24.10. A heat balance around Part 1 in Figure 24.11 gives (assuming the specific heat capacity of the cooling water to be constant):

$$F_1(T_{PINCH} - T_1) = F_2(T_{PINCH} - T_{in,max}) \qquad (24.13)$$

Moving the mixing junction to the inlet of the operation, as shown in Figure 24.11, and carrying out a heat balance

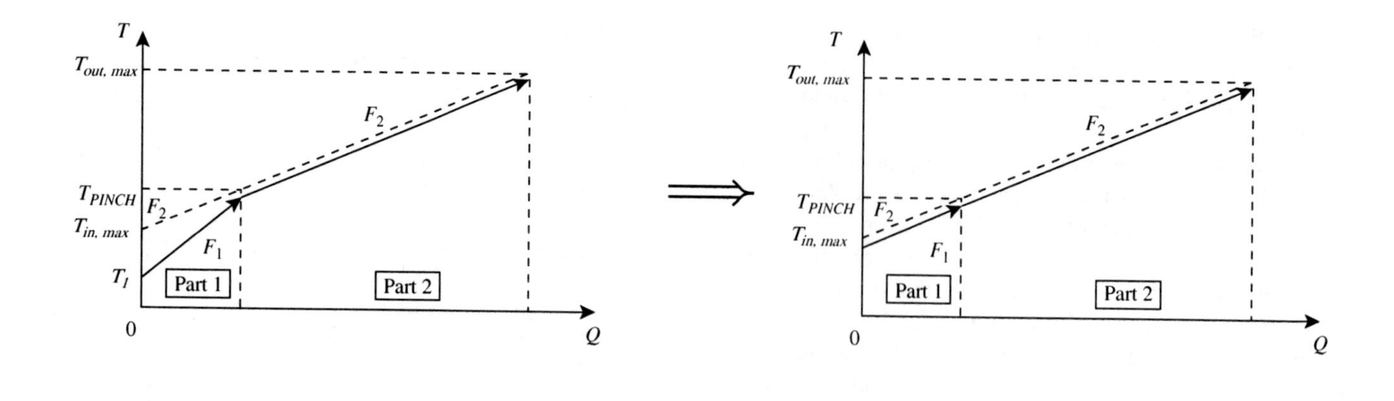

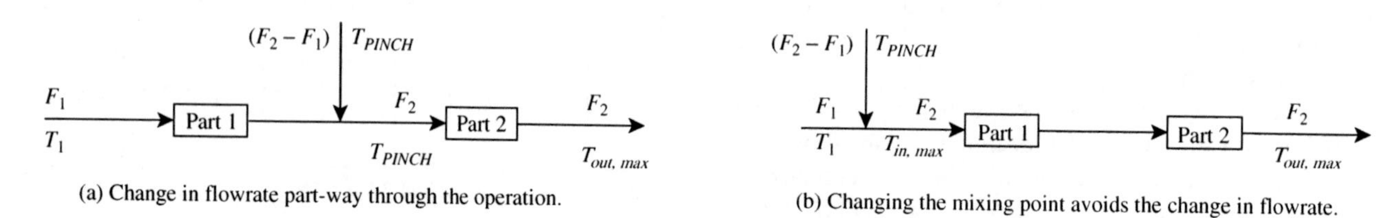

(a) Change in flowrate part-way through the operation.

(b) Changing the mixing point avoids the change in flowrate.

Figure 24.11 Changing the mixing arrangement can avoid the change in cooling water flowrate.

at the new mixing junction gives:

$$T_{in} = \frac{(F_2 - F_1)T_{PINCH} + F_1 T_1}{F_2}$$

$$= \frac{F_2 T_{PINCH} - F_1(T_{PINCH} - T_1)}{F_2} \quad (24.14)$$

Substituting Equation 24.13 into Equation 24.14 gives:

$$T_{in} = \frac{F_2 T_{PINCH} - F_2(T_{PINCH} - T_{in,max})}{F_2}$$

$$= T_{in,max} \quad (24.15)$$

In other words, if the mixing junction is moved from the middle of the operation to the beginning of the operation, there is a constant flowrate throughout the operation corresponding with the maximum inlet temperature, after mixing.

Figure 24.12 shows the corresponding grid diagram to correct the change in flowrate. The change in flowrate for Stream 3 that previously occurred at the pinch temperature mains is now added from the pinch temperature mains to the inlet of Stream 3. This provides a design for the cooling water network that achieves the target minimum flowrate of $CP = 90$ kW·K^{-1}. The arrangement shown in Figure 24.12 involves reuse of cooling water from Streams 1 and 2 into Streams 3 and 4 via a cooling water main at pinch temperature 40°C. An alternative way to arrange the design is to make the connection directly rather than through an intermediate cooling water main. If the intermediate cooling water main is removed, then there are basically two sources of cooling water from Streams 1 and 2 at 40°C and two sinks for cooling water in Streams 3 and 4 at 40°C. These sources and sinks can be connected together in different ways. If the intermediate water main at 40°C is removed from the design, then the streams can be connected directly together. Figure 24.13 shows a flowsheet for one possible arrangement. In Figure 24.13, there is reuse from Heat Exchanger 1 to Heat Exchanger 3 and from Heat Exchanger 2 to Heat Exchanger 4. Some of the cooling water from the cooling tower goes through Heat Exchanger 1 and some bypasses to be mixed in

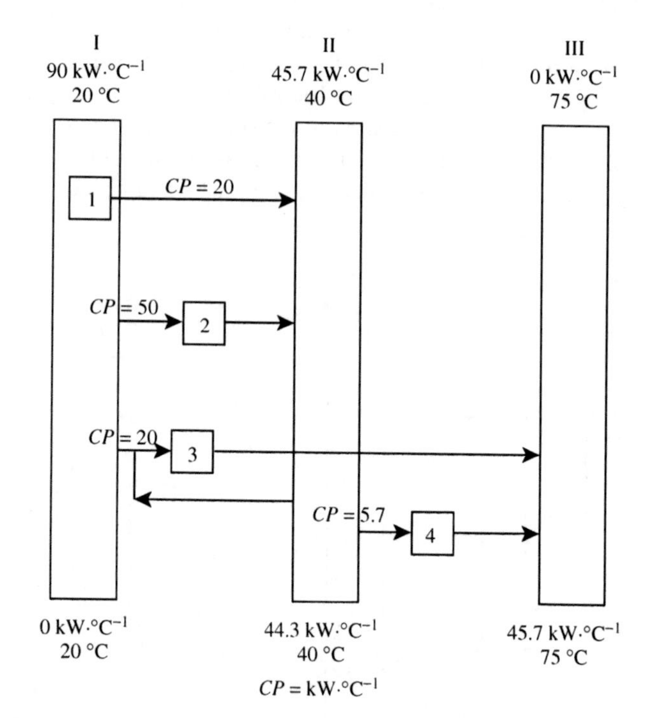

Figure 24.12 The final grid design after changing the mixing arrangement to avoid flowrate changes.

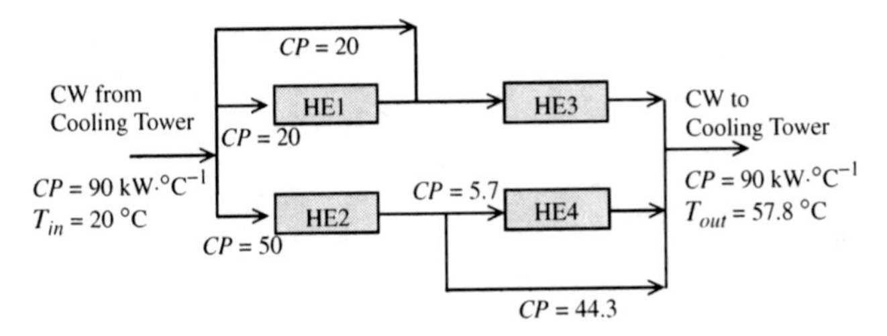

Figure 24.13 Flowsheet of cooling water network with maximum reuse of cooling water. (From Kim J-K and Smith R, 2001, *Chem Eng Sci*, **56**: 3641, reproduced by permission of Elsevier Ltd.)

before entering Heat Exchanger 3. This is necessary to comply with the inlet constraints for Heat Exchanger 3. The flowsheet in Figure 24.13 then needs to be assessed for its practicality and operability. The design can then be evolved. For example, the bypass around Heat Exchanger 1 in Figure 24.13 can be eliminated and all of the $CP =$ 40 kW·K^{-1} can be put through Heat Exchanger 1, before entering Heat Exchanger 3. The essential feature of the design to achieve the target is the reuse of cooling water between exchangers. Different arrangements are possible than the one shown in Figure 24.13 to achieve the target by connecting the sources and sinks at 40°C differently.

Parallel cooling water use that is minimized to achieve ΔT_{min} at the outlet of each heat exchanger leads to a flowrate for the network of 25.4 kg·s^{-1}, with a cooling water return temperature of 52°C. Maximizing the reuse reduces this flowrate to 21.5 kg·s^{-1}, and the cooling water return temperature increases to 57.8°C.

A number of complications need to be addressed that are likely to be encountered.

1. Figure 24.14 shows the stream grid for another problem, which shows a feature not present in the previous example. In this stream grid, Streams 1 and 3 both have maximum outlet temperatures corresponding with one of the cooling water mains. However, Stream 2 has a maximum outlet temperature below the final water main temperature. It might be suspected that the outlet of Stream 2 could be discharged directly to the final cooling water main. However, if this is done, the design will not be feasible and meet the target. The heat balance in Figure 24.8 that determines the cooling water usage from each cooling water main is based on the assumption that all of the cooling water reaches the temperature of the next cooling water main, before it is discharged to that cooling water main. If the water from the outlet of Stream 2 in Figure 24.14 is discharged to the final water main directly, it is not complying with this energy balance and the design will be infeasible. The water from the exit of Stream 2 must therefore find another reuse, in this case to the inlet of Stream 3 before being discharged to the cooling water main at the appropriate cooling water temperature.

2. Another complication that can occur is that the number of design regions can be greater than two. Figure 24.8 shows a design problem involving just two design regions: below the pinch and above the pinch. The cooling water composite curve in Figure 24.15 has three design

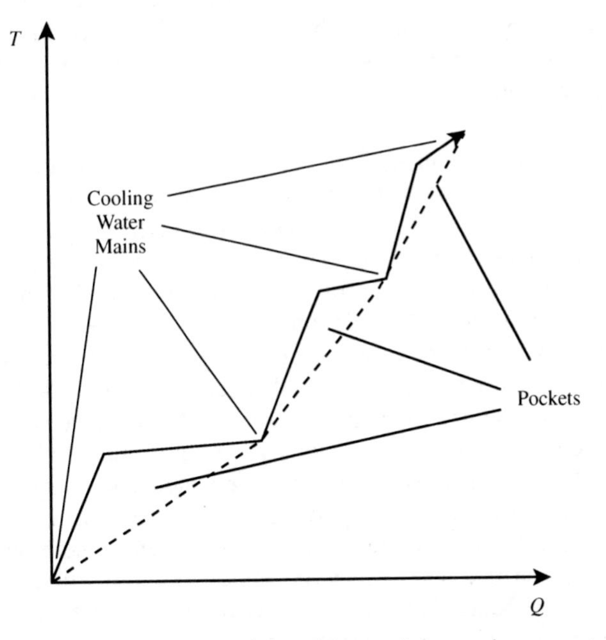

Figure 24.15 More complex problems might require a greater number of cooling water mains.

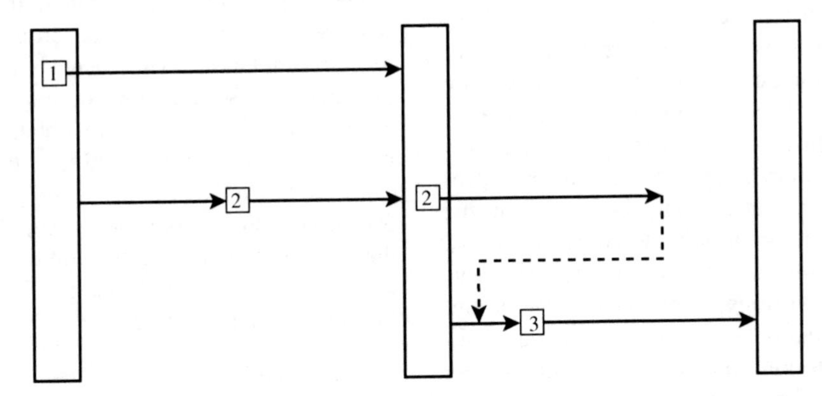

Figure 24.14 Operation outlets must be reused until reaching mains temperature.

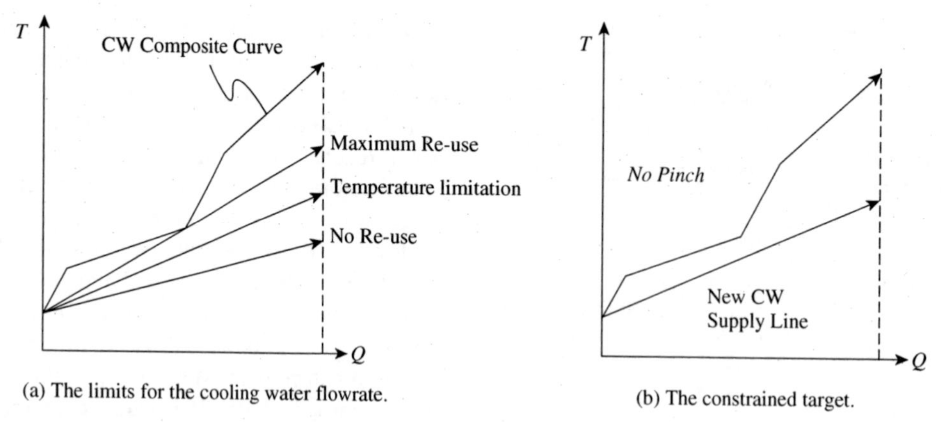

Figure 24.16 Cooling water targets limited by constraints on the return temperature.

regions, rather than two. The design regions are identified by drawing straight lines between the extreme convex points of the cooling water composite curve, as illustrated in Figure 24.15. If more design regions exist, then this requires additional cooling water mains. However, the procedure is basically the same as in the example described above, taking into account the additional mains.

3. Constraints might limit the cooling water return temperature. Figure 24.16a shows a cooling water composite curve with cooling water supply lines matched against it. The highest flowrate corresponds with no reuse (a parallel arrangement). The steepest slope corresponds with minimum flowrate and maximum reuse. It might be the case though that there is a constraint restricting the cooling water return temperature, as illustrated in Figure 24.16a. In this case, the constraint restricting the return temperature prevents the target from achieving minimum flowrate, and hence maximum reuse. Higher temperatures increase the corrosion and fouling potential in cooling systems. The corrosion rate increases with increase in temperature, and corrosion rate approximately doubles for every 10°C rise in temperature. Fouling is also related to temperature. For example, calcium carbonate, which is the most common scaling problem in cooling water systems, has inverse solubility characteristics with temperature. Chemical dosing of the cooling water can to some extent counteract increases in corrosion and fouling. The temperature limitation for the cooling water return might also result from temperature limitations for the packing material in the cooling tower.

The target incorporating such a constraint is straightforward. The maximum return temperature is set, and providing the entire cooling water supply line is below the cooling water composite curve, then the target is feasible and the corresponding flowrate can be calculated. However, this leads to a problem in design. The basis of the design method was to divide the problem into design regions corresponding with the pinch point of ΔT_{min}. In Figure 24.16b, there is only one point for the cooling water supply line featuring ΔT_{min} at the initial temperature of the cooling water. There is no pinch in the problem. A design strategy to overcome this problem is to shift the cooling water composite curve by a *temperature shift*, to artificially create a pinch for the system[2]. The problem now though is how to modify the composite curve with a temperature shift and how to find the new pinch. For the example in Table 24.1, suppose there is a limitation on the cooling water return temperature to be less than 55°C. The cooling water composite curve needs to be shifted down to create the pinch for the system, as illustrated in Figure 24.17. A simple energy balance allows the new pinch temperature to be calculated:

$$\frac{T^*_{PINCH} - 20}{55 - 20} = \frac{1800}{3400}$$

$$T^*_{PINCH} = 38.5^\circ\text{C} \qquad (24.16)$$

The new pinch temperature is 38.5°C, involving a temperature shift of 1.5°C. Cooling water Streams 1, 2 and 3 all take part in creating the original pinch, which means that these streams are candidates for the corresponding temperature shift. The limiting cooling water modifications are in two stages. This is illustrated in Figure 24.18. Starting from the original limiting cooling water profile in Figure 24.18a, the calculated temperature shift is applied (1.5°C for this example), Figure 24.18b. Modified profiles might cross the cooling water supply line, and thus another step is needed. In the second stage, the flowrate of the limiting water profile is increased when the shifted profile is restricted by temperature limitations. This increase in flowrate is illustrated in Figure 24.18c. For the example in Table 24.1, Streams 2 and 3 can be modified to obtain new limiting cooling water profiles simply by shifting the temperatures. However, for Stream 1, it is necessary to increase the flowrate, because the 20°C cooling water supply temperature restricts the temperature shift of the limiting data. A heat balance determines the required increase in flowrate and the new limiting exit temperature:

$$CP^* = CP\frac{T_{PINCH} - T_{CW,in}}{T^*_{PINCH} - T^*_{CW,in}} \qquad (24.17)$$

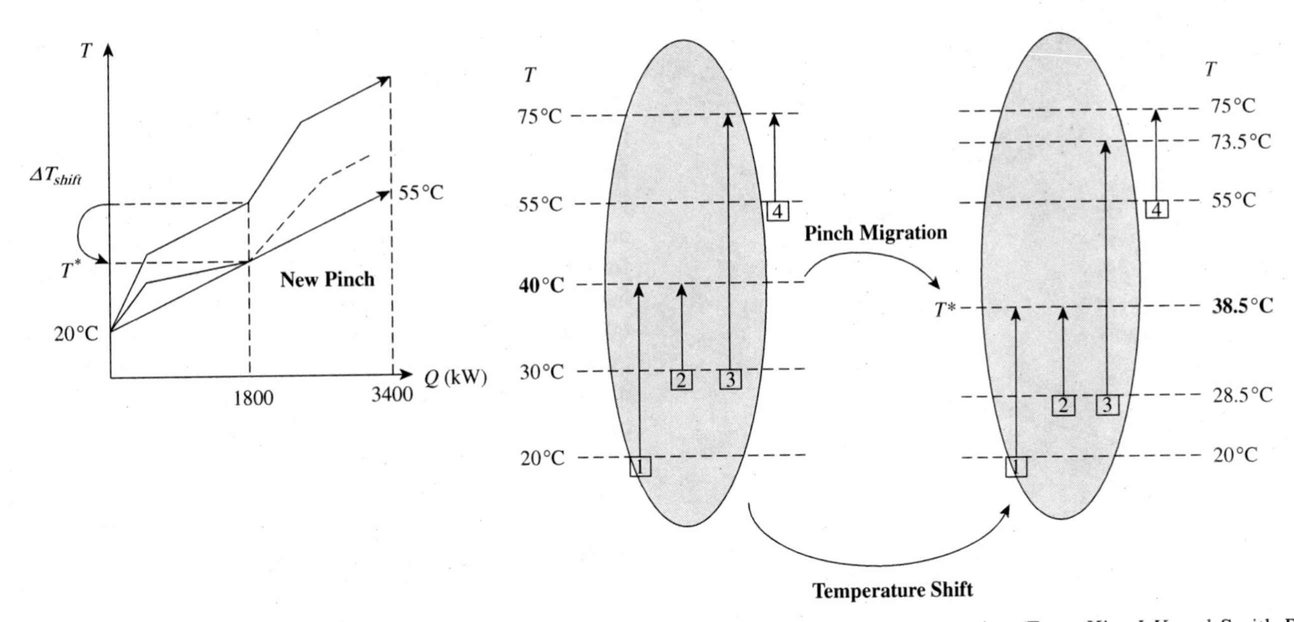

Figure 24.17 Temperature shift and pinch migration from cooling water return temperature constraint. (From Kim J-K and Smith R, 2001, *Chem Eng Sci*, **56**: 3641, reproduced by permission of Elsevier Ltd.)

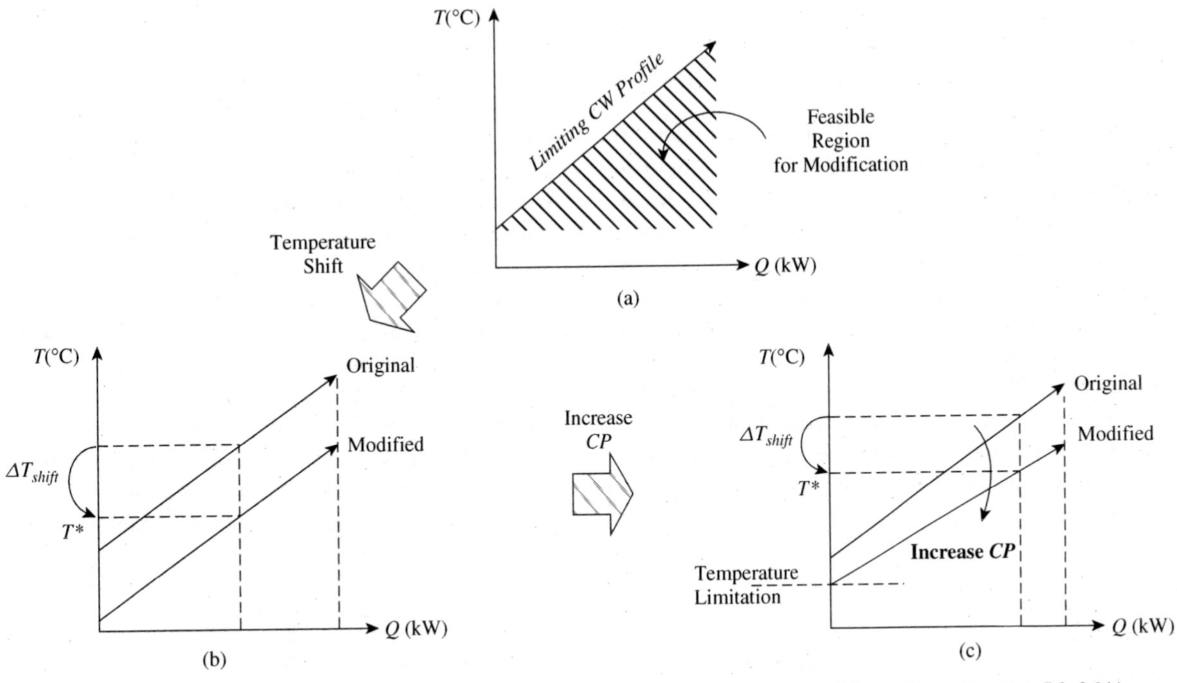

Figure 24.18 Stream modification from the temperature shift. (From Kim J-K and Smith R, 2001, *Chem Eng Sci*, **56**: 3641, reproduced by permission of Elsevier Ltd.)

where CP^* = modified heat capacity flowrate
T^*_{PINCH} = modified pinch temperature
$T^*_{CW,in}$ = modified cooling water inlet temperature

$$T^*_{CW,out} = CP\frac{T_{CW,out} - T_{CW,in}}{CP^*} + T^*_{CW,in}$$

where $T^*_{CW,out}$ = modified cooling outlet temperature

The shifted limiting cooling water data are given in Table 24.2.

Table 24.2 Temperature-shifted limiting cooling water data.

Heat exchanger	$T_{CW,in}$ (°C)	$T_{CW,out}$ (°C)	CP (kW·K^{-1})	Q (kW)
1*	20	38.5	21.6	400
2*	28.5	38.5	100	100
3*	28.5	73.5	40	1800
4	55	75	10	200

(* modified)

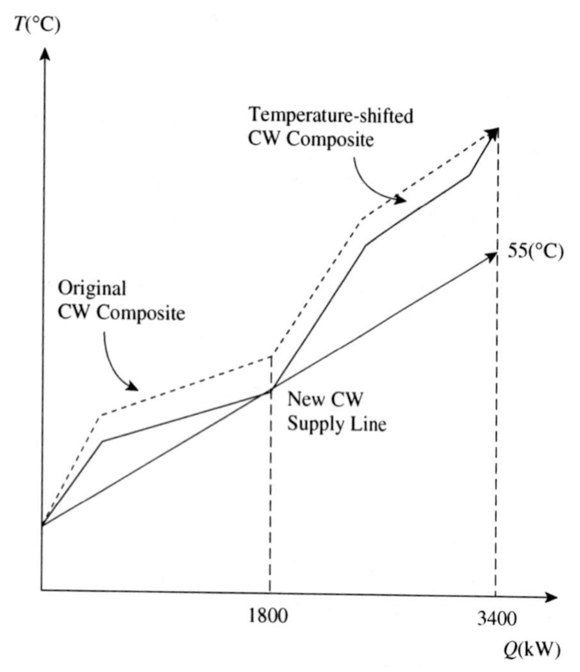

Figure 24.19 Temperature-shifted cooling water composite curve. (From Kim J-K and Smith R, 2001, *Chem Eng Sci*, **56**: 3641, reproduced by permission of Elsevier Ltd.)

For Stream 1 in Table 24.2, the *CP* has increased from 20 $kW \cdot K^{-1}$ to 21.6 $kW \cdot K^{-1}$. The cooling water composite curve corresponding with the data in Table 24.2 is shown in Figure 24.19. This compares the original cooling water composite curve with the temperature-shifted curve. The cooling water supply line now touches the cooling water composite curve at 38.5°C. A design for this condition, developed in the same way as described previously, is shown in Figure 24.20. Again, this is only one of a number of possible options for network structures to achieve the target. The network structures in Figures 24.13 and 24.20 are slightly different, corresponding with the different design objectives.

It should be emphasized that the network structures created here can always be evolved to simplify the structure. The simplest structure of all is the parallel structure, but this brings penalties in performance of the cooling tower. The parallel structure is also the easiest to control. Evolution of the maximum reuse network toward a parallel arrangement might bring penalties in performance in return for network simplicity. The designer must strike a balance between network simplicity and system performance and take account of the change in pressure drop. Also, the temperature difference in the heat exchangers will decrease, decreasing the performance of the heat exchangers. The optimum design requires all of these issues to be traded off to obtain the minimum cost design. The most important thing to be aware of is that cooling water can be used in parallel or series, and a few critical reuse opportunities might be extremely effective in improving the design.

24.5 RETROFIT OF COOLING WATER SYSTEMS

Situations are often encountered when cooling water networks need to increase the heat load of individual coolers, or a new heat exchanger is introduced into an existing system. In such situations, cooling water systems can become bottlenecked. This might call for a new cooling tower to be built or a new cell added to the existing tower. However, this is only one of many possible options to debottleneck the system.

As the increase of cooling load influences the cooling tower performance, and there are interactions between cooling water networks and the cooling tower, the best solution might be obtained by modifying the cooling water network. Changing inlet cooling water conditions from high flowrate and low temperature to low flowrate and high temperature can increase the heat removal of the cooling tower. This will, in general, require changing the cooling

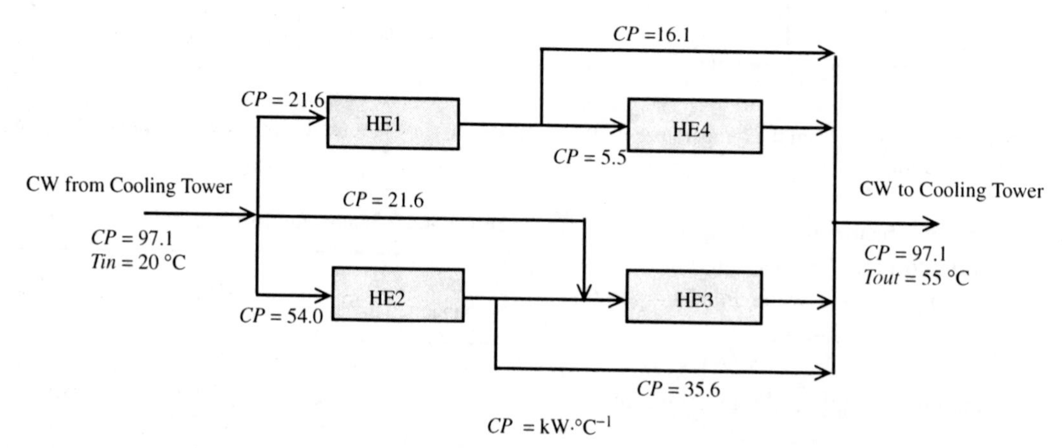

Figure 24.20 Cooling water network design with target return temperature of 55°C. (From Kim J-K and Smith R, 2001, *Chem Eng Sci*, **56**: 3641, reproduced by permission of Elsevier Ltd.)

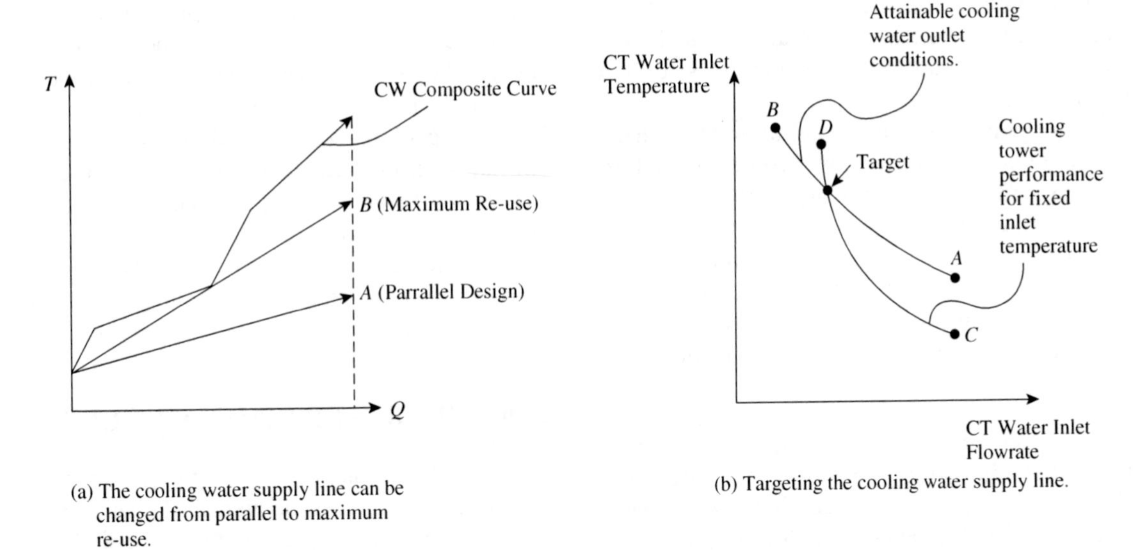

Figure 24.21 Targeting cooling water flowrate in retrofit. (From Kim J-K and Smith R, 2001, *Chem Eng Sci*, **56**: 3641, reproduced by permission of Elsevier Ltd.)

water network design from parallel to series or mixed parallel/series arrangements with reuse of cooling water, decreasing the flowrate of cooling water and increasing the return temperature. By changing from parallel arrangements to cooling water reuse designs, the heat removal of the cooling tower can be increased, without any energy penalty, and perhaps without investment in a new cooling tower or cooling tower cell.

A design procedure for debottlenecking cooling water systems can readily be developed[2]. The cooling water composite curve can first be constructed from the limiting cooling water data at the new conditions, as described previously. The cooling water network performance can be changed within a feasible region that is bounded by the maximum reuse supply line (or maximum cooling tower return temperature) and the parallel design supply line (i.e. the flowrate corresponding with parallel design). In Figure 24.21a, *AB* represents the attainable outlet conditions by changing the cooling water network design configuration from parallel to maximum reuse. It is necessary to know how the inlet conditions affect the cooling tower performance.

The cooling water supply line has the same heat load from the viewpoint of the cooling water network. But the heat removal of the cooling water system is changed as the inlet conditions to the cooling tower are changed. The heat removal of the cooling tower increases as the design configuration changes from parallel to maximum reuse (or maximum return temperature).

The next stage is to find the target supply conditions for the cooling tower. This can be achieved by constructing a simulation model of the cooling tower to simulate the conditions of the exit water and air for given inlet air and water conditions[2]. The model must also consider the influence of the cooling tower makeup and blowdown on the overall heat balance for the cooling system[2]. The feasible cooling water supply line can move from *A* to *B* in Figure 24.21a. This gives the corresponding profile of cooling tower inlet temperature versus inlet flowrate shown in Figure 24.21b. The cooling tower performance (including the makeup and blowdown streams) is then simulated for fixed cooling tower outlet temperature (the desired supply temperature) for a given flowrate to the tower, and the corresponding outlet temperature calculated. By varying the inlet flowrate and fixing the cooling tower inlet temperature, the corresponding outlet temperature can be calculated for a range of conditions. This is line *CD* in Figure 24.21b. The target, where the heat removal of the cooling system is the same as the heat load of the cooling water network, is where the two curves cross in Figure 24.21b. At this point, the inlet temperature to the cooling water network is satisfied for the new conditions.

At target conditions, the cooling demand of the network at the new conditions is satisfied without additional cooling tower capacity. Below the target temperature, the current cooling systems cannot operate for the required cooling demand. If the target conditions can be achieved at the new conditions, then the cooling water network can be designed with the target conditions. The reconfigured cooling water network should, in principle, be able to cope with the increased load on the cooling system without investment in a new cooling capacity. However, there will need for investment in piping changes and control systems to reconfigure the network. The heat exchangers will need to be checked to see if they can operate under the changed heat transfer conditions (cooling water temperature and flowrate). Also, the cooling water pumps will need to be checked to ensure that they can operate with the new conditions of flowrate and pressure drop. The retrofit of cooling water systems thus involves trade-offs, including cooling tower costs, pressure drops, piping costs, design complexity, and so on.

The debottlenecking procedure for cooling water systems maintains high temperature and low flowrate of return cooling water to increase the heat removal capacity of the cooling tower. In some situations, this might not be adequate to satisfy increased cooling demands, or the required network changes might not be practical or economic. But before investment in new cooling tower capacity is considered, other options should be evaluated, perhaps in conjunction with cooling water network changes.

1. *Hot blowdown extraction.* If the cold blowdown is changed to hot blowdown (i.e. blowdown taken from the cooling tower return rather than the cooling tower basin), the heat load of the cooling tower is reduced because the flowrate of the hot cooling tower is decreased. The required hot blowdown might exceed the original cold blowdown flowrate. This will result in an increase of makeup water and decrease of cycles of concentration. Also, the resulting increase in effluent temperature might be unacceptable.

2. *Introduction of air-cooled heat exchangers.* If the temperature of the hot return cooling water can be decreased, the heat duty on the cooling tower will also be decreased. To decrease the cooling water return temperature, air-cooled heat exchangers can be installed between the cooling tower and the cooling water network. Air-cooled heat exchangers have been discussed in Chapter 15. The flowrate of the return hot cooling water does not change as a result of the air-cooled heat exchanger, but the temperature is decreased. Hot blowdown tends to be more effective than air-cooled heat exchangers. The capacity of the cooling tower is used more effectively in the case of hot blowdown, because the inlet conditions of the cooling tower favor high temperature and low flowrate. However, hot blowdown incurs penalties from an increase in makeup water and effluent temperature.

Alternatively, air-cooled heat exchangers can be installed within the cooling water network to replace cooling water coolers. This has the effect of reducing the cooling load on the cooling tower and, in principle, allows the flowrate of cooling water to be decreased.

3. *Other debottlenecking design options.* The use of cold seawater is yet another way to decrease the return temperature of the hot cooling water. When cold seawater is available to use as a cooling medium, the air-cooled heat exchanger may be replaced with a cooler using cold seawater.

It should also not be forgotten that the cooling tower itself leaves room to improve the cooling tower performance. For example, the packing can be changed to one with a higher efficiency, to provide greater surface area between the air and water. Also, improving the water distribution system across the cooling tower packing to provide a more uniform distribution pattern can improve the performance. Finally, the performance of the air fan can be improved to increase the induced/forced air flowrate. The driving force for cooling is increased when the ratio of water flowrate to air flowrate in the cooling tower is decreased. An increase in the air flowrate is therefore an alternative way to increase the driving force for cooling, and consequently the heat removal of the cooling tower. Thus, greater cooling can be obtained by upgrading the water- and air-distribution systems.

24.6 REFRIGERATION CYCLES

A refrigeration system is a heat pump with the purpose of providing cooling at temperatures below that which can normally be achieved using cooling water or air cooling[4–6]. For most applications, such cooling is required below ambient temperature. Thus, the removal of heat using refrigeration leads to its rejection at a higher temperature. This higher temperature might be to an external cooling utility (e.g. cooling water), a heat sink within the process or to another refrigeration system. Refrigeration is important in chemical engineering operations where low temperatures are required to condense gases[6], crystallize solids, control reactions, and so on, and for the preservation of foods, air conditioning, and so on. Generally, the lower the temperature of the cooling required to be serviced by the refrigeration system and the larger its duty, the more complex the refrigeration system. Processes involving the liquefaction and separation of gases (e.g. ethylene, liquefied natural gas, etc.) usually have complex refrigeration systems.

There are two broad classes of refrigeration system:

- compression refrigeration
- absorption refrigeration.

Compression refrigeration cycles are by far the most common and absorption refrigeration is only applied in special circumstances.

Consider first a simple ideal compression refrigeration cycle using a pure component as the refrigerant fluid, as illustrated in Figure 24.22a. Process cooling is provided by a cold liquid refrigerant in the *evaporator*. A mixture of vapor and liquid refrigerant at Point 1 enters the evaporator where the liquid vaporizes and produces the cooling effect before exiting the evaporator at Point 2. The refrigerant vapor is then compressed to Point 3. At Point 3, the vapor is not only at a higher pressure but is also superheated by the compression process. After the compressor, the vapor refrigerant enters a *condenser* where it is cooled and condensed to leave as a saturated liquid at Point 4 in the cycle. The liquid is then expanded to a lower pressure to Point 1 in the cycle. The expansion process partially vaporizes the liquid refrigerant across

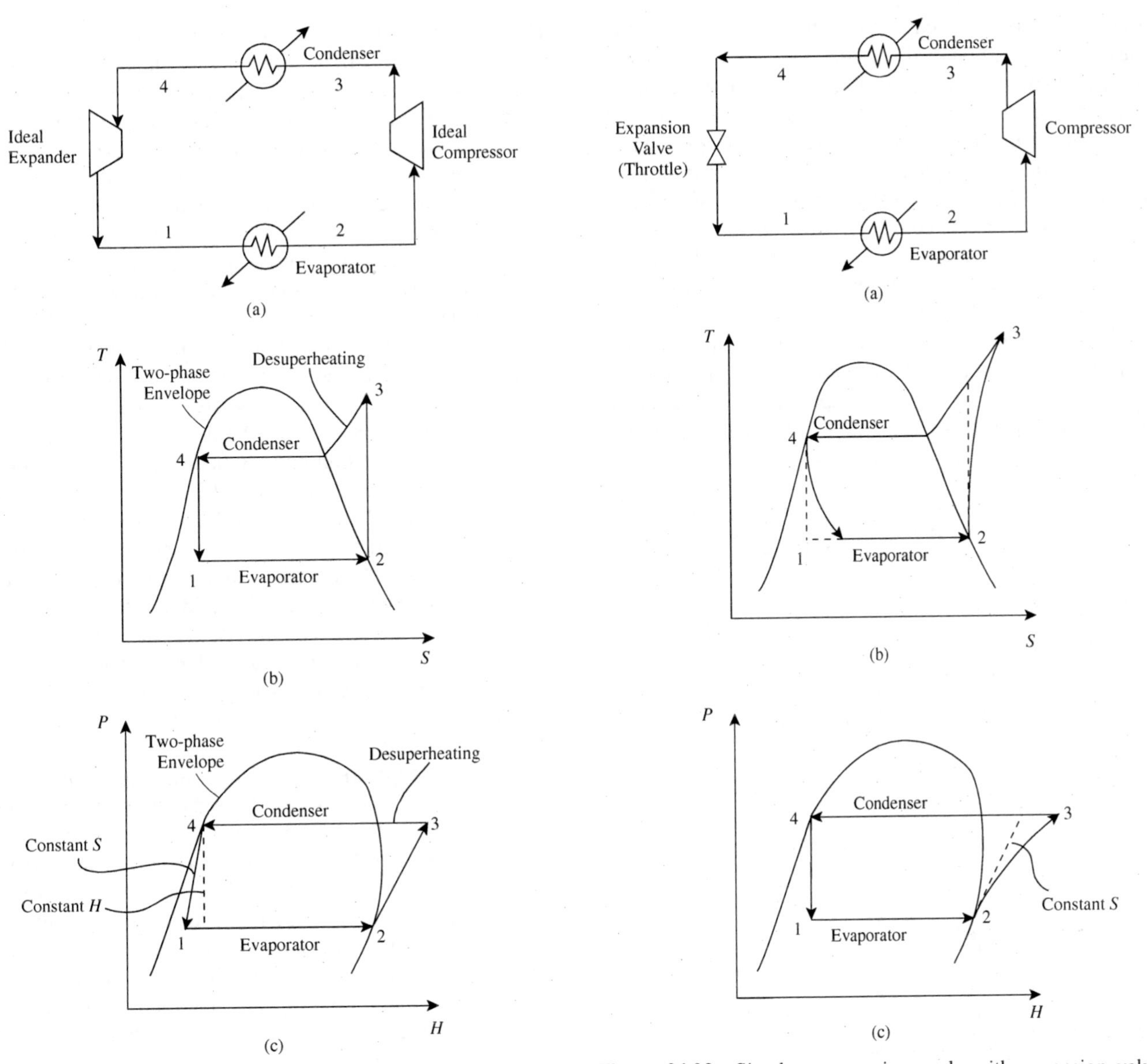

Figure 24.22 A simple (ideal) compression refrigeration cycle.

Figure 24.23 Simple compression cycle with expansion valve and nonideal compression.

the expander, producing a cooling effect to provide refrigeration, and the cycle continues. Figure 24.22b shows the cycle on a temperature-entropy diagram. The diagram shows a two-phase envelope, inside of which the refrigerant is present as both vapor and liquid. To the left of the two-phase envelope, the refrigerant is liquid, and to the right of the two-phase envelope, the refrigerant is vapor. Starting at Point 1 with a two-phase mixture, the refrigerant enters the evaporator where there is an isothermal vaporization to Point 2. From Point 2 to Point 3, there is a compression process that, at this stage, is assumed to be ideal. This is an isentropic process finishing at Point 3. From Point 3, the refrigerant is first cooled to remove the superheat and is then condensed to Point 4, where there is a saturated liquid. The refrigerant is expanded from Point 4 to Point 1, which is assumed here to be an ideal expansion with no increase in entropy. In Figure 24.22c, the same cycle is represented on a pressure-enthalpy diagram. Again the two-phase envelope is evident. From Point 1 to Point 2, the enthalpy of the refrigerant increases in the evaporator. From Point 2 to Point 3, the pressure is increased in the compressor with a corresponding increase in the enthalpy. From Point 3 to Point 4, there is no change in pressure but a decrease in the enthalpy in the desuperheating and the condensation. Between Point 4 and Point 1, there is a decrease in pressure in the expander. The constant entropy assumption leads to a decrease in the enthalpy to Point 1. The cycle in Figure 24.22 assumed ideal isentropic compression and expansion processes. Next consider nonideal compression and expansion processes.

Figure 24.23a again shows a simple compression cycle involving a pure refrigerant fluid, but now with an *expansion valve* or *throttle*, rather than an ideal expander. It also features a nonideal compression. On a temperature-entropy diagram as shown in Figure 24.23b, the compression process now features an increase in entropy. The expansion also features an increase in entropy. On the pressure-enthalpy diagram in Figure 24.23c, the nonideal compression features a greater enthalpy change between Points 2 and 3 when compared with an isentropic compression. The expansion between Points 4 and 1 is now assumed to be isenthalpic, leading to a vertical line on the pressure-enthalpy diagram.

Figure 24.24a shows the corresponding simple compression cycle, but with subcooled condensate. The cycle is basically the same as before, except that the liquid leaving the condenser is now subcooled rather than being saturated.

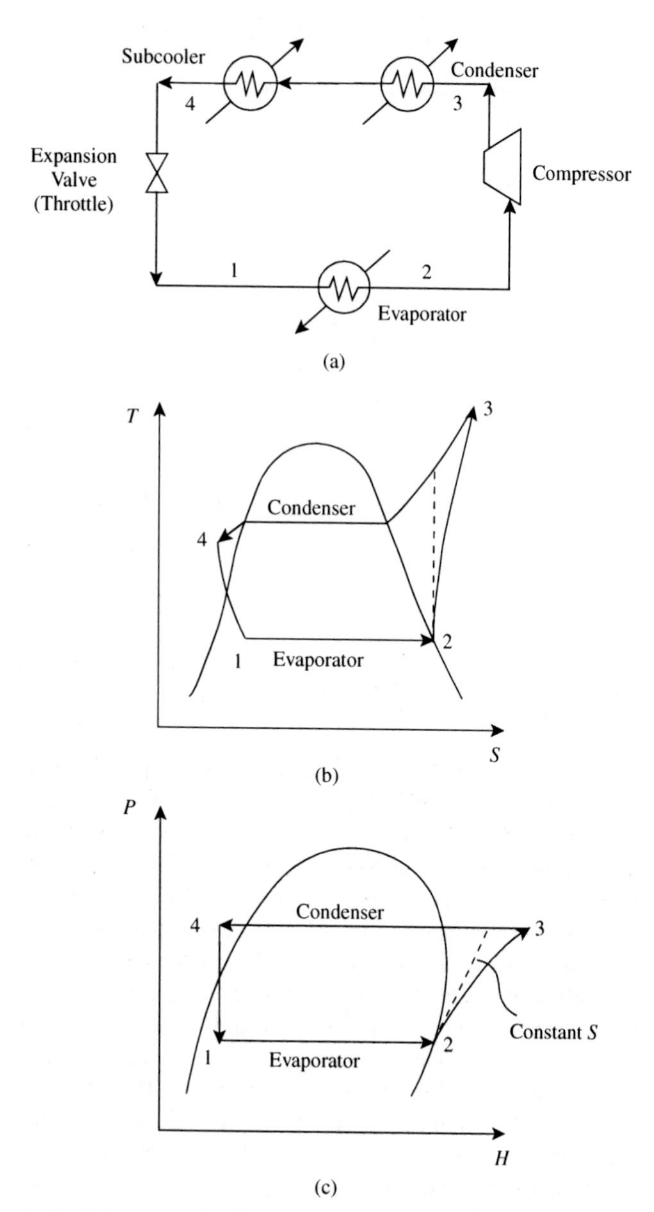

Figure 24.24 Simple compression cycle with subcooled condensate.

The subcooling effect can be seen on the temperature-entropy in Figure 24.24b and pressure-enthalpy diagram in Figure 24.24c. Subcooling the liquid in the way shown in Figure 24.24 brings a benefit to the cycle. The fact that the liquid is subcooled when entering the expansion process means that there is less vaporization in the expansion process to produce refrigerant at the required temperature. This leads to an increase in the proportion of the liquid refrigerant at Point 1 entering the evaporator. The liquid refrigerant provides the cooling and the vapor present does not provide a cooling effect as the vapor enters and leaves at the same temperature without change in enthalpy. If a greater proportion of the refrigerant entering the evaporator is liquid, it means that the flowrate around the cycle for a given refrigeration duty can be decreased. This, in turn, reduces the compression duty. Subcooling the liquid in this way therefore increases the overall efficiency of the cycle.

In Figures 24.22a, 24.23a and 24.24a, the expansion of the liquid refrigerant produces a two-phase vapor–liquid mixture that is fed directly to the evaporator to provide process cooling. For a vapor–liquid mixture of a pure refrigerant at saturated conditions, the vapor provides no cooling, as it enters and leaves the evaporator at the same temperature without change in enthalpy (neglecting any pressure drop). Only the vaporization of the liquid produces cooling. Some designs of refrigeration use a vapor–liquid separation drum after the expansion of the refrigerant. The vapor from the separation drum is then fed directly to the compressor. The liquid from the separation drum is fed to the evaporator and then to the compressor after vaporization. This arrangement with a vapor–liquid separator is thermodynamically equivalent to the design without a separator (neglecting pressure drops). The separation drum adds capital cost and complexity to the design without bringing any thermodynamic benefit. Whether such a separation drum is used largely depends on the type of heat exchanger used for the evaporator. For example, if the evaporation takes place on the shell-side of a kettle reboiler, then feeding a two-phase vapor–liquid mixture does not present any problems and a vapor–liquid separator would not normally be justified. However, if the evaporation takes place in the channels of a plate-fin heat exchanger, then feeding a two-phase vapor–liquid mixture would present problems, as it is difficult to distribute a two-phase mixture evenly across a manifold of channels. Under such circumstances, a vapor–liquid separation drum would normally be used.

The performance of refrigeration cycles is measured as a *coefficient of performance* (COP_{REF}), as illustrated in Figure 24.25. The coefficient of performance is the ratio of cooling duty performed per unit power required.

$$COP_{REF} = \frac{Q_C}{W} \tag{24.18}$$

where Q_C = cooling duty
W = refrigeration power required

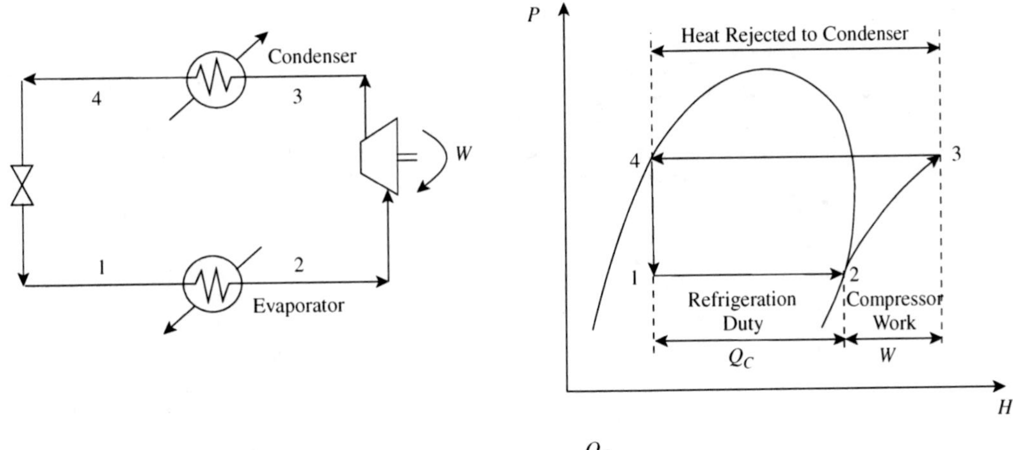

$$\text{Coefficient of Performance } (COP) = \frac{Q_C}{W}$$

Figure 24.25 Performance of practical refrigeration cycles.

The higher the coefficient of performance, the more efficient is the refrigeration cycle. An ideal coefficient of performance can be defined by:

$$\text{Ideal } COP_{REF} = \frac{Q_C}{W} = \frac{T_{EVAP}}{T_{COND} - T_{EVAP}} \tag{24.19}$$

where T_{EVAP} = evaporation temperature (K)
T_{COND} = condensing temperature (K)

Actual performance is typically 0.6 of ideal, but can be much lower for complex cycles operating at extremely low temperatures. Thus:

$$\text{Approximate } COP_{REF} = \frac{Q_C}{W} = \frac{0.6T_{EVAP}}{T_{COND} - T_{EVAP}} \tag{24.20}$$

Example 24.1 Estimate the coefficient of performance and power requirement for a refrigeration cycle with $T_{EVAP} = -30°\text{C}$, $T_{COND} = 40°\text{C}$ and $Q_{EVAP} = 3$ MW.

Solution

$$COP_{REF} \approx 0.6\frac{243}{(313 - 243)} = 2.08$$

$$W \approx \frac{3}{2.08} = 1.4 \text{ MW}$$

It is obvious from Equation 24.20 that the larger the temperature difference across the refrigeration cycle ($T_{COND} - T_{EVAP}$), the lower will be the coefficient of performance and the higher will be the power requirements for a given cooling duty.

For the performance of practical refrigeration cycles, rather than assuming 0.6 of the ideal performance, the refrigeration cycle can be followed using thermodynamic properties of the refrigerant to obtain a more accurate calculation of the coefficient of performance. This will be discussed later.

Simple cycles like the ones discussed so far can be used to provide cooling as low as typically −40°C. For lower temperatures, complex cycles are normally used.

One way to reduce the overall power requirement of a refrigeration cycle is to introduce multistage compression and expansion, as shown in Figure 24.26a. The expansion is carried out in two stages with a vapor–liquid separator between the two stages, often called an *economizer*. Vapor from the economizer passes directly to the high-pressure compression stage. Liquid from the economizer passes to the second expansion stage. The introduction of an economizer reduces the vapor flow in the low-pressure compression stage. This reduces the overall power requirement. Figure 24.26a also shows an intercooler for the vapor between the low-pressure and the high-pressure compression stages. This intercooler reduces further the compressor power in the high-pressure compression stage, reducing the overall power requirement. The cycle is shown as a pressure-enthalpy diagram in Figure 24.26b.

Figure 24.27a shows another way to reduce the overall power requirement of a refrigeration cycle by introducing multistage compression and expansion. Again the expansion is carried out in two stages with a vapor–liquid separator between the two. However, this time the cooled liquid and vapor from the first expansion stage is contacted directly with the compressed vapor from the low-pressure compression stage in a *presaturator*. The presaturator cools the vapor from the low-pressure compression stage by direct contact and acts as a vapor–liquid separator between the stages. The vapor leaving the presaturator being passed to the high-pressure compression stage is saturated. Liquid from the presaturator passes to the second expansion stage. Again, the introduction of a presaturator reduces the vapor flow in the low-pressure compression stage, reducing the overall power requirement. The cycle is shown in Figure 24.27b as a pressure-enthalpy diagram.

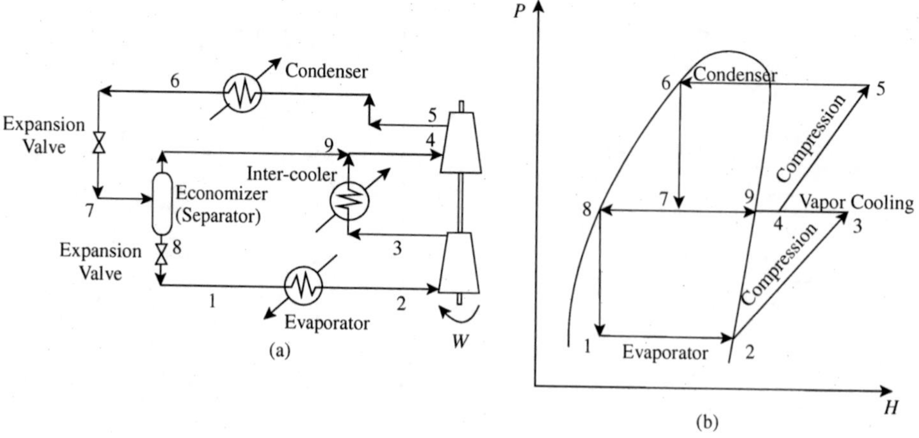

Figure 24.26 Multistage compression and expansion with an economizer.

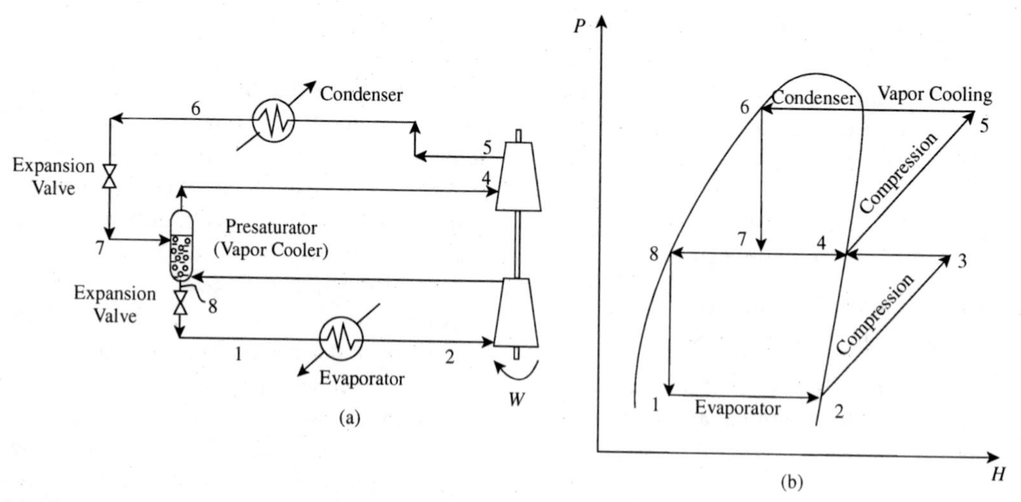

Figure 24.27 Multistage compression and expansion with a presaturator.

Figure 24.28a shows a *cascade cycle*. This involves two refrigerant cycles linked together. Each cycle will use a different refrigerant fluid. The low-temperature cycle provides the process cooling in the evaporator and rejects its heat to the other cycle, which rejects its heat in the condenser to cooling water. Cascade cycles are used to provide very low temperature refrigeration where a single refrigerant fluid would not be suitable to operate across such a wide temperature range. In the cascade in Figure 24.28a, each cycle comprises a simple cycle. The cascade cycle is represented in Figure 24.28b in a pressure-enthalpy diagram. The temperature of the interface between the two cycles is normally specified in terms of the temperature of the evaporator of the upper (high-temperature) cycle and is known as the *partition temperature*. The partition temperature is an important degree of freedom.

Complex refrigeration cycles can be built up in different ways. Figure 24.29 shows a cascade cycle with multistage compression and expansion of the high-temperature cycle.

Figure 24.30a shows a *two-level* refrigeration cycle. Level 1 and Level 2 provide process cooling at different temperatures. This reduces the refrigerant flow through the low-pressure compression stage and reduces the overall power requirements. It is represented in Figure 24.30b in a pressure-enthalpy diagram.

To illustrate one of the many ways the different features can be put together, Figure 24.31a shows a two-level cascade system with presaturators. The evaporators in Level 1 and Level 2 provide process cooling at different temperatures. Two cycles are cascaded together, each separate cycle having multistage compression involving presaturators. The cycle is shown in a pressure-enthalpy diagram in Figure 24.31b. This is only one of many possible complex arrangements that combine the various features.

24.7 PROCESS EXPANDERS

If a gas or vapor process stream is available at a high pressure and downstream conditions do not require this high pressure, the stream can be expanded to provide useful cooling. The cooling might allow partial condensation for recovery of the less volatile components in a mixture, or to

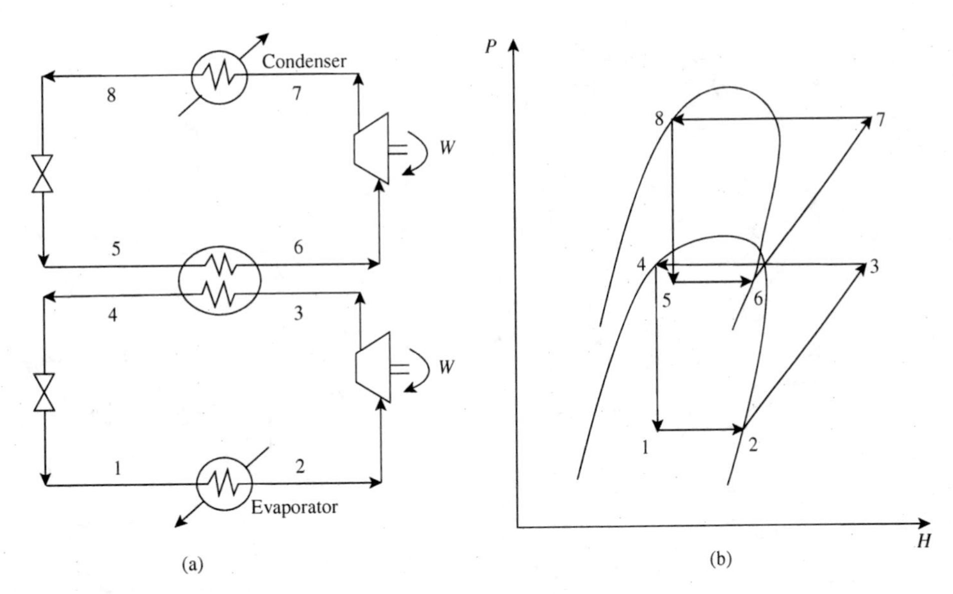

Figure 24.28 A cascade cycle.

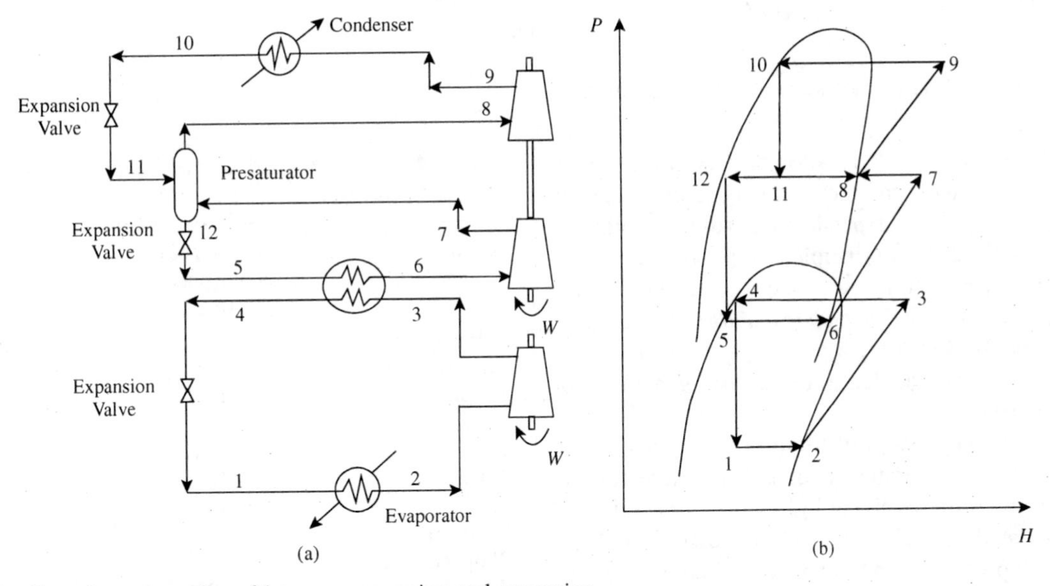

Figure 24.29 Cascade cycle with multistage compression and expansion.

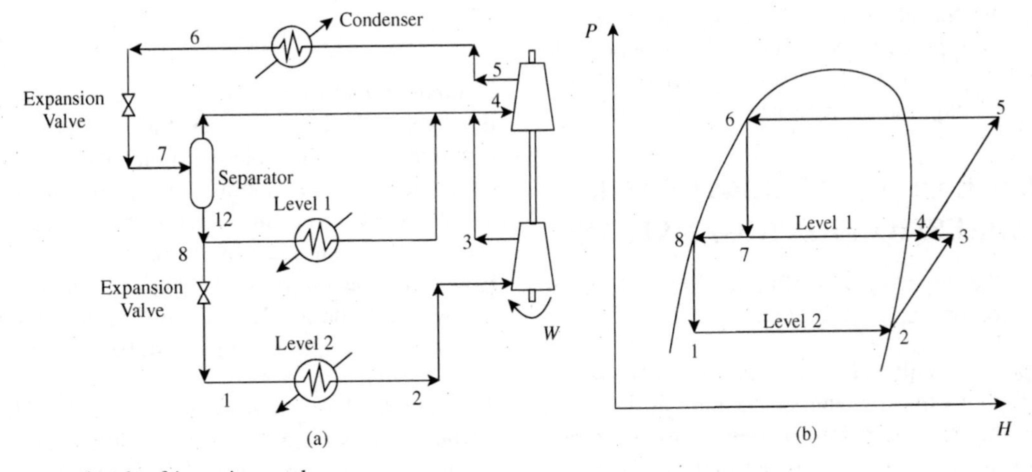

Figure 24.30 A two-level refrigeration cycle.

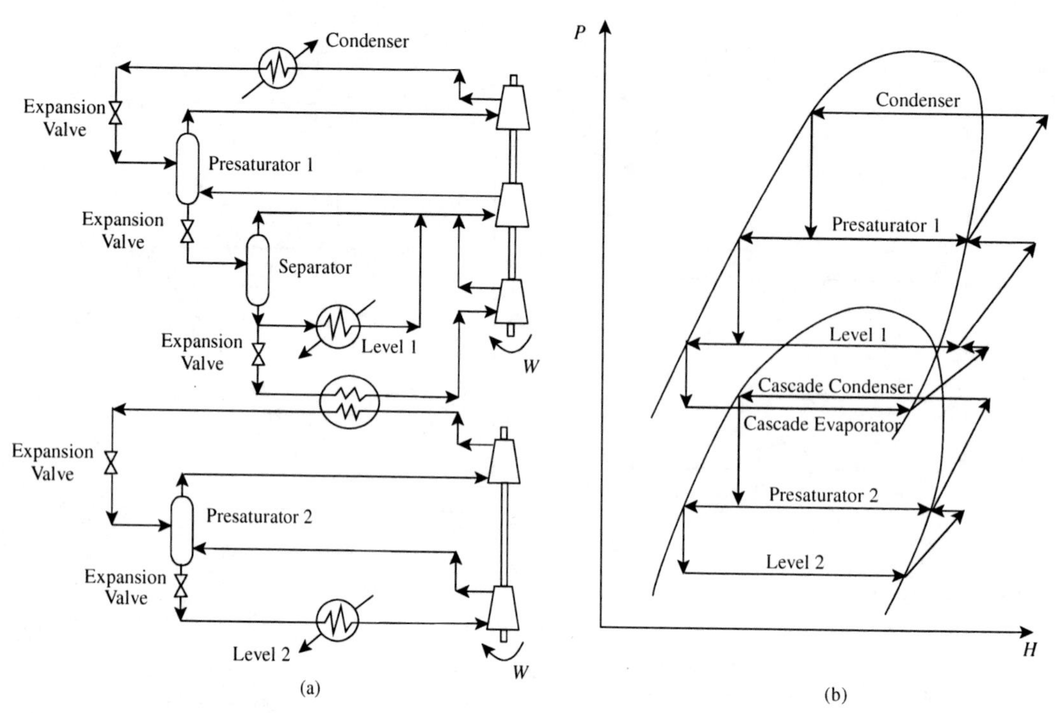

Figure 24.31 A two-level cascade system with presaturators.

provide a cold stream for refrigeration purposes. Expansion across a throttle valve provides cooling only, whereas expansion across a process expander can provide cooling as well as power recovery. In principle, the process expander can be a reciprocating piston and cylinder design but is most often a radial flow turbine or *turboexpander*. The process expander can be used to drive a process machine (e.g. gas compressor) or can be coupled to an electric generator for the production of electricity.

As an example, suppose a distillation is required to be carried out at low temperature and high pressure for the separation of a mixture involving light gases. Uncondensed light gases from the condenser at high pressure, and not required at high pressure, can be expanded to cool the gases, which in turn can be used, for example, to precool the feed to the distillation.

Substituting the throttle valve in a compression refrigeration cycle by a process expander both increases the efficiency of the cycle and allows power recovery. However, this is at the expense of increased capital cost.

24.8 CHOICE OF REFRIGERANT FOR COMPRESSION REFRIGERATION

Now consider the main factors affecting the choice of refrigerant for compression refrigeration.

1. *Freezing point*. Firstly, the evaporator temperature should be well above the freezing temperature at the operating pressure; the freezing points of some common refrigerants at atmospheric pressure are given in Table 24.3.

Table 24.3 Freezing and normal boiling points for some common refrigerants.

Refrigerant	Freezing point at atmospheric pressure (°C)	Boiling point at atmospheric pressure (°C)
Ammonia	−78	−33
Chlorine	−101	−34
n-butane	−138	0
i-butane	−160	−12
Ethylene	−169	−104
Ethane	−183	−89
Methane	−182	−161
Propane	−182	−42
Propylene	−185	−48
Nitrogen	−210	−196

2. *Vacuum operation*. The second consideration, illustrated in Figure 24.32, is that at the evaporator temperature, evaporator pressure below atmospheric pressure should be avoided. An evaporator pressure above atmospheric avoids potential problems with the ingress of air into the cycle, which can cause performance and safety problems. However, special designs can use evaporator pressures below atmospheric. The boiling points of some common refrigerants are given in Table 24.3 at atmospheric pressure.

3. *Latent heat of vaporization*. It is desirable to have a refrigerant with a high latent heat. A high latent heat will lead to a lower flowrate of refrigerant around the loop and reduce

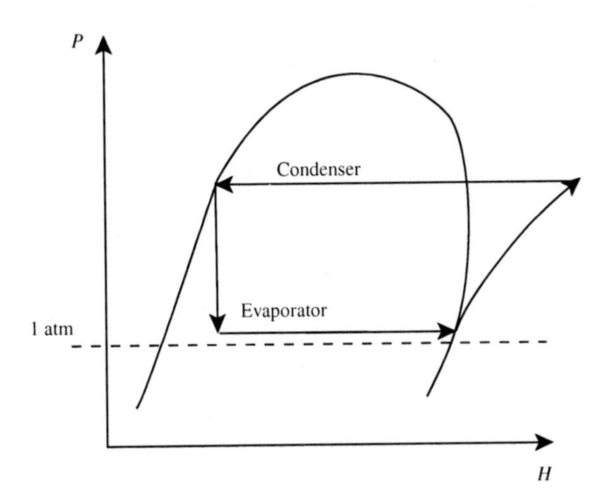

Figure 24.32 Choice of refrigerant for pressure.

the power requirements correspondingly. The shape of the two-phase envelope is such that as the temperature of the refrigerant increases, the latent heat decreases as the critical temperature is approached. It is therefore desirable to have evaporator and condenser temperatures significantly below the critical temperature. Also, as the condenser temperature approaches the critical point, a greater portion of the heat is extracted in the desuperheating, relative to condensation, as illustrated in Figure 24.33. This reduces the coefficient of performance, as it increases the average heat rejection temperature. It also increases the heat transfer area in the condenser. Given that it is desirable to operate away from the critical point where the latent heat of vaporization is high, for a given refrigerant fluid, how close would it be desirable to go to the critical temperature? The change of latent heat with temperature can be correlated using the Watson Equation[7]:

$$\frac{\Delta H_{VAP2}}{\Delta H_{VAP1}} = \left(\frac{T_C - T_2}{T_C - T_1}\right)^{0.38} \tag{24.21}$$

where $\Delta H_{VAP1}, \Delta H_{VAP2}$ = heat of vaporization at temperatures T_1 and T_2 respectively

T_1, T_2 = absolute temperature (K)

T_C = critical temperature (K)

Writing Equation 24.21 between the normal boiling point T_{BPT} and temperature T:

$$\frac{\Delta H_{VAP}}{\Delta H_{VAP,BPT}} = \left(\frac{T_C - T}{T_C - T_{BPT}}\right)^{0.38} \tag{24.22}$$

where $\Delta H_{VAP,BPT}$ = heat of vaporization at the normal boiling point

Let λ be the ratio of the heats of vaporization, such that:

$$\lambda = \left(\frac{T_C - T}{T_C - T_{BPT}}\right)^{0.38} \tag{24.23}$$

Rearranging Equation 24.23 gives:

$$T = T_C - \lambda^{2.63}(T_C - T_{BPT}) \tag{24.24}$$

Thus, for a given refrigerant with normal boiling point T_{BPT} and critical temperature T_C, Equation 24.24 allows the maximum temperature to be fixed if the minimum value of λ is specified. For example, for ethylene, $T_C = 282$ K and $T_{BPT} = 169$ K. If the minimum heat of vaporization in the evaporator is specified to be no lower than 50% of that at the normal boiling point, then the maximum temperature is given by:

$$T = 282 - 0.5^{2.63}(282 - 169)$$
$$= 264 \text{ K}$$

The value of λ is at the discretion of the designer and more conservative values of 60% or 70% could be taken. A value of 50% is probably as low as would be desirable for many applications.

4. *Shape of the two-phase envelope.* Another factor affecting the choice of refrigerant relates to the shape of the two-phase region on a temperature-entropy diagram, as illustrated in Figure 24.34. Figure 24.34 shows two different two-phase regions. The important difference is the slope of the saturated vapor line to the right of the two-phase region. Figure 24.34b shows a two-phase region with a steep slope relative to the one in Figure 24.34a. If the slope is steep, as is the case with the two-phase region in Figure 24.34b,

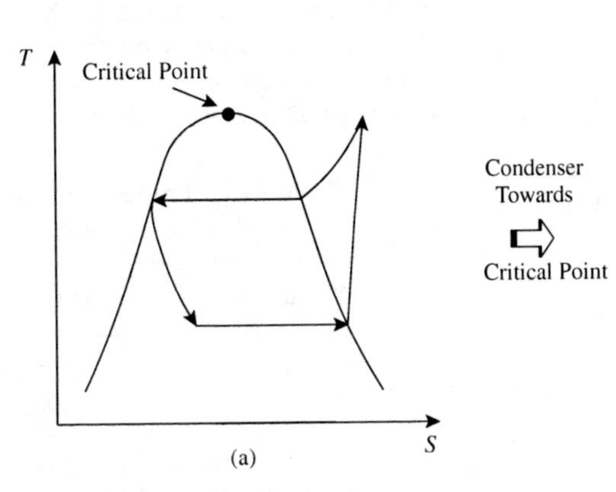

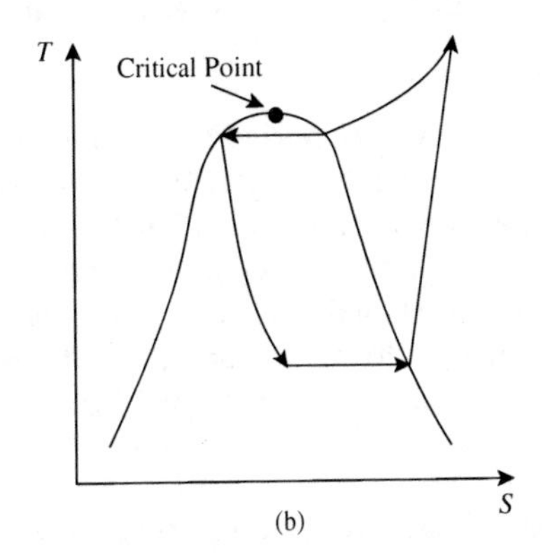

Figure 24.33 Choice of refrigerant – critical point.

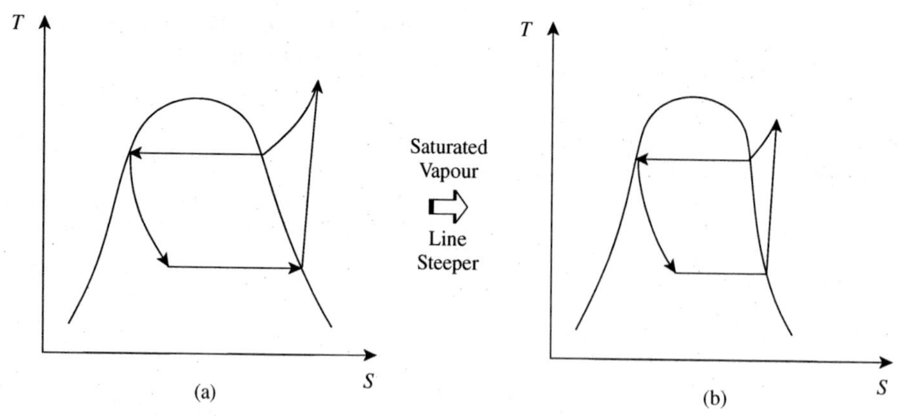

Figure 24.34 Choice of refrigerant-saturated vapor line.

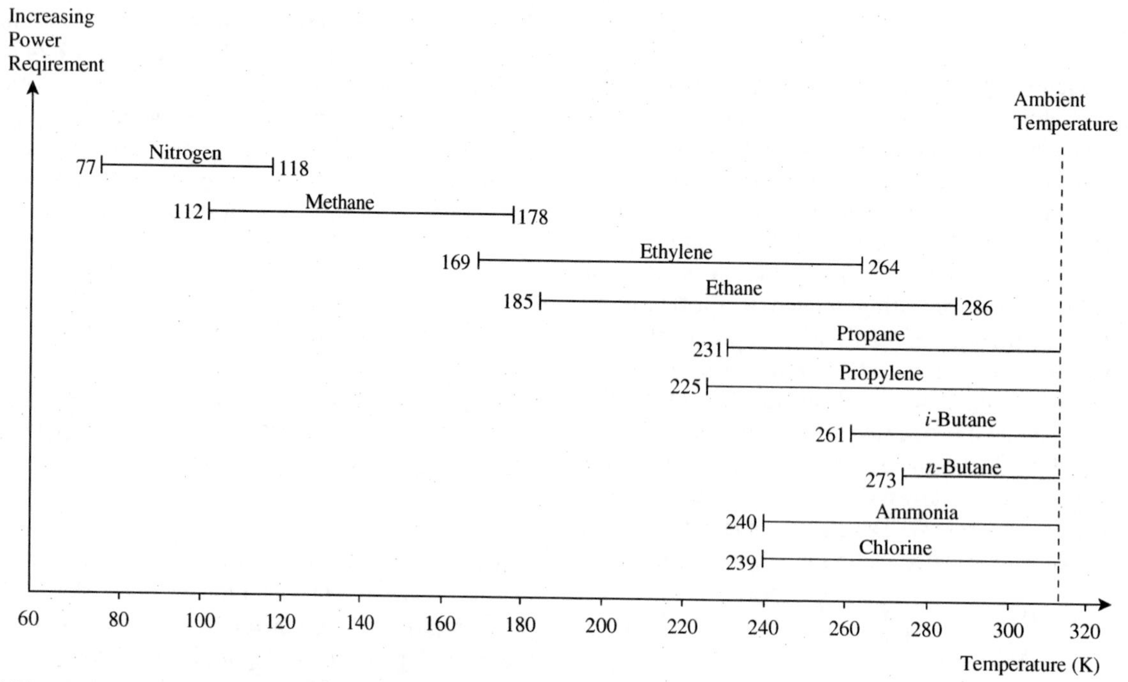

Figure 24.35 Operating ranges of refrigerants.

this reduces superheat and decreases heat extracted in desuperheating relative to condensation. In turn, this increases the coefficient of performance, as it decreases the average heat rejection temperature. It also decreases the heat transfer area.

5. *General considerations.* The refrigerant should, as much as possible, be nontoxic, nonflammable, noncorrosive and have low ozone depletion potential and a low global warming potential (see Chapter 25). Fluids that are already present in the process should be used where possible. Introducing new materials for refrigeration introduces new storage requirements and new safety and environmental problems.

Example 24.2 Figure 24.35 gives the operating ranges of a number of refrigerants. The upper limit of the operating range in Figure 24.35 has been set to ambient temperature if the critical temperature is well above ambient temperature, or to a temperature corresponding with a heat of vaporization of 50% of that at atmospheric pressure. The lower temperature in Figure 24.35 has been limited by the normal boiling point to avoid vacuum conditions in the refrigeration loop. It should be noted that the operating range of the refrigerants can be extended

Table 24.4 Process streams to be cooled by refrigeration.

Stream no.	Temperature (K)
1	175
2	200
3	230
4	245

at the lower limit by operation under vacuum conditions and at the upper limit by relaxing the restriction that the latent heat should not be lower than 50% of that at atmospheric pressure. The refrigerants in Figure 24.35 have been placed in approximate order of power requirement for a given refrigeration duty. Four process streams given in Table 24.4 require refrigeration cooling. Given the information in Figure 24.35, make an initial choice of refrigerants for each of the streams in Table 24.4.

Solution

Stream 1. At 175 K, the only suitable refrigerant from Figure 24.35 is ethylene. Methane is too close to its upper limit. If ethylene is chosen as a refrigerant and heat rejection is required to ambient at 313 K, then it needs to be cascaded with another refrigerant for heat rejection to ambient. In principle, ethylene could be cascaded with chlorine, ammonia, *i*-butane, propylene or propane. However, the overlap with *i*-butane is small and will require a small temperature difference in the heat exchanger linking the two cycles. The choice of refrigerant for the cascade will depend on a number of factors. Choosing chlorine will minimize the power for the system, but introduces significant safety problems. Choosing a component already in the process would be desirable in order to avoid introducing new safety problems (e.g. propylene).

Stream 2. At 200 K, either ethane or ethylene would be suitable refrigerants, with ethane being slightly better from the point of view of the power requirements. As for Stream 1, a cascade system would be required for heat rejection to ambient temperature.

Stream 3. At 230 K, propylene, ethane and ethylene would all be suitable refrigerants. Given that propylene requires the lowest power and does not require a cascade for heat rejection to ambient, this would be a suitable choice.

Stream 4. At 245 K, chlorine, ammonia, propylene and propane could all be chosen. In principle, ethane and ethylene could also have been included but at 245 K they are too close to their critical temperature and would require significantly higher refrigeration power than the other options. The safety problems associated with chlorine are likely to be greater than ammonia. Thus, ammonia might be a suitable choice of refrigerant. Choosing a component already in the process would be desirable.

24.9 TARGETING REFRIGERATION POWER FOR COMPRESSION REFRIGERATION

To evaluate the impact of each design option, a quick and reliable method for estimating refrigeration power requirements can be very useful[8,9]. The benefits from setting targets for refrigeration power are to

- evaluate refrigeration power requirements before complete design
- assess the performance of the whole process prior to detailed design
- allow many alternative design options to be screened quickly and estimated reliably
- assess energy and capital costs.

1. *Simple cycles.* Consider a simple cycle as shown in Figure 24.24. The problem is how to estimate the actual power requirement, given the cooling duty (Q_{EVAP}), condensing temperature (T_{COND}) and evaporating temperature (T_{EVAP}).

The calculation procedure is performed as follows[8,9]:

(1) Find the vapor pressure of the refrigerant in the condenser P^{SAT}_{COND} at T_{COND} and the vapor pressure in the evaporator P^{SAT}_{EVAP} at T_{EVAP} using a correlation for the saturated liquid – vapor pressure, such as the Antoine equation (see Chapter 4):

$$\ln P^{SAT} = A - \frac{B}{C + T} \tag{4.19}$$

where A, B, C = constants determined experimentally for each refrigerant
T = temperature (K)

Once these pressures have been set, this sets the pressure difference across the compressor and the throttle, if the pressure drops in the heat exchangers, flash drums and the pipework are neglected.

(2) An energy balance around the evaporator gives the mass flowrate of refrigerant. Referring to Figure 24.23:

$$m = \frac{Q_{EVAP}}{H_2 - H_1} = \frac{Q_{EVAP}}{H_2 - H_4} \tag{24.25}$$

where m = refrigerant flowrate (kg·s^{-1})
Q_{EVAP} = evaporator heat duty (J·s^{-1}, kJ·s^{-1})
H_1 = specific enthalpy at the evaporator inlet (J·kg^{-1}, kJ·kg^{-1})
H_2 = specific enthalpy at the evaporator outlet (saturated vapor enthalpy at the evaporator pressure (J·kg^{-1}, kJ·kg^{-1})
H_4 = specific enthalpy at the condenser outlet (saturated liquid enthalpy at the condenser pressure (J·kg^{-1}, kJ·kg^{-1})

The enthalpy can be calculated for most refrigerant fluids using the departure function of Peng – Robinson Equation of State (see Chapter 4). The enthalpy calculations are complex and usually carried out using commercial physical property software packages. However, given the capability to determine the enthalpy, Equation 24.25 allows the refrigerant mass flowrate to be determined from a knowledge of the evaporator and condenser pressures. Once the mass flowrate is known, the volumetric flowrate into the compressor can be obtained from the vapor density, which can again be obtained from an equation of state such as the Peng – Robinson Equation. Thus:

$$F = \frac{m}{\rho_V} \tag{24.26}$$

where F = volumetric flowrate ($m^3 \cdot s^{-1}$)
ρ_V = vapor density ($kg \cdot m^{-3}$)

(3) Estimate the compressor efficiency. Compressor manufacturer's data should be used where possible. If this is not available, then for a reciprocating compressor, the isentropic efficiency can be estimated from Equation 13.17:

$$\eta_{IS} = 0.1091(\ln r)^3 - 0.5247(\ln r)^2 + 0.8577 \ln r + 0.3727 \quad 1.1 < r < 5 \tag{13.17}$$

where η_{IS} = isotropic efficiency
r = pressure ratio P_{out}/P_{in}

For a centrifugal compressor, the polytropic efficiency can be estimated from Equation 13.18:

$$\eta_P = 0.017 \ln F + 0.7 \tag{13.18}$$

where η_P = polytropic efficiency
F = inlet flowrate of gas ($m^3 \cdot s^{-1}$)

(4) Estimate the polytropic coefficient n. For reciprocating compressors (see Appendix B):

$$n = \frac{\ln\left(\dfrac{P_{COND}}{P_{EVAP}}\right)}{\ln\left[\dfrac{\eta_{IS}\left(\dfrac{P_{COND}}{P_{EVAP}}\right)}{\eta_{IS} - 1 + \left(\dfrac{P_{COND}}{P_{EVAP}}\right)^{\frac{\gamma-1}{\gamma}}}\right]} \tag{B.28}$$

For centrifugal compressors (see Appendix B):

$$n = \frac{\gamma\eta_P}{\gamma\eta_P - \gamma + 1} \tag{B.38}$$

(5) Find the outlet temperature of the compressor (T_{out}).

$$T_{out} = T_{EVAP}\left(\frac{P_{COND}^{SAT}}{P_{EVAP}^{SAT}}\right)^{\frac{n-1}{n}} \tag{B.26}$$

(6) An energy balance around the compressor gives

$$W = m \cdot (H_2 - H_3) \tag{24.27}$$

where W = power required for compression ($kJ \cdot s^{-1} = kW$)
m = refrigerant flowrate ($kg \cdot s^{-1}$)
H_2 = specific enthalpy at the compressor inlet P_{EVAP}, T_{EVAP} ($kJ \cdot kg^{-1}$)
H_3 = specific enthalpy at the compressor outlet P_{COND}, T_{out} ($kJ \cdot kg^{-1}$)

Again, the compressor inlet and outlet enthalpies can be calculated using the departure function of an equation of state (see Chapter 4) using a commercial physical property software package. Thus, the refrigeration power requirement of the compressor for this simple vapor-compression cycle is obtained. Alternatively, rather than perform an energy balance around the compressor, the refrigeration power can be calculated directly for a reciprocating compressor from (see Appendix B):

$$W = \left(\frac{\gamma}{\gamma - 1}\right)\frac{P_{EVAP}F_{in}}{\eta_{IS}}\left[1 - \left(\frac{P_{COND}}{P_{EVAP}}\right)^{\frac{\gamma-1}{\gamma}}\right] \tag{B.15}$$

where W = power required for compression ($N \cdot m \cdot s^{-1} = J \cdot s^{-1} = W$)
P_{EVAP}, P_{COND} = inlet and outlet pressures for the compressor ($N \cdot m^{-2}$)
F_{in} = inlet volumetric flowrate ($m^3 \cdot s^{-1}$)
γ = ratio of heat capacities C_P/C_V (–)
η_{IS} = isentropic efficiency (–)

For a centrifugal or axial compressor, the refrigeration power can be calculated directly from (see Appendix B):

$$W = \frac{n}{n-1}\frac{P_{EVAP}F_{in}}{\eta_P}\left[1 - \left(\frac{P_{COND}}{P_{EVAP}}\right)^{\frac{n-1}{n}}\right] \tag{B.35}$$

where η_P = polytropic efficiency (ratio of polytropic power to actual power)

2. *Multistage cycles.* The calculation procedure for targeting refrigeration power for simple cycles can be extended to multistage cycles, such as the one shown in Figure 24.36a. In order to be able to apply the procedure, the temperature after the mixing junction between Levels 1 and 2 (T_{MIX} in Figure 24.36a) must be determined. T_{MIX} is not at saturated conditions, as a result of the superheat from the low-pressure (Level 2) compressor. The effect of the mixing between stages in a multistage cycle, leading to superheated conditions at the inlet to the high-pressure compression stage, can be estimated by correcting the temperature between the stages by weighting according to mass flows[8,9]. An energy balance around the mixing junction in Figure 24.36a assuming the heat capacity of the vapor to be constant gives[8,9]:

$$(m_1 + m_2) \cdot T_{MIX} = m_2 \cdot T_{out} + m_1 \cdot T_{EVAP1} \tag{24.28}$$

where m_1, m_2 = mass flowrate for the refrigerant in Levels 1 and 2 ($kg \cdot s^{-1}$)
T_{MIX} = temperature after the mixing junction (K)

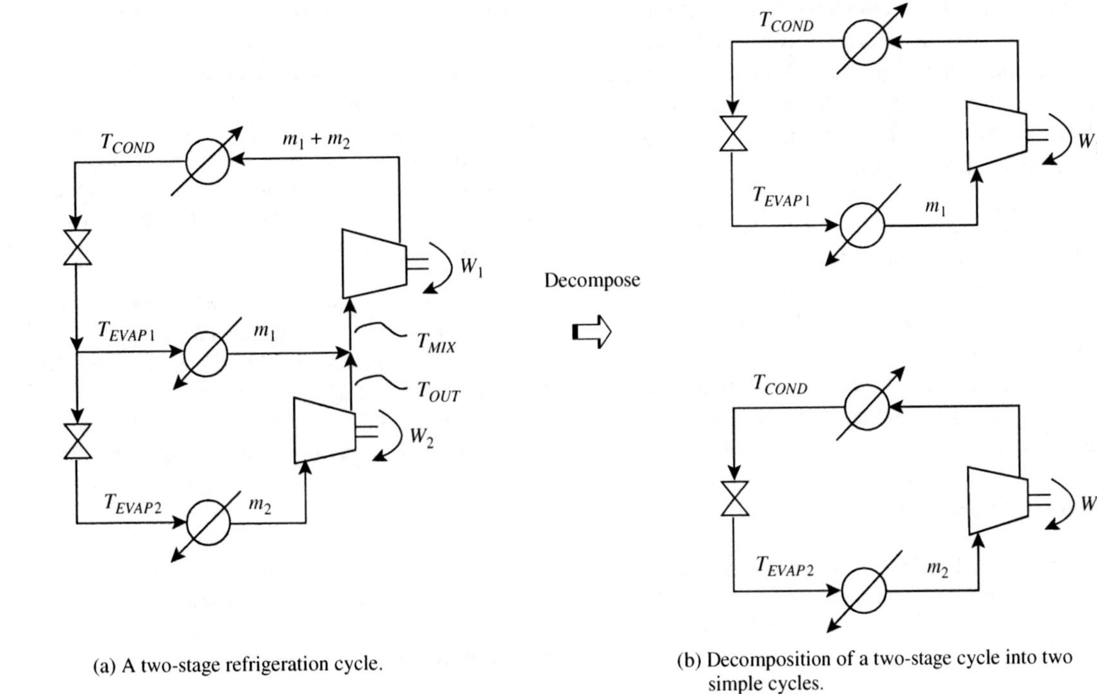

(a) A two-stage refrigeration cycle.

(b) Decomposition of a two-stage cycle into two simple cycles.

Figure 24.36 Targeting refrigeration work for a multistage cycle.

T_{out} = temperature exit the compressor for Level 2 (low-pressure compressor) (K)

T_{EVAP1} = evaporator temperature for Level 1 (K)

Rearranging Equation 24.28 gives:

$$T_{MIX} = \frac{m_2 \cdot T_{out} + m_1 \cdot T_{EVAP1}}{m_1 + m_2} \tag{24.29}$$

The outlet temperature of the high-pressure compressor in Figure 24.36a is found by using Equation B.26 with T_{EVAP} replaced by T_{MIX}. Thus, mass flow weighting of flows around mixing junctions can be used to extend the procedure for refrigeration power targeting to complex cycles.

However, attempting to target the power requirements for complex problems in this way is very restrictive. It requires the configuration of the refrigeration system to be specified. The more complex the problem, the more options available in the design of the refrigeration cycle (or cycles). What is required is an approach that allows the interactions between the design of the process and the refrigeration to be explored without restricting options.

A multistage cycle, such as the one shown in Figure 24.36a, can be decomposed into an assembly of simple vapor-compression cycles in order to estimate the power requirements. Figure 24.36b shows the two-stage cycle being modeled as two simple cycles operating in parallel to estimate the power requirements. One simple cycle operates between Level 1 and ambient with cooling load given by the evaporator for Level 1. The other simple cycle operates between Level 2 and ambient with cooling load given by the evaporator for Level 2.

Thus, for the purposes of refrigeration power targeting for screening options, the power requirements can be estimated from an assembly of simple cycles. Rather than reject heat to ambient, heat rejection from the refrigeration cycle to the process or to another refrigeration cycle can also be allowed for, as will be discussed later. As the design evolves, the simple cycles used for targeting can be merged together to produce complex cycles. As the simple cycles are merged, a number of effects can change the power of the merged design to differ from that predicted by an assembly of simple cycles.

- Complex cycles can introduce nonisothermal mixing in the cycle, which increases the power requirements. In multistage cycles such as the one shown in Figure 24.36a, the high-pressure compressor is subject to superheated inlet conditions, which is a result of mixing of the saturated stream from the low-pressure (Level 1) evaporator with the superheated compressor outlet from the low-pressure (Level 2) compression stage. The superheated inlet conditions to the high-pressure compression stage lead to an overall increase in the power requirements when compared with two simple cycles operating in parallel, as in Figure 24.36b.
- The introduction of vapor–liquid separators (economizers) acts to reduce the power requirements. For example, if a vapor–liquid separator is introduced into the cycle in Figure 24.36a, as in Figure 24.30, vapor from the first (Level 1) throttle is not passed to the low-pressure

(Level 2) compressor. This reduces the load on the low-pressure compressor and the overall power requirements.
- If compressor loads can be merged, such that a larger compressor is required, then the larger compressor is likely to have a higher efficiency than the smaller compressors used in simple cycles.

This means that the refrigeration power targets calculated by decomposing the refrigeration system into an assembly of simple cycles might be lower or higher than the final refrigeration design, depending on the problem and the options chosen by the designer. Complex cycles are likely to be cheaper in capital cost than a collection of simple cycles. Thus, in determining the final design for the refrigeration, there are difficult trade-offs to be considered, involving operating cost, capital cost, complexity, operability and safety.

Example 24.4 Following Example 24.1, calculate the target for refrigeration power for a cooling duty of 3 MW with an evaporator temperature of −30°C and refrigeration condenser temperature of 40°C using

a. ammonia
b. propane
c. propylene

Assume that reciprocating compressors are to be used. Enthalpies can be calculated from the Peng – Robinson Equation of State.

Solution

a. From saturated liquid – vapor pressure data, for an evaporator temperature of −30°C and a condenser temperature of 40°C,

	Pressure (bar)		
	Ammonia	Propylene	Propane
Evaporator	1.2009	2.1191	1.6815
Condenser	15.548	16.431	13.718
Pressure ratio	12.95	7.75	8.16

The values of the heat capacity ratio γ can be obtained from physical property data at average conditions.

	Ammonia	Propylene	Propane
γ	1.30	1.13	1.16

From the values of γ and the pressure ratios, it is clear that the pressure ratio for ammonia is very high for a single-stage compression. The guidelines for gas compression given in Chapter 13 indicate that propylene and propane compressions should be able to be achieved in a single compression stage, but the ammonia would seem to require two compression stages with intercooling. However, in this case, the gas entering the compressor is at a low temperature, which results in a correspondingly lower outlet temperature from the compressor. This would need to be examined in some detail. Indeed, if a two-stage compression is introduced, it would be sensible to consider using an economizer, together with intercooling, to reduce vapor flow in the low-pressure compression stage. In this case, the ammonia compression will be assumed to be carried out in a single stage so as to compare with the other refrigerants on a common basis.

Calculation of the power requirements for the three refrigerants requires the flowrate to be calculated for a duty of 3 MW. This can be calculated from the enthalpy difference across the evaporator $(H_2 - H_1)$. The enthalpy difference across the evaporator is assumed to be the difference between the saturated vapor enthalpy at the evaporator pressure and the saturated liquid enthalpy at the condenser pressure. This assumes no subcooling of the refrigerant.

	Ammonia	Propylene	Propane
$H_2 - H_1\,(\text{kJ}\cdot\text{kg}^{-1})$	1047	245.8	229.33
$m = \dfrac{Q_{EVAP}}{(H_2 - H_1)}(\text{kg}\cdot\text{s}^{-1})$	2.865	12.21	13.10
$\rho_V\,(\text{kg}\cdot\text{m}^{-3})$	1.007	4.605	3.850
$F_{in} = \dfrac{m}{\rho_V}(\text{m}^3\cdot\text{s}^{-1})$	2.845	2.650	3.402

The compressor efficiency (in this case, isentropic efficiency) can now be calculated and in turn the polytropic coefficient. This allows the outlet temperature to be calculated, together with the compressor power.

	Ammonia	Propylene	Propane
$\eta_{IS}\,(-)$ (from Equation 13.11)	0.960	0.866	0.870
$n(-)$ (from Equation B.28)	1.312	1.150	1.184
T_{out} (K) (from Equation B.26)	447	317	337
W (MW) (from Equation B.15)	1.24	1.50	1.60

From these results, the high temperature of the ammonia from the compressor outlet should be noted. This reinforces the points made earlier regarding ammonia compression.

In principle, ammonia is the best refrigerant fluid in terms of power requirement. However, this conclusion disregards the potential practical problems associated with compression. There is little to choose between propylene and propane in terms of the power requirements.

It is also interesting to compare the more detailed calculations performed here with the approximate calculation in Example 24.1. Example 24.1 estimated the power to be 1.4 MW on the basis of the *COP* for an ideal cycle with an assumed 60% efficiency.

24.10 HEAT INTEGRATION OF COMPRESSION REFRIGERATION PROCESSES

Figure 24.37a shows an example of a refrigeration cycle matched against a grand composite curve. Heat is rejected from the process into the refrigerant at the lowest temperature. In Figure 24.37a, the heat rejection from the refrigeration cycle is to cooling water. Closer inspection of the grand composite curve in Figure 24.37a shows that it has a large requirement for heat at a temperature below cooling water temperature. Thus, Figure 24.37b shows a refrigeration cycle that rejects its heat into the process to reduce the power requirement with a lower temperature lift and to save hot utility.

Figure 24.38 shows the same grand composite curve as in Figure 24.37, but cooled below ambient by a two-level refrigeration system. The grand composite curve is used to determine the heat rejection at each refrigeration level. Using a two-level refrigeration system, as shown in Figure 24.38, reduces the power requirement compared with the single-level arrangement shown in Figure 24.37. This results from the smaller temperature lift required by the higher temperature refrigeration level.

Figure 24.39 shows a much more complex refrigeration cycle matched against the same grand composite curve. Heat is rejected between Points 1 and 2 into the lowest level of refrigeration. Another level of refrigeration is placed between Points 3 and 4. So far, this is the same solution as shown in Figure 24.38. Figure 24.39 shows an additional feature in which some heat is rejected back into the process between Points 5 and 6 in Figure 24.39. Rejecting the heat in the way shown in Figure 24.39 at an intermediate temperature saves power for the rejection of heat that would otherwise have to be lifted above pinch temperature before rejection. However, the intermediate heat rejection between Points 5 and 6 in Figure 24.39 disturbs the energy balance in the pocket of the grand composite curve, requiring a third level of refrigeration to be introduced between Points 7 and 8 for heat rejection from the process. Going from the single-level refrigeration in Figure 24.37 to the two-level refrigeration in Figure 24.38 and onto the three-level refrigeration system with intermediate heat rejection in Figure 24.39 saves power with each additional level of complexity. However, the capital cost increases as the complexity increases. Thus, whether it is worthwhile to introduce a complex solution as the one shown in Figure 24.39 depends very much on the economy of scale and the trade-off between energy costs and capital costs.

When refrigeration profiles are matched against the grand composite curve, either heat removal from the process or heat rejection to the process, the selection of temperature is not always straightforward. Sometimes, the refrigeration

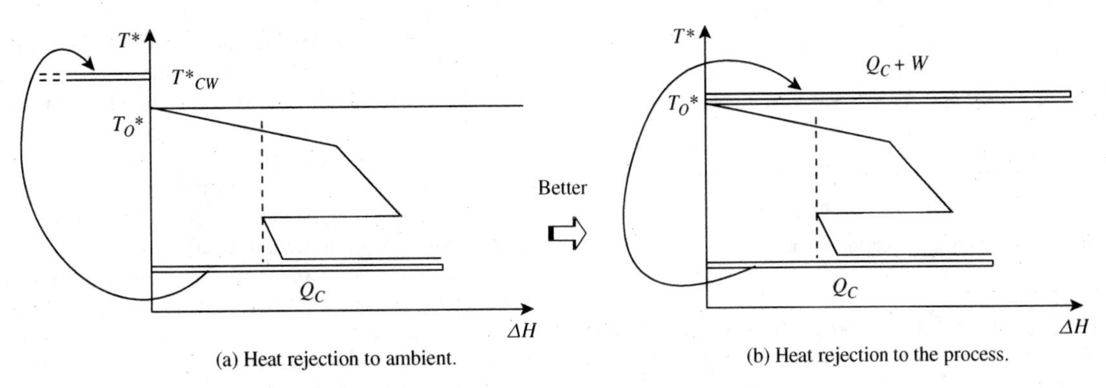

(a) Heat rejection to ambient. (b) Heat rejection to the process.

Figure 24.37 Heat integration of a single-level refrigeration system.

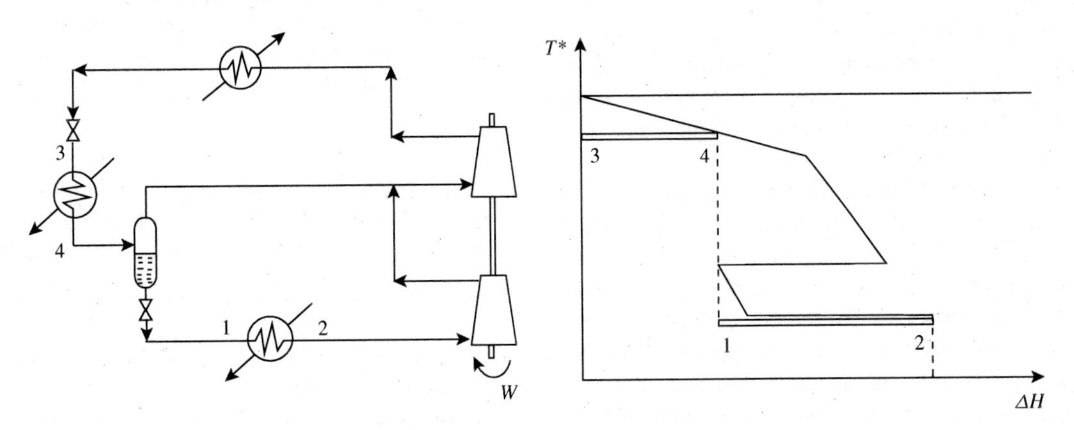

Figure 24.38 Heat integration of a two-level refrigeration system.

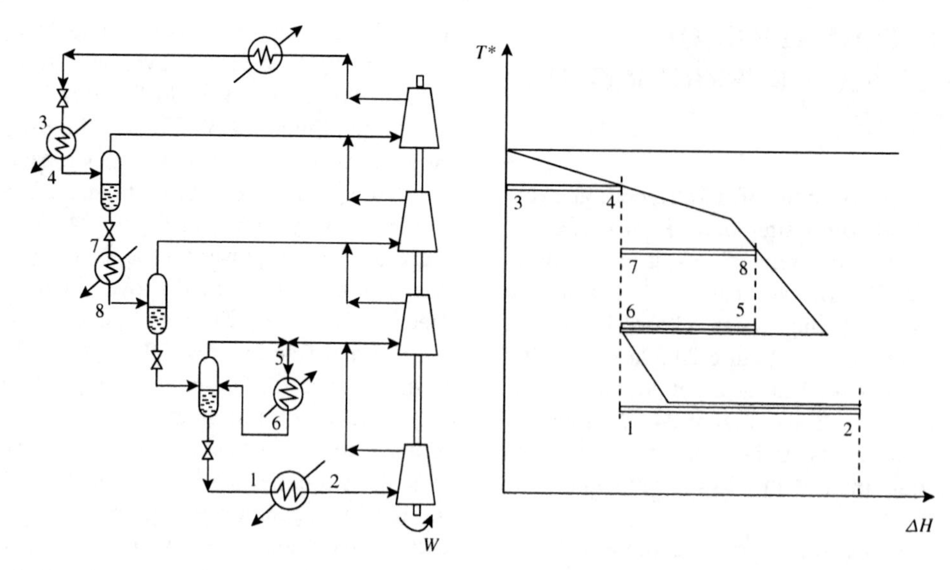

Figure 24.39 Exploiting the pockets in the grand composite curve reduces the overall temperature lift and the shaftwork requirements.

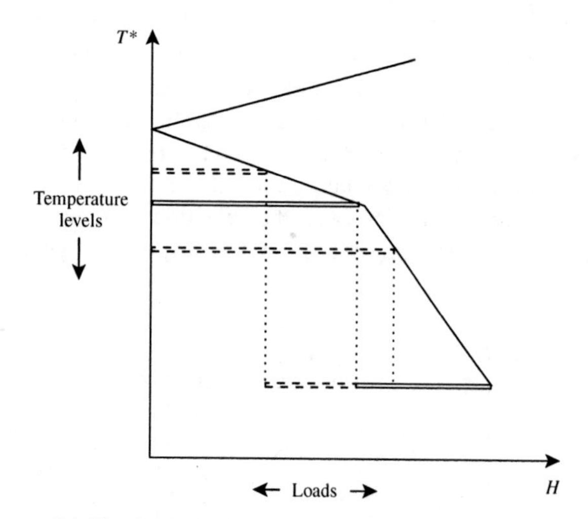

Figure 24.40 Optimization of refrigeration levels.

level fits neatly against a flat portion of the grand composite curve (e.g. Figure 24.37). Sometimes, the refrigeration level needs to be fitted against a sloping section of the grand composite curve, as shown in Figure 24.40. In this case, there is a degree of freedom to choose the level of the refrigeration. For the example in Figure 24.40, there are two levels of refrigeration. As the higher temperature refrigeration level is increased, the heat load and power requirement per unit of heat load both decrease. This will decrease the cost of the power requirements for the higher temperature refrigeration level. However, as the heat load on the higher temperature refrigeration level decreases, the heat load on the lower temperature refrigeration level increases, increasing its power requirement. As the temperature of the higher-level refrigeration decreases, these trends reverse. There is thus a degree of freedom that needs to be optimized to determine the heat duty allocated to each of the two levels of refrigeration. To analyze such a trade-off requires a quantitative prediction of the power required for given refrigeration evaporation and condensing temperatures.

If the refrigeration system involves a cascade cycle, the partition temperature is an important additional degree of freedom to be optimized.

Example 24.5 Determine the refrigeration requirements for the low-temperature distillation process from Example 16.1 shown in Figure 16.19 for $\Delta T_{min} = 5°C$.

a. Plot the grand composite curve from the heat cascade given in Figure 16.21b and determine the temperature and duties of the refrigeration if two levels of refrigeration are to be used. Assume that both vaporization and condensation of the refrigerant occur isothermally.
b. Calculate the power requirements for the refrigeration for heat rejection to cooling water operating between 20 and 25°C approximated by Equation 24.20.
c. Repeat the calculation from Part b using refrigeration power targeting, assuming propylene as the refrigerant and a reciprocating compressor.
d. Heat rejection from the refrigeration system into the process can be used to reduce the refrigeration power requirements. Calculate the power using Equation 24.20.
e. Repeat the calculation from Part d using refrigeration power targeting, assuming propylene as the refrigerant and a reciprocating compressor.

Solution

a. Figure 24.41a shows a plot of the grand composite curve from the problem table cascade given in Figure 16.21b. Also shown in Figure 24.41a are two refrigeration profiles.

	T^*(°C)	T(°C)	Q_{EVAP} (MW)
Level 1	−24.5	−25	1.04
Level 2	−42.5	−45	(1.84 − 1.04) = 0.8

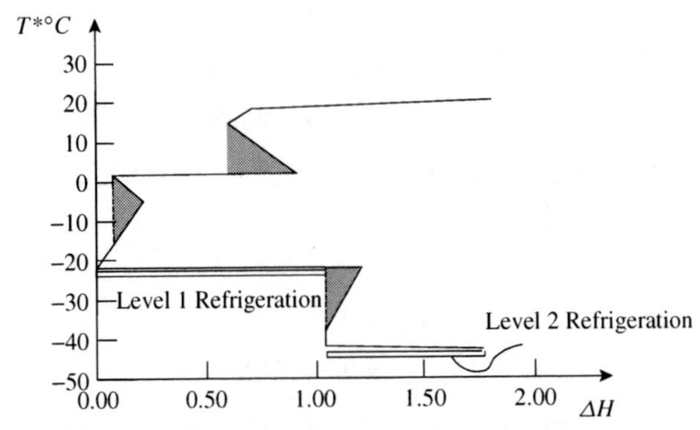

(a) Two refrigeration levels can be used to provide process cooling.

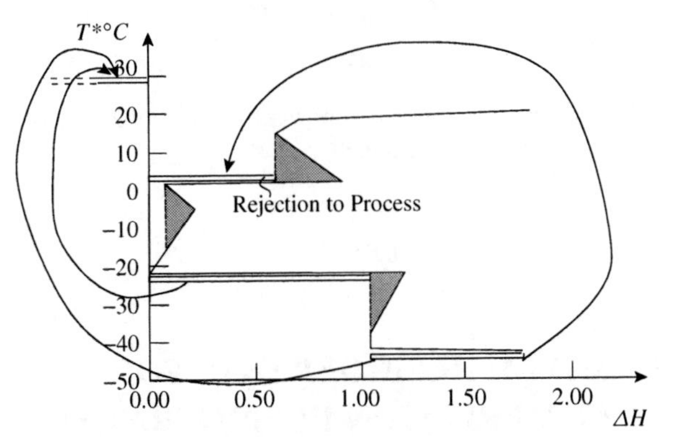

(b) Heat rejection to the process allows the power requirements to be reduced.

Figure 24.41 Refrigeration system for Example 6.6.

b. For heat rejection to cooling water,

$$T_H = 25 + 5 = 30^\circ\text{C} = 303\text{ K}$$

$$W_1 = \frac{1.04}{0.6}\left[\frac{303 - 248}{248}\right] = 0.38\text{ MW}$$

$$W_2 = \frac{0.8}{0.6}\left[\frac{303 - 228}{228}\right] = 0.44\text{ MW}$$

Total power for heat rejection to cooling water = 0.38 + 0.44 = 0.82 MW

c. Repeat the calculation from Part c but now using refrigeration targeting, assuming propylene as the refrigerant. Designate Cycle 1 to operate between −25°C and 30°C and Cycle 2 to operate between −45°C and 30°C.

	Pressure (bar)	
	Cycle 1	Cycle 2
Evaporator	2.5239	1.1291
Condenser	13.0081	13.0081
Pressure ratio	5.15	11.52

The pressure ratio is a little high for Cycle 2 to be carried out in a single compression stage. However, single-stage compression will be assumed for the sake of comparison between different options. The heat capacity ratio γ will be assumed to be constant with a value of 1.13.

	Cycle 1	Cycle 2
Q_{EVAP} (MW)	1.04	0.8
$H_2 - H_1$ (kJ·kg^{-1})	279.4	258.2
$m = \dfrac{Q_{EVAP}}{(H_2 - H_1)}$ (kg·s^{-1})	3.722	3.098
ρ_V (kg·m^{-3})	5.492	2.593
$F_{in} = \dfrac{m}{\rho_V}$ (m^3·s^{-1})	0.6778	1.195

The compressor efficiency (in this case, isentropic efficiency) can now be calculated and in turn the polytropic coefficient. This allows the outlet temperature to be calculated.

	Cycle 1	Cycle 2
η_{IS} (−) (from Equation 13.17)	0.849	0.928
n(−) (from Equation B.28)	1.154	1.140
T_{out} (K) (from Equation B.26)	309	308
W (MW) (from Equation B.15)	0.36	0.41

Total power for heat rejection to cooling water = 0.36 + 0.41 = 0.77 MW

d. Now assume that part of Level 2 can be rejected to the process above the pinch as shown in Figure 24.41b. The heat exchangers represented in Figure 24.41b might be several exchangers in practice. The rejection load is fixed at 0.54 MW.

$$W = \frac{Q_{EVAP}}{0.6}\left[\frac{T_{COND} - T_{EVAP}}{T_{EVAP}}\right]$$

$$W = \frac{Q_{COND} - W}{0.6}\left[\frac{T_{COND} - T_{EVAP}}{T_{EVAP}}\right]$$

$$W = \frac{0.54 - W}{0.6}\left[\frac{(5 + 273) - 248}{248}\right]$$

$$W = 0.091\text{ MW}$$

Process cooling by Level 2 by this arrangement across the pinch

$$= 0.54 - 0.091$$

$$= 0.45\text{ MW}$$

The balance of the cooling demand on Level 2 (0.8 − 0.45) = 0.35 MW, together with the load from Level 1 must be either rejected to the process at a higher temperature above the pinch or to cooling water.

The process has a heating demand at 20°C, which means that heat could be rejected at 25°C. However, there seems little advantage in such an arrangement since the heat can be rejected to cooling water at 30°C, and rejection to the process would add to the complexity of both design and operation. Assume that the rest of the rejection heat goes to cooling water.

$$W_1 = \frac{1.04}{0.6}\left[\frac{303 - 248}{248}\right] = 0.38 \text{ MW (as before)}$$

$$W_2 = \frac{0.8 - 0.45}{0.6}\left[\frac{303 - 228}{228}\right] = 0.19 \text{ MW}$$

Total refrigeration power for part rejection of Level 2 to the process

$$= 0.38 + 0.19 + 0.091$$
$$= 0.66 \text{ MW}$$

Thus, the saving in refrigeration power by integration with the process

$$= 0.82 - 0.66$$
$$= 0.16 \text{ MW}$$

e. Repeat the above calculation using refrigeration power targeting. Designate the cycle operating between – 45°C and 5°C to be Cycle 3.

	Pressure (bar) Cycle 3
Evaporator	1.1291
Condenser	6.7033
Pressure ratio	5.15

For Cycle 3, the refrigerant flowrate and power are dictated by the heat load on the condenser, $Q_{COND} = Q_{EVAP} + W$. Since W is unknown, some trial and error is required. Assume initially that $Q_{COND} = 0.4$ MW for Cycle 3.

Q_{EVAP} (MW)	0.4	0.38	0.388
$H_2 - H_1$ (kJ·kg^{-1})	258.2	258.2	258.2
$m = \dfrac{Q_{EVAP}}{(H_2 - H_1)}$ (kg·s^{-1})	1.549	1.472	1.503
ρ_V (kg·m^{-3})	2.593	2.593	2.593
$F_{in} = \dfrac{m}{\rho_V}$ (m^3·s^{-1})	0.5974	0.5676	0.5795
η_{IS} (–) (from Equation 13.11)	0.852	0.852	0.852
n(–) (from Equation B.28)	1.153	1.153	1.153
T_{out} (K) (from Equation B.26)	289	289	289
W (MW) (from Equation B.15)	0.156	0.149	0.152
Q_{COND} (MW)	0.556	0.529	0.540

The duty on Cycle 1 remains unchanged from Part c, but the duty on Cycle 2 must be reduced from Part c to compensate by the duty on Cycle 3. For Cycle 2:

$$Q_{EVAP} = 0.8 - 0.388$$
$$= 0.412 \text{ MW}$$

Adjusting the calculation in Part c to reduce Q_{EVAP} to 0.412 MW:

$$m = 1.596 \text{ kg·s}^{-1}$$
$$F_{in} = 0.6153 \text{ m}^3\text{·s}^{-1}$$
$$W = 0.211 \text{ MW}$$

Total refrigeration power for part rejection of Level 2 to the process from refrigeration targeting

$$= 0.36 + 0.21 + 0.15$$
$$= 0.72 \text{ MW}$$

Thus, the saving in refrigeration power by integration with the process from refrigeration targeting

$$= 0.77 - 0.72$$
$$= 0.05 \text{ MW}$$

24.11 MIXED REFRIGERANTS FOR COMPRESSION REFRIGERATION

It was discussed above how pure refrigerants have a restricted temperature range over which they can operate. The working range of refrigerant fluids can be extended and modified by using a mixture for the refrigerant rather than a pure component. Mixed refrigerants can then be applied in simple, multistage or cascade refrigeration systems. However, unlike pure refrigerants and azeotropic mixtures, the temperature and vapor and liquid compositions of nonazeotropic mixtures do not remain constant at constant pressure as the refrigerants evaporate or condense. Use of mixed refrigerants can be particularly effective if the cooling duty involves a significant change in temperature. The composition of the mixture is selected such that the liquid refrigerant evaporates over a temperature range similar to that of the process cooling demand. Figure 24.42 compares a pure refrigerant (or an azeotropic mixture)

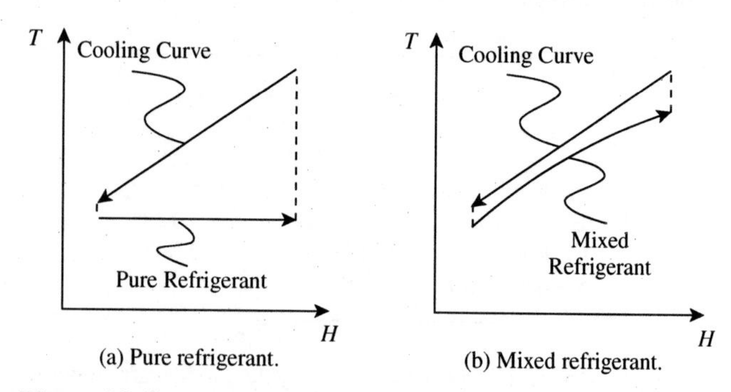

Figure 24.42 Pure and mixed refrigerants.

and a mixed refrigerant, both cooling a low-temperature heat source that changes temperature significantly. In Figure 24.42a, it can be seen that a pure refrigerant (or azeotropic mixture) evaporates at a constant temperature, leading to a small difference in temperature at one end of the heat exchanger, but a large difference in temperature at the other end. By contrast, a mixed refrigerant, as shown in Figure 24.42b, evaporates over a range of temperature and follows more closely the low-temperature cooling profile. Small temperature differences throughout the heat exchange in low-temperature systems lead to lower refrigeration power requirements.

The design of a mixed refrigerant system requires the degrees of freedom to be manipulated simultaneously. These are as follows[10].

1. *Pressure level*. The choice of pressure level for the mixed refrigerant evaporation affects the temperature difference between the process cooling curve and refrigerant evaporation curve. Increasing the overall temperature difference will increase the refrigeration power requirements.

2. *Refrigerant flowrate*. Increasing the refrigerant flowrate can widen the temperature difference between the process cooling curve and refrigerant evaporation curve and vice versa. Increasing the refrigerant flowrate increases the refrigeration power requirement. However, if the refrigerant flowrate is too low, it might cause temperature crosses (negative temperature differences) in the heat exchange. Equally, if the refrigerant flowrate is too high, there might be some wetness in the inlet stream to the compressor, which should be avoided. Therefore, the refrigerant flowrate can be changed only within a feasible range.

3. *Refrigerant composition*. The composition of the mixture can be varied to achieve the required characteristics. Introduction of new components or replacing existing components by new ones provides additional freedom to achieve better performance. Unlike adjusting the refrigerant pressure level and flowrate, optimization of the composition does not inevitably mean that increasing the temperature difference in some part of the heat exchange will result in higher refrigeration power requirements. Since pressure levels and refrigerant flowrate can only be changed within certain ranges, the refrigerant composition is the most flexible and significant variable when designing mixed refrigerant systems.

The major difficulty in selection of refrigerant composition for mixed refrigerants comes from the interactions between the variables, and the small temperature difference between the process cooling curve and refrigerant evaporation curve. Any change in the refrigerant composition, evaporating pressure or the refrigerant flowrate will change the shape and position of the refrigerant evaporation curve.

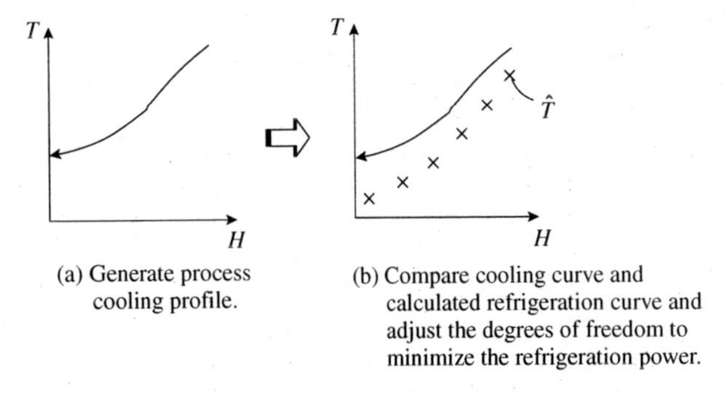

Figure 24.43 Optimization of refrigerant composition.

Optimization can be used to vary the refrigerant composition, flowrate and pressure level to obtain the desired refrigerant properties. Ideal matching between the process cooling curve and refrigerant evaporation curve would lead to the evaporation curve tending to follow the general shape of the cooling curve, but with the two curves not necessarily being parallel with each other. The procedure to adjust the degrees of freedom is illustrated in Figure 24.43. The method starts by first generating the process cooling profile for the given process operating conditions. An initial setting must be made for the refrigerant composition, flowrate and evaporator pressure. The cooling and evaporation curves are divided into enthalpy intervals. Within each Interval i, a material and energy balance calculates the actual refrigerant evaporation temperature $\hat{T}_i$[10]. The difference between the cooling curve and the calculated refrigeration evaporation curve can then be compared to identify any infeasibilities such that $(T_i - \hat{T}_i) \leq \Delta T_{min}$, Figure 24.43. The objective function of the optimization should be to minimize refrigeration power requirements. A nonlinear programming method (e.g. SQP, see Chapter 3) can be used to perform the optimization by manipulating the composition, flowrate and pressure to minimize the refrigeration power requirement[10]. However, this approach is vulnerable to the highly nonlinear characteristics of the optimization. Alternatively, a stochastic method can be used at the expense of the additional computational requirements[11].

One major application of mixed refrigerant systems is in the liquefaction of natural gas[12–14]. A mixture of hydrocarbons (usually with carbon numbers in the C_1 to C_5 range) and nitrogen is normally used as refrigerant. The refrigerant characteristics are chosen such that there is a close match between the cooling and heating profiles with small temperature differences throughout the whole temperature range for a specific refrigeration demand (but not necessarily parallel). The simplest form of process for liquefaction of natural gas using a mixed refrigerant is illustrated in Figure 24.44. The function of such the process is to convert natural gas to the liquid state for transportation and/or storage. The natural gas enters the heat exchanger at ambient temperature, and high pressure and is to be liquefied by the mixed refrigerant flowing countercurrently

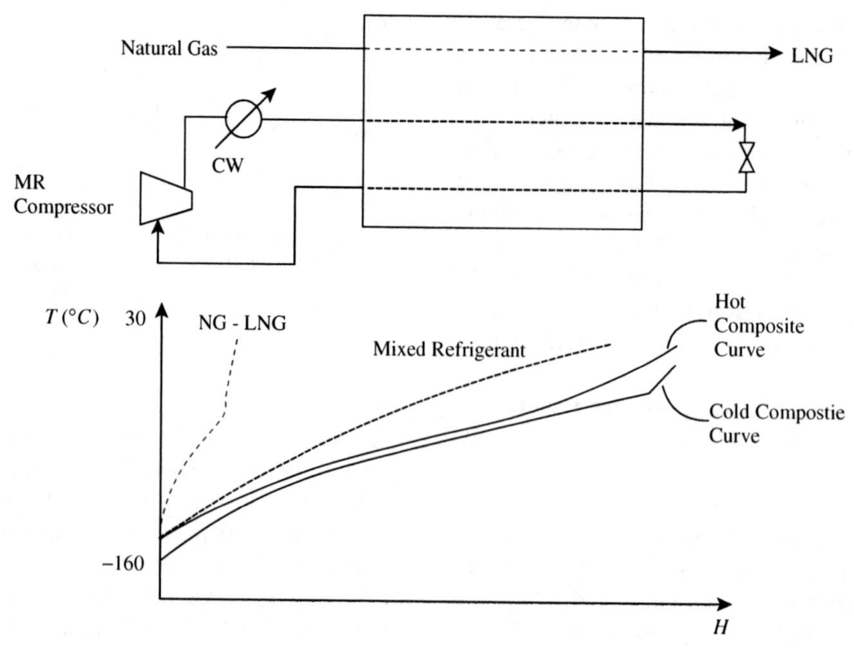

Figure 24.44 A simple liquefied natural gas (LNG) process. (From Lee GC Smith R and Zhu XX, 2002, Ind Eng Chem Res, 41: 5016, reproduced by permission of the American Chemical Society.)

through the main heat exchanger. The main heat exchanger could typically be a plate-fin design (see Chapter 15). The mixed refrigerant is then recompressed and partially condensed, normally by cooling water. Because there is no other heat sink in the process that can totally condense the mixed refrigerant, it must be condensed by the cold refrigerant itself. Thus, the mixed refrigerant passes through the main heat exchanger, where it is condensed. It is then expanded across a valve and returns countercurrently through the heat exchanger to the compressor.

Figure 24.44 shows the temperature profiles of the three streams in the main heat exchanger: the natural gas stream, the warm refrigerant stream (before the expansion valve) and the cold refrigerant stream (after the expansion valve)[10]. The natural gas stream and the warm refrigerant stream need to be combined as a hot composite curve, as described in Chapter 16, to give the total cooling demand. The mixed refrigerant evaporates and condenses at two constant pressure levels, while both the evaporating and condensing temperatures vary over a wide range. The choice of the two pressure levels affects the temperature difference between the hot composite curve and the refrigeration evaporation profile. It should be noted that, in this case, as the degrees of freedom are manipulated, this alters the shape of both the cooling curve and the refrigerant evaporation curve. This means that the cooling curve (hot composite curve) must be recalculated, as the optimization progresses[10].

Rather than use the simple cycle shown in Figure 24.44 for the liquefaction of natural gas, much more complex arrangements using multiple cycles (with both pure and mixed refrigerants) and cascade systems can be used.

24.12 ABSORPTION REFRIGERATION

Consider now absorption refrigeration. Compression refrigeration is powered by a compressor compressing refrigerant vapor, Figure 24.45a. The basic problem with this is the high compression costs. Figure 24.45b illustrates an alternative way to bring about the compression. A refrigerant vapor is first absorbed in a solvent. The resulting liquid solution can have its pressure increased using a pump. The compressed refrigerant is then separated from the solvent in a stripper (regenerator). The pump requires significantly less power to bring about the increase in pressure compared with the corresponding gas compression. The overall effect is to increase the pressure of the refrigerant with far less power. The drawback is that a heat supply is needed for the stripper (regenerator).

Figure 24.46 shows a typical absorption refrigeration arrangement. To the left of the cycle in Figure 24.46 is the absorber and stripper (regenerator) arrangement. Low-pressure refrigerant vapor from the evaporator is first absorbed in a solvent, which is then increased in pressure and then increased in temperature in a heat exchanger. The refrigerant then enters the vapor generator where the refrigerant is stripped from the solvent. Heat is input to the vapor generator and the solvent is cooled in the heat exchanger, decreased in pressure and returned to the absorber. The high-pressure vapor from the vapor generator is condensed in the condenser, expanded in the expansion valve to produce the cooling effect and then enters the evaporator to provide process cooling.

The features of absorption refrigeration are that there is a low power requirement relative to compression

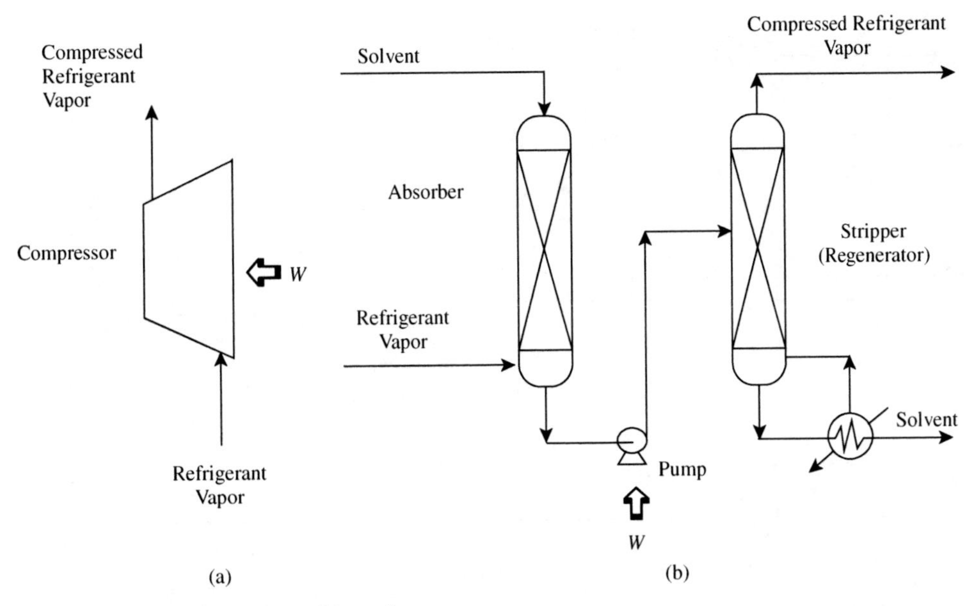

Figure 24.45 Compression versus absorption refrigeration.

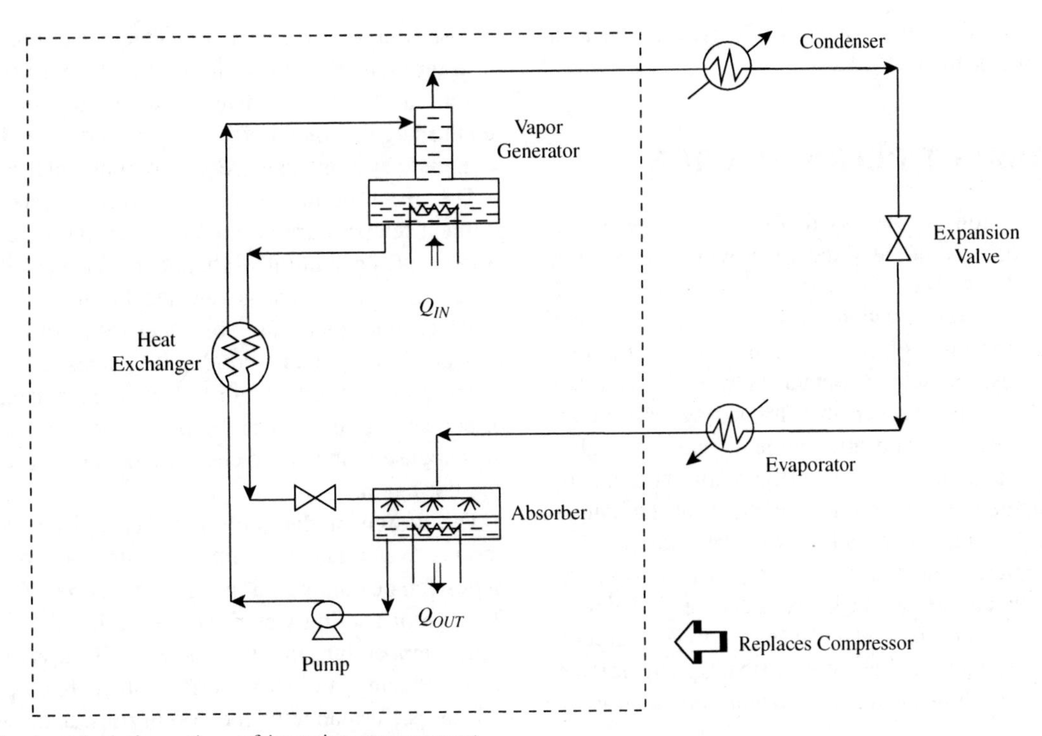

Figure 24.46 A typical absorption refrigeration arrangement.

refrigeration, but heat input to the vapor generator (stripper) is required and heat output from the absorber is required, usually to cooling water. Absorption refrigeration is only useful for moderate temperature refrigeration.

The most common working fluids for absorption refrigeration are given in Table 24.5, together with the working range.

When should absorption refrigeration be used rather than compression refrigeration? There are two important criteria. The first is that absorption refrigeration can only be used when moderate levels of refrigeration are required. Second, it should only be used when there is a large source of waste

Table 24.5 Common working fluids for absorption refrigeration.

Refrigerant	Solvent	Lower temperature limit (°C)
Water	Lithium bromide	5
Ammonia	Water	−40

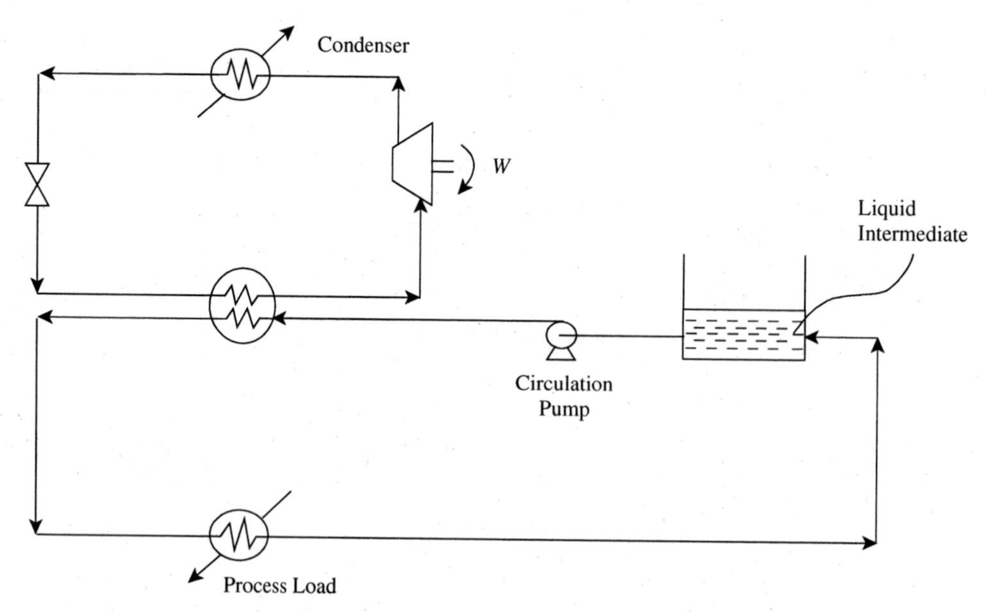

Figure 24.47 Indirect refrigeration.

heat available for the vapor generator. This must be at a temperature greater than 95°C.

24.13 INDIRECT REFRIGERATION

Indirect refrigeration is often used. Figure 24.47 shows a liquid intermediate being recycled to provide the cooling to a process load. Heat is removed from the liquid intermediate by a refrigeration cycle. This arrangement is used for distribution of refrigeration across a number of process loads, or when contact between refrigerant and process fluids is unacceptable for safety or product contamination reasons. The arrangement leads to higher power requirements than direct refrigeration because of the extra temperature lift required in the heat exchanger between the refrigerant fluid and intermediate liquid.

The liquid intermediates used are either water solutions of various concentrations such as salts (e.g. calcium chloride, sodium chloride), glycols (e.g. ethylene glycol, propylene glycol) and alcohols (e.g. methanol, ethanol), or pure substances such as acetone, methanol and ethanol.

24.14 COOLING WATER AND REFRIGERATION SYSTEMS – SUMMARY

Rejection of waste heat is a feature of most chemical processes. Once the opportunities for recovery of heat to other process streams and to the utilities system (e.g. steam generation) have been exhausted, then waste heat must be rejected to the environment. The most direct way to reject heat to the environment above ambient temperature is by the use of air-cooled heat exchangers. Cooling water using once-through cooling systems based on river water, and so on, can be used. However, such once-through systems require large amounts of water, and the waste heat rejected to aquatic systems can create environmental problems.

The most common way to reject waste heat above ambient temperature is through recirculating cooling water systems. Both natural draft and mechanical draft cooling towers are used. Heat is rejected through the evaporation of water, but other losses of water result from the need for blowdown to prevent buildup of undesired contaminants in the recirculation and drift losses. Increasing the cycles of concentration reduces the blowdown losses, probably at the expense of increased chemical dosing to prevent fouling and corrosion.

The design of the cooling tower and the design of the cooling water network interact with each other. Cooling duties can be configured in parallel or series. Decreasing the flowrate of cooling water and increasing the cooling tower return temperature increase the effectiveness of the cooling tower. Placing coolers in series, rather than parallel, benefits the performance of the cooling tower through reduced flowrate and increasing return temperature. However, this is at the expense of reduced temperature differences in the coolers and increased pressure drop for the cooling water. Placing coolers in series for a few critical reuse opportunities might be extremely effective in improving the design.

Situations are often encountered when cooling water networks need to increase the heat load of individual coolers, which requires investment in new cooling tower capacity. Such situations demand a more complete analysis of the whole cooling water system.

Refrigeration is required to produce cooling below ambient temperature. There are two broad classes of refrigeration:

- compression refrigeration
- absorption refrigeration.

Compression refrigeration is by far the most common. Simple cycles can be employed to provide cooling typically as low as −40°C. For lower temperatures, complex cycles are used. Economizers, presaturators and the introduction of multiple refrigeration levels allow multistage compression and expansion to be used to reduce the power requirements for refrigeration. Very low temperature systems require the use of cascade cycles with two refrigerant cycles linked together, with each cycle using a different refrigerant fluid.

The choice of refrigerant fluid depends on a number of issues. Use of an evaporator pressure below atmospheric pressure is normally avoided. It is desirable to have a high latent heat at the evaporator conditions in order to reduce the flowrate around the refrigeration loop to reduce the power requirements correspondingly. Various other factors relating to the shape of the two-phase region for the refrigerant fluid also affect the choice of refrigerant, as well as the toxicity, flammability, corrosivity and environmental impact.

Refrigeration cycles offer many opportunities for heat integration with the process. These can be explored using the grand composite curve.

Refrigeration power can be targeted for specific refrigerant fluids by performing an energy balance around the cycle. Simple, multistage and complex cycles can be targeted for minimum refrigeration power.

Mixed refrigerant systems use a mixture as refrigerant instead of pure refrigerants. Evaporation of the refrigerant mixture over a range of temperature allows a better match between the refrigerant duty and refrigerant evaporation if the refrigeration duty varies significantly in temperature. When designing mixed refrigerant systems, important degrees of freedom include

- composition of the refrigerant mixture
- pressure levels
- refrigeration flowrate.

Absorption refrigeration is much less common than compression refrigeration. Absorption refrigeration powers by compressing the refrigerant fluid dissolved in a solvent using a pump.

24.15 EXERCISES

1. Cooling water is being circulated at a rate of 20 $m^3 \cdot min^{-1}$ to the cooling network. The cooling water from the cooling tower is at a temperature of 25°C and is returned at 40°C. Measurements on the concentrations of the feed and circulating water indicate that there are five cycles of concentration. Assuming that the heat capacity of the water is 4.2 $kJ \cdot kg^{-1} \cdot K^{-1}$ and the makeup water is at a temperature of 10°C, calculate:
   a. the rate of evaporation assuming the latent heat of vaporization to be 2423 $kJ \cdot kg^{-1}$ at the conditions in the cooling tower.
   b. the makeup water requirement, assuming drift losses are negligible.
2. A cooling tower supplies water to a cooling water network at 25°C. The maximum inlet and outlet temperatures for the three cooling duties together with their flowrates are given in Table 24.6.

Table 24.6 Cooling duties for Exercise 2.

Operation	T_{in} (°C)	T_{out} (°C)	Flowrate ($t \cdot h^{-1}$)
1	25	40	20
2	30	40	50
3	50	70	40

   a. Calculate the minimum cooling water flowrate. How big is the reduction in flowrate relative to a parallel design?
   b. Design a cooling water network to achieve the target.
3. Using the data for Exercise 2,
   a. Calculate the minimum cooling water flowrate assuming that the cooling water return temperature must be limited at 50°C.
   b. Design a cooling water network to achieve the target for a 50°C return temperature.
4. An existing cooling water system is shown in Figure 24.48. The cooling tower is performing with an outlet temperature of 20°C for an inlet temperature of 30°C, a wet bulb temperature of 15°C and an inlet flowrate of 100 $t \cdot h^{-1}$. A new cooling duty is to be added with a duty of 280 kW, with a maximum cooling water inlet temperature of 20°C. The heat capacity of water can be assumed to be 4.2 $kJ \cdot kg^{-1} \cdot K^{-1}$. Data for the performance of the cooling tower indicate that it can be modeled by the empirical relationship:

$$T_{out} = 0.8422(T_{in} + 17.8)^{0.38}(T_{WBT} + 17.8)^{0.405}F^{0.2} - 17.8 \quad (24.30)$$

where T_{in}, T_{out} = cooling tower inlet and outlet temperatures (°C).

T_{WBT} = wet bulb temperature (°C)

F = flowrate to the cooling tower ($t \cdot h^{-1}$)

After the new cooling duty is added, the temperature of the cooling water flow to the cooling network must be maintained to be 20°C.

   a. Confirm that if the new cooler is simply added in parallel to the existing coolers, the temperature to the cooling water network will be too high if the inlet to the cooling tower is maintained to be 30°C.
   b. Plot the cooling water composite curve. Assume that the maximum inlet and outlet temperatures for Cooler C1 are the existing ones from Figure 24.48 but that for Cooler C2 the maximum inlet and outlet temperatures can be increased to 30°C and 45°C.

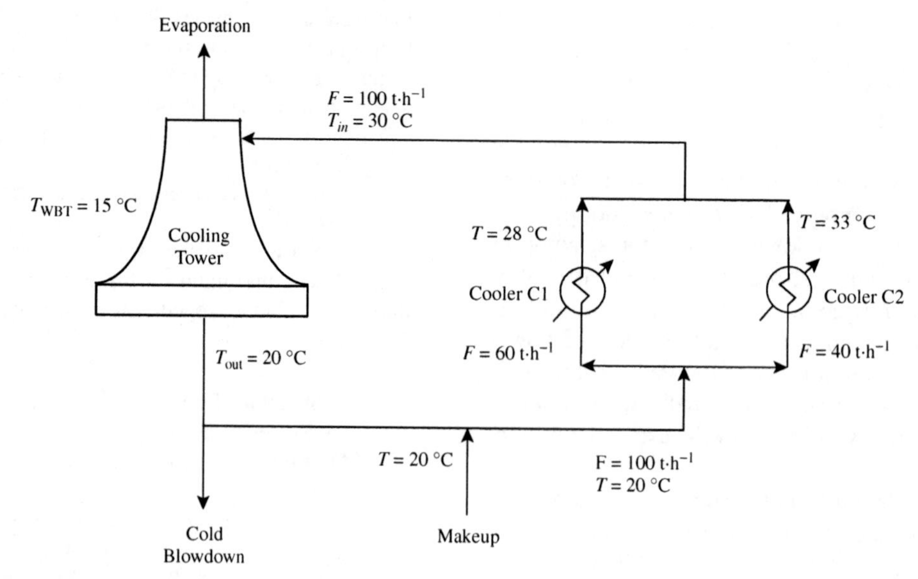

Figure 24.48 An existing cooling water system.

c. Determine the minimum cooling water flowrate and the corresponding cooling water return temperature. Confirm that this is above the maximum desirable return temperature for this system of 35°C.
d. If the return temperature is maintained to be 35°C, confirm using Equation 24.33 that the outlet temperature from the cooling tower will be acceptable.
e. Shift the limiting cooling data to allow for the maximum return temperature of 35°C and plot the shifted cooling water composite curve.
f. Design a cooling water network to achieve the target conditions of inlet and outlet cooling tower temperatures of 35°C and 20°C respectively (Hint: there are more than two design regions).

5. Rather than reconfigure the cooling water network in Exercise 4, what other options might have been considered to accommodate the new cooling duty without having to invest in a new cooling tower?
6. For cooling duties at
   a. −40°C
   b. −60°C

   Estimate the power required to reject 1 MW of cooling duty to cooling water with a return temperature of 40°C. Assume $\Delta T_{min} = 5$°C for the refrigeration evaporator and $\Delta T_{min} =$ 10°C for the condenser. Suggest suitable refrigeration fluids for these two duties using Figure 24.35.
7. The stream data for a low-temperature process are given in Table 24.7.

Table 24.7 Stream data for Exercise 7.

Stream	T_S (°C)	T_T (°C)	ΔH (MW)
1. Hot	65	−35	−15
2. Hot	−54	−55	−4
3. Cold	−85	55	7
4. Cold	−6	−5	12

Assume $\Delta T_{min} = 10$°C. Steam is available at 120°C and cooling water at 25°C to be returned at 35°C.
a. Carry out a problem table analysis and plot the grand composite curve.
b. Match a single-level refrigeration system using a pure refrigerant against the process. Determine the power required if heat is rejected from the refrigeration cycle to cooling water.
c. Match a two-level refrigeration system using a pure refrigerant against the process. Determine the power required if heat is rejected from the refrigeration cycle to the process.
d. For the two-level refrigeration system, devise a scheme to reduce the power requirement by heat rejection to the process rather than to cooling water. All of the heat from the refrigeration should be rejected to the process.

8. The data for a heat recovery problem are given in Table 24.8.

Table 24.8 Stream data for Exercise 8.

Stream		T_S (°C)	T_T (°C)	Stream heat duty (MW)
No	Type			
1	Hot	−20	−20	1.0
2	Hot	−40	−40	1.0
3	Hot	20	0	0.8
4	Cold	20	20	1.0
5	Cold	0	0	1.0
6	Cold	0	20	0.2
7	Cold	−40	20	0.6

Carry out a problem table analysis on the data for $\Delta T_{min} = 5$°C and plot the grand composite curve.
a. To reduce power requirements to a minimum, two levels of refrigeration are to be used. What are the temperatures and duties on the two levels?

b. Both the vaporization and condensation of the refrigerant occur isothermally in the refrigeration cycle. Assuming heat is to be rejected from the refrigeration cycle to cooling water, calculate the power requirements. Cooling water operates between 20°C and 25°C. The power required for the refrigeration system is given by Equation 24.20.

c. Rather than reject heat from the refrigeration cycle to cooling water, part can be rejected back into the process. Calculate the power requirement if the maximum heat possible is rejected to the process. The balance is to be rejected into cooling water.

REFERENCES

1. Kröger DG (1998) *Air-cooled Heat Exchangers and Cooling Towers*, Dept Mech Engg, University of Stellenbosch, South Africa.
2. Kim J-K and Smith R (2001) Cooling Water System Design, *Chem Eng Sci*, **56**: 3641.
3. Kim J-K and Smith R (2003) Automated Retrofit Design of Cooling Water Systems, *AIChE J*, **49**: 1712.
4. Gosney WB (1982) *Principles of Refrigeration*, Cambridge University Press.
5. Dossat RJ (1991) *Principles of Refrigeration*, 3rd Edition, Prentice Hall.
6. Isalski WH (1989) *Separation of Gases*, Oxford Science Publications.
7. Poling BE, Prausnitz JM and O'Connell JP (2001) *The Properties of Gases and Liquids*, McGraw-Hill.
8. Lee G-C (2001) *Optimal Design and Analysis of Refrigeration Systems for Low-temperature Processes*, PhD Thesis, UMIST, UK.
9. Lee G-C, Zhu XX and Smith R (2000) *Synthesis of Refrigeration Systems by Shaftwork Targeting and Mathematical Optimisation, ESCAPE 10*, Florence, Italy.
10. Lee G-C, Smith R and Zhu XX (2003) Optimal Synthesis of Mixed-refrigerant Systems for Low-temperature Processes, *Ind Eng Chem Res*, **41**: 5016.
11. Wang J (2004) *Synthesis and Optimization of Low Temperature Gas Separation Processes*, PhD Thesis, UMIST, UK.
12. Kinard GE and Gaumer LS (1973) Mixed Refrigerant Cascade Cycles for LNG, *Chem Eng Prog*, **69**: 56.
13. Bellow EG, Ghazal FP and Silverman AJ (1997) Technology Advances Keeping LNG Cost-Competitive, *Oil Gas J*, **June**: 74.
14. Finn AJ, Johnson GL and Tomlinson T (1999) Developments in Natural Gas Liquefaction, *Hydrocarbon Process*, **April**: 47.

25 Environmental Design for Atmospheric Emissions

25.1 ATMOSPHERIC POLLUTION

There are many types of emissions to atmosphere, and these can be characterized as particulate (solid or liquid), vapor and gaseous. Overall, the control of atmospheric emissions is difficult because the majority of emissions come from small sources that are difficult to regulate and control. Legislators therefore control emissions from sources that are large enough to justify monitoring and inspection. Industrial emissions of major concern are as follows.

1. PM_{10} – Particulate material less than 10 μm diameter is formed as a byproduct of combustion processes through incomplete combustion and through the reaction between gaseous pollutants in the atmosphere. PM_{10} is a particular problem as it causes damage to the human respiratory system.
2. $PM_{2.5}$ – Particulate material less than 2.5 μm diameter forms in the same way as PM_{10}, but it can penetrate deeper into the respiratory system than PM_{10}.
3. O_3 – Ozone is a very reactive compound present in the upper atmosphere (stratosphere) and the lower atmosphere (troposphere). Whilst ozone is vital in the stratosphere, its presence at ground levels is a danger to human health and contributes to the formation of other pollutants.
4. *VOCs* – A VOC is any compound of carbon, excluding carbon monoxide, carbon dioxide, carbonic acid, metal carbides or carbonates and ammonium carbonate, which participate in atmospheric photochemical reactions[1]. VOCs are precursors to ground-level ozone production and various photochemical pollutants and are major components in the formation of smog through photochemical reactions[2,3]. There are many sources of VOCs, as will be discussed later.
5. SO_x – Oxides of sulfur (SO_2 and SO_3) are formed in the combustion of fuels containing sulfur and as a byproduct of chemicals production.
6. NO_x – Oxides of nitrogen (principally NO and NO_2) are formed in combustion processes and as a byproduct of chemicals production.
7. *CO* – Carbon monoxide is formed by the incomplete combustion of fuel and as a byproduct of chemicals production.
8. CO_2 – Carbon dioxide is formed principally by the combustion of fuel but also as a byproduct of chemicals production.

Chemical Process Design and Integration R. Smith
© 2005 John Wiley & Sons, Ltd ISBNs: 0-471-48680-9 (HB); 0-471-48681-7 (PB)

Atmospheric emissions are controlled by legislation because of their damaging effect on the environment and human health. These effects can be categorized into their local effects and global effects. There are four main problems associated with atmospheric emissions[2,3].

1. *Urban smog*. Urban smog is commonly found in modern cities especially where air is trapped in a basin. It is observable as a brownish colored air. The formation of urban smog is through complex photochemical reactions that can be characterized by:

$$\text{VOCs} + NO_x + O_2 \xrightarrow{hf} O_3 + \text{other photochemical pollutants} \quad (25.1)$$

Photochemical pollutants such as ozone, aldehydes and peroxynitrates such as peroxyethanoyl (or peroxyacetyl) nitrate (PAN) are formed. Ozone and other photochemical pollutants have harmful effects on living organisms and on building structures. High levels of these pollutants can cause breathing difficulties and bring on asthma attacks in humans. Warm weather and still air exacerbate the problem. In addition to VOCs and NO_x, the problem of urban smog is made worse by particulate emissions and carbon monoxide from incomplete combustion of fuel.

2. *Acid rain*. Natural (unpolluted) precipitation is naturally acidic with a pH often in the range of 5 to 6 caused by carbonic acid from dissolved carbon dioxide and sulfurous and sulfuric acids from natural emissions of SO_x and H_2S. Human activity can reduce the pH very significantly down to the range 2 to 4 in extreme cases, mainly caused by emissions of oxides of sulfur. Because atmospheric pollution and clouds travel over long distances, acid rain is not a local problem. The problem may manifest itself a long way from the source. Problems associated with acid rain include:

- damage to plant life, particularly in forests
- acidification of water, leading to dead lakes and streams, loss of aquatic life and possible damage to human water supply
- corrosion of buildings, particularly those made of marble and sandstone.

3. *Ozone layer destruction*. The upper atmosphere contains a layer rich in ozone. Whilst ozone in the lower levels of the atmosphere is harmful, ozone in the upper levels

of the atmosphere is essential as it absorbs considerable amounts of ultraviolet light that would otherwise reach the Earth's surface. The destruction of ozone is catalyzed by oxides of nitrogen in the upper atmosphere:

$$NO\cdot + O_3 \longrightarrow NO_2^{\cdot} + O_2$$

$$NO_2^{\cdot} + O \longrightarrow NO\cdot + O_2$$

$$NO_2^{\cdot} \xrightarrow{hf} NO\cdot + O$$

Destruction is also initiated by certain halocarbon compounds such as:

$$CCl_2F_2 \xrightarrow{hf} \cdot CClF_2 + Cl\cdot$$

$$Cl\cdot + O_3 \longrightarrow ClO\cdot + O_2$$

$$ClO\cdot + O \longrightarrow Cl\cdot + O_2$$

The Cl· can then react with further ozone. Destruction of ozone has led to the appearance of ozone holes over the North and South Poles where the ozone layer is thinnest. The size of the holes varies during the year, but their growth has led to the need to reduce pollutants that destroy the ozone layer. The result of ozone layer destruction is increased ultraviolet light reaching the Earth, potentially increasing skin cancer and endangering polar species. This is a global effect that requires global solutions.

4. *The Greenhouse effect*. Gases such as CO_2, CH_4 and H_2O are present in low concentrations in the Earth's atmosphere. These gases reduce the Earth's emissivity and reflect some of the heat radiated by the Earth. Thus, the effect is to create a "blanket" to keep the Earth warmer than it would otherwise be. The problem arises mainly from burning fossil fuels and clearing forests by burning. The result is that global temperatures increase, leading to melting of the polar ice caps and glaciers, rising sea levels, desertification of areas, thawing of permafrost, increased weather disruptions and changes to ocean currents. This is a global problem requiring global solutions.

When applying legislation to atmospheric emissions, the regulatory authorities can either control emissions from individual points of release or combine all of the releases together as a combined release from a "bubble" around the manufacturing facility.

25.2 SOURCES OF ATMOSPHERIC POLLUTION

As noted above, one of the major problems with atmospheric emissions is the number of potential sources. Solid emissions arise from:

- incomplete combustion or fuel ash from furnaces, boilers and thermal oxidizers
- incomplete combustion in flares
- solids drying operations
- kilns used for high temperature treatment of solids
- metal manufacture
- crushing and grinding operations for solids
- any solids handling operation open to the atmosphere, and so on.

Vapor emissions are even more difficult as they have even more sources such as:

- condenser vents
- venting of pipes and vessels
- inert gas purging of pipes and vessels
- process purges to atmosphere
- drying operations
- incomplete combustion of fuel in furnaces, boilers and thermal oxidizers
- incomplete combustion in flares
- application of solvent-based surface coatings
- open operations such as filters, mixing vessels, and so on, leading to evaporation of volatile organic materials
- drum emptying and filling, leading to evaporation of volatile organic materials
- spillages of volatile organic material
- process plant ventilation of buildings processing volatile organic materials
- storage tank loading and cleaning
- road, rail and barge tank loading and cleaning
- fugitive emissions through gaskets and shaft seals
- fugitive emissions from sewers and effluent treatment
- fugitive emissions from process sampling points, and so on.

In larger plants, significant reductions in VOC emissions can usually be made by controlling major sources, such as tank venting, condensers and purges, and by inspection and maintenance of gaskets and shaft seals.

The largest volume of atmospheric emissions from process plants occurs from combustion producing gaseous emissions. Such emissions are created from:

- furnaces, boilers and thermal oxidizers
- gas turbine exhausts
- flares
- process operations where coke needs to be removed from catalysts (e.g. fluid catalytic cracking regeneration in refineries), and so on.

In addition to the gaseous emissions from the combustion of fuel, gaseous emissions are also produced by chemical production, for example, SO_x from sulfuric acid production, NO_x from nitric acid production, HCl from chlorination reactions, and so on.

Example 25.1 A storage tank with a vent to atmosphere is to be filled at 25°C with a mixture containing benzene with a mole fraction of 0.2 and toluene with a mole fraction of 0.8. Estimate the concentration of benzene and toluene in the tank vent:

a. at 25°C
b. corrected to standard conditions of 0°C and 1 atm

Assume that the mixture of benzene and toluene obeys Raoult's Law and the molar mass in kilograms occupies 22.4 m^3 in the vapor phase at standard conditions. The molar masses of benzene and toluene are 78 and 92 respectively. The vapor pressures of benzene and toluene at 25°C are 0.126 bar and 0.0376 bar respectively.

Solution

a.

$$y_{BEN} = 0.2 \times \frac{0.126}{1.013}$$
$$= 0.0249$$
$$y_{TOL} = 0.8 \times \frac{0.0376}{1.013}$$
$$= 0.0297$$

The balance is inert gas (air). Thus, the concentration at 25°C is given by:

$$C_{BEN} = y_{BEN} \times \frac{1}{22.4\left(\frac{T}{273}\right)} \times 78$$
$$= 0.0249 \times \frac{1}{22.4\left(\frac{298}{273}\right)} \times 78$$
$$= 0.0794 \text{ kg·m}^{-3}$$
$$= 79{,}400 \text{ mg·m}^{-3}$$
$$C_{TOL} = y_{TOL} \times \frac{1}{22.4\left(\frac{T}{273}\right)} \times 92$$
$$= 0.0297 \times \frac{1}{22.4\left(\frac{298}{273}\right)} \times 92$$
$$= 0.112 \text{ kg·m}^{-3}$$
$$= 112{,}000 \text{ mg·m}^{-3}$$

b. Corrected to standard conditions of 0°C and 1 atm:

$$C_{BEN} = y_{BEN} \times \frac{1}{22.4} \times 78$$
$$= 0.0249 \times \frac{1}{22.4} \times 78$$
$$= 0.0867 \text{ kg·m}^{-3}$$
$$= 86{,}700 \text{ mg·m}^{-3}$$
$$C_{TOL} = y_{TOL} \times \frac{1}{22.4} \times 92$$
$$= 0.0297 \times \frac{1}{22.4} \times 92$$
$$= 0.122 \text{ kg·m}^{-3}$$
$$= 122{,}000 \text{ mg·m}^{-3}$$

These approximate calculations reveal that the concentrations are many orders of magnitude greater than those permitted for such compounds in most situations (permitted levels are typically less than 10 mg·m^{-3}). More precise calculations could be performed with an equation of state (e.g. Peng–Robinson Equation of State, see Chapter 4).

25.3 CONTROL OF SOLID PARTICULATE EMISSIONS TO ATMOSPHERE

The selection of equipment for the treatment of solid particle emissions to atmosphere depends on a number of factors[4–8]:

- size distribution of the particles to be separated
- particle loading
- gas throughput
- permissible pressure drop
- temperature.

There is a wide range of equipment available for the control of emissions of solid particles, as discussed in Chapter 8. These methods are classified in broad terms in Table 25.1[5].

1. *Gravity settlers*. Gravity settlers have already been discussed in Chapter 8 and illustrated in Figures 8.1 and 8.3. These are used to collect coarse particles and may be used as prefilters. Only particles in excess of 100 μm can reasonably be removed[5–8].

2. *Inertial collectors*. Inertial collectors were also discussed in Chapter 8 and illustrated in Figure 8.4. The particles are given a downward momentum to assist the settling. Only particles in excess of 50 μm can be reasonably

Table 25.1 Methods of control of emissions of solid particles.

Equipment	Approximate particle size range (μm)
Settling chambers	>100
Inertial separators	>50
Cyclones	>5
Scrubbers	>3
Venturi scrubbers	>0.3
Bag filters	>0.1
Electrostatic precipitators	>0.001

removed[5–8]. Like gravity settlers, inertial collectors are widely used as prefilters.

3. *Cyclones.* Cyclones are also primarily used as prefilters. These have already been discussed in Chapter 8 and illustrated in Figure 8.6. The particle-laden gas enters tangentially and spins downwards and inwards, ultimately leaving the top of the unit. Particles are thrown radially outwards to the wall by the centrifugal force and leave at the bottom.

Cyclones can be used under conditions of high particle loading. They are cheap, simple devices with low maintenance requirements. Problems occur when separating materials, which have a tendency to stick to the cyclone walls.

4. *Scrubbers.* Scrubbers are designed to contact a liquid with the particle-laden gas and entrain the particles with the liquid. They offer the obvious advantage that they can be used to remove gaseous as well as particulate pollutants. The gas stream may need to be cooled before entering the scrubber. Some of the more common types of scrubbers are shown in Figure 8.11.

Packed columns are widely used in gas absorption, but particulates are also removed in the process, Figure 8.11a. The main disadvantage of packed columns is that the solid particles will often accumulate in the packing and require frequent cleaning. Some designs of packed column allow the packing to self-clean through inducing movement of the bed from the up flow of the gas. Spray scrubbers are less prone to fouling and use a tangential inlet to create a swirl to enhance the separation, as illustrated in Figure 8.11b. A considerable reduction in particle size separation can be achieved at the expense of increased pressure drop using a venturi scrubber, Figure 8.11c.

5. *Bag filters.* Bag filters, as discussed in Chapter 8 and illustrated in Figure 8.10b, are probably the most common method of separating particulate materials from gases. A cloth or felt filter material is used, which is impervious to the particles. Bag filters are suitable for use in very high dust load conditions. They have a high efficiency but suffer from the disadvantage that the pressure drop across them may be high.

6. *Electrostatic precipitators.* Electrostatic precipitators are used where collection of fine particles at a high efficiency coupled with a low-pressure drop is necessary. The arrangement is illustrated in Figure 8.9. Particle-laden gas enters a number of tubes or passes between parallel plates. The particulates are charged and are deposited on the earthed plates or tube walls. The walls are mechanically *rapped* periodically to remove the accumulated dust layer.

25.4 CONTROL OF VOC EMISSIONS TO ATMOSPHERE

Release limits for VOCs are set for either specific components (e.g. benzene, carbon tetrachloride), or as VOCs for organic compounds with a lower environmental impact and classed together and reported, for example, as toluene.

The hierarchy appropriate for control of VOC emissions is[9] as follows:

1. Eliminate or reduce VOC emissions at source
2. Recover the VOC for reuse
3. Recover the VOC for treatment and disposal
4. Treatment and disposal of the VOC-laden gas stream.

The problem with eliminating or reducing VOC emissions at source is that the sources of VOC emissions can be many and varied.

Tank loading can be a significant source of VOC emissions from atmospheric storage tanks used for the storage of organic liquids. Figure 25.1 shows an atmospheric storage tank first being emptied. As the liquid level falls, air or inert gas must be drawn into the tank to prevent the tank from collapsing from the pressure of the atmosphere. The air or nitrogen drawn into the vapor space will become saturated in organic vapor from evaporated liquid in the tank. As the tank is filled, the liquid level rises, displacing the air or nitrogen from the vapor space in the tank that is now saturated in VOC material. There are a number of ways in which such VOC emissions can be prevented from being omitted from storage tanks. Figure 25.2 shows a simple technique involving a balancing line. The atmospheric storage tank in Figure 25.2 is fitted with a vacuum/pressure relief system to guard against over-pressure or under-pressure of the storage tank. Now as the atmospheric storage tank is emptied and the road tanker filled in Figure 25.2, the displaced vapor from the road tanker is pushed back into the atmospheric storage tank. In this way, displaced vapor from the road tanker is prevented from being released to the atmosphere. The same system works if the road tanker is being emptied into the atmospheric storage tank, rather than being filled from the atmospheric storage tank as shown in Figure 25.2. If the atmospheric storage tank is being filled, then the displaced vapor from the atmospheric storage tank enters the road tanker vapor space and is not released to the atmosphere. Clearly, the same technique works for rail cars and barges.

Another way storage tanks can "breathe" to atmosphere results from the expansion and contraction of the tank contents as the temperature changes between day and night. However, such effects tend to be smaller than those associated with tank filling and emptying.

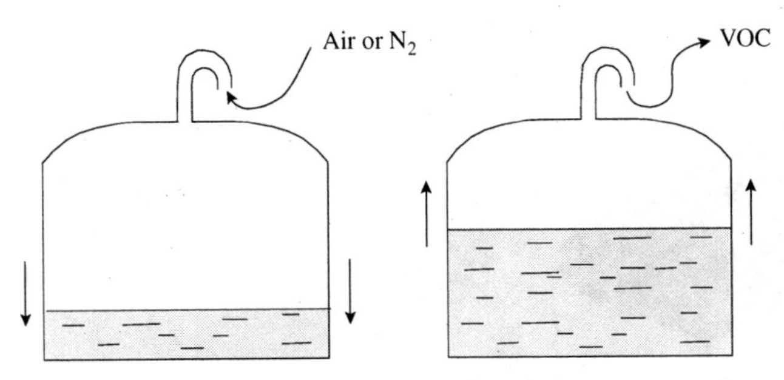

Figure 25.1 Tank loading causes VOC emissions to atmosphere.

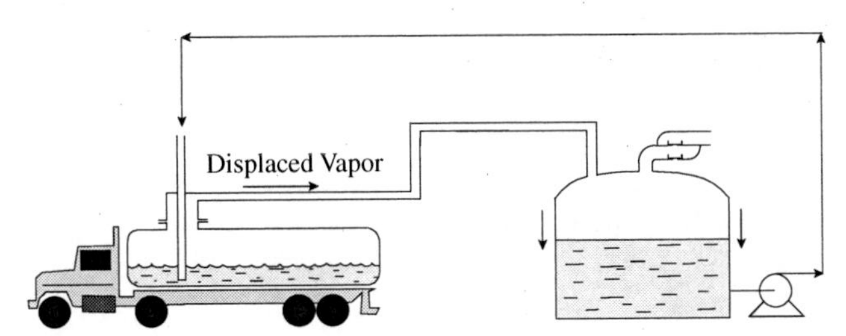

Figure 25.2 A balancing line can prevent VOC emissions when filling and emptying tankers.

Floating roofs and flexible membranes can be used, as illustrated in Figure 25.3:

1. *External floating roof.* In this arrangement, the roof floats directly on the surface of the stored liquid and is open to the atmosphere (Figure 25.3a). The floating roof has a seal system for the gap between the roof and the tank wall. In principle, this eliminates breathing of the tank to atmosphere, but maintaining a reliable seal between the edge of the floating roof and the tank wall can be problematic.
2. *Internal floating roof.* In this arrangement, the tank has a fixed roof in addition to the floating roof, and the floating roof is not open to the atmosphere (Figure 25.3b). The space between the floating and fixed roofs should normally be purged with an inert gas. The inert gas would be vented to atmosphere after treatment. Internal floating roofs are used when there is likely to be a heavy accumulation of rainwater or snow on the floating roof.
3. *Flexible membrane.* In this arrangement, the roof changes shape to stop the vapor space breathing to atmosphere, as shown in Figure 25.3c.

Effluent systems can be significant sources of the VOC emissions. If organic material is sent to effluent, along with aqueous effluent, then evaporation from open drains can create significant VOC emissions. If this is the case, then having separate sewers for organic and aqueous waste, as illustrated in Figure 25.4, can eliminate the problem. The organic waste is collected in a closed system, typically draining to a sump tank, from which the organic material can be recovered or treated before disposal. The segregation system shown in Figure 25.4 has many other benefits

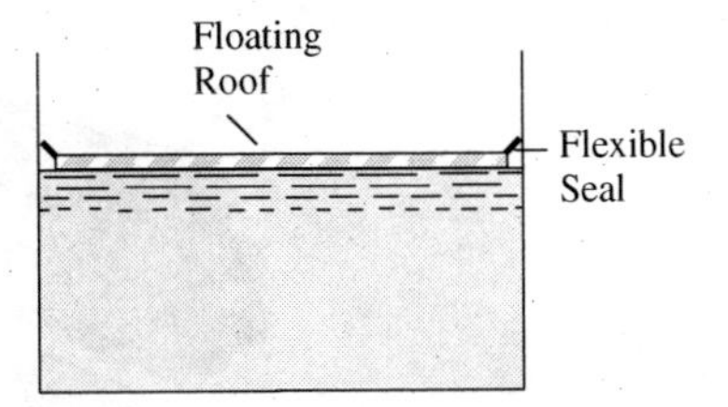

(a) External floating roof storage tank.

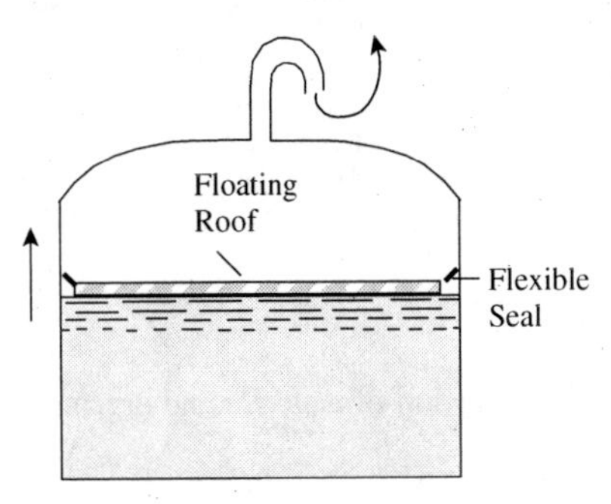

(b) Internal floating roof storage tank.

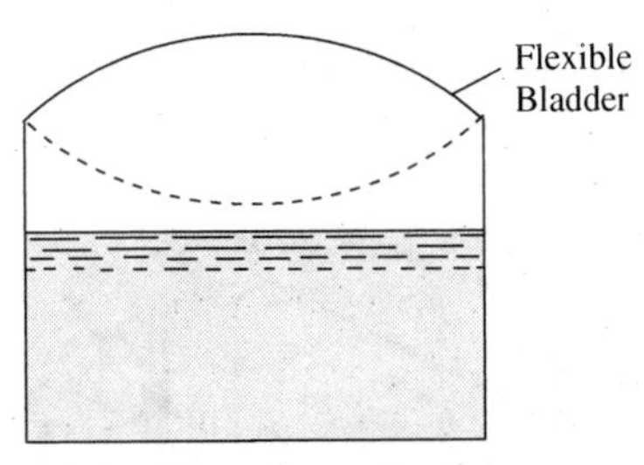

(c) Flexible roof storage tank.

Figure 25.3 Floating roof and flexible membranes can be used to prevent the release of material. (From Smith R and Patela EA, 1992, *The Chemical Engineer*, No 517: 9 April, reproduced by permission of the Institution of Chemical Engineers).

than simply eliminating VOC emissions from drains. Segregation of the effluents, as illustrated in Figure 25.4, is also useful from the point of view of the treatment of liquid waste from the process, as will be discussed in the next chapter.

Sampling of organic liquids for analysis and quality control can also be a significant source of VOC emissions. Figure 25.5 shows a traditional sampling technique involving a sample point with a valve. The sampling technique might start with flushing the sample point to drain to obtain a representative sample. The sample vessel may then be rinsed out with the organic material, again to obtain a representative sample, and the rinsed material rejected. The sample vessel would then be taken to the laboratory, the required sample taken and the remainder or the contents of the sample vessel sent to waste. The material going to drain from this procedure can create VOC emissions from evaporation from the sewers as discussed above. It might be suspected that such procedures would only contribute a small amount of waste and, therefore, VOC emissions. However, in some chemical processes where a considerable

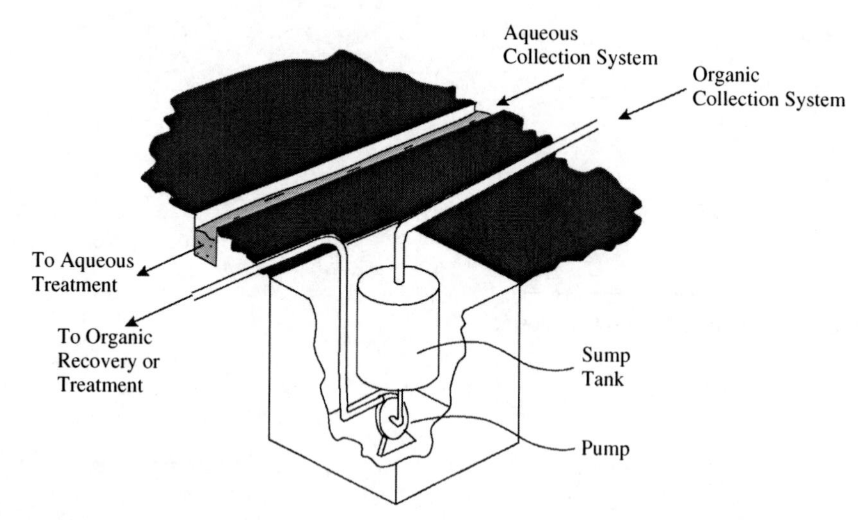

Figure 25.4 Segregation of aqueous and organic waste can prevent VOC emissions from open drains.

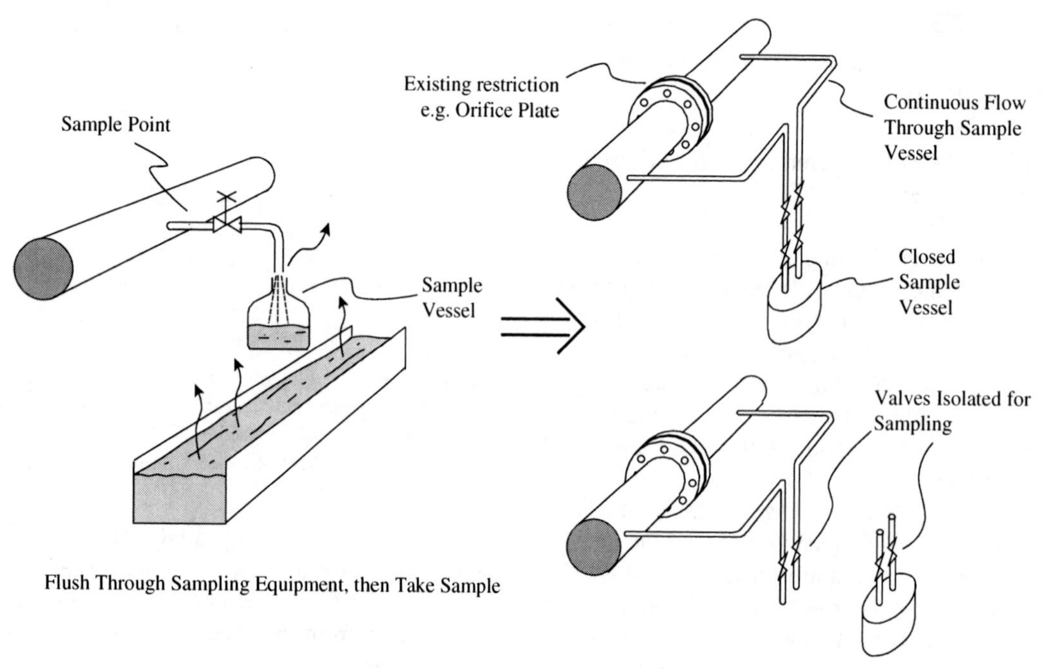

Figure 25.5 Prevention of emissions from sampling.

amount of analysis is required, sampling waste can be one of the biggest sources of waste sent to liquid effluent. Figure 25.5 shows a method to eliminate sampling waste and, therefore, the VOC emissions associated with sampling waste. A sample vessel is connected directly to the pipe from which the sample needs to be taken by two tubes. The two tubes are connected to the pipe from which the sample needs to be taken on either side of a restriction, such as an orifice plate or valve. If the valves in the tubes connecting the sample to the pipe are open, then the pressure drop across the restriction will create a flow through the sample vessel. This continuous flow will ensure that a representative sample is obtained. When the sample needs to be taken, valves are closed to isolate the closed sampling vessel. This can then be taken to the laboratory for analysis. Any unused material in the sample vessel is returned to the plant and reconnected, eliminating the waste of sampling material from the laboratory.

Slow leaks from gaskets and from compressor, pump and valve seals can create significant fugitive emissions of VOCs. Such emissions can be reduced at source by better maintenance. However, more sophisticated sealing arrangements are desirable to eliminate the problem at source.

Once VOC emissions have been eliminated or reduced at source, then recovery of the VOC for reuse should be considered. Figure 25.6 shows a vapor recovery system associated with an atmospheric storage tank. The storage tank is fitted with a vacuum-pressure relief valve, which

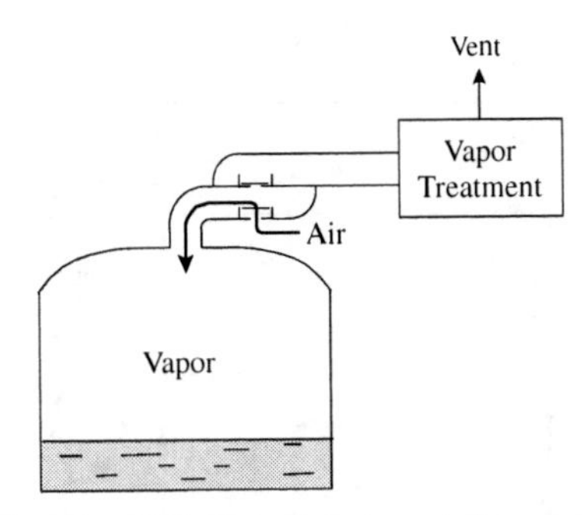

(a) As liquid level falls, air or N_2 is drawn in from atmosphere.

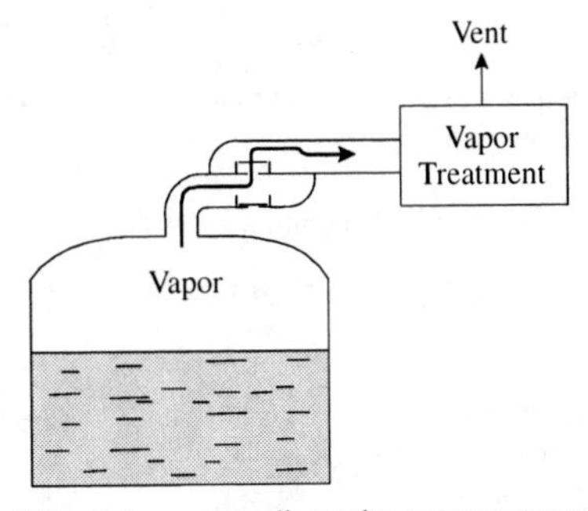

(b) As liquid level rises, vapor diverted to a vapor recovery system.

Figure 25.6 Storage tank fitted with a vapor treatment system. (From Smith R and Patela EA, 1992, *The Chemical Engineer*, No 517: 9 April, reproduced by permission of the Institution of Chemical Engineers).

allows air (or nitrogen) into the vapor space when the liquid level falls (Figure 25.6a) but forces the vapor through a recovery system when the tank is filled (Figure 25.6b).

Many processes involve open operations (e.g. filters, drum handling, etc.) that create VOC emissions. If this is the case, it is often not practical to enclose all such operations, in which case, a ventilation system needs to draw a continuous flow across the operation into a duct and then a vapor recovery system, before being released to the atmosphere.

The methods of VOC recovery are[1,10]:

- condensation
- membranes
- absorption
- adsorption.

1. *Condensation.* Condensation can be accomplished by increase in pressure or decrease in temperature. Most often, decrease in temperature is preferred. Figure 25.7 illustrates two ways in which VOC recovery can be accomplished from an airstream used in a process. Figure 25.7a shows an open loop in which air enters the process, picks up VOCs and enters a condenser for recovery of the VOCs before being vented directly or passed to secondary treatment before venting. An alternative way is shown in Figure 25.7b. This shows a recirculation system in which the VOC is recovered from recycled air. If this is possible, it is a better arrangement than Figure 25.7a, as the exhaust is eliminated.

Unfortunately, condensation of VOCs usually requires refrigeration. Figure 25.8a shows an arrangement with refrigerated condensation using the primary refrigerant in a simple refrigeration loop. Figure 25.8b shows an arrangement in which a secondary refrigerant is used in the VOC condenser. The use of a secondary refrigerant

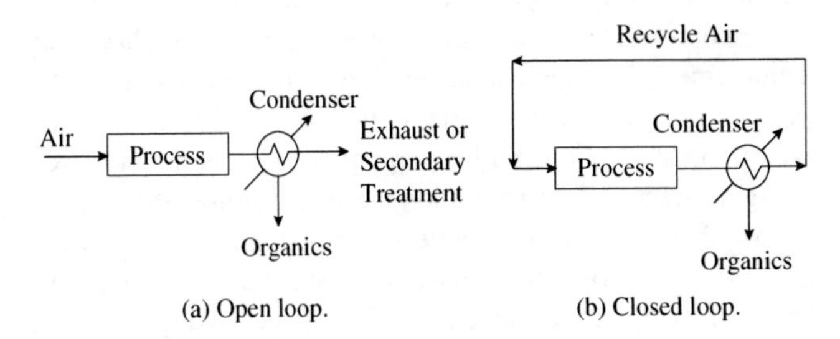

(a) Open loop. (b) Closed loop.

Figure 25.7 Condensation of volatile material.

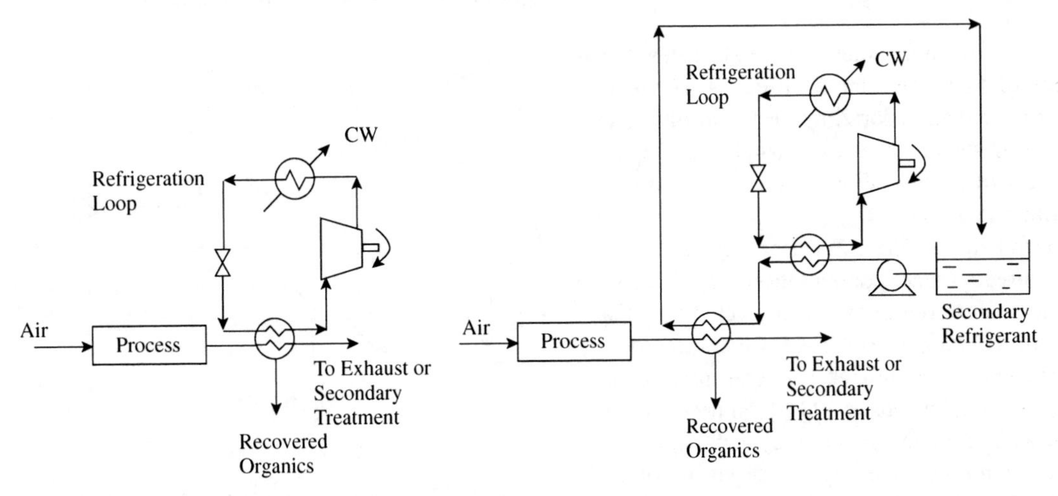

(a) Use of primary refrigerant.

(b) Use of secondary refrigerant (if a number of cooling duties required).

Figure 25.8 Refrigeration is usually required for the condensation of vapor emissions.

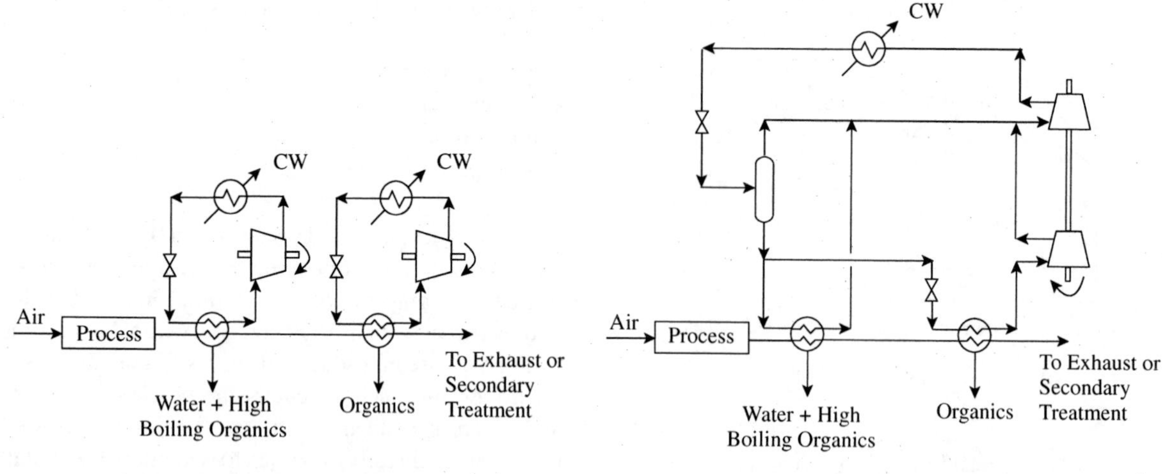

Figure 25.9 If a gas stream contains water vapor or organic compounds with high freezing points, then precooling is necessary.

is less efficient than using the primary refrigerant directly. However, if a number of cooling duties are required, then a single refrigeration loop cooling a secondary refrigerant can be used for a number of condenser duties.

A further complication of refrigeration is that, if the gas stream contains water vapor or organic compounds with high freezing points, then these will freeze in the condenser, causing it to block up. It is therefore usual to precool the gas stream before condensing the VOCs. Figure 25.9a shows an arrangement in which a refrigeration loop is first used to precool the gas stream to the region of 2 to 4°C to remove water vapor and high-boiling organics. A second condenser using a second refrigeration loop then condenses the VOCs. Figure 25.9b shows the same basic arrangement but with a two-level refrigeration system, instead of two separate loops. In both cases in Figure 25.9, it is likely that the VOC condenser will have to be taken off-line periodically to remove frozen water vapor and organic material that will accumulate over time.

If the site uses nitrogen as inert gas and stores the nitrogen as a liquid, then this can be used as a source of refrigeration. Rather than vaporizing the liquid nitrogen by low-pressure steam or an air heat exchanger, the liquid nitrogen can be used to cool a VOC-laden gas stream to condense the organic material.

The problem of freezing on the condenser surface can be avoided by using direct contact condensation, as shown in Figure 25.10. A secondary refrigerant is contacted directly with the vent stream containing VOC. A refrigeration loop is used to cool the secondary refrigerant. The mixture of secondary refrigerant and condensed VOC then needs to be separated and the secondary refrigerant recycled.

It is usually uneconomic to compress a gas in order to bring about the condensation of VOC. Figure 25.11 shows a flowsheet in which the VOC-laden vent gas is compressed in order that condensation can take place using

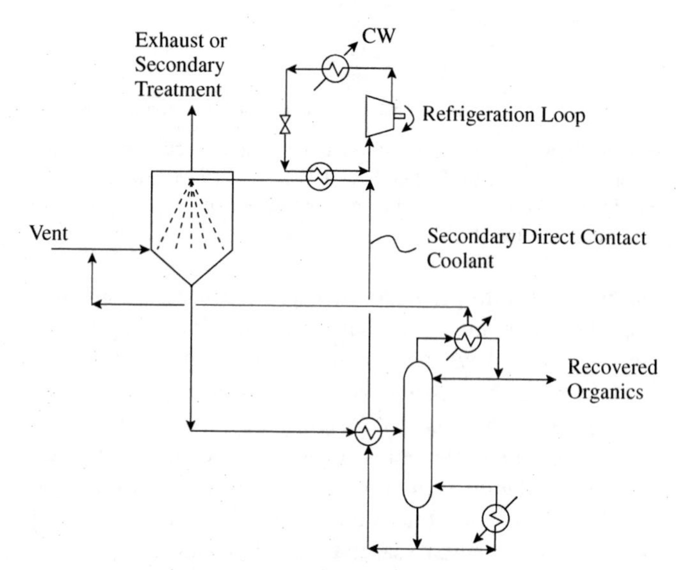

Figure 25.10 Direct contact condensation.

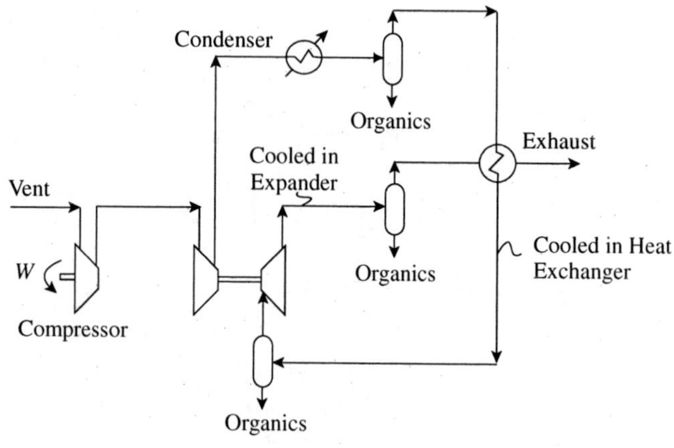

Figure 25.11 Compression of the vent with recovery of compression power.

cooling water rather than refrigeration. The arrangement in Figure 25.11 uses an expander on the same shaft as the high-pressure compression stage in order to recover part of the work of compression. Around 60% of the work of compression can be recovered in an arrangement like the one in Figure 25.11.

Example 25.2 An airstream contains acetone vapor. Estimate the following.

a. The outlet concentration that can be achieved by cooling to 35°C at 1 atm pressure (cooling water temperature).
b. The outlet concentration that can be achieved by cooling to −20°C at 1 atm pressure.
c. The outlet temperature to which the air stream must be cooled to obtain an outlet concentration of 20 $mg{\cdot}m^{-3}$.

The concentrations should be quoted at standard conditions of 0°C and 1 atm. It can be assumed that the molar mass in kilograms occupies 22.4 m^3 at standard conditions. The molar mass of acetone can be taken to be 58. The vapor pressure of acetone is given by:

$$\ln P^{SAT} = 10.0310 - \frac{2940.46}{T - 35.93}$$

where P^{SAT} = saturated liquid vapor pressure (bar)
T = absolute temperature (K)

Solution

a. Acetone will start to condense when the partial pressure equals the vapor pressure. At 35°C (308 K), the vapor pressure is given by:

$$P^{SAT} = \exp\left[10.0310 - \frac{2940.46}{308 - 35.93}\right]$$
$$= 0.460 \text{ bar}$$

For a partial pressure of 0.608 bar, the mole fraction is given by:

$$y = \frac{0.460}{1.013}$$
$$= 0.454$$
$$C = y \times \frac{1}{22.4} \times 58$$
$$= 0.454 \times \frac{1}{22.4} \times 58$$
$$= 1.18 \text{ kg} \cdot \text{m}^{-3}$$
$$= 1.18 \times 10^6 \text{ mg} \cdot \text{m}^{-3}$$

b. At −20°C (253 K):

$$P^{SAT} = 0.0297 \text{ bar}$$
$$C = \frac{0.0297}{1.013} \times \frac{1}{22.4} \times 58$$
$$= 0.0759 \text{ kg}{\cdot}\text{m}^{-3}$$
$$= 75{,}900 \text{ mg}{\cdot}\text{m}^{-3}$$

c. For a concentration of 20 $mg{\cdot}m^{-3}$ (20×10^{-6} $kg{\cdot}m^{-3}$):

$$y = \frac{20 \times 10^{-6}}{58} \times 22.4$$
$$= \frac{P^{SAT}}{1.013}$$
$$P^{SAT} = 1.013 \times \frac{20 \times 10^{-6}}{58} \times 22.4$$
$$= 7.825 \times 10^{-6} \text{ bar}$$

Rearranging the vapor pressure correlation,

$$T = 35.93 - \frac{2940.46}{\ln P^{SAT} - 10.310}$$
$$= 35.93 - \frac{2940.46}{\ln 7.825 \times 10^{-6} - 10.310}$$
$$= 169 \text{ K}$$
$$= -104°\text{C}$$

This temperature is outside of the range of the vapor pressure correlation and below the freezing point of acetone (−95°C). However, it illustrates just how low temperatures must be to achieve very low concentrations of VOC.

Example 25.2 illustrates that low temperature is required for a high recovery rate of VOCs. However, extremely low temperatures are required to achieve high environmental standards. Therefore, condensation is often used in conjunction with another process (e.g. adsorption) to achieve high environmental standards.

When the VOC-laden gas stream contains a mixture of VOCs, then the calculations must be performed using the methods described for single-stage equilibrium calculations in Chapter 4. The temperature at the exit of the condenser must be assumed, together with a condenser pressure. The vapor fraction is then solved by trial and error using the methods described in Chapter 4, and the complete mass balance can be determined on the basis of the assumption of equilibrium.

The major advantage of condensation is that VOCs are recovered without contamination (by contrast with absorption and adsorption with steam or hot gas regeneration). The significant disadvantages of refrigeration are that there is a relatively low efficiency of recovery (typically less than 95%) and refrigeration systems have high operating costs.

2. *Membranes.* VOCs can be recovered using an organic selective membrane that is more permeable to organic vapors than permanent gases[10]. Figure 25.12 shows one possible arrangement for recovery of VOCs using a membrane[10]. A vent gas is compressed and enters a condenser in which VOCs are recovered. The gases from the condensers then enter a membrane unit, in which the VOC permeates through the membrane. A VOC-enriched permeate is then recycled back to the compressor inlet. Recovery rates for such an arrangement can be as high

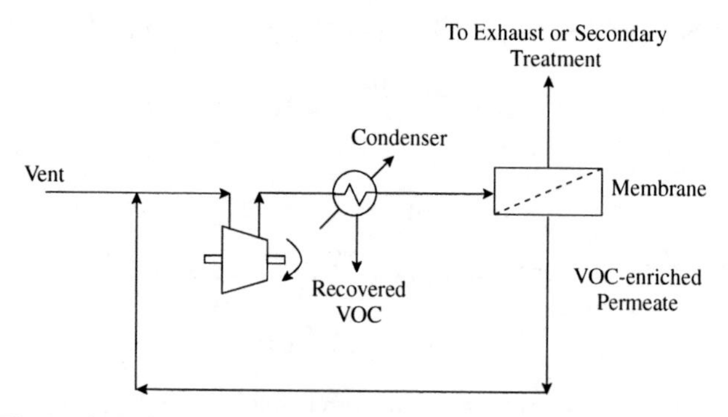

Figure 25.12 Recovery of VOCs using membranes.

as 90 to 99%. Other arrangements than the one shown in Figure 25.12 are possible, again using a combination of membrane and condensation.

3. *Absorption*. Physical absorption, as discussed in Chapter 10, can be used for the recovery of VOCs. The solvent used should be regenerated and recycled, with the VOC being recovered if at all possible. If the VOC is water-soluble (e.g. formaldehyde), then water can be used as the solvent. However, high-boiling organic solvents are most often used for VOC absorption. Efficiency of recovery depends on the VOC, solvent and absorber design, but efficiencies of recovery can be as high as 95%. The efficiency of recovery increases with decreasing temperature and increases with increasing pressure. When using an organic solvent, care must be exercised in the selection of the solvent that vaporization of the solvent into the gas stream does not create a new environmental problem to replace the recovered VOC.

4. *Adsorption*. Adsorption of VOCs is most often carried out using activated carbon with in-situ regeneration of the carbon using steam. The details have been discussed in Chapter 10. A number of different arrangements are used for adsorption. One possible arrangement is shown in Figure 25.13. This involves a three-bed system. The first bed encountered by the VOC-laden gas stream is the primary bed. The gas from the primary bed enters the secondary bed, which will have just been regenerated using steam and is now cooling. The third bed is off-line being regenerated using steam. The steam from the regenerating bed is condensed and the condensate separated to recover the organics. The vent from the condenser receiver will normally be too concentrated to vent directly to atmosphere and therefore will normally be connected back to the inlet of the primary bed. Once the primary bed has become saturated and breakthrough occurs (see Chapter 10), the beds are switched. The current secondary bed becomes the new primary bed. The regenerated bed becomes the secondary bed, and the previous primary bed is now regenerated. The switching cycle is normally based on a timing arrangement.

Systems with one bed can be used for small duties (typically less than 5 kg per day discharge of organic material) on the basis of disposable cartridges. Two-bed arrangements can be used when environmental standards are not too demanding. One bed is absorbing, whilst the other is being regenerated, with constant switching between the two. More complex cycles than the one in Figure 25.12 can be used involving four beds, instead of three, for more demanding duties.

The efficiency of recovery for adsorption:

- increases with molar mass of the VOC (carbon adsorbers are not good for controlling low-boiling VOCs with molar mass less than around 40)
- increases with decreasing adsorption temperature

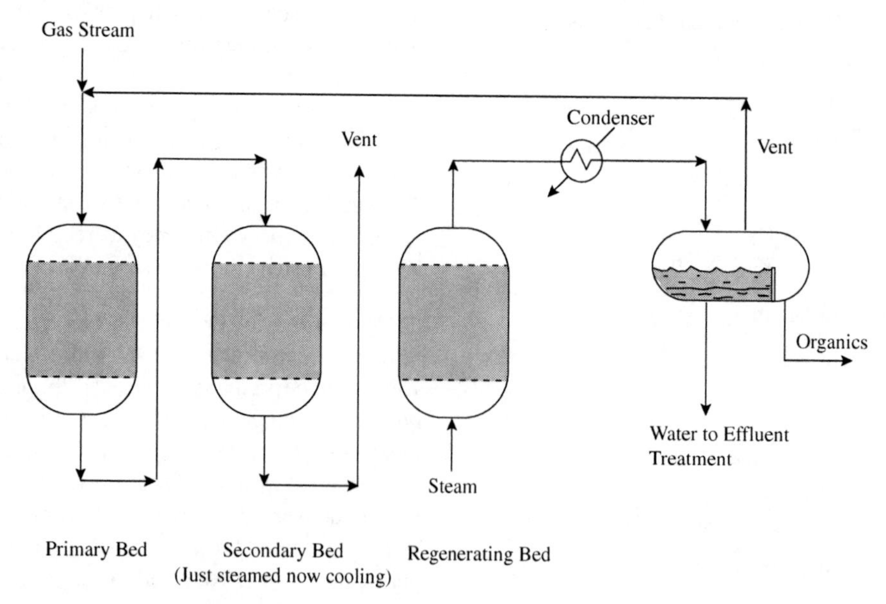

Figure 25.13 Adsorption with a three-bed system.

- increases with increasing pressure
- increases with decreasing concentration of water vapor in the gas stream (due to competition of water for the adsorption sites).

In summary, condensation and absorption are usually the simplest methods of VOC recovery. Recovery methods can be used in combination effectively (but at a cost). Adsorption is usually the only method capable of recovery to achieve very low concentrations of VOC. If the gas stream contains a mixture of VOCs, then the recovered liquid might not be suitable for reuse and will need to be separated by distillation or destroyed by thermal oxidation.

Once the potential for minimizing VOC emissions at source has been exhausted and the recovery of VOCs exhausted, then any remaining VOC must be destroyed. There are two methods of VOC destruction[1]:

- thermal oxidation
- biological treatment.

Consider now the various methods of thermal oxidation.

1. *Flare stacks*. Flares use an open combustion process with oxygen for the combustion provided by air around the flame. Good combustion depends on the flame temperature, residence time in the combustion zone and good mixing to complete the combustion. The flare stack can be dedicated to a specific vent stream or for a combination of streams via a header. The flare can be used for normal process releases or emergency upsets. Flares can be categorized by the:

- height of the flare tip (elevated versus ground flares)
- method of enhancing mixing at the flare tip (steam assisted, air assisted, unassisted).

The most common type used in the process industries is the steam-assisted elevated flare. This is illustrated in Figure 25.14. The vent stream in Figure 25.14 first enters a knockout drum to remove any entrained liquid. Liquids must be removed as they can extinguish the fame or lead to irregular combustion in the flame. Also, there is a danger that the liquid might not be completely combusted, which can result in liquid reaching the ground and creating hazards. The vent stream then passes through a *flame arrester* or a *gas barrier*. The flame arrester prevents the flame passing down into the vent system when the vent stream flowrate is low. The vent stream then enters the flare tip where the combustion takes place. The flare tip requires pilot burners, an ignition device and fuel gas to maintain the flame. In steam-assisted flares, steam nozzles around the perimeter of the flare tip are used to provide the mixing. Flare performance can achieve a destruction efficiency of up to 98% with steam-assisted flares. Halogenated and sulfur-containing compounds should not be flared. Flares are applicable to almost any VOC-laden stream and can be used when there are large variations in the composition, heat content and flowrate of the vent stream. The use of flares is mostly restricted to emergency release.

2. *Thermal oxidation*. Figure 25.15 illustrates three designs of thermal oxidizer. Figure 25.15a shows a vertical thermal oxidizer. The vent enters a refractory-lined combustion chamber fed with auxiliary air and auxiliary fuel. Figure 25.15b shows the corresponding arrangement mounted horizontally. Figure 25.15c shows a fluidized bed arrangement. In this arrangement, the vent enters a bed of fluidized inert material (e.g. sand), along with supplementary fuel. The bed is fluidized by an airstream. The fluidized bed provides good mixing in the combustion zone. It also

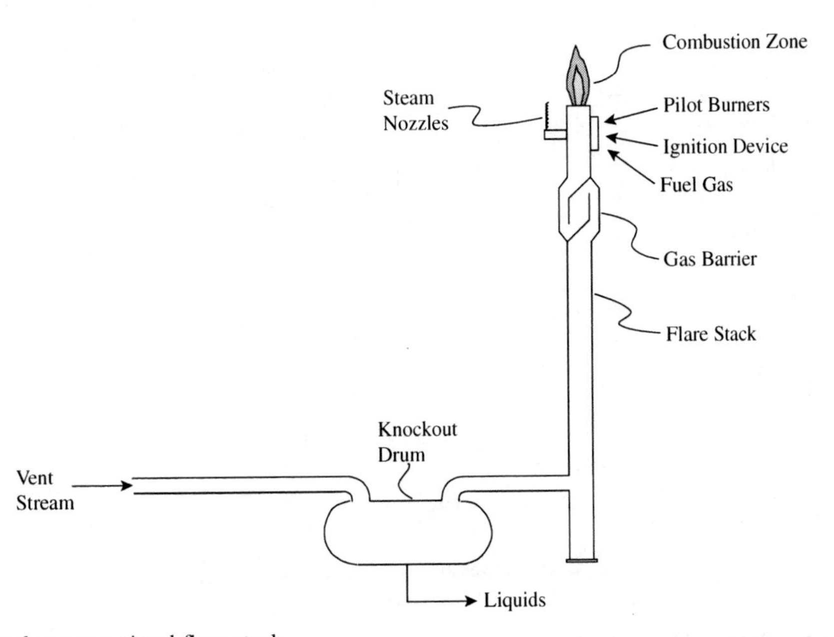

Figure 25.14 An elevated steam-assisted flare stack.

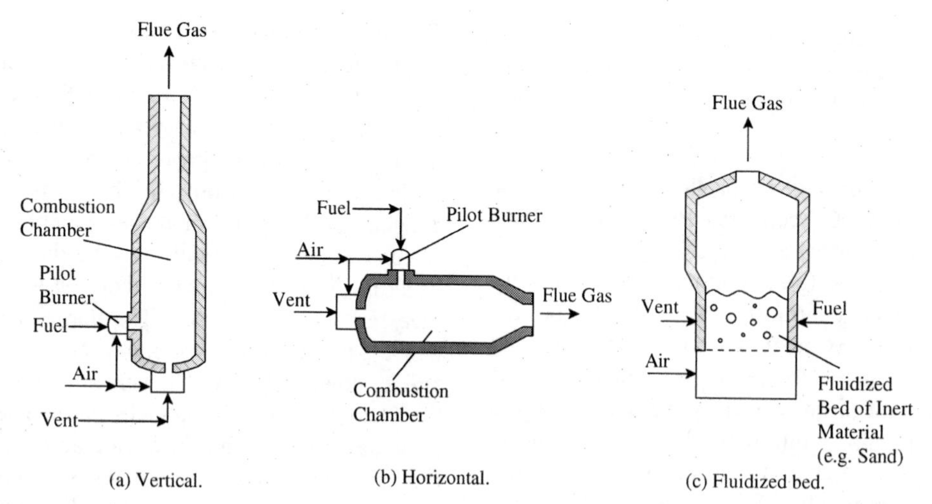

Figure 25.15 Thermal oxidation.

provides some heat storage that can compensate for variations in the entering vent stream.

Conditions in the thermal oxidizer are typically 25% oxygen over and above stoichiometric requirements. Minimum temperatures required for VOCs comprising carbon, hydrogen and oxygen are around 750°C, but 850 to 900°C are more typical. The residence time of the VOCs in the combustion zone is typically 0.5 to 1 second. When thermally oxidizing halogenated organic compounds, the minimum temperature used is in the range of 1100 to 1300°C, with a residence time of up to two seconds.

Supplementary fuel is required for start-up and required if the feed concentration of organic material is low or the feed concentration varies. Supplementary fuel is usually required for vent streams, as processed vents are normally designed to operate below the lower flammability limit in the case of destruction of VOCs in air, or below the minimum oxygen concentration for VOCs in an inert gas mixture (see Chapter 27). The VOC should normally be less than 25 to 30% of the lower flammability limit for air mixtures or less than 40% of the minimum oxygen concentration for inert gas mixtures. These limits can be increased if on-line analysis equipment is installed to monitor VOC concentrations continuously. If the vent stream is above this limit, the airflow in the vent is normally increased accordingly. Otherwise, nitrogen can be introduced, but this is more expensive.

Heat recovery is an important feature of many designs. Figure 25.16a shows thermal oxidizer with recuperative heat recovery. In this arrangement, fuel and air enter directly into

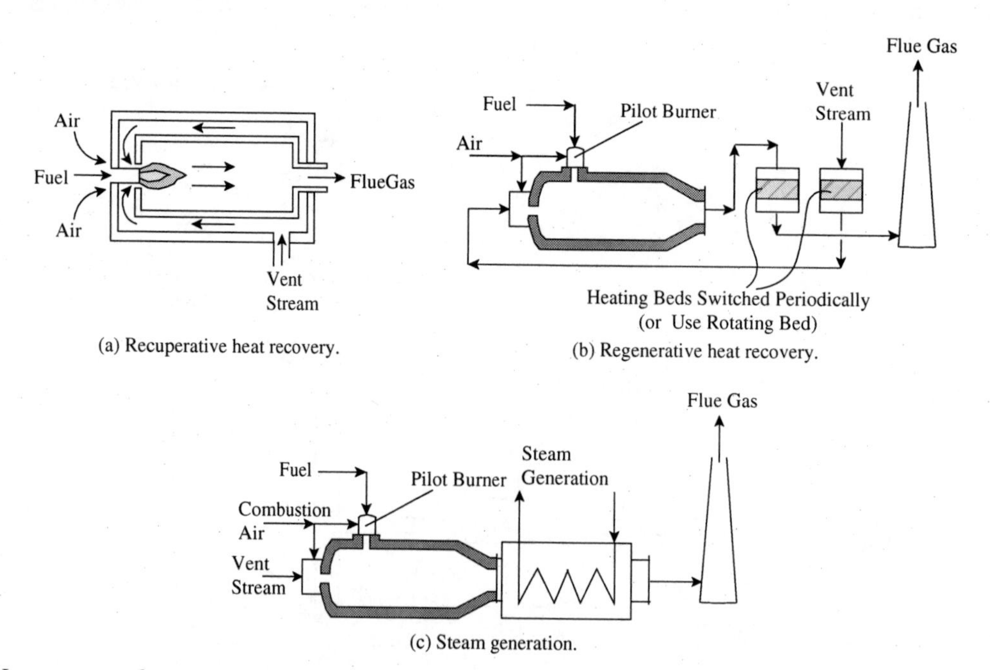

Figure 25.16 Heat recovery from thermal oxidation.

the combustion zone. However, the vent stream is preheated by heat transfer through the walls of the combustion chamber. Many different schemes for recuperative heat recovery through indirect exchange of heat in heat transfer equipment are possible. Figure 25.16b shows a thermal oxidizer with regenerative heat recovery. Fuel and air enter the combustion zone directly. However, the vent stream enters via a hot bed that preheats the vent stream prior to entering the combustion chamber. The scheme in Figure 25.16b shows two regenerative beds that are switched periodically. As the bed used to heat the vent stream cools, the other bed is being heated by the hot exhaust gases before being vented. Once the bed that is preheating the vent stream cools, then the beds are switched such that the cool bed is then preheated by the hot flue gas and the new hot bed is used to preheat the incoming vent stream. The material used in the regenerative beds can be refractory material or metal. It should be noted that with a two-bed regenerative system, when one bed is switched from heating mode to cooling mode, that bed will contain vapor feed that has not yet been treated. This untreated vapor within the bed will thus be vented to atmosphere directly untreated. This problem can be overcome by the introduction of an additional bed in a three-bed system. However, this introduces additional capital cost and complexity of operation. Alternatively, the regeneration can be carried out continuously by a rotating bed rather than having beds being switched. The rotating bed is in the form of a wheel that slowly rotates. The ducting arrangement is such that the incoming vent stream and the hot exhaust gas both flow axially across the wheel, with different parts of the wheel being exposed to each stream. Different parts of the wheel are thus in heating and cooling mode at the same time. This achieves the same purpose as the switching beds shown in Figure 25.16b but operates continuously. This continuous operation is preferable to switching beds, but it is mechanically more complex and introduces some challenging sealing requirements for the ducting of streams around the wheel. Lastly, Figure 25.16c shows an arrangement whereby the heat recovery steam generator is placed at the outlet of the combustion zone to generate steam from the waste heat.

If the vent is to be preheated prior to combustion, it should be lower than the flammability limit or the minimum oxygen concentration (see Chapter 27). Recuperative heat recovery above 70% is usually not economic for gas-to-gas heat exchange. Heat recovery increases to typically 80% if steam generation is used. Regenerative heat recovery can be as high as 95%.

If the vent stream contains halogenated material or sulfur, the exhaust gases must be scrubbed before release. Figure 25.17 shows a typical arrangement in which the cooled exhaust gases are scrubbed first with water and then with sodium hydroxide solution before being vented. The water and sodium hydroxide scrubbers shown in Figure 25.17 are likely to be on a recirculation arrangement with a purge stream to liquid effluent. The arrangement in Figure 25.17 shows the exhaust gases from the thermal oxidation being cooled by steam generation before scrubbing. If no steam generation is required, then the hot flue gases must be cooled by quenching with direct water injection before entering the scrubbers.

The stack must be maintained above 80°C to avoid a visible plume, which might be a legislative requirement. If this is the case, the stack might need to be reheated before release.

The performance of thermal oxidizers usually gives destruction efficiency greater than 98%. When designed to do so, destruction efficiency can be virtually 100% (depending on temperature, residence time and mixing in the thermal oxidizer). Thermal oxidizers can be designed to handle minor fluctuations but are not suitable for large fluctuations in flowrate.

3. *Catalytic thermal oxidation.* A catalyst can be used to lower the combustion temperature for thermal oxidation and to save fuel[1]. Figure 25.18 shows a schematic of a catalytic thermal oxidation arrangement. Combustion air and supplementary fuel, along with preheated VOC-laden stream, enter the preheater before the catalyst bed. Heat recovery, as illustrated in Figure 25.18, is not always used. The catalysts are typically platinum or palladium on an alumina support or metal oxides such as chromium, manganese or cobalt. The vent stream should be well below the lower flammability limit (see Chapter 27, typically 25 to 30% of the lower flammability limit). The catalyst can be deactivated by compounds containing sulfur, bismuth, phosphorus, arsenic, antimony, mercury, lead, zinc, tin or halogens. Catalysts are gradually deactivated

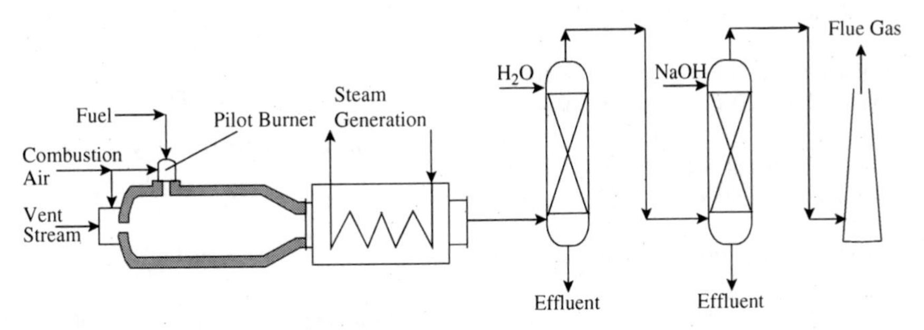

Figure 25.17 Flue gas scrubbing after thermal oxidation.

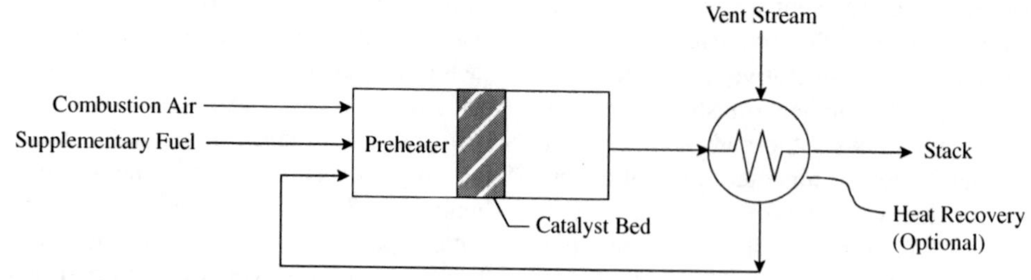

Figure 25.18 Catalytic thermal oxidation.

over time, and their life is generally two to four years. Operating temperature for the catalyst should not normally exceed 500 to 600°C; otherwise the catalyst can suffer from sintering. Typical operating conditions are between 200 and 480°C with a residence time of 0.1 seconds. Catalytic thermal oxidation is not suitable for halogenated compounds because of the lower combustion temperature used. If the heat release in the combustion is such that the temperature rise in the catalyst is too high, then part of the cooled stack gas can be recycled and mixed with the incoming vent stream. Recuperative heat recovery is possible up to 70 to 80% and regenerative heat recovery up to 95%. The destruction efficiency is up to 99.9%, but the arrangement is not suitable for large fluctuations in flowrate.

The advantage of catalytic thermal oxidation is that the lower temperature of operation can lead to fuel savings (although effective heat recovery without a catalyst can offset this advantage). The major disadvantages of catalytic thermal oxidation are that the catalyst needs to be replaced every two to four years and the capital cost tends to be higher than thermal oxidation without a catalyst. Catalytic thermal oxidation also tends to increase the pressure drop through the system.

4. *Gas turbines.* If the vent flow is a high flowrate of air with a significant VOC loading, then a gas turbine can be used for thermal oxidation. A few gas turbine models have been designed specifically for such applications. This allows power to be generated directly from the vent flow. Such applications should be restricted to nonhalogenated, sulfur-free VOCs. If the VOC loading is high, then the application can be achieved without the use of supplementary fuel, otherwise supplementary will be required. Destruction efficiency can be up to 99.9%.

Example 25.3 A stream of air containing 0.4 vol % butane at 20°C is to be thermally oxidized at 800°C. A minimum excess of air of 25% is to be used. The heat of combustion of butane is 2.66×10^6 kJ·kmol^{-1}, and the mean heat capacity of the combustion gases can be taken to be 37 kJ·kmol^{-1}·K^{-1}.

a. Is there sufficient oxygen in the inlet air for efficient combustion?
b. What efficiency of heat recovery would be necessary to sustain combustion without auxiliary fuel at steady state?

Solution

a. The combustion requirements are given by:

$$C_4H_{10} + 6\tfrac{1}{2}O_2 \longrightarrow 4CO_2 + 5H_2O$$

Thus $6\frac{1}{2}$ kmol O_2 required for combustion per kmol of butane

$$= 0.004 \times 6\tfrac{1}{2}$$
$$= 0.026$$

kmol O_2 in steam

$$= 0.21 \times (1 - 0.004)$$
$$= 0.2092$$

Excess oxygen

$$= \frac{0.2092 - 0.026}{0.2092}$$
$$= 88\%$$

Thus, there is plenty of excess air for the combustion.

b. Temperature rise from combustion

$$= \frac{0.004 \times 2.66 \times 10^6}{37}$$
$$= 288°\text{C}$$

Inlet temperature required before combustion

$$= 800 - 288$$
$$= 512°\text{C}$$

Feed preheat duty

$$= (512 - 20) \times 37$$
$$= 492 \times 37 \text{ kJ} \cdot \text{kmol}^{-1} \text{ gas}$$

Heat available in exhaust gases

$$= (800 - 20) \times 37$$
$$= 780 \times 37$$

Efficiency of heat recovery to avoid supplementary fuel

$$= \frac{492}{780}$$
$$= 63\%$$

5. *Biological treatment.* Microorganisms can be used to oxidize VOCs[10]. The VOC-laden stream is contacted

with microorganisms. The organic VOC is food for the microorganisms. Carbon in the VOC is oxidized to CO_2, hydrogen to H_2O, nitrogen to nitrate and sulfur to sulfate. Biological growth requires an ample supply of oxygen and nutrients. Figure 25.19 shows a *bio-scrubber* arrangement. The VOC-laden gas stream enters a chamber and rises up through a packing material, over which flows water. Nutrients (phosphates, nitrates, potassium and trace elements) need to be added to promote biological growth. The microorganisms grow on the surface of the packing material. The water from the exit of the tower needs to be adjusted for pH before recycle to prevent excessively low pH. As the film of microorganism grows on the surface of the packing through time, eventually it becomes too thick and detaches from the packing. This excess *sludge* must be removed from the recycle and disposed of.

An alternative arrangement, as illustrated in Figure 25.20, replaces the packing with soil or compost in a *bio-filter*. One advantage of the bio-filter is that the compost usually used for such systems contains a reservoir of nutrients to sustain biological growth, and additional nutrients do not normally need to be added. However, biological activity eventually starts to decrease, and the bed will need to be replaced after typically five years.

Destruction efficiency for biological treatment is typically up to 95%, but some VOCs are very difficult to degrade. Biological treatment is limited to low concentration streams (typically less than 1000 ppm) with flowrates typically less than 100,000 $m^3 \cdot h^{-1}$.

25.5 CONTROL OF SULFUR EMISSIONS

Sources of atmospheric sulfur emissions from the process industries are:

- chemical production, for example, sulfuric acid production, sulfonation reactions, and so on
- smelting processes, for example, production of copper
- fuel processing operations, for example, fuel desulfurization
- combustion of fuel, for example, steam generation.

Before considering treatment of sulfur emissions to atmosphere, their minimization at source should be considered. Sulfur emissions can be minimized at source through:

- increase in process yield in chemical production
- increase in energy efficiency leading to less fuel burnt
- switch to a fuel with lower sulfur
- desulfurizing the fuel prior to combustion.

Sulfur emissions from combustion processes can be reduced by switching to a fuel with lower sulfur. The sulfur content of fuels varies significantly. Generally, the sulfur content is gas < liquid < solid. This order arises mainly from the relative ease with which the fuels can be desulfurized. Desulfurizing coal prior to combustion is extremely difficult. It must be first ground to very fine particles (of the order of 100 μm) to liberate the mineral (inorganic) sulfur. The light coal can then be

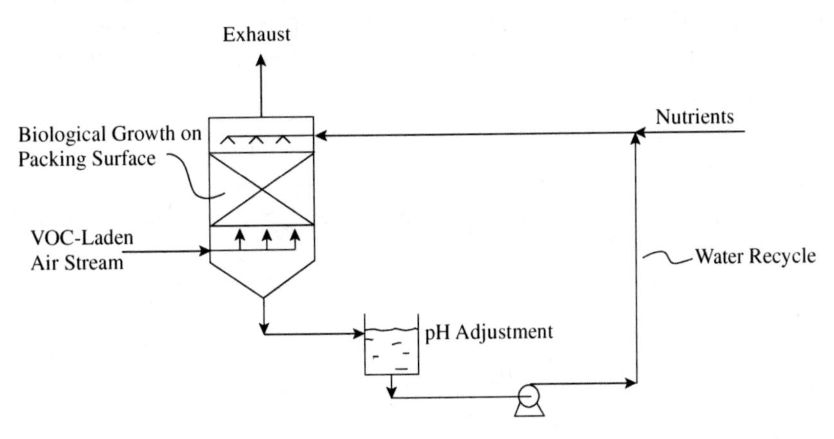

Figure 25.19 A bio-scrubber for treatment of VOC.

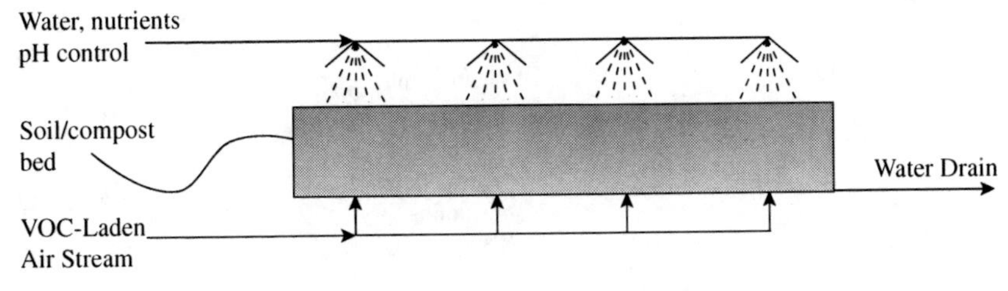

Figure 25.20 A bio-filter for treatment of VOC.

separated from the heavy mineral in a flotation process. Unfortunately, even this desulfurization only removes 30 to 60% of the sulfur, as the remaining sulfur is organic. Desulfurizing a liquid fuel is much more straightforward and is commonly practiced in refinery operations. The sulfur is present in the form of mercaptans, thiols, thiophenes, and so on. Desulfurizing the liquid fuel requires reaction with hydrogen over a catalyst at elevated temperature and pressure. The organic sulfur in the liquid fuel is reacted to H_2S. Desulfurizing gaseous fuel is the most straightforward. The sulfur in gaseous fuel is mostly present as H_2S. This can be removed straightforwardly, for example, in an amine separation absorption, as discussed in Chapter 10.

Removal of sulfur from gas streams is generally removal of H_2S or removal of SO_2[11]. Generally, removal of the sulfur in the form of H_2S is much more straightforward than in the form of SO_2. H_2S can be removed by absorption, as already discussed. Chemical absorption using amines is the most commonly used method. However, other solvents can be used for chemical absorption, for example, potassium carbonate. Physical absorption is also possible using solvents such as propylene carbonate and methanol.

When it is necessary to combust a fuel containing a high sulfur content, gasification is the preferred route. Gasification is the partial combustion of the fuel using pure oxygen (or air) in the presence of steam. Temperatures up to 1600°C and pressures up to 150 bar are used. The product of the gasification is a mixture of CO, CO_2, H_2, H_2O, CH_4, H_2S, COS, N_2 and NH_3. The sulfur is converted mostly to H_2S (typically 95%), with the balance to COS (carbonyl sulfide). Hence, gasification has advantages over conventional combustion when dealing with fuels with high sulfur content. Figure 25.21 shows a flowsheet for an integrated gasification combined cycle. Fuel containing the sulfur, along with steam and oxygen, enters the gasification process. The gasification products are then treated to remove particulate material. COS may then be converted to H_2S by reaction over a catalyst, followed by H_2S removal. This gas can then be used as a chemical feedstock. Alternatively, it can be combusted in a gas turbine combined cycle, as shown in Figure 25.21. In Figure 25.21, the compressor for the gas turbine provides not only the compressed gas for the gas turbine cycle but also compressed air for the air separation unit that produces the oxygen. The nitrogen from the air separation unit, under pressure, is then expanded in the gas turbine to recover the pressure energy from the nitrogen. If the nitrogen is passed to the combustion chamber of the gas turbine, it can be used also to reduce NO_x formation through lowering the peak flame temperature (see later in Chapter 25). The hot exhaust gases from the gas turbine are used to generate steam before being vented. The steam is used to generate further power in a steam turbine. Many variations around the basic theme in Figure 25.21 are possible. The important thing about gasification is that it is an effective method of producing chemical feedstock, power and heat from fuel that has high sulfur content as the sulfur is not oxidized all the way to SO_2 before being treated. Having removed sulfur in the form of H_2S, it cannot be vented to atmosphere. The sulfur from H_2S is normally recovered as elemental sulfur by partial oxidation in a Claus Process. Figure 25.22 shows the outline of a Claus Process. Hydrogen sulfide in air enters a reaction furnace where a partial combustion takes place. After cooling the gases by steam generation, the gases enter a converter. The

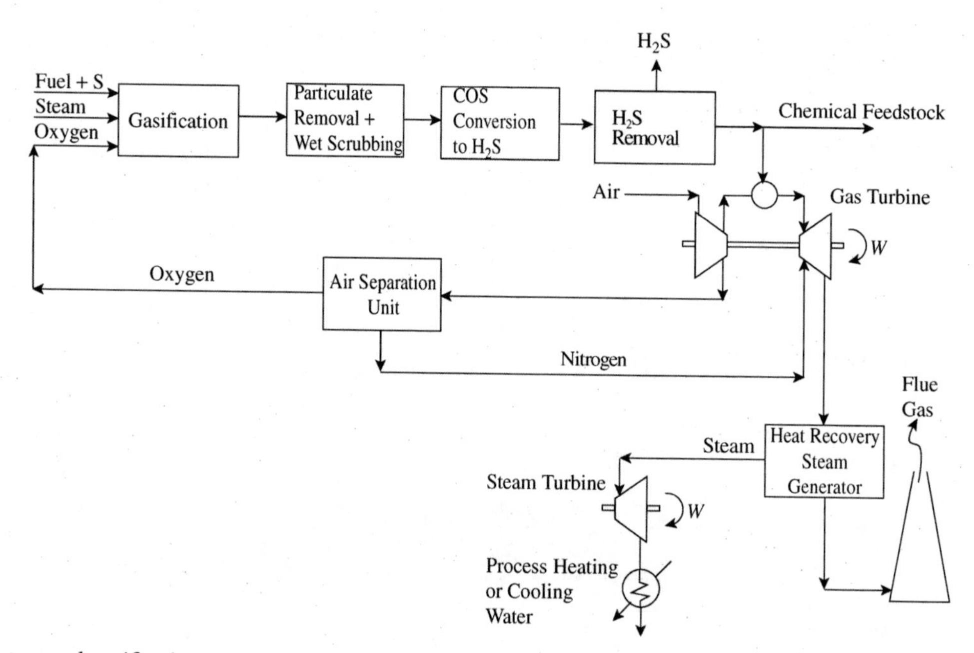

Figure 25.21 Integrated gasification combined cycle.

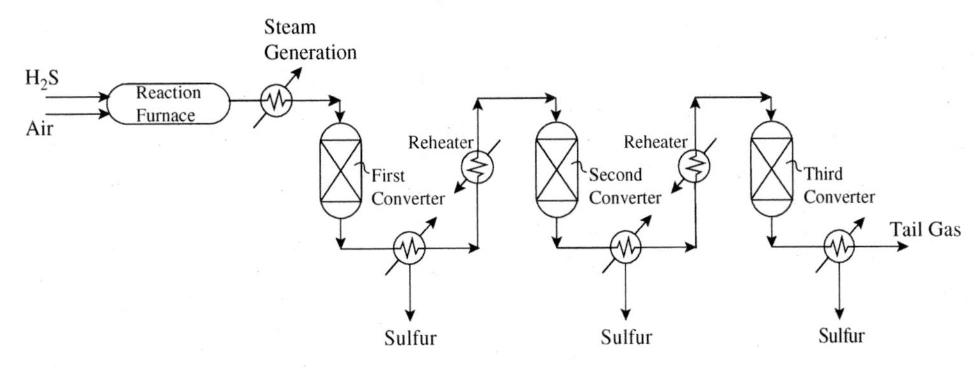

Figure 25.22 Removal of H_2S using the Claus process.

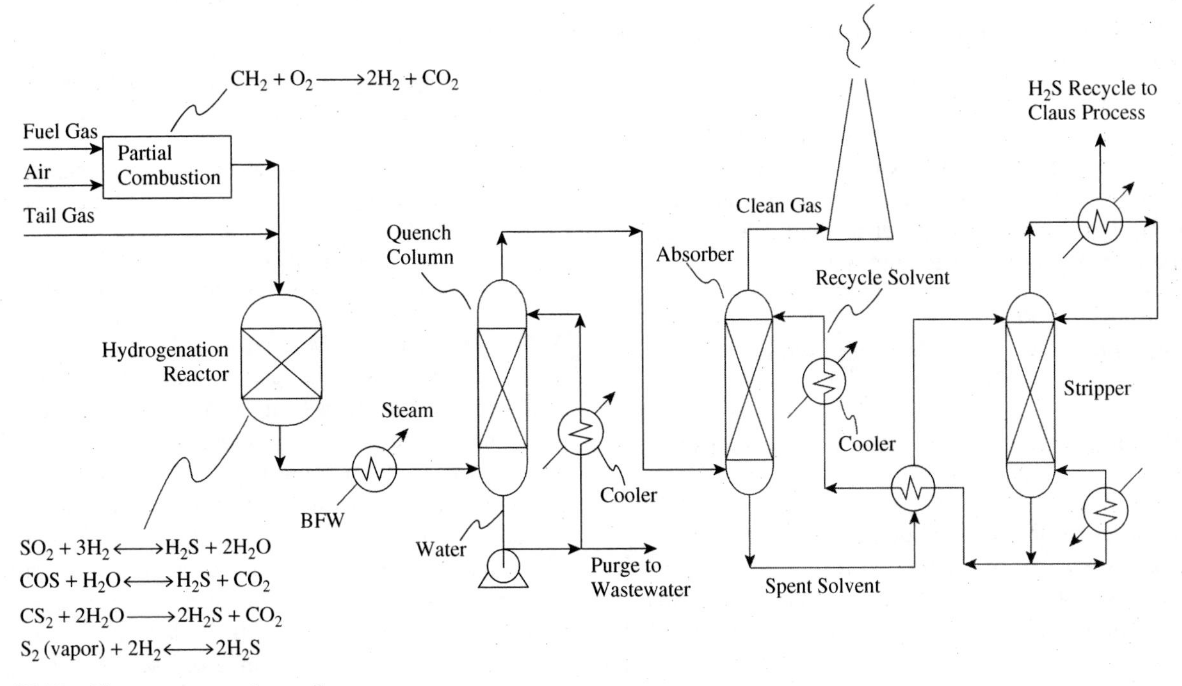

Figure 25.23 Claus process tail gas cleanup.

chemistry is:

$$\text{Burner } 2H_2S + 3O_2 \longrightarrow 2H_2O + 2SO_2$$

$$\text{Converter } 2H_2S + SO_2 \longrightarrow 2H_2O + 3S$$

$$\text{Overall } H_2S + \tfrac{1}{2}O_2 \longrightarrow S + H_2O$$

After the converter in Figure 25.22, the sulfur is recovered by condensation. Conversion is limited by reaction equilibrium to about 60%, so several (usually three) stages of conversion are used, as shown in Figure 25.22. After each conversion stage, the sulfur is condensed and the gas is reheated before going to the next conversion stage. The tail gas contains CO_2, H_2S, SO_2, CS_2, COS and H_2O and cannot be released directly to atmosphere. Various processes are available for the cleanup of Claus Process tail gas. Figure 25.23 shows a process in which the sulfur compounds are converted to H_2S by reaction with hydrogen. Hydrogen is generated by partial oxidation of fuel gas. Sulfur and sulfur compounds are hydrogenated to hydrogen sulfide in the hydrogenation reactor. Gases from the reactor are cooled by generating steam and then in a quench column by recirculating water, Figure 25.23. Hydrogen sulfide is then separated from the other gases by absorption (e.g. in an amine solution). The solvent from the absorber is regenerated in the stripper and the hydrogen sulfide recycled to the Claus Process.

An alternative process for the removal of H_2S by partial oxidation is shown in Figure 25.24[10]. This uses an iron chelate to partially oxidize the H_2S. As shown in Figure 25.24, the gas is contacted in an absorber with the iron chelate solution. Fe^{3+} reacts with H_2S to produce Fe^{2+} and elemental sulfur. Reduced iron chelate solution is oxidized by sparging with air and returned to the absorber. This process can operate over a wide range of conditions.

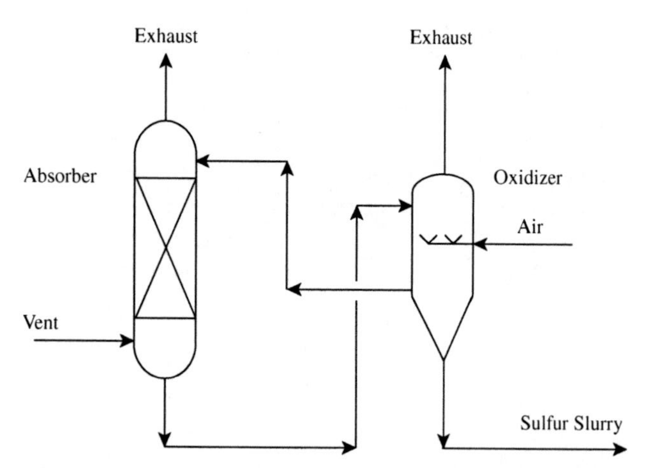

Figure 25.24 Removal of H_2S by partial oxidation of H_2S using iron chelate.

So far, the techniques to recover sulfur have all related to sulfur in the form of H_2S. Consider now how sulfur in the form of SO_2 can be dealt with. Sulfur dioxide can be reacted with limestone to produce solid calcium sulfate (gypsum), according to the equations[11]:

$$\underset{\text{Limestone}}{CaCO_3} + SO_2 \longrightarrow \underset{\text{Calcium sulfite}}{CaSO_3} + CO_2$$

$$CaSO_3 + \tfrac{1}{2}O_2 + 2H_2O \longrightarrow \underset{\text{Gypsum}}{CaSO_4 \cdot 2H_2O}$$

The flowsheet for the removal of SO_2 using wet limestone scrubbing is shown in Figure 25.25[11]. The gas stream containing SO_2 enters a scrubbing chamber into which a limestone slurry is sprayed. The slurry is then collected, separated by thickening and filtration. This creates a solid material (gypsum) that does have some potential for use as building material. However, the solid product is usually of low value and often needs to be disposed of.

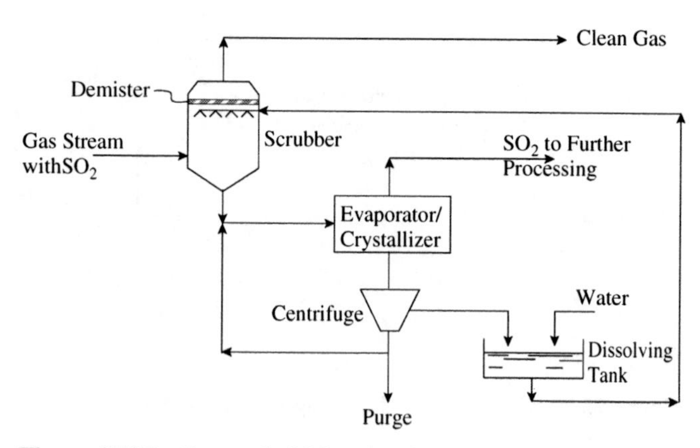

Figure 25.26 Removal of SO_2 using the Wellman–Lord process.

Another process for removal of SO_2 is the Wellman–Lord process[11]. The reactions are:

$$SO_2 + \underset{\text{Sodium sulfite}}{Na_2SO_3} + H_2O \longrightarrow \underset{\text{Sodium bisulfite}}{2NaHSO_3}$$

$$2NaHSO_3 \xrightarrow{\text{crystallize}} Na_2SO_3 + H_2O + SO_2$$

A flowsheet for the Wellman–Lord process is shown in Figure 25.26. Again the gas stream with SO_2 enters a scrubber into which is sprayed a sodium sulfite solution. This then goes to an evaporator/crystallizer to crystallize out the resulting sodium bisulfite, which converts the sodium bisulfite back to sodium sulfate, releasing the SO_2. The crystals are dissolved in water and recycled to the scrubber. The effect of the Wellman–Lord process is to produce a concentrated SO_2 stream from a dilute SO_2 stream. The resulting concentrated SO_2 still needs to be treated.

The SO_2 from the Wellman–Lord process or any other concentrated SO_2 stream (e.g. gases from copper smelting) can be oxidized to SO_3 to produce sulfuric acid as discussed in Chapter 6.

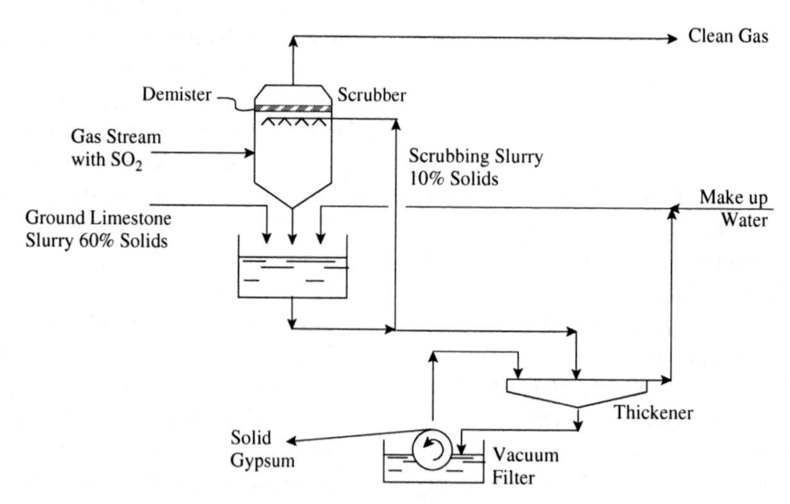

Figure 25.25 Removal of SO_2 using wet limestone scrubbing.

Rather than oxidize the SO_2, it can be reduced in a process similar to the one illustrated in Figure 25.23. The SO_2 is reduced to H_2S. The resulting H_2S can then be separated and fed to a sulfur recovery process.

Example 25.4 A power plant produces 500 standard $m^3 \cdot s^{-1}$ of exhaust gas with 0.1% SO_2 by volume. It is required to remove 90% of the SO_2 before the gas is discharged to atmosphere. Three methods of removal of SO_2 are to be evaluated.

a. The first option is absorption at atmospheric pressure in water taken from a local river at a temperature of 10°C. Assume the concentration of water at the exit of the absorber to be 80% of equilibrium. The equilibrium can be assumed to follow Henry's Law:

$$p_i = H_i x_i$$

where p_i = partial pressure of Component i
H_i = Henry's Law Constant (determined experimentally)
x_i = mole fraction of Component i in the liquid phase

Assuming ideal gas behavior ($p_i = y_i P$),

$$x_i = \frac{y_i^* P}{H_i}$$

where y_i^* = mole fraction of Component i in vapor phase in equilibrium with the liquid
P = total pressure
At 10°C, H_{SO_2} = 22 atm

How much water would be required for the removal of the SO_2?

b. The second option is absorption in sodium hydroxide solution. Assume that the sodium hydroxide and SO_2 react in an oxidizing environment to produce sodium sulfate. Calculate the quantity of material required for the absorption.
c. The third option is absorption in a slurry of calcium carbonate. Calculate the quantity of material required for the absorption.

Assume the molar mass in kilograms occupies 22.4 m^3 at standard conditions and the plant operates for 8600 $h \cdot y^{-1}$.

Solution

a. $$x_{SO2} = \frac{0.8 y_{SO_2}^* P}{H_{SO_2}} = \frac{0.8 \times 0.001 \times 1}{22} = 3.6 \times 10^{-5}$$

Now calculate the gas flowrate G:

$$G = 500 \times \frac{1}{22.4} = 22.32 \text{ kmol} \cdot \text{s}^{-1}$$

Assuming gas and liquid flowrates constant:

$$G(y_{in} - y_{out}) = L(x_{out} - x_{in})$$

$$L = \frac{G(y_{in} - y_{out})}{(x_{out} - x_{in})} = \frac{22.32(0.001 - 0.1 \times 0.001)}{(3.6 \times 10^{-5} - 0)} = 558 \text{ kmol} \cdot \text{s}^{-1} = 558 \times 18 \times \frac{1}{10^3} \text{t} \cdot \text{s}^{-1} = 10 \text{ t} \cdot \text{s}^{-1}$$

that is, an impractically large flowrate

b. $$2NaOH + SO_2 + \tfrac{1}{2}O_2 \longrightarrow Na_2SO_4 + H_2O$$

$$SO_2 \text{ in exhaust} = 500 \times \frac{1}{0.0224} \times \frac{1}{10^3} \times 0.001 = 0.0223 \text{ kmol} \cdot \text{s}^{-1}$$

NaOH required to remove 90%

$$= 2 \times 0.0223 \times 0.9 \text{ kmol} \cdot \text{s}^{-1} = 2 \times 0.0223 \times 0.9 \times 40 \text{ kg} \cdot \text{s}^{-1} = 1.606 \text{ kg} \cdot \text{s}^{-1} = \frac{1.606 \times 3600 \times 8600}{10^3} \text{t} \cdot \text{y}^{-1} = 49{,}722 \text{ t} \cdot \text{y}^{-1}$$

c.

$$CaCO_3 + SO_2 + \tfrac{1}{2}O_2 + 2H_2O \longrightarrow CaSO_4 \cdot 2H_2O + CO_2$$

$$SO_2 \text{ in exhaust} = 0.0223 \text{ kmol} \cdot \text{s}^{-1}$$

$CaCO_3$ required to remove 90%

$$= 0.0223 \times 0.9 \text{ kmol} \cdot \text{s}^{-1} = 0.0223 \times 0.9 \times 100 \text{ kg} \cdot \text{s}^{-1} = 2.007 \text{ kg} \cdot \text{s}^{-1} = \frac{2.007 \times 3600 \times 8600}{10^3} = 62{,}137 \text{ t} \cdot \text{y}^{-1}$$

25.6 CONTROL OF OXIDES OF NITROGEN EMISSIONS

Nitrogen forms eight oxides, but the principal concern is with the two most common ones:

- nitric oxide (NO)
- nitrogen dioxide (NO_2).

These are collectively referred to as NO_x. NO_x emissions are produced from:

- chemicals production (e.g. nitric acid production, nitration reactions, etc.)
- use of nitric acid in metal and mineral processing
- combustion of fuels.

NO_x from the combustion of fuels is formed initially as NO, which subsequently oxidizes to NO_2. There are three sources of NO production from the combustion of fuels[11,12].

1. *Fuel NO*. Organically bound nitrogen in fuel reacts with oxygen to form NO.

$$[\text{Fuel N}] + \tfrac{1}{2}\,O_2 \rightleftharpoons NO$$

   Organic nitrogen is more reactive than nitrogen in air. Fuel NO formation is weakly dependent on temperature.
2. *Thermal NO*.
   Thermal NO is formed when molecular nitrogen in the combustion air reacts with oxygen to form NO, according to the reaction:

$$N_2 + O_2 \rightleftharpoons 2NO$$

   The reaction mechanism is via free radicals and is highly temperature dependent.
3. *Prompt (rapidly forming) NO*. Prompt or rapidly forming NO is formed when molecular nitrogen reacts with hydrocarbon radicals in the flame. The formation only occurs in the flame and is weakly dependent on temperature.

Figure 25.27 illustrates the contributions of the three mechanisms to NO formation. It can be seen from Figure 25.27 that both the fuel and prompt NO are weakly dependent on temperature. Below around 1300°C thermal NO formation is negligible. However, at the highest temperatures thermal NO is the most important. Once NO has been formed, it can then oxidize to NO_2 according to:

$$NO + \tfrac{1}{2}\,O_2 \rightleftharpoons NO_2$$

NO_x in flue gases is predominantly (of the order 90%) NO. NO oxidizes in the atmosphere to NO_2.

NO_x formation in combustion depends on the fuel and the type of combustion device. Generally, coal produces the highest NO_x concentrations, then fuel oil, with natural gas producing the lowest concentrations. Thus, one way of reducing NO_x formation is to change fuel. For example, a large steam boiler fired with fuel oil might produce a flue gas with 150 ppm, which might be reduced to 100 ppm by switch to natural gas. However, it should be noted that the actual NO_x concentration depends on both the fuel and the combustion device, and these are only illustrative figures.

Minimization of NO_x emissions at source means:

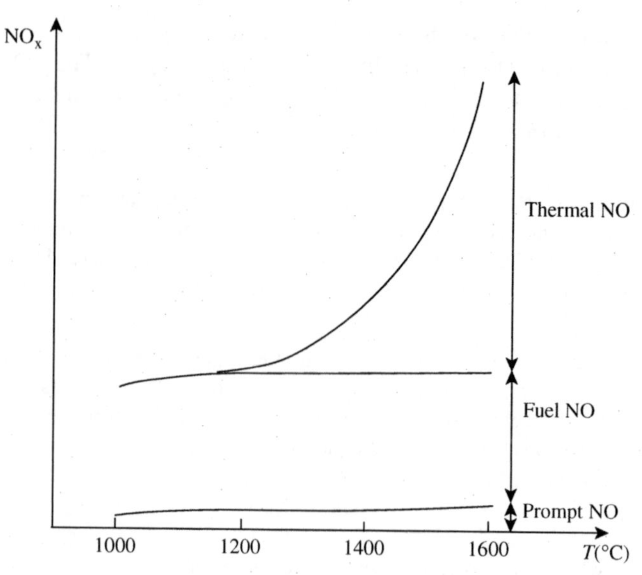

Figure 25.27 The contributions to NO_x formation.

- increase the process yield in chemicals production
- increase in energy efficiency leading to less fuel burnt
- switch to a fuel with lower nitrogen content
- minimize NO_x formation in combustion processes.

Switching to a fuel with lower nitrogen content typically means switching from coal to fuel oil, or fuel oil to natural gas. It should be noted that nitrogen can be a high percentage of natural gas (e.g. 5%), but the nitrogen will be molecular nitrogen and, therefore, behave like nitrogen in the combustion air and be relatively unreactive.

NO_x formation in combustion processes can be reduced by[12,13] the following.

1. *Reduced air preheat*. Preheating the air to combustion processes increases the flame temperature and furnace efficiency (see Chapter 15). However, the increase in flame temperature increases NO_x formation. Thus, reducing preheat lowers the thermal NO_x formation. However, this lowers the furnace efficiency at the same time.
2. *Flue gas recirculation*. Recirculating part of the flue gas, as shown in the Figure 25.28, also reduces the peak flame temperature and reduces the thermal NO_x formation. Usually 10 to 20% of the combustion air is recirculated.
3. *Steam injection*. Steam injection reduces combustion temperature by adding an inert to the combustion process. In principle, water or steam could be injected, but dry superheated steam is usually used. This reduces the furnace efficiency as energy is used to produce the steam, and the latent heat in the steam cannot be recovered. It is only used for moderate NO_x reduction. Steam injection can also be used in gas turbines.

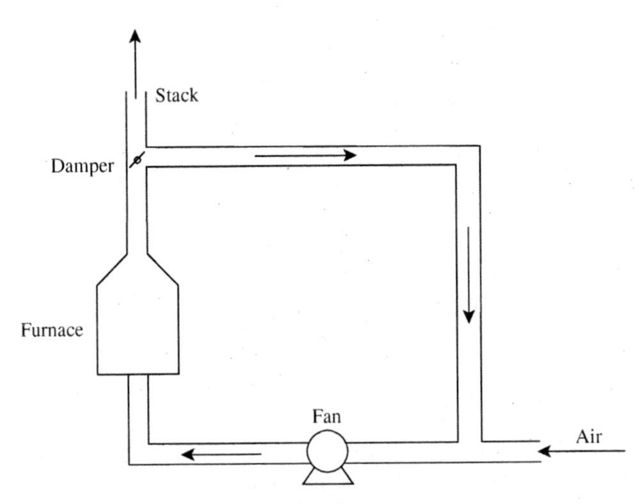

Figure 25.28 Flue gas recirculation.

4. *Reduced excess air.* Reducing the amount of excess air also reduces the NO_x formation.
5. *Air staging.* Air staging in the burner can be used to reduce NO_x formation. The principle is illustrated in Figure 25.29a. Fuel and primary air is injected into the flame with a substoichiometric amount of oxygen. Secondary air is then injected to complete the combustion to give overall greater than stoichiometric requirements. The effect is to reduce the fuel NO and the thermal NO by reducing the peak flame temperature. Air staging can be used both in furnaces and gas turbines.
6. *Fuel staging.* Fuel staging is illustrated in Figure 25.29b. The primary fuel and primary air are first combusted at greater than stoichiometric requirements. In the staging zone, secondary fuel (typically 10 to 20% of the primary fuel) is burned under substoichiometric (reducing) conditions. The temperature should be at least 1000°C in the staging zone. In the final combustion zone, air is added to complete the combustion. Low temperatures (less than 1000°C) are preferred in the final combustion zone to minimize NO formation.

Table 25.2 Performance of the NO_x minimization techniques[12].

Technique	NO_x reduction
Reduced air preheat	25–65%
Flue gas recirculation	40–80%
Water/steam injection	40–70%
Reduced excess air	1–15%
Air staging	30–60%
Fuel staging	30–50%

Table 25.2 compares the performance of the NO_x minimization techniques[12].

Once NO_x formation has been minimized at source, if the NO_x levels still do not achieve the required environmental standards, then treatment must be considered.

The first option that might be considered is absorption into water. The problem is that NO_2 is soluble in water, but NO is only sparingly soluble. To absorb NO, a complexing agent or oxidizing agent must be added to the solvent. One method that can be used is to absorb into a solution of hydrogen peroxide according to:

$$2NO + 3H_2O_2 \longrightarrow 2HNO_3 + 2H_2O$$

$$2NO_2 + H_2O_2 \longrightarrow 2HNO_3$$

This allows nitric acid to be recovered from the oxides of nitrogen. It can be a useful technique, for example, in metal finishing operations where nitric acid is used. More commonly, NO_x is removed using reduction. Ammonia is usually used as the reducing agent, according to the reactions:

$$6NO + 4NH_3 \longrightarrow 5N_2 + 6H_2O$$

$$4NO + 4NH_3 + O_2 \longrightarrow 4N_2 + 6H_2O$$

$$2NO_2 + 4NH_3 + O_2 \longrightarrow 3N_2 + 6H_2O$$

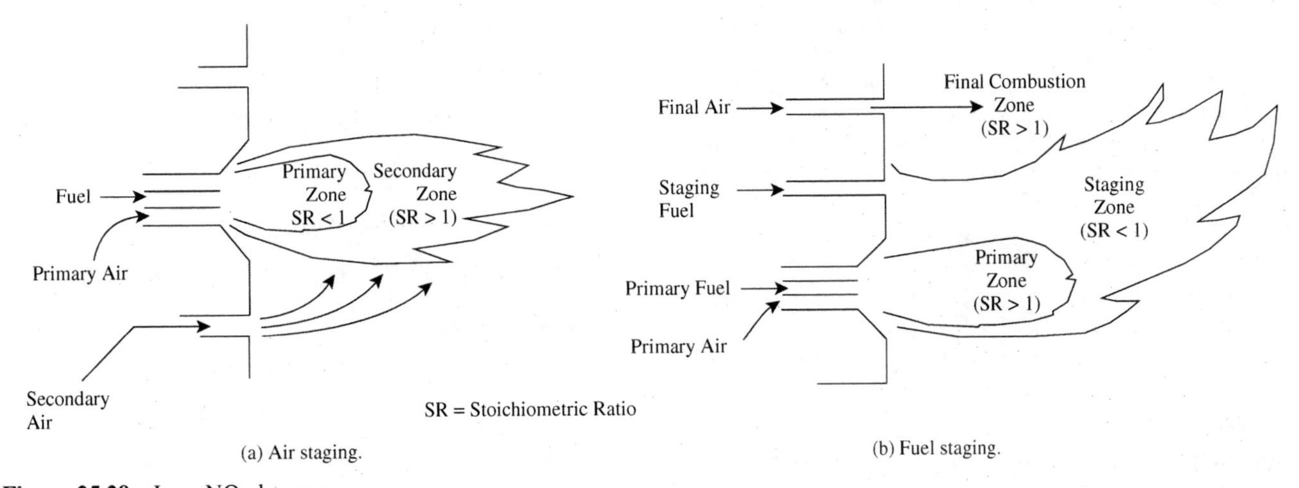

Figure 25.29 Low NO_x burners.

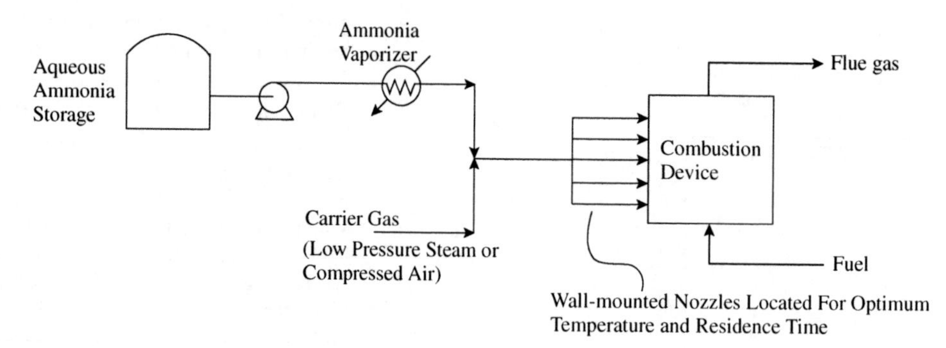

Figure 25.30 Removal of NO_x using selective non-catalytic reduction.

This can be carried out without a catalyst in narrow temperature range (850 to 1100°C). This is known as *selective non-catalytic reduction*[10]. Below 850°C, the reaction rate is too slow. Above 1100°C, the dominant reaction becomes:

$$NH_3 + O_2 \longrightarrow NO + \tfrac{3}{2}\, H_2O$$

Thus, the technique can become counterproductive. A typical arrangement for selective non-catalytic reduction is shown in Figure 25.30. Aqueous ammonia is vaporized and mixed with a carrier gas (low-pressure steam or compressed air) and injected into nozzles located in the combustion device for optimum temperature and residence time[10]. NO_x reduction of up to 75% can be achieved. However, *slippage* of excess ammonia must be controlled carefully.

Rather than selective non-catalytic reduction, the reduction can be carried out over a catalyst (e.g. zeolite) at 150 to 450°C. This is known as selective catalytic reduction. Figure 25.31 shows a typical selective catalytic reduction arrangement[10]. Either anhydrous or aqueous ammonia can be used. This is mixed with air and injected into the flue gas stream upstream of the catalyst. Removal efficiency of up to 95% is possible. Again, slippage of excess ammonia needs to be controlled.

Example 25.5 A gas turbine exhaust is currently operating with a flowrate of 41.6 $kg{\cdot}s^{-1}$ and a temperature of 180°C after a heat recovery steam generator. The exhaust contains 200 ppmv NO_x to be reduced to 60 $mg{\cdot}m^{-3}$ (expressed as NO_2) at 0°C and 1 atm. The NO_x is to be treated in the exhaust using low temperature selective catalytic reduction. Ammonia slippage must be restricted to be less than 10 $mg{\cdot}m^{-3}$, but a design basis of 5 $mg{\cdot}m^{-3}$ will be taken. Aqueous ammonia is to be used at a cost of 300 $\${\cdot}t^{-1}$ (dry NH_3 basis). Estimate the cost of ammonia if the plant operates 8000 $h{\cdot}yr^{-1}$. Assume exhaust gas composition to be similar to that of air. Molar mass data are given in Table 25.3. Assume the molar mass in kilograms occupies 22.4 m^3 at standard conditions.

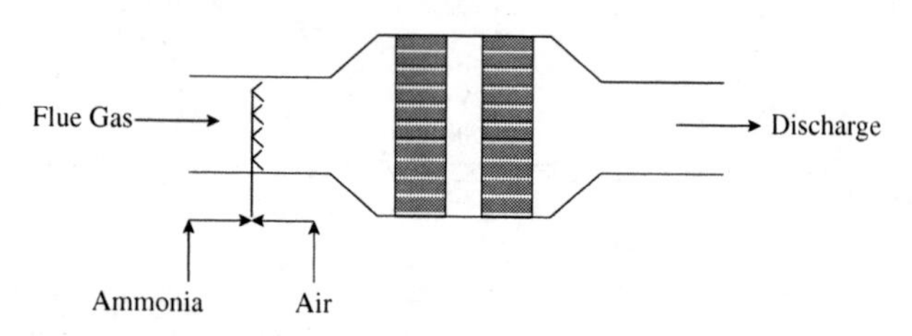

Figure 25.31 Removal of NO_x using selective catalytic reduction.

Table 25.3 Molar mass data.

	Molar mass ($kg{\cdot}kmol^{-1}$)
Air	28.9
NO_2	46
NH_3	17

Solution Volume of exhaust at standard conditions

$$= 41.6 \times 22.4 \times \frac{1}{28.9}$$

$$= 32.2\ m^3{\cdot}s^{-1}$$

Concentration of NO_2 at standard conditions

$$= y \times \frac{1}{22.4} \times 46$$

$$= 0.0002 \times \frac{1}{22.4} \times 46$$

$$= 0.0004107\ kg{\cdot}m^{-3}$$

$$= 410.7\ mg{\cdot}m^{-3}$$

NO_2 to be removed

$$= 32.2(410.7 - 60) \times 10^{-6}\ kg{\cdot}s^{-1}$$

$$= 0.01129\ kg{\cdot}s^{-1}$$

The reduction reaction is given by:

$$2NO_2 + 4NH_3 + O_2 \longrightarrow 3N_2 + 6H_2O$$

NH_3 required for NO_2

$$= 0.01129 \times \frac{17 \times 4}{46 \times 2}$$

$$= 8.345 \times 10^{-3}\ kg \cdot s^{-1}$$

$$NH_3 \text{ slippage } = 32.2 \times 5 \times 10^{-6}$$

$$= 0.161 \times 10^{-3} \text{ kg}\cdot\text{s}^{-1}$$

$$\text{Cost of NH}_3 = (8.345 + 0.161) \times 10^{-3} \times 3600 \times 8000 \times 300 \times 10^{-3}$$

$$= \$73{,}500 \text{ y}^{-1}$$

25.7 CONTROL OF COMBUSTION EMISSIONS

The major emissions from the combustion of fuel are CO_2, SO_x, NO_x and particulates[14]. The products of combustion are best minimized by making the process efficient in its use of energy through efficient heat recovery and avoiding unnecessary thermal oxidation of waste through minimization of process waste. Flue gas emissions can be minimized at source by:

- increased energy efficiency at the point of use (e.g. better heat integration)
- increased energy efficiency of the utility system (e.g. increased cogeneration)
- improvements to combustion processes (e.g. low NO_x burners)
- changing fuel (e.g. changing from fuel oil to natural gas).

For a given energy consumption, fuel change is the only way to reduce CO_2 and SO_x emissions at source. Fuel switch from, say, coal to natural gas reduces the CO_2 emissions for the same heat release because of the lower carbon content of natural gas. Fuel change can also be useful for reducing NO_x emissions. Once emissions have been minimized at source, then treatment can be considered to solve any residual problems.

1. *Treatment of SO_x*. The various techniques that can be considered for the treatment of SO_x are:
   - absorption into NaOH, $CaCO_3$, water, and so on
   - oxidation and conversion to H_2SO_4
   - reduction by conversion to H_2S and then sulfur.
2. *Treatment of NO_x*. The techniques used for treatment of NO_x are:
   - absorption into an NO-complexing agent or oxidizing agent
   - reduction using selective non-catalytic or catalytic processes.
3. *Treatment of particulates*. The various methods for treatment of particulates were reviewed in Chapter 8. These include:
   - scrubbers
   - inertial collectors
   - cyclones
   - bag filters
   - electrostatic precipitators.
4. *Treatment of CO_2*. If CO_2 needs to be separated from the other combustion products for the purpose of reducing greenhouse gas emissions, there are four ways in which this can, in principle, be brought about:
   - absorption (e.g. using amine)
   - adsorption (e.g. using alumina, zeolite, etc.)
   - membrane
   - cryogenic separation.

The problem faced after separation of the CO_2 is what to do with it. The uses of CO_2 are so small relative to the volumes produced by combustion that the only realistic solution *sequestration*. This involves finding a stable location for the CO_2 (e.g. by injection into oil wells or underground caverns). Thus, whilst it is technically feasible to recover CO_2, it is both expensive and difficult to dispose of the CO_2.

Example 25.6 A medium fuel oil is to be burnt in a furnace with 10% excess air. Ambient temperature is 10°C and 60% relative humidity. Saturated vapor pressure of water at 10°C is 0.0123 bar. The analysis of the fuel is given in Table 25.4:

Table 25.4 Fuel analysis for Example 25.6.

	Molar mass ($kg\cdot kmol^{-1}$)	Fuel analysis (wt%)
C	12	83.7
H	1	12.0
S	32	4.3

Estimate the composition of the flue gas.

Solution First, calculate the mole fraction of water vapor in the combustion air. The partial pressure of water in the combustion air is given by:

$$p^{SAT}(10°\text{C}) = 0.0123 \text{ bar}$$

$$p_{H_2O} = 0.6 \times 0.0123 = 0.00738 \text{ bar}$$

Mole fraction water in combustion air:

$$y_{H_2O} = \frac{p_{H_2O}}{P} = 0.00738$$

	Fuel				Stoichiometric flue gas from fuel		
	Molar mass ($kg\cdot kmol^{-1}$)	m (kg)	N (kmol)	N_{O_2} (kmol)	N_{CO_2} (kmol)	N_{SO_2} (kmol)	N_{H_2O} (kmol)
C	12	83.7	6.975	6.975	6.975		
H	1	12.0	12.000	3.000			6.000
S	32	4.3	0.134	0.134		0.134	
		100.0		10.109	6.975	0.134	6.000

$$N_2 \text{ from combustion air} = 10.109 \times \frac{79}{21} = 38.03 \text{ kmol}$$

$$H_2O \text{ from combustion air} = (10.109 + 38.03)\left(\frac{y_{H_2O}}{1 - y_{H_2O}}\right)$$

$$= 48.139\left(\frac{0.00738}{1 - 0.00738}\right)$$

$$= 0.358 \text{ kmol}$$

$$\text{Stoichiometric moist air} = 10.109 + 38.03 + 0.358$$

$$= 48.497 \text{ kmol}$$

$$\text{Stoichiometric flue gas} = 6.975 + 0.134 + 6.000 + 38.03 + 0.358$$

$$= 51.497 \text{ kmol}$$

Assume 10% excess air:

$$N_{O_2} = 0.1 \times 10.109 = 1.011 \text{ kmol}$$

$$N_{N_2} = 0.1 \times 38.03 + 38.03 = 41.833 \text{ kmol}$$

$$N_{H_2O} = 0.1 \times 0.358 + 0.358 + 6.000 = 6.394 \text{ kmol}$$

$$\text{Total flue gas flowrate} = 1.011 + 41.833 + 6.394 + 6.975 + 0.134$$

$$= 56.347 \text{ kmol}$$

$$y_{CO_2} = \frac{6.975}{56.347} = 0.1238$$

$$y_{SO_2} = \frac{0.134}{56.347} = 0.0024$$

$$y_{N_2} = \frac{41.833}{56.347} = 0.7424$$

$$y_{H_2O} = \frac{6.394}{56.347} = 0.1135$$

$$y_{O_2} = \frac{1.011}{56.347} = 0.0179$$

In addition to these components, some NO_x will be formed. For fuel oil, this will be of the order of 300 ppm.

$$y_{NO_x} \approx 0.0003$$

25.8 ATMOSPHERIC DISPERSION

The objective must be to reduce atmospheric emissions to a minimum or at least below legislative requirements. However, there is inevitably some residual emission and this must be safely dispersed in the environment. The factors that affect the dispersion of gases to atmosphere are[3]:

- temperature
- wind
- turbulence.

Temperature is a critical factor. Generally the temperature of the atmosphere decreases with height, and the actual change of temperature with height is known as the *environmental lapse rate*. Air originating at the surface of the Earth, if rising, will cool owing to expansion as the pressure changes. The rate of cooling is known as the *dry adiabatic lapse rate* and is about 9.8°C per kilometer, until condensation occurs. The environmental lapse rate (ELR) will determine what happens to *pockets* of air if they are forced to rise. Figure 25.32a shows the situation where the ELR has a greater change of temperature with height than the dry adiabatic lapse rate. This means that a small volume of air displaced upward will become less dense than the surroundings and would continue its upward motion. This is a desirable condition for atmospheric dispersion, termed *unstable conditions*. Figure 25.32b shows a situation in which the ELR and dry adiabatic lapse rate are roughly equal, giving *neutral conditions*. In this case, there is no tendency for a displaced volume to gain or lose buoyancy. The third situation is shown in Figure 25.32c, in which the ELR is such that the temperature increases with height, known as an *inversion*. These are termed *stable conditions* and provide a strong resistance to vertical motion of a displaced volume. Such stable conditions are problematic from the point of view of gas dispersion.

In the lower levels of the atmosphere, the ELR changes with the time of day. Figure 25.33 shows a typical daily variation of atmospheric stability. Starting before sunrise, the minimum temperature is at the surface. This is caused

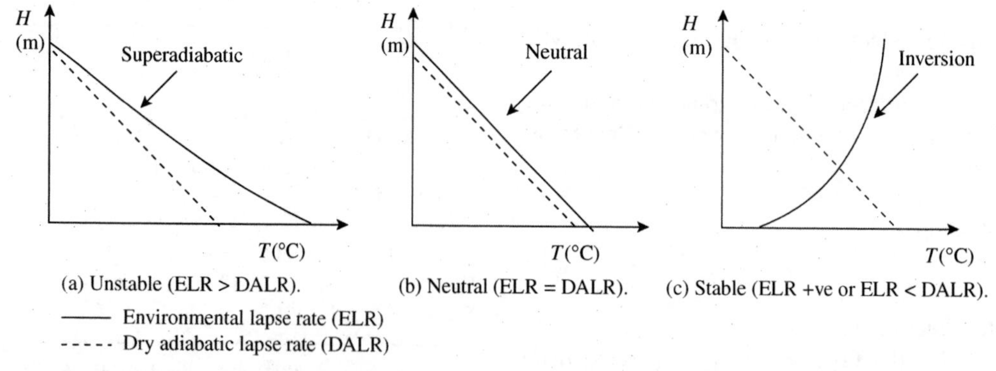

Figure 25.32 Temperature and atmospheric stability.

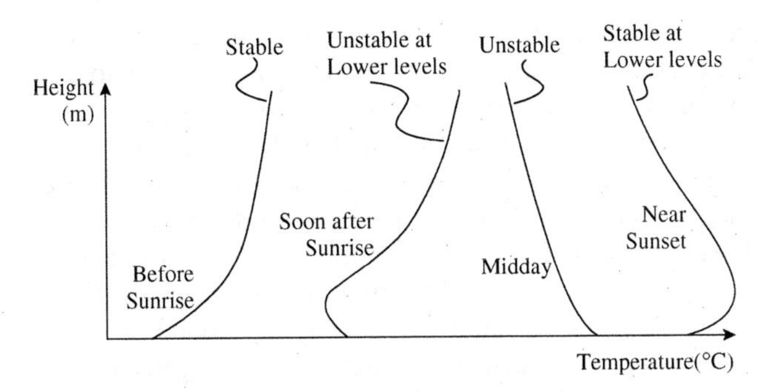

Figure 25.33 Typical daily variation of atmospheric stability.

by heat loss from long-wave radiation. It creates an inversion (increase of temperature with height) up to perhaps 100 meters. Soon after sunrise in Figure 25.33, warming of the surface layer occurs, but the inversion remains at higher levels. Around midday in Figure 25.33, warming has extended from the ground, and now there are unstable conditions (decrease of temperature with height) throughout the lower atmosphere. Near sunset in Figure 25.33, there is radiation from the ground, and an inversion begins to extend from the surface.

Not only does wind direction change but also the wind speed increases with height above the ground to a maximum at a height at which speed is equal to the *free air* or *geostrophic* wind speed. The rate of change of wind speed with height is affected by the atmospheric conditions. It is also affected by the terrain. The buildings in urban areas, for example, slow the air close to the ground, meaning that the maximum speed occurs at a greater height than, for example, over level country.

The third factor affecting dispersion is turbulence. Mechanical turbulence is caused by the roughness of the Earth's surface. Away from the surface, convective turbulence (heated air rising and cooler air falling) becomes increasingly important. The amount of turbulence and the height to which it operates depends on the surface roughness, wind speed and atmospheric stability.

The problem for the designer is to determine the appropriate stack height. This is illustrated in Figure 25.34. This shows that the effective stack height is a combination of the *actual stack height* and the *plume rise*. The plume rise is a function of discharge velocity, temperature of emission and atmospheric stability[3].

The emissions from the stack itself must comply with environmental regulations relating to concentration and flowrate of pollutants. However, the stack must also be high enough such that any pollutant reaching the ground must be lower than ground level concentrations specified by the regulatory authorities. Pollution concentration at ground level depends on many factors, the most important being:

- height of the emitting stack
- velocity and temperature of the emission from the stack
- wind speed
- atmospheric stability
- nature of the surrounding terrain.

Calculation of pollution concentration at ground level requires specialized modeling techniques that are outside the scope of this text.

25.9 ENVIRONMENTAL DESIGN FOR ATMOSPHERIC EMISSIONS – SUMMARY

Emissions to atmosphere can be categorized according to their phase (gas, vapor, liquid and solid). Industrial emissions of major concern are:

- VOCs
- NO_x
- SO_x
- CO
- CO_2
- Particulates
- PM_{10}
- $PM_{2.5}$

Urban smog, acid rain, ozone layer disruption and the greenhouse effect are environmental problems caused by atmospheric emissions.

Solid (particulate) emissions are produced from incomplete combustion of fuels, solids, drying operations, crushing and grinding operations, solids handling operations, and so on. The largest volume of emissions is from products of combustion (CO_2, CO, NO_x, SO_x and particulates). Acid

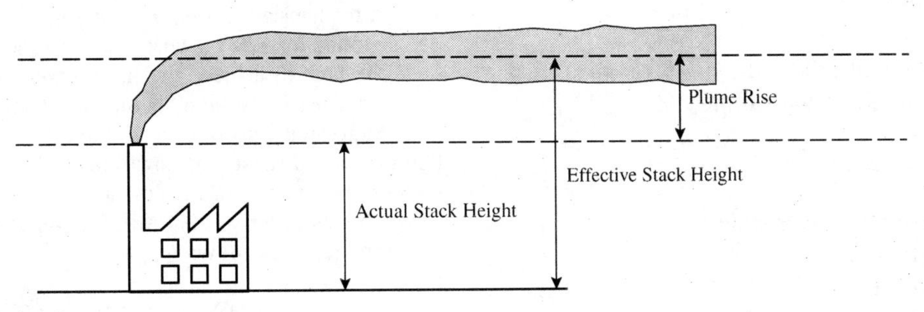

Figure 25.34 Stack height.

gases are produced by chemical production (CO_2, NO_x, SO_x, H_2S, HCl). Vapor emissions are often more difficult to deal with because of the variety of sources.

Remediation processes to deal with atmospheric emissions include:

- removal of particulates
- condensation
- membranes
- adsorption
- physical absorption
- chemical absorption
- thermal oxidation
- gas dispersion.

The hierarchy of control of VOC emissions is:

- eliminate or reduce its source
- recover for reuse
- recover for treatment and disposal
- treatment and disposal of the VOC-laden gas stream.

Four methods can be used for VOC recovery:

- condensation
- membranes
- absorption
- adsorption.

Once VOCs have been minimized at source and recovery possibilities have been exhausted, then any residual VOC needs to be destroyed using:

- flares
- thermal oxidation
- catalytic thermal oxidation
- gas turbines
- bioscrubbers
- biofiltration.

Atmospheric sulfur emissions can be minimized at source by improving the process yields and desulfurization of fuels prior to combustion. The difficulty of fuel desulfurization is solid > liquid > gas. Sulfur can be removed from emissions either as SO_2 or H_2S. Removal of the H_2S can be by:

- physical absorption
- chemical absorption
- gasification of material followed by H_2S removal
- partial oxidation to elemental sulfur.

Removal of SO_2 can be by:

- absorption using sodium hydroxide
- wet limestone scrubbing
- Wellman–Lord process
- or many other methods.

NO_x (principally NO and NO_2) are produced in chemical production, metal and mineral processing and combustion of fuels. NO_x formed in combustion processes is by three mechanisms (fuel, thermal and prompt NO). NO_x emissions can be minimized at source by increasing process yields in chemical production, switching to a fuel with lower nitrogen content or minimizing NO_x formation in combustion processes. After minimizing the production of NO_x, removal by either oxidation or reduction can be considered. NO_x can be absorbed in hydrogen peroxide, which forms nitric acid. Reduction is usually carried out using ammonia:

- without a catalyst in a narrow temperature range (850 to 1100°C)
- with a catalyst with higher removal efficiency (up to 95%) at lower temperatures (120 to 480°C).

Flue gas emissions can be minimized at source by:

- increased energy efficiency at the point of use
- increased energy efficiency of the utility system
- improvements to combustion processes
- changing fuel.

Increased energy efficiency at the point of use can be achieved effectively by improved heat integration. Increased energy efficiency of the utility system can be improved through better matching between processes and the utility system, improved cogeneration, and so on. Improvements to combustion processes are effective for NO_x reduction.

25.10 EXERCISES

1. A storage tank is filled with toluene at 40°C and has a vent that is open to the atmosphere:
   a. Estimate the concentration of toluene in the vent from the tank. Is this above a legislative limit of 80 $mg \cdot m^{-3}$?
   b. The concentration of toluene in the vent from the tank is to be reduced using a refrigerated condensation system. Is it possible to achieve the legislative requirement of 80 $mg \cdot m^{-3}$ by cooling the vent to −20°C?
   c. Sketch a flowsheet to recover any volatile organic compounds from the vent by refrigerated condensation.
   d. What problems are likely to be encountered in using such a refrigerated condensation system?
   e. If the legislative limit of 80 $mg \cdot m^{-3}$ cannot be achieved by cooling to −20°C, what would be the required pressure of the tank to achieve the limit, given the same refrigerated condensation system? Is this a realistic proposal?
   f. Suggest at least three alternatives to refrigerated condensation that would satisfy the environmental requirement. .

   It can be assumed that the molar mass in kilograms at standard conditions occupies 22.4 m^3. The vapor pressures of toluene can be represented by:

$$\ln P^{SAT} = 9.3935 - \frac{3096.52}{T - 53.67}$$

where P^{SAT} = saturated liquid vapor pressure (bar)
T = absolute temperature (K)

The freezing point of toluene is 95°C and its molar mass is 92 kg·kmol^{-1}.

2. A covered tank with a volume of 50 m^3 is used for the storage of acetone. The tank has a vent to atmosphere. The temperature of the storage area changes from 5°C during the night to 25°C during the day. The tank breathes to atmosphere as a result of the change in temperature. Assume that the vapor space above the liquid reaches equilibrium with the temperature of the surrounding air, the vapor follows ideal gas behavior and that the change in volume of the acetone can be neglected.
   a. Estimate the concentration of acetone in the tank vent (corrected to standard conditions of 0°C and 1 atm) as the temperature begins to rise.
   b. Estimate the concentration of acetone in the tank vent (corrected to standard conditions of 0°C and 1 atm) at the maximum temperature.
   c. Estimate the change in volume if the tank is on average 20% full. Assume ideal gas behavior.
   d. Estimate the mass of acetone vented to atmosphere from the expansion assuming a vapor pressure at the average temperature.
   e. Regulations dictate that the concentration and mass of volatile organic compounds (VOCs) should be related to the toluene equivalent. Express the concentration and mass released as toluene equivalent (according to the ratio of the molar masses). The regulations dictate that the vent should be a concentration of less than 80 mg·m^{-3} (toluene equivalent) if the mass vented is greater than 5000 kg·y^{-1} toluene equivalent. If the temperature change is assumed to take place 365 days per year, does this emission violate the regulations?
   f. One way of reducing the mass released would be to operate the tank with higher average level. What would the average level need to be in order to reduce the mass released by half?
   g. Would you consider the estimate in Part d. above to be high or low in terms of the concentration and the mass vented?
   h. How would you estimate the mass vented more accurately than taking the vapor pressure at the average temperature?
   i. Without using abatement equipment, how would you reduce the breathing losses resulting from temperature variations with fixed level in the tank?

   It can be assumed that the molar mass in kilograms occupies 22.4 m^3 at standard conditions. The molar mass of acetone can be taken to be 58 kg·kmol^{-1} and toluene 92 kg·kmol^{-1}. The vapor pressure of acetone is given by:

$$\ln P^{SAT} = 10.0310 - \frac{2940.46}{T - 35.93}$$

where P^{SAT} = saturated liquid vapor pressure (bar)
T = absolute temperature (K)

3. Toluene is used as a solvent for the application of surface coatings. The solvents evaporate as a result of the application, creating a problem with for the emission of volatile organic compounds (VOCs). The legislative framework for the emission of VOCs requires that the mass load of VOC emissions allowed to be released to atmosphere should be less than 60% of the mass of solids deposited during the coating process. The coating operations are currently depositing 50 t·y^{-1} of solids. The concentration of the solids in the coating material is 20%, the remainder being toluene.
   a. Calculate the mass of toluene released during the coating operations and the mass that needs to be recovered or destroyed as a result of the legislation.
   b. It is proposed to solve the emissions problem by installing a ventilation system using air to collect the vapors and to destroy these using thermal oxidation. For safety reasons, the concentration of the flammable material in the ventilation system must be less than 30% of the lower flammability limit. The lower flammability limit for toluene in air is 1.2% by volume. The release of VOCs can be assumed to be evenly distributed over 8000 hours per year of operation. Calculate the air flowrate to the thermal oxidizer in m^3·h^{-1} required to collect the VOCs that needs to be destroyed.
   c. Toluene is also used for cleaning purposes and is stored on the site in a covered tank with a volume of 10 m^3 that is maintained on average to be half full. The tank has a vent to atmosphere. The temperature of the storage area changes from 5°C during the night to 30°C during the day. The tank breathes to atmosphere as a result of the change in temperature. Assume that the vapor space above the liquid reaches equilibrium with the temperature of the surrounding air, the vapor follows ideal gas behavior and that the change in volume of the toluene can be neglected. Estimate the concentration of toluene in the tank vent (corrected to standard conditions of 0°C and 1 atm) and the mass released as a result of the expansion, using the vapor pressure at the mean temperature.
   d. Regulations dictate that the vent should be a concentration of less than 80 mg·m^{-3} if the mass vented is greater than 5000 kg·y^{-1}. If the temperature change is assumed to take place 365 days per year, does this emission violate the regulations?

   It can be assumed that the molar mass in kilograms at standard conditions occupies 22.4 m^3. The vapor pressures of toluene can be represented by:

$$\ln P^{SAT} = 9.3935 - \frac{3096.52}{T - 53.67}$$

where P^{SAT} = saturated liquid vapor pressure (bar)
T = absolute temperature (K)

The freezing point of toluene is 95°C and its molar mass 92.

4. A flue gas with a flow of 10 Nm^3·s^{-1} (Nm^3 = normal m^3) contains 0.1% vol NO_x (expressed as NO_2 at 0°C and 1 atm) and 3% vol oxygen. It is proposed to remove NO_x by absorption in water to 100 ppmv for discharge. Whilst NO_2 is highly soluble, NO is only sparingly soluble and there is a reversible reaction in the gas phase between the two according to:

$$NO + \tfrac{1}{2}O_2 \rightleftharpoons NO_2$$

The equilibrium relationship for the reaction is given by:

$$K_a = \frac{p_{NO_2}}{p_{NO}\, p_{O_2}^{0.5}}$$

where K_a is the equilibrium constant for the reaction and p the partial pressure. At 25°C, the equilibrium constant is 1.4×10^6 and at 725°C is 0.14.

a. Calculate the mole ratio of NO_2 to NO assuming chemical equilibrium at 25°C and at 725°C.
b. Calculate the flowrate of NO_2 and NO assuming chemical equilibrium at 25°C and at 725°C. Assume the kilogram molar mass occupies 22.4 m^3 at standard conditions.
c. It is proposed to remove the NO_x by absorption in water at 25°C. Do you consider this to be a good option if chemical equilibrium is assumed?
d. In the worst case, if equilibrium is not attained, none of the NO would react to NO_2. Calculate the flowrate of water to remove the NO to a concentration of 100 ppmv by absorption in water at 25°C and 1 atm. The solubility of NO in water can be assumed to follow Henry's Law:

$$x_i = \frac{y_i{}^* P}{H}$$

where x_i = mole fraction of Component i in the liquid phase

y_i^* = mole fraction of Component i in the vapor phase in equilibrium with the liquid

P = total pressure

H = Henry's Law constant

At 25°C, $H = 28$ atm for NO. If the concentration in the gas phase at the exit of the absorber is assumed to be 80% of equilibrium and gas phase is assumed to be ideal, calculate the flowrate of water required.

e. Suggest other methods to treat the NO_x.

The molar masses of NO and NO_2 are 30 kg·kmol^{-1} and 46 kg·kmol^{-1} respectively.

5. Coal with the analysis in Table 25.5 is to be burnt in 20% excess air in a boiler.

Table 25.5 Fuel analysis for Exercise 5.

	Per cent by wt
C	78
O	7
H	3
S	1
Ash	6
H_2O	6

The ash contains calcium (0.15% of the coal) and sodium (0.18% of the coal). Assume that the calcium reacts to calcium sulfate $CaSO_4$ and the sodium to sodium sulfate Na_2SO_4. Assume all the remaining sulfur reacts to SO_2. Regulations on the flue gas are to be based on a dry gas (free of water vapor) at 0°C and 1 atm.

a. What proportion of the sulfur in the coal is trapped in the ash and what proportion is oxidized to SO_2?
b. What is the mass of the ash after combustion as a result of sulfate formation?
c. Calculate the flowrate of the flue gases per kilogram of fuel for 20% excess air on a dry basis, neglecting any ash carried from the boiler. Assume the kilogram molar mass occupies 22.4 m^3 at standard conditions.
d. Calculate the composition of the flue gas.

Atomic and molar masses are given in Table 25.6.

Table 25.6 Atomic and molar masses.

	Atomic/molar mass (kg·kmol^{-1})
H	1
C	12
O	16
N	14
Na	23
S	32
Ca	40
CO_2	44
H_2SO_4	98
$CaCO_3$	100
$CaSO_4$	136
Na_2SO_4	142

6. Coal with the analysis in Table 25.7 is to be burnt in 20% excess air in a boiler.

Table 25.7 Fuel analysis for exercise 6.

	% by wt
C	75
O	7
H	3
S	2
Ash	7
H_2O	5

The ash contains calcium (0.2% of the coal) and sodium (0.24% of the coal). Assuming the calcium reacts to form solid calcium sulfate $CaSO_4$ and the sodium to form solid sodium sulfate $Na_2 SO_4$,

a. what proportion of the sulfur in the coal is trapped in the ash and what proportion is oxidized to SO_2?
b. one method to prevent the emission of the remaining sulfur as gaseous SO_2 is to carry out the combustion in a fluidized bed with the addition of limestone ($CaCO_3$) to react the sulfur to calcium sulfate ($CaSO_4$). What mass of limestone must be added per mass of coal to prevent the emission of the remaining SO_2, assuming a 20% excess of limestone is added?
c. in what other way could limestone be used to prevent the emission of the SO_2? Write down the key reactions and steps in the process. What advantages could the alternative method have?
d. an alternative to the react of SO_2 with limestone is to recover the SO_2 in a Wellman–Lord process. If the resulting SO_2 is converted to sulfuric acid, how much sulfuric acid (as 100% pure) would be produced per mass of coal combusted?

Atomic and molar masses are given in Table 25.6.

7. A gas turbine with power output of 10.7 MW and an efficiency of 32.5% burns natural gas. In order to reduce the NO_x emissions to the environmental limits, 0.6 kg steam is injected into the combustion per kg of fuel. The airflow through the gas turbine is 41.6 $kg \cdot s^{-1}$. The composition of the natural gas can be assumed to be effectively 100% methane with a molar mass of 16 $kg \cdot kmol^{-1}$. The kilogram molecular volume can be assumed to occupy 22.4 m^3 at standard conditions.
   a. Calculate the mass flowrate of natural gas. The fuel value of the natural gas is 34.9 $MJ \cdot m^{-3}$ at the standard conditions.
   b. Calculate the flowrate of steam for NO_x abatement.
   c. Without the steam flow, the turbine produces 10.7 MW power. Assuming the power output is proportional to the mass flow through the turbine, estimate the power output with NO_x abatement.
   d. Assume that the steam used for NO_x abatement was raised in a boiler burning natural gas (assumed to be 100% methane) with an efficiency of 90%. The condition of the steam is 20 bar and 350°C with an enthalpy of 3138 $kJ \cdot kg^{-1}$ and is raised from boiler feedwater at 100°C with an enthalpy of 420 $kJ \cdot kg^{-1}$. Estimate the total CO_2 emissions from the boiler and the gas turbine.
   e. Instead of using a gas turbine, a steam turbine could have been used to generate power. For an ideal expansion of the 20 bar steam to 0.075 bar in a steam turbine, a flowrate of 3.7 kg of steam is required per kWh of power production. The steam turbine can be assumed to have an efficiency of 0.75. Calculate the CO_2 emissions using the power output calculated from Part c as the basis.
   f. Compare the emissions from the gas turbine and steam turbine for power generation. Suggest why one process is better than the other.

REFERENCES

1. US Environmental Protection Agency (1992) *Hazardous Air Pollution Emissions from Units in the Synthetic Organic Chemical Industry – Background Information for Proposed Standards, Vol 1B Control Technologies*, US Department of Commerce, Springfield, VA.
2. Harrison RM (1992) *Understanding Our Environment: An Introduction to Environmental Chemistry and Pollution*, Royal Society of Chemistry, Cambridge, UK.
3. De Nevers N (1995) *Air Pollution Control Engineering*, McGraw Hill, New York.
4. Svarovsky L (1981) *Solid-Gas Separation*, Elsevier Scientific, New York.
5. Stenhouse JIT (1981) *Pollution Control, in Teja AS, Chemical Engineering and the Environment*, Blackwell Scientific Publications.
6. Rousseau RW (1987) *Handbook of Separation Process Technology*, Wiley, New York.
7. Dullien FAL (1989) *Introduction to Industrial Gas Cleaning*, Academic Press.
8. Schweitzer PA (1997) *Handbook of Separation Process Techniques for Chemical Engineers*, 3rd Edition, McGraw-Hill, New York.
9. Hui C-W and Smith R (2001) Targeting and Design for Minimum Treatment Flowrate for Vent Streams, *Trans IChemE*, **79A**: 13.
10. Hydrocarbon Processing's Environmental Processes '98, *Hydrocarbon Process*, **August**: 71 (1998).
11. Crynes BL (1977) *Chemical Reactions as a Means of Separation - Sulfur Removal*, Marcel Dekker Inc.
12. Wood SC (1994) Select the Right NO_x Control Technology, *Chem Eng Prog*, **Jan**: 32.
13. Garg A (1994) Specify Better Low-NO_x Burners for Furnaces, *Chem Eng Prog*, **Jan**: 46.
14. Glassman J (1987) *Combustion*, 2nd Edition, Academic Press.

26 Water System Design

In the past, water has been assumed to be a limitless low-cost resource. However, there is now increasing awareness of the danger to the environment caused by the overextraction of water. In some locations, future increases in water use are being restricted. At the same time, the imposition of ever-stricter discharge regulations has driven up effluent treatment costs, requiring capital expenditure with little or no productive return. Figure 26.1 shows a schematic of a greatly simplified water system for a processing site. Raw water enters the processing environment and might need some raw water treatment. This might be something as simple as sand filtration. In many instances, the raw water supply is good enough to enter the processes directly. Water is used in various operations as a reaction medium, solvent in extraction processes, for cleaning, and so on. Water becomes contaminated and is discharged to effluent. Also, as shown in Figure 26.1, some of the raw water is required for the steam system. It first requires upgrading in boiler feed water treatment to remove suspended solids, dissolved salts and the dissolved gases before being fed to the steam boiler, as discussed in Chapter 23. Deionized water might also be required for process use. The steam from steam boilers is distributed and some of the steam condensate is not returned to boilers, but lost to effluent. The boiler requires blowdown to remove the build up of solids, as discussed in Chapter 23. Also, as discussed in Chapter 23, the ion-exchange beds used to remove dissolved salts and soluble ions need to be regenerated by saline solutions or acids and alkalis and this goes to effluent. Finally, as shown in Figure 26.1, water is used in evaporative cooling systems to make up for the evaporative losses and blowdown from the cooling water circuit, as discussed in Chapter 24. All of the effluents tend to be mixed together, along with contaminated storm water, treated centrally in a wastewater treatment system and discharged to the environment. It is usual for most of the water entering the processing environment to leave as wastewater. If the use of water can be reduced, then this will reduce the cost of water supplied. However, it will also reduce the cost of effluent treatment, as a result of most of the water entering the processing environment leaving as effluent. There is thus considerable incentive to reduce both freshwater consumption and wastewater generation.

Figure 26.2a shows three operations, each requiring freshwater and producing wastewater. By contrast, Figure 26.2b shows an arrangement where there is a *reuse* of water from Operation 2 to Operation 1. Reusing water in this way reduces both the volume of the freshwater and the volume of wastewater, as the same water is used twice. However, for such an arrangement to be feasible, any contamination level at the outlet of Operation 2 must be acceptable at the inlet of Operation 1. Not all operations require the highest quality of water. For example, the extraction process described in Section 11.7 for desalting crude oil prior to distillation does not need the highest quality water. Some level of contamination is acceptable, but certain specific contaminants (e.g. hydrogen sulfide and ammonia in the case of crude oil desalting) can cause problems. Another example would be a multistage washing operation. Low quality water could be used in the initial stages, and high-quality water used in the final stages. There are many examples when water with some level of certain contaminants is acceptable for use rather than using the highest quality water.

Figures 26.2c and 26.2d both show arrangements involving *regeneration*. Regeneration is a term used to describe any treatment process that regenerates the quality of water such that it is acceptable for further use. Figure 26.2c shows *regeneration reuse* where the outlet water from Operation 2 is too contaminated to be used directly into Operation 3. A regeneration process between the two allows reuse to take place. Regeneration reuse reduces both the volume of the freshwater and the volume of the wastewater, as with reuse, but also removes part of the *effluent load* (i.e. kilograms of contaminant). The regeneration, in addition to allowing a reduction in the water volume, also removes part of the contaminant load that would have to be otherwise removed in the final effluent treatment before discharge.

A third option is shown in Figure 26.2d where a regeneration process is used on the outlet water from the operations and the water is *recycled*. The distinction between the regeneration reuse shown in Figure 26.2c and the *regeneration recycling* shown in Figure 26.2d is that in regeneration reuse the water only goes through any given operation once. Figure 26.2c shows that the water goes from Operation 2 to regeneration, then to Operation 3 and then discharge. By contrast, in Figure 26.2d, the water can go through the same operation many times. Regeneration recycling reduces the volume of freshwater and wastewater and also reduces the effluent load by virtue of the regeneration process taking up part of the required effluent treatment load.

Regeneration reuse and regeneration recycling are similar in terms of their outcomes. Regeneration recycling allows larger reductions in the freshwater use and wastewater generation than in the regeneration reuse. However, problems can be encountered with regeneration recycling. Regeneration costs can be high. What is usually a greater problem

Chemical Process Design and Integration R. Smith
 ISBNs: 0-471-48680-9 (HB); 0-471-48681-7 (PB)

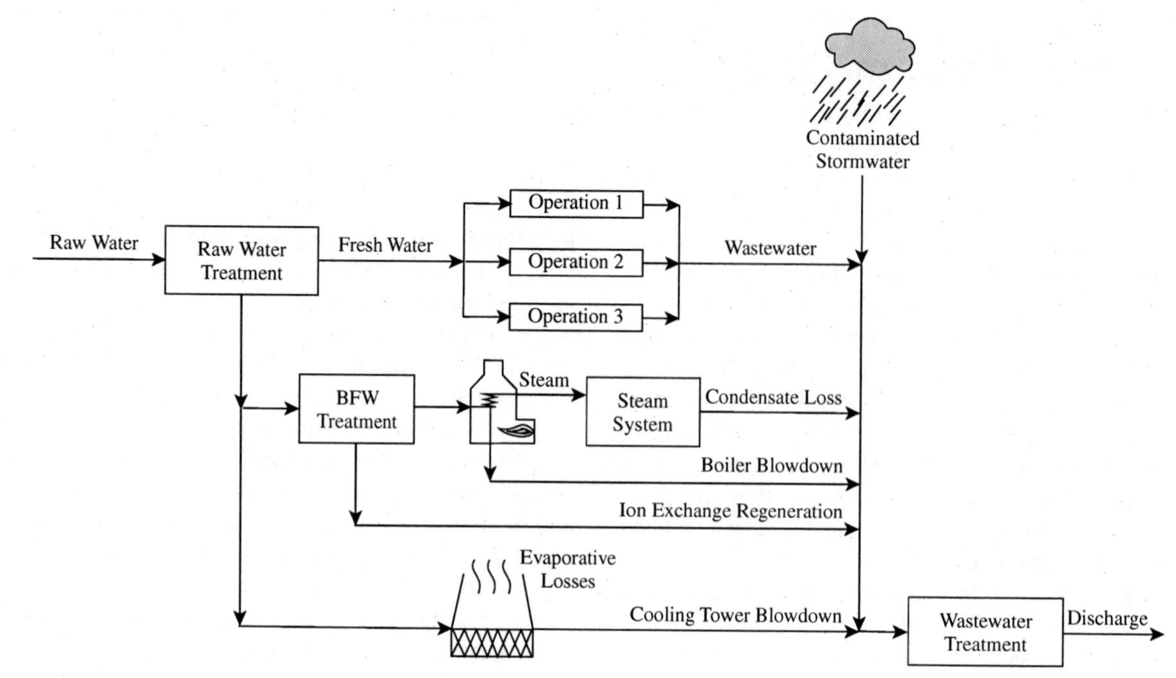

Figure 26.1 A typical water and effluent treatment system.

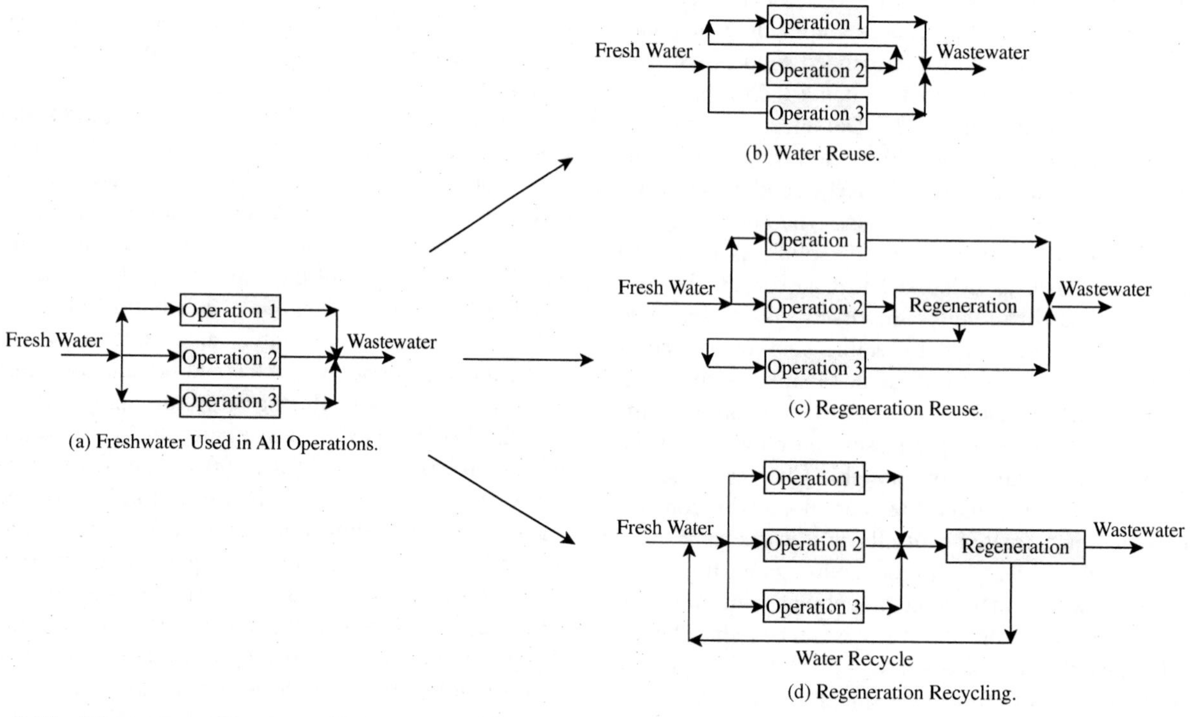

Figure 26.2 Water reuse and regeneration.

though, is that the recycling can allow the build up of undesired contaminants in the recycle, such as microorganisms or products of corrosion. These contaminants not removed in the regeneration might build up to the extent of creating problems to the process.

As shown in Figure 26.1, all of the wastewater was collected together and treated centrally before discharge. Another way to deal with the effluent treatment is by *distributed effluent treatment* or *segregated effluent treatment*. The basic idea is illustrated in Figure 26.3. In addition to some reuse of water between Operation 2 and Operation 1, some local treatment (distributed treatment) is taking place on the outlet of Operation 1 and on the outlet of Operation 3 before going to final wastewater treatment and discharge.

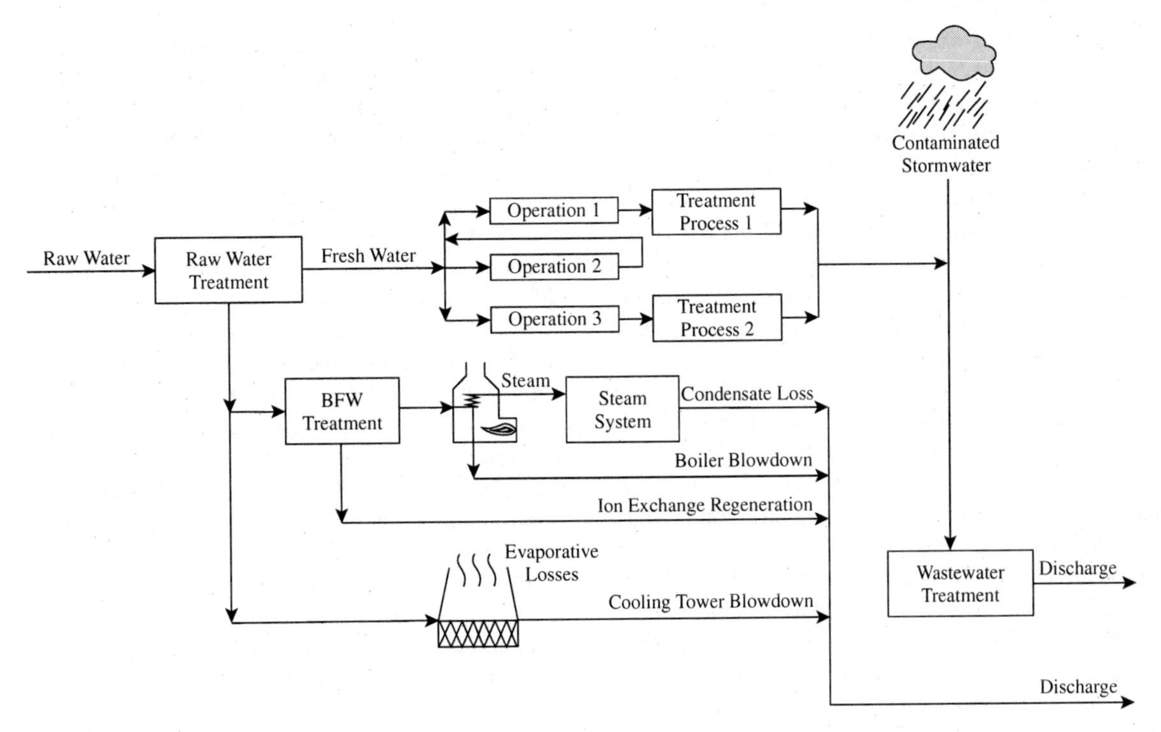

Figure 26.3 Distributed effluent treatment.

Another feature of the arrangement in Figure 26.3 is that a number of the effluents are no longer being treated before discharge. The arrangement in Figure 26.1 is such that the more contaminated effluents from the outlet of Operations 1, 2 and 3 need to be diluted by the less contaminated streams before being treated and discharged. By contrast, as shown in Figure 26.3, the pretreatment on the outlet of Operation 1 and Operation 3 is such that the dilution is no longer required before final effluent treatment and discharge. This means that the streams with only light contamination now no longer need to be treated.

The capital cost of generally aqueous waste treatment operations is generally proportional to the total flow of wastewater and the operating cost generally increases with decreasing concentration for a given mass of contaminant to be removed. Thus, if two streams require different treatment operations, it makes no sense to mix them and treat the two streams in both treatment operations. This will increase both capital and operating costs. The concept of distributed effluent treatment is one that tends to treat effluents before they are mixed together. Treatment is made specific to individual (or small numbers of) contaminants while still concentrated. The benefit is that, by avoiding mixing, this increases the potential to recover material, leading to less waste and lower cost of raw materials. However, the overriding benefit is usually that the effluent volume to be treated is reduced significantly, leading to lower effluent treatment costs overall.

Before developing more systematic ways of designing water systems, first consider the general issues of water contamination and treatment.

26.1 AQUEOUS CONTAMINATION

There are two significant reasons why water contamination needs to be considered. The first is that aqueous effluent must comply with environmental regulations before discharge. The concentration, and perhaps load, of contamination of various specified contaminants must be less than the regulatory requirements. The second reason is that contaminant levels will affect the feasibility of reuse and recycling of water, as shown in Figure 26.2. If water is to be reused or recycled, then the level of inlet contamination to the operation receiving reused or recycled water must be acceptable. What types of contamination need to be considered?

Consider first aqueous emissions of organic waste material. When this is discharged to the receiving water, bacteria feed on the organic material. This organic material will eventually be oxidized to stable end products. Carbon in the molecules will be converted to CO_2, hydrogen to H_2O, nitrogen to NO_3^-, sulfur to SO_4^{2-} and so on. As an example, consider the degradation of urea:

$$\underset{\text{urea}}{CH_4N_2O} + \underset{\text{oxygen}}{\tfrac{9}{2}O_2} \longrightarrow \underset{\text{carbon dioxide}}{CO_2} + \underset{\text{water}}{2H_2O} + \underset{\text{nitrate}}{2NO_3^-}$$

This equation indicates that every molecule of urea requires 9/2 molecules of oxygen for complete oxidation. The oxygen required for the reactions depletes receiving water of oxygen, causing the death of aquatic life. Standard tests have been developed to measure the amount of oxygen required to degrade a sample of wastewater[1–3].

1. *Biochemical oxygen demand (BOD).* A standard test has been devised to measure biological oxygen demand (BOD) in which the oxygen utilized by microorganisms in contact with the wastewater over a five-day period at 20°C is measured (usually termed BOD_5). The period of the test can be extended to a much longer period (in excess of 20 days) to measure the *ultimate demand.* While the BOD_5 test gives a good indication of the effect the effluent will have on the environment, it requires five days to carry out (or longer for the ultimate BOD). Other tests have been devised to accelerate the oxidation process.

2. *Chemical oxygen demand (COD).* In the chemical oxidation demand (COD) test, oxidation with acidic potassium dichromate is used. A catalyst (silver sulfate) is required to assist oxidation of certain classes of organic compounds. Chemical oxygen demand results are generally higher than BOD_5, since the COD test oxidizes materials that are only slowly biodegradable. Although the COD test provides a strong oxidizing environment, certain organic compounds are oxidized only slowly, or not at all.

The ratio of BOD_5 to COD varies according to the contamination. The ratio can vary between 0.05 and 0.8, depending on the chemical species[2]. Domestic sewage has a value of typically around 0.37[4]. An average value across all types of contaminants is around 0.35.

3. *Total oxygen demand (TOD).* The total oxygen demand (TOD) test measures the oxygen consumed when a sample of wastewater is oxidized in a stream of air at high temperature (900°C) in a furnace. Under these harsh conditions, all the carbon is oxidized to CO_2. The oxygen demand is calculated from the difference in oxygen content of the air before and after oxidation. The resulting value of TOD embraces oxygen required to oxidize both organic and inorganic substances present.

The relationship between BOD_5, COD and TOD for the same organic waste is in the order,

$$BOD_5 < COD < TOD$$

4. *Total organic carbon (TOC).* The total organic carbon (TOC) test measures the carbon dioxide produced when a sample of wastewater is subjected to a strongly oxidizing environment. One option is to oxidize the sample in a stream of air at a high temperature (800 to 900°C) in a furnace, similar to the TOD test, but measuring the change in CO_2 rather than the change in O_2. Rather than using high temperature in a furnace, other strongly oxidizing environments (e.g. chemical oxidation) can also be used. To obtain the TOC requires that inorganic carbon compounds must be removed prior to test, or results corrected for their presence. The inorganic carbon can be removed prior to test by adding acid to convert the inorganic carbon to carbon dioxide that must be stripped from the sample by a sparge carrier gas.

A chemical process is designed from knowledge of physical concentrations, whereas aqueous effluent treatment systems are designed from knowledge of BOD_5 and COD. Thus, the relationship between BOD_5 and COD and the concentration of waste streams leaving the process needs to be established. Without measurements, relationships can only be established approximately. The relationship between BOD_5 and COD is not easy to establish, since different materials will oxidize at different rates. To compound the problem, many wastes contain complex mixtures of oxidizable materials, perhaps together with chemicals that *inhibit* the oxidation reactions.

If the composition of the waste stream is known, then the *theoretical oxygen demand (ThOD)* can be calculated from the appropriate stoichiometric equations. As a first level of approximation, it can be assumed that the ThOD would be equal to the COD. The following example will help to clarify these relationships.

Example 26.1 A process produces an aqueous waste stream containing 0.1 mol % acetone. Estimate the COD and BOD_5 of the stream.

Solution First, calculate the ThOD from the equation that represents the overall oxidation of the acetone:

$$\underset{\text{acetone}}{(CH_3)_2CO} + \underset{\text{oxygen}}{4O_2} \longrightarrow \underset{\substack{\text{carbon}\\\text{dioxide}}}{3CO_2} + \underset{\text{water}}{3H_2O}$$

Approximating the molar density of the waste stream to be that of pure water (i.e. 56 $kmol{\cdot}m^{-3}$), then theoretical oxygen demand

$$= 0.001 \times 56 \times 4 \text{ kmol } O_2{\cdot}m^{-3}$$

$$= 0.001 \times 56 \times 4 \times 32 \text{ kg } O_2{\cdot}m^{-3}$$

$$= 7.2 \text{ kg } O_2{\cdot}m^{-3}$$

Thus

$$COD \cong 7.2 \text{ kg}{\cdot}m^{-3}$$

$$BOD_5 \cong 7.2 \times 0.35$$

$$\cong 2.5 \text{ kg}{\cdot}m^{-3}$$

Effluent treatment regulations might specify a level of BOD_5, COD or both. Increasingly, the tendency is toward the specification of *toxicity*. This measures the toxicity of an effluent to some kind of living species. Other contaminants that might be specified are:

- specifically nominated contaminants (e.g. phenol, benzene, etc),
- heavy metals (e.g. chromium, cobalt, vanadium, etc.),
- halogenated organic compounds,
- organic nitrogen,
- organic sulfur,
- nitrates,

- phosphates,
- suspended solids,
- pH, and so on.

In addition to the levels of contamination, typically specified as parts per million (ppm) or milligrams per liter ($mg \cdot l^{-1}$), there might be regulations for the total discharge of a contaminant in kilograms per day ($kg \cdot d^{-1}$) or tons per year ($t \cdot y^{-1}$). There might also be a specification for the maximum effluent temperature.

Water treatment processes can be classified in order as[1]:

- Primary (or pretreatment)
- Secondary (or biological)
- Tertiary (or polishing)

26.2 PRIMARY TREATMENT PROCESSES

Primary or *pretreatment* of wastewater can involve either physical or chemical treatment, depending on the nature of the contamination, and serves three purposes:

- Allows reuse or recycling of water;
- Recovers useful material where possible;
- Prepares the aqueous waste for biological treatment by removing excessive load or contaminants that will inhibit the biological processes in biological treatment.

The pretreatment processes may be most effective when applied to individual waste streams from particular processes or process steps before effluent streams are combined for biological treatment.

A brief review of the primary treatment methods will now be given[1–3]. Pretreatment usually starts with phase separation if the effluent is a heterogeneous mixture.

1. *Solids separation.* Methods used for solids separation include most of the commonly used techniques:

- Sedimentation
- Centrifugal separation
- Filtration.

Clarifiers can be used, as discussed in Chapter 8 and as illustrated in Figure 8.2. As an example of the effectiveness of clarifiers, in petroleum refinery applications, they are capable of removing typically 50 to 80% of suspended solids, together with 60 to 95% of dispersed hydrocarbon (which rises to the surface), 30 to 60% of BOD_5 and 20 to 50% of COD[5]. The performance depends on the design and the effluent being treated. The performance can be enhanced by the use of chemical additives. The chemical additives neutralize the electric charges on the particles that cause them to repel each other and remain dispersed.

Settling to remove solids from aqueous effluent can be enhanced by the use of centrifugal forces in hydrocyclones. Their design was discussed in Chapter 8. For solids removal, the solids are driven toward the wall of the hydrocyclone and removed from the base. Centrifugation can also be used for solids removal, but restricted to smaller volumes. Solids removal by centrifugation can typically be in the range 50 to 80%, increasing to typically 80 to 95% with chemical addition.

Filtration can be used to remove solid particles down to around 10 μm. Both cake filtration and depth filtration can be used, as discussed in Chapter 8.

2. *Coalescence.* Coalescence by gravity in simple settling devices can often be used to separate immiscible liquid–liquid mixtures. The coalescence can be enhanced by the use of mesh pads and centrifugal forces, as discussed in Chapter 8.

In petroleum and petrochemical plants, a common device used for the separation of dispersed hydrocarbon liquids from aqueous effluent is the American Petroleum Institute (API) separator, as illustrated in Figure 26.4. This is a simple settling device in which the effluent enters a large volume. The resulting low velocity allows light particles of hydrocarbon to rise to the surface and any heavy solid particles to settle to the base of the device. Rakes might be employed to move the light material along the surface and the heavy material along the base for collection. The light material can be removed using a *scum skimmer.* This, for example, could be as simple as a horizontal pipe with a slot in its upper surface. Rotation of the device, such that the slot is just below the surface of the liquid allows the light material (hydrocarbon) to overflow into the pipe for collection. In petroleum refinery applications, API separators typically remove 60 to 99% of dispersed hydrocarbon liquids, together with 10 to 50% of suspended solids, 5 to 40% of BOD_5 and 5 to 30% of COD[5]. It is a very simple device used for the first stage of treatment. The performance depends on the design and the effluent being treated, and the performance can be enhanced by the use of chemical additives.

Centrifugal forces can also be exploited in hydrocyclones. As discussed in Chapter 8, designs are different for the removal of solids and dispersed oil. In contrast to solids

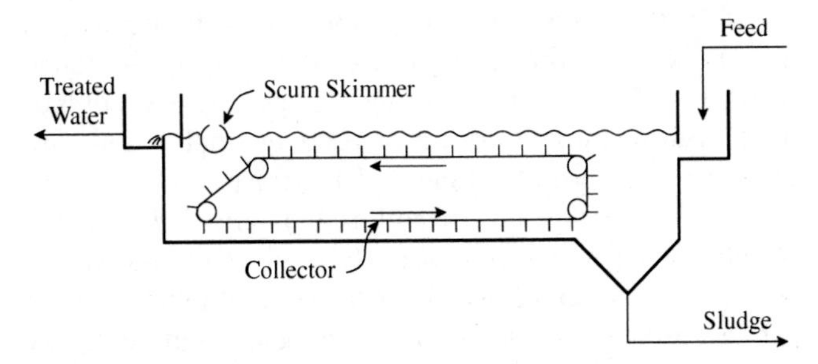

Figure 26.4 A typical API (American Petroleum Institute) separator.

removal, for oil removal, the water is driven to the wall of the hydrocyclone and the treated water removed from the base. Performance for the removal of oil from an aqueous effluent is typically in the range 70 to 90%.

3. *Flotation.* Flotation can be used to separate solid or immiscible liquid particles from aqueous effluent, as discussed in Chapter 8. These particles should have a lower density than water and be naturally hydrophobic. Typical applications include removal of dispersed hydrocarbon liquids from petroleum and petrochemical effluents and removal of fibers from pulp and paper effluents. Direct air injection is usually not the most effective method of flotation. Dissolved-air flotation is usually superior. In dissolved-air flotation, air is dissolved in the effluent under a pressure of several atmospheres and then liberated in the flotation cell by reducing the pressure, as discussed in Chapter 8. The fine air bubbles attach to the particles and rise to the surface. Once the particles have floated to the surface, they are collected by skimming. Flotation is particularly effective for the separation of very small and light particles. As an example of the performance of such units, when applied to effluent from a petroleum refinery, dissolved-air flotation can typically remove 70 to 85% of dispersed oil and 50 to 85% of suspended solids, together with 20 to 70% of BOD_5 and 10 to 60% of COD[5]. The performance depends on the equipment design and the effluent being treated. In such applications, it is normally used after an API separator has been used for preliminary treatment.

4. *Membrane processes.* Conventional filtration processes can separate particles down to a size of around 10 μm. If smaller particles need to be separated, a porous polymer membrane can be used. Microfiltration retains particles down to a size of around 0.05 μm. A pressure difference across the membrane of up to 4 bar is used. The two most commonly used arrangements are spiral wound and hollow fiber, as discussed in Chapter 8.

In ultrafiltration, the effluent is passed across a semipermeable membrane (see Chapter 10). Water passes through the membrane, while submicron particles and large molecules are rejected from the membrane and concentrated. The membrane is supported on a porous medium for strength, as discussed in Chapter 10. Ultrafiltration is used to separate very fine particles (typically in the range 0.001 to 0.02 μm), microorganisms and organic components with molar mass down to 1000 $kg \cdot kmol^{-1}$. Pressure drops are usually in the range 1.5 to 10 bar.

Reverse osmosis and nano-filtration are high-pressure membrane separation processes (typically 10 to 50 bar for reverse osmosis and 5 to 20 bar for nano-filtration), which can be used to reject dissolved inorganic salt or heavy metals. The processes were discussed in Chapter 10 and are particularly useful for removal of ionic species, such as sodium, magnesium, nitrate, sulfate, chloride ions, and so on. Depending on the membrane system design and the ionic species, these can be removed with an efficiency of up to 85 to 99%.

Configurations used include tubes, plate-and-frame arrangements and spiral wound modules. Spiral wound modules should be treated to remove particles down to 20 to 50 μm, while hollow fiber modules require particles down to 5 μm to be removed. If necessary, pH should be adjusted to avoid extremes of pH. Also, oxidizing agents such as free chlorine must be removed. Because of these restrictions, reverse osmosis is only useful if the wastewater to be treated is free of heavy contamination. The concentrated waste material produced by membrane processes should be recycled if possible but might require further treatment or disposal.

Electrodialysis can be an alternative to reverse osmosis, as discussed in Chapter 10.

5. *Stripping.* Volatile organic compounds and dissolved gases can be stripped from wastewater. The usual arrangement would involve wastewater being fed down through a column with packing or trays and the stripping agent (usually steam or air) fed to the bottom of the column.

If steam is used as a stripping agent, either live steam or a reboiler can be used. The use of live steam increases the effluent volume. The volatile organic materials are taken overhead, condensed and, if possible, recycled to the process. If recycling is not possible, then further treatment or disposal is necessary. A common application of steam stripping is the stripping of hydrogen sulfide and ammonia from water in petroleum refinery operations to regenerate contaminated water for reuse. If significant quantities of both H_2S and NH_3 need to be stripped, then this can be a problem. At pH less than 7, the ammonia is present predominantly as NH_4^+ and it will not readily strip. At a pH of 12, the ammonia is present in solution predominantly as free NH_3 and will strip much more readily. If H_2S also needs to be stripped, then below a pH of 6.5, H_2S is stripped preferentially. Thus, if significant quantities of both H_2S and NH_3 need to be stripped, then it might be necessary to use a two-stage stripping operation. One is carried out at low pH (e.g. by the addition of sulfuric acid before stripping) to strip the H_2S. The other is carried out after adjusting to high pH (e.g. by the addition of sodium hydroxide before stripping) to strip the NH_3.

Steam stripping is capable of removing typically 90 to 99% H_2S, 90 to 97% NH_3 and 75 to 99% organic materials. It should be noted though, that some organic materials are resistant to steam stripping and it is thus not a universal solution to contamination with organic materials.

If air is used as a stripping agent, further treatment of the stripped material will be necessary. The gas might be fed to a thermal oxidizer or some attempt made to recover material by use of adsorption.

6. *Crystallization.* If contamination in wastewater has a solubility that varies significantly with temperature, then cooling might allow crystallization of a significant proportion of the contamination. Crystallization was discussed in Chapter 10. Unfortunately, cooling crystallization to treat wastewater streams might require the use of refrigeration to cool below cooling water temperature and this can be expensive to install and incur a high cost of power to operate. An alternative way to create supersaturation is to use evaporation, as discussed in Chapter 10. This can be expensive to operate in terms of the heat required for evaporation, unless the heat required can be supplied by heat recovery, or the latent heat available in the evaporated water can be recovered. An additional benefit is the production of relatively clean water from the evaporated water.

7. *Evaporation.* An extreme case of the use of evaporative crystallization is to use evaporation to simply concentrate the contamination as a concentrated waste stream. This will generally only be useful if the wastewater is low in volume and the waste contamination is nonvolatile. The relatively pure evaporated water might still require treatment after condensation if it is to be disposed of. The concentrated waste can then be recycled or sent for further treatment or disposal. As with evaporative crystallization, the cost of energy for such operations can be prohibitively expensive, unless the heat required for evaporation can be supplied by heat recovery, or the heat available in the evaporated water can be recovered.

8. *Liquid–liquid extraction.* With liquid–liquid extraction, wastewater containing organic waste is contacted with a solvent in which the organic waste is more soluble. The waste is then separated from the solvent by evaporation or distillation and the solvent recycled.

One common application of liquid–liquid extraction is the removal and recovery of phenol and compounds of phenol from wastewaters. Although phenol can be removed by biological treatment, only limited levels can be treated biologically. Variations in phenol concentration are also a problem with biological treatment, since the biological processes take time to adjust to the variations.

9. *Adsorption.* Adsorption can be used for the removal of organic compounds (including many toxic materials) and heavy metals (especially when complexed with organic compounds). Activated carbon is primarily used as the adsorbent, although synthetic resins are also used. Both fixed and moving beds can be used, but fixed beds are by far the most commonly used arrangement. For activated carbon, the removal of organic compounds depends, amongst other things, on the molar mass and the polarity of the molecule. A general trend is that nonpolar molecules (e.g. benzene) tend to adsorb more readily than polar molecules (e.g. methanol) on activated carbon. As an example, adsorption with activated carbon when applied to petroleum refinery effluent can remove typically 70 to 95% BOD_5 and 70 to 90% COD, depending on the effluent being treated[5].

As the adsorbent becomes saturated, regeneration is required. When removing organic materials, activated carbon can be regenerated by steam stripping or heating in a furnace. Stripping allows recovery of material, whereas thermal regeneration destroys the organic material. Thermal regeneration requires a furnace with temperatures above 800°C to oxidize the adsorbates. This causes a loss of carbon of around 5 to 10% per regeneration cycle.

10. *Ion exchange.* Ion exchange is used for selective ion removal and finds some application in the recovery of specific materials from wastewater, such as heavy metals. As with adsorption processes, regeneration of the medium is necessary. Resins are regenerated chemically, which produces a concentrated waste stream requiring further treatment or disposal.

11. *Wet oxidation.* In wet oxidation, an aqueous mixture is heated in the liquid phase under pressure in the presence of air or pure oxygen, which oxidizes the organic material. The efficiency of the oxidation process depends on reaction time and pressure. Temperatures of 150 to 300°C are used, together with pressures of 3 to 200 bar, depending on the process and the nature of the waste being treated and whether a catalyst is used. Oxidation at low temperatures and pressures is only possible if a catalyst is used. Carbon is oxidized to CO_2, hydrogen to H_2O, chlorine to Cl^-, nitrogen to NH_3 or N_2, sulfur to SO_4^{2-}, phosphorous to PO_4^{2-}, and so on.

Wet oxidation is particularly effective in treating aqueous wastes containing organic contaminants with a COD up to 2% prior to biological treatment. Chemical oxygen demand can be reduced by up to 95% and organic halogen compounds by up to 95%. Organic halogen compounds are particularly resistant to biological degradation. If designed for a specific organic contaminant (e.g. phenol), removal can be 99% or greater. Wet oxidation is often used prior to biological treatment to pretreat wastes that would otherwise be resistant to biological oxidation.

A basic flowsheet for a wet oxidation process is shown in Figure 26.5. Although the oxidation reactions release heat,

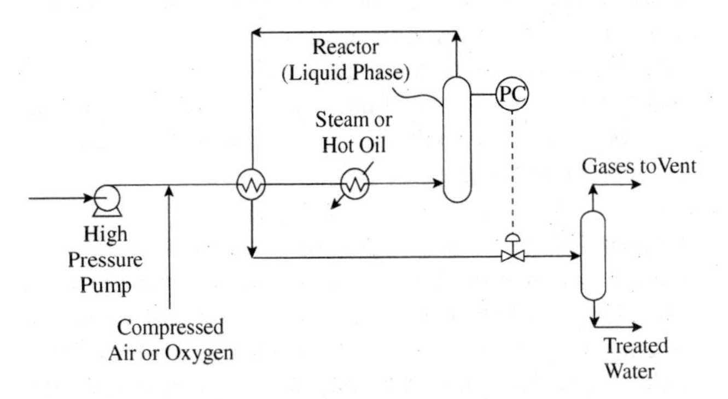

Figure 26.5 A typical wet oxidation process.

the process might still require a net input of heat from an external source (e.g. steam or hot oil). In some cases, the heat release can be high enough to avoid the need for an external source of heat. The largest cost associated with the process is the capital cost of the high-pressure reactor, which normally has an internal titanium cladding.

12. *Chemical oxidation.* Chemical oxidation can be used for the oxidation of organic contaminants that are difficult to treat biologically. It can be used to kill microorganisms by oxidation. When used before biological treatment, organic pollutants that are difficult to treat biologically can be oxidized to simpler, less refractory organic compounds. Chlorine (as gaseous chlorine or hypochlorite ion) can be used. In solution, chlorine reacts with water to form hypochlorous acid (HOCl) according to:

$$Cl_2 + H_2O \longrightarrow HOCl + HCl$$

$$HOCl \rightleftharpoons H^+ + OCl^-$$

Then, for example, the reaction for nitrogen removal is:

$$2NH_3 + 3HOCl \longrightarrow N_2 + 3H_2O + 3HCl$$

Heterogeneous solid catalysts can enhance the performance. Chlorination has the disadvantage that small quantities of undesirable halogenated organic compounds can be formed.

Hydrogen peroxide is another common oxidizing agent. It is used as a 30 to 70% solution for the treatment of:

- cyanides
- formaldehyde
- hydrogen sulfide
- hydroquinone
- mercaptans
- phenol
- sulfites and so on.

Ozone can also be used as an oxidizing agent, but because of its instability, it must be generated on-site. It is a powerful oxidant for some organic materials, but others are oxidized only slowly or not at all. Ozone is only suitable for low concentrations of oxidizable materials. A common use is for the sterilization of water.

The effectiveness of these chemical oxidizing agents is enhanced by the presence of ultraviolet (UV) light and solid catalysts. Chemical oxidation is also sometimes applied after biological treatment.

13. *Sterilization.* In some processes, such as food and beverage and pharmaceutical processes, water might need to be sterilized before it is reused or recycled. Chemical oxidation (e.g. ozonation) can be used. Ultraviolet light is an alternative for lightly contaminated water. Alternatively, a combination of chemical oxidation and UV light can be used. Also, heat treatment (pasteurization) can be used. Heat treatment of water to 80°C provides sterilization that is often good enough for many purposes.

14. *pH adjustment.* The pH of the wastewater often needs adjustment prior to reuse, discharge or biological treatment. For biological treatment, the pH is normally adjusted to between 8 and 9. Bases used include sodium hydroxide, calcium oxide and calcium carbonate. Acids used include sulfuric acid and hydrochloric acid.

15. *Chemical precipitation.* Chemical precipitation followed by solids separation is particularly useful for separating heavy metals. The heavy metals of particular concern in the treatment of wastewaters include cadmium, chromium, copper, lead, mercury, nickel and zinc. This is a particular problem in the manufacture of dyes and textiles and in metal processes such as pickling, galvanizing and plating.

Heavy metals can often be removed effectively by chemical precipitation in the form of carbonates, hydroxides or sulfides. Sodium carbonate, sodium bisulfite, sodium hydroxide and calcium oxide can be used as precipitation agents. The solids precipitate as a *floc* containing a large amount of water in the structure. The precipitated solids need to be separated by thickening or filtration and recycled if possible. If recycling is not possible, then solids are usually disposed of to landfill.

The precipitation process tends to be complicated when a number of metals are present in solution. If this is the case, then the pH must be adjusted to precipitate out the individual metals, since the pH at which precipitation occurs depends on the metal concerned.

26.3 BIOLOGICAL TREATMENT PROCESSES

In *secondary* or *biological treatment*, a concentrated mass of microorganisms is used to break down organic matter into stabilized wastes. The degradable organic matter in the wastewater is used as food by the microorganisms. Biological growth requires supplies of oxygen, carbon, nitrogen, phosphorus and inorganic ions such as calcium, magnesium, and potassium. Domestic sewage satisfies the requirements, but industrial wastewaters may lack nutrients and this can inhibit biological growth. In such circumstances, nutrients may need to be added. As the waste treatment progresses, the microorganisms multiply producing an excess of this *sludge*, which cannot be recycled.

There are three main types of biological process[1-3]:

1. *Aerobic.* Aerobic reactions take place only in the presence of free oxygen and produce stable, relatively inert end products such as carbon dioxide and water. Aerobic reactions are by far the most widely used. The oxidation

reactions are of the type:

$$\begin{bmatrix} C \\ O \\ H \\ N \\ S \end{bmatrix} + O_2 + [\text{Nutrients}] \longrightarrow CO_2 + H_2O$$

Organic matter

$$+ NH_3 + [\text{Cells}] + [\text{Other end products}]$$

Endogenous respiration reactions also occur, which reduce the sludge formation:

$$[\text{Cells}] + O_2 \longrightarrow CO_2 + H_2O + NH_3 + [\text{Energy}]$$

Nitrification reactions can occur, in which organic nitrogen and ammonia are converted to nitrate:

$$\text{Organic Nitrogen} \longrightarrow NH_4^+$$

$$NH_4^+ + CO_2 + O_2 \longrightarrow NO_2^-$$
$$+ H_2O + [\text{Cells}] + [\text{Other end products}]$$

$$NO_2^- + CO_2 + O_2 \longrightarrow NO_3^-$$
$$+ [\text{Cells}] + [\text{Other end products}]$$

The nitrification reactions are inhibited by high concentrations of ammonia.

2. *Anaerobic*. Anaerobic reactions function without the presence of free oxygen and derive their energy from organic compounds in the waste. Anaerobic reactions proceed relatively slowly and lead to end products that are unstable and contain considerable amounts of energy, such as methane and hydrogen sulfide.

$$\begin{bmatrix} C \\ O \\ H \\ N \\ S \end{bmatrix} + [\text{Nutrients}] \longrightarrow CO_2 + CH_4$$

Organic matter

$$+ [\text{Cells}] + [\text{Other end products}]$$

3. *Anoxic*. Anoxic reactions also function without the presence of free oxygen. However, the principal biochemical pathways are not the same as in anaerobic reactions, but are a modification of aerobic pathways and hence termed anoxic. Anoxic reactions are used for *denitrification* to convert nitrate to nitrogen:

$$NO_3^- + BOD \longrightarrow N_2 + CO_2 + H_2O + OH^- + [\text{Cells}]$$

Various methods are used to contact the microorganisms with the wastewater. In a completely mixed system, the hydraulic residence time of the wastewater and the solids residence time of the microorganisms would be the same. Thus, the minimum hydraulic residence time would be defined by the growth rate of the microorganisms. Since the crucial microorganisms can take considerable time to grow, this would lead to hydraulic residence times that would be prohibitively long. To overcome this, a number of methods have been developed to decouple the hydraulic and solids residence times.

1. *Aerobic digestion*. The *suspended growth* or *activated sludge* method is illustrated in Figure 26.6. Biological treatment takes place in a tank where the waste is mixed with a flocculated biological sludge. To maintain aerobic conditions, the tank must be aerated. Sludge separation from effluent is normally achieved by gravity sedimentation. Part of the sludge is recycled and excess sludge is removed. The hydraulic flow pattern in the aeration tank can vary between extremes of mixed-flow and plug-flow. For mixed-flow reactors, the wastewater is rapidly dispersed throughout the reactor and its concentration is reduced. This feature is advantageous at sites where periodic discharges of more concentrated waste are received. The rapid dilution of the waste means that the concentration of any toxic compounds present will be reduced, and thus the microorganisms within the reactor may not be affected by the toxicant. Thus, mixed-flow reactors produce an effluent of uniform quality in response to fluctuations in the feed.

Plug-flow reactors have a decreasing concentration gradient from inlet to outlet, which means that toxic compounds in the feed remain undiluted during their passage along the reactor, and this may inhibit or kill many of the microorganisms within the reactor. The oxygen demand along the reactor will also vary. On the other hand, the increased concentration means that rates of reaction are increased, and for two reactors of identical volume and hydraulic retention time, a plug-flow reactor will show a greater degree of BOD_5 removal than a mixed-flow reactor.

The biochemical population can be specifically adapted to particular pollutants. However, in the majority of cases, a wide range of organic materials must be dealt with and mixed cultures are used.

Nitrification–denitrification reactions can be carried out in suspended growth by separating the reactor into different cells. A typical arrangement would control the first cell under anoxic conditions to carry out the denitrification reactions. These reactions require organic carbon and this

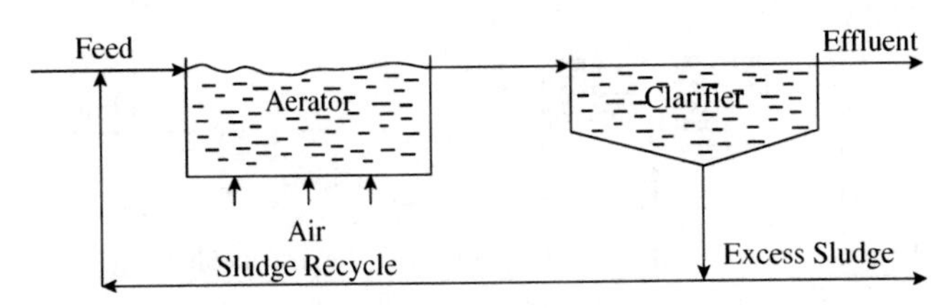

Figure 26.6 Suspended growth aerobic digestion.

will already be present in mixed effluents. If there is insufficient organic carbon in the feed, then this must be added (e.g. adding methanol). The water from the anoxic cell would then overflow into a cell maintained under aerobic conditions in which the nitrification reactions are carried out, along with the oxidation of organic carbon. The nitrification reactions produce nitrate and this requires a recycle back to the anoxic cell.

In aerobic processes, the mean sludge residence time is typically 5 to 10 days. The hydraulic residence time is typically 0.2 to 0.3 days. Suspended growth aerobic processes are capable of removing up to 95% of BOD. The inlet concentration is restricted to a maximum BOD of around 1 $kg \cdot m^{-3}$ (1000 $mg \cdot liter^{-1} \approx$ 1000 ppm) or COD of around 3.5 $kg \cdot m^{-3}$ (3500 $mg \cdot liter^{-1} \approx$ 3500 ppm). Chloride (as Cl^-) should be less than 8 $kg \cdot m^{-3}$ to 15 $kg \cdot m^{-3}$. The performance of the process can be enhanced by the use of pure oxygen, rather than air. This reduces the plant size and increases the allowable inlet concentration of COD to around 10 $kg \cdot m^{-3}$, but does not have a significant effect on the overall removal. If powdered activated carbon is added, this increases the overall removal slightly. More importantly, though, it helps to treat refractory organic compounds by adsorption and evens out fluctuations in the performance caused by fluctuations in the feed concentration.

Attached growth (film) methods. Here the wastewater is trickled over a packed bed through which air is allowed to percolate. A *biological film* or *slime* builds up on the packing under aerobic conditions. Oxygen from the air and biological matter from the wastewater diffuses into the slime. As the biological film grows, it eventually breaks its contact with the packing and is carried away with the water. Packing material varies from pieces of stone to preformed plastic packing.

Figure 26.7 shows a typical attached growth arrangement. Alternatively, the wastewater can be used to fluidize a bed of carbon or sand onto which the film is attached.

Attached growth processes are capable of removing up to 90% of BOD_5 and are thus less effective than suspended growth methods. Nitrification–denitrification reactions can also be carried out in attached growth processes.

2. *Anaerobic digestion.* With wastewaters containing a high organic content, the oxygen demand may be so high that it becomes extremely difficult and expensive to maintain aerobic conditions. In such circumstances, anaerobic processes can provide an efficient means of removing large quantities of organic material. Anaerobic processes tend to be used when BOD_5 levels exceed 1 $kg \cdot m^{-3}$ (1000 $mg \cdot liter^{-1}$). However, they are not capable of producing very high quality effluents and further treatment is usually necessary.

The inability to produce high-quality effluents is one significant disadvantage. Another disadvantage is that anaerobic processes must be maintained at temperatures between 35 and 40°C to get the best performance. If low-temperature waste heat is available from the production process, then this is not a problem. One advantage of anaerobic reactions is that the methane produced can be a useful source of energy. This can be fed to steam boilers or burnt in a heat engine to produce power.

Suspended growth methods. The contact type of anaerobic digester is similar to the activated sludge method of aerobic treatment. The feed and microorganisms are mixed in a tank (this time closed). Mechanical agitation is usually required since there is no air injection. The sludge is separated from the effluent by sedimentation or filtration, part of the sludge recycled and excess sludge removed.

Another method is the *upflow anaerobic sludge blanket* illustrated in Figure 26.8. Here the sludge is contacted by upward flow of the feed at a velocity such that the sludge is not carried out of the top of the digester.

A third method of contact known as an *anaerobic filter* also uses upward flow but keeps the sludge in the digester by a physical barrier such as a grid.

Attached growth (film) methods. As with aerobic digestion, the microorganisms can be encouraged to grow attached to a support medium such as plastic packing or sand. In anaerobic attached growth digestion, the bed is usually fluidized rather than a fixed-bed arrangement, as shown in Figure 26.9.

Anaerobic processes typically remove 75 to 85% of COD[1,2].

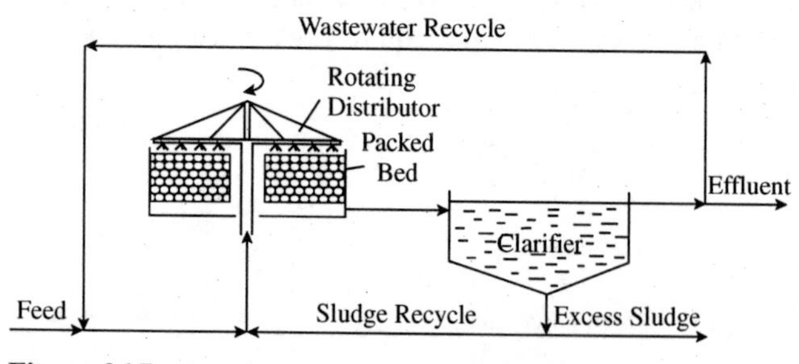

Figure 26.7 Attached growth aerobic digestion.

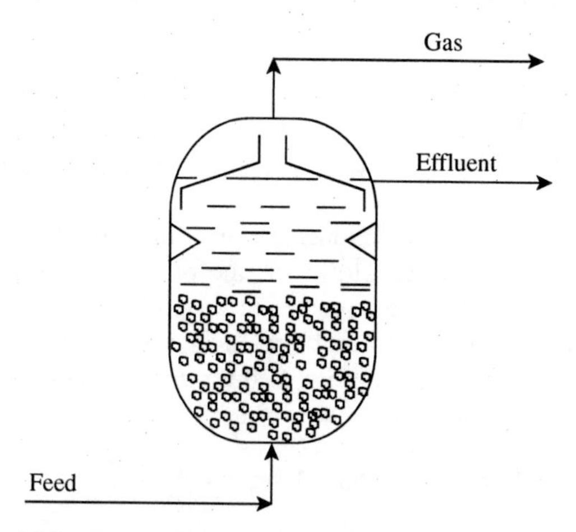

Figure 26.8 Suspended growth anaerobic digestion using an upward flow anaerobic sludge blanket.

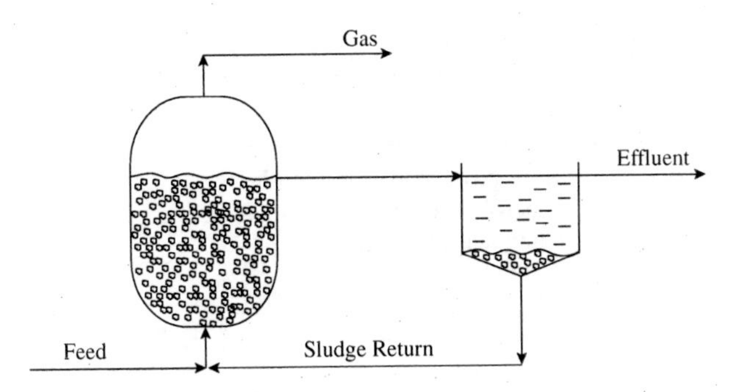

Figure 26.9 Attached growth anaerobic digestion using a fluidized anaerobic bed.

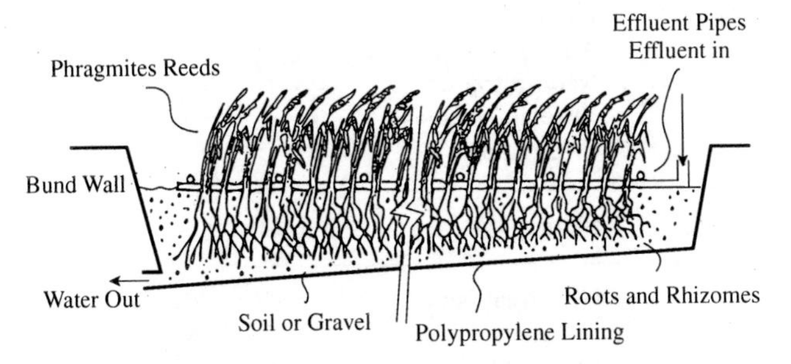

Figure 26.10 A reed bed.

3. *Reed beds.* Reed bed processes require reeds (phragmites) to be planted in soil or gravel in the wastewater. Figure 26.10 shows a typical reed bed arrangement. Oxygen from air is transported through the leaves, stems and rhizomes to high concentrations of microorganisms in the root zone. Aerobic treatment takes place in the region of the rhizomes, with anoxic and anaerobic treatment in the surrounding soil. Different flow arrangements other than that shown in Figure 26.10 can also be used. Reed beds are capable of removing 60 to 80% of BOD_5[6]. They are also capable of removing 25 to 50% of total nitrogen and 20 to 40% of total phosphorous[6]. They have the advantage that there is no sludge disposal. The maximum inlet concentration can be as high as 3 $kg \cdot m^{-3}$ BOD_5. A significant disadvantage of reed beds is the time taken for them to become fully established and they need between six months and two years for this. The difference between winter and summer performance needs to be considered for such arrangements.

Excess sludge is produced in most biological treatment processes, which must be disposed of. The treatment and disposal of sludge is a major problem that can be costly to deal with. Anaerobic processes have the advantage here, since they produce considerably less sludge than aerobic processes (of the order of 5% of aerobic processes for the same throughput). Sludge disposal can typically be responsible for 25 to 40% of the operating costs of an aerobic biological treatment system. Treatment of sludge is primarily aimed at reducing its volume. This is because the sludge is usually 95 to 99% water and the cost of disposal is closely linked to its volume. The water is partly free, partly trapped in the flocs and partly bound in the microorganisms. Anaerobic digestion of the sludge can be used, followed by dewatering. The dewatering can be by filtration or centrifugation. Alternatively, filtration or centrifugation can be used directly to carry out the dewatering. For centrifugation, the dewatering process can be enhanced by the addition of clay. Adding powdered activated carbon to aerobic suspended growth processes can also helps with sludge dewatering. The resulting water content after these processes is reduced to typically 60 to 85%. The water content can be reduced to perhaps 10% by drying. The sludge may finally be used for agricultural purposes (albeit a poor fertilizer) or thermally oxidized.

Large sites might require their own biological treatment processes for final treatment before discharge. Smaller sites might rely on local municipal treatment processes, which treat a mixture of industrial and domestic effluent, for final effluent discharge.

Table 26.1 provides a summary of the main features of biological wastewater treatment.

Table 26.2 summarizes the treatment processes that can be used for various types of contamination.

26.4 TERTIARY TREATMENT PROCESSES

Tertiary treatment or *polishing treatment* prepares the aqueous waste for final discharge. The final quality of the effluent depends on the nature and flow of the receiving water. Table 26.3 gives an indication of the final quality required[7].

Table 26.1 Comparison of biological wastewater treatments.

Aerobic	Anaerobic	Reed beds
$BOD_5 < 1\ kg \cdot m^{-3}$ (higher if O_2 used)	$BOD_5 > 1\ kg \cdot m^{-3}$	$BOD_5 < 3.5\ kg \cdot m^{-3}$
Stable end products (CO_2, H_2O, etc.)	Unstable end products (CH_4, H_2S, etc.)	Stable and unstable end products
BOD_5 removal up to 95%	BOD_5 removal 75–85%	BOD_5 removal 60–80%
High sludge formation	Low sludge formation	No sludge disposal or centrifugation

Table 26.2 Summary of treatment processes for some common contaminants.

Suspended solids	Dispersed oil	Dissolved organic
Gravity separation Centrifugal separation Filtration Membrane filtration	Coalescence Centrifugal separation Flotation Wet oxidation Thermal oxidation	Biological oxidation (aerobic, anaerobic, reed beds) Chemical oxidation Activated carbon Wet oxidation Thermal oxidation
Ammonia	**Phenol**	**Heavy metals**
Steam stripping Air stripping Biological nitrification Chemical oxidation Ion exchange	Solvent extraction Biological oxidation (aerobic) Wet oxidation Activated carbon Chemical oxidation	Chemical precipitation Ion exchange Adsorption Nano-filtration Reverse osmosis Electrodialysis
Dissolved solids	**Neutralization**	**Sterilization**
Ion exchange Reverse osmosis Nano-filtration Electrodialysis Crystallization Evaporation	Acid Base	Heat treatment UV light Chemical oxidation

Table 26.3 Typical effluent quality for various receiving waters[7].

Receiving water	Typical effluent	
	BOD_5 ($mg \cdot l^{-1}$)	Suspended solids ($mg \cdot l^{-1}$)
Tidal estuary	150	150
Lowland river	20	30
Upland river	10	10
High quality river with low dilution	5	5

Aerobic digestion is normally capable of removing up to 95% of the BOD. Anaerobic digestion is capable of removing less, in the range 75 to 85%. With municipal treatment processes, which treat a mixture of domestic and industrial effluent, some disinfection of the effluent might be required to destroy any disease-causing organisms before discharge to the environment. Tertiary treatment processes vary, but constitute the final stage of effluent treatment to ensure that the effluent meets specifications for disposal. Tertiary processes used include:

1. *Filtration*. Examples of such processes are microstrainers (a fine screen with openings) and sand filters. They are designed to improve effluents from biological treatment processes by removing suspended material, and with it, some of the remaining BOD_5. Sand filtration can remove effectively all of the remaining BOD_5 in many circumstances.
2. *Ultrafiltration*. Ultrafiltration was described under pretreatment methods. It is used to remove finely divided suspended solids and when used as a tertiary treatment can in many circumstances remove virtually all the BOD_5 remaining after biological treatment.
3. *Adsorption*. Some organic materials are not removed in biological systems operating under normal conditions. Removal of residual organic material can be achieved by adsorption. Both activated carbon and synthetic resins are used. As described earlier under pretreatment methods, regeneration of activated carbon in a furnace can cause carbon losses of perhaps 5 to 10%.
4. *Nitrogen and phosphorous removal*. Since nitrogen and phosphorous are essential for growth of the microorganisms, the effluent from secondary treatment will contain some nitrogen and phosphorous. The amount that is discharged to receiving waters can have a considerable effect on the growth of algae. If discharge is to a high-quality receiving water and/or dilution rates are low, then removal may be necessary. Nitrogen principally occurs as ammonium (NH_4^+), nitrate (NO_3^-) and nitrite (NO_2^-). Phosphorous principally occurs as orthophosphate (PO_4^{3-}). A variety of biological and chemical processes are available for the removal of nitrogen and phosphorous[1,2]. These

processes produce extra biological and inorganic sludge that requires disposal.

5. *Disinfection.* Chlorine, as gaseous chlorine or as the hypochlorite ion, is widely used as a disinfectant. However, its use in some cases can lead to the formation of toxic organic chlorides and the discharge of excess chlorine can be harmful. Hydrogen peroxide and ozone are alternative disinfectants that lead to products that have a lower toxic potential. Treatment is enhanced by ultraviolet light. Indeed, disinfection can be achieved by ultraviolet light on its own.

26.5 WATER USE

Water is used for a wide variety of purposes in process operations:

- reaction medium (vapor or liquid)
- extraction processes
- steam stripping
- steam ejectors for production of vacuum
- equipment washing
- hosing operations, and so on.

The one thing all of these operations have in common is that the water comes into contact with process materials and becomes contaminated. Figure 26.11 shows the quantitative representation of this on the plot of concentration versus mass load of contaminant. The water starts with no contamination and its level of contamination increases as a result of the mass transfer.

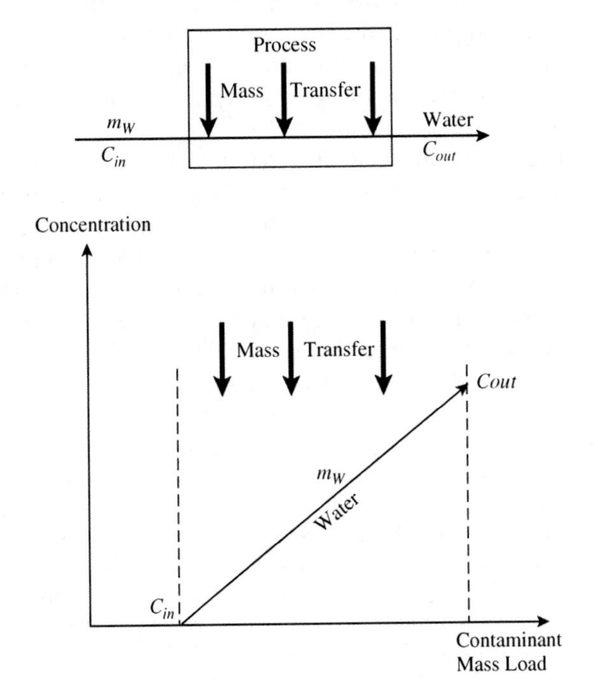

Figure 26.11 Water use representation.

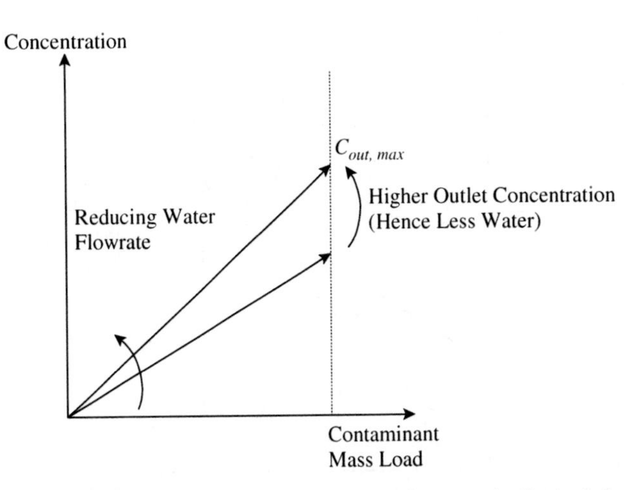

Figure 26.12 Reduction in the water flowrate is limited by minimum flowrate or maximum outlet concentration limits.

If the flowrate of water to an operation is decreased by some change to the operation, then for the same mass load transferred, the reduction in the water flowrate will lead to a steeper line and higher outlet concentration as shown in Figure 26.12. The reduction in the water usage will be limited by either the operation requiring some minimum flowrate, below which it cannot operate, or the outlet concentration from the operation goes to a maximum value. The maximum outlet concentration might be set by a number of considerations:

- maximum solubility
- corrosion limitations
- fouling limitations
- minimum of mass transfer driving force
- minimum flowrate requirements
- maximum inlet concentration for downstream treatment

If all operations use clean water, then reducing the flowrate to its minimum value can be used to minimize the water consumption, as illustrated in Figure 26.12. However, such an approach misses the opportunity to reuse water. To open up the opportunity to reuse water between operations, some level of inlet contamination must be accepted. Figure 26.13 shows a water-using profile such that both the inlet concentration and outlet concentration have been set to their maximum values. This particular setting, where both the inlet and outlet concentrations are set to their maximum values, can be used to define the *limiting water profile*[4,8]. As shown in Figure 26.14, the limiting water profile is used to define a boundary between feasible and infeasible concentrations. The concentration of a water profile is considered to be feasible as long as it is below the limiting water profile, as illustrated in Figure 26.14. This approach will be used later to identify reuse opportunities. The approach has a number of advantages:

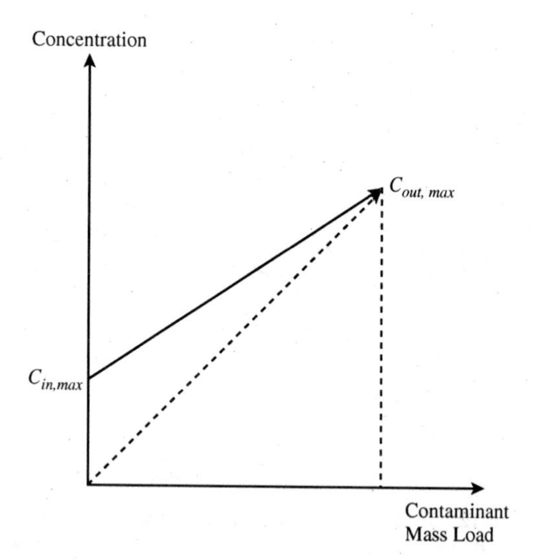

Figure 26.13 An alternative water profile uses more water but accepts slightly contaminated water.

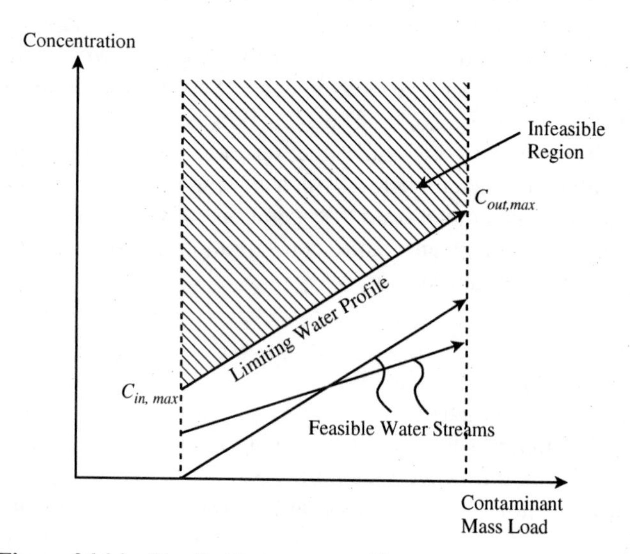

Figure 26.14 The limiting water profile.

- Operations with different characteristics can be compared on a common basis (e.g. water used in an extraction process versus a hosing operation).
- It does not require a model of the operation to represent the mass transfer.
- It does not depend on any particular flow pattern (countercurrent versus cocurrent).
- It works on any type of water-using operation (e.g. firewater makeup, cooling tower makeup, and so on).

26.6 TARGETING MAXIMUM WATER REUSE FOR SINGLE CONTAMINANTS

As already noted, if freshwater is being used for all operations, then minimizing the flowrates to individual operations will minimize the water consumption. However, this misses opportunities for reuse. If some level of inlet contamination is allowed, then, in principle, water can be reused between operations. To illustrate how the overall minimum water consumption can be targeted when allowing reuse, consider the data for the simple problem given in Table 26.4. This specifies maximum inlet and outlet concentrations or *limiting concentrations* for a single contaminant. The single contaminant might be a specific component (e.g. phenol, acetone, starch) or an *aggregate property* (e.g. total organic material, suspended solids, dissolved solids, COD). Later the approach will be extended to systems where limiting concentrations for multiple contaminants are specified.

A number of points need to be noted regarding the data in Table 26.4. First, the concentration is specified on the basis of the mass flowrate of water and not the mass flowrate of the mixture.

$$C = \frac{m_C}{m_W} \quad not \quad C = \frac{m_C}{m_W + m_C} \tag{26.1}$$

where C = concentration (ppm)
m_C = mass flowrate of contaminant ($g\cdot h^{-1}$, $g\cdot d^{-1}$)
m_W = mass flowrate of water ($t\cdot h^{-1}$, $t\cdot d^{-1}$)

In most problems, the concentration of contaminant is so small that there is virtually no difference between the concentration based on the mass flowrate of water and the mass flowrate of the mixture. However, it is important to be consistent and follow the convention given in Equation 26.1. The other point to note is regarding the units. It is convenient to define the flowrate in terms of metric tons (typically tons per hour or tons per day). It is also convenient to define the concentration in terms of parts per million (ppm). If the basic unit of flowrate is taken to be tons and concentration to be parts per million, then the mass load is measured in grams (typically grams per hour or grams per day).

The flowrate in Table 26.4 refers to the *limiting water flowrate*. The limiting water flowrate is the flowrate required if the specified mass of contaminant is picked up by the water between the maximum inlet and outlet concentrations. If an operation has a maximum inlet contaminant concentration greater than zero and it is fed by water with zero concentration, then for the specified mass load, a lower flowrate than the limiting water flowrate could be used.

Table 26.4 Data for a problem with four operations.

Operation number	Contaminant mass ($g\cdot h^{-1}$)	C_{in} (ppm)	C_{out} (ppm)	Limiting water flowrate ($t\cdot h^{-1}$)
1	2000	0	100	20
2	5000	50	100	100
3	30,000	50	800	40
4	4000	400	800	10

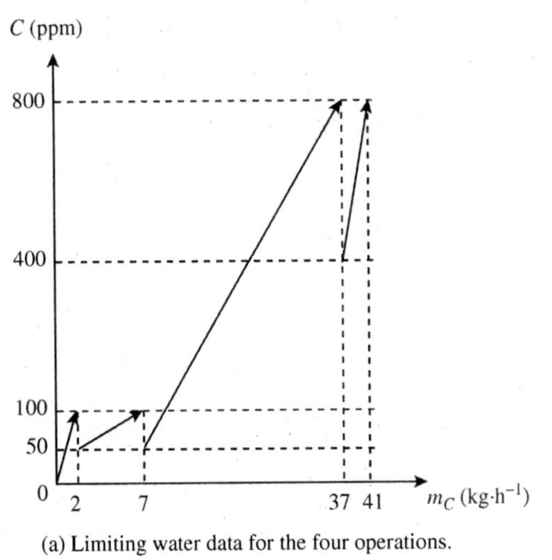

(a) Limiting water data for the four operations.

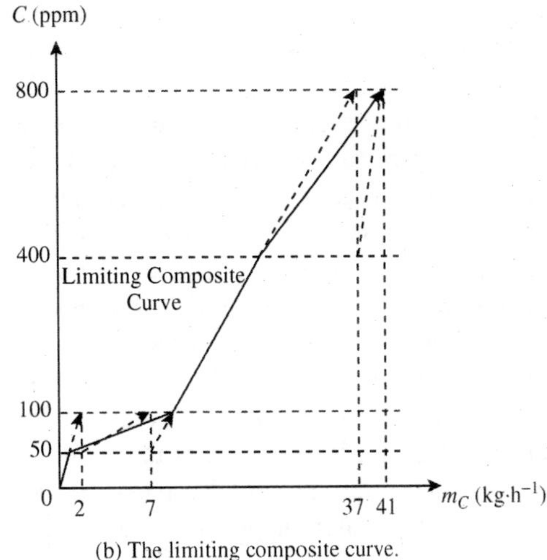

(b) The limiting composite curve.

Figure 26.15 Construction of the limiting composite curve for the simple example from Table 26.4. (From Wang YP and Smith R, 1994, *Chem Eng Sci*, **49**: 981, reproduced by permission of Elsevier Ltd.)

Analysis of the data in Table 26.4 can be started by calculating the flowrate that would be required for each of these operations, if each operation was fed by freshwater with zero concentration. The relationship between mass pickup of contaminant, mass flowrate of water and concentration change is given by:

$$\Delta m_C = m_W \Delta C \qquad (26.2)$$

where Δm_C = mass pickup of contaminant (g·h^{-1}, g·d^{-1})
m_W = flowrate of water (t·h^{-1}, t·d^{-1})
ΔC = concentration change (ppm)

For freshwater feed and minimum flowrate for the streams from Table 26.4:

$$m_{W1} = \frac{2000}{100 - 0} = 20 \text{ t·h}^{-1}$$

$$m_{W2} = \frac{5000}{100 - 0} = 50 \text{ t·h}^{-1}$$

$$m_{W3} = \frac{30{,}000}{800 - 0} = 37.5 \text{ t·h}^{-1}$$

$$m_{W4} = \frac{4000}{800 - 0} = 5 \text{ t·h}^{-1}$$

Total flowrate of freshwater

$$= 20 + 50 + 37.5 + 5$$

$$= 112.5 \text{ t·h}^{-1}$$

Consider now the possibility of reuse. To determine the maximum potential for reuse, Figure 26.15a shows the four operations plotted on axes of concentration versus mass load. Note the concentrations are maximum inlet and outlet concentrations (limiting concentrations). Figure 26.15b shows a *limiting composite curve* of the four water streams[4]. The construction of the limiting composite curve is analogous to the composite curves for energy developed in Chapter 16. To construct the limiting composite curve in Figure 26.15b, the diagram is divided into concentration intervals and the mass load within each concentration interval combined to obtain the limiting composite curve[4]. This represents a quantitative profile of the single-stream equivalent to the four separate streams. It is a combined boundary between feasible and infeasible concentrations. To target for the minimum water flowrate, a *water supply line* is drawn to represent the water supply, as shown in Figure 26.16[4]. The water supply line starts at zero concentration in this case (as it must for this problem to satisfy the concentration requirements). In other cases, the water supply might not necessarily start from zero, depending on the quality of the water fed to the process. In principle any slope can be drawn, as long as it is below the limiting composite curve. If the minimum water flowrate

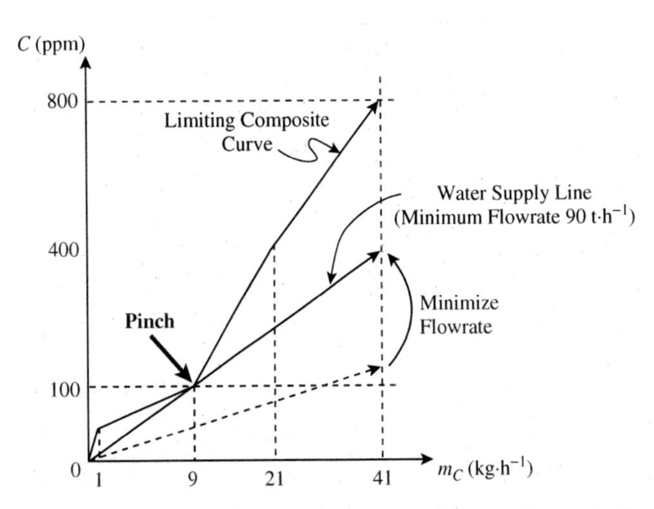

Figure 26.16 Targeting minimum water flowrate for a single contaminant.

is to be obtained, the steepest line possible must be drawn for the water supply. This is shown in Figure 26.16, where the steepest slope corresponds with the water supply line touching the limiting composite curve at the *pinch point*. Knowing the mass load from zero concentration up to the pinch point (9000 $g \cdot h^{-1}$ in this case) and the concentration of the pinch (100 ppm in this case), the flowrate for the system can be calculated from Equation 26.2 to be 90 $t \cdot h^{-1}$. This represents the minimum flowrate target unless there are constraints that there should be a minimum flowrate[9].

A number of points should be noted regarding Figure 26.16. If a steeper water supply line was drawn than the one shown in Figure 26.16, it would cross the limiting composite curve and concentrations will be infeasible at some point. The fact that the water supply line touches the limiting composite curve does not imply, for example, zero concentration difference in a mass exchange operation. It must be remembered that any allowances for driving forces have already been included in the limiting data. Where the water supply line touches the limiting composite curve implies that the water concentration goes to its maximum value at that point. This could correspond with minimum mass exchange driving force, maximum solubility limit, and so on. At points in Figure 26.16 other than zero concentration and the pinch point, the concentrations of the water will all be below their maximum values. The point where the water concentration goes to its maximum at the pinch point dictates the minimum flowrate for the system.

An alternative graphical representation has been suggested involving a plot of concentration (or purity) versus volumetric flowrate of water[10]. Using this approach, a plot is developed involving the water sinks (inlet concentration and flowrate) and water sources (outlet concentration and flowrate). While this plot is an alternative representation, it does not allow the minimum water flowrate to be targeted in a similar way to that developed in Figure 26.16[11].

26.7 DESIGN FOR MAXIMUM WATER REUSE FOR SINGLE CONTAMINANTS

Having seen how to set a target for the data in Figure 26.16, consider now how to achieve that target in design[12]. Figure 26.17 illustrates the basis of the design strategy. First, the minimum water requirements in each region of the design are identified. In this simple example, there are two design regions: above the pinch and below the pinch. Below the pinch in Figure 26.17, by definition, the full amount of the target minimum flowrate is needed (90 $t \cdot h^{-1}$ in this example). Above the pinch, not all 90 $t \cdot h^{-1}$ are required and the process could, in principle, operate with a lower flowrate than the target. The minimum flowrate required above the pinch is determined by a simple mass balance (45.7 $t \cdot h^{-1}$ in this example). The design regions are identified by straight lines drawn between the convex points, to identify the *pockets* in the problem. The basis of the design strategy is to use the target flowrate below the pinch and then to use only the required amount above the pinch with the balance going to effluent. This design strategy serves two purposes:

a. The mass balance is tightly defined, allowing very specific rules to be applied in the design.
b. The strategy also is compatible with minimizing effluent treatment costs overall through distributed effluent treatment, to be discussed later.

Having formulated the basic design strategy, a grid can then be set up as shown in Figure 26.18[12]. The design grid starts by setting up three *water mains*, corresponding with freshwater concentration, pinch concentration and the maximum concentration from Figure 26.17. The flowrate required by each water main is shown at the top of the main

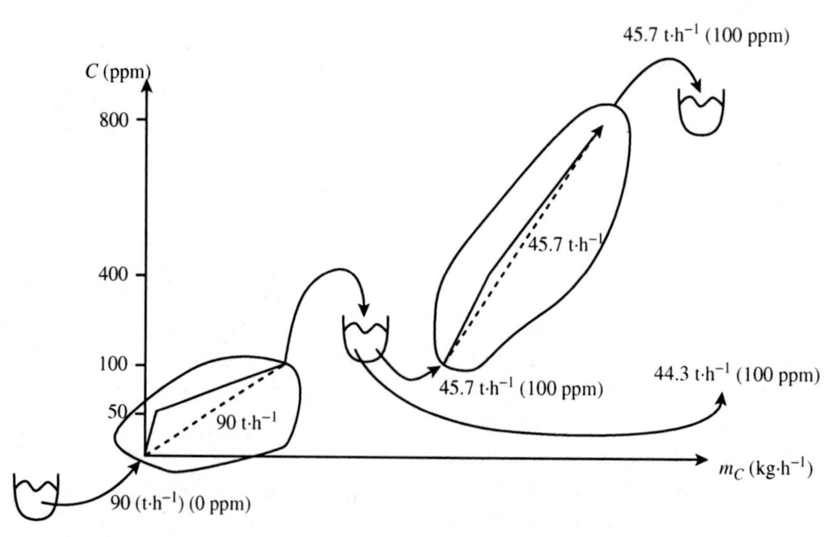

Figure 26.17 The basis of the design strategy.

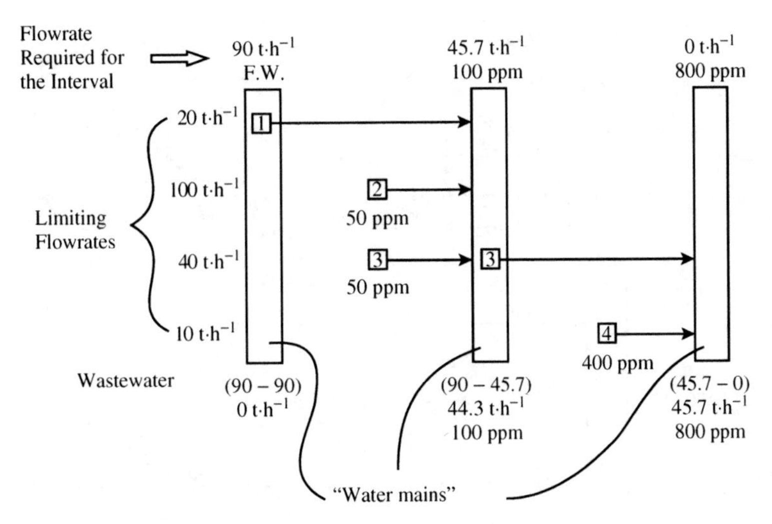

Figure 26.18 The design grid for the water system.

and the wastewater generated by the main at the bottom. Streams representing the individual operation requirements are superimposed on the water mains. Operation 1 has limiting data corresponding with freshwater at the inlet and 100 ppm at the outlet and is therefore shown between the freshwater main and the pinch concentration water main. Operation 2 starts at 50 ppm and terminates at 100 ppm and is therefore shown between the freshwater and pinch concentration mains. Operation 3 is broken into two parts as it features both below and above-pinch concentrations. Operation 4 starts at 400 ppm and ends at 800 ppm and therefore features between the pinch concentration and final concentration mains. The operations are then connected to the appropriate water mains, as illustrated in Figure 26.19. Figure 26.19 also shows the concentrations and limiting concentrations versus mass load in the individual operations. This, in principle, is a working design. However, there is a problem created by Operation 3, in that, below the pinch it receives a flowrate of 20 $t{\cdot}h^{-1}$ of freshwater, but at the pinch it receives a flowrate of 40 $t{\cdot}h^{-1}$ of water at 100 ppm. This change in flowrate in the middle of Operation 3 would be impractical for most operations. It could be practical if Operation 3 involved, for example, an operation with multiple stages of washing. If this was the case, then the different stages could be fed with different flowrates and qualities of water. However, for most situations this change in flowrate would be completely impractical.

The change in flowrate for Operation 3 is in fact readily corrected. Consider Figure 26.20, which shows the concentration versus mass load for an operation with a change in flowrate similar to Operation 3 in Figure 26.19.

A mass balance around Part 1 in Figure 26.20 gives:

$$m_{W1}(C_{PINCH} - C_0) = m_{W2}(C_{PINCH} - C_{in,max}) \qquad (26.3)$$

Moving the mixing junction to the inlet of the operation, as shown in Figure 26.20, and carrying out a mass balance on the new mixing junction gives:

$$\begin{aligned} C_{in} &= \frac{(m_{W2} - m_{W1})C_{PINCH} + m_{W1}C_0}{m_{W2}} \\ &= \frac{m_{W2}C_{PINCH} - m_{W1}(C_{PINCH} - C_0)}{m_{W2}} \end{aligned} \qquad (26.4)$$

Substituting Equation 26.3 into Equation 26.4 gives:

$$\begin{aligned} C_{in} &= \frac{m_{W2}C_{PINCH} - m_{W2}(C_{PINCH} - C_{in,max})}{m_{W2}} \\ &= C_{in,max} \end{aligned} \qquad (26.5)$$

In other words, if the mixing junction is moved from the middle of the operation to the beginning of the operation, then there is a constant flowrate throughout the operation corresponding with an inlet concentration after mixing of the maximum inlet concentration. Also, by definition, the flowrate will be the limiting water flowrate.

In Figure 26.21, the change in flowrate for Operation 3 that previously occurred at the pinch concentration water mains is now added at that concentration to the inlet of Operation 3. The design now features a constant flowrate in all of the operations and achieves the target minimum flowrate of 90 $t{\cdot}h^{-1}$. The arrangement shown in Figure 26.21 involves reuse of water from Operations 1 and 2 into Operations 3 and 4 via a water main at the pinch concentration of 100 ppm. An alternative way to arrange the design is to make the connections directly, rather than through an intermediate water main. If the intermediate water main is removed, then there are basically two sources of water from Operations 1 and 2 at 100 ppm and two sinks for water for Operations 3 and 4 at 100 ppm, as illustrated in Figure 26.22. Figure 26.22 shows a direct connection from Operation 1 to Operation 3 and another from Operation 2 to Operation 4 with 44.3 $t{\cdot}h^{-1}$ going to wastewater from Operation 2. The arrangement shown in Figure 26.22 is the only one possible

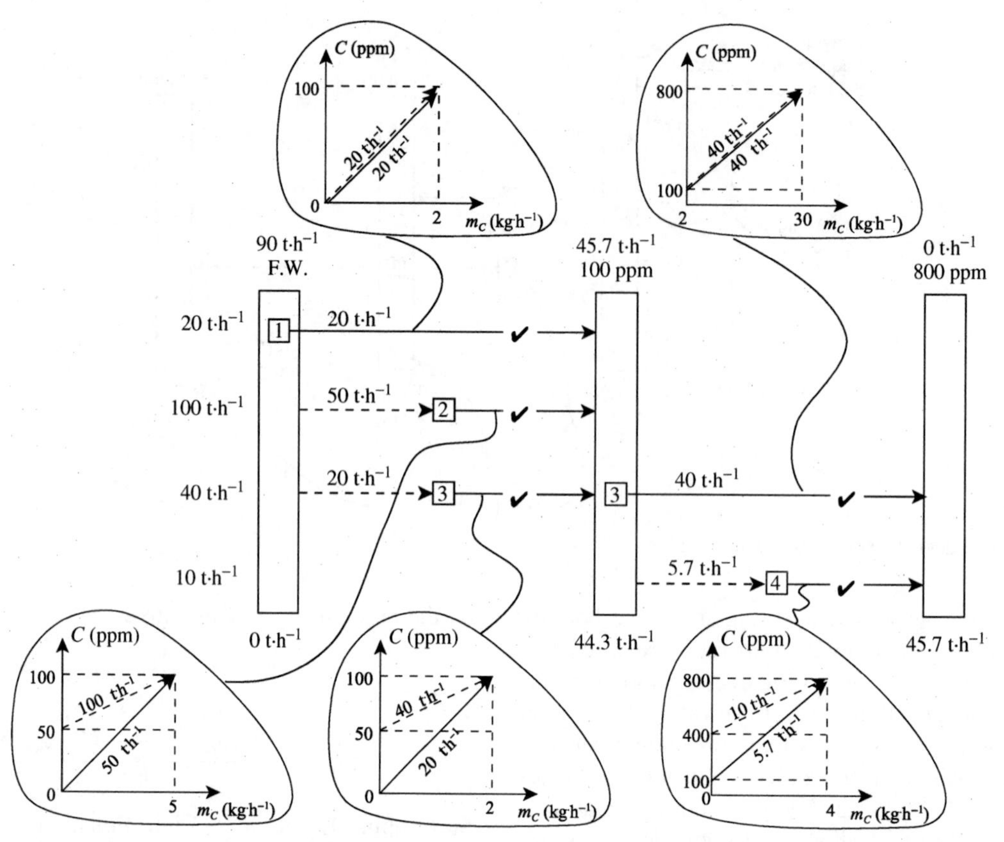

Figure 26.19 The streams are connected with the water mains.

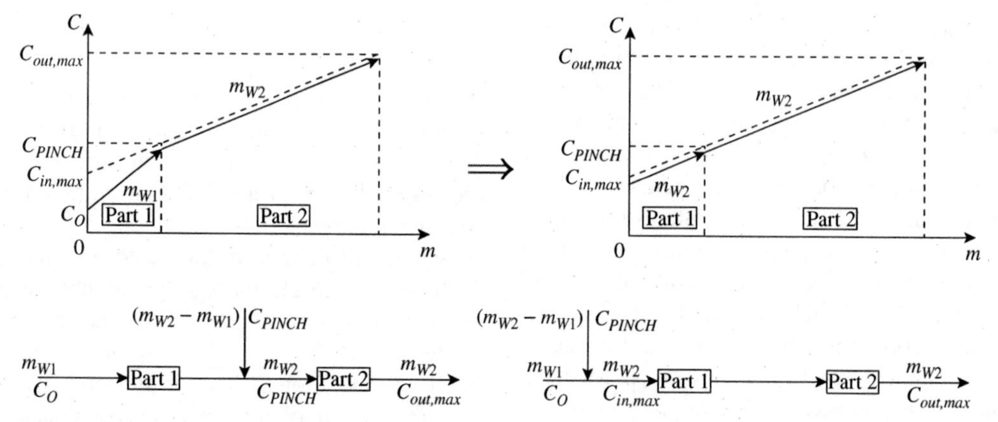

Figure 26.20 An operation involving a change in flowrate. (From Kuo W-CJ and Smith R, 1998, *Trans IChemE*, **76A**: 287, reproduced by permission of the Institution of Chemical Engineers.).

arrangement of connections between the sources and the sinks. Figure 26.23 shows this arrangement in the grid diagram, and Figure 26.24 shows the final design as a conventional flowsheet. The design in Figure 26.24 can be evolved to produce alternative networks. For example, rather than splitting the flow of water at the inlet of Operation 1 with 20 t·h^{-1} going through the operation and 20 t·h^{-1} bypassing, all 40 t·h^{-1} could be put through Operation 1. Then, the outlet of Operation 1 could be fed directly to the inlet of Operation 3. Other options are possible.

This simple example illustrates the basic principles of water network design for maximum reuse for a single contaminant. A number of issues need to be considered that would apply to more complex examples. Consider Figure 26.25 involving three water mains and three operations. Operation 2 above the pinch terminates at a concentration less than the concentration for the high concentration water main. The outlet of Operation 2 must not be fed directly into this final water main. The basis of the mass balance from Figure 26.17 dictates that all streams must achieve the concentration of the water mains into

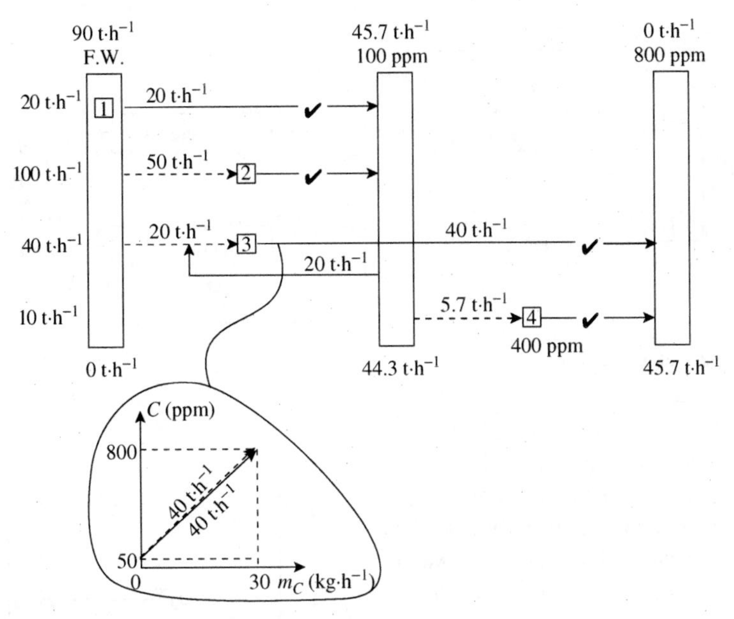

Figure 26.21 Correcting the change in flowrate in the design grid.

which they are being fed; otherwise the mass balance will be violated. Thus, in Figure 26.25a, the outlet of Operation 2 must not be fed to the final water mains, but must be used again and brought to the concentration of the water main into which it is being fed, Figure 26.25b.

Figure 26.26 shows a more complex limiting composite curve that involves three design regions, rather than two. The design regions are identified by straight lines drawn between the convex points, to identify the pockets in the problem. A water main is required at each of the extreme points. In Figure 26.26, there would be four water mains required in the design grid. However, the basic principles are exactly the same as those described so far, but with the additional water main.

It is often the case that there is more than one source of freshwater available. There may be raw water available

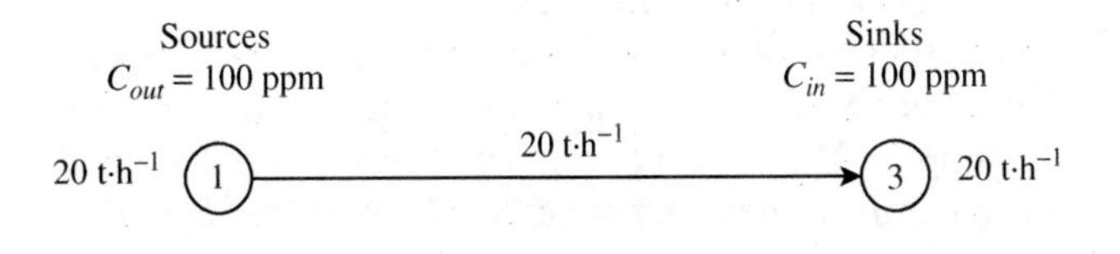

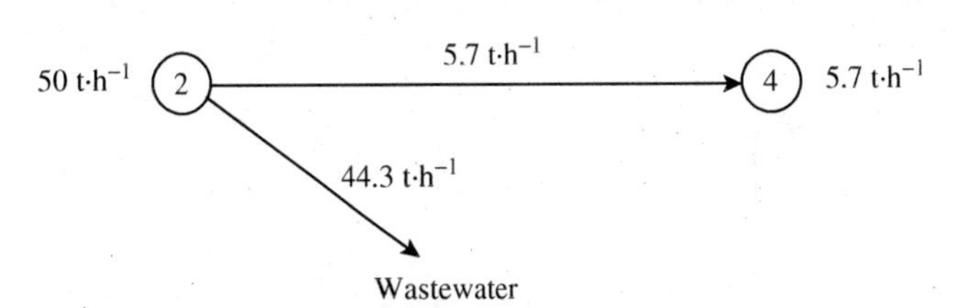

Figure 26.22 Removing the intermediate water main allows the connections to be made directly.

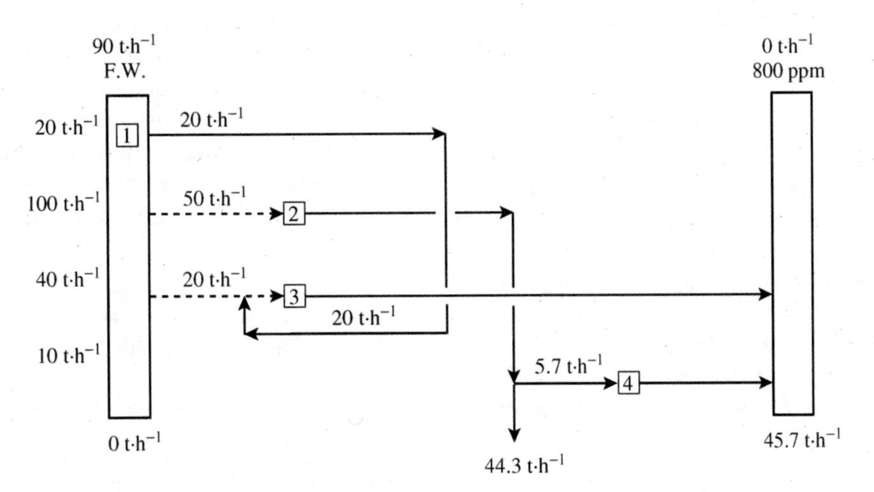

Figure 26.23 Water network without the intermediate water main.

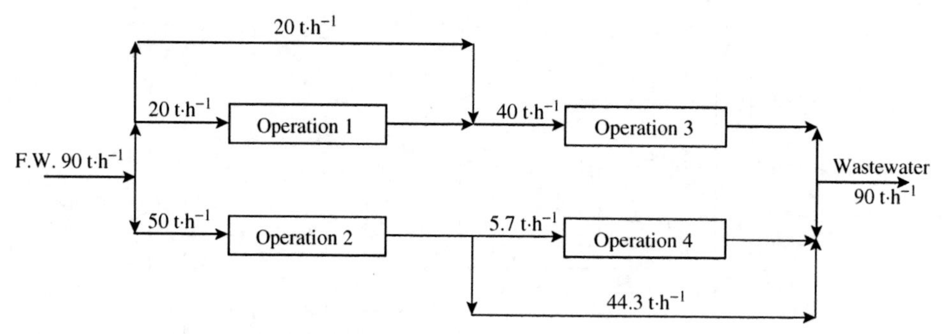

Figure 26.24 Flowsheet for the water network.

from river, lake and so on. It would also be usual to have potable water available. In addition, there may be water available from a borehole and demineralized water (see Chapter 23). The order of water quality is usually also the order of the unit cost of the water. Demineralized water will normally be the most expensive, then potable water, and so on. Consider a simple example to illustrate how to deal with multiple water sources. Limiting data for three operations are given in Table 26.5.

Two sources of freshwater are available; WSI with a concentration of 0 ppm and WSII with a concentration of 25 ppm. The limiting composite curve for the three operations is plotted in Figure 26.27a. Matched against it is a water supply line that is a composite of the two water sources. Between 0 ppm and 25 ppm only WSI can satisfy

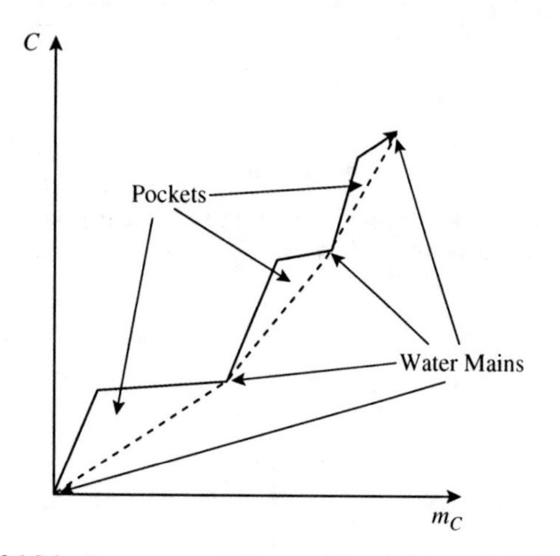

Figure 26.26 In more complex problems there might be more design intervals.

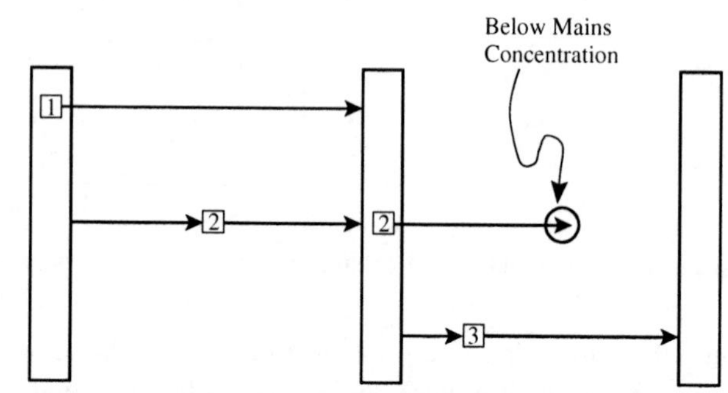

(a) The final concentration of a stream might be below that of a water main.

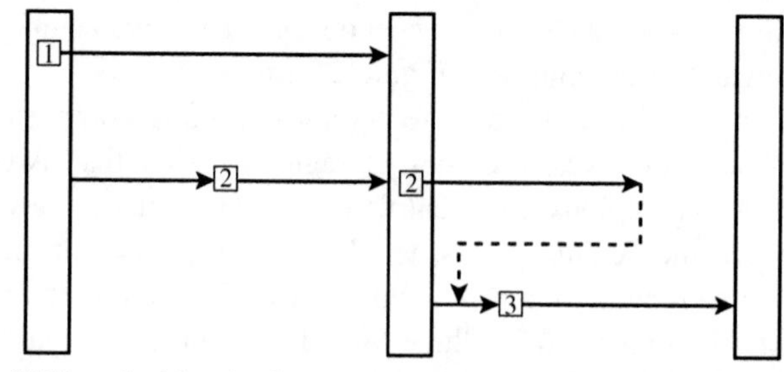

(b) Water should not be discharged to water mains until it has reached mains concentration.

Figure 26.25 Connecting to water mains if the final concentration of a steam is below mains concentration.

the requirements of the process. Here the flowrate of WSI has been minimized to 20 t·h^{-1}. This is dictated by the slope of the limiting composite curve between 0 ppm and 25 ppm. The requirements of the process above 25 ppm are being satisfied by the continued use of the 20 t·h^{-1} of WSI and the balance by the introduction of WSII at 25 ppm. The amount of WSII required can be determined by a mass

Table 26.5 Limiting data for three operations.

Operation	Contaminant mass (kg·h^{-1})	$C_{in,max}$ (ppm)	$C_{out,max}$ (ppm)	Limiting flowrate (t·h^{-1})
1	2	0	100	20
2	5	50	100	100
3	30	50	800	40

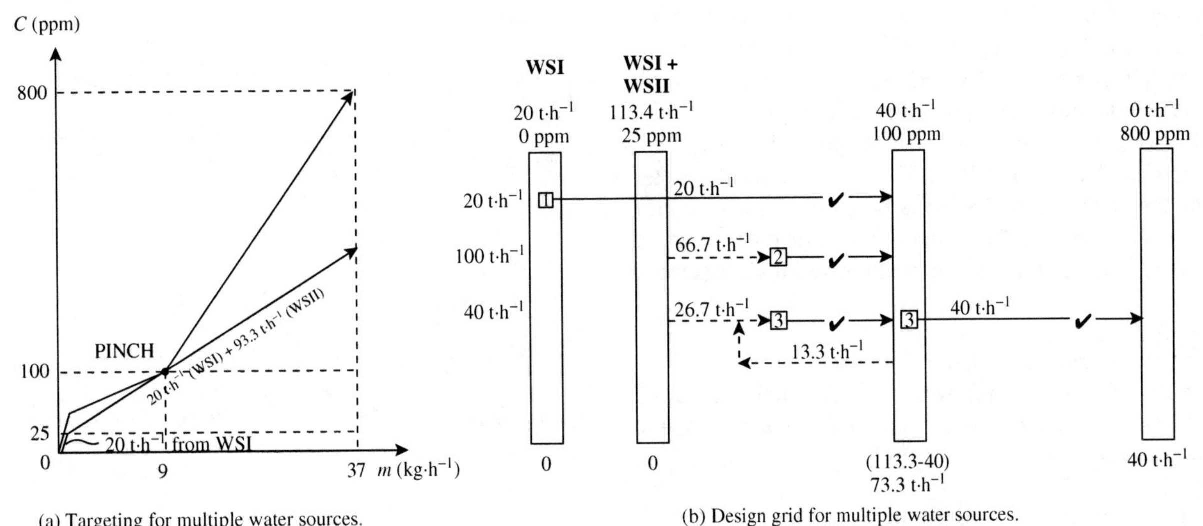

(a) Targeting for multiple water sources. (b) Design grid for multiple water sources.

Figure 26.27 Targeting and design for multiple water sources.

balance below the pinch.

$$9000 = m_{WSI}\,\Delta C_{WSI} + m_{WSII}\,\Delta C_{WSII}$$
$$= 20 \times (100 - 0) + m_{WSII}(100 - 25)$$
$$m_{WSII} = 93.3 \text{ t·h}^{-1}$$

Figure 26.27b shows the corresponding design grid for the two water sources. An additional water main is introduced into the design grid to account for the second water source.

Another complication that often arises is loss of water from the system. This could be, for example, the loss of water to effluent from a hosing operation or the evaporative loss to atmosphere from a cooling tower, neither of which becomes available for reuse. To illustrate how water losses can be accounted for, suppose that an operation is added to those in Table 26.5 with a maximum inlet concentration of 80 ppm and a flowrate of 10 t·h^{-1}, all of which is lost. Figure 26.28a shows the limiting composite curve for the three operations from Table 26.5 (i.e. excluding the flowrate loss). Matched against the limiting composite curve is a water supply line that shows a change in slope where the flowrate loss occurs. The target flowrate can be determined by a mass balance below the pinch.

$$9000 = m_W(\Delta C\mathit{before\ loss}) + (m_W - \mathit{Loss}) \times (\Delta C\mathit{after\ loss})$$
$$= m_W(80 - 0) + (m_W - 10) \times (100 - 80)$$
$$m_W = 92.0 \text{ t·h}^{-1}$$

Figure 26.28b shows the corresponding design grid for the loss. Note that no additional water main is required to account for the water loss. It should also be noted that the freshwater target for the three operations in Table 26.5 with a single source of freshwater at 0 ppm is 90 t·h^{-1},

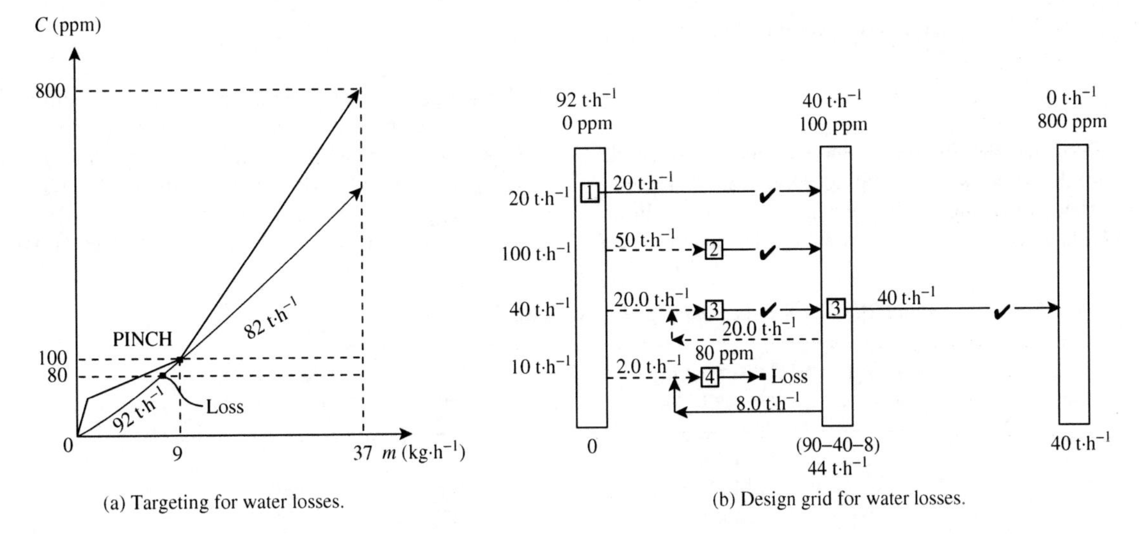

(a) Targeting for water losses. (b) Design grid for water losses.

Figure 26.28 Targeting and design for water losses.

which increases to 92 t·h^{-1} with a water loss of 10 t·h^{-1} at 80 ppm, rather than 100 t·h^{-1}. This is because the water loss can be partially fed by reuse. Had the same loss occurred above the pinch concentration, then there would have been no increase in flowrate as a result of the introduction of the water loss. Finally, it should be noted that the target in Figure 26.28a is based on the assumption that the loss is being fed at its maximum inlet concentration.

The design procedure can be summarized in four steps[12]:

1. Set up the grid.
2. Connect operations with water mains.
3. Correct for changes in flowrate in individual operations.
4. Remove intermediate water mains and connect operations directly.

The final step in the procedure may or may not be appropriate. A design with an intermediate water main at the pinch concentration might in some circumstances be convenient. Having a water main on the plant corresponding with the one introduced for the design procedure might be a convenient way of distributing slightly contaminated water around the plant and provides flexibility that direct connections between operations do not provide. Also, in batch operation, the demands on the water system and the production of water from the outlet of the operations will vary through time. This normally requires buffering capacity to bridge between the times when the water becomes available and when it is required[13]. Thus, the intermediate water main in a batch environment could easily be envisaged to be a storage tank providing buffer capacity between operations producing water and consuming water during different time intervals.

Example 26.2 Table 26.6 presents water-use data for a simple example involving three operations.

a. Target the minimum water consumption for the system through maximum reuse
b. Design a network for the target water consumption

Solution

a. Figure 26.29 shows the limiting composite curve for Example 26.2. Figure 26.30 shows the limiting composite curve with the appropriate water supply line pinched at 150 ppm. From Table 26.6, the mass load up to 150 ppm can be calculated as:

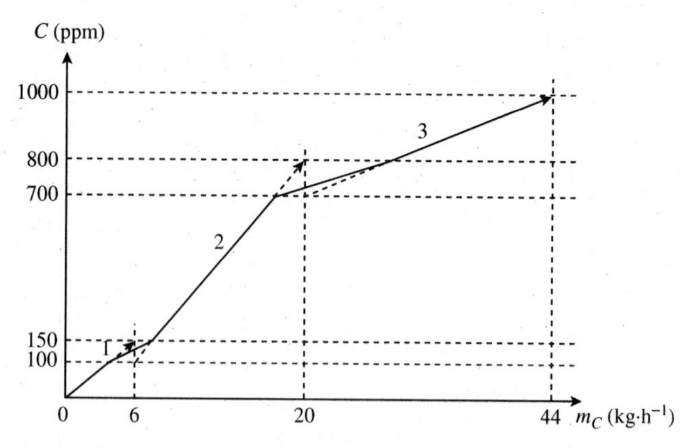

Figure 26.29 The limiting composite curve for Example 26.2.

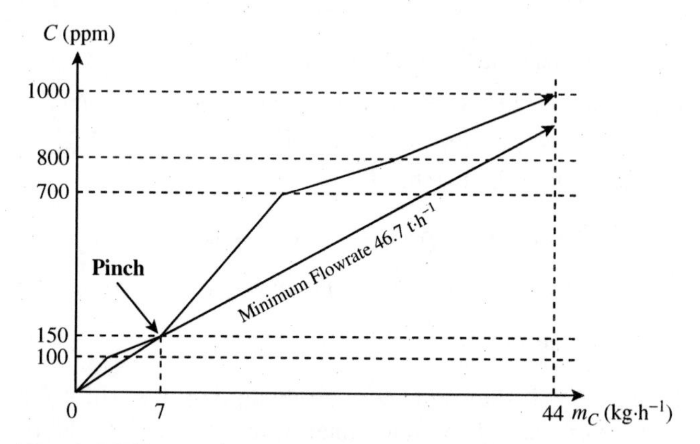

Figure 26.30 The minimum water target for Example 26.2.

Mass load up to pinch concentration

$$= 40(150 - 0) + 20(150 - 100)$$

$$= 7000 \text{ g·h}^{-1}$$

$$\text{Minimum flowrate} = \frac{\text{Mass load up to pinch concentration}}{\text{Concentration change of water to pinch}}$$

$$= \frac{7000}{150 - 0}$$

$$= 46.7 \text{ t·h}^{-1}$$

b. The design to achieve the target first must identify how many regions are involved. From the limiting composite curve in Figure 26.30, there are two design regions between zero and 150 ppm and 150 and 1000 ppm. Below the pinch, the first design region requires 46.7 t·h^{-1} by definition. Above the pinch, the concentration change extends from 150 to 1000 ppm with a

Table 26.6 Water-use data for Example 26.2.

Operation number	Contaminant mass (g·h^{-1})	C_{in} (ppm)	C_{out} (ppm)	Limiting water flowrate (t·h^{-1})
1	6000	0	150	40
2	14,000	100	800	20
3	24,000	700	1000	80

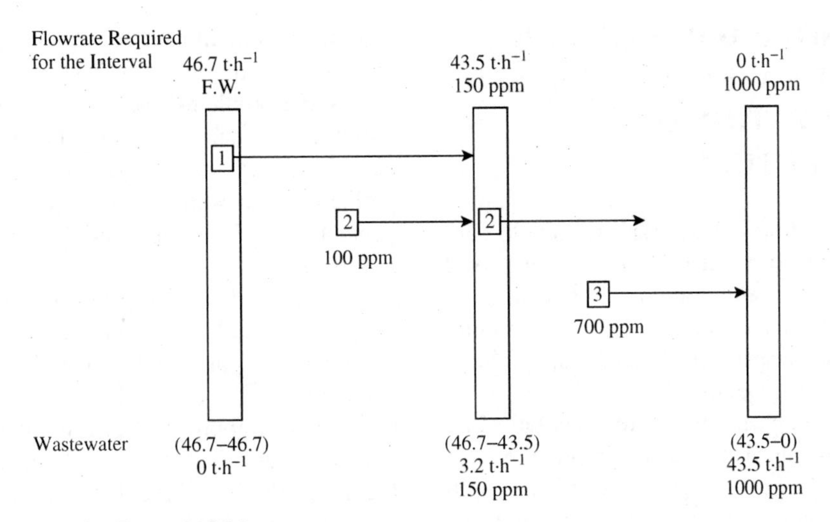

Figure 26.31 The design grid for Example 26.2.

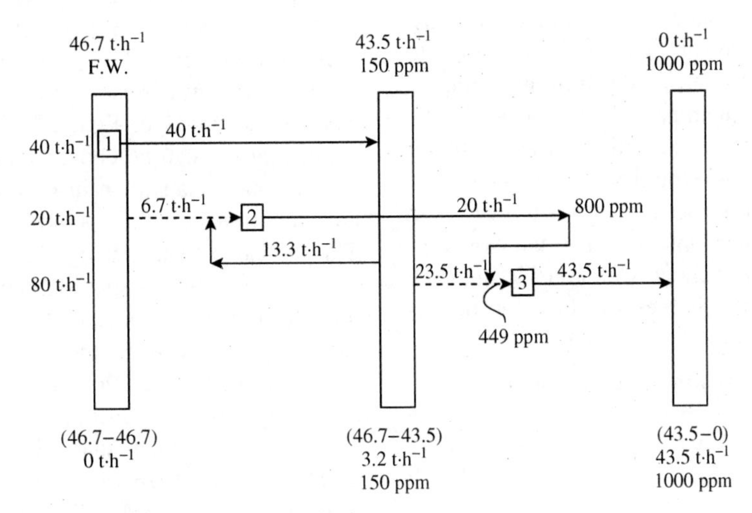

Figure 26.32 The completed network design for Example 26.2.

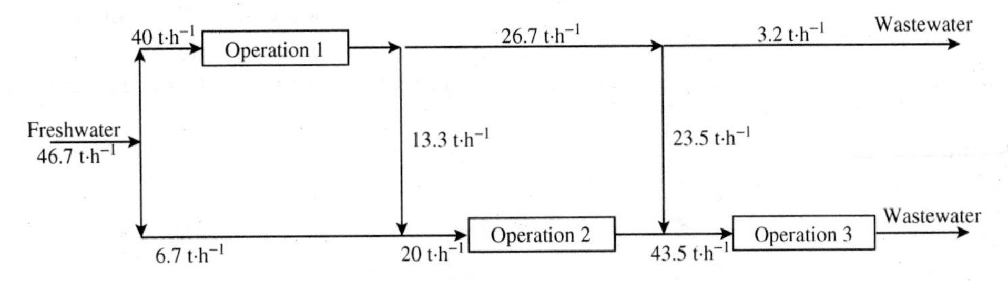

Figure 26.33 Flowsheet for the water network for Example 26.2.

mass load of 37,000 g·h^{-1}. This corresponds with a minimum flowrate requirement above the pinch of 43.5 t·h^{-1}. Figure 26.31 shows the design grid for the example with the streams in place. Figure 26.32 shows the completed network design, achieving the target of 46.7 t·h^{-1}. Note that the outlet water from Operation 2 at 800 ppm has not been put directly into the water main at 1000 ppm. If this had been done, the mass balance would have been violated. Instead, further use of the water from the outlet of Operation 2 requires it to be fed to the inlet of Operation 3, finally discharging to the 1000 ppm main. Figure 26.33 shows a flowsheet for the completed water network.

Even for a small problem involving three operations this turns out to be a solution that would be difficult to be achieved by inspection. Of course, the practicality and operability of a network would have to be explored before this design was accepted. It might be that some design features are considered to be inoperable. In which case, the design would have to be simplified. However, the consequence might be an increase in the water consumption. The procedure simply allows the maximum potential for the reuse to be identified, thereafter the practicality must be explored and the design evolved if necessary.

26.8 TARGETING AND DESIGN FOR MAXIMUM WATER REUSE BASED ON OPTIMIZATION OF A SUPERSTRUCTURE

The methods presented so far are adequate for single contaminants (e.g. total solids, suspended solids, total dissolved solids, organic concentration, etc), but it is often required to deal with problems in which the concentration limits require multiple contaminants to be specified. Consider the problem in Table 26.7 involving two operations.

This time, maximum concentrations are specified separately for two Contaminants A and B. A simple approach to such a problem might be to target Contaminant A in isolation from Contaminant B, then Contaminant B to be targeted in isolation from Contaminant A, and the worst case chosen. Figure 26.34a shows the limiting composite curve and target for Contaminant A in isolation to be 115 $t \cdot h^{-1}$. Figure 26.34b shows the limiting composite curve and target for Contaminant B in isolation to be 112 $t \cdot h^{-1}$. On the basis of these calculations, it would therefore be suspected that the target would be the worst case and would be 115 $t \cdot h^{-1}$. In fact, the true target for this problem is 126.5 $t \cdot h^{-1}$ by considering Contaminant A and Contaminant B simultaneously. The problem with considering each contaminant in isolation is that it does not allow for the fact that when Contaminant A is picked up, Contaminant B is picked up at the same time and vice versa.

Table 26.7 Data for a problem with two contaminants.

Operation number	Limiting water flowrate ($t \cdot h^{-1}$)	Contaminant	C_{in} (ppm)	C_{out} (ppm)
1	90	A	0	120
		B	25	85
2	75	A	80	220
		B	30	100

In some problems, the simple approach of targeting the individual contaminants and taking the worse case can give the correct answer. The target for a multicontaminant problem might well be the largest target of the individual contaminant. However, it might be a larger value, as it is in this case.

A more sophisticated approach is therefore required when dealing with multiple contaminants. It is possible to extend the graphical approach described here to deal with multiple contaminants, but it gets significantly more complex[4]. Also, the graphical approach has a number of other limitations. Sometimes different mass transfer models might be required. Figure 26.35 gives three different models that are useful in different circumstances for the design of water networks. Figure 26.35a shows the model that has been used so far involving a fixed mass load, but allowing the flowrate to vary. However, it is often the case that an operation requires a fixed flowrate, irrespective of the inlet concentration. This is illustrated in Figure 26.35b, where different inlet concentrations can be chosen, but the slope must be fixed. A third option is shown in Figure 26.35c where the outlet concentration from the operation is fixed, as is the flowrate. This last case can happen if water is coming into contact with a contaminant that is sparing soluble and can only dissolve up to a maximum concentration, the maximum solubility. In these circumstances, the mass load will vary according to the inlet concentration, as illustrated in Figure 26.35c. As an example of how different mass transfer models might be required, consider again the example of the crude oil desalter described in Section 11.7. This is a simple extraction process in which crude oil is mixed with water to extract salt from the crude oil into the water. The function of the desalter is to extract salt. Thus, the load of salt is specified. The flowrate of water can be varied to some extent, but there will be limits on this. At the same time that the water

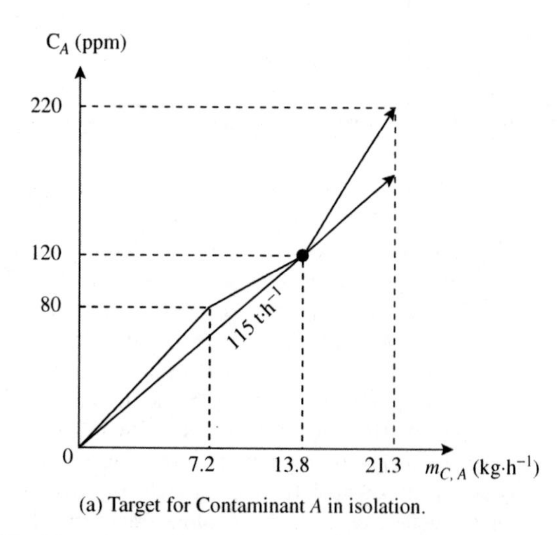

(a) Target for Contaminant A in isolation.

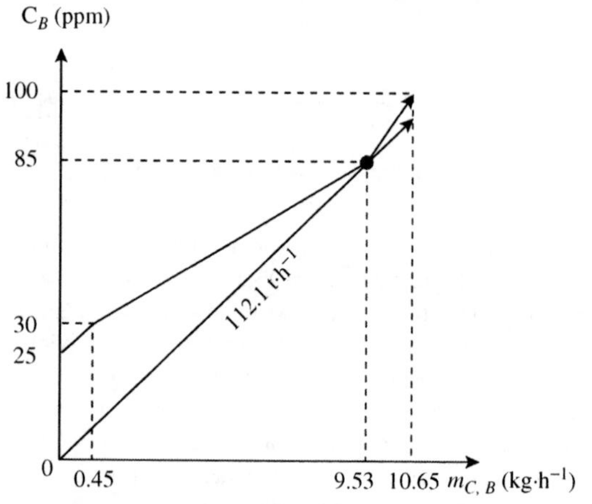

(b) Target for Contaminant B in isolation.

Figure 26.34 Targeting individual contaminants for a multiple contaminant problem from Table 26.7.

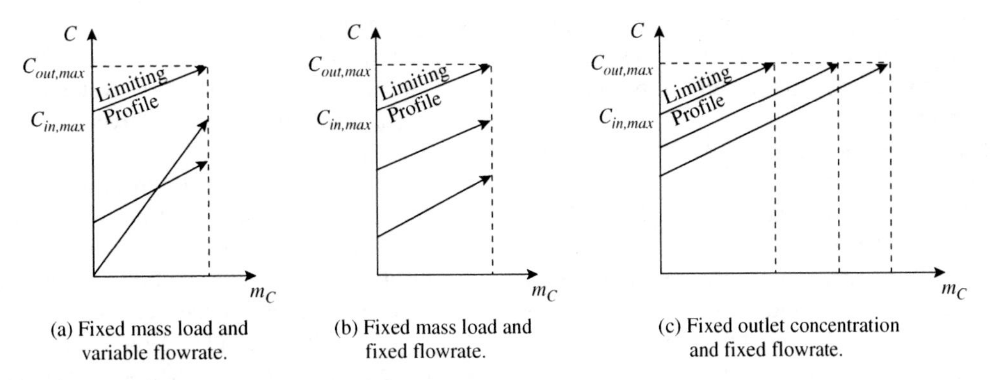

Figure 26.35 Different mass transfer models.

extracts the salt from the crude oil, other contaminants will be transferred from the crude oil to the water. These might typically be hydrocarbons with a maximum solubility and thus a maximum concentration might wish to be specified for such contaminants. Thus, not only might it be necessary to specify multiple contaminants, it also might be necessary to specify different mass transfer models in an individual operation. Also, the flowrate of water might be fixed, or allowed to vary only within a specified range.

In addition to these complexities, there may be other issues that need to be included in the analysis that are not readily included using the graphical approach. These might include forbidden matches (e.g. because a long pipe run may be necessary), compulsory matches (e.g. for operability) and capital cost issues (e.g. the cost of running pipes between operations).

To include all of these complexities requires a different approach from the one described so far. The design approach based on the optimization of a superstructure can be used to solve such problems[14]. Figure 26.36 shows the superstructure for a problem involving two operations and a single source of fresh water[14]. The superstructure allows for reuse from Operation 1 into Operation 2, reuse from Operation 2 to Operation 1, local recycles around both operations, fresh water supply to both operations and

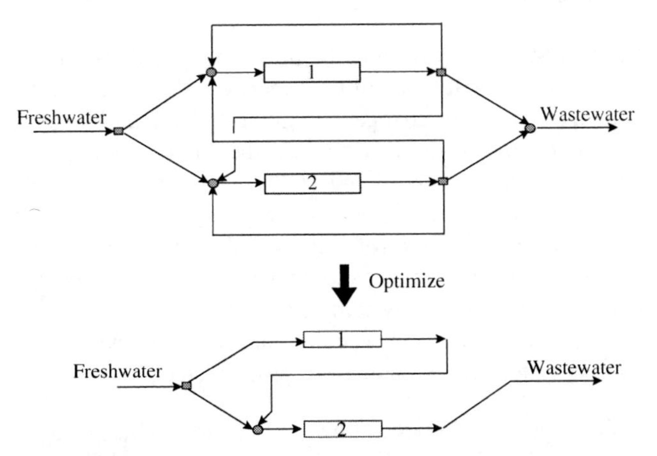

Figure 26.36 Water network design based on the optimization of a superstructure.

wastewater discharge from both operations. All structural features that are candidates for the final design have been included in the superstructure. The mass balances need to be modeled, cost models set up and the superstructure can then be optimized, as illustrated in Figure 26.36. The objective functions could be as simple as minimizing the supply of fresh water, or could be more sophisticated and minimize cost. Minimum cost could include features such as[15]:

- Multiple water sources with different costs
- Effluent treatment costs
- Piping costs.

The cost of water from a given source will be proportional to its flowrate. Effluent treatment costs can also be formulated as function of flowrate. The economics of water reuse is, however, often critically dependent on piping costs. Piping costs can be included in the optimization by providing information on the approximate length of pipe required to connect a water source and a water sink, together with cost information for the piping. During the course of the optimization, the flowrate of water will be calculated. Assuming a reasonable velocity of 1 to 2 $m \cdot s^{-1}$ allows the pipe diameter to be calculated. From knowledge of the length and diameter, the cost of a pipe between a source and sink can be estimated. Given this information, the optimization will tend to produce solutions avoiding long and expensive pipe connections[15].

The nature of the optimization problem can turn out to be linear or nonlinear depending on the mass transfer model chosen[14]. If a model based on a fixed outlet concentration is chosen, the model turns out to be a linear model (assuming linear cost models are adopted). If the outlet concentration is allowed to vary, as in Figure 26.35a and Figure 26.35b, then the optimization turns out to be a nonlinear optimization with all the problems of local optima associated with such problems. The optimization is in fact not so difficult in practice as regards the nonlinearity, because it is possible to provide a good initialization to the nonlinear model. If the outlet concentrations from each operation are initially assumed to go to their maximum outlet concentrations, then this can then be solved by a linear optimization. This usually provides a good initialization for the nonlinear optimization, as the network design will tend, in most instances, to try to force concentrations to their maximum outlet concentrations to maximize water reuse[14].

The advantages of an optimization-based approach are that it provides automation for complex problems, provides a design as well as a target and allows all kinds of constraints and costs to be included. One disadvantage is that there is no longer a graphical representation allowing conceptual insights. This shortcoming can be overcome by displaying the output from the optimization in graphical form. This is most conveniently done, one contaminant at a time. The optimization provides a network design with inlet and outlet concentrations that can be taken and combined together in a composite plot, one contaminant at a time. By superimposing the flowrate predicted by the optimization onto this composite curve, conceptual insights can be gained.

26.9 PROCESS CHANGES FOR REDUCED WATER CONSUMPTION

So far the discussion on water minimization has restricted consideration to identify opportunities for water reuse. Maximizing water reuse minimizes both fresh water consumption and wastewater generation. However, the process constraints for inlet concentrations, outlet concentrations and flowrates have so far been fixed. Often there is freedom to change the conditions within the operation. Typical process changes that might be contemplated include:

- Introducing a local recycle around a scrubber or washing operation.
- Removing water from the operation completely if it is extraneous (e.g. replace the scrubbing operation used to separate an organic contaminant from a vapor stream by condensation, or replacing an extraction process using water by cooling crystallization).
- Increasing the number of stages in an extraction process to reduce the water demand for a given separation.
- Increasing the number of stages in absorption and scrubbing operations that use water to reduce the water consumption for the same separation.
- Replacing live steam injection for distillation operations with a reboiler using indirect steam heating.
- Introducing multiple stages for equipment cleaning, such that the initial stages are carried out using slightly contaminated water (e.g. to remove residual solids), followed by high-quality water for the final stages.
- In multiproduct plants, scheduling operations to minimize product changeovers and thereby reduce the cleaning requirements between product changeovers.
- Introducing mechanical cleaning of vessels (e.g. wall wipers) and pipes (e.g. pipeline pigging) that process viscous materials and require periodic cleaning with water.
- For cleaning agitated vessels, rather than filling up the vessel and use agitation for the cleaning, introducing a local recycle using less water but with a spray system inside the vessel to give effective cleaning with less water.
- Fixing triggers to hoses to prevent unattended running of water from the hoses when not in use, so that as soon as the trigger is released by the operator, the flow is stopped.
- Introducing local recycles around liquid ring pumps.
- Replacing direct contact condensation using a water spray by indirect condensation using cooling water.
- Introducing local recycles around quench operations that use a water spray with some indirect cooling in the recycle loop.
- For batch operations, introducing solenoid valves in water scrubbers, direct contact condensation and the flushes used to protect pump and agitator seals, such that if the operation is not in use, the flow of water is stopped.
- Improving the energy efficiency to reduce steam demand and hence reduce the wastewater generated by the steam system through boiler blowdown, boiler feedwater treatment and condensate loss (see Chapter 23).
- Increasing condensate return for steam systems to reduce makeup water requirements, reduce aqueous waste from boiler feedwater treatment and boiler blowdown (see Chapter 23).
- Improving control of cooling tower blowdown (see Chapter 24) for evaporative cooling water circuits to increase the cycles of concentration and reduce the cooling tower blowdown rate.
- Decreasing the heat duty on evaporative cooling tower circuits by better energy efficiency to decrease the evaporation from the cooling tower and hence, decrease the makeup requirements.
- Introducing air cooling to replace evaporative cooling towers (see Chapter 24).
- Considering reducing the water mains pressure or isolating areas when not in use, if leaks and unaccountable losses are a problem from underground piping systems and so on

Listed above are just some of the many possible changes that can be made to operations to reduce their inherent water demand, but which changes are going to have an influence on the overall system? Consider the limiting composite curves in Figures 26.16 and 26.30. It is clear that only changes at or below the pinch for the system will reduce the demand for water for the overall system. A process located above the pinch should have more than enough water available in the system through the appropriate reuse opportunities. Thus, in general, the ways in which the overall system can benefit from process changes are

- For a given volumetric flowrate for the processes below the pinch, shift mass load of contaminant from below the pinch to above the pinch by increasing concentrations.

- For a fixed mass load of contaminant for the processes below the pinch, decrease the volumetric flowrate of the processes below the pinch, thus increasing concentrations.
- A combination of the above two measures.

The location of an operation relative to the pinch is one criterion for deriving a benefit from process changes. The other criterion is one of scale. The larger the flowrate to an operation, the larger is likely to be the overall effect of the process changes. To understand the resulting magnitude of suggested process changes on the overall water consumption, the targeting methods described above should be used to screen options.

26.10 TARGETING MINIMUM WASTEWATER TREATMENT FLOWRATE FOR SINGLE CONTAMINANTS

Wastewater treatment in the process industries is most often carried out in a central treatment facility. Effluent streams are collected together in a common sewer system and passed to a central treatment facility, or in some instances, given some pretreatment and sent off-site to a municipal treatment facility. The problem with centralized treatment is that combining waste streams that require different treatment technologies leads to a cost of treating the combined streams that is likely to be more expensive than the individual treatment of the separate streams. This situation results from the fact that the capital cost of most waste treatment processes is related to the total flowrate of wastewater. Also, operating costs for treatment usually increase with decreasing concentration for a given mass load of contaminant to be treated. On the other hand, if two waste streams require exactly the same treatment then it makes sense to combine them for treatment to obtain economies of scale for the equipment.

A philosophy of design is required that will allow the designer to combine effluents for treatment when it is appropriate, but also segregate them if it is appropriate. Consider now the targeting of minimum treatment flowrate for a given set of effluent streams that need to be treated to bring the concentration to below an environmental discharge limit.

Figure 26.37a shows a system involving four effluent streams with different inlet concentrations that all need to be treated to remove mass load and bring down the concentration to an acceptable level for environmental discharge C_e. To obtain an overall picture, rather than deal with four separate effluent streams, the streams can be combined together to produce a *composite effluent stream*[16]. The construction is analogous to that for the limiting composite curve. The diagram is divided into concentration intervals and the mass load of the streams within each concentration interval combined together, Figure 37b. This provides a picture of the overall effluent treatment problem and what is required to happen to the effluent streams.

The mass load can be removed in a number of different network arrangements. The objective in designing an effluent treatment network is to minimize its cost. In the first instance, this means minimizing the flowrate through the treatment process. Figure 26.38a shows a composite effluent curve. Superimposed are *treatment lines* representing the actual performance of the effluent treatment process[16]. The effluent treatment process is required to remove the mass load in order to bring down the effluent concentration to its environmental discharge limit. The steeper the treatment line, the lower will be the flowrate through the effluent treatment. The maximum slope of the treatment line in Figure 26.38a is dictated at the low concentration end by a minimum outlet concentration, below which the treatment process cannot operate, and

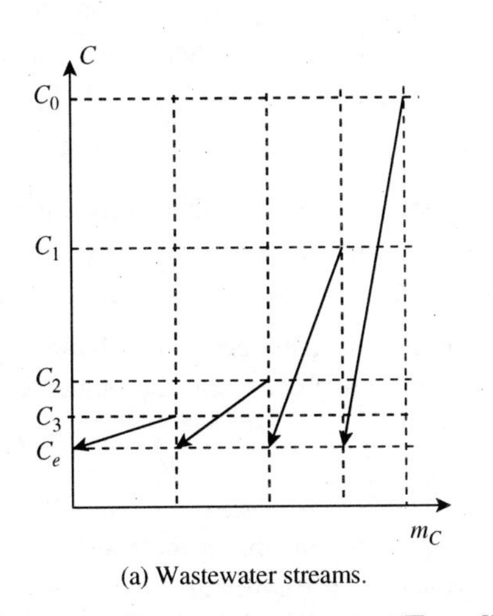

(a) Wastewater streams.

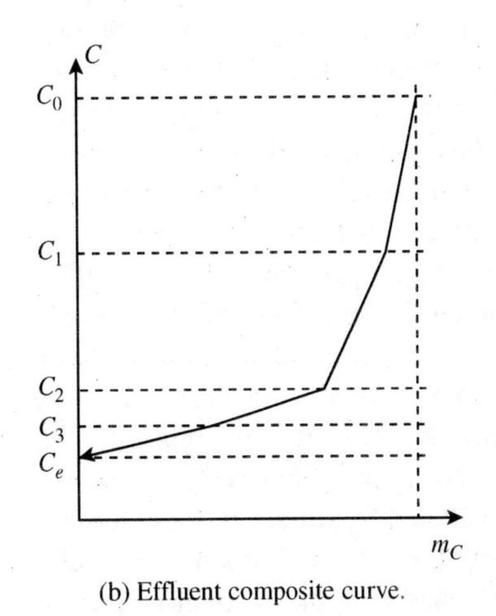

(b) Effluent composite curve.

Figure 26.37 A composite curve of wastewater streams. (From Wang YP and Smith R, 1994, *Chem Eng Sci*, **49**: 3127, reproduced by permission of Elsevier Ltd.)

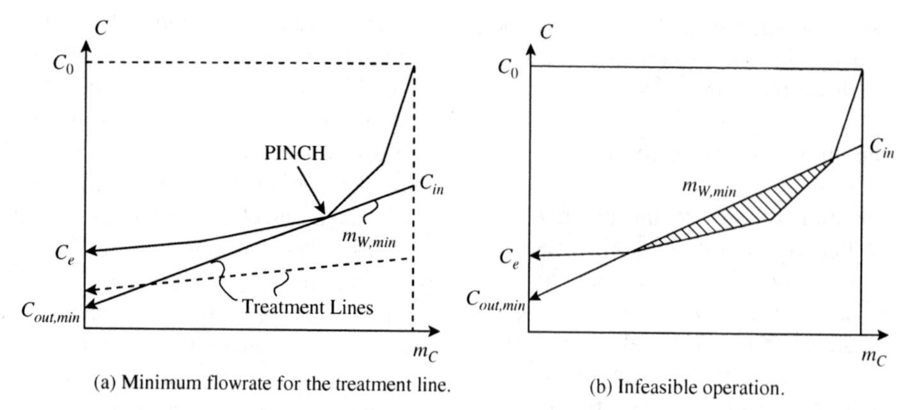

(a) Minimum flowrate for the treatment line. (b) Infeasible operation.

Figure 26.38 Targeting minimum wastewater treatment flowrate. (From Wang YP and Smith R, 1994, *Chem Eng Sci*, **49**: 3127, reproduced by permission of Elsevier Ltd.

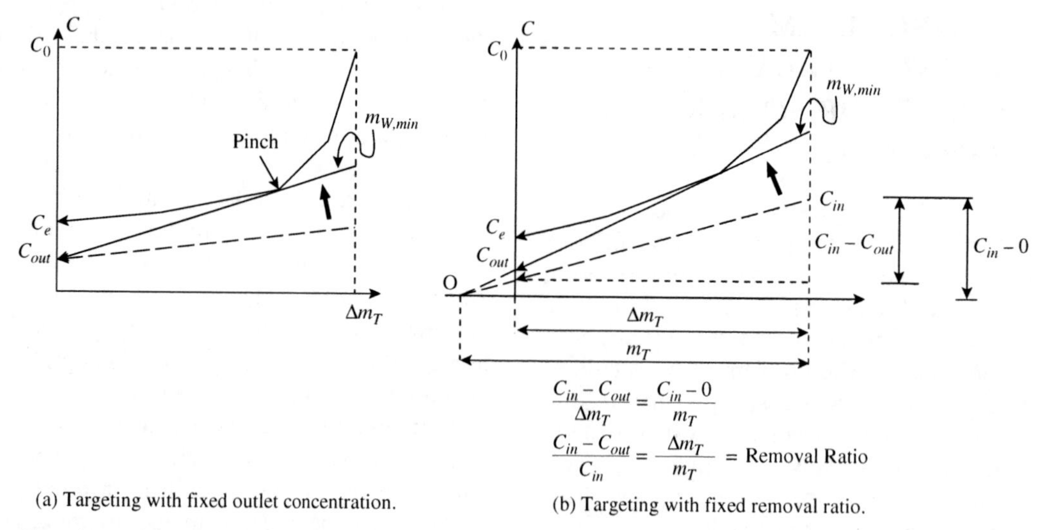

(a) Targeting with fixed outlet concentration. (b) Targeting with fixed removal ratio.

Figure 26.39 Targeting minimum wastewater treatment flowrate. (From Wang YP and Smith R, 1994, *Chem Eng Sci*, **49**: 3127, reproduced by permission.)

the *pinch point* between the effluent composite curve and treatment line[16]. The treatment line cannot be made any steeper than shown in Figure 26.38a. If the slope is increased, as shown in Figure 26.38b, then the treatment line crosses the effluent composite curve. This is infeasible, as it requires mass to be treated in concentration ranges where it is not available. Any attempt to design this flowrate will inevitably lead to an infeasible mass balance in the design of the effluent treatment system.

In specifying effluent treatment processes, two specifications are used in conceptual design. The simplest case is where the concentration at the outlet of an effluent treatment process is specified to be a fixed value. Targeting with a fixed outlet concentration is illustrated in Figure 26.39a. The effluent composite curve is first constructed. The outlet concentration for the effluent treatment is specified and the slope of the treatment line pivoted around this value until the treatment line pinches with the effluent composite curve. The slope of the line will then dictate the minimum treatment flowrate.

The performance of the treatment process is more likely to be specified by a removal ratio:

$$R = \frac{\text{Mass of contaminant removed}}{\text{Mass of contaminant fed}} = \frac{m_{W,in}C_{in} - m_{W,out}C_{out}}{m_{W,in}C_{in}} \tag{26.6}$$

where

R = removal ratio
$m_{W,in}, m_{W,out}$ = inlet and outlet flowrates
C_{in}, C_{out} = inlet and outlet concentrations

In many instances, the change in flowrate through the treatment process can be neglected, giving:

$$R = \frac{C_{in} - C_{out}}{C_{in}} \tag{26.7}$$

Figure 26.39b shows the construction for targeting minimum wastewater treatment flowrate when removal ratio has been specified. If the initial effluent treatment line shown dotted in Figure 26.39b is considered, then there is a point of origin corresponding with a specified removal ratio. Applying the geometry of similar triangles, as shown in Figure 26.39b and rearranging indicates that the ratio $\Delta m_T/m_T$ equals the removal ratio. Thus, a given removal ratio corresponds with a given ratio of $\Delta m_T/m_T$ and, therefore, a fixed point of origin for the treatment line. If the treatment line is then pivoted around this point of origin as shown in Figure 26.39b, the ratio $\Delta m_T/m_T$ and, therefore, the removal ratio are both fixed. Finding the steepest slope where the treatment line is pinched against the composite effluent curve corresponds with the minimum treatment flowrate.

Example 26.3 The data for a wastewater treatment problem are given in Table 26.8. Centralized treatment would correspond with a flowrate to be treated of 75 $t \cdot h^{-1}$. For an environmental discharge limit of 20 ppm, determine the minimum treatment flowrate.

a. For a process with a fixed outlet concentration of 10 ppm
b. For a treatment process with a removal ratio of 95%

Table 26.8 Data for an effluent treatment problem.

Stream	C_{in} (ppm)	Water flowrate ($t \cdot h^{-1}$)
1	250	40
2	100	25
3	40	10

Solution

a. Figure 26.40 shows the effluent composite curve plotted for the three streams in Table 26.8. The end of the effluent treatment line is fixed with a concentration of 10 ppm and the steepest line drawn until it pinches with the effluent composite curve at 100 ppm. The removal flowrate is then given by:

$$m_{W,min} = \frac{\Delta m_T}{\Delta C} = \frac{5400}{100 - 10} = 60 \text{ t} \cdot \text{h}^{-1}$$

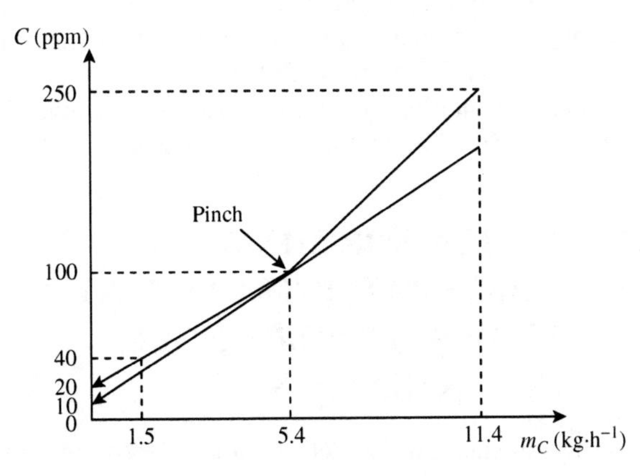

Figure 26.40 Targeting for a fixed outlet concentration of 10 ppm in Example 26.4.

b. For a fixed removal ratio with no change in flowrate of water:

$$R = 0.95 = \frac{C_{in} - C_{out}}{C_{in}}$$

Rearranging gives:

$$C_{in} = 20\ C_{out} \tag{26.8}$$

The total mass load is given by:

$$\Delta m_T = m_{W,min} \Delta C$$

$$11{,}400 = m_{W,min}(C_{in} - C_{out}) \tag{26.9}$$

For minimum flowrate, the treatment line will pass through the pinch point at 100 ppm. Writing the mass balance below the pinch point:

$$5400 = m_{W,min}(100 - C_{out}) \tag{26.10}$$

Dividing Equation 26.9 by Equation 26.10 gives:

$$\frac{11{,}400}{5400} = \frac{C_{in} - C_{out}}{100 - C_{out}}$$

Substituting Equation 26.8 gives:

$$C_{out} = 10 \text{ ppm}$$

Thus:

$$C_{in} = 200 \text{ ppm}$$

$$m_{W,min} = 60 \text{ t} \cdot \text{h}^{-1}$$

By coincidence the answer is the same in both cases, that is, 10 ppm outlet concentration corresponds with a removal ratio of 0.95.

One complication not considered so far that is very commonly encountered in effluent systems is where some of the effluents are already below their environmental discharge limit. Table 26.9 presents a problem involving four effluent streams. The environmental discharge limit is 20 ppm.

Note from Table 26.9 that Stream 4 with a concentration of 10 ppm is already below the environmental limit of 20 ppm. Suppose that this is to be treated with an operation capable of producing 5 ppm outlet concentration. How can the target minimum flowrate be set under such circumstances?

In fact, the availability of the stream with a concentration below the environmental limit, removes part of the

Table 26.9 A problem where one of the effluents is below the environmental discharge limit.

Stream	C_{in} (ppm)	Water flowrate ($t \cdot h^{-1}$)
1	200	40
2	100	30
3	30	20
4	10	80

environmental load to be taken up by the effluent treatment system through its dilution effect. The construction of the effluent composite to include this effect is shown in Figure 26.41. It can be seen that as the stream starting at 10 ppm increases in concentration to the specified environmental limit of 20 ppm, it takes up part of the effluent load from the other streams that would otherwise have to be removed by the effluent treatment process. The dilution effect of the stream below the environmental limit effectively creates a new environmental limit greater than 20 ppm. The target for the problem is shown in Figure 26.42. In this case, the specification for the treatment process was based on an outlet concentration. If the specification had been based on removal ratio, then the basic approach would be the same. The only change would be the specification of the effluent treatment line.

Sometimes the effluent treatment cannot be accomplished in a single effluent treatment process. For example, different treatment processes might be required to treat contaminants over different concentration ranges. It might also be cheaper

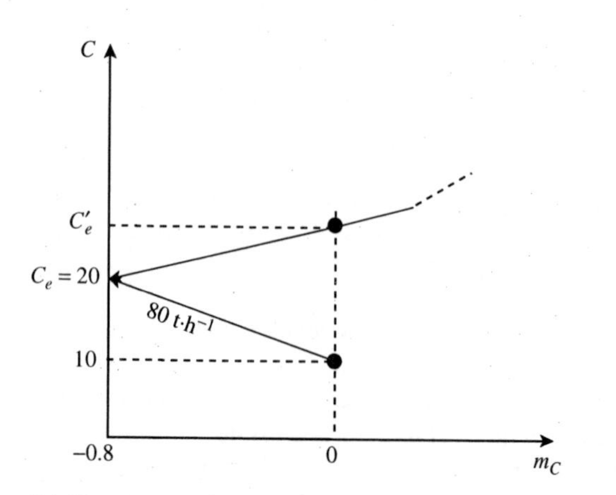

Figure 26.41 A stream below the environmental limit defines a new effective environmental limit.

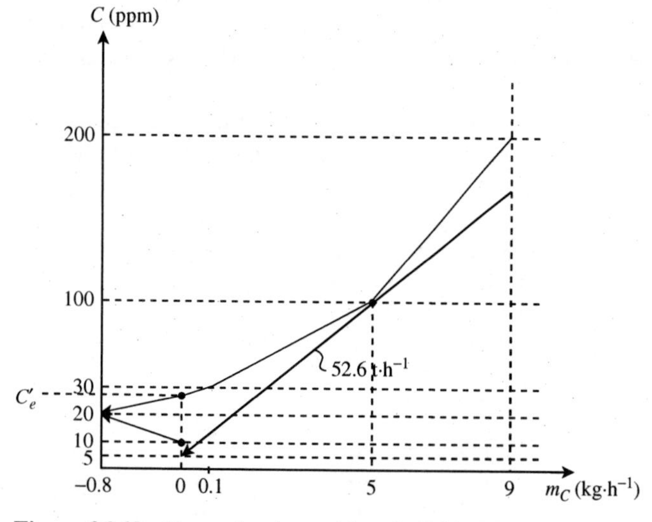

Figure 26.42 Target for the problem in Table 26.6.

to use multiple treatment processes, rather than a single process. For example, when treating petroleum refinery effluent, it is usually cheap to start the treatment with an API Separator and then complete the treatment with other treatment processes, such as dissolved-air flotation (see Chapter 8) and biological treatment. The API Separator is a cheap process to remove part of the contamination before passing the effluent to more expensive treatment processes. Thus, limitations in the performance or the costs associated with effluent treatment processes might dictate the requirement for multiple treatment processes.

The graphical approach to targeting minimum treatment flowrate can be extended to multiple treatment processes, as shown in Figure 26.43[17]. In Figure 26.43a, the effluent composite curve is divided into parts of the effluent mass load being treated by the different processes (m_{TPI} and m_{TPII} in Figure 26.43a). The effluent composite curve is then decomposed into two parts at this point of division in the mass load, as shown in Figure 26.43b. Decomposition is the opposite procedure from creating a composite. In Figure 26.43b, the mass balance division is such that the effluent composite curve between C_1 and C_e needs to be decomposed. In this way, two separate effluent composite curves are created for treatment by Treatment Process I (TPI) and Treatment Process II (TPII). Treatment lines for TPI and TPII are then matched against the respective decomposed effluent composite curves, as shown in Figure 26.43c. The matching of the treatment lines against the decomposed effluent composite curve is now the same as that for single treatment processes, as described previously. Specifications can be based on outlet concentration or removal ratio. Figure 26.43d shows that decomposed effluent composite curves and the treatment lines combined to obtain composites for each. It should be noted that the treatment line for TPII in Figure 26.43c would actually cross the original effluent composite curve and would appear to be infeasible. However, a composite of the two treatment lines does not cross the original effluent composite curve. The combination of the two treatment lines leads to a feasible solution[17].

Finally, the relative contaminant mass load on the two treatment processes can usually be varied within reasonable limits. As load is shifted from TPI to TPII, and vice versa, their relative capital and operating costs will change. This is a degree of freedom that can be optimized.

26.11 DESIGN FOR MINIMUM WASTEWATER TREATMENT FLOWRATE FOR SINGLE CONTAMINANTS

Having established how to set the target for the minimum treatment flowrate, the next question is how to design to achieve the target. The design principles to achieve the

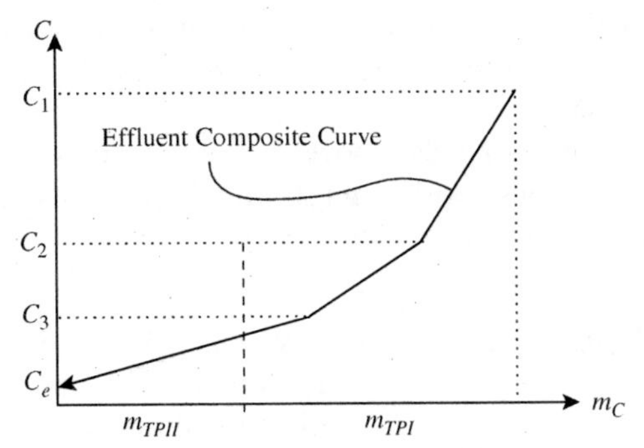

(a) Division of mass balance for the multiple treatment processes.

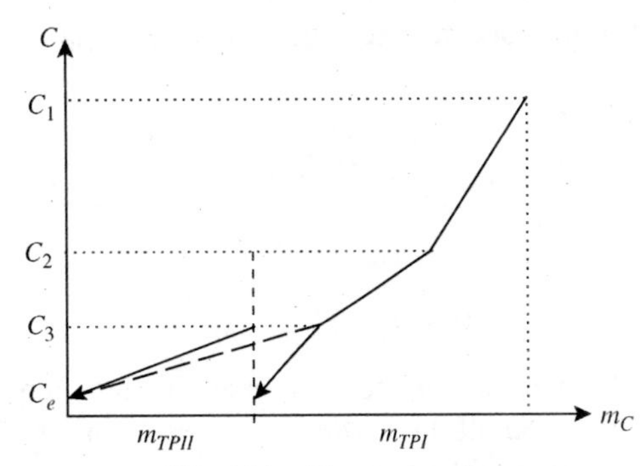

(b) Decomposition of the effluent composite curve.

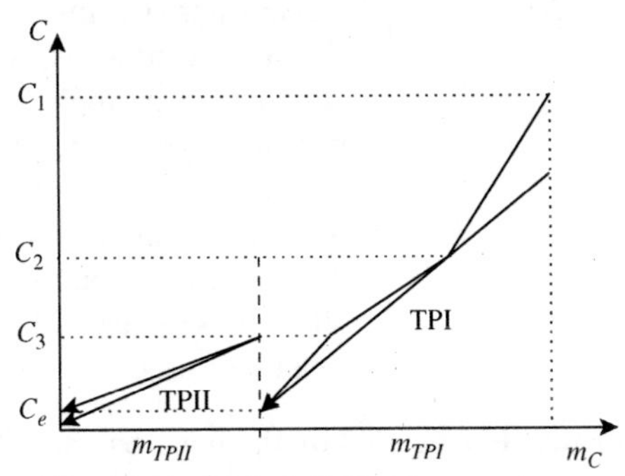

(c) Specification of the individual treatment processes.

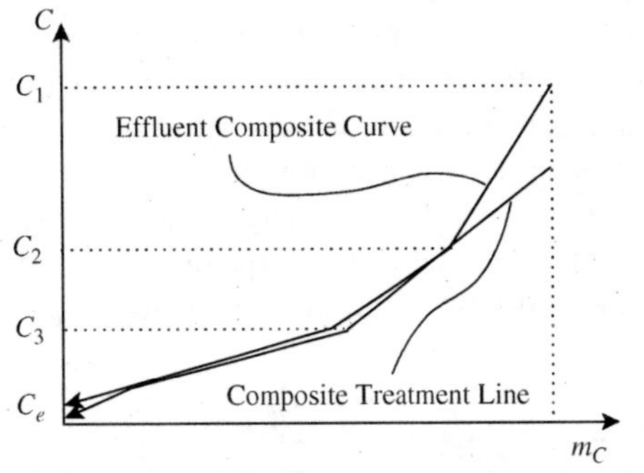

(d) Composites of the effluent streams and treatment lines.

Figure 26.43 Targeting for multiple treatment processes.

target are based on the starting concentrations of the streams relative to the pinch concentration.

$$\textbf{Group I: } C_{I,j} > C_{PINCH} \quad (26.11)$$

$$\textbf{Group II: } C_{II,j} = C_{PINCH} \quad (26.12)$$

$$\textbf{Group III: } C_{III,j} < C_{PINCH} \quad (26.13)$$

where $C_{I,j}, C_{II,j}, C_{III,j}$ = concentration of Stream j in Groups I, II and III respectively

If $m_{W,above}$ and $m_{W,below}$ are defined to be the total wastewater flowrates in the concentration intervals immediately above and below the pinch concentration

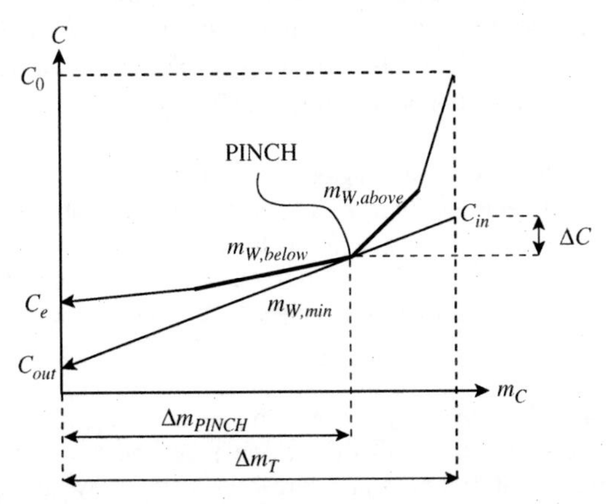

Figure 26.44 Achieving the treatment flowrate target. (From Wang YP and Smith R, 1994, *Chem Eng Sci*, **49**: 3127, reproduced by permission of Elsevier Ltd.)

respectively, as shown in Figure 26.44, then:

$$m_{W,above} = \sum_{j}^{S_I} m_{WI,j} \quad (26.14)$$

and:

$$m_{W,below} = \sum_{j}^{S_I} m_{WI,j} + \sum_{j}^{S_{II}} m_{WII,j} \quad (26.15)$$

where $m_{W,above}, m_{W,below}$ = water flowrate immediately above and below the pinch

$m_{WI,j}, m_{WII,j}$ = water flowrate for Stream j in Group I and II respectively

S_I = total number of streams in Group I

S_{II} = total number of streams in Group II

Because the treatment line with a flowrate m_{WT} is pinched with the composite curve at this point, the following relation holds:

$$m_{W,above} \leq m_{W,min} \leq m_{W,below} \quad (26.16)$$

which means that:

$$\sum_{j}^{S_I} m_{WI,j} \leq m_{W,min} \leq \sum_{j}^{S_I} m_{WI,j} + \sum_{j}^{S_{II}} m_{WII,j} \quad (26.17)$$

Therefore, the minimum treatment flowrate target $m_{W,min}$ is given by:

$$m_{W,min} = \sum_{j}^{S_I} m_{WI,j} + \theta \sum_{j}^{S_{II}} m_{WII,j} \qquad (26.18)$$

where:

$$0 \le \theta \le 1$$

Equation 26.18 reveals that the treatment flowrate target can be achieved when all the wastewater streams in Group I are treated, together with some of the wastewater streams from Group II and none from Group III.

Having identified the streams to be treated in order to achieve the flowrate target, it is still necessary to prove that the mean inlet concentration of those streams, $C_{mean,in}$, is equal to the targeted inlet concentration for the treatment process, C_{in}. If the treatment flowrate target $m_{W,min}$ consists of the wastewater streams defined by Equation 26.18, then the mean inlet concentration of the streams treated, $C_{mean,in}$, is given by:

$$C_{mean,in} = \frac{\sum_j^{S_I} m_{WI,j} C_{I,j} + \theta \sum_j^{S_{II}} m_{WII,j} C_{PINCH}}{\sum_j^{S_I} m_{WI,j} + \theta \sum_j^{S_{II}} m_{WII,j}} \qquad (26.19)$$

From Figure 26.44, the targeted inlet concentration, C_{in} for the treatment flowrate target is given by:

$$C_{in} = C_{PINCH} + \Delta C = C_{PINCH} + \frac{\Delta m_T - \Delta m_{PINCH}}{m_{W,min}} \qquad (26.20)$$

where ΔC is the concentration change through the treatment process above the pinch. Then knowing that:

$$\Delta m_T - \Delta m_{PINCH} = \sum_{j}^{S_I} m_{WI,j}(C_{I,j} - C_{PINCH}) \qquad (26.21)$$

where Δm_T = total mass treatment load
Δm_{PINCH} = mass treatment load below the pinch

Equations 26.21 and 26.18 can be substituted into Equation 26.20 to give the targeted inlet concentration, C_{in}.

$$C_{in} = C_{PINCH} + \frac{\sum_j^{S_I} m_{WI,j}(C_{I,j} - C_{PINCH})}{\sum_j^{S_I} m_{WI,j} + \theta \sum_j^{S_{II}} m_{WII,j}}$$

$$= \frac{\sum_{j}^{S_I} m_{WI,j} C_{I,j} - \sum_{j}^{S_I} m_{WI,j} C_{PINCH} + \sum_{j}^{S_I} m_{WI,j} C_{PINCH} + \theta \sum_{j}^{S_{II}} m_{WII,j} C_{PINCH}}{\sum_j^{S_I} m_{WI,j} + \theta \sum_j^{S_{II}} m_{WII,j}}$$

$$= \frac{\sum_j^{S_I} m_{WI,j} C_{I,j} + \theta \sum_j^{S_{II}} m_{WII,j} C_{PINCH}}{\sum_j^{S_I} m_{WI,j} + \theta \sum_j^{S_{II}} m_{WII,j}}$$

$$= C_{mean,in} \qquad (26.22)$$

Equation 26.22 proves that the mean inlet concentration of all the streams treated is equal to the targeted inlet concentration.

Summarizing the above discussion gives the following design rules, which must be observed to achieve the treatment flowrate target.

Design Rule I: If wastewater streams start above-pinch concentration, then all of the flowrate of all of these streams must be treated.

Design Rule II: If wastewater streams start at the pinch concentration, then these streams are partially treated and partially bypass the effluent treatment.

Design Rule III: If wastewater streams start below the pinch concentration, then all of these streams bypass the treatment process completely.

Design Rule II is dictated by the mass balance. The amount of the wastewater streams from Group II to be treated can be calculated from:

$$m_{WII} = \frac{\Delta m_T - \Delta m_{CI}}{C_{PINCH} - C_{out}} \qquad (26.23)$$

where

$$\Delta m_{CI} = \sum_{j}^{S_I} m_{WI,j}(C_{I,j} - C_e) \qquad (26.24)$$

These three basic rules create a segregation policy to ensure that the target is achieved in design.

If the effluent treatment involves multiple treatment processes, as shown in Figure 26.43, then the same basic approach can be followed as for single treatment processes. The problem is decomposed, as shown in Figure 26.43. This provides the target and the basis of the design. Each treatment process has its own pinch and the network design for each can be developed separately. The network designs are then simply connected together[17].

Example 26.4 Consider again Example 26.3. The target for the three streams in Table 26.8 was determined to be 60 $t \cdot h^{-1}$. This corresponded both with a treatment process achieving 10 ppm at its outlet and also one with a removal ratio of 95%. Design an effluent treatment network to achieve the target of 60 $t \cdot h^{-1}$.

Solution To achieve the target in practice requires that the starting concentrations relative to the pinch be first identified. The pinch is at 100 ppm and Stream 1 starts above the pinch. Stream 2 starts at the pinch and Stream 3 starts below the pinch. Thus, Stream 1 must be completely treated, Stream 2 partially treated and partially bypassed and Stream 3 totally

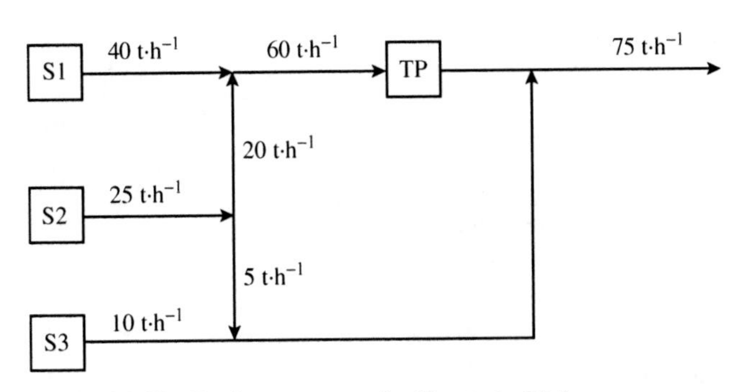

Figure 26.45 Design structure for Example 26.4.

bypasses the effluent treatment. The question remains as to how much of Stream 2 goes through the effluent treatment and how much bypasses. If Stream 1 with a flowrate of 40 $t{\cdot}h^{-1}$ must go through the effluent treatment and the target for effluent treatment is 60 $t{\cdot}h^{-1}$, then 20 $t{\cdot}h^{-1}$ from Stream 2 must go through the effluent treatment process. Figure 26.45 shows the design structure for the effluent treatment, together with the flowrates.

The design to achieve the target is therefore a matter of following three simple rules. It is worth highlighting a few points though. Firstly, streams starting at the pinch must be partially treated and partially bypassed in order to achieve the target. This splitting of the stream, while it is necessary to solve the problem as posed, is often not acceptable in practice. It provides design complexity in the need to split the stream. Also, regulatory authorities often find such an arrangement to be unacceptable, even though it solves the problem according to their regulations. Environmental regulations often have requirements for criteria such as *best practice* or *best environmental option*. It is therefore often considered to be not the best environmental practice to partially treat a stream. If part of the stream needs to be treated, then why not the whole stream? If the partial bypass is eliminated and all of the flow for the stream starting at the pinch is fed through the treatment process, then the performance of the system will exceed that of the regulatory requirements. While this might seem to be a drawback in small systems, in practice it is not such a problem. In practice, rather than dealing with three streams, it is likely that the number of streams from a site might be of the order of 100 or more. In this situation, changing perhaps one stream that is at the pinch to fully go through the treatment process, rather than partially bypass, will have a very small effect on the overall system performance[18].

Another point regarding the design process is the effect of the streams that are already below the environmental limit. In practice, for a large site, many of the streams will already be below the environmental limit. These are dealt with as shown in Figure 26.42. In design terms, these should not be treated, but bypassed along with all other streams below the pinch concentration[18].

26.12 REGENERATION OF WASTEWATER

The basic principle of regeneration was introduced in Figure 26.2. Regeneration is a treatment process but applied with the objective of reusing or recycling the water, rather than discharge to the environment. Regeneration of wastewater is most likely to be economic if

- it allows process materials to be recovered;
- it substitutes investment in treatment for discharge by taking up part of the effluent load, as well as reducing the overall demand for water;
- limits are being imposed on freshwater consumption.

Consider the data in Table 26.10 for three water-using operations. It is desired to minimize the consumption of freshwater and generation of wastewater. A regeneration process is available with a performance that can achieve an outlet concentration of 5 ppm.

Start by considering the performance of the system for reuse only. Figure 26.46 shows the target for reuse only turns out to be 60 $t{\cdot}h^{-1}$.

When targeting for regeneration, whether regeneration reuse or regeneration recycling, the objective is to divide the water-using operations into two groups: those using freshwater and those using regenerated water[19]. The grouping strategy depends on whether it is desirable

Table 26.10 Water-using data for a problem requiring the use of regeneration.

Stream number	Limiting water flowrate ($t{\cdot}h^{-1}$)	C_{in} (ppm)	C_{out} (ppm)	Contaminant mass flowrate ($g{\cdot}h^{-1}$)
1	20	0	200	4000
2	50	100	200	5000
3	30	100	400	9000

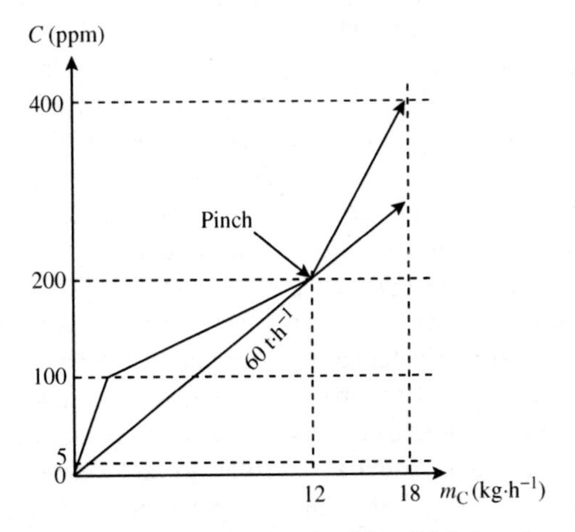

Figure 26.46 Target for the data in Table 26.6 for reuse only.

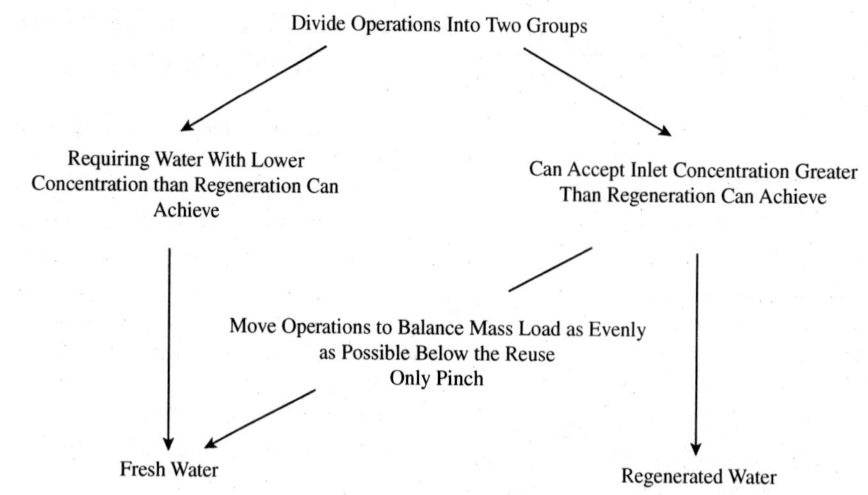

Figure 26.47 Algorithm to provide an initial grouping for regeneration reuse.

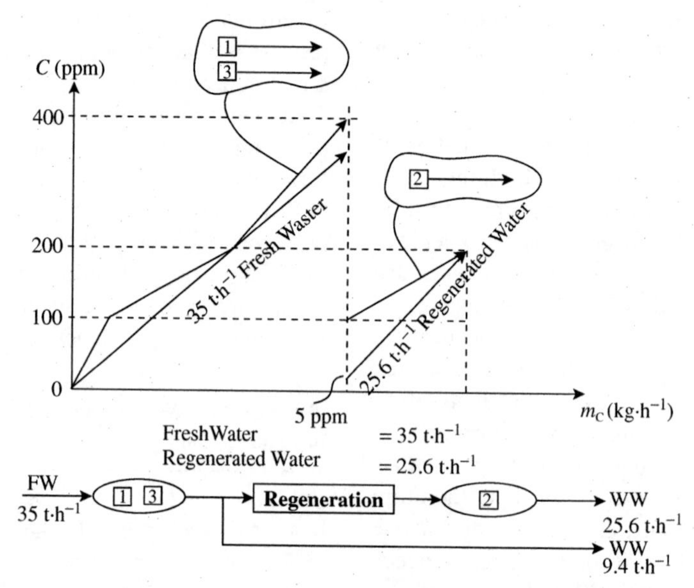

Figure 26.48 The targets for the data from Table 26.10 for regeneration reuse exploiting a regeneration process with an outlet concentration of 5 ppm.

to include or avoid recycling. Take first the case of regeneration reuse, in which no recycling is allowed.

Figure 26.47 gives a simple algorithm to provide an initial grouping for problems exploiting regeneration reuse in which recycling is to be avoided[19]. The initial grouping is such that the streams are first divided into those that require a lower concentration than the regeneration can achieve, and those that can accept a concentration greater than regeneration can achieve. In Figure 26.47, some of the operations that can accept a concentration greater than regeneration can achieve are also put to freshwater, such that the balance of the mass loads between freshwater and regeneration water below the reuse pinch in Figure 26.46 is as evenly distributed as possible. In this example, from the data given in Table 26.10, Operation 1 must be fed with freshwater. It is then a matter of deciding whether Operations 2 and 3 are fed by freshwater or regenerated water. The mass loads below the reuse pinch are:

Stream 1	4000 $g \cdot h^{-1}$
Stream 2	5000 $g \cdot h^{-1}$
Stream 3	3000 $g \cdot h^{-1}$

Given that Operation 1 belongs to the freshwater group, then an arrangement with Operations 1 and 3 in the freshwater group and Operation 2 in the regenerated water group balances the mass loads below the reuse pinch as evenly as possible. This arrangement is shown in Figure 26.48. The streams being fed by freshwater (Operations 1 and 3) are used to construct a limiting composite curve. For these two streams the target freshwater is 35 $t \cdot h^{-1}$. Only Operation 2 is included in the group using regenerated water and the target for regenerated water starting at 5 ppm is 26.6 $t \cdot h^{-1}$.

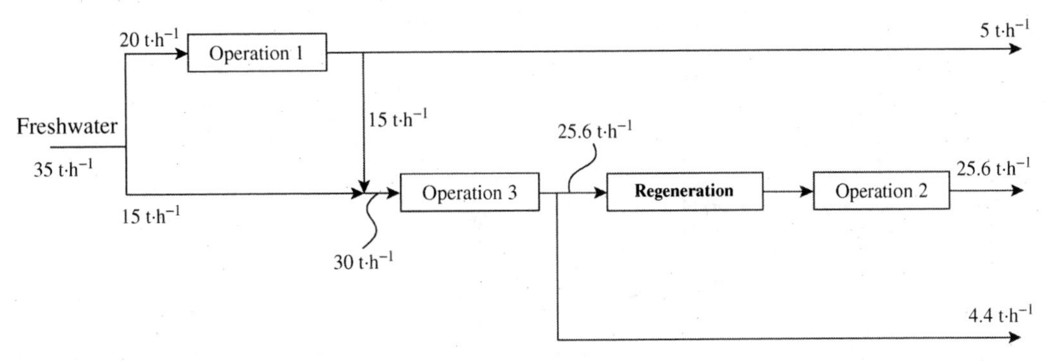

Figure 26.49 Design for regeneration reuse for the problem in Table 26.10.

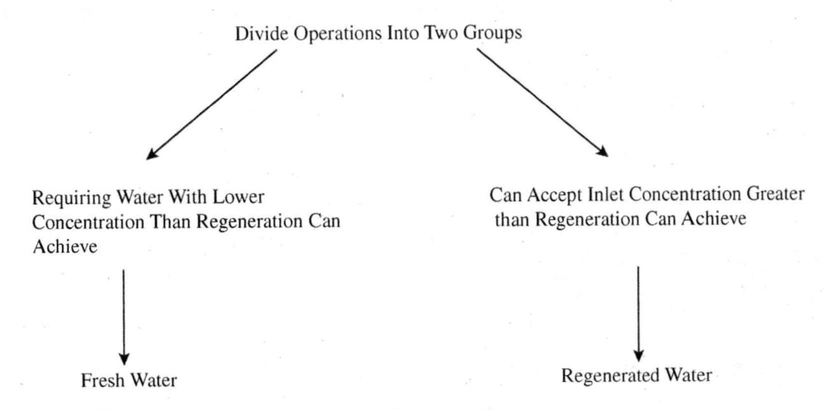

Figure 26.50 Algorithm to provide initial grouping for generation recycling.

Given that the target for freshwater of 35 $t \cdot h^{-1}$ is greater than the target for regenerated water of 26.6 $t \cdot h^{-1}$, some of the water from the exit of Operations 1 and 3 goes directly to effluent, the remainder being regenerated and used in Operation 2. While Figure 26.47 provides an initial grouping, there is sometimes scope to improve the targets through migration of streams from the freshwater group into the regenerated water group and vice versa. This scope to improve arises from the fact that there are now two pinches for the system: the pinch for freshwater and the pinch for regenerated water. In some problems, exploiting the interactions between the pinches allows the targets to be improved. There are three basic approaches to refine the grouping of streams by such migration:

1. A conceptual approach based on the freshwater and regeneration pinches[19]
2. A combinatorial search
3. Mixed integer programming.[15]

In fact, in this simple example there is no scope to improve the targets by migration and the final target is that given in Figure 26.48. Design to achieve the target uses exactly the same water network design procedures as described so far but carried out separately on the two groups of streams. The two designs are then connected together via a regeneration process to comply with the outline flow diagram shown in Figure 26.48. Figure 26.49 shows a design for regeneration reuse. Compared with a freshwater target for reuse only of 60 $t \cdot h^{-1}$, regeneration reuse allows the freshwater to be reduced to 35 $t \cdot h^{-1}$.

As with regeneration reuse, the objective of regeneration recycling is to divide the water-using operations into two groups, those requiring freshwater and those requiring regenerated water[19]. Figure 26.50 presents an algorithm to provide initial grouping for regeneration recycling[19]. This time the algorithm is as simple as dividing the streams into those that require water with a lower concentration than regeneration can achieve and those that can accept a concentration greater than regeneration can achieve. The targets based on this initial grouping are shown in Figure 26.51. Again, there is sometimes scope to improve the initial targets by migration. This again arises from the fact that there are two pinches in the problem, and the interactions between the two pinches can in some cases be exploited. In contrast to regeneration reuse, however, the migration will only move streams from regenerated water onto freshwater in order to reduce regenerated water costs[19]. This migration again can be based on a conceptual approach[19], a combinatorial search or mixed integer programming[15]. Once the final targets have been accepted, design is a matter of designing the network for the two groups separately and then joining the two designs together via the regeneration process and the recycle. The final design to achieve the target is given in Figure 26.52. The target for freshwater consumption is now reduced to

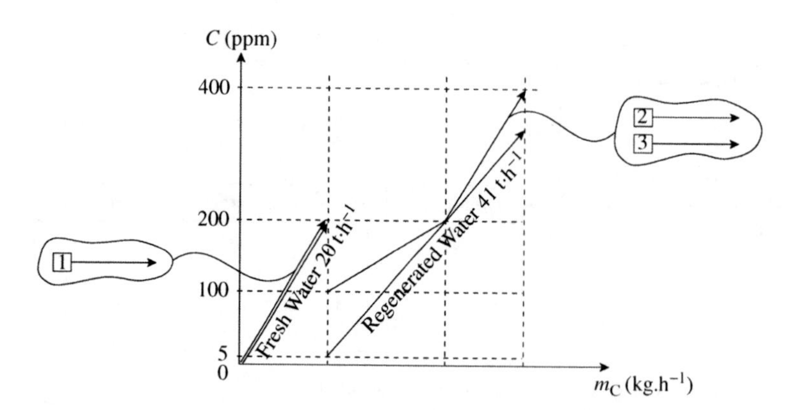

Figure 26.51 The targets for the data from Table 26.10 for regeneration recycling exploiting a regeneration process with an outlet concentration of 5 ppm.

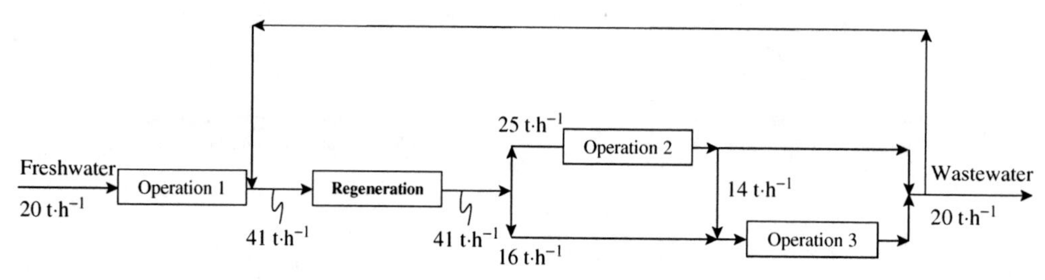

Figure 26.52 Design for regeneration recycling for the problem in Table 26.10.

20 $t \cdot h^{-1}$, in contrast with 35 $t \cdot h^{-1}$ for regeneration reuse and 60 $t \cdot h^{-1}$ for reuse only.

26.13 TARGETING AND DESIGN FOR EFFLUENT TREATMENT AND REGENERATION BASED ON OPTIMIZATION OF A SUPERSTRUCTURE

As with the case for water minimization, the graphical methods used for effluent treatment and regeneration have some severe limitations. As before, multiple contaminants are difficult to handle, constraints, piping and sewer costs, multiple treatment processes and retrofit are all difficult to handle. To include all of these complications requires an approach based on the optimization of the superstructure.

Figure 26.53 presents a superstructure for the design of an effluent treatment system involving three effluent streams and three treatment processes[17]. The superstructure allows for all possibilities. Any stream can go to any effluent process and potential bypassing options have been included. Also, the connections toward the bottom of the superstructure allow for the sequence of the treatment processes to be changed. To optimize such a superstructure requires a mathematical model to be developed for the various material balances for the system and costing correlations included. Such a model then allows the superstructure to be optimized and a final design obtained, as shown in Figure 26.53. The advantages of the superstructure approach are that targeting for large complex problems is facilitated, the design process is automated, it provides a design as well as a target, constraints can be readily included, piping and sewer costs can be included and the cost trade-offs explored[15,20]. The major disadvantage of the approach is that the optimization is more difficult than the corresponding optimization for water reuse. The general problem is nonlinear and the best approach is to provide an initial solution for the nonlinear optimization via a simplified linear model[15,20].

A disadvantage of the automated approach based on the optimization of a superstructure is that conceptual insights are lost. However, as with the case of optimization of water reuse networks, graphical representations based on concentration versus mass load can be constructed from the output of the optimization. The actual concentrations from the design are used to plot the effluent composite curve and the composite of the effluent treatment processes. This can be done on a single component basis, as illustrated in Figure 26.54. Figure 26.54 shows the output from an optimization involving the treatment of two contaminants. The composite treatment line involves multiple treatment processes and pinches with the effluent composite curve for Contaminant *A*. Also shown in Figure 26.54 is the corresponding representation for Contaminant *B*. The effluent composite curve is different. Also, the composite treatment line shows some interesting features. Firstly, it is

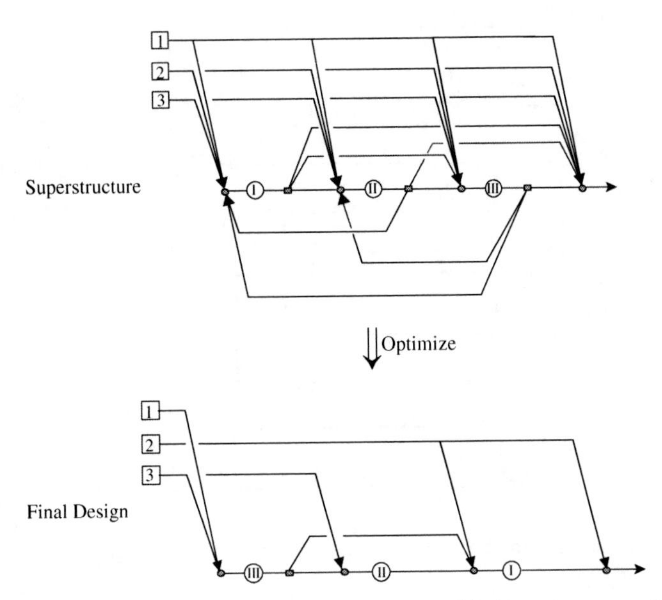

Figure 26.53 Superstructure for the design of an effluent treatment process with three streams and three treatment processes.

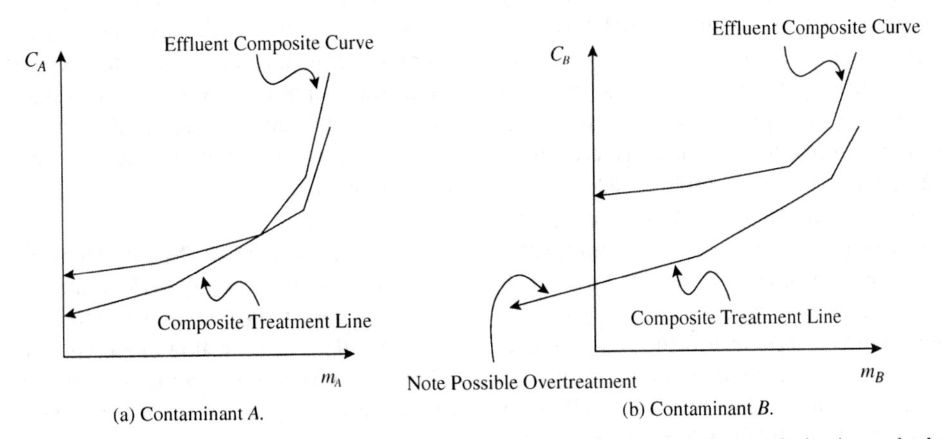

Figure 26.54 Plots of composition versus mass load can be developed from the output of an optimization calculation.

not pinched. Secondly, the composite treatment line extends beyond the effluent composite curve. This indicates that Contaminant *B* has been overtreated. This presumably has arisen as a result of a necessity to fulfill the discharge requirements for Contaminant *A*. Contaminant *A* and *B* are treated simultaneously in the various treatment processes and Contaminant *A* was obviously limiting in this particular problem.

A further option for treatment processes is that, rather than used for effluent discharge, they can be used for regeneration of wastewater for reuse or recycling. In principle, any treatment process can be used for effluent discharge or wastewater regeneration. Figure 26.55 shows a superstructure for two operations and a single treatment process[15]. The superstructure allows for all of the basic reuse options between Operations 1 and 2, together with the use of the treatment process for regeneration reuse or regeneration recycling. Alternatively, the same treatment process can be used for treatment and discharge of effluent. In this way, the analyses for reuse only, regeneration reuse, regeneration recycling and effluent treatment for discharge can all be examined simultaneously by the optimization of the superstructure in Figure 26.55.

26.14 DATA EXTRACTION

When considering heat exchanger and network design, the important issue of data extraction was highlighted, as lost opportunities can result from poor data extraction in heat exchanger and network design. Similarly, there are fundamental issues associated with data extraction for the design of water systems.

1. *Water balance.* Before an existing system can be studied from the point of view of water consumption and effluent treatment, the first step must be to establish a water balance. For a new design, this is a question simply of extracting the information relating to water streams from the flowsheets

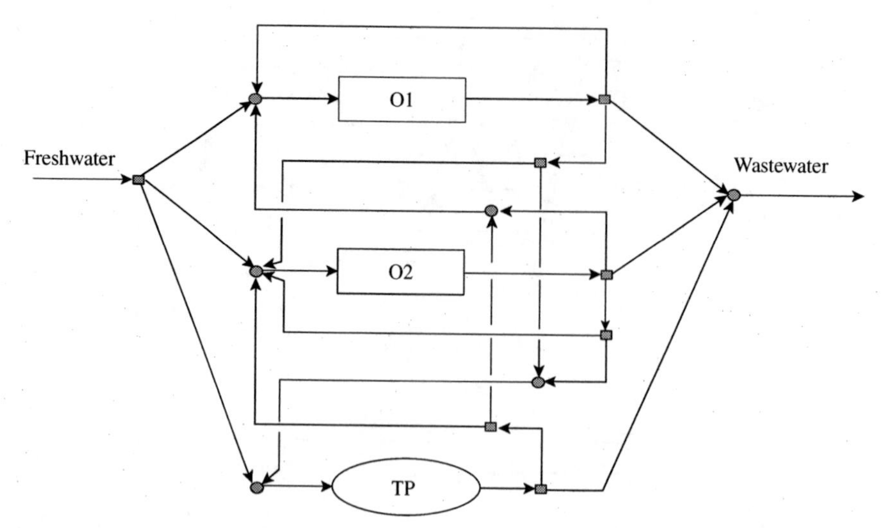

Figure 26.55 A superstructure allowing for a treatment process to be used for either regeneration or discharge.

and material, and energy balance for the processes on a site and the associated utility systems. However, it is rarely the case that the designer is faced with a completely new site. The most common situation is retrofit of an existing site, with all of the existing units in place, or the addition of new units to an existing site, or shutdown of units that modifies the basic site configuration. The objective could be to meet changes in the site load through addition or shutdown of units, improve the existing environmental and economic performance or to meet changes in environmental regulations. In retrofit situations, it is often very difficult to produce a water balance, yet it is a necessity. General lack of information, poor records and poor instrumentation mean that it is usually not possible to close a water balance to better than 90% accountability. This certainly applies to overall sites, and will often apply to individual processes. A water balance that accounts for at least 90% of the water in a process, or on a site is a necessary starting point.

2. *Contaminants.* The next issue to address is which contaminants should be included in any analysis? The answer is, as few contaminants as absolutely necessary. Contaminants should be lumped together where possible, for example, total organic material, salt, suspended solids, total solids and so on. In some instances, it will be necessary to include individual components (e.g. phenol), if a particular constraint relating specifically to that component must be included. Care should be taken not to include contaminants into an analysis, simply because analytical data is available from process outlets. Contaminants should only be included where there is an inlet restriction. A good strategy in developing an analysis is to start by assuming that the whole problem can be analyzed with a single contaminant, for example, dissolved solids. Additional contaminants should then only be added where necessary. It should be noted that the methods for studying water systems described here are not attempting to simulate the water system. All that is of concern at this stage in the design of water-using systems is whether the contaminant specifications reflect the constraints that allow or prevent reuse or recycling of water. For effluent treatment systems, the study can often be performed by putting in the most important contaminants related to discharge regulations. Of course, when the design is completed, it must be checked to make sure that all environmental regulations are complied with.

3. *Flowrate constraints.* Some processes require fixed flowrates, for example, cooling tower makeup, hosing operations, steam ejectors and so on. In extracting the data, the designer needs to identify such flowrate constraints. If the objective of the study is to make a major impact in the consumption of freshwater and the wastewater generation, then it makes sense not to include small streams in the analysis. Reuse or recycling of small streams is unlikely in most cases to be economic. Only the largest streams should be studied.

4. *Data accuracy.* Water systems are usually complex and there is usually a lack of data for a study. It is important to avoid collecting a lot of unnecessary data early in a study. The most accurate data requirements will be around the area of the pinch and below the pinch for the system. By definition, this is the data that is limiting. A good approach therefore is to collect approximate data first and to carry out a preliminary analysis. An initial target and analysis of the data will then reveal where the design is most sensitive to data errors. Collecting better data in these areas can be carried out to refine the target. This approach should avoid collecting too much of unnecessary data.

5. *Limiting conditions.* When determining the opportunities for reuse, regeneration reuse and regeneration recycling, the starting point is the limiting water data that determines the most contaminated water acceptable to an operation. But how are limiting conditions determined? There is

no straightforward answer to this and many factors can influence the setting of the limiting conditions:

- judgment and experience
- corrosion limitations
- maximum solubility
- simulation studies
- laboratory studies
- plant trials
- sensitivity analysis.

Sensitivity analysis is a particularly useful tool when determining the limiting concentrations. Starting with initial limiting conditions, that might well be the existing concentrations, an initial target can be set as illustrated in Figure 26.56. The data can then be adjusted to increase the inlet concentrations and retargeted. It might be that the result is sensitive or insensitive to changes in the inlet concentration. Operation 2 in Figure 26.56 shows that a small increase in the inlet concentration initially brings a large reduction in the water requirement. Further increases in the inlet concentration gradually bring diminishing returns. The point where flowrate becomes insensitive to changes in the inlet concentrations should then be examined more closely for practicality using simulation, laboratory studies and so on. By contrast, large increases in the inlet concentration to Operation 1 only bring a modest reduction in the overall flowrate. Carrying out such a sensitivity analysis allows the critical design variables to be identified in order that they can be studied more closely.

6. *Treatment data.* When carrying out a study of an effluent treatment system, as with a water reuse study, as few contaminants as possible should be included. Again, contaminants should be lumped together where possible, for example, suspended solids, COD, BOD_5. When studying the performance of an existing biological treatment process, great care should be exercised if a mixed effluent from different processes is being treated. For example, suppose a site has an existing biological treatment unit that has an existing performance that removes 90% of the COD. Suppose now that this performance is no longer acceptable and an increase in the performance is required, either on COD removal or individually nominated contaminants. Additional capacity could be added to the biological treatment unit, distributed-treatment applied upstream, or waste minimization applied at source (see Chapter 28). It might be extremely deceptive simply to assume that the existing performance of the biological treatment can be improved by adding extra capacity. It might be that when the performance of the treatment on the individual effluents from the different processes is examined more closely, the effect is quite different on the different effluent streams. It might be that some effluent streams, because the contaminants are easily digested, are treated almost completely in the biological treatment unit. Other effluents from other processes might be largely unchanged by the biological treatment, because the materials are refractory. In other words, some effluent streams are overtreated, others are undertreated and the overall effect is to give a removal ratio for COD of 90%. Such situations can occur on chemical processing sites involving a diverse range of chemical processes. To study an improvement in the system for such circumstances would require the effluent to be broken down into different categories of COD, depending on the origin, rather than just an overall COD. Each process would be given a COD category characterized by how effective the biological treatment was on that effluent. To identify how the individual streams are degraded requires biological digestion tests to be carried out on the individual streams (e.g. BOD_5). The relative rates of digestion can then be modeled to give the overall removal ratio of the mixed effluent in the existing biological treatment unit. The options of waste minimization at source, pretreatment or additional biological treatment capacity can then be evaluated with greater confidence.

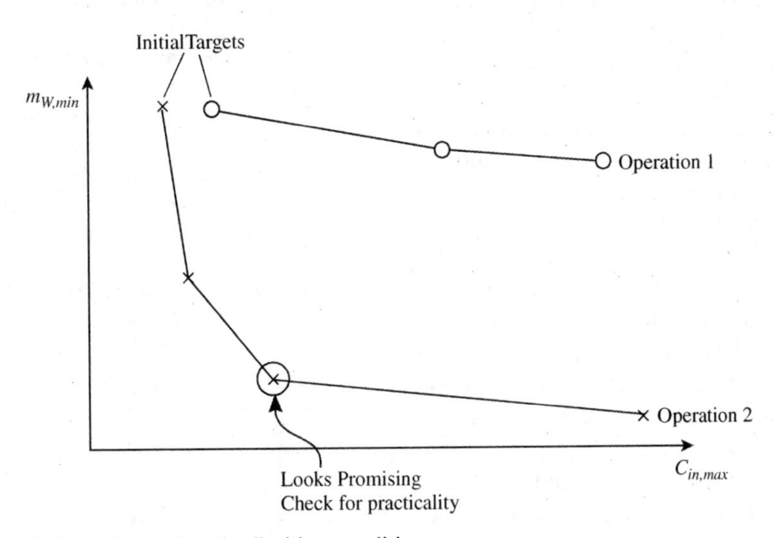

Figure 26.56 Sensitivity analysis to determine the limiting conditions.

7. *Environmental limits*. Each site will have environmental discharge limits specified by regulatory authorities. These might relate to concentrations, total load (i.e. total kilograms of contaminant discharged) or a combination of both. It is usual to have the major contaminants specified as concentrations (especially BOD_5 and COD). Given environmental *discharge limits* or *consent limits* from the regulatory authorities, it would be a bad practice to design precisely for those concentrations. The problem is, as with steam systems, water systems are almost never at steady state. Upset conditions cause surges in both the concentration and volume of effluent. For example, effluent is often collected in effluent pits until the pit is full, when the control system starts a pump and empties it. This causes both an increase in the flowrate and the concentration of the effluent. Given the dynamic nature of water systems, it is good practice to design for discharge concentrations significantly below the consent limits. Design for 60% of the consent limit is usually reasonable to allow for upset conditions.

26.15 WATER SYSTEM DESIGN – SUMMARY

Increasing awareness regarding the problems of overextraction of water and increasingly strict discharge regulations mean that the design of water systems requires care. Water consumption and wastewater generation can be reduced through reuse, regeneration reuse and regeneration recycling.

Distributed effluent treatment requires that a philosophy of design be adopted that segregates effluent for treatment wherever appropriate, and combines it for treatment where appropriate. Various primary, secondary and tertiary treatment processes are available to achieve the required discharge concentrations.

Maximum water reuse can be identified from limiting water profiles. These identify the most contaminated water that is acceptable in an operation. A composite curve of the limiting water profiles can be used to target the minimum water flowrate. While this approach is adequate for simple problems, it has some severe limitations. A more mathematical approach using the optimization of a superstructure allows all of the complexities of multiple contaminants, constraints, enforced matches, capital and operating costs to be included. A review of this area has been given by Mann and Liu[21].

Creating a composite of the effluent streams and matching a treatment line against this composite, with an appropriate specification, can target the minimum flowrate of effluent to be treated. Targeting and design for simple problems is straightforward. Regeneration of wastewater requires that the streams for regeneration be separated into two groups: those that require fresh water and those that require regenerated water. Various methods are available to divide the operations into two groups. Targeting and design for effluent treatment and regeneration can be based on the optimization of a superstructure.

Water systems can be very complex to study and potentially require an enormous amount of data if a whole site is to be studied. Understanding the data requirements more fully can prevent collection of unnecessary data and avoid missed opportunities.

When viewing effluent treatment methods, it is clear that the basic problem of disposing safely of waste material is, in many cases, not so much solved but moved from one place to another. If a method of treatment can be used that allows material to be recycled, then the waste problem is truly solved. However, if the treatment simply concentrates the waste as concentrated liquid, slurry or solid in a form, which cannot be recycled, then it will still need to be disposed of. Landfill disposal of such waste is increasingly unacceptable and thermal oxidation causes pollution through products of combustion and liquors from scrubbing systems. The best method for dealing with effluent problems is to solve the problem at source by waste minimization, as will be discussed in Chapter 28.

26.16 EXERCISES

1. A process involves four water-using operations as detailed in Table 26.11.

Table 26.11 Limiting water data for Exercise 1.

Operation	Limiting water flowrate ($t\cdot h^{-1}$)	$C_{in,max}$ (ppm)	$C_{out,max}$ (ppm)
1	20	0	300
2	46	100	300
3	14	100	600
4	40	400	600

   a. Sketch the limiting water composite curve for the operations.
   b. Target the water flowrate for reuse only.
   c. Generate the grid diagram and design the water-using operation network to achieve the target.

2. Four operations in a process are required to be supplied with water. Table 26.12 below gives the maximum inlet and outlet concentrations for the water and the mass pickup of contamination in each operation.

Table 26.12 Limiting water data for Exercise 2.

Operation	$C_{in,max}$ (ppm)	$C_{out,max}$ (ppm)	Mass pickup ($kg\cdot h^{-1}$)
1	0	400	16
2	200	400	16
3	200	1000	32
4	700	1000	6

a. If freshwater with no inlet contamination is used in each of the four processes, calculate the total water requirements in $t \cdot h^{-1}$.
b. If water is reused up to its maximum potential, calculate the corresponding water requirements in $t \cdot h^{-1}$.
c. Maximum water reuse alone does not satisfy the requirements for water reduction; hence, process changes are to be considered in conjunction with water reuse. What changes to the process conditions would be required to the operations in Table 26.12 to reduce the target, assuming the mass pickup is fixed.

3. Consider a water-using network with a single contaminant. The limiting data for the problem are given in Table 26.13.

Table 26.13 Limiting water data for Exercise 3.

Operation	C_{in} (ppm)	C_{out} (ppm)	Limiting water flowrate ($t \cdot h^{-1}$)	Flowrate loss ($t \cdot h^{-1}$)
1	0	100	30	
2	25	100	50	
3	50	200	40	
4		70		10

a. Construct the limiting water composite curve without Operation 4.
b. Target the minimum freshwater and wastewater flowrate for maximum reuse considering all operations.
c. Design a network to achieve the target. There is one constraint to design a water network. Water from only Operation 1 can be reused for other operations. Show the design steps and represent the water network as a conventional flowsheet.
d. Suppose that Operations 1, 2 and 3 could be operated at fixed flowrate at limiting conditions. Show the modifications of the network design given in Part c.
   (i) When local recycling around operations is accepted.
   (ii) When local recycling around operations in not accepted.
e. Water for Operation 4 is a cooling tower makeup. There are several process changes that could reduce the water requirement for cooling tower operation. Suggest process changes.

4. Consider a water-using network with a single contaminant, represented by the COD. The limiting data for the problem are given in Table 26.14.

Table 26.14 Limiting water data for Exercise 4.

Operation	C_{in} (ppm)	C_{out} (ppm)	m_C ($g \cdot h^{-1}$)	Limiting water flowrate ($t \cdot h^{-1}$)	Water loss ($t \cdot h^{-1}$)
1	0	100	1000	10	–
2	50	100	1000	20	–
3	100	400	6000	20	–
4	0	10	200	20	–
5	100	200	4000	40	20

a. Target the minimum fresh water and wastewater flowrate for maximum reuse only.
b. Design a network to achieve the target.
c. Show the design steps and represent the water network as a conventional flowsheet.
d. Suggest a process change that could reduce the water requirement.

5. The effluent data for a plant are given in Table 26.15.

Table 26.15 Effluent data for Exercise 5.

Stream	C (ppm)	Water flowrate ($t \cdot h^{-1}$)
1	1000	20
2	600	53.4
3	200	22.6

For an environmental discharge limit of 50 ppm,
a. Sketch the composite curve for the effluent streams;
b. Target the treatment flowrate for a fixed outlet concentration $C_{OUT} = 20$ ppm;
c. Target for treatment flowrate removal ratio $RR = (C_{in} - C_{out})/C_{in} = 0.97$;
d. Determine the stream groups for treatment system design;
e. Sketch flow pattern.

6. A chemical process has three major water-using operations that are contaminated mainly by NH_3. The effluent data are given in Table 26.16.

Table 26.16 Effluent data for Exercise 6.

Stream	C (ppm)	Water flowrate ($t \cdot h^{-1}$)
1	300	50
2	150	30
3	80	20

The effluent streams are currently combined and sent to a steam stripper to remove NH_3 from wastewater. The environmental discharge limit for NH_3 is 30 ppm.
a. Construct the effluent composite curve for this process.
b. What is the minimum treatment flowrate for the steam stripper (removal ratio of steam stripper = 90%)?
c. Design a distributed-treatment network that achieves the minimum treatment flowrate target in Part b. Draw the final design as a conventional flowsheet, giving flowrate and concentration levels for all wastewater streams.
d. The steam stripper is used to separate NH_3 from wastewater. If the wastewater stream contains NH_3 as well as H_2S, describe process modifications or operating changes in the steam stripper that can remove both contaminants.

7. A water network produces the 6 effluent streams in Table 26.17. The environmental discharge limit for the contaminant concentration is 50 ppm. There are two treatment processes available on the site, which are capable of reducing

Table 26.17 Effluent streams for Exercise 7.

Stream	Water flowrate ($t \cdot h^{-1}$)	C (ppm)
1	25	200
2	40	350
3	50	200
4	90	150
5	85	120
6	40	75

Table 26.18 Effluent treatment processes for Exercise 7.

Treatment process	C_{out} (ppm)	Unit treatment cost ($\$ \cdot t^{-1}$)	Maximum flowrate capacity ($t \cdot h^{-1}$)
TP1	20	2.25	700
TP2	40	1.5	500

the contaminant concentration respectively to 20 ppm and 40 ppm (Table 26.18).

a. Draw the effluent composite curve for the problem.
b. Draw a superstructure for the problem.

8. In a country with severe water scarcity, the government has placed limits on the amount of fresh water that may be used by industry. A company has three main water-using operations. In each operation, the water picks up contamination in the form of suspended solids (SS). The existing concentrations of SS are given in Table 26.19.

Table 26.19 Water data for Exercise 8.

Operations	C_{in} (ppm)	C_{out} (ppm)	Water flowrate ($t \cdot h^{-1}$)
1	0	100	20
2	0	100	30
3	0	330	50

The company has now been told that it must reduce the total water demand by at least 10 per cent. Carry out a water minimization study to determine the scope for reusing water. Discussions with the operations managers have concluded that not all the operations need to be fed with freshwater. The maximum concentrations that can be tolerated at the inlet of the operations are given in Table 26.20.

a. What is the contaminant mass load and limiting flowrate of each operation? Assume that the outlet concentrations in Table 26.19 are the maximum allowable ones.
b. What is the minimum fresh water flowrate that is required if water is reused?
c. Can the company achieve the required water reduction by reusing water?
d. Design a network that will achieve the minimum water target. Draw the final design as a conventional flowsheet, giving flows and concentrations for all water streams.

Table 26.20 Maximum inlet concentrations for Exercise 8.

Operation	$C_{in,max}$ (ppm)
1	0
2	50
3	30

9. Table 26.21 lists the limiting data for a water network.

Table 26.21 Water data for Exercise 9.

Operation	Limiting water flow ($t \cdot h^{-1}$)	m_C ($g \cdot h^{-1}$)	C_{in} (ppm)	C_{out} (ppm)
1	10	1000	0	100
2	55	8250	100	250
3	40	4000	300	400

The environmental discharge limit for the contaminant is 50 ppm. Freshwater is available with a concentration of 0 ppm. A treatment process is available capable of achieving a contaminant outlet concentration of 20 ppm.

a. Construct the limiting water composite curve and obtain the target for minimum fresh water flowrate for reuse only.
b. Design a water-use network, which achieves this target. Draw the network as a conventional flowsheet.
c. From the streams directed to effluent, devise a distributed treatment problem. Construct the effluent composite curve. Obtain the target for minimum treatment flowrate.
d. Design a distributed effluent treatment network, which achieves the minimum treatment target. Draw the network as a conventional flowsheet.

10. The data for five water-using operations are shown in Table 26.22.

Table 26.22 Water data for Exercise 10.

Operation	Mass load of contaminant ($kg \cdot h^{-1}$)	C_{in} (ppm)	C_{out} (ppm)	Limiting water flowrate ($t \cdot h^{-1}$)
1	8	0	200	40
2	5	100	200	50
3	6	100	400	20
4	6	300	400	60
5	8	400	600	40

Targeting indicates that for reuse only the process requires 80 $t \cdot h^{-1}$. It is proposed to reduce this target by introducing regeneration with a removal ratio of 0.95.

a. The streams are divided into two groups for regeneration. The first group contains Operations 1 and 2 and the second group Operations 3, 4 and 5. Determine the targets for freshwater and regenerated water.

b. An alternative grouping has only Operation 1 in the first group with the remainder in the second group. Again, determine the targets for freshwater and regenerated water.

c. A third grouping has Operations 1 and 4 in the first group with the remainder in the second group. Determine the targets for freshwater and regenerated water.

Which group would you recommend? In each case, sketch the flow pattern of how the water flows would be distributed between the two groups. It is not necessary to determine the flow pattern within each group.

REFERENCES

1. Tchobanoglous G and Burton FL (1991) *Metcalf & Eddy Wastewater Engineering – Treatment, Disposal and Reuse*, McGraw-Hill.
2. Eckenfelder WW and Musterman JL (1995) *Activated Sludge Treatment of Industrial Wastewater*, Technomic Publishing.
3. Berne F and Cordonnier J (1995) *Industrial Water Treatment*, Gulf Publishing.
4. Wang YP and Smith R (1994) Wastewater Minimization, *Chem Eng Sci*, **49**: 981.
5. Betz Laboratories (1991) *Betz Handbook of Industrial Water Conditioning*, 9th Edition.
6. Schierup H-H and Brix H (1990) Danish Experience With Emergent Hydrophyte Treatment Systems (EHTS) and Prospects in the Light of Future Requirements on Outlet Water Quality, *Water Sci Tech*, **22**: 65.
7. Tebbutt THY (1990) *BASIC Water and Wastewater Treatment*, Butterworths.
8. Takama N, Kuriyama T, Shiroko K and Umeda T (1980) Optimal Water Allocation in a Petroleum Refinery, *Comp Chem Eng*, **4**: 251.
9. Wang YP and Smith R (1995) Wastewater Minimization with Flowrate Constraints, *Trans IChemE*, **A73**: 889.
10. Dhole VR, Ramchandani N, Tainsh R and Wasilewski M (1996) Make Your Process Water Pay for Itself, *Chem Eng*, **103**: 100.
11. Polley GT and Polley HL (2000) Design Better Water Networks, *Chem Eng Prog*, **Feb**: 47.
12. Kuo W-CJ and Smith R (1998) Designing for the Interactions Between Water Use and Effluent Treatment, *Trans IChemE*, **A76**: 287–301.
13. Wang YP and Smith R (1995) Time Pinch Analysis, *Trans IChemE*, **A73**: 905.
14. Doyle SJ and Smith R (1997) Targeting Water Reuse with Multiple Contaminants, *Trans IChemE*, **B75**: 181.
15. Alva-Argaez A (1999) *Integrated Design of Water Systems*, PhD Thesis, UMIST, UK.
16. Wang YP and Smith R (1994) Design of Distributed Effluent Treatment Systems, *Chem Eng Sci*, **49**: 3127.
17. Kuo W-CJ and Smith R (1998) Design of Water-Using Systems Involving Regeneration, *Trans IChemE*, **B76**: 94.
18. Smith R, Petela EA and Howells J (1996) Breaking a Design Philosophy, *Chem Eng*, **606**: 21.
19. Kuo W-CJ and Smith R (1997) Effluent Treatment System Design, *Chem Eng Sci*, **52**: 4273.
20. Galen B and Grossmann IE (1998) Optimal Design of Distributed Wastewater Treatment Networks, *Ind Eng Chem Res*, **37**: 4036.
21. Mann JG and Liu YA, (1999), *Industrial Water Reuse and Waste water Minimization*, McGraw-Hill.

27 Inherent Safety

Although safety considerations have been left until late in this text, they should not be left to the final stages of the design. Consideration has been left late here to allow the many issues in process design that have an impact on safety to be dealt with first.

Early decisions made purely for process reasons can often lead to problems of safety, health and environment that require complex solutions. It is far better to consider them early as the design progresses. Designs that avoid the need for hazardous materials, use less of them, use them at lower temperatures and pressures or dilute them with inert materials will be *inherently safe* and will not require elaborate safety systems[1].

Here, safety and health considerations will be restricted to features of inherent safety that can be included as the design is developing, rather than the detailed *hazard and operability* studies that take place in the later stages of design when the design is nearing completion.

The three major hazards in process plant are fire, explosion and toxic release[2].

27.1 FIRE

The first major hazard in process plant is fire, which is usually regarded as having a disaster potential lower than both explosion and toxic release[2]. However, fire is still a major hazard and can under the worst conditions approach explosion in its disaster potential. Fire requires a combustible material (gas or vapor, liquid, solid, solid in the form of a dust dispersed in a gas), an oxidant (usually oxygen in air) and usually, but not always, a source of ignition. Consider now the important factors in assessing fire as a hazard.

1. *Autoignition temperature.* The autoignition temperature of a gas or vapor is the temperature at which it will ignite spontaneously in air, without any external source of ignition.

2. *Flammability limits.* A flammable gas or vapor will burn in air only over a limited range of composition. Below a certain concentration of the flammable gas, the *lower flammability limit*, the mixture is too "lean" to burn. Above a certain concentration, the *upper flammability limit*, it is too "rich" to burn. Concentrations between these limits constitute the *flammable range.*

Combustion of a flammable gas–air mixture occurs if the composition of the mixture lies in the flammable range *and* if there is a source of ignition. Alternatively, the combustion of the mixture occurs without a source of ignition if the mixture is heated up to its autoignition temperature.

The most flammable mixture is usually the stoichiometric mixture for combustion. It is often found that the concentrations of the lower and upper flammability limits are approximately one-half and twice that of the stoichiometric mixture, respectively[2,3].

Flammability limits are affected by pressure. The effect of pressure changes is specific to each mixture. In some cases, decreasing the pressure can narrow the flammable range by raising the lower flammability limit and reducing the upper flammability limit until eventually the two limits coincide and the mixture becomes nonflammable. Conversely, an increase in pressure can widen the flammable range. However, in other cases, increasing the pressure has the opposite effect of narrowing the flammable range[2,3].

Flammability limits are also affected by temperature. An increase in temperature usually widens the flammable range[2,3].

3. *Minimum oxygen concentration.* Whereas the lower flammability limit measures the lowest concentration that will allow combustion of a vapor-air mixture, sometimes inert gas (usually nitrogen, but sometimes carbon dioxide or steam) is added to the mixture to prevent combustion. The *minimum oxygen concentration* is a limit below which the reaction cannot generate enough energy for the mixture (including inerts) to allow self propagation of a flame. The minimum oxygen is stated as a percent of oxygen in air plus combustible material.

4. *Flash point.* The *flash point* of a liquid is the lowest temperature at which it gives off enough vapor to form an ignitable mixture with air. The flash point generally increases with increasing pressure.

5. *Limiting oxygen concentration of a dust.* The *limiting oxygen concentration* of a dust is the minimum concentration of oxygen (displaced by an inert gas such as nitrogen) capable of supporting combustion of dust that is dispersed in the form of a cloud. A mixture with oxygen concentration below the limiting oxidant concentration is not capable of supporting combustion and hence cannot support a subsequent dust explosion.

6. *Minimum ignition temperature of a dust.* The *minimum ignition temperature* of a dust is the lowest temperature at

Chemical Process Design and Integration R. Smith
 ISBNs: 0-471-48680-9 (HB); 0-471-48681-7 (PB)

which dust that is dispersed in the form of a cloud can ignite. The minimum ignition temperature is an important factor in evaluating the sensitivity of dust to ignition sources such as hot surfaces. Decreasing particle size of dust and decreasing moisture content both lower the minimum ignition temperature.

Flammable liquids are potentially much more dangerous than flammable gas mixtures, if the flammable liquids are processed or stored under pressure at a temperature above their atmospheric boiling point. Gases leak at a lower mass flowrate than liquids through an opening of a given size. Flashing liquids leak at about the same rate as a subcooled liquid but then turn into a mixture of vapor and spray on release. The spray, if fine, is just as hazardous as the vapor and can be spread as easily by the wind. Thus, the leak of a flashing liquid through a hole of a given size produces a much greater hazard than the corresponding leak of gas[1].

When synthesizing a process, the occurrence of flammable gas and dust mixtures should be avoided, rather than relying on the elimination of sources of ignition. This can be achieved in the first instance by changing the process conditions such that dust mixtures are below their limiting oxygen concentration and gas mixtures are outside their flammable range. If this is not possible, inert material such as nitrogen, carbon dioxide or steam should be introduced. The use of flammable liquids held under pressure above their atmospheric boiling points should be avoided. Adopting atmospheric subcooled conditions or vapor conditions in the process will be much safer. In addition, sources of ignition such as flames, sparks from electrical equipment, sparks from static electricity, sparks from impact or friction, and so on, should also be eliminated wherever possible. Sources of ignition from static electricity tend to be less obvious than other sources of ignition. Electrical charge can accumulate on a powder or any plant item if electrically isolated from the ground. The resulting spark discharges can be avoided by electrically grounding all conductive items. Where high-surface charging processes exist (e.g. when processing dusty materials), the use of electrically insulating materials (e.g. plastic pipes) should be avoided, as this allows the accumulation of electrical charge. For the particular case of dusts, high relative humidity can reduce the resistivity of some powders, increasing the rate of charge decay and decreasing the accumulation of static charge.

27.2 EXPLOSION

The second of the major hazards is explosion, which has a disaster potential usually considered to be greater than fire but lower than toxic release[2]. Explosion is a sudden and violent release of energy. The energy released in an explosion on a process plant is either of the following.

1. *Chemical energy.* Chemical energy derives from a chemical reaction. The source of the chemical energy is exothermic chemical reactions or combustion of flammable material (dust, vapor or gas). Explosions based on chemical energy can be either *uniform* or *propagating*. An explosion in a vessel will tend to be a uniform explosion, while an explosion in a long pipe will tend to be a propagating explosion. For a dust, the *minimum explosible concentration* is the lowest concentration in $g \cdot m^{-3}$ in air that will give rise to flame propagation on ignition.

2. *Physical energy.* Physical energy may be pressure energy in gases, thermal energy, strain energy in metals or electrical energy. An example of an explosion caused by release of physical energy would be fracture of a vessel containing high-pressure gas.

Thermal energy is generally important in creating the conditions for explosions rather than a source of energy for the explosion itself. In particular, as already mentioned, superheat in a liquid under pressure causes flashing of the liquid if it is accidentally released to the atmosphere.

There are two basic kinds of explosions involving the release of chemical energy:

1. *Deflagration.* In a *deflagration*, the flame front travels through the flammable mixture relatively slowly.
2. *Detonation.* In a *detonation*, the flame front travels as a shock wave followed closely by a combustion wave that releases the energy to sustain the shock wave. The detonation front travels with a velocity greater than the speed of sound in the unreacted medium.

A detonation generates greater pressures and is more destructive than a deflagration. The peak pressure caused by the deflagration of a hydrocarbon–air mixture or a dust mixture at atmospheric pressure is of the order of 8 to 10 bar. However, a detonation may give a peak pressure of the order of 20 bar. A deflagration may turn into a detonation particularly if traveling down a long pipe.

Just as there are two basic kinds of explosions, they can occur in two different conditions:

1. *Confined explosions. Confined explosions* are those that occur within vessels, pipework or buildings. The explosion of a flammable mixture in a process vessel or pipework may be a deflagration or a detonation. The conditions for a deflagration to occur are that the gas mixture is within the flammable range *and* that there is a source of ignition. Alternatively, the deflagration can occur without a source of ignition if the mixture is heated to its autoignition temperature. An explosion starting as a deflagration can make the transition into a detonation. This transition can occur in a pipeline but is unlikely to happen in a vessel.

2. *Unconfined explosions.* Explosions that occur in the open air are *unconfined explosions*. An unconfined vapor cloud explosion is one of the most serious hazards in the

process industries. Although a large toxic release may have a greater disaster potential, unconfined vapor explosions tend to occur more frequently[2]. Most unconfined vapor cloud explosions have been the result of leaks of flashing flammable liquids.

The problem of the explosion of an unconfined vapor cloud is not only that it is potentially very destructive but also that it may occur some distance from the point of vapor release and may thus threaten a considerable area.

If the explosion occurs in an unconfined vapor cloud, the energy in the blast wave is generally only a small fraction of the energy theoretically available from the combustion of all the material that constitutes the cloud. The ratio of the actual energy released to that theoretically available from the heat of combustion is referred to as the *explosion efficiency*. Explosion efficiencies are typically in the range of 1 to 10%. A value of 3% is often assumed.

The hazard of an explosion should in general be minimized by avoiding flammable gas–air mixtures in the process. Again, this can be done either by changing process conditions or by adding an inert material. It is bad practice to rely solely on elimination of sources of ignition.

27.3 TOXIC RELEASE

The third of the major hazards and the one with the greatest disaster potential is the release of toxic chemicals[2]. The hazard posed by toxic release depends not only on the chemical species but also on the conditions of exposure. The high disaster potential from toxic release arises in situations in which large numbers of people are briefly exposed to high concentrations of toxic material. However, the long-term health risks associated with prolonged exposure at low concentrations over a working life also present serious hazards.

For a chemical to affect health, a substance must come into contact with an exposed body surface. The three ways in which this happens are by inhalation, skin contact and ingestion.

In preliminary process design, the primary consideration is contact by inhalation. This happens either through accidental release of toxic material to the atmosphere or the *fugitive emissions* caused by slow leakage from pipe flanges, valve glands, pump and compressor seals. Tank filling also causes emissions when the rise in liquid level causes vapor in the tank to be released to atmosphere, as discussed in Chapter 25.

The acceptable limits for toxic exposure depend on whether the exposure is brief or prolonged. *Lethal concentration* for airborne materials and *lethal dose* for nonairborne materials are measured by tests on animals. The limits for brief exposure to toxic materials that are airborne are usually measured by the concentration of toxicant that is lethal to 50% of the test group over a given exposure period, usually four hours. It is written as LC_{50} (lethal concentration for 50% of the test group). The test gives a comparison of the absolute toxicity of a compound in a single concentrated dose, *acute exposure*. For nonairborne materials, lethal dose LD_{50} refers to the quantity of material administered, which results in death of 50% of the test group. It should be emphasized that it is extremely difficult to extrapolate tests on animals to human beings.

The limits for prolonged exposure are expressed as the *threshold limit values*. These are essentially acceptable concentrations in the workplace. There are three categories of threshold limit values:

1. *Time weighed exposure*. This is the time weighted average concentration for a normal 8-hour workday or 40-hour workweek to which nearly all workers can be exposed, day after day, without adverse effects. Excursions above the limit are allowed if compensated by other excursions below the limit.

2. *Short-term exposure*. This is the maximum concentration to which workers can be exposed for a period of up to 15 min continuously without suffering from (1) intolerable irritation, (2) chronic of irreversible tissue change or (3) narcosis of sufficient degree to increase accident proneness, impair self-rescue or materially reduce efficiency, provided that no more than four excursions per day are permitted, with at least 60 min between exposure periods, and provided the daily time weighted value is not exceeded.

3. *Ceiling exposure*. This is the concentration that should not be exceeded, even instantaneously.

When synthesizing a flowsheet, it is obviously best to try and avoid, where possible, the use of toxic materials altogether. However, this is often just not possible. In this case, the designer should take particular care to avoid processing and storing toxic liquids under pressure at temperatures above their atmospheric boiling points. As with flammable materials, if a leak occurs, whether large or small, the mass flowrate through a hole of a given size is far greater for a liquid than for a gas. Release of a flashing liquid will result in higher levels of exposure than the release of a subcooled liquid or a gas.

The best way to avoid fugitive emissions is by using leak-tight equipment (e.g. changing from packing to mechanical seals or even using sealless pumps, etc.). If this is not possible, then regular maintenance checks can reduce fugitive emissions. If all else fails, the equipment can be enclosed and ventilated. The air would then be treated before finally passing to atmosphere. To reduce emissions from tank filling, the vapor space must be prevented from breathing to the atmosphere. This can be achieved through vapor treatment, use of floating roofs or use of membranes in the tank roof, as discussed in Chapter 25.

27.4 INTENSIFICATION OF HAZARDOUS MATERIALS

The best way of dealing with a hazard in a flowsheet is to remove it completely. The provision of safety systems to control the hazard is much less satisfactory. One of the principal approaches to making a process inherently safe is to limit the inventory of hazardous material, *intensification* of hazardous material. The inventories to be avoided most of all are flashing, flammable liquids or flashing, toxic liquids.

Once the process route has been chosen, it may be possible to synthesize flowsheets that do not require large inventories of materials in the process. The design of the reaction and separation system is particularly important in this respect but heat transfer, storage and pressure relief systems are also important.

1. *Reactors.* Perhaps the worst safety problem that can occur with reactors occurs when an exothermic reaction generates heat at a faster rate than the cooling system can remove it. Such *runaway reactions* are usually caused by coolant failure, perhaps for a temporary period, or reduced cooling capacity due to perhaps a pump failure in the cooling water circuit. The runaway happens because the rate of reaction, and hence the rate of heat generation, increases exponentially with temperature, whereas the rate of cooling increases only linearly with temperature. Once heat generation exceeds available cooling capacity, the rate of temperature rise becomes progressively faster[4]. If the energy release is large enough, liquids will vaporize, and overpressurization of the reactor follows.

Clearly, the potential hazard from runaway reactions is reduced by reducing the inventory of material in the reactor.

Batch operation requires a larger inventory than the corresponding continuous reactor. Thus, there may be a safety incentive to change from batch to continuous operation. Alternatively, the batch operation can be changed to semibatch in which one (or more) of the reactants is added over a period. The advantage of semibatch operation is that the feed can be switched off in the event of a temperature (or pressure) excursion. This minimizes the chemical energy stored up for a subsequent exotherm.

For continuous reactors, plug-flow designs require smaller volumes and hence smaller inventories than mixed-flow designs for the same conversion, as discussed in Chapter 5.

Reaction rates may often be improved by the use of more extreme operating conditions. More extreme conditions may reduce inventory appreciably. However, more extreme conditions bring their own problems, as will be discussed later. A very small reactor operating at a high temperature and pressure may be inherently safer than one operating at less extreme conditions because it contains a much lower inventory. A large reactor operating close to atmospheric temperature and pressure may be safe for different reasons. Leaks are less likely, and if they do happen, the leak will be small because of the low pressure. Also, little vapor is produced from the leaking liquid because of the low temperature. A compromise solution employing moderate pressure, temperature and medium inventory may combine the worst features of the extremes. The compromise solution may be such that the inventory is large enough for a serious explosion or serious toxic release if a leak occurs, the pressure will ensure the leak is large and the high temperature results in the evaporation of a large proportion of the leaking liquid[1].

2. *Distillation.* There is a large inventory of boiling liquid, sometimes under pressure, in a distillation column both in the base and held up in the column. If a sequence of columns is involved, then, as discussed in Chapter 11, the sequence can be chosen to minimize the inventory of hazardous material. If all materials are equally hazardous, then choosing the sequence that tends to minimize the flowrate of nonkey components will also tend to minimize the inventory. Use of the partition column or dividing-wall column shown in Figure 11.14c will reduce considerably the inventory relative to two simple columns. Partition columns or dividing-wall columns are inherently safer than conventional arrangements since they not only lower the inventory but also the number of items of equipment is fewer, and hence there is a lower potential for leaks.

The column inventory can also be reduced by the use of low hold-up column internals, including low hold-up in the column base. As the design progresses, other features can be included to reduce the inventory. Thermosyphon reboilers have a lower inventory than kettle reboilers. Peripheral equipment such as reboilers can be located inside the column[1].

3. *Heat transfer operations.* Heat transfer fluids other than steam and cooling water utilities are sometimes introduced into the design of the heat exchange system. These heat transfer media are sometimes liquid hydrocarbons used at high pressure. When possible, higher boiling liquids should be used. Better still, the flammable material should be substituted with a nonflammable medium such as water or molten salt.

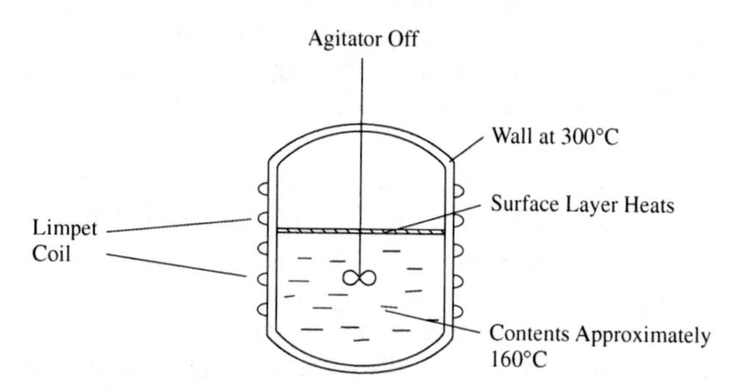

Figure 27.1 Schematic of the Seveso reaction system.

The use of an unnecessarily high-temperature hot utility or heating medium should be avoided. This may have been a major factor that led to the runaway reaction at Seveso in Italy in 1976, which released toxic material over a wide area. The reactor was liquid-phase and operated in a stirred tank, Figure 27.1. It was left containing an uncompleted batch at around 160°C, well below the temperature at which a runaway reaction could start. The temperature required for a runaway reaction was around 230°C[5]. The reaction was normally carried out under vacuum at about 160°C in a reactor heated by steam at about 300°C. The temperature of the liquid could not rise above its boiling point of 160°C at the operating pressure.

In this accident, the steam was isolated from the reactor containing the unfinished batch, and the agitator was switched off. The reactor walls below the liquid level fell to the same temperature as the liquid, around 160°C. The reactor walls above the liquid level remained hotter because of the high-temperature steam that had been used (but now isolated). Heat then passed by conduction and radiation from the walls to the top layer of the stagnant liquid, which became hot enough for a runaway reaction to start, Figure 27.1. Once started in the upper layer, the reaction then propagated throughout the reactor. If the steam had been cooler, say 170°C, the runaway could not have occurred[1].

Some operations need to be carried out at low temperature that needs refrigeration. The refrigeration fluid might, for example, be propylene and present a major hazard. Operation of the process at a higher pressure on the one hand brings increased hazards in the process equipment but, on the other hand, might allow a less hazardous refrigeration fluid.

4. *Storage.* Some of the largest inventories of hazardous materials tend to be held up in the storage of raw materials and products and intermediate (buffer) storage.

The most obvious way of reducing the inventory in storage is by locating producing and consuming plants near each other so that hazardous intermediates do not have to be stored or transported[1]. It may also be possible to reduce storage requirements by making the design more flexible. Adjusting the capacity could then be used to cover delays in the arrival of raw material, upsets in one part of the plant, and so on, and thus reduce the need for storage[1].

Large quantities of toxic gases such as chlorine and ammonia and flammable gases such as propane and ethylene oxide can be stored either under pressure or at atmospheric pressure under refrigerated conditions. If there is a leak from atmospheric refrigerated storage, the quantity of hazardous material that is discharged will be less than the corresponding pressurized storage at atmospheric temperature. For large storage tanks, refrigeration is safer. However, this might not be the case with small-scale storage, since the refrigeration equipment provides sources for leaks. Thus, in small-scale storage, pressurized storage may be safer[1,2].

5. *Relief systems.* Emergency discharge from relief valves can be dealt with in a number of ways:

a. Direct discharge to atmosphere under conditions leading to rapid dilution.
b. Total containment in a connected vessel, with ultimate disposal being deferred.
c. Partial containment, in which some of the discharge is separated either physically through gravitational, centrifugal means, and so on, or chemically through absorption, and so on, and contained.
d. Combustion in a flare. Flare systems might include a catchpot that collects liquids and passes gases to flare (see Chapter 25).

Relief systems are expensive and bring considerable environmental problems.

Sometimes it is possible to dispense with relief valves and all that comes after them by using stronger vessels, strong enough to withstand the highest pressures that can be reached[1].

For example, if the vessel can withstand the pump delivery pressure, then a relief valve for overpressurization by the pump may not be needed. However, there may still be need for a small relief device to guard against overpressurization in the event of a fire. It may be possible to avoid the need for a relief valve on a distillation column by making it strong enough to withstand the pressure developed if cooling is lost but heat input and feed pumping continues[1].

At first sight it might seem that making vessels strong enough to withstand the possible overpressurization would be an expensive option. However, it must be remembered that a comparison is not being made between one vessel with a thick wall versus one vessel with a thin wall protected by a relief valve. Material discharged through the relief valve might need to be partially contained, in which case the comparison might be between Figures 27.2a and 27.2b[1].

Similarly, instead of installing vacuum relief valves, the vessels can be made strong enough to withstand vacuum. In addition, if the vessel contains flammable gas or vapor, vacuum relief valves will often need to admit nitrogen to avoid flammable mixtures. A stronger vessel may often be safer and cheaper[1].

6. *The overall inventory.* The optimization of reactor conversion was considered in Chapter 13. As the conversion increased, the size (and cost) of the reactor increased but that of separation, recycle and heat exchanger network systems decreased. The same trends also occur with the inventory of material in these systems. The inventory in the reactor increases with increasing conversion, but the inventory in the other systems decreases. Thus, in some processes, it is possible to optimize for minimum overall

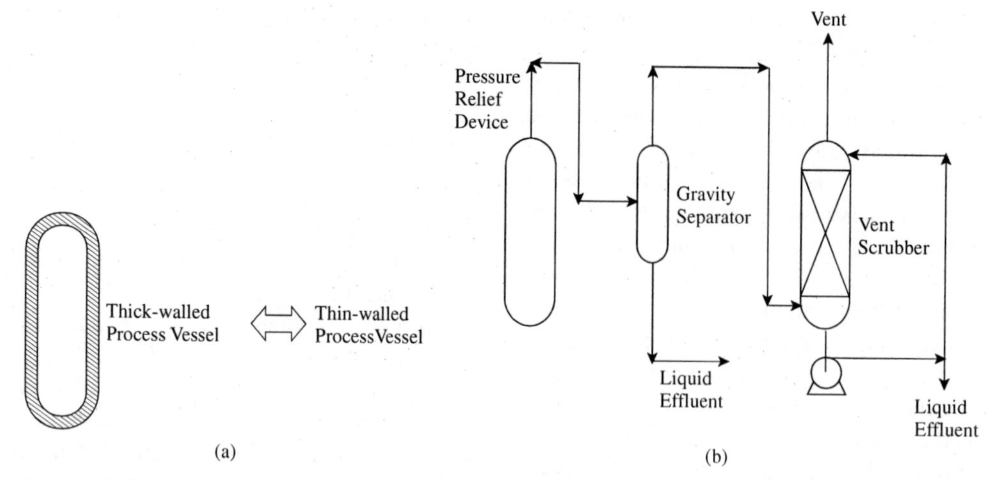

Figure 27.2 A thick-walled pressure vessel might be economic when compared with a thin-walled vessel with its relief and venting system.

inventory[6]. In the same way as reactor conversion can be varied to minimize the overall inventory, the recycle inert concentration can also be varied.

It might be possible to reduce the inventory significantly by changing reactor conversion and recycle inert concentration without a large cost penalty if the cost optimization profiles are fairly flat.

Intensification of hazardous material results in a safer process. The minimization of waste will most often bring environmental benefits.

27.5 ATTENUATION OF HAZARDOUS MATERIALS

So far, the emphasis has been on substituting hazardous materials or using less intensification. Now consider use of hazardous material under less hazardous conditions of less extreme temperatures or pressures, as a vapor rather than superheated liquid, or diluted, in other words, *attenuation*[1].

Operation at extremes of pressure and temperature brings a number of safety problems:

1. *High pressure.* Most process plants operate at pressures below 250 bar, but certain processes, such as high-pressure polyethylene plants, operate at pressures up to about 3000 bar. The use of high pressure greatly increases the stored energy in the plant. Although high pressures in themselves do not pose serious problems in materials of construction, the use of high temperatures, low temperatures or corrosive chemicals together with high pressure does[2].

With high-pressure operation, the problem of leaks becomes much more serious, since this increases the mass flowrate of fluid that can leak out through a given hole. This is particularly so when the fluid is a flashing liquid.

2. *Low pressure.* Low pressures are not in general as hazardous as the other extreme operating conditions. However, one particular hazard that does exist in low-pressure plant handling flammable materials is the ingress of air with consequent formation of a flammable mixture.

3. *High temperature.* The use of high temperatures in combination with high pressures greatly increases stored energy in the plant. The heat required to obtain a high temperature is often provided by furnaces. These have a number of hazards, including possible rupture of the tubes carrying the process fluid and explosions in the radiant zone.

There are materials-of-construction problems associated with high-temperature operation. The main problem is creep.

4. *Low temperature.* Low-temperature processes (below 0°C) contain large amounts of fluids kept in the liquid state by pressure and/or low temperature. If for any reason it is not possible to keep them under pressure or keep them cold, then the liquids will begin to vaporize. If this happens, impurities in the fluids are liable to precipitate from solution as solids, especially if equipment is allowed to boil dry. Deposited solids may not only be the cause of blockage but also in some cases the cause of explosions. It is necessary, therefore, to ensure that the fluids entering a low-temperature plant are purified. A severe materials-of-construction problem in low-temperature processes is low-temperature embrittlement. Also, in low temperature as in high-temperature operations, the equipment is subject to thermal stresses, especially during start-up and shutdown.

When synthesizing a flowsheet, the designer should consider carefully the problems associated with operation under extreme conditions. Attenuation will result in a safer plant, providing the attenuation does not increase the inventory of hazardous material. If the inventory does not increase, then not only will attenuation make the process safer but also it will be cheaper since cheaper materials of construction and thinner vessel walls can be used, and it is not necessary to add on so much protective equipment[1].

27.6 QUANTITATIVE MEASURES OF INHERENT SAFETY

Although, for example, higher reactor temperature might lead to safer plant if the inventory can be reduced, it is necessary to be able to assess such changes quantitatively. Lowering the inventory makes the plant safer, but raising the temperature makes it less safe. Which effect is more significant?

Safety indices, such as the Dow Index, have been suggested as measures of safety[1,2]. In these indices, the hazard associated with each material in the process is assessed and given a number largely based on judgment and experience. The numbers are weighed and combined to give an overall index for the process. The indices have no significance in an absolute sense but can be used to compare the relative hazards between two alternative designs. However, they are intended more for use in the later stages of design when more information is available.

The major hazard from the release of flammable or toxic material arises from material that, having been released to atmosphere, enters the vapor phase. A simple quantitative measure of inherent safety for fire and explosion hazards is the energy released by combustion of material that enters the vapor phase upon release from containment. The combustion energy releases associated with two process alternatives can be compared and some judgment made of the relative inherent safety of the two options as far as fire and explosion hazards are concerned. However, the difficulty is one of defining the mode of release. In the worst case, catastrophic failure involving release of all the materials could be assumed and the energy release calculated from the part that would vaporize. On the other hand, the release could be assumed to occur from a standard-sized hole in the equipment. This would be a less hazardous scenario than catastrophic failure but more likely to occur. Comparing process options on the basis of these two alternative modes of release will not necessarily lead to the same conclusions when comparing two process alternatives. Judgment is required as to which mode of release is most appropriate.

On the other hand, if the hazard is toxicity, process alternatives can be compared by assessing the mass of toxic material that would enter the vapor phase on release from containment, weighting the components according to their lethal concentration.

Example 27.1 A process involves the use of benzene as a liquid under pressure. The temperature can be varied over a range. Compare the fire and explosion hazard of operating with a liquid process inventory of 1000 kmol at 100°C and 150°C, on the basis of the theoretical combustion energy resulting from catastrophic failure of the equipment. The normal boiling point of benzene is 80°C, latent heat of vaporization 31,000 $kJ \cdot kmol^{-1}$, specific heat capacity 150 $kJ \cdot kmol^{-1} \cdot K^{-1}$ and heat of combustion 3.2×10^6 $kJ \cdot kmol^{-1}$.

Solution The fraction of liquid vaporized on release is calculated from a heat balance[3]. The sensible heat above saturated conditions at atmospheric pressure provides the heat of vaporization. The sensible heat of the superheat is given by:

$$mC_P(T_{SUP} - T_{BPT})$$

where m = mass of liquid
C_P = heat capacity
T_{SUP} = temperature of the superheated liquid
T_{BPT} = normal boiling point

If the mass of liquid vaporized is m_V then:

$$m_V = \frac{mC_P(T_{SUP} - T_{BPT})}{\Delta H_{VAP}}$$

where m_V = mass of liquid vaporized
ΔH_{VAP} = latent heat of vaporization

Thus, the vapor fraction (VF) is given by:

$$VF = \frac{m_V}{m} \frac{C_P(T_{SUP} - T_{BPT})}{\Delta H_{VAP}}$$

For operation at 100°C:

$$VF = \frac{50(100 - 80)}{31{,}000}$$

$$= 0.097$$

$$m_V = 0.097 \times 1000$$

$$= 97 \text{ kmol}$$

Theoretical combustion energy

$$= 97 \times 3.2 \times 10^6$$

$$= 310 \times 10^6 \text{ kJ}$$

For operation at 150°C,

$$VF = 0.339$$

$$m_V = 339 \text{ kmol}$$

Theoretical combustion energy

$$= 1085 \times 10^6 \text{ kJ}$$

Thus, against this measure of inherent safety the fire hazard is 3.5 times larger for operation at 150°C compared to operation at 100°C.

In fact, the true potential fire load will be greater than the energy release calculated in Example 27.1. In practice, such a release of superheated liquid generates large amounts of fine spray in addition to the vapor. This can double the energy release based purely on vaporization.

If the material in two process alternatives is both flammable and highly toxic, then they can be compared on both bases separately. If the assessments of the relative flammability and toxicity are in conflict, then a safety index can be used for a measure.

27.7 INHERENT SAFETY – SUMMARY

Designs that avoid the need for hazardous materials or use less of them or use them at lower temperatures and pressures or dilute them with inert materials will be inherently safe and will not require elaborate safety systems.

When synthesizing a process, the occurrence of flammable gas mixtures should be avoided, rather than relying on the elimination of sources of ignition.

One of the principal approaches to making a process inherently safe is to limit the inventory of hazardous material. The inventories to be avoided most of all are flashing flammable or toxic liquids, that is, liquids under pressure above their atmospheric boiling points.

The following changes should be considered to improve safety[1–3]:

Reactors

- batch to continuous
- batch to semibatch
- mixed-flow reactors to plug-flow
- reduce the inventory in the reactor by increasing temperature or pressure, by changing catalyst or by better mixing
- lower the temperature of a liquid-phase reactor below the normal boiling point, or dilute it with a safe solvent
- substitute a hazardous solvent
- externally heated/cooled to internally heated/cooled.

Distillation

- choose the distillation sequence to minimize the inventory of hazardous material
- use partition or dividing-wall columns to reduce the inventory relative to two simple columns and reduce the number of items of equipment and hence lower the potential for leaks
- use of low hold-up internals.

Heat Transfer Operations

- use water or other nonflammable heat transfer media
- use a lower temperature utility or heat transfer medium
- use a liquid heat transfer medium below its atmospheric boiling point if flammable or toxic
- if refrigeration is required, consider higher pressure if this allows a less hazardous refrigerant to be used

Storage

- locate producing and consuming plants near to each other so that hazardous intermediates do not have to be stored and transported
- reduce storage by increasing design flexibility
- store in a safer form (less extreme pressure, temperature or in a different chemical form).

Relief Systems

- consider strengthening vessels rather than relief systems

Overall Inventory

- consider changes to reactor conversion and recycle inert concentration to reduce the overall inventory. This might be possible without significant cost if the cost optimization profiles are fairly flat.

When synthesizing a flowsheet, the designer should consider carefully the problems associated with operation under extreme conditions. Attenuation will result in a safer plant, providing the attenuation does not increase the inventory of hazardous material.

What you don't have, can't leak[1]. If plants can be designed so that they use safer raw materials and intermediates or not so much of the hazardous ones or use the hazardous ones at lower temperatures and pressures or diluted with inert materials, then many problems later in the design can be avoided[1].

As the design progresses, it is necessary to carry out hazard and operability studies[2]. These are generally only meaningful when the design has been progressed very far and are outside the scope of this text.

27.8 EXERCISES

1. When processing flammable and toxic materials, why are superheated liquids above their atmospheric boiling point to be most avoided?
2. a. Compare the flammability hazard of storing 1000 kmol of cyclohexane at 100°C and 200°C, using catastrophic failure of the vessel as a basis for comparison. The atmospheric boiling point of cyclohexane is 81°C, its latent heat of vaporization is 30,000 $kJ{\cdot}kmol^{-1}$, liquid heat capacity is 210 $kJ{\cdot}kmol^{-1}{\cdot}K^{-1}$ and its heat of combustion is 3.95×10^6 $kJ{\cdot}kmol^{-1}$.
   b. How much of the theoretical heat of combustion would you expect to be released in the event of an explosion?
3. A chemical process uses as raw material hazardous Component A. Component A is fed as liquid under pressure to a continuous stirred tank vessel in which it performs a first-order reaction to useful Product B. The reaction in the vessel is isothermal first order with a rate of reaction given by $-r_A = k_0 \exp(-E/RT)$, where r_A is the rate of reaction ($kmol{\cdot}min^{-1}$), $k_0 = 1.5 \times 10^6$ min^{-1}, $E = 67{,}000$ $kJ{\cdot}kmol^{-1}$,

$R = 8.3145$ kJ·K^{-1}·kmol^{-1} and T is the reactor temperature (K). T is also the supply temperature of A whose yet unknown inventory m_A is in the form of a superheated liquid. The total amount of B to be produced is 1000 kmol. T and m_A are to be selected with the additional consideration of safety. The normal boiling point of A is 70°C, its latent heat of vaporization is 25,000 kJ·kmol^{-1}, the liquid specific heat capacity is 140 kJ·kmol^{-1}·K^{-1}, and its heat of combustion is 2.5×10^6 kJ·kmol^{-1}. The residence time of the reactor is 1 min, and the safety is measured in terms of fire and explosion hazards on the basis of the theoretical combustion energy resulting form catastrophic failure of the equipment.

a. An initial choice of 80°C has been made for T. Calculate the required inventory of A, the vapor fraction of A in the feed and the theoretical combustion energy at this temperature.
b. Sketch the combustion energy against T and determine the least safe temperature in terms of this measure. Does the shape of the curve in the sketch depend on the problem data? Comment on the selection made in Part a and explain whether the selection of $T = 130$°C makes a better choice.

REFERENCES

1. Kletz TA (1984) *Cheaper, Safer Plants*, IChemE Hazard Workshop, 2nd Edition, Institution of Chemical Engineers, UK.
2. Lees FP (1989) *Loss Prevention in the Process Industries*, Vol. 1, Butterworths.
3. Crowl DA and Louvar JF (1990) *Chemical Process Safety – Fundamentals with Applications*, Prentice Hall.
4. Tharmalingam S (1989) Assessing Runaway Reactions and Sizing Vents, *Chem Eng*, **Aug**: 33.
5. Cardillo P and Girelli A (1981) The Seveso Runaway Reaction: A Thermoanalytical Study, *IChemE Symp Ser*, **68**, 3/N: 1.
6. Boccara K (1992) *Inherent Safety for Total Processes*, MSc Dissertation, UMIST, UK.

28 Clean Process Technology

28.1 SOURCES OF WASTE FROM CHEMICAL PRODUCTION

As with safety, environmental considerations are usually left to a late stage in the design. However, like safety, early decisions can often lead to difficult environmental problems that later require complex solutions. Again, it is better to consider effluent problems as the design progresses in order to avoid complex waste treatment systems.

The effects of pollution can be direct such as toxic emissions providing a fatal dose of toxicant to fish, animal life and even human beings. The effects can also be indirect. Toxic materials that are nonbiodegradable such as insecticides and pesticides, if released to the environment, are absorbed by bacteria and enter the food chain. These compounds can remain in the environment for long periods of time, slowly being concentrated at each stage in the food chain until ultimately they prove fatal, generally to predators at the top of the food chain such as fish or birds.

Thus, emissions must not exceed levels where they are considered to be harmful. There are two approaches to deal with emissions:

1. Treat the effluent using thermal oxidation, biological digestion, and so on, to a form suitable for discharge to the environment, the so-called *end-of-pipe* treatment.
2. Reduce or eliminate the production of the effluent at source through *clean process technology* by *waste minimization*.

The problem with relying on end-of-pipe treatment is that once waste has been created, it cannot be destroyed. The waste can be concentrated or diluted, its physical or chemical form can be changed, but it cannot be destroyed. Thus, the problem with end-of-pipe effluent treatment systems is that they do not so much solve the problem but move it from one place to another. For example, aqueous solutions containing heavy metals can be treated by chemical precipitation to remove the metals. If the treatment system is designed and operated correctly, the aqueous stream can be passed on for further treatment or discharged to the receiving water. But what about the precipitated metallic sludge? This is usually disposed of to landfill[1].

The whole problem is best dealt with by not making the waste in the first place through clean process technology. If waste can be minimized at source, this brings the dual benefit of reducing waste treatment costs and reducing raw materials costs.

Two classes of waste from chemical processes can be identified[1]:

1. The two inner layers of the onion diagram in Figure 1.7 (the reaction and separation and recycle systems) produce *process waste*. The process waste is waste byproducts, purges, and so on.
2. The outer layers of the onion (the utility system) produce *utility waste*. The utility waste is products of fuel combustion, waste from the production of boiler feedwater for steam generation, and so on. However, the design of the utility system is closely tied together with the design of the heat exchanger network. Hence, in practice, the three outer layers should be considered as being the source of utility waste.

There are three sources of process waste[1]:

1. *Reactors*. Waste is created in reactors through the formation of waste byproducts, and so on.
2. *Separation and recycle systems*. Waste is produced from separation and recycle systems through the inadequate recovery and recycling of valuable materials from waste streams.
3. *Process operations*. The third source of process waste can be classified under the general category of process operations. Operations such as start-up and shutdown of continuous processes, product changeover, equipment cleaning for maintenance, tank filling, and so on, all produce waste.

The principal sources of utility waste are associated with hot utilities (including cogeneration), cold utilities and the water system. Furnaces, steam boilers, gas turbines and diesel engines all produce waste as gaseous combustion products. These combustion products contain carbon dioxide, oxides of sulfur and nitrogen and particulates, which contribute in various ways to the greenhouse effect, acid rain and the formation of smog. In addition to gaseous waste, steam generation creates aqueous waste from boiler feedwater treatment, and so on. The water system also produces waste from the use of water for extraction, scrubbing and washing operations, and so on.

Consider now how clean process technology can be developed by minimizing waste from each of these sources. Since one of the themes running throughout the design philosophy has been minimizing waste through high processed yields, elimination of extraneous materials, and

Chemical Process Design and Integration R. Smith
 ISBNs: 0-471-48680-9 (HB); 0-471-48681-7 (PB)

so on, much of the discussion to follow will summarize the arguments already presented. However, here the arguments will be drawn together into an overall philosophy of clean process technology. Since the reactor is at the heart of the process, this is where to start when considering clean process technology[1]. The separation and recycle system comes next, followed by process operations and finally utility waste.

28.2 CLEAN PROCESS TECHNOLOGY FOR CHEMICAL REACTORS

Under normal operating conditions, waste is produced in reactors in the following ways[2]:

1. If it is not possible, for some reason, to recycle unreacted feed material to the reactor inlet, then low conversion will lead to waste of that unreacted feed.
2. The primary reaction can produce waste byproducts, for example:

$$FEED1 + FEED2 \longrightarrow PRODUCT + WASTE\ BYPRODUCT \quad (28.1)$$

3. Secondary reactions can produce waste byproducts, for example:

$$FEED1 + FEED2 \longrightarrow PRODUCT$$
$$PRODUCT \longrightarrow WASTE\ BYPRODUCT \quad (28.2)$$

In Chapter 5, the objective set was to maximize selectivity for a given conversion. This will also minimize waste generation in reactors for a given conversion.

4. Impurities in the feed materials can undergo reaction to produce waste byproducts.
5. Catalyst is either degraded and requires changing or is lost from the reactor and cannot be recycled.

Consider each of these in turn and how clean chemical reactor technology can be developed.

1. *Reducing waste when recycling is difficult..*

a. *Increasing conversion for single irreversible reactions.* If unreacted feed material is difficult to separate and recycle, it is necessary to force as high conversion as possible. If the reaction is irreversible, then the low conversion can be forced to higher conversion by longer residence time in the reactor, a more effective catalyst, higher temperature, higher pressure or a combination of these.

b. *Increasing conversion for single reversible reactions.* The situation becomes worse if the fact that unreacted feed material is difficult to separate and recycle coincides with the reaction being reversible. Chapter 6 considered what can be done to increase equilibrium conversion.

- *Excess reactants.* An excess of one of the reactants can be used as shown in Figure 6.8a.
- *Product removal during reaction.* Separation of the product before completion of the reaction can force a higher conversion as discussed in Chapter 6. Figure 6.6 showed how this is done in sulfuric acid processes. Sometimes, the product (or one of the products) can be removed continuously from the reactor as the reaction progresses, for example, by allowing it to vaporize from a liquid-phase reactor or incorporating a membrane in the reactor to allow it to permeate through the membrane.
- *Inert material concentration.* The reaction might be carried out in the presence of an inert material. This could be a solvent in a liquid-phase reaction or an inert gas in a gas-phase reaction. Figure 6.8b shows that if the reaction involves an increase in the number of moles, then adding inert material will increase equilibrium conversion. On the other hand, if the reaction involves a decrease in the number of moles, then inert concentration should be decreased, Figure 6.8b. If there is no change in the number of moles during reaction, then inert material has no effect on equilibrium conversion.
- *Reactor temperature.* For endothermic reactions, Figure 6.8c shows that the temperature should be set as high as possible, consistent with materials of construction limitations, catalyst life and safety. For exothermic reactions, the ideal temperature is continuously decreasing as conversion increases, Figure 6.8c.
- *Reactor pressure.* In Chapter 6, it was deduced that vapor-phase reactions involving a decrease in the number of moles should be set to as high a pressure as practicable, taking into account that the high pressure might be expensive to obtain through compressor power, mechanical construction might be expensive and high pressure brings safety problems, Figure 6.8d. Reactions involving an increase in the number of moles should ideally have a pressure that is continuously decreasing as conversion increases, Figure 6.8d. Reduction in pressure can be brought about either by a reduction in the absolute pressure or by the introduction of an inert diluent.

If the separation and recycle of unreacted feed material is not a problem, then squeezing extra conversion from the reactor is usually not a problem.

2. *Reducing waste from primary reactions, which produce waste byproducts.* If a waste byproduct is formed from the reaction, as in Equation 28.1 above, then it can only be avoided by different reaction chemistry, that is, a different reaction path.

3. *Reducing waste from multiple reactions producing waste byproducts.* In addition to the losses described above

for single reactions, multiple reaction systems lead to further waste through the formation of waste byproducts in secondary reactions. Let us briefly review from Chapters 5 to 7 what can be done to minimize byproduct formation.

a. *Reactor type.* Firstly, make sure that the correct reactor type has been chosen to maximize selectivity for a given conversion in accordance with the arguments presented in Chapter 5.
b. *Reactor concentration.* Selectivity can often be improved by one or more of the following actions[2]:
   - Use an excess of one of the feeds when more than one feed is involved.
   - Increase the concentration of inert material if the byproduct reaction is reversible and involves a decrease in the number of moles.
   - Decrease the concentration of inert material if the byproduct reaction is reversible and involves an increase in the number of moles.
   - Separate the product part way through the reaction before carrying out further reaction and separation.
   - Recycle waste byproducts to the reactor if byproduct reactions are reversible.

   Each of these measures can, in the appropriate circumstances, minimize waste, Figure 6.9.
c. *Reactor temperature and pressure.* If there is a significant difference between the effects of temperature or pressure on primary and byproduct reactions, then temperature and pressure should be manipulated to improve selectivity and minimize the waste generated by byproduct formation.
d. *Catalysts.* Catalysts can have a major influence on selectivity. Changing the catalyst can change the relative influence on the primary and byproduct reactions. This might result directly from the reaction mechanisms at the active sites, or the relative rates of diffusion in the support material, or a combination of both.

4. *Reducing waste from feed impurities that undergo reaction.* If feed impurities undergo reaction, it causes waste of feed material, products or both. Avoiding such waste is most readily achieved by purifying the feed. Thus, increased feed purification costs are traded off against reduced raw materials, product separation and waste disposal costs, Figure 28.1.

5. *Reducing waste by upgrading waste byproducts.* Waste byproducts can sometimes be upgraded to useful materials by subjecting them to further reaction in a different reaction system. An example was quoted in Chapter 13 in which hydrogen chloride, which is a waste byproduct of chlorination reactions, can be upgraded to chlorine and then recycled to a chlorination reactor.

6. *Reducing catalyst waste.* Both homogeneous and heterogeneous catalysts are used. In general, heterogeneous catalysts should be used whenever possible, rather than homogeneous catalysts, since separation and recycling of homogeneous catalysts can be difficult, leading to waste.

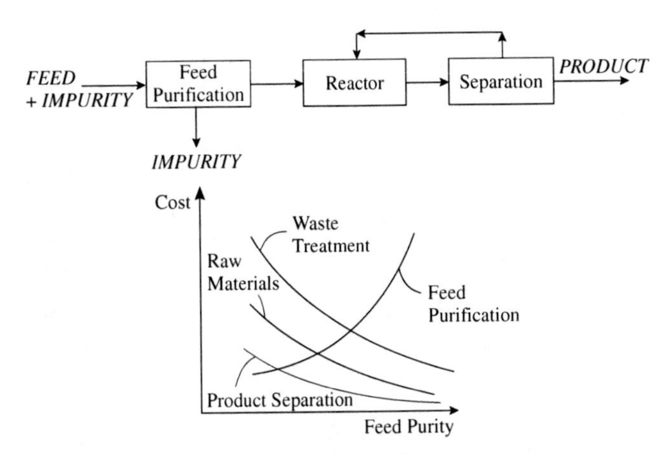

Figure 28.1 If feed impurity undergoes a reaction, then there is an optimum feed purity.

Heterogeneous catalysts are more common in large-scale processes. However, they degrade and need replacement. If contaminants in the feed material or recycle shorten catalyst life, then extra separation to remove those contaminants before entering the reactor might be justified. If the catalyst is sensitive to extreme conditions, such as high temperature, then some measures can help avoid local hot spots and extend catalyst life:

- better flow distribution
- better heat transfer
- introduction of catalyst diluent
- better instrumentation and control.

Fluidized-bed catalytic reactors tend to generate loss of catalyst through attrition of the solid particles, causing fines to be generated, which are lost. More effective separation of catalyst fines from the reactor product and recycling the fines will reduce catalyst waste up to a point. Improving the mechanical strength of the catalyst is likely to be the best solution in the long term.

Now consider the separation and recycle system.

28.3 CLEAN PROCESS TECHNOLOGY FOR SEPARATION AND RECYCLE SYSTEMS

Clean process technology requires that useful materials can be separated and recycled more effectively.

Waste from the separation and recycle system can be minimized in five ways[3]:

- Recycling waste streams directly.
- Reduction of feed impurities by purification of the feed.

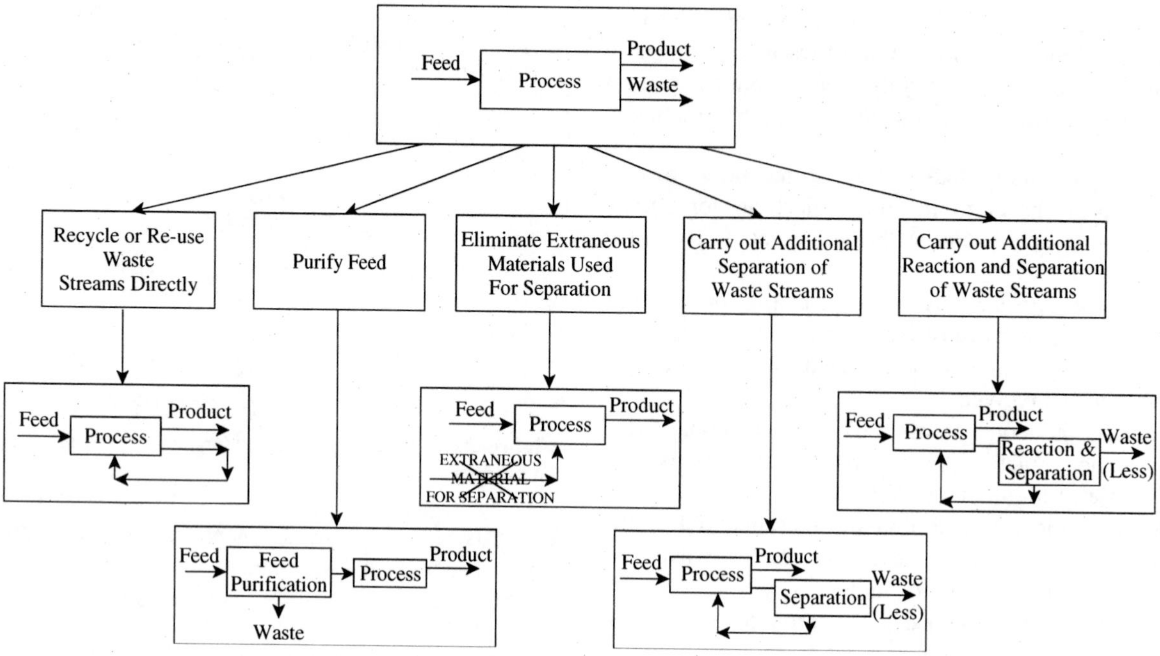

Feed → Process → Product / Waste

Recycle or Re-use Waste Streams Directly | Purify Feed | Eliminate Extraneous Materials Used For Separation | Carry out Additional Separation of Waste Streams | Carry out Additional Reaction and Separation of Waste Streams

Feed → Process → Product

Feed → Process → Product; EXTRANEOUS MATERIAL FOR SEPARATION

Feed → Process → Product; Reaction & Separation → Waste (Less)

Feed → Feed Purification → Process → Product; Waste

Feed → Process → Product; Separation → Waste (Less)

Figure 28.2 Waste minimization in separation and recycle systems.

- Elimination of extraneous materials used for separation.
- Additional separation of waste streams to allow increased recovery.
- Additional reaction and separation of waste streams to allow increased recovery.

Although this is generally the sequence in which the five actions would be considered, this sequence will not necessarily be always correct. The best sequence to consider the five actions will depend on the process. The magnitude each action will have will vary for different processes.

Figure 28.2 illustrates the basic approach for reducing waste from separation and recycle systems.

1. *Recycling waste streams directly*. Sometimes, waste can be reduced by recycling waste streams directly. This is the first option in Figure 28.2. If this can be done, it is clearly the simplest way to reduce waste and should be considered first. Most often, the waste streams that can be recycled directly are aqueous streams, which, although contaminated, can substitute part of the freshwater feed to the process, as discussed in Chapter 26.

Figure 28.3a shows a simplified flowsheet for the production of isopropyl alcohol by the direct hydration of propylene[3]. Different reactor technologies are available for the process and separation and recycle systems vary, but Figure 28.3a is representative. Propylene containing propane as an impurity is reacted with water according to the reaction:

$$\underset{\text{propylene}}{C_3H_6} + \underset{\text{water}}{H_2O} \longrightarrow \underset{\text{isopropyl alcohol}}{(CH_3)_2CHOH}$$

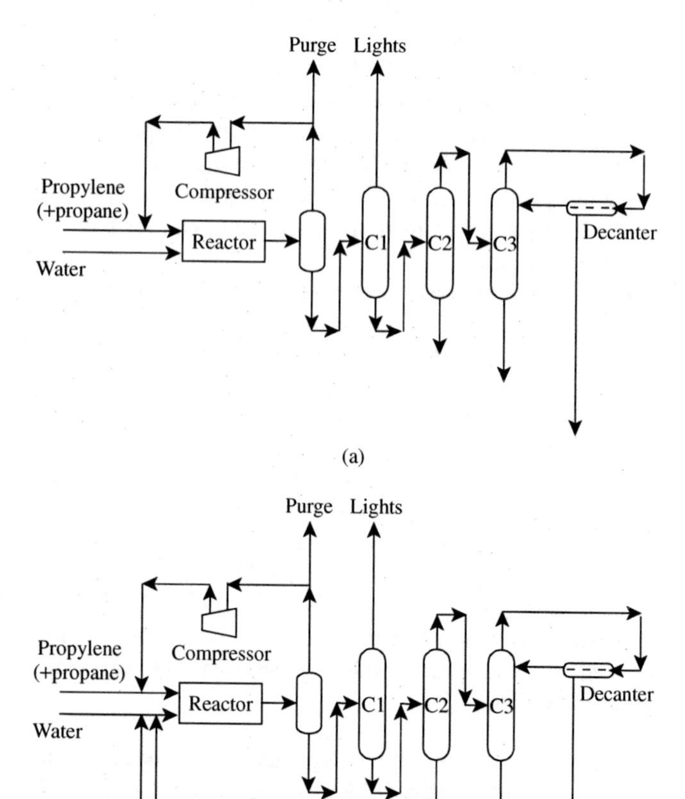

Figure 28.3 Outline flowsheet for the production of isopropyl alcohol by direct hydration of propylene. (From Smith R and Petela EA, 1991, *Chem Eng*, **513**: 24, reproduced by permission of the Institution of Chemical Engineers).

Some small amount of byproduct formation occurs. The principal byproduct is di-isopropyl ether. The reactor product is cooled and a phase separation of the resulting vapor–liquid mixture produces a vapor containing predominantly propylene and propane and a liquid containing predominantly the other components. Unreacted propylene is recycled to the reactor and a purge taken to prevent the buildup of propane. The first distillation in Figure 28.3a (Column C1) removes light ends (including the di-isopropyl ether). The second (Column C2) removes as much water as possible to approach the azeotropic composition of the isopropyl alcohol–water mixture. The final column in Figure 28.3a (Column C3) is an azeotropic distillation using an entrainer. In this case, one of the materials already present in the process, di-isopropyl ether, can be used as entrainer.

Wastewater leaves the process from the bottom of the second column and the decanter of the azeotropic distillation column. Although both of these streams are essentially pure water, they will nevertheless contain small quantities of organics and must be treated before final discharge. This treatment can be avoided altogether by recycling the wastewater to the reactor inlet to substitute part of the freshwater feed, Figure 28.3b.

Sometimes, waste streams can be recycled directly, but between different processes. Waste streams from one process can become the feedstock for another. The scope for such *industrial symbiosis* is often not fully realized since it often means waste being transferred between different business units within a company or between different companies.

If waste streams can be recycled directly, this is clearly the simplest method for reducing waste. However, most often, additional separation is required.

2. *Feed purification.* Impurities that enter with the feed inevitably cause waste. If feed impurities undergo reactions, then this causes waste from the reactor, as already discussed. If the feed impurity does not undergo reaction, then it can be separated out from the process in a number of ways as discussed in Section 13.1. The greatest source of waste occurs when a purge is used. Impurity builds up in the recycle, and it is desirable for it to build up to a high concentration to minimize waste of feed materials and product in the purge. However, two factors limit the extent to which the feed impurity can be allowed to build up:

a. High concentrations of inert material can have an adverse effect on reactor performance.
b. As more and more feed impurity is recycled, the cost of the recycle increases (e.g. through increased recycle gas compression costs etc.) to the point where that increase outweighs the savings in raw materials lost in the purge.

In general, the best way to deal with a feed impurity is to purify the feed before it enters the process. This is the second option shown in Figure 28.2.

Returning to the isopropyl alcohol process from Figure 28.3, propylene is fed to the process containing propane as a feed impurity. In Figure 28.3, the propane is removed from the process using a purge. This causes waste of propylene, together with a small amount of isopropyl alcohol. The purge can be eliminated if the propylene is purified before entering the process. In this case, the purification can be achieved by distillation. Examples of where similar schemes can be implemented are plentiful.,

Many processes are based on an oxidation step for which air would be the first obvious source of oxygen. A partial list would include acetic acid, acetylene, acrylic acid, acrylonitrile, carbon black, ethylene oxide, formaldehyde, maleic anhydride, nitric acid, phenol, phthalic anhydride, sulfuric acid, titanium dioxide, vinyl acetate and vinyl chloride[4]. Because the nitrogen in the air is not required by the reaction, it must be separated and leave the process at some point. Because gaseous separations are difficult, the nitrogen is normally removed from the process using a purge if a recycle is used. Alternatively, the recycle is eliminated from the design and the reactor forced to as high a conversion as possible to avoid recycling. The nitrogen will carry with it process materials, both feeds and products, and will probably require treatment before final discharge. If the air for the oxidation is substituted by pure oxygen, then, at worst, the purge will be very much smaller. At best it can be eliminated altogether. Of course, this requires an air separation plant upstream of the process to provide the pure oxygen. However, despite this disadvantage, very significant benefits can be obtained, as the following example shows.

Consider vinyl chloride production (see Example 5.1). In the "oxychlorination" reaction step of the process ethylene, hydrogen chloride and oxygen are reacted to form dichloroethane:

$$\underset{\text{ethylene}}{C_2H_4} + \underset{\text{hydrogen chloride}}{2HCl} + \underset{\text{oxygen}}{1/2O_2} \longrightarrow \underset{\text{dichloroethane}}{C_2H_4Cl_2} + \underset{\text{water}}{H_2O}$$

If air is used, then a single pass with respect to each feedstock is used and no recycle to the reactor, Figure 28.4a. Thus, the process operates at near stoichiometric feedrates to achieve high conversions. Typically, between 0.7 and 1.0 kg vent, gases are emitted per kilogram of dichloroethane produced[5].

If the air is substituted by pure oxygen, then the problem of the large flow of inert gas is eliminated, Figure 28.4b. Unreacted gases can be recycled to the reactor. This allows oxygen-based processes to be operated with an excess of ethylene, thereby enhancing the HCl conversion without sacrificing ethylene yield. Unfortunately, this introduces a safety problem downstream of the reactor. Unconverted ethylene can create explosive mixtures with the oxygen.

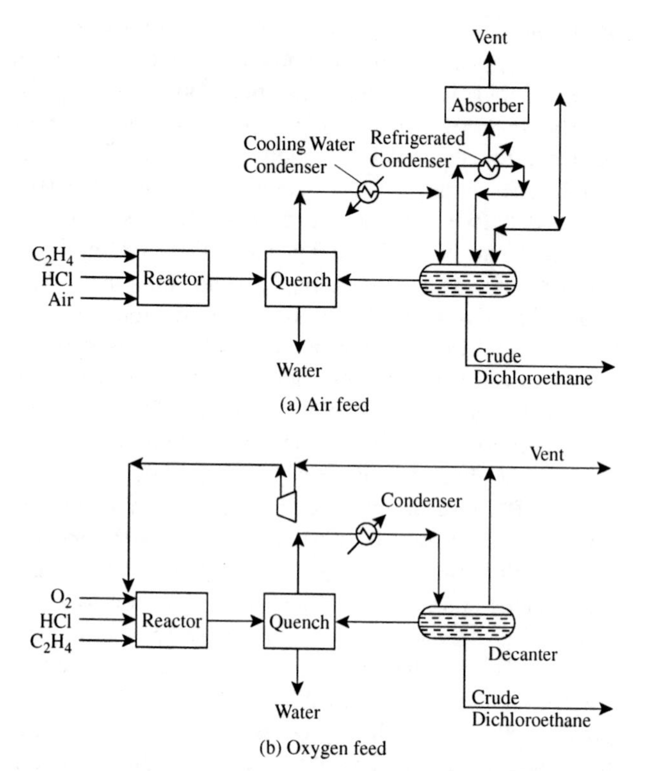

Figure 28.4 The oxychlorination step of the vinyl chloride process. (From Smith R and Petela EA, 1991, *Chem Eng*, No 513:24, reproduced by permission of the Institution of Chemical Engineers).

To avoid explosive mixtures, a small amount of nitrogen can be introduced.

Since nitrogen is drastically reduced in the feed and essentially all ethylene is recycled, only a small purge is required to be vented. This results in a 20- to 100-fold reduction in the size of the purge[5].

3. *Elimination of extraneous materials for separation.* The third option from Figure 28.2 is to eliminate extraneous materials added to the process to carry out separation. The most obvious example would be addition of a solvent, either organic or aqueous. Also, acids or alkalis are sometimes used to precipitate other materials from solution. If these extraneous materials used for separation can be recycled with a high efficiency, there is not a major problem. Sometimes, however, they cannot be recycled efficiently. If this is the case, then waste is created by the discharge of that material. To reduce this waste, alternative methods of separation are needed, such as use of evaporation instead of precipitation.

As an example, consider again the manufacture of vinyl chloride. In the first step of this process, ethylene and chlorine are reacted to form dichloroethane:

$$\underset{\text{ethylene}}{C_2H_4} + \underset{\text{chlorine}}{Cl_2} \longrightarrow \underset{\text{dichloroethane}}{C_2H_4Cl_2}$$

A flowsheet for this part of the vinyl chloride process is shown in Figure 28.5[6]. The reactants (ethylene and chlorine) dissolve in circulating liquid dichloroethane and react in solution to form more dichloroethane. Temperature is maintained between 45 and 65°C, and a small amount of ferric chloride is present to catalyze the reaction. The reaction generates considerable heat.

In early designs, the reaction heat was typically removed by cooling water. Crude dichloroethane was withdrawn from the reactor as a liquid, acid-washed to remove ferric chloride, then neutralized with dilute sodium hydroxide, and purified by distillation. The material used for separation of the ferric chloride can be recycled up to a point, but a purge must be taken. This creates waste streams contaminated with chlorinated hydrocarbons that must be treated prior to disposal.

The problem with the flowsheet shown in Figure 28.5 is that the ferric chloride catalyst is carried from the reactor with the product. This is separated by washing. If a design of reactor can be found that prevents the ferric chloride leaving the reactor, the effluent problems created by the

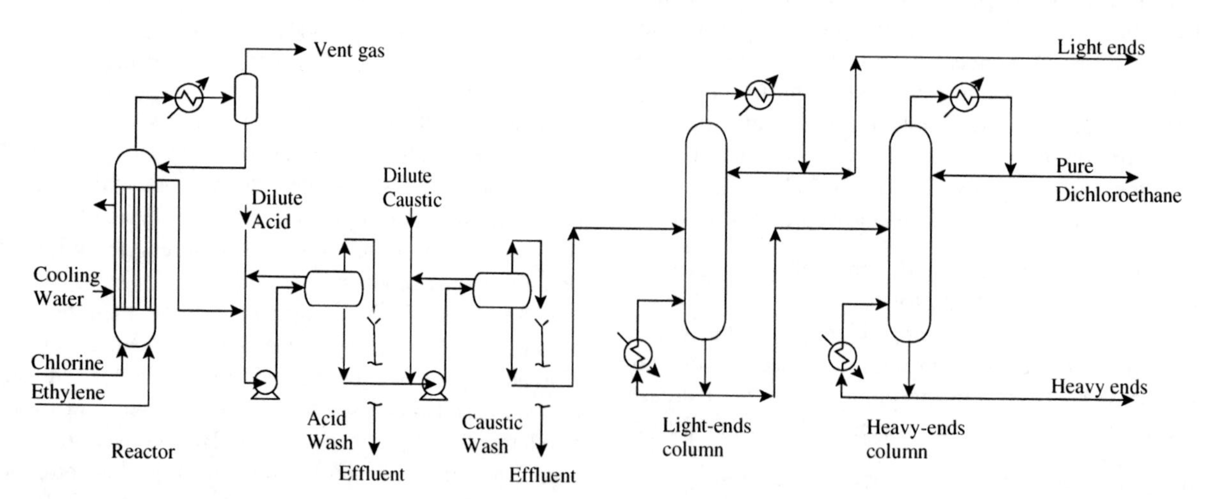

Figure 28.5 The direct chlorination step of the vinyl chloride process using a liquid-phase reactor. (From McNaughton KJ, 1983, *Chem Eng*, **12**: 54, reproduced by permission of McGraw Hill).

washing and neutralization are avoided. Because the ferric chloride is nonvolatile, one way to do this would be to allow the heat of reaction to raise the reaction mixture to boiling point and remove the product as a vapor, leaving the ferric chloride in the reactor. Unfortunately, if the reaction mixture is allowed to boil, then there are two problems:

- ethylene and chlorine are stripped from the liquid phase, giving a low conversion;
- excessive byproduct formation occurs.

This problem is solved in the reactor shown in Figure 28.6[6]. Ethylene and chlorine are introduced into circulating liquid dichloroethane. They dissolve and react to form more dichloroethane. No boiling takes place in the zone where the reactants are introduced, or in the zone of reaction. As shown in Figure 28.6, the reactor has a U-leg in which dichloroethane circulates as a result of gas lift and thermosyphon effects. Ethylene and chlorine are introduced at the bottom of the up-leg, which is under sufficient hydrostatic head to prevent boiling.

The reactants dissolve and immediately begin to react to form further dichloroethane. The reaction is essentially complete at a point only two-thirds up the rising leg. As the liquid continues to rise, boiling begins and finally the vapor–liquid mixture enters the disengagement drum. A slight excess of ethylene ensures essentially 100% conversion of chlorine.

As shown in Figure 28.6, the vapor from the reactor flows into the bottom of a distillation column and high purity dichloroethane is withdrawn as a sidestream, several trays from the column top[6]. The design shown in Figure 28.6 is elegant in that the heat of reaction is conserved to run the separation and no washing of the reactor products is required. This eliminates two aqueous waste streams, which would inevitably carry organics with them, requiring treatment and causing loss of materials.

It is often possible to use the energy release inherent in the process to drive the separation system by improved heat recovery and in so doing carry out the separation at little or no increase in operating costs.

4. *Additional separation and recycling*. Once the possibilities for recycling streams directly have been exhausted, feed purification and extraneous materials for separation eliminated that cannot be recycled efficiently, attention is turned to the fourth option in Figure 28.2, the degree of material recovery from the waste streams that are left. It should be emphasized that once the waste stream is rejected, any valuable material turns into a liability as an effluent material. The level of recovery for such situations needs careful consideration. It may be economic to carry out additional separation of valuable material with a view to recycling that additional recovered material, particularly when the cost of downstream effluent treatment is taken into consideration.

Perhaps the most extreme situation is encountered with purge streams. Purges are used to deal with both feed impurities and byproducts of reaction. In the previous section, reduction of the size of purges by feed purification was considered. However, if it is impractical or uneconomic to reduce the purge by feed purification, or the purge is required to remove a byproduct of reaction, then the additional separation can be considered.

Figure 28.7 shows the basic trade-off to be considered, as additional feed and product material is recovered from waste streams and recycled. As the fractional recovery increases, the cost of the separation and recycle increases. On the other hand, the cost of the lost material decreases. It should be noted that the raw materials cost is a *net* cost, which means that the cost of lost material should be adjusted to either:

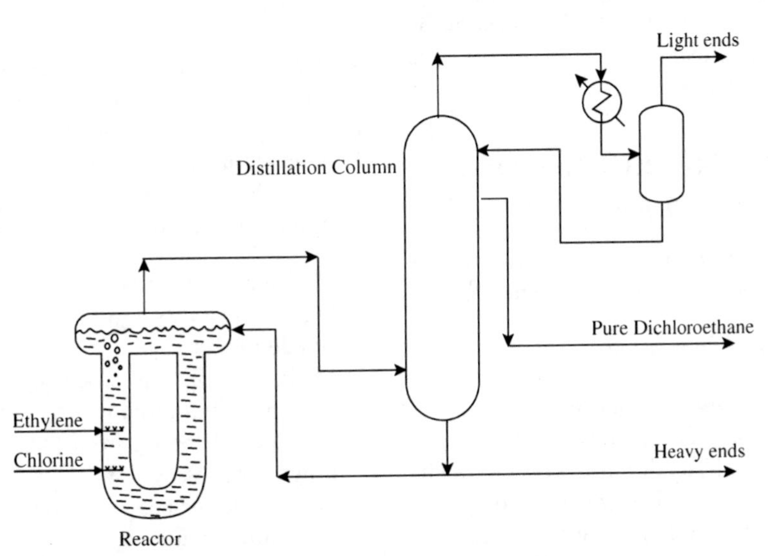

Figure 28.6 The direct chlorination step of the vinyl chloride process using a boiling reactor eliminates the washing and neutralization steps and the resulting effluents.

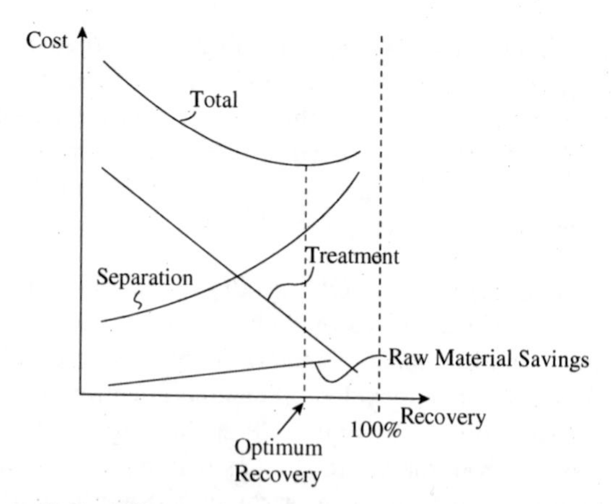

Figure 28.7 Effluent treatment costs should be included with raw materials costs when traded off against separation costs to obtain the optimum recovery. (From Smith R and Petela EA, 1991, *Chem Eng*, No 513:24, reproduced by permission of the Institution of Chemical Engineers).

a. add the cost of waste treatment for unrecovered material, or
b. deduct the fuel value if the recovered material is to be burnt to provide useful heat in a furnace or boiler.

Figure 28.7 shows that the trade-off between separation and net raw materials cost gives an economic optimum recovery. It is possible that significant changes in the degree of recovery can have a significant effect on costs, other than those shown in Figure 28.7 (e.g. reactor costs). If this is the case, then these must also be included in the trade-offs.

It must be emphasized that any energy costs for the separation in the trade-offs shown in Figure 28.7 must be taken within the context of the overall heat integration problem. The separation might after all be driven by heat recovery.

5. *Additional reaction and separation of waste streams.* Sometimes it is possible to carry out further reaction as well as separation on waste streams. Some examples have already been discussed in Chapter 13.

28.4 CLEAN PROCESS TECHNOLOGY FOR PROCESS OPERATIONS

The third source of process waste after the reactor and separation and recycle systems is process operations[7].

1. *Sources of waste in process operations.*

a. *Start-up/shutdown in continuous processes.*
   - Reactors give lower than design conversions.
   - Reactors at nonoptimal conditions produce (additional) unwanted byproducts. This might not only lead to loss of material through additional byproduct formation, but also might prevent the recycling of material produced during the start-up.
   - Separators working at unsteady conditions produce intermediates with compositions that do not allow them to be recycled. Alternatively, if the intermediate can be recycled, a nonoptimal recycle might produce (additional) unwanted byproducts in the reactor.
   - Process intermediates are generated that, because the downstream process is not operational, cannot be processed further.
   - When working at unsteady conditions, separators that normally split useful material from waste streams might lose material unnecessarily to the waste streams.
   - Separators working at unsteady conditions produce products that do not meet the required sales specification.

b. *Product changeover.*
   - In continuous processes, all those sources of process waste associated with start-up and shutdown also apply to product changeover in multiproduct plants.
   - In both batch and continuous processes, it may be necessary to clean equipment to prevent contamination of new product. Materials used for equipment cleaning often cannot be recycled, leading to waste.

c. *Equipment cleaning for maintenance, tank filling and fugitive emissions.* Equipment needs to be cleaned and made safe for maintenance.
   - When process tanks, road tankers, rail tank cars or barges are filled, material in the vapor space is forced out of the tank and lost to atmosphere, as discussed in Chapter 25.
   - Material transfer requires pipework, valves, pumps and compressors. Fugitive emissions occur from pipe flanges, valve glands and pump and compressor seals, as discussed in Chapter 25.

Consider now what can be done, particularly in design, to overcome such waste.

2. *Process operation for clean process technology.* Many of the problems associated with waste from process operations can be mitigated if the process is designed for low inventories of material in the process. This is also compatible with design for inherent safety. Other ways to minimize waste from process operation are[7]:

- Minimize the number of shutdowns by designing for high availability. Install more reliable equipment or standby equipment.
- Design continuous processes for flexible operation, for example, high turndown rate rather than shutdown.
- Consider changing from batch to continuous operation. Batch processes, by their very nature, are always at unsteady-state, and thus difficult to maintain at optimum conditions.
- Install enough intermediate storage to allow reworking of off-specification material.

- Changeover between products causes waste since equipment must be cleaned. Such waste can be minimized by scheduling operation to minimize product changeovers (see Chapter 14).
- Install a waste collection system for equipment cleaning and sampling waste, which allows waste to be segregated and recycled where possible. This normally requires separate sewers for organic and aqueous waste, collecting to sump tanks and recycle or separate and recycle if possible. This was discussed in Chapter 25. If equipment is steamed out during the cleaning process, the plant should allow collection and condensation of the vapors and recycling of materials where possible.
- Reduce losses from fugitive emissions and tank breathing as discussed in Chapter 25.

There are many other sources of waste associated with process operations that can only be taken care of in the later stages of design, or after the plant has been built and became operational. For example, poor operating practice can mean that the process operates under conditions for which it was not designed, leading to waste. Such problems might be solved by an increased level of automation or better management of the process[7]. These considerations are outside the scope of this text.

28.5 CLEAN PROCESS TECHNOLOGY FOR UTILITY SYSTEMS

1. *Utility systems as sources of waste*. The principal sources of utility waste are associated with hot utilities (including cogeneration systems) and cold utilities[8]. Furnaces, steam boilers, gas turbines, diesel engines and gas engines all produce waste from products of combustion. The principal problem here is the emission of carbon dioxide, oxides of sulfur and nitrogen and particulates (metal oxides, unburnt carbon and hydrocarbon). As well as gaseous waste, the combustion of coal produces solid waste as ash. Steam systems and cooling water systems also produce aqueous waste.

The waste streams created by utility systems tend, on the whole, to be less environmentally harmful than process waste. Unfortunately, complacency would be misplaced. Even though utility waste tends to be less harmful than process waste, the quantities of utility waste tend to be larger than process waste. This sheer volume can then result in greater environmental impact than process waste. Gaseous combustion products contribute in various ways to the greenhouse effect, acid rain and can produce a direct health hazard because of the formation of smog. The aqueous waste generated by utility systems can also be a major problem if it is contaminated.

2. *Energy efficiency of the process*. If the process requires a furnace or steam boiler to provide hot utility, then any excessive use of hot utility will produce excessive utility waste through excessive generation of CO_2, NO_x, SO_x and particulates. Improved heat recovery will reduce the overall demand for utilities and hence reduce utility waste.

3. *Local and global emissions*. When considering utility waste, it is tempting to consider only the *local* emissions from the process and its utility system, Figure 28.8a. However, this only gives part of the picture. The emissions generated from central power generation associated with any power import are just as much part of the process as those emissions generated on-site, Figure 28.8b. These emissions should be included in the assessment of utility waste. Thus, *global* emissions are defined to be[9]:

Global emissions = (Emissions from on-site utilities)
+ (Emissions from central power generation corresponding with the amount of electricity imported)
− (Emissions saved at central power generation corresponding with the amount of electricity exported from the site)

This is particularly important when considering the effect that cogeneration has on utility waste.

4. *Cogeneration*. Cogeneration can have a very significant effect on the generation of utility waste. However, great care must be taken to assess the effects on the correct basis. Assessing only the local effects of cogeneration is misleading. Cogeneration increases the local utility emissions since, besides the fuel burnt to supply the heating demand, additional fuel must be burnt to generate the power. It is only when the emissions are viewed on a global

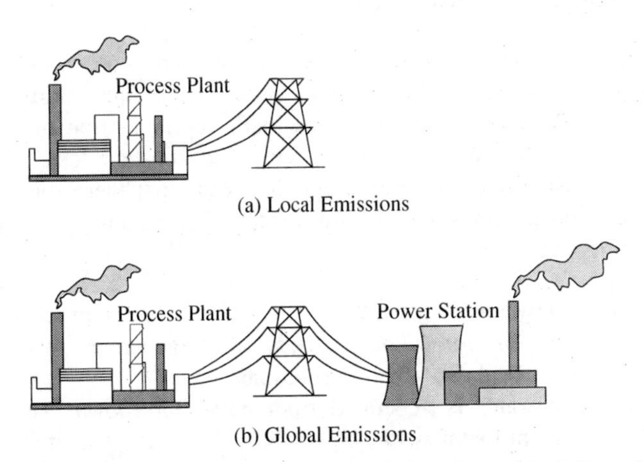

Figure 28.8 Local and global emissions. (From Smith R and Petela EA, 1992, *Chem Eng*, **523**: 32, reproduced by permission of the Institution of Chemical Engineers).

basis and the emissions from central power generation are viewed that the true picture is obtained. Once these are included, on-site cogeneration can make major reductions in global utility waste. The reason for this is that even the most modern central power stations have a poor efficiency of power generation compared to a cogeneration system. Once the other inefficiencies associated with centralized power generation are taken into account, such as distribution losses, the gap between the efficiency of cogeneration systems and centralized power generation widens.

As an example, consider a process that requires a furnace to satisfy its hot utility requirements. Suppose it is a state-of-the-art furnace with a thermal efficiency of 90% producing 300 kg $CO_2 \cdot h^{-1}$ for each megawatt of heat delivered to the process. Power is being imported from centralized generation via the grid. If, instead of the furnace, a gas turbine is installed, this produces 500 kg $CO_2 \cdot h^{-1}$ for each megawatt of heat delivered to the process, an increase in local emissions of 200 kg $CO_2 \cdot h^{-1}$ per megawatt of heat. However, the gas turbine also generates 400 kW of power, replacing that much in centralized generation. If the same power was generated centrally to supplement the furnace, 450 kg $CO_2 \cdot h^{-1}$ would be released from centralized generation, giving a global emission of 750 kg $CO_2 \cdot h^{-1}$ for the furnace plus power from the grid[9].

5. *Fuel switch.* The choice of fuel used in furnaces and steam boilers has a major effect on the gaseous utility waste from products of combustion. For example, a switch from coal to natural gas in a steam boiler can lead to a reduction in carbon dioxide emissions of typically 40% for the same heat released[9]. This results from the lower carbon content of the natural gas. In addition, it is likely that the switch from coal to natural gas will also lead to a considerable reduction in both SO_x and NO_x emissions.

Such a fuel switch, while being desirable in reducing emissions, might be expensive. If the problem is SO_x and NO_x emissions, there are other ways to combat these, as discussed in Chapter 25.

6. *Waste from steam systems.* If steam is used as hot utility, then inefficiencies in the steam system itself cause utility waste. Figure 23.2 shows a schematic representation of a boiler feedwater treatment system. The constant loss of condensate from the steam system means that there must be a constant makeup with freshwater. This makeup causes utility waste:

(a) Wastewater is generated in the deionization process when the ion-exchange beds are regenerated with saline, acid and alkaline solutions.
(b) Wastewater is generated from *boiler blowdown*. The main problem with boiler blowdown is that it is contaminated with water treatment chemicals.
(c) The lost condensate does not create a direct problem since it is likely to be contaminated only with perhaps a few parts per million of amines added to prevent corrosion in the condensate system. The major problems are indirect. The heat loss caused by the condensate loss must ultimately be made up by burning extra fuel and the generation of extra products of combustion.

These sources of waste from the steam system can be reduced by increasing the percentage of condensate returned and by reducing steam generation through increased heat recovery.

7. *Waste from cooling water systems.* Cooling water systems also give rise to wastewater generation. Most cooling water systems recirculate water rather than using "once-through" arrangements. Water is lost from recirculating systems in the cooling tower mainly through evaporation but also, to a much smaller extent, through drift (see Chapter 24). The buildup of solids is prevented by cooling tower blowdown. Cooling tower blowdown is the source of the largest volume of wastewater on many sites. The blowdown will contain corrosion inhibitors, polymers to prevent solid deposition and biocides to prevent the growth of microorganisms.

Cooling tower blowdown can be reduced by improving the energy efficiency of processes, thus reducing the thermal load on cooling towers and by increasing the cycles of concentration. Alternatively, cooling water systems can be switched to air-coolers, which eliminates the problem altogether.

28.6 TRADING OFF CLEAN PROCESS TECHNOLOGY OPTIONS

When tackling effluent treatment problems, it is usually better to remove the problem at source through waste minimization to give clean process technology rather than add on a waste treatment system. Given that many problems can be solved either through waste minimization, or effluent treatment or a combination or both, how can such trade-offs be made? Figure 28.9 shows the superstructure from Figure 26.53, but modified to include options for in-plant clean process technology options. The superstructure allows for the options to be included or bypassed and therefore eliminated from the solution.

Combustion in a thermal oxidizer is the only practical way to deal with many waste streams. This is particularly true of solid and concentrated waste and toxic wastes such as those containing halogenated hydrocarbons, pesticides, herbicides, and so on. Many of the toxic substances encountered resist biological degradation and persist in the natural environment for a long period. Unless they are in dilute aqueous solution, the most effective treatment is usually thermal oxidation.

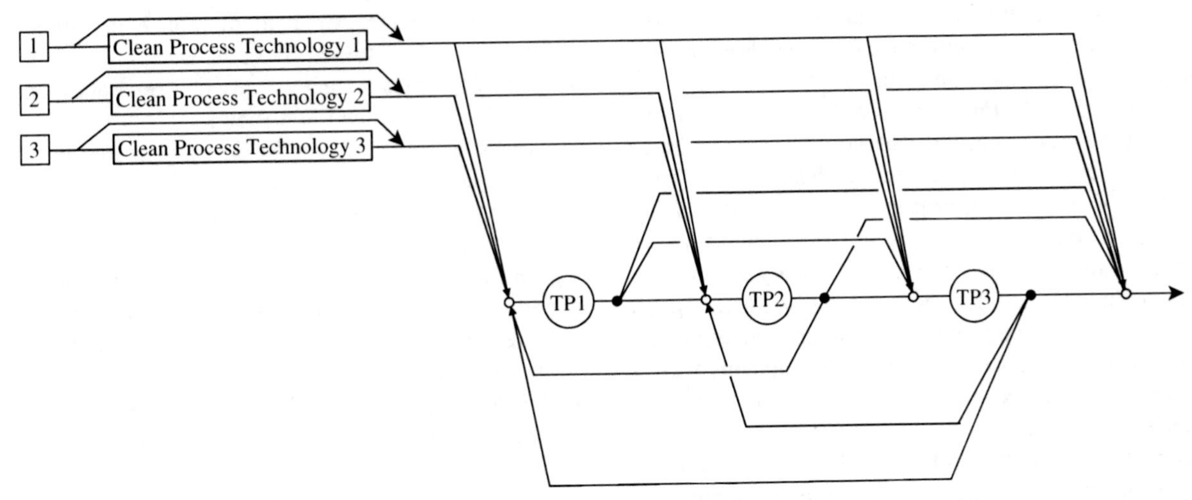

Figure 28.9 Waste minimization projects can be traded off against treatment options by optimization of a superstructure.

Thermal oxidation of toxic materials such as halogenated hydrocarbons, pesticides, herbicides, and so on, requires a sustained temperature of 1100 to 1300°C in an excess of oxygen. The incinerator stack gases will contain acid gases such as hydrogen chloride, oxides of sulfur and oxides of nitrogen, depending on the waste being thermally oxidized. These acid gases require scrubbers to treat the gaseous waste stream. This scrubbing, in turn, produces an aqueous effluent.

Depending on the mix of waste being burnt, the thermal oxidizer may or may not require auxiliary firing from fuel oil or natural gas. The various designs of thermal oxidizer for gaseous waste were discussed in Chapter 25. The designs discussed in Chapter 25 are capable of oxidizing both gaseous and liquid waste, but are not suited to solid waste. If solid waste needs to be oxidized, then two classes are suited to such duties:

1. *Rotary kilns*. Rotary kilns involve a cylindrical refractory-lined shell mounted at a small angle to horizontal and rotated at low speed. Solid material, sludges and slurries are fed at the higher end and flow under gravity along the kiln. Liquids can also be oxidized. Rotary kilns are ideal for treating solid waste but have the disadvantage of high capital and maintenance costs.
2. *Hearth thermal oxidizers*. This type of thermal oxidizer is primarily designed to oxidize solid waste. Solids are moved through the combustion chamber mechanically using a rake.

The policy for waste heat recovery from the flue gas varies between different operators. Thermal oxidizers located on the waste producer's site tend to be fitted with waste heat recovery systems, usually steam generation, which is fed into the site steam mains. *Merchant* operators, who treat other people's waste and operate in isolation remotely from the waste producers, tend not to fit heat recovery systems. Instead, the flue gases tend to be cooled prior to scrubbing and the waste heat used to reheat the flue gases after scrubbing to avoid a visible steam plume from the stack.

While thermal oxidation is the preferred method of disposal for wastes containing high concentrations of organics when their production cannot be avoided, it becomes expensive for aqueous wastes with low concentrations of organics, since auxiliary fuel is required, making the treatment expensive. Weak aqueous solutions of organic material that are still too strong to be treated by biological treatment are better treated by wet oxidation (see Section 26.2).

28.7 LIFE CYCLE ANALYSIS

When utility waste was considered, it was found that to obtain a true picture of the flue gas emissions associated with a process, both the local on-site emissions and those generated by centralized power generation corresponding with the amount of power imported (or exported) need to be included. In the limit, this basic idea can be extended to consider the total emissions (process and utility) associated with the manufacture of a given product in a *life cycle analysis*[10–12]. In life cycle analysis, a cradle-to-grave view of a particular product is taken. The analysis starts with the extraction of the initial raw material from natural resources. The various transformations of the raw materials are followed through to the manufacture of the final consumer product, the distribution and use of the consumer product, recycling of the product, if this is possible, and finally to its eventual disposal. Each step in the life cycle creates waste. Waste generated by transportation and the manufacture and maintenance of processing equipment should also be included.

There are three components in a life cycle analysis[11]:

1. The life cycle is first defined and the complete resource requirements (materials and energy) quantified. This

allows the total environmental emissions associated with the life cycle to be quantified by putting together the individual parts. This defines the *life cycle inventory*.
2. Once the life cycle inventory has been quantified, the effects of the environmental emissions are characterized and assessed in a *life cycle impact analysis*. While the life cycle inventory can, in principle at least, be readily assessed, the resulting impact is far from straightforward to assess. Environmental impacts are usually not directly comparable. For example, how can the production of a kilogram of heavy metal sludge waste be compared with the production of a ton of contaminated aqueous waste? A comparison of two life cycles is required to pick the preferred life cycle.
3. Having attempted to quantify the life cycle inventory and impact, a *life cycle improvement analysis* suggests environmental improvements.

Life cycle analysis, in principle, allows an objective and complete view of the impact of processes and products on the environment[10]. For a manufacturer, life cycle analysis requires an acceptance of responsibility for the impact of manufacturing in total. This means not just the manufacturers' operations, and the disposal of waste created by those operations, but those of raw materials suppliers and those of product users.

To the process designer, life cycle analysis is useful because in focusing exclusively on waste minimization at some point in the life cycle, sometimes problems are created elsewhere in the cycle. The designer can often obtain useful insights by changing the boundaries of the system under consideration to be wider than those of the process being designed.

28.8 CLEAN PROCESS TECHNOLOGY – SUMMARY

The best solution to effluent problems is to not produce the waste in the first place through clean process technology. If waste can be minimized at source, not only are effluent treatment costs reduced but also raw materials costs.

There are three sources of process waste:

1. reactor
2. separation and recycle system
3. process operations

Since the reactor is at the heart of the process, this is where to start when considering clean process technology. The separation and recycle system is next and finally process operations are considered.

Process waste minimization in general terms is a question of:

- changing the reaction path to reduce or eliminate the formation of unwanted byproducts;
- increasing reactor conversion when separation and recycle of unreacted feed is difficult;
- increasing process yields of raw materials through improved selectivity in the reactor;
- reducing catalyst waste by changing from homogeneous to heterogeneous catalysts;
- protecting heterogeneous catalysts from contaminants and extreme conditions that will shorten their life;
- increasing process yields through improved separation and recycling;
- increasing process yields through feed purification to reduce losses in the reactor, separation and recycle system;
- reducing the use of extraneous materials that cannot be recycled with high efficiency;
- reducing process inventories;
- allowing enough immediate storage to rework off-specification material;
- designing for minimum number of shutdowns and product changeovers;
- reducing the use of fluids (aqueous or organic) used for equipment cleaning;
- segregating waste to maximize the potential for recycling;
- reducing losses from fugitive emissions and tank breathing.

The utility system also creates waste through and products of combustion from boilers and furnaces, and so on, and wastewater from water treatment, boiler blowdown, and so on. Utility waste minimization is in general terms a question of

- reducing products of combustion from furnaces, steam boilers and gas turbines by making the process more energy efficient through improved heat recovery;
- reducing wastewater associated with steam generation by both reducing steam usage through improved heat recovery and by making the steam system itself more efficient;
- reducing wastewater associated with cooling water systems.

28.9 EXERCISES

1. A reaction between organic compounds is carried out in the liquid phase in a stirred-tank reactor in the presence of excess formaldehyde. The organic reactants are nonvolatile in comparison with the formaldehyde. The reactor is vented to atmosphere via an absorber to scrub any organic material carried from the reactor. The absorber is fed with freshwater and the water from the absorber rejected to effluent. The major contaminant in the aqueous waste from the absorber is formaldehyde.

a. If the absorption system is kept, how can the volume of aqueous waste be reduced from the system?

b. How might the organic waste to effluent be reduced at source?

2. A chemical manufacturing site has a large aqueous effluent flowrate that passes through biological treatment before discharge to a river. Although the outlet concentrations of pollutants from the biological treatment are within permitted limits, the temperature at the outlet is too high. The maximum temperature permitted for discharge is 30°C, whereas the current outlet is 40°C. The inlet temperature to biological treatment is 36°C. A temperature rise of 4°C occurs across biological treatment. Heat is generated within the biological treatment from the reactions but is also lost to the environment directly. The longer the residence time in biological treatment, the greater the heat loss. It has been proposed to solve the problem by installing a cooling tower downstream biological treatment. A better solution would be to solve the problem at source.

a. What factors could be changed at the inlet to the biological treatment to alleviate the problem?

b. The processes that create the effluent are batch in nature and involve various washing operations at different temperatures. What changes should be sought to solve the problem?

3. A chemical production site producing a variety of specialty chemicals has a problem with its aqueous effluent. The site produces aqueous effluent that is currently discharged without treatment. The effluent has a high load of organic material and has a low pH. The regulatory authorities have demanded a reduction in the organic load before discharge of 90%, together with neutralization. An effluent treatment system has been designed and costed. The treatment system consists of collecting all of the effluent steams together, followed by neutralization using lime and then biological treatment. The cost of both the neutralization and biological treatment operations are proportional to the volume of effluent to be treated. The cost of biological treatment also increases with the load of organic material. The cost of the treatment system is unacceptable, and the company is prepared to consider an alternative solution based on waste minimization.

a. The worst effluent stream, in terms of its organic load, comes from Operation 1 of Plant A. This effluent is created when an organic product is washed free of salts in an extraction operation. This is done by mixing the product with water in a tank followed by separation of the water from the organic product by settling in a decanter. The washing operation picks up organic product as well as the salts. The salts are extremely soluble, whereas the organic product is sparingly soluble. The effluent leaving this operation is saturated with organic contaminants, but well below saturation for the salts. Taking this operation in isolation, what can be done to reduce the effluent volume and organic load?

b. Another operation, Operation 2, in Plant A uses water in a cooling circuit. The water is used for condensation of organic vapor by direct contact. In this operation, the organic vapor is passed through a vessel into which is sprayed the cooling water. The resulting two-phase mixture is again separated by settling in a decanter. The water becomes saturated with organic contaminants and is recirculated through a cooling tower. A purge must be taken from the cooling circuit, which is sent to effluent. This purge is another highly contaminated effluent stream from Plant A. What can be done to reduce the effluent form Plant A as a whole by integrating Operations 1 and 2? Explain what effect your suggestions are likely to have on the volume and the load.

c. Plant B uses water to scrub hydrogen chloride from a vent. The resulting water is highly acidic but not contaminated with organic material. The other effluents on the site are essentially neutral. Can anything be done to reduce the cost of treating the effluent?

d. Plant C produces an aqueous effluent contaminated with organic contaminants. In addition, there is no policy for recovery of steam condensate resulting from the use of steam for heating. Large quantities of steam condensate are sent to drain. There is also a large cooling tower on-site that requires a large water makeup to compensate for evaporative losses. The blowdown from the cooling tower is sent to drain and contains no organic contaminants. Taking the site as a whole, what, in addition to the measures suggested so far, can be done to reduce effluent treatment costs?

REFERENCES

1. Smith R and Petela EA (1991) Waste Minimization in the Process Industries Part 1 The Problem, *Chem Eng*, **506**: 31.
2. Smith R and Petela EA (1991) Waste Minimization in the Process Industries Part 2 Reactors, *Chem Eng*, **509/510**: 12.
3. Smith R and Petela EA (1991) Waste Minimization in the Process Industries Part 3 Separation and Recycle Systems, *Chem Eng*, **513**: 24.
4. Chowdhury J and Leward R (1984) Oxygen Breathes More Life Into CPI Processing, *Chem Eng*, **19**: 30.
5. Reich P (1976) Air or Oxygen for VCM? *Hydrocarbon Process*, **March**: 85.
6. McNaughton KJ (1983) Ethylene Dichloride Process, *Chem Eng*, **12**: 54.
7. Smith R and Petela EA (1992) Waste Minimization in the Process Industries Part 4 Process Operations, *Chem Eng*, **517**: 9.
8. Smith R and Petela EA (1992) Waste Minimization in the Process Industries Part 5 Utility Waste, *Chem Eng*, **523**: 16.
9. Smith R and Delaby O (1991) Targeting Flue Gas Emissions, *Trans IChemE*, **69A**: 492.
10. Hindle P and Payne AG (1991) Value-Impact Assessment, *Chem Eng*, **28**: 31.
11. Curran MA (1993) Broad-Based Environmental Life Cycle Assessment, *Environ Sci Technol*, **27**: 430.
12. Guinee JB, Udo HA and Huppes G (1993) Quantitative Life Cycle Assessment of Products, *J Cleaner Production*, **1**: 3.

29 Overall Strategy for Chemical Process Design and Integration

29.1 OBJECTIVES

The purpose of chemical processes is not to make chemicals: the purpose is to make money. However, the profit must be made as part of a sustainable industrial activity. Chemical processes should be designed as part of a sustainable industrial activity that retains the capacity of ecosystems to support both industrial activity and life into the future. Sustainable industrial activity must meet the needs of the present without compromising the needs of future generations. For chemical process design, this means that processes should use raw materials as efficiently as is economic and practicable, both to prevent the production of waste that can be environmentally harmful and to preserve the reserves of raw materials as much as possible. Processes should use as little energy as economic and practicable, both to prevent the buildup of carbon dioxide in the atmosphere from burning fossil fuels and to preserve reserves of fossil fuels. Water must also be consumed in sustainable quantities that do not cause deterioration in the quality of the water source and the long-term quantity of the reserves. Aqueous and atmospheric emissions must not be environmentally harmful, and solid waste to landfill must be avoided.

Relying on methods of waste treatment is usually not adequate, since waste treatment processes tend not so much to solve the waste problem but simply to move it from one place to another. Chemical processes also must not present significant short-term or long-term hazards, either to the operating personnel or to the community.

When developing a chemical process design, it helps if it is recognized that there is a hierarchy that is intrinsic to chemical processes. Design starts at the reactor. The reactor design dictates the separation and recycle problem. The reactor design and separation problem together dictate the heating and cooling duties for the heat exchanger network. Those heating and cooling duties that cannot be satisfied by heat recovery dictate the need for external heating and cooling utilities. All of these issues combine to dictate the need for water and effluent treatment. This hierarchy is represented by the layers in the onion diagram.

Following this hierarchy, all too often, safety, health and environmental considerations are left to the final stages of design. This approach leaves much to be desired, since early decisions made purely for process reasons often can lead to problems of safety, health and environment that require complex solutions. It is better to consider them as the design progresses. Designs that avoid the need for hazardous materials, or use less of them, or use them at lower temperatures and pressures, or dilute them with inert materials will be inherently safe and not require elaborate treatment systems. These considerations need to be addressed as the design progresses.

29.2 THE HIERARCHY

1. *Choice of reactor.* The first and usually most important decisions to be made are those for the reactor type and its operating conditions. In choosing the reactor, the overriding consideration is usually raw materials efficiency (bearing in mind materials of construction, safety, etc.). Raw materials costs are usually the most important costs in the whole process. Also, any inefficiency in raw materials use is likely to create waste streams that become an environmental problem.

The design of the reactor usually interacts strongly with the rest of the flowsheet. Hence, a return must be made to the reactor design when the process design has progressed further.

2. *Choice of separator.* For a heterogeneous mixture, separation usually can be achieved by phase separation. Such phase separation normally should be carried out before any homogeneous separation. Phase separation tends to be easier and usually should be done first.

Distillation is by far the most commonly used method for the separation of homogeneous fluid mixtures. No attempt should be made to optimize pressure, reflux ratio or feed condition of distillation in the early stages of design. The optimal values will almost certainly change later once heat integration with the overall process is considered.

The most common alternative to distillation for the separation of low-molecular-weight materials is absorption. Liquid flowrate, temperature and pressure are important variables to be set, but no attempt should be made to carry out any optimization at this stage. Other commonly used separation methods are adsorption and membranes.

As with distillation and absorption, when evaporators and dryers are chosen, then no attempt should be made to carry out any optimization at this stage in the design.

Chemical Process Design and Integration R. Smith

ISBNs: 0-471-48680-9 (HB); 0-471-48681-7 (PB)

3. *Distillation sequencing*. The separation of homogeneous nonazeotropic mixtures using distillation usually offers the degree of freedom to choose the distillation sequence. The choice between different sequences can be made on the basis of total vapor load, energy consumption, refrigeration shaft power for low-temperature systems, or total cost. However, there is often little to choose between the best few sequences in terms of such measures of system performance if simple distillation columns are used.

Complex column arrangements, such as thermally coupled designs, offer large potential savings in energy when compared with sequences of simple columns. The partition column or dividing-wall column also offers large potential savings in capital cost. However, it is recommended that complex column arrangements should only be considered on a second pass through the design after first establishing a complete design with simple columns. Once this first complete design is established, then thermally coupled arrangements can be evaluated in the context of the overall heat-integrated design.

If the system forms azeotropes, then the azeotropic mixtures can be separated by exploiting the change in azeotropic composition with pressure, or the introduction of an entrainer or membrane to change the relative volatility in a favorable way. If an entrainer is used, then efficient recycle of the entrainer material is necessary for an acceptable design. In some cases, the formation of two liquid phases can be exploited in heterogeneous azeotropic distillation.

4. *The synthesis of reaction-separation systems*. The recycling of material is an essential feature of most chemical processes. The use of excess reactants, diluents, solvents or heat carriers in the reactor design has a significant effect on the flowsheet recycle structure. Sometimes, the recycling of unwanted byproduct to the reactor can inhibit its formation at the source.

Batch processes can be synthesized by first synthesizing a continuous process and then converting it to batch operation. A Gantt (time-event) diagram can be used to identify the scope for improved equipment utilization and the need for intermediate storage.

5. *Heat exchanger network and utilities targets*. Having established a design for the reaction and separation and recycle, the material and energy balance is known. This allows the hot and cold streams for the heat recovery problem to be defined. Energy targets can then be calculated directly from the material and energy balance. It is not necessary to design a heat exchanger network in order to establish the energy costs. Alternative utility scenarios and cogeneration schemes can be screened quickly and conveniently using the grand composite curve.

Targets also can be set for total heat exchange area, number of units, and number of shells for 1–2 shell-and-tube heat exchangers. These can be combined to establish targets for capital costs, taking into account mixed materials of construction, pressure rating and equipment type. Furthermore, the targets for energy and capital cost can be optimized to produce an optimal setting for the capital/energy trade-off before any network design is carried out.

Although it is not necessary to design the heat exchanger network in order for the design to progress, it is sometimes desirable to carry out a preliminary design to ensure that there are no significant features of the design that are unacceptable. If there are unacceptable features, the targets will have to be modified by inclusion of constraints.

6. *Process changes for improved heat integration*. Process changes can be exploited to allow the energy targets to be reduced. The ultimate reference in guiding process changes is the plus/minus principle. The appropriate placement of the major items of equipment in relation to the heat recovery pinch is as follows:

- Exothermic reactors should be above the pinch.
- Endothermic reactors should be below the pinch.
- Distillation columns, evaporators and dryers should be above the pinch, below the pinch but not across the pinch.

The grand composite curve can be used to quantitatively assess the appropriate placement of reactors and separators. If reactors and separators are not appropriately placed, then the plus/minus principle can be used to direct changes to bring about improvements through, for example, pressure change. If a reactor is not appropriately placed, then it is more likely that the rest of the process would be changed to bring about appropriate placement rather than change the reactor design.

The sequence of distillation columns should be addressed again at this stage and the possibility of introducing complex configurations considered. Prefractionator arrangements (both with and without thermal coupling) can be used to replace direct or indirect distillation pairings. Alternatively, direct pairings can be replaced by side-rectifiers and indirect pairings replaced by side-strippers. Partition or dividing-wall columns can make significant reductions in both capital and operating costs when compared to a conventional distillation column pairings.

7. *Heat exchanger network design*. Having explored the major degrees of freedom, the material and energy balance is fixed, and the hot and cold streams that define the heat exchanger network are fixed. The heat exchanger network can then be defined.

The pinch design method is a step-by-step approach that allows the designer to interact as the design progresses. For more complex network designs, especially those involving many constraints, mixed equipment specifications, and so on, design methods based on the optimization of a superstructure can be used.

8. *Economic trade-offs*. Interactions between the reactor and the rest of the process are extremely important. Reactor conversion is often the most significant optimization variable, because it tends to influence most operations through the process. Also, when inert material is present in the recycle, the concentration of inert material is another important optimization variable, again influencing operations throughout the process.

In carrying out these optimizations, targets can be used for the energy and capital cost of the heat exchanger network. This is often the only practical way to carry out these optimizations, since changes in reactor conversion and recycle inert concentration change the material and energy balance of the process, which in turn changes the heat recovery problem. Each change in the material and energy balance, in principle, calls for a different heat exchanger network design. Furnishing a new heat exchanger network design for each setting of reactor conversion and recycle inert concentration is just not practical. On the other hand, targets for energy and capital cost of the heat exchanger network are, by comparison, easily generated.

9. *Steam and cogeneration systems*. Most process heating is provided from the distribution of steam around sites at various pressure levels. Normally, two or three steam mains will distribute steam at different pressures around the site. The steam system is not only important from the point of view of process heating but is also used to generate a significant amount of power on the site. Steam turbines are used to convert part of the energy of the steam into power, and can be configured in different ways. Gas turbines are available in a wide range of sizes but are restricted to standard frame sizes. The hot exhaust gas is useful for raising steam, and the temperature of the exhaust gas can be increased by using supplementary firing. The highest-pressure steam main is normally used for power generation, rather than process heating. Steam is expanded from high to lower levels by either steam turbines or expansion valves.

The site power-to-heat ratio is very important in determining the most appropriate cogeneration system for the site.

Complex steam systems usually feature many important degrees of freedom to be optimized. To establish the steam costs for retrofit of site processes requires an optimization model to be developed. This allows the steam loads for process heating to be gradually decreased and the steam system reoptimized at each setting. The result in cost savings establishes the true cost of steam for retrofit projects aiming to reduce steam consumption on the site.

Drivers are required for many different types of process machine on a site. Power can be generated and distributed to drive electric motors or direct drives can be used. A combination of direct drives and electric motors is usually the best solution for the site as a whole.

10. *Water and effluent treatment*. Increasing awareness regarding the problems of overextraction of water and increasingly strict discharge regulations mean that water consumption and wastewater generation should be reduced through reuse, regeneration reuse and regeneration recycling.

Distributed effluent treatment requires that a philosophy of design be adopted that segregates effluent for treatment wherever appropriate and combines it for treatment where appropriate.

When viewing effluent treatment methods, it is clear that the basic problem of safely disposing of waste material is, in many cases, not so much solved but moved from one place to another. If a method of treatment can be used that allows material to be recycled or reused in someway, then the waste problem is truly solved. However, if the treatment simply concentrates the waste as concentrated liquid, slurry or solid in a form that cannot be recycled, then it will still need to be disposed of. Landfill disposal of such waste is increasingly unacceptable, and thermal oxidation causes pollution through products of combustion and liquors from scrubbing systems. The best method for dealing with effluent problems is to solve the problem at source by waste minimization.

29.3 THE FINAL DESIGN

Although the sequence of the design tends to follow the onion diagram in Figure 1.7, the design rarely can be taken to a successful conclusion by a single pass. More often, there is a flow of the design process in both directions. This follows from the fact that decisions are made in the early stages of design on the basis of incomplete information. As more detail is added to the design with a more complete picture emerging, the decisions might need to be readdressed, moving back to early decisions.

As the flowsheet becomes more firmly defined, the detailed process and mechanical design of the equipment can progress. The control scheme must be added and detailed hazard and operability studies carried out. All this is beyond the scope of the present text. However, all these considerations might require the flowsheet to be readdressed if problems are uncovered.

Appendix A Annualization of Capital Cost

Derivation of Equation 2.7 is as follows[1]. Let

P = present worth of estimated capital cost

F = future worth of estimated capital cost

i = fractional interest rate per year

n = number of years

After the first year, the future worth F of the capital cost present value P is given by:

$$F(1) = P + Pi = P(1 + i) \quad \text{(A.1)}$$

After the second year, the worth is:

$$F(2) = P(1 + i) + P(1 + i)i = P(1 + i)^2 \quad \text{(A.2)}$$

After the third year, the worth is:

$$F(3) = P(1 + i)^2 + P(1 + i)^2 i = P(1 + i)^3 \quad \text{(A.3)}$$

After year n, the worth is:

$$F(n) = P(1 + i)^n \quad \text{(A.4)}$$

Equation A.4 is normally written as:

$$F = P(1 + i)^n \quad \text{(A.5)}$$

Take the capital cost and spread it as a series of equal annual payments A made at the end of each year, over n years. The first payment gains interest over $(n-1)$ years, and its future value after $(n-1)$ years is:

$$F = A(1 + i)^{n-1} \quad \text{(A.6)}$$

The future worth of the second annual payment after $(n-2)$ years is:

$$F = A(1 + i)^{n-2} \quad \text{(A.7)}$$

The combined worth of all the annual payments is:

$$F = A[(1 + i)^{n-1} + (1 + i)^{n-2} + (1 + i)^{n-3} + \cdots + (1 + i)^{n-n}] \quad \text{(A.8)}$$

Multiplying both sides of this equation by $(1 + i)$ gives:

$$F(1 + i) = A[(1 + i)^n + (1 + i)^{n-1} + (1 + i)^{n-2} + \cdots + (1 + i)] \quad \text{(A.9)}$$

Subtracting the Equations A.8 and A.9 gives:

$$F(1 + i) - F = A[(1 + i)^n - 1] \quad \text{(A.10)}$$

which on rearranging gives:

$$F = \frac{A[(1 + i)^n - 1]}{i} \quad \text{(A.11)}$$

Combining Equation A.11 with Equation A.5 gives:

$$A = \frac{P[i(1 + i)^n]}{(1 + i)^n - 1} \quad \text{(A.12)}$$

Thus, Equation 2.7 is obtained.

REFERENCES

1. Holland FA, Watson FA and Wilkinson, JK (1983) *Introduction to Process Economics*, 2nd Edition, Wiley, New York.

Chemical Process Design and Integration R. Smith
 ISBNs: 0-471-48680-9 (HB); 0-471-48681-7 (PB)

Appendix B Gas Compression

B.1 RECIPROCATING COMPRESSORS

Reciprocating compressors compress gases by a piston moving backwards and forwards in a cylinder. Valves control the flow of low-pressure gas into the cylinder and high-pressure gas out of the cylinder. The mechanical work to compress a gas is the product of the external force acting on the gas and the distance through which the force moves. Consider a cylinder with cross-sectional area A containing a gas to be compressed by a piston. The force exerted on the gas is the product of the pressure (force per unit area) and the area A of the piston. The distance the piston travels is the volume V of the cylinder divided by the area A. Thus:

$$W = \int_{V_1}^{V_2} PAd\left(\frac{V}{A}\right)$$

$$= \int_{V_1}^{V_2} PdV \tag{B.1}$$

where W = work for gas compression
P = pressure of the gas
V = volume of the gas
A = area of the cylinder and piston

For gas compression, the final volume V_2 is less than the initial volume V_1 and the work of compression is negative. A compressor adds energy to the gas by doing work. A simple ideal compression process is shown in Figure B.1. In compressing the gas from pressure P_1 and volume V_1 to pressure P_2 and volume V_2, the work as defined by Equation B.1 is the area under the graph. The integral in Equation B.1 can be transformed from integration over V to integration over P by considering the areas in Figure B.1:

$$W = \int_{V_1}^{V_2} PdV$$

$$= -\int_{V_2}^{V_1} PdV$$

$$= -\left[\int_{P_1}^{P_2} VdP - P_2V_2 + P_1V_1\right]$$

$$= P_2V_2 - P_1V_1 - \int_{P_1}^{P_2} VdP \tag{B.2}$$

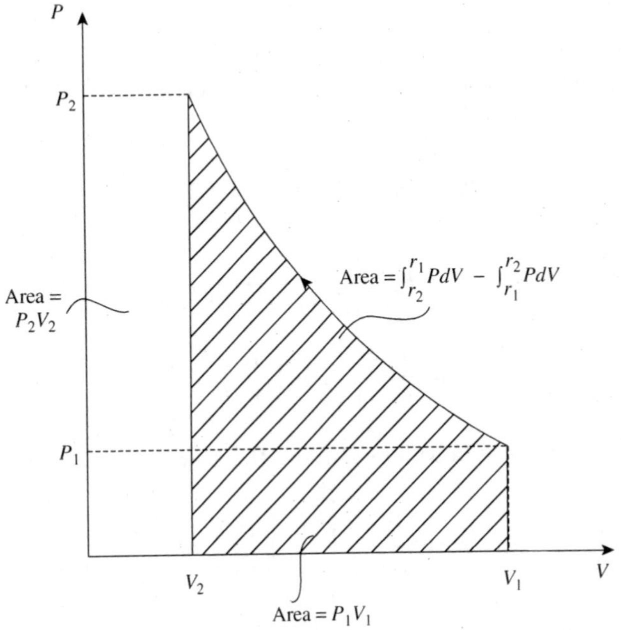

Figure B.1 A simple ideal compression process.

Thus:

$$\int_{V_1}^{V_2} PdV + P_1V_1 - P_2V_2 = -\int_{P_1}^{P_2} VdP \tag{B.3}$$

At this stage, the compression process is considered to be frictionless.

To evaluate the integral in Equation B.1 requires the pressure to be known at each point along the compression path. In principle, compression could be carried out either at constant temperature or adiabatically. Most compression processes are carried out close to adiabatic conditions. Adiabatic compression of an ideal gas along a thermodynamically reversible (isentropic) path can be expressed as:

$$PV^{\gamma} = \text{constant} \tag{B.4}$$

where $\gamma = C_P/C_V$
$= \dfrac{C_P}{(C_P - R)}$ for an ideal gas
C_P = heat capacity at constant pressure
C_V = heat capacity at constant volume

Thus, from Equation B.4 for an adiabatic ideal gas (isentropic) compression:

$$\frac{P_1}{P_2} = \left(\frac{V_2}{V_1}\right)^{\gamma} \tag{B.5}$$

Chemical Process Design and Integration R. Smith
 ISBNs: 0-471-48680-9 (HB); 0-471-48681-7 (PB)

where P_1, P_2 = initial and final pressures
V_1, V_2 = initial and final volumes

A general ideal gas adiabatic (isentropic) compression process is given by:

$$P = \frac{P_1 V_1^\gamma}{V^\gamma} \tag{B.6}$$

where P and V are the pressure and volume, starting from initial conditions of P_1 and V_1. Substituting Equation B.6 into Equation B.1 gives:

$$\begin{aligned} W &= P_1 V_1^\gamma \int_{V_1}^{V_2} \left(\frac{1}{V^\gamma}\right) dV \\ &= P_1 V_1^\gamma \left[-\frac{1}{(\gamma-1)V^{\gamma-1}}\right]_{V_1}^{V_2} \\ &= \frac{P_1 V_1}{\gamma - 1}\left[1 - \left(\frac{V_1}{V_2}\right)^{\gamma-1}\right] \end{aligned} \tag{B.7}$$

Combining Equations B.7 and B.5 gives:

$$W = \frac{P_1 V_1}{\gamma - 1}\left[1 - \left(\frac{P_2}{P_1}\right)^{\frac{\gamma-1}{\gamma}}\right] \tag{B.8}$$

Equation B.8 only considers the work accompanying a change of state. In a reciprocating compressor, these changes form only one step in a cycle of changes. Figure B.2 represents the pressure and volume changes that occur in the cylinder of an ideal reciprocating compressor.

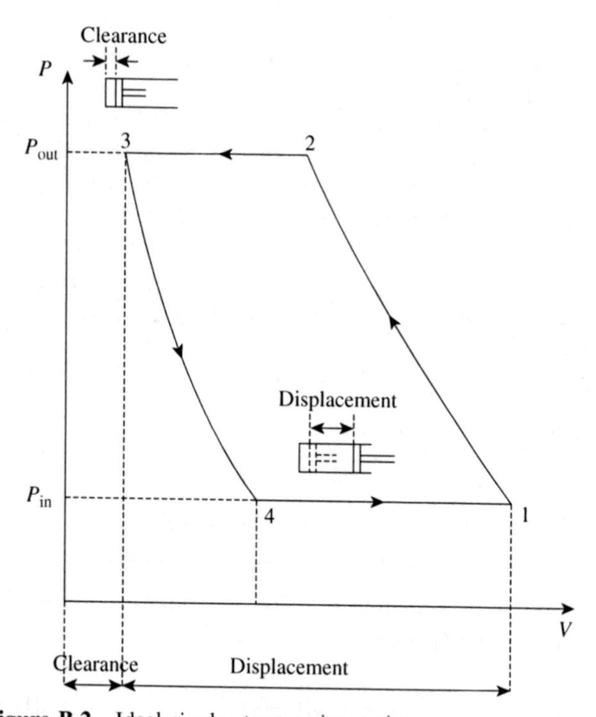

Figure B.2 Ideal single-stage reciprocating gas compressor.

Between Points 1 and 2 in Figure B.2, the intake and discharge valves are closed and the gas in the cylinder is compressed from P_1 to P_2. When the pressure reaches P_2, the discharge valve opens and the gas is pushed from the cylinder between Points 2 and 3 in Figure B.2. Between Points 3 and 4, the intake and discharge valves are closed and any gas remaining in the cylinder is expanded to the intake pressure of P_1. Between Points 4 and 1, the intake valve opens and the suction stroke draws gas into the cylinder at pressure P_1. The total work for the cycle is the sum of the work for the four steps. The work required by the compression is often termed *shaft work* W_S. Thus:

$$W_S = W_{12} + W_{23} + W_{34} + W_{41} \tag{B.9}$$

where

$$W_{12} = \int_{V_1}^{V_2} P\,dV$$

$$W_{23} = \int_{V_2}^{V_3} P\,dV = P_3V_3 - P_2V_2$$

$$W_{34} = \int_{V_3}^{V_4} P\,dV$$

$$W_{41} = \int_{V_4}^{V_1} P\,dV = P_1V_1 - P_4V_4$$

Substituting in Equation B.9 gives:

$$\begin{aligned} W_S &= \int_{V_1}^{V_2} P\,dV + P_3V_3 - P_2V_2 \\ &\quad + \int_{V_3}^{V_4} P\,dV + P_1V_1 - P_4V_4 \\ &= \left[\int_{V_1}^{V_2} P\,dV + P_1V_1 - P_2V_2\right] \\ &\quad + \left[\int_{V_3}^{V_4} P\,dV + P_3V_3 - P_4V_4\right] \end{aligned} \tag{B.10}$$

Combining Equations B.10 and B.3 gives:

$$\begin{aligned} W_S &= -\int_{P_1}^{P_2} V\,dP - \int_{P_3}^{P_4} V\,dP \\ &= -\int_{P_1}^{P_2} V\,dP + \int_{P_4}^{P_3} V\,dP \end{aligned} \tag{B.11}$$

For an adiabatic ideal gas (isentropic) compression or expansion:

$$\begin{aligned} \int_{P_1}^{P_2} V\,dP &= P_1^{1/\gamma} V_1 \int_{P_1}^{P_2} \left(\frac{1}{P^{1/\gamma}}\right) dP \\ &= P_1^{1/\gamma} V_1 \left[-\frac{1}{(1/\gamma - 1)P^{1/\gamma - 1}}\right]_{P_1}^{P_2} \end{aligned}$$

$$= \frac{\gamma}{1-\gamma} P_1 V_1 \left[1 - \left(\frac{P_2}{P_1}\right)^{\frac{\gamma-1}{\gamma}}\right] \quad \text{(B.12)}$$

Thus, from Equations B.11 and B.12:

$$W_S = -\frac{\gamma}{1-\gamma} P_1 V_1 \left[1 - \left(\frac{P_2}{P_1}\right)^{\frac{\gamma-1}{\gamma}}\right] + \frac{\gamma}{1-\gamma} P_4 V_4 \left[1 - \left(\frac{P_3}{P_4}\right)^{\frac{\gamma-1}{\gamma}}\right] \quad \text{(B.13)}$$

Given that $P_1 = P_4 = P_{in}$, $P_2 = P_3 = P_{out}$ and $(V_1 - V_4) = V_{in}$:

$$W_S = \left(\frac{\gamma}{\gamma-1}\right) P_{in} V_{in} \left[1 - \left(\frac{P_{out}}{P_{in}}\right)^{\left(\frac{\gamma-1}{\gamma}\right)}\right] \quad \text{(B.14)}$$

Equation B.14 is the work required for an ideal adiabatic (isentropic) compression. To account for inefficiencies in the compression process and the mechanical inefficiency, the isentropic compression efficiency is introduced:

$$W = \left(\frac{\gamma}{\gamma-1}\right) \frac{P_{in} V_{in}}{\eta_{IS}} \left[1 - \left(\frac{P_{out}}{P_{in}}\right)^{\left(\frac{\gamma-1}{\gamma}\right)}\right] \quad \text{(B.15)}$$

where W = work required for gas compression
P_{in}, P_{out} = inlet and outlet pressures
V_{in} = inlet gas volume
η_{IS} = isentropic efficiency
γ = heat capacity ratio C_P/C_V

The work for a real adiabatic compression can also be calculated from the difference between the total enthalpy of the outlet and inlet flows:

$$W = H_{in} - H'_{out} = \frac{H_{in} - H_{out}}{\eta_{IS}} \quad \text{(B.16)}$$

where H_{in} = total enthalpy of the inlet stream
H_{out} = total enthalpy of the outlet stream for an isentropic compression
H'_{out} = total enthalpy of the outlet stream for a real compression

If the heat capacity is constant, then:

$$W = \frac{CP(T_{in} - T_{out})}{\eta_{IS}} \quad \text{(B.17)}$$

where CP = total heat capacity (product of mass flowrate and specific heat capacity)
T_{in} = temperature of the inlet stream
T_{out} = temperature of the outlet stream for an isentropic compression

By definition, the isentropic efficiency η_{IS} is given by:

$$\eta_{IS} = \frac{H_{in} - H_{out}}{H_{in} - H'_{out}} \quad \text{(B.18)}$$

If the heat capacity is assumed to be constant:

$$\eta_{IS} = \frac{CP(T_{in} - T_{out})}{CP(T_{in} - T'_{out})} = \frac{T_{in} - T_{out}}{T_{in} - T'_{out}} \quad \text{(B.19)}$$

where T_{in} = temperature of the inlet stream
T_{out} = temperature of the outlet stream for an isentropic compression
T'_{out} = temperature of the outlet stream for a real compression

To obtain the temperature rise for an ideal gas isentropic compression, substitute $P = RT/V$ in Equation B.5 to give:

$$\frac{T_{in}}{T_{out}} = \left(\frac{V_{out}}{V_{in}}\right)^{\gamma-1} \quad \text{(B.20)}$$

Combining Equations B.5 and B.20 gives

$$\frac{T_{out}}{T_{in}} = \left(\frac{P_{out}}{P_{in}}\right)^{\frac{\gamma-1}{\gamma}} \quad \text{(B.21)}$$

Equation B.21 assumes the compression to be adiabatic ideal gas (isentropic) compression. In practice, the compression will be neither perfectly adiabatic nor ideal. To allow for this, the gas compression can be assumed to follow a *polytropic* compression represented by the empirical expression:

$$PV^n = \text{constant} \quad \text{(B.22)}$$

where n = polytropic coefficient

For $n = \gamma$, Equation B.22 reduces to the expression for an adiabatic ideal gas. Thus, for a polytropic compression:

$$\frac{P_1}{P_2} = \left(\frac{V_2}{V_1}\right)^n \quad \text{(B.23)}$$

A polytropic compression is neither adiabatic nor isothermal, but specific to the physical properties of the gas and the design of the compressor. The polytropic coefficient n must therefore be determined experimentally. If the initial and final conditions for a given compression process are known, then n can be determined from a rearrangement of Equation B.23:

$$n = \frac{\ln(P_2/P_1)}{\ln(V_1/V_2)} \quad \text{(B.24)}$$

For an isentropic compression:

$$T_{out} = T_{in}\left(\frac{P_{out}}{P_{in}}\right)^{\frac{\gamma-1}{\gamma}} \tag{B.25}$$

If the real compression is assumed to follow a polytropic compression:

$$T_{out} = T_{in}\left(\frac{P_{out}}{P_{in}}\right)^{\frac{n-1}{n}} \tag{B.26}$$

Substituting Equation B.25 and B.26 in Equation B.19 gives:

$$\eta_{IS} = \frac{1-\left(\dfrac{P_{out}}{P_{in}}\right)^{\frac{\gamma-1}{\gamma}}}{1-\left(\dfrac{P_{out}}{P_{in}}\right)^{\frac{n-1}{n}}} \tag{B.27}$$

Equation B.27 relates the isentropic compression efficiency, heat capacity ratio γ and polytropic coefficient n to the inlet and outlet pressures and can be rearranged to give:

$$n = \frac{\ln\left(\dfrac{P_{out}}{P_{in}}\right)}{\ln\left[\dfrac{\eta_{IS}\left(\dfrac{P_{out}}{P_{in}}\right)}{\eta_{IS}-1+\left(\dfrac{P_{out}}{P_{in}}\right)^{\frac{\gamma-1}{\gamma}}}\right]} \tag{B.28}$$

Equation B.28 is useful to estimate the polytropic coefficient n if the inlet and outlet pressures are known, along with an estimate of the isentropic efficiency. Knowing the polytropic coefficient allows the outlet temperature for a real gas compression to be estimated from Equation B.26.

B.2 CENTRIFUGAL COMPRESSORS

Centrifugal compressors increase gas pressure by accelerating the gas as it flows radially out from a rotating impeller. The increase in velocity is then converted to increase in pressure as the gas leaves the compression stage. Unlike reciprocating compressors, centrifugal compressors involve a constant flow through the compressor.

Consider a volume of gas V_1 flowing into the compressor. Compression work W_1 is required to force the gas into the system. The constant force exerted on the gas is P_1A_1, where A_1 is the cross-sectional area of the inlet duct. The distance through which the gas is forced as it enters the system is $-V_1/A_1$. The negative value of $-V_1/A_1$ results from the force acting from the surroundings on the system. Thus:

$$\begin{aligned} W_1 &= P_1A_1\left(-\frac{V_1}{A_1}\right) \\ &= -P_1V_1 \end{aligned} \tag{B.29}$$

Similarly, for the outlet from the compressor, the system must do work on the surroundings to force the gas out, such that:

$$W_2 = P_2V_2 \tag{B.30}$$

The work done by the compression is described by Equation B.1. Thus:

$$\int_{V_1}^{V_2} PdV = W_S + W_1 + W_2$$

Substituting Equations B.20 and B.21:

$$\int_{V_1}^{V_2} PdV = W_S - P_1V_1 + P_2V_2 \tag{B.31}$$

Rearranging Equation B.31:

$$W_S = -\int_{V_1}^{V_2} PdV + P_1V_1 - P_2V_2 \tag{B.32}$$

which from Equation B.3 gives:

$$W_S = -\int_{P_1}^{P_2} VdP \tag{B.33}$$

For an adiabatic ideal gas compression, from Equation B.12:

$$W_S = \frac{\gamma}{\gamma-1}P_{in}V_{in}\left[1-\left(\frac{P_{out}}{P_{in}}\right)^{\frac{\gamma-1}{\gamma}}\right] \tag{B.34}$$

Introducing the isentropic compression efficiency gives Equation B.15, the same result for a reciprocating compressor. In a reciprocating compressor, the net effect of the cycle is a flow process, even though intermittent nonflow steps are involved.

Whereas reciprocating compressors are normally designed on the basis of adiabatic work (together with an isentropic efficiency), centrifugal compressors are usually designed on the basis of polytropic work. By analogy with Equation B.15, the work required for a polytropic compression is given by:

$$W = \frac{n}{n-1}\frac{P_{in}V_{in}}{\eta_P}\left[1-\left(\frac{P_{out}}{P_{in}}\right)^{\frac{n-1}{n}}\right] \tag{B.35}$$

where η_P = polytropic efficiency (ratio of polytropic work to actual work)

For centrifugal compressors, the value of n is usually greater than γ. The inefficiencies caused by frictional losses, and so on, keep the operation from being truly adiabatic.

The isentropic and polytropic efficiencies can be related by taking the ratio of Equations B.15 and B.35 for the same compression process:

$$\frac{\eta_{IS}}{\eta_P} = \frac{\left(\frac{\gamma}{\gamma-1}\right)\left[1-\left(\frac{P_{out}}{P_{in}}\right)^{\frac{\gamma-1}{\gamma}}\right]}{\left(\frac{n}{n-1}\right)\left[1-\left(\frac{P_{out}}{P_{in}}\right)^{\frac{n-1}{n}}\right]} \quad \text{(B.36)}$$

Substituting Equation B.27 into Equation B.36 gives:

$$\eta_P = \frac{\left(\frac{n}{n-1}\right)}{\left(\frac{\gamma}{\gamma-1}\right)} \quad \text{(B.37)}$$

Rearranging Equation B.37 gives:

$$n = \frac{\gamma\eta_P}{\gamma\eta_P - \gamma + 1} \quad \text{(B.38)}$$

Equation B.38 is a useful expression to estimate the polytropic coefficient. The heat capacity ratio is a function of physical properties and, given an estimate of the polytropic efficiency for a compressor, the polytropic coefficient can be estimated from Equation B.38.

B.3 STAGED COMPRESSION

The temperature rise accompanying single-stage gas compression might be unacceptably high because of the properties of the gas, materials of construction of the compressor or the properties of the lubricating oil used in the machine. If this is the case, the overall compression can be broken down into a number of stages with intermediate cooling. Also, intermediate cooling will reduce the volume of gas between stages and reduce the work of compression for the next stage. On the other hand, the intercoolers will have a pressure drop that will increase the work, but this effect is usually small compared with the reduction in work from gas cooling.

Consider a two-stage compression in which the intermediate gas is cooled down to the initial temperature. The total work for a two-stage adiabatic gas compression of an ideal gas is given by:

$$W_S = \frac{\gamma}{\gamma-1}P_1V_1\left[1-\left(\frac{P_2}{P_1}\right)^{\frac{\gamma-1}{\gamma}}\right] + \frac{\gamma}{\gamma-1}P_2V_2\left[1-\left(\frac{P_3}{P_2}\right)^{\frac{\gamma-1}{\gamma}}\right] \quad \text{(B.39)}$$

where P_1, P_2, P_3 = initial, intermediate and final pressures

V_1, V_2 = initial and intermediate gas volumes

For an ideal gas with intermediate cooling to the initial temperature:

$$P_1V_1 = P_2V_2 \quad \text{(B.40)}$$

Combining Equations B.39 and B.40:

$$W_S = \frac{\gamma}{\gamma-1}P_1V_1\left[2-\left(\frac{P_2}{P_1}\right)^{\frac{\gamma-1}{\gamma}} - \left(\frac{P_3}{P_2}\right)^{\frac{\gamma-1}{\gamma}}\right] \quad \text{(B.41)}$$

The intermediate pressure P_2 can be chosen to minimize the overall work of compression. Thus:

$$\frac{dW_S}{dP_2} = 0 = \frac{\gamma}{\gamma-1}P_1V_1\left[\left(\frac{1}{P_1}\right)^{\frac{\gamma-1}{\gamma}}\left(\frac{\gamma-1}{\gamma}\right)P_2^{-1/\gamma} - P_3^{\frac{\gamma-1}{\gamma}}\left(\frac{\gamma-1}{\gamma}\right)P_2^{\frac{1-2\gamma}{\gamma}}\right] \quad \text{(B.42)}$$

Simplifying and rearranging Equation B.42 gives:

$$P_2^{\frac{2\gamma-2}{\gamma}} = (P_1P_3)^{\frac{\gamma-1}{\gamma}}$$

or

$$P_2 = \sqrt{P_1P_3} \quad \text{(B.43)}$$

Rearranging Equation B.43 gives:

$$\frac{P_2}{P_1} = \frac{P_3}{P_2} = \left(\frac{P_3}{P_1}\right)^{1/2} \quad \text{(B.44)}$$

Thus, for minimum shaft work, each stage should have the same compression ratio, which is equal to the square root of the overall compression ratio. This result is readily extended to N stages. The minimum work is obtained when the compression ratio in each stage is equal:

$$\frac{P_2}{P_1} = \frac{P_3}{P_2} = \frac{P_4}{P_3} = \ldots\ldots = r \quad \text{(B.45)}$$

where r = compression ratio

Since

$$\left(\frac{P_2}{P_1}\right)\cdot\left(\frac{P_3}{P_2}\right)\cdot\left(\frac{P_4}{P_3}\right)\ldots\ldots = r^N = \frac{P_{N+1}}{P_1} \quad \text{(B.46)}$$

The pressure ratio for minimum work for N stages is given by:

$$r = \sqrt[N]{\frac{P_{OUT}}{P_{IN}}} \quad \text{(B.47)}$$

The total shaft work for N stages is then given by:

$$W_S = \frac{\gamma}{\gamma-1}P_{in}V_{in}N\left[1-(r)^{\frac{\gamma-1}{\gamma}}\right] \quad \text{(B.48)}$$

Introducing the isentropic compression efficiency gives:

$$W = \frac{\gamma}{\gamma - 1} \frac{P_{in} V_{in} N}{\eta_{IS}} \left[1 - (r)^{\frac{\gamma-1}{\gamma}} \right] \qquad \text{(B.49)}$$

In principle, the isentropic efficiency might change from stage to stage. However, if the isentropic efficiency for a reciprocating compressor is assumed to be only a function of the pressure ratio and the pressure ratio is constant between stages, then it is legitimate to use a single value as in Equation B.49. It should be noted that these results for staged compression are based on adiabatic ideal gas compression and are therefore not strictly valid for real gas compression. It is also assumed that intermediate cooling is back to inlet conditions, which might not be the case with real intercoolers. For fixed inlet conditions and outlet pressure, the overall power consumption is usually not sensitive to minor changes in the intercooler temperature.

The corresponding equation for a polytropic compression is given by:

$$W = \frac{n}{n - 1} \frac{P_{in} V_{in} N}{\eta_P} \left[1 - (r)^{\frac{n-1}{n}} \right] \qquad \text{(B.50)}$$

If the polytropic efficiency of a centrifugal or axial compressor is assumed to be a function of volumetric flowrate, then the efficiency, in principle, will change from stage to stage. This is because the density changes between stages, even if the gas is cooled back to the same temperature as a result of the pressure increase. However, such effects are not likely to have a significant influence on the predicted power.

Consider now a two-stage compression in which the intermediate gas is cooled to a defined temperature T_2 different from the inlet temperature T_1. For an ideal gas:

$$P_2 V_2 = \frac{T_2}{T_1} P_1 V_1 \qquad \text{(B.51)}$$

Substituting Equation B.51 into Equation B.39 and differentiating with respect to intermediate pressure P_2 (assuming T_2 to be constant) gives:

$$\frac{dW_S}{dP_2} = 0 = \frac{\gamma}{\gamma - 1} P_1 V_1 \left[\left(\frac{1}{P_1}\right)^{\frac{\gamma-1}{\gamma}} \left(\frac{\gamma - 1}{\gamma}\right) P_2^{-1/\gamma} - \frac{T_2}{T_1} P_3^{\frac{\gamma-1}{\gamma}} \left(\frac{\gamma - 1}{\gamma}\right) P_2^{\frac{1-2\gamma}{\gamma}} \right] \qquad \text{(B.52)}$$

Rearranging Equation B.52 gives:

$$P_2 = \left(\frac{T_2}{T_1}\right)^{\frac{\gamma}{2\gamma-2}} (P_1 P_3)^{1/2} \qquad \text{(B.53)}$$

Thus, for a specified intercooler outlet temperature T_2 for the gas and fixed inlet and outlet pressures P_1 and P_3, Equation B.53 predicts the intermediate pressure for minimum shaft work for compression of an ideal gas. The corresponding expression for a polytropic compression is given by replacing γ by n in Equation B.53. Although Equation B.53 changes the intermediate pressure for an intercooler temperature different from the inlet temperature, the effect on the overall shaft work for compression is often insensitive to modest deviations of the intercooler temperature from the inlet temperature.

Appendix C Heat Transfer Coefficients and Pressure Drop in Shell-and-tube Heat Exchangers

At the conceptual stage for heat exchanger network synthesis, the calculation of heat transfer coefficient and pressure drop should depend as little as possible on the detailed geometry. Simple models will be developed in which heat transfer coefficient and pressure drop are both related to velocity[1]. It is thus possible to derive a correlation between the heat transfer coefficient, pressure drop and the surface area by using velocity as a bridge between the two[1].

Total pressure drop for a stream can be obtained by adding the pressure losses incurred in individual exchangers for each stream, allowing estimation of the pressure drops for both single heat exchangers and streams within a network.

C.1 PRESSURE DROP AND HEAT TRANSFER CORRELATIONS FOR THE TUBE-SIDE

The total pressure drop for the tube-side includes the pressure drop in the tube, sudden contractions, sudden expansions and flow reversals. There are two major sources of pressure losses on the tube-side of a shell-and-tube exchanger:

a. friction loss in a tube, which can be calculated as[2,3]:

$$\Delta P = 4c_f \frac{L}{d_I}\frac{\rho v_T^2}{2} \tag{C.1}$$

where ΔP = pressure drop
c_f = Fanning friction factor
L = length of tube
d_I = inside diameter of tube
ρ = fluid density
v_T = fluid velocity inside the tubes

b. losses due to the sudden contractions, expansions and flow reversals through the tube arrangement, which can be estimated per tube-pass as[4]:

$$\Delta P = 2.5\frac{\rho v_T^2}{2} \tag{C.2}$$

Thus, the total pressure drop for the tube-side is:

$$\Delta P_T = N_{TP}\left(4c_f\frac{L}{d_I} + 2.5\right)\frac{\rho v_T^2}{2} \tag{C.3}$$

where ΔP_T = tube-side pressure drop
N_{TP} = number of tube passes

The friction factor for turbulent flow ($Re > 4000$) can be approximated by[5]:

$$c_f = 0.046Re^{-0.2} \tag{C.4}$$

where Re = Reynolds number
$= \dfrac{\rho v_T d_I}{\mu}$
μ = fluid viscosity

Substituting this into Equation C.3:

$$\Delta P_T = N_{TP}4 \times 0.046\left(\frac{d_I v_T \rho}{\mu}\right)^{-0.2}\frac{L}{d_I}\frac{\rho v_T^2}{2} + N_{TP}\frac{2.5}{2}\rho v_T^2$$

$$= K_{PT}N_{TP}Lv_T^{1.8} + 1.25N_{TP}\rho v_T^2 \tag{C.5}$$

$$\text{where } K_{PT} = 0.092\rho^{0.8}\mu^{0.2}d_I^{-1.2} \tag{C.6}$$

Now, the relationship between velocity (v_T) and the heat transfer coefficient (h_T) needs to be determined to relate pressure drop to h_T.

The tube-side heat transfer coefficient can be calculated from[3]:

$$h_T = C\frac{k}{d_I}Pr^{\frac{1}{3}}Re^{0.8}\left(\frac{\mu}{\mu_W}\right)^{0.14} \tag{C.7}$$

where h_T = tube-side heat transfer coefficient
C = 0.021 for gases
= 0.023 for nonviscous liquids
= 0.027 for viscous liquids
k = fluid thermal conductivity
Pr = Prandtl number
$= \dfrac{C_P\mu}{k}$
C_P = fluid heat capacity
μ = fluid viscosity at the bulk fluid temperature
μ_W = fluid viscosity at the wall

Assuming $\mu/\mu_W = 1$ and rearranging Equation C.7 gives:

$$h_T = K_{hT}v_T^{0.8} \text{ or } v_T = \left(\frac{h_T}{K_{hT}}\right)^{\frac{1}{0.8}} \tag{C.8}$$

Chemical Process Design and Integration R. Smith
 ISBNs: 0-471-48680-9 (HB); 0-471-48681-7 (PB)

$$\text{where } K_{hT} = C\left(\frac{k}{d_I}\right)Pr^{\frac{1}{3}}\left(\frac{d_I\rho}{\mu}\right)^{0.8} \tag{C.9}$$

Now, consider the relationship between the pressure drop and the surface area for the tube-side. The heat transfer surface area A, based on the outside tube surface area, is given by:

$$A = N_T\pi\, d_O L \tag{C.10}$$

where A = heat transfer surface area
N_T = number of tubes
d_O = outside diameter of tube

Volumetric flowrate on the inside (tube-side) F_I is given by:

$$F_I = \frac{\pi\, d_I^2}{4}\frac{N_T}{N_{TP}}v_T \tag{C.11}$$

where F_I = volumetric flowrate on the inside (tube-side)

This equation can be rearranged to give an expression for the number of tubes N_T:

$$N_T = \frac{4F_I N_{TP}}{\pi\, d_I^2 v_T} \tag{C.12}$$

Substituting N_T into the expression for surface area gives:

$$A = \frac{4F_I N_{TP}}{d_I^2 v_T}d_O L \tag{C.13}$$

or

$$L = \frac{Ad_I^2 v_T}{4F_I N_{TP}\, d_O} \tag{C.14}$$

Thus, Equation C.5 can be rearranged to give:

$$\Delta P_T = K_{PT}\frac{Ad_I^2}{4F_I\, d_O}v_T^{2.8} + 1.25N_{TP}\rho v_T^2 \tag{C.15}$$

Substituting v_T from Equation C.8:

$$\begin{aligned}\Delta P_T &= K_{PT}\frac{Ad_I^2}{4F_I\, d_O}\left(\frac{h_T}{K_{hT}}\right)^{\frac{2.8}{0.8}} + 1.25N_{TP}\rho\left(\frac{h_T}{K_{hT}}\right)^{\frac{2}{0.8}}\\ &= K_{PT1}Ah_T^{3.5} + K_{PT2}h_T^{2.5}\end{aligned} \tag{C.16}$$

where

$$\begin{aligned}K_{PT1} &= K_{PT}\frac{d_I^2}{4F_I\, d_O}\left(\frac{1}{K_{hT}}\right)^{3.5}\\ &= \frac{0.023\rho^{0.8}\mu^{0.2}\, d_I^{0.8}}{F_I\, d_o}\left(\frac{1}{K_{hT}}\right)^{3.5}\end{aligned} \tag{C.17}$$

$$K_{PT2} = 1.25N_{TP}\rho\left(\frac{1}{K_{hT}}\right)^{2.5} \tag{C.18}$$

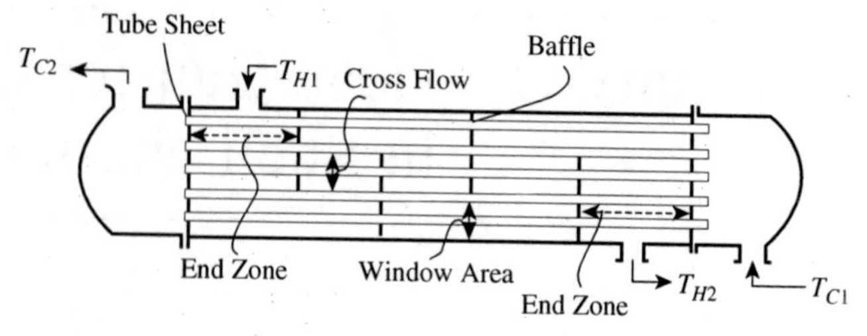

Figure C.1 Zone definition of the shell-side of shell-and-tube heat exchangers.

C.2 PRESSURE DROP AND HEAT TRANSFER CORRELATIONS FOR THE SHELL-SIDE

The total pressure drop for the shell-side includes those in the two ends, cross sections and window sections (see Figure C.1). The resulting pressure drop is defined from the inlet to the outlet of an exchanger. Figure 15.2a illustrates idealized flow on the shell-side involving a combination of idealized axial and cross flow. Ideal cross flow gives the higher heat transfer coefficients than axial flow. Figure 15.2b illustrates the actual flow pattern. Clearances between the tubes and baffles allow leakage (bypassing) of some of the fluid. This acts to reduce the outside heat transfer coefficient. The heat transfer coefficient and pressure drop is first estimated for ideal cross flow. Then the effects of leakage, bypassing and flow in the window zone are considered by applying correction factors.

The pressure drop for the shell-side has three components from inlet to outlet (see Figure C.1). These are the pressure drop for the ends (ΔP_e), the pressure drop for the cross-flow sections (ΔP_c), and the pressure drop for the window sections (ΔP_w).

The total pressure drop for the shell-side is given by summing the pressure drops over all the zones in series from inlet to outlet:

$$\Delta P_S = 2\Delta P_e + \Delta P_c(N_B - 1) + \Delta P_w N_B \tag{C.19}$$

where ΔP_S = shell-side pressure drop
ΔP_e = pressure drop for the end
ΔP_c = pressure drop for the cross-flow section
ΔP_w = pressure drop for the window section
N_B = number of baffles

a. *Pressure drop in the two ends.* There will be only one baffle window in the end zones. The total number of restrictions in the end zone will be the sum of the number of tubes in the cross-flow and window sections. The zone between tube sheet and the baffle (the end zone) pressure drop (ΔP_e) is given by:

$$\Delta P_e = F_{Pb}\Delta P_{IC}\left(1 + \frac{N_w}{N_c}\right) \tag{C.20}$$

where F_{Pb} = bypass correction factor for pressure drop to allow for flow between the tube bundle and the shell wall and is a function of the shell-to-bundle clearance. Typically, F_{Pb} lies between 0.5 and 0.8, depending on construction of the exchanger and the sealing arrangements. $F_{Pb} = 0.8$ can be used as a reasonable assumption for the clean condition. Fouling will tend to reduce bypassing and increase the pressure drop. Fouling will tend to increase the value to approach 1.0 in the worst case.
ΔP_{IC} = for ideal flow across the tubes based on the number of tubes in the cross-flow section
N_w = number of tube rows in the window section
N_c = number of tube rows in the cross-flow section

The ideal cross-flow pressure drop ΔP_{IC} can be expressed as:

$$\Delta P_{IC} = 8 j_f N_c \frac{\rho v_S^2}{2}\left(\frac{\mu}{\mu_W}\right)^{-0.14} \tag{C.21}$$

where j_F = cross-flow friction factor
v_S = shell-side fluid velocity

The shell-side fluid velocity is normally based on the area of flow for a hypothetical row of tubes across the diameter of the shell between two baffles. Thus:

$$v_S = \frac{F_O}{A_S} \tag{C.22}$$

where F_O = volumetric flowrate on the outside (shell-side) ($m^3 \cdot s^{-1}$)
A_S = mass cross-flow area (m^2)
= number of tubes × space between the tubes × baffle spacing
$= \frac{D_S}{p_T}(p_T - d_o)L_B$
D_S = shell diameter (m)
p_T = tube pitch, that is, center to center distance between adjacent tubes (m)
L_B = distance between baffles (m)

Assuming $\mu/\mu_W = 1$ in Equation C.21 and substituting into Equation C.20 gives:

$$\Delta P_e = F_{Pb}\left(1 + \frac{N_w}{N_c}\right) 8 j_f N_c \frac{\rho v_S^2}{2}$$
$$= F_{Pb}(N_w + N_c) 8 j_f \frac{\rho v_S^2}{2} \tag{C.23}$$

The friction factor j_f for cross flow can be correlated for turbulent flow ($Re > 4000$) as[1]:

$$j_f = 0.3245 Re^{-0.17} \tag{C.24}$$

The number of restrictions for cross flow in the window zones N_W can be calculated as:

$$N_w = \frac{B_C D_S}{p_T} \tag{C.25}$$

where N_w = number of tube rows in the window section (–)
B_C = baffle cut (–)

The number of tube rows in the cross-flow section N_c is:

$$N_c = \frac{D_S}{p_T} - 2 \times \frac{B_C D_S}{p_T} = \frac{D_S(1 - 2B_C)}{p_T} \tag{C.26}$$

Substituting N_w, N_c and j_f into Equation C.23 gives:

$$\Delta P_e = F_{Pb}\frac{D_S(1 - B_C)}{p_T} 8 j_f \frac{\rho v_S^2}{2}$$
$$= F_{Pb}\frac{D_S(1 - B_C)}{p_T} 8 \times 0.3245 Re^{-0.17} \frac{\rho v_S^2}{2}$$
$$= F_{Pb}\frac{D_S(1 - B_C)}{p_T} 8 \times 0.3245 \left(\frac{\rho d_O}{\mu}\right)^{-0.17} \frac{\rho}{2} v_S^{1.83}$$
$$= K_{PS1} v_S^{1.83} \tag{C.27}$$

where

$$K_{PS1} = 1.298 \frac{F_{Pb} D_S (1 - B_C) \rho^{0.83} \mu^{0.17}}{p_T d_O^{0.17}} \tag{C.28}$$

b. *Pressure drop for the cross sections.* The pressure drop in the cross flow zones between the baffle tips is calculated from the correlation for ideal tube banks, and corrected for leakage and bypassing:

$$\Delta P_c = \Delta P_{IC} F_{Pb} F_{PL} \tag{C.29}$$

where ΔP_c = pressure drop in cross-flow section
ΔP_{IC} = pressure drop for ideal cross flow
F_{PL} = leakage correction factor to allow for leakage through the tube-to-baffle clearance and the baffle-to-shell clearance. Typically, F_{PL} varies between 0.4 and 0.5. A value of $F_{PL} = 0.5$ can be used as a reasonable assumption for the clean condition. Fouling will tend to reduce leakage and increase the shell-side pressure drop. Fouling will tend to increase the value to approach 1.0 in the worst case.

The pressure drop for the cross-flow sections can be calculated as:

$$\Delta P_c(N_B - 1) = (N_B - 1)8 j_f N_c \frac{\rho v_S^2}{2} F_{Pb} F_{PL} \quad \text{(C.30)}$$

where N_B = number of baffles
The surface area A is given by:

$$A = N_T \pi d_O L \quad \text{(C.31)}$$

For a square pitch, each one contains four quarter tubes. Thus, for a square pitch, each tube is contained in an area of p_T^2. The number of tubes can then be approximated as:

$$N_T = \frac{\frac{\pi}{4} D_S^2}{p_T^2} \quad \text{(C.32)}$$

For a triangular pitch, each one with sides p_T, having an area of $0.5 p_T^2 \sin 60^o$ contains half a tube. Thus, a single tube is contained in an area of $p_T^2 \sin 60^o = 0.866 p_T^2$. The number of tubes can then be generalized as:

$$N_T = \frac{\frac{\pi}{4} D_S^2}{p_C p_T^2} \quad \text{(C.33)}$$

where p_C = pitch configuration factor
= 1 for square pitch
= 0.866 for triangular pitch

It should be noted that Equations C.32 and C.33 will tend to overestimate the number of tubes that can be contained in a given shell diameter. The larger the diameter shell, the smaller will be the error in the tube count. Substituting Equation C.33 into Equation C.31 gives:

$$A = \frac{\pi D_S^2}{4 p_C p_T^2} \pi d_O L = \frac{\pi D_S^2}{4 p_C p_T^2} \pi d_O L_B (N_B + 1) \quad \text{(C.34)}$$

The area for cross flow A_S is given by:

$$A_S = \frac{p_T - d_O}{p_T} D_S L_B \quad \text{(C.35)}$$

Rearranging Equation C.35 gives:

$$L_B = \frac{A_S p_T}{(p_T - d_O) D_S} \quad \text{(C.36)}$$

Substituting L_B into Equation C.34 gives:

$$A = \frac{\pi D_S^2}{4 p_C p_T^2} \pi d_O \frac{A_S p_T}{(p_T - d_O) D_S} (N_B + 1)$$
$$= \frac{\pi}{4 p_C p_T^2} \pi d_O \frac{A_S p_T}{(p_T - d_O)} D_S (N_B + 1) \quad \text{(C.37)}$$

Rearranging Equation C.37 gives:

$$D_S(N_B + 1) = \frac{A}{\frac{\pi}{4 p_C p_T^2} \pi d_O \frac{A_S p_T}{(p_T - d_O)}}$$
$$= \frac{A}{\frac{\pi}{4 p_C p_T^2} \pi d_O \frac{F_O p_T}{(p_T - d_O)}} v_S \quad \text{(C.38)}$$

Substituting the friction factor j_f and N_C into Equation C.30 gives:

$$\Delta P_c(N_B - 1) = (N_B - 1)8 \times 0.3245 \left(\frac{\rho d_O v_S}{\mu}\right)^{-0.17} \times \frac{D_S(1 - 2B_C)}{p_T} \frac{\rho v_S^2}{2} F_{Pb} F_{PL} \quad \text{(C.39)}$$

Combining Equations C.38 and C.39 gives:

$$\Delta P_c(N_B - 1) = \frac{8 \times 0.3245 \left(\frac{\rho d_O}{\mu}\right)^{-0.17}}{\frac{\pi}{4 p_C p_T^2} \pi d_O \frac{F_O p_T}{(p_T - d_O)}} \times \frac{(1 - 2B_C)}{p_T} \frac{\rho}{2} A v_S^{2.83} F_{Pb} F_{PL} - 2 \times 8 \times 0.3245 \left(\frac{\rho d_O v_S}{\mu}\right)^{-0.17} \times \frac{(1 - 2B_C) D_S}{p_T} \frac{\rho v_S^2}{2} F_{Pb} F_{PL}$$
$$= K_{PS2} A v_S^{2.83} - K_{PS3} v_S^{1.83} \quad \text{(C.40)}$$

where

$$K_{PS2} = \frac{0.5261 F_{Pb} F_{PL} p_C (1 - 2B_C)(p_T - d_O) \rho^{0.83} \mu^{0.17}}{d_O^{1.17} F_O} \quad \text{(C.41)}$$

$$K_{PS3} = \frac{2.596 F_{Pb} F_{PL} (1 - 2B_C) D_S \rho^{0.83} \mu^{0.17}}{p_T d_O^{0.17}} \quad \text{(C.42)}$$

c. *Pressure drop for the window sections.* The pressure drop calculation for the window zone is less accurate than that for the cross-flow sections. One correlation is[6]:

$$\Delta P_w = F_{PL}(2 + 0.6 N_w) \frac{\rho v_S^2}{2} \quad \text{(C.43)}$$

The pressure drop for the window sections can be calculated as:

$$N_B \Delta P_w = N_B F_{PL}(2 + 0.6 N_w) \frac{\rho v_S^2}{2}$$
$$= N_B F_{PL} \left(2 + 0.6 \frac{B_C D_S}{p_T}\right) \frac{\rho v_S^2}{2} \quad \text{(C.44)}$$

In order to remove N_B from the correlation, the following approximation can be made:

$$N_B + 1 \approx N_B \quad \text{(C.45)}$$

The bigger the exchanger, the better this assumption becomes. Using Equation C.45 and C.38 gives:

$$\begin{aligned} N_B \Delta P_w &= F_{PL}\left(\frac{2}{D_S} + 0.6\frac{B_C}{p_T}\right) D_S(N_B+1)\frac{\rho v_S^2}{2} \\ &= F_{PL}\left(\frac{2}{D_S} + 0.6\frac{B_C}{p_T}\right) \frac{A}{\dfrac{\pi}{4p_C p_T^2}\pi d_O \dfrac{F_O p_T}{(p_T - d_O)}} \\ &\quad \times v_S \frac{\rho v_S^2}{2} \\ &= K_{PS4} A v_S^3 \end{aligned} \quad \text{(C.46)}$$

where

$$K_{PS4} = \frac{0.2026 F_{PL} p_C p_T (p_T - d_O)\rho}{d_O F_O}\left(\frac{2}{D_S} + \frac{0.6 B_C}{p_T}\right) \quad \text{(C.47)}$$

d. *Shell-side heat transfer coefficient*. The shell-side heat transfer coefficient is given by:

$$h_S = h_{IS} F_{hn} F_{hw} F_{hb} F_{hL} \quad \text{(C.48)}$$

where h_S = shell-side heat transfer coefficient

h_{IS} = shell-side heat transfer coefficient for ideal cross flow

F_{hn} = correction factor to allow for the effect of the number of tube rows crossed. The basic heat transfer coefficient is based on ten rows of tubes. For turbulent flow, F_{hn} is close to 1.0.

F_{hw} = the window correction factor. This allows for flow through the baffle window and is a function of the heat transfer area in the window zones and the total heat transfer area. A typical value for a well-designed exchanger is near 1.0.

F_{hb} = the bypass stream correction factor. This allows for flow between the tube bundle and the shell wall and is a function of the shell-to-bundle clearance. Typical values are in the range 0.7 to 0.9 for clean exchangers with effective sealing arrangements. Fouling will tend to reduce bypassing and increase the shell-side heat transfer coefficient by increasing the cross flow. A conservative assumption would be to assume a value of 0.8 both for the clean and fouled condition. However, fouling will tend to increase the value to approach 1.0.

F_{hL} = the leakage correction factor. This allows for leakage through the tube-to-baffle clearance and the baffle-to-shell clearance. Typical values are in the range 0.7 to 0.8 for clean exchangers. Fouling will tend to reduce leakage and also increase the shell-side heat transfer coefficient by increasing the cross flow. A conservative assumption would be to assume a value of 0.8 both for the clean and fouled condition. However, fouling will tend to increase the value to approach 1.0.

The heat transfer coefficient for ideal cross flow over a tube bank is given as[3]:

$$h_{IS} = j_h C_P Pr^{-\frac{2}{3}} \rho v_S \quad \text{(C.49)}$$

The heat transfer factor j_h can be correlated for turbulent flow ($Re > 4000$) as[1]:

$$j_h = 0.24 Re^{-0.36} \quad \text{(C.50)}$$

Substituting j_h into h_{IS} gives:

$$h_{IS} = 0.24 C_P Pr^{-\frac{2}{3}} \rho \left(\frac{\rho d_O}{\mu}\right)^{-0.36} v_S^{0.64} \quad \text{(C.51)}$$

From Equation C.48:

$$\begin{aligned} h_S &= F_{hn} F_{hw} F_{hb} F_{hL} 0.24 C_P Pr^{-\frac{2}{3}} \rho \left(\frac{\rho d_O}{\mu}\right)^{-0.36} v_S^{0.64} \\ &= K_{hS} v_S^{0.64} \end{aligned} \quad \text{(C.52)}$$

where

$$K_{hS} = \frac{0.24 F_{hn} F_{hw} F_{hb} F_{hL} \rho^{0.64} C_P^{1/3} k^{2/3}}{\mu^{0.307} d_O^{0.36}} \quad \text{(C.53)}$$

Reasonable assumptions for the clean and fouled condition are $F_{hn} = F_{hw} = 1$ and $F_{hb} = F_{hL} = 0.8$. By rearranging Equation C.52:

$$v_S = \left(\frac{h_S}{K_{hS}}\right)^{1/0.64} \quad \text{(C.54)}$$

e. *Pressure drop correlation for the shell-side*. Substituting Equations C.27, C.40, C.46 and C.54 into Equation C.19 gives

$$\begin{aligned} \Delta P_S &= 2\Delta P_e + 2\Delta P_c(N_B - 1) + \Delta P_w N_B \\ &= 2K_{PS1} v_S^{1.83} + K_{PS2} A v_S^{2.83} \\ &\quad - K_{PS3} v_S^{1.83} + K_{PS4} A v_S^3 \\ &= 2K_{PS1}\left(\frac{h_S}{K_{hS}}\right)^{\frac{1.83}{0.64}} + K_{PS2} A \left(\frac{h_S}{K_{hS}}\right)^{\frac{2.83}{0.64}} \end{aligned}$$

$$- K_{PS3}\left(\frac{h_S}{K_{hS}}\right)^{\frac{1.83}{0.64}} + K_{PS4}A\left(\frac{h_S}{K_{hS}}\right)^{\frac{3}{0.64}}$$

$$= K_{S1}h_S^{2.86} + K_{S2}Ah_S^{4.42} + K_{S3}Ah_S^{4.69} \quad (C.55)$$

where

$$K_{S1} = \frac{2K_{PS1} - K_{PS3}}{K_{hS}^{2.86}} \quad (C.56)$$

$$K_{S2} = \frac{K_{PS2}}{K_{hS}^{4.42}} \quad (C.57)$$

$$K_{S3} = \frac{K_{PS4}}{K_{hS}^{4.69}} \quad (C.58)$$

The constants in the heat transfer and pressure drop correlations are functions of the fluid physical properties, volumetric flowrate, tube size and pitch. In preliminary design, it is reasonable to assume either 20 mm outside diameter tubes with a 2 mm wall thickness or 25 mm outside diameter tubes with 2.6 mm wall thickness. The tube pitch is normally taken to be $p_T = 1.25d_O$. A square tube pitch configuration can be assumed as a conservative assumption. Baffle cut can be assumed to be 0.25 in preliminary design.

It should be noted that the pressure drop calculation is much less accurate than the film transfer coefficient calculation. Moreover, calculations for the shell-side are less reliable than those for the tube-side. The pressure drop correlations for the shell-side should be treated with great caution. Even experimental data tends to show considerable scatter when correlated[7].

REFERENCES

1. Nie X-R (1998) *Optimisation Strategies for Heat Exchanger Network Design Considering Pressure Drop Aspects*, PhD Thesis, UMIST, UK.
2. Kern DQ (1950) *Process Heat Transfer*, McGraw-Hill.
3. Sinnott RK (1996) *Chemical Engineering*, *Vol 6 Chemical Engineering Design*, 2nd Edition, Butterworth Heinemann.
4. Frank O (1978) Simplified Design Procedure for Tubular Exchangers in Practical Aspects of Heat Transfer, *Chem Eng Prog Tech Manual*, AIChE.
5. Hewitt GF, Shires GL and Bott TR (1994) *Process Heat Transfer*, CRC Press Inc.
6. Bell KJ (1963) Final Report of the Cooperative Research Program on Shell and Tube Heat Exchangers, University of Delaware, Eng Expt Sta Bull 5.
7. Taborek J (1992) Calculation of Shell-side Heat Transfer Coefficient and Pressure Drop in GF, *Hewitt Handbook of Heat Exchanger Design*, Hemisphere.

Appendix D The Maximum Thermal Effectiveness for 1–2 Shell-and-tube Heat Exchangers

The derivation of Equation 15.53 is as follows[1]. From Bowman *et al*[2]:

When $R \neq 1$:

$$F_T = \frac{\sqrt{R^2+1}\ln\left[\dfrac{(1-P)}{(1-RP)}\right]}{(R-1)\ln\left[\dfrac{(2-P(R+1-\sqrt{R^2+1}))}{(2-P(R+1+\sqrt{R^2+1}))}\right]} \quad \text{(D.1)}$$

When $R = 1$:

$$F_T = \frac{\left[\dfrac{\sqrt{2}P}{(1-P)}\right]}{\ln\left[\dfrac{(2-P(2-\sqrt{2}))}{(2-P(2+\sqrt{2}))}\right]} \quad \text{(D.2)}$$

The maximum value of P, for any R, occurs as F_T tends to $-\infty$. From the F_T functions above, for F_T to be determinate:

1. $P < 1$
2. $RP < 1$
3. $\dfrac{2-P(R+1-\sqrt{R^2+1})}{2-P(R+1+\sqrt{R^2+1})} > 0$

Condition 3 applies to Equation D.2 when $R = 1$. Both Conditions 1 and 2 are always true for a feasible heat exchange with positive temperature differences.

Taking Condition 3, either:

a. $P < \dfrac{2}{R+1-\sqrt{R^2+1}}$ and $P < \dfrac{2}{R+1+\sqrt{R^2+1}}$ (D.3)

or

b. $P > \dfrac{2}{R+1-\sqrt{R^2+1}}$ and $P > \dfrac{2}{R+1+\sqrt{R^2+1}}$ (D.4)

but not both. Consider Condition b in more detail. For positive values of R, $R+1-\sqrt{R^2+1}$ is a continuously increasing function of R, and

- as R tends to 0, $R+1-\sqrt{R^2+1}$ tends to 0
- as R tends ∞, $R+1-\sqrt{R^2+1}$ tends to 1

For Condition b to apply, for positive values of R, $P > 2$. However, $P < 1$ for feasible heat exchange. Thus, Condition b does not apply.

Now consider Condition a. Because

$$R+1+\sqrt{R^2+1} > R+1-\sqrt{R^2+1} \quad \text{(D.5)}$$

both inequalities for Condition a are satisfied when

$$P < \frac{2}{R+1-\sqrt{R^2+1}} \quad \text{(D.6)}$$

Thus, the maximum value of P for any value of R, P_{max} is given by:

$$P_{max} = \frac{2}{R+1+\sqrt{R^2+1}} \quad \text{(D.7)}$$

REFERENCES

1. Ahmad S (1985) *Heat Exchanger Networks: Cost Trade-offs in Energy and Capital*, PhD Thesis, UMIST, UK.
2. Bowman RA, Mueller AC and Nagle WM (1940) Mean Temperature Differences in Design, *Trans ASME*, **62**: 283.

Chemical Process Design and Integration R. Smith

ISBNs: 0-471-48680-9 (HB); 0-471-48681-7 (PB)

Appendix E Expression for the Minimum Number of 1–2 Shell-and-tube Heat Exchangers for a Given Unit

The derivation of Equations 15.62 to 15.64 is as follows[1]. From Bowman[2], the value of P over N_{SHELLS} number of 1–2 shells in series, P_{N-2N}, can be related to the value of P for each 1–2 shell, P_{1-2}, according to:

$R \neq 1$:

$$P_{N-2N} = \frac{1 - \left[\dfrac{1 - P_{1-2}R}{1 - P_{1-2}}\right]^{N_{SHELLS}}}{R - \left[\dfrac{1 - P_{1-2}R}{1 - P_{1-2}}\right]^{N_{SHELLS}}} \tag{E.1}$$

$R = 1$:

$$P_{N-2N} = \frac{P_{1-2}N_{SHELLS}}{P_{1-2}N_{SHELLS} - P_{1-2} + 1} \tag{E.2}$$

Now, the maximum possible value of P_{1-2} in a 1–2 shell is (see Appendix D):

$$P_{max\ 1-2} = \frac{2}{R + 1 + \sqrt{R^2 + 1}} \tag{E.3}$$

The value of P_{1-2} required in each 1–2 shell to satisfy a chosen value of X_P is defined by:

$$P_{1-2} = X_P P_{1-2\max} \tag{E.4}$$

This therefore requires that:

$R \neq 1$:

$$P_{N-2N} = \frac{1 - \left[\dfrac{1 - \dfrac{2X_P R}{R + 1 + \sqrt{R^2 + 1}}}{1 - \dfrac{2X_P}{R + 1 + \sqrt{R^2 + 1}}}\right]^{N_{SHELLS}}}{R - \left[\dfrac{1 - \dfrac{2X_P R}{R + 1 + \sqrt{R^2 + 1}}}{1 - \dfrac{2X_P}{R + 1 + \sqrt{R^2 + 1}}}\right]^{N_{SHELLS}}} \tag{E.5}$$

$R = 1$:

$$P_{N-2N} = \frac{\dfrac{2X_P N_{SHELLS}}{2 + \sqrt{2}}}{\dfrac{2X_P N_{SHELLS}}{2 + \sqrt{2}} - \dfrac{2X_P}{2 + \sqrt{2}} + 1} \tag{E.6}$$

These expressions define P_{N-2N} for N_{SHELLS} number of 1–2 shells in series in terms of R and X_P in each shell. The expressions can be used to define the number of 1–2 shells in series required to satisfy a specified value of X_P in each shell for a given R and P_{N-2N}. Hence, the relationship can be inverted to find that value of N that satisfies X_P exactly in each 1–2 shell in the series:

$R \neq 1$:

$$N_{SHELLS} = \frac{\ln\left[\dfrac{1 - RP_{N-2N}}{1 - P_{N-2N}}\right]}{\ln W} \tag{E.7}$$

where

$$W = \frac{R + 1 + \sqrt{R^2 + 1} - 2X_P R}{R + 1 + \sqrt{R^2 + 1} - 2X_P} \tag{E.8}$$

$R = 1$:

$$N_{SHELLS} = \left(\frac{P_{N-2N}}{1 - P_{N-2N}}\right)\left(\frac{1 + \dfrac{\sqrt{2}}{2} - X_P}{X_P}\right) \tag{E.9}$$

Choosing the number of 1–2 shells in series to be the next largest integer above N_{SHELLS} ensures a practical exchanger design satisfying X_P.

REFERENCES

1. Ahmad S (1985) *Heat Exchanger Networks: Cost Trade-offs in Energy and Capital*, PhD Thesis, UMIST, UK.
2. Bowman RA (1936) Mean Temperature Difference Correction in Multipass Exchangers, *Ind Eng Chem*, **28**: 541.

Chemical Process Design and Integration R. Smith

ISBNs: 0-471-48680-9 (HB); 0-471-48681-7 (PB)

Appendix F Algorithm for the Heat Exchanger Network Area Target

Figure F.1 shows a pair of composite curves divided into vertical enthalpy intervals. Also shown in Figure F.1 is a heat exchanger network for one of the enthalpy intervals, which will satisfy all of the heating and cooling requirements. The network shown in Figure F.1 for the enthalpy interval is in grid diagram form. Hot streams are at the top, running left to right. Cold streams are at the bottom, running right to left. A heat exchange match is represented by a vertical line joining two circles on the streams being matched. The network arrangement in Figure F.1 has been placed such that each match experiences the ΔT_{LM} of the interval. The network also uses the minimum number of matches ($S - 1$). Such a network can be developed for any interval, provided each match within the interval:

a. completely satisfies the enthalpy change of a stream in the interval, and
b. achieves the same ratio of CP's as exists between the composite curves (by stream splitting if necessary).

As each such match is successively placed in the interval, the minimum number of matches can be achieved because there is one fewer stream to match *and* the CP-ratio of the remaining streams (that is, ratio of ΣCP_H and ΣCP_C of the remaining streams) in the interval still satisfies the CP-ratio between the composite curves.

It is, thus, always possible to achieve the interval design with $S - 1$ matches, with each match operating with the log mean temperature differences of the interval.

Now, consider the heat transfer area required by enthalpy interval k, in which the overall heat transfer coefficient is allowed to vary between individual matches.

$$A_k = \frac{1}{\Delta T_{LM_k}} \sum_{ij} \frac{Q_{ij,k}}{U_{ij,k}} \tag{F.1}$$

where A_k = network area based on vertical heat exchange in enthalpy interval k

ΔT_{LM_k} = log mean temperature difference for enthalpy interval k

$Q_{ij,k}$ = heat load on match between hot stream i and cold stream j in enthalpy interval k

$U_{ij,k}$ = overall heat transfer coefficient between hot stream i and cold stream j in enthalpy interval k

Introducing individual film transfer coefficients:

$$A_k = \frac{1}{\Delta T_{LM_k}} \sum_{ij} Q_{ij,k} \left[\frac{1}{h_i} + \frac{1}{h_j} \right] \tag{F.2}$$

where h_i, h_j are film transfer coefficients for hot stream i and cold stream j (including wall and fouling resistances). From Equation F.2:

$$A_k = \frac{1}{\Delta T_{LM_k}} \left[\sum_{ij} \frac{Q_{ij,k}}{h_i} + \sum_{ij} \frac{Q_{ij,k}}{h_j} \right] \tag{F.3}$$

Since enthalpy interval k is in heat balance, then summing overall cold stream matches with hot stream i gives the stream duty on hot stream i:

$$\sum_{j}^{J} Q_{ij,k} = q_{i,k} \tag{F.4}$$

where $q_{i,k}$ = stream duty on hot stream i in enthalpy interval k

J = total number of cold streams in enthalpy interval k

Similarly, summing overall hot stream matches with cold j gives the cold stream duty on cold stream j:

$$\sum_{i}^{I} Q_{ij,k} = q_{j,k} \tag{F.5}$$

where $q_{j,k}$ = stream duty on cold stream j in enthalpy interval k

I = total number of hot streams in enthalpy interval k

Thus, from Equation F.4:

$$\sum_{ij} \frac{Q_{ij,k}}{h_i} = \sum_{i}^{I} \frac{q_{i,k}}{h_i} \tag{F.6}$$

and from Equation F.5:

$$\sum_{ij} \frac{Q_{ij,k}}{h_j} = \sum_{j}^{J} \frac{q_{j,k}}{h_j} \tag{F.7}$$

Chemical Process Design and Integration R. Smith
 ISBNs: 0-471-48680-9 (HB); 0-471-48681-7 (PB)

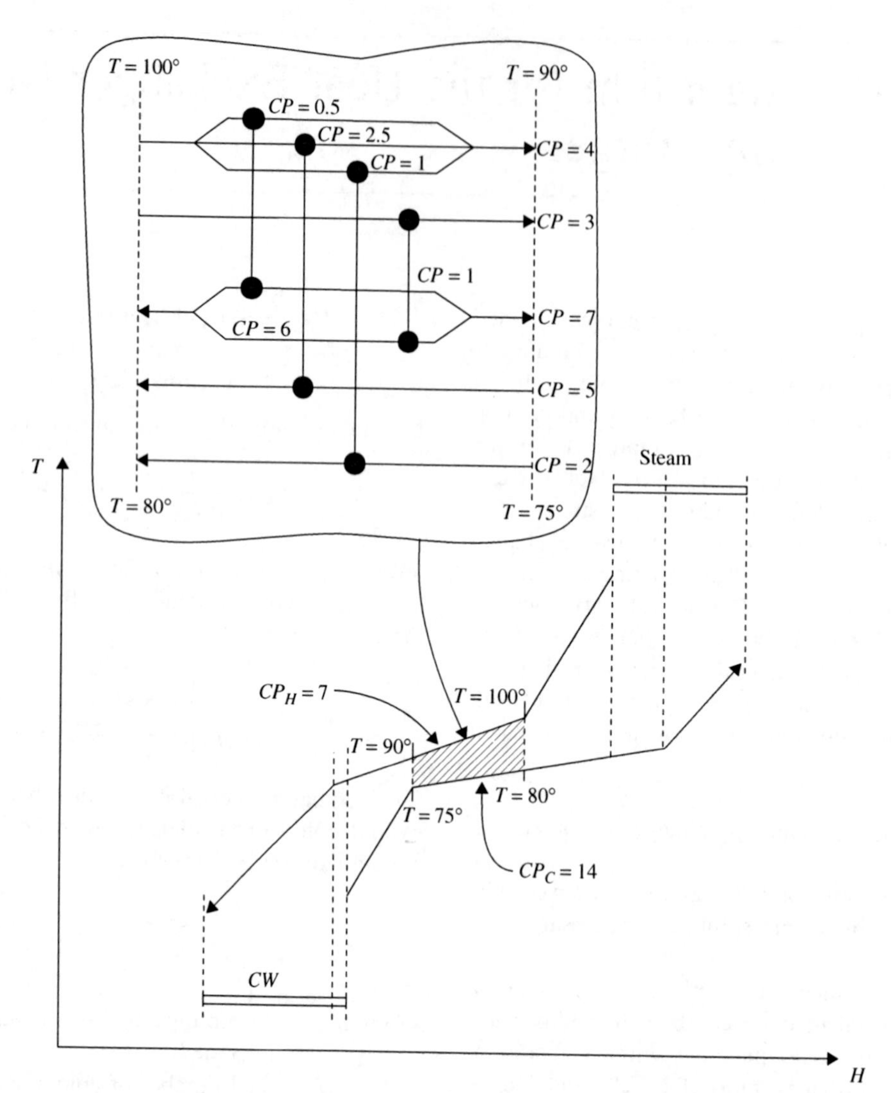

Figure F.1 Within each enthalpy interval it is possible to design a network in $(S - 1)$ matches. (From Ahmad S and Smith R (1989) Targets and Design for Minimum Number of Shells in Heat Exchanger Networks, *IChemE, ChERD*, **67**: 481, reproduced by permission of the Institution of Chemical Engineers.)

Substituting these expressions in Equation F.3 gives:

$$A_k = \frac{1}{\Delta T_{LM_k}} \left(\sum_{i}^{I} \frac{q_{i,k}}{h_i} + \sum_{j}^{J} \frac{q_{j,k}}{h_j} \right) \tag{F.8}$$

Extending this equation to all enthalpy intervals in the composite curves gives:

$$A_{NETWORK} = \sum_{k}^{INTERVALS\ K} \frac{1}{\Delta T_{LM_k}} \times \left(\sum_{i}^{HOT\ STREAMS\ I} \frac{q_{i,k}}{h_i} + \sum_{j}^{COLD\ STREAMS\ J} \frac{q_{j,k}}{h_j} \right) \tag{F.9}$$

Appendix H Algorithm for Heat Exchanger Network Capital Cost Targets

The area target, Equation 17.6, sums the area contributions from each enthalpy interval. This equation can be rearranged to an equivalent expression that sums the area contribution of each stream[1,2]:

$$A_{NETWORK} = \sum_{k}^{INTERVALS\ K} \frac{1}{\Delta T_{LM_k}} \times \left[\sum_{i}^{HOT\ STREAMS\ I} \frac{q_{i,k}}{h_i} + \sum_{j}^{COLD\ STREAMS\ J} \frac{q_{j,k}}{h_j} \right]$$

$$= \sum_{i}^{HOT\ STREAMS\ I} \left[\sum_{k}^{INTERVALS\ K} \frac{q_{i,k}}{\Delta T_{LM_k} h_i} \right] + \sum_{j}^{COLD\ STREAMS\ J} \left[\sum_{k}^{INTEVALS\ K} \frac{q_j}{\Delta T_{LM_k} h_j} \right]$$

$$= \sum_{i}^{HOT\ STREAMS\ I} A_i + \sum_{j}^{COLD\ STREAMS\ J} A_j \quad \text{(H.1)}$$

where A_i = contribution to area target from hot stream i
A_j = contribution to area target from cold stream j

Equation H.1 shows that each stream makes a contribution to total heat transfer area defined only by its duty, position in the composite curves and its film transfer coefficient. This contribution to area means a contribution to capital cost also. If, for example, a corrosive stream requires special materials of construction, it will have a greater contribution to capital cost than a similar noncorrosive stream. If only one cost law is to be used for a network comprising mixed materials of construction, the area contribution of streams requiring special materials must somehow increase. One way of doing this is by weighting the heat transfer coefficient to reflect the cost of the material the stream requires.

To develop the approach, first consider a single exchanger whose cost may be represented as:

$$\textit{Installed capital cost of reference exchanger} = a_1 + b_1 A^{c_1} \quad \text{(H.2)}$$

where a_1, b_1 and c_1 are cost law coefficients for the reference exchanger.

If, instead, the heat exchanger is made to a different specification, its cost may be represented as:

$$\textit{Installed capital cost of special exchanger} = a_2 + b_2 A^{c_2} \quad \text{(H.3)}$$

where a_2, b_2 and c_2 are cost law coefficients for a special exchanger.

In the approach to be developed, the cost of the special exchanger can be determined from the reference cost law by using a modified area A^*:

$$\textit{Installed capital cost of special exchanger} = a_2 + b_2 A^{*c_1} \quad \text{(H.4)}$$

Heat exchanger cost data can usually be manipulated such that fixed costs, represented by the Coefficient a in Equations H.2 to H.4, do not vary with exchanger specification[2]. Equations H.3 and H.4 can now be rearranged to give the modified heat exchanger area A^* as a function of actual area A and the cost law coefficients.

$$A^* = \left(\frac{b_2}{b_1}\right)^{\frac{1}{c_1}} A^{\frac{c_2}{c_1} - 1} A \quad \text{(H.5)}$$

The relationship between heat exchanger area and overall heat transfer coefficient U is given by:

$$A = \frac{Q}{\Delta T_{LM} U} \quad \text{(H.6)}$$

where Q is exchanger heat load. The ratio $Q/\Delta T_{LM}$ is a constant for a given heat exchanger and, hence, the modified overall heat transfer coefficient U^* can be related to actual overall heat transfer coefficient U:

$$\frac{1}{U^*} = \left(\frac{b_2}{b_1}\right)^{\frac{1}{c_1}} A^{\frac{c_2}{c_1} - 1} \frac{1}{U} \quad \text{(H.7)}$$

The overall heat transfer coefficient in a single heat exchanger comprises resistance contributions from both streams. Each contribution contains allowances for film, wall and fouling resistances. In practice, the overall heat transfer coefficient will depend, to some extent, on the exchanger flow arrangement. It is not possible to specify such details at the targeting stage, hence the overall heat transfer coefficient must be assumed independent of the flow arrangement:

$$\frac{1}{U} = \frac{1}{h_H} + \frac{1}{h_C} \quad \text{(H.8)}$$

Equation H.7 may be split streamwise to obtain an expression for the modified film transfer coefficient h_j^* of

Chemical Process Design and Integration R. Smith
 ISBNs: 0-471-48680-9 (HB); 0-471-48681-7 (PB)

either stream in the match:

$$h_j{}^* = \left(\frac{b_1}{b_2}\right)^{\frac{1}{c_1}} A^{1-\frac{c_2}{c_1}} h_j \tag{H.9}$$

A stream-specific cost-weighting factor ϕ_j to be applied to the film transfer coefficient of a special stream j can now be defined. This is the ratio of weighted-to-actual stream film transfer coefficients:

$$\phi_j = \frac{h_j{}^*}{h_j} = \left(\frac{b_1}{b_2}\right)^{\frac{1}{c_1}} A^{1-\frac{c_2}{c_1}} \tag{H.10}$$

The same philosophy of weighting area contributions in a single heat exchanger can be extended to weighting stream area contributions for a whole network. Some additional error in the targets is incurred by this extension, resulting from the fact that a stream may pass through several exchangers, all with different areas. At the targeting stage, heat exchangers are assumed to be all the same size. The special stream cost-weighting factor is then expressed as:

$$\phi_j = \left(\frac{b_1}{b_2}\right)^{\frac{1}{c_1}} \left(\frac{A_{NETWORK}}{N}\right)^{1-\frac{c_2}{c_1}} \tag{H.11}$$

Once the ϕ-factor has been evaluated for each stream, a weighted network area target $A^*_{NETWORK}$ can be calculated:

$$A_{NETWORK}{}^* = \sum_k^{INTERVALS\ K} \frac{1}{\Delta T_{LM_k}} \times \left(\sum_i^{HOT\ STREAMS\ I} \frac{q_{i,k}}{\phi_i h_i} + \sum_j^{COLD\ STREAMS\ J} \frac{q_{j,k}}{\phi_j h_j} \right) \tag{H.12}$$

REFERENCES

1. Ahmad S (1985) *Heat Exchanger Networks: Cost Trade-Offs in Energy and Capital*, PhD Thesis, UMIST, UK.
2. Hall SG, Ahmad S and Smith R (1990) Capital Cost Targets for Heat Exchanger Networks Comprising Mixed Materials of Construction, Pressure Ratings and Equipment Types, *Comput Chem Eng*, **14**: 319.

Index

Chemical Process Design and Integration R. Smith
 ISBNs: 0-471-48680-9 (HB); 0-471-48681-7 (PB)

Lightning Source UK Ltd.
Milton Keynes UK
UKOW03f1550050615

252973UK00002B/14/P